GEOENVIRONMENTAL ENGINEERING

GEOENVIRONMENTAL ENGINEERING:

Site Remediation, Waste Containment, and Emerging Waste Management Technologies

Hari D. Sharma, Ph.D., P.E., G.E.
GeoSyntec Consultants
Oakland, California

Krishna R. Reddy, Ph.D., P.E.
University of Illinois at Chicago
Chicago, Illinois

WILEY

JOHN WILEY & SONS, INC.

With Love and Affection
Dedicated to
Jaya, Samir, and Lena Sharma
and
Hema, Nishith, and Navya Reddy

Library of Congress Cataloging in Publication Data:

Sharma, Hari D.,
   Geoenvironmental Engineering: Site Remediation, Waste Containment,
and Emerging Waste Management Technologies / Hari D. Sharma, Krishna R. Reddy.
         p. cm.
Includes bibliographical references and index.
   ISBN 0-471-21599-6 (Cloth)
   1. Geoenvironmental engineering   I. Reddy, Krishna R.   II. Title.

Printed in the United States of America

10 9 8 7 6 5

CONTENTS

PART I BASIC PRINCIPLES

PART II REMEDIATION TECHNOLOGIES

PART III LANDFILLS AND SURFACE IMPOUNDMENTS

15
**SOURCES AND CHARACTERISTICS OF
WASTES**   **605**

PART IV   EMERGING TECHNOLOGIES

PREFACE

Rapid industrialization and economic growth, especially during the twentieth century, resulted in the production and disposal of various chemicals and their by-products into the environment. This resulted in the contamination of our air, water, and earth materials. These contaminants affected our ecosystem and, eventually, had an adverse impact on human health and the environment.

The response to environmental degradation resulted in public action and awareness that began in the 1960s and 1970s. This led to the passage of numerous environmental laws and regulations. In the United States, the federal environmental laws are passed by the U.S. Congress. These laws define goals in general terms and do not provide technical details on how to accomplish the goals. The U.S. Congress, however, authorizes certain government agencies, such as the U.S. Environmental Protection Agency (USEPA), to develop full details, including technical details, on how to achieve these goals. In response, the USEPA has developed various technical guidance documents and has sponsored various applied research topics to solve technical challenges related to environmental protection. This effort has been complemented by state government agencies and various private organizations, resulting in a large volume of technical information on various ways to investigate, design, remediate, and construct facilities with a goal to protect human health and the environment.

This information encompasses various disciplines of science, engineering, and public policy, consisting of environmental laws and regulations, geochemistry, geology and hydrogeology, geotechnical engineering, environmental engineering, fluid mechanics, and statics and dynamics, including earthquake engineering, plastics and geotextiles, and other fields of engineering: thus, the title *Geoenvironmental Engineering*. The book presents a concise, systematic, and complete treatment of the subject for senior-level undergraduate and graduate students; practicing civil, geotechnical, and environmental engineers; project managers; and regulatory agency professionals.

The book is divided into four major parts: Part I, Basic Principles; Part II, Remediation Technologies; Part III, Landfills and Surface Impoundments; and Part IV, Emerging Technologies. In Part I we present environmental laws and regulations, chemical and geochemistry background, geotechnical background, groundwater flow, and contaminant fate and transport. In Part II we discuss sources of contamination, contaminated site characterization, risk assessment, in-situ waste containment, and soil and groundwater remediation technologies. In Part III we present information on the design and evaluation of landfills and surface impoundments. These include sources and characteristics of wastes, regulations, and siting requirements, liner systems, leachate collection and removal systems, final cover systems, gas management, and groundwater monitoring requirements. Finally, in Part IV we outline various emerging technologies, such as beneficial uses of closed landfills, recycling, bioreactors, and in-situ capping of subaquatic waste sediments.

The book has been used in university-level geoenvironmental courses and therefore includes basic principles, example problems, case histories, and questions/problems with lists of comprehensive up-to-date references at the end of each chapter. Various parts of the book have also been used in short courses and seminars to practicing engineers, students, and regulatory professionals. Practical and academic experiences of the authors have helped to prepare this book such that its content will be useful to both practitioners and academic professionals.

Thanks go to various associations, organizations, individuals, and journal and technical document publishers for permitting the use of materials from their publications. The authors would like to thank their families for providing continued support and long patience during the completion of this manuscript. We would also like to thank Ala Prasanth and Adam Urbanek for their invaluable support in typing the manuscript and preparing illustrations. We are also grateful to Kranti Maturi, Jeff Adams, Supraja Chinthamreddy, Rich Saichek, Robin Semer, Andy Jazdanian, Jeff Schuh, Chien Wu, and many other former and current students and colleagues for their encouragement, confidence, and support in the preparation of this book.

Hari D. Sharma
Oakland, California

Krishna R. Reddy
Chicago, Illinois

PART I
BASIC PRINCIPLES

1

INTRODUCTION

1.1 EMERGENCE OF GEOENVIRONMENTAL ENGINEERING

Traditionally, geotechnical engineers have been responsible for (1) investigating subsurface conditions; (2) designing foundations for roads, buildings, machines, storage tanks, and offshore structures; (3) designing earthworks for dams, levees, roads, tunnels, and underground structures; and (4) investigating landmass problems such as landslides, slope stability, and subsidence (Taylor, 1948; Terzaghi and Peck, 1948; Lambe and Whitman, 1969; Peck et al., 1974). Geotechnical engineering was initially a loosely structured practice relying primarily on a few rules of thumb. After the 1950s, the practice of soil mechanics and foundation engineering grew tremendously throughout the world. The main contributors to this growth were Coulomb, Rankine, Darcy, Terzaghi, Casagrande, Taylor, Skempton, Bishop, and Peck. Through the work of these investigators, the practice of soil mechanics and foundation engineering became more rigorous, based on rational design approaches. As development of soil mechanics and foundation engineering continued, the practice of rock mechanics and engineering geology also evolved. Because all these topics proved interrelated, they were combined and renamed *geotechnical engineering* in the early 1980s. Since then, geotechnical engineering has developed into an important and necessary specialty of the civil engineering profession.

Simultaneous with the growth of geotechnical engineering, the post–World War II economic boom of the 1950s led to rapid industrialization. In particular, chemical industries grew in number and increased the standard of living by producing a wide variety of products. Manufacture of these products required the production of tremendous amounts of organic chemicals while increasing the use of heavy metals. Mass production of these products also increased the quantity of wastes to be disposed of. Improper disposal practices and accidental spills of these chemicals have created numerous contaminated sites in the United States and throughout the world.

The onset of nuclear power plants and nuclear waste generated awareness of contamination problems and perhaps initiated the involvement of geotechnical engineers in environmental matters (Daniel, 1993). According to the National Environmental Policy Act of 1970, environmental impact assessments were required for any federal project that could affect the environment, particularly those involving the selection of nuclear power plant sites. Geotechnical engineers lead the detailed site investigations for these impact assessments. Another concern was the ultimate disposal of high-level radioactive waste, which can remain lethal for thousands of years. Geotechnical engineers played an important role in the investigation and characterization of suitable host soil and rocks for waste repositories. Geotechnical engineers also determined long-term performance of earth materials under realistic temperature and pressure, probable groundwater impacts, and potential risks. With increased attention, environmental concerns became a new aspect of geotechnical engineering.

One specific environmental event that increased attention was the widely publicized contamination at Love Canal in upstate New York. At this site, chemical wastes were buried in an old canal and covered with clayey soils. The chemicals seeped out slowly, contaminating the soil and groundwater. The health of the residents was adversely affected and the entire area had to be evacuated. This incident drew national attention to the effects of improper disposal and management of chemical waste. As a result, in the late 1970s and early 1980s, discussions began on the prevention and mitigation of such improper waste disposal practices. Also, as a result of this incident, new U.S. regulations for remediation of contaminated sites, in addition to the design of effective waste containment systems for newly created wastes, were promulgated. In 1970, the Resource Conservation and Recovery Act (RCRA) and its subsequent amendments addressed issues of disposal of newly generated waste. In 1980, the U.S. Congress passed the Comprehensive Environmental Response, Compensation and Liability Act (CERCLA), also known as the Superfund, to clean up contaminated sites in cases where the responsible polluting parties could not be identified or were incapable of paying for the cleanup. The geotechnical engineer's knowledge of earth material and groundwater became vital to the investigation, design, and actual cleanup of contaminated sites as well as the design of containment facilities.

Waste containment and remediation problems require an understanding of the physical characteristics of the subsurface and the ability to engineer it using the skills of classical geotechnical engineering. However, these problems also require an understanding of the chemical characteristics of the subsurface and the ability to engineer pollution control or removal using the skills of environmental engineering. A combined expertise of geotechnical engineering and environmental engineering is needed to address various aspects of such problems. In addition, knowledge of environmental regulations, hydrogeology, environmental chemistry, geochemistry, and microbiology is needed. Recognizing this fact, a new specialty of civil engineering known as *geoenvironmental engineering* (also known as *environmental geotechnology* and *environmental geotechnics*) emerged in the early 1990s. Geoenvironmental engineering encompasses the behavior of soils, rocks, and groundwater when they interact with contaminants and addresses problems of hazardous and nonhazardous waste management and contaminated sites.

Geoenvironmental engineering is evolving rapidly. Environmental laws and regulations that have a significant impact on geoenvironmental engineering are constantly changing. Environmental problems are numerous in all industrial and developing countries and will continue to grow with increased chemical waste generation and handling. For these reasons, geoenvironmental engineers will play a vital role in pollution control strategies, particularly in the design of effective and economical waste containment and remediation systems.

1.2 TYPES OF GEOENVIRONMENTAL PROBLEMS

Geoenvironmental problems may be grouped into three categories:

1. *Contaminated site remediation:* remediation of already contaminated soils and groundwater using in-situ barriers and in-situ or ex-situ treatment methods
2. *Waste containment:* safe disposal of newly generated wastes in engineered impoundments and landfills
3. *Waste minimization by recycling:* minimization of waste generation and disposal by recycling and using waste materials in various civil engineering applications, and beneficial use of closed waste disposal sites

A general overview of these problems with examples is provided in this section.

1.2.1 Contaminated Site Remediation

Until the early 1970s, environmental laws and regulations did not exist; therefore, chemicals were used and wastes were disposed of without proper consideration of potential impacts on public health and the environment. As a result, numerous sites have been contaminated by toxic chemicals. The U.S. Environmental Protection Agency (USEPA) estimated in 1997 that over 200,000 contaminated sites, where toxic chemicals pose unacceptable risk to public health and the environment exist in the United States.

The Wide Beach development site in Brant, New York, a 55-acre lakeside community development, is an example of contamination resulting from improper use of chemicals (USEPA, 1998b). From 1964 until 1978, waste oil containing polychlorinated biphenyls (PCBs) was applied to roadways in the community to control dust. In 1980, during the installation of a 1-mile sanitary sewer trench, soil from the roadways was excavated and used as fill in several residential yards. An odor complaint in the community led the regulators to discover drums containing waste oil with PCBs. Further investigation revealed that PCBs were present in soils from roadways and residential yards, in vacuum cleaner dust from residential homes, and in water from residential wells. An extensive remedial action, including the removal of drums and treatment of contaminated soils, cost more than $15 million.

The highly publicized Love Canal site, located in Love Canal near Niagara Falls in New York, demonstrated the consequences of improper disposal of wastes (*www.usepa.gov*). Between 1942 and 1953, over 20,000 tons of chemical waste was disposed of in an abandoned canal by a chemical manufacturing company. In 1953, the Niagara school board bought the property, despite a warning regarding the chemical waste present. A school was built and opened in 1955, with some buildings atop the waste-filled canal. By 1972, several homes with basements were built surrounding the school. In 1976, heavy rainfall in the area caused groundwater to rise. This rise caused subsidence of the waste-fill area, resulting in contamination of surface water. In addition, seepage of groundwater transported toxic chemicals into the basements of the surrounding homes. When children in the area fell sick in 1977 and 1978, the contamination was discovered. The Love Canal site was then evacuated and a state of emergency was declared. This incident drew national attention, initiating many environmental laws and regulations. Extensive remedial action at this site has been undertaken, at a total cost exceeding $200 million.

In addition to site contamination resulting from improper chemical use or waste disposal practices, spillage of toxic chemicals during handling, transportation, and storage have polluted soils and groundwater. Leaking underground storage tanks that contain petroleum products and other toxic chemicals are common occurrences. In 1995, the USEPA estimated that there are over 400,000 sites where soils and groundwater have been contaminated by leaking underground storage

tanks. An example of such sites is the Fairfield Semiconductor Corporation site in San Jose, California, which operated from 1977 until its closure in1983 (USEPA, 1998c). In 1981, an underground storage tank containing organic solvents failed, resulting in both soil and groundwater contamination by a mixture of solvents. An estimated 60,000 gal of waste solvents was released. Extensive remedial action was necessary, including removal of tanks, installation of slurry walls around the site perimeter to contain contaminated groundwater, extraction of contaminated groundwater, and treatment of contaminated soils. The remedial cost was over $4 million.

The role of geoenvironmental engineers in the remediation of contaminated sites, especially in dealing with contaminated soil and groundwater, is critical. Knowledge of soil composition, soil stratigraphy, groundwater hydraulics, and geochemistry can be applied to assess, develop, and implement effective remedial methods. In particular, geoenvironmental engineers have the capacity to lead subsurface investigations for the design of in-situ remedial strategies.

1.2.2 Waste Containment

Wastes are created that require disposal despite the best waste management practices. Such wastes include household garbage, mine refuse, highly toxic industrial by-products, and nuclear wastes. Proper disposal of these wastes in engineered waste containment facilities is crucial to protect public health and the environment.

Containment facilities for liquid wastes are known as *surface impoundments,* or more commonly, *lagoons* and *ponds.* These impoundments have to be lined properly at the bottom to prevent infiltration into the subsurface of their chemical constituents. In the past, because regulations on such linings did not exist, linings were not provided, resulting in several contaminated sites. An example of such a site is the Anderson Development Company (ADC) site in Adrian, Michigan (USEPA, 1998a). The ADC occupied approximately 12.5 acres of land, surrounded by residential areas. Between 1970 and 1979, the site was used for the manufacture of 4,4-methyl-bis(2-chloroamile) (MBOCA), a hardening agent used in plastics manufacturing. Process wastewaters were discharged to an unlined 0.5-acre lagoon. Later, contamination was found in the soils surrounding the lagoon. Because of the potential health hazard, the lagoon was closed and the contaminated soil was treated. Approximately $2.7 million was spent for completion of this project.

Containment facilities for solid wastes are known as *landfills*. In the early 1970s, wastes were disposed of in open ditches or pits, or piled above the ground surface. With no lining at the bottom, such disposal practices led to several contaminated sites. For instance, at a southern Illinois coal mine, liquid waste impounds from coal processing were created within massive surface piles of coarse tailings (Reddy and Schuh, 1994). Waste constituents from the piles and impoundments have infiltrated the subsurface, causing groundwater pollution. A community in close proximity to the site drew the groundwater for drinking purposes. Because of the public health concerns, control of contamination at the source and the cleanup of groundwater became necessary, and remedial action, which includes closing the piles and impoundments and implementing remedial action, is expected to cost over $1 million.

Several Army bases also reportedly contain numerous unlined pits where toxic chemical wastes have been dumped, and soil and groundwater contamination has occurred as a result. For instance, at McClellan Air Force Base (AFB) near Sacramento, California, fuel and solvents were disposed of in several pits from the early 1940s until the mid-1970s (USEPA, 1998a,c). Later, an impermeable cap was constructed over the pits to reduce rainwater infiltration and subsequent leaching of contaminants into the groundwater. However, contaminants seeped out of the pits, contaminating the soil as well as the groundwater more than 100 ft below the ground surface. Corrective action involving groundwater treatment and soil remediation cost exceeded $3.8 million.

Newly generated liquid and solid wastes are required to be disposed of in engineered impoundments and landfills. All these containment facilities require liner systems that perform as both hydraulic and chemical barriers. In addition, these facilities must be located where hydrogeologic conditions are favorable.

Upon reaching their waste storage capacity, the containment facilities should be covered properly to isolate the waste and to prevent infiltration of precipitation. The mechanical stability of liner and cover systems should also be ensured. The role of geoenvironmental engineers is crucial in the selection of a suitable site for locating waste containment facilities as well as for the design and construction of liner and cover systems for effective containment of wastes.

1.2.3 Waste Minimization by Recycling

During the past few years, intense efforts have been made to prevent or at least minimize generation of wastes through the use of environmentally benign materials and innovative manufacturing processes. Recycling efforts have been increased to divert wastes from disposal in containment facilities. The recycled materials include domestic wastes such as paper, plastics, glass, and tires, as well as industrial wastes such as fly ash and bottom ash. Recycling efforts have been fairly successful, and large quantities of materials have been collected. However, except for paper, markets that can utilize such materials are limited or can consume only minimal quantities. Therefore, new applications that have potential to consume large quantities of recycled materials are urgently sought. Large-scale beneficial use of recycled and waste materials is possible for civil engineering applications. In addition, in several cases, closed waste containment sites have been used for recreational, industrial, and commercial purposes, such as parks, golf courses, and buildings.

Geoenvironmental engineers are involved in evaluating the feasibility of using recycled materials as soil substitutes in various large-scale civil engineering applications, including roadways, embankments, and retaining structures. In addition, large-scale environmental engineering use of recycled and waste materials, including soil substitutes in the design of waste containment systems and reactive media substitute in soil and groundwater remediation systems, is growing rapidly. These applications involve evaluation not only of their mechanical properties and behavior, but also environmental aspects, including chemical compatibility and durability.

1.3 BOOK ORGANIZATION

The book is divided into four parts:

- Part I: Basic Principles
- Part II: Remediation Technologies
- Part III: Landfills and Surface Impoundments
- Part IV: Emerging Technologies

Background on environmental regulations, inorganic and organic chemistry, soil composition and properties, geochemistry, groundwater, and contaminant fate and transport, all essential to an understanding of geoenvironmental problems, are covered in Part I. In Part II, assessment and remediation of contaminated sites are presented, including a general overview of sources of subsurface contamination, methods to characterize the contaminated sites, methods to quantify the risk posed by the site contamination, and various remedial methods: in-situ barriers, soil remediation technologies, and groundwater remediation technologies. In Part III, waste characterization and waste containment systems are described. Various sources and types of wastes and their properties are described first. Siting, permitting and design of landfills are then described. Finally, design of surface impoundments is discussed. In Part IV, several different emerging waste management technologies are described, including waste material recycling, end use of closed landfills, bioreactor landfills, and subaqueous sediment containment.

1.4 SUMMARY

Originally, geotechnical engineering was a loosely designed discipline focused primarily on the use of earth materials and the roles they play in construction. Since the 1950s, however, geotechnical engineering has undergone a tremendous amount of growth. A shift in job involvement began to occur in the 1970s as a series of high-profile environmental disasters occurred, leading to increased public awareness. As environmental problems have become more prevalent, geotechnical engineers have played an increasing role in finding solutions. Geoenvironmental engineering has evolved by utilizing combined geotechnical and environmental engineering technologies to address the problems of contaminated sites as well as hazardous and nonhazardous waste management. This new field of engineering is evolving rapidly, due to changed regulations and advanced knowledge through research and innovative practice methods. In this book we present the current state of practice in geoenvironmental engineering.

QUESTIONS/PROBLEMS

1.1. Name and briefly discuss three notable incidents that have had an adverse impact on the environment in the past 50 years.

1.2. How do geotechnical engineering and geoenvironmental engineering differ?

1.3. Discuss briefly how accidental releases of hazardous chemicals into the subsurface would affect the environment. How may an engineer prevent releases by containment facilities?

1.4. Why are geoenvironmental engineers uniquely qualified to address subsurface contamination problems?

1.5. Discuss how recycled materials may be used in engineering projects. Discuss why it would be important to analyze the engineering properties of such materials.

1.6. A former unlined landfill that accepted municipal and industrial waste is located 500 ft from a drinking water well in a rural town in Illinois. Investigation indicated the presence of benzene and lead in groundwater immediately downgrade of the landfill, and the groundwater plume is found to migrate in the direction of the drinking water well in the town. A potential remedial strategy considered in this situation includes (1) capping the landfill, (2) installing a leachate collection system, (3) constructing an impermeable slurry wall, (4) installing a groundwater extraction well system, (5) air stripping of extracted leachate and groundwater, and (6) discharge of treated leachate and groundwater to nearby streams. Identify various geoenvironmental engineering issues and/or tasks involved in this project. Explain them briefly.

REFERENCES

Daniel, D. E., *Geotechnical Practice for Waste Disposal,* Chapman & Hall, London, 1993.

Lambe, T. W., and Whitman, R. V., *Soil Mechanics,* Wiley, New York, 1969.

Peck, R. B., Hanson, W. E., and Thornburn, T. H., *Foundation Engineering,* 2nd ed., Wiley, New York, 1974.

Reddy, K. R., and Schuh, J. C., Groundwater contamination due to seepage from surface impoundments, pp. 319–329 in *Proceedings of the 4th Annual WERC Technology Development Conference,* Las Cruces, NM, 1994.

Taylor, D. W., *Fundamentals of Soil Mechanics,* Wiley, New York, 1948.

Terzaghi, K., and Peck, R. B., *Soil Mechanics in Engineering Practice,* Wiley, New York, 1948.

USEPA (U.S. Environmental Protection Agency), *Remediation Case Studies: In-Situ Soil Treatment Technologies (Soil Vapor Extraction, Thermal Processes),* EPA/542/R-98/012, Federal Remediation Technologies Roundtable, USEPA, Washington, DC, Vol. 8, 1998a.

———, *Remediation Case Studies: Groundwater Pump and Treat (Chlorinated Solvents),* EPA/542/R-98/013, Federal Remediation Technologies Roundtable, USEPA, Washington, DC, Vol. 9, 1998b.

———, *Remediation Case Studies: Groundwater Pump and Treat (Nonchlorinated Contaminants),* EPA/542/R-98/014, Federal Remediation Technologies Roundtable, USEPA, Washington, DC, Vol. 8, 1998c.

2

RELEVANT ENVIRONMENTAL LAWS AND REGULATIONS

2.1 INTRODUCTION

Increased public concern for the protection of human health and the environment has led to the passage of numerous environmental laws and regulations. The purpose of these laws and regulations is to protect human health and the environment without putting an unreasonable financial burden on industry or on taxpayers. The federal, state, and local governments all play a part in the development of these regulations. The federal government sets certain conditions, and state and local agencies may either adopt the federal laws and regulations, or promulgate more stringent laws and regulations. Under no circumstances, however, are state and local agencies allowed to impose laws and regulations that contradict federal laws and regulations.

The impact of environmental laws and regulations on the geoenvironmental engineering practice cannot be understated. Geoenvironmental studies dealing with the characterization and remediation of contaminated sites as well as the design, construction, and monitoring of waste containment facilities (landfills and impoundments) must conform to the procedures and minimum design criteria set forth in environmental laws and regulations. Environmental laws and regulations have actually made it necessary to advance the field of geoenvironmental engineering through research and the development of innovative practices.

In this chapter, the general process followed to develop environmental legislation within the United States is presented first. The evolution of major federal environmental laws and regulations is then described.

Finally, specific regulatory requirements dealing with various geoenvironmental aspects are outlined.

2.2 DEVELOPMENT OF LAWS AND REGULATIONS

Federal environmental *laws* are passed by the U.S. Congress. They define goals in general terms but do not provide technical details on how to accomplish the goals. The creation of a law involves the following three steps:

1. A member of Congress proposes a bill.
2. If both houses of Congress approve a bill, it goes to the President, who has the option to either approve it or veto it. If approved, the bill is called a law or *act,* and the text of the act is known as a *public statute.*
3. Once an act is passed, the House of Representatives standardizes the text of the law and publishes it in the *United States Code* (USC).

Congress authorizes certain government agencies, primarily the U.S. Environmental Protection Agency (USEPA), to develop full details, including technical details, on how to achieve the goals stated in environmental laws. These details, known as *regulations,* become legal requirements in compliance with the law. The following procedure is followed in the development of regulations:

1. The USEPA decides that a regulation may be needed, does the research, and if necessary, proposes a regulation. The proposal is listed in the

Federal Register for a specified period to allow the opportunity for public comment.

2. The USEPA may then revise the proposal based on these comments, and publish a legally binding regulation.

3. The regulation is then published in the *Code of Federal Regulations* (CFR). The CFR is the official record of all regulations created by the federal government. It is divided into 50 volumes called *titles,* each title focusing on a particular area. All environmental regulations are printed in *Title 40.* Further, the regulations are subdivided as *parts* and *sections.* For example, 40 CFR 264.300 refers to Title 40, Part 264, Section 300. The CFR is revised and/or printed yearly.

Since environmental regulations are as binding as the law, practitioners make no distinction between the law and the regulations. Together, they are just referred to either as laws or regulations. The USEPA develops *technical guidance documents,* which provide technical details to help practitioners conform to the regulations. The USEPA may also issue *policies,* which serve as decision-making guidance to meeting regulatory or environmental objectives.

Many states and local governments have adopted federal laws and regulations directly, and some states and local agencies have enacted laws and regulations that are *stricter* than federal laws and regulations. Some states have also promulgated additional laws and regulations to address problems specific to a particular state. The procedures followed in the development of state and local laws and regulations are similar to those of federal laws and regulations. The legislative branch passes the laws, and a state agency (generally, the state EPA) both prepares the regulations and enforces them. The laws and regulations are then published and made available to the public.

2.3 FEDERAL ENVIRONMENTAL LAWS AND REGULATIONS

Numerous federal environmental laws (or acts) and regulations have been promulgated to protect human health and the environment. Table 2.1 provides a chronology of major environmental laws and regulations. Complete authoritative information on these environmental laws and regulations may be found in the *United States Code Database* (1994) and the *Code of Federal Regulations* (CFR, 2002). Several other references also describe environmental laws and regulations in detail (e.g., USEPA, 1990; Wagner, 1991; Tchobanoglous et al., 1993; LaGrega et al., 1994; Sharma and Lewis, 1994; Wentz, 1995). An overview of these laws and regulations is given in the following sections.

2.3.1 Rivers and Harbors Act (RHA) (1899)

The Rivers and Harbors Act (RHA) prohibited the placement of any objects into waterways that may create a hazard to navigation. This was the only law with an impact on the environment before the 1950s. The main provisions of this law included the following:

- Construction of a wharf, pier, breakwater, jetty, or similar structure was illegal without federal permission.
- Federal approval was required to alter the course, location, condition, or capacity of a waterway.
- A permit was required for dredge and fill projects and similar construction activities in waterways.

2.3.2 Atomic Energy Act (AEA) (1946, 1954)

The Atomic Energy Act (AEA) regulates the atomic energy industry. Listed below are the main features of this act.

- The Atomic Energy Commission (AEC) was created to regulate the atomic energy industry.
- Civilian participation in research and development of nuclear power was allowed.
- Regulations on mining and milling of uranium, construction of nuclear power plants, and medical and scientific research were created.
- Through amendments, the AEC was divided into two agencies: (1) the Nuclear Regulatory Com-

Table 2.1 **Development of major federal environmental laws and regulations**

Year	Title	Purpose
1899	River and Harbors Act	Prohibits the disposal of solid objects and placement of any objects in waterways that would create a hazard to navigation.
1946, 1954	Atomic Energy Act	Regulates the atomic industry.
1965	Solid Waste Disposal Act	Reduces the volume of waste.
1969	National Environment Policy Act	Creates a council on environmental quality. This law requires environmental impact statements (EISs) for any federal project that affected the environment.
1970	Occupational Safety and Health Act	Ensures the health and safety of employees in the workplace.
1972	Marine Protection, Research and Sanctuaries Act	Prevents or limits ocean dumping of any wastes that would adversely affect human health or the marine environment.
1970, 1977, 1990	Clean Air Act	Ensures that emissions of hazardous air pollutants from both existing and new emission sources are controlled to the maximum extent possible.
1977, 1981, 1987	Clean Water Act	Ensures the protection of U.S. waters by regulating point-source discharges to surface waters within the United States.
1974, 1977, 1986	Safe Drinking Water Act	Requires the USEPA to set national primary drinking water standards.
1976	Toxic Substances Control Act	Controls the manufacture of toxic materials required for industrial processes.
1976, 1980	Resource Conservation and Recovery Act	Manages nonhazardous and hazardous wastes and underground storage tanks, with emphasis on recovery of reusable materials, as opposed to disposal.
1980	Comprehensive Environmental Response, Compensation and Liabilities Act (Superfund)	Ensures immediate cleanup of abandoned hazardous waste sites.
1972, 1982, 1988	Federal Insecticide, Fungicide and Rodenticide Act	Regulates the storage and disposal of pesticides, requiring informative and accurate labeling.
1984	Hazardous and Solids Waste Amendments	Includes six amendments to the RCRA passed in 1976: restrictions on liquid wastes; new requirements for waste management and treatment; new regulations on USTs; new technology standards; specific requirements for treatment, storage, or disposal facilities; USEPA responsibility for inspection and enforcement.
1986	Superfund Amendments	Ensures further hazardous site cleanup and establishes requirements for current landowners to be responsible for cleaning the site.
1990	Pollution Prevention Act	Prevents the production of wastes.
2002	Small Business Liability Relief and Brownfields Revitalization Act	Provides relief from liability for small businesses, promotes cleanup of brownfields, and provides financial assistance.

Source: Tchobanoglous et al. (1993); Sharma and Lewis (1994); Wentz (1995).

mission (NRC) and (2) the Department of Energy (DOE). The NRC is responsible for enforcing laws related to civilian uses of nuclear power, mainly nuclear power plants. The DOE is responsible for the nuclear weapons program. The DOE is also responsible for remediation of sites that are contaminated with nuclear wastes that result from uranium mining and processing operations.

2.3.3 Solid Waste Disposal Act (SWDA) (1965, 1970)

The Solid Waste Disposal Act (SWDA) was the first federal legislation aimed at regulating municipal solid waste. The main provisions of this law are:

- Emphasis on reducing the volume of solid waste to protect human health and the environment
- Emphasis on the improvement of solid waste disposal practices
- Providing funds to the states to better manage their solid wastes

This law was amended in 1970 and is now known as the Resource Recovery Act. This amendment encouraged waste reduction and resource recovery and created a system of national disposal sites for hazardous wastes.

2.3.4 National Environmental Policy Act (NEPA) (1969)

The National Environmental Policy Act (NEPA) was the major environmental legislation representing the nation's commitment to protect and maintain environmental quality. The provisions of this law included:

- The creation of the Council on Environmental Quality (CEQ), a new executive branch agency. Subsequently, the Environmental Protection Agency was created by presidential action.
- Requirement of an environmental impact statement (EIS) for any federal project that might have

a significant effect on the environment. The EIS is required to address:

- The potential impact on the environment of the proposed project
- Alternatives to the proposed action
- Short-term usefulness and long-term productivity of the project
- Resources that would be depleted irreversibly

Unfortunately, the NEPA provisions were used by various parties for numerous lawsuits. Individuals, environmental groups, state governments, and business groups filed numerous environmental lawsuits against various federal agencies for failing to prepare an EIS or for preparing an inadequate EIS. A total of 146 and 94 lawsuits were filed in 1983 and 1990, respectively (Wentz, 1995).

2.3.5 Occupational Safety and Health Act (OSHA) (1970)

The Occupational Safety and Health Act (29 U.S.C. 651) was passed to ensure worker and workplace safety. The goal is to make sure that employers provide their workers a place of employment free of recognized safety and health hazards, including exposure to toxic chemicals. The National Institute for Occupational Safety and Health (NIOSH) was established as the research institution to establish standards for workplace health and safety; the Occupational Safety and Health Administration (OSHA) was established to enforce the standards of the act. As a result of this act, the following regulations concerning waste sites were developed (29 CFR 1910.120):

- Stringent health and safety programs are required at all hazardous waste sites. Included are (1) a site-specific health and safety plan, (2) a medical surveillance program, (3) continuous air monitoring, (4) 40-hour health and safety training for workers, and (5) a decontamination program.
- All employers are required to retain health records of employees working on hazardous waste remediation projects.

- Four levels of personal protection are specified for workers:
- Level A, the highest level of protection, requires workers to wear fully enclosed positive-pressure suits and to breathe supplied air.
- Level B requires workers to wear protective clothing such as coveralls, gloves, hard hats, and steel-toed boots and to breathe supplied air.
- Level C is similar to level B, except that workers wear air-purifying respirators rather than breathing supplied air.
- Level D, the lowest level of protection, does not require workers to use respiratory protection but simply to wear the protective clothing described in level B.

2.3.6 Marine Protection, Research and Sanctuaries Act (MPRSA) (1972)

The Marine Protection, Research and Sanctuaries Act (MPRSA) was passed to prevent or limit ocean dumping of any wastes that would adversely affect human health or the marine environment. The primary provisions of this law include the following:

- The sources of waste regulated under this act include runoff from land via rivers and streams, atmospheric fallout, point-source discharges, dumping of dredged materials, operational discharges from ships and offshore platforms, and accidental spills.
- Certain types of wastes are prohibited from ocean dumping, including high-level radioactive wastes, biological, chemical, or radiological warfare materials; incompletely characterized materials; and persistent inert materials that float or remain suspended in the ocean.
- A permit is required for the transportation of all wastes that are to be dumped at sea.
- States are not permitted to enact their own regulations regarding marine environmental issues controlled under this law.

2.3.7 Federal Insecticide, Fungicide and Rodenticide Act (FIFRA) (1972, 1982, 1988)

As a result of the extensive use of pesticides for agricultural purposes, groundwater contamination problems have become a major concern. The Federal Insecticide, Fungicide and Rodenticide Act (FIFRA) (7 U.S.C.s/s 136) was created to regulate the storage and disposal of pesticides. In accordance with this law:

- Informative and accurate labeling of all pesticides is required.
- Users (farmers, utility companies, and others) are required to register when purchasing pesticides and must take exams for certification as applicators of pesticides.
- All pesticides used must be registered by the USEPA to assure that they are labeled properly and will not cause significant harm to the environment.
- Tolerance levels are specified for certain pesticides to prevent unreasonable hazards.

The USEPA and U.S. Department of Agriculture have undertaken major collaborative efforts to further minimize the effects of pesticides on the environment.

2.3.8 Clean Air Act (CAA) (1970, 1977, 1990)

The Clean Air Act (42 U.S.C.s/s 7401) is a comprehensive law that regulates air emissions from area, stationary, and mobile sources. Some of the major provisions of this law are:

- The establishment of National Ambient Air Quality Standards (NAAQSs) for certain pollutants, known as *criteria pollutants*, which must not be exceeded anywhere in the United States (Table 2.2).
- Development of standards for other hazardous air pollutants, such as asbestos, beryllium, vinyl chloride, mercury, benzene, radionuclides, and arsenic, for which NAAQSs are not specified but which can contribute to or cause adverse effects on human health and the environment.

Table 2.2 **National ambient air quality standards**

Pollutant	Primary Standards (Health-Related)
Ozone	0.12 ppm (1-h average)
Carbon monoxide	9 ppm (8-h average)
	35 ppm (1-h average)
Particulate matter (PM_{10}^*)	150 $\mu g/m^3$ (24-h average)
	50 $\mu g/m^3$ (annual arithmetic mean)
Sulfur dioxide	0.140 ppm (24-h average)
	0.03 ppm (annual arithmetic mean)
Nitrogen dioxide	0.053 ppm (annual arithmetic mean)
Lead	1.5 $\mu g/m^3$ (arithmetic mean averaged over a calendar quarter)

PM_{10}^* is the particulate matter that contains particles with an aerodynamic diameter less than or equal to a nominal 10 micrometers.

Source: 40 CFR 51.151. (as of July, 2003)

- Division of the United States into regions known as *air quality control regions.* Regions meeting NAAQS are considered *in attainment,* and regions not meeting these standards are considered *in nonattainment.* Most regions in the United States are in attainment, except for ozone and carbon monoxide standards.
- The 1990 amendments mandate an extensive operating permit program for all major emission sources. These amendments are designed for ozone attainment through reduction in volatile organic compounds (VOCs) and oxides of nitrogen emissions, and for carbon monoxide (CO) attainment through tighter motor vehicle emissions.

2.3.9   Clean Water Act (CWA) (1977, 1981, 1987)

The Clean Water Act (CWA), known initially as the Federal Water Pollution Control Act (FWPCA), (33 U.S.C.s/s 1251), stipulates the basic structure for regulating the discharge of pollutants into U.S. waters. U.S. waters include navigable waters, their tributaries, adjacent wetlands, and other waters or wetlands where degradation or destruction could affect interstate or foreign commerce. Some major provisions of this law include:

- A total of 129 priority pollutants are identified as hazardous wastes.
- Wastewater can be discharged into surface water only if effluent standards based on the best available technology are met (40 CFR 125.3; 40 CFR 122.44).
- It is unlawful for any person to discharge any pollutant from a point source into navigable waters unless a National Pollutant Discharge Elimination System (NPDES) permit is obtained.
- Discharge of dredged and fill materials into U.S. waters, including wetlands, is allowed only if a permit is obtained (33 CFR 323).
- Discharges from publicly owned treatment works (POTWs) must meet pretreatment standards.

The U.S. Congress is currently discussing amending of the Clean Water Act's effluent limitation guidelines, such as control on toxic discharges, limits on non-point-source pollution, and restrictions on development at wetland sites.

2.3.10   Safe Drinking Water Act (SDWA) (1974, 1977, 1986)

The Safe Drinking Water Act (42 U.S.C.s/s 300f) was passed to protect the quality of drinking water in the United States. This law focuses on all waters actually or potentially designated for drinking use, and whether their sources are above or below ground. The main provisions of this law are:

- Establishment of drinking water standards that consist of (1) maximum contaminant level goals (MCLGs), (2) maximum contaminant levels (MCLs), and (3) secondary maximum contaminant levels (SMCLs). MCLGs are the concentration levels at which no known or anticipated adverse health effects are predicted. A list of MCLGs is given in 40 CFR 141. MCLs are the primary drinking water standards that must be met by all water suppliers. A list of MCLs, as given in 40 CFR 141, is provided in Table 2.3. SMCLs include the compounds listed in 40 CFR 143 that

Table 2.3 **Maximum contaminant levels (MCLs) for drinking water**

Chemical	MCL (µg/L)	Chemical	MCL (µg/L)
Organic Chemicals			
Adipates [di(2-ethylhexyl)adipate]	400	Glyphosate	700
Alachlor	2	Heptachlor	0.4
Aldicarb	3	Heptachlor epoxide	0.2
Aldicarb sulfoxide	4	Hexachlorobenzene	1
Aldicarb sulfone	2	Hexachlorocyclopentadiene (HEX)	50
Atrazine	3	Lindane	0.2
Benzene	5	Methoxychlor	40
Benzo[a]pyrene	0.2	Monochlorobenzene	100
Carbonfuran	40	Oxamyl (vydate)	200
Chlorodane	2	Polychlorinated biphenyls PCBs	0.5
Dalapon	200	Pentachlorophenol	1
Dibromochloropropane (DBCP)		Picloram	500
Dichloromethane 5 µg/L	0.2	Simazine	4
o-Dichlorobenzene	600	Styrene	100
p-Dichlorobenzene	75	2,3,7,8-TCDD (dioxin)	3×10^{-5}
1,2-Dichloroethane	5	Tetrachloroethylene	5
1,1-Dichloroethylene	7	1,2,4-Trichlorobenzene	70
cis-1,2-Dichloroethylene	70	1,1,1-Trichloroethane	200
trans-1,2-Dichloroethylene	100	1,1,2-Trichloroethane	5
1,2-Dichloropropane	5	Trichloroethylene (TCE)	5
2,4-Dichlorophenoxyacetic acid (2,4-D)	70	2-(2,4,5-Trichlorophenoxy)propionic acid (2,4,5-TP, or silvex)	50
Di(2-ethylhexyl)phthalate	6		
Dinoseb	7	Toluene	1,000
Diquat	20	Toxaphene	3
Endothall	100	Vinyl chloride	2
Endrin	2	Xylenes (total)	10,000
Ethylbenzene	700		
Ethylene dibromide (EDB)	0.05		
Inorganic Chemicals			
Antimony	6	Cyanide	200
Arsenic	10	Fluoride	4000
Asbestos (fibers per liter)	7×10^{6}	Mercury	2
Barium	2,000	Nitrate (as N)	10,000
Beryllium	4	Nitrite (as N)	1,000
Cadmium	5	Selenium	50
Chromium	100	Thallium	2
Radionuclides			
Beta particle and photon radioactivity	4 millirem/yr (mrem/yr)	Combined radium −226 and −228	5 pCi/L

Source: 40 CFR 141. (as of July, 2002)

are not a health hazard but cause an offensive taste, color, or odor in drinking water.

- Protection of underground water supplies through regulation of the injection of hazardous wastes into injection wells (40 CFR 144 through 148).
- An aquifer may be designated as a *sole source aquifer* if it is the sole or principal drinking water source for an area. This designation prevents activities that could contaminate the aquifer.

2.3.11   Toxic Substances Control Act (TSCA) (1976)

The Toxic Substance Control Act (TSCA) (15 U.S.C.s/ s 2601) was enacted to regulate the introduction and use of new hazardous chemicals. Provisions in this law include:

- Requirement of industries to report or test chemicals that may pose an environmental or human-health hazard.
- Banning of the manufacture and import of chemicals that pose an unreasonable risk.
- Requirement of premanufacture notification by companies to the USEPA 90 days prior to production.
- Banning of polychlorinated biphenyls (PCBs). All storage or disposal of PCB-contaminated products and wastes must be performed in accordance with the USEPA regulations developed under this law.
- The management of asbestos is also regulated under this law.

2.3.12   Resource Conservation and Recovery Act (RCRA) (1976, 1980)

The Solid Waste Disposal Act of 1965 regulated municipal solid waste. Other laws, such as the Clean Air Act and Clean Water Act, regulated disposal of hazardous waste into the atmosphere or into surface waters. To circumvent these regulations, disposal of hazardous waste in roadside ditches, vacant lots, abandoned industrial sites, and other inappropriate lo-

cations became common occurrences. Later, land disposal and deep well injection became common practices. All of these practices still led to the pollution of air, water, and land. In response to these problems, the Resource Conservation and Recovery Act (42 U.S.C.s/ s 6901) was passed to manage nonhazardous and hazardous wastes and underground storage tanks, with an emphasis placed on the recovery of reusable materials (recycling) as an alternative to their disposal.

As a result of this law, comprehensive regulations were developed by the USEPA, and their main features included the following:

- *Subtitle C regulations* were developed to control hazardous waste from "cradle to grave." These regulations provide criteria for defining hazardous waste, generators' responsibilities, transporters' requirements, manifest systems, and treatment, storage, and disposal facility (TSDF) requirements (40 CFR 264).
- *Subtitle D regulations* were set forth for the management of nonhazardous wastes (40 CFR 257 and 258).
- *Subtitle I regulations* were set forth to address environmental problems that could result from underground tanks storing petroleum and other hazardous substances (40 CFR 264.190 through 264.200).

2.3.13   Hazardous and Solid Waste Amendments (HSWA) (1984)

The Congress passed several amendments to the RCRA to protect groundwater. These amendments, known as the Hazardous and Solid Waste Amendments (HSWA), include the following:

- Restrictions were imposed on the land disposal of untreated liquid and solid hazardous wastes. Initially, free liquids were banned. However, disposal of containerized liquids is now permitted but only if USEPA determines that no reasonable alternatives are available and that disposal is environmentally acceptable. HSWA also prohibits land disposal of specific chemicals (e.g., halogenated organic compounds with concentrations greater than 1000 mg/kg).

- New requirements were developed for the management and treatment of small quantities of waste (auto repair shops, dry cleaners, etc.). Under the RCRA, a small-quantity generator was defined in terms of 1000 kg/month; however, in HSWA, the definition was changed to less than 100 kg/month. Small-quantity generators are exempted from manifest systems and are allowed to dispose of hazardous wastes in landfills that are permitted to accept them.

- New regulations on underground storage tanks located in urban areas were developed. These regulations require tanks to have leak detection systems or inventory control with regular tank testing. The amendments also state that owners are responsible for record keeping and are liable for any third-party damage due to leakage.

- New technology standards for land disposal facilities (e.g., new liner systems) were developed. These regulations require double liners, leachate collection systems, groundwater monitoring systems, and leak detection systems. Removal efficiencies for incinerators are also specified.

- Specific requirements are imposed for treatment, storage, or disposal facilities (TSDF), including corrective action procedures, plans for accidental spills, bans on disposal, and five-year permit reviews. These regulations are specifically applicable to solid waste management units (SWMUs or "Smoos") that are currently inactive but were used formally as hazardous waste disposal sites located within an RCRA facility.

- The USEPA is authorized to inspect and enforce the regulations. Penalties are established for violating the regulations.

2.3.14 Comprehensive Environmental Response, Compensation and Liabilities Act (CERCLA) or Superfund (1980)

The RCRA and HSWA addressed newly generated hazardous and nonhazardous wastes but failed to address serious historical pollution problems such as the Love Canal case noted in Chapter 1. The Comprehensive Environmental Response, Compensation and Liabilities Act (CERCLA), also commonly known as the Superfund (42 U.S.C.s/s 9601), was promulgated to address abandoned or uncontrolled hazardous waste sites, which required immediate cleanup to ensure protection of human health and the environment. The main features of this law include the following:

- A $1.6 billion fund was created based on taxes levied on the chemical and petroleum industries and was made available to clean up hazardous waste sites. After the cleanup had begun, additional funds were used for litigation to recover costs from potentially responsible parties (PRPs). PRPs include present and past owners of a site, operators at the time of disposal, generators, and transporters.

- A hazard ranking system (HRS) was established to rank sites and prioritize cleanup. This system takes into account the degree of risk to human health and the environment posed by each suspect site, and also considers population, contaminants, and potential pathways. The National Priorities List (NPL) is prepared based on hazard ranking scores (HRSs) of these sites.

- A systematic remedial response is required, which includes remedial investigation (RI) and a feasibility study (FS). RI provides a description of the current situation and analysis of air, surface water, groundwater, and other wastes. The FS provides an evaluation of alternative remedial solutions based on the following criteria: short- and long-term effectiveness; implementability; reduction in toxicity, mobility, and volume; cost-effectiveness; compliance with standards; human health protection; and concurrence with state and local requirements.

The USEPA has identified 36,000 potential Superfund sites, and as of 1993, 1200 sites were on the NPL, with about 100 sites added each year. Cleanups, averaging 11 years and $25 million per site, have been completed at 183 NPL sites and are in progress at over 1000 others (USEPA, 1997).

2.3.15 Superfund Amendments and Reauthorization Act (SARA) (1986)

Contrary to predictions under CERCLA, not all contaminated sites were cleaned up within five years. In 1986, there was considerable debate in the Congress regarding reauthorization of the Superfund. The debate resulted in the Superfund Amendments and Reauthorization Act (SARA) (42 U.S.C. 9601). The main provisions of this law are:

- The creation of an $8.5 billion fund for the remediation of abandoned sites and a $500 million fund for cleanup of leaking underground storage tank (LUST) sites.
- In view of the Union Carbide–Bhopal tragedy in India, community right-to-know provisions were included. The Union Carbide–Bhopal tragedy involved the release of 45 tons of methyl isocyanate at the Union Carbide pesticide plant in Bhopal, India, that killed over 3000 people and injured over 200,000 people in, and in the vicinity of, the plant. The right-to-know provisions require the release of detailed emergency plans and information to the public regarding hazardous substances at a particular site.
- The determination of cleanup standards was a major problem in CERCLA. SARA specifies that remediation must meet applicable or relevant and appropriate requirements (ARARs), which were adopted from other regulations, such as RCRA. These specific requirements are based on chemical-specific, action-specific, and location-specific criteria, or alternatively, on quantitative risk assessment. As a result, harsh cleanup liability was created and two other important requirements were included: liability for innocent purchases, and the disclosure of the annual release of hazardous substances.
- Under this law, current landowners became liable for cleanup; therefore, in real estate transactions

banks and lending institutions were forced to be concerned about contamination and any ramifications. This situation has led to environmental audits, known as *environmental site assessments*, which are performed in three phases:

- *Phase I* includes a visual inspection of property and a review of the following information: (a) prior ownership and use, (b) agency records and known contamination sites, (c) aerial photographs, and (d) interviews with neighbors and/or past employees. This phase determines whether or not the purchaser has reason to believe that contamination could exist at the site. If problems are detected in phase I, phase II is performed, with the purpose of determining the extent of the problem.
- *Phase II* involves soil testing and groundwater monitoring to determine the magnitude of soil and groundwater contamination.
- *Phase III* involves remediation effort. Generally, closure of a real estate deal becomes difficult when soil and groundwater contamination is detected. Buyers and lenders require full cleanup and certification of closure.

The Superfund reauthorization expired at the end of fiscal year 1994 (September 30, 1994). Despite significant debate in the Congress, it has not yet been reauthorized as of 2003, except for special activities. Major issues being resolved include risk assessment, cleanup standards, innovative technologies, and the "polluter pays" system. Emphasis on actual remediation rather than litigation will be expected.

2.3.16 Pollution Prevention Act (PPA) (1990)

The Pollution Prevention Act (PPA) (42 U.S.C. 13101 and 13102), promulgated to prevent and/or minimize the production of wastes, included the following features:

- A pollution prevention office was created within the USEPA.
- The act focused industry, government, and public attention on reducing the amount of pollution

through cost-effective changes in production, operation, and raw material use.

- Pollution prevention includes other practices that increase efficiency in the use of energy, water, or other natural resources, and protect our resource base through conservation.
- Practices include recycling, source reduction, and sustainable agriculture.

Opportunities for source reduction are often not realized because of existing regulations and because the industrial resources required for compliance focus on treatment and disposal. Source reduction is fundamentally different and more desirable than waste management or pollution control.

2.3.17 Small Business Liability Relief and Brownfields Revitalization Act (SBLR&BRA) (2002)

The Small Business Liability Relief and Brownfields Revitalization Act (SBLR&BRA) (P.L. 107-118) aims to accomplish the following:

- Providing certain relief for small businesses from liability imposed under the CERCLA
- Promoting the cleanup and reuse of brownfields
- Providing financial assistance for brownfields revitalization
- Enhancing state response programs

According to this act, the term *brownfield site* means "real property, the expansion, redevelopment, or reuse of which may be complicated by the presence or potential presence of hazardous substance, pollutant, or contaminant." Under CERCLA, owners and operators of a contaminated property can be held liable for the cost of cleanup, regardless of whether they actually caused any of the contamination. This potential liability creates a strong incentive for businesses to avoid redeveloping brownfield sites. The SBLR&BRA provides liability protection for prospective purchasers, contiguous property owners, and innocent landowners, and authorizes increased funding for state and local programs that assess and clean up brownfields. This

law provides relief from liability under CERCLA for small business owners who sent small amounts of waste, protecting innocent small businesses while ensuring that polluted sites continues to be cleaned up by those most responsible for the contamination.

2.4 STATE AND LOCAL LAWS AND REGULATIONS

As stated earlier, state and local agencies may adopt federal laws and regulations or may develop more stringent laws and regulations. Under no circumstances may state and local agencies develop laws and regulations that contradict the federal laws and regulations. As an example, Table 2.4 shows major environmental laws and regulations that have been developed in the state of Illinois. Similar information may be obtained for other states by contacting the state's environmental protection agency or department of environment. Local regulations or requirements should not be taken for granted. For example, landfill siting cannot proceed without the approval of local governments, such as a city council or county board of directors.

2.5 IMPACT OF REGULATIONS ON GEOENVIRONMENTAL PRACTICE

It is impossible to think of a geoenvironmental project that does not involve a careful consideration of environmental laws and regulations. All environmental laws and regulations influence the management of hazardous and nonhazardous wastes. Of these, the environmental laws and regulations that have profound effects on geoenvironmental practice are the RCRA, HSWA, CERCLA, SARA, PPA, and SBLR&BRA. Table 2.5 shows some of the regulatory requirements that must be addressed for various geoenvironmental problems. These regulatory requirements are discussed in more detail in later chapters of the book. The characterization and remediation of contaminated sites is dictated significantly by the regulations developed under RCRA, HSWA, CERCLA, SARA, and SBLR&BRA. The design, construction, and monitoring of landfills and impoundments must conform to the minimum de-

Table 2.4 **Major environmental laws in Illinois**

Year	Title	Purpose
1970	Illinois Environmental Protection Act	Establishes a unified, statewide program supplemented by private remedies to restore, protect, and enhance the quality of the environment.
1971	Anti-Pollution Bond Act	Provides grant assistance to local communities for water projects.
1972	Federal Clean Water Act adopted	
1972	Illinois Asbestos Abatement Act	Controls the amount of asbestos fiber released into the environment from the major sources of emission.
1974	Federal Safe Drinking Water Act adopted	
1976	RCRA adopted	
1976	Illinois Uniform Hazardous Substance Act	Prohibits the disposal of hazardous substances.
1977	Environmental Training Resource Center	Provides training courses for wastewater and drinking water treatment plants.
1980	CERCLA (Superfund) adopted	
1980	Underground injection control program draft	Regulates disposal of liquid wastes into underground formations.
1980	Degradable Plastic Act	Evaluates the environmental impact of degradable plastics and develops and proposes ways to reduce the impact.
1981	Legislation to ban landfilling of liquid hazardous wastes	Prohibits landfilling of any hazardous wastes.
1983	Vehicle emission testing program	Reduces emissions from mobile sources.
1983	Illinois Toxic Substances Disclosure to Employees Act	Ensures that employees be given information concerning the nature of the toxic substances with which they work.
1984	Illinois Toxic Pollution Prevention Act	Reduces the disposal and release of toxic substances that have adverse effects on health and the environment.
1984	National Municipal Policy Act	Regulates wastewater treatment standards; all facilities to be in compliance by July 1, 1988.
1985	Illinois Pesticide Act	Controls the purchase and use of pesticide pertaining to the production, protection, care, storage, or transportation of agricultural commodities.
1986	Solid Waste Management Act adopted	
1987	Underground Storage Tank Fund	Assists underground tank owners in meeting the requirements for safe operation.
1987	Illinois Groundwater Protection Act	Ensures expansion of groundwater protection programs.
1987	Illinois Emergency Planning and Community Right to Know Act	Regulates toxic air pollutants.
1987	Poison Control System Act	Establishes a comprehensive, statewide regional poison control system so that any person may obtain expert hazard, toxicology, and treatment information on a 24-h basis.
1988	Illinois Responsible Property Transfer Act	Ensures that parties involved in certain real estate transactions are made aware of the existing environmental liabilities associated with ownership of such properties, as well as the past use and environmental status of such properties.
1990	Clean Air Act amendments	Provides plans for reduction of hazardous air pollutants and acid rain.
1991	Waste Tire Program	Regulates storage of waste tires.

Source: http://legis.state.il.us/icls/.

Table 2.5 **Impact of federal environmental laws and regulations on geoenvironmental practice**

Geoenvironmental Problem	Relevant Laws	Selected Regulations	Issue to Address
Cleanup of contaminated sites	CERCLA, SARA	40 CFR 300, Appendix A	Hazard ranking system
		40 CFR 300.430(e)	Feasibility investigation
		40 CFR 300.415(d)	Urgent removal actions
	RCRA, HSWA	40 CFR 261.20 to 261.24, 40 CFR 261.31 to 261.33	Hazardous waste definition
		40 CFR 262	Generator requirements
Waste containment systems	RCRA, HSWA	40 CFR 264.300 to 264.317, 40 CFR 257 and 258	Design of landfills
		40 CFR 264.220 to 264.232	Design of surface impoundments
		40 CFR 268.30 to 268.32	Wastes subject to land disposal restrictions
		40 CFR 268.41 to 268.42	Treatment standards before disposal
		40 CFR 264.95 to 264.97	Groundwater program and monitoring

sign specifications set forth in environmental laws and regulations, particularly RCRA and HSWA. The beneficial use of waste and scrap materials for civil engineering construction involves addressing not only geotechnical issues, but also a wide range of regulatory issues covered under various laws, including RCRA and PPA.

2.6 SUMMARY

In the 1970s, a great deal of federal legislation was passed with the increase in environmental and health concerns. The legislation enacted included the National Environmental Policy Act, the Solid Waste Disposal Act, the Clean Air Act, the Occupational Safety and Health Act, the RCRA, the CERCLA, and the Pollution Prevention Act. Although RCRA and CERCLA provided a new basis for environmental regulation, it became apparent that some changes were necessary. As

a result, amendments in the form of the HSWA and SARA were created to help facilitate the changes. All these forms of legislation have proven helpful in addressing a wide range of geoenvironmental problems, including the remediation of contaminated sites, the design of waste containment systems, and the beneficial use of waste materials.

State and local laws and regulations may be stricter than federal laws and regulations. Therefore, one must carefully assess and conform to these regulations. Information on state and local laws and regulations may be obtained by contacting local enforcing agencies, such as the state environmental protection agency or a city's department of environment.

It should be noted that existing laws and regulations may undergo frequent changes, and new laws and regulations may evolve. Therefore, one must keep abreast of the latest laws and regulations that may affect geoenvironmental engineering practices.

QUESTIONS/PROBLEMS

2.1. What is the main objective of environmental laws and regulations?

2.2. What are the major laws that relate to remediation of contaminated sites?

2.3. What are the major laws that relate to design of landfills?

2.4. Which environmental laws are aimed at protecting groundwater? Explain them briefly.

2.5. Which environmental laws control the underground injection of wastes? Explain them briefly.

2.6. Name the environmental laws and regulations that relate to the disposal of contaminated dredged sediments.

2.7. What is CERCLA (Superfund)?

2.8. Explain the hazard ranking system established in the CERCLA. What hazard ranking score is needed in order to include a contaminated site on the National Priorities List (NPL)?

2.9. What is the current status of CERCLA (Superfund)?

2.10. What is meant by "cradle to grave" in RCRA? Explain with examples.

2.11. A site is suspected to be contaminated with toxic chemicals. As an engineer in charge, what initial actions would you undertake?

2.12. Where and how can you get regulations related to leaking underground storage tanks (LUSTs)?

2.13. Explain the role of state and local governments in the development and implementation of environmental protection measures.

2.14. Discuss the similarities and differences between RCRA and CERCLA.

2.15. You are a geoenvironmental engineer working on a design for a new landfill. What subtitle sections of the *Federal Register* would you consult for design guidelines?

2.16. List five examples of possible small-quantity generators of waste.

2.17. Describe the specific tasks included in phase I, II, and III site investigations performed as a result of SARA.

2.18. A city lot is purchased by J. Berger Enterprises, in order to build a shopping mall. The lot has been abandoned for 20 years. Once it was the site of Saichek Plating Works, which went out of business 20 years ago. During construction, a site investigation revealed high amounts of chromium, nickel, and cadmium in the soil. Who is financially responsible for the cleanup?

2.19. The U.S. Department of Agriculture plans on building a major research facility in Grand Forks, North Dakota. To do this, the Red River needs to be diverted from its present course. Chadams Consultants, Inc. has been hired to perform an environmental impact study for the project. Name three possible topics that would be applicable for the study.

2.20. CERCLA allows the USEPA to clean up NPL sites and then recover costs from potentially responsible parties (PRPs). What practical limitations does the USEPA experience in being able to clean up all NPL sites?

2.21. Summarize the amendments (if any) to the following environmental laws and regulations since 2000: (*a*) RCRA; (*b*) CERCLA; (*c*) Clean Water Act; (*d*) Clean Air Act.

2.22. Summarize the evolution of laws and regulations in the state in which you live that are relevant to geoenvironmental studies (preferably in tabular form).

2.23. Select a foreign country of your choice. Explain the similarities and differences in environmental laws in that country to those in the United States in dealing with (*a*) solid waste; (*b*) hazardous waste; (*c*) groundwater protection.

2.24. Identify potential regulatory requirements involved in addressing the contamination problem stated in Problem 1.6.

2.25. What are brownfields? List advantages of brownfield redevelopment. Compare and contrast the cleanup process of brownfield sites with that of Superfund sites.

REFERENCES

CFR (*Code of Federal Regulations*), Title 40, U.S. Government Printing Office, Washington, DC, 2002; *www. access.gpo.gov/nara/cfr/*.

LaGrega, M. D., Buckingham, P. L., and Evans, J. C., *Hazardous Waste Management*, McGraw-Hill, New York, 1994.

Sharma, H. D., and Lewis, S. P., *Waste Containment Systems, Waste Stabilization, and Landfills Design and Evaluation*, Wiley, 1994.

Tchobanoglous, G., Theisen, H., and Vigil, S., *Integrated Solid Waste Management: Engineering Principles and Management Issues*, McGraw-Hill, New York, 1993.

U.S. Code Database, U.S. Government Printing Office, Washington, DC, 1994.

USEPA (U.S. Environmental Protection Agency), *RCRA Orientation Manual*, Office of Solid Waste, USEPA, Washington, DC, 1990.

————*Cleaning Up the Nation's Waste Sites: Market and Technology Trends*, 1996 ed., EPA/542/R-96/005, Office of Solid Waste and Emergency Response, USEPA, Washington, DC, 1997.

Wagner, T. P., *Hazardous Waste Regulations*, Van Nostrand Reinhold, New York, 1991.

Wentz, C. A., *Hazardous Waste Management*, McGraw-Hill, New York, 1995.

3

CHEMICAL BACKGROUND

3.1 INTRODUCTION

Geoenvironmental engineering covers remediation of polluted sites as well as the design of containment facilities for future waste. In dealing with already contaminated sites, there are two ways of addressing the problem. The first is to treat the contaminated material to immobilize the spread of contaminants (toxic chemicals); the second is to recover or remove the contaminants from the soil or groundwater. When designing waste containment facilities, the task is to develop chemical-resistant barriers that form a part of the containment lining system. In each of these situations, an understanding is needed of the behavior of various contaminants in soil and groundwater. The sources of hazardous and nonhazardous wastes and the methods to characterize these wastes are presented in Chapter 15.

In this chapter, various chemicals that are of environmental concern are presented first. Then a brief review of the chemistry of these chemicals in water or aqueous solutions is provided. This background is essential in understanding the behavior of chemicals in soils and groundwater, also known as geochemistry, which is presented in Chapter 6. The information presented in this chapter has been obtained from various sources (e.g., Wertheim and Jeskey, 1956; O'Connor, 1974; Snoeynik and Jenkins, 1980; Solomons, 1980; Stumm and Morgan, 1981; Brady and Holum, 1984; Perry and Green, 1984; USEPA, 1986; Arnikar, 1987; Lide and Frederikse, 1994; Sawyer et al., 1994; Droste, 1997). The information has been summarized and organized here to provide a chemical background and quick reference on various chemical terms and concepts. For further details, readers are encouraged to read the original references.

3.2 TOXIC CHEMICALS

The U.S. Environmental Protection Agency (USEPA) identified several toxic chemicals based on their known or suspected adverse effects on human health and the environment (e.g., USDHHS, 1985; USEPA, 1997; CFR, 2002). These are also known as *priority pollutants*. All these chemicals can be grouped under one of three categories: toxic inorganic chemicals, toxic organic chemicals, and radionuclides.

3.2.1 Toxic Inorganic Chemicals

The toxic inorganic chemicals include heavy metals such as Cd, Ni, Pb, Cr, and Hg (USEPA, 1997). Nonmetals such as arsenic and selenium are also grouped under this category. Table 3.1 lists typical toxic inorganic chemicals found at contaminated sites and in waste leachates. The study of these chemicals is termed *inorganic chemistry*. A brief background of inorganic chemistry is provided in Section 3.3.

3.2.2 Toxic Organic Chemicals

The toxic organic chemicals include chemicals such as benzene, toluene, trichloroethylene, endrin, and lindane (USEPA, 1997). Table 3.1 lists typical organic chemicals that are common at contaminated sites and waste leachates. As will be discussed later, toxic organic chemicals are further subdivided based on structure and functional groups. The study of these chemicals is known as *organic chemistry*. A brief background of organic chemistry is provided in Section 3.4.

Table 3.1 **Typical toxic inorganic, organic, and radionuclide chemicals**

Inorganic Chemicals	Organic Chemicals	Radionuclides
Antimony	Benzene	Cesium
Arsenic	Endrin	Cobalt
Asbestos	4,4'-DDT	Krypton
Beryllium	Ethylbenzene	Plutonium
Cadmium	Napthalene	Radium
Chromium	Pentachlorophenol	Ruthenium
Copper	Phenanthrene	Strontium
Cyanide	Phenol	Thorium
Lead	Polychlorinated biphenyls	
Mercury	Pyrene	
Nickel	Toluene	
Selenium	Toxaphene	
Silver	Trichloroethylene	
Thallium	Vinyl chloride	
Zinc	o-Xylene	

Source: USDHHS (1985), USEPA (1997), 40 CFR 261, 40 CFR 141.

3.2.3 Toxic Radionuclides

The radionuclides include chemicals such as uranium and thorium (USEPA, 1997). Table 3.1 contains a listing of typical radionuclides found at contaminated sites. Nuclear chemistry is the study and use of these substances. A brief background of nuclear chemistry is provided in Section 3.5.

3.3 INORGANIC CHEMISTRY BACKGROUND

In this section we present basic definitions and concepts, types of chemical reactions, chemical kinetics, and gas laws.

3.3.1 Atoms, Elements, and the Periodic Table

An *atom* is the smallest subdivision or particle of an element. An *element* is defined as a substance that is composed of only one kind of atom, all of which have the identical number of protons in each nucleus. An element is also defined as a substance that cannot be

decomposed into other pure substances. For example, silver is an element because it is entirely made up of silver atoms, containing 47 protons in each nucleus. So no chemical or physical process could separate silver into any other pure substance.

As shown in Figure 3.1, the *periodic table* was created to classify and arrange all the elements. The periodic table contains the symbols, atomic numbers, and atomic weights of the elements, with the atomic weights determining the order. For each element, the atomic number refers to the number of protons in the atom that makes up the element. The number of protons and neutrons in the atom gives the atomic weight.

The rows in the periodic table are named *periods*, while the columns of the table are named *groups*. Elements in each group possess similar properties. Some common names of groups or elements in the periodic table are:

group IA	alkali metals
group IIA	alkaline earth metals
group VIIA	halogens
group 0	noble or inert gases
elements 57–71	lanthanides or rare earths
elements 89–106	actinides
groups IB–VIIB	transition metals

3.3.2 Molecules and Chemical Compounds

A *molecule* is the smallest subdivision or particle of a chemical compound. Bonding of atoms of the same element, or of two or more elements together, forms a molecule. For example, bonding of two atoms of the element oxygen (O) results in a molecule of oxygen (O_2); bonding of an atom of carbon and an atom of oxygen results in a molecule of carbon monoxide (CO); and bonding of several atoms of the elements carbon, hydrogen and, oxygen results in a molecule of ordinary table sugar ($C_{12}H_{22}O_{11}$). A *chemical compound* is a substance that consists of the same type of molecules. Thus, oxygen gas consists entirely of O_2 molecules, carbon monoxide consists entirely of CO molecules, and sugar consists entirely of molecules of $C_{12}H_{22}O_{11}$.

Periodic table of elements.

IA	IIA	IIIB	IVB	VB	VIB	VIIB	VIIIB	VIIIB	VIIIB	IB	IIB	IIIA	IVA	VA	VIA	VIIA	O
1 H 1.0079																	2 He 4.0026
3 Li 6.941	4 Be 9.0122											5 B 10.811	6 C 12.011	7 N 14.007	8 O 15.999	9 F 18.998	10 Ne 20.180
11 Na 22.990	12 Mg 24.305											13 Al 26.982	14 Si 28.086	15 P 30.974	16 S 32.065	17 Cl 35.453	18 Ar 39.948
19 K 39.098	20 Ca 40.078	21 Sc 44.956	22 Ti 47.867	23 V 50.942	24 Cr 51.996	25 Mn 54.938	26 Fe 55.845	27 Co 58.933	28 Ni 58.693	29 Cu 63.546	30 Zn 65.409	31 Ga 69.723	32 Ge 72.64	33 As 74.922	34 Se 78.96	35 Br 79.904	36 Kr 83.798
37 Rb 85.468	38 Sr 87.62	39 Y 88.906	40 Zr 91.224	41 Nb 92.906	42 Mo 95.94	43 Tc (98)	44 Ru 101.07	45 Rh 102.91	46 Pd 106.42	47 Ag 107.87	48 Cd 112.41	49 In 114.82	50 Sn 118.71	51 Sb 121.76	52 Te 127.60	53 I 126.90	54 Xe 131.29
55 Cs 132.91	56 Ba 137.33	57-71 *	72 Hf 178.49	73 Ta 180.95	74 W 183.84	75 Re 186.21	76 Os 190.23	77 Ir 192.22	78 Pt 195.08	79 Au 196.97	80 Hg 200.59	81 Tl 204.38	82 Pb 207.2	83 Bi 208.98	84 Po (209)	85 At (210)	86 Rn (222)
87 Fr (223)	88 Ra (226)	89-103 #	104 Rf (261)	105 Db (262)	106 Sg (266)	107 Bh (264)	108 Hs (277)	109 Mt (268)	110 Ds (281)	111 Uuu (272)	112 Uub (285)		114 Uuq (289)				

* Lanthanide series

57 La 138.91	58 Ce 140.12	59 Pr 140.91	60 Nd 144.24	61 Pm (145)	62 Sm 150.36	63 Eu 151.96	64 Gd 157.25	65 Tb 158.93	66 Dy 162.50	67 Ho 164.93	68 Er 167.26	69 Tm 168.93	70 Yb 173.04	71 Lu 174.97

Actinide series

89 Ac (227)	90 Th 232.04	91 Pa 231.04	92 U 238.03	93 Np (237)	94 Pu (244)	95 Am (243)	96 Cm (247)	97 Bk (247)	98 Cf (251)	99 Es (252)	100 Fm (257)	101 Md (258)	102 No (259)	103 Lr (262)

Figure 3.1 Periodic table of elements.

26

There are two types of chemical bonds that make up a molecule: ionic bonds and covalent bonds. An *ionic bond* occurs whenever there is complete transfer of one or more electrons from one atom to the other. An example is the molecule NaCl, which is formed by an ionic bond between Na^+ and Cl^-. The force holding the two particles together is the electrostatic attraction between the oppositely (positive and negative) charged ions. Conversely, a *covalent bond* occurs whenever one or more electrons are shared between two atoms. Examples of covalent bonding are found in molecules such as H_2 and H_2O. The force holding the atoms together is the electrostatic attraction between the shared electrons and the positively charged nuclei.

The composition of a molecule (or chemical compound formed with the molecules) is described by a molecular formula that defines the number of atoms of each element in the molecule. For example, $C_{12}H_{22}O_{11}$ is the formula that represents a molecule of sugar formed by binding of 12 atoms of carbon element, 22 atoms of hydrogen element, and 11 atoms of oxygen element. The *molecular weight* of a compound is the sum of the atomic weights of the combined elements.

In a large number of molecules or chemical compounds, certain groups of atoms exist together as a unit. These groups, called *radicals,* contain one or more unpaired electrons. A few examples of radicals are:

hydroxyl (OH^-)   sulfite (SO_3^{2-})
carbonate (CO_3^{2-})   ammonium (NH_4^+)
bicarbonate (HCO_3^-) nitrate (NO_3^-)
sulfate (SO_4^{2-})   nitrite (NO_2^-)

The *oxidation number* (*oxidation state* or *valence*) of an element in a molecule or chemical compound refers to the total number of electrons transferred or partly transferred to or from an atom of the element in the bonds it forms with other atoms of different elements. If electrons are transferred from the atom (the atom loses the electrons), the oxidation number of the element is positive. If electrons are transferred to the atom (the atom gains the electrons), the oxidation number of the element is negative. Rules for assigning oxidation numbers are as follows:

1. The algebraic sum of the oxidation numbers of all atoms in a neutral compound is zero. Otherwise, the sum must be equal to the charge on the ionic species. For example, for NaCl, the sum of oxidation numbers of Na and Cl should be equal to zero, and for CO_3^{2-}, the sum of oxidation numbers of C and O should be equal to -2.

2. The oxidation number of all elements in the free state is zero. For example, the oxidation number of O in O_2 and H in H_2 is equal to zero.

3. The alkali metals (i.e., group IA elements: H, Li, Na, K, Rb, Cs, and Fr) assume a $+1$ oxidation number, while the alkaline earth metals (Group IIA elements: Be, Mg, Ca, Sr, Ba, and Ra) assume a $+2$ oxidation number in chemical compounds.

4. The oxidation number of oxygen is -2, except in peroxide, where the oxidation number is -1. This means that only in hydrogen peroxide (H_2O_2) is the oxidation number of oxygen -1; in all other chemical compounds containing oxygen, such as CO_2 and $KMnO_3$, it is -2.

5. Hydrogen has an oxidation number of $+1$, except when bonded to metallic hydride, when its oxidation number is -1.

The rules above can be applied to determine the oxidation numbers of all elements in a chemical compound. For example, to determine the oxidation numbers of Na and Br in NaBr, the first rule implies that the sum of the oxidation numbers of Na and Br must be equal to zero. Na being a group IA element, its oxidation number according to the rule 3 is $+1$. Therefore, the oxidation number of Br must be -1. Similarly, for $K_2Cr_2O_7$, the sum of oxidation numbers of K, Cr, and O must be equal to zero. According to rule 3, the oxidation number of K is $+1$, and according to rule 4, the oxidation number of O is -2. Two atoms of K contribute a charge of $+2$ and seven atoms of O contribute a charge of -14; therefore, the net balance charge of $+12$ must be derived from Cr. Since there are two atoms of Cr, the oxidation number of each Cr atom must be $+6$.

3.3.3   Concentration of Chemical Compounds

The concentration of ions[1] or chemicals in solution is commonly expressed as (1) molar concentration, (2) mole fraction, or (3) mass concentration, as defined below.

- The *molar concentration* (*M*) is defined as the number of moles[2] per volume of solution. For example, 1 *M* solution equals 1 mole per liter (mol/L).
- The *mole fraction* (x_i) refers to the number of moles of a substance (n_i) divided by the total number of moles. This can be expressed mathematically as $x_i = n_i/\Sigma\, n_j$, where $\Sigma\, n_j$ is the total number of moles.
- The *mass concentration* is expressed as the mass of the element or compound in milligrams per liter of solution (mg/L). Often, the term *parts per million* (ppm) is used. If the specific gravity of solution (SG) = 1, then 1 L of solution = 1000 g. Then 1 mg/L = (1 g/1000 mg) × (1 L/1000 g) = $1/10^6$ = 1 ppm. Therefore, mg/L is the same as ppm.

Occasionally, concentration is expressed as meq/L. This expression is the equivalent weight of an element in milligrams in 1 L of solution. The *equivalent weight* of an element is defined as the atomic weight divided by the oxidation number. For example, Ca^{2+}, with an atomic weight of 40 g/mol, has an equivalent weight = 40/2 = 20 g.

3.3.4   Chemical Reaction

A *chemical reaction* occurs whenever a chemical compound is formed from the elements or from other chemical compounds. This process involves transformation of the arrangement of atoms in molecules of the initial substances, known as *reactants,* into a new arrangement of atoms in the molecules of the final substances, known as *products.* A chemical reaction is expressed as

$$A + B = C + D$$

where A and B are the reactants and C and D are the products. For example, in the following reaction, Zn and HCl are the reactants and $ZnCl_2$ and H_2 are the products.

$$Zn + 2HCl = ZnCl_2 + H_2$$

3.3.5   Chemical Reaction Balance

The expression for a chemical reaction must be balanced such that mass is conserved. In other words, the total number of atoms in the molecules of an element must be equal on both sides of the reaction. For example,

$$HNO_3 + H_2S \rightarrow NO + S + H_2O$$

In this reaction, the total numbers of atoms of H, N, O, and S on the left-hand side are 3, 1, 3, and 1, respectively. The total numbers of atoms of H, N, O, and S on the right-hand side are 2, 1, 2, and 1, respectively. This shows that the total number of atoms of H and O are not balanced; consequently, the reaction is not balanced. The reaction as modified below contains the same number of atoms of all elements on both sides of the reaction; therefore, the reaction is balanced:

$$2HNO_3 + 3H_2S \rightarrow 2NO + 3S + 4H_2O$$

In addition to mass conservation in a reaction, the charge should be conserved. Occasionally, electrons are present in chemical reactions. Although their mass is insignificant to mass balance consideration, the electron (e) charge must be conserved. The total charge on both sides of the reaction must be equal. For example,

---

[1] An *ion* is defined as an atom or a group of atoms that has acquired a net electric charge by gaining or losing one or more electrons.

[2] A *mole* is the amount of a substance that contains as many atoms, molecules, ions or other elementary units as the number of atoms in 0.012 kg of carbon. The number is 6.0225 × 10^{23}, *Avogadro's number.* They are also known as *gram molecules.*

the following reaction is unbalanced because of the net charge of -2 on the left side and -1 on the right side.

$$MnO_4^- + H_2O + e^- \rightarrow MnO_2 + OH^-$$

This reaction as modified below is balanced because it has the same charge (-4) on both sides of the reaction:

$$MnO_4^- + 2H_2O + 3e^- \rightarrow MnO_2 + 4OH^-$$

3.3.6 Equilibrium

A chemical reaction is said to be in *equilibrium* if the concentrations of all species (elements, molecules, etc.) are constant. A general expression for an equilibrium reaction is as follows:

$$aA + bB \Leftrightarrow cC + dD$$

where a, b, c, and d are stoichiometric coefficients used to balance the reaction. For example, the following reaction is said to be in equilibrium:

$$NH_3 + H_2O \Leftrightarrow NH_4^+ + OH^-$$

Under equilibrium conditions, an equilibrium constant, K_{eq}, is commonly used. The equilibrium constant is expressed as

$$K_{eq} = \frac{\{C\}^c\{D\}^d}{\{A\}^a\{B\}^b}$$

where $\{\cdot\}$ represents the chemical activity of the substance. However, chemical activity is related to concentration, $[\cdot]$:

$$\{A\} = \gamma_A[A]$$

where γ is known as the *activity coefficient*. This coefficient depends mainly on the ionic strength or concentration of ions in solution. For dilute solutions, $\gamma = 1$ and $\{A\} = [A]$. Therefore, activity and concentration are approximately the same, and K_{eq} can be rewritten:

$$K_{eq} = \frac{[C]^c[D]^d}{[A]^a[B]^b}$$

For the example reaction considered above, K_{eq} is given by

$$K_{eq} = \frac{[NH_4^+][OH^-]}{[NH_3][H_2O]}$$

3.3.7 Types of Chemical Reactions

There are four different types of chemical reactions that are important in environmental studies: acid–base reactions, precipitation–dissolution reactions, complexation reactions, and oxidation–reduction reactions. Each type is described briefly in the following subsections.

Acid–Base Reactions. An *acid–base reaction* is a chemical reaction that involves either the gain or loss of a proton (H^+) or the gain or loss of a hydroxyl (OH^-). In general, acid is a proton donor and base is a proton acceptor. Acid–base reactions affect the pH and chemistry of soils and groundwater; therefore, these reactions may have a major effect on remediation and waste containment. The dissociation of water, acids and bases as well as typical acid–base reactions, are explained in this section. When a compound is dissolved in water, the ions that are packed tightly together in the compound become separated; we call this *dissociation*.

Dissociation of Water and pH. The dissociation of water molecules is expressed as

$$H_2O \Leftrightarrow H^+ + OH^-$$

The equilibrium constant for this reaction is given as

$$K_{eq} = \frac{[H^+][OH^-]}{[H_2O]}$$

The concentration of water does not change significantly; therefore, instead of K_{eq}, K_w is used and defined as

$$K_w = [H^+][OH^-]$$

K_w is temperature dependent. At 25°C, $K_w = 1 \times 10^{-14}$ = $[H^+][OH^-]$. Based on this equation, if the concentration of H^+ increases, the concentration of OH^- must decrease, and vice versa.

Because the magnitude of $[H^+]$ can vary over a wide range, it is convenient to use the log scale; p notation is used for further simplicity: $p = -\log_{10}$. Taking $-\log_{10}$ of both sides of the equation for K_w gives $14 = -\log_{10}[H^+] - \log_{10}[OH^-]$, or $14 = pH + pOH$. Usually, only the value of pH is used because pOH is always $(14 - pH)$. In the case of pure water, $[H^+] \simeq [OH^-]$ and pH = 7. This situation is called the *neutral condition*. If the pH lies between 0 and 7, *acidic conditions* exist, where $[H^+] > [OH^-]$. If the pH lies between 7 and 14, *basic* or *alkaline conditions* exist, where $[H^+] < [OH^-]$.

Dissociation of Acids and Bases. The dissociation of acids and bases varies with strength. *Strong acids* dissociate completely. For example, $HCl \rightarrow H^+ + Cl^-$. *Strong bases* also dissociate completely, as seen in this example: $NaOH \rightarrow Na^+ + OH^-$. Strong acid and strong base are relative terms that indicate any acid or base that dissociates completely in the pH range of interest.

Weak acids and *bases* are those acids and bases that dissociate only partially. For example, acetic acid is a weak acid because it does not dissociate completely. This dissociation can be given by the reaction

$$CH_3COOH \Leftrightarrow CH_3COO^- + H^+$$

The equilibrium constant is given by $K_{eq} = [CH_3COO^-][H^+]/[CH_3COOH] = 1.8 \times 10^{-5}$. This constant can also be expressed using p notation: $pK_{eq} = -\log_{10}[1.8 \times 10^{-5}] = 4.7$. The pK_{eq} values for chemical reactions involving weak acids and bases are useful to determine the extent of dissociation or the strengths of the acids and bases. Table 3.2 provides pK_{eq} values for selected acids. The lower the pK_{eq} value, the higher will be the dissociation of the compound. For strong acids, pK_{eq} can be a negative value, while for strong bases, pK_{eq} values can be higher than 20.

Typical Acid–Base Reactions. Acid-base reactions involve reaction between an acid and a base and is given by the general reaction

$$A_1 + B_2 = A_2 + B_1$$

In the acid–base reaction above, base B_2 gains the proton lost by acid A_1, whereas base B_1 gains the proton lost by acid A_2. For example, salts are formed by a combination of an acid (HA) and a base (BOH) as given by

$$HA + BOH = H_2O + BA$$

In this reaction, the acid HA loses H^+ and gives rise to a potential H^+ acceptor A^- (A^- is called the *conjugate base* of HA). The base BOH loses OH^- and gives rise to a potential OH^- acceptor B^+ (B^+ is called the *conjugate acid* of BOH). Acid–base reactions involving carbonate species (e.g., HOC_3^- and CO_3^{2-}) in waters can significantly affect the pH and must be considered in remedial designs.

Precipitation–Dissolution Reactions. When a solid chemical compound (e.g., chloride and sulfate salts and metal hydroxides) is added to water, it dissolves to some extent. This reaction, known as *dissolution*, is expressed as

$$A_aB_b(s) \Leftrightarrow aA^{b+} + bB^{a-}$$

where (s) indicates a solid. The reaction proceeding from left to right is called *dissolution*, while the reaction proceeding from right to left is called *precipitation*. For example, the precipitation–dissociation of NaCl is expressed as

$$NaCl(s) \Leftrightarrow Na^+ + Cl^-$$

The dissolution–precipitation of metal hydroxide, in general, is expressed as

$$MOH(s) \Leftrightarrow M^+ + OH^-$$

where M represents any cationic metal.

Table 3.2 pK_{eq} **Values for selected acids**

Acid	Reaction	$pK_{eq} = -log_{10}[K_{eq}]$
Perchloric acid ($HClO_4$)	$HClO_4 \Leftrightarrow ClO_4^- + H^+$	-7
Hydrochloric acid (HCl)	$HCl \Leftrightarrow Cl^- + H^+$	~ -3
Sulfuric acid (H_2SO_4)	$H_2SO_4 \Leftrightarrow HSO_4^- + H^+$	~ -3
Nitric acid (HNO_3)	$HNO_3 \Leftrightarrow NO_3^- + H^+$	0
Hydronium ion ($H3O^+$)	$H_3O^+ \Leftrightarrow H_2O + H^+$	0
Iodic acid (HIO_3)	$HIO_3 \Leftrightarrow IO_3^- + H^+$	0.8
Bisulfate ion (HSO_4^-)	$HSO_4^- \Leftrightarrow SO_4^{2-} + H^+$	2
Phosphoric acid (H_3PO_4)	$H_3PO_4 \Leftrightarrow H_2PO_4^- + H^+$	2.1
Ferric ion [$Fe(H_2O)_6^{3+}$]	$Fe(H_2O)_6^{3+} \Leftrightarrow Fe(H_2O)_5OH^{2+} + H^+$	2.2
Hydrofluoric acid (HF)	$HF \Leftrightarrow F^- + H^+$	3.2
Nitrous acid (HNO_2)	$HNO_2 \Leftrightarrow NO_2^- + H^+$	4.5
Acetic acid (CH_3COOH)	$CH_3COOH \Leftrightarrow CH_3COO^- + H^+$	4.7
Aluminum ion [$Al(H_2O)_6^{3+}$]	$Al(H_2O)_6^{3+} \Leftrightarrow Al(H_2O)_5OH^{2+} + H^+$	4.9
Carbon dioxide and carbonic acid (H_2CO_3)	$H_2CO_3^- \Leftrightarrow HCO_3^- + H^+$	6.3
Hydrogen sulfide (H_2S)	$H_2S \Leftrightarrow HS^- + H^+$	7.1
Dihydrogen phosphate ($H_2PO_4^-$)	$H_2PO_4^- \Leftrightarrow HPO_4^{2-} + H^+$	7.2
Hypochlorous acid (HOCl)	$HOCl \Leftrightarrow OCl^- + H^+$	7.5
Hydrocyanic acid (HCN)	$HCN \Leftrightarrow CN^- + H^+$	9.3
Boric acid (H_3BO_3)	$H_3BO_3 \Leftrightarrow B(OH)_4^- + H^+$	9.3
Ammonium ion (NH_4^+)	$NH_4^+ \Leftrightarrow NH_3 + H^+$	9.3
Orthosilicic acid (H_4SiO_4)	$H_4SiO_4 \Leftrightarrow H_3SiO_4^- + H^+$	9.5
Phenol (C_6H_5OH)	$C_6H_5OH \Leftrightarrow C_6H_5O^- + H^+$	9.9
Bicarbonate ion (HCO_3^-)	$HCO_3^- \Leftrightarrow CO_3^{2-} + H^+$	10.3
Monohydrogen phosphate (HPO_4^{2-})	$HPO_4^{2-} \Leftrightarrow PO_4^{3-} + H^+$	12.3
Trihydrogen silicate (H_3SiO4^-)	$H_3SiO4^- \Leftrightarrow H_2SiO_4^{2-} + H^+$	12.6
Bisulfide ion (HS^-)	$HS^- \Leftrightarrow S^{2-} + H^+$	14
Water (H_2O)	$H_2O \Leftrightarrow OH^- + H^+$	14
Ammonia (NH_3)	$NH_3 \Leftrightarrow NH_2^- + H^+$	~ 23
Hydroxide ion (OH^-)	$OH^- \Leftrightarrow O^{2-} + H^+$	~ 24

Source: Snoeyink and Jenkins (1980).

The equilibrium constant for a precipitation–dissolution reaction can be given by

$$K_{eq} = \frac{[A^{b+}]^a[B^{a-}]^b}{[A_aB_b(s)]}$$

or for reactions involving metal hydroxides,

$$K_{eq} = \frac{[M^+][OH^-]}{[MOH(s)]}$$

The concentrations of solids do not change; therefore, instead of the equilibrium constant K_{eq}, the *solubility product constant, K_{sp}*, is used:

$$K_{sp} = [A^{b+}]^a[B^{a-}]^b \quad \text{or} \quad K_{sp} = [M^+][OH^-]$$

The values of K_{sp} [or $pK_{sp} = -log_{10}(K_{sp})$] of different chemical compounds will help to determine the extent of solubility. Table 3.3 provides pK_{sp} values for selected solid substances. The lower the pK_{sp} value, the higher will be the dissolution of the solid chemical compound. Such information is very useful for geoenvironmental studies, particularly those dealing with remedial system design.

Complexation Reactions. Some ions or molecules, dissolved in water, will combine with a variety of other ions or molecules to form several different species, known as *complexes*. This process is known as *com-*

Table 3.3 **Solubility constants for selected solid substances**

Solid	Reaction	$pK_{sp} = -log(K_{sp})$
Fe(OH)$_3$ (amorph.)	Fe(OH)$_3$ $\Leftrightarrow$ Fe^{3+} + 3OH$^-$	38
FePO$_4$	FePO$_4$ $\Leftrightarrow$ Fe$^+$ + PO$_4^-$	17.9
Fe$_3$(PO$_4$)$_2$	Fe$_3$(PO$_4$)$_2$ $\Leftrightarrow$ 3Fe$_3^{2:+}$ + 2PO$_4^{3-}$	33
Fe(OH)$_2$	Fe(OH)$_2$ $\Leftrightarrow$ Fe^{2+} + 2OH$^-$	14.5
FeS	FeS $\Leftrightarrow$ Fe^{2+} + S^{2-}	17.3
Fe$_2$S$_3$	Fe$_2$S$_3$ $\Leftrightarrow$ 2Fe^{3+} + 3S^{2-}	88
Al(OH)$_3$ (amorph.)	Al(OH)$_3$ $\Leftrightarrow$ Al^{3+} + 3OH$^-$	33
AlPO$_4$	AlPO$_4$ $\Leftrightarrow$ Al^{3+} + PO$_4^{3-}$	21.0
CaCO$_3$ (calcite)	CaCO$_3$ $\Leftrightarrow$ Ca$^+$ + CO$_3^-$	8.34
CaF$_2$	CaF$_2$ $\Leftrightarrow$ Ca^{2+} + 2F$^-$	10.3
Ca(OH)$_2$	Ca(OH)$_2$ $\Leftrightarrow$ Ca^{2+} + 2OH$^-$	5.3
Ca$_3$(PO$_4$)$_2$	Ca$_3$(PO$_4$)$_2$ $\Leftrightarrow$ 3Ca^{2+} + 2PO$_4^{3-}$	26.0
CaSO$_4$	CaSO$_4$ $\Leftrightarrow$ Ca^{2+} + SO$_4^{2-}$	4.59
BaSO$_4$	BaSO$_4$ $\Leftrightarrow$ Ba^{2+} + SO$_4^{2-}$	10
Cu(OH)$_2$	Cu(OH)$_2$ $\Leftrightarrow$ Cu^{2+} + 2OH$^-$	19.3
PbCl$_2$	PbCl$_2$ $\Leftrightarrow$ Pb^{2+} + 2Cl$^-$	4.8
Pb(OH)$_2$	Pb(OH)$_2$ $\Leftrightarrow$ Pb^{2+} + 2OH$^-$	14.3
PbSO$_4$	PbSO$_4$ $\Leftrightarrow$ Pb^{2+} + SO$_4^{2-}$	7.8
PbS	PbS $\Leftrightarrow$ Pb$^+$ + S$^-$	27.0
MgCO$_3$	MgCO$_3$ $\Leftrightarrow$ Mg^{2+} + CO$_3^{2-}$	5.0
Mg(OH)$_2$	Mg(OH)$_2$ $\Leftrightarrow$ Mg^{2+} + 2OH$^-$	10.74
Mn(OH)$_2$	Mn(OH)$_2$ $\Leftrightarrow$ Mn^{2+} + 2OH$^-$	12.8
AgCl	AgCl $\Leftrightarrow$ Ag$^+$ + Cl$^-$	10.0
Ag$_2$CrO$_4$	Ag$_2$CrO$_4$ $\Leftrightarrow$ 2Ag$^+$ + CrO$_4^{2-}$	11.6
Ag$_2$SO$_4$	Ag$_2$SO$_4$ $\Leftrightarrow$ 2Ag$^+$ + SO$_4^{2-}$	4.8
Zn(OH)$_2$	Zn(OH)$_2$ $\Leftrightarrow$ Zn^{2+} + 2OH$^-$	17.2
ZnS	ZnS $\Leftrightarrow$ Zn^{2+} + S^{2-}	21.5

Source: Snoeyink and Jenkins (1980).

plexation. Generally, the complex consists of a central atom or *cation* (typically, one of the metals) surrounded by two to nine other atoms, anions (negatively charged ions), or small molecules. These surrounding atoms, anions, and molecules, called *ligands,* can include inorganic species such as OH$^-$, Cl$^-$, SO$_4^-$, NH$_3$, and PO$_4^{3-}$ and organic molecules such as amino acid and ethylenediaminetetraacetic acid (EDTA). Complexes may be charged or uncharged. Also, depending on the complex, the solubility may be increased or decreased.

A simple example of a complexation reaction is

$$Mn^{2+} + Cl^- = MnCl^+ \qquad K_{eq} = \frac{[MnCl^+]}{[Mn^{2+}][Cl^-]}$$

Complexation reactions can also occur stepwise, with equilibrium constants known as *stepwise formation constants.* For example, complexation of metal, Cr^{3+}, with ligand, OH$^-$, can occur in the following forms:

$$Cr^{3+} + OH^- = Cr(OH)^{+2} \qquad K_1 = \frac{Cr(OH)^{2+}}{[Cr^{3+}][OH^-]}$$

$$Cr(OH)^{2+} + OH^- = Cr(OH)_2^+ \qquad K_2 = \frac{[Cr()H)_2^+]}{[Cr(OH)^{2+}][OH^-]}$$

$$Cr(OH)_2^+ + OH^- = Cr(OH)_3 \qquad K_3 = \frac{[Cr(OH)^3]}{[Cr(OH)_2^+][OH^-]}$$

Cr(OH)$^{2+}$, Cr(OH)$_2^+$, and Cr(OH)$_3$ are the complexes, and K_1, K_2, and K_3 are the stepwise formation constants. Such stepwise complexation can occur for other metals, such as mercury, lead, copper, and zinc with various ligands. Table 3.4 provides stability constants in p notation, pK, for selected metals with selected ligands. The lower pK values indicate a preference for such complexes to occur.

The properties of metal complexes may be quite different than those of the metal itself; for example, the solubility of a metal complex may be much higher than that of the metal itself. Complexation reactions

Table 3.4 **Stability constants for selected metal–ligand complexes[a]**

Complex	pK	Complex	pK
$AgCl^0$	3.36	FeF^{2+}	5.25
$AgCl_2^-$	5.2	FeF_2^+	9.25
$AgCl_3^{2-}$	5.85	FeF_3^0	12.25
$AgCl_4^{3+}$	5.2	$FeOH^{2+}$	-2.73
$AlOH^{2+}$	-5.5	$Fe(OH)_2^+$	-6.50
$Al(OH)_2^+$	-10.13	$Fe(OH)_3^0$	-12.83
$Al(OH)_3^0$	-15.83	$Fe(OH)_4^-$	-22.13
$Al(OH)_4^+$	-23.53	$HgCl^+$	6.72
$CdCl^+$	1.35	$HgCl_2^0$	13.23
$CdCl_2^0$	1.70	$HgCl_3^-$	14.2
$CdCl_3^+$	1.50	$HgCl_4^{2-}$	15.3
$CdNH_3^{2+}$	2.51	MgF^+	1.82
$Cd(NH_3)_2^{2+}$	4.47	$NiNH_3^{2+}$	2.67
$Cd(NH_3)_3^{2+}$	5.77	$Ni(NH_3)_2^{2+}$	4.79
$Cd(NH_3)_4^{2+}$	6.56	$Ni(NH_3)_3^{2+}$	6.4
$CuCl^0$	3.36	$Ni(NH_3)_4^{2+}$	7.47
$CuCl_2^-$	5.2	$PbCl^+$	1.6
$CuCl_3^{2-}$	5.85	$PbCl_2^0$	1.78
$CuCl_4^{3-}$	5.2	$PbCl_3^-$	1.68
$CuNH_3^{2+}$	3.99	$PbCl^{2-}$	1.38
$Cu(NH_3)_2^{2+}$	7.33	$ZnCl^+$	-0.5
$Cu(NH_3)_3^{2+}$	10.06	$ZnCl_2^0$	-1.0
$Cu(NH_3)_4^{2+}$	12.03	$ZnCl_3^-$	0
$CrOH^{2+}$	-4.54	$ZnCl_4^{2+}$	-1.0
$Cr(OH)^{2+}$	-10.53	ZnF^+	1.26
$Cr(OH)_3^0$	-18.83	$ZnNH_3^{2+}$	2.67
$Cr(OH)_4^-$	-27.93	$Zn(NH3)_2^{2+}$	4.43
$FeCl^{2+}$	1.48	$Zn(NH_3)_3^{2+}$	6.74
$FeCl_2^+$	2.13	$Zn(NH_3)_4^{2+}$	8.7
$FeCl_3^0$	1.13		

Sources: Stumm and Morgan (1996); Snoeyink and Jenkins (1980).

[a] General equilibrium reaction: $M^{+n} + iL^{a-} \Leftrightarrow ML_i^{n-a}$ (M, metal; L, ligand, ML, complex).

are considered carefully when designing remedial systems to either enhance contaminant removal or immobilize the contaminants in soils and groundwater. The stability constants provide information on the preference of contaminants to complex with various ligands.

Oxidation–Reduction Reactions. An *oxidation–reduction reaction,* or *redox reaction,* is a chemical reaction in which an atom or ion loses electrons to another atom or ion. In this type of reaction, one compound loses electrons and another compound gains them. The overall reaction can be formulated from two half-reactions: one producing electrons, the other receiving them. For example, in the half-reaction $4Fe^{2+} = 4Fe^{3+} + 4e^-$, Fe^{2+} is losing electrons, thereby increasing its charge to +3. Fe^{2+} is oxidized, or has become a reducing agent. In the other half-reaction, $O_2 + 4H^+ + 4e^- = 2H_2O$, O_2 is taking electrons, decreasing its charge from 0 to −2. O_2 is reduced or has become an oxidizing agent. The overall balanced reaction is simply the sum of the two half-reactions:

$$4Fe^{2+} + O_2 + 4H^+ = 4Fe^{3+} + 2H_2O$$

A redox reaction is influenced by pH if H^+ ions appear in the reaction. If H^+ is present on the left-hand side of the reaction, a high H^+ concentration, or low pH, will favor the formation of products. If H^+ is present on the right-hand side of the reaction, a low H^+ concentration, or high pH, will favor the formation of products.

In writing redox reactions, half-reactions are first written. Then both reactions are combined to obtain the overall redox reaction. The following rules are followed when writing half-reactions:

1. Write the core reactants and products on opposite sides of an equal sign. Balance the core reaction with respect to the atom that is changing oxidation number. For example, consider the half-reaction involving $KMnO_4$. The core reactant will be MnO_4^- and the core product will be MnO_2. This means that Mn is changing its oxi-

dation number from $+7$ in MnO_4^- to $+4$ in MnO_2. Using the electrons (e^-), the balanced reaction can be written as:

$$MnO_4^- + 3e^- = MnO_2$$

2. Balance the reaction with respect to oxygen by adding H_2O where needed. For the example reaction above, this step requires adding two molecules of H_2O to the right side of the reaction, resulting in

$$MnO_4^- + 3e^- = MnO_2 + 2H_2O$$

3. Balance H by adding H^+ ions. For the example reaction, this rule requires adding four H^+ ions on the left side of the reaction, leading to

$$MnO_4^- + 4H^+ + 3e^- = MnO_2 + 2H_2O$$

4. Balance the charge by adjusting the number of electrons. For the example reaction, the charge is balanced; therefore, no changes are required.

5. Add any extraneous ions that do not participate in the redox reaction. For the example reaction, suppose that K^+ is an extraneous ion. Adding this to the reaction on both sides provides a balanced half-reaction:

$$K^+ + MnO_4^- + 4H^+ + 3e^-$$
$$= MnO_2 + 2H_2O + K^+$$

In general, a half-reaction can be represented by the following equation:

$$Ox + ne^- = Red$$

where Ox represents the oxidant and Red represents the reductant. The equilibrium constant for this general equation is as follows:

$$K_{eq} = \frac{\{Red\}}{\{Ox\}\{e^-\}^n}$$

Solving for $\{e^-\}$, the electron activity is given by

$$\{e^-\} = \left(\frac{\{Red\}}{\{Ox\}K_{eq}}\right)^{1/n}$$

Assuming that the activity is approximately equal to the concentration, the concentration of electrons can be expressed using the p notation as

$$p^e = -\log_{10}[e^-]$$

A large p^e value indicates low $[e^-]$, meaning that primarily oxidized species are present. A small p^e value indicates high $[e^-]$, meaning that primarily reduced species are present. Three redox regions are defined based on these p^e values, at a pH value of 7. When p^e is greater than 7, the region is called *oxic*. A value of p^e between 2 and 7 indicates the *suboxic* region. When p^e is less than 2, the region is called *anoxic*.

Direct measurement of p^e is not practical. Instead, the redox potential, Eh, is used to characterize the reactions. Eh is given in units of volts and measured using an electrochemical cell. Eh and p^e are related by $Eh = (2.3RT/F)p^e$, where F is Faraday's constant, R the universal gas constant, and T the temperature in Kelvin (K). At 25°C, $2.3RT/F = 0.059$ V, thereby making $Eh = (0.059$ V$)p^e$. This equation displays a proportional relationship between Eh and p^e. Knowing Eh and pH, the potential for redox reactions occurring may be assessed.

Table 3.5 provides selected half-reactions. In this table, values of standard free energy change (ΔG_0) and standard electrode potential (E_0) are given for each half-reaction. The redox reactions are important for chemicals that can exist in different valence states (e.g., carbon, arsenic, chromium, copper, iron, manganese, nitrogen, and sulfur). The form of a chemical can significantly affect a remedial process. For example, chromium exists as trivalent chromium under reducing conditions and as hexavalent chromium under oxidizing conditions. Since the properties of these two chromium

Table 3.5 **Selected half-reactions**

Half-Reaction	ΔG_0 (kcal/mol)	E_0 (V)
$\frac{1}{2} Br_2(aq) + e^- = Br^-$	−25.2	1.09
$\frac{1}{2} BrCl + e^- = \frac{1}{2} Br^- + \frac{1}{2} Cl^-$	−31.1	1.35
$Ce^{4+} + e^- = Ce^{3+}$	−33.2	1.44
$\frac{1}{4} CO_3^{2-} + \frac{7}{8} H^+ + e^- = \frac{1}{8} CH_3COO^- + \frac{1}{4} H_2O$	−1.73	0.075
$\frac{1}{4} CO_3^{2-} + H^+ + e^- = \frac{1}{24} C_6H_{12}O_6 + \frac{1}{4} H_2O$	0.35	−0.0015
$\frac{1}{2} Cl_2(aq) + e^- = Cl^-$	−32.1	1.391
$ClO_2 + e^- = ClO_2^-$	−26.6	1.15
$ClO_3^- + 2H^+ + e^- = ClO_2 + H_2O$	−26.6	1.15
$\frac{1}{2} OCl^- + H^+ + e^- = \frac{1}{2} Cl^- + \frac{1}{2} H_2O$	−39.9	1.728
$\frac{1}{8} ClO_4^- + H^+ + e^- = \frac{1}{8} Cl^- + \frac{1}{2} H_2O$	−31.6	1.37
$\frac{1}{6} Cr_2O_7^{2-} + \frac{7}{3} H^+ + e^- = \frac{1}{3} Cr^{3+} + \frac{7}{6} H_2O$	−30.7	1.33
$\frac{1}{2} Cu^{2+} + e^- = \frac{1}{2} Cu$	−7.78	0.337
$\frac{1}{2} Fe^{2+} + e^- = \frac{1}{2} Fe$	9.45	−0.409
$\frac{1}{3} Fe^{3+} + e^- = Fe^{2+}$	−17.78	0.770
$Fe^{3+} + e^- = \frac{1}{3} Fe$	0.84	−0.0036
$H^+ + e^- = \frac{1}{2} H_2(g)$	0.00	0.00
$\frac{1}{2} H_2O_2 + H^+ + e^- = H_2O$	−40.8	1.77
$\frac{1}{2} Hg^{2+} + e^- = \frac{1}{2} Hg$	−19.7	0.851
$\frac{1}{2} I_2(aq) + e^- = I^-$	−14.3	0.62
$\frac{1}{5} IO_4^- + \frac{6}{5} H^+ + e^- = \frac{1}{10} I_2(g) + \frac{3}{5} H_2O$	−27.6	1.197
$\frac{1}{2} MnO_2 + 2H^+ + e^- = \frac{1}{2} Mn^{2+} + H_2O$	−27.9	1.208
$\frac{1}{5} MnO_4^- + \frac{8}{5} H^+ + e^- = \frac{1}{5} Mn^{2+} + \frac{4}{5} H_2O$	−34.4	1.491
$\frac{1}{3} MnO_4^- + \frac{4}{3} H^+ + e^- = \frac{1}{3} MnO_2 + \frac{2}{3} H_2O$	−39.2	1.695
$\frac{1}{6} No_2^- + \frac{4}{3} H^+ + e^- = \frac{1}{6} NH_4^+ + \frac{1}{3} H_2O$	−20.75	0.898
$\frac{1}{8} NO_3^- + \frac{5}{4} H^+ + e^- = \frac{1}{8} NH_4^+ + \frac{3}{8} H_2O$	−20.33	0.88
$\frac{1}{3} NO_2^- + \frac{4}{3} H^+ + e^- = \frac{1}{6} N_2(g) + \frac{2}{3} H_2O$	−35.16	1.519
$\frac{1}{5} NO_3^- + \frac{6}{5} H^+ + e^- = \frac{1}{10} N_2(g) + \frac{3}{5} H_2O$	−28.73	1.244
$\frac{1}{4} O_2(aq) + H^+ + e^- = \frac{1}{2} H_2O$	−29.32	1.23
$\frac{1}{2} O_3(g) + H^+ + e^- = \frac{1}{2} O_2(g) \frac{1}{2} H_2O$	−47.8	2.07
$\frac{1}{6} SO_4^{2-} + \frac{1}{3} H^+ + e^- = \frac{1}{6} S + \frac{2}{3} H_2O$	−8.24	0.357
$\frac{1}{8} SO_4^{2-} + \frac{5}{4} H^+ + e^- = \frac{1}{8} H_2S(aq) + \frac{1}{2} H_2O$	−7.00	0.303
$\frac{1}{4} SO_4^{2-} + \frac{5}{4} H^+ + e^- = \frac{1}{8} S_2O_3^{2-} + \frac{5}{8} H_2O$	−7.00	0.303
$\frac{1}{2} SO_4^{2-} + H^+ + e^- = \frac{1}{2} SO_3^{2-} + \frac{1}{2} H_2O$	0.93	−0.039
$\frac{1}{2} Zn^{2+} + e^- = \frac{1}{2} Zn$	17.6	−0.763

Source: Droste (1997). This material is used by permission of John Wiley & Sons, Inc.

forms are quite different, the remedial processes may have to be different.

3.3.8 Chemical Kinetics

All reactions have thus far been assumed to be in equilibrium (concentrations remaining constant). In reality, however, concentrations often change with time. The rate of reaction depends on the concentration of species in a reaction. The general reaction rate equation is

$$\frac{dC_A}{dt} = \pm\, kC_A^a C_B^b \cdots C_N^n$$

where C_A, C_B, . . . , C_N are the concentrations of chemical substances A, B, . . . , N, respectively; a, b, . . . , n are the constants; and k is the reaction rate constant. A positive value of k implies increasing concentrations, while a negative k value implies decreasing concentrations of the chemical species or substance.

The order of an entire reaction is defined as the sum of the exponents a, b, . . . , n. The order of a reaction for a particular substance, A, is equal to a. A *zero-order reaction* is independent of the concentration of other species. The rate of the reaction is given by $dC_A/dt = \pm k$. A *first-order reaction,* where the rate $dC_A/dt = \pm kC_A$, is the most commonly used reaction order. There is also a *second-order reaction,* whose rate can be written in two different ways: (1) $dC_A/dt = \pm kC_A^2$, where A is independent of other chemical species; and (2) $dC_A/dt = \pm kC_A^a C_B^b$ where species B affects the reaction, and $a + b = 2$.

The rate of reaction provides a real-time prediction of the concentrations of contaminants. The rates can also indicate the controlling reactions. These kinetic reactions may include precipitation–dissolution reactions and redox reactions. For instance, the time required to dissolve a metal salt may be predicted based on the rate of precipitation–dissolution reactions. Similarly, if a dissolved contaminant is to be precipitated due to redox reactions, the rates of redox reactions could provide an estimate of the time required for formation of the precipitates.

3.3.9 Gas Laws

When substances exist in gaseous states, their behavior is controlled by the following gas laws:

- Boyle's law
- Charles's law
- Ideal gas law
- Dalton's law of partial pressures
- Henry's law of gas solubility

Boyle's law states that at constant temperature, the volume of a fixed amount of gas (V) is inversely proportional to the pressure (P). In equation form this can be written as $PV = k$, where k is a constant.

Charles's law states that volume at a fixed amount of gas (V) at a constant pressure is directly proportional to the absolute temperature (T). In equation form, this can be written as $V = kT$, where k is a constant.

The *ideal gas law* states that PV is proportional to nT, where T is the absolute temperature and n is the mass in terms of moles. In equation form, it can be written as $PV = nRT$, where R is the ideal gas constant.

Dalton's law of partial pressures states that the total pressure of a gas mixture (P_t) equals the sum of the partial pressures exerted by each individual component gas. In equation form, it can be expressed as $P_t = P_1 + P_2 + \cdots + P_n$, where P_1, P_2, . . . are the partial pressures of components 1, 2, . . . , respectively. The partial pressure of an individual component gas (i) is given by the equation $P_i = (n_i/n)P_t$. The value n_i/n is known as the *mole fraction.*

Henry's law describes the solubility of gas in a liquid. The weight of a gas that can dissolve in a given volume of liquid is directly proportional to the partial pressure that the gas exerts on the surface of the liquid. Under equilibrium conditions, $G_{(g)} \Leftrightarrow G_{(aq)}$, where $G_{(g)}$ and $G_{(aq)}$ are the gas and dissolved phases of a gas, respectively. The equilibrium constant (K_{eq}) for this situation is called *Henry's constant,* K_H, and is defined as $K_H = [G_{(aq)}]/[G_{(g)}]$. The concentration of the gas in the dissolved phase, $[G_{(aq)}]$, is given in moles or milligrams per liter, while the concentration of the gas in the gas phase, $[G_{(g)}]$, is given in units of atmospheres. Assuming equilibrium, a simplified relationship is $C_A = K_H P_A$ where C_A is the concentration of gas A in a liquid (in mg/L), P_A is the partial pressure of gas A in a gas mixture (in atmospheres), and K_H is the Henry's law constant (in mg/L · atm). Table 3.6 provides Henry's law constants for selected gases. Henry's law is used to determine the volatilization of dissolved gases during remedial operations; this is explained further in Chapter 6.

3.4 ORGANIC CHEMISTRY BACKGROUND

Organic chemistry is the study of compounds that contain carbon. A few exceptions, however, are CO_2 and

Table 3.6 **Henry's constants for selected compounds**[a]

Compound	K_H (mol/L · atm)
$Cl_2(g)$	1.38×10^{-2}
$CH_4(g)$	1.29×10^{-3}
$CH_2O(g)$	6.3×10^3
$CH_3COOH(g)$	7.66×10^2
$CO(g)$	9.55×10^{-4}
$CO_2(g)$	3.23×10^{-2}
$H_2(g)$	7.8×10^{-4}
$H_2O_2(g)$	1.0×10^5
$HNO_2(g)$	49
$H_2S(g)$	1.05×10^{-1}
$N_2(g)$	6.61×10^{-4}
$NO(g)$	1.9×10^{-3}
$NO_2(g)$	1.00×10^{-2}
$N_2O(g)$	2.57×10^{-2}
$O_2(g)$	1.26×10^{-3}
$O_3(g)$	9.4×10^{-3}
$SO_2(g)$	1.3
Alkanes	
n-Octane (C_8H_{18})	3.14×10^{-4}
1-Hexane (C_6H_{12})	2.86×10^{-3}
Aromatic substances	
Benzene	0.184
Toluene	0.151
Naphthalene	2.495
Biphenyl	0.654
Pecticides	
p,p'-DDT ($C_{14}H_9Cl_5$)	1.06×10^2
Lindane	3.1×10^2
Dieldrin	8.59×10^1
Polychlorinated biphenyls (PCBs)	
2,2'4,4'-CBP ($C_{12}H_6Cl_4$)	3.23
2,2'3,3'4,4'-CBP ($C_{12}H_4Cl_6$)	8.59
Dimethylsulfide (C_2H_6S)	0.572
Mercury (Hg^0)	0.0859
Water	1.759×10^3

Source: Stumm and Morgan (1996).

[a] General equilibrium reaction: $G_{(g)} \Leftrightarrow G_{(aq)}$.

related compounds, such as carbonic acids, bicarbonates, and carbonates, as well as cyanides and cyanates. These all contain carbon but are not considered organic compounds. Living systems are made up of organic molecules. In addition, numerous organic compounds are produced synthetically.

3.4.1 Characteristics of Organic Compounds

Carbon is a unique atom for three main reasons. First, carbon forms four bonds. These bonds can be with carbon atoms or with other atoms near carbon in the periodic table. Most commonly, these other atoms are hydrogen and oxygen. However, nitrogen, phosphorus, sulfur, and certain metals also bond with carbon. Second, carbon bonds are covalent; the electrons are shared between atoms. Third, carbon bonds are stable, allowing chains to be formed. Because of these unique characteristics, carbon can form numerous organic compounds.

Organic compounds possess the following properties, which make them significantly different from inorganic compounds:

1. They are combustible.

2. They generally possess lower melting and boiling points than do inorganic compounds.

3. They are less soluble in water than are inorganic compounds.

4. They can have several isomers (same chemical formulas but a different arrangement of atoms).

5. Their reactions are molecular, leading to generally slower reactions (inorganic compounds have ionic reactions).

6. The valency or oxidation state of carbon allows for multiple bonding.

7. Organic compounds are biodegradable; they serve as a source of food for bacteria and other microorganisms.

The specific properties of organic compounds depend on three main factors: the chain structure, the groups attached, and the presence of multiple bonds. For simplicity, the bond between carbon atoms is denoted as C—C. In general, there are three types of carbon chain structures:

1. *Basic chain structure*

$$-\overset{|}{C}-\overset{|}{C}-\overset{|}{C}-\overset{|}{C}-\overset{|}{C}-$$

2. *Branched chain structure*

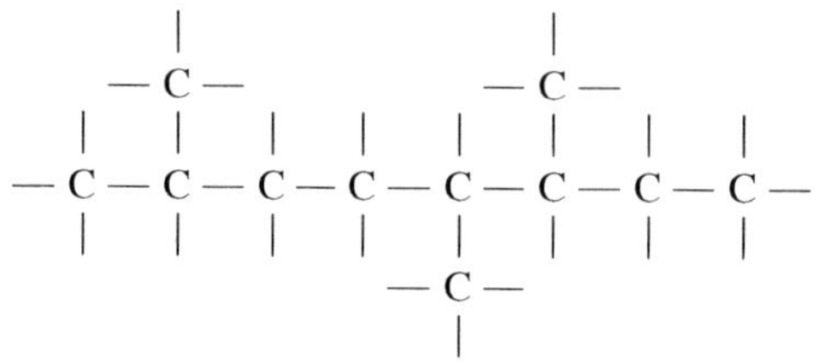

3. *Cyclic structure*

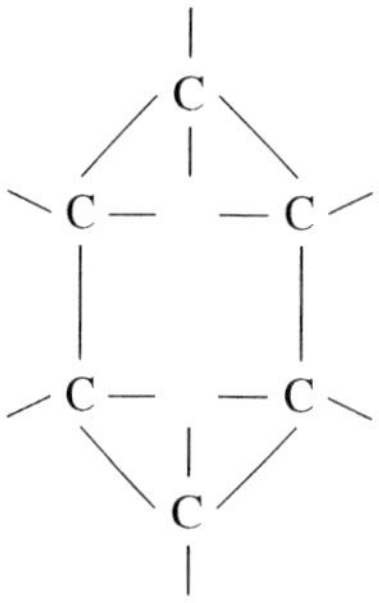

Different functional groups may be attached to the carbon chain. Depending on the functional group attached, different groups of organic compounds are formed.

Functional Group	Organic Compound
—H	hydrocarbons
—OH	alcohols
O	aldehydes
|	
—C—H	
—COOH	carboxylic acids
O	ketones
|	
—C—	
—O—	ethers
—NH$_2$	amines
—SH	thiols
—NO$_2$	nitro compounds

When a carbon atom bonds with another carbon atom, it is possible to form a single, double, or triple bond. If only single bonds exist, the compound is known as an *alkane*. If at least one double bond is present, the compound is known as an *alkene*. The double bond is denoted C=C. If at least one triple bond is present, the compound is known as an alkyne. The triple bond is denoted C≡C.

3.4.2 Classification of Organic Compounds

It is important for geoenvironmental engineers to be familiar with the terminology involved in the classification of organic compounds. There are two common methods of classifying these compounds: classification based on structure, and classification based on functional groups.

Classification by Structure. Based on structure, organic compounds are classified into aliphatic, aromatic, and heterocyclic compounds. The *aliphatic compounds* are molecules that consist of simple straight chains or branched straight chains. Compounds such as octane, 3-methyl-5-propyloctane, acetylene, and propene are classified as aliphatic compounds.

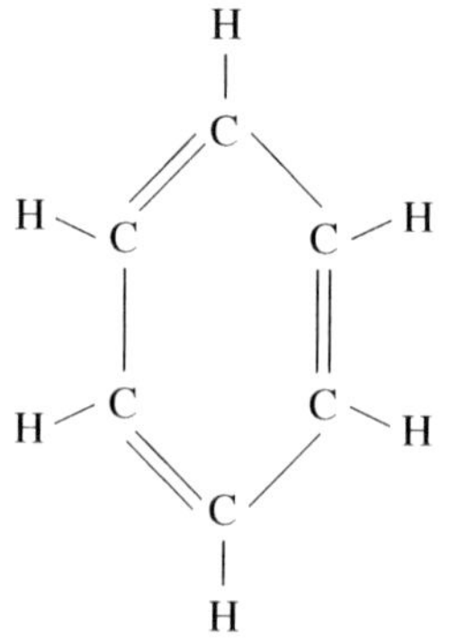

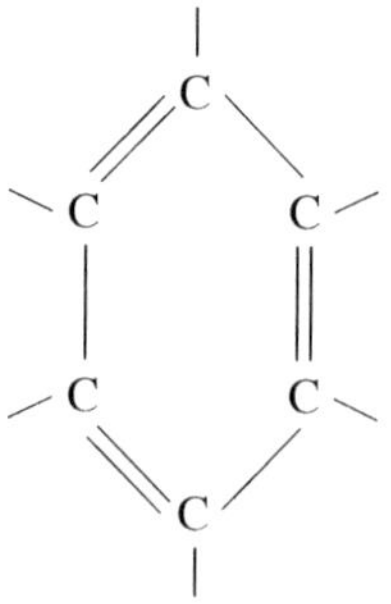

The *aromatic compounds* are molecules that consist of cyclic chains. The basic building block for aromatic compounds is the benzene skeleton or benzene ring, with the formula C_6H_6. The benzene molecule, can be expressed as

or more simply as

Multiple benzene rings result in different compounds, such as, phenanthrene:

Similar to aromatic compounds, *heterocyclic compounds* are molecules that contain a ring structure, but some hydrogens are substituted with other atoms (e.g., Cl) or functional groups (e.g., OH, NO$_2$). Phenol, nitrobenzene, and pentachlorophenol are examples of heterocyclic compounds.

Classification by Functional Group. As stated above, many different functional groups may be attached to the carbon chain, leading to different organic compound types. For simplicity and based on common occurrence at contaminated sites, organic compounds are divided into two groups: hydrocarbons, and all other organic compounds. The hydrocarbons, which consist of C and H, can be divided into three subgroups: aliphatic, aromatic, and halogenated.

Aliphatic hydrocarbons consist of a straight or branched-chain structure with single, double, or triple bonds. Examples of alkanes, or single-bonded hydrocarbons include methane, ethane, propane, and hexane. Ethene and propene are examples of alkenes, or double-bonded hydrocarbons. Alkynes, or triple-bonded hydrocarbons, are not commonly found.

Aromatic hydrocarbons are based on benzene rings. Examples include benzene, naphthalene, and others with an increasing number of rings. These hydrocarbons are also known as *polycyclic aromatic hydrocarbons* (PAHs).

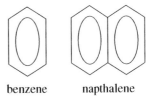

benzene   napthalene

It is also possible for a benzene ring to bond to a functional group, creating isomers. An example is toluene:

CH$_3$

An additional type of aromatic hydrocarbon is the *polychlorinated biphenyl* (PCB), which consists of a benzene ring joined to another functional group, such as phenyl, in combination with chlorine. Phenol is also an aromatic ring, but with an attached hydroxyl group.

Halogenated hydrocarbons consist of both aliphatic and aromatic hydrocarbons but contain one or more halogen atoms (Cl, Br, F) as well. A classic example is trichloroethene, a common solvent used in industry:

$$Cl\diagdown \qquad \diagup Cl$$
$$C = C$$
$$Cl \diagup \qquad \diagdown H$$

A second example is vinyl chloride:

$$H\diagdown \qquad \diagup Cl$$
$$C = C$$
$$H \diagup \qquad \diagdown H$$

Halogenated hydrocarbons are a common problem at contaminated sites. In fact, all three types of hydrocarbons account for most of the organic contaminants encountered.

The other organic compounds, formed with different functional groups, are not as common a problem. Examples of these are phosphorus-based compounds (found in pesticides), ketones, aldehydes, and alcohols.

3.4.3 Properties of Organic Compounds

The properties of organic compounds that are important from a geoenvironmental engineering point of view are:

- Aqueous solubility
- Vapor pressure
- Henry's law constant
- Octanol–water partition coefficient
- Rate of biodegradation

Detailed explanation of these properties is provided in Chapters 6 and 8. These properties are measured by experiments or calculated using theoretical equations and correlations. For many organic compounds, these properties have been reported in standard chemistry handbooks (e.g., Perry and Green, 1984; Lide and Frederikse, 1994; Agteren et al., 1998). Quantitative values of these properties for different organic compounds will be cited in succeeding chapters, as applicable.

3.5 NUCLEAR CHEMISTRY BACKGROUND

Nuclear chemistry is a speciality subject and is not commonly dealt with by environmental professionals. However, when dealing with radioactive wastes or with sites contaminated with radionuclides, an understanding of nuclear chemistry is essential. Basic concepts and terminology of nuclear chemistry are provided in this section; more detailed information can be found in other references (e.g., Choppin and Rydberg, 1980; Arnikar, 1987).

Some elements have nuclei with unfavorable proton/neutron ratios. These elements spontaneously break down to achieve a more stable form. During this breakdown process, energy is released and the unstable nuclei emit ionizing radiation in the form of α particles, β particles, and γ rays. This process is known as *radioactive decay.*

All elements with an atomic number greater than 83 (termed *radionuclides*) are unstable. Examples of common natural radionuclides include uranium and thorium. Other examples of radionuclides are plutonium, strontium, cesium, and cobalt. All of these radionuclides are used extensively for nuclear weapons and nuclear power generation.

Radioactive decay is described by a first-order reaction, $dN/dt = -kN$ where N is the number or mass of the undecomposed nucleide and k is the decay constant. More commonly, the radioactivity of a substance is described by its half-life, $t_{1/2}$. The half-life of a substance is the time required for 50% of a given number of radioactive atoms to decay. $t_{1/2}$ and k are related by the equation $t_{1/2} = 0.693/k$. The unit of radioactivity is the curie (Ci; 1 Ci = 3.7×10^{10} disintegrations per second).

Radioactive contamination is common at Department of Energy (DOE) sites and nuclear power plant sites. These sites are limited in number, but their remediation is complicated and challenging and requires special expertise in nuclear chemistry.

3.6 CHEMICAL ANALYSIS METHODS

Accurate determination of the nature of chemicals and their concentrations is critical in geoenvironmental studies. Our focus is on toxic inorganic and organic contamination, as this is the most common contamination at the majority of sites. The following two standard references are used routinely to measure the inorganic and organic contaminants:

1. *Standard Methods for the Examination of Water and Wastewater,* published by the American

Public Health Association (APHA), Washington, DC

2. *Test Methods for Evaluating Solid Wastes,* Vols. IA and IB, *Laboratory Manual, Physical/Chemical Methods,* also known as SW-846, prepared by the USEPA

Different types of equipment and analysis procedures are available that can detect chemical concentrations in the range parts per billion (10^{-9}) to parts per trillion (10^{-12}). A strict quality assurance and quality control (QA/QC) is required to ensure accuracy of the analysis results. Generally, a QA/QC plan is prepared, conformed to this plan during testing, and documentation prepared for each project (USEPA, 1986).

3.6.1 Methods Used to Measure Toxic Inorganic Chemicals

The toxic inorganic chemicals in water are commonly measured using the following methods (USEPA, 1986; APHA, 1992): (1) atomic absorption spectrophotometry (AAS) and (2) inductively coupled plasma–mass spectrometry (ICP-MS). AAS is suitable for quantifying the concentration of inorganic chemicals, particularly metals, in water. In this procedure, a sample is aspirated and atomized in a flame. A light beam from a hollow cathode lamp is directed through the flame, into a monochromator, and onto a detector that measures the amount of absorbed light. Absorption depends on the presence of free unexcited ground-state atoms in the flame. Because the wavelength of the light beam is characteristic only of the metal being determined, the light energy absorbed by the flame is a measure of the concentration of that metal in the sample.

ICP-MS is suitable for detection and quantification of a large number of inorganic chemicals simultaneously in water. In this method, a sample is ionized at a high temperature in a *plasma* (a high-temperature ionized gas) and then sent to the spectrometer, which detects and quantifies individual atoms by matching the signatures of various chemicals with those of the known chemicals.

3.6.2 Methods Used to Measure Toxic Organic Chemicals

The toxic organic chemicals in water are commonly measured using the following methods (USEPA, 1986; APHA, 1992): (1) gas chromatography (GC), (2) gas chromatography–mass spectrometry (GC-MS), and (3) high-performance liquid chromatography (HPLC). GC is suitable for the analysis of organic compounds that can be volatilized without changing chemical structure. It can be used to determine large numbers of volatile hydrocarbons. In this method, a liquid sample is injected into the instrument, which vaporizes the chemicals. They then travel at different speeds through a GC column and exit at different times. By recording the chromatograph (i.e., the pattern that is exhibited by the chemicals as they exit), the nature and quantities of organic compounds present in the sample can be determined.

GC-MS is used to identify and quantify a large number of organic compounds, including volatile and semivolatile organics. In this method, a small sample is injected into the instrument, where the sample vaporizes instantly. Individual molecules travel through a narrow column (tube) in the instrument at different speeds, exit out of the column, and enter the mass spectrometer, where an electronic beam breaks them into fragments. By studying the pattern formed by the broken molecules and their abundance, the nature and quantity of the chemicals present in the sample can be determined.

HPLC is suitable for analyzing less volatile compounds that are not suitable to be tested with GC. It is used to identify and quantify organic compounds such as polycyclic aromatic hydrocarbons (PAHs). In this method, a sample is injected into the instrument along with a solvent. The detector in the instrument identifies the compounds.

3.7 SUMMARY

Basic concepts such as chemical concentration, equilibrium reactions, acids and bases, kinetics, redox reactions, and gas laws are key parts of understanding inorganic contaminants. The characterization, classification, and properties of organic chemicals are important aspects of understanding organic contaminants. Knowledge of radioactivity at the atomic level is essential in understanding radionuclide contamination.

A background in inorganic, organic, and nuclear chemistry is essential for an understanding of the cause of contamination and to develop technologies for its removal. An in-situ remediation method known as chemical oxidation, for example, is based on redox re-actions between organic/inorganic soil contaminants and a given oxidant. It is important to be familiar with the chemical processes that determine whether the oxidant will alter the contaminant successfully and which new substances will be produced by the reaction. Nuclear chemistry is important if one is dealing with radioactive contamination, which is far less common than organic and inorganic contamination. When the situation does arise, however, it is imperative that the problem be resolved promptly and efficiently by someone who is well trained in that field.

QUESTIONS/PROBLEMS

3.1. What are the oxidation numbers for each of the atoms in the following compounds? **(a)** $Al_2(SO_4)_3$; **(b)** $C_6H_{12}O_6$; **(c)** NO_3^-.

3.2. Calculate the formula weight of **(a)** NaCl and **(b)** H_3AsO_4.

3.3. Calculate the number of moles in 100 g of the following samples: **(a)** NaCl; **(b)** $KMnO_4$; **(c)** $PbCrO_4$; **(d)** $C_6H_{12}O_6$.

3.4. Balance the following reaction:

$$MnO_2 + NaCl + H_2SO_4 \rightarrow MnSO_4 + H_2O + Cl_2 + Na_2SO_4$$

3.5. The pH of a solution is 5.0. What are the pOH, the hydrogen ion concentration, and hydroxyl ion concentration at 25°C?

3.6. Calculate the molar solubility of $PbCl_2$ in water. (*Hint:* Obtain the K_{sp} value from Table 3.3.)

3.7. What is a ligand?

3.8. Write the half-reaction for the oxidation of NH_4^+ to N_2.

3.9. State Henry's law and explain its significance.

3.10. What are the major functional groups, and how do they affect organic molecules?

3.11. What are the general chemical formulas for alcohols and ketones?

3.12. What properties of organic compounds are important from a geoenvironmental engineering point of view? Summarize these properties for the following compounds: **(a)** benzene; **(b)** trichloroethylene; **(c)** phenanthrene.

REFERENCES

Agteren, M. H., Keuning, S., and Janssen, D. B., *Handbook on Biodegradation and Biological Treatment of Hazardous Organic Compounds,* Kluwer Academic, Norwell, MA, 1998.

APHA (American Public Health Association), *Standard Methods for the Examination of Water and Wastewater,* 16th ed., APHA, Washington, DC, 1992.

Arnikar, H. J., *Essentials of Nuclear Chemistry,* Wiley, New York, 1987.

Brady, J. E., and Holum, J. R., *Fundamentals of Chemistry,* Wiley, New York, 1984.

Choppin, G. R., and Rydberg, J., *Nuclear Chemistry: Theory and Applications,* Pergamon Press, New York, 1980.

CFR (*Code of Federal Regulations*), U.S. Government Printing Office, Washington, DC, 2002

Droste, R. L., *Theory and Practice of Water and Wastewater Treatment,* Wiley, New York, 1997.

Lide, D. R., and Frederikse, H. P. R., *Handbook of Chemistry and Physics,* 75th ed., CRC Press, Boca Raton, FL, 1994.

O'Connor, R., *Fundamentals of Chemistry,* Harper & Row, New York, 1974.

Perry, R. H., and Green, D. W., *Perry's Chemical Engineers' Handbook,* 6th ed., McGraw-Hill, New York, 1984.

Sawyer, C. N., McCarty, P. L., and Parkin, G. F., *Chemistry for Environmental Engineering,* 5th ed., McGraw-Hill, New York, 1994.

Snoeynik, V. L., and Jenkins, D., *Water Chemistry,* Wiley, New York, 1980.

Solomons, T. W. G., *Organic Chemistry,* 2nd ed., Wiley, New York, 1980.

Stumm, W., and Morgan, J. J., *Aquatic Chemistry,* 2nd ed., Wiley, New York, 1981.

USDHHS (U.S. Department of Health and Human Services), *Occupational Safety and Health Guidance Manual for Hazardous Waste Site Activities,* USDHHS, Washington, DC, 1985.

USEPA (U.S. Environmental Protection Agency), *Test Methods for Evaluating Solid Wastes,* Vols. IA and IB, *Laboratory Manual, Physical/Chemical Methods,* SW-846, USEPA, Washington, DC, 1986.

———, *Cleaning Up the Nation's Waste Sites: Markets and Technology Trends,* EPA/542/R-96/005, USEPA, Washington, DC, 1997.

Wertheim, E., and Jeskey, H., *Introductory Organic Chemistry,* 3rd ed., McGraw-Hill, New York, 1956.

4

COMPOSITION OF SOILS

4.1 INTRODUCTION

Classical geotechnical engineering is concerned primarily with the mechanical behavior of soils. Soil is considered to be a three-phase material that consists of solid, liquid, and gas. The liquid phase is assumed to be water and the gas phase is assumed to be air. The behavior and properties, such as permeability, compressibility, shrinkage and swelling, and shear strength of soils depend on the interaction of these three phases. However, when dealing with contaminated soils and chemical transport through the soils, it is important to characterize the composition of the three phases accurately. The liquid phase may contain water as well as chemicals and dissolved gases, while the gas phase may contain air as well as chemical vapors. The presence of chemicals may influence the soil properties and behavior. Therefore, detailed understanding of the different phases in soils is critical when determining the chemical compatibility of materials used in waste containment systems and in developing effective soil remediation methods. This chapter provides a brief description of how soils are formed, the composition of soils, and methods of determining the soil composition. The chemical and mechanical properties of soils are presented in Chapter 5.

4.2 SOIL FORMATION

To understand the composition of soils, it is important to know how they are formed. Earth scientists have postulated that the Earth was formed 4.5 billion years ago from a huge molten ball of cosmic gases and debris. Our point of interest is *Earth's crust,* which extends approximately 10 to 15 km below the ground surface and consists of bedrock and overlying a layer of unconsolidated material. The unconsolidated layer consists of several different layers of soils that are composed mostly of weathered rock. The composition of Earth's crust is shown in Table 4.1. The elements listed exist in combinations, as minerals.

The soils in the unconsolidated material layer are formed due to the weathering of three types of rocks: igneous, sedimentary, and metamorphic. *Igneous rocks* are parent rocks formed by the cooling of molten magma. Granite and basalt are examples of igneous rocks. The typical composition of these rocks is given in Table 4.2. The weathering and lithification of igneous rocks forms *sedimentary rocks.* The weathering process reduces rock mass to fragments, which are transported by wind, water, and ice. When these fragments are settled, compacted, and lithified, a new rock is formed from the fragments. Examples of sedimentary rocks include shale, sandstone, limestone, and dolomite. *Metamorphic rocks* are formed by metamorphism through high temperatures and pressures acting on either sedimentary or igneous rocks. The original rock undergoes both chemical and physical alterations. Examples of metamorphic rocks include slate, quartzite, and marble. Detailed information on the various rock types can be found in Goodman (1993).

The weathering of rocks that results in soils can occur due to mechanical (or physical) processes, chemical processes, or a combination of mechanical and chemical processes. The *mechanical processes* include the following (Mitchell, 1993):

- *Unloading,* which causes cracks and joints in rocks to open up, in turn causing reduction of the effective confining pressure; this creates uplift, erosion, or changes in fluid pressures.

Table 4.1 **Composition of Earth's crust**

Element	Percent by Volume	Percent by Weight
Oxygen (O)	93.8	47.3
Silicon (Si)	0.9	27.7
Aluminum (Al)	0.5	7.8
Iron (Fe)	0.4	4.5
Calcium (Ca)	1.0	3.5
Sodium (Na)	1.3	2.5
Potassium (K)	1.8	2.5
Magnesium (Mg)	0.3	2.2
Others (titanium, hydrogen, carbon, phosphorus, sulfur, etc.)	<0.1	2.0

Source: Mitchell (1993).

Table 4.2 **Chemical composition of basaltic and granitic rocks**

Chemical	Percent by Weight	
	Basalt	Granite
SiO_2	50.4	69.5
Al_2O_3	16.9	15.7
Fe_2O_3	2.8	2.6
FeO	6.9	0.7
MgO	7.5	0.9
CaO	8.2	1.0
MnO	0.1	—
K_2O	1.6	2.8
Na_2O	3.3	2.6
TiO_2	1.4	0.9
H_2O	0.8	1.0

Source: Tan (1994).

- *Thermal expansion and contraction,* which create planes of weakness.
- *Crystal growth* due to the freezing of water, causing disintegration.
- *Colloid plucking,* which refers to shrinkage of colloidal materials upon drying, can exert tensile stresses on adjacent surfaces.
- *Organic activity,* which represents plant roots in fractures, worms, rodents and humans, can cause disintegration.

The *chemical processes* include the following (Mitchell, 1993):

- *Hydration* of adsorbed water.
- *Oxidation* of rocks, rich with iron, in the presence of dissolved oxygen in water (rainwater).
- *Hydrolysis:* reaction between minerals and H^+ and OH^- of water. The H^+ ion is small in size; hence, it can enter the crystal lattice of a mineral.
- *Chelation:* complexation and removal of metal ions.
- *Cation exchange*[1]: ions held by Al_2O_3 and SiO_2 influence the type of clay mineral.
- *Carbonation:* combination of carbonate or bicarbonate ions with rock.

A detailed explanation of both physical and chemical processes may be found in Mitchell (1993).

Based on the method of deposition, soils may be classified as either residual soils or transported soils. The movement of weathered rock to another site results in *transported soils.* The transporting agents may be water, glaciers, wind, or gravity and the soils formed are called *alluvial deposits, glacial deposits, aeolian deposits* (loess and dune sands), and *talus* or *colluvial deposits,* respectively. The thickness and composition of transported soil layers vary significantly. *Residual soils* are formed in place through weathering of the parent rock. These soils are commonly found in tropical areas of the southeastern and southwestern United States.

Weathering and leaching[2] cause changes in soil with depth. Topsoil and organic matter can exist from a few inches in depth to up to several feet below the ground surface. Underlying this top layer, soil layers possessing carbonates, sulfates, and chlorides exist with thicknesses varying up to several feet. A transitional zone containing weathered parent rock generally overlies the parent rock.

[1] A chemical process in which cations of like charge are exchanged equally between two media.
[2] Removing soluble and other constituents by the action of a percolating liquid.

4.3 SOIL COMPOSITION

All soils consist of a solid phase, which is a mixture of inorganic and organic materials, making up the skeletal framework. Enclosed within this framework is a system of pores, which are shared jointly by the liquid and gaseous phases. In classic geotechnical engineering, the liquid phase is considered to be water and the gaseous phase is considered to be air. However, for geoenvironmental studies, the composition of liquid and gaseous phases may need more detailed evaluation than is generally required for understanding conventional geotechnical behavior. In addition, the exact composition of solid phase should be known. In this section we detail the three phases of soils: solid phase, liquid phase, and gas phase.

4.3.1 Solid Phase

In geotechnical engineering, the soil solid phase is classified simply according to particle sizes as boulders, cobbles, gravel, sand, silt, clay and colloids, as shown in Table 4.3. This classification is not universal; different agencies use different grade boundaries. Boulders and cobbles are rock fragments and are generally separated. Gravels and sands are the result of weathering rock and can exist in different shapes: rounded, subrounded, angular, and subangular. Silt, clay, and colloidal particles are invisible to the naked eye. Generally, grain size less than 0.002 mm is called *clay fraction.* Gravels and sands are nonplastic and may be poorly graded or well graded. Silts could be

Table 4.3 Soil classification based on grain sizes according to ASTM D 422

Soil Fraction	Grain Size (mm)
Boulders	>300
Cobbles	75–300
Gravel	4.75–75
Sand	0.075–4.75
Silt	0.005–0.075
Clay	0.001–0.005
Colloids	<0.001

Source: ASTM (2003).

either plastic or nonplastic, and clays and colloids exhibit different degrees of plasticity.

For geoenvironmental studies, the physical and chemical composition of the soil solid phase should be known. The solid-phase composition can be divided into inorganic constituents and organic constituents. The *inorganic constituents* are derived from the weathering of rocks and consist of minerals of varying sizes and composition. In addition to these minerals, the inorganic fraction may consist of oxides, hydroxides and other amorphous, noncrystalline constituents. The *organic constituents* are derived from biomass and consist of both living and dead organic matter. The inorganic and organic constituents and the various methods to determine these constituents are described in the following subsections.

Inorganic Constituents. The inorganic fraction of the solid phase consists mainly of soil minerals. Free oxides and hydroxides may also be present in the form of discrete particles, coatings on soil particles, or cementing agents between soil particles. Chemical analyses provide the quantitative estimates of these oxides and hydroxides. Soil minerals are, by definition, naturally occurring, homogeneous, solid inorganic substances that have definite physical properties and chemical compositions. Soil minerals originate from parent rocks, which are composed mostly of the elements O, Si, Al, Fe, Ca, Mg, Na, and K, as shown in Table 4.2. Soil minerals are therefore made up predominantly of these elements. Soil minerals may be grouped into (1) carbonates, which are predominantly calcite ($CaCO_3$) and dolomite [$CaMg(CO_3)_2$], and (2) silicates, such as quartz, feldspar, mica, and clay minerals (kaolinite, illite, montmorillonite, etc.). The carbonate and silicate minerals that are commonly found in soils are described briefly in the following sections. For more detailed information, readers should refer to Grim (1968), Moore and Reynolds (1989), and Blackburn and Dennen (1994).

Carbonate Minerals. The carbonate minerals are derived from limestone or dolomite rocks or reefs. These minerals exhibit perfect cleavage planes; therefore, they break down readily into smaller sizes. If the addition of dilute hydrochloric acid (HCl) to the soil results in effervescence, it is indicative that carbonates are present. Several chemical analysis methods are available to determine the amount of carbonates present in soils (Black, 1965). These methods include:

- *Vacuum distillation and titration method.* In this method, carbon dioxide is liberated in vacuum at low HCl and low temperature (50 to 55°C) in the presence of $SnCl_2$ (antioxidant). This decreases the tendency for CO_2 to evolve from soil organic matter. Evolved CO_2 is swept by water vapor into an upper flask, where it is absorbed by $Ba(OH)_2$. The absorption occurs concurrently with the evolution to minimize loss of CO_2. Titration of the $Ba(OH)_2$ provides an accurate index of carbonate content in the sample.

- *Gravimetric method.* In this method, carbon dioxide is released by treating the soil sample with $2 N H_2SO_4$ containing $FeSO_4$ as an antioxidant to prevent release of CO_2 from the organic matter. The air stream containing evolved CO_2 is passed through a series of traps to remove water and any extraneous constituents. The evolved CO_2 is then absorbed into a Nesbitt absorption bulb. Weighing the bulb before and after absorption provides the amount of CO_2, and this value determines the carbonate content.

- *Acid neutralization method.* In this method, soil is treated with hot dilute HCl and all carbonates are decomposed. Few other soil constituents may also react with the acid. Thus, the acid neutralized by the soil is a rough index of the carbonate content of the soil samples.

- *Gravimetric method for loss of carbon dioxide.* When carbonates are decomposed with acid, CO_2 is released to the atmosphere. The weight of CO_2 lost is an index of the carbonate content of the soil sample.

- *Volumetric calcimeter method.* When carbonates are treated with acid in a closed system at constant pressure and temperature, the increase in volume is a direct measure of CO_2 if no other gases are evolved. The apparatus used to measure this increase in volume is called a *calcimeter.*

- *Pressure calcimeter method.* When carbonates are treated with acid in a closed system at constant

volume and temperature, the increase in pressure observable on a manometer is a direct measure of CO_2. The apparatus used to measure the pressure increase is termed a *pressure calcimeter.*

All the methods above essentially involve measuring the CO_2 released from the carbonates when an acid is added. These measurements should be taken carefully, however, so that CO_2 released from the decomposition of soil organic matter is not included in the measurement.

Silicate Minerals. Silicate minerals are the most common minerals found in soils. These minerals are derived mainly from the mechanical and chemical weathering of igneous and metamorphic rocks. Silicates possess crystalline structures and are formed by various combinations of the two basic building units of silicates: (1) the silica tetrahedron, and (2) the aluminum or magnesium octahedron. The *silica tetrahedron* $[(SiO_4)^{4-}]$ consists of silicon (Si^{4+}) bonded to four oxygen atoms (O^{2-}), which results in a strong bond, as shown in Figure 4.1(*a*). A tetrahedral layer or sheet is composed of silica tetrahedra, all sitting on one of their triangular faces, with their apexes pointing in the same direction. At the base, the tetrahedra are connected to each other by sharing oxygen atoms, as shown in Figure 4.1(*b*). Schematically, this sheet is represented as shown in Figure 4.1(*c*) and (*d*). The *octahedron,* on the other hand, consists of six hydroxyl radicals surrounding Al^{3+}, Fe^{3+}, Fe^{2+}, or Mg^{2+}, as shown in Figure 4.2(*a*). An octahedral layer or sheet is composed of octahedra, all lying on one of their triangular faces, where each corner is shared by three octahedra, as shown in Figure 4.2(*b*). A schematic representation of this layer is shown in Figure 4.2(*c*) and (*d*). If the octahedron consists of Al, it is called a *Gibbsite sheet,* or if the octahedron consists of Mg, it is called a *Brucite sheet.* Different combinations of tetrahedral and octahedral units result in different types of silicate minerals.

Common Silicate Minerals in Coarse-Grained Soils. Coarse-grained soils consist predominantly of silicate minerals such as quartz, feldspar, and mica. These various minerals are described below.

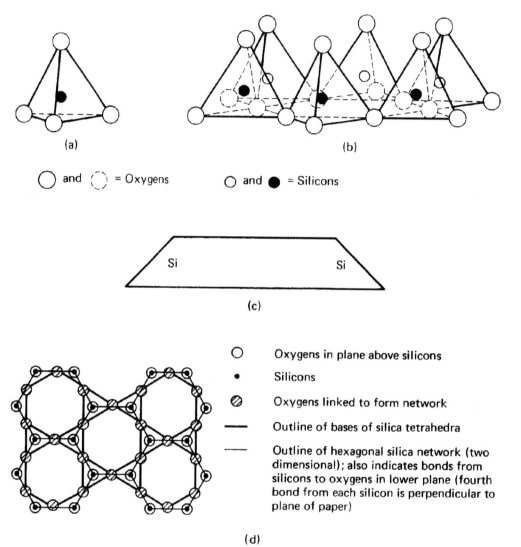

Figure 4.1 (a) *Single silica tetrahedron from Grim (1959); (b) isometric view of the tetrahedral or silica sheet from Grim (1959); (c) schematic representation of the silica sheet from Lambe (1953); (d) top view of the silica sheet from Warshaw and Roy (1961). (From Holtz and Kovacs, © 1981. Reprinted by permission of Pearson Education, Inc., Upper Saddle River, N.J.)*

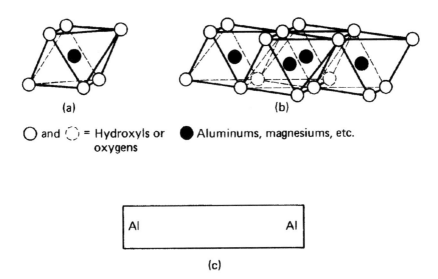

(a)

(b)

○ and ◌ = Hydroxyls or oxygens   ● Aluminums, magnesiums, etc.

Al	Al

(c)

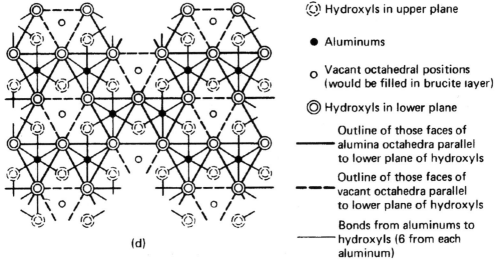

(d)

◎ Hydroxyls in upper plane

● Aluminums

○ Vacant octahedral positions (would be filled in brucite layer)

◎ Hydroxyls in lower plane

—— Outline of those faces of alumina octahedra parallel to lower plane of hydroxyls

- - - Outline of those faces of vacant octahedra parallel to lower plane of hydroxyls

—— Bonds from aluminums to hydroxyls (6 from each aluminum)

Figure 4.2 (a) *Single aluminum (or magnesium) octahedron from Grim (1959);* (b) *isometric view of the octahedral sheet from Grim (1959);* (c) *schematic representation of the octahedral or alumina (or) magnesia) sheet from Lambe (1953);* (d) *top view of the octahedral sheet from Warshaw and Roy (1961). (From Holtz and Kovacs, © 1981. Reprinted by permission of Pearson Education, Inc., Upper Saddle River, NJ.)*

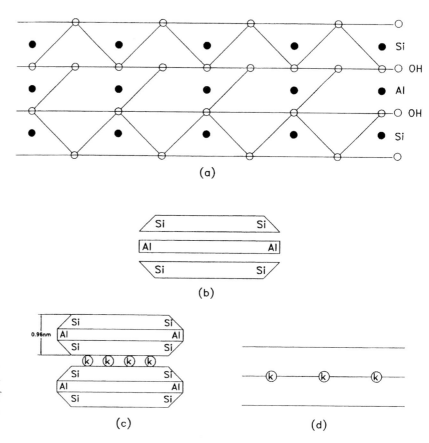

Figure 4.3 (a) *Atomic structure of mica;* (b) *schematic diagram of the structure of mica;* (c) *K-bonds;* (d) *schematic diagram of k-bond.*

- *Quartz.* Quartz has a chemical composition of SiO_2. This mineral is most commonly found in soils. In quartz, silica tetrahedra are arranged in a three-dimensional network. Each tetrahedron is firmly and equally braced in all directions. As a result, the mineral has no cleavage[3] and is very hard and insoluble in all common acids.

- *Feldspar.* Feldspar has a crystalline structure similar to quartz with one exception: aluminum (Al^{3+}) atoms replace some of the silicon atoms (Si^{4+}) in the tetrahedron. This structure results in a negative charge, balanced by cations such as K^+, Na^+, and Ca^{2+}. Depending on the type of

cation balancing the charge, the following types of feldspars will result: (a) orthoclase ($KAlSi_3O_8$) and (b) plagioclase, which can be either albite ($NaAlSi_3O_8$) or anorthite ($CaAl_2Si_2O_8$). The presence of cations in the tetrahedra gives feldspars cleavage; hence, they are less resistant to mechanical and chemical weathering.

- *Mica.* Mica is made up of one octahedral layer sandwiched between two tetrahedral layers, as shown in Figure 4.3. Some of the silicon atoms (Si^{4+}) in the tetrahedral layer are replaced by Al^{3+} or Mg^{2+}, which causes a deficiency in positive charge balanced by the adsorption of potassium cations. Shared by adjacent sheets, the potassium (K^+) bonds the sheets to each other, as shown in Figure 4.3. Such bonding is also known as the *K-bond*. A particle of mica is formed from

---

[3] *Cleavage* is the tendency of a mineral to break parallel to a specific atomic plane due to weaker bonding.

many sheets connected to each other by the K-bond. The two most common types of mica are muscovite, $K_2Al_4(Al_2Si_6)O_{20}(OH)_4$, and biotite, $K_2Mg_6(Al_2Si_6)O_{20}(OH)_4$. Micas are nonswelling, moderately plastic, and relatively resistant to physical and chemical weathering.

Common Silicate Minerals in Fine-Grained Soils. Fine-grained soils consist of silicate minerals, predominantly clay minerals. Clay minerals are formed with different combinations of silicon tetrahedral and octahedral sheets. If the atomic structure of the mineral consists of one tetrahedral sheet and one octahedral sheet, the mineral is known as *1:1 type.* If the atomic structure consists of a combination of two tetrahedral sheets and one octahedral sheet, the mineral is known as *2:1 type.* The most common clay minerals encountered in fine-grained soils are kaolinite, illite, and montmorillonite.

- *Kaolinite.* One tetrahedral layer and one octahedral layer form a sheet of kaolinite. Particles of kaolinite are formed by the piling up of sheets on top of one another, with the octahedral face positioned toward the tetrahedral face. The strong bonding between the sheets is through O to OH, also known as the *hydrogen bond,* as shown in Figure 4.4.
- *Illite.* Illite is also known as *decomposed mica* or *hydrous mica.* A basic sheet of illite clay mineral is made up of one octahedral layer and two tetrahedral layers. The sheets of illite are connected to each other by the K-bond, as shown in Figure 4.3.
- *Montmorillonite.* In montmorillonite, one octahedral layer is sandwiched between two tetrahedral layers, as depicted in Figure 4.5. There are isomorphous substitutions in both the tetrahedral (Al^{3+} for Si^{4+}) and octahedral (Mg^{2+} for Al^{3+}) layers. No apparent secondary bonding exists to connect these sheets together; as a result, water entering between these sheets causes high swelling.

Kaolinite, which occurs commonly in soils, is very stable, with little tendency for volume change when exposed to water. Illite is commonly present in marine deposits and soils derived from micaceous rock and is more plastic than kaolinite. Illite clays do not expand when exposed to water unless there is a deficiency in potassium. Montmorillonite is commonly found in soils derived from ferromagnesium rocks and volcanic ashes. Because of negative charge resulting from isomorphous substitutions, montmorillonite clays readily absorb water between layers and undergo a large volume increase. The chemical and mechanical properties of these clay minerals are explained further in Chapter 5.

It is rare to find pure clay minerals in practice. However, relatively pure clay minerals may be obtained from the American Clay Minerals Society Repository and are often used for research purposes. The typical chemical composition of clay minerals is shown in Table 4.4.

Methods to Identify Soil Minerals. The determination of the mineralogy of soils is important when investigating the fundamental effects of soil composition and explaining various physicochemical phenomena. The "real-world" soils consist of different minerals in different proportions. The major types of minerals present in sand, silt, and clay are shown in Table 4.5. Methods to determine the amount of carbonate minerals in soils are described above; the rest of the inorganic fraction consists of silicate minerals. Even if the amounts of carbonate and silicate minerals are known, it may be necessary to identify the specific types of minerals that are present in the soils. Methods to determine the different minerals in soils are detailed below.

For gravel and coarse sand, minerals can be identified by visual observation of the physical characteristics of the particles and comparing them with the standard minerals (Blackburn and Dennen, 1994). For fine sands and silts, optical methods are used to identify the minerals; these methods are also useful to study the size, shape, and texture of the particles. The common optical techniques used are (1) the optical microscope, (2) the petrographic microscope, and (3) the electron microscope. Electron microscopy involves

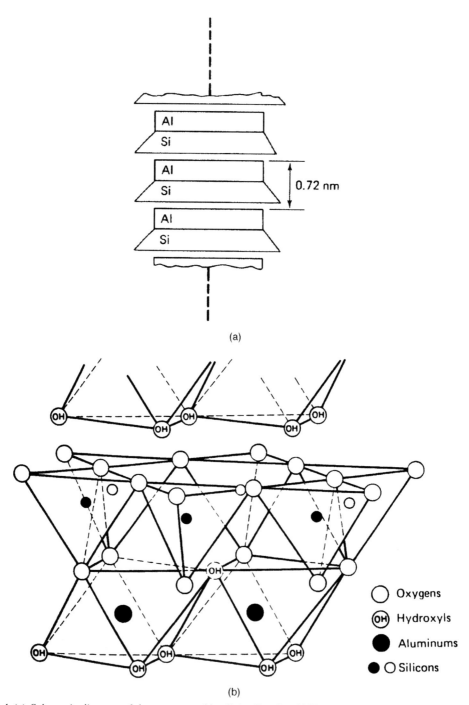

(a)

(b)

Figure 4.4 (a) *Schematic diagram of the structure of kaolinite (Lambe, 1953); (b) atomic structure of kaolinite (Grim, 1959). (From Holtz and Kovacs, © 1981. Reprinted by permission of Pearson Education, Inc., Upper Saddle River, NJ.)*

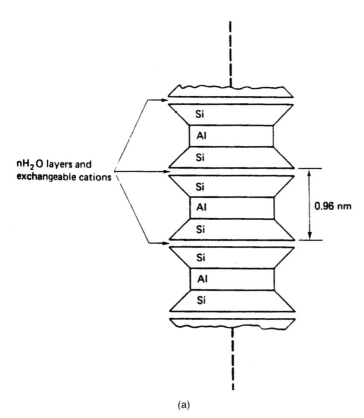

nH₂O layers and
exchangeable cations

0.96 nm

(a)

Figure 4.5 (a) *Schematic diagram of the structure of montmorillonite* (*Lamb, 1953*); (b) atomic structure of montmorillonite (*Grim, 1959*). (*From Holtz and Kovacs,* © *1981. Reprinted by permission of Pearson Education, Inc., Upper Saddle River, NJ.*)

either scanning electron microscopy (SEM) or transmission electron microscopy (TEM) with magnifications of 20 to 150,000 times on the prepared soil specimens. The method of specimen preparation for this testing can affect the results significantly; therefore, it must be performed carefully. More details on these can be found in Black (1965).

SEM and TEM are also used to identify clay minerals. In addition, two other techniques, x-ray diffraction analysis and differential thermal analysis (DTA), are commonly used for mineral identification. The *x-ray diffraction analysis* technique is suited especially for the analysis of minerals that possess crystal structures (Moore and Reynolds, 1989). X-ray radiation has wavelengths ranging from 0.01 to 100 Å.[4] When an x-

ray of wavelength λ strikes a crystal at an angle θ to the parallel atomic planes spaced at a distance d, the parameter n (also known as the *order of diffraction*) must be a whole number in order for the Bragg equation ($2d \sin \theta = n\lambda$) to be satisfied. This relationship forms the basis for identifying different crystals based on their observed x-ray diffraction patterns.

For x-ray diffraction testing, the coarse fraction from the soil is separated first. The fine fraction is then treated with glycerol or other solutions to dry the sample. The dry, powdered soil specimen is subjected to a collimated beam of parallel x-rays. The diffraction beams of various intensities are then recorded. A record of reflection intensity versus θ data is obtained. Typical x-ray diffraction records are shown in Figure 4.6. Each peak in this type of x-ray diffraction record represents a reflection, or a θ value that satisfies the Bragg equation. Mineral types are identified based on

[4] Å is the angstrom unit $= 10^{-10}$ m.

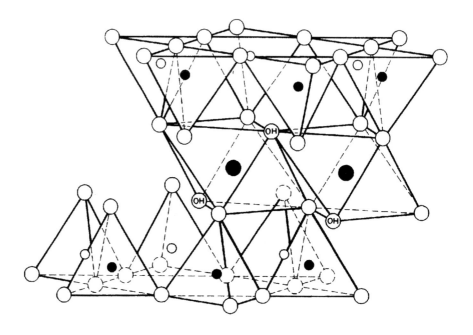

nH$_2$O layers and exchangeable cations

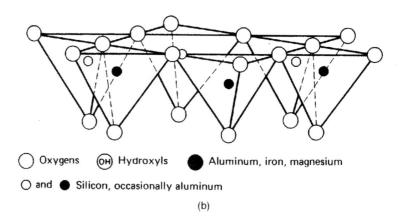

○ Oxygens   (OH) Hydroxyls   ● Aluminum, iron, magnesium

○ and ● Silicon, occasionally aluminum

Figure 4.5 (*continued*)   (b)

a comparison of the measured diffraction record with standard diffraction records for common minerals found in soils.

X-ray diffraction testing does not determine the exact amounts of different minerals present in a given soil sample. Rather, this testing will provide semiquantitative information on the minerals present. X-ray diffraction results can be influenced by a number of fac-

tors, such as sample weight, surface texture, particle orientation, crystallinity, hydration, and mass absorption. Therefore, careful sample preparation is important and x-ray diffraction results must be interpreted carefully.

In the *differential thermal analysis* (DTA) method, a test soil sample of approximately 1 g is placed in a ceramic container and heated in a furnace at a constant

Table 4.4 **Typical chemical composition of clay minerals**

Chemical	Kaolinite (wt %)	Illite (wt %)	Montmorillonite (wt %)
SiO_2	44.2	49.3	62.9
Al_2O_3	39.7	24.25	19.6
TiO_2	1.39	0.55	0.090
Fe_2O_3	0.13	7.32	3.35
FeO	0.08	0.55	0.32
MnO	0.002	0.03	0.006
MgO	0.03	2.56	2.05
CaO	—	0.43	1.68
Na_2O	0.013	0	1.53
K_2O	0.05	7.83	0.53
P_2O_5	0.013	—	0.111
P_{25}	0.034	0.08	0.049
S	—	—	0.05

Source: Van Olphen and Fripiat (1979); Clay Mineral Society, Aurora, Colorado.

rate of approximately 10°C per minute to a maximum temperature of approximately 1000°C. Simultaneously, a thermally inert substance weighing approximately 1 g is also heated in the same way as the test soil. During the heating, the actual temperatures of the test soil and the inert substance are measured using thermocouples and recorded continuously. The plot of the difference in temperature between the test soil and the inert substance (ΔT) versus the temperature is known as a *thermogram*. Typical thermograms of soils are shown in Figure 4.7.

Soil undergoes different reactions during heating. If ΔT is negative, this indicates that heat is taken up by the sample, and this is called an *endothermic reaction;* whereas if ΔT is positive, this indicates that heat is liberated and the reaction is exothermic. When the temperature reaches about 100°C, the free pore water

in the soil is evaporated. With an increase in temperature to 300°C, the diffuse double layer and adsorbed water in the soil is removed. With a further increase in temperature to 1000°C, the interlayer water in montmorillonite minerals will be removed. Above 1000°C, water in the crystal lattice is removed, and as a result, the crystalline structure changes. These changes in crystalline structure occur at specific temperatures, depending on the mineral, and are helpful in soil mineral identification. The amplitude and area of the reaction peak provide estimates of the type and amount of the specific mineral. The mineral type is identified using reference thermograms for pure minerals (Black, 1965).

X-ray diffraction and differential thermal analysis methods are complex, costly, and time-consuming and are not performed routinely. In cases where it is sufficient simply to know the predominant type of minerals present in the soil, Casagrande's plasticity chart with mineral correlation, shown in Figure 4.8, is helpful. Knowing the liquid limit and plasticity index (as explained in Chapter 5), the chart can be used to identify the type of clay mineral that is predominant in the soil.

Organic Constituents. The organic fraction of the solid phase, also known as the *soil organic matter,* is derived from soil biomass. It includes nonhumified and humified compounds (Tan, 1994). The nonhumified compounds are released by the decay of plant, animal, and microbial tissue and include carbohydrates, amino acids, proteins, lipids, nucleic acids, lignins, pigments, hormones, and a variety of organic acids. The humified compounds are products that have been synthesized from the nonhumified substances by a process called *humification* and include complex substances such as

Table 4.5 **Most common minerals found in natural soils**

Mineral	Sand	Silt	Clay
Carbonates	Dominant in beach sand	Occasionally dominant	Present
Quartz	Most common	Very common	Generally present in small quantities
Feldspar	Very common	Less common	—
Mica	—	Very common in small quantities	Common
Clay minerals	—	—	Dominant

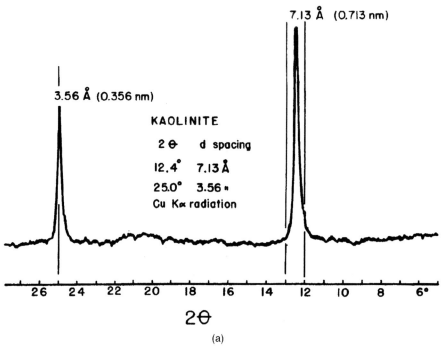

Figure 4.6 (a) *Interpretation and identification of kaolinite using its x-ray diffraction pattern. A 12.4 value for 2θ corresponding to a d spacing of 7.13 Å (0.713 nm), which is the diagnostic d spacing of kaolinite. (b) Interpretation and identification of illite and montmorillonite using x-ray diffraction pattern. (Top) A 8.7 value for 2θ corresponds to a d spacing of 10.1 Å (1.01 nm), which is diagnostic for illite. (Bottom) A 7.2 value of 2θ reading corresponds to a d spacing of 12.3 Å, which is characteristic for air-dry montmorillonite. (From Tan, 1982, by courtesy of Marcel Dekker, Inc.)*

humic and fulvic acids. Soil organic matter affects the physical, chemical, and biological conditions in soils. An increase in organic carbon causes soils (1) to become lightweight, darker in color, and less plastic; (2) to increase in adsorption of water and chemicals; and (3) to experience biochemical reactions.

The quantity of organic matter in soils is determined by either direct or indirect methods. These methods are described in detail in Jackson (1958), Black (1965), and ASTM (2003). Direct methods target the destruction of all organic matter, after which the loss in weight of the soil is taken as the organic content. Two methods are commonly used to achieve destruction of organic matter: (1) oxidation of the organic matter with hydrogen peroxide (H_2O_2) (Black, 1965), and (2) ignition of the soil at high temperature

(ASTM D 2974). The H_2O_2 method may not result in the oxidation of all organic matter, thereby underestimating the organic content of the soil. On the other hand, the high-temperature ignition method may also destroy inorganic constituents of the soil, such as hydrated aluminosilicates, thereby overestimating the organic content of the soil. To eliminate these problems, an alternative method is used which consists of pretreating the soil with 37% hydrochloric acid (HCl) and 48% hydrofluoric acid (HF) to remove the hydrated mineral matter, followed by determining the loss on ignition. This loss on ignition gives the most accurate organic matter content of the soil (Black, 1965).

With indirect methods, organic carbon is measured experimentally (Black, 1965). The amount of organic carbon is multiplied by an empirical factor of 1.724 to

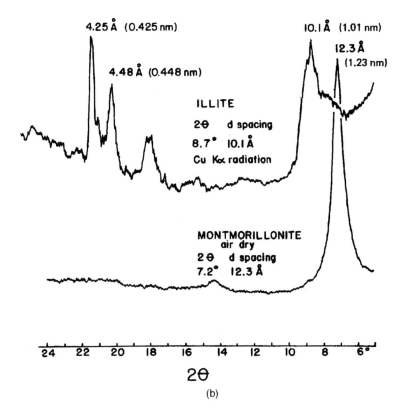

4.25 Å (0.425 nm)

4.48 Å (0.448 nm)

10.1 Å (1.01 nm)

12.3 Å (1.23 nm)

ILLITE

2θ   d spacing

8.7°   10.1 Å

Cu Kα radiation

MONTMORILLONITE
air dry

2θ   d spacing

7.2°   12.3 Å

24 22 20 18 16 14 12 10 8   6°

2θ

(b)

Figure 4.6 (*continued*)

estimate the total amount of organic matter. The organic carbon quantity is determined by either wet or dry combustion procedures with measurement of the CO_2 evolved, or by determining the extent of reduction of a strong oxidizing agent.

4.3.2 Liquid Phase

The liquid phase is generally composed of water and is therefore commonly known as *soil water* or *pore water*. In contaminated soil, however, chemicals may also be present in dissolved and free (or pure) phases. In addition, dissolved gases may also be present in this liquid phase. Therefore, the liquid phase may also be termed *soil solution*. In the saturated zone, the soil pores are completely filled with liquid (or groundwater, as detailed in Chapter 7), while in the unsaturated zone, the soil pores are partially filled with liquid. In dry soils, the soil pores do not contain any liquid. Due to the complex and variable nature of the liquid phase in contaminated soils, we focus first on a water-only liquid phase. The soil water may exist as adsorbed water, double-layer water, and free water as detailed below.

Adsorbed Water. Because of their unique dipolar molecular structure, water molecules can form relatively strong bonds with many substances, including the surfaces of soil solids (or the solid phase). This water is known as *adsorbed water.* Mitchell (1993) described the various mechanisms contributing to the adsorption of water to the soil solids. These mechanisms include hydrogen bonding, hydration of exchangeable cations, attraction by osmosis, charged surface–dipole attraction, and attraction by London dispersion forces. The thickness of the adsorbed water can be up to three to

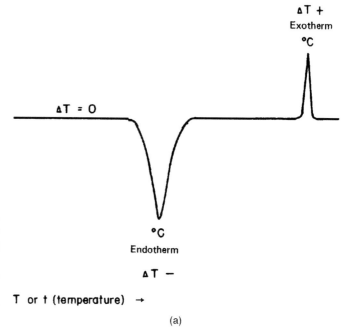

Figure 4.7 (a) Idealized DTA curve; (b) differential thermal analysis (DTA) curves of composite sand (2.0 to 0.05 mm), fine sand (0.25 to 0.10 mm), silt (0.050 to 0.002 mm), coarse clay (0.002 to 0.0002 mm), and fine clay fractions (<0.2 μm) of a Cecil soil. (From Tan, 1982, by courtesy of Marcel Dekker, Inc.)

four molecular layers of water (2.5 to 3 Å each layer). The significance of adsorbed water in soil depends on the type of soil mineral. For example, when a 10-Å-thick adsorbed water layers on the surfaces of quartz and a montmorillonite sheet are compared, the thickness of the adsorbed water film on the quartz particle is insignificant, whereas with the montmorillonite, a substantial surface area is adsorbed by water. The properties of the adsorbed water are shown to be slightly different from those of ordinary water. For instance, density and viscosity may be higher than those of ordinary water. However, for practical purposes, these differences in properties are not significant (Mitchell, 1993).

Double-Layer Water. Clay minerals generally carry a negative charge on their surfaces; this is discussed in Chapter 5. The negative charge is balanced by cations (positively charged) such as Na^+, Ca^{2+}, Mg^{2+}, and so on, which are present in a dissolved form in pore water. These cations are not attached to the surface of the clay minerals; therefore, they are called *exchangeable cations*. On one hand, the cations are electrostatically attracted to the negatively charged clay minerals; these cations have thermal energy and tend to move around. Furthermore, the cations repel each other, and water molecules try to surround them (hydrate the cations). For these reasons, the cations do not crowd the surface of the clay; rather, they form a *diffused layer* around the minerals as depicted in Figure 4.9. The charged surface and the distributed charge in the adjacent phase are together termed the *diffuse double layer*, and water within this layer is known as *double-layer water*. The concentration of cations, also known as *salt concentration* or *electrolyte concentration*, decreases exponentially with distance from the surface of the clay. The concentration of cations is high, but finite, near the clay surface, and it decreases to the concentration of cations in the free pore water, as shown in Figure 4.9.

Several mathematical theories have been proposed to predict ion distribution in the diffuse double layer. The two most popular among these theories are (1) Gouy–Chapman theory and (2) Stern's theory. These

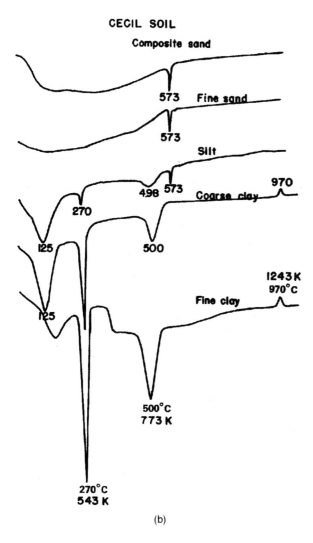

(b)

Figure 4.7 (*continued*)

theories have been described in detail in Grim (1959, 1968) and Mitchell (1993). Based on the Gouy–Chapman theory, the diffuse double-layer thickness (t) is given by

$$t = \sqrt{\frac{\varepsilon_0 DkT}{2n_0 e^2 v^2}}$$

where ε_0 is the permittivity of vacuum (8.8542×10^{-12} $C^2/J \cdot m$), D the dielectric constant of pore liquid (unit

less)[5], k the Boltzmann constant (1.38×10^{-23} J/K), T the temperature (K), n_0 the electrolyte concentration (ions/m^3)[6], e the unit electronic charge (1.602×10^{-19}

[5] Dielectric constant of water = 80.
[6] The electrolyte concentration is commonly given as the mass per unit volume (e.g., mg/L) or as molarity (M). This concentration is converted first as mol/m^3 and then multiplied by Avogadro's number (6.02×10^{23}) to express concentration in terms of ions/m^3.

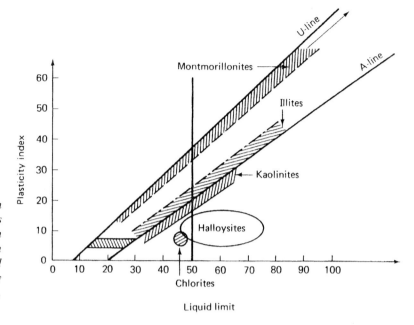

Figure 4.8 *Location of common clay minerals on Casagrade's plasticity chart developed from Casagrande (1948) and data in Mitchell (1976). (From Holtz and Kovacs, © 1981. Reprinted by permission of Pearson Education, Inc., Upper Saddle River, NJ.)*

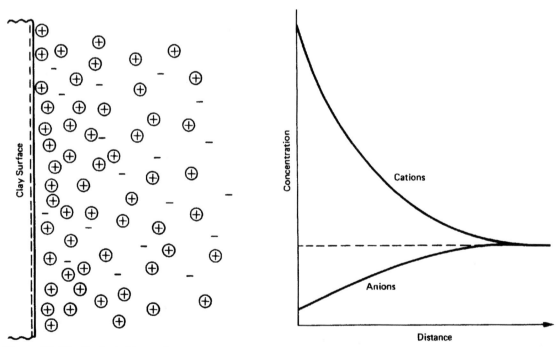

Figure 4.9 *Distribution of ions adjacent to a clay surface according to the concept of the diffuse double layer. (From Mitchell, 1993. This material is used by permission of John Wiley & Sons, Inc.)*

C), and v the cation valance. This equation demonstrates that the diffuse double-layer thickness increases with an increase in D, k, and T, whereas the thickness, t, decreases with increases in n_0 and v. Several other important factors, such as secondary energy considerations (effects of electrical field strengths on D, coulombic interaction, etc.), adsorbed water, ion size, pH, and anion adsorption, are not accounted for in deriving this equation. This equation is helpful in understanding the effects of various system variables rather than defining the exact double-layer thickness. The thickness of the double-layer water ranges from 20 to 500 Å. Like adsorbed water, the effect of double-layer water on soil behavior depends on the mineral type. A 300-Å-thick double-layer water, for example, would completely immerse a sheet of montmorillonite; however, it is less significant to a particle of kaolinite, as shown in Figure 4.10.

Free Water. Water surrounding the diffuse double-layer water is known as *free water.* Different dissolved materials exist in this water; many of them are metallic ions. Since metallic ions cannot exist by themselves,

these ions react with water molecules to form hydration shells. In addition to metallic ions, dissolved gases may also be present in pore water. The two most common gases present in soils are oxygen and carbon dioxide. The amounts of these gases dissolved in pore water can be calculated using Henry's law (refer to Chapter 3). Dissolved CO_2 affects many biological and chemical reactions in soil water. For example, when CO_2 is dissolved in water, it forms carbonic acid, which is a weak acid that can disassociate some of its protons (H^+ ions) in soil water, thereby increasing the acidity. Dissolved oxygen can promote different biochemical reactions.

In contaminated soils or during the transport of contaminants (chemicals) through soils, the liquid phase may undergo significant change in its composition. Depending on the chemical or mixture of chemicals, the adsorbed water, diffuse double-layer water, and free water properties may be affected. Liquid chemicals may exist in the dissolved phase as well as the free (or pure) phase. Gaseous chemicals may be dissolved in water. In Chapter 5 we discuss the effects on the behavior and properties of soils of chemicals present in the liquid phase.

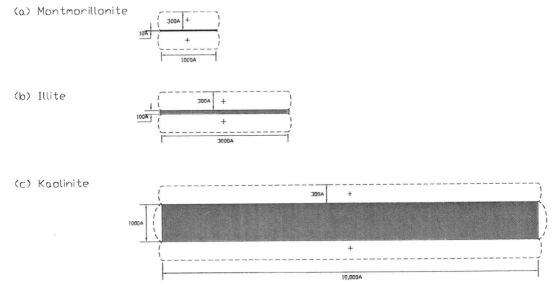

Figure 4.10 Double diffuse layer on clay minerals.

Methods to Determine the Amount and Composition of Liquid Phase. The volume of liquid phase may be calculated using the mass–volume relationships (or using the phase diagram) presented in Chapter 5.[7] In most cases, the amount of pore water present in soil is determined in accordance to the ASTM standard water content test procedure ASTM D 2216 (ASTM, 2003). For this testing, a representative, preweighed soil sample is heated at a constant temperature of 105°C for 24 h. The weight loss is equal to the weight of the pore water in the soil. This procedure does not remove the adsorbed water and interlayer water. Therefore, water driven off during testing represents the amount of *free water* present in the pores. In addition, this method is not valid for granular soils that easily drain during the sampling process, and as a result, in-situ moisture conditions are difficult to determine for granular soils.

The liquid phase may be extracted by various methods as specified in ASTM D 4542 (ASTM, 2003). These methods include (1) collection of drainage water, (2) displacement by a fluid immiscible with water, and (3) direct extraction by vacuum, applied pressure, or centrifugation. The collection of *drainage water,* which represents a soil's liquid phase, may be achieved in-situ under saturated conditions. For this, piezometers or monitoring wells are constructed to allow the collection of water samples (ASTM D 4448), or direct push and sampling methods (ASTM D 6001) are used; these are detailed in Chapter 10. This approach is applicable for saturated and highly permeable soils. Care must be taken in sampling and storing the collected samples so that the liquid composition is not altered. The *displacement method* is used in the laboratory. The soil sample is mixed with a dense, nonreactive organic liquid of low solubility, such as trichlorotrifluoroethane, and then centrifuged. The soil solution floats on top of the displacement solution because of its lower density and can then be separated. The *vacuum ex-*

traction method is used routinely in-situ under unsaturated conditions (vadose zone) and in the laboratory for withdrawal of a soil's liquid phase through a filter under high-vacuum application. For in-situ extraction, lysimeters may be constructed. A lysimeter essentially consists of a ceramic porous cup attached to a small-diameter PVC tube placed in the soil, making sure that there is good contact with the soil. A one-hole rubber plug is placed in the other end of the tube and small-diameter tubing at the base of the ceramic cup runs through the hole to the surface. A vacuum is applied to the small tubing and the soil solution is drawn into a small flask. Various other samplers may also be used to collect soil's liquid phase (ASTM D 4696). Care should be taken to use sampling equipment in the field as well as in the laboratory, so that the extracted liquid represents the composition of a soil's true liquid phase. Instead of vacuum extraction, extraction by applying pressure or centrifuging may also be used in the laboratory. All of these extraction methods can be used for different soil types, including low-permeability soils (ASTM D 4542).

The composition of liquid phase extracted from soil, including the concentrations of dissolved constituents, pH, and electrolytic conductivity, may be determined using standard methods for water and wastewater, as defined in Chapter 3 (Black, 1965; USEPA, 1986; APHA, 1992; ASTM D 596).

4.3.3 Gas Phase

The gas phase, also known as *soil air,* exists in unsaturated soils (or the vadose zone). The amount of gas in pore spaces is usually controlled by the liquid phase. When the liquid phase increases, air is pushed out of the pore spaces. When the liquid phase in the pore spaces decreases, gas content increases by mass flow and diffusion. In surficial soils, the gas phase is composed of the same type of gases as those commonly found in the atmosphere. The major gaseous components and their relative amounts in atmospheric air are listed in Table 4.6. Many other types of gases may be present in minute amounts in a soil's gas phase. The

---

[7] Assumes static conditions. For flowing conditions, the hydraulic gradient and the hydraulic conductivity of the soil are used to calculate the flow rate using Darcy's law.

Table 4.6 **Typical composition of atmospheric air**

Component	Percent by Volume
Nitrogen (N_2)	78.0
Oxygen (O_2)	21.0
Carbon dioxide (CO_2)	0.031
Argon (Ar)	0.93
Helium (He)	0.0005
Methane (CH_4)	0.0002
Hydrogen (H_2)	0.00005
Nitrous oxide (NO_2)	0.00002

Source: Tan (1994), by courtesy of Marcel Dekker, Inc.

quality of air is usually measured by its O_2 and CO_2 content. Most biological reactions in soils consume O_2 and produce CO_2. No biochemical reactions exist in soils that can generate O_2. Therefore, the O_2 content decreases and CO_2 content increases with soil depth.

In contaminated soils or during the transport of contaminants through soils, volatile contaminants may exist in vapor phase. In such cases, the gas phase of the soil consists of a mixture of atmospheric gas and contaminant vapors. The composition of the gas phase affects various geochemical processes in the soils.

Methods to Determine the Amount and Composition of Gas Phase. The volume of soil gas may be calculated using mass–volume relationships[8] (or using a phase diagram), as presented in Chapter 5. The composition of soil gas may be approximated by carefully placing the soil sample in a vial immediately upon retrieval from the subsurface and measuring the headspace in the vial for volatile organics using a portable analyzer called a *photoionization detector* (PID). However, a sample of gas is required to determine its composition accurately. The ASTM standard procedure D 5314 provides guidance on soil gas sampling and analysis procedures. Soil gas samples can be obtained using passive sampling or grab sampling. *Passive sampling* involves placing a sorbent material (such as

activated carbon) in the soil zone being monitored for a period of days to weeks. Contaminants in the soil gas are sorbed onto the sorbed material. The sorbed material is then retrieved, sealed, and shipped to the laboratory for desorption and analysis of the contaminants. *Grab sampling* involves obtaining a small volume of the soil gas that is present at the time of sampling. Grab samples are classified as *static* (the sample is obtained from a body of gas that is more or less immobile) or *dynamic* (the sample is obtained from an actively moving volume). To obtain a grab sample, a small probe is inserted to the depth to be tested, then vacuum is applied to extract the soil gas sample. Grab samples can also be obtained from a lysimeter installed in a boring or well.

Soil gas can be analyzed on-site using gas detector tubes or portable analyzers. A gas detector tube is filled with a solid reagent and when the gas is passed through it, the reagent color changes. These tubes are available for different types of gas contaminants and are only suitable to detect the presence of a particular gas contaminant. Portable analyzers, such as PID, are used to indicate gross levels of volatile organics. The PID probe is inserted into the gas sample to measure the total volatiles. For the identification of individual components of gas, gas chromatography (GC) equipment is used. A portable GC may be used to analyze soil air samples in the field. However, laboratory analysis of soil gas using GC or GC-MS is performed to determine the types and concentrations of soil gas constituents (refer to Chapter 3).

4.4 SOIL FABRIC

In addition to knowing the different phases of the soil, it is very important to know the arrangement of solids and pore spaces in a particular soil, often known as *soil fabric*. The terms *soil fabric* and *soil structure* are used interchangeably; however, *structure* refers to the fabric of the soil plus interparticle forces. The understanding of soil fabric and structure is important for understanding the mechanical behavior of soils as well for geoenvironmental studies, especially when deline-

---

[8] Assumes static conditions. Under flowing conditions, the pressure gradient and soil gas permeability are used to calculate gas flow rate using Darcy's law.

ating the form and distribution of contaminants in the soil. This information on contaminant distribution will help in assessing the fate of the contaminants for long-term conditions and in assessing methods to effectively remove the contaminants from the soil.

The fabric of *coarse-grained soils* such as gravels, sands, and in some cases, silts, consists of a single-grained structure, as shown in Figure 4.11(*a*). Depending on the grain size distribution and the arrangement of these particles, loose to dense structures could result. Under certain conditions, these soils could form a honeycombed structure, as shown in Figure 4.11(*b*). The fabric of the soils can be characterized using simple optical methods such as petrography. Loose and honeycomb structures have larger pore volumes than those of dense structures. A structure with larger pores will have higher permeability and compressibility behavior. This helps us to understand the contaminant transport and retention in the soil fabric and can help in designing remediation measures.

The fabric of *fine-grained soils* is complex, and single-grained structures rarely exist in these soils. The fabric of fine-grained soils is generally presented on different scale levels: microfabric, minifabric, and macrofabric (Mitchell, 1993). *Microfabric* refers to regular aggregations of particles and the very small pores between them. *Minifabric* refers to aggregations of the microfabric and the interassemblage pores between them. *Macrofabric* refers to cracks, fissures, root holes, laminations, and so on, that correspond to the transassemblage pores.

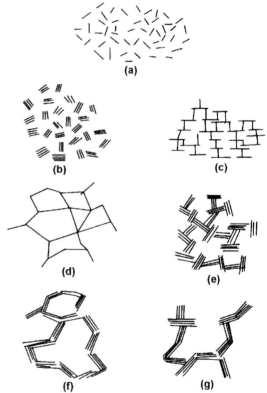

Figure 4.12 *Modes of particle association in clay suspensions and terminology; (a) dispersed and deflocculated; (b) aggregated but deflocculated (face-to-face association, or parallel or oriented aggregation); (c) edge-to-face flocculated but dispersed; (d) edge-to-edge flocculated but dispersed; (e) edge-to-face flocculated and aggregated; (f) edge-to-edge flocculated and aggregated; (g) edge-to-face and edge-to-edge flocculated and aggregated (van Olphen, 1977). (From Mitchell, 1993. This material is used by permission of John Wiley & Sons, Inc.)*

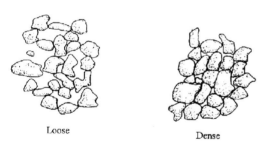

(a) Single grained soil structures

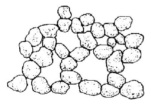

(b) Honeycomb structure

Figure 4.11 *Soil structure. (From Holtz and Kovacs, © 1981. Reprinted by permission of Pearson Education, Inc., Upper Saddle River, NJ.)*

Fabric of fine-grained soils can result in *flocculent* or *dispersed structures,* as depicted in Figure 4.12. The dispersed structure involves clay particles oriented in a more parallel arrangement. Because the equilibrium of forces between particles is disturbed, soils with dispersed structure possess low strengths. The flocculent structure involves the attraction and contact between many of the clay particles through an edge-to-face arrangement. Soils with a flocculent structure are strong and resistant to external forces because of the attraction between particles.

The fabric of fine-grained soils is characterized by electron microscopy using either a transmission electron microscope (TEM) or a scanning electron microscope (SEM), x-ray diffraction, or transmission x-ray (or radiography) (Mitchell, 1993). Careful preparation of the sample is extremely important when performing any of these analyses. Sample preparation includes removing pore fluid by various methods, such as air drying, oven drying, or freeze drying, then replacing pore fluid with Carbowax or other resins, and finally, preparing surfaces to expose the original fabric for electron microscope characterization.

The fabric of real-world soils is extremely complex. They generally contain a mixture of sand, silt, and clay, and the fabric depends on particle surface forces, the adsorption of clays on nonclays, and the chemical reactions of clay surfaces.

4.5 SUMMARY

Unlike classic geotechnical engineering, the detailed chemical characterization of three phases of soil—the solid phase, the liquid phase, and gas phase—is essential when addressing geoenvironmental problems. The solid phase may consist of minerals, free oxides and hydroxides, and organic matter. The liquid phase may consist of water, dissolved chemicals and gases, and free-phase chemicals. The gas phase may consist of air and chemical vapors. Various techniques may be used to determine the amounts and composition of these phases. In addition to the phases, an understanding of the arrangement of solids and pores, called the soil fabric, is also very valuable in assessing the fate and transport of chemicals in soils.

QUESTIONS/PROBLEMS

4.1. Explain the differences between mechanical and chemical weathering of rocks.

4.2. Why is detailed soil composition information needed for geoenvironmental studies?

4.3. What are the components of the solid phase of soils?

4.4. Explain the differences between the carbonate minerals and silicate minerals.

4.5. Distinguish between 1:1- and 2:1-type silicate minerals.

4.6. Explain the most important characteristics of the following minerals: **(a)** quartz; **(b)** kaolinite; **(c)** montmorillonite.

4.7. Which minerals are commonly found in the sand, silt, and clay fractions of soils? Why?

4.8. Does mica have stronger or weaker resistance to mechanical and chemical weathering than quartz? Explain why.

4.9. How do you determine the mineralogical composition of a soil?

4.10. Explain the basis for identifying soil minerals using x-ray diffraction analysis.

4.11. Explain how a small amount of organic matter can have a profound effect on soil properties.

4.12. What is humification?

4.13. What is the diffuse double layer? Why is it important in geoenvironmental studies?

4.14. Explain the factors affecting the diffuse double layer.

4.15. Kaolinite clay has a specific surface of 25 m^2/g, a cation exchange capacity of 10 meq per 100 g, and a constant surface charge. Calculate the thickness of the diffusive double layer if the pore water consists of **(a)** water only; **(b)** 0.01 M NaOH only, **(c)** 0.1 M NaOH only. Assume a temperature of 20°C, a dielectric constant of water (D) of 80, $k = 1.38 \times 10^{-23}$ J/K, $e = 1.602 \times 10^{-19}$ C, and $\varepsilon_0 = 8.8542 \times 10^{-12}$ C^2/ J · m. Discuss the results.

4.16. Estimate the water content of a sodium montmorillonite when mixed with a sufficient solution of 0.001 M NaCl such that the average spacing between the particles is twice the thickness of the double layer as defined by Gouy–Chapman theory. Assume that the dielectric constant for the solution is 80 and the temperature is 20°C. Boltzmann's constant is 1.38×10^{-16} erg/K and the unit electrical charge is 4.80 $\times 10^{-10}$ esu.

4.17. Repeat Problem 4.16 with kaolinite in the same solution.

4.18. Calculate the diffusive double-layer thickness in kaolin with the following different pore water composi-tions: **(a)** 0.1 M NaOH; **(b)** 1 M NaOH; **(c)** 0.1 M $CaCl_2$; **(d)** 1 M $CaCl_2$. Explain the results briefly. Clearly state any assumptions you made.

4.19. Discuss advantages and disadvantages of various methods of soil liquid (pore water) sampling.

4.20. Discuss advantages and disadvantages of various methods of soil gas sampling.

4.21. Let's consider that you are the responsible geoenvironmental engineer dealing with remediation of a contaminated site. Explain your approach, preferably using a flowchart, to determining the composition of the site soil.

REFERENCES

APHA (American Public Health Association), *Standard Methods for the Examination of Water and Wastewater,* 16th ed., APHA, Washington, DC, 1992.

ASTM (American Society for Testing and Materials), *Annual Book of ASTM Standards,* ASTM, West Conshohocken, PA, 2003.

Black, C. A. (ed.), *Methods of Soil Analysis,* American Society of Agronomy, Madison, WI, 1965.

Blackburn, W. H., and Dennen, W. H., *Principles of Mineralogy,* 2nd ed., Wm. C. Brown, Dubuque, IA, 1994.

Casagrande, A., Classification and identification of soils, *Trans. ASCE,* Vol. 113, pp. 901–930, 1948.

Goodman, R. E., *Engineering Geology: Rock in Engineering Construction,* Wiley, New York, 1993.

Grim, R. E., Physio-chemical properties of soils: clay minerals, *J. Soil Mech. Found. Div., ASCE,* Vol. 85, No. SM2, pp. 1–17, 1959.

———, *Clay Mineralogy,* 2nd ed., McGraw-Hill, New York, 1968.

Holtz, R. D., and Kovacs, W. D., *An Introduction to Geotechnical Engineering,* Prentice Hall, Upper Saddle River, NJ, 1981.

Jackson, M. L., *Soil Chemical Analysis,* Prentice Hall, Upper Saddle River, NJ, 1958.

Lambe, T. W., The structure of inorganic soil, *Proc., ASCE,* Vol. 79, Sep. No. 315, 1953.

Mitchell, J. K., *Fundamentals of Soil Behavior,* 1st and 2nd eds., Wiley, New York, 1976, 1993.

Moore, D. M., and Reynolds, R. C., Jr., *X-ray Diffraction and the Identification and Analysis of Clay Minerals,* Oxford University Press, New York, 1989.

Tan, K. H., *Principles of Soil Chemistry,* Marcel Dekker, New York, 1982.

———, *Environmental Soil Science,* Marcel Dekker, New York, 1994.

USEPA (U.S. Environmental Protection Agency), *Test Methods for Evaluating Solid Wastes,* Vols. IA and IB, *Laboratory Manual, Physical/Chemical Methods,* SW-846, USEPA, Washington, DC, 1986.

Van Olphen, H., *An Introduction to Clay Colloid Chemistry,* 2nd ed., Wiley, New York, 1977.

Van Olphen, H., and Fripiat, J. J., *Data Handbook for Clay Minerals and Other Non-metallic Materials,* Pergamon Press, New York, 1979.

Warshaw, G. B., and Roy, R., Classification and a scheme for the identification of layer silicates, *Geol. Soc. Am. Bull.,* Vol. 72, pp. 1455–1492, 1961.

5

SOIL PROPERTIES

5.1 INTRODUCTION

In this chapter we present geotechnical and chemical properties of soils. The geotechnical properties consist of index properties, compaction characteristics, hydraulic characteristics, consolidation characteristics, and shear strength. The chemical properties include pH, surface charge, cation exchange capacity, anion exchange capacity, and specific surface. As discussed in other chapters, these properties are important in understanding geoenvironmental behavior of soils.

5.2 GEOTECHNICAL PROPERTIES

The most important geotechnical properties of soils used in the analysis and design of waste containment and waste remediation, are reviewed in this section. Standard geotechnical engineering publications provide detailed information on this subject (e.g., Lambe and Whitman, 1969; USACE, 1970; Holtz and Kovacs, 1990; Mitchell, 1993; Das, 1994; Terzaghi et al., 1996; Coduto, 1999; ASTM, 2003).

5.2.1 Mass–Volume Relationships

Figure 5.1(a) shows a soil mass. For illustration purposes, various components of soil mass can be identified in a phase diagram, as shown in Figure 5.1(b). The figure is used to represent three phases of the soil. Using the volumes and masses of these phases, several useful mass–volume ratios are defined, as shown in Table 5.1. Typical values of porosity and density for soils are provided in Table 5.2.

5.2.2 Index Properties and Soil Classification

Index properties of soils, specifically grain size distribution and plasticity or consistency represented by Atterberg limits, are used for classification of soils. In addition, these can be correlated with various engineering properties of soils.

Grain Size Distribution. In general, soil particles are irregularly shaped. The general practice is to represent them by equivalent particle diameter. This particle diameter is determined by sieve analysis and hydrometer analysis of the soil (ASTM D 422). *Sieve analysis* involves passing a known amount of dry soil through a series of sieves, stacked with the larger sieve sizes atop the smaller sieve sizes. The sieve sizes commonly used for sand and silt soils include No. 4 (4.75 mm), No. 10 (2 mm), No. 20 (0.85 mm), No. 40 (0.425 mm), No. 60 (0.25 mm), No. 100 (0.15 mm), and No. 200 (0.075 mm). After the soil is placed in the top sieve, the stack is shaken in a sieve shaker. The amount of soil retained on each sieve is then measured. Following the sieve analysis, the grain size distribution of the soil passing through a No. 200 sieve can be determined by performing a hydrometer analysis. *Hydrometer analysis* starts with a mixture of soil, water, and a dispersing agent placed in a graduated glass jar, and the particle size is measured based on the settling velocity of those particles using a hydrometer. This relationship is determined using *Stokes' law:*

$$D = \sqrt{\frac{18\eta}{(G_s - 1)\rho_w}} \sqrt{\frac{L}{t}} \qquad (5.1)$$

where D is the particle diameter, L the length of fall, t the time of fall, η the viscosity of water, and G_s the

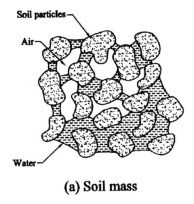

(a) Soil mass

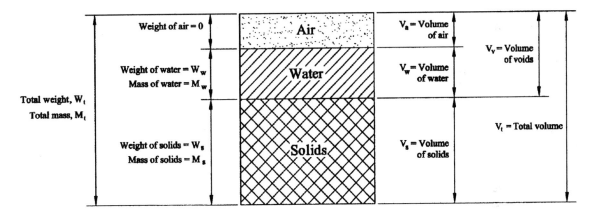

(b) Phase diagram

Figure 5.1 Phase distribution of soils and phase diagram.

specific gravity of the soil. A grain size distribution can then be plotted, based on information from the sieve analysis and the hydrometer analysis; an example of this is shown in Figure 5.2.

Two parameters can be calculated based on grain size distribution: the coefficient of uniformity (C_u) and coefficient of curvature (C_c), defined as $C_u = D_{60}/D_{10}$ and $C_c = D_{30}^2/D_{10}D_{60}$, where D_{10}, D_{30}, and D_{60} are the equivalent grain diameters corresponding to 10, 30, and 60% passing the sieve sizes, respectively. Using these coefficients, a soil may be defined as well graded or poorly graded. A soil is called *well graded* if the

coefficient of curvature falls between 1 and 3, whereas the coefficient of uniformity is greater than 4 for gravelly soils and 6 for sandy soils. Figure 5.2 exhibits a numerical example of calculating C_c and C_u values.

Atterberg Limits. The *Atterberg limits* of a soil relate a change from solid state to fluid state or to semisolid state with a change in water content. As shown in Figure 5.3, a soil changes from a solid state to a fluid state with increasing water content. The water content at which a soil changes from a solid to a semisolid is referred to as the *shrinkage limit* (SL). The *plastic limit*

Table 5.1 Mass–volume relationships of soils

Ratio	Definition	Units
Void ratio, e	V_v/V_s	Nondimensional
Porosity, n	$V_v/V_t \times 100$	%
Degree of saturation, S	$V_w/V_v \times 100$	%
Water content,[a] w	$M_w/M_s \times 100$	%
Wet or total density, ρ_t	M_t/V_t	g/cm³
Dry density, ρ_d	M_s/V_t	g/cm³
Saturated density, ρ_{sat}	M_t/V_t when $S = 100\%$	g/cm³
Submerged density, ρ_{sub}	$\rho_{sat} - \rho_w$	g/cm³
Density of solids, ρ_s	M_s/V_s	g/cm³
Specific gravity, G_s	ρ_s/ρ_w	Nondimensional

[a] In some geoenvironmental applications, the water content is also defined based on volumes and is called the *volumetric moisture content* (θ) $= V_w/V_t$. At 100% saturation, $\theta = n$.

Table 5.2 Porosity and density of typical soils[a]

Soil Type	Porosity, n (%)	Dry Density, ρ_d (g/cm³)	Saturated Density, ρ_{sat} (g/cm³)
Gravels	25–40	1.5–2.0	1.8–2.4
Sands	25–50	1.4–1.8	1.8–2.3
Silts	35–50	0.6–1.8	1.4–2.1
Clays	40–70	0.8–2.3	1.4–2.4
Organic clays	60–80	0.1–0.3	0.5–1.5

Source: Holtz and Kovacs (1981); Lambe and Whitman (1969); Terzaghi et al. (1996); Coduto (1999).
[a] Void ratio (e) can be calculated using $e = n/(1 - n)$.

(PL) refers to the water content at which a soil changes from a semisolid state to a plastic state. The water content at which a soil changes from a plastic state to a fluid state is called the *liquid limit* (LL) of the soil. The plastic limit and the liquid limit are commonly used soil properties in engineering practice.

Standard test methods for determining Atterberg limits have been developed (ASTM D 4318). The plastic limit test defines the PL as the water content (w) at which a thread of soil crumbles when it is carefully rolled out to a diameter of 3 mm ($\frac{1}{8}$ in.) and breaks up into segments 3 to 10 mm ($\frac{1}{8}$ to $\frac{3}{8}$ in.) long. The liquid limit is the water content (w) at which a standard-sized groove, which has been cut with a standard grooving tool into remolded soil that has been placed into a standard cup, will close. The width of the groove is a dis-

tance of 13 mm ($\frac{1}{2}$ in.) and must close at 25 blows of the liquid limit cup. For each blow, the liquid limit cup must fall a distance of 10 mm onto a hard rubber or Micarta plastic base. In practice, several liquid limit tests are performed at various water contents and the results are plotted as displayed in Figure 5.4. The liquid limit is then taken to be the water content corresponding to 25 blow counts.

The *plasticity index* (PI) is defined as the LL minus the PL. The PI indicates the range of water contents at which soil is plastic. It is also a useful property for soil classification and for correlation purposes when related with engineering properties.

Soil Classification. Many systems exist for the classification of soils. The most common system is the *Unified Soil Classification System* (USCS). This system (ASTM D 2487) uses the index properties primarily to classify soils into three main groups: (1) coarse-grained soils, (2) fine-grained soils, and (3) peat.

Coarse-grained soils are composed of material with less than 50% of particles passing a No. 200 sieve. These soils may be further identified as gravelly soils (G), with greater than 50% retained on a No. 4 sieve, or sandy soils (S), with greater than 50% passing a No. 4 sieve. The final classification is then made based on the amount of fines present in the soil. Table 5.3 displays the possible classifications of coarse-grained soils.

Fine-grained soils are soils with greater than 50% of particles passing a No. 200 sieve, defined as silt (M), clay (C), and organic soil (O). These soils are further defined based on their plasticity: low plasticity (L) and high plasticity (H). Fine-grained soils can be classified using a plasticity chart, as shown in Figure 5.5.

The final group of soils consists of *Peat* (Pt), which is characterized by a dark color, strong odor, and high water content. Soils with peat may have low to high plasticity and therefore are classified as OL and OH.

5.2.3 Compaction Characteristics

By definition, compaction is the densification of an unsaturated soil by reducing the volume of voids filled

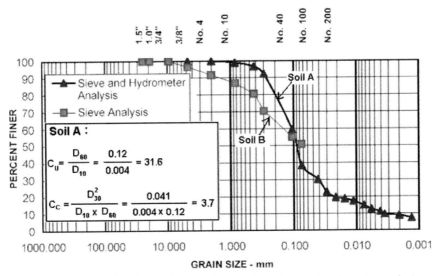

Figure 5.2 *Grain size distribution of soils based on sieve and hydrometer analysis.*

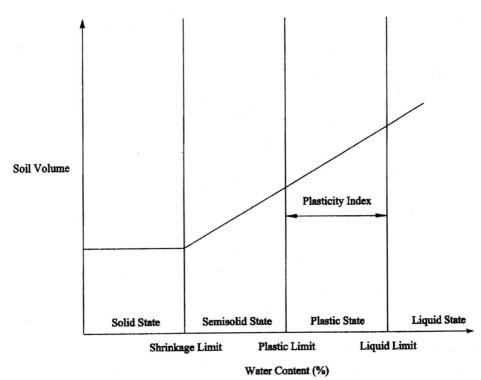

Figure 5.3 *Various states of soil with increasing water content and Atterberg limits.*

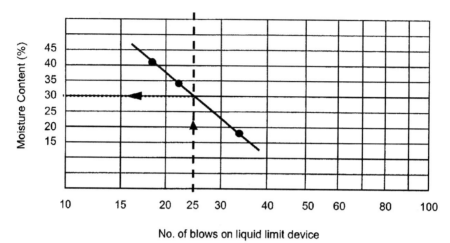

No. of blows on liquid limit device

Figure 5.4 *Determination of liquid limit.*

Table 5.3 Classification of coarse-grained soils

Percent Fines	Type of Fines	Symbols
<5	Silt or clay	GW, GP, SW, SP (W = well graded, P = poorly graded)
5–12	Silt (M)	GW-GM, GP-GM, SW-SM, SP-SM
5–12	Clay (C)	GW-GC, GP-GC, SW-SC, SP-SC
>12	Silt	GM, SM
>12	Clay	GC, SC

Source: ASTM D 2487.

with air. Compaction generally improves soil properties by increasing shear strength and decreasing compressibility. This process can be accomplished on both coarse- and fine-grained soils, and depends on the dry density, water content, compaction effort, and soil type.

R. R. Proctor studied compaction of soils extensively when dealing with the construction of earthen dams. He established compaction tests, known as *Proctor tests,* to provide a relationship between dry density and water content. There are two different methods of compaction tests: the standard Proctor (ASTM D 698) and the modified Proctor (ASTM D 1557). The *standard Proctor test* involves filling a standard mold (with a volume of 944 cm^3 or $\frac{1}{30}$ ft^3) with soil in three layers.

Each layer is compacted with 25 blows of a standard 5.5-lb hammer falling from a height of 12 in. After this preparation, the total mass and water content of the soil are determined. Then the total and dry density (ρ_t and ρ_d, respectively) are calculated using: $\rho_t = M_t/V_t$ and $\rho_d = \rho_t/(1 + w)$. The test is repeated four or five times at different water contents, and w versus ρ_d is plotted, forming a compaction curve. As depicted in Figure 5.6, the compaction curve gives the maximum dry density (ρ_{dmax}) and the optimum water content (OMC). The *theoretical zero air voids curve* (ZAVC), indicating the ρ_d–w relationship under 100% saturation, is always plotted along with the compaction curve. The ZAVC is plotted as ρ_d versus w, based on $\rho_d = \rho_w S/(w + S/G_s)$ with $S = 1$ to serve as the theoretical upper bound on compaction effort. Similar relationships can be obtained from *modified Proctor tests,* where the same mold is used but the soil is compacted in five layers, with each layer compacted with 25 blows using a 10-lb hammer falling from a height of 18 in.

Properties of soils depend on the level of compaction (also called *relative compaction,* ρ_d/ρ_{dmax}) that has been achieved. For illustration purposes, Figure 5.7 depicts a typical variation of shrinkage, shear strength, and hydraulic conductivity corresponding to a typical compaction curve. These properties are defined in the following sections.

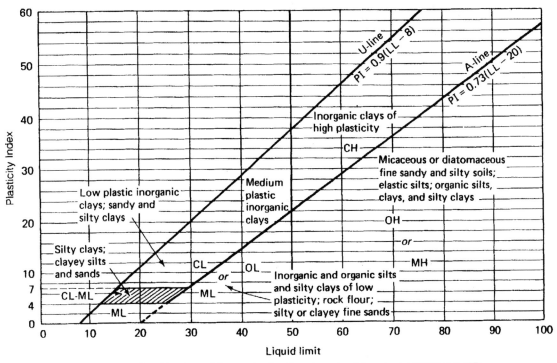

Figure 5.5 *Classification of fine-grained soils using plasticity chart (ASTM D 2487).*

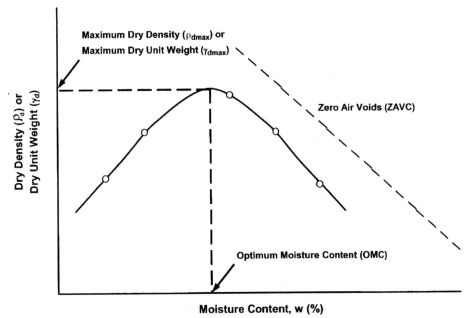

Figure 5.6 *Typical compaction test results showing relationship between the moisture content and dry density.*

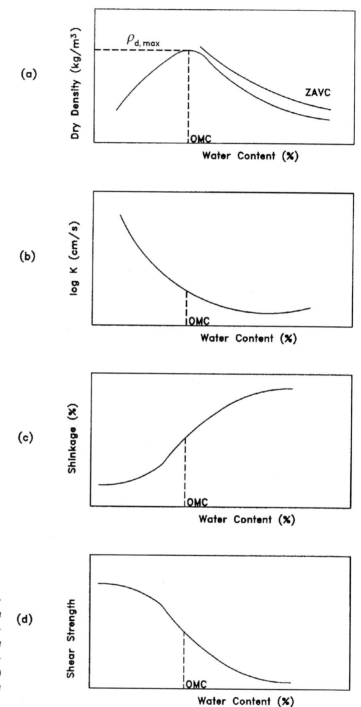

Figure 5.7 Effect of compaction on properties of fine-grained soils: (a) *compaction curve;* (b) *water content and hydraulic conductivity relation with respect to compaction curve;* (c) *water content and shrinkage relation with respect to compaction curve;* (d) *water content and shear strength relation with respect to compaction curve.*

5.2.4 Hydraulic Characteristics

The presence of water in soils can affect their behavior significantly. The water in soils can exist under static or flowing conditions. Under static conditions, the important hydraulic characteristics of soils include capillarity, shrinkage and swelling, and frost action. Under flowing conditions, seepage rate is significant. Seepage rate is dependent on the hydraulic conductivity of soils.

Capillarity. *Capillarity* refers to a rise in water level, due to surface tension (soil suction), that occurs between a soil and the pore water. It is given by

$$h_c = \frac{\sigma}{r\gamma_w} \cos\theta \qquad (5.2)$$

where h_c is the capillary rise, r the mean pore radius, σ the surface tension, γ_w the unit weight of water, and θ the contact angle. The capillary rise is inversely proportional to the pore size of the soil. In fine-grained soils, capillary rise is high, whereas it is negligible in gravelly soils. Empirically, the capillary rise is given as (Holtz and Kovacs, 1981)

$$h_c = \frac{-0.03}{0.2D_{10}} \qquad (5.3)$$

where h_c is reported in meters and D_{10} in mm. The capillary pressure (u_c) is defined as $u_c = h_c\rho_w g$. The capillarity rise and pressure changes significantly affect the spreading of contaminants at the interface of the vadose zone and the groundwater, otherwise known as the *capillary zone.*

Shrinkage and Swelling. In general, as shown in Figure 5.3, soil volume decreases with decreasing water content (or drying) until the *shrinkage limit* (SL) is reached. After the SL is reached, the volume remains constant with further decrease in water content. *Swelling,* on the other hand, is referred to as an increase in soil volume with an increase in water content. This behavior is predominant in soils that contain montmorillinite minerals and are near a water source.

Frost Action. *Frost action* occurs in frost-susceptible soils (primarily fine-grained soils) when the temperature is below freezing and a source of water is close to the frostline. Freeze–thaw effects can alter soil properties significantly. Although clays are also susceptible to frost action, silts are affected most by freeze–thaw action, due to their higher hydraulic conductivity.

Hydraulic Conductivity. *Hydraulic conductivity* refers to the ease with which water can flow through a soil. This property is defined using Darcy's law: $q = KiA$ where q the flow rate, i the hydraulic gradient, and K the hydraulic conductivity. The K value depends on the density and viscosity of water, the soil's specific surface, porosity, and tortuosity (the irregularity of the flow path), and the degree of saturation. The hydraulic conductivity value is required for seepage or groundwater flow calculations, as detailed in Chapter 7.

There are three different ways to determine hydraulic conductivity: laboratory tests, field tests, and empirical correlations. The laboratory tests consist of permeameter tests (ASTM D 2434) and triaxial tests (ASTM D 5084). Further details on these are provided in Chapter 17. The field tests are covered in a discussion of site characterization methods in Chapter 10. From numerous empirical correlations, the two most common methods for hydraulic conductivity estimation are:

1. The *Hazen equation* for coarse-grained soils, given by

$$K(\text{cm/s}) = CD_{10}^2$$

   where D_{10} is expressed in cm and C varies from 100 to 150 for fairly uniform sand in a loose state.

2. *Taylor's equation,* which relates different hydraulic conductivities of the same coarse-grained soil to the void ratio, given by

$$\frac{K_1}{K_2} = \frac{e_1^2}{e_2^2} \qquad (5.4)$$

Typical values of K for different soils are shown in Table 5.4. These values may be used as a guide only and may be used at the feasibility design level. For detailed design and evaluation, hydraulic conductivity should be obtained by performing tests on representative soils.

Total Stress, Pore Water Pressure, and Effective Stress. Total stress (σ), pore water pressure (u), and effective stress (σ_e) are related by $\sigma = \sigma_e + u$. The total stress and pore water pressure are calculated using the densities and thicknesses of the soil layers and location of the groundwater table. Total stress is given by: $\sigma = \Sigma \rho g h$, where ρ is the total (wet) density of the soil, g the acceleration due to gravity, and h the thickness of the soil layer. Every soil layer above the point where stress is being calculated is considered in this calculation. The pore water pressure is given by: $u = \rho_w g h_w$, where ρ_w is the density of water and h_w is the height of the water above the point where stress is being calculated. The effective stress, therefore, can be calculated by subtracting pore water pressure from the total stress.

5.2.5 Consolidation (Compressibility) Characteristics

When saturated soil is loaded, excess pore water pressure is generated. This pressure is slowly dissipated,

Table 5.4 Typical hydraulic conductivity values

Soil Type	Hydraulic Conductivity[a] (cm/s)
Gravels	10^2–10^{-2}
Sands	1–10^{-5}
Silts	10^{-3}–10^{-7}
Clays	10^{-5}–10^{-11}

Source: Lambe and Whitman (1969); Holtz and Kovacs (1981); Terzaghi et al. (1996); Coduto (1999).

[a] The range indicates the grain size distribution primarily for gravels, sands, and silts. For silts and clays, the range of hydraulic conductivity, in addition to grain size distribution, depends on plasticity, degree of saturation, degree of compaction, and placement moisture contents.

due to the squeezing out of excess pore water, and as a result, the soil grains rearrange into a more stable and dense configuration. With this rearrangement, a decrease in volume occurs with time, causing ground surface settlement. This time-dependent compression of saturated soil, called *consolidation,* is mainly a phenomenon of practical importance for low-permeability fine-grained soils. In coarse-grained soils, this excess pore water pressure dissipates in a short period of time, mostly as a load is applied.

Figure 5.8 displays the changes in stress, pore water pressure (both initial and excess), effective stress, and settlement with change in time upon application of a load at the ground surface, or at a known level from the top of the compressible soil layer. The excess pore water pressure (u_e) that is generated dissipates with time, causing an increase in effective stress. The compression or consolidation occurs as the effective stress increases. For practical purposes, under field conditions, consolidation is assumed to be in one dimension (i.e., the vertical direction).

The total primary consolidation settlement (Figure 5.8) is calculated using the following equation:

$$S_C = \frac{C_r}{1 + e_0} H \log \frac{\sigma_{evm}}{\sigma_{ev0}} + \frac{C_C}{1 + e_0} H \log \frac{\sigma_{evf}}{\sigma_{evm}}$$

(5.5)

where H is the thickness of the compressible soil layer, e_0 the initial void ratio, σ_{ev0} the initial effective vertical stress given by $\sigma_{ev0} = \sigma_{v0} - u_0$ (where σ_{v0} is the initial total vertical stress and u_0 is the initial pore water pressure), σ_{evf} is the final effective vertical stress given by $\sigma_{evf} = \sigma_{ev0} + \Delta\sigma_v$ (where $\Delta\sigma_v$ is the increase in vertical stress). The increase in vertical stress can be calculated using: $\Delta\sigma_v = I\Delta P$, where ΔP is the applied pressure at the ground surface or at any known location. The *influence factor* (I) is calculated using the theory of elasticity, which considers the two-dimensional effect of loaded area and location where $\Delta\sigma$ is applied. If the loading area is very large when compared to the depth at which $\Delta\sigma$ is needed, $I \approx 1.0$. For further information on influence factors, readers

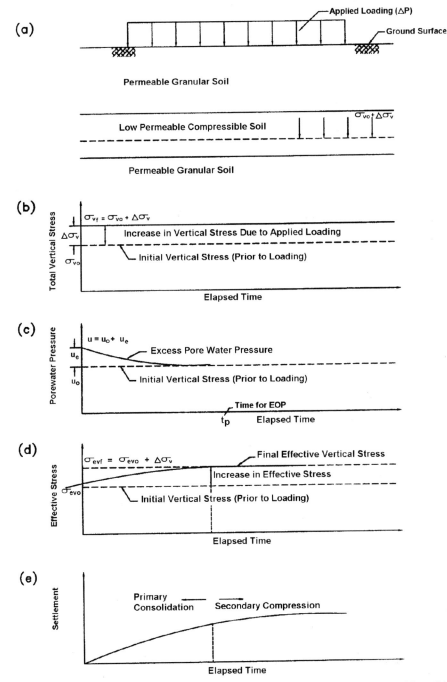

Figure 5.8 Consolidation of fine-grained soils: (a) hypothetical areal loading on a compressible soil layer; (b) total vertical stress; (c) porewater pressure; (d) effective stress; (e) settlement.

are referred to standard textbooks on geotechnical engineering, such as Holtz and Kovacs (1981), Das (1994), and Terzaghi et al. (1996).

The recompression index (C_r), the compression index (C_c), and the maximum past pressure or preconsolidation pressure (σ_{vm}) are known as the *consolidation properties* of a soil. These values are found by conducting a one-dimensional laboratory consolidation test (ASTM D 2435) on the soil. The *consolidation test* involves placing a soil specimen in a ring with porous stones on either end. It is then submerged in water to ensure saturated conditions. A constant load is applied, and a dial gauge reading of vertical deformation is recorded with time. Figure 5.9 is an example of a plot of the change in deformation with time. The point of a change in slope, as indicated, is known as the *end of primary consolidation* (EOP). Beyond this point, *secondary consolidation*, or *creep*, occurs, where

$u_e = 0$. The slope of the line beyond the EOP is used to calculate the coefficient of secondary compression (C_α), which is defined by $\Delta e / \Delta \log t$. By applying different stress increments, change in strain or void ratio at EOP as a function of effective stress can be plotted, such as in Figure 5.10, displaying the loading, unloading, and reloading processes. C_c and C_r are determined as the slopes on this plot.

To calculate the maximum past pressure, *Casagrande's construction method* is used, as shown in Figure 5.10. As outlined in Figure 5.10, an estimate point of sharpest curvature is first identified and a horizontal line is drawn through it. After a tangent line is drawn at this point, a bisecting line is drawn between the horizontal and tangent lines. The consolidation line is extended upward, intersecting the bisecting line at σ_{vm}. σ_{vm} is also expressed in terms of the *over-consolidation ratio* (OCR), defined as OCR $= \sigma_{vm}/\sigma_{v0}$. If the OCR

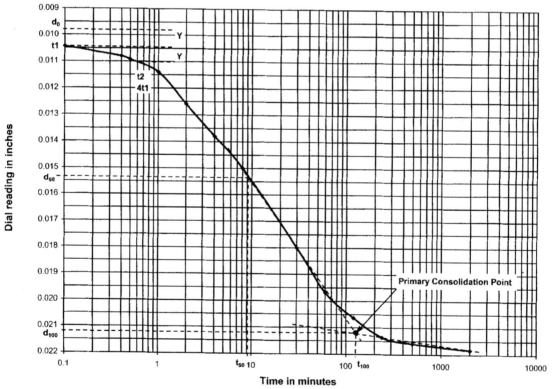

Figure 5.9 *Compression of soil specimen under constant vertical stress in consolidation.*

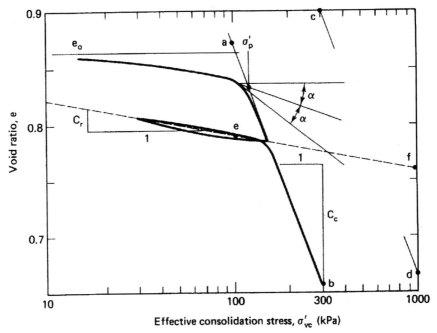

Figure 5.10 *Vertical effective stress versus void ratio at the end of primary consolidation. (From Holtz and Kovacs,* © *1981. Reprinted by permission of Pearson Education, Inc., Upper Saddle River, NJ.)*

= 1, the soil is normally consolidated. If the OCR > 1, the soil is overconsolidated.

Consolidation tests are the preferred method of determining C_c, C_r, and σ_{vm}. However, the tests do require undisturbed soil samples and a large amount of time. Therefore, for feasibility studies, empirical correlations, as summarized in Table 5.5, may be used.

In certain situations, it may be important to know the rate of consolidation [t versus u_e and t versus S_c, as depicted in Figure 5.8(c) and (e)]. This calculation is made through the use of *Terzaghi's one-dimensional consolidation theory*. This theory, expressed in the partial differential equation, is as follows:

$$C_v \frac{\partial u_e}{\partial t} = \frac{\partial^2 u_e}{\partial z^2} \qquad (5.6)$$

where u_e is the excess pore water pressure, z the depth from the top of the compressible layer, and C_v the coefficient of consolidation. The solution to this equation

(with simple boundary and initial conditions) yields (Holtz and Kovas, 1981)

$$u_e = \sum_{m=0}^{\infty} \frac{2u_{ie}}{(\pi/2)(2m + 1)} \sin\left[\frac{\pi(2m + 1)}{2} \frac{z}{H_d}\right]$$

$$e^{-\left(\frac{\pi(2m + 1)}{2}\right) T} \qquad (5.7)$$

where T is the time factor, equal to $C_v t / H_d^2$. H_d, defined as the maximum drainage length, depends on the K value of soil or bedrock layers that overlie and underlie the compressive clay layer that is being considered. If a double drainage condition exists (i.e., if high-permeability soil/rock layers overlie and underlie the compressive clay layer), the maximum drainage length is equal to half the thickness of the soil layer. If a single drainage condition exists (i.e., only one layer, either the top or bottom layer, possesses a higher K value than the compacted clay layer), the maximum drainage length is equal to the thickness of the soil

Table 5.5 **Typical correlations to estimate consolidation characteristics of cohesive soils**

Property	Correlation
Compression index (C_c)	$C_c = 0.009(LL - 10)$ LL = liquid limit
	$C_c = 1.15(e_0 - 0.35)$ e_0 = initial void ratio
	$C_c = 1.15 \times 10^{-2}w_n$ w_n = natural water content
Recompression index (C_r)	$C_r = 0.1C_c$
	$C_r = 0.0463(LL)G_s$ G_s = specific gravity of solids
Maximum past pressure or preconsolidation pressure (σ_{vm})	$\sigma_{vm} = 3.78C_u - 2.9$ (kPa)
	$\sigma_{vm} = (q_u/2)/(0.11 + 0.0037PI)$
	$\log \sigma_{vm} = (5.97 - 5.32w_n)/(LL - 0.25 \log \sigma_v)$ (kPa)
	q_u = unconfined compressive strength
	C_u = undrained shear strength ($\approx q_u/2$)
	σ_v = effective overburden pressure
Coefficient of consolidation (C_v)	See Figure 5.12 for LL vs. C_v

Source: U.S. Navy (1971); Holtz and Kovacs (1981).

layer. C_v is the coefficient of consolidation, another consolidation property of the soil. Defining $Z = z/H_d$ and $U_z = (1 - u_e/u_{ie})$, a nondimensional chart of Z versus U_z versus T, as shown in Figure 5.11, is developed. This chart is commonly used to determine the u_e. Note that u_{ie} is the initial excess pore water pressure (at $t = 0$), generally equal to $\Delta\sigma$.

To develop a relationship between t and S_c, the average degree of consolidation (U_{ave}) is defined as

$$U_{avg} = \frac{S_{c,\text{at any time}}}{S_{c,\text{total}}} \quad (5.8)$$

S_c total is calculated as explained previously, and U_{ave} is related to T. For $T < 0.3$,

$$U_{avg} = 2\sqrt{\frac{T}{\pi}} \quad (5.9)$$

and for $T \geq 0.3$,

$$U_{avg} = 1 - \frac{8}{\pi^2} e^{-\pi^2 T/4} \quad (5.10)$$

Charts defining these relationships are also included in various standard geotechnical textbooks, such as Holtz and Kovacs (1981) and Das (1994).

C_v can also be calculated using one-dimensional laboratory consolidation test data. A value of t_{50} is found using Casagrande's log t method, as shown in Figure 5.9. For this, a time t_1 is first selected, and a time t_2 is calculated as four times the value of t_1. The vertical distance between t_1 and t_2 is added to t_1, producing a dial reading d_0. On the lower portion of the figure, two linear segments are identified whose intersection is designated with a dial reading of d_{100}. D_{50} is calculated as the average of d_0 and d_{100}, with a corresponding time t_{50}. Then C_v is calculated using $C_v = T_{50}H_d^2/t_{50}$, where $T_{50} = 0.197$. In the absence of consolidation test data, C_v may be estimated using charts such as that shown in Figure 5.12.

For organic soils, secondary consolidation (S_s) may be significant. S_s may be calculated using

$$S_S = \frac{C_\alpha}{1 + e_p} H \log \frac{t}{t_p} \quad (5.11)$$

where e_p is the void ratio at the end of primary consolidation settlement and t_p is the time for the entire primary consolidation settlement to occur. The total settlement is then given as $S_t = S_c + S_s$. For feasibility studies, C_α may be assumed equal to $0.05C_c$ (Holtz and Kovacs, 1981).

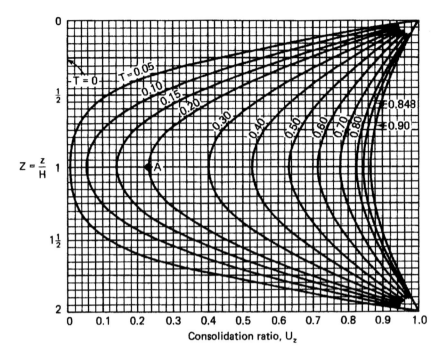

$Z = \dfrac{z}{H}$

Consolidation ratio, U_z

Figure 5.11 Nondimensional chart showing U_z vs. Z and T to calculate excess porewater pressure (u_e) distribution in a compressible soil during consolidation. (From Holtz and Kovacs, © 1981. Reprinted by permission of Pearson Education, Inc., Upper Saddle River, NJ.)

5.2.6 Shear Strength

The shear resistance to failure is called the *shear strength* (τ) of the soil. In other words, the failure condition is based on excessive deformation that cannot be tolerated in a specific design. The Mohr–Coulumb failure criterion of materials, given by $\tau = c + \sigma \tan \phi$, is used to define the shear strength of soils. If normal stress, σ, is known, the shear strength parameters c and ϕ are needed to define the shear strength. In general, these parameters have no physical meaning. They depend on the test methods and test drainage conditions. c is commonly known as the cohesion parameter, and ϕ is known as the angle of internal friction or the friction parameter.

Under *drained conditions,* the loading rate is slow and u_e is dissipated. Volume change occurs during shearing. The drained shear strength parameters are denoted c' and ϕ'. In most cases (e.g., for cohesionless soils and for normally consolidated cohesive soils), $c' \approx 0$. Under *undrained conditions,* the loading rate is fast and excess pore water pressure is generated; vol-

ume change does not occur. The undrained shear strength parameters are denoted c and ϕ. A special case of undrained conditions occurs when soil is not allowed to consolidate either prior to or during shearing which is referred to as *unconsolidated undrained condition* (or $\phi = 0$ *condition*).

Several methods exist to determine the shear strength of soils: laboratory tests, field tests, and empirical correlations. Lab tests include the direct shear test (ASTM D 3080), the triaxial test (e.g., ASTM D 4767), and the unconfined compression test (ASTM D 2166). In special cases, the hollow cylinder test, the plane-strain test, the true-triaxial test, the torsional shear test, and the direct simple shear test can be conducted. The triaxial test can be conducted under three conditions: UU (unconsolidated, undrained) per ASTM D 2850, CU (consolidated, undrained) per ASTM D 4767, and CD (consolidated, drained). Field tests include the standard penetration test (ASTM D 1586), the cone penetration test (ASTM D 3441), and the vane shear test (ASTM D 2573). Empirical correlations to

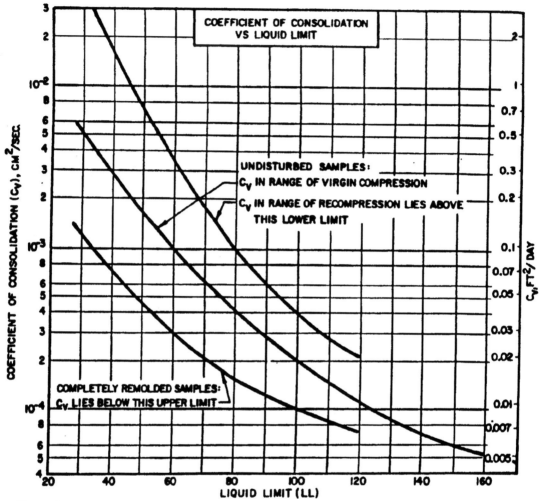

Figure 5.12 *Estimation of coefficient of consolidation* (C_v) *using liquid limit (LL). (From U.S. Navy, 1971.)*

estimate shear strength are based on relationships between the index properties and c, ϕ, or c', ϕ'. Figure 5.13 shows correlations between the SPT blow count and ϕ' for granular soils. Figure 5.14(a) correlates undrained shear strength parameter (c) of fine-grained soils with PI, while Figure 5.14(b) correlates the drained shear strength parameter (ϕ') with PI for fine-grained soils. For detailed design, tests must be per-

formed on representative site soils to determine their shear strength.

5.3 CHEMICAL PROPERTIES

The chemical soil properties that are generally of practical use in geoenvironmental engineering are pH, sur-

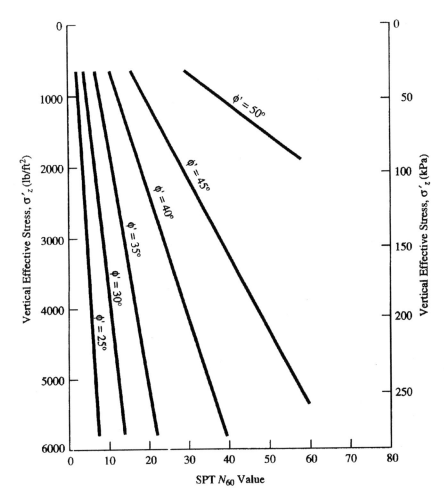

Figure 5.13 Empirical correlation between N_{60} and ϕ' for uncemented sands. [Note: N_{60} is STP N-value corrected for field procedure; see Coduto (1999) for details.] (From Coduto, © 1999. Reprinted by permission of Pearson Education, Inc., Upper Saddle River, NJ.)

face charge, cation exchange capacity, anion exchange capacity, and specific surface. These properties are explained briefly in this section.

5.3.1 pH or Hydrogen-Ion Activity

The most important chemical property of soil is the hydrogen-ion activity, or pH. *pH* is the negative logarithm of hydrogen-ion activity. pH values range from 0 to14. A pH value below 7 indicates acidic soil; a pH value greater than 7 indicates basic or alkaline soil; a pH value of 7 indicates neutral soil.

The pH of soils is measured using a glass electrode pH meter following standard procedures such as ASTM D 4972 and USEPA 9045C (USEPA, 1986; ASTM, 2003). Generally, the testing procedures essentially involve mixing one part of soil with one part of distilled water in a beaker, stirring the suspension for 30 minutes, allowing the suspended clay to settle for 1 hour, immersing the glass electrode partly into the settled suspension, and measuring the pH.

The measured soil pH is influenced by (1) the soil-to-water ratio (or dilution), (2) the soluble salts, and (3) the CO_2 in the air. The soil-to-water ratio ranges from 1:1 to 1:10, with 1:1 being the most commonly

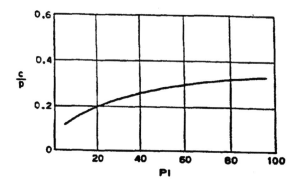

(a) c/p versus plasticity index for
normally consolidated soils

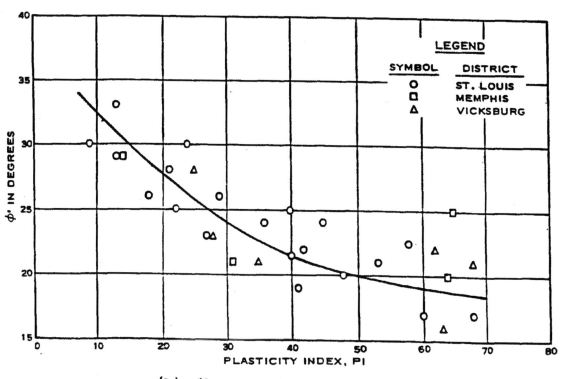

(b) Ø' versus plasticity index

Figure 5.14 *Example correlations of strength characteristics for fine-grained soils. (From U.S. Navy, 1971.)*

used. Highly plastic soils may require higher dilution to keep soil particles in suspension. In general, the more dilute the soil suspension, the higher the soil pH value found, whether the soil is acidic or alkaline. The rise in soil pH with dilution may be on the order of 0.2 to 0.5 pH unit, but can be 1 or more pH units in certain neutral and alkaline soils.

The pH of a soil suspension may decrease due to the solubilization of salts, if present, in the soil. To mask the variability of salt content in soils, the soil pH is measured using one part soil and two parts 0.01 M $CaCl_2$ solution, prepared with distilled water. This soil suspension promotes accurate soil pH and is independent of the soil-to-water ratio (dilution) as well as the dissolved salts. The presence of CO_2 decreases soil pH, but at the common atmospheric partial pressure of CO_2, this effect is very small.

As explained in Chapter 3, pH is the chemical property that affects various chemical processes, such as adsorption–desorption, precipitation–dissolution, and oxidation–reduction. These processes, in turn, control the fate and transport of chemicals in soils. Therefore, determination of pH value is very important in understanding various geochemical reactions.

5.3.2 Surface Charge and Point of Zero Charge

Coarse-grained soils such as gravel, sand, and silt are chemically inert. Clay mineral surfaces generally carry electronegative charges. The negative charges on clay surfaces are a result of two occurrences: (1) isomorphous substitution and (2) the disassociation of exposed hydroxyl groups. Isomorphous substitution is the substitution of atoms for other atoms without affecting the crystal structure. This substitution is possible in both the silica tetrahedra and aluminum octahedra of the clay mineral. For example, in the absence of isomorphous substitution, kaolinite is electrically balanced. However, an isomorphic replacement of one octahedral Al^{3+} by Mg^{2+} yields one unbalanced negative charge in the crystal. This is because Mg^{2+} is divalent and contributes only two positive charges to the neutralization of the crystal. A similar substitution can also

occur in a Si tetrahedron, where Si^{4+} can be replaced by Al^{3+}, resulting in one negative charge that has not been neutralized. These types of charges, called *permanent negative charges*, are independent of pH. Isomorphous substitution will occur only between atoms of almost equal size when the difference in valance does not exceed one unit.

The negative charge on a clay surface is also due to the presence of exposed hydroxyls (OH) on the surface of Al octahedral sheets. Prevalent in 1:1-type clay minerals, these OH are in contact with the soil solution and tend to dissociate, releasing their protons (H^+). This dissociation of the H^+ leaves one negative charge, in the octahedron, which is not neutralized. Such a dissociation reaction is dependent on pH. The dissociation reaction occurs at high pH and decreases at low pH. Therefore, the magnitude of negative charge also increases and decreases with change in pH. This type of negative charge is called *pH-dependent charge* or *variable charge*.

Sometimes, clay mineral surfaces carry positive charges. Positive charges generally result from the protonation of exposed OH groups. This involves the addition of H^+ ions to the exposed OH of clay minerals. The H^+ is adsorbed with a relatively weak bond. Because of this addition of H^+, the OH group is oversaturated with protons and the clay surface becomes positively charged. Protonation of exposed OH groups occurs only at low pH values, because acidic conditions are required for the supply of extra protons.

The permanent charge resulting from isomorphic substitution is negative and is independent of pH, as shown in Figure 5.15. The pH-dependent or variable surface charge resulting from dissociation of OH at high pH is negative, while the surface charge resulting from protonation of OH at low pH is positive. This charge is also depicted in Figure 5.15. The net total charge on the soil surface is the summation of the permanent charge and the pH-dependent charge, as shown in Figure 5.15. The pH at which the net surface charge is equal to zero is called the *point of zero charge* (PZC) or *zero point of charge* (ZPC). The PZC is a specific characteristic of each clay mineral. Typical values of PZC for various minerals containing water as the soil solution (liquid phase) are listed in Table 5.6.

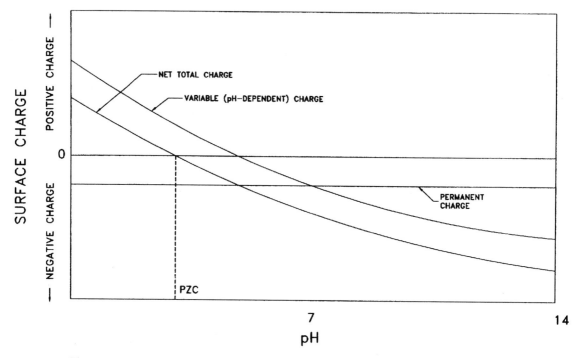

Figure 5.15 *Surface charge of soils as a function of pH and the point of zero charge (PZC).*

Table 5.6 **Point of zero charge of different minerals**

Mineral	Chemical Formula	ZPC
Quartz	SiO_2	2–3
Birnessite	$(Na,Ca)Mn_7O_{14} \cdot (2 \cdot 8H_2O)$	1.5–2.5
Kaolinite	$Si_4Al_4O_{10}(OH)_8$	4.0–5.0
Geothite	$FeO(OH)$	7.0–8.0
Hematite	Fe_2O_3	8.0–8.5
Gibbsite	$Al(OH)_3$	8.0–9.0
Montmorillonite	$Na_x[(Al_{2-x}Mg_x)Si_4O_{10}(OH)_2]$	<2.5

Source: Sposito (1989); Drever (1997).

The surface charge may have major implications on the distribution of ionic contaminants in soils. If the surface charge is negative, cationic metals such as lead (Pb^{2+}) will be bonded to the soil surfaces. However, if the surface charge is positive, anionic metal complexes, such as chromate, (CrO_4^{2-}) will be bonded to the soil surfaces. The PZC is useful in determining the pH range in which soil surfaces are positively or negatively charged.

5.3.3 Cation Exchange Capacity

When the surface charge of a soil is negative, the negative charges are balanced by cations such as Na^+, Ca^{2+}, Mg^{2+}, and others, which form the diffuse double layer. These cations can be replaced rather easily by one another; therefore, they are called *exchangeable cations*. The sum of exchangeable cations is called the *cation exchange capacity* (CEC). CEC is expressed as milliequivalent per 100 g of dry soil and is denoted as meq/100 g. As explained in Chapter 3, the milliequivalent of an ion is the atomic weight of the ion in milligrams divided by the valence of the ion. For example, 1 meq of H^+ is equal to 1 mg of H; 1 meq of Na^+ is equal to 23 mg of Na; 1 meq of K^+ is equal to 39 mg of K, and 1 meq of Ca^{2+} is equal to $40/2 =$ 20 mg of Ca.

Many methods have been reported for the determination of CEC of soils (Black, 1965). These methods essentially involve replacing exchangeable cations by saturating the soil with a selected cation. The saturating cation may be generated using one of three reagents: 1 N ammonium acetate (pH 7.0), 1 N sodium acetate (pH 8.2), or 0.5 N barium chloride plus 0.2 N triethanolamine solution (pH 8.2). The first two reagents are used for calcareous and noncalcareous soils, and the third reagent is used for soils where it is desired to determine both exchange capacity and the amount of exchangeable hydrogen. The test procedure essentially consists of mixing known amounts of dry soil with the reagent solution, shaking thoroughly, and filtering or centrifuging to separate the supernatant. This process is repeated three or four times to ensure that all of the exchangeable cations are replaced by the cation in the reagent selected (e.g., ammonium or Na). Finally, the amount of adsorbed reagent cations in the soil (ammonium or Na) is equal to the CEC and is extracted and determined by standard methods. The detailed procedures can be found in publications such as Black (1965) and USEPA (1986).

The CEC differs from soil to soil depending on (1) clay content, (2) clay types, and (3) organic content. The CEC is higher in soils that contain high clay content and high organic content. The CEC values for various clay minerals are given in Table 5.7.

Adsorption of contaminants in soils depends on soil CEC values. The higher the CEC, the higher the adsorption of cationic contaminants on soil surfaces, increasing the difficulty of removal during remediation process implementation.

5.3.4 Anion Exchange Capacity

The *anion exchange capacity* (AEC) is the capacity of soil to adsorb and exchange anions. The soil must be positively charged to adsorb negatively charged ions. Positive charges in soils occur only in low-pH or acidic conditions, when the soil pH is below the PZC of the

Table 5.7 Cation exchange capacity of clay minerals and soil types

Clay Minerals / Soil Types	CEC (meq / 100 g)
Chlorite	10–40
Illite	10–40
Kaolinite	3–15
Montmorillonite	80–150
Oxides and oxyhydroxides	2–6
Saponite	80–120
Vermiculite	100–150
Soil organic matter	>200
Sand	2–7
Sandy loam	2–18
Loam	8–22
Silt loam	9–27
Clay loam	4–32
Clay	5–60

Source: USEPA (2000).

soil. A positive charge can also develop from of broken bonds on broken surfaces of clay minerals. In general, the positive charge, hence the anion exchange capacity of soils, is considered smaller than the CEC.

The procedure used to determine the AEC of a soil is similar to the CEC procedure, except that different reagent solutions are used to replace all of the exchangeable anions in the soil (Jackson, 1958). Most commonly used, and preferred, reagents contain Cl^-, due to its nonspecific adsorption characteristics.

Data on AEC of soils are scarce. Generally, AEC is negligible at the natural pH (5 to 8) of most soils. AEC values commonly range from 1 to 10 mmol/kg, but can be as high as 1 mol/kg in soils with high organic matter and metal oxide contents.

Similar to CEC, adsorption of anionic contaminants on soil surfaces depends on the AEC. Higher AEC results in higher adsorption of anionic contaminants and may be an important consideration in the design of remedial processes.

5.3.5 Specific Surface

The *specific surface* of minerals is the ratio of surface area to either volume or mass. Specific surface is determined by the ethylene glycol, glycerol, or ethylene glycol monoethyl ether (EGME) adsorption procedure (Black, 1965; Mitchell, 1993). These amounts provide a quantitative determination of clay minerals and an estimate of specific surface area. Specific surface values for typical minerals are given in Table 5.8.

Particles smaller than 1 μm possess significant surface area, and the surface properties of such particles will have significant consequences. For example, larger

surface area may result in higher unbalanced surface charge, which in turn may cause greater sorption of contaminants. Such circumstances influence remedial processes and waste leaching conditions.

5.4 SUMMARY

The geotechnical properties of soils that are determined routinely for various geotechnical projects, such as foundations and earth structures, are also important for geoenvironmental studies. Standard properties such as hydraulic conductivity, compaction, consolidation, and shear strength are required for the assessment and design of earth barriers in waste containment systems. Chemical properties of soils are also very important for geoenvironmental studies. Among these are soil pH, surface charge, point of zero charge, CEC, AEC, and specific surface. The chemical properties are useful in the design of waste containment systems as well as remedial processes.

Table 5.8 **Specific surface of various minerals**

Mineral	Specific Surface (m^2/g)
Quartz	0.14
Gibbsite	120
Hematite	1.8
Kaolinite	10–38
Illite	65–100
Montmorillonite	600–800

Source: Langmuir (1997).

QUESTIONS/PROBLEMS

5.1. The mass of a moist soil sample collected from the field is 580 g, and its oven-dry mass is 525 g. The specific gravity of the soil solids as determined in the laboratory is 2.65. The void ratio of the soil in the natural state is 0.60. Draw the phase diagram and calculate **(a)** water content; **(b)** degree of saturation; **(c)** wet density; **(d)** dry density.

5.2. An undisturbed moist soil sample obtained from the field has a volume of 110 cm³ and weighs 210 g. When it is dried in a drying oven, it weighs 168 g. The specific gravity of soil solids is 2.65. Draw the phase diagram and calculate the water content, void ratio, porosity, degree of saturation, wet density, and dry density.

5.3. A soil sample has a void ratio of 0.48, water content of 12%, and specific gravity of soil solids of 2.6. Draw the phase diagram and calculate the porosity, degree of saturation, wet density, and dry density.

5.4. A soil sample has a moist unit weight of 120 lb/ft³. Its water content and specific gravity are 23% and 2.6, respectively. Draw the phase diagram and calculate the void ratio, degree of saturation, and dry unit weight.

5.5. Sieve analysis test results for a soil are given below. Draw the grain size distribution curve and classify the soil according to the Unified Soil Classification System (USCS).

U.S. Sieve Number	Sieve Size Opening (mm)	Weight Retained (g)
4	4.75	8.8
10	2.00	1.9
20	0.85	3.3
40	0.425	7.3

U.S. Sieve Number	Sieve Size Opening (mm)	Weight Retained (g)
60	0.300	57.3
140	0.106	369.4
200	0.075	26.1
Pan	—	56.6

5.6. Sieve analysis test results for three soils are provided below. Draw the grain size distribution curves and determine the coefficient of uniformity and the coefficient of curvature for these soils. Classify these soils according to the USCS.

Mass of Soil Retained (g)	U.S. Sieve Size							
	No. 4	No. 10	No. 20	No. 40	No. 60	No. 140	No. 200	Pan
Soil A	72.8	57.8	29.5	19.4	84.6	253	19.7	39.6
Soil B	0	1.5	1.8	2.0	66.9	556.9	44.3	42.2
Soil C	4.2	4.6	3.4	6.1	19.7	485.6	26.5	37

5.7. Determine the liquid limit based on the following results obtained from the liquid limit test.

Number of blows	8	18	29	34
Wt. of container + wet soil (g)	19.8	18.4	19.4	14.6
Wt. of container + dry soil (g)	19.0	17.8	18.8	14.0
Wt. of container (g)	15.9	15.2	15.9	11.1

5.8. The following results were obtained from a liquid limit test on a clayey soil.

Number of blows	17	22	26	39
Water content (%)	38.2	36.0	34.4	30.6

Two tests for plastic limit were conducted which gave values of 24.1 and 23.8, respectively. Determine the liquid limit, plastic limit, and plasticity index. Classify the soil according to the USCS.

5.9. The results of sieve analysis are given below. The Atterberg limit tests for the soil are liquid limit of 35% and plastic limit of 20%. Classify the soil according to the USCS.

U.S. Sieve Size	Percent Passing
$\frac{3}{8}$ in.	100
No. 4	100
No. 10	100
No. 40	94.6
No. 200	87.0

5.10. The following data were obtained from Proctor's compaction test on a soil.

Wt. of mold + wet soil (g)	3457.2	3721.1	3909.0	3782.5	3715.2
Moisture content (%)	8.0	11.0	12.8	15.7	17.0

The weight of the mold is 1850 g and volume is 945 cm³. The specific gravity of soil solids is 2.65.

(a) Plot the compaction curve and determine the maximum dry density and optimum moisture content.

(b) Plot the zero air voids curve (ZAVC).

(c) Calculate the degree of saturation and the void ratio at the maximum dry density.

5.11. A series of laboratory compaction tests on a soil provided the following data:

Water content (%)	8.3	10.2	12.9	14.6
Wet unit weight (lb/ft³)	116	131	138	135

Plot the moisture–unit weight curve and determine the maximum dry unit weight and optimum moisture content (OMC). Plot the zero air voids curve assuming that the specific gravity of soil solids is 2.60. Determine the range of moisture content at which 95% or more of the maximum dry unit weight can be obtained.

5.12. Soil with a void ratio of 0.52 as it exists in a borrow pit is to be excavated and transported to a landfill site, where it will be used as a 3-ft-thick compacted soil liner with a void ratio of 0.4. Find the volume of soil that must be excavated from the borrow pit to provide the required volume of liner over a 1-acre area.

5.13. Estimate the capillary rise in soils with D_{10} equal to **(a)** 0.001 mm; **(b)** 0.075 mm; **(c)** 4.75 mm; **(d)** 75 mm. Comment on the validity of such results.

5.14. A constant-head permeability test was conducted on a clean sand sample. The diameter and length of the test specimen were 10 and 15 cm, respectively. The head difference across the sample was 30 cm during the test. If it took 100 s for 750 mL of water to discharge, determine the hydraulic conductivity of the soil.

5.15. In a constant-head permeameter, a sample of coarse sand 15 cm high and 6.5 cm in diameter was tested. Water was allowed to pass through the soil under a constant hydraulic head of 30 cm for a period of 20 s and the discharge water collected was 455 mL. Calculate the hydraulic conductivity of the soil.

5.16. In a falling-head permeameter, a sample of clayey soil was tested and the following data were obtained: internal diameter of permeameter, 65.5 mm; length of sample, 150 mm; standpipe inside diameter, 10.0 mm; initial water level in standpipe, 950 mm; water level in standpipe after 20 min, 700 mm. Using these data, calculate the hydraulic conductivity of the soil.

5.17. A falling-head permeability test was conducted on a silty clay sample. The diameter and length of the test specimen were 8.5 and 15.5 cm, respectively. The cross-sectional area of the standpipe was 1.95 cm². If it took 45 minutes for the water in the standpipe to drop from a height of 85 cm at the beginning of the test to 60 cm at the end, determine the hydraulic conductivity of the soil.

5.18. A sample of sand has a void ratio of 0.4 and a hydraulic conductivity of 1×10^{-2} cm/s. Estimate the hydraulic conductivity for the same soil at a void ratio of 0.3.

5.19. A soil profile at a site consists of 5 ft of fill ($\gamma = 100$ lb/ft³) over 10 ft of fine sand ($\gamma = 110$ lb/ft³) over 20 ft of silty clay ($\gamma = 115$ lb/ft³) over 10 ft of sand/gravel ($\gamma = 120$ lb/ft³). The water level is 10 ft below the ground surface. Calculate the **(a)** total stress distribution; **(b)** pore water pressure distribution; **(c)** effective stress distribution.

5.20. A soil profile consists of 5 ft of gravel fill ($\gamma = 120$ lb/ft³) over 5 ft of sand ($\gamma = 115$ lb/ft³) over 10 ft of silt ($\gamma = 110$ lb/ft³) over 8 ft of clay ($\gamma = 100$ lb/ft³) over intact bedrock. The water level is 7 ft below the ground surface. Calculate total stress, pore water pressure, and effective stress distribution with depth.

5.21. Below the landfill base, the soil profile consists of 20 ft of silty clay ($\gamma = 100$ lb/ft³, $e_0 = 0.54$, $C_c = 0.4$, OCR = 1.2, $C_r/C_c = 0.1$, $C_\alpha/C_c = 0.05$, $C_v = 0.002$ ft²/day) over a very dense sand and gravel layer. The water table is 5 ft below the base of the landfill. A 30-ft-high waste layer with a unit weight of 60 lb/ft^3 is placed in the landfill. Calculate the (a) total primary settlement and time it takes for this settlement to occur; (b) settlement and excess pore water pressure distribution after two years of loading; (c) settlement (including secondary compression, if any) after 30 years and 100 years of waste placement.

5.22. A 50-ft-high landfill with an average unit weight of 65 lb/ft^3 was constructed at a site. The soil profile below the landfill base consists of 5 ft of sand (γ = 120 lb/ft^3) over 20 ft of silty clay (γ = 100 lb/ft^3, e_0 = 0.4, C_c = 0.32, OCR = 1.4, C_r/C_c = 0.15, C_α/C_c = 0.05, C_v = 0.006 ft^2/day) over 10 ft of marine clay (γ = 100 lb/ft^3, e_0 = 0.6, C_c = 0.4, OCR = 1, C_r/C_c = 0.1, C_α/C_c = 0.06, C_v = 0.004 ft^2/day) over coarse sand. The water table is at the interface between the sand and silty clay layers. Calculate the (a) total settlement resulting from primary consolidation of the silty clay and marine clay; (b) time required for a total settlement of 1 in. to occur.

5.23. A direct shear test was performed on a sand and the results are shown in Figure P.5.23. Using these data, determine the shear strength parameters of the soil.

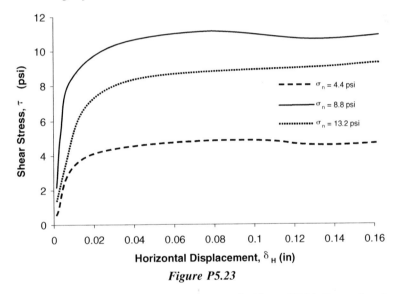

Figure P5.23

5.24. A series of triaxial UU tests performed on a soil are shown in Figure P5.24. Determine the shear strength parameters of the soil.

5.25. A soil has an unconfined compressive strength of 4.5 tons/ft^2. Determine the UU shear strength parameters of the soil.

5.26. A series of triaxial CU tests performed on a soil are shown in Figure P5.26. Determine the shear strength parameters of the soil.

5.27. A series of triaxial CD tests performed on a soil are shown in Figure P5.27. Determine the shear strength parameters of the soil.

5.28. Explain the difference between permanent and pH-dependent charge.

5.29. Which functional groups commonly cause pH-dependent charge in soils?

5.30. What is the zeta potential of soils?

5.31. What is the zero point of charge (ZPC) of soil?

5.32. What are the major exchangeable cations in soils?

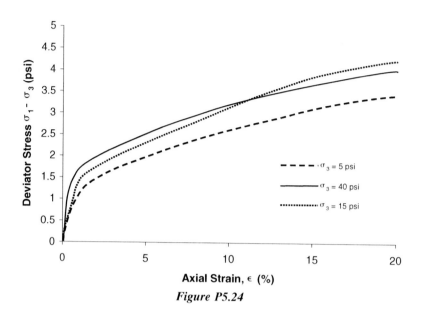

Figure P5.24

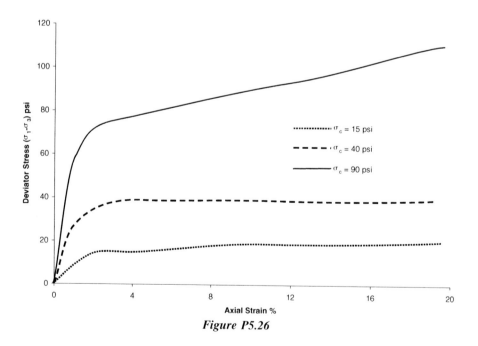

Figure P5.26

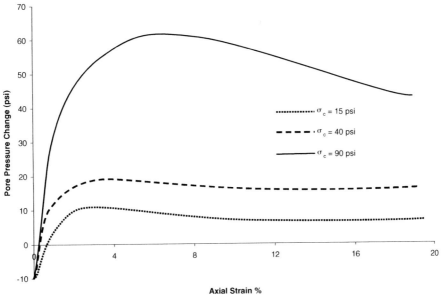

Figure P5.26 (*continued*)

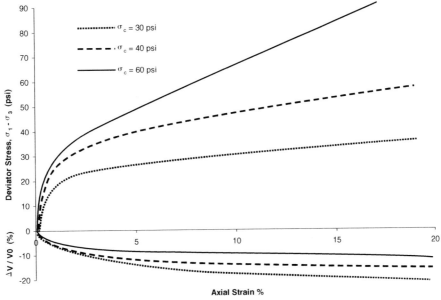

Figure P5.27

5.33. Estimate the CEC of a soil containing 60% kaolinite, 30% illite, and 10% montmorillonite.

5.34. A soil has a pH of 4.50 and CEC of 10.0 meq/100 g. It is titrated with OH^- to a pH of 6.5. Plot the variation of CEC with pH that is expected during this process.

REFERENCES

Black, C. A. (ed.), *Methods of Soil Analysis,* American Society of Agronomy, Madison, WI, 1965.

ASTM (American Society for Testing and Materials), *Annual Book of ASTM Standards,* ASTM, West Conshohocken, PA, 2003.

Coduto, D. P., *Geotechnical Engineering: Principles and Practice,* Prentice Hall, Upper Saddle River, NJ, 1999.

Das, B. M., *Principles of Geotechnical Engineering,* 3rd ed., PWS Publishing, Boston, 1994.

Drever, J. I., *The Geochemistry of Natural Waters: Surface and Groundwater Environments,* 3rd ed., Prentice Hall, Upper Saddle River, NJ, 1997.

Holtz, R. D., and Kovacs, W. D., *An Introduction to Geotechnical Engineering,* Prentice Hall, Upper Saddle River, NJ, 1981.

Jackson, M. L., *Soil Chemical Analysis,* Prentice Hall, Upper Saddle River, NJ, 1958.

Lambe, T. W., and Whitman, R. V., *Soil Mechanics,* Wiley, New York, 1969.

Langmuir, D., *Aqueous Environmental Geochemistry,* Prentice Hall, Upper Saddle River, NJ, 1997.

Mitchell, J. K., *Fundamentals of Soil Behavior,* 2nd ed., Wiley, New York, 1993.

Sposito, G., *The Surface Chemistry of Soils,* Oxford University Press, Oxford, 1989.

Terzaghi, K., Peck, R. B., and Mesri, G., *Soil Mechanics in Engineering Practice,* 3rd ed., Wiley, New York, 1996.

USACE (U.S. Army Corps of Engineers), *Laboratory Soil Testing,* EM 1110-2-1906, U.S. Department of the Army, 1970.

U.S. Navy, *Soil Mechanics, Foundations, and Earth Structures,* DM-7, Naval Facilities and Engineering Command, Washington, DC, 1971.

USEPA (U.S. Environmental Protection Agency), *Test Methods for Evaluating Solid Wastes,* Vols. IA and IB, *Laboratory Manual, Physical/Chemical Methods,* SW-846, USEPA, Washington, DC, 1986.

——, *In Situ Treatment of Soil and Groundwater Contaminated with Chromium,* EPA/625/R-00/005, Office of Research and Development, USEPA, Washington, DC, 2000.

6

GEOCHEMISTRY BACKGROUND

6.1 INTRODUCTION

In Chapter 3 we discussed the behavior of chemicals in water. In this chapter we study the behavior and reactions of chemicals in soils. The chemical aspects of soil are generally studied in geochemistry courses, which apply the principles of chemistry to the study of chemical reactions and processes that are associated with earth materials. We are particularly interested in environmental low-temperature geochemistry, which deals with the distribution and behavior of chemicals in soils and groundwater.

In this chapter, we review briefly the geochemistry of inorganic and organic chemicals. This background is essential in addressing geoenvironmental problems, particularly issues dealing with the soil and groundwater remediation process and the chemical-resistant barriers. The following sections cover basic concepts of inorganic and organic geochemistry. For detailed information, readers should consult references such as Lindsay (1979), Kramer and Allen (1988), Sposito (1989), Huling and Weaver (1991), McLean and Bledose (1992), Schwarzenbach et al. (1993), Newell et al. (1995), Drever (1997), and Langmuir (1997).

6.2 INORGANIC GEOCHEMISTRY

One aspect of inorganic chemistry is the behavior of inorganic chemicals, such as heavy metals, in soils. In this section we present various types of toxic metals found at contaminated sites. We then explain briefly the distribution of metals in soils, geochemical processes affecting the distribution of metals in soils, and mathematical models that can predict the metal distribution in soils. We then describe various laboratory chemical analysis methods used to determine metal concentrations in soils.

6.2.1 Metal Contamination

Of the many different inorganic chemicals, we will concern ourselves only with those that are identified as toxic by the USEPA. The toxic inorganic chemicals most commonly found at contaminated sites and in waste leachates include lead (Pb), chromium (Cr), arsenic (As), cadmium (Cd), nickel (Ni), zinc (Zn), copper (Cu), mercury (Hg), silver (Ag), and selenium (Se) (USEPA, 1997). These chemicals are also commonly referred to as metals, toxic metals, heavy metals, trace metals, transition metals, or micronutrients. Strictly speaking, a *metal* is any element that has a metallic luster and is a good conductor of heat and electricity. Although As and Se have both metallic and nonmetallic properties (metalloids), they are also commonly called metals. Unlike organic compounds, the unique characteristic of all metals is that they are not biodegradable.

6.2.2 Distribution of Metals in Soils

When metals are introduced into soils, they may be distributed in one or more of the following forms:

1. Dissolved in soil solution (pore water)
2. Occupying exchange sites on inorganic soil constituents
3. Specifically adsorbed on inorganic soil constituents
4. Associated with insoluble soil organic matter

5. Precipitated as pure or mixed solids

6. Present in the structure of minerals

In situations where metals have been introduced into the environment through human activities, they only exist in the first five forms. The last fraction generally exists if metals occur in natural soils that have been subjected to various geologic processes. For simplicity, the metal forms in soils can be grouped under three phases: (1) aqueous phase, representing the first form; (2) adsorbed phase, representing the second, third, and fourth forms; (3) solid phase, representing the fifth form.

The *aqueous phase* represents the metals existing in soil solution as free (uncomplexed) metal ions (e.g. Cd^{2+}, Ni^{2+}, Zn^{2+}, Cr^{3+}), and as various soluble complexes (e.g. $CdSO_4^0$, $ZnCl^+$, $CdCl_3^-$). Metals can form soluble complexes with inorganic and organic ligands. Common inorganic ligands are SO_4^{2-}, Cl^-, OH^-, PO_4^{3-}, NO_3^-, and CO_3^{2-}, and organic ligands include soluble constituents of fulvic acids, and low-molecular-weight aliphatic, aromatic, and amino acids (see Chapter 3 for details). The potential for migration (or mobility) of metals in soils will be higher if higher amounts of metals are present in the aqueous phase; this is because metals in the aqueous phase are readily transported by processes such as advection, dispersion, and diffusion, as explained in detail in Chapter 8.

The *adsorbed phase* represents the accumulation of metal ions at the interface between soil solids and the aqueous phase. This phase is associated with the surfaces of soil solids such as organic matter, soil minerals, iron and manganese oxides, and hydroxides, carbonates, and amorphous aluminosilicates.

The *solid phase* represents metals present in precipitate form. These precipitates exist in three-dimensional solid phase and may consist of pure solids [e.g., $CdCO_3$, $Pb(OH)_2$, ZnS] or mixed solids [e.g., $(Fe_xCr_{1-x})(OH)_3$, $Ba(CrO_4,SO_4)$]. Mixed solids are formed when various elements coprecipitate. The solid phase may be present in soils when metal concentrations are significantly higher than their solubility limits. The metals that exist in solid phase are immobile in soils, meaning that the solid-phase metals remain inplace and cannot migrate. Certain remediation tech-

nologies such as stabilization (discussed in Chapters 13 and 14) aim to convert all of the metals in aqueous and adsorbed phases into the solid phase. The conversion of metals into the solid phase eliminates the potential for migration or spreading of the contaminants in soils, thereby reducing the risk to public health and the environment.

6.2.3 Geochemical Processes Controlling the Distribution of Metals in Soils

The amounts of metals present in different soil phases (i.e., the aqueous, adsorbed, and solid phases) are controlled by the following interdependent geochemical processes: (1) adsorption and desorption, (2) redox reactions, (3) complex formations, (4) precipitation and dissolution of solids, and (5) acid–base reactions. The processes and the factors affecting them are described briefly in the following subsections.

Adsorption and Desorption. *Adsorption* is defined as the accumulation of ions at the interface between a solid phase and an aqueous phase. *Desorption* is the opposite of adsorption, defined as the decrease of ions at the interface of solid and aqueous phases. In soils, the solid phase may consist of organic matter, soil minerals, iron and manganese oxides and hydroxides, carbonates, and amorphous aluminosilicates.

As discussed in Chapter 4, soil organic matter consists of biochemicals and humic substances with different functional groups (e.g., carboxylic, phenolic, alcoholic, and amino groups) that provide sites for metal adsorption. Soil organic matter can be the main source of soil cation exchange capacity (>200 meq/100 g at reactive sites). So this type of adsorption is significant in highly organic surface soils.

As discussed in Chapter 5, soils composed of clay minerals carry either a net negative or positive charge on their surface, depending on the nature of the surface and the soil pH. The net negative or positive surface charge is balanced by adsorption of cationic or anionic metals, respectively. A surface complexation model is often used to describe adsorption behavior (Sposito, 1989). Several types of surface complexes can form between soil surface functional groups and a metal, depending on the extent of bonding between the metal ion and the surface (Figure 6.1). As discussed in Chapter 5, metals in a diffuse ion association or in an outer sphere complex are not bonded directly to the soil surface. In response to electrostatic forces, these ions (often called *exchangeable ions*) accumulate at the charged surfaces.

With inner sphere complexation, the metal is bound directly to the soil surface; no waters of hydration are involved. It is distinguished from the exchangeable state in that the bonding between the metal and the particle surface has ionic and/or covalent character. The bonding energy involved with inner sphere complexation is much higher than that in exchange reactions, and the bonding depends on the electron configuration of both the surface group and the metal. The adsorption mechanism associated with inner sphere complexation is often termed *specific adsorption*. The word *specific* implies that there are differences in the energy of adsorption among cations, such that other ions, including major cations (Na, Ca, and Mg), do not compete effectively for specific surface sites. Specifically, adsorbed metal cations are relatively immobile and unaffected by high concentrations of the major

cations, due to the large difference in their energies of adsorption.

At low concentrations, metals are adsorbed at specific adsorption sites, and adsorbed metals are not removed by the input of major cations. With increasing concentration of the metal, specific sites become saturated and the exchange sites are filled. Metals associated with these nonspecific sites are exchangeable with other metal cations and thus are potentially mobile.

Although soil particles are generally negatively charged, they may also carry positive charge in certain conditions (e.g., low pH). In such situations, the adsorption of metals in anionic form becomes important. Common anionic metal contaminants of concern include arsenic (AsO_4^{3-} and AsO_2^-), selenium (SeO_3^{2-} and SeO_4^{2-}), and chromium in its hexavalent oxidation state (CrO_4^{2-}). The adsorption capacity for anions is small relative to the adsorption capacity of cations.

The adsorption capacity (both exchange and specific adsorption) of a soil is determined by the number and type of exchange sites available. Adsorption of metal cations has been correlated with soil composition, as well as soil chemical properties such as pH and redox potential. The amounts and types of clay minerals, organic matter, Fe and Mn oxides, and calcium carbonates determine the surface area, surface charge, and cation exchange capacity of soils, and hence affect the adsorption capacity. As the soil pH increases beyond PZC, adsorption of all cationic metals increases. This is because with an increase in pH, the number of negative sites for cationic adsorption increases. Figure 6.2 shows the adsorption behavior of selected cationic metals at various pH conditions. The adsorption of cations increases with an increase in soil pH. The *redox potential* of soils indicates whether the metals are in an oxidized or reduced state. For instance, depending on the redox conditions, Fe can exist as Fe(III) solid or soluble Fe(II). If Fe(III) solids exist, they provide additional surfaces for adsorption of cation metals.

In addition, the type of metal cation, the presence of competing cations, and the formation of complexes affect the adsorption capacity. The adsorption of a cation metal depends on its relative preference to form

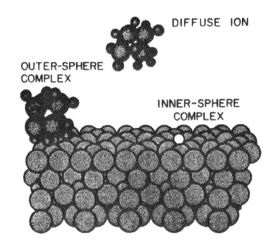

Figure 6.1 Surface complexation model to describe adsorption of metals in soils. (From Sposito, 1989, copyright 1989 by Oxford University Press, Inc. Used by permission of Oxford University Press, Inc.)

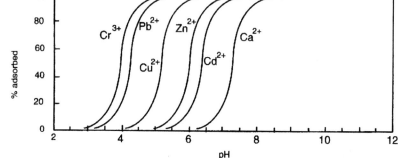

Figure 6.2 Typical adsorption behavior of metal cations on hydrous ferric oxide as a function of pH. (From Drever, © 1997. Reprinted by permission of Pearson Education, Inc., Upper Saddle River, NJ.)

strong bonds (inner sphere complexes) with the soil surface. For example, the general order of preference for cations to adsorb onto soil surfaces is found to be Pb > Cu > Zn > Cd > Ni (i.e., Pb has a higher order of preference for adsorption onto soil than do Cu, Zn, Cd, and Ni).

The adsorption of metal anions has been correlated with soil composition, pH, and redox potential. The presence of Fe and Mn oxides and low-pH conditions (below PZC) increase the adsorption of metal anions. Figure 6.3 shows the adsorption behavior of selected anionic metal complexes at various pH conditions. Contrary to the adsorption of cations, the adsorption of anions decreases with an increase in soil pH.

The use of published data that do not reflect the complex mixture of metals and the type of soils present at a specific contaminated site may not be useful in understanding or accurately characterizing the adsorption capacity. A careful assessment of adsorption capacity of site-specific metals and soils should be made

based on laboratory or field testing, as described in Chapter 8.

Redox Reactions. As explained in Chapter 3, oxidation–reduction reactions involve the transfer of electrons and are important for metals that possess more than one oxidation state. Important toxic inorganic chemicals that possess multiple oxidation states are U, Cr, As, Mo, V, Se, Sb, W, Cu, Au, Ag, and Hg. Redox reactions also involve other major elements that possess more than one oxidation state, such as H, O, C, S, N, Fe, and Mn. As explained in Chapter 3, Eh is used instead of p^c (or p^E) to express redox potential. p^c is the negative common logarithm of electron activity or electron concentration, $p^c = -\log_{10}[e^-]$. Eh and p^c are related by Eh = $0.059p^c$. Eh is measured in aqueous samples using the standard hydrogen electrode (SHE), which is formed by bubbling hydrogen gas at 1 bar pressure over a platinum electrode in a 1 *N* HCl solution.

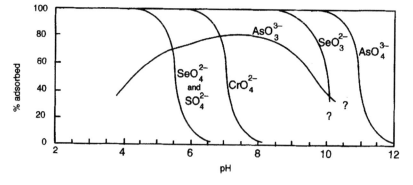

Figure 6.3 Typical adsorption behavior of metal anions in hydrous ferric oxide as a function of pH. (From Drever, © 1997. Reprinted by permission of Pearson Education, Inc., Upper Saddle River, NJ.)

Eh–pH diagrams are used to describe and visualize theoretical relationships among redox-sensitive elements (Pourbaix, 1974). The theoretical stability of a field of water in Eh and pH terms encompasses all theoretical redox reactions taking place in water. The upper bound represents water in equilibrium with O_2 gas, and the lower bound represents water in equilibrium with H_2 gas, both at 1 bar pressure. Based on thermodynamic considerations, speciation of redox-sensitive elements at different Eh–pH conditions can be determined under the assumption that equilibrium conditions exist. A typical theoretical Eh-pH relationship for chromium, including the lower and upper bounds, is shown in Figure 6.4. Depending on the soil Eh and pH, chromium could exist in neutral, cationic or anionic form, and as explained above, the form of Cr will affect its adsorption behavior in the soil.

In soils, organic matter is a common reducing agent. Lowering of Eh can also be caused by oxidation of minerals containing ferrous iron or reduced sulfur species. Conversely, the presence of oxidized manganese, increases Eh in soils. These conditions affect the chemical form (or speciation) of redox-sensitive metals. Speciation of metals affects adsorption processes. For instance, if Cr is present in an oxidation state of 3 as Cr^{3+}, it behaves as a cation, whereas if Cr is present in an oxidation state of 6 as CrO_4^{-2}, it behaves as an anion. As explained above, the adsorption behavior of cations is drastically different from that of anions.

Complex Formations. Depending on the inorganic and organic ligands present in soil, metals may form various complexes, as explained in detail in Chapter 3. These complexes may be cationic, anionic, or neutral. This will then affect the surface reactions, and hence adsorption and desorption. For instance, Cr^{3+} may complex with hydroxyl to form $Cr(OH)^{2+}$, $Cr(OH)_2^+$, or $Cr(OH)_3$. The adsorption of complexes is expected to be reduced and becomes negligible when $Cr(OH)_3$ solids are formed.

Precipitation and Dissolution of Solids. Precipitation and dissolution of solids can significantly influence the phase distribution of metals. The dissolution of solid minerals and metal oxides and hydroxides increases the aqueous phase, while precipitation of the

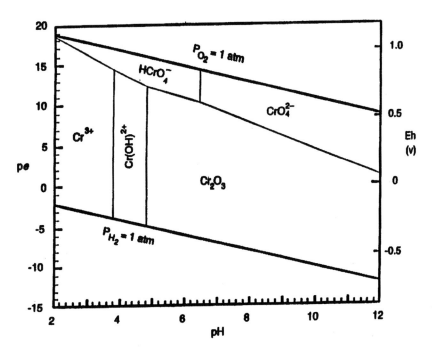

Figure 6.4 p^e (*or Eh*)–*pH diagram for chromium.* (*From Pourbaix, 1974.*)

solids decreases the aqueous phase. As explained in Chapter 3, the solubility property of the solids controls the extent of dissolution. The solubility of solids is influenced by the soil solution composition, including ionic strength.

Acid–Base Reactions. As explained in Chapter 3, acid–base reactions affect the pH and the ion chemistry of soil solution. CO_2 and water in soil solution generate H^+ and carbonate species (H_2CO_3, HCO_3^-, and CO_3^{2-}) through a series of acid–base reactions that increase the pH. This increase in pH affects the adsorption and desorption of metals.

6.2.4   Predictive Methods for Metal Distribution in Soils

Several geochemical models have been reported that calculate the distribution of metals in different phases. Assuming equilibrium conditions, these models use thermodynamic data to model possible adsorption–desorption, oxidation–reduction, complexation, precipitation–dissolution, and acid–base reactions that are likely to occur under a given set of conditions. These models calculate the speciation of free metal ions and metal complexes in soil solution (aqueous phase) and are helpful in predicting the fate of metals added to soil. The effect of changing soil solution parameters such as pH, redox, ligand concentration, or metal concentration can also be evaluated.

Models that are commonly used are GEOCHEM (Mattigod and Sposito, 1979), MINTEQA2 (USEPA, 1987), MINEQL (Westall et al., 1976; Schecher and McAvoy, 1992), and PHREEQC (Parkhurst, 1995). The required input data for these models includes concentration of the metal of interest, the inorganic and organic ligands, the major cations and other metal ions, and pH. In specific cases, the redox potential and pCO_2 also may be required. These models include thermodynamic data such as formation constants for metal–ligand complexes; therefore, such data need not be specified. The output from these models consists of an estimation of the concentration of free metals and com-

plexed metals at equilibrium for the specified conditions.

It should be remembered that these geochemical models are equilibrium models and thus do not consider the kinetics of the reactions (refer to Chapter 3 for more information on chemical kinetics). These models are also limited by the accuracy of the thermodynamic data incorporated into the models.

6.2.5   Chemical Analysis of Metals in Soils

Types of Chemical Analysis. As explained in Chapter 3, metal concentrations in water and wastewater are determined by using atomic absorption (AA) spectrophotometers and inductively coupled plasma emission (ICP) spectrophotometers. In the case of soils, an extraction (digestion) procedure is first used to remove metals from a known mass of dry soil into solution, and then the AA or ICP is used to determine the metal concentrations. Using the concentration and volume of solution, the mass of metal is calculated. The mass of metal divided by the dry mass of soil gives the metal concentration expressed as mg/kg or μg/g, both denoted as ppm. Expressed as μg/kg, the concentration is denoted as ppb.

AA and ICP spectrophotometers measure the total metal concentration in solution without distinguishing metal speciation or oxidation state. Free metal concentrations, complexed metal ion concentrations, and concentration of metals in different oxidation states can be determined using special methods such as ion-selective electrodes, polarography, colorimetric procedures, gas chromatography–AA, and high-performance liquid chromatography (HPLC) (Kramer and Allen, 1988). However, these methods are not performed routinely by commercial laboratories and are not standardized by the USEPA.

The following three types of chemical analysis procedures are commonly performed on metal-contaminated soils: (1) total concentrations of metals in soils, (2) toxicity characteristic leaching procedure (TCLP), and (3) sequential extractions of metals in soils. The extraction (digestion) procedure is different in each of the above, but the analysis of metals in resulting solution by AA or ICP is the same.

Total Concentrations of Metals in Soils. In this method, the total concentration of metal in soil is determined, and no consideration is given to the chemical form of metals. A rigorous digestion procedure is used to dissolve all solid-phase components in the soil. This involves using either a heated mixture of nitric acid, sulfuric acid, hydrofluoric acid, and perchloric acid, or a fusion of the soil with sodium carbonate (Black, 1965).

A more commonly used procedure is the USEPA acid digestion method 3050B, which involves digestion of the soil using hot nitric acid–hydrogen peroxide (USEPA, 1986; 1996). The metal concentrations in the extract solution are determined using AA or ICP, and the concentration of metals per unit dry mass of soil is then calculated. Figure 6.5 shows a flowchart describing the various steps involved in this analysis procedure.

TCLP. The TCLP simulates the leaching that a waste might undergo when disposed of in a landfill. The TCLP is performed in accordance with the USEPA method 1311 (USEPA, 1986, 1992). This method is frequently used to determine the leaching potential of cationic metals in landfill situations where due to microbial degradation of the waste under anaerobic conditions, acetic acid is produced. In this procedure, metals in soils are extracted using 0.1 M acetic acid. The extract solution is analyzed using AA or ICP similar to total metal concentration. Figure 6.6 shows a flowchart describing the different steps involved in TCLP metal analysis procedure. The TCLP is also used to characterize a waste as hazardous or nonhazardous, as detailed in Chapter 15, and to determine the necessary level of treatment to hazardous waste using stabilization or immobilization remedial technologies, as explained in Chapters 13 and 14.

Sequential Extractions of Metals in Soils. In this method, a series of extraction procedures are used to remove metals selectively from various geochemical forms. Although these procedures cannot be used to identify the actual form of a given metal in a soil, they are useful in categorizing the metals into several operationally defined geochemical fractions: exchangeables, metals associated with carbonates, metals associated with Mn and Fe oxides, metals associated with organic matter, and residual metals.

Numerous sequential extraction procedures have been developed for metal ions. The extraction procedures include the reaction of a soil sample with chemical solutions of increasing strengths. Typically, water or a salt solution (e.g., KNO_3, $CaCl_2$) is the first extractant used. Following this first extraction, mild acids, bases, chelating agents, and oxidizing solutions are used. Table 6.1 summarizes the wide variety of extractants that have been used for the sequential extraction of metal cations. Typical results of sequential extractions presented in Figure 6.7 show how each metal present in the soil is distributed. An extraction procedure developed by Tessier et al. (1979) is commonly used, and the various steps involved in this extraction procedure are shown in Table 6.2.

The sequential extraction procedures have several limitations. These methods do not always isolate a specific geochemical fraction of the soil, so the extractant may remove additional metals that are associated with other fractions. Second, adsorption of extracted metals to the remaining solid phase of soil may occur. This may result in concentration measurements that are lower than the actual concentrations of the metal associated with that fraction. Finally, no single extraction procedure is universal for all metals, both cations and anions, and all soils. As a result, the results of sequential extractions should be interpreted with caution.

6.3 ORGANIC GEOCHEMISTRY

Organic geochemistry deals with the behavior and reactions of organic chemicals in soils. In this section we present the type and characteristics of organic contaminants commonly found at contaminated sites. The distribution of organic contaminants in soils and processes affecting this distribution are also discussed. Finally, in

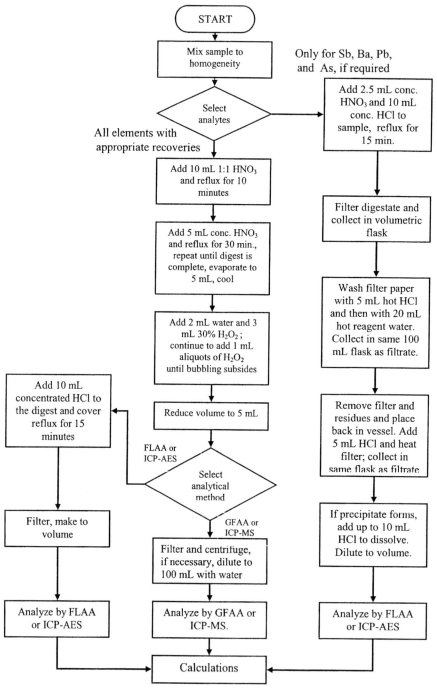

Figure 6.5 *Method 3050B, acid digestion of sediments, sludges, and soils. (From USEPA, 1986, 1996.)*

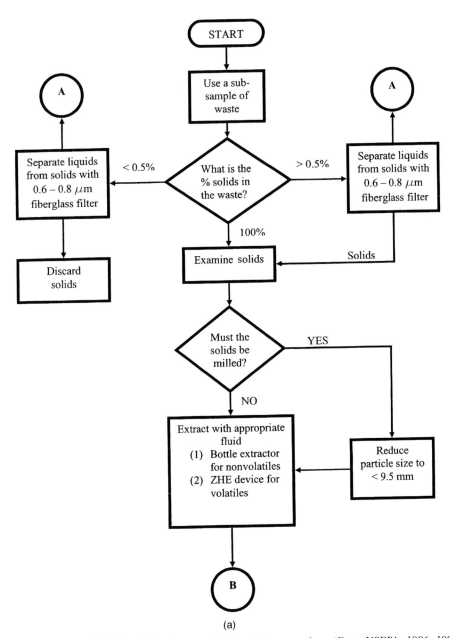

Figure 6.6 *Method 1311, toxicity characteristic leaching procedure.* (*From USEPA, 1986, 1996*)

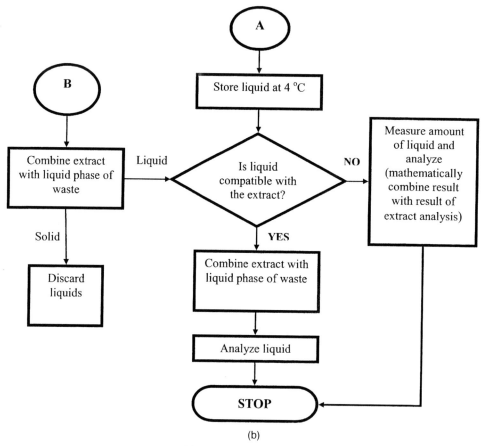

(b)

Figure 6.6 (*continued*)

the section we explain laboratory chemical analysis methods to determine concentrations of organic contaminants in soils.

6.3.1 Organic Contamination

Accidental spillage or improper disposal of products such as gasoline, diesel fuel, fuel oil, jet fuels, coal tars, motor oil and waste oil, and chlorinated solvents and degreasers has caused a variety of organic chemicals to enter into the subsurface. Organic contaminants may exist in liquid or solid form in soils and may exhibit volatile or semivolatile characteristics. The

most common types of organic contaminants found at contaminated sites include:

- Hydrocarbons such as gasoline components (e.g., benzene, toluene, ethylbenzene, and xylene)
- Chlorinated hydrocarbons such as PCE and TCE
- Polycyclic aromatic hydrocarbons (PAHs) such as naphthalene, phenanthrene, and pyrene
- Polychlorinated biphenyls (PCBs) such as aroclor
- Pesticides such as aldrin and endrin

As stated in Chapter 3, these organic compounds are classified in different ways. The organic compounds classified as hydrocarbons are commonly en-

Table 6.1 **Different methods of sequential extraction of heavy metals present in soils**

Method	Exchangeable	Fraction Associated with Carbonates	Fraction Associated with Oxides	Fraction Associated with Organic Matter	Total Amount and Residual Fraction
1	$0.05N$ $CaCl_2$	2.5% CH_3COOH	0.1 M $(COOH)_2$ + 0.175 M $(COONH_4)_2$ at pH 3.5	1 M $K_4P_2O_7$	HF
2	1 M KNO_3 + NaF	—	0.1 M EDTA at pH 6.5	0.1 M $Na_4P_2O_7$	1 M HNO_3
3	—	1 M CH_3COONH_4 + 1 M CH_3COOH at pH 4.5	0.1 M NH_2OH + 1 M CH_3COONH_4 at pH 4.5	30% H_2O_2	HF–$HClO_4$
4	—	1 M CH_3COOH	0.25 M NH_2OH, HCl in 25% (v/v) CH_3COOH	Acidified 30% H_2O_2	HNO_3–HF–$HClO_4$
5	1 M $MgCl_2$ or 1 M CH_3COONa at pH 8.2	1 M CH_3COONa + 1 M CH_3COOH at pH 5.0	0.04 M NH_2OH, HCl in 25% (v/v) CH_3COOH at 96 ± 3°C or 0.3 M $Na_2S_2O_4$ + 0.175 M Na-citrate + 0.025 M citric acid	0.02 M HNO_3 + 30% H_2O_2, pH 2 at 85 ± 2°C, 2 h +30% H_2O_2 + HNO_3, pH 2 at 85 ± 2°C, 3 h 3.2 M CH_3COONH_4 in 20% HNO_3	HF–$HClO_4$
6	0.2 M $BaCl_2$	—	0.1 M NH_2OH,HNO_3 + 25% (v/v) CH_3COOH + HCl	30% H_2O_2 + NH_4OH	HF–$HClO_4$
7	1 M KNO_3	0.5 M NaF at pH 6.5	0.1 M EDTA at pH 6.5 double extraction	0.1 M $Na_4P_2O_5$	1 M HNO_3
8	1 N $CaCl_2$	2.5% CH_3COOH	0.05 M EDTA at pH 7	0.1 N $Na_4P_2O_5$	HF
9	—	1 M CH_3COONH_4 + 0.5 M $(CH_3COO)_2Mg$	0.25 M NH_2OH, HCl, pH 2	30% H_2O_2 + 1 M CH_3COONH_4	—
10	1 M CH_3COONH_4 at pH 7	—	(1) 0.1 M NH_2OH, HCl + 0.01 M HNO_3, pH 2 (2) 0.2 M $(COONH_4)_2$ + 0.2 M $(COOH)_2$, pH 3	30% H_2O_2, HNO_3, pH 2 at 85°C extraction with 1 M CH_3COONH_4	HNO_3 at 180°C
11	—	Resin-H^+	(1) $(COONa)_2$ (2) $(COONa)_2$ + UV	30% H_2O_2 at 40°C	—
12	0.5 M KNO_3	—	0.5 M $Na_2 \cdot$ EDTA	0.5 M NaOH	4 M HNO_3 at 80°C
13	—	1 M CH_3COONH_4 at pH 4.5	(1) 0.1 M NH_2OH, HCl (2) 0.2 M $(COONH_4)_2$ $(HCOOH)_2$, pH 3.3 obscurite (3) Same as (2) + UV	—	HNO_3–HF–HCl
14	1 M $MgCl_2$	—	(1) $(COONa)_2$ (2) Citrate dithionite bicarbonate	6% $NaClO_4$ at 85°C	HNO_3–$HClO_4$
15	1 M $BaCl_2$	1 M CH_3COOH + 0.6 M CH_3COONa	0.1 M NH_2OH + 25% (v/v) CH_3COOH	30% H_2O_2 + 0.02 M HNO_3 + 3.2 M CH_3COONH_4	HF–HCl

Table 6.1 (*continued*)

Method	Exchangeable	Fraction Associated with Carbonates	Fraction Associated with Oxides	Fraction Associated with Organic Matter	Total Amount and Residual Fraction
16	1 M Mg(NO$_3$)$_2$ at pH 7	—	(1) 0.1 M NH$_2$OH, HCl, pH 2 (2) 0.2 M (COONH$_4$)$_2$ + 0.2 M (COOH)$_2$, pH 3 (3) Same as (2) + ascorbic acid	0.7 M NaOCl, pH 8.5	HF–HNO$_3$–HCl
17	1 M CH$_3$COONH$_4$ at pH 7	1 M CH$_3$COONa at pH 5	(1) 0.1 M NH$_2$OH, HCl + 0.01 M HNO$_3$ (2) 1 M NH$_2$OH,ClH in 25% (v/v) CH$_3$COOH	30% H$_2$O$_2$ + 0.02 M HNO$_3$ at 85°C	Aqua regia + HF

Source: USEPA (1992).

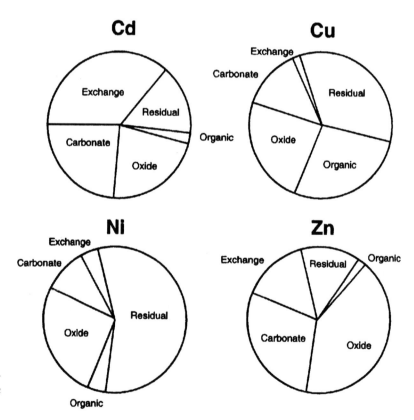

Figure 6.7 *Typical sequential extraction results. (From McLean and Bledsoe, 1992.)*

Table 6.2 **Sequential extraction procedure for speciation of heavy metals**

Fraction	Extraction Procedure (per 1-g Dry Soil Sample)
Exchangeable	Add 8 mL of 1 M sodium acetate solution (pH 8.2) and mix continuously for 1 h.
Carbonates bound	To the residue from above, add 8 mL of 1 M sodium acetate (pH adjusted to 5.0 with acetic acid) and mix continuously for 5 h.
Fe–Mn oxides bound	To the residue from above, add 20 mL of 0.04 M hydroxylamine hydrochloride ($NH_2OH \cdot HCl$) in 25% (v/v) acetic acid, and heat to 96°C with occasional stirring for 6 h.
Organic bound	To the residue from above, add 3 mL of 0.02 M nitric acid (HNO_3) and 5 mL of 30% hydrogen peroxide (H_2O_2) (pH adjusted to 2.0 with nitric acid), and heat to 85°C for 2 h with occasional agitation. Then add 3 mL of 30% H_2O_2 (pH adjusted to 2.0 with HNO_3) and mix continuously for 3 h. Cool the mixture and add 5 mL of 3.2 M ammonium acetate (NH_4OAc) in 20% (v/v) HNO_3. Finally, dilute to 20 mL and mix continuously for 30 min.
Residual	Sum up the four fractions above and subtract it from the total concentration.

Source: Tessier et al. (1979).

countered at contaminated sites. As explained in Chapter 3, these hydrocarbons are classified as aliphatic hydrocarbons, aromatic hydrocarbons, and halogenated hydrocarbons. A group of these hydrocarbons are often known as non-aqueous phase liquids (NAPLs; pronounced "naples") have been encountered at numerous contaminated sites. This term is used because NAPLs exist primarily as a separate, immiscible phase when in contact with water and/or air. Differences in the physical and chemical properties of water and NAPLs result in the formation of a physical interface between the liquids that prevents the two fluids from mixing (or solubilizing).

Classification of NAPLs. NAPLs are classified into light non-aqueous phase liquids (LNAPLs) and dense non-aqueous phase liquids (DNAPLs). LNAPLs are NAPLs that have densities less than that of water, and DNAPLs are NAPLs that have densities greater than that of water. Because of differences in density alone, the fate and migration of LNAPLs and DNAPLs in soils can be significantly different.

Typical NAPLs (LNAPLs and DNAPLs) found at contaminated sites, along with their important properties, are presented in Table 6.3. LNAPLs consist primarily of petroleum products such as gasoline, kerosene, and diesel, which are all associated with spills and accidental releases during production, refinement, and distribution. DNAPLs consist primarily of chlori-

nated solvents, polycyclic aromatic hydrocarbons (PAHs), and pentachlorobiphenyls (PCBs) resulting from a wide variety of industrial activities such as degreasing, metal stripping, chemical manufacturing, pesticide manufacturing, wood-treating operations, and manufactured gas plant operations.

Characteristics of NAPLs. The most important characteristics that affect the behavior of NAPLs in the subsurface are (1) density, (2) viscosity, (3) solubility, (4) vapor pressure, (5) volatility, (6) interfacial tension, (7) wettability, and (8) capillary pressure. A brief explanation of each of these characteristics is provided below. A detailed review of these characteristics can be found in Bear (1972), Lyman et al. (1982), and Mercer and Cohen (1990).

Density is defined as the mass of a substance per unit volume. Density is also expressed as *specific gravity* (SG), which is the ratio of the mass of a given volume of substance at a specified temperature to the mass of the same volume of water at the same temperature. LNAPLs have a SG less than 1 and will float on water. DNAPLs have SG greater than 1 and will sink in water. Table 6.3 shows density values for selected LNAPLs and DNAPLs.

Viscosity is a fluid's resistance to flow. Dynamic or absolute viscosity is expressed in units of mass per unit length per unit time [in centipoise (cP)]. Table 6.3 presents viscosity values of selected NAPLs. Both density

Table 6.3 **Typical NAPLs and their properties**

NAPL Type	Chemical	Density (g/cm³)	Dynamic Viscosity (cP)	Water Solubility (mg/L)	Vapor Pressure (mmHg)	Henry's Law Constant (atm · m³/mol)
LNAPL	Benzene	0.8765	0.6468	1,780	76	5.43×10^{-3}
	Ethyl benzene	0.867	0.678	152	7	7.9×10^{-3}
	Toluene	0.8669	0.58	515	22	6.61×10^{-3}
	o-Xylene	0.880	0.802	170	7	4.94×10^{-3}
	Methyl ethyl ketone	0.805	0.40	26,800	71.2	2.74×10^{-5}
	Vinyl chloride	0.910		2,670	2,660	2.78
DNAPL	Trichloroethylene	1.462	0.570	1,100	72.6	0.0091
	Tetrachloroethylene	1.625	0.890	150	20	0.0153
	Methylene chloride	1.3250	0.430	13,200	3,789	0.010
	Naphthalene	1.162	—	3.1	0.23	4.6×10^{-4}
	Phenanthrene	1.20	—	1.3	6.8×10^{-4}	2.56×10^{-5}
	Pyrene	1.271	—	0.148	6.85×10^{-7}	1.87×10^{-5}
	Anthracene	1.25	—	0.075	1.95×10^{-4}	6.51×10^{-5}
Water		1.0	1.14	—	—	—

Source: Mercer and Cohen (1990); Huling and Weaver (1991); Newell et al. (1995).

and viscosity of fluids generally decrease as temperature increases. The permeability (K) of soils depends on density and viscosity of fluid and is given by the expression

$$K = \frac{k\rho g}{\mu} \qquad (6.1)$$

where k is the intrinsic permeability, K the hydraulic conductivity, ρ the fluid mass density, g the acceleration due to gravity, and μ the dynamic (absolute) viscosity. From the relationship (6.1) it can be seen that as density increases, k increases; however, as viscosity increases, k decreases.

Solubility is the equilibrium concentration of NAPLs in water. The solubility of NAPLs varies considerably from infinitely miscible, for compounds such as ethanol and methanol, to extremely low solubility, for compounds such as polycyclic aromatic compounds (PAHs). Tables 6.3 and 8.7 present solubility of selected NAPLs.

Vapor pressure is the pressure exerted by the vapor above a liquid, it is the partial pressure exerted by the free molecules of the NAPL. Vapor pressure determines how readily vapors volatilize or evaporate from pure phase liquids. Vapor pressure increases with in-

crease in temperature. Table 6.3 provides vapor pressure values for selected NAPLs.

Volatility is a measure of the transfer of the NAPL from the aqueous phase to the gaseous phase. The *Henry's law constant*, defined as the vapor pressure divided by the aqueous solubility, is used to help evaluate the volatilization of a NAPL from the water. Tables 6.3 and 8.7 provide Henry's law constants for selected NAPLs. More details on the volatilization process and Henry's law are provided in Section 6.3.3.

Interfacial tension is the surface energy at the interface that results from differences in the forces of molecular attraction within two immiscible fluids. It is expressed in units of energy per unit area. In general, the greater the interfacial tension, the greater the stability of the interface between the liquids. Interfacial tension decreases with increasing temperature and may be affected by pH, surfactants, and dissolved gases. When this force is present between a liquid and a gaseous phase, the same force is called *surface tension.* Low interfacial tension between a NAPL and water allows the NAPL to exist mostly in pore spaces, thus facilitating greater migration in the subsurface.

Wettability is generally defined as the tendency of one fluid to spread on or adhere to a solid surface (i.e., preferentially coat) in the presence of another fluid with which it is immiscible. This concept has been used to describe fluid distribution at the pore scale. In a multiphase system, the wetting fluid will preferentially coat (wet) the solid surfaces and tend to occupy smaller pore spaces. The nonwetting fluid will generally be restricted to the largest interconnected pore spaces. Water is generally the wetting fluid displacing NAPL from pore spaces, while air is a nonwetting fluid that allows NAPL to adhere to soil surfaces. This means that unsaturated (or dry) soils have more susceptibility than saturated soils for the NAPL to adhere to a solid surface.

Capillary pressure is the pressure across the interface between the wetting and nonwetting phases and is often expressed as the height of an equivalent water column. It determines the size of the pores in which an interface can exist. Capillary pressure is a measure of the relative attraction of the molecules of a liquid (cohesion) to each other and for a solid surface (adhesion), represented by the tendency of the porous medium to attract the wetting fluid and repel the nonwetting fluid. The capillary pressure of the largest pore spaces must be exceeded before the nonwetting fluid (NAPL) can enter the porous medium. The minimum pressure required for the NAPL to enter the medium is termed the *entry pressure*. In general, capillary pressure increases with decreasing pore size, decreasing initial moisture content, and increasing interfacial tension. In general, when the NAPL pressure head is greater than the capillary pressure, the NAPLs can enter into the soil pores, including the smaller pores. When sufficient NAPL volume has been released and the pressure head of it exceeds the water capillary pressure at the capillary fringe (entry pressure), the NAPL will enter the groundwater. If the NAPL pressure head is less than the water capillary pressure at the capillary fringe, NAPLs tend to migrate laterally on the top of the water table. Therefore, the extent of migration, and residual distribution of NAPLs in the subsurface, are controlled by the capillary forces.

6.3.2 Distribution of NAPLs in Soils

NAPLs are comprised of either a single organic compound or a complex mixture of several organic compounds. The exact constituents of NAPLs depend on the source of contamination and may include LNAPLs, DNAPLs, or a combination. NAPLs may exist in any of four phases in soils: (1) dissolved phase, (2) adsorbed phase, (3) gaseous phase, and (4) free NAPL phase. In unsaturated soils, all four phases exist, as depicted in Figure 6.8(a). In saturated soils, NAPLs are present in three phases: the dissolved, adsorbed, and free NAPL phases, as depicted in Figure 6.8(b).

Among these different phases, free NAPL phase is most significant. The free NAPL phase may exist in soil pores as continuous slugs of NAPL near contamination source locations or as small individual NAPL blobs (or ganglia). NAPL ganglia form due to capillary forces when continuous slugs of NAPL become discontinuous during migration.

The amount of NAPL present in soils is often expressed by saturation (S_i), which is the relative fraction of total pore space containing NAPL in a representative volume of soil:

$$S_i = \frac{V_i}{V_v} \tag{6.2}$$

where V_i is the volume of NAPL in a general fluid i, and V_v is the volume of voids in soil. When NAPL migrates through soils, some of it may be trapped in soil pores, due to capillary forces, and exist in the form of blobs. The amount of this trapped NAPL is expressed as the residual saturation (S_r). S_r is defined as the ratio of the volume of residual NAPL and the volume of pore space. S_r is generally higher in low-permeability soils than in highly permeable soils. S_i and S_r describe mobile and immobile amounts of NAPLs, respectively, and are often used in assessing the potential for contaminant migration and the need for remediation.

6.3.3 Processes Controlling Distribution of NAPLs in Soils

NAPLs may change from one phase to another, depending on the following processes: (1) volatilization, (2) dissolution, (3) adsorption, and (4) biodegradation. These processes are interrelated, as depicted in Figure

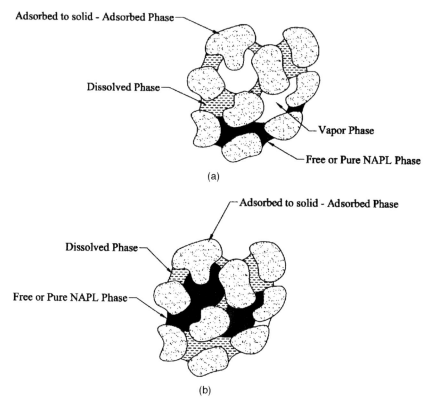

Adsorbed to solid - Adsorbed Phase

Dissolved Phase

Vapor Phase

Free or Pure NAPL Phase

(a)

Adsorbed to solid - Adsorbed Phase

Dissolved Phase

Free or Pure NAPL Phase

(b)

Figure 6.8 (a) *Phase distribution of NAPL in unsaturated soils;* (b) *phase distribution of NAPL in saturated soils.*

6.9(*a*) and (*b*), and are described briefly in the following subsections.

Volatilization. *Volatilization* refers primarily to the partitioning of NAPLs from dissolved and free phases into the gaseous phase. Henry's law describes the partitioning of an organic compound between the aqueous and gaseous phases. For a dilute solution (concentrations less than approximately 10^{-3} mol/L), the ideal gas vapor pressure of a volatile organic is proportional to its mole fraction in solution. Simply stated, the escaping tendency of the organic molecules from the dissolved phase to the gas phase is proportional to the dissolved organic concentrations. This relationship assumes local equilibrium between dissolved and gaseous phases. Given mathematically,

$$C_{aq} = HP_g \qquad (6.3)$$

where C_{aq} is the concentration in water, H is Henry's constant, and P_g is the vapor pressure of the organic component. Tables 6.3 and 8.7 provide the values of Henry's constant for selected organic compounds of environmental concern.

For a more concentrated dissolved phase or free NAPL phase, which contains multiple organic compounds, Raoult's law describes the volatilization of an organic compound. Raoult's law states that the vapor pressure over a solution is equal to the mole fraction of the solute times the vapor pressure of the pure phase liquid. This law is given mathematically as

$$P_a = X_a P_0^a \qquad (6.4)$$

where P_a is the vapor pressure of the NAPL mixture (atm), P_0^a the vapor pressure of the organic compound a as a pure-phase component (atm), and X_a the mole

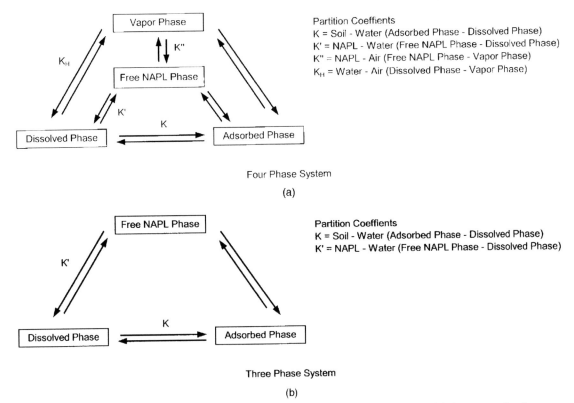

Figure 6.9 (a) *Partitioning of NAPL in unsaturated soils;* (b) *partitioning of NAPL in saturated soils.*

fraction of a hydrocarbon *a* in the NAPL mixture. This relationship is valid only for equilibrium conditions. Tables 6.3 and 8.7 present the values of vapor pressure of selected organic compounds that are of environmental concern. To represent mass-based concentrations, equation (6.4) is written using the ideal gas law ($PV = nRT$) as

$$C_a(\text{g/L}) = \frac{X_a P_0^a \cdot \text{MW}}{RT} \qquad (6.5)$$

where X_a is the mole fraction, P_0^a is represented in atm, MW is the molecular weight in g/mol, $R = 0.0821$ (atm · L/mol · K), and T is the temperature in kelvin. The MW values of selected organic compounds are provided in Table 8.7.

Although Henry's law and Raoult's law assume equilibrium conditions, in reality transient nonequilibrium conditions may exist, especially during the implementation of remedial processes. This problem must be evaluated based on site-specific conditions.

Dissolution. *Dissolution* refers to the partitioning of free NAPL phase into the dissolved phase. An NAPL in physical contact with soil solution or groundwater will dissolve (solubilize, partition) into the aqueous phase (dissolved phase). The solubility of a single organic compound is the equilibrium concentration of that compound in water at a specified temperature and pressure. In other words, the solubility represents the maximum possible concentration of that compound in water. Tables 6.3 and 8.7 provide solubility data for selected organic compounds of environmental concern.

EXAMPLE 6.1

A NAPL of pure TCE is released into the unsaturated zone. Calculate the theoretical concentration of TCE in the gaseous phase (expressed in mg/L).

Solution:

Use equation (6.5) with values as follows: $X_{TCE} = 1.0$, $P_0^{TCE} = 72.6$ mmHg, $MW_{TCE} = 131$ g/mol (from Table 8.7), and

$$C_{TCE} = \frac{[(1.0)(72.6 \text{ mmHg})(1 \text{ atm}/760 \text{ mmHg})(131.39 \text{ g/mol})]}{[0.0821 \text{ atm} \cdot \text{L/mol} \cdot \text{K})(298K)]} = 0.409 \text{ g/L} = 409 \text{ mg/L}$$

EXAMPLE 6.2

A NAPL consisting of benzene and TCE in mole fractions of 0.7 and 0.3, respectively, exists in the unsaturated zone. Calculate the theoretical concentration of benzene and TCE in the gaseous phase (expressed in mg/L). Assume that adsorption and biodegradation are negligible.

Solution:

Use equation (6.5) with values as follows: $X_{BENZENE} = 0.7$, $P_0^{BENZENE} = 76$ mmHg, $MW_{BENZENE} = 78$ g/mol, $X_{TCE} = 0.3$, $P_0^{TCE} = 72.6$ mmHg, $MW_{TCE} = 131$ g/mol,

$$C_{BENZENE} = \frac{[(0.7)(76 \text{ mmHg})(1 \text{ atm}/760 \text{ mmHg})(78 \text{ g/mol})]}{[(0.0821 \text{ atm} \cdot \text{L/mol} \cdot \text{K})(298K)]} = 0.223 \text{ g/L} = 223 \text{ mg/L}$$

and

$$C_{TCE} = \frac{[(0.3)(72.6 \text{ mmHg})(1 \text{ atm}/760 \text{ mmHg})(131 \text{ g/mol})]}{[(0.0821 \text{ atm} \cdot \text{L/mol} \cdot \text{K})(298K)]} = 0.155 \text{ g/L} = 155 \text{ mg/L}$$

Ranging over several orders of magnitude, the solubility of organic compounds is affected by temperature, pH, cosolvents, dissolved organic matter, and dissolved inorganic compounds.

For a multicomponent NAPL in contact with water, the equilibrium dissolved-phase concentrations may be estimated using the solubility of the pure liquid in water and its mole fraction in the NAPL mixture. The maximum concentration in such cases is referred to as the *effective solubility*. This value is expressed mathematically as

$$S_i^e = X_i S_i \tag{6.6}$$

where S_i^e is the effective aqueous solubility of compound i in the NAPL mixture, X_i the mole fraction of compound i in the NAPL mixture, and S_i the aqueous solubility of the pure-phase compound. The effective solubility represents the concentration that may occur at equilibrium under ideal conditions. However, this approach does not account for the tendency of certain cosolvents (e.g., alcohols) to increase the solubility of organic compounds. Dissolution rates may also increase with higher groundwater velocity, higher NAPL saturation, increased contact between NAPL and water, and increased fraction of soluble components. Many studies report that dissolution kinetics affects solubility as well and that the dissolution rate is limited under certain conditions, such as high groundwater velocity.

Adsorption. *Adsorption* refers to the partitioning of NAPL from various phases, primarily from the dissolved phase, into adsorbed phase. On the other hand, *desorption* refers to the partitioning of adsorbed phase into, generally, dissolved phase. Adsorption and sorption are used interchangeably to include NAPLs accumulated at the interface with soil solids, as well as those partitioned into soil organic carbon.

The extent of adsorption and desorption depends on parameters such as solubility, polarity, ionic charge, pH, redox potential, and the octanol–water partition coefficient (K_{ow}). This is discussed in detail in Chapter 8.

EXAMPLE 6.3

Consider a two-component NAPL with equal mole fractions (0.5), where the solubilities of the pure-phase compounds are 1000 and 10 mg/L, respectively. Calculate the effective solubilities and explain the implication on the transport of these organic compounds in the subsurface.

Solution:
Use equation (6.6), with values as follows: $X_{NAPL\ A} = 0.5$, $S_{NAPL\ A} = 1000$ mg/L, and $S^e_{NAPL\ A} = (0.5)(1000$ mg/L$) = 500$ mg/L; $X_{NAPL\ B} = 0.5$, $S_{NAPL\ B} = 10$ mg/L, and $S^e_{NAPL\ B} = (0.5)(10$ mg/L$) = 5$ mg/L. The greater the effective solubility, the greater the amount of NAPL present in the dissolved phase. Therefore, a larger amount will be transported in the subsurface.

EXAMPLE 6.4

Consider a gasoline containing numerous compounds where the mole fractions of benzene are 0.0076 and 0.0021 for fresh and weathered gasoline, respectively. The solubility of benzene is 1780 mg/L. Calculate the effective solubility of benzene and explain its implications on the transport in the subsurface.

Solution:
Use equation (6.6), with values as follows: For fresh gasoline, $X_{BENZENE} = 0.0076$; for weathered gasoline, $X_{BENZENE} = 0.0021$; and $S_{BENZENE} = 1780$ mg/L (Table 6.3). So for fresh gasoline;

$$S^e_{BENZENE} = (0.0076)(1780 \text{ mg/L}) = 13.53 \text{ mg/L}$$

and for weathered gasoline;

$$S^e_{BENZENE} = (0.0021)(1780 \text{ mg/L}) = 3.738 \text{ mg/L}.$$

With a fresh gasoline spill, more benzene will be dissolved and transported in the subsurface.

Biodegradation. Many NAPLs, especially LNAPLs, may undergo biological degradation. The dissolved-phase NAPLs are particularly amenable to biological degradation if the right, naturally occurring microorganisms are present in the subsurface. However, free-NAPL-phase biodegradation may not be conducive, because the free NAPL phase may hinder microbial activity. The biodegradation process essentially involves oxidation–reduction reactions (this is discussed in detail in Chapter 8).

6.3.4 Chemical Analysis of NAPLs in Soils

To understand the need for remediation and for design of a remediation system, we should know the constit-uents of NAPL and their concentration. It may also be valuable to determine the distribution of NAPLs in dissolved, gaseous, adsorbed, and free NAPL phases. In addition, we should also know if the mobile and residual NAPLs are present in the system. Accurate determination of each of the preceding factors under in-situ conditions is quite complex. However, methods to estimate these values are available, as presented in the following paragraphs.

Commonly, soil, soil solution/groundwater, and soil gas samples are collected from the field and analyzed in the laboratory. For the analysis of soils, an extraction is done first to recover all of the organics from the soil sample by dissolving them into solution. The USEPA Method 3540C Soxhlet extraction procedure is the most commonly used extraction procedure for this pur-

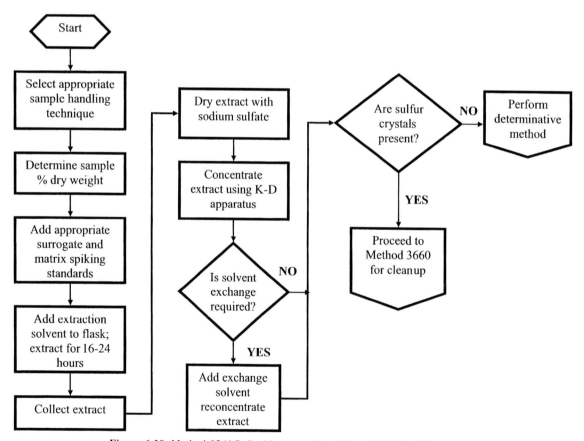

Figure 6.10 *Method 3540C, Soxhlet extraction. (From USEPA, 1986, 1996.)*

pose (USEPA, 1986; 1996). Figure 6.10 shows a flowchart describing different steps involved in this extraction procedure. The extract solution is then analyzed using GC, GC-MS, or HPLC, as described in Chapter 3, to identify individual compounds and their concentrations. The soil solution/groundwater may be analyzed using the samples directly, or they may require filtration and/or concentration prior to analysis using GC, GC-MS, or HPLC. Soil gas analysis may also be performed by injecting the sample directly into the GC or GC-MS.

The chemical analyses above provide the constituents in NAPL and their total concentrations but do not distinguish directly between dissolved, adsorbed, gaseous, and free NAPL phases. However, using these data, an estimate of these phases may be made assuming that equilibrium conditions are valid. This procedure consists of the following steps:

1. Calculate the effective solubility of organic compound of interest (S_i^e) as explained in Section 6.3.3.

EXAMPLE 6.5

By accident, 1000 gal of liquid waste containing benzene and TCE, with equal mole fractions, was released into unsaturated subsurface. The contaminated zone is 100 ft long by 100 ft wide by 20 ft deep. The soil in the contaminated zone consists of silty sand with a porosity of 0.3 and a degree of saturation of 60%. Calculate the theoretical amounts of benzene and TCE that exist in gaseous phase.

Solution:
Using equation (6.5), we obtain the following: $\rho_{BENZENE} = 0.876$ g/cm^3, $\rho_{TCE} = 1.462$ g/cm^3, $X_{BENZENE} = 0.5$, $X_{TCE} = 0.5$, $P_0^{BENZENE} = 76$ mmHg, $P_0^{TCE} = 72.6$ mmHg, $MW_{BENZENE} = 78$ g/mol; $MW_{TCE} = 131$ g/mol, $T = 25°C$, porosity (n) = 0.3, and degree of saturation (S) = 60%.

Total mass of benzene spilled = $(0.876$ g/cm$^3)(0.5)(1000$ cm^3/L$)(3.78$ L/gal$)(1000$ gal$)(1$ kg/1000 g$)$

$$= 1655 \text{ kg}$$

Total mass of TCE spilled = $(1.462$ g/cm$^3)(0.5)(1000$ cm^3/L$)(3.78$ L/gal$)(1000$ gal$)(1$ kg/1000 g$)$

$$= 2763 \text{ kg}$$

Total pore volume = $(20$ ft$)(100$ ft$)(100$ ft$)(0.3) = 60,000$ ft^3

Volume of air phase = $(0.4)(60,000$ ft$^3) = 24,000$ ft^3

$$C_{TCE} = \frac{(0.5)(72.6 \text{ mmHg})(1 \text{ atm}/760 \text{ mmHg})(131 \text{ g/mol})}{(0.0821 \text{ L} \cdot \text{atm}/\text{K} \cdot \text{mol})(298\text{K})} = 0.256 \text{ g/L}$$

Mass$_{TCE\ VAPOR}$ = $(0.256$ g/L$)(24,000$ ft$^3)(28.317$ L/ft$^3)(1$ kg/1000 g$) = 174$ kg

$$C_{BENZENE} = \frac{(0.5)(76 \text{ mmHg})(1 \text{ atm}/760 \text{ mmHg})(78 \text{ g/mol})}{(0.0821 \text{ L} \cdot \text{atm}/\text{K} \cdot \text{mol})(298\text{K})} = 0.159 \text{ g/L}$$

Mass$_{BENZENE\ VAPOR}$ = $(0.159$ g/L$)(24,000$ ft$^3)(28.317$ L/ft$^3)(1$ kg/1000 g$) = 108$ kg

2. Calculate concentration in gas phase using vapor pressures and mole fractions (C_a) as explained in Section 6.3.3.

3. Estimate adsorbed concentration of organic chemicals. This is discussed in detail in Chapter 8.

4. The difference between the total measured concentration and the sum of concentrations of dissolved, vapor, and adsorbed phases (from steps 1 to 3) provides the amount of free NAPL phase.

This method requires the determination of soil moisture content, organic carbon content, porosity, NAPL composition, adsorption parameters, effective solubilities, and total constituent concentrations in soil samples. When NAPLs are present in large quantities (significantly higher than the solubility limits), the method is most useful. The results of such analysis should be used with caution, due to the fact that equilibrium conditions were assumed.

The free NAPL phase may be identified visually in soil cores and groundwater samples. The thickness of floating LNAPL in a groundwater monitoring well may be used to estimate the amount of LNAPL in the free phase. In addition, noninvasive geophysical methods may also help identify the presence of free NAPL phase in subsurface (refer to Chapter 10).

6.4 SUMMARY

Understanding the various aspects of geochemistry that deal with the behavior of chemicals in soils is essential for an understanding of the fate and transport of contaminants in subsurface environments. Inorganic geochemistry deals with the study of inorganic contaminants, particularly toxic metals, while organic geochemistry includes the study of organic contaminants such as hydrocarbons. An understanding of the phase distribution of contaminants introduced into the subsurface is essential for the design of strategies that can effectively remediate contaminated soils. Various interdependent geochemical processes control the distribution of contaminants in different phases. Mathematical models or simple mass-balance analysis methods are often useful to assess the amount of contaminants in different phases under varying environmental conditions, such as pH, solution chemistry, and temperature. Laboratory analysis methods to determine contaminant concentrations involve extraction procedures that must be selected carefully in order to attain accurate and useful information.

QUESTIONS/PROBLEMS

6.1. What is geochemistry? Why is it important for geoenvironmental studies?

6.2. List typical toxic metals and their major chemical properties (molecular weight, solubility, etc.).

6.3. Explain the different phases that metals could exist in soils.

6.4. List and explain different geochemical processes that affect the various phases of metal contamination in soils.

6.5. What are the advantages and disadvantages of geochemical models to study metals in soils?

6.6. What are the different laboratory test methods to determine metal concentrations in soils?

6.7. What is TCLP? What is its significance?

6.8. Explain situations when sequential extraction of metals in soils is needed.

6.9. What are the major organic contaminants in soils? How are they commonly classified for geoenvironmental studies?

6.10. How are the organic contaminants distributed in the soils?

6.11. What processes govern the distribution of organic contaminants in soils?

6.12. Explain the common laboratory methods to determine concentration of organic contaminants in soils.

6.13. Consider a two-component NAPL with mole fractions of 0.4 and 0.6, where the solubilities of the pure-phase compounds are 1000 and 10 mg/L, respectively. Calculate the effective solubilities and explain which of these organic compounds has a greater potential for transport in the subsurface.

6.14. Consider a chemical spill containing numerous compounds where the mole fractions of TCE are 0.01 and 0.005 for fresh and old spill conditions, respectively. The solubility of TCE is 1100 mg/L. Calculate the effective solubility of TCE and explain its implications on the transport in the subsurface.

6.15. A NAPL of pure benzene is released into the unsaturated zone. Calculate the theoretical concentration of benzene in the gaseous phase (expressed in mg/L).

6.16. A NAPL consisting of benzene, toluene, TCE, and phenanthrene in mole fractions of 0.3, 0.2, 0.4, and 0.1, respectively, exists in the unsaturated zone. Calculate the theoretical concentration of these organic compounds in the gaseous phase (expressed in mg/L).

6.17. By accident, 100 gal TCE was released into unsaturated subsurface. The contaminated zone is 100 ft long by 100 ft wide by 20 ft deep. The soil in the contaminated zone consists of silty sand with porosity 0.3 and degree of saturation 60%. Calculate the theoretical amount of TCE that exist in the gaseous phase.

6.18. At an industrial plant, 30,000 gal of TCE was spilled accidentally into the subsurface. The subsurface consists primarily of sandy soil. The sandy soil is fully saturated and has a porosity of 0.35. Site characterization revealed that a dissolved-phase TCE plume approximately 100 ft long, 10 ft wide, and 3 ft deep exists with an average concentration of 1000 mg/L. Assuming that adsorption and biodegradation are negligible, calculate the amount of free-phase NAPL present in the aquifer.

REFERENCES

Bear, J., *Dynamics of Fluids in Porous Media*, American Elsevier, New York, 1972.

Black, C. A. (ed.), *Methods of Soil Analysis*, American Society of Agronomy, Madison, WI, 1965.

Drever, J. I., *The Geochemistry of Natural Waters: Surface and Groundwater Environments*, 3rd ed., Prentice Hall, Upper Saddle River, NJ, 1997.

Huling, S. G., and Weaver, J. W., *Dense Nonaqueous Phase Liquids, Ground Water Issue*, EPA/540/4-91/002, R.S. Kerr Environmental Research Laboratory, USEPA, Ada, OK, 1991.

Kramer, J. R., and Allen, H. E., *Metal Speciation: Theory, Analysis and Application*, Lewis Publishers, Chelsea, MI, 1988.

Langmuir, D., *Aqueous Environmental Geochemistry*, Prentice Hall, Upper Saddle River, NJ, 1997.

Lindsay, W. L., *Chemical Equilibria in Soils*, Wiley, New York, 1979.

Lyman, W. J., Reehl, W. F., and Rosenblatt, D. H., *Handbook of Chemical Property Estimation Methods*, McGraw-Hill, New York, 1982.

Mattigod, S. V., and Sposito, G., Chemical modeling of trace metals equilibrium in contaminated soil solutions using the computer program GEOCHEM, in *Chemical Modeling in Aqueous Systems*, Jenne, E. A. (ed.), ACS No. 93, American Chemical Society, Washington, DC, 1979.

McLean, J. E., and Bledsoe, B. E., *Behavior of Metals in Soils, Ground Water Issue*, EPA/540/S-92/018, R. S. Kerr Environmental Research Laboratory, USEPA, Ada, OK, 1992.

Mercer, J. W., and Cohen, R. M., A review of immiscible fluids in the subsurface: properties, models, characterization and remediation, *J. Contam. Hydrol.*, Vol. 6, pp. 107–163, 1990.

Newell, C. J., Acree, S. D., Ross, R. R., and Huling, S. G., *Light Nonaqueous Phase Liquids, Ground Water Issue,* EPA/540/S-95/500, R. S. Kerr Environmental Research Laboratory, USEPA, Ada, OK, 1995.

Parkhurst, D. L., *Users Guide to PHREEQC: A Computer Program for Speciation, Reaction-Path, Advective-Transport, and Inverse Geochemical Calculations,* USGS Water-Resources Investigations Report 95-4227, U.S. Geological Survey, Washington, DC, 1995.

Pourbaix, M., *Atlas of Electrochemical Equilibria,* in *Aqueous Solutions,* Pergamon, Elmsford, N.Y. 1974.

Schecher, W. D., and McAvoy, D. C., MINEQL+: a software environment for chemical equilibrium modeling, *Comput., Environ. Urban Syst.,* Vol. 16, No. 1, pp. 65–76.

Schwarzenbach, R. P., Gschwend, P. N., and Imboden, D. M., *Environmental Organic Chemistry,* Wiley, New York, 1993.

Sposito, G., *The Chemistry of Soils,* Oxford University Press, New York, 1989.

Tessier, A., Campbell, P. G. C., and Bisson, M., Sequential extraction procedure for the speciation of particular trace metals, *Anal. Chem.,* Vol. 51, pp. 844-850, 1979.

USEPA (U.S. Environmental Protection Agency), *Test Methods for Evaluating Solid Waste: Physical/Chemical Methods,* SW-846, Office of Solid Waste and Emergency Response, USEPA, Washington, DC, 1986, 1992 (rev.), 1996 (rev.).

———, *MINTEQ2: An Equilibrium Metal Speciation Model: User's Manual,* EPA/600/3-87/012, USEPA, Athens, GA, 1987.

———, *Cleaning Up the Nation's Waste Sites: Markets and Technology Trends,* EPA/542/R-96/005, Office of Solid Waste and Emergency Response, USEPA, Washington, DC, 1997.

Westall, J. C., Zachary, J. L., and Morel, F. M. M., *MINEQL: A Computer Program for Calculation of Chemical Equilibrium Composition of Aqueous Systems,* Technical Note 18, Department of Civil Engineering, MIT, Cambridge, MA, 1976.

7

GROUNDWATER FLOW

7.1 INTRODUCTION

Groundwater is the main source of drinking water in the United States as well as worldwide. Therefore, protecting groundwater from contamination is of utmost importance to protect public health and the environment. Any groundwater that has already been contaminated due to improper waste disposal or handling practices needs to be remediated. At the same time, future occurrence of groundwater contamination must be prevented by designing the waste disposal facilities that can effectively contain the wastes that are being disposed. As a geoenvironmental professional, an understanding of occurrence and movement of groundwater is essential to design remedial systems and to evaluate potential for contamination at waste disposal facilities.

In this chapter, basic groundwater flow terminology and concepts are introduced first. Then groundwater flow behavior in aquifers and toward pumping wells is described. In addition, the basis for each of the common field tests to determine aquifer hydraulic properties is presented. Finally, mathematical modeling of groundwater flow is explained briefly.

7.2 HYDROLOGIC CYCLE AND GROUNDWATER

The Earth's water circulatory process is known as the *hydrologic cycle*. As shown in Figure 7.1, in the hydrologic cycle water evaporates from the oceans into the atmosphere continuously. Water vapor moves through the atmosphere and eventually returns to the ocean through one or more routes. A portion of this water precipitates over the oceans and another portion falls over the land. As soon as it falls on the land, part of the precipitation becomes runoff that flows back to the oceans in the form of streams and rivers. Some of the precipitation infiltrates into the subsurface.

Water in the subsurface can exist in two zones: the *unsaturated zone* (also known as the *vadose zone*) and the *saturated zone*, which underlies the unsaturated zone. The unsaturated zone consists of soil pores that are filled to a varying degree with air and water, whereas the saturated zone consists of water-filled pores that are assumed to be at hydrostatic pressure. The water in the saturated zone which is under a pressure equal to or greater than atmospheric pressure is known as *groundwater*. The water level in the saturated zone at which the pressure is equal to the atmospheric pressure is known as *groundwater table*. Generally, a zone of soil may be saturated just above the groundwater table due to capillary rise, and this soil zone is called *capillary zone* or *capillary fringe*. When a hydraulic gradient (change in head per unit distance) exists within the saturated zone, groundwater flow occurs. We focus on groundwater flow in this chapter; the reader should refer to Tindall and Kunkel (1999), Guymon (1994), and Fredlund and Rahardjo (1993) for details on flow in unsaturated zone.

7.3 AQUIFER, AQUICLUDE, AND AQUITARD

An *aquifer* is a geologic unit that can store and transmit water. A variety of geologic formations can act as aquifers. The most common aquifers occur in permeable soils such as sand and gravel as well as porous and permeable rocks such as sandstones. Aquifers are generally categorized into two types, depending on the

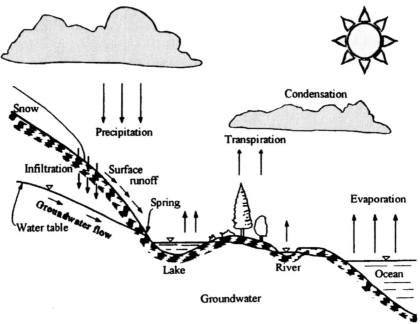

Figure 7.1 Hydrologic cycle and groundwater. (From USACE, 1999.)

geologic environment in which they occur: unconfined and confined. Figure 7.2 depicts these basic aquifer types.

An *unconfined aquifer* is a permeable formation that contains a phreatic surface (water table) as an upper boundary that fluctuates in response to recharge and discharge (such as from a pumping well). Unconfined aquifers are generally close to the ground surface, with the water table in direct contact with the atmosphere. A special type of unconfined aquifer is a *perched aquifer*, which occurs when the precipitation moving downward through the unsaturated zone is intercepted by a layer of relatively low permeability (e.g., a layer of clay within sand). This layer has a limited areal extent. The water is accumulated on the top of a restricting layer and forms a perched water table. Perched aquifers generally yield temporary or small quantities of water.

A *confined aquifer* is a permeable zone between two geologic formations of relatively low permeability. The aquifer is completely saturated and it does not

have a free water table. Water in a confined aquifer is trapped under pressure. The pressure conditions are characterized by the potentiometric surface (see Figure 7.2). The potentiometric surface represents the total head in an aquifer; that is, it represents the height above a datum plane at which the water level stands in tightly cased wells that penetrate the aquifer. In other words, the water level will rise above the bottom of the overlying confining layer to an elevation at which it is equal to atmospheric pressure if the aquifer is penetrated by a well. In some geologic settings, the potentiometric surface may be above the top of the ground surface, causing water to flow from springs or wells associated with the confined aquifer. This type of confining aquifer is also known as an *artesian aquifer*.

The two geologic units separating an aquifer or forming the boundaries of an aquifer are known as *confining layers* (see Figure 7.2). Confining layers are generally subdivided into aquicludes and aquitards. An *aquiclude* is a confining layer that is essentially im-

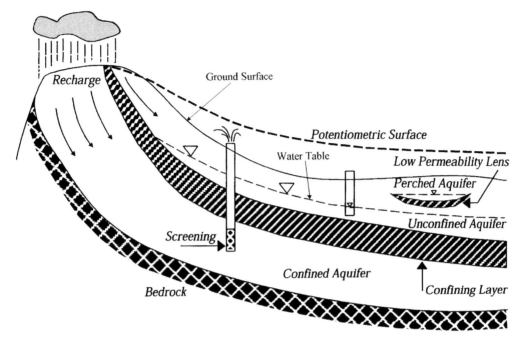

Figure 7.2 *Aquifer formation. (From USACE, 1999.)*

permeable. An *aquitard* is a confining layer of low permeability that can store groundwater and transmit slowly from one aquifer to another. An aquifer bounded by an aquitard is often called a *leaky* or *semi-confined aquifer.*

7.4 HYDRAULIC HEAD AND AQUIFER PROPERTIES

The groundwater flow depends on the slope of the hydraulic head (hydraulic gradient) as well as on the aquifer properties, specifically porosity, specific yield, hydraulic conductivity, transmissivity, and storativity. All of these are explained briefly below.

7.4.1 Hydraulic Head

Groundwater moves through the subsurface from areas of greater hydraulic head to areas of lower hydraulic head. The hydraulic head at any location within the subsurface is defined by Bernoulli's equation as

$$h = z + \frac{p}{\rho g} + \frac{v^2}{2g} \qquad (7.1)$$

where h is the hydraulic head, z the elevation above datum, p the fluid pressure with constant density ρ, g the acceleration due to gravity, and v the fluid velocity. Pressure head (or fluid pressure), h_p, is defined as

$$h_p = \frac{p}{\rho g} \qquad (7.2)$$

Below the water table, in the saturated zone, pressure head is greater than atmospheric pressure ($h_p > 0$). Because groundwater velocities are usually very low, the velocity component of head is neglected. Thus, hydraulic head is expressed as

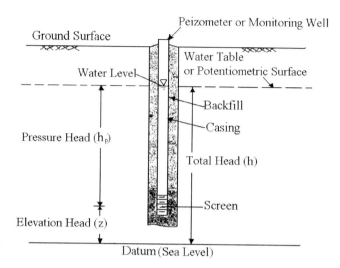

Figure *7.3 Relationship between total head, pressure head, and elevation head within a well.*

$$h = z + h_p \qquad (7.3)$$

Figure 7.3 shows the relationship between hydraulic head, pressure head, and elevation head within a well.

7.4.2 Porosity

As explained in Chapter 5, porosity (*n*) is defined as the ratio of volume of void space to the total volume of medium:

$$n = \frac{V_V}{V_T} \qquad (7.4)$$

where V_V is the volume of void space and V_T is the total volume. Porosity is governed by the grain packing, grain shape, and grain size distribution. Typical values of porosity of different types of soils are given in Chapter 5. Porosity is an index of how much total groundwater can be stored in the saturated material; however, it does not indicate how much water the porous material will yield. The effective porosity (n_e) is actually needed for seepage calculations. The effective porosity is the ratio of the volume of the void space through which flow can occur to the total volume. Since part of the total porosity is occupied by static fluid being held to the mineral surface by surface ten-

sion, it follows that effective porosity will be less than total porosity. Typical values of porosity for different types of soils are given in Table 7.1.

7.4.3 Specific Yield

During the drainage of water from unconfined aquifers by gravity forces, only a part of the total volume stored in void space can be released. This drainage behavior

Table 7.1 **Typical Relationship between specific yield, specific retention, and total porosity for various soil types**

Soil Type	Specific Yield (%)	Specific Retention (%)	Total Porosity (%)
Clay and silt	5	45	50
Fine sand	20	25	45
Medium sand	30	10	40
Coarse sand	5	30	35
Fine gravel	5	28	33
Medium gravel	5	20	25
Coarse gravel	4	15	19
Limestone	18	2	20
Sandstone	6	5	11
Granite	0.09	0.01	0.1
Basalt	8	3	11

Source: USEPA (1990); USACE (1999).

is defined by specific yield and specific retention of an aquifer. The *specific yield* (S_y) of an aquifer is the ratio of the volume of water that drains from saturated material due to the attraction of gravity to the total volume of the material. The ratio of volume of water that is retained against the force of gravity to the total volume is called the *specific retention* (S_r).

The porosity (n) of the media can be related to the specific yield and specific retention of the media as follows:

$$n = S_y + S_r \qquad (7.5)$$

For most practical applications, the value of effective porosity (n_e) can be considered equivalent to specific yield (S_y). Table 7.1 shows typical values of specific yield and specific retention of different soil types.

7.4.4 Hydraulic Conductivity

As explained in Chapter 5, *hydraulic conductivity* or *permeability* is defined as the property of a porous media that permits the transmission of water through it. The groundwater flow in porous media is described using Darcy's law as

$$Q = -KA \frac{dh}{dL} = -KAi \qquad (7.6)$$

where Q is the volumetric flow rate, K the hydraulic conductivity, A the cross-sectional area of flow, h the hydraulic head, and L the distance between two points. The negative sign on the right-hand side is used by convention to indicate a downward-trending flow gradient. The hydraulic conductivity (K) is a function of the properties of the medium and the properties of the fluid and can be given by

$$K = \frac{k\rho g}{\mu} \qquad (7.7)$$

where k is the intrinsic permeability of porous medium, ρ the fluid density, μ the dynamic viscosity of fluid, and g the acceleration due to gravity. The ranges of K

values for water flow through different soil types are given in Chapter 5.

7.4.5 Transmissivity

Transmissivity (T) is a measure of the amount of water that can be transmitted horizontally through a unit width by the fully saturated thickness of an aquifer under a hydraulic gradient equal to 1. Transmissivity is equal to the hydraulic conductivity multiplied by the saturated thickness of the aquifer and is given by

$$T = Kb \qquad (7.8)$$

where K is the hydraulic conductivity and b is the saturated thickness of the aquifer. Since transmissivity depends on hydraulic conductivity and saturated thickness, its value will differ at different locations within aquifers if the thickness of the confined aquifer varies or the saturated thickness varies with the water table.

7.4.6 Specific Storage and Storativity

In any aquifer, when the hydraulic head is reduced due to lowering of water table or potentiometric surface (as commonly occurs during pumping a well), water is released (or expelled) from the aquifer. This occurs for the following two reasons (Heath, 1989):

1. *Compressibility of aquifer solids skeleton.* Lowering the hydraulic head causes pore water pressure to decrease. This causes the effective stress to increase. The increase in effective stress causes compression of the solids skeleton (the reduction of soil pores as a result of consolidation).
2. *Compressibility of water.* The decrease in pore water pressure also causes the pore water to expand.

Specific storage, S_s, defines the amount of water released (or expelled) from an aquifer (for these two reasons) per *unit volume* per unit change in hydraulic head. It can be expressed mathematically by

$$S_S = \rho_w g(\alpha + n\beta) \qquad (7.9)$$

where ρ_w is the density of water, g the acceleration due to gravity, α the compressibility of the aquifer skeleton, n the porosity, and β the compressibility of water. The specific storage of sands and gravels is about 1×10^{-6} ft^{-1} and of clays and silts is about 3.5×10^{-6} ft^{-1} (USACE, 1999).

Storativity (also known as *storage coefficient*), S, is used to define the amount of water released from an aquifer per *unit surface area* per unit change in hydraulic head. Within a confined aquifer the full thickness of the aquifer remains saturated when water is released (or expelled). Therefore, the storativity of confined aquifers is given by

$$S = bS_s \qquad (7.10)$$

where b is the thickness of the aquifer. Note that S is dimensionless and is generally less than 0.005 for confined aquifers (USACE, 1999).

It should be noted that in unconfined aquifers, when hydraulic head is reduced, the water table gets lowered. This means that water is released from the aquifer due to the compression of aquifer solid skeleton and expansion of water similar to confined aquifer, but additional water is also released, due to gravity drainage. The water released due to compression of solids skeleton and expansion of water is given by specific storage (S_s) as given in equation (7.9), while the additional water released due to gravity drainage is given by the specific yield (S_y) as explained in Section 7.3.3. Therefore, the storativity of an unconfined aquifer is given by the sum of the specific yield, and the specific storage as given by

$$S - hS_s + S_y \qquad (7.11)$$

where h is the saturated head in unconfined aquifer. In general, the value of specific storage (S_s) of aquifers is very small, less than 1×10^{-4} ft^{-1} (USACE, 1999), whereas the specific yield (S_y) is usually several orders of magnitude greater than specific storage. Therefore, the storativity of an unconfined aquifer is approximately equal to its specific yield: $S \approx S_y$. The S_y values range from 0.01 to 0.30 (USACE, 1999).

In both confined and unconfined aquifers, the volume of water released (or expelled) due to a lowering of the hydraulic head can be calculated from

$$V_w = SA \, \Delta h \qquad (7.12)$$

where V_w is the volume of water drained from aquifer, S the storativity (or storage coefficient), A the surface area overlying the drained aquifer, and Δh the average decline in hydraulic head. As stated earlier, $S \approx bS_s$ for confined aquifers and $S \approx S_y$ for unconfined aquifers.

7.5   GROUNDWATER FLOW IN AQUIFERS

The groundwater flow may occur in three dimensions (x, y, and z), and the aquifers may be anisotropic and heterogeneous. *Anisotropic conditions* imply that the hydraulic conductivity varies with the direction of measurement at a point within a formation. *Heterogeneous conditions* imply that within the same direction, the hydraulic conductivity varies. General governing equations describing the groundwater flow in anisotropic, heterogeneous confined and unconfined aquifers are presented in this section. The general flow equations can then be simplified for isotropic and homogeneous aquifer conditions as detailed in this section.

7.5.1   Flow in Confined Aquifer

The general governing flow equation for confined aquifers is derived from the application of the law of mass conservation (continuity equation) to the elemental volume shown in Figure 7.4. Continuity is given by: rate of mass accumulation = rate of mass inflow − rate of mass outflow. Integrating the conservation of mass (under constant density) with Darcy's law, the following general flow equation in three dimensions for a heterogeneous anisotropic material is derived (USACE, 1999):

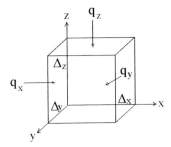

Figure 7.4 *Elemental control volume.*

$$\frac{\partial}{\partial x}\left(K_x \frac{\partial h}{\partial x}\right) + \frac{\partial}{\partial y}\left(K_y \frac{\partial h}{\partial y}\right) + \frac{\partial}{\partial z}\left(K_z \frac{\partial h}{\partial z}\right) = S_S \frac{\partial h}{\partial t}$$

$$(7.13)$$

Equation (7.13) is the general flow equation in three dimensions for a heterogeneous anisotropic material. Discharge (from a pumping well, etc.) or recharge to or from the control volume is represented as volumetric flux per unit volume:

$$\frac{\partial}{\partial x}\left(K_x \frac{\partial h}{\partial x}\right) + \frac{\partial}{\partial y}\left(K_y \frac{\partial h}{\partial y}\right) + \frac{\partial}{\partial z}\left(K_z \frac{\partial h}{\partial z}\right)$$

$$= S_S \frac{\partial h}{\partial t} + W \qquad (7.14)$$

where W is the volumetric flux per unit volume. Assuming that the material is homogeneous (i.e., K does not vary with position), equation (7.13) can be written as

$$K_x \frac{\partial}{\partial x}\left(\frac{\partial h}{\partial x}\right) + K_y \frac{\partial}{\partial y}\left(\frac{\partial h}{\partial y}\right) + K_z \frac{\partial}{\partial z}\left(\frac{\partial h}{\partial z}\right) = S_S \frac{\partial h}{\partial t}$$

$$(7.15)$$

If the material is both homogeneous and isotropic (i.e., $K_x = K_y = K_z = K$, then equation (7.15) becomes

$$K \left(\frac{\partial^2 h}{\partial x^2} + \frac{\partial^2 h}{\partial y^2} + \frac{\partial^2 h}{\partial z^2}\right) = S_S \frac{\partial h}{\partial t} \qquad (7.16)$$

Using the definitions for storativity (or storage coefficient) given by $S = bS_s$ and transmissivity given by $T = Kb$, where b is the aquifer thickness, equation (7.16) becomes

$$\frac{\partial^2 h}{\partial x^2} + \frac{\partial^2 h}{\partial y^2} + \frac{\partial^2 h}{\partial z^2} = \frac{S}{T}\frac{\partial h}{\partial t} \qquad (7.17)$$

If the flow is steady state, the hydraulic head does not vary with time and equation (7.17) becomes

$$\frac{\partial^2 h}{\partial x^2} + \frac{\partial^2 h}{\partial y^2} + \frac{\partial^2 h}{\partial z^2} = 0 \qquad (7.18)$$

Equation (7.18) is known as the *Laplace equation*.

7.5.2 Flow in Unconfined Aquifer

In an unconfined aquifer, the saturated thickness of the aquifer changes with time as the hydraulic head changes. Therefore, the ability of the aquifer to transmit water (the transmissivity) is not constant. For such conditions, the flow equation can be expressed as (USACE, 1999)

$$\frac{\partial}{\partial x}\left(K_x h \frac{\partial h}{\partial x}\right) + \frac{\partial}{\partial y}\left(K_y h \frac{\partial h}{\partial y}\right) + \frac{\partial}{\partial z}\left(K_z h \frac{\partial h}{\partial z}\right)$$

$$= S_y \frac{\partial h}{\partial t} \qquad (7.19)$$

where S_y is the specific yield. For a homogeneous, isotropic aquifer, the general equation governing unconfined flow is known as the *Boussinesq equation* and is given by

$$\frac{\partial}{\partial x}\left(h \frac{\partial h}{\partial x}\right) + \frac{\partial}{\partial y}\left(h \frac{\partial h}{\partial h}\right) + \frac{\partial}{\partial z}\left(h \frac{\partial h}{\partial z}\right) = \frac{S_y}{K}\frac{\partial h}{\partial t}$$

$$(7.20)$$

If the change in the elevation of the water table is small in comparison to the saturated thickness of the aquifer, the variable thickness, h, can be replaced with an average thickness, b, that is assumed to be constant over the aquifer. Equation (7.20) can then be linearized to the form

$$\frac{\partial^2 h}{\partial x^2} + \frac{\partial^2 h}{\partial y^2} + \frac{\partial^2 h}{\partial z^2} = \frac{S_y}{Kb} \frac{\partial h}{\partial t} \qquad (7.21)$$

For steady-state condition, equation (7.21) reduces to the *Laplace equation* as given by equation (7.18).

7.5.3 Methods to Solve Groundwater Flow Equation

The governing flow equations (7.13–7.21) can be solved using appropriate boundary conditions by using either analytical or numerical methods. The following flow problems are commonly addressed in practice:

- One-dimensional steady flow; this is presented in Section 7.6.
- Radial flow toward a pumping well; this is described in Section 7.7.
- Two and three-dimensional flow; this is described in Section 7.9. For two-dimensional steady flow conditions, the governing flow equation, the Laplace equation [equation (7.18)], can be solved mathematically or by a graphical construction of a flow net. A brief description of flow nets is also given in Section 7.9.

7.6 ONE-DIMENSIONAL STEADY FLOW

7.6.1 Uniform Aquifer

Darcy's law is used to describe the rate of flow through a porous medium and is given by

$$Q = -KA \frac{dh}{dl} = -KAi \qquad (7.22)$$

The volumetric flow velocity (V) (also known as *specific discharge* or *Darcy flux*), can be determined by dividing the volumetric flow rate (Q) by the cross-sectional area of the flow as

$$V = \frac{Q}{A} = -K \frac{dh}{dl} = -Ki \qquad (7.23)$$

The volumetric flow velocity represents the velocity at which water would move through an aquifer if the aquifer was an open conduit. However, the cross-sectional area is not entirely available for flow due to the presence of the porous matrix. The average pore water velocity or seepage velocity (V_s) is derived by dividing volumetric flow velocity by effective porosity (n_e) to account for the actual open space available for the flow. This velocity represents the average rate at which water moves between two points and is given by

$$V_S = \frac{Q}{n_e A} = -\frac{K}{n_e} \frac{dh}{dl} = -\frac{Ki}{n_e} \qquad (7.24)$$

As shown in Figure 7.5(a) and (b), the hydraulic gradient (i) is the difference in hydraulic head between two points, ($h_1 - h_2$), divided by the corresponding distance (L) as

$$i = \frac{h_1 - h_2}{L} \qquad (7.25)$$

The quantity of flow can be calculated using equation (7.22). If transmissivity (T) is known, K can be calculated by dividing the transmissivity by the thickness of the aquifer (b).

7.6.2 Stratified Aquifer

If the aquifer is stratified as shown in Figure 7.6(a) and (b), the flow through it may be described by calculating an equivalent hydraulic conductivity. Horizontal flow through the stratified aquifer is given by Darcy's law as (USACE, 1999)

$$Q_x = K_x A_x \frac{\Delta h_T}{X} \qquad (7.26)$$

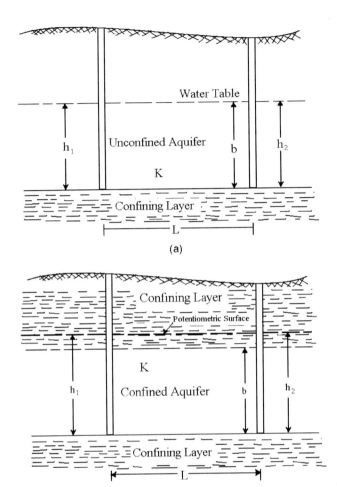

(a)

(b)

Figure 7.5 (a) *Steady flow through an unconfined aquifer;* (b) *steady flow through a confined aquifer.*

where Δh_T is the total hydraulic head drop across flow distance X. The horizontal equivalent hydraulic conductivity (K_x) is given by

$$K_x = \frac{\Sigma\, K_{xi} X_i}{X} \qquad (7.27)$$

Similarly, vertical flow is given by Darcy's law as

$$Q_z = K_z A_z \frac{\Delta h_T}{L} \qquad (7.28)$$

where Δh_T is the total hydraulic head drop across flow distance L. The vertical equivalent hydraulic conductivity (K_z) is given by

$$K_z = \frac{L}{\Sigma\, b_i/K_{zi}} \qquad (7.29)$$

7.6.3 Spatial Variation of Hydraulic Conductivity

Considering spatial variations in two dimensions, Gutjahr et al. (1978), Dagan (1979), and Gelhar (1993)

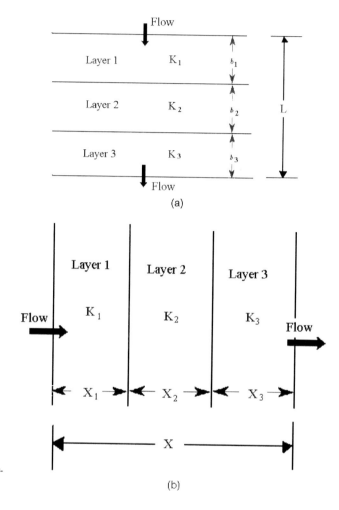

Figure 7.6 *Flow through stratified media:* (a) *horizontal flow;* (b) *vertical flow.*

show that the effective hydraulic conductivity (K_e) of the porous media can be obtained by geometric mean of various spatial hydraulic conductivity values K_1, K_2, K_3, . . . , K_n and is given by the following relationship:

$$K_e = \sqrt[n]{K_1 \times K_2 \times K_3 \times \cdots \times K_n} \quad (7.30)$$

7.7 FLOW TOWARD A PUMPING WELL

Pumping wells are used for determining the aquifer properties as well as for remediating contaminated groundwater. Therefore, flow of water toward a pumping well in both confined and unconfined aquifers

should be understood. Figure 7.7 depicts the groundwater flow toward a well. When pumping is started, the water level in the vicinity of the pumping well is lowered. The greatest amount of lowering or drawdown occurs at the well. Drawdown is less at greater distances from the well, and at some distance away, aquifer lowering is nonexistent. The force or pressure that drives the water toward the well is the hydraulic head, given by the difference between the water level inside the well and water level at any place outside the well, $H - h$. The velocity of the groundwater increases as it approaches the well. According to Darcy's law, with increasing velocity, the hydraulic gradient increases as flow converges toward the well. As a result,

EXAMPLE 7.1

Consider the cross section of the confined aquifer that has two observation wells, as shown in Figure E7.1. Determine the flow rate for a unit width of aquifer.

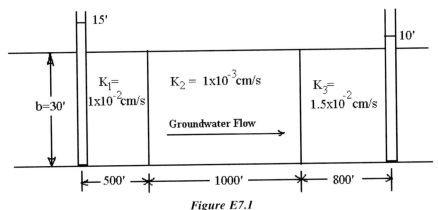

Figure E7.1

Solution:

$$Q_x = K_x A_x \frac{\Delta h_T}{X} \qquad K_x = \frac{\Sigma \, k_{xi} X_i}{X}$$

$$K_x = \frac{(1 \times 10^{-2})(500) + (1 \times 10^{-3})(1000) + (1.5 \times 10^{-2})(800)}{2300}$$

$$= 7.83 \times 10^{-3} \text{ cm/s} = 22.1 \text{ ft/day}$$

$$Q_x = 22.1(1 \times 30)\left(\frac{(15-10)}{2300}\right) \qquad Q = 1.45 \text{ ft}^3/\text{day}$$

the lowered water surface develops a continually steeper slope toward the well, referred to as the *cone of depression*. The distance from the center of the well to the limit of the cone of depression is called the *radius of influence* (R_0).

The shape and size of the cone of depression is dependent on the pumping rate, pumping period, and aquifer characteristics. The aquifer characteristics that are important are hydraulic conductivity and specific yield for unconfined aquifers, and transmissivity and storativity for confined aquifers. A number of well formulas have been developed to describe both equilib-

rium (steady state) and nonequilibrium flow conditions in confined and unconfined aquifers. All of these formulas are based on the assumption that the aquifer is homogeneous and isotropic and flow is radial. A brief review of these formulas is presented below.

7.7.1 Confined Aquifer

Equilibrium (Steady) Flow. Figure 7.8 shows the drawdown and cone of depression in a confined aquifer under steady flow conditions. The formula that describes the flow is given by (Freeze and Cherry, 1979)

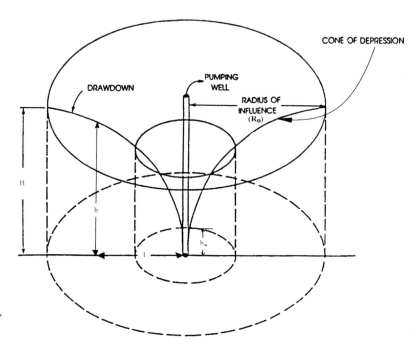

Figure *7.7 Formation of cone depression for a pumped well.*

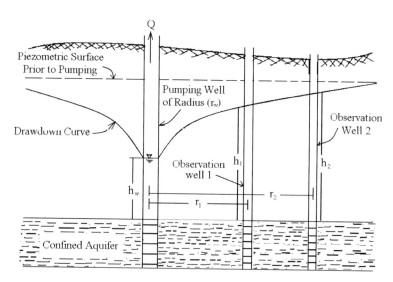

Figure *7.8 Radial flow to a well penetrating a confined aquifer.*

$$Q = 2\pi T \frac{h - h_w}{\ln(r/r_w)} \qquad (7.31)$$

or

$$Q = 2\pi T \frac{h_2 - h_1}{\ln(r_2/r_1)} \qquad (7.32)$$

Example 7.2 shows how T and h calculations can be made using the equation above.

Nonequlibrium (Unsteady-State) Flow. A long pumping time is required to reach equilibrium or steady-state conditions represented above. Therefore, to represent a point in time prior to steady-state conditions, it is often necessary to use a nonequilibrium or unsteady-state formula. The governing flow equa-

tion for such condition is given by (Freeze and Cherry, 1979)

$$\frac{\partial^2 h}{\partial r^2} + \frac{1}{r}\frac{\partial h}{\partial r} = \frac{S}{T}\frac{\partial h}{\partial t} \qquad (7.33)$$

where h is the head along the drawdown curve at a distance r from the pumping well, r the radial distance, S the storativity, and T the transmissivity. The two solutions commonly used to solve the above governing flow equation are the Theis (1935) solution and the Cooper–Jacob (1946) solution.

The *Theis solution* is given by the following equation (Freeze and Cherry, 1979):

$$s' = \frac{Q}{4\pi T}\int_u^\infty \frac{e^{-u}\,du}{u} = \frac{Q}{4\pi T}W(u) \qquad (7.34)$$

EXAMPLE 7.2

An 18-in.-diameter well was used to pump water from a confined aquifer at 100 gal/min. The drawdowns recorded at observation wells located at 50 and 200 ft from the pumping well were 10 ft and 8 ft, respectively. Calculate the drawdown at the pumping well.

Solution:
Given $Q = 100$ gal/min $= 100 \times 2.23 \times 10^{-3} = 0.22$ ft³/s (1 gal/min $= 2.23 \times 10^{-3}$ ft³/s), let h_0 be the initial head, so the measured heads at $r_1 = 50$ ft and $r_2 = 200$ ft are $h_1 = (h_0 - 10)$ ft and $h_2 = (h_0 - 8)$ ft, respectively. Using equation (7.32) yields

$$Q = 2\pi T \frac{h_2 - h_1}{\ln(r_2/r_1)} \Rightarrow 0.22 = 2\pi T \frac{(h_0 - 8) - (h_0 - 10)}{\ln(200/50)}$$

$$T = 0.0245 \text{ ft}^2/\text{s}$$

At the well location, $r = r_w =$ radius of the well $= 0.75$ ft; assume that the drawdown at the well is x. Substituting these in equation (7.32) with respect to an observation well at 50 ft distance and substituting $T = 0.0245$ ft²/s, we get

$$0.22 = 2\pi(0.0245)\frac{(h_0 - 10) - (h_0 - x)}{\ln(50/0.75)}; \qquad x = 16 \text{ ft}$$

Therefore, the drawdown at the pumping well is 16 ft.

where s' is drawdown in the well under a pumping rate of Q, and u is given by

$$u = \frac{r^2 S}{4Tt} \tag{7.35}$$

$W(u)$ is known as the exponential integral or well function which can be expanded as a series:

$$W(u) = -0.5772 - \ln(u) + u - \frac{u^2}{2 \cdot 2!}$$

$$+ \frac{u^3}{3 \cdot 3!} + \cdots \tag{7.36}$$

Tables showing the values of $W(u)$ as a function of u, such as Table 7.2, are often used for this solution.

In the *Cooper–Jacob solution*, the $W(u)$ infinite series above is approximated and s' is given by (USBR, 1977)

$$s' = \frac{Q}{4\pi T}\left(-0.5772 - \ln\frac{r^2 S}{4Tt}\right) \tag{7.37}$$

Equation (7.37) can be rewritten as

$$s' = \frac{2.3Q}{4\pi T}\log\frac{2.25Tt}{r^2 S} \tag{7.38}$$

Example 7.3 shows how to use Theis and Cooper–Jacob solutions to describe flow in confined aquifer under pumping conditions.

7.7.2   Unconfined Aquifer

Equilibrium (Steady) Flow. Figure 7.9 shows the drawdown and cone of depression in an unconfined aquifer under steady flow conditions. The formula that describes the flow is given by (Freeze and Cherry, 1979)

$$Q = \pi K\frac{h_2^2 - h_1^2}{\ln(r_2/r_1)} \tag{7.39}$$

Nonequlibrium (Unsteady-State) Flow. Long time of pumping is required in order to reach equilibrium or steady-state conditions; therefore, it is often necessary to use nonequilibrium or unsteady-state formula. As mentioned earlier, flow in unconfined aquifer results from gravity drainage as well as a result of com-

Table 7.2 **Tabulation of $W(u)$ values for use in the Theis equation**

| u | \multicolumn{9}{c}{Values of W(u) for values of u} | | | | | | | | |
	1.0	2.0	3.0	4.0	5.0	6.0	7.0	8.0	9.0
$\times 1$	0.219	0.049	0.013	0.0036	.0011	0.00036	0.00012	0.000038	0.000012
$\times 10^{-1}$	1.82	1.22	0.91	0.70	0.56	0.45	0.37	0.31	0.26
$\times 10^{-2}$	4.04	3.35	2.96	2.68	2.47	2.30	2.15	2.03	1.92
$\times 10^{-3}$	6.33	5.64	5.23	4.95	4.73	4.54	4.39	4.26	4.14
$\times 10^{-4}$	8.63	7.94	7.53	7.25	7.02	6.84	6.89	6.55	6.44
$\times 10^{-5}$	10.94	10.24	9.84	9.55	9.33	9.14	8.99	8.86	8.74
$\times 10^{-6}$	13.24	12.55	12.14	11.85	11.63	11.45	11.29	11.16	11.04
$\times 10^{-7}$	15.54	14.85	14.44	14.15	13.93	13.75	13.60	13.46	13.34
$\times 10^{-8}$	17.84	17.15	16.74	16.46	16.23	16.05	15.90	15.76	15.65
$\times 10^{-9}$	20.15	19.45	19.05	18.76	18.54	18.35	18.20	18.07	17.95
$\times 10^{-10}$	22.45	21.76	21.35	21.03	20.84	20.66	20.50	20.37	20.25
$\times 10^{-11}$	24.75	24.06	23.65	23.36	23.14	22.96	22.81	22.67	22.55
$\times 10^{-12}$	27.05	26.36	25.96	25.67	25.44	25.26	25.11	24.97	24.86
$\times 10^{-13}$	29.35	28.66	28.26	27.97	27.75	27.56	27.41	27.28	27.16
$\times 10^{-14}$	31.66	30.97	30.56	30.27	30.05	29.87	29.71	29.58	29.46
$\times 10^{-15}$	33.96	33.27	32.86	32.58	32.35	32.17	32.02	31.88	31.76

Source: USACE (1999).

EXAMPLE 7.3

Two wells (A and B) are located 100 m apart in a confined aquifer with transmissivity 1×10^{-4} m²/s and a storativity of 1×10^{-5}. Well A is pumped at a rate of 100 gal/day and the well B at a rate of 200 gal/day. Find (a) the drawdown at the midway of the wells one day after the pumping starts, and (b) the time required to cause a total drawdown of 10 cm at the midway of the wells.

Solution:
(a) Consider the midway location, this means that $r = 50$ m from the wells. For well A, $Q = 100$ gal/day $= (100)(4.3 \times 10^{-8}) = 4.3 \times 10^{-6}$ m³/s (1 gal/day $= 4.3 \times 10^{-8}$ m³/s). Using equation (7.35) yields

$$u = \frac{r^2 S}{4Tt} = \frac{(50)^2(1 \times 10^{-5})}{4(1 \times 10^{-4})(24 \times 3600)} = 7.23 \times 10^{-4} \Rightarrow W(u) = 6.58 \quad \text{(using Table 7.2)}$$

$$s' = \frac{Q}{4\pi T} \times W(u) = \frac{(4.3 \times 10^{-6})6.58}{4\pi(1 \times 10^{-4})} = 2.25 \text{ cm}$$

For well B, $Q = 200$ gal/day $= (200)(4.3 \times 10^{-8}) = 8.6 \times 10^{-6}$ m³/s.

$$s' = \frac{Q}{4\pi T} W(u) = \frac{(8.6 \times 10^{-6})6.58}{4\pi(1 \times 10^{-4})} = 4.5 \text{ cm}$$

Total drawdown at midlocation $= 2.25 + 4.5 = 6.75$ cm.

(b) Assuming that the drawdown is proportional to the pumping rate, well A has a drawdown of 3.33 cm. Calculate the time required for this drawdown to occur.

$$0.033 = \frac{(4.3 \times 10^{-6})W(u)}{4\pi(1 \times 10^{-4})} \Rightarrow W(u) = 9.74 \Rightarrow u = 3.34 \times 10^{-5} \quad \text{(using Table 7.2)}$$

$$u = \frac{r^2 S}{4Tt} \Rightarrow 3.34 \times 10^{-5} = \frac{(50)^2(1 \times 10^{-5})}{4(1 \times 10^{-4})t} \Rightarrow t = 21 \text{ days}$$

We know that within this time, the drawdown from well B should be 6.67 cm. We will check this as follows:

$$u = \frac{r^2 S}{4Tt} = \frac{(50)^2(1 \times 10^{-5})}{4(1 \times 10^{-4})(21 \times 24 \times 3600)} \Rightarrow u = 3.34 \times 10^{-5} \Rightarrow W(u) = 9.74$$

$$s' = \frac{(8.6 \times 10^{-6})(9.74)}{4\pi(1 \times 10^{-4})} \Rightarrow s' = 6.67 \text{ cm}$$

Total drawdown in 21 days $\rightarrow 6.67 + 3.33 = 10$ cm. Therefore, 10 cm drawdown occurs in 21 days.

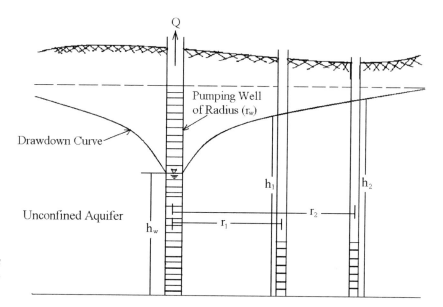

Figure 7.9 *Radial flow to a well penetrating an unconfined aquifer.*

paction of the skeleton and expansion of the pore water. The governing flow equation for such conditions is given by (Fetter, 1987)

$$T = \frac{Q}{4\pi s} W(u_A, u_B, \Gamma) \qquad (7.40)$$

where $W(u_A, u_B, \Gamma)$ is the well function for the unconfined aquifer, and

$$u_A = \frac{r^2 S}{4Tt} \qquad (7.41a)$$

$$u_B = \frac{r^2 S_y}{4Tt} \qquad (7.41b)$$

$$\Gamma = \frac{r^2 K_v}{b^2 K_h} \qquad (7.42)$$

where s is the the drawdown, Q the pumping rate, T the transmissivity, r the radial distance from the pumping well, S the storativity, S_y the specific yield, t the time, K_h the horizontal hydraulic conductivity, K_v the vertical hydraulic conductivity, and b the initial saturated thickness of the aquifer.

The formulation above assumes that during the initial stage of pumping, the aquifer behaves as a confined aquifer and the time–drawdown data follows the Theis nonequilibrium curve [equation (7.41a) with S equal to the storativity of the aquifer]. Flow is horizontal during this period, as the water is being derived from the entire aquifer thickness. Following this initial stage, water is derived primarily from the gravity drainage of the aquifer, and there are both horizontal and vertical flow components. The drawdown–time relationship is a function of the ratio of horizontal to vertical conductivities of the aquifer, the distance to the pumping, and the thickness of the aquifer. As the time progresses, the rate of drawdown decreases and the contribution of the particular annular region to the overall well discharge diminishes. Flow is again essentially horizontal, and the time–discharge data again follow a Theis nonequilibrium flow with storativity equal to the specific yield of the aquifer.

7.7.3 Effects of Multiple Wells and Boundaries

Several wells are often placed at selected spacing such that their cones of depression overlap and no flow occurs between the wells. For confined aquifers, the composite pumping cone can be calculated by summing the individual drawdowns caused by each well as shown in Figure 7.10. This can be accomplished by graphical superposition or by calculations using either equilibrium or nonequilibrium solutions. Theoretically, the method of cumulative drawdown cannot be applied to unconfined aquifers because transmissivity changes with drawdown. Refer to Freeze and Cherry (1979), Driscoll (1986), and Fetter (1987) for composite drawdowns for unconfined aquifers under equilibrium conditions.

Typically, wells are not located in aquifers that have infinite areal extent; therefore, drawdown cones extend until they intercept a recharge boundary or a barrier boundary. Recharge boundaries (e.g., streams) are areas where aquifers are replenished with water, and barrier boundaries (e.g., impermeable zones) are areas where aquifers terminate. The actual total drawdown and the rate of drawdown are less than theoretical predictions using the equations given in Sections 7.7.1 and 7.7.2 in situations where recharge boundaries or barrier boundaries are present. To predict hydraulic head drawdown when recharge or barrier boundary conditions exist, the methods of images are used. Further details on these methods are available in Bouwer

(1978), Freeze and Cherry (1979), Todd (1980), Driscoll (1986), and Domenico and Schwartz (1990).

7.8 PUMPING AND SLUG TESTING

Different field testing methods can be employed to characterize site's hydrogeologic conditions as described in Chapter 10. Basically, these methods include installation of wells in various geologic formations. The groundwater elevations are measured in these wells over time. This information is then used to construct groundwater contour maps. These maps provide the hydraulic gradients and direction of the groundwater flow. In addition to the groundwater gradients, hydraulic properties of different geologic formations must be determined. In particular, the hydraulic conductivity (K) and specific yield (S_y) of unconfined aquifers and the transmissivity (T) and storativity (S) of confined aquifers are required. These properties are often determined by conducting aquifer tests in the wells. The most common aquifer tests include pumping tests and slug tests. The detailed testing procedures are provided in Chapter 10. The theoretical background for determining aquifer characteristics (K, S_y, T, S) based on the field measured data is presented in this section.

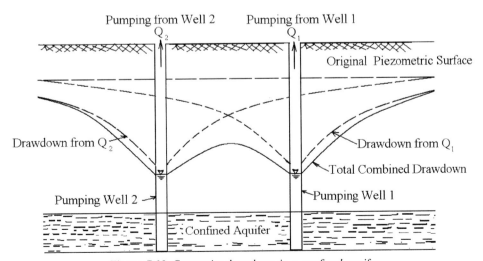

Figure 7.10 *Composite drawdown in a confined aquifer.*

7.8.1 Pumping Tests

Pumping tests generally consist of pumping a well at a constant discharge rate for a set period of time (usually, 24 or 72 h) and monitoring the water levels in at least one observation well located at a certain distance from the pumping well. The number and location of observation wells is dependent on the type of aquifer. These pumping test data can be used to determine transmissivity and storativity of confined aquifers and hydraulic conductivity and specific yield of unconfined aquifers using the applicable well hydraulic solutions as described below.

Pumping Test in Confined Aquifer. For confined aquifers, the nonequilibrium flow solutions (i.e., the Theis solution, the Cooper–Jacob solution, or the Jacob distance-drawdown solution) can be used. The Theis solution requires solving the following two equations with four unknowns (Theis, 1935; Fetter, 1987):

$$s' = \frac{Q}{4\pi T} W(u) \qquad (7.43)$$

$$\frac{r^2}{t} = \frac{4T}{S} u \qquad (7.44)$$

It is recognized that the relation between $W(u)$ and u is the same as that between s' and r^2/t because all other terms are constants in the equations. Based on this, Theis suggested a graphical superposition method to determine T and S using time–drawdown data from the pumping test. A plot of $W(u)$ versus $1/u$ is drawn first and this plot is called a *type curve*. Figure 7.11(*a*) shows such a curve. Then, time versus drawdown data are plotted and superimposed over the type curve while keeping the coordinate axes parallel. The two plots are adjusted until a position is found by trial such that most of the data observed fall on a segment of the type curve. Any convenient point is selected, and the values of $W(u)$, u, s', and r^2/t are recorded and these values then substituted in the equations above to determine S and T.

To use the *Cooper–Jacob solution*, drawdown (s') versus time (t) are plotted on semilog paper (Cooper and Jacob, 1946). A best-fit straight line is drawn and the line is projected to $s' = 0$, the intercepting t value (t_0) is determined. Substituting this in the Cooper–Jacob solution as given in equation (7.38) discussed in Section 7.7.1 can be written as

$$0 = \frac{2.3Q}{4\pi T} \log \frac{2.25 T t_0}{r^2 S} \qquad (7.45)$$

The above will be true if

$$\frac{2.25 T t_0}{r^2 S} = 1 \qquad (7.46)$$

or

$$S = \frac{2.25 T t_0}{r^2} \qquad (7.47)$$

T can be calculated by rewriting the Cooper–Jacob solution equation in the form

$$T = \frac{2.3Q}{4\pi \, \Delta s'} \qquad (7.48)$$

where $\Delta s'$ is the drawdown difference of data per log cycle of t, which can readily be obtained from the $\Delta s'$–$\log t$ plot. After plotting $\log t$–s', t_0 and $\Delta s'$ are determined. Substituting $\Delta s'$ in equation (7.48), T is calculated first. Then, substituting T and t_0 in equation (7.47), S is calculated.

To use the *Jacob distance-drawdown solution*, drawdowns at a particular time, from at least three observation wells located at different distances from the pumping well, are plotted on a drawdown (arithmetic) and distance (logarithmic) axis. A best-fit straight line is drawn. The slope of this line is proportional to transmissivity and pumping rate. Storativity can then be calculated as a function of transmissivity, time, and the value of the intercept at the point of zero drawdown using the following equations:

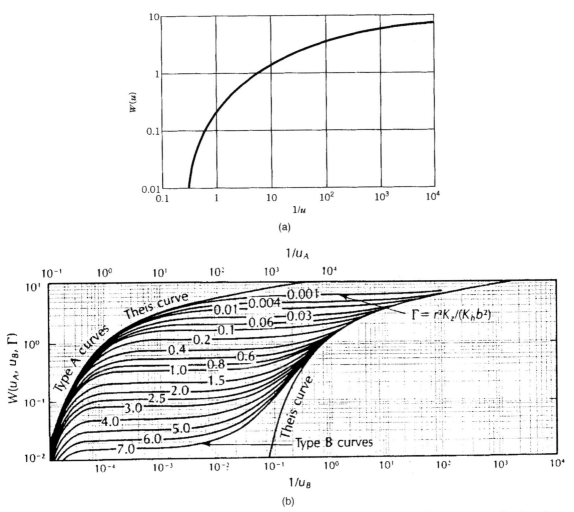

Figure 7.11 (a) *Values of* W(u) *plotted against values of 1/u for a confined aquifer;* (b) *type curves for drawdown data from fully penetrating wells in an unconfined aquifer.* [(a) *From Domenico and Schwartz, 1990. This material is used by permission of John Wiley & Sons, Inc.* (b) *From Neuman, 1974.*]

$$T = \frac{2.3Q}{4\pi \, \Delta s'} \qquad (7.49)$$

$$S = \frac{2.25Tt_0}{r_0^2} \qquad (7.50)$$

where Q is the pumping rate, t the time at which the drawdowns were measured, $\Delta s'$ the drawdown across one log cycle, and r_0 the radial distance from the pumping well to the point where there is zero drawdown.

Pumping Test in Unconfined Aquifer. Based on the formulation described in Section 7.7.2, two sets of type curves are generated as shown in Figure 7.11(b). Type

EXAMPLE 7.4

A fully penetrating well in a 30-ft-thick confined aquifer is pumped at a rate of 44.5 gal/min for 400 min. Drawdown measured at an observation well located 200 ft away is given below. Calculate the transmissivity, hydraulic conductivity, and storativity of the aquifer using (a) the Theis method and (b) the Cooper–Jacob method.

Elapsed Time (min)	Drawdown (ft)	Elapsed Time (min)	Drawdown (ft)
1	0.158	30	0.505
2	0.205	40	0.536
3	0.268	50	0.536
4	0.282	60	0.568
5	0.315	70	0.568
6	0.347	80	0.583
7	0.347	90	0.583
8	0.363	100	0.599
9	0.378	200	0.646
10	0.394	300	0.678
20	0.473	400	0.710

Solution:

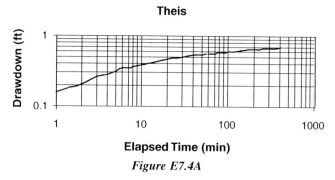

Theis

Figure E7.4A

(*a*) The curve-matching graphical procedure is used. First, plot the time versus drawdown data (Figure E7.4A) on logarithmic paper of the same scale as the type curve [Figure 7.11(*a*)]. The field curve is superimposed on the type curve, with the coordinate axes of the two curves kept parallel while matching the field data to the type curve. Any point on the sheets is selected arbitrarily (the point does not have to be on the matched curves). For the point selected, find $W(u)$ and s and $1/u$ and t. By doing this, for this problem, the following were obtained (matching graph not shown): $W(u) = 4$; $1/u = 100$; $t = 240$ s; $s' = 0.4$ ft. We are given $Q = 44.5(2.23 \times 10^{-3}) = 0.099$ ft^3/s (1 gal/min $= 2.23 \times 10^{-3}$ ft^3/s). Substituting in equation (7.34) yields

$$s' = \frac{Q}{4\pi T} W(u) \Rightarrow 0.4 = \frac{0.099}{4\pi T} 4 \Rightarrow T = 0.078 \text{ ft}^2/\text{s}$$

Using equation (7.35) gives us

$$u = \frac{r^2 S}{4Tt} \Rightarrow S = \frac{4Ttu}{r^2} = \frac{4(0.078)(240)(1/100)}{200^2} \Rightarrow S = 1.87 \times 10^{-5}$$

T and K are related by equation (7.8):

$$T = Kb \Rightarrow K = \frac{T}{b} \Rightarrow K = \frac{0.078}{30} \Rightarrow K = 2.6 \times 10^{-3} \text{ ft/s}$$

(b) Plot the logarithm of time versus drawdown (Figure E7.4B). Draw a best-fit straight line through the data. Based on this best-fit straight line, determine drawdown per log cycle ($\Delta s'$). Also, determine t_0, which is the time intercept where the drawdown line intercepts the zero drawdown axis. For these data; $\Delta s' = 0.2$ ft and $t_0 = 12$ s.

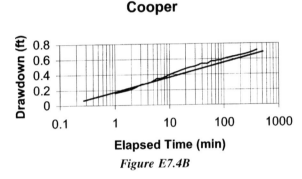

Cooper

Figure E7.4B

Using equations (7.48), (7.47), and (7.8) gives us

$$T = \frac{2.3Q}{4\pi \, \Delta s'} \Rightarrow T = \frac{2.3(0.099)}{4\pi(0.20)} \Rightarrow T = 0.09 \text{ ft}^2/\text{s}$$

$$S = \frac{2.25Tt_0}{r^2} \Rightarrow S = \frac{(2.25)(0.09)(12)}{(200)^2} \Rightarrow S = 6.07 \times 10^{-5}$$

$$T = Kb \Rightarrow K = \frac{T}{b} \Rightarrow K = \frac{0.09}{30} \Rightarrow K = 3.0 \times 10^{-3} \text{ ft/s}$$

A curves account for both instantaneous release of water from storage and gravity drainage. Type B curves account for late drawdown data when effects of gravity drainage are smaller. These type curves and values of $W(u_A, \Gamma)$ and $W(u_B, \Gamma)$ can be found in many standard references (Driscoll, 1986; Fetter, 1987; Walton, 1987; Dawson and Istok, 1991). These type curves and the drawdown versus time data from pumping test are used to determine hydraulic properties of aquifers following the procedure given below.

1. Time–drawdown data are plotted on logarithmic paper of the same scale as the type curve.

2. Superimpose the latest time–drawdown data on the type B curves. The axes of the graph papers should be parallel and the data matched to the

curve with the best fit. At any match point, the values of $W(u_B,\Gamma)$, u_B, t, and s are determined. The value of Γ comes from the type curve. The value of T is calculated using

$$T = \frac{Q}{4\pi s} W(u_A, u_B, \Gamma) \qquad (7.51)$$

The specific yield is then calculated using

$$u_B = \frac{r^2 S_y}{4Tt} \qquad (7.52)$$

3. The early drawdown data are then superimposed on the type A curve for Γ value of the previously matched type B curve. A new set of match points is determined. The value of T calculated in step 2 should be approximately equal to that computed from the type B curve. Then calculate storativity using

$$u_A = \frac{r^2 S}{4Tt} \qquad (7.53)$$

4. The value of the horizontal hydraulic conductivity is calculated using

$$K_h = \frac{T}{b} \qquad (7.54)$$

5. The value of the vertical hydraulic conductivity can also be calculated using

$$K_V = \frac{\Gamma b^2 K_h}{r^2} \qquad (7.55)$$

The procedure above assumes that the drawdown is small as compared to the saturated thickness of the aquifer. If the drawdown is significant, the same procedure may be used using the corrected drawdown values, which are calculated using

$$s' = s - \frac{s^2}{2b} \qquad (7.56)$$

7.8.2 Slug Tests

Slug tests are performed routinely to determine hydraulic conductivity of geologic formations. Slug tests are applicable to a wide range of geologic settings as well as small-diameter piezometers or wells and in areas of low permeability where it would be difficult to conduct a pumping test. Slug testing basically involves injecting or withdrawing a known volume of water from a well and measuring the aquifer's response by the rate at which the water level returns to equilibrium. A detailed slug testing procedure is provided in Chapter 10.

There are three analytical methods that are commonly used to calculate hydraulic conductivity based on the slug test data: (1) Hvorslev method, (2) Bouwer and Rice method, and (3) Cooper, Bredehoeft, and Papadopulos method. These three methods are described briefly in this section.

Hvorslev Method. Hvorslev (1951) assumed the geometry of the test conditions as shown in Figure 7.12. The method is applicable for confined isotropic or anisotropic aquifers, and the horizontal hydraulic conductivity (K_h) is given by the following equations:

$$K_h = \frac{\pi r_c^2}{Ft_L} \qquad (7.57)$$

where r_c is the effective casing radius and t_L is the time lag, defined as the time required for the initial excess head (H_0) to dissipate if the initial flow rate is maintained and is given by

$$t_L = \frac{t}{\ln(H_w/H_0)} \qquad (7.58)$$

where H_w is the excess head at time t. The F in equation (7.57), called the *shape factor,* is given by three empirical equations for three different well and aquifer geometries, called cases A, B, and C, as shown in Table 7.3. In these equations, a_k is a correction factor that accounts for anisotropy and is given by

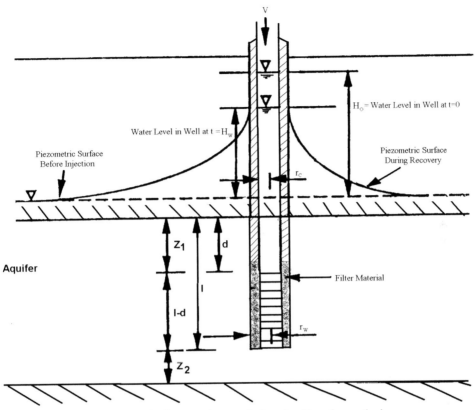

Figure 7.12 *Slug test data analysis using Hvorslev method.*

$$a_k = \sqrt{\frac{K_h}{K_v}} \qquad (7.59)$$

It should be noted that for case C, radius of influence R is required. Although R can be calculated using empirical equations, the Bouwer and Rice method, in which R is estimated based on electrical resistance analogy, as described in the next section, is more appropriate to use in such a case.

The Hvorslev slug test data analysis consists of the following steps:

1. Calculate the shape factor (F) that represents the test condition. For this calculation, a_k may be estimated based on the site geologic information; otherwise, assume that $a_k = 1$.

2. Plot $\ln(H_w/H_0)$ versus t. Draw a best-fit straight line.

3. Calculate time lag, (t_L), using one of the following methods:

   a. Inverse of the slope of the straight line fitted to the data, which can be expressed as

   $$\frac{1}{t_L} = \frac{\ln(H_w/H_0)_1 - \ln(H_w/H_0)_2}{t_1 - t_2} \qquad (7.60)$$

   b. t_L is the time to where $1/\ln(H_w/H_0) = 1$ or $(H_w/H_0) = 0.37$.

4. Calculate K_h using

   $$K_h = \frac{\pi r_c^2}{F t_L} \qquad (7.61)$$

Table 7.3 **Selected equations for the shape factor, F^a**

Case	Conditions of Development	Shape Factor
A Injection well	$z_1 = 0$, $z_2 = \infty$ Flow emanates from a line source for which the equipotential surface is semiellipsoids.	$F = \begin{cases} \dfrac{2\pi(1-d)}{\ln\left[a_k L_s/r_w + \sqrt{1 + (a_k L_s/r_w)^2}\right]} \\[2mm] \dfrac{2\pi L_s}{\ln(2a_k L_s/r_w)} \quad \text{for } \dfrac{a_k L_s}{r_w} > 4 \end{cases}$
B	$z_1 = \infty$, $z_2 = \infty$ Flow emanates from a line source. Flow lines are symmetrical with respect to a horizontal plane through the center of the intake.	$F = \begin{cases} \dfrac{2\pi L_s}{\ln\left[a_k L_s/2r_w + \sqrt{1 + (a_k L_s/2r_w)^2}\right]} \\[2mm] \dfrac{2\pi L_s}{\ln(2a_k L_s/r_w)} \quad \text{for } \dfrac{a_k L_s}{2r_w} > 4 \end{cases}$
C	$z_1 = 0$, $z_2 = 0$ Flow emanates from a line source and moves radially outward.	$F = \dfrac{2\pi L_s}{\ln(R/r_w)}$

Source: Dawson and Istok (1991).

[a] See Figure 7.12 for definitions of various terms.

5. Calculate K_v using

$$K_v = \frac{K_h}{a_k^2} \tag{7.62}$$

Bouwer and Rice Method. The Bouwer and Rice (1976) (see also Bouwer, 1989) method is used most commonly for analysis of slug test data. Although this method was originally developed for unconfined aquifers, as shown in Figure 7.13, it is also applicable for confined aquifers if the top of the well screen is located below the upper confining aquifer. This method is based on the following equation:

$$K = \frac{r_c^2 \ln(R/r_w)}{2L_s} \frac{1}{t} \ln \frac{H_0}{H_w} \tag{7.63}$$

Or using the definition of time lag (t_L) as used in the Hvorslev method, the governing equation can be rewritten as

$$K = \frac{r_c^2 \ln(R/r_w)}{2L_s} \frac{1}{t_L} \tag{7.64}$$

where $t_L = t/\ln(H_w/H_0)$. Bouwer and Rice determined the radius of influence R for different values of r_w, L_s, H_w, and m using measurements made with an electrical resistance analogy model and developed the following empirical equation for R:

$$\ln \frac{R}{r_w} = \left\{ \frac{1.1}{\ln(l/r_w)} + \frac{A + B \ln[(m-l)/r_w]}{L_s r_w} \right\}^{-1} \tag{7.65}$$

EXAMPLE 7.5

A slug test was performed in a piezometer screened in a confined silty sand aquifer. The well and aquifer geometry is shown in Figure E7.5A. When a slug was introduced into the well, the water level was raised by 0.61 ft. The time versus drawdown data obtained from slug testing is given below. Assume that $k_h/k_v = 50$ and calculate the hydraulic conductivity of the aquifer.

Elapsed Time (min)	Drawdown (ft)	Elapsed Time (min)	Drawdown (ft)
1	0.48	9	0.19
2	0.44	10	0.19
3	0.4	15	0.15
4	0.38	20	0.11
5	0.34	25	0.07
6	0.29	30	0.02
7	0.25	35	0
8	0.24		

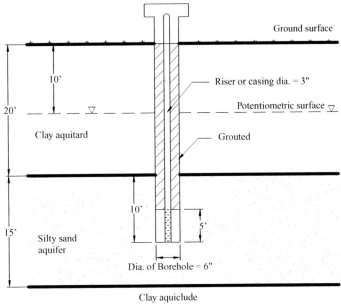

Figure E7.5A

Solution:

Given data: $H_0 = 0.61$ ft, $k_h/k_v = 50$, $r_w = 3$ in., $r_c = 1.5$ in., and $L_s = 5$ ft. To solve for hydraulic conductivity, the time lag, t_L, and an appropriate shape factor, F, must be determined. From Figure E7.5B, the time lag can be determined from the slope of the fitted (dashed) line.

Elapsed Time (min)	Drawdown (ft)	H_w/H_0	Elapsed Time (min)	Drawdown (ft)	H_w/H_0
1	0.48	0.79	9	0.19	0.31
2	0.44	0.72	10	0.19	0.31
3	0.4	0.66	15	0.15	0.25
4	0.38	0.62	20	0.11	0.18
5	0.34	0.56	25	0.07	0.11
6	0.29	0.48	30	0.02	0.03
7	0.25	0.41	35	0	0.00
8	0.24	0.39			

$$\frac{1}{t_L} = \frac{\ln(H_w/H_0)_1 - \ln(H_w/H_0)_2}{t_1 - t_2} = \frac{\ln(0.2) - \ln(0.35)}{15 - 10} = 0.11$$

Therefore, $t_L = 8.93$ min.

$$a_k = \sqrt{\frac{k_h}{k_v}} = \sqrt{50} = 7.07$$

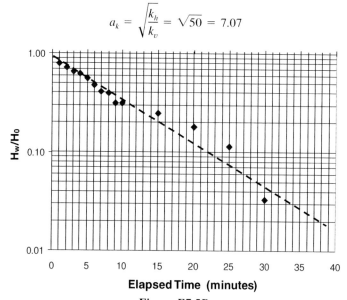

Figure E7.5B

Consider case B for shape factor F. Check

$$\frac{a_k L_s}{2r_w} = \frac{7.07 \times 5}{2 \times 0.25} = 70.7 > 4$$

Therefore,

$$F = \frac{2\pi L_s}{\ln(a_k L_s / r_w)} = \frac{2(3.14)5}{\ln(7.07 \times 5)/0.25} = 6.34$$

$$K_h = \frac{\pi r_c^2}{F t_L} = \frac{3.14(0.125)^2}{6.34(8.93)} = 8.67 \times 10^{-4} \text{ ft/min} = 4.4 \times 10^{-4} \text{ cm/s}$$

$$K_v = 4.4 \times 10^{-4} \times 50 = 2.2 \times 10^{-2} \text{ cm/s}$$

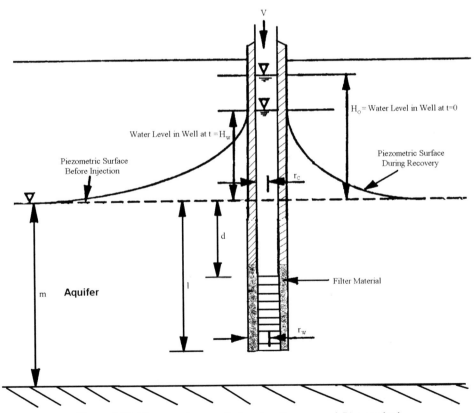

Figure 7.13 *Slug test data analysis using Bouwer and Rice method.*

where A and B are dimensionless coefficients that are functions of L_s/r_w, as shown in Figure 7.14. If the well fully penetrates the aquifer ($l = m$), equation (7.65) cannot be used and the following equation is employed:

$$\ln \frac{R}{r_w} = \left\{ \frac{1.1}{\ln(l/r_w)} + \frac{C}{L_s/r_w} \right\} \qquad (7.66)$$

where C is a dimensionless coefficient, as shown in Figure 7.14.

To determine K from the slug test data, the following steps may be followed:

1. Plot $\ln(H_w/H_0)$ versus t. Draw the best-fit straight line through the data points.

2. Determine t_L using either the slope of the best-fit straight line or determining t where $H_w/H_0 = 0.37$ (this is similar to Hvorslev procedure).

3. Calculate R using the empirical equations as given above.

4. Calculate K using

$$K = \frac{r_c^2 \ln(R/r_w)}{2L_s t_L} \qquad (7.67)$$

Cooper, Bredehoeft, and Papadopulos Method. Cooper et al. (1967) and Papadopoulous et al. (1973) developed a slug test analysis method applicable for confined aquifers, as shown in Figure 7.15. This method is similar to the Theis method for pumping test analysis, in that a curve-matching procedure is used to obtain the transmissivity (T) and storativity (S) of the aquifer. A family of type curves H_w/H_0 versus Tt/r_c^2, as shown in Figure 7.16, was developed. To calculate T and S, the following step-by-step procedure is used:

1. Plot $\ln(H_w/H_0)$ versus t to the same scale as the type curves.

2. Match the data with type curves by horizontal translation until the best match is achieved. Determine the α value of the best-fit curve. The value of S can then be calculated using

$$S = \alpha \frac{r_c^2}{r_s^2} \qquad (7.68)$$

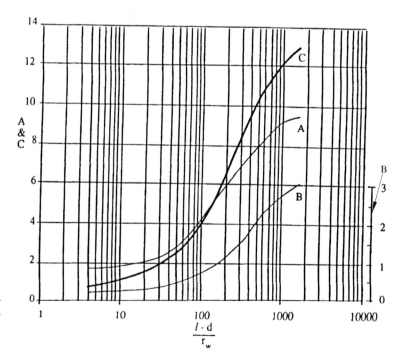

Figure 7.14 Values of the coefficients A, B, and C for use in estimating the radius of influence, R. (From Bouwer and Rice, 1976.)

3. The horizontal time axis, t, which overlays the horizontal axis for $\beta = Tt/r_c^2 = 1.0$ is selected and a value of T can then be calculated:

$$T = 1.0 \frac{r_c^2}{t} \qquad (7.69)$$

7.9 TWO- AND THREE-DIMENSIONAL GROUNDWATER FLOW

Two-dimensional steady flow problems can be solved either by graphical construction of groundwater flow nets or by using groundwater flow models, while two-dimensional unsteady flow problems as well as three-

EXAMPLE 7.6

A slug test was performed in an unconfined aquifer with a piezometer. The geometry of the piezometer and the aquifer is shown in Figure E7.6A. When a slug was introduced, the water level rose by 0.7 ft. The drawdown–time data are given below and in Figure E7.6B. Calculate the hydraulic conductivity of aquifer using the Bouwer and Rice Method.

Elapsed Time (s)	Drawdown (ft)
20	0.64
45	0.58
75	0.37
101	0.31
138	0.24
164	0.20
199	0.18

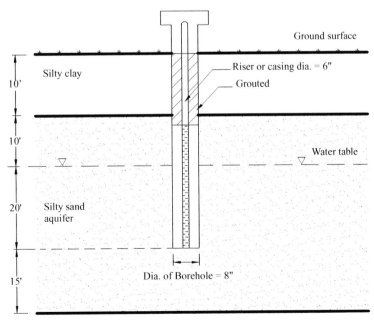

Figure E7.6A

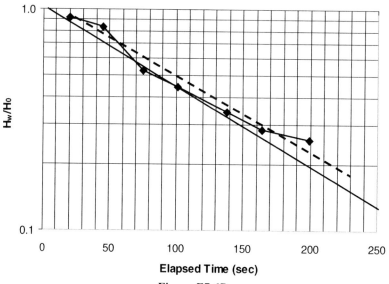

Figure E7.6B

Solution:

Given data: r_w = 4 in., r_c = 3 in., m = 20 + 15 = 35 ft, L_s = 20 ft, l = 20 ft, and H_0 = 0.7 ft.

Elapsed Time (s)	Drawdown (ft)	H_w/H_0
20	0.64	0.91
45	0.58	0.83
75	0.37	0.53
101	0.31	0.44
138	0.24	0.34
164	0.2	0.29
199	0.18	0.26

$$\frac{L_s}{r_w} = \frac{20}{0.25} = 60 \rightarrow c = 2.9$$

$$\ln \frac{R}{r_w} = \frac{1.1}{\ln(l/r_w)} + \frac{c}{L_s/r_w} = \frac{1.1}{\ln(60)} + \frac{2.9}{60} = 0.317$$

$$\frac{R}{r_w} = 1.373 \Rightarrow R = 1.373 \times 0.25 = 0.458 \text{ ft}$$

$$K = \frac{r_c^2 \ln(R/r_w)}{2 \times L_s \times t_L}$$

From the graph, t_L = 0.0076. Therefore,

$$K = \frac{(0.23)^2 0.317}{2(20)(0.0076)} = 0.0652 \text{ ft/s} = 1.98 \text{ cm/s}$$

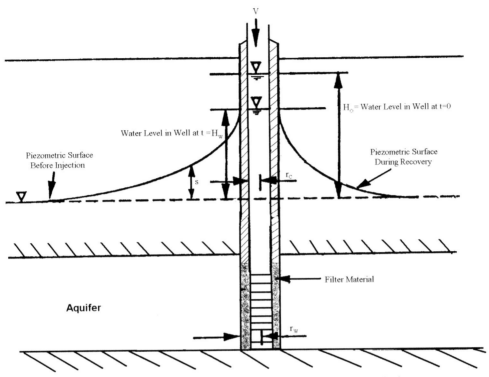

Figure 7.15 Slug test data analysis using Cooper et al. method.

dimensional steady and unsteady flow problems are addressed through groundwater flow models. Groundwater flow nets followed by groundwater flow modeling are described in this section.

7.9.1 Groundwater Flow Nets

Flow nets have been used extensively in geotechnical engineering practice to determine seepage in earth structures (e.g., earth dams, levees). A flow net consists of two sets of lines, flow lines and equipotential lines (Cedergren, 1997). A *flow line* is an imaginary line that traces the path that a particle of groundwater would follow as it flows through an aquifer. An *equipotential line* represents locations of equal potentiometric head. A flow net in an isotropic aquifer is a family of equipotential lines with sufficient orthogonal flow lines drawn so that a pattern of square elements results.

More detailed explanation and construction of flow nets can be found in Cedergren (1997).

In groundwater investigations, water level or potentiometric surface contour maps are produced to study the direction in which groundwater is flowing and provide an estimate of the hydraulic gradient, which controls velocity. Figure 7.17 shows a typical potentiometric surface map. A potentiometric surface map can then be used to develop a flow net by constructing flow lines that intersect the equipotential lines at right angles. Darcy's law can be used to estimate the quantity of water flowing through any "square" element using:

$$q = Kbw \frac{dh}{dl} \qquad (7.70)$$

and the total flow through any set of "squares" is given by:

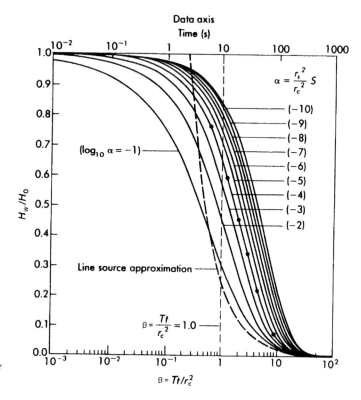

Figure 7.16 Cooper et al. slug test type curves. (From Papadopulos et al., 1973.)

$$Q = nq \qquad (7.71)$$

where K is the hydraulic conductivity, b the aquifer thickness at the midpoint between equipotential lines, w the distance between flow lines, dh the difference in head between equipotential lines, dl the distance between equipotential lines, and n the number of squares through which the flow occurs.

Figure 7.18 shows a flow net in both plan view and cross section for an area underlain by an unconfined sand aquifer. The aquifer overlies a horizontal confining layer. As shown in Figure 7.18(a), flow lines originate in the central part of the interstream divide (recharge area) and terminate at the streams (discharge line). As shown in Figure 7.18(b), the curved flow lines illustrate that the groundwater is flowing in the same direction but not in the same manner as implied from the plan view.

Groundwater flows not only through aquifers, but across confining units as well. Owing to the great differences in hydraulic conductivity between aquifers and confining units, most of the flow occurs through aquifers where the head loss per unit of distance is far less than in a confining unit. As a result, flow lines tend to parallel aquifer boundaries; they are less dense and trend nearly perpendicular through confining units (Figure 7.19). Consequently, lateral flow in units of low hydraulic conductivity is small compared to aquifers, but vertical leakage through them can be significant. Where an aquifer flow line intersects a confining unit, the flow line is refracted to produce the shortest path. The degree of refraction is proportional to the differences in hydraulic conductivity. Cedergren

EXAMPLE 7.7

A slug test was performed in a fully penetrating well screened in a 10-ft thick confined aquifer. The well consisted of a casing and screen radius of 4 in. and a well radius of 8 in. When the slug was introduced into the well, the water level rose by 4.76 ft. The slug test data are given below and in Figure E7.7. Calculate the T, K, and S values of the aquifer using the Cooper et al. method.

Elapsed Time (min)	Drawdown (ft)	Elapsed Time (min)	Drawdown (ft)
0.5	4.2	7	3.11
1	4.05	8	2.93
2	3.9	10	2.59
3	3.76	15	2.02
4	3.53	20	1.64
5	3.4	30	1.28
6	3.27	40	1.1

Solution:

Given data: $H_0 = 4.76$ ft, $r_w = 4$ in., $r_c = 2$ in., and $m = 10$ ft.

Elapsed Time (min)	Drawdown (ft)	H_w/H_0	Elapsed Time (min)	Drawdown (ft)	H_w/H_0
0.5	4.2	0.88	7	3.11	0.65
1	4.05	0.85	8	2.93	0.62
2	3.9	0.82	10	2.59	0.54
3	3.76	0.79	15	2.02	0.42
4	3.53	0.74	20	1.64	0.34
5	3.4	0.71	30	1.28	0.27
6	3.27	0.69	40	1.1	0.23

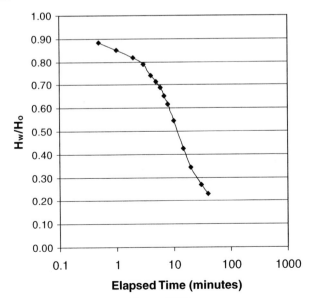

Figure E7.7

151

The data above, are plotted on a semilogarithmic scale and overlaid on a family of type curves (Figure 7.16). Based on this, the shape of the data curve mostly matches the $\alpha = 10^{-3}$ type curve. With the curves aligned, the following match-point values were selected: $\beta = 1$, $t = 10$ min, and $\alpha = 1 \times 10^{-3}$. Therefore,

$$S = \alpha \frac{r_c^2}{r_w^2} = 10^{-3} \times \frac{2^2}{4^2} = 2.5 \times 10^{-4} \text{ in}^2/\text{min} = 2.69 \times 10^{-5} \text{ cm}^2/\text{s}$$

$$T = \frac{\beta r_c^2}{t} = \frac{1(2^2)}{10} = 0.4 \text{ in}^2/\text{min} = 0.043 \text{ cm}^2/\text{s}$$

$$K = \frac{T}{m} = \frac{0.4}{10} = 0.04 \text{ in.}/\text{min} = 1.69 \times 10^{-3} \text{ cm/s}$$

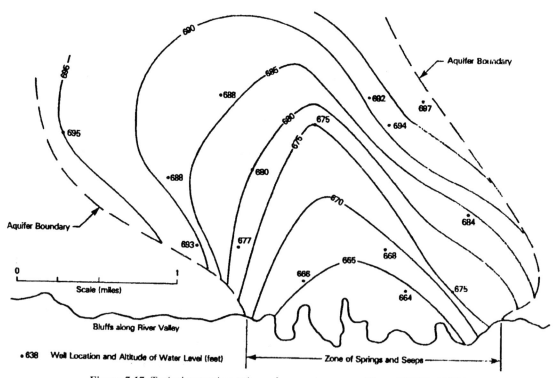

Figure 7.17 Typical potentiometric surface contour map. (*From USEPA, 1991.*)

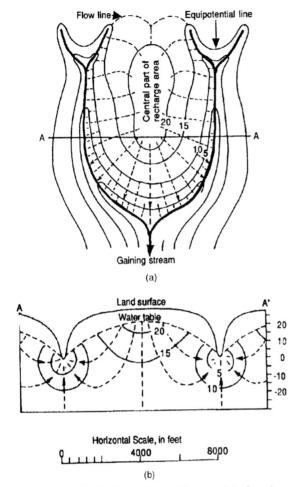

Flow line ➤
Equipotential line
Central part of recharge area
A — A
20 15
10 5
Gaining stream

(a)

Land surface
A'
Water table
20
20
10
15
0
5
-10
10
-20

Horizontal Scale, in feet
0   4000   8000

(b)

Figure 7.18 *Typical groundwater flow net:* (a) *plan view;* (b) *cross-sectional view.* (*From USEPA, 1991.*)

(1997) provides detailed information on construction of flow nets under heterogeneous soil conditions.

7.9.2 Groundwater Flow Modeling

Mathematical Models. Two- and three-dimensional flow problems are often solved by using a groundwater flow model. A groundwater flow model is an application of a mathematical model to represent a site-specific flow system. A groundwater model can

be as simple as a construction of saturated sand packed in a glass container or as complex as a three-dimensional mathematical representation requiring solving a large number of equations using a computer. The governing equations for groundwater flow modeling are explained in Section 7.5. The different methods that can be used to solve the governing flow equation can be categorized as analytical methods and numerical methods.

The *analytical methods,* also called *closed-form solutions,* use classical mathematical approaches to solve differential equations into exact solutions. The solution is continuous in space and time. The analytical methods provide exact solutions but employ many simplifying assumptions. One must review these assumptions before using these models for any groundwater investigation. Analytical solutions require assumption of homogeneity and are limited to one- and two-dimensional problems. For instance, the Theis equation can be used to estimate long-term drawdown resulting from pumping in a confined aquifer and is an analytical method.

The *numerical methods* are capable of addressing more complicated problems, do not require simplifying assumptions, and provide approximate solutions. However, these methods require significantly more data. These methods are most frequently based on one of the following numerical solution techniques: finite-difference methods and finite element methods.

The *finite difference method* approximates the solution of partial differential equations by using finite difference equivalents, whereas the *finite element method* approximates differential equations by an integral approach. The finite difference methods solve the partial difference equations describing the system by using algebraic equations to approximate the solution at discrete points in a rectangular grid. The grid can be one-, two-, or three-dimensional. The points in the grid, called *nodes,* represent the average for the surrounding rectangular block, called a *cell.* The finite element methods differ from finite difference methods in that the area (or volume) between adjacent nodes forms an element over which exact solution values are defined everywhere by means of basis functions. The finite element solution requires a more labor-intensive grid setup than that of a finite difference solution. Fig-

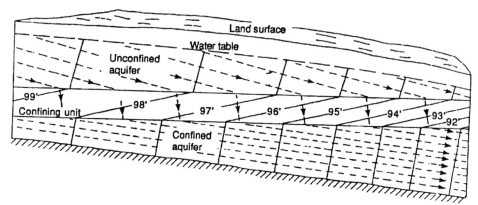

Figure 7.19 *Typical groundwater flow net for aquifer and confining layer system. (From USEPA, 1991.)*

ure 7.20 illustrates the mathematical and computational differences in the two approaches. Table 7.4 compares the relative advantages and disadvantages of the two methods. In general, finite difference methods are best suited for relatively simple hydrogeologic settings, whereas finite element methods are required where hydrogeology is complex.

Several numerical models have been reported for groundwater flow simulations (USEPA, 1988). MODFLOW, a finite difference solution method (McDonald and Harbaugh, 1988), is the most commonly used model and is described further below.

MODFLOW   Overview. MODFLOW (Modular Three-Dimensional Finite-Difference Groundwater Flow Model) is a finite difference model with modular structure that simulates flow in three dimensions in saturated porous media. The model is developed by the U.S. Geological Survey (McDonald and Harbaugh, 1984). The model is based on the following governing equation, which describes three-dimensional movement of groundwater of constant density through porous media:

$$\frac{\partial}{\partial x}\left(K_x \frac{\partial h}{\partial x}\right) + \frac{\partial}{\partial y}\left(K_y \frac{\partial h}{\partial y}\right) + \frac{\partial}{\partial z}\left(K_z \frac{\partial h}{\partial z}\right) - W$$

$$= S_s \frac{\partial h}{\partial t} \qquad (7.72)$$

where K_x, K_y, and K_z represent hydraulic conductivity along the x, y, and z coordinates axes, which are assumed to be parallel to the major axes of hydraulic conductivity; h is the potentiometric head; W a volumetric flux per unit volume and represents sources and/or sinks of water; S_s the specific storage of the porous material; and t the time. In general, S_s, K_x, K_y, and K_z may be functions of space and time and W may be a function of space and time and equation (7.72) can describe groundwater flow under nonequilibrium conditions in a heterogeneous and anisotropic medium. The governing equation, together with specification of flow and/or head conditions at the boundaries of an aquifer system and specification of initial head conditions, constitutes a mathematical representation of a groundwater flow system. The governing equation is solved by the finite difference method in order to determine the head distribution in space with time, $h(x,y,z,t)$.

The modular structure of MODFLOW consists of a main program and a series of highly independent subroutines called *modules*. The modules are grouped into *packages*. Each package deals with a specific feature of the hydrologic system that is to be simulated. Table 7.5 shows the packages available in MODFLOW. More packages are being added continuously. The modular structure allows examination of specific hydrologic features of the model independently and also facilitates development of additional capabilities because new

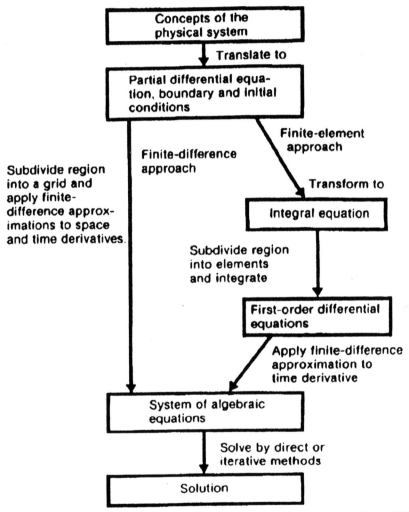

Figure 7.20 *Model development using finite difference and finite element methods. (From USEPA, 1991.)*

Table 7.4 **Advantages and disadvantages of numerical methods**

Method	Advantages	Disadvantages
Finite difference method	Intuitive basis Easy data entry Efficient matrix techniques Programming changes easy	Low accuracy for some problems Regular grids required
Finite element method	Flexible grid geometry High accuracy possible Evaluates cross-product terms better	Complex mathematical basis Difficult data input Difficult programming

Source: USEPA (1991).

Table 7.5 **Selected packages in MODFLOW**

Package Name	Abbreviation	Package Description
Basic	BAS	Handles those tasks that are part of the model as a whole. Among those tasks are specification of boundaries, determination of time-step length, establishment of initial conditions, and printing of results.

Flow Component Packages

Stress packages		
Block-Centered Flow	BCF	Calculates terms of finite difference equations that represent flow within porous medium; specifically, flow from cell to cell and flow into storage.
Well	WEL	Adds terms representing areally distributed recharge to the finite difference equations.
Recharge	RCH	Adds terms representing flow to wells to the finite difference equations
River	RIV	Adds terms representing flow to rivers to the finite difference equations.
Drain	DRN	Adds terms representing flow to drains to the finite difference equations.
Evapotranspiration	EVT	Adds terms representing ET to the finite difference equations.
General-Head Boundaries	GHB	Adds terms representing general-head boundaries to the finite difference equations.
Solver packages		
Strongly Implicit Procedure	SIP	Iteratively solves the system of finite difference equations.
Slice-Successive Overrelaxation	SOR	Iteratively solves the system of finite difference equations.

Source: McDonald and Harbaugh (1988).

packages can be added to the program without modifying the existing packages.

Figure 7.21 shows a spatial discretization of an aquifer system with a mesh of blocks called *cells,* the locations of which are described in terms of rows, columns, and layers using an (i,j,k) indexing system. Layers correspond to the geologic units and can be simulated as confined, unconfined, or a combination of confined and unconfined. For unconfined or a combination of unconfined and confined aquifer conditions, the elevations of the top and bottom of the aquifer and the hydraulic properties are required as input. For a confined aquifer, thickness is not required; rather, transmissivity values are required. Confining layers are generally not included in discretization, but their effect is included through the use of conductance terms be-

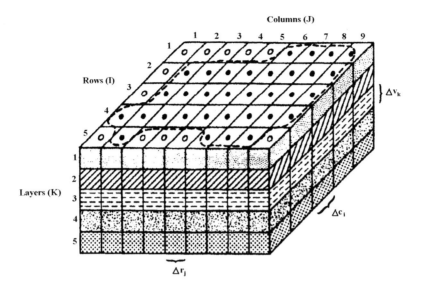

Columns (J)

Rows (I)

Layers (K)

Explanation

– – – – Aquifer Boundary

● Active Cell

○ Inactive Cell

Δr_J Dimension of Cell Along the Row Direction. Subscript (J) Indicates the Number of the Column

Δc_I Dimension of Cell Along the Column Direction. Subscript (I) Indicates the Number of the Row

Δv_K Dimension of the Cell Along the Vertical Direction. Subscript (K) Indicates the Number of the Layer

Figure 7.21 Discretized hypothetical aquifer system. (*From McDonald and Harbaugh, 1988.*)

tween the aquifers. Simulation is divided into stress periods, which are in turn divided into time steps. A *stress period* is a time during which aquifer stresses such as pumping and recharge rates remain constant.

The model assumes that the flow in aquifers is horizontal and the flow through confining units is vertical. The model can be used to simulate two-dimensional flow in a single layer. Within each cell, there is a point called a *node* (at the center of the cell if block-centered grid is used) at which the head is to be calculated. A set of finite difference equations is formulated at each time step, and these equations are solved using iterative methods to determine the heads. Constant-head and no-flow cells are used in the model to represent con-

ditions along various hydrologic boundaries. Other boundary conditions, such as areas of constant inflow or areas where inflow varies with head, can be simulated through the use of external source terms or through a combination of no-flow cells and external source terms. Flow associated with external stresses, such as wells, areal recharge, evapotranspiration, drains, and streams, can also be simulated. The finite difference equations can be solved using iterative methods such as the strongly implicit procedure, slice-successive overrelaxation, or preconditioned gradient.

MODFLOW is written in the Fortran 77 programming language. The model may be used for either two- or three-dimensional applications. MODFLOW is simple to use, can be modified readily, can be executed on a variety of computers with minimal changes, and is relatively efficient with respect to computer memory and execution time. It can be complemented easily with transport models. All of these features are responsible for the widespread use of the model for groundwater flow systems.

Modeling Process. Irrespective of the type of mathematical model, the groundwater flow modeling requires a systematic approach that typically involves the following tasks (USEPA, 1988):

- Determination of modeling objectives
- Data gathering and organization
- Development of a conceptual model
- Numerical code selection
- Assignment of properties and boundary conditions
- Calibration and sensitivity analysis
- Model execution and interpretation of results
- Report preparation

More details on each of these tasks can be found in Mercer and Faust (1981), Wang and Anderson (1982),

USEPA (1988, 1992, 1993), and Anderson and Woessner (1992), The accuracy of a model is dependent on the level of understanding of the system the model is to represent. Thus, a complete site investigation and accurate conceptualization of the site hydrogeology are necessary precursors to a successful modeling.

7.10   SUMMARY

Protection of groundwater from contamination is of utmost importance to protect public health and the environment. Groundwater investigations are needed in remediating the sites that have already been contaminated. These investigations are also needed in assessing the potential for causing groundwater contamination by newly planned landfills and impoundments. Groundwater flow velocity can be determined by knowing the hydraulic heads and hydraulic properties of aquifers. In-situ aquifer tests such as pumping tests and slug tests provide the hydraulic properties of aquifers. Simple analytical models to comprehensive numerical models are available to simulate groundwater flow conditions at project sites. A systematic selection and use of mathematical models is essential to better define groundwater flow conditions at any site.

QUESTIONS/PROBLEMS

7.1. Calculate the groundwater flow rate in the confined aquifer as shown in Figure P7.1. Also calculate the height of the potentiometric surface 50 ft from each well in the groundwater flow direction.

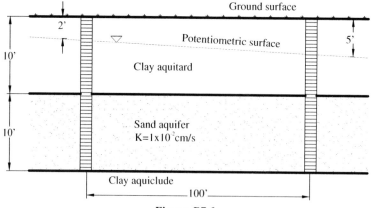

Figure P7.1

7.2. Calculate the groundwater flow rate in the unconfined aquifer as shown in Figure P7.2. Also calculate the height of the water table midway between the piezometers.

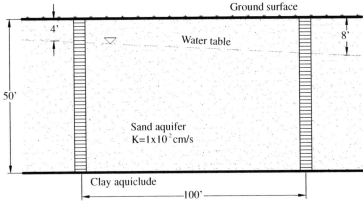

Figure P7.2

7.3. Calculate the groundwater flow rate between the drainage ditch and the river as shown in Figure P7.3.

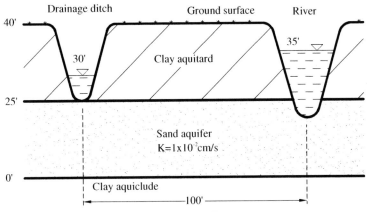

Figure P7.3

7.4. Calculate the groundwater flow between the drainage ditch and the river for the conditions shown in Figure P7.4. If a monitoring well is placed at the interface between the sand and the silty sand aquifers, what would be the potentiometric surface height?

7.5. Calculate the groundwater flow between the drainage ditch and the river for the conditions shown in Figure P7.5.

7.6. Calculate the groundwater flow between the pond and the confined gravel aquifer as shown in Figure P7.6. A steady vertical flow condition exists across the soil formations. Calculate the hydraulic head at the top and bottom of the silt layer.

7.7. A 12-in.-diameter pumping well was constructed in an unconfined aquifer. The water table is located 50 ft below the ground surface. Two observation wells, OW-1 and OW-2, were constructed 50 and 200 ft from the pumping well location. When the pumping well is pumped at 50 gal/min, the drawdowns

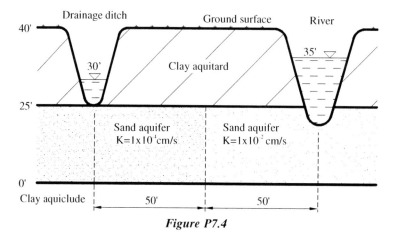

Figure P7.4

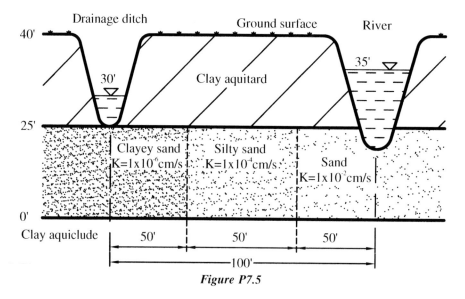

Figure P7.5

observed in the observation wells were 10 ft and 4 ft in wells OW-1 and OW-2, respectively. Calculate the hydraulic conductivity of the aquifer.

7.8. A pumping well (PW) with a diameter of 18 in. was constructed in an unconfined aquifer that is 100 ft thick and has a hydraulic conductivity of 1000 gal/day per square foot. The water table is located 80 ft below the ground surface. Two observation wells, OW-1 and OW-2, were located at 100 ft and 200 ft, respectively, from the PW. As a result of steady pumping, the measured drawdowns in OW-1 and OW-2 were 15 ft and 10 ft, respectively. What is the pumping rate?

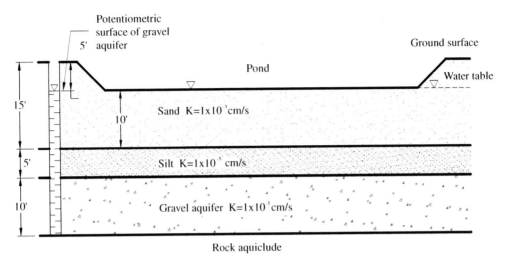

Figure P7.6

7.9. A pumping well was constructed in a confined aquifer that is 30 ft thick. Two observation wells, OW-1 and OW-2, were constructed at distances of 50 ft and 500 ft, respectively, from the pumping well. Water is pumped from the pumping well at a steady rate of 1000 gal/day. Under steady-state conditions, the measured drawdown was 20 ft in OW-1 and 5 ft in OW-2. Determine the hydraulic conductivity and transmissivity of the aquifer.

7.10. A 18-in.-diameter well was used to pump water from a confined aquifer at 100 gal/min. The drawdowns recorded at observation wells located at 50 and 200 ft from the pumping well were 10 and 8 ft, respectively. Calculate the drawdown at the pumping well.

7.11. A confined aquifer has the transmissivity of 100,000 gal/day per foot and a storativity of 0.00005. Determine the following:

(a) The drawdown at 1000 ft from a pumping well if the well is pumped at 100 gal/min for 30 days.

(b) The maximum pumping rate at which a well could be pumped for 45 days so that drawdown at a distance of 1000 ft from the pumping well would not exceed 5 ft.

(c) The time required to induce 2 ft drawdown at a distance of 1000 feet from a pumping well discharging at 125 gal/min.

(d) The distance after one day of pumping at which drawdown would equal 1 foot if the well produced 80 gal/min.

7.12. A 12-in.-diameter pumping well was constructed in a confined aquifer. The aquifer transmissivity is 1×10^{-3} m^2/s, and the storativity is 0.0005. Determine the following:

(a) The drawdown at 50 m from the pumping well 30 min and 24 h after the beginning of pumping with the rate of 10 L/s.

(b) The time after which the drawdown at 50 m from the pumping well would be 2 m if the pumping rate was 5 L/s.

(c) The pumping rate needed to produce 2 m of drawdown at 50 m from the pumping well after 10 days of pumping.

(d) The distance at which the drawdown is 2 m after one week of pumping at a rate of 5 L/s.

7.13. A confined aquifer has a transmissivity of 100 m^2/day and a storativity of 1×10^{-5}. A 12-in.-diameter well is used to pump continuously at 1000 m^3/day. Determine the following:

(a) The drawdown at distances of 10, 100, and 1000 m from the pumping well after one day of pumping.

(b) The time to reach steady state.

(c) The steady-state drawdown in the well.

7.14. Two wells are located 100 m apart in a confined aquifer with a transmissivity of 1×10^{-4} m^2/s and storativity of 1×10^{-5}. One well is pumped at a rate of 100 gal/day and the other at a rate of 200 gal/day. Find **(a)** the drawdown at the midway of the wells one day after the pumping starts, and **(b)** the time required to cause a total drawdown of 0.1 m at the midway of the wells.

7.15. A fully penetrating well in a 5-m-thick confined aquifer is pumped at a rate of 1000 L/min for a period of 24 h. Time–drawdown data for an observation well located 100 m away are given below. Calculate the values of T, K, and S of the aquifer using **(a)** the Theis method; **(b)** the Cooper–Jacob method.

Time (min)	Drawdown (ft)	Time (min)	Drawdown (ft)
1	0.80	70	2.68
2	1.04	80	2.76
3	1.20	90	2.83
4	1.32	100	2.89
5	1.42	200	3.31
6	1.50	300	3.57
7	1.57	400	3.79
8	1.63	500	3.935
9	1.68	600	4.11
10	1.73	700	4.2
20	2.05	800	4.34
30	2.24	900	4.44
40	2.38	1000	4.495
50	2.49	1440	4.805
60	2.59		

7.16. A fully penetrating well in a 30-ft-thick confined aquifer is pumped at a rate of 44.5 gal/min for 400 min. Drawdown measured at an observation well located 200 ft away is given below. Calculate the transmissivity, hydraulic conductivity, and storativity of the aquifer using **(a)** the Theis method; **(b)** the Cooper–Jacob method.

Elapsed Time (min)	Drawdown (ft)	Elapsed Time (min)	Drawdown (ft)
1	0.158	30	0.505
2	0.205	40	0.536
3	0.268	50	0.536
4	0.282	60	0.568
5	0.315	70	0.568

Elapsed Time (min)	Drawdown (ft)	Elapsed Time (min)	Drawdown (ft)
6	0.347	80	0.583
7	0.347	90	0.583
8	0.363	100	0.599
9	0.378	200	0.646
10	0.394	300	0.678
20	0.473	400	0.710

7.17. A slug test was performed in a piezometer installed in a silty clay layer. The piezometer and the soil conditions are shown in Figure P7.17. Water was added to the piezometer to cause the initial head of 35.57 ft. The drawdown versus time is given in the table below. Calculate the hydraulic conductivity of silty clay using the Hvorslev method.

Elapsed Time (min)	Drawdown (ft)	Elapsed Time (min)	Drawdown (ft)
0	35.57	71.83	33.12
0.5	35.42	95.68	32.68
1.0	35.33	106.76	32.48
2.48	35.02	119.93	32.24
4.44	34.61	125.60	32.14
11.30	34.39	130.67	32.06
23.66	34.10	143.97	31.84
34.46	33.88	153.64	31.65
47.60	33.61	167.71	31.41
59.26	33.39	192.0	31.01

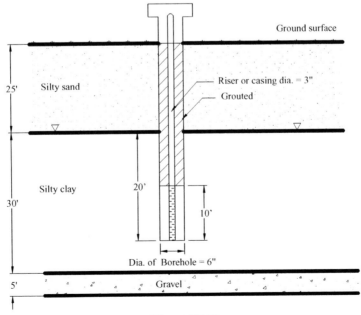

Figure P7.17

7.18. A 8-in.-diameter fully penetrating well was installed in a 10-ft-thick confined aquifer. A screen with a radius of 2 in. was installed in the casing. When a slug of water was injected, the water level rose by 1.25 ft. The measured drawdown data are given below. Using these data, calculate the values of T, K, and S for the aquifer using the Cooper et al. method.

Time (s)	Drawdown (ft)	Time (s)	Drawdown (ft)
0.2	1.18	3	0.60
0.3	1.15	4	0.46
0.4	1.12	5	0.30
0.5	1.00	10	0.18
1.0	0.94	20	0.05
2.0	0.75		

7.19. A slug test was performed in a monitoring well installed in an unconfined sand aquifer. The well geometry and aquifer conditions are shown in Figure P7.19. A slug of water was injected that raised the water level by 0.4 m. The change in water level with time is given below. Calculate the hydraulic conductivity of the aquifer using the Bouwer and Rice method.

Elapsed Time (min)	Drawdown (ft)
200	0.38
400	0.29
600	0.22
1000	0.15

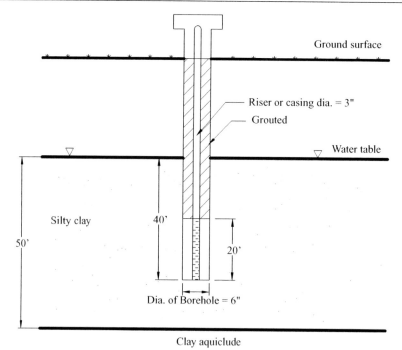

Figure P7.19

REFERENCES

Anderson, M. P., and Woessner, W. W., *Applied Groundwater Modeling: Simulation of Flow and Advective Transport,* Academic Press, San Diego, CA, 1992.

Bouwer, H., *Groundwater Hydrology,* McGraw-Hill, New York, 1978.

———, The Bouwer and Rice slug test: an update, *Ground Water,* Vol. 27, No. 3, pp. 304–309, 1989.

Bouwer, H., and Rice, R. C., A slug test for determining hydraulic conductivity of unconfined aquifers with completely or partially penetrating wells, *Water Resour. Res.,* Vol. 12, No. 3, pp. 423–428, 1976.

Cedergren, H. R., *Seepage, Drainage, and Flow Nets,* 3rd ed., Wiley, New York, 1997.

Cooper, H. H., and Jacob, C. E., A generalized graphical method for evaluating formation constants and summarizing well-field history, *Trans., Am. Geophys. Union,* Vol. 27, 1946.

Cooper, H. H., Bredehoeft, J. D., and Papadopulos, I. S., Response to a finite diameter well to an instantaneous charge of water, *Water Resour. Res.,* Vol. 3, pp. 263–269, 1967.

Dagan, G., Models of groundwater flow in statistically homogeneous porous formations, *Water Resour. Res.,* Vol. 15, No. 1, pp. 47–63, 1979.

Dawson, K. J., and Istok, J. D., *Aquifer Testing: Design and Analysis of Aquifer Tests,* Lewis Publishers, Chelsea, MI, 1991.

Domenico, P. A., and Schwartz, F. W., *Physical and Chemical Hydrogeology,* Wiley, New York, 1990.

Driscoll, F. G., *Groundwater and Wells,* 2nd ed., Johnson Screens, Wheelabrator Water Technologies, Inc., St. Paul, MN, 1986.

Fetter, C. W., *Applied Hydrogeology,* Charles E. Merrill, Columbus, OH, 1987.

Fredlund, D. G., and Rahardjo, H., *Soil Mechanics for Unsaturated Soils,* Wiley, New York, 1993.

Freeze, R. A., and Cherry, J. A., *Groundwater,* Prentice Hall, Upper Saddle River, NJ, 1979.

Gelhar, L. W., *Stochastic Subsurface Hydrology,* Prentice Hall, Upper Saddle River, NJ, 1993.

Gutjahr, A. L., Gelhar, L. W., Bakr, A., and MacMillan, J. R., Stochastic analysis of spatial variability in subsurface flows, *Water Resour. Res.,* Vol. 14, No. 5, pp. 953–959, 1978.

Guymon, G. L., *Unsaturated Zone Hydrology,* Prentice Hall, Upper Saddle River, NJ, 1994.

Heath, R. C., *Basic Ground-Water Hydrology,* U.S. Geological Survey Water-Supply Paper 2220, U.S. Government Printing Office, Washington, DC, 1989.

Hvorslev, M. J., *Time Lag and Soil Permeability in Ground Water Observations,* Bulletin 36, U.S. Army Corps of Engineers Waterways Experiment Station, Vicksburg, MS, 1951.

McDonald, M. G., and Harbaugh, A. W., *A Modular Three-Dimensional Finite Difference Ground-Water Flow Model,* U.S. Geological Survey Open-File Report 83-875, U.S. Government Printing Office, Washington, DC, 1988.

Mercer, J. W., and Faust, C. R., *Ground-Water Modeling,* National Water Well Association, Columbus, OH, 1981.

Neuman, S. P., Effect of partial penetration on flow in unconfined aquifers considering delayed gravity response, *Water Resour. Res.,* Vol. 10, No. 2, pp. 303–312, 1974.

Papadopulos, I. S., Bredehoeft, J. D., and Cooper, H. H., On the analysis of slug test data, *Water Resour. Res.,* Vol. 9, pp. 1087–1089, 1973.

Theis, C. V., The relationship between the lowering of the piezometric surface and the rate and duration of discharge of a well using ground-water storage, *Transactions of the American Geophysical Union 16th Annual Meeting,* Part I, pp. 519–524, 1935.

Tindall, J. A., and Kunkel, J. R., *Unsaturated Zone Hydrology for Scientists and Engineers,* Prentice Hall, Upper Saddle River, NJ, 1999.

Todd, D. K., *Ground Water Hydrology,* 2nd ed., Wiley, New York, 1980.

USACE (U.S. Army Corps of Engineers), *Groundwater Hydrology,* EM 1110-2-1421, U.S. Department of the Army, Washington, DC, 1999.

USBR (U.S. Bureau of Reclamation), *Groundwater Manual,* U.S. Government Printing Office, Washington, DC, 1977.

USEPA (U.S. Environmental Protection Agency), *Groundwater Modeling: An Overview and Status Report,* EPA/600/2-89/028, by Van der Heijde, P. K. M., Al-Kadi, A. I., and Williams, S. A., for the R. S. Kerr Environmental Research Laboratory, Office of Research and Development, USEPA, Ada, OK, 1988.

———, *Handbook: Groundwater,* Vol. I, *Groundwater and Contamination,* EPA/625/6-90/016a, Office of Research and Development, USEPA, Washington, DC, 1990.

———, *Handbook: Groundwater,* Vol. II, *Methodology,* EPA/625/6-90/016b, Office of Research and Development, USEPA, Washington, DC, 1991.

———, *Quality Assurance and Quality Control in the Development and Application of Ground-Water Models,* EPA/600/R-93/011, R. S. Kerr Environmental Research Laboratory, USEPA, Ada, OK, 1992.

———, *Compilation of Ground-Water Models,* EPA/600/R-93/118, R. S. Kerr Environmental Research Laboratory, USEPA, Ada, OK, 1993.

Walton, W. C., *Groundwater Pumping Tests: Design and Analysis,* Lewis Publishers, Chelsea, MI, 1987.

Wang, H. F., and Anderson, M. P., *Introduction to Groundwater Modeling: Finite Difference and Finite Element Methods,* W.H. Freeman, San Francisco, 1982.

8

CONTAMINANT TRANSPORT AND FATE

8.1 INTRODUCTION

In traditional geotechnical engineering, the flow of water through soils due to hydraulic gradient or pressure gradient is studied. In geoenvironmental engineering, one must also study the contaminant transport and fate. It is important to understand various processes that affect the transport (migration) and fate (chemical form and concentration) of the contaminants in soils and groundwater. The major processes affecting the contaminants in subsurface can be grouped in three categories:

1. *Transport processes:* include advection, diffusion, and dispersion
2. *Chemical mass transfer processes:* include sorption and desorption, dissolution and precipitation, oxidation and reduction, acid–base reactions, complexation, ion exchange, volatilization, and hydrolysis
3. *Biological process* (or biodegradation)

These various processes occur under different hydraulic and chemical conditions, which either exist naturally or are applied externally in the soil and groundwater systems (USEPA, 1989). Contaminant transport modeling incorporating these processes is performed to determine contaminant concentrations at exposure locations while performing the risk assessment (described in Chapter 11) and to evaluate and/or design effective waste containment systems (such as slurry walls and landfill liner systems, as detailed in Chapters 12 and 17) and site remediation systems (such as soil flushing, soil vapor extraction and bioremediation, as detailed in Chapters 13 and 14).

In this chapter we describe the major transport, chemical, and biological processes that affect the fate and transport of contaminants in subsurface. In addition, mathematical models that incorporate these processes and predict contaminant transport and fate in the subsurface are described. Finally, examples of application of contaminant transport modeling to address various geoenvironmental problems are presented.

8.2 TRANSPORT PROCESSES

In this section we present various contaminant transport processes: advection, diffusion, and dispersion. These processes control the extent of contaminant migration in the subsurface. These are the only processes that one must consider in dealing with the transport of nonreactive contaminants in the subsurface. Nonreactive contaminants are dissolved contaminants that are not influenced by chemical reactions or microbiological processes. For reactive contaminants, these transport processes are considered along with various chemical mass transfer and microbial degradation processes.

8.2.1 Advection

Advection, also known as *advective transport* or *convection,* refers to the contaminant movement by flowing water in response to a hydraulic gradient. As described in Chapter 7, the groundwater velocity for one-dimensional steady-state flow is given by Darcy's law as

$$V = Ki \qquad (8.1a)$$

where V is the discharge velocity or Darcy's velocity, K the hydraulic conductivity, and i the hydraulic gradient. The seepage or actual velocity is calculated using

$$V_s = \frac{Ki}{n} \qquad (8.1b)$$

where V_s is the seepage velocity and n is the porosity. To be precise, the average seepage or flow velocity is given by

$$V_s = \frac{Ki}{n_e} \qquad (8.1c)$$

where n_e is the effective porosity. *Effective porosity* is defined as the volume of void space that conducts most of the flow, divided by the total volume of the soil. The effective porosity excludes the noninterconnected and dead-end pores; thus, $n_e < n$.

Under one-dimensional steady-state flow condition, the contaminant mass flux (i.e., mass per unit area per unit time) due to advection (F_v) is given by

$$F_v = VC = n_e V_s C \qquad (8.2)$$

where C is the dissolved contaminant concentration. Under uniform flow conditions, the advective transport is described by

$$\frac{dC}{dt} = -V_s \frac{dC}{dx} \qquad (8.3)$$

The solution to this equation implies that if a contaminant of concentration C_0 is introduced in the flowing groundwater, then in time t it will be transported a distance $x = V_s t$ as a plug, due to advection. This implies a sharp concentration interface, as shown in Figure 8.1(a).

For two- and three-dimensional steady-state or unsteady-state flow conditions, the groundwater flow velocities can be determined by performing ground-water flow modeling (using models such as MODFLOW) as described in Chapter 7.

In general, the following porous media parameters must be known to perform advective transport analyses:

- *Hydraulic gradients.* The potentiometric surface or water table contour maps are used to calculate hydraulic gradients, as explained in Chapter 7.

- *Hydraulic conductivity and transmissivity.* The hydraulic conductivity of unconfined aquifers and the transmissivity of confined aquifers are required (as detailed in Chapter 7). These properties can be determined based on laboratory methods and field test methods. The *laboratory test methods* include a constant-head permeameter test, variable-head permeameter test, triaxial test, and consolidation test. The laboratory methods are discussed briefly in Chapters 5 and 17. The *field test methods* include pumping tests and slug tests and generally consist of creating a head difference by displacing water in wells and measuring the drawdown or recovery of the water levels with time and then using the data to calculate the K value using an appropriate analysis method. All of these field test methods are discussed in detail in Chapters 7 and 10.

- *Specific yield and storativity.* For transient flow simulations, specific yield for unconfined aquifers and storativity for confined aquifers are required (as described in Chapter 7). These properties are determined by performing pumping tests or slug tests, as described in Chapter 7.

- *Porosity and effective porosity.* Porosity and effective porosity are calculated based on laboratory testing and using the phase relationships presented in Chapter 5.

Typical hydraulic conductivity values of various geologic formations and typical values of porosity, effective porosity (same as specific yield), and storativity of different soils are given in Chapters 7 and 17.

8.2.2 Diffusion

Diffusion, also known as *molecular diffusion*, refers to the movement of contaminants under a chemical con-

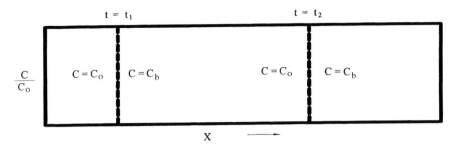

(a) ADVECTION ONLY (Co=INITIAL CONCENTRATION
Cb=BACKGROUND CONCENTRATION)

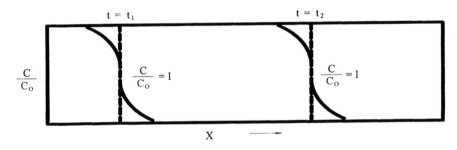

(b) DIFFUSION ONLY

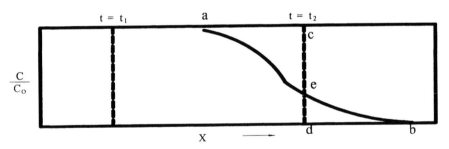

(c) ADVECTION AND DISPERSION

Figure 8.1 *Contaminant transport processes.*

centration gradient (i.e., from an area of greater concentration toward an area of lower concentration). Diffusion can occur even when the fluid is not flowing or is flowing in the direction opposite to contaminant movement. Diffusion will cease only if there is no concentration gradient. Diffusion is characterized using *Fick's first law.* For one-dimensional conditions, this law can be expressed as

$$F_d = -D^* \frac{dC}{dx} \qquad (8.4a)$$

where F_d is the diffusive mass flux per unit area per unit time, D^* the effective diffusion coefficient, and dC/dx the concentration gradient. The values of D^* do not vary significantly for different soils and contaminant combinations and they range from 1×10^{-9} to 2

EXAMPLE 8.1

At a landfill site, dissolved chloride in a concentration of 1000 mg/L is being transported along with the leachate through a 5-ft-thick natural silty sand layer into an underlying aquifer. The flow velocity is 0.03 m/day and the effective porosity of silty sand is 0.1. What is the mass flux of chloride into the aquifer per unit base area of the landfill due to advection alone?

Solution:

Calculate the flux per 1 m × 1 m base area of the landfill, using equation (8.2).

$$F_v = n_e V_s C$$

$$= 0.1 \left(\frac{0.03 \text{ m}}{\text{day}} \right) \left(\frac{1000 \text{ mg}}{\text{L}} \right) \left[\frac{1000 \text{ L}}{1 \text{ m}^3} \left(\frac{1 \text{ g}}{1000 \text{ mg}} \right) \right]$$

$$= 3 \frac{\text{g}}{\text{day}} \frac{1}{\text{m}^2}$$

$\times 10^{-9}$ m^2/s (Mitchell, 1976). The D^* in equation (8.4a) also accounts for tortuosity in the soils and can be related to self-diffusion coefficients (D_0) of chemicals as follows:

$$D^* = \tau D_0 \tag{8.4b}$$

τ is the tortuosity coefficient, which has a value less than 1. The D_0 values for chemicals may be found in standard chemistry or environmental handbooks, and Table 8.1 presents D_0 values for selected chemicals. The value of τ ranges from 0.01 and 0.5 (Freeze and Cherry, 1979). The values of τ (or D^*) can be determined by laboratory tests using the steady-state method, time-lag method, and transient method. The transient method is subdivided further into the column method and the half-cell method. A description of these test methods and the determination of D^* are given by Rowe et al. (1988), Shackelford and Daniel (1991), and Sharma and Lewis (1994).

From Fick's first law and the equation of continuity, the rate at which contaminant can diffuse in soils can be given by

$$\frac{dC}{dt} = D^* \frac{d^2 C}{dx^2} \tag{8.5}$$

This is known as *Fick's second law*. For the following initial condition, $C(x,0) = 0$ (assuming that the porous medium is contaminant free initially) and boundary conditions, $C(0,t) = C_0$ and $C(\infty,0) = 0$, the solution to equation (8.5) is given by (Crank, 1956)

$$C(x,t) = C_0 \text{ erfc} \left(\frac{x}{2\sqrt{D^* t}} \right) \tag{8.5a}$$

where $C(x,t)$ is the concentration at distance x from the source at time t; C_0 the initial contaminant concentration, which is assumed to be constant; and erfc the complementary error function. A typical transient diffusive transport of a contaminant as predicted by equation (8.5a) is depicted in Figure 8.1(b).

The complementary error function (erfc) is given by

$$\text{erfc}(u) = 1 - \text{erf}(u) \tag{8.5b}$$

$$\text{erf}(u) = \frac{2}{\sqrt{\pi}} \int_0^u e^{-\eta^2} \, d\eta \tag{8.5c}$$

Table 8.1 Self-diffusion coefficients (D_0) for ions at infinite dilution in water

Chemical	$D_0 \times 10^{-10}\ (m^2/s)$	Chemical	$D_0 \times 10^{-10}\ (m^2/s)$
Anions		*Cations*	
OH^-	52.8	H^+	93.1
F^-	14.7	Li^+	10.3
Cl^-	20.3	Na^+	13.3
Br^-	20.8	K^+	19.6
I^-	20.4	Rb^+	20.7
HCO_3^-	11.8	Cs^+	20.5
NO_3^-	19.0	Be^{2+}	5.98
SO_4^{2-}	10.6	Mg^{2+}	7.05
CO_3^{2-}	9.22	Ca^{2+}	7.92
		Sr^{2+}	7.90
Organics		Ba^{2+}	8.46
Acetone	10.24	Pb^{2+}	9.25
N-Butyl alcohol	8.98	Cu^{2+}	7.13
Chlorobenzene	8.42	Fe^{2+}	7.19
Ethylbenzene	7.54	Cd^{2+}	7.17
Tertachlorothylene	8.4	Zn^{2+}	7.02
Methyl alcohol	14.65	Ni^{2+}	6.79
1,1,1-Trichloroethane	8.53	Fe^{3+}	6.07
Trichloroethylene	9.89	Cr^{3+}	5.94
Carbon disulfide	11.53	Al^{3+}	5.95

Source: Sharma and Lewis (1994); Fetter (1999)

Tables of error function and complementary error function are available in mathematics handbooks. Error function values can also be obtained from other sources, including Microsoft Excel and Matlab. [Table 8.2 provides erf(u) and erfc(u) values for selected u values.] Sometimes, it may be necessary to find erfc of a negative number that can be calculated using the following relationship:

$$\text{erfc}(-u) = 1 + \text{erf}(u) \qquad (8.5d)$$

8.2.3 Dispersion

At the macroscale level, the contaminant transport is defined by the average groundwater velocity. However, at the microscale level, the actual velocity of water may vary from point to point and can be either lower or higher than the average velocity. This difference in microscale water velocities arises due to:

- *Pore size.* The water velocity is inversely proportional to the pore size, which implies that lower velocities exist in larger pore spaces, while greater velocities exist in smaller pore spaces for the same quantity of flow.
- *Path length.* The greater the flow path, the lower the velocity; the flow path can be longer or shorter depending on the particle size and distribution.
- *Friction in pores.* Water flowing close to the soil solids possesses lower velocities due to friction, while the water away from the soil solids possesses higher velocity.

Because of these differences in velocities, mixing occurs along the flow path. This mixing is called *mechanical (or hydrodynamic) dispersion*, or simply *dispersion*. The mixing that occurs along the direction of the flow path is called *longitudinal dispersion*. The contaminants will also spread in directions normal to the direction of the flow path, called *transverse dis-*

EXAMPLE 8.2

At a landfill site, leachate accumulated over a 0.3-m-thick clay liner contains chloride concentration of 1000 mg/L. If the tortousity is equal to 0.5, what would be the concentration of chloride at a depth of 3 m after 100 years of diffusion? Neglect the effects of advection.

Solution:

Given: initial concentration, $C_0 = 1000$ mg/L; tortuosity (τ) = 0.5. For chloride, $D_0 = 20.3 \times 10^{-10}$ m^2/s (from Table 8.1). Therefore,

$$D^* = \tau D_0 = (0.5)(20.3 \times 10^{-10}) = 1.015 \times 10^{-9} \text{ m}^2/\text{s} = 0.032 \text{ m}^2/\text{yr}$$

Using equation (8.5a) yields

$$C(x,t) = C_0 \text{erfc} \left(\frac{x}{2\sqrt{D^*t}} \right)$$

$$C(3100) = 1000 \text{ erfc} \left[\frac{3}{2\sqrt{(0.032)(100)}} \right]$$

$$= 1000 \text{ erfc}(0.8385) = 230 \text{ mg/L}$$

---

persion. Both longitudinal dispersion and transverse dispersion are a function of the average velocity. The longitudinal dispersion is defined by

$$D_L = \alpha_L V_s \tag{8.6}$$

where D_L is the coefficient of longitudinal mechanical dispersion, α_L the longitudinal dispersivity, and V_s the average seepage velocity. Similarly, the transverse dispersion is defined by

$$D_T = \alpha_T V_s \tag{8.7}$$

where D_T is the coefficient of transverse mechanical dispersion and α_T is the transverse dispersivity. Generally, the molecular diffusion and mechanical dispersion are combined and the dispersion coefficients D_L^* and D_T^* are defined as

$$D_L^* = \alpha_L V_S + D^* \tag{8.8}$$

$$D_T^* = \alpha_T V_S + D^* \tag{8.9}$$

where D_L^* and D_T^* are called the longitudinal hydrodynamic dispersion coefficient and the transverse hydrodynamic dispersion coefficient, respectively.

As depicted in Figure 8.1(c), the one-dimensional dispersion of the contaminant results in a dilution of the contaminants at the advancing edge of flow. This one-dimensional condition considers only longitudinal hydrodynamic dispersion coefficient (D_L^*). However, in the two-dimensional case, both D_L^* and D_T^* are considered.

To describe the dispersive transport of the contaminants, α_L and α_T must first be known. If molecular diffusion is incorporated, D^* must also be known. Methods to determine D^* were discussed in Section 8.2.2. The dispersivities (α_L and α_T) can be determined from laboratory tests and/or field tests as detailed below.

Laboratory Testing. In laboratory testing, a column of soil with height L and cross-sectional area A is first prepared and saturated. The contaminant solution with known concentration is then injected from one end and

Table 8.2 **Error and complementary error function values**

u	erf(u)	erfc(u)
0.00	0.0	1.0
0.05	0.0563720	0.9436280
0.10	0.1124629	0.8875371
0.15	0.1679960	0.8320040
0.20	0.2227026	0.7772974
0.25	0.2763264	0.7236736
0.30	0.3286268	0.6713732
0.35	0.3793821	0.6206179
0.40	0.4283924	0.5716076
0.45	0.4754817	0.5245183
0.50	0.5204999	0.4795001
0.55	0.5633234	0.4366766
0.60	0.6038561	0.3961439
0.65	0.6420293	0.3579707
0.70	0.6778012	0.3221988
0.75	0.7111554	0.2888446
0.80	0.7421008	0.2578992
0.85	0.7706679	0.2293321
0.90	0.7969081	0.2030919
0.95	0.8208907	0.1791093
1.00	0.8427007	0.1572993
1.10	0.8802050	0.1197950
1.20	0.9103140	0.0896860
1.30	0.9340079	0.0659921
1.40	0.9522851	0.0477149
1.50	0.9661051	0.0338949
1.60	0.9763484	0.0236516
1.70	0.9837905	0.0162095
1.80	0.9890905	0.0109095
1.90	0.9927904	0.0072096
2.00	0.9953223	0.0046777
2.10	0.9970205	0.0029795
2.20	0.9981372	0.0018628
2.30	0.9988568	0.0011432
2.40	0.9993115	0.0006885
2.50	0.9995930	0.0004070
2.60	0.9997640	0.0002360
2.70	0.9998657	0.0001343
2.80	0.9999250	0.0000750
2.90	0.9999589	0.0000411
3.00	0.9999779	0.0000221

Source: Generated using MS Excel software (2003).

the volume and concentration of the effluent from the other end is recorded with time. A constant flow rate Q, hence a constant flow velocity V, condition is created. V_s can be calculated using Q/tAn. One pore volume of the column is equal to ALn. Unit discharge rate is equal to $V_s nA$; therefore, the total discharge is equal to $V_s nAt$. The total number of pore volumes U is calculated by dividing the total discharge by the volume of a single pore volume. Mathematically,

$$U = \frac{V_s nAt}{ALn} = \frac{V_s t}{L} = t_r \qquad (8.10)$$

where t_r is known as the dimensionless time. This means that the number of pore volumes is equal to dimensionless time t_r. The column test results are plotted as U versus C, where C is the concentration of the contaminant in effluent. These data are then used to calculate α_L based on analytical solution to a one-dimensional continuous injection (see Section 8.5.1) given by

$$C = \frac{C_0}{2} \, \text{erfc}\left(\frac{L - V_s t}{2\sqrt{D_L^* t}}\right) \qquad (8.11)$$

In nondimensional form, the solution above can be given by (Fetter, 1999)

$$\frac{C}{C_0} = 0.5 \, \text{erfc}\left(\frac{1 - U}{2 \sqrt{U D_L^* / V_s L}}\right) \qquad (8.12)$$

The solution of this equation for different values of D_L^* is obtained and compared with the test data of U versus C. The value of D_L^* that best fits the data is then determined. As $D_L^* = \alpha_L V_s + D^*$, α_L is then equal to $(D_L^* - D^*)/V_s$.

Field Testing. In field testing, a tracer is introduced into a well and its concentration is monitored in observation wells located at different locations from the injection well. The spatial and temporal variation of the tracer concentrations are then compared with the predicted two-dimensional advective–dispersive solution (see Section 8.5), and the values of α_L and α_T that best describe the observed test results are selected.

Unfortunately, both laboratory testing and field testing do not provide accurate values of dispersivities. Gelhar (1986) assessed the dispersivity values determined from several reported laboratory and field studies and found that these values range widely and depend significantly on the scale (i.e., flow length). As shown in Figure 8.2, a correlation is found to exist between the flow length and the dispersivities. A best-fit straight line provides an approximate quantitative relationship between the longitudinal dispersivity and flow length and is expressed as $\alpha_L = 0.1X$, where X is the flow distance. The transverse dispersivity α_T is generally assumed to be 10% of the α_L. These correlations are often used initially to estimate α_L and α_T values, and these values are then refined through mathematical model calibration with the site data.

8.3 CHEMICAL MASS TRANSFER PROCESSES

The chemical mass transfer processes that are important in evaluating contaminant fate and transport in

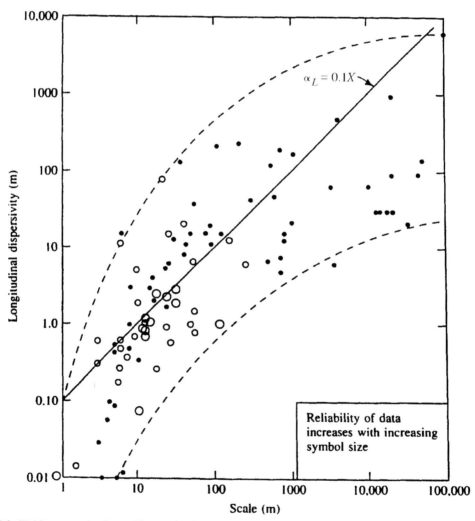

Figure 8.2 *Field-measured values of longitudinal dispersivity as a function of the scale of measurement. The largest circles represent the most reliable data. (From Gelhar, 1986. Copyright 1986 American Geophysical Union. As presented in Fetter, 1999.)*

subsurface include sorption and desorption, dissolution and precipitation, oxidation and reduction, acid–base reactions, complexation, ion exchange, volatilization, and hydrolysis. These chemical mass transfer processes are described in this section. Some of these processes are also described in further detail in Chapters 3 and 6.

8.3.1 Sorption and Desorption

Sorption refers to contaminants attached to mineral grains and organic matter in the soil. Sorption encompasses various processes, such as adsorption, chemisorption, and absorption. *Adsorption* refers to contaminants clinging to a solid surface. *Chemisorption* refers

---

EXAMPLE 8.3

A one-dimensional column test was conducted on a soil sample. Deionized water was flushed through the column to saturate the soil sample. Then chloride solution at a concentration of 1000 mg/L was passed through the column. The length of the column was 0.5 m. The seepage velocity was 0.0005 m/s. The concentrations of the effluent at various time periods are given below. Calculate the value of longitudinal dispersivity (α_L) for the soil in the column.

Time (s)	Concentration (mg/L)
200	2.5
400	66.8
700	290
900	450

Solution:

Given data: chloride solution concentration, $C_0 = 1000$ mg/L; length of column, $L = 0.5$ m; average seepage velocity, $V_s = 0.0005$ m/s. Assume that $D_L^* = 1.0 \times 10^{-5}$ m^2/s and calculate C for specified t values using equation (8.11):

Time (s)	$U = V_s t/L$	$x = \dfrac{L - V_s t}{2\sqrt{D_L^* t}}$	erf(x)	erfc(x) = 1 − erf(x)	0.5 erfc(x)	C (mg/L) [Eq. (8.11)]
200	0.2	4.472	1.000	0.000	0.000	0.000
400	0.4	2.372	0.999	0.001	0.000	0.398
700	0.7	0.896	0.795	0.205	0.102	102.447
900	0.9	0.264	0.291	0.709	0.355	354.694

Plot U versus C and compare the calculated values with the test data. Repeat this procedure for other D_L^* values. The results are shown in Figure E8.3. $D_L^* = 5.5 \times 10^{-5}$ m^2/s results match closely with the test data. For chloride (Cl$^-$), $D_0 = 20.3 \times 10^{-10}$ m^2/s (from Table 8.1) and assume that $\tau = 0.5$.

$$D^* = \tau D_0 \Rightarrow D^* = (0.5)20.3 \times 10^{-10} \text{ m}^2/\text{s}$$

$$\Rightarrow D^* = 1.015 \times 10^{-9} \text{ m}^2/\text{s}$$

Substitute D^* and D_L^* in $D_L^* = \alpha_L V_s + D^*$:

$$0.000055 = \alpha_L (0.0005 \text{ m/s}) + 1.015 \times 10^{-9} \Rightarrow \alpha_L = 0.11 \text{ m}$$

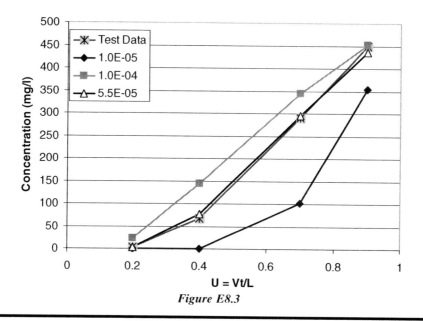

Figure E8.3

to contaminants incorporating on soil surface by a chemical reaction. *Absorption* refers to contaminants diffusing into the particle and being adsorbed onto interior surfaces. Generally, we are interested in the combined effect of all these processes, collectively known as sorption. *Sorption* defines the contaminant distribution between the solution and solid phase; therefore, it is also known as *partitioning*. *Desorption,* the opposite of sorption, refers to contaminants becoming detached from mineral grains and organic matter and entering into the pore fluid.

Many factors affect the sorption of contaminants in soils. These factors include (1) contaminant characteristics, such as water solubility, polar–ionic character, and octanol–water partition coefficient; (2) soil characteristics, such as mineralogy, permeability, porosity, texture, homogeneity, organic carbon content, surface charge, and surface area; and (3) fluid media characteristics, such as pH, salt content, and dissolved organic carbon content. The relationship between the contaminant sorbed onto soil solids (S) and that present in soil pore water (C) at equilibrium is known as the *isotherm.* The term *isotherm* is used to refer to constant-temperature conditions. There are three dif-

ferent types of sorption isotherms: (1) the linear isotherm, (2) the Freundlich isotherm, and (3) the Langmuir isotherm. These are depicted in Figure 8.3 and described briefly below.

Linear Isotherm. The linear isotherm is expressed mathematically as

$$S = K_d C \qquad (8.13)$$

where S is the mass of contaminant sorbed per unit dry mass of solid (e.g., mg/kg), C the concentration of contaminant in solution at equilibrium (e.g., mg/L), and K_d is known as the distribution coefficient or partitioning coefficient with such units as L/kg or cm^3/g. The K_d is used to define the *retardation coefficient, R* (see Section 8.5):

$$R = 1 + \frac{\rho_d}{n} K_d \qquad (8.14)$$

where n and ρ_d are porosity and dry density of the transport medium, respectively. Using this retardation

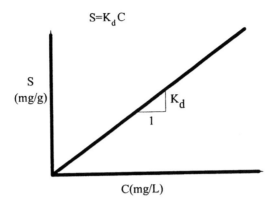

(a) LINEAR SORPTION ISOTHERM

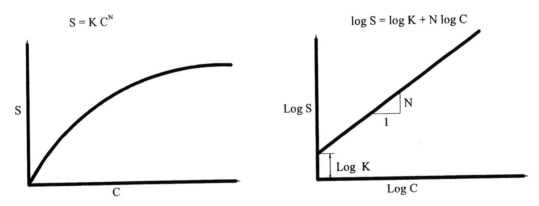

(b) FREUNDLICH SORPTION ISOTHERM

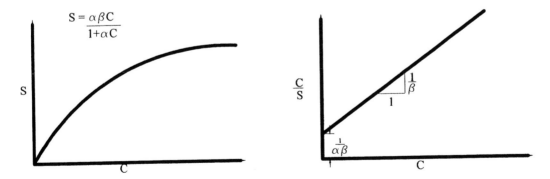

(c) LANGMUIR SORPTION ISOTHERM

Figure 8.3 *Various types of sorption isotherms.*

factor, the velocity of contaminant V_{cont} is given by (see Section 8.5)

$$V_{cont} = \frac{V_S}{R} \tag{8.15}$$

R is generally greater than 1; therefore, equation (8.15) shows that the contaminant migration is lower than the flow of water. The linear isotherm is the most commonly used isotherm in contaminant migration analysis because of its simplicity. However, the linear sorption isotherm is generally valid at low contaminant concentration ranges.

Freundlich Isotherm. For higher concentrations, the Freundlich sorption isotherm as given below is used:

$$S = KC^N \tag{8.16}$$

where K and N are constants. To calculate K and N, equation (8.16) is linearized by plotting log C versus log S as shown in Figure 8.3(b). The slope of this straight line is N and the intercept is equal to log K. According to this isotherm, the retardation factor R is given by

$$R = 1 + \frac{\rho_d}{n} KNC^{N-1} \tag{8.17}$$

The contaminant velocity can be calculated using equation (8.15) and is dependent on the contaminant concentration. If $N = 1$, equation (8.17) reduces to linear sorption.

Langmuir Isotherm. Another nonlinear relationship between S and C is the Langmuir sorption isotherm, which has been developed based on the concept that a solid surface possesses a finite number of sorption sites. Sorption does not occur if all the sorption sites are filled. Mathematically, this isotherm is expressed as

$$S = \frac{\alpha\beta C}{1 + \alpha C} \tag{8.18}$$

where α is the adsorption constant related to the binding energy (mg^{-1}), and β is the maximum amount of contaminant that can be adsorbed by the solid (mg/kg). To determine these two parameters, a plot of C versus C/S is plotted and a best-fit straight line is drawn as shown in Figure 8.3(c). The slope of the straight line is $1/\beta$ and the intercept is equal to $1/\alpha\beta$. Using these values, α and β can be calculated. The retardation factor per this isotherm is given by

$$R_R = 1 + \frac{\rho_d}{n} \frac{\alpha\beta}{(1 + \alpha C)^2} \tag{8.19}$$

The velocity of the contaminant migration, dependent on the concentration C, can again be calculated using equation (8.15).

All of the isotherms above are valid for equilibrium conditions. For most situations, this assumption is valid; however, when the rate of change of concentrations due to sorption is significantly higher than that of other transport processes, specifically, advection and dispersion, and depends on time, nonequilibrium (kinetic) sorption models are needed. More details on this may be found in Fetter (1999).

Linear sorption isotherm is the most commonly used isotherm in practice; therefore, K_d is often needed to define sorption. Three different approaches may be followed to determine K_d: (1) empirical methods, (2) laboratory methods, and (3) field methods. These methods are described briefly below.

Empirical Methods. The most common empirical methods to determine sorption are (1) the use of empirical field data, (2) the use of a contaminant property such as the octanol–water partition coefficient (K_{ow}) and water solubility (S_w), and (3) use of the surface area of soil solids. All these methods assume linear sorption; therefore, they provide an approximate value of K_d.

Use of Empirical Field Data. For this method, the contaminant concentration in groundwater (C) and the contaminant concentration in soil based on dry soil weight (S) at the same location are used. The ratio of S and C gives the value of K_d. It is assumed that the concentration in soil water is assumed to be the same as that in groundwater.

Use of Contaminant Property. These methods are generally applicable for organic contaminants and assume that the sorption of organic contaminants in soils is due primarily to sorption onto the soil's organic carbon fraction (f_{oc}). The sorption of the contaminant onto the soil is directly proportional to f_{oc}. Krickhoff et al. (1979) found a correlation between K_d and f_{oc}:

$$K_d = K_{oc} f_{oc} \qquad (8.20)$$

where K_{oc} is the partitioning coefficient for the organic contaminant onto a hypothetical pure organic carbon phase. As explained in Chapter 4, the organic matter (OM) of soil can be determined using either the wet combustion or dry combustion method, and the f_{oc} can then be calculated approximately using the relationship between f_{oc} and OM (Olsen and Davis, 1990):

$$f_{oc} = \frac{OM}{1.724} \qquad (8.21)$$

Generally, K_{oc} is calculated using its correlation with some of the standard properties of organic chemicals, such as K_{ow} and S_w (Morel, 1983; Lyman, 1982). Krickhoff (1981) used an analogy between the sorption of an organic contaminant onto soil organic matter and the sorption of an organic contaminant onto other organic compounds, such as octanol. Partitioning of the latter combination is represented by the octanol–water partitioning coefficient (K_{ow}), which represents an organic contaminant in octanol (organic matter) and water when they are in contact with each other at equilibrium conditions as

$$K_{OW} = \frac{\text{conc. in octanol phase}}{\text{conc. in aqueous phase}} \qquad (8.22)$$

The K_{ow} values for many organic chemicals have either been measured in a laboratory or calculated, and are published in standard organic chemistry books; these

values vary by several orders of magnitude, ranging from 10^{-3} to 10^7, depending on the chemical type. Using K_{ow}, K_{oc} can be calculated using various correlations, shown in Table 8.3. To use this method, the particular organic contaminant should first be known. The K_{ow} of that organic chemical can be found from standard organic chemistry references, K_{oc} is calculated using an appropriate correlation between K_{ow} and K_{oc} listed in Table 8.3, and K_d is obtained from equation (8.20).

Correlations between the contaminant solubility (S) and K_{oc} have also been proposed by a few investigators. These correlations are summarized in Table 8.4 and can be used in a manner similar to K_{ow} correlations to calculate K_d.

The methods above assume a linear relationship between f_{oc} and K_d which may not be valid if f_{oc} is very low (<0.001) for large amounts of swelling clays or for polar organic compounds. Therefore, caution must be exercised when using these methods to calculate K_d.

Use of Surface Area of Soil Solids. This method is useful for both organic and inorganic contaminants, when the organic fraction in the soil is below the critical level and when the mineral surface area is high. In this method, sorption is assumed to occur onto the inorganic mineral surfaces. Assuming linear sorption,

Table 8.3 Correlation between K_{oc} and K_{ow}[a]

Equation	Chemical Class Represented
$\log K_{oc} = 0.544 \log K_{ow} + 1.377$	Wider variety, mostly pesticides
$\log K_{oc} = 0.937 \log K_{ow} - 0.006$	Aromatics, polynuclear aromatics, triazines, and dinitroaniline herbicides
$\log K_{oc} = 1.00 \log K_{ow} - 0.21$	Mostly aromatic or polynuclear aromatics; two chlorinated
$\log K_{oc} = 0.94 \log K_{ow} + 0.02$	s-Triazines and dinitroaniline herbicides
$\log K_{oc} = 1.029 \log K_{ow} - 0.18$	Variety of insecticides, herbicides, and fungicides
$\log K_{oc} = 0.524 \log K_{ow} + 0.855$	Substituted phenylureas and alkyl-N-phenylcarbamates

Source: Bedient et al. (1994).

[a] K_{oc}, soil (or sediment) adsorption coefficient; K_{ow} octanol–water partition coefficient.

relationships between K_d and surface area of the mineral (SA) have been developed by various investigators; two representative equations are:

$$\log K_d = 0.061(\text{SA}) + 2.89 \qquad (8.23a)$$

$$K_d = \frac{\text{SA}}{(K_{ow})^{0.16}} \qquad (8.23b)$$

where SA is given in m^2/g.

Laboratory Test Methods. Laboratory sorption tests can be either (1) batch tests or (2) column tests. In *batch tests,* a known volume of solution (V_w) containing an initial concentration (C_0) of a contaminant is placed into a container. A known mass of dry soil (M_s) is then added and the mixture is shaken and allowed to equilibrate. The soil is then separated from the solution by centrifuging, and an aliquot of the supernatant is sampled. The concentration of the contaminant in this aliquot (C) is measured and the concentration on the soil (S) is calculated by

$$S = \frac{V_w(C_0 - C)}{M_s} \qquad (8.24)$$

The test is run several times with different initial contaminant concentrations or different masses of soil. The result is a series of contaminant concentrations (S) with corresponding aqueous-phase concentrations (C), which can then be used to plot the isotherm. Based on the data, the most appropriate isotherm is determined. The results of batch tests depend on the equilibrium time and the soil/solution ratio used for testing; therefore, the effect of these variables must be evaluated. Care must also be taken to prevent volatilization of the organic contaminants during this testing. In addition, if nonsettling particles are present in aqueous samples, the concentrations may be overestimated.

In *column testing,* a column of soil is prepared, and a solution containing a nonadsorbing tracer and the contaminant of interest is run through the column (Figure 8.4). The concentrations of the tracer and contaminant can be measured in the water that passes through the column. The retardation factor is then the ratio of the time (or volume) for the center of mass of the

contaminant to break through the column, to the time (or volume) for the center of mass of the nonreactive tracer to break through the column (Figure 8.4). This method provides an R value directly; however, it is also well suited for those contaminants that have relatively low retardation factors (<10). The K_d value, if desired, can be calculated using equation (8.14). The disadvantages of column tests are that slow flow rates in fine-grained soils require a long test duration, and the destruction of soil structure by soil repacking may affect the test results.

Field Test Methods. Site-specific field testing provides an estimate of the retardation factor. In this method, reactive and nonreactive tracers are introduced into an injection well and the concentrations of both tracers are monitored in downgradient wells (see Figure 8.5). The ratio of velocities provides the retardation factor (R). One of the potential disadvantages of this method is that other processes that are not included in the data analysis are occurring within the soil. Ignoring these processes can result in poor estimates of the retardation factor. The temporal variation of the contaminant plume migration can also be used to calculate the retardation factor (see Figure 8.6).

Each method of estimating sorption has advantages and disadvantages. A general approach to choosing an isotherm is to perform laboratory experiments to determine S versus C and then determine which isotherm best fits the data. Example 8.4 shows how to select an appropriate isotherm based on the experimental data.

8.3.2 Precipitation and Dissolution

Precipitation and dissolution reactions involve chemicals in solid phases and are described by an equilibrium constant (K), as explained in Chapter 3. In many reactions where the activity of the reacting solid is equal to 1, a comparison of the relative size of the equilibrium constant K provides an indication of the solid solubility in pure water (Table 8.5). Chloride and sulfate salts tend to be the most soluble phase, and sulfide and hydroxide groups to be the least soluble. Minerals in the carbonate and silicate and aluminum silicate groups have low solubility.

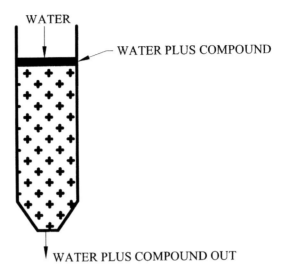

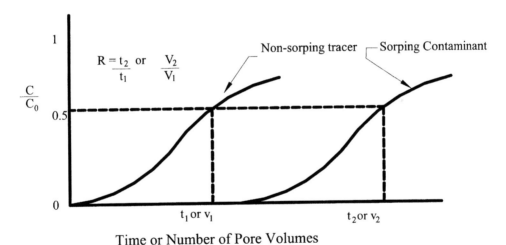

Figure 8.4 *Column testing to determine R and K_d.*

When other ions are present in a solution, the solubility of a solid can differ from its value in pure water. Solubility increases due to solution nonidealities and decreases due to the common-ion effect. Generally, the solubility of a solid increases with increasing ionic strength, because other ions in solution reduce the activity of the ion involved in the reaction.

The common-ion effect occurs when a solution contains the same ions that will be released when the solid dissolves. The presence of the common ion means that less solid is able to dissolve before the solution reaches saturation with respect to that ion. Thus, the solubility of a solid is less in a solution containing a common ion than it would be in water alone.

8.3.3 Oxidation–Reduction (Redox) Reactions

As described in Chapter 3, redox reactions involve the transfer of electrons. The redox reactions are meas-

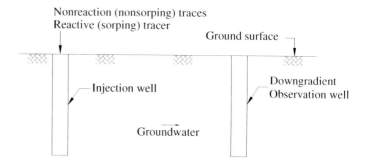

$R=t_1/t_2$
t_1=Time for arrival of reactive traces in observation well
t_2=Time for arrival of nonreactive tracer in observation well

Figure 8.5 *Field testing to determine* R.

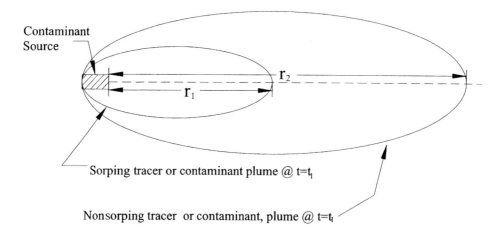

$$R= r_1 / r_2$$

Figure 8.6 *Determination of* R *based on contaminant plume migration.*

ured in terms of p^e or Eh. Large values of p^e indicate low electron activity, which favors the existence of electron-poor (oxidized) species. Small values of p^e indicate high electron activity, which favors the existence of electron-rich (reduced) species. For a pH of 7.0, depending on p^e, the redox regions are defined as $p^e > 7$, oxic; $2 < p^e < 7$, suboxic; and $p^e < 2$, anoxic. The redox reaction is also characterized by the redox potential (Eh), which has units of volts.

The redox reactions can be caused by naturally occurring microorganisms, as shown in Figure 8.7. In this case, oxidation of organic matter occurs by reduction of oxygen. This is followed by reduction of NO_3^- and NO_2^-. The succession of these reactions follows the decreasing p^e level. Reduction of MnO_2, if present, should occur at about the same p^e level as that of nitrate reduction, followed by reduction of FeOOH(s) to Fe^{2+}. When sufficiently negative p^e levels have been reached, fermentation reactions and reduction of SO_4^{2-} and CO_2 may occur almost simultaneously.

8.3.4 Acid–Base Reactions

As discussed in Chapter 3, acid–base reactions involve either the gain or loss of a proton (H^+) or the gain or loss of a hydroxyl (OH^-). The strength of an acid or base refers to the extent to which protons are lost or gained, respectively. Consider the following generalized ionization reaction for an acid HA in water:

$$HA + H_2O = H_3O^+ + A^-$$

where A^- is an anion like Cl^-. The strength of the acids and bases depends on whether equilibrium in the reaction is established to the right or left side of the equation. Table 8.6 shows a few weak acid–base re-

---

EXAMPLE 8.4

The batch sorption test results performed using Na-montmorillonite soil and hexavalent chromium [Cr(VI)] are given below. Which adsorption isotherm fits the experimental test results best? Determine the appropriate coefficients for the isotherm selected.

Equilibrium Concentration (mg/L)	Cr(VI) Sorbed (mg/g)	Equilibrium Concentration (mg/L)	Cr(VI) Sorbed (mg/g)
0.842	0.0168	36.13	0.7227
8.701	0.1740	65.67	1.3134
12.27	0.2454	90.989	1.8198
17.00	0.3400	138.574	2.7702
22.49	0.4499	183.79	3.6758
31.27	0.6285		

Solution:

Based on the data given:

C (mg/L)	S (mg/g)	C/S	Log C	Log S
0.842	0.017	50.1190	−0.0747	−1.7747
8.701	0.174	50.0057	0.9396	−0.7595
12.270	0.245	50.0000	1.0888	−0.6101
17.000	0.340	50.0000	1.2304	−0.4685
22.490	0.449	50.0891	1.3520	−0.3478
31.270	0.629	49.7534	1.4951	−0.2017
36.130	0.723	49.9931	1.5579	−0.1410
65.670	1.313	50.0000	1.8174	0.1184
90.980	1.820	49.9945	1.9589	0.2600
138.570	2.770	50.0217	2.1417	0.4425
183.790	3.676	50.0000	2.2643	0.5654

The data above are plotted in Figures E8.4A, E8.4B, and E8.4C, and the parameters for each isotherm are determined by fitting a straight line. As can be seen from these results, the linear sorption isotherm predicts the results accurately. Because of the simplicity, a linear isotherm is recommended.

Linear Isotherm: $S = K_d\,C$

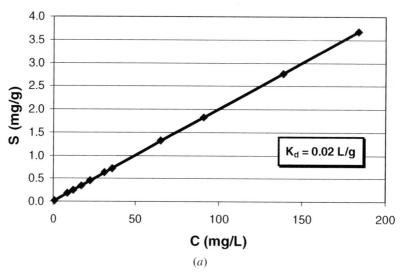

(*a*)

Langmuir Isotherm:
$C/S = 1/\alpha\beta + 1/\beta\ (C)$

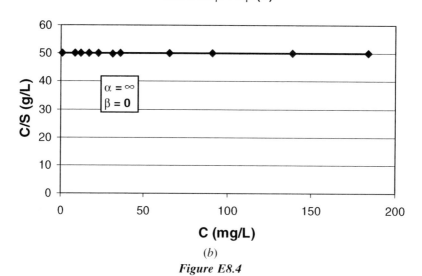

(*b*)

Figure E8.4

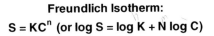

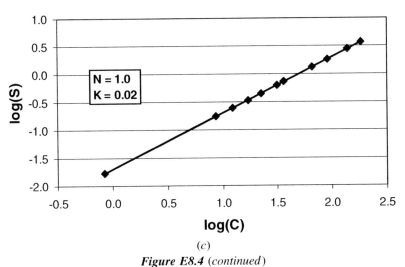

Freundlich Isotherm:
$$S = KC^n \text{ (or log } S = \log K + N \log C)$$

$N = 1.0$
$K = 0.02$

(c)
Figure E8.4 (continued)

EXAMPLE 8.5

Estimate the travel time for the following contaminants from the source area to a water well located 100 m away in an aquifer with bulk density of soil = 1.5 g/cm³, n = 40%, organic carbon = 1%: (a) chloride ion (nonsorbing); (b) benzene; (c) TCE; (d) pyrene. The solubilities of benzene, TCE, and pyrene are given as 1780, 1100, and 0.148 mg/L. The groundwater velocity is 10 m/yr.

Solution:

$$V_{cont} = \frac{V_S}{R} \qquad R = 1 + \frac{\rho_d}{n} K_d \qquad K_d = K_{oc} f_{oc} \qquad \log K_{oc} = -0.55S + 3.64$$

Use equations (8.15), (8.14), and (8.20) and the equation from Table 8.4, respectively.

Chemical	Solubility (mg/L)	K_{oc}	K_d	R	V_{cont} (m/yr)	Time (yr) = 100 m/V_{cont}
Chloride[a]	—	—	—	1	10	10
Benzene	1780	0	0	1	10	10
TCE	1100	0	0	1	10	10
Pyrene	0.148	3619.095	36.19	136.72	0.00731	1367.16

[a] Chloride ion is nonsorbing, so R = 1.0.

Table 8.4 Correlation between K_{oc} and S^a

Equation	Chemical Class Represented
$\log K_{oc} = -0.55 \log S + 3.64$ (S in mg/L)	Wide variety, mostly pesticides
$\log K_{oc} = -0.54 \log S + 0.44$ (S in mole fraction)	Mostly aromatic or polynuclear aromatics; two chlorinated
$\log K_{oc} = -0.557 \log S + 4.277$ (S in μmol/L)	Chlorinated hydrocarbons

Source: Bedient et al. (1994).

$^a K_{oc}$, soil (or sediment) adsorption coefficient; S, water solubility.

actions that are important in natural groundwater. The reactions involving CO_2 show that CO_2 dissolved in water partitions among H_2CO_3, HCO_3^-, and CO_3^{2-}. When the pH of a solution is fixed, the mass law equation can be used to determine the concentrations of the species.

Acid–base reactions affect the pH and ion chemistry of soils and groundwater. In groundwater, the pH and carbonate speciation are interdependent and are a function of ionization equilibria for the carbonate species and water, and strong bases added through the dissolution of carbonate and silicate minerals. The *alkalinity* is defined as the net concentration of strong base in excess of strong acid with a pure CO_2–water system as the point of reference:

$$\text{Alk} = \Sigma(i^+)_{sb} - \Sigma(i^-)_{sa}$$
$$= -(H^+) + (OH^-) + (HCO_3^-) + 2(CO_3^{2-})$$

When the alkalinity is zero, the equation above becomes the charge-balance equation for a pure CO_2–water system. The equation above shows that an increase in alkalinity increases the net positive charge on the left side of the equation. This increase is not balanced simply by an increase in one of the negative ions on the right side because equilibrium relationships

Table 8.5 Some common mineral dissolution reactions and associated equilibrium constants

Mineral or Solid	Reaction	$\log K(25°C)$
Chlorides and sulfates		
Halite	$NaCl = Na^+ + Cl^-$	1.54
Sylvite	$KCl = K^+ + Cl^-$	0.98
Gypsum	$CaSO_4 \cdot 2H_2O = Ca^+ + SO_4^{2-} + 2H_2O$	−4.62
Carbonates		
Magnesite	$MgCO_3 = Mg^{2+} + CO_3^{2-}$	−7.46
Aragonite	$CaCO_3 = Ca^{2+} + CO_3^{2-}$	−8.22
Calcite	$CaCO_3 = Ca^{2+} + CO_3^{2-}$	−8.35
Siderite	$FeCO_3 = Fe^{2+} + CO_3^{2-}$	−10.7
Dolomite	$CaMg(CO_3) = Ca^{2+} + Mg^{2+} + 2CO_3^{2-}$	−16.7
Hydroxides		
Brucite	$Mg(OH)_2 = Mg^{2+} + 2OH^-$	−11.1
Ferrous hydroxide	$Fe(OH)_2 = Fe^{2+} + 2OH^-$	−15.1
Gibbsite	$Al(OH)_3 = Al^{3+} + 3OH^-$	−33.5
Sulfides		
Pyrrhotite	$FeS = Fe^{2+} + S^{2-}$	−18.1
Sphalerite	$ZnS = Zn^{2+} + S^{2-}$	−23.9
Galena	$PbS = Pb^{2+} + S^{2-}$	−27.5
Silicates and aluminum silicates		
Quartz	$SiO_2 + H_2O = H_2SiO_3$	−4.00
Na-montmorillonite	$3Na\text{-mont} + 11\frac{1}{2}H_2O = 3\frac{1}{2}\text{kaoline} + 4H_4SiO_4 + Na^+$	−9.1
Kaolinite	$\text{Kaoline} + 5H_2O = 2Al(OH)_3 + 2H_4SiO_4$	−9.4

Source: Domenico and Schwartz (1998). This material is used by permission of John Wiley & Sons, Inc.

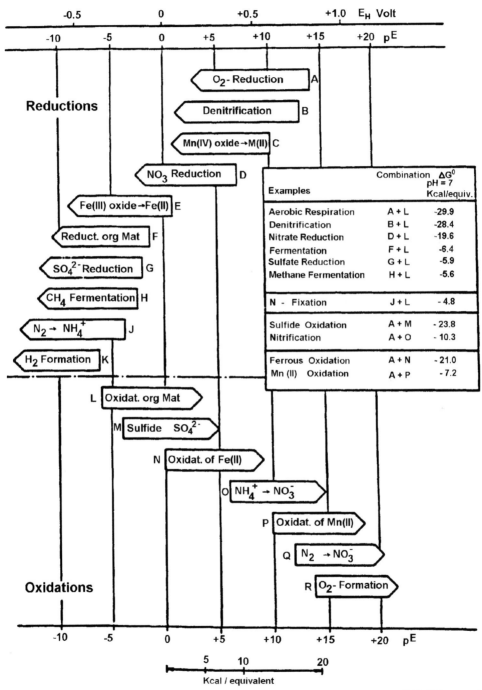

Figure 8.7 Sequence of microbially mediated redox processes. (*From Stumm and Morgan, 1981. This material is used by permission of John Wiley & Sons, Inc.*)

Table 8.6 **Important weak acid–base reactions in natural water systems**

Reaction	Mass Law Equation	$-\log K(25°C)$
$H_2O = H^+ + OH^-$	$K_w = (H^+)(OH^-)$	14.0
$CO_2(g) + H_2O = H_2CO_3$	$K_{co_2} = \dfrac{H_2CO_3}{P_{CO_2}(H_2O)}$	1.46
$H_2CO_3^* = HCO_3^- + H^+$	$K_1 = \dfrac{(HCO_3^-)(H^+)}{H_2CO_3^-}$	6.35
$HCO_3^- = CO_3^{2-} + H^+$	$K_2 = \dfrac{(CO_3^{2-})(H^+)}{HCO_3^-}$	10.33
$H_2SiO_3 = HSiO_3^- + H^+$	$K = \dfrac{(HSiO_3^-)(H^+)}{H_2SiO_3}$	9.86
$HSiO_3^- = SiO_3^{2-} + H^+$	$K = \dfrac{(SiO_3^{2-})(H^+)}{HSiO_3^-}$	13.1

Source: Domenico and Schwartz (1998). This material is used by permission of John Wiley & Sons, Inc.

in the solution must be maintained. The increase in alkalinity is matched by an increase in the concentration of negatively charged species, generated from the ionization of HCO_3^- to CO_3^{2-} and an increase in pH. Thus, increasing the alkalinity with a strong base leads ultimately to an increase in pH. This behavior is commonly observed in groundwater as it dissolves minerals along a flow path.

8.3.5 Complexation

In Chapter 3, complexation is described as the process of ion or molecule combination in the dissolved phase. The ligands, or surrounding atoms or ions, can include organic and inorganic species. As discussed, complexes may be charged or uncharged, and the solubility of the complex can be affected.

8.3.6 Ion Exchange

Ion exchange is a specific category of adsorption called *adsorbent-motivated sorption*. Accumulation occurs due to an affinity of the solid surface for the chemical. Ion exchange is important for clays. The exchange of a multivalent cation (B^{n+}) for a monovalent cation (A^+) on a natural mineral surface (R^-) could occur:

$$nR^-A^+ + B^{n+} \Leftrightarrow R_n^-B^{n+} + nA^+$$

For example, $2R^-Na^+ + Ca^{2+} \Leftrightarrow R_2^-Ca^{2+} + 2Na^+$ shows the exchange of Ca for Na ions. The most common ions occurring naturally in soil are as follows in decreasing order of occurrence: cations: Ca^{2+}, Mg^{2+}, Na^+, and K^+ and anions: SO_4^{2-}, Cl^-, PO_4^{3-}, and NO_3^-. Valocchi (1984) suggested the use of an "effective" partitioning coefficient approach for describing the transport of ion-exchanging contaminants. Generally, this coefficient is also included in sorption.

8.3.7 Volatilization

As explained in Chapter 6, volatilization is a process of liquid- or solid-phase evaporation that occurs when contaminants (nonaqueous or dissolved) contact a gas phase. At equilibrium, *Raoult's law* describes the equilibrium partial pressure of volatile organics in the atmosphere above an ideal solvent (e.g., benzene):

$$P_{org} = X_{org} P_{org}^0 \qquad (8.25)$$

where P_{org} is the partial pressure of the vapor in the gas phase, X_{org} the mole fraction of the organic solvent, and P_{org}^0 the vapor pressure of the pure organic solvent.

Volatilization of the dissolved organic solutes from water is described by Henry's law:

$$C_a = K_H C_w \qquad (8.26)$$

where K_H is Henry's constant and C_a and C_w are concentrations in gas and dissolved phases. Table 8.7 lists vapor pressure and K_H (or simply H) of typical organic contaminants.

Table 8.7 **Properties of selected organic compounds**

Chemical	M (g/mol)	Solubility mg/L	Solubility mol/m³	Vapor Pressure at 20°C mm Hg	Vapor Pressure at 20°C atm	Henry's Law Constant atm · m³/mol	Henry's Law Constant Nondim.
Acetylene tetrabromide	344	650	1.9	0.3	3.9×10^{-4}	2.1×10^{-4}	8.9×10^{-3}
Aldrin	365	0.2	5.5×10^{-4}	6×10^{-6}	7.9×10^{-9}	1.4×10^{-5}	6.1×10^{-4}
Aroclor 1242	254	0.24	9.5×10^{-4}	4.1×10^{-4}	5.3×10^{-7}	5.6×10^{-4}	2.4×10^{-2}
Benzene	78	1780	22.8	95.2	1.25×10^{-1}	5.5×10^{-3}	0.24
Biphenyl	154	7.5	0.05	0.06	7.5×10^{-5}	1.5×10^{-3}	6.8×10^{-2}
m-Bromonitrobenzene	170	104	58.8	0.07	9.2×10^{-5}	1.6×10^{-6}	7.4×10^{-5}
3-Bromo-f-propanol	139	1.7×10^{-5}	1233	0.1	1.3×10^{-4}	1.1×10^{-7}	4.6×10^{-6}
4-t-Butylphenol	150	1×10^3	6.7	0.046	6.1×10^{-5}	9.1×10^{-6}	3.8×10^{-4}
Carbon tetrachloride	154	800	5.2	91	0.12	2.3×10^{-2}	0.97
Chlorobenzene	113	472	4.2	11.8	1.6×10^{-2}	3.7×10^{-3}	0.165
Chloroform	119	8×10^3	67	246	0.32	4.8×10^{-3}	0.2
Cumene (isopropyl benzene)	120	50	0.416	4.6	6.1×10^{-3}	1.5×10^{-2}	0.62
DDT	354.5	1.2×10^{-3}	3.4×10^{-6}	1×10^{-7}	1.3×10^{-10}	3.8×10^{-5}	1.7×10^{-3}
Dieldrin	381	0.25	6.6×10^{-4}	1×10^{-7}	1.3	2×10^{-7}	8.9×10^{-6}
Ethyl benzene	106	152	1.43	9.5	1.25×10^{-2}	8.7×10^{-3}	0.37
Ethyl bromide	109	900	8.3	460	6.1×10^{-1}	7.3×10^{-2}	3.1
Ethylene	28	131	4.7	—	>40	>8.6	~360
Ethylene dibromide	188	4.3×10^3	22.9	11.6	1.5×10^{-2}	6.6×10^{-4}	2.8×10^{-2}
Ethylene dichloride	99	8.0×10^3	80.8	67	0.09	1.1×10^{-3}	4×10^{-2}
Fluorotrichloromethane	137	—	—	—	—	5	—
Lindane	290.9	7.3	2.5×10^{-2}	9.4×10^{-6}	1.2×10^{-8}	4.8×10^{-7}	2.2×10^{-5}
Mercury	201	3×10^{-2}	1.5×10^{-4}	1.3×10^{-3}	1.7×10^{-6}	1.1×10^{-2}	0.48
Methyl bromide	95	1.3×10^4	137	1.4×10^3	1.8	1.3×10^{-2}	0.56
Methyl chloride	50.5	7.4×10^3	146	3.6×10^3	4.74	2.4×10^{-2}	0.36
Methylene chloride	85	1.3×10^4	155	349	0.46	3×10^{-3}	0.13
Naphthalene	128	33	0.026	0.23	3×10^{-4}	1.15×10^{-3}	4.9×10^{-2}
Nitrobenzene	123	2×10^3	16.3	0.27	3.5×10^{-4}	2.2×10^{-5}	9.3×10^{-4}
n-Octane	114	0.66	5.8×10^{-3}	14.1	1.85×10^{-2}	3.2	136
Pentachlorophenol	266	14	5.3×10^{-2}	1.4×10^{-4}	1.8×10^{-7}	3.4×10^{-6}	1.5×10^{-4}
Perchloroethylene	166	400	2.4	14.3	2×10^{-2}	8.3×10^{-3}	0.34
Phenanthrene	178	1.29	7×10^{-3}	2.1×10^{-4}	2.8×10^{-7}	3.9×10^{-5}	1.7×10^{-3}
Toluene	92	515	5.6	28.4	3.7×10^{-2}	6.6×10^{-3}	0.28
Trichloroethylene	131	1×10^3	7.6	60	8×10^{-2}	1×10^{-2}	0.42
1,1,1-Trichloroethane	133	950	7.1	100	0.13	1.8×10^{-2}	0.77
2,2,4-Trimethyl pentane	114	2.44	2.1×10^{-2}	49.3	6.5×10^{-2}	3.1	129
Vinyl chloride	62.5	90	1.44	2580	3.4	2.4	99
o-Xylene	106	175	1.7	6.6	8.7×10^{-3}	5.1×10^{-3}	0.22

Source: Bedient et al. (1994).

8.3.8 Hydrolysis

Hydrolysis is a chemical transformation process in which an organic molecule, RX, reacts with water. The reaction results in the introduction of a hydroxyl group (OH) into the organic compound:

$$RX + H_2O = ROH + H^+ + X^-$$

An example of this reaction is

$$CH_3CH_2CH_2Br + H_2O = CH_3CH_2CH_2OH + H^+$$
$$+ Br^-$$

Hydrolysis is likely to be the most important reaction of organic compounds with water in an aqueous environment. Table 8.8 shows the typical organic compounds that are potentially susceptible to hydrolysis, as well as those generally resistant to hydrolysis.

The hydrolysis reactions are first order with respect to the concentration of the chemical reactant:

Table 8.8(a) Types of organic functional groups that are potentially susceptible to hydrolysis

Alkyl halides	Nitriles
Amides	Phosphonic acid esters
Amines	Phosphoric acid esters
Carbamates	Sulfinic acid esters
Carboxylic acid esters	Sulfuric acid esters
Epoxides	

Table 8.8(b) Types of organic functional groups that are generally resistant to hydrolysis

Alkanes	Aromatic amines
Alkenes	Alcohols
Alkynes	Aldehydes
Aromatic nitro compounds	Carboxylic acids
Benzenes/biphenyls	Ethers
Dieldrin/aldrin and related halogenated	Glycols
hydrocarbon pesticides	Ketones
Polycyclic aromatic hydrocarbons	Phenols
Halogenated aromatics/PCBs	Sulfonic acids
Heterocyclic polycyclic aromatic	
hydrocarbons	

Source: Domenico and Schwartz (1998). This material is used by permission of John Wiley & Sons, Inc.

$$-\frac{d[RX]}{dt} = K_T[RX]$$

where K_T is known as the hydrolysis rate constant. This first-order dependence implies that the hydrolysis half-life is independent of the RX concentration. Reaction half-lives range from several seconds to thousands of years. First-order rate equation is an oversimplification for most organic hydrolysis reactions. K_T can be expressed as

$$K_T = K_H[H^+] + K_0 + K_{OH}[OH^-]$$

where K_H is the rate constant for specific acid-catalyzed hydrolysis or catalysis by hydronium ion, H^+; K_0 is the rate constant for neutral hydrolysis; and K_{OH} is the rate constant for specific acid-catalyzed hydrolysis or catalysis by hydroxide ion, OH^-.

8.4 BIOLOGICAL PROCESS (BIODEGRADATION)

Biodegradation, an oxidation–reduction reaction that is mediated by microorganisms, is the biological process that must be considered for organic contaminants. In general, an organic compound is oxidized (loses electrons) by an electron acceptor, which is reduced (gains electrons). This can occur in the following two ways: aerobic biodegradation and anaerobic biodegradation. *Aerobic biodegradation* occurs under aerobic or oxic environmental conditions in which oxygen, when present, commonly acts as the electron acceptor. The oxidation of organic compounds, coupled with the reduction of molecular oxygen, is termed *aerobic heterotrophic respiration.* *Anaerobic biodegradation* occurs under anaerobic or anoxic conditions (O_2 not present). Microorganisms can use organic chemicals or inorganic anions as alternative electron acceptors.

Anaerobic biodegradation can occur under fermentative, denitrifying, iron-reducing, sulfate-reducing, or methanogenic conditions. Fermenting organisms utilize their substrata as both an electron donor and an acceptor. In this process an organic compound is metabolized (metabolism = degradation of organic compounds to provide energy and carbon for growth), with a portion of that compound becoming a reduced end product and another becoming an oxidized product:

$$starch \rightarrow CO_2 \ (oxidized \ product)$$

$$+ \ ethanol \ (reduced \ product)$$

Denitrifying bacteria utilize NO_3^- as the electron acceptor and reduce it to NO_2^-, N_2O, and N_2. Iron- or sulfate-reducing bacteria utilize ferric iron or SO_4^{2-} as electron acceptors and reduce them to ferrous iron and H_2S. Methanogens or methane-producing bacteria utilize CO_2 as an electron acceptor.

The organisms are divided into two types on the basis of their ability to function: (1) *oligotrophs*— active in the presence of low concentrations of organic carbon and effective for low-contaminant concentrations; (2) *eutrophs*—active under conditions of high organic carbon (do not function at low organic carbon levels) and effective for high contaminant concentrations. On the basis of nutrition, organisms are grouped into three types: (1) *chemotrophs*, which capture energy from the oxidation of organic or inorganic materials; (2) *autotrophs*, which are capable of synthesizing their cell carbon from simple compounds such as CO_2; and (3) *heterotrophs*, which require a fixed organic source of carbon.

Six basic requirements must be met for biodegradation to occur (Bedient, et al 1994):

1. Presence of the appropriate organisms: bacteria and indigenous microorganisms to degrade specific contaminants. Organisms can be natural or genetically engineered.

2. Energy source—organic carbon is required as an energy source and is used by the organisms for cell maintenance and growth. The organic carbon is transformed into inorganic carbon, energy, and electrons.

3. Carbon source—about 50% of the dry weight of bacteria is carbon. Organic chemicals serve as both carbon and energy sources. As a carbon source, organic carbon is used in conjunction with energy to generate new cells.

4. Electron acceptor—some chemicals must accept the electrons released from the energy source.

Typically, acceptors include O_2, NO_3^-, SO_4^{2-}, and CO_2 as follows: $e^- + O_2 = H_2O$; $e^- + NO_3^- = N_2$; $e^- + SO_4^{-2} = H_2S$; $e^- + CO_2 = CH_4$.

5. Nutrients—the required nutrients include nitrogen, phosphorus, calcium, magnesium, iron, and trace elements.

6. Acceptable environmental conditions—these include temperature, pH, salinity, hydrostatic pressure, radiation, and low levels of heavy metals, or other toxic materials.

Microbial population in soils and groundwater can be determined based on direct microscopic, cultivation, metabolic, or biochemical methods. Based on these investigations, the natural microorganism populations in soils and groundwater are generally found in the range 1×10^5 to 1×10^7 cells per gram dry weight in soils or cells per milliliter in groundwater. In most situations, we want to know if the microorganisms are metabolically active, the diversity of their metabolism, the factors affecting stimulation and/or limiting their growth and activity, and the advantage of their metabolism for use in soil or groundwater remediation. These questions are answered by field and laboratory investigations. In all such investigations, the first task is to collect representative field soil. The sampling of the soils must be such that no cross-contamination occurs due to drilling equipment, surface soils, and so on. After collecting the soil, the top and bottom portions, as well a few outer centimeters of the soil, are discarded and the central portion is used for analysis. The samples are dissected as soon as possible with a sterilized paring device. For anaerobic conditions, the samples are dissected inside plastic anaerobic glove bags. Samples prepared this way are termed *asceptically obtained* and are suitable for microbiological analysis.

Numerous methods are reported to detect microorganisms and to estimate their biomass and metabolic activities. These methods include:

- *Microscopic methods:* direct light and electron microscopy
- *Cultural methods:* standard procedures, which include plate counts, etc.

- *Biochemical indicators:* indicators of metabolic activity
- *Radioscopic methods:* measure metabolic activity and growth of microorganisms
- *Microcosms*

Microcosm study is an attempt to bring an intact, minimally disturbed piece of an ecosystem into the laboratory for study in its natural state. It is the establishment of a physical model or simulation of part of the ecosystem in the laboratory within definable physical and chemical boundaries under a controlled set of experimental conditions. Microcosms are used extensively to determine the biodegradability of organic contaminants at the laboratory scale and under the influence of the site-specific physical, chemical, and hydrogeologic characteristics. Microcosms range from simple batch incubation systems to large, complex flow-through devices. The microcosms help to (1) identify biodegradable pollutants, and (2) determine the metabolic pathways of biotic or abiotic transformations. Decay of a particular pollutant in a microcosm may define the rate of biotransformation in situ without relying on indirect measures of microbial biodegradation for rate prediction. The advantages of microcosms are the appropriate controls and time efficiency as compared to field tests. The limitations include the structure of the ecosystem and scale effects. Studies have shown that most hydrocarbons, such as gasoline, crude oil, and heating oil, are biodegradable.

During aerobic biodegradation, the removal of the hydrocarbons, the consumption of oxygen in the process, and the growth of microbes in the aquifer can be described by the following equations, which are modifications of the Monod function (also known as the Michaelis–Menten function) (Borden and Bedient, 1986):

$$\frac{dH}{dt} = -M_t h_u \frac{H}{K_h + H} \frac{O}{K_o + O} \qquad (8.27a)$$

$$\frac{dO}{dt} = -M_t h_u G \frac{H}{K_h + H} \frac{O}{K_o + O} \qquad (8.27b)$$

$$\frac{dM_t}{dt} = M_t h_u Y \frac{H}{K_h + H} \frac{O}{K_o + O} + k_c Y C_{oc} - bM_t$$

$$(8.27c)$$

where H is the hydrocarbon concentration in pore fluid, O the oxygen concentration in pore fluid, M_t the total aerobic microbial concentration, h_u the maximum hydrocarbon utilization rate per unit mass of aerobic microorganisms, Y the microbial yield coefficient, K_h the hydrocarbon half-saturation constant, K_o the oxygen half-saturation constant, C_{oc} the natural organic carbon concentration, b the microbial decay rate, and G the rate of oxygen to hydrocarbon consumed.

Anaerobic decomposition of hydrocarbons can be described by another variation of the Monod function, which describes two-step catalyst chemical reactions (Bouwer and McCarty, 1984). This function is

$$\frac{dH}{dt} = -h_{ua}M_a \frac{H}{K_a + H} \qquad (8.28)$$

where M_a is the total mass of anaerobic microbes, h_{ua} the maximum hydrocarbon utilization rate per unit mass of anaerobic microbes, and K_a the half-maximum rate concentration of the hydrocarbon for anaerobic decay.

8.5   CONTAMINANT TRANSPORT AND FATE MODELING

Contaminant transport and fate modeling is performed for risk analysis as well as for evaluation and design of waste containment and site remediation systems. Incorporating all of the transport processes, chemical mass transfer processes, and biodegradation into a mathematical model to predict the contaminant form and concentration in space and time is a difficult task. Different geochemical models (e.g., MINTEQA2 and PHREEQC) have been developed that take into account various chemical mass transfer processes and are useful to determine speciation and distribution of inorganic contaminants. On the other hand, different multiphase and multispecies (multiple contaminants) models (e.g., UTCHEM and TOUGH2) have been developed that take into account various geochemical processes to determine the distribution of multiple organic contaminants in different phases (e.g., dissolved, sorbed, gases). These models may also incorporate transport processes; therefore, they also describe the transport of the contaminants in the subsurface. Geochemical and multiphase models are not presented here because these are not used routinely in engineering practice. However, simple mathematical models that incorporate advection, diffusion, dispersion, sorption, and biodegradation are commonly used in engineering practice to assess the fate and transport of individual dissolved contaminants and are described in this section.

Most models are based on the following governing partial differential equation derived from the conservation of mass principle (continuity equation), whereby the net rate of change of contaminant mass within a volume of porous media is equal to the difference between the flux of contaminant into and out of the volume, adjusted for the loss or gain of contaminant due to chemical and biological reactions (Freeze and Cherry, 1979; Zheng and Bennett, 2002):

$$R \frac{\partial C}{\partial t}$$

$$= \left[\frac{\partial}{\partial x} \left(D_x \frac{\partial C}{\partial x} \right) + \frac{\partial}{\partial y} \left(D_y \frac{\partial C}{\partial y} \right) + \frac{\partial}{\partial z} \left(D_z \frac{\partial C}{\partial z} \right) \right]$$

$$- \left[\frac{\partial}{\partial x} (V_{sx}C) + \frac{\partial}{\partial y} (V_{sy}C) + \frac{\partial}{\partial z} (V_{sz}C) \right] \pm \lambda RC$$

$$(8.29)$$

where D_x, D_y, and D_z are dispersion coefficients and V_{sx}, V_{sy}, and V_{sz} are seepage velocities in x, y, and z directions, respectively; C is dissolved contaminant concentration; t is time; R is retarding coefficient; and λ is rate of decay coefficient for both the dissolved and absorbed phases.

The general equation (8.29) assumes linear sorption and first-order chemical–biological transformation in both dissolved and adsorbed phases. The governing equation can be solved by either analytical or numerical methods. Analytical methods involve the solution of the partial differential equations using calculus based on the initial and boundary conditions. They are limited to simple geometry and in general require that the aquifer be homogeneous. Numerical methods involve solution of the partial differential equation by numerical methods of analysis. The most common analytical and numerical methods are described in the following sections.

8.5.1 Analytical Methods

To obtain a unique analytical solution, a complete set of boundary and initial conditions must be specified. *Initial conditions* are used to define the contaminant concentration in porous media just prior to the beginning of contaminant transport. The *boundary conditions* specify the interaction between the area under investigation and its external environment. The following three types of boundary conditions are used:

- The first type (or Dirichlet) boundary specifies the value of the concentration along a section of the flow system boundary.
- The second type (or Neumann) boundary condition specifies the gradient in contaminant concentration across a section of the boundary.
- The third type (or Cauchy) boundary condition is applied where the flux of contaminant across the boundary is dependent on the difference between a specified concentration value on one side of the boundary and the contaminant concentration on the opposite side of the boundary.

These three types of boundary conditions are used to describe conditions at the inflow and outflow ends of the flow system (one-dimensional case) and also along the lateral boundaries of two- and three-dimensional systems. Several one-, two-, and three-dimensional analytical solutions have been reported, depending on the transport, chemical, and biological processes included as well as initial and boundary conditions assigned (Van Genuchten and Alves, 1982; Javandel et al., 1984; Wexler, 1992; Fetter, 1999; Choy and Reible, 2000); however, a few selected analytical solutions are presented below.

One-Dimensional Solutions. Assuming uniform average linear velocity (V_s), linear sorption, and first-order decay, the governing equation for one-dimensional contaminant transport in a homogeneous, isotropic porous media is given by (Fetter, 1999)

$$\frac{\partial C}{\partial t} = D_L \frac{\partial^2 C}{\partial x^2} - V_s \frac{\partial C}{\partial x} - \frac{\rho_d}{n} \frac{\partial S}{\partial t} + \left(\frac{\partial C}{\partial t}\right)_{rxn} \quad (8.30)$$

where C is the concentration of contaminant in the liquid phase, t the time, D_L the longitudinal dispersion coefficient, V_s the average linear groundwater velocity, ρ_d the bulk density of aquifer, n the porosity for saturated media, S the amount of contaminant sorbed per unit weight of solid, and rxn a subscript indicating a biological or chemical reaction of the contaminant (other than sorption). Neglecting reactions and with linear isotherm ($S = K_d C$), equation (8.30) becomes

$$\frac{\partial C}{\partial t} = \frac{D_L}{R} \frac{\partial^2 C}{\partial x^2} - \frac{V_s}{R} \frac{\partial C}{\partial x} \quad (8.31)$$

where

$$R = 1 + \frac{\rho_d}{n} K_d \quad (8.32)$$

Neglecting both sorption and reactions, equation (8.30) becomes

$$\frac{\partial C}{\partial t} = D_L \frac{\partial^2 C}{\partial x^2} - V_S \frac{\partial C}{\partial x} \quad (8.33)$$

Several analytical solutions have been reported in the literature for equations (8.30), (8.31), and (8.33) with different initial and boundary conditions, and a few selected examples are presented below.

1. Assume the following initial and boundary (first-type) conditions:

$$C(x,0) = 0 \qquad x \geq 0 \quad \text{initial condition}$$

$$\left.\begin{array}{l} C(0,t) = C_0 \quad t \geq 0 \\[2mm] \dfrac{\partial C(\infty,t)}{\partial x} = 0 \quad t \geq 0 \end{array}\right\} \text{boundary conditions} \quad (8.34)$$

The solution to equation (8.33) for these conditions is (Ogata and Banks, 1961)

$$C(x,t) = \frac{C_0}{2}\left[\text{erfc}\left(\frac{x - V_s t}{2\sqrt{D_L t}}\right) \right.$$
$$\left. + \exp\left(\frac{V_s x}{D_L}\right) \text{erfc}\left(\frac{x + V_s t}{2\sqrt{D_L t}}\right) \right] \quad (8.35)$$

If the initial condition $C(x,0) = C_i$ and for the same boundary conditions as above,

$$C(x,t) = C_i + \frac{C_0 - C_i}{2}\left[\text{erfc}\left(\frac{x - V_s t}{2\sqrt{D_L t}}\right) \right.$$
$$\left. + \exp\left(\frac{V_s x}{D_L}\right) \text{erfc}\left(\frac{x + V_s t}{2\sqrt{D_L t}}\right) \right] \quad (8.35a)$$

2. Assume the following initial and boundary (second-type) conditions:

$$\begin{array}{ll} C(x,0) = 0, & -\infty < x < +\infty \\ & \text{initial condition} \end{array}$$
$$\left.\begin{array}{ll} \displaystyle\int_{-\infty}^{+\infty} n_e C(x,t)\,dx = C_0 n_e V_S t, & t > 0 \\ C(\infty,t) = 0, & t \geq 0 \end{array}\right\} \begin{array}{l}\text{boundary} \\ \text{condition}\end{array}$$

$$(8.36)$$

The first boundary condition states that the injected mass of contaminant over the domain from $-\infty$ to $+\infty$ is proportional to the length of time of the injection. The solution to equation (8.33) (Sauty, 1980) is

$$C = \frac{C_0}{2}\left[\text{erfc}\left(\frac{x - V_S t}{2\sqrt{D_L t}}\right) \right.$$
$$\left. - \exp\left(\frac{V_S x}{D_L}\right) \text{erfc}\left(\frac{x + V_S t}{2\sqrt{D_L t}}\right) \right] \quad (8.37)$$

Note that the difference between equations (8.35) and (8.37) is that second term is subtracted rather than added in equation (8.37).

3. Assume the following initial and boundary (third-type) conditions (Van Genuchten, 1981):

$$C(x,0) = 0 \qquad \text{initial condition}$$

$$\left. \left(-D\frac{\partial C}{\partial x} + V_s C \right) \right|_{x=0} = V_s C_0 \left. \vphantom{\frac{\partial C}{\partial x}} \right\} \text{boundary}$$

$$\left. \frac{\partial C}{\partial x} \right|_{x\to\infty} = \text{(finite)} \left. \vphantom{\frac{\partial C}{\partial x}} \right\} \text{conditions} \qquad (8.38)$$

The second boundary condition specifies that as x approaches infinity, the concentration gradient will still be finite. Under these conditions, the solution to equation (8.33) is

$$
\begin{aligned}
C = \frac{C_0}{2} &\left[\text{erfc}\left(\frac{x - V_s t}{2\sqrt{D_L t}} \right) \right. \\
&+ \left(\frac{V_s^2 t}{\pi D_L} \right)^{1/2} \exp\left[-\frac{(x - V_s t)^2}{4 D_L t} \right] \\
&- \frac{1}{2}\left(1 + \frac{V_s x}{D_L} + \frac{V_s^2 t}{D_L} \right) \\
&\left. \exp\left(\frac{V_s x}{D_L} \right) \text{erfc}\left(\frac{x - V_s t}{2\sqrt{D_L t}} \right) \right]
\end{aligned}
\qquad (8.39)
$$

The second term on the righthand side of Equations (8.35) and (8.37) and both the second and the third terms on the right hand side of Equation (8.39) are commonly neglected in practical applications for ease for calculations, and the approximate solution for all of the boundary conditions above becomes

$$C = \frac{C_0}{2}\left[\text{erfc}\left(\frac{x - V_s t}{2\sqrt{D_L t}} \right) \right] \qquad (8.40)$$

It should be noted that this approximate solution introduces errors that can be large. The Peclet number (Pe), is often used to assess this error:

$$\text{Pe} = \frac{V_s x}{D_L} \qquad (8.41)$$

The Peclet number is useful in determining the relative importance of advection and dispersion. A Peclet number greater than 1000 guarantees errors of less than 2% in concentration and less than 1% in velocity (Ogata and Banks, 1961; Binning, 2000). Ideally, full analytical solution given in Equations (8.35), (8.37), and (8.39) should be used to avoid any errors in calculating concentrations and velocities.

Two-Dimensional Solutions. In a homogeneous medium with *a uniform velocity field*, equation (8.29) for two-dimensional flow with the direction of *flow parallel to x axis* becomes

$$D_L \frac{\partial^2 C}{\partial x^2} + D_T \frac{\partial^2 C}{\partial y^2} - V_S \frac{\partial C}{\partial x} = \frac{\partial C}{\partial t} \qquad (8.42)$$

where D_L is the longitudinal hydrodynamic dispersion and D_T is the transverse hydrodynamic dispersion. Note that only advection, diffusion and dispersion transport processes are considered and both sorption and reactions are neglected for this case. Several solutions are presented in the literature for different initial and boundary conditions, and a few examples of such solutions are presented below.

Continuous Injection into a Uniform Two-Dimensional Flow Field. If a tracer is injected continuously into a uniform field from a single point that fully penetrates the aquifer, a two-dimensional plume will form. It will spread along the axis of flow due to longitudinal dispersion and normal to the axis of flow due to transverse dispersion. This is the type of contamination that would spread from the use of an injection well, which would be a point source.

Assume that the well is located at the origin ($x = 0$, $y = 0$) and there is a uniform flow velocity at a rate V_s parallel to the x axis. There is a continuous injection at the origin of a contaminant with a concentration C_0 at a rate Q over the aquifer thickness, b. The solution to equation (8.42) under steady-state conditions is as follows (Fetter, 1999):

$$C(x,y) = \frac{C_0(Q/b)}{2\pi\sqrt{D_L D_T}} \exp\left(\frac{V_s x}{2D_L}\right) K_0$$

$$\left[\left(\frac{V_s^2}{4D_L}\left(\frac{x^2}{D_L} + \frac{y^2}{D_T}\right)\right)^{1/2}\right] \quad (8.43)$$

where K_0 is the modified Bessell function of the second kind and zero order, Q the rate that the contaminant is injected, and b the thickness of the aquifer over which the contaminant is injected. K_0 values as a function of x are given in Hantush (1956) and Fetter (1999).

Equation (8.43) can also be solved for a specific value of time, so that the spread of a two-dimensional plume with time can be determined. This solution is given by (Hantush, 1956; Fetter, 1999)

$$C(x,y,t) = \frac{C_0(Q/b)}{4\pi\sqrt{D_L D_T}} \exp\left(\frac{V_s x}{2D_L}\right)[W(0,B)$$

$$- W(t_D,B)] \quad (8.44)$$

where t_D and B are defined as

$$t_D = \frac{V_s^2 t}{D_L} \quad (8.44a)$$

$$B = \sqrt{\frac{V_s^2 x^2}{4D_L^2} + \frac{V_s^2 y^2}{4D_L D_T}} \quad (8.44b)$$

t_D is a dimensionless form of time. Values for $W[t_D,B]$, known as the well function, are given in Hantush (1956) and Fetter (1999).

Slug Injection into a Uniform Two-Dimensional Flow Field. If a slug of contamination is injected over the full thickness of a two-dimensional uniform flow field in a short period of time, it will move in the direction of flow and spread with time. If a tracer with concentration C_0 is injected into a two-dimensional flow field over an area A at a point (x_0, y_0), the concentration at a point (x,y) at time t after the injection is given by (Fetter, 1999)

$$C(x,y,t) = \frac{C_0 A}{4\pi t\sqrt{D_L D_T}}$$

$$\exp\left[-\frac{[(x - x_0) - V_s t]^2}{4D_L t} - \frac{(y - y_0)^2}{4D_T t}\right] \quad (8.45)$$

Three-Dimensional Solutions. The solution obtained for slug injection into a uniform two-dimensional flow field [equation (8.45)] can also be extended into a uniform three-dimensional flow field $(x, y,$ and $z)$. For such three-dimensional conditions, the solution is given by (Bedient et al., 1994)

$$C(x,y,z,t) = \frac{C_0 V_0}{8(\pi t)^{3/2}(D_L D_T D_Z)^{1/2}}$$

$$\exp\left[-\frac{[(x - x_0) - V_s t]^2}{4D_L t} - \frac{(y - y_0)^2}{4D_T t} - \frac{(z - z_0)^2}{4D_z t}\right]$$

$$(8.46)$$

where V_0 is the original volume of slug injected, D_Z the dispersion coefficient in the z-direction, and all other terms are as defined above. Note that Equation (8.46) assumes advection, diffusion, and dispersion and ignores sorption and reactions.

Incorporation of Sorption and Reactions. It should be noted that the analytical solutions above assume that sorption and reactions are negligible and the contaminant transport is governed by advection, diffusion, and dispersion. Several analytical solutions have been reported that account for sorption (using a retardation coefficient of R) and reactions (using the first-order decay constant, λ). More details may be found in Section 8.6 and references such as Van Genuchten and Alves (1982), Javandel et al. (1984), Wexler (1992), Fetter (1999), Choy and Reible (2000), and Zheng and Bennett (2002).

8.5.2 Numerical Methods

Numerical methods are used when the flow system possesses irregular geometry and nonuniform aquifer properties. Recently, many numerical contaminant transport models have been developed. Similar to the numerical models for groundwater flow, numerical transport models use numerical methods (commonly, the finite difference method and finite element method) to solve governing equations with appropriate initial and boundary conditions.

Numerical models that utilize the flow predictions based on MODFLOW have received wide attention from practitioners. These models use the same grid system as that used for MODFLOW and perform contaminant transport calculations with assigned sources and sinks. MT3D and RT3D are the two numerical models used extensively in the practice for contaminant transport and fate analysis and are explained briefly below.

MT3D, a modular three-dimensional transport model, complements the MODFLOW flow simulation model (Zheng, 1990). MT3D, based on the finite difference method, solves the advection–dispersion equation with sorption and a first-order reaction. The reaction model controls radioactive decay or biodegradation that is treated as first-order decay, and sorption using the linear, Freundlich, or Langmuir model. The model uses the same grid as MODFLOW and uses flow fields calculated during a corresponding MODFLOW model run. MT3D can simulate the same two- and three-dimensional scenarios as those simulated in MODFLOW.

RT3D, a reactive transport in three dimensions model, was developed at Battelle Laboratories by Clement (1997). RT3D is based on an MT3D finite difference numerical scheme and provides additional multispecies modeling. RT3D can simulate multiple sorbed and aqueous-phase species with any one of seven predefined reaction frameworks or any other novel framework that the user may define. This allows simulation of natural attenuation and different active remedial processes. Most of the reactions modeled by RT3D are for simulation of various biodegradation processes. RT3D preprogrammed reaction packages include:

- Instantaneous aerobic decay of BTEX
- Instantaneous degradation of BTEX using multiple electron acceptors

- Kinetically limited hydrocarbon biodegradation using multiple electron acceptors
- Rate-limited sorption reactions
- Double Monod method
- Sequential decay reactions and an aerobic–anaerobic model for PCE/TCE degradation

A systematic approach to the application of the numerical contaminant transport model selected is shown in Figure 8.8. A site investigation must precede modeling at the site of concern. A modeling work plan should be prepared with the following components:

- *Modeling goals.* The goals and reasons for the modeling should be specific and measurable.
- *Conceptual site model for modeling.* Based on available data, a geologic and hydrogeologic conceptual site model should be prepared.
- *Technical approach.* Technical considerations to achieve the stated goal such as numerical or analytical, modeling dimensions (one-, two-, or three-dimensional), saturated or unsaturated flow, multiphase or single phase, reactive or nonreactive, dispersion, retardation, and/or degradation.
- *Computer model.* Select the computer model. Discussion of the abilities and limitations of the model in general, and also in terms of the site-specific conceptual model, should be included. Site-specific input parameters are preferred. Boundary and initial conditions should be described.
- *Sensitivity analysis.* To model sensitivity with the variations in calibrated input parameters within reasonable ranges to find more sensitive parameters.
- *Model calibration.* Calibrate the model to reproduce appropriate field-measured parameters. The parameters will be different for different models and for different transporting media. Calibration establishes that the model can reproduce field conditions. Models have no predictive value if they cannot reproduce observed concentrations.
- *Model verification.* Plan to reproduce a second set of field data using the set of calibrated parameter

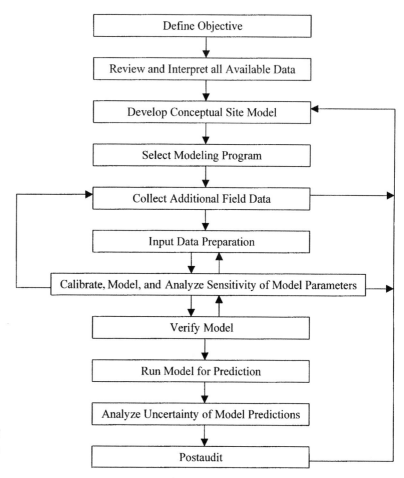

Figure 8.8 *Steps in contaminated site model application. (From ADEC, 1998.)*

values of a calibrated model may not represent the contaminated site conditions. If the parameter values established during calibration change during verification, the model should be calibrated again with the changed parameter values. By reproducing a second set of field data, it may be possible to verify a model. Collection of a second set of field data is often possible at sites with multiple observation wells.

- *Uncertain analysis.* Describe the uncertainty associated with the modeling of a given problem. The uncertainty in modeling exists because of uncertainties in the transport mechanism, sink/ source within the transporting media, temporal and spatial variation of model parameters, initial and boundary conditions, and matrix heterogeneity.

- *Postaudit of modeling.* Plan to conduct a postaudit several years after the modeling study is completed. Collect new field data to determine whether the prediction was correct. Analyze what went wrong with modeling and means to improve it.

- *Modeling report.* A final modeling report should describe model predictions, technical analysis, input parameters, results of parameter sensitivity, and model uncertainty analysis. Critical input and output files should be in the report's appendix as well as on portable computer disks. The report should have sufficient information for an independent reviewer to duplicate model runs.

8.6 APPLICATIONS

Contaminant transport and fate modeling is performed to assess and/or design geoenvironmental engineering systems such as landfills, impoundments, in-situ barriers, and groundwater remedial systems. More details on these systems are provided in other chapters of this book. A few applications of contaminant transport and fate modeling for these geoenvironmental systems are presented below.

8.6.1 Landfills and Surface Impoundments

As shown in Figure 8.9, landfills and impoundments are designed with liner systems to protect groundwater quality from the disposed wastes. The liner systems consist of low-permeable materials (compacted clay, geomembrane, geosynthetic clay liner) and must contain the disposed wastes effectively. While designing the liner systems, contaminant transport modeling is performed to determine potential breakthrough of the waste constituents (contaminants). Regulations also require that a liner system be designed such that at no time should the concentrations of a selected set of contaminants (see Chapter 17) exceed the values specified in the uppermost aquifer at the point of compliance or the property boundary, whichever is less.

Because of the low permeability, the primary contaminant transport mechanism in liner systems is diffusion. In addition to diffusion, advection and sorption effects are commonly taken into account in assessing contaminant transport through liner systems. Because of the large areal extent of a landfill compared to the liner and subsurface profile to the upper aquifer, an assumption of one-dimensional contaminant transport (in the vertical direction or z direction) is valid. Thus, based on equation (8.31), the governing equation to describe contaminant transport through the liner systems of landfills and impoundments is given by

$$\frac{\partial C}{\partial t} = \frac{D^*}{R}\frac{\partial^2 C}{\partial z^2} - \frac{V_s}{R}\frac{\partial C}{\partial z} \qquad (8.47)$$

where C is the contaminant concentration, t the elapsed time, D^* the effective diffusion coefficient, z the distance in the direction of flow, V_s the seepage velocity, and R the retardation coefficient. Since a linear sorption isotherm ($S = K_d C$) is assumed, the retardation coefficient, R is:

$$R = 1 + \frac{\rho_d}{n}K_d \qquad (8.47a)$$

where n and ρ_d are porosity and dry density of the transport medium, respectively.

Equation 8.47 assumes no reactions. As explained in Section 8.5.1, equation (8.47) can be solved using different initial and boundary conditions. The following initial and boundary conditions are commonly applicable for liner systems in new landfills:

$$C(z,0) = 0 \qquad z \geq 0 \text{ initial condition}$$

$$\left.\begin{array}{ll} C(0,t) = C_0 & t \geq 0 \\ C(\infty,t) = 0 & t \geq 0 \end{array}\right\} \text{ boundary conditions} \quad (8.47b)$$

For these initial and boundary conditions, the solution to equation (8.47) given by Ogata and Banks (1961) is as follows:

$$C(z,t) = \frac{C_0}{2}\left[\operatorname{erfc}\left(\frac{Rz - V_s t}{2\sqrt{D^* t R}}\right)\right.$$
$$\left. + \exp\left(\frac{V_s z}{D^*}\right)\operatorname{erfc}\left(\frac{Rz + V_s t}{2\sqrt{D^* t R}}\right)\right] \quad (8.48)$$

For a Peclet number [as defined in equation (8.41) > 1000, the solution above is approximated by

$$C(z,t) = \frac{C_0}{2}\operatorname{erfc}\left(\frac{Rz - V_s t}{2\sqrt{D^* t R}}\right) \qquad (8.49)$$

In diffusion-dominated systems (Pe < 1), Fick's second law is used to describe the transient diffusion, which is given mathematically by

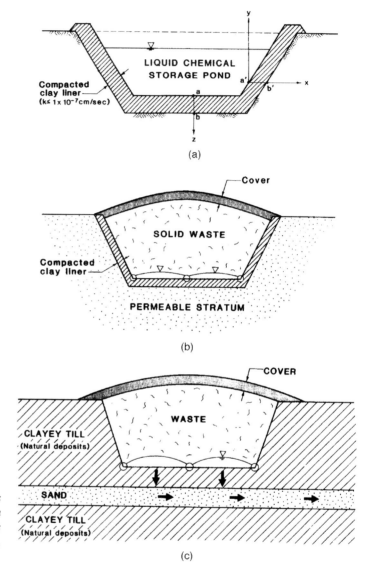

Figure 8.9 *Contaminant transport analysis of landfills and impoundments. (From Sharma and Lewis, 1994. This material is used by permission of John Wiley & Sons, Inc.)*

$$\frac{dC}{dt} = \frac{D^*}{R}\frac{d^2C}{dz^2} \qquad (8.50)$$

$$C(z,t) = C_0\,\mathrm{erfc}\left(\frac{z}{2\sqrt{D^*t/R}}\right) \qquad (8.51)$$

With initial and boundary conditions $C(z,0) = 0$, and $C(0,t) = C_0$ and $C(\infty,0) = 0$, the solution to the equation above is given by

where $C(z,t)$ is the concentration at distance z from the source at time t, C_0 the initial contaminant concentration, which is assumed to be constant, and erfc the complementary error function.

The analytical solutions above for equations (8.48), (8.49), and (8.51) are valid for a single homogeneous layer. When we need to simulate contaminant transport through a multilayer system, a numerical model such as POLLUTE is used (Rowe et al., 1984). In addition, in situations where vertical transport of contaminants through the liner system and the underlying low-permeability soils, followed by lateral migration in the uppermost aquifer, must be simulated, a combination of numerical models may be used. Models such as MIGRATE have been developed specifically to address such problems (Rowe and Booker, 1988).

8.6.2 In-Situ Barriers

In-situ barriers are constructed to prevent migration of contaminants in the subsurface. More details on in-situ barriers are given in Chapter 12. In-situ barriers, which include horizontal barriers (bottom barriers), and vertical barriers (cutoff walls or slurry walls) are commonly used at contaminated sites. Vertical barriers (Figure 8.10) are used commonly. These barriers are constructed with soil–bentonite or soil–cement–bentonite, which possess low permeability; therefore, contaminant transport through these walls due to advection is less significant, although diffusion is the primary transport mechanism.

Figure 8.11 shows bottom barrier and vertical barrier configurations. With uniform seepage velocity (V_x), linear sorption, and first-order decay, the governing equation for contaminant transport through these barrier systems is given by

$$\frac{\partial C}{\partial t} = D^* \frac{\partial^2 C}{\partial x^2} - V_x \frac{\partial C}{\partial x} - \frac{\rho_d}{n} \frac{\partial S}{\partial t} + \left(\frac{\partial C}{\partial t}\right)_{rxn} \quad (8.52)$$

where C is the concentration of solute in the liquid phase, t the time, D^* the effective molecular diffusion coefficient, V_x the seepage velocity, ρ_d the bulk density of barrier material, n the porosity of the barrier material, S the amount of solute sorbed per unit weight of solid, and rxn a subscript indicating a biological or chemical reaction of the solute (other than sorption). The seepage velocity is given by

$$V_x = -\frac{K \Delta h}{n_e L} \quad (8.52a)$$

where K is the effective hydraulic conductivity of the barrier material in the direction normal to the plane of the barrier, L the barrier thickness, and Δh the difference in piezometric head across the barrier wall. For inward seepage, V_x would be negative.

Neglecting reactions, equation (8.52) becomes the same as equation (8.47). With initial condition $C(x,0) = 0$ and boundary conditions $C(0,t) = C_0$ (contamination side) and $C(x \to \infty, t) = 0$ (assuming that the contaminant transported out of the wall is removed), the analytical solution is given by equation (8.48), and if only diffusive flux is assumed, the solution is given by equation (8.51).

It is important to recognize that the one-dimensional system above assumes the flow media to be infinite. Instead of infinite system, finite systems may be considered in modeling. In finite systems such as cutoff wells, water containing a known concentration of the contaminant enters the system at the origin (at $x = 0$). Water and contaminant exit at opposite ends of the system (at $x = L$), which could represent the outer boundary of the wall. For such conditions, the governing equation (8.47) can be solved with the following initial and boundary conditions: $C(x,0) = 0$, $C(0,t) = C_0$, and $C(x = L,t) = 0$. The second boundary condition considers the finite length of the material. The solution given by Rabideau and Khandelwal (1998) is as follows:

$$C_{x,t} = C_0 \exp\left(\frac{xV_s}{2D}\right) \left[\frac{\sinh\{[(L - x)V_s]/2D\}}{\sinh LV_s/2D} \right.$$

$$+ \frac{2\pi}{L^2} \sum_{m=1}^{\infty} \frac{(-1)^m \, m \sin [m\pi(L - x)/L]}{(V_s^2/4D^2) + (m^2\pi^2/L^2)}$$

$$\left. \exp\left[-\left(\frac{V_s^2}{4D} + \frac{Dm^2\pi^2}{L^2}\right)\left(\frac{t}{R}\right) \right] \right]$$

$$(8.53)$$

Using the analytical solutions above [equations (8.48), (8.49), (8.51), and (8.53)], contaminant trans-

EXAMPLE 8.6

A 1-ft leachate head is accumulated over a 3-ft-thick clay liner. The leachate contained trichloroethylene (TCE) at a concentration of 100 mg/L. Calculate the distribution of TCE within the liner after 10 years for (a) advective–diffusion transport, and (b) diffusion transport only. Assume that $k = 5 \times 10^{-8}$ cm/s, $n_e = 0.4$, $D^* = 1 \times 10^{-6}$ cm^2/s, $\rho_d = 1.2$ g/cm^3, and $K_d = 8.0$ cm^3/g. Determine with and without sorption effects.

Solution:

Given $h = 1$ ft, $T = 3$ ft, $D^* = 1 \times 10^{-6}$ cm^2/s, $K = 5 \times 10^{-8}$ cm/s, $n_e = 0.4$, TCE $C_0 = 100$ mg/L, $\rho_d = 1.2$ g/cm^3, and $K_d = 8.0$ cm^3/g:

$$i = \frac{h + T}{T} = \frac{1 + 3}{3} = 1.33$$

$$V_s = \frac{Ki}{n_e} = \frac{(5 \times 10^{-8})(1.33)}{0.4} = 1.66 \times 10^{-7} \text{ cm/s}$$

Case (a), with no sorption: For $t = 10$ years, contaminant distribution due to advective–diffusion transport is calculated using equations (8.48) (exact solution) and (8.49) (approximate solution), and contaminant distribution due to diffusion only is calculated using equation (8.51). The results are summarized below.

	TCE Concentration (mg/L)		
Depth, Z (ft)	Advective-Diffusion [Exact Solution, Eq. (8.48)]	Advective-Diffusion [Approximated Solution, Eq. (8.49)]	Diffusion Only [Eq. (8.51)]
0	100.00	98.14	100.00
0.5	97.49	93.02	54.40
1.0	88.47	80.81	22.49
1.5	69.73	60.41	6.87
2.0	44.56	36.59	1.52
2.5	21.91	17.11	0.24
3.0	8.00	5.98	0.03

Case (b), with sorption:

$$R = 1 + (\rho_d/n_e)K_d = 25.0$$

For $t = 10$ years, contaminant distribution due to advective–diffusion transport is calculated using equations (8.48) (exact solution) and (8.49) (approximate solution), and contaminant distribution due to diffusion only is calculated using equation (8.51). The results are summarized below.

	TCE Concentration (mg/L)		
Depth Z (ft)	Advective-Diffusion [Exact Solution, Eq. (8.48)]	Advective-Diffusion [Approximated Solution, Eq. (8.49)]	Diffusion Only [Eq. (8.51)]
0	100.00	66.16	100.00
0.5	0.79	0.44	0.24
1.0	0	0	0
1.5	0	0	0
2.0	0	0	0
2.5	0	0	0
3.0	0	0	0

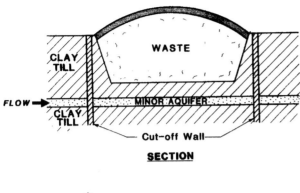

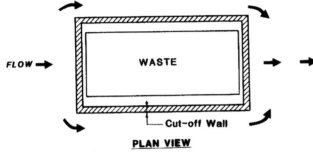

Figure 8.10 *Cutoff wall used to divert ground-water flow from beneath a natural clay barrier. (From Sharma and Lewis, 1994. This material is used by permission of John Wiley & Sons, Inc.)*

port through the barrier systems can be computed. It should be noted that (1) neglecting reactions and/or sorption in the analysis will normally yield conservative estimates of contaminant flux, and (2) in a finite-length system, the outflow boundary is close enough that it will have an effect on the magnitude of concentrations within the area of interest. If the outflow boundary is far away to have a negligible effect on contaminant concentrations in the area of interest, solutions for infinite system can be used and are generally easier to evaluate.

The flux calculated from the one-dimensional models can be used as source terms for two- and three-dimensional models for the fate and transport of contaminants outside the containment zone. These models are generally numerical models, as described in Section 8.5.2.

8.6.3 Groundwater Contamination

Contaminant transport modeling is very helpful in groundwater contamination assessment and remedia-

tion studies. Contaminant transport modeling is performed to determine the contaminant concentrations at exposure locations during risk assessment (refer to Chapter 11 for details). Contaminant transport modeling is also performed to assess the fate of contamination under different remedial scenarios, including a scenario in which no remedial action is proposed.

One-dimensional analytical solutions can be used to study some simple groundwater contamination conditions. Figure 8.12 shows an example using one-dimensional analytical solutions. This example shows horizontal groundwater flow from river *A*, which has been contaminated, to river *B*. It should be noted that the two-dimensional analytical solutions can be used for homogeneous aquifers with no vertical flow component. In two dimensions, the governing equation in two dimensions (*x* and *y* directions) is solved with relatively simple initial and boundary conditions. Selected examples of such analytical solutions are presented in Section 8.5.1.

Analytical solutions require constant parameters, simple geometries, and well-defined boundary condi-

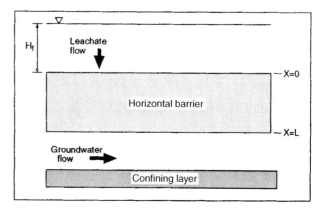

(a) bottom barrier

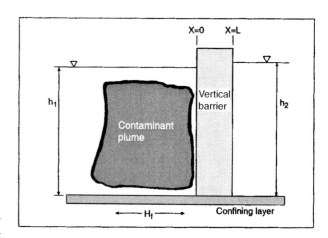

Figure 8.11 Schematic diagram of barrier configuration. (*From Mitchell and Rumer, 1998. Reproduced by permission of ASCE.*)

(b) vertical barrier wall

tions. For many site conditions, groundwater flow and
contaminant transport are complicated and two- and
three-dimensional numerical models may be required.

8.7 SUMMARY

In this chapter we presented various transport, chemical, and biological processes that affect contaminant
migration, transfer, and transformation in natural soil
and groundwater systems. An understanding of these
processes is essential for the design of effective waste
containment and remedial systems. For instance, if one

wants to evaluate the potential for leachate migration
through liner systems in a landfill, the contaminant
transport analysis must include the processes that affect
the site-specific materials and conditions. Similarly, in
the development or selection of a remediation method
for contaminated soils and groundwater, the processes
that effectively separate or transform the contaminants
under the site-specific contaminant, soil, and groundwater combinations must be assessed carefully. The
concepts presented here are explained further by citing
practical examples of landfill, surface impoundment,
containment barrier, and groundwater systems.

EXAMPLE 8.7

A 3-ft-thick cutoff wall is used to contain groundwater contaminated with benzene at a concentration of 1000 mg/L. Assume that $K = 1 \times 10^{-7}$ cm/s, $D^* = 1 \times 10^{-5}$ cm^2/s, $n_e = 0.4$, $\rho_d = 1.3$ g/cm^3, and $K_d = 3.0$ cm^3/g. The groundwater surface elevations inside and outside the wall are 500 ft and 501 ft, respectively (Figure E8.7). Calculate breakthrough contaminant concentration for a period of five years for (a) advection–diffusion transport, and (b) diffusive transport only. Calculate with and without sorption effects.

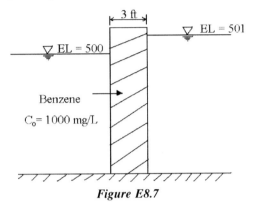

Figure E8.7

Solution:

Given $x = 3$ ft, $D^* = 1 \times 10^{-5}$ cm^2/s, benzene $C_0 = 1000$ mg/L, $K = 1 \times 10^{-7}$ cm/s, $n_e = 0.4$, $\rho_d = 1.2$ g/cm^2, and $K_d = 3.0$ cm^3/g:

$$i = \frac{h}{T} = \frac{500 - 501}{3} = -0.33$$

Negative hydraulic gradient implies that the flow is inward:

$$V_s = \frac{Ki}{n_e} = \frac{(1 \times 10^{-7})(-0.33)}{0.4} = -8.25 \times 10^{-8} \text{cm/s}$$

Case (a), without sorption effect: For $t = 5$ years, the breakthrough concentration (C_e) is calculated using equations (8.48), (8.49), and (8.51) by substituting $z = x = L = 3$ ft:

- Advective–diffusion transport [exact method, equation (8.48)], $C_e = 69.66$ mg/L
- Advective–diffusion transport (approximate method, equation (8.49)], $C_e = 31.45$ mg/L
- Diffusion transport only [equation (8.51)], $C_e = 103.46$ mg/L

The results show that the approximate method is not appropriate to use. In addition, because of the inward gradient, which is in the direction opposite to the diffusion, the mass flux is reduced. Assuming only diffusive transport will be very conservative.

Case (b), with sorption: $R = 1 + (\rho_d/n_e)K_d = 1 + (1.2/0.4)(3) = 10$. For $t = 5$ years, the breakthrough concentration (C_e) is calculated using equations (8.48), (8.49), and (8.51) by substituting $z = x = L = 3$ ft:

- Advective–diffusion transport [exact method, equation (8.48)], $C_e = 0$ mg/L
- Advective–diffusion transport [approximated method, equation (8.49)], $C_e = 0$ mg/L
- Diffusion transport only [equation (8.51)], $C_e = 0$ mg/L

These results show that there will be no breakthrough in a period of five years.

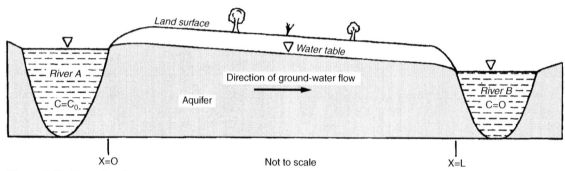

Figure 8.12 *Example of contaminant movement in field settings that can be simulated as one-dimensional solute-transport system.* (*From Wexler, 1992.*)

EXAMPLE 8.8

A chemical with a concentration of 1000 mg/L leaks continuously from an underground storage tank into an underlying homogeneous aquifer. Assume that $i = 0.05$, $K = 0.01$ cm/s, $n_e = 0.1$, and $\alpha_L = 100$ cm. A drinking water well is located 100 m away from the tank location. How long will it take for the chemical concentration to reach 100 mg/L at the drinking water well location? Assume that advection and dispersion control the mass transport.

Solution:
Seepage velocity v_x is calculated from Darcy's law:

$$V_x = \frac{Ki}{n_e} = \frac{0.01(0.05)}{0.1} = 0.005 \text{ cm/s}$$

$$D_x = \alpha_L v_x = 100(0.005) = 0.5 \text{ cm}^2/\text{s}$$

Using equation (8.49) (with $R = 1$) yields

$$100 = \frac{1000}{2} \, \text{erfc}\left(\frac{10,000 - 0.005t}{2\sqrt{0.5t}}\right)$$

$$0.2 = \text{erfc}\left(\frac{10,000 - 0.005t}{2\sqrt{0.5t}}\right)$$

Solving the equation above by trial and error, $t = 1,669,000$ s $= 19.3$ days.

EXAMPLE 8.9

A spill occurred over an area of 20 ft $\times$ 20 ft at a machine shop. Contaminant at a concentration of 1000 mg/L leaked into the underlying homogeneous aquifer. Assume that $K = 0.5$ ft/day, $n = 0.2$, $i = 1.0$, $D_L = 5$ ft^2/day, and $D_T = 0.5$ ft^2/day. Assuming that advection and dispersion control the transport, calculate the peak contaminant concentration that would occur at a drinking well located $x = 100$ ft and $y = 0$ ft from the spill location.

Solution:

Calculate the time required for a contaminant to reach 100 ft using

$$V_s = \frac{Ki}{n} = \frac{(0.5)(1)}{0.2} = 2.5 \text{ ft/day}$$

Assume no retardation (i.e., $R = 1$):

$$t = \frac{\text{flow distance}}{\text{flow velocity}} = \frac{100}{2.5} = 40 \text{ days}$$

Using equation (8.45) gives

$$C(x,y,t) = \frac{C_0 A}{4\pi t \sqrt{D_L D_T}} \, \exp\left[-\frac{[(x - x_0) - V_s t]^2}{4D_L t} - \frac{(y - y_0)^2}{4D_T t}\right]$$

Substitute $(x - x_0) = 100$ ft, $(y - y_0) = 0$, and $t = 40$ days; then the peak concentration at the well location (C_{max}) is calculated:

$$C_{max} = \frac{(1000)(20 \times 20)}{4\pi(40)\sqrt{(5)(0.5)}} \, \exp\left[-\frac{(100 - 2.5 \times 40)^2}{4(5)(40)} - \frac{(0)^2}{4(0.5)(40)}\right]$$

$$= 503 \text{ mg/L}$$

QUESTIONS/PROBLEMS

8.1. Dissolved chloride in a concentration of 1000 mg/L is being advected with flowing groundwater at a velocity of 0.3 m/day in an aquifer with a porosity of 0.25. Groundwater from the aquifer discharges into a river. What is the mass flux of chloride into the river if the aquifer is 2 m thick and 100 m wide where it discharges into the river?

8.2. Assuming a tortuosity factor of 0.5, determine the effective diffusion coefficients of OH^-, Cl^-, H^+, Na^+, and Cr^{3+}. Find the value of the concentration ratio (C_i/C_0) of these ionic species at a distance of 5 m in a silty clay layer after 100 years of diffusion.

8.3. At a landfill site, leachate accumulated over a very thick clay layer contains a chloride concentration of 1000 mg/L. If the tortousity is 0.5, what would be the concentration of chloride at a depth of 1 and 3 m after 10, 100, and 1000 years of diffusion? Neglect the effects of advection.

8.4. A one-dimensional column test was conducted on a soil sample. Deionized water was flushed through the column to saturate the soil sample. Then chloride solution at a concentration of 1000 mg/L was passed through the column. The length of the column was 0.5 m. The seepage velocity was 0.0005 m/s. The concentrations of effluent at different time periods are given below. Assume an effective molar diffusion coefficient of 1×10^{-10} m^2/s. Calculate the value of D_1 and α_1 for the soil in the column.

Time (s)	Concentration (mg/L)
500	200
700	400
1400	700
2000	960

8.5. Estimate the dispersivity and dispersion coefficient for groundwater flow between a contaminant source area and a water well 100 m downgradient that is potentially at risk. Assume a soil porosity of 40%, an effective molecular diffusivity coefficient of 10^{-5} cm^2/s, and a Darcy velocity of 10 m/year.

8.6. Chloride was injected as a continuous source into a one-dimensional column 50 cm long at a seepage velocity of 10^{-3} cm/s. The concentration measured after 1800 s from the beginning of the test was 0.3 of the initial concentration, and after 2700 seconds it reached 0.4 of the initial concentration. Assume that $\tau = 0.5$. Find the coefficient of dispersion and longitudinal dispersivity.

8.7. The batch sorption test results performed using kaolin and Cr(VI) are given below. Which adsorption isotherm fits the experimental test results best? Determine the appropriate coefficients for the isotherm selected.

Equilibrium Concentration (mg/L)	Cr(VI) Sorbed (mg/g)
0.7	0.0035
2.4	0.0258
7.4	0.0260
58.3	0.4168
453.0	0.4702
954.0	0.4603

8.8. Batch sorption test results performed using glacial till and Cr(VI) are given below. Which adsorption isotherm best fits the experimental test results? Determine the appropriate coefficients for the isotherm selected.

Equilibrium Concentration (mg/L)	Cr(VI) Sorbed (mg/gm)
0.75	0.00252
4.33	0.0067
8.73	0.0127
11.78	0.0322
15.68	0.0435
24.84	0.0511
32.7	0.073
40.84	0.0916
87.26	0.1274
215.52	0.3448
478.55	0.21446
977.8	0.222

8.9. Estimate the travel time for the following contaminants from the source area to a water well located 100 m away in an aquifer with bulk density of soil = 1.5 g/cm^3, n = 40%, organic carbon = 1%: **(a)** chloride ion (nonsorbing); **(b)** benzene; **(c)** TCE; **(d)** pyrene. The groundwater velocity is 10 m/yr.

8.10. An accidental spill from a point source introduced 10 kg of contaminant mass to an aquifer. The seepage velocity in the aquifer is 0.1 ft/day in the x-direction. The longitudinal dispersion coefficient D_L = 0.01 ft^2/day; the lateral and vertical dispersion coefficient $D_y = D_z$ = 0.001 ft^2/day. Calculate **(a)** the maximum concentration at x = 100 ft and t = 5 years; **(b)** concentration at a point x = 200 ft, y = 5 ft, z = 2 ft, five years after the spill.

8.11. A slug of contaminant was injected into a well for a tracer test (two dimensions). If the initial contaminant concentration is 1000 mg/L, the background seepage velocity in the aquifer is 0.022 m/day, the well radius is 0.05 m, the longitudinal dispersion coefficient is 0.034 m^2/day, and the transverse dispersion coefficient is 0.01 of the longitudinal dispersion coefficient, **(a)** what is the maximum concentration reached after 24 h? **(b)** Plot the concentration versus distance at t = 24 h.

8.12. For a 3-ft-thick clay liner with a hydraulic conductivity of 1×10^{-7} cm/s, estimate the breakthrough time for **(a)** advective transport only and **(b)** advective–dispersive transport for concentration of 2%. Assume a gradient = 1.0, a dispersion D_L = 1×10^{-6} cm^2/s, and R = 1.

8.13. A waste deposit is designed to be placed over a clay liner that has a hydraulic conductivity of 1×10^{-7} cm/s. Estimate the liner thickness so that the exit concentration at the base of the liner does not exceed 40% of the original contaminant concentration at the top of the liner. Assume that D = 5×10^{-6} cm^2/s, R = 1, hydraulic gradient = 1, and design life = 40 years.

8.14. For a 3-ft clay barrier with a 1-ft head, estimate transit times for C/C_0 = 0.5 and for **(a)** advection only, **(b)** diffusion only, and **(c)** advective–dispersion transport for the following clay properties: k = 5×10^{-8} cm/s, n = 0.5, and D = 6×10^{-6} cm^2/s.

8.15. Calculate the distribution of the contaminant through a 3-ft-thick clay liner of a surface impoundment at times 100, 1000 and 10,000 years for the following conditions: i = 1, k = 1×10^{-8} cm/s, n_e = 0.5, ρ_d = 1.25 g/cm^3, D^* = 3×10^{-6} cm^2/s, K_d = 500 mL/g.

REFERENCES

ADEC (Alaska Department of Environmental Conservation), *Guidance for Fate and Transport Modeling*, CSRP-98-001, Division of Spill Prevention and Response, ADEC, Juneau, AK, 1998.

Bedient, P. B., Rifai, H. S., and Newell, C. J., *Ground Water Contamination, Transport, and Remediation,* 1st and 2nd eds., Prentice Hall, Upper Saddle River, NJ, 1994, 1999.

Binning, P., Discussion: on the misuse of the simplest transport model, *Ground Water,* Vol. 38, No. 1, pp. 4–5, 2000.

Borden, R. C., and Bedient, P. B., Transport of dissolved hydrocarbons influenced by oxygen-limited biodegradation: 1. Theoretical development, *Water Resour. Res.,* Vol. 22, No. 13, pp. 1973–1983, 1986.

Bouwer, E. J., and McCarty, P. L., Modeling of trace organics biotransformation in the subsurface, *Ground Water,* Vol. 22, No. 4, pp. 433–440, 1984.

Crank, J., *The Mathematics of Diffusion,* Oxford University Press, New York, 1956.

Choy, B., and Reible, D. D., *Diffusion Models of Environmental Transport,* Lewis Publishers, Chelsea, MI, 2000.

Clement, T. P., *RT3D: A Modular Computer Code for Simulating Reactive Multi-species Transport in Three-Dimensional Groundwater Aquifers,* Pacific Northwest National Laboratory, Richmond, WA, 1997.

Domenico, P. A., and Schwartz, F. W., *Physical and Chemical Hydrogeology,* 2nd ed., Wiley, New York, 1998.

Fetter, C. W., *Contaminant Hydrogeology,* 2nd ed., Prentice Hall, Upper Saddle River, NJ, 1999.

Freeze, R. A., and Cherry, J. A., *Groundwater,* Prentice Hall, Upper Saddle River, NJ, 1979.

Gelhar, L. W., Stochastic subsurface hydrology from theory to applications, *Water Resour. Res.,* Vol. 22, No. 9, pp. 135S–145S, 1986.

Hantush, M. S., Analysis of data from pumping tests in leaky aquifers, *Trans. Am. Geophys. Union,* Vol. 37, pp. 702–714, 1956.

Javandel, I., Doughty, C., and Tsang, C. F., *Groundwater Transport: Handbook of Mathematical Models,* Water Resources Monograph 10, American Geophysical Union, Washington, DC, 1984.

Krickhoff, S. W., Semi-empirical estimation of sorption of hydrophobic pollutants on natural sediments and soils, *Chemosphere,* Vol. 10, No. 8, pp. 833–846, 1981.

Krickhoff, S. W., Brown, D. S., and Scott, T. A., Sorption of hydrophobic pollutants on natural sediments, *Water Res.,* Vol. 13, pp. 241–248, 1979.

Lyman, W. J., Adsorption coefficients for soils and sediments, in *Handbook of Chemical Property Estimation Methods,* Lyman, W. J., Reehl, W. F., and Rosenblatt, D. H. (eds.), McGraw-Hill, New York, 1982.

Mitchell, J. K., *Fundamentals of Soil Behavior,* Wiley, New York, 1976.

Mitchell, J. K., and Rumer, R. R., Waste containment barriers: evaluation of the technology, pp. 1–25 in *In-Situ Remediation of Geoenvironment,* ASCE Special Publication, ASCE, Reston, VA, 1998.

Morel, F. M. M., *Principles of Aquatic Chemistry,* Wiley, New York, 1983.

Ogata, A., and Banks, R. B., *A Solution of the Differential Equation of Longitudinal Dispersion in Porous Media,* U.S. Geological Survey Professional Paper 411-A, U.S. Government Printing Office, Washington, DC, 1961.

Olsen, R. L., and Davis, A., Predicting the fate and transport of organic compounds in groundwater, Part 1, pp. 40–64 in *Proceedings of the Hazardous Materials Conference,* Hazardous Materials Resource Control, Greenbelt, MD, 1990.

Rabideau, A., and Khandelwal, A., Boundary conditions for modeling transport in vertical barriers, *J. Environ. Eng.,* Vol. 124, No. 11, pp. 1135–1139, 1998.

Rowe, R. K., and Booker, J. R., *MIGRATE: Finite Layer Analysis Program for 2D Analysis,* Geotechnical Research Center, University of Western Ontario, London, Ontario, Canada, 1988.

Rowe, R. K., Booker, J. R., and Caers, C. J., *POLLUTE-1D Pollute Migration through a Non-homogeneous Soil: Users Manual,* SACDA 84-13, Systems Analysis Control and Design Activity, Faculty of Engineering Science, University of Western Ontario, London, Ontario, Canada, 1984.

Rowe, R. K., Caers, C. J., and Barone, F., Laboratory determination of diffusion and distribution coefficients of contaminants using undisturbed clayey soils, *Can. Geotech. J.,* Vol. 25, pp. 108–118, 1988.

Sauty, J.-P., Analysis of hydrodispersive transfer in aquifers, *Water Resour. Res.,* Vol. 16, No. 1, pp. 145–158, 1980.

Shackelford, C. D., and Daniel, D. E., Diffusion in saturated soil: II. Results for compacted clay, *J. Geotech. Eng.,* Vol. 117, No. 3, pp. 485–506, 1991.

Sharma, H. D., and Lewis, S. P., *Waste Containment Systems, Waste Stabilization, and Landfills: Design and Evaluation,* Wiley, New York, 1994.

Stumm, W., and Morgan, J. J., *Aquatic Chemistry,* 2nd ed., Wiley, New York, 1981.

USEPA (U.S. Environmental Protection Agency), *Transport and Fate of Contaminants in the Subsurface,* EPA/625/4-89/019, USEPA, Washington, DC, 1989.

Valocchi, A. J., Describing the transport of ion-exchanging contaminants using an effective K_d approach, *Water Resour. Res.,* Vol. 20, No. 4, pp. 499–503, 1984.

Van Genuchten, M. T., Analytical solutions for chemical transport with simultaneous adsorption, zero-order production, and first order decay, *J. Hydrol.,* Vol. 9, pp. 213–233, 1981.

Van Genuchten, M. T., and Alves, W. J., *Analytical Solutions of the One-Dimensional Convective–Dispersive Solute Transport Equation,* Technical Bulletin 1661, U.S. Department of Agriculture, Washington, DC, 1982.

Wexler, E. J., Analytical solutions for one-, two-, and three-dimensional solute transport in ground-water systems with uniform flow, in *Techniques of Water-Resources Investigations of the United States Geological Survey,* Chap. B7, Book 3, USGS, Washignton, DC, 1992.

Zheng, C., *MT3D: A Modular Three-Dimensional Transport Model for Simulation of Advection, Dispersion and Chemical Reactions of Contamination in Groundwater Systems,* Report to USEPA, Ada, OK, 1990.

Zheng, C., and Bennett, G. D., *Applied Contaminant Transport Modeling,* 2nd ed., Wiley, New York, 2002.

PART II
REMEDIATION TECHNOLOGIES

9

SUBSURFACE CONTAMINATION: SOURCES, CONTAMINANTS, REGULATIONS, AND REMEDIAL APPROACH

9.1 INTRODUCTION

As stated in Chapter 1, geoenvironmental problems involve remediation of contaminated sites, design of waste containment systems, and reuse of waste and recycled materials for engineering applications. In Part II of the book, we present general information on subsurface contamination (this chapter), methods to characterize contaminated sites (Chapter 10), assessment of risk posed by the site's contamination (Chapter 11), and various methods to remediate contaminated sites (Chapters 12 to 14).

Subsurface contamination is a widespread problem in the United States and worldwide (Bredehoeft, 1994; Page, 1997; USEPA, 1997). Before the passage of environmental regulations, wastes have been disposed of in different ways, including burning, placement in streams, and storage on and placement in the ground. These disposal methods were primarily reflections of convenience, expedience, expense, or best available technology at the time of disposal. We are now realizing that these past disposal practices have contaminated our valuable groundwater resources, affecting or posing serious threats to both human health and the surrounding environment. The contamination problems vary from simple inconveniences such as taste, odor, color, hardness, or foaming, to serious health hazards due to pathogenic organisms, flammable or explosive substances, or toxic chemicals and their by-products.

In this chapter we present an overview of various sources of subsurface contamination, typical contaminants encountered at contaminated sites, relevant environmental regulations concerning subsurface remediation, and a general remedial approach to contaminated sites.

9.2 SOURCES OF CONTAMINATION

A variety of sources can cause subsurface contamination, as depicted in Figure 9.1 (USEPA, 1990, 1998a–c). From a geoenvironmental point of view, various sources of contamination may be divided into the following three groups: (1) sources originating on the ground surface, (2) sources originating above the water table (vadose zone), and (3) sources originating below the water table (saturated zone).

9.2.1 Sources Originating on the Ground Surface

When stored or spread on the ground surface, various water-soluble products cause subsurface contamination:

- *Infiltration of contaminated surface water.* During pumping, many wells are recharged by infiltration from surrounding surface water bodies. In these

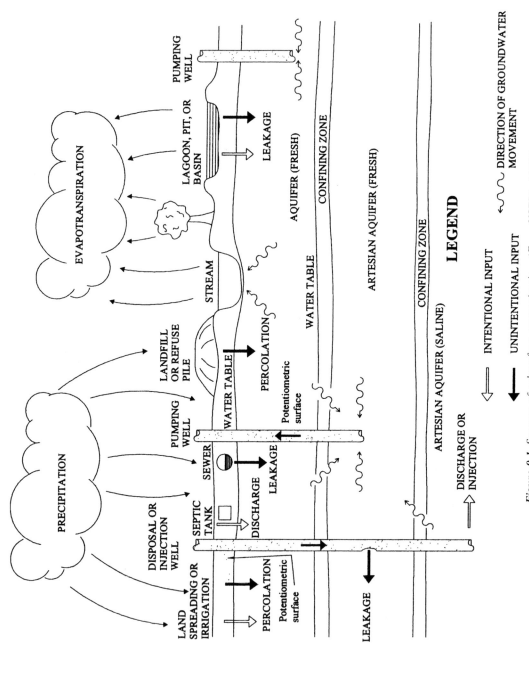

Figure 9.1 Sources of subsurface contamination. (From USEPA, 1990.)

situations, if the surface water is contaminated, such infiltration causes contamination of aquifer and drinking water.

- *Land disposal of solid and liquid wastes.* Many types of wastes, such as manure sludges, garbage, industrial wastes, and mine tailings, have been spread directly onto the ground surface over large areas or formed as individual mounds. The infiltration of soluble substances from these wastes causes subsurface contamination.

- *Accidental spill.* A large volume of toxic materials are transported throughout the country by truck, rail, and aircraft, transferred at handling facilities, and stored in tanks. Accidental spills of these materials were and are common occurrences. When spilled, these materials may either infiltrate or flow into a stream.

- *Fertilizers and pesticides.* Fertilizers and pesticides, many of which are highly toxic, are used in agricultural applications. These substances may infiltrate or be carried by surface runoff.

- *Disposal of sewage and water treatment plant sludges.* Sludge is the residue of treatment of municipal and industrial waste and may contain decomposed organic matter, inorganic salts, heavy metals, and so on. Land application of sludge is a common practice, resulting in infiltration of contaminants and hence, subsurface contamination.

- *Salt storage and spreading on roads.* Unprotected deicing salt piles dissolve readily, allowing infiltration or runoff. Salts spread on the roads also dissolve and infiltrate, causing subsurface contamination.

- *Animal feedlots.* Feedlots used for cattle, hogs, sheep, and poultry cover relatively small areas but provide a huge volume of wastes. These wastes, and seepage from lagoons that store any liquid wastes, cause subsurface contamination.

- *Particulate matter from airborne sources.* The fallout of particulate matter originating from smoke, flue dust, aerosols, and automobile emissions may deposit water-soluble and toxic chem-

icals on the ground surface. Infiltration of these chemicals result in subsurface contamination.

9.2.2 Sources Originating above the Water Table (Vadose Zone)

A variety of substances deposited or stored in the ground above the water table (vadose zone) lead to subsurface contamination. Such typical conditions include:

- *Waste disposal in excavations.* Open excavations are often used as unregulated dumps. They are used to dispose of a variety of wastes, including liquid wastes. Infiltration from such dumps has caused the contamination of numerous sites.

- *Landfills.* Municipal and industrial wastes were disposed of in unlined open dumps. Leachates generated from waste decomposition and infiltration of precipitation and surface runoff spread into the subsurface, causing contamination.

- *Surface impoundments.* Surface impoundments, including ponds and lagoons, consist of shallow excavations and are used to dispose of agricultural, municipal, and industrial wastes. Many of these impoundments have been located in permeable soils, causing extensive infiltration of soluble waste constituents into the subsurface.

- *Leakage from underground storage tanks.* Underground storage tanks are used to store billions of gallons of liquids for municipal, industrial, and agricultural purposes. Leakage, due to corrosion of tanks, causes contamination. Gasoline leakage is the most widespread problem.

- *Leakage from underground pipelines.* Thousands of miles of buried pipelines exist to transport liquids. Leakage of these pipelines, generally difficult to detect, can cause extensive contamination.

- *Septic tanks.* Septic tanks are abundant in areas not served by municipal or privately owned sewage treatment systems. Effluent discharges from these tanks can cause high levels of biological contamination into the subsurface, especially if these tanks are located in permeable soils.

9.2.3 Sources Originating below the Water Table (Saturated Zone)

Numerous situations exist where materials were stored or disposed of below the water table (saturated zone), leading to serious groundwater contamination problems. Some of such situations include:

- *Waste disposal in wet excavations.* Generally, mining excavations extend to the level of groundwater and are filled with water. When abandoned, they serve as dumps for both solid and liquid wastes. Because of the direct connection to the aquifer, extensive groundwater contamination occurs.

- *Deep well injection.* For many years, liquid wastes were disposed of by pumping into deep wells. The injection of highly toxic wastes has led to extensive subsurface contamination.

- *Mines.* Many mines are deeper than the water table. Pumping of mine waters to the surface, leaching of the spoil material, milling wastes, and so on, caused numerous contamination problems.

- *Agricultural drainage wells and tiles.* To enhance drainage in swampy areas, field tiles and drainage wells are used to allow the water to drain into deeper, more permeable soils. Such drainage may contain agricultural chemicals that cause groundwater contamination.

- *Abandoned or improperly constructed wells.* When a well is abandoned, the casing is pulled. In other situations, the casing may become so corroded that holes develop. Under both conditions, fluids under higher pressure are permitted to migrate and contaminate adjacent aquifers. Improperly constructed water supply wells may either contaminate an aquifer or produce contaminated water.

9.3 TYPES OF CONTAMINANTS

The U.S. Environmental Protection Agency (USEPA) estimated that there are thousands of sites that have been contaminated in the United States, and over 217,000 of these sites require urgent remedial action

(USEPA, 1997). These sites are divided into the following groups:

- National Priorities List (NPL) or Superfund sites
- Resource Conservation and Recovery Act (RCRA) corrective action sites
- Underground storage tank (USTs) sites
- Department of Defense (DOD) sites
- Department of Energy (DOE) sites
- Civilian federal agency sites
- State and private parties (including brownfield) sites

Table 9.1 shows a breakdown of these sites. Groundwater and soils are the most prevalent contaminated media at these sites. In addition, large quantities of other contaminated material, such as sediments, landfill waste, and slag, are present. About 70% of NPL, RCRA, DOD, and DOE sites have contaminated soils or groundwater, or both. The contaminants encountered include organic compounds, heavy metals, and radionuclides. DOE sites contain mixed wastes, including radioactive wastes, while DOD sites contain explosives and unexploded ordnance. The cost to clean up these sites is estimated to be $187 billion dollars (Table 9.1).

Table 9.2 summarizes the most common contaminants found at the contaminated sites. This table also shows the chemical characteristics and toxicity of the contaminants as well as the major sources and pathways leading to subsurface contamination. Because of

Table 9.1 **Estimated number of contaminated sites and associated remediation cost**

Group	Number of Sites	Cost ($\times 10^9$ dollars)
Superfund	547	7
RCRA corrective action	3,000	39
RCRA-USTs	165,000	21
DOD	8,336	29
DOE	10,500	63
Civilian federal agencies	>700	15
States and brownfields	29,000	13
Total	>217,083	$187

Source: USEPA (1997).

Table 9.2 Common contaminants found at contaminated sites

Contaminant Group	Most Common Contaminants in the Group	Major Chemical Characteristics of Contaminants	Toxic Effects	Major Sources of Subsurface Contamination	Causes / Pathways of Subsurface Contamination
Heavy metals	Chromium (Cr), cadmium (Cd), nickel (Ni), lead (Pb)	Malleable, ductile, good conductors; cationic forms precipitate under high-pH conditions	Cr, Cd and Ni can be carcinogenic with long-term exposure; Pb, Cr, Cd and Ni are toxic with short-term exposure to large doses	Metal reclamation facilities, electroplating facilities and other metallurgical applications, car exhaust	Atmospheric deposition, urban runoff, municipal and industrial discharge, landfill leachate
Arsenic	Arsenic, plus various inorganic forms and some organic forms	Solid at standard conditions, gray metallic color, pure As is insoluble in water, melts at 817°C and 28 atm, sublimes at 613°C, density of 5.727 g/cm^3, 74.92 atomic mass, 5.73 specific gravity, and a vapor pressure of 1 mmHg at 373°C	Carcinogen; high dosages will cause death	Earth's crust, some seafood, volcanoes, geological process, industrial waste, and arsenical pesticides	Weathering of soil or rocks, minerals in the soil, mining operations, coal power plants, wastewater, etc.
Radionuclides	Uranium (U), radium (Ra), radon (Rn)	U: radioactive metal, Ra: radioactive metal, Rn: radioactive gas	U: lung disease; Ra: leukopenia, tumors in the brain and lungs; Rn: pneumonia, cancer	U: nuclear weapons, power plants, accidental spills; Ra: mineral deposits, rocks, soils; Rn: mineral deposits, rocks, soils	U: dismantled nuclear weapons; Ra: groundwater contamination, gas escapes from water and fills the air; Rn: groundwater contamination, gas escapes from water and fills the air
Chlorinated solvents	Perchloroethylene (PCE), trichloroethylene (TCE), trichloroethane (TCA), methylene chloride (MC)	Volatile, nonflammable; have low viscosity and high surface tension	Causes dermatitis, anesthetic, and poisonous	Dry cleaners, pharmaceuticals, chemical plants, electronics, etc.	Improper handling and disposal, spills, leaks from storage tanks

219

Table 9.2 (continued)

Contaminant Group	Most Common Contaminants in the Group	Major Chemical Characteristics of Contaminants	Toxic Effects	Major Sources of Subsurface Contamination	Causes/Pathways of Subsurface Contamination
Polycyclic aromatic hydrocarbons (PAHs)	Anthracene, Benzo[a]pyrene, naphthalene	Made of carbon and hydrogen, formed through incomplete combustion, colorless, pale yellow, or white solid	Carcinogen	Coal, aerosols, soot, air, creosote	Direct input, coal tar plants, manufactured gas plants, spills, garbage dumps, car exhausts
PCBs	Aroclor 1016, Aroclor 1221, Aroclor 1232, Aroclor 1242, Aroclor 1248, Aroclor 1254, Aroclor 1260, Aroclor 1262, Aroclor 1268,	Water solubility (mg/L), 1.50×10 to 2.7×10^{-3}	Cancer and noncancer effects, including immune, reproductive, nervous, and endocrine systems	PCB fluid containing electrical equipment and appliances	Manufacture, use disposal, and accidental spills; can travel long distances in the air; once in surface waters they are taken up by aquatic animals and thus enter the food chain
Pesticides	Organochlorines: DDT, dieldrin, chlordane, aldrin Organophosphates: parathion, malathion, dianzinon Carbamates: aldicarb, carbofuran	Organochlorines: low VP, low solubility, high toxicity, high persistence Organophosphates: less stable and more readily broken down than organochlorines	Chronic: cancer, liver toxicity Acute: central nervous system, respiration	Agriculture	Adsorption to soil, then soil leaching, runoff to surface waters, contaminated soil comes in contact with GW table, resides in crops or livestock
Explosives	Trinitrotoluene (TNT), cyclotrimethylene–trinitramine (RDX)	TNT: density 1.65 g/mL; melting point: 82°C; boiling point: 240°C; water solubility: 130 g/L at 20°C; vapor pressure: 0.0002 mmHg at 20°C	Inhalation or ingestion: gastrointestinal disturbance, toxic hepatitis, anemia, cyanosis, fatigue, lassitude, headache, delirium, convulsions, coma	Military training manufacturing and testing	Impact craters, improper design of settling lagoons containing manufacturing process waters

Source: Reddy (2002).

the distinctly different properties, as well as the complex distribution and behavior of the contaminants in the subsurface, the remediation of contaminated sites has been a daunting task to many environmental professionals. For example, when heavy metals are present in soils, they may be distributed in one or more of the following forms: (1) dissolved in soil solution (pore water), (2) occupying exchange sites on inorganic soil constituents, (3) specifically adsorbed on inorganic soil constituents, (4) associated with insoluble soil organic matter, (5) precipitated as pure or mixed solids, and (5) present in the structure of the minerals. The amounts of metals present in these different phases are controlled by the interdependent geochemical processes such as (1) adsorption and desorption, (2) redox reactions, (3) complex formations, (4) precipitation and dissolution of solids, and (5) acid–base reactions. On the other hand, organic compounds may exist in four phases in soils: (1) dissolved phase, (2) adsorbed phase, (3) gaseous phase, and (4) free or pure phase. The organic compounds may change from one phase to another phase depending on the following processes: (1) volatilization, (2) dissolution, (3) adsorption, and (4) biodegradation. An in-depth understanding of the various geochemical processes that control the phase distribution of the contaminants in soils is critical for the assessment and remediation of contaminated sites. Refer to Chapter 6 for more information on the geochemistry of contaminants.

9.4 RELEVANT REGULATIONS

As presented in Chapter 2, the assessment and remediation of previously contaminated sites and the proper management of newly created hazardous wastes have been regulated through the passage of major environmental laws and regulations. Specifically, the most relevant laws and regulations dealing with the remediation of contaminated sites include:

- Comprehensive Environmental Response, Compensation and Liabilities Act
- Resource Conservation and Recovery Act
- State voluntary site remediation programs

A brief overview of the laws and regulations above related to the assessment and remediation of contaminated sites is provided below.

9.4.1 Comprehensive Environmental Response, Compensation and Liabilities Act

In 1980, the U.S. Congress established the Superfund program, also known as the Comprehensive Environmental Response, Compensation and Liabilities Act (CERCLA), to provide financial assistance for the remediation of abandoned hazardous waste sites. It was determined that these sites pose a serious risk to the health and safety of the public as well as the welfare of the environment. The Superfund program is administered by the USEPA in cooperation with regional governmental agencies. To determine which sites are eligible to receive federal aid under the Superfund program, a ranking system has been established to allow for a quantitative rating of sites across the United States. Sites that score high enough on the USEPA's hazard ranking system are placed on the National Priorities List (NPL). The National Priorities List is a published list of hazardous sites that require extensive, long-term remediation and are deemed eligible to receive funding from the Superfund program. Superfund sites must comply with the stringent remediation codes, liability standards, and documentation required by the Superfund program. According to this program, the purchasers of contaminated sites may be held responsible for the damage caused by the previous owners, even if the activities that caused the damage in the past were legal at that time. Additionally, Superfund regulations require that a contaminated site be remediated to very low contaminant levels, so that the risk to public health is minimized.

As shown in Figure 9.2, the CERCLA requires nine stages of response to the hazardous waste sites (USEPA, 1991):

- Discovery and notification
- Preliminary assessment
- Removal action
- NPL determination

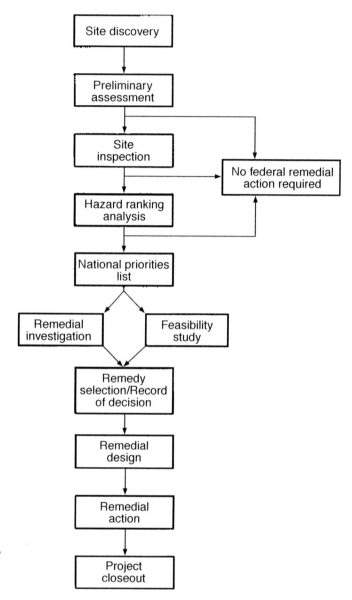

Figure 9.2 *Various steps involved in remedial action according to CERCLA. (From USEPA, 1991.)*

- Remedial investigation
- Feasibility study
- Record of decision
- Remedial design
- Remedial action

Discovery and Notification. CERCLA specifies that the responsible party (RP) must immediately report the release of a hazardous substance to the National Response Center (800-424-8802) if the release quantity (RQ) is exceeded (refer to 40 CFR 302.4 and 40 CFR 117.3 for RQs). RQs range from 1 to 500 lb per 24 hours. For mixtures and solutions, there is a "release" when a component hazardous substance exceeds its RQ.

Preliminary Assessment (PA). This stage consists of a review of existing data to determine if there is a need for a removal or remedial action. PA involves a review of available regulatory history, site operations, demographic data, aerial photos, and local topography. The review may be followed up with a site investigation (SI) that involves limited sampling.

Removal Action (RA). This stage is undertaken when the USEPA determines that there is an immediate threat to human health or the environment. This response usually does not lead to a complete cleanup. EPA must terminate a removal action after the expenditure of $2 million or the lapse of 12 months from the date of initial removal unless circumstances warrant continued action. Examples of removal actions include installation of a security fence, installation of warning signs, removal of wastes to an approved RCRA facility, drainage controls to prevent runoff, capping of contaminated soils, segregation and storage of drummed wastes, removal of highly contaminated soil or drums, provision of alternative water supplies, incineration or treatment of waste on-site, and stabilization of impoundments.

NPL Determination. If a site is not listed on the NPL, it is eligible for a Superfund-financed removal action but not a remedial action. The decision to list is based on the hazard ranking system score. The score is sensitive to the potential for release and the proximity of people. A score of 28.5/100 makes a site eligible for the NPL.

Remedial Investigation (RI). A RI is a detailed site investigation by the USEPA or responsible parties (RPs) which includes sampling and exposure assessment to determine the nature and extent of contamination of air, soil, and water. It also includes a risk assessment. A RP can be fined $25,000 per day if the RP does not comply with an information request.

Feasibility Study (FS). A FS is an evaluation of remdial alternatives. Off-site transport and disposal without treatment is the least desired remedy. The average cost of an RI/FS per site is $1.3 million. A RI/FS may take 18 to 30 months to complete. The FS must consider nine factors:

1. Protection of human health and the environment
2. Compliance with applicable and relevant and appropriate requirements (ARARs) (The exact level of cleanup has not been specified by the Congress, the USEPA, or the courts. Drinking water standards and lifetime cancer risk $< 10^{-4}$ are commonly used.)
3. Long-term effectiveness and permanence
4. Reduction in volume, toxicity, and mobility
5. Short-term effectiveness
6. Implementability
7. Cost
8. State acceptance
9. Community acceptance

Record of Decision (ROD). The USEPA selects a remedy, prepares a remedial plan, and seeks state, community, and RP acceptance during a public review and comment period. The ROD can be reopened and amended.

Remedial Design (RD). The RD provides engineering and construction specifications for the selected remedy and may take 12 to 18 months to complete.

Remedial Action (RA). There are several prerequisites for a remedial action: (1) listing on the NPL; (2) the state must contribute 10% when no financially feasible RP can be located (50% if the site is state owned); and (3) RI/FS, ROD, and RD must be complete.

9.4.2 Resource Conservation and Recovery Act

In 1976, the U.S. Congress promulgated the Resource Conservation and Recovery Act (RCRA) to control newly created hazardous waste from "cradle to grave." These regulations provide criteria for defining hazardous waste, generator responsibilities, transporter requirements, manifest systems, and treatment, storage, and disposal facility (TSDF) requirements. The regulations also address problems that could result from underground tanks storing petroleum and other hazardous substances. Although the terminology is different, the steps required for the remediation for contaminated sites under RCRA are similar to those required under CERCLA (as detailed in Section 9.4.1). Figure 9.3 and Table 9.3 show the comparison of steps and terminology related to the remediation of contaminated sites according to RCRA and CERCLA. More details on RCRA are provided in Chapter 2.

9.4.3 State Voluntary Site Remediation Programs

Many state governments are assisting in the cleanup of contaminated sites. Nearly half of the states in the United States offer some type of voluntary remediation program. The purpose of such programs is to encourage the remediation of sites with possible contamination, while preventing any increased liability for the participating parties. When a remediation project is completed, many states will issue a statement releasing the participants from state liability for any contamination that may exist at the site. Moreover, state agencies will often offer assistance to project participants if they are subject to federal liability.

Typically, sites that meet the following criteria are eligible to participate in state's voluntary remediation programs (Birkel et al., 1998):

- Sites not listed or proposed for listing on the National Priority List of federal Superfund sites
- Sites that are not the subject to current environmental enforcement actions or orders
- Sites that are not the subject of a RCRA corrective action
- Sites that do not pose an immediate and significant risk of harm to human health or the environment

- Sites that do not pose a significant threat to public or private drinking water supplies

One should contact the state's environmental agency (the state in which the site is located) to obtain specific criteria for a voluntary site remediation program. One should be careful to learn the eligibility requirements, voluntary participation agreements, site investigation/characterization procedures, response action requirements, cleanup certifications and inspections, and a final agency sign-off, release, and covenant not to sue. A comparison of salient features of voluntary site remediation programs in six midwestern states is given in Table 9.4.

9.5 OTHER CONSIDERATIONS

In addition to the regulations above, many other interrelated factors affect the remediation of contaminated sites. These factors include:

- *End use of the site.* The proposed future use of the site after the site has been remediated will dictate the need for remediation and the cleanup levels if remediation is required.
- *Cost of cleanup.* The cost of remediation depends on the site conditions and applicable regulations. The more stringent the regulations, the higher will be the cost of the remediation.
- *Health and safety.* Federal regulations require stringent safety measures at contaminated sites. These regulations include OSHA requirements stipulated in 29 CFR 1910.120: Protection of Workers in Hazardous Waste Operations. State regulations also require stringent safety measures. A site-specific health and safety plan is prepared and strictly followed. All persons who work at the site or who visit the site are required to follow the safety measures.
- *Environmental liability.* Who is responsible for contamination and who will pay for the remediation are contentious questions to be answered. CERCLA uses the court system to assign specific liability for the cleanup of contaminated sites. CERCLA defines four classes of "potentially re-

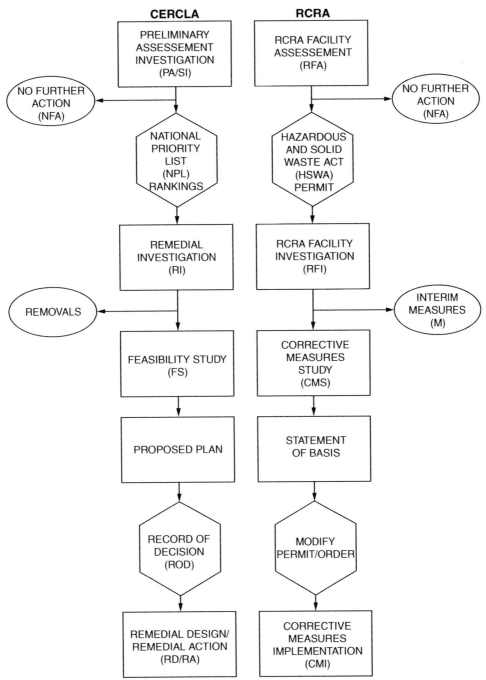

Figure 9.3 *Comparison of steps involved in corrective action according to CERCLA and RCRA. (From USACE, 1994.)*

Table 9.3 **Comparison of terminology used in corrective action according to RCRA and CERCLA**

CERCLA Process	RCRA Process	Objectives
Preliminary assessment (PA)	RCRA facility assessment (RFA)	Determine the potential for a present or the past release, based primarily on historical records.
Site investigation (SI)	a	Provide sufficient information to determine the need for a full remedial investigation, based on the preliminary site data and field sampling for contamination.
Remedial investigation (RI)	RERA facility investigation (RFI)	Characterize the nature, extent, direction, rate, movement, and concentration of releases.
Feasibility study (FS)	Corrective measures study (CMS)	Evaluate potential remedial actions and provide sufficient information to decision makers to allow an informed decision.

Source: USACE (1994).

[a] There is no direct RCRA equivalent for the SI. The RFA may have many of the field investigation aspects of the SI.

sponsible parties": (1) the current owner or operator of the site, (2) any person who formerly owned or operated the site at the time of disposal of any hazardous waste, (3) any person who arranged for disposal or treatment of hazardous waste at the site, and (4) any transporter of hazardous waste to the site. This implies that almost anyone involved with the site is a potentially responsible party and liable for the cost of cleanup.

9.6   REMEDIATION APPROACH

A systematic approach for the assessment and remediation of contaminated sites is necessary to facilitate the remediation process and avoid undue delays. The most important aspects of the approach include (1) site characterization, (2) risk assessment, and (3) selection of an effective remedial action. Figure 9.4 outlines this systematic approach (Reddy et al., 1999; Reddy, 2002). Innovative integration of various tasks can often lead to a faster, cost-effective remedial program.

9.6.1   Site Characterization

Site characterization is often the first step in a contaminated site remediation strategy. It consists of the collection and assessment of data representing contaminant type and distribution at a site under investigation.

The results of a site characterization form the basis for decisions concerning the requirements of remedial action. Additionally, the results serve as a guide for design, implementation, and monitoring of the remedial system. Detailed procedures of site characterization are presented in Chapter 10.

Each site is unique; therefore, site characterization must be tailored to meet site-specific requirements. An inadequate site characterization may lead to the collection of unnecessary or misleading data, technical misjudgment affecting the cost and duration of possible remedial action, or extensive contamination problems resulting from inadequate or inappropriate remedial action. Site characterization is often an expensive and lengthy process, so it is advantageous to follow an effective site characterization strategy to optimize efficiency and cost.

An effective site characterization includes the collection of data pertaining to (1) site geologic data, including site stratigraphy and important geologic formations; (2) hydrogeologic data, including major water-bearing formations and their hydraulic properties; and (3) site contamination data, including type, concentration, and distribution. Additionally, surface conditions both at and around the site must be taken into consideration.

Because little information regarding a particular site is often known at the beginning of an investigation, it is often advantageous to follow a phased approach for

Table 9.4 Comparison of salient features of voluntary site remediation programs in six midwestern states

Comparison Criterion	Illinois Site Remediation Program	Iowa Land Recycle Program	Kansas Voluntary Cleanup and Property Redevelopment Program	Minnesota Voluntary Investigation and Cleanup Program	Missouri Voluntary Cleanup Program	Wisconsin Contaminated Land Recycling Program
Eligibility	Owner, tenant, or anyone with interest in cleaning up site, as long as the owner grants permission.	Anyone with a signed acknowledgment of access/control of the property.	Any person or entity with access to property (potential developers, owners, operators, municipalities, or other units of government).	Property owners, potential property buyers or developers, insurers, environmental consultants, attorneys, or local governmental units.	Any individual, partnership, co-partnership, firm, company, public or private corporation, association, joint stock company, trust, estate, political subdivision, or any agency, board, department, or bureau of the state or federal government, or any other legal entity that is recognized by the law as the subject of rights and duties, or any person acquiring, disposing of, or possessing a lien holder interest in real property that is known to be or suspected to be contaminated by hazardous substance.	Anyone who did not intentionally or recklessly cause release of a hazardous substance on the property.
Participation	SRP application, service agreement, minimum deposit of $500 for oversight costs.	LRP application, $750 application fee, participation agreement, reimbursement of oversight costs.	VCPRP application, $200 application fee, past assessment reports, nonbinding voluntary agreement, reimbursement of oversight costs.	VIC application, reimbursement of oversight costs. No application fee.	Intent to participate agreement and consent for access to property agreement, $200 application fee, reimbursement of oversight costs.	VPLE application, $250 application fee, and reimbursement of oversight costs.
Agency involvement	IEPA Bureau of Land or independent RELPE review and approval. Authority to assess oversight fees up to $5000.	IDNR Solid Waste Section review and approval. Authority to assess oversight fees up to $7500.	KDOH Bureau of Environmental Remediation review and approval. Authority to assess oversight fees up to $5000.	MPCA Ground Water and Solid Waste Division Site Response Section review and approval. Oversight average $85 to $95 per hour.	MDNR Hazardous Waste Program review and approval. Authority to bill participant on a fee-for-service (average $40 per hour) basis.	WDNR Bureau for Remediation and Redevelopment review and approval. Authority to assess oversight fees up to $1,000 if property < 1 acre or $3000 if property > 1 acre.

Table 9.4 (*continued*)

Comparison Criterion	Illinois Site Remediation Program	Iowa Land Recycle Program	Kansas Voluntary Cleanup and Property Redevelopment Program	Minnesota Voluntary Investigation and Cleanup Program	Missouri Voluntary Cleanup Program	Wisconsin Contaminated Land Recycling Program
Scope of response	Site investigation, development of remedial objectives, remedial action, and completion report. In accordance with 35 IAC 740 and 742.	Site assessment, risk evaluation/response action report.	Site classification, site assessment and/or site investigation, cleanup plan, public notification, verification sampling program.	Phase I and II investigation, community relations, focused feasibility study and/or response action plan, implementation report.	Site assessment/characterization, site classification, ecology risk assessment develop cleanup levels, remedial action, final CALM report.	Phase I and II environmental assessment, site investigation, remedial action, construction documentation, and summary report. In accordance with NR700 rule series.
Future liability after site closure	No further remediation letter. The NFR letter must be recorded against deed of each property. NFR release participant for future responsibility for issues addressed.	No further action certificate. Provides liability protection related to state authority to participate and future property owners.	No further action letter. Adjacent property owners may receive protection. Memorandum of agreement to provide voluntary parties with assurance of relief from future liability at a participant's property.	No further action letter. Provide assurance regarding future Superfund enforcement actions by MPCA. Assurances apply to parties identified in request for assistance form and to lenders and successive owners.	Certificate of completion letter. Provides a measure of protection for both MDNR and EPA, through a memorandum of agreement, against future environmental liability related to the property.	Certificate of completion releases participant from future liability for past contamination. Liability exemption applies to any future owner, as long as they maintain and monitor the property, and do not engage in any activities that are inconsistent with the maintenance of the property.

Source: Birkel et al. (1998). Used with permission from Gas Technology Institute, Des Plaines, Illinois.

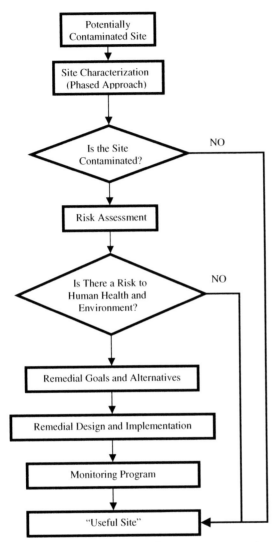

Figure 9.4 *Systematic approach to assessment and remediation of contaminated sites. (From Reddy, 2002.)*

site characterization. A phased approach may also minimize the financial impact by improving the planning of the investigation and ensuring the collection of relevant data. Phase I consists of the definition of investigation purpose and the performance of a preliminary site assessment. A preliminary assessment provides the geographical location, background information, regional hydrogeologic information, and potential

sources of contamination pertaining to the site. The preliminary site assessment consists of two tasks, a literature review and a site visit.

Based on the results of the phase I activities, the purpose and scope of the phase II exploratory site investigation need to be developed. If contamination was detected at the site during the course of the preliminary investigation, the exploratory site investigation must be used to confirm such findings as well as to obtain further data necessary for the design of a detailed site investigation program. A detailed work plan should be prepared for the site investigations describing the scope of related field and laboratory testing. The work plan should provide details about sampling and testing procedures, sampling locations and frequency, a quality assurance/quality control (QA/QC) plan, a safety and health (S&H) plan, a work schedule, and a cost assessment.

Phase III includes a detailed site investigation in order to define site geology, hydrogeology, and the contamination profile. The data obtained from the detailed investigation must be adequate to properly assess the risk posed at the site as well as to allow for effective designs of possible remedial systems. As with the exploratory investigations, a detailed work plan, including field and laboratory testing programs as well as QA/QC and S & H plans, should be outlined. Depending on the size, accessibility, and proposed future purpose of a site, this investigation may last anywhere from a few weeks to a few years. Because of the time and effort required, this phase of the investigation is very costly. If data collected after the first three phases are determined to be inadequate, phase IV should be developed and implemented to gain additional information. Additional phases of site characterization must be performed until all pertinent data has been collected.

Depending on the logistics of the project, site characterization may require regulatory compliance and/or approval at different stages of the investigation. Thus, it is important to review the applicable regulations during the preliminary site assessment (phase I). Meetings with regulatory officials may also be beneficial to ensure that investigation procedures and results conform to regulatory standards. This proactive approach may prevent delays in obtaining the required regulatory permits and/or approvals.

Innovative site characterization techniques are increasingly being used to collect relevant data in an efficient and cost-effective manner. Recent advances in cone penetrometer and sensor technology have enabled contaminated sites to be characterized rapidly using vehicle-mounted direct push probes. Probes are available for directly measuring contaminant concentrations in-situ, in addition to measuring standard stratigraphic data, to provide flexible, real-time analysis. The probes can also be reconfigured to expedite the collection of soil, groundwater, and soil gas samples for subsequent laboratory analysis. Noninvasive geophysical techniques such as ground-penetrating radar, cross-well radar, electrical resistance tomography, vertical induction profiling, and high-resolution seismic reflection produce computer-generated images of subsurface geological conditions and are qualitative at best. Other approaches, such as chemical tracers, are used to identify and quantify contaminated zones, based on their affinity for a particular contaminant and the measured change in tracer concentration between wells employing a combination of conservative and partitioning tracers.

9.6.2   Risk (Impact) Assessment

Once site contamination has been confirmed through the course of a thorough site characterization, a risk assessment is performed. A *risk assessment,* also known as an *impact assessment,* is a systematic evaluation used to determine the potential risk posed by the detected contamination to human health and the environment under the present and possible future conditions (Figure 9.4). If the risk assessment reveals that an unacceptable risk exists due to the contamination, a remedial strategy is developed to assess the problem. If corrective action is deemed necessary, the risk assessment will assist in the development of remedial strategies and goals necessary to reduce the potential risks posed at the site. The details of impact assessment methods are provided in Chapter 11.

The USEPA and the American Society for Testing and Materials (ASTM) have developed comprehensive risk assessment procedures. The USEPA procedure was originally developed by the U.S. Academy of Sciences in 1983. It was adopted with modifications by the USEPA for use in Superfund feasibility studies and RCRA corrective measure studies (USEPA, 1989). This procedure provides a general, comprehensive approach for performing risk assessments at contaminated sites. It consists of four steps: (1) hazard identification, (2) exposure assessment, (3) toxicity assessment, and (4) risk characterization. ASTM Standard E 1739, the *Guide for Risk-Based Corrective Action* (RBCA), is a tiered assessment originally developed to help assess sites that contained leaking underground storage tanks containing petroleum (ASTM, 2002). Although the standard is geared toward such sites, many regulatory agencies use a slightly modified version for non-UST sites. This approach integrates risk and exposure assessment practices with site assessment activities and selection of the remediation technique. The RBCA process allows corrective action activities to be tailored for site-specific conditions and risks and assures that the chosen course of action will protect both human health and the environment.

9.6.3   Remedial Action

When the results of a risk assessment reveal that a site does not pose risks to human health or the environment, no remedial action is required. In some cases, however, monitoring of a site may be required to validate the results of the risk assessment. Corrective action is required when risks posed by the site are deemed unacceptable. When action is required, a remedial strategy must be developed to ensure that the intended remedial method complies with all technological, economic and regulatory considerations.

The costs and benefits of various remedial alternatives are often weighed by comparing the flexibility, compatibility, speed, and cost of each method. A remedial method must be flexible in its application to ensure that it is adaptable to site-specific soil and groundwater characteristics. The method selected must be able to address site contamination while offering compatibility with the geology and hydrogeology of the site.

Generally, remediation methods are divided into two categories: in-situ remediation methods and ex-situ remediation methods. In-situ methods treat contaminated soils and/or groundwater in-place, eliminating the need to excavate the contaminated soils and extract groundwater. In-situ methods are advantageous because they are less expensive, cause less site disturbance, and provide increased safety to both on-site workers and the general public within the vicinity of the remedial project. Successful implementation of in-situ methods requires a thorough understanding of subsurface conditions. In-situ containment, using bottom barriers, vertical walls, and caps, may be a feasible strategy to minimize the risk posed by the contamination at some sites. The in-situ barriers may serve as interim remedial action. The details on in-situ barriers are provided in Chapter 12. Ex-situ remediation methods are used to treat excavated soils and/or extracted groundwater. Surface treatment may be performed either on-site or off-site, depending on site-specific conditions. Ex-situ treatment methods are attractive because consideration does not need be given to subsurface conditions. Ex-situ treatment also offers easier control and monitoring during remedial activity implementation.

Remedial technologies are classified into two groups based on their scope of application: (1) vadose zone technologies and (2) saturated zone technologies. The *vadose zone* is the geological profile extending from the ground surface to the upper surface of the principal water-bearing formation. The financial impact of the remediation program may be substantially reduced if the source of pollution is identified and remediated while it is still in the vadose zone, before the onset of groundwater contamination. A number of remedial technologies are suitable for vadose zone treatment; however, many of these options are not capable of treating contaminated groundwater. Details on the vadose zone remediation technologies are given in Chapter 13. In the case of saturated zone contamination, other technologies must be considered for possible implementation as detailed in Chapter 14. To remediate subsurface contamination properly, it is

essential to understand the operation, applicability, advantages, and drawbacks of available subsurface remedial technologies.

Using just one technology may not be adequate to remediate some contaminated sites where different types of contaminants exist (e.g., heavy metals combined with volatile organic compounds) and/or when the contaminants are present within a complex geological environment (e.g., a heterogeneous soil profile consisting of lenses or layers of low-permeability zones surrounded by high-permeability soils). Under these situations, different remediation technologies can be used sequentially to achieve the remedial goals. The use of such multiple remediation technologies is often referred to as *treatment trains*. Alternatively, different remediation technologies can be used concurrently.

To confirm adequate remediation, a monitoring program is implemented during and after remedial operations. Once the site is declared a clean site, it may be then reused or redeveloped.

9.7 SUMMARY

Numerous contaminated sites exist throughout the United States that pose a threat to human health and the environment. These sites require the characterization of potential contamination and the possible implementation of remedial action. It is of the utmost importance to characterize the site properly, and such a characterization includes defining the site's geology, hydrology, and contamination, potential releases to the environment, and the locations and demographics of nearby populations. Once the site has been characterized, a risk assessment of hazards at the site is performed and a suitable remedial action may be selected. To perform these different tasks in a financially responsible manner, it is important that a rational remedial strategy is followed, from the initial site characterization efforts until the completion of site cleanup.

If contamination has been detected and the risk posed by the contamination is unacceptable, an appropriate remedial technology must be selected and implemented properly. This requires not only a thorough understanding of the conditions within the subsurface, but the advantages and drawbacks of the available remedial options must also be understood. Such an understanding is necessary because improper implementation can often exacerbate site contamination. By possessing knowledge of the available technologies, remediation professionals will be better equipped to utilize proper judgment for the decisions regarding the remediation of contaminated sites.

QUESTIONS/PROBLEMS

9.1. Explain the need for remediation of contaminated sites.

9.2. What are the major sources of subsurface contamination?

9.3. How are the contaminated sites classified by the USEPA?

9.4. Why is a systematic approach needed to remediate a contaminated site? What are its components?

9.5. Who is a potentially responsible party in accordance with the CERCLA?

9.6. Explain why in-situ remediation approach is preferred to an ex-situ remediation approach.

9.7. Explain situations in which in-situ barriers could become the only remedial action required.

9.8. Contact the state environmental agency and obtain relevant information on voluntary site remediation and/ or brownfields program. Discuss the differences between the state's voluntary site remediation approach to that of typically followed for Superfund sites.

9.9. On what basis are chemicals classified as carcinogens and noncarcinogens? List five chemicals that are carcinogens.

9.10. Explain, with examples, how a proposed site's use after remediation could affect the actual remedial design and implementation.

9.11 What are the major obstacles to remediation of contaminated sites?

REFERENCES

ASTM (American Society for Testing and Materials), *Annual Book of ASTM Standards,* ASTM, West Conshohocken, PA, 2002.

Birkel, C., English, D., Shefchek, G., and Hargens, D., Steps to obtaining closure of contaminated sites under state regulatory programs, in *Proceedings of the International Symposium on Environmental Biotechnologies and Site Remediation Technologies,* Institute of Gas Technology, Des Plaines, IL, 1998.

Bredehoeft, J. D., Hazardous waste remediation: a 21st century problem, *Ground Water Monitor. and Remed.,* Vol. 14, pp. 95–100, 1994.

Page, G. W., *Contaminated Sites and Environmental Cleanup: International Approaches to Prevention, Remediation, and Reuse,* Academic Press, San Diego, CA, 1997.

Reddy, K. R., Assessment and remediation of contaminated sites, in *Proceedings of the Indian Geotechnical Conference: Keynote Presentations,* Allahabad, India, 2002.

Reddy, K. R., Adams, J. A., and Richardson, C., Potential technologies for remediation of brownfields, *Pract. Per. Hazard. Toxic Radioact. Waste Manag.,* Vol. 3, No. 2, pp. 61–68, 1999.

USACE (U.S. Army Corps of Engineers), *Technical Guidelines for Hazardous and Toxic Waste Treatment and Cleanup Activities,* EM 1110-1-502, U.S. Department of the Army, Washington, DC, 1994.

USEPA (U.S. Environmental Protection Agency), *Risk Assessment Guidance for Superfund,* Vol. I, *Human Health Evaluation Manual,* Part A, EPA/540/1-89/002, Office of Emergency and Remedial Response, USEPA, Washington, DC, 1989.

————, *Ground Water,* Vol. I, *Ground Water and Contamination,* EPA/625/6-90/016a, Office of Research and Development, USEPA, Washington, DC, 1990.

————, *Conducting Remedial Investigations/Feasibility Studies for CERCLA Municipal Landfill Sites,* EPA/540/P-91/001, Office of Emergency and Remedial Response, USEPA, Washington, DC, 1991.

————, *Cleaning Up the Nation's Waste Sites: Markets and Technology Trends,* EPA/542/R-96/005, Office of Solid Waste and Emergency Response, USEPA, Washington, DC, 1997.

————, *Remediation Case Studies: In-Situ Soil Treatment Technologies (Soil Vapor Extraction, Thermal Processes),* EPA/542/R-98/012, Federal Remediation Technologies Roundtable, USEPA, Washington, DC, Vol. 8, 1998a.

————, *Remediation Case Studies: Groundwater Pump and Treat (Chlorinated Solvents),* EPA/542/R-98/013, Federal Remediation Technologies Roundtable, USEPA, Washington, DC, Vol. 9, 1998b.

————, *Remediation Case Studies: Groundwater Pump and Treat (Nonchlorinated Contaminants),* EPA/542/R-98/014, Federal Remediation Technologies Roundtable, USEPA, Washington, DC, Vol. 8, 1998c.

10

CONTAMINATED SITE CHARACTERIZATION

10.1 INTRODUCTION

Site characterization is a systematic investigation aimed at obtaining appropriate and adequate data in order to define the type and extent of contamination as well as to assess the fate and transport of contaminants under various scenarios. The site characterization will require data related to the (1) geology, (2) hydrogeology, and (3) contamination information for the site under investigation. The geologic data provide site stratigraphy and properties of the various geologic formations existing at the site. The hydrogeologic data identify the major water-bearing formations and their hydraulic properties. The contamination data define the nature and distribution of the chemicals encountered at the site. Each site is unique; therefore, the site characterization process must be tailored to meet the site-specific requirements.

Site characterization data are necessary to (1) describe the occurrence and movement of contaminants at the site; (2) assess the background concentrations in the soil and water for the purpose of describing their precontamination conditions; (3) determine the impact of contamination on the subsurface; (4) evaluate the risk posed by the contamination to the public and the environment; (5) predict future contaminant trends under various conditions, including conditions created under selected remedial action implementation; (6) design and implement efficient and cost-effective remediation methods; and (7) design a monitoring program to verify the effectiveness of the site remediation technology selected (USEPA, 1991a,b).

In this chapter, a general methodology for site characterization is presented first. Following this, various

methods of obtaining soil and rock data, hydrogeologic data, and chemical data are described briefly. Finally, expedited or accelerated site characterization approaches are discussed.

10.2 GENERAL METHODOLOGY

A systematic phased approach for site characterization is shown in Figure 10.1. A phased approach is recommended because of the unknown nature of the site conditions at the beginning of the investigation. By facilitating the planning of the investigation to collect the pertinent data, a phased approach minimizes costs. The following phases of investigations are, therefore, generally planned.

The first phase, also known as the *preliminary site assessment,* includes the collection and review of available or published site-specific or regional data as well as conducting site reconnaissance. This is explained in detail in Section 10.3. The second phase consists of performing an exploratory site characterization to obtain preliminary information on the site conditions. This information can facilitate development of a detailed site investigation program, if necessary. The field investigation for this phase is performed with a stringent safety and health (S&H) plan, assuming that the worst-case contaminant conditions exist at the site (USEPA, 1984). A quality assurance and quality control (QA/QC) plan is also implemented to obtain accurate and reliable data. The results obtained from this phase will help to revise the safety and health plan to meet the site-specific conditions and to develop the scope of work for the third phase, detailed site inves-

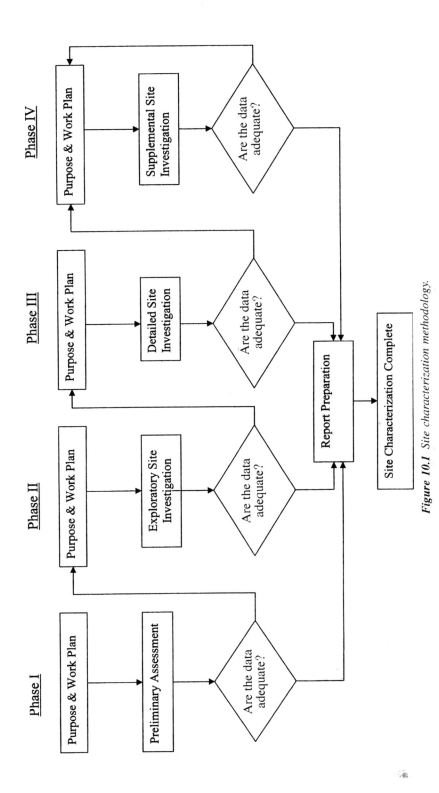

Figure 10.1 Site characterization methodology.

tigation, if necessary. This is explained further in Section 10.4. The third phase of the site characterization process involves performing a detailed site investigation to define the site geology, site hydrogeology, and the nature and extent of the contamination. During this phase of investigation, S&H and QA/QC plans must be carefully implemented to obtain accurate and reliable data. The detailed methods are described in Section 10.5.

Generally, the first three phases should provide sufficient site characterization information. However, if the data collected from these phases is determined to be inadequate, the purpose and scope of an additional investigation should be developed and implemented. This phased approach is repeated until all pertinent data have been collected. Depending on the type and location of a project, site characterization may require regulatory compliance and/or approval at different stages of the investigation. Thus, it is important to review the applicable regulations at the beginning of the project. Project meetings with the regulatory personnel are desirable so that the investigation procedures and results conform to regulatory standards. This proactive approach can avoid delays in obtaining the required regulatory permits and/or approvals.

This general site characterization methodology is comprehensive and is typically followed; however, it may be lengthy and expensive to execute. In some situations, expedited or accelerated procedures are applied to obtain adequate site characterization data within the shortest time period. These procedures are reviewed in Section 10.6.

10.3   PRELIMINARY SITE ASSESSMENT

A preliminary assessment of the project site is a crucial step in the site characterization process. It provides geographical location of the project site, background information on past and current activities at the site, regional hydrogeological information, and potential sources of contamination. The preliminary assessment involves two important tasks: (1) literature review and (2) site visit. The results of these two tasks form the basis for the second phase: exploratory site investigation.

10.3.1   Literature Review

A serious effort should be made to collect all relevant literature related to a project site. This information should include but not be limited to the following:

1. *Site records,* including engineering design drawings of various facilities that exist or once existed at the site, geotechnical investigation reports, environmental audits, and records of operations at the site.

2. *Site personnel interviews* with the plant manager, safety officer, and/or site engineer.

3. *Site permits,* obtainable from regulatory agencies, for the subject site and/or neighboring sites.

4. *Geological maps and reports,* obtainable from agencies such as the U.S. Geological Survey (USGS) and state geological surveys, which provide regional geological and hydrological information.

5. *Water well logs and records,* obtainable from state agencies, to approximate geological conditions and water yield at well locations.

6. *Topographic maps,* of the site prepared before and after development of the site. USGS topographic maps may also provide information on regional topography and features such as ponds, lakes, and rivers.

7. *Aerial photos,* possibly obtainable from local commercial sources, taken at different times throughout the sites development.

8. *Soil survey maps,* obtainable from the U.S. Department of Agriculture, which provide characteristics of surface soils.

9. *Other sources,* such as satellite imagery, local fire and emergency service records, and interviews with local communities.

10.3.2   Site Visit

After reviewing the information obtained from the sources listed above, the project engineer should visit the site to observe and record all potentially important surface features that exist at the site. This information

includes any abandoned objects suspected of being potential sources of contamination as well as surface features, including surface soils and surface water. Samples of surface water and near-surface soils may be
obtained during the site visit. Proper procedures should
be followed in sampling, labeling, and handling the
samples collected in the field. Tests may be performed
on these samples to determine their chemical composition, suspected contaminants, or chemical indicators
by an approved analytical laboratory.

10.4 EXPLORATORY SITE INVESTIGATION

Based on the results of the preliminary site assessment,
the purpose and scope of exploratory field investigation
should be developed. The purpose of this investigation
is to confirm findings in the preliminary assessment.
The exploratory investigation is also necessary to obtain preliminary site-specific data in order to design a
detailed site investigation program, including the
health and safety plan.

A written work plan is prepared describing the
scope of field and laboratory investigation. The work
plan should provide details on the sampling and testing
procedures, the sampling locations and frequency, a
quality assurance/quality control plan, a health and
safety plan, and a schedule and cost assessment. The
scope of work is limited in the exploratory field investigation, but the procedures followed for this investigation are the same as those required for the detailed
site investigation, as explained in Section 10.5.

10.5 DETAILED SITE INVESTIGATION

The detailed site investigation is performed to characterize the geology and hydrogeology of the site as
well as the nature and distribution of the contamination. The data must be adequate to assess the fate and
transport of the contamination and to design an optimal
remedial system, if necessary. Along with S&H and
QA/QC plans, a detailed work plan that includes a
comprehensive field and laboratory test program
should be prepared. Typically, this investigation may

take a few weeks to a few years to complete, depending on the purpose, size, and accessibility of the site.

A variety of techniques are available to characterize
the site geology, site hydrogeology, and site contamination. These methods are described briefly in the following sections.

10.5.1 Methods of Obtaining Soil and Rock Data

Site geology defines the principal stratigraphic units
underlying a site. The data include soil layer thicknesses and the soil's lateral continuity and physical
properties. The methods used to obtain the data are the
same as those used for conventional geotechnical subsurface investigations. However, because the soils will
also be tested to determine their chemical constituents,
stringent decontamination procedures for sampling
equipment must be followed between samples to prevent cross-contamination of soils.

The methods employed to determine site geology
may be categorized into three types: (1) direct methods, (2) geophysical methods, and (3) drive methods.
The selection of a particular type of method depends
on the project objective, project area, and the cost.
Prior to any fieldwork, local utility authorities should
be notified so that the existing buried utilities within
the footprint of the investigation area can be located
and marked.

Direct Methods. Direct methods involve soil or rock
sampling through the use of hand augers, trenches, or
boreholes. The soil and rock samples are subsequently
tested for their physical properties and chemical constituents. To characterize the entire project area adequately, the locations and methods of sampling should
be selected carefully. The various methods of sampling
are described in the following paragraphs.

Near-surface soils may be sampled by using spoons,
scoops, and shovels, as depicted in Figure 10.2. Stainless steel or Teflon-coated tools should be used. A
shovel is used to remove the soil to the desired depth,
and spoons or scoops are used to collect the soil. Similar procedures can be followed to sample in open
trenches either horizontally or vertically. The proce-

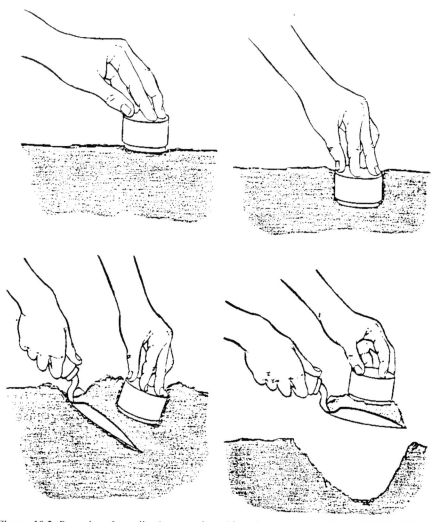

Figure 10.2 *Procedure for collecting samples with soil moisture tin. (From USEPA, 1993a.)*

dure is simple and inexpensive; however, it is limited to near-surface sampling only.

Hand-held augers may be used to sample soils at shallow depths. These augers consist of an auger bit, a solid or tubular rod, and a T handle, as shown in Figure 10.3. The drill rods are threaded; extensions can be added, or auger bits may be interchanged. The auger tip bites into the soil as the handle is rotated, and the soil retained on the auger tip is brought to the surface for use as the soil sample. The auger bits can be screw-

type, bucket-type, or spiral-type, as shown in Figure 10.3. Generally, two people are required to operate hand-held augers. This type of sampling is simple and inexpensive; however, it is difficult to know exact locations of the samples. There is also a possibility that cross-contamination could occur due to caving-in or sloughing of soil.

For sampling at deeper depths, it is common practice to drill boreholes and collect soil samples continuously or at certain intervals with depth. When the

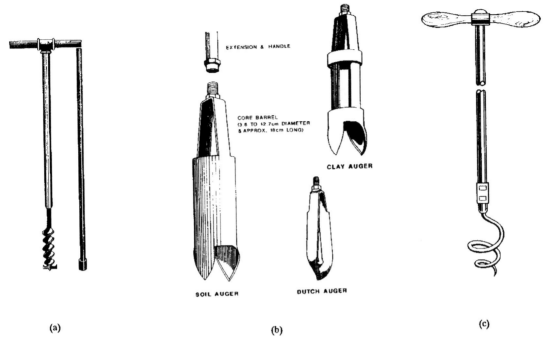

EXTENSION & HANDLE

CORE BARREL
(3.8 TO 12.7cm DIAMETER
& APPROX. 18cm LONG)

CLAY AUGER

SOIL AUGER

DUTCH AUGER

(a) (b) (c)

Figure 10.3 Handheld augers: (a) *screw auger;* (b) *examples of bucket augers;* (c) *spiral or ram's horn auger.* (*From USEPA, 1993a.*)

bedrock is encountered, rock coring may be performed to collect rock cores to the desired depths. In addition to obtaining soil and rock samples, the boreholes or coreholes may be used for performing borehole geophysical tests, installing piezometers, or installing monitoring wells. A wide variety of methods are available to drill boreholes/coreholes. The selection of a particular method is made based on the availability, cost, type of geological materials, and sampling requirements.

Drilling. Generally, a power-operated drill rig, mounted on either a truck or an all-terrain vehicle (ATV), is used for drilling boreholes/coreholes (see Figure 10.4). The ATV-mounted drill rigs are used when drilling in soft areas or in marshlands. The boreholes are advanced by any of the following common methods: solid-stem auger drilling, hollow-stem auger drilling, wet rotary drilling, air rotary drilling, rotary diamond drilling, or directional drilling. Any one or

combination of these methods may be selected based on the required drilling depth, nature of geologic formation, and the purpose of drilling. The purpose of drilling may be to sample for chemical testing, well installation, and so on. Hollow-stem auger drilling is typically used for environmental studies; the other methods are not typically used, due to difficulties in obtaining representative samples at different depths.

In *solid-stem auger drilling,* auger sections with a solid stem and flight, as shown in Figure 10.5, are connected in a continuous string to the lowermost section; the cutting head is set approximately 2 inches larger than the flighting. Cuttings are rotated upward to the surface by moving along the continuous flightings as the cutting head advances into the ground. Unfortunately, this process makes it difficult to obtain reliable depth-specific soil samples from the cuttings that are brought to the surface. In cohesive soils, though, drilling can be stopped at the desired depth. The augers are then removed from the borehole, and a sample is

(a)

(b)

Figure 10.4 Drill rigs: (a) *rubber tire all-terrain vehicle mounted;* (b) *track vehicle mounted. (Courtesy of Diedrich Drill, Inc.)*

taken from the bottom flight. The way to collect representative samples is to remove the auger string, attach a split-spoon or thin-walled sampler to the end of the drill rod, and put the entire string back into the borehole. The details on split-spoon and thin-walled samplers are presented later in this section. Note that recovery of the cohesionless soil samples from saturated zones using this method is not possible.

In *hollow-stem auger drilling,* augers used consist of a hollow pipe with a continuous ramp of upward-spiraling flight welded around them, as shown in Figure 10.6. During drilling, a center rod equipped with a pilot bit and plug is lowered inside the auger; the center rod and hollow stem are rotated together. The plug prevents the soil from entering into the hollow

stem, causing the soil to ride up the outer ramp of auger flights. When the sample interval is reached, the center rod, the plug, and the bit are removed to allow lowering of a soil-sampling tool. During sampling, the hollow stem of the auger prevents collapse of the borehole wall. After sampling is completed, an in-situ testing or monitoring well installation may be performed. This method also provides the uppermost level of the water table, as there is no introduction of fluids.

Wet rotary drilling is used to drill boreholes for installing piezometers or monitoring wells at greater depths. Figure 10.7 shows the major elements of a wet rotary drilling system. Drilling fluid, consisting of bentonite and potable water, is pumped down hollow rotary drill rods and through a bit attached at their lower end. The fluid circulates back to the surface by moving up the annular space between the drill rods and the borehole wall. The fluid is discharged at the surface through a pipe or ditch into a sedimentation tank, pond, or pit. Cuttings settle in the pond, and the fluid overflows into a suction pit, where a pump recirculates it back through the drill rods. The drilling fluid serves to cool and lubricate the bit, stabilize the borehole wall, and prevent the inflow of formation fluids. As a result, the drilling fluid also serves to minimize cross-contamination of aquifers. Samples can be obtained directly from the circulated fluid by placing a sample-collecting device in the discharge flow before the settling pit. For accurate sampling, the flow of drilling fluid is interrupted, and a split-spoon, thin-walled, or core sampler is inserted down the drill rods. The sample is then taken ahead of the bit. Although this drilling method allows drilling boreholes in hard clays, cemented sandstone, or shale to depths of 500 ft, the invasion by drilling fluid makes it difficult to identify aquifers.

Air rotary drilling is similar to wet rotary drilling, but the circulation medium is air instead of water or drilling mud. Figure 10.8 shows the basic rig setup for air rotary drilling with a tricone bit. Compressed air is circulated through the drill rods to cool the bit, carrying cuttings up the open hole to the surface. A cyclone separator slows the air velocity and allows the cuttings to fall into a container. A roller cone drill bit is used for unconsolidated and hard-to-soft consoli-

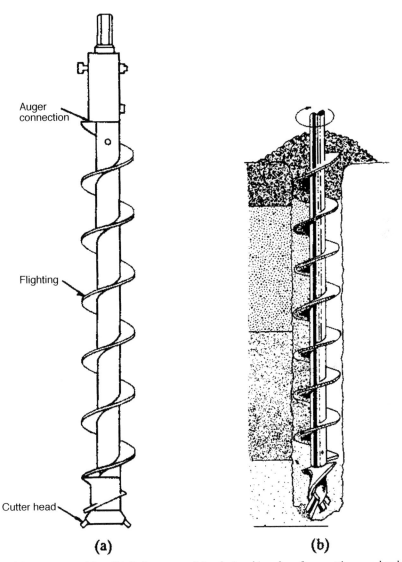

(a)   **(b)**

Figure 10.5 *Power-driven augers:* (a) *solid-flight auger;* (b) *relationship of surface cuttings and subsurface.* (*From USEPA, 1993a.*)

dated rock. This drilling method is often used for installing piezometers or monitoring wells in consolidated material, and for deeper unconsolidated materials that form a stable hole. Drilling is fast, using no drilling fluids, and is well suited for highly fractured rock. This method is relatively expensive, however, and uncontrolled cuttings blown during drilling can pose a hazard to drilling personnel if toxic contaminants are present.

Rotary diamond drilling is used to drill and core in consolidated rock. A rotating bit is used, consisting of a tube 10 to 20 ft long with a diamond-studded ring

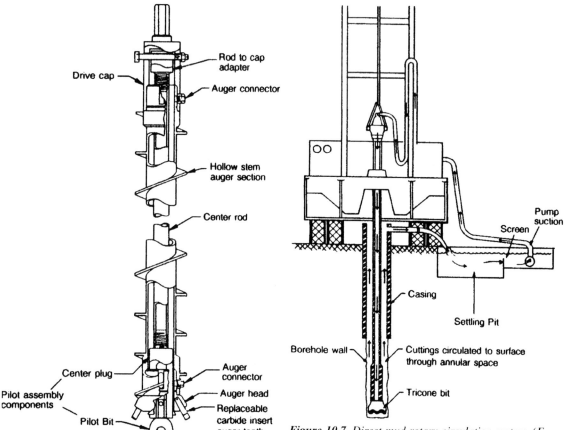

Figure 10.6 *Typical components of a hollow-stem auger.* (*From USEPA, 1993a.*)

Figure 10.7 *Direct mud rotary circulation system.* (*From USEPA, 1993a.*)

fitted to the end of the core barrel. This bit may also be attached to either an air or a mud rotary rig. Figure 10.9 shows a typical diamond drilling rig setup. Typically, water circulates through the bit to cool the cutting surface. The diamond bit cuts through rock while retaining a solid core in the tube. In soft and medium formations, sawtooth or carbide tips may be used. This method allows drilling to any depth and provides continuous cores of rock formations; however, this method is limited to drilling through consolidated rocks. Several negative aspects of rotary diamond drilling include possible alteration of the chemistry of groundwater samples due to water from the drill, expensive diamond bits, and comparatively slow drilling.

Directional drilling is used in certain cases where vertical drilling may not be possible, due to the inaccessibility of the drilling location. In these instances it is advantageous to employ directional drilling (USACE, 1996; Kaback, 2002). Directional drilling utilizes a slant drill in combination with systems to monitor the inclination and general location of the drilling bit. A schematic of the method is shown in Figure 10.10. Another advantage of this system is its cost compared to auger drilling.

Sampling. During drilling, soil samples are commonly obtained at desired depth intervals by employing either (1) split-spoon sampling or (2) thin-walled Shelby tube sampling. Rock sampling, on the other hand, is performed using the core barrel in the rotary diamond drilling method, as described in Section 10.4.

Split-spoon sampling is widely used during drilling for stratigraphy characterization. Split spoons are tubes constructed of high-strength alloy steel with a tongue-and-groove arrangement running the length of the tube. This arrangement allows the tubes to be split in half. A schematic of a split-spoon sampler is shown in Figure 10.11. The two halves are held together by a threaded drive head assembly at the top and a hardened shoe at the bottom with a beveled cutting tip. The sampler is driven by a 140-lb weight dropped through a 30-in. interval, known in classic geotechnical engineering as *standard penetration testing* (SPT). The number of blows required to drive the sampler for every 6 in. of penetration recorded and the sum of the blow count for the last two 6 in. of penetration, known as the SPT N value, provides an indication of the density and strength of the formation being sampled. When the split spoon is brought to the surface, it is dissembled and the core is removed. A basket or spring retainer may be placed near the inside tip of the tube to reduce loss of sample material to the borehole as the sampler is being withdrawn. Standard geotechnical investigation sampling involves penetrating an 18-in. interval for every 5 ft penetrated. Continuous samples can be taken by augering or drilling to the bottom of the previously sampled interval and repeating the sampling operation.

Thin-walled Shelby tube sampling is employed to collect undisturbed cohesive soil samples. Typical Shelby tube dimensions are shown in Figure 10.12(a). The sample collection procedure is similar to split-spoon sampling except that the tube is pushed into the soil as opposed to being driven. Use of a continuous thin-walled sampler, as shown in Figure 10.12(b), with a hollow-stem auger, avoids the time delays involved in the collection of continuous cores from conventional thin-walled samplers. The 5-ft thin-walled tube is placed down the stem of the auger. The tube is then attached to a nonrotating sampling rod or a wireline assembly that allows the auger to rotate while the tube remains stationary. Undisturbed material then enters the tubes, and the auger flights advance. The sample is collected every 5 ft before a new auger flight is added.

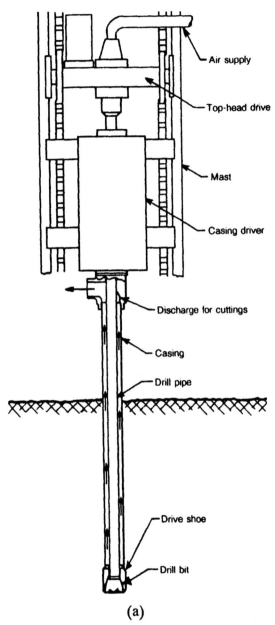

(a)

Figure 10.8 Drill-through methods: (a) rotary drill-through casing drive; (b) dual rotary method. (From USEPA, 1993a.)

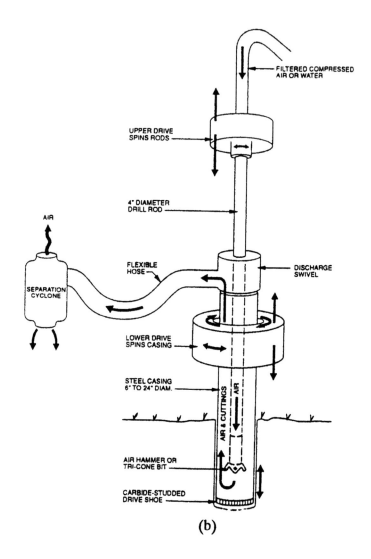

Figure 10.8 (*continued*)   **(b)**

Decontamination Procedures. The drilling tools, augers, samplers, and other equipment used for drilling and sampling should be decontaminated using a steam-generated pressure washer between each boring. The logging tools should be washed in a solution of Alconox (or other such washing solution) and clean water, and then rinsed in clean water. Rock coring and grouting at all locations should be performed using clean water.

Logging. All field sampling is generally performed under the supervision of an experienced geotechnical engineer or geologist. The field engineer or geologist maintains the daily drilling records and logs the soil samples and rock cores. The logging is generally performed according to the soil description terminology and the Unified Soil Classification System (USCS), ASTM D 2487. The soil samples and rock cores are labeled, sealed in proper containers, stored, and transported to the laboratory for testing.

Field Testing. In situations where geotechnical properties of soils are required, field testing on the split-spoon samples or in-situ borehole tests are performed

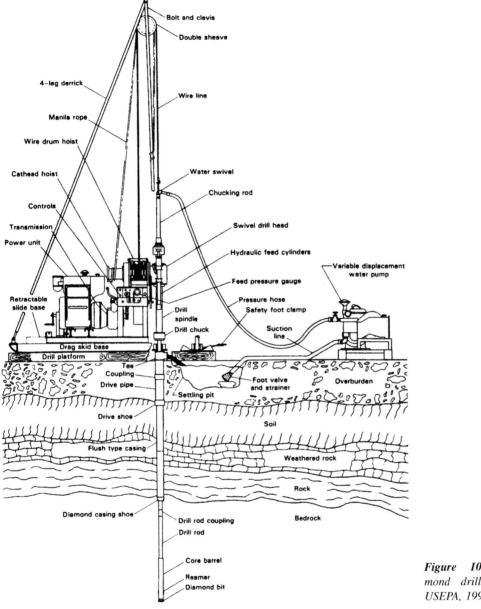

Figure 10.9 *Typical diamond drilling rig. (From USEPA, 1993a.)*

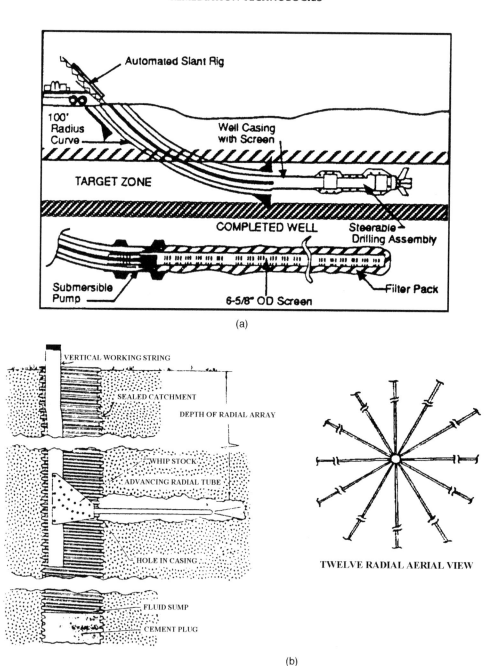

Figure 10.10 Directional drilling methods: (a) Eastman–Christensen slant rig; (b) Petrolphysics rig with a shallow radial system. (From USEPA, 1993a.)

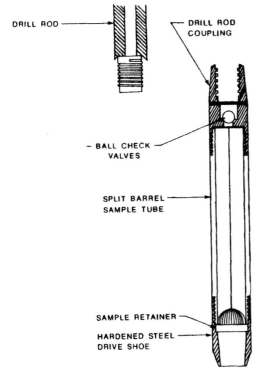

Figure 10.11 Split-spoon sampler. (From USEPA, 1993a.)

to determine properties such as hydraulic conductivity, shear strength, and so on. However, such testing is not commonly performed in site characterization for environmental purposes. For environmental site characterization, generally, soil classification and hydraulic conductivity tests are performed. However, project-specific testing requirements should be identified based on project needs.

Surveying. The location and ground surface elevation of each boring are surveyed using surveying equipment, including GPS.

Borehole Abandonment. The boreholes that are not converted into piezometers or monitoring wells are abandoned in accordance with applicable standards. Generally, these boreholes are grouted from the bottom of the borehole to the ground surface with bentonite

grout. Documentation of the borehole abandonment must be prepared carefully, as improperly abandoned boreholes may serve as potential pathways for contaminant migration.

Laboratory Testing. Laboratory testing is performed on soil and rock samples to determine their physical properties and to aid in characterizing different geological units present at a project site. The common laboratory tests performed on the samples include moisture content tests, Atterberg limits tests, grain size analysis, specific gravity, unit weight, hydraulic conductivity tests, cation exchange capacity tests, and organic content tests. These tests are performed in accordance with the ASTM standards or the applicable regulatory standards. In addition, determination of mineralogy of soils may be required.

Geophysical Methods. Geophysical methods to define site geology are useful in reducing costs when investigating large sites (USACE, 1979; Haeni, 1988; USEPA, 1993a,b, 1998, 2000a,b; ITRC, 2000). These methods are also used in preliminary investigations and provide verification of direct method results. However, the interpretation of the geophysical test data is difficult and requires special expertise. The geophysical methods may be grouped under two categories: (1) borehole geophysical methods and (2) surface geophysical methods.

When using *borehole geophysical methods,* a probe is lowered into the borehole using a cable. The probe transmits signals to surface instruments that generate logs or charts, which relate changes in the parameter being measured with depth. The parameter is then correlated to the types of geologic formations in the area. Borehole geophysical methods can be classified into three categories: (1) electrical/electromagnetic methods, which measure resistivity and conductivity of fluids and surrounding rocks; (2) nuclear methods, which use natural or artificial sources of radiation and radiation detectors to characterize rock and fluid properties; and (3) acoustic/seismic methods, which measure the elastic response of subsurface rock to seismic sources. The selection of a particular type of method depends on the type of borehole (cased or uncased)

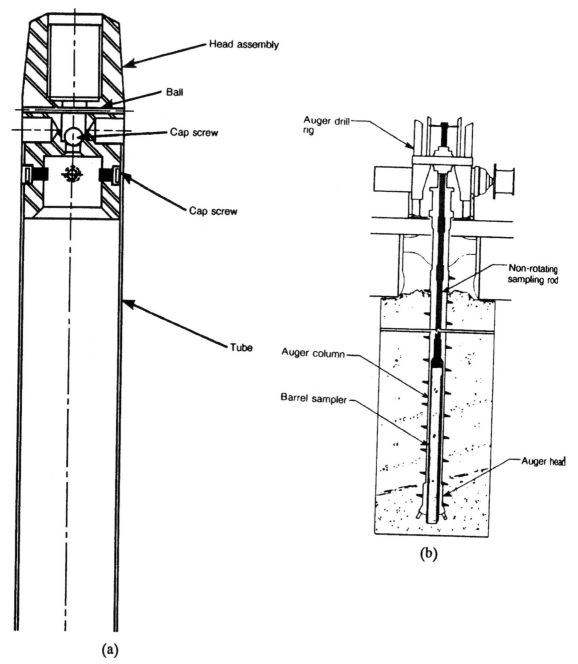

Figure 10.12 (a) *Thin-walled Shelby tube sampler;* (b) *continuous thin-walled sampler.* (*From USEPA, 1993a.*)

and whether it is filled with fluid or is dry. For example, most electrical methods require uncased boreholes and either drilling fluid or water present. Logs are generated in terms of spontaneous potential, single-point resistance or normal resistivity, natural gamma, neutron, caliper, fluid conductivity, temperature, or acoustic velocity, and are used for stratigraphic interpretations. More details on these methods can be found in USEPA (1993a,b, 2000a,b).

Surface geophysical methods, on the other hand, do not require boreholes. The most common surface geophysical techniques include electric, seismic, and electromagnetic surveys as well as the use of ground-penetrating radar. A brief description of these methods is given below.

Electric (or *resistivity*) *surveys* use the measured resistivity of the soil to determine the material type, water content, and porosity of a soil. In this process, the soil acts as a component in a circuit. The apparatus typically incorporates two sets of electrodes for application of current and monitoring of voltage differences in the soil.

Seismic surveys utilize an energy impulse to determine the properties of a soil. The surveys are performed through measurement of the wave velocity created by an impact made on the ground surface. The resulting waves are monitored by remote sensing devices such as geophones, which are placed at various distances from the impact. By analyzing the various times at which the impact wave reaches the remote sensor, soil properties are determined.

Electromagnetic surveys induct an electromagnetic field into the ground by way of a transmitter antenna. The strength of the magnetic field is measured at various locations within a site using a receiving antenna. The electromagnetic conductivity is then interpreted to determine soil properties and may be used to distinguish subsurface objects such as metallic drums. One person may typically operate this apparatus.

Ground penetrating radar utilizes radio-frequency microwaves to penetrate the soil and identify layers possessing contrasting electron properties. The returned signal is dependent on geologic variations in porosity and water content. This method is interpreted to determine soil properties as well as to identify other subsurface resources or hazards.

Drive Methods. The most common drive method is cone penetrometry. In this method, a cone-shaped instrument is pushed into the soil hydraulically and the resistance to penetration is measured. A schematic is shown in Figure 10.13. The resistance is measured by sensitive strain gauges that transmit electronic data to the acquisition system. The resistances measured are correlated with different soil types. Cone penetrometry is very useful in initial site characterization. Cone penetration soil characterization must be calibrated with properties obtained from drilling and sampling data for the site.

In lieu of cone penetrometry, environmental professionals commonly use other direct push methods such as the Geoprobe system for sampling soils. More details on such methods can be found in the ASTM Standard Guide D 6282 (ASTM, 2003). The direct push methods use percussion hammers and static vehicle weight combined with hydraulic cylinders to advance tools to depth. The direct push methods save time and money, provide representative soil samples, and minimize waste generation and exposure hazards. Many regulatory agencies permit using direct push methods to characterize and monitor contaminated sites.

10.5.2 Methods of Obtaining Hydrogeologic Data

Hydrogeologic data are essential to determine groundwater flow conditions, including the rate and direction of flow at a site. The data consist of hydraulic properties of geologic formations as well as the hydraulic gradients. The hydraulic properties of importance are hydraulic conductivity (or transmissivity) and specific yield (or storativity). Using the water levels measured in nested piezometers or monitoring wells at a particular location for vertical gradients, hydraulic gradients are calculated. Measurements for horizontal gradients require the use of water-level measurements in piezometers/wells at different locations throughout the site.

The methods used to obtain hydrological data can be grouped under either direct methods or drive methods. The direct methods involve constructing piezometers or monitoring wells and conducting tests; drive

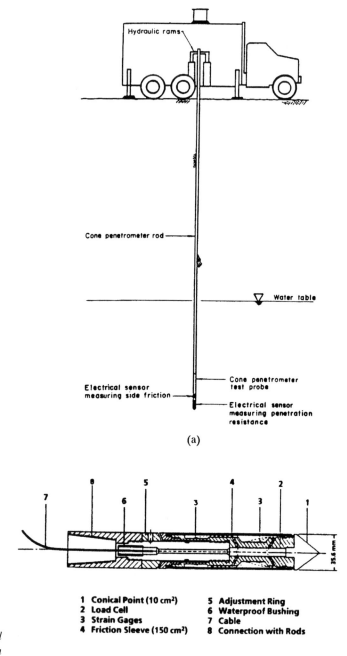

(a)

(b)

1 Conical Point (10 cm²)   **5 Adjustment Ring**
2 Load Cell   **6 Waterproof Bushing**
3 Strain Gages   **7 Cable**
4 Friction Sleeve (150 cm²) **8 Connection with Rods**

Figure 10.13 Cone penetrometry: (a) typical cone penetrometer test rig; (b) electric friction cone pentrometer tip. (From USEPA, 1993a.)

methods involve one-time sampling and testing at selected locations by driving a cone or similar device into the subsurface.

Direct Methods. Direct methods essentially involve the construction and development of piezometers and monitoring wells and the measurement of water levels at different time periods. In addition, in-situ tests can be conducted using these piezometers/wells to determine hydraulic conductivity of geologic formations.

Piezometers and Monitoring Wells. Piezometers and monitoring wells are used to monitor saturated geologic formations. Piezometers are generally used for measuring water levels and conducting hydraulic conductivity tests. In addition to these uses, monitoring wells are used to collect groundwater samples to determine the chemical constituents and their concentrations (groundwater quality). As such, the monitoring wells are constructed with extreme care and are composed of chemical-resistant materials such as stainless steel or Teflon. Strict decontamination procedures must also be followed to prevent cross-contamination problems. Installation procedures must comply with all applicable OSHA rules and other regulations to ensure safe drilling conditions and to minimize potential exposure to contaminants.

Construction Procedures. A schematic diagram of a piezometer is shown in Figure 10.14. Installation begins with drilling a borehole to the required well depth using a hollow-stem auger drill. Once drilling is complete, the drill rods are removed, the auger is pulled up 1 ft, and a weighted tape measure is used to determine the finished depth and to check for caving of the borehole walls. Silica sand is then added to fill the 1-ft annulus below the auger; if water is present, time is allowed for settlement. A 5- or 10-ft-long slotted pipe is connected to the solid pipe sections and lowered into the borehole. Upon lowering, a 2- to 3-ft section of riser should be left above the ground level. The process of removing the auger and filling the resulting annulus with sand should be continued until the sand level is approximately 1 ft above the top of the well screen section. After placing the sand filter pack, bentonite

pellets are placed above the filter pack to form a 2- to 3-ft-thick seal. The pellets are added in the same manner as the sand and allowed approximately 20 minutes to hydrate. The remaining section of the borehole is filled with grout, comprised of either volclay or Portland cement using the tremie method.

Important factors in determining the placement of piezometers include selection of the location, depth, and materials to be used. The locations selected should be accessible and distributed across the site. The screen depths should be selected based on the geologic formation to be monitored. The material used for construction should be chosen based on the soil conditions, nature of potential contaminants, and overall cost effectiveness. Polyvinyl chloride (PVC) piping is typically used to construct piezometers.

A schematic diagram of a monitoring well is shown in Figure 10.15. The construction procedure for monitoring wells is similar to the construction of piezometers. Generally, either stainless steel or Teflon is used as the construction material (McCaulou et al., 1995). By decontaminating the drilling and sampling equipment prior to use, cross-contamination can be prevented. For security purposes, all monitoring wells should be capped and locked with a steel plate.

When water levels in different geologic formations are needed in order to calculate vertical hydraulic gradient, clustered piezometers or monitoring wells should be constructed as depicted in Figure 10.16. These piezometers or monitoring wells are constructed within the same borehole or constructed in different boreholes located close to each other, generally within a 10-ft distance. Each piezometer or monitoring well should be screened independently at selected depths in different formations.

Development Procedures. Postinstallation conditions in piezometers or monitoring wells typically do not reflect the in-situ conditions, as excessive sediments may enter the piezometer or well during formation. Physicochemical changes may also occur in the groundwater. The situation must be rectified through piezometer/well development in order to obtain representative groundwater samples.

The process of piezometer/well development primarily removes excess sediment from within the well

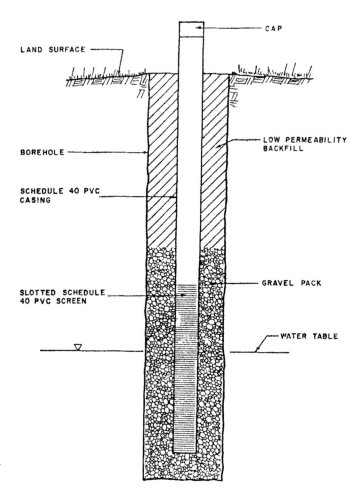

Figure 10.14 Typical piezometer design. *(From USEPA, 1993a.)*

casing and the surrounding geologic formation near the well intake. Different development procedures are shown in Figure 10.17. The particular procedure used is determined by the characteristics of the geologic formation in which the well was installed. Typically, clearing of formation material adjacent to the well screen is accomplished by pumping out water using the pump mounted on the drill rig and/or bailing water using a decontaminated stainless steel bailer. The piezometer/well development should be continued until stabilized conditions exist. Stabilized conditions are typically inferred by monitoring the pH, conductivity, and temperature of extracted water.

Water-Level Measurements. Water levels in piezometers and monitoring wells are measured to determine the groundwater flow direction and gradients at a site. The depth to water from the top of casing is typically measured using a weighted tape measure, an electronic water-level indicator probe, or a pressure transducer (see Figure 10.18). Electronic water-level indicator probes are commonly used for water-level measurements.

When using an electronic water-level indicator, the probe should be lowered into the well slowly until the light goes on and/or the buzzer sounds. This procedure should be repeated several times, until the exact static

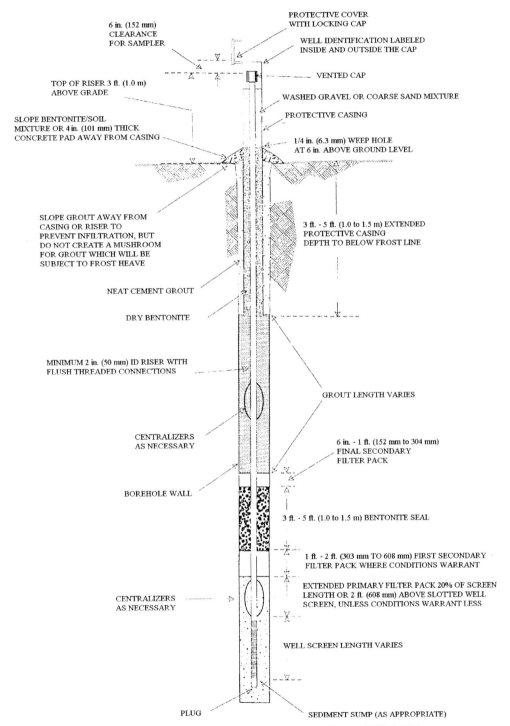

6 in. (152 mm)
CLEARANCE
FOR SAMPLER

PROTECTIVE COVER
WITH LOCKING CAP

WELL IDENTIFICATION LABELED
INSIDE AND OUTSIDE THE CAP

VENTED CAP

TOP OF RISER 3 ft. (1.0 m)
ABOVE GRADE

WASHED GRAVEL OR COARSE SAND MIXTURE

PROTECTIVE CASING

SLOPE BENTONITE/SOIL
MIXTURE OR 4 in. (101 mm) THICK
CONCRETE PAD AWAY FROM CASING

1/4 in. (6.3 mm) WEEP HOLE
AT 6 in. ABOVE GROUND LEVEL

SLOPE GROUT AWAY FROM
CASING OR RISER TO
PREVENT INFILTRATION, BUT
DO NOT CREATE A MUSHROOM
FOR GROUT WHICH WILL BE
SUBJECT TO FROST HEAVE

3 ft. - 5 ft. (1.0 to 1.5 m) EXTENDED
PROTECTIVE CASING
DEPTH TO BELOW FROST LINE

NEAT CEMENT GROUT

DRY BENTONITE

MINIMUM 2 in. (50 mm) ID RISER WITH
FLUSH THREADED CONNECTIONS

GROUT LENGTH VARIES

CENTRALIZERS
AS NECESSARY

6 in. - 1 ft. (152 mm to 304 mm)
FINAL SECONDARY
FILTER PACK

BOREHOLE WALL

3 ft. - 5 ft. (1.0 to 1.5 m) BENTONITE SEAL

1 ft. - 2 ft. (303 mm TO 608 mm) FIRST SECONDARY
FILTER PACK WHERE CONDITIONS WARRANT

EXTENDED PRIMARY FILTER PACK 20% OF SCREEN
LENGTH OR 2 ft. (608 mm) ABOVE SLOTTED WELL
SCREEN, UNLESS CONDITIONS WARRANT LESS

CENTRALIZERS
AS NECESSARY

WELL SCREEN LENGTH VARIES

PLUG

SEDIMENT SUMP (AS APPROPRIATE)

Figure 10.15 Typical monitoring well design. (From ASTM, 2003.)

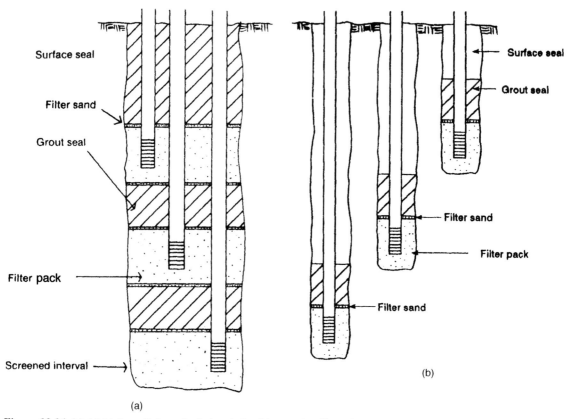

Figure 10.16 (a) *Multiple wells in a single borehole; (b) nested wells with multiple boreholes. (From USEPA, 1993a.)*

water level has been determined. The water levels are then recorded, generally to the nearest 0.01 ft. The exact reference location for water-level measurement, such as the top of casing pipe, should be recorded clearly. The depth to the bottom of the well should also be measured using the electric water-level probe or a weighted stainless steel tape measure. Preferably, the same person should perform all water-level measurements on the same day. If the water-level measurements are to be taken in several monitoring wells, the water-level indicator probe and cable should be decontaminated between each well.

At gasoline spill sites, immiscible light-phase organic liquids or hydrocarbons may exist, floating on the groundwater surface. To calculate the corrected water level, the thickness of a free-phase hydrocarbon

layer may be measured using an interface probe. The interface probe should be lowered into the well carefully until the probe sends a distinct signal indicating that the probe is in an organic liquid, and another distinct signal indicating that the probe is in water. The probe should be raised and lowered slowly, as well, so that the exact top and bottom of the light-phase immiscible layer (if present) can be determined.

In-Situ Hydraulic Conductivity Testing. Three types of field tests are commonly performed to determine the hydraulic conductivity of a geologic formation: packer tests, slug tests, and pumping tests. Packer tests are performed during borehole drilling (in corehole) to determine hydraulic conductivity of rock formations. Slug tests and pump tests are both performed using

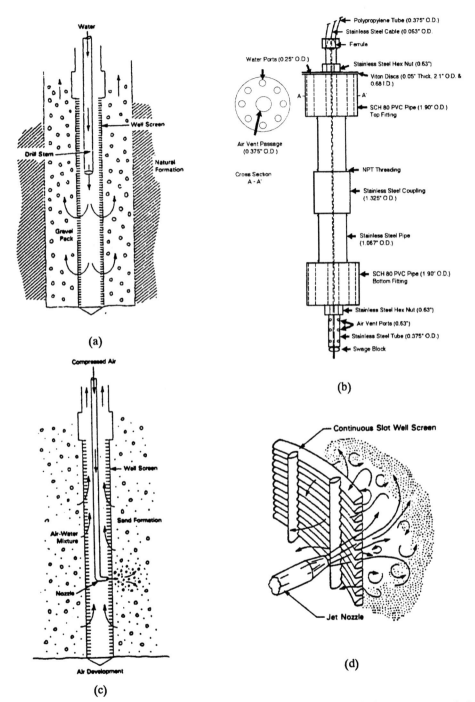

Figure 10.17 *Well development methods:* (a) *backwashing;* (b) *specialized surge block;* (c) *compressed air;* (d) *high-velocity jetting. (From USEPA, 1993a.)*

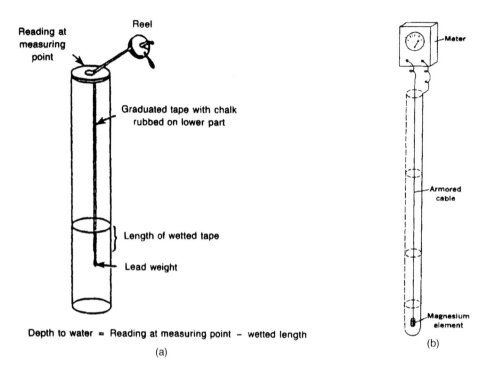

Depth to water = Reading at measuring point − wetted length

(a)

(b)

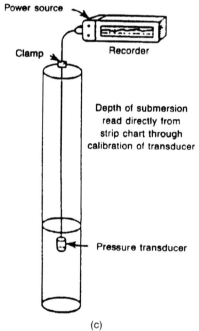

(c)

Figure 10.18 Methods for measuring water levels: (a) *steel tape method;* (b) *electric probe;* (c) *pressure transducer.* (*From USEPA, 1993a.*)

piezometers and monitoring wells to determine hydraulic conductivity of soils. The test procedures and data analysis methods are described in this section.

Packer Tests. Packer tests, also known as *pressure tests,* are performed in boreholes to determine the hydraulic conductivity of rock formations. A schematic of testing conditions is shown in Figure 10.19. These tests are performed using an NX-size inflatable packer setup. For each test, each packer is inflated under nitrogen gas pressure to hydraulically isolate a portion of the borehole. Single and double packer tests may be conducted. A single-packer test utilizes one packer to seal off the borehole, so that the interval below the packer to the bottom of the borehole may be tested. Water is then pumped in a pipe through the packer to the zone between the packer and the bottom of the borehole. Double-packer tests utilize two packers to isolate a specific length of test zone. The two packers are separated by a perforated conductor pipe, which permits water to be pumped into the zone between them.

TEST PROCEDURE. The typical procedure for performing a packer test consists of the following steps:

1. Measure the static water level and record the test configuration details. Prepare an as-built model of the test.
2. Connect the drill rods, with the tool joints covered with Teflon tape to prevent leakage.
3. Determine the maximum allowable inflation pressure for two packers.
4. Attach the packer assembly to the predetermined length of drill rods and lower it into the boring.
5. Inflate the packers. Monitor the nitrogen pressure gauge for fluctuations that would result in leakage or fracturing of rock.
6. Pump water into the test zone at a desired pressure (injection pressure) through the drill rods. Monitor the pressure gauge on the instrument string for pressure loss, which would indicate a poor seal in the test zone.
7. Measure the flow, in rate versus time, using a water meter fixed to the aboveground instrument string.
8. Repeat steps 6 and 7 for various injection pressures.

Each test zone is tested initially at a small constant pressure, and increasing pressures are tested in one or two steps. If a significant flow into the test zone is observed, a stepped pressure test should be performed. The purpose of step tests is to evaluate the impact of increased pressure on the calculated hydraulic conductivity.

DATA ANALYSIS PROCEDURE. The field data are analyzed in accordance with the method described in the USBR (1981) and summarized in Figure 10.20. As shown in Figure 10.20, two methods of analysis exist. Method 1 is used for the analysis of single-packer test data, whereas method 2 is used for the analysis of double-packer test data.

Slug Tests. Slug tests are performed routinely because they are easy, economical, acceptably accurate, and repeatable over a short period of time. Slug tests involve introduction or withdrawal of a slug to displace water in a well and measurement of displacement of water vs. time until equilibrium conditions are reached. The slug is generally made of a sealed cylindrical PVC or stainless steel pipe filled with sand. An alternative method involves initial water displacement in the well created by introducing water as shown in Figure 10.21(*a*) or pneumatically as shown in Figure 10.21(*b*). The data are then analyzed to calculate the hydraulic conductivity.

TEST PROCEDURE. The equipment used for slug tests includes a (1) slug, (2) water-level indicator, (3) pressure transducer, (4) data logger, and (5) laptop computer. In low-permeability soils, the water displacement may be very slow and water displacement vs. time data may be obtained by using the water-level indicator alone. However, for medium- to high-permeability soils, the water displacement vs. time data can only be measured accurately using the pressure

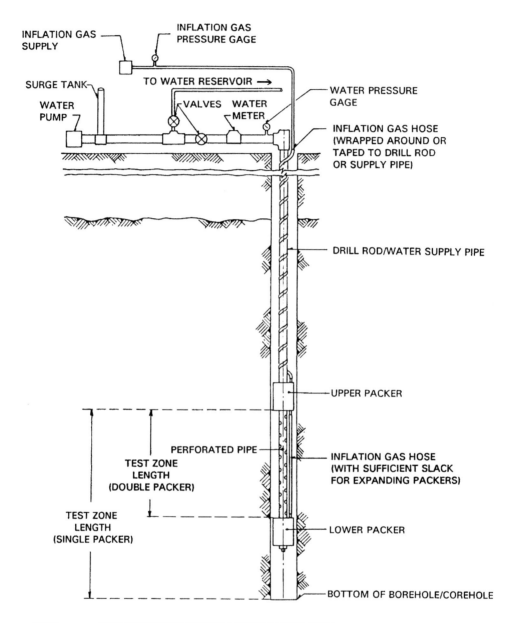

NOTE: THIS SCHEMATIC SHOWS DOUBLE PACKER TEST APPARATUS. IN CASE
            OF SINGLE PACKER TEST APPARATUS, THE LOWER PACKER IS NOT
            PROVIDED AND THE INFLATION GAS HOSE AT THE BOTTOM OF UPPER PACKER
            IS SEALED. THE TEST ZONE LENGTH FOR SINGLE PACKER TEST,
            THEREFORE, IS THE DISTANCE FROM BOTTOM OF UPPER PACKER TO THE
            BOTTOM OF BOREHOLE/COREHOLE.

Figure 10.19 Schematic of packer test.

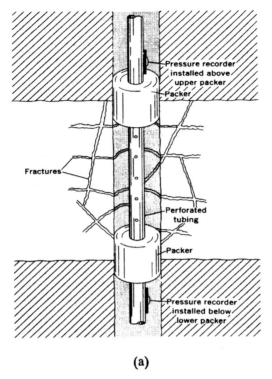

(a)

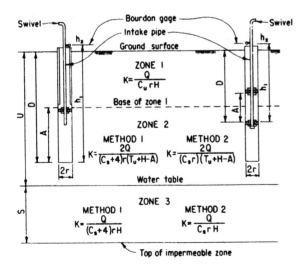

K = coefficient of permeability, feet per second under a unit gradient
Q = steady flow into well, ft³/s
H = h, + h₂ - L = effective head, ft
h, (above water table) = distance between Bourdon gage and bottom of hole for method 1 or distance between gage and upper surface of lower packer for method 2, ft
h, (below water table) = distance between gage and water table, ft
h₂ = applied pressure at gage, 1 lb/in² = 2.307 ft of water
L = head loss in pipe due to friction, ft; ignore head loss for Q<4 gal/min in 1¼ - inch pipe; use length of pipe between gage and top of test section for computations
X = H/U (100) = percent of unsaturated stratum
A = length of test section, ft
r = radius of test hole, ft
C_u = conductivity coefficient for unsaturated materials with partially penetrating cylindrical test wells
C_s = conductivity coefficient for semi-spherical flow in saturated materials through partially penetrating cylindrical test wells
U = thickness of unsaturated material, ft
S = thickness of saturated material, ft
T_u = U - D + H = distance from water surface in well to water table, ft
D = distance from ground surface to bottom of test section, ft
a = surface area of test section, ft²; area of wall plus area of bottom for method 1; area of wall for method 2
Limitations:
Q/a ≤ 0.10, S ≥ 5A, A ≥ 10r, thickness of each packer must be ≥ 10r in method 2

(b)

Figure 10.20 *Packer test: (a) typical straddle-packer installation; (b) single- and double-packer permeability tests for use in saturated or unsaturated consolidated rock. (From USEPA, 1993a.)*

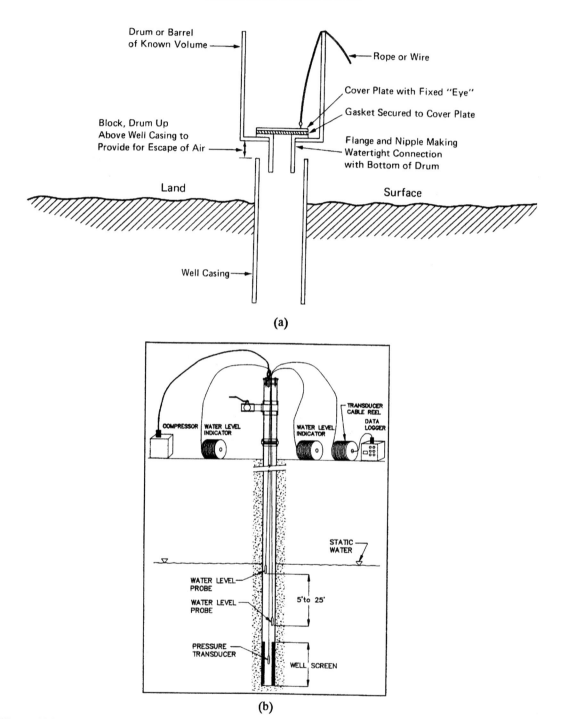

Figure 10.21 *Slug tests:* (a) *apparatus for performing water injection slug tests;* (b) *equipment setup for conducting a pneumatic rising-head slug test. (From USEPA, 1993a.)*

transducer and data logger, which may be connected to a computer for data processing. The typical slug test procedure consists of the following steps:

1. Open the well.

2. Use the water-level indicator to measure the distance from the top of well casing to the water surface.

3. Insert the pressure transducer into the well to about 10 ft from the water surface, to avoid damage due to contact with the slug (when inserted later), yet close enough to the water surface to provide accurate water-level measurements. Secure the transducer and its cable so that it will stay in the desired position.

4. Connect the transducer to the data logger and program the data logger with water-level information and transducer calibration coefficients.

5. Measure the length of the slug and its rope so that the slug will be completely submerged but safely above the transducer when inserted into the well.

6. Insert the slug and start the data logger simultaneously to collect the water-level information.

7. After the water level reaches equilibrium, extract the slug from the well and record the water-level data simultaneously using the data logger.

8. Repeat the test procedure above to confirm repeatability.

9. Transfer the data onto a computer for storage and analysis.

DATA ANALYSIS PROCEDURE. The drawdown or rise of water in the well vs. time data is used to determine hydraulic conductivity, using the Hvorslev (1951) method, the Cooper et al. (1967) method, or the Bouwer and Rice (1976) and Bouwer (1989) method. These methods are described, along with examples, in Section 7.8.2.

Pumping Tests. Pumping tests are performed to determine the hydraulic properties of an aquifer. A pumping test consists of pumping a well at a known rate and measuring the drawdown in both the pumping well and one or more nearby observation wells (Figure 10.22). Pumping continues until equilibrium is reached in the pumping well and in each of the observation wells. Once equilibrium is reached, the pump is turned off. Water-level measurements should be taken in the pumping well and observation wells at specified time intervals during the pumping and recovery phases of the test. Water levels may be measured using a weighted tape measure, electric water-level indicator, or commercially available data logger and pressure transducers. There are a variety of commercially available data recorders that can monitor water levels in all wells simultaneously during the test (Osborne, 1993).

Prior to the pumping test, preliminary estimates of the hydraulic conductivity of the aquifer should be made. Simple analyses should also be performed to estimate pumping rates, radius of influence from the pumping well, and drawdown in the pumping well and observation wells. The pumping well and observation wells may then be installed and developed. If monitoring wells or water supply wells are to be used during the pump test, all equipment placed in the wells should be decontaminated. All field personnel involved in setting up and running the pumping test should have some understanding of the basic purpose and operation of the tests.

TEST PROCEDURE. The typical pumping test procedure consists of the following steps:

1. All pumps within the anticipated zone of influence or any of the observation wells must be shut off until water levels are allowed to reach static equilibrium condition.

2. Static water levels in the pumping well and observation wells are measured immediately prior to starting the test to confirm static conditions.

3. Depths to the bottom of all wells are measured. The volume of water within each well is calculated.

4. A submersible pump is installed in the pumping well.

5. Pressure transducers are installed in the pumping well and in observation wells to the preselected depths below the static water level. The transducers are set below the anticipated

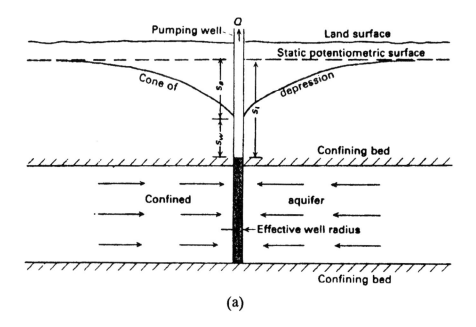

(a)

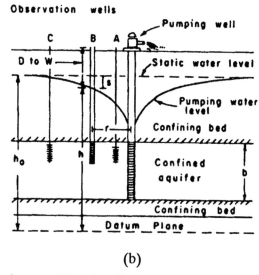

(b)

Figure 10.22 *Pumping tests:* (a) *single-well test;* (b) *multiple-well test.* (*From USEPA, 1993a.*)

drawdowns in the wells to ensure that the transducers will remain fully submerged during the test. The transducers are connected to a data logger.

6. The water-level measurements are entered into the data logger and the data logger is programmed to collect water levels at frequent time intervals.

7. The water levels in the wells are remeasured to ensure that water levels are consistent with the static water levels.

8. Pumping is started and the data logger is simultaneously activated to record water levels.

9. The flow rate is calculated by measuring the volume of water pumped out in a known time interval and is checked periodically to verify that a constant-flow condition exists.

10. After pumping for the required amount of time (generally, 24 hours), the pump is shut off and water-level recovery is started immediately using the data logger.

11. After the water level reaches the initial static water level or there is no significant change in water level, the test is considered complete. The pump and the pressure transducers are removed from the wells.

The pumping test may be conducted as a series of "steps." During each step, the pump should be operated at a constant rate of discharge until equilibrium conditions are observed in all observation wells. In the next step, the pump discharge is increased, and data are collected until a new state of equilibrium is achieved. This may be repeated several times during the test.

DATA ANALYSIS PROCEDURE. Several methods are available to analyze the pumping test data to determine the hydraulic properties of the aquifers. The most commonly used methods, for confined aquifers, are the (1) Theis method, (2) Cooper–Jacob method, (3) Papadopulos–Cooper method, and (4) Theis recovery method. Many assumptions are made in the data analysis methods. These assumptions include:

1. The aquifer is of infinite extent in area and has constant thickness.

2. The aquifer is homogeneous and isotropic.

3. The aquifer is fully confined.

4. The potentiometric surface in the aquifer is initially horizontal.

5. The well fully penetrates the aquifer.

6. The pumping rate is constant.

7. The flow is horizontal and radial toward the pumping well during the unsteady-state conditions, which occur during initial pumping.

8. Well storage effects are negligible.

9. Water is released instantaneously from storage.

The Theis method and the Cooper–Jacob method are commonly used and are described, along with examples, in Section 7.8.1. For a pumping test in an unconfined aquifer, the Neuman type curve matching procedure is used (details on this procedure are given in Section 7.8.1). For details on these analysis methods, the reader should refer Walton (1989) and Dawson and Isotok (1991).

Drive Methods. An alternative to installing piezometers and monitoring wells for assessing hydrogeology is the use of drive methods (Kram et al., 2001). In these methods, a cone-type device is caused to penetrate the soil to the desired depth using either a hydraulic or a hammering mechanism. The device may then be controlled to obtain a groundwater sample and/or perform hydraulic conductivity tests (ASTM D 6001). Recently, direct push methods have been standardized and have been used extensively for installation of monitoring wells [more details on these may be found in ASTM D 6724 and ASTM D 6725 (ASTM, 2003)]. The small-diameter direct push installed wells and groundwater samplers have also been used to determine the hydraulic conductivity of soils. As stated earlier, direct push methods are simple, save time and money, and minimize waste generation and exposure hazard; therefore, these methods are commonly used at contaminated sites.

Two types of drive methods, the hydropunch and the BAT (given the name of its inventor, Bengt Arne Torstensson), are commonly employed. The hydropunch is comprised of shield and well screen mechanisms as shown in Figure 10.23. The hydropunch is pushed or hammered into the ground, and the screen is shielded until the desired depth is reached. At the desired depth, the entire mechanism is raised 1.5 ft, and the screen fills with water due to hydrostatic pressure. Once the chamber is filled, the hydropunch is raised to the surface, and the sample is transferred to a sample vial.

The BAT consists of a cone with a porous filter, as shown in Figure 10.24 (Mines and Stauffer, 1992). The cone may be fitted with one of two types of probes: a probe with an exposed filter or a probe with a shielded filter. In either case, the BAT has the distinct advantage over the hydropunch of allowing hermetically sealed samples to be collected at the in-situ hydrostatic pressure. This is accomplished by inserting an evacuated test tube through the drilling rods to connect to one side of a double-ended needle. The other end of the needle is fitted to a porous filter. After sufficient sampling time, the test tube is removed and stored in a refrigerated container. The BAT may also be fitted to determine the depth of the water table and the horizontal hydraulic conductivity of the soil, and to perform tracer tests. A penetrometer rig may be used to drive the BAT probes into the ground.

Limitations of the BAT lie in its inability to be driven through soils with large amounts of gravel or cobbles. The BAT is also limited in the relatively small sample sizes (35 mL) that it obtains. Studies have shown improved results in sampling volatile organics using BAT in comparison to using peristaltic pumps in monitoring wells. Peristaltic pumps have the tendency to move the volatile organics from the liquid to the gaseous phase. This, however, does not suggest that the BAT is suitable to replace the long-term monitoring capability of monitoring wells.

10.5.3   Methods of Obtaining Chemical Data

The conventional geotechnical site characterization studies culminate with the characterization of site geology and hydrogeology. However, when dealing with contaminated sites, additional characterization of the type and extent of contamination must be performed. For this purpose, soils in the vadose and saturated zones as well as the groundwater must be tested to determine their chemical constituents. The soil and groundwater samples should be collected and tested such that the extent of contamination in horizontal and vertical directions can be delineated. Although less common, sampling of soil moisture and soil gas samples in the vadose zone may also be required at some project sites. The in-situ soil moisture samples in the vadose zone may be collected using different methods, including lysimeters, as described in ASTM D4696, whereas, in-situ soil gas samples may be obtained using different methods described in ASTM D 5314 and in Lutenegger and DeGroot (2000).

To represent the site contamination, extreme care should be taken in collecting and handling the soil and groundwater samples. Stringent quality assurance and quality control protocols must be implemented. In addition, standard methods of chemical analysis must be performed on these samples by a certified analytical laboratory.

In the following sections, details on the sampling, handling, and testing of soil and groundwater samples are presented.

Sampling Procedures. Various methods to collect soil samples have been discussed in Section 10.5.1. In this section, methods to collect groundwater samples from monitoring wells are presented. The generalized methodology for groundwater sampling is shown in Figure 10.25. The quantity of sample and method of preservation depend on the type of analyses to be performed (Puls and Barcelona, 1989; USEPA, 1993a; Yeskis and Zavala, 2002). Prior to sampling, the depth to the water level is measured and the volume of water in the well is calculated (Thornhill, 1989). The well should then be pumped or bailed either to remove a

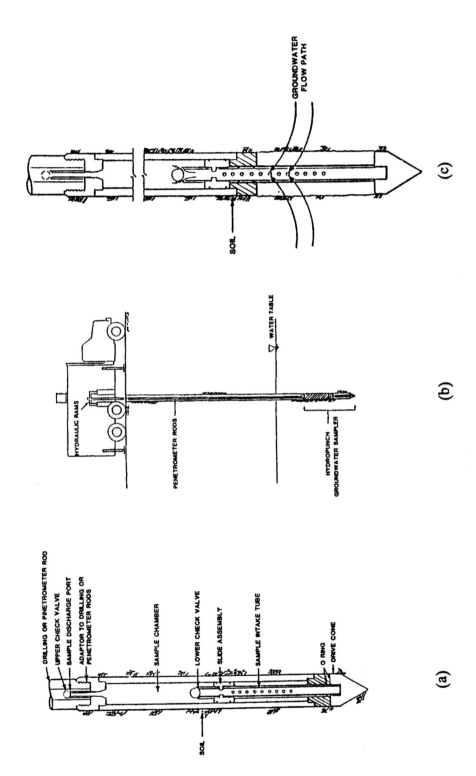

Figure 10.23 Hydropunch: (a) schematic; (b) operation; (c) sample collection. (From USEPA, 1993a.)

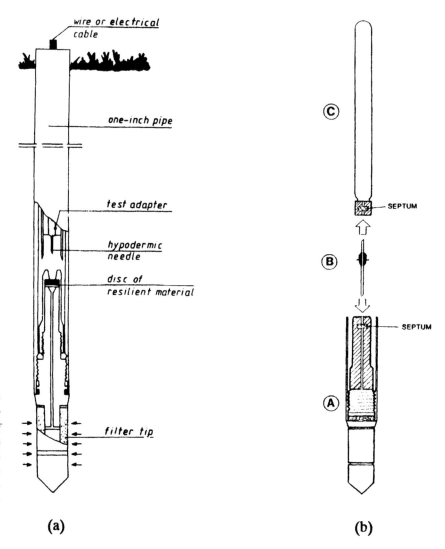

Figure 10.24 BAT system: (a) *schematic of test adaptor for sampling of groundwater and gas;* (b) *filter tip attached to cone penetration ring (A) with double-ended hypodermic needle (B) and sample vial (C). (From USEPA, 1993a.)*

minimum of three casing volumes or until it is dry (Puls and Barcelona, 1996). Two types of pumps commonly used are bladder and submersible pumps, shown in Figures 10.26 and 10.27, respectively. If bailers are used, they should be made of stainless steel or Teflon. The physical appearance of the water, including the color, turbidity, odor, and so on, is recorded. Its pH, conductivity, and temperature may also be measured. Once water conditions are stabilized, samples are col-

lected in prelabeled containers and stored on ice for transportation to a laboratory.

Decontamination Procedures. All equipment that comes into contact with soil samples should be decontaminated before use and between sampling points. The decontamination of soil sampling equipment can be accomplished by directing high-pressure hot wash water over the outside and inside surfaces of the equip-

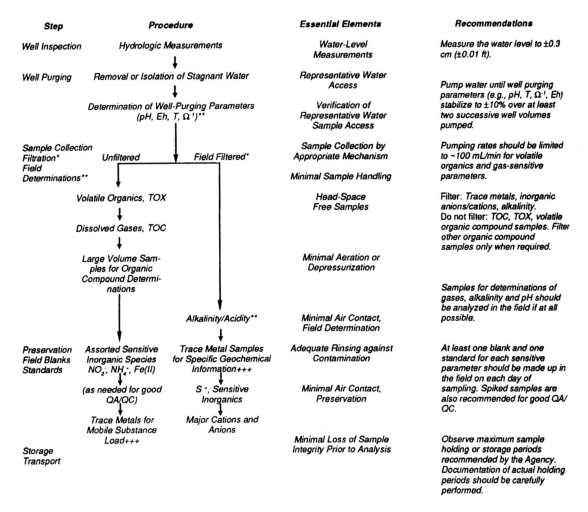

Step	Procedure	Essential Elements	Recommendations
Well Inspection	Hydrologic Measurements	Water-Level Measurements	Measure the water level to ±0.3 cm (±0.01 ft).
Well Purging	Removal or Isolation of Stagnant Water	Representative Water Access	Pump water until well purging parameters (e.g., pH, T, Ω^{-1}, Eh) stabilize to ±10% over at least two successive well volumes pumped.
	Determination of Well-Purging Parameters (pH, Eh, T, Ω^{-1})**	Verification of Representative Water Sample Access	
Sample Collection Filtration* Field Determinations**	Unfiltered / Field Filtered*	Sample Collection by Appropriate Mechanism / Minimal Sample Handling	Pumping rates should be limited to ~100 mL/min for volatile organics and gas-sensitive parameters.
	Volatile Organics, TOX	Head-Space Free Samples	Filter: Trace metals, inorganic anions/cations, alkalinity. Do not filter: TOC, TOX, volatile organic compound samples. Filter other organic compound samples only when required.
	Dissolved Gases, TOC		
	Large Volume Samples for Organic Compound Determinations	Minimal Aeration or Depressurization	
	Alkalinity/Acidity**	Minimal Air Contact, Field Determination	Samples for determinations of gases, alkalinity and pH should be analyzed in the field if at all possible.
Preservation Field Blanks Standards	Assorted Sensitive Inorganic Species NO_2^-, NH_4^+, Fe(II) / Trace Metal Samples for Specific Geochemical Information+++	Adequate Rinsing against Contamination	At least one blank and one standard for each sensitive parameter should be made up in the field on each day of sampling. Spiked samples are also recommended for good QA/QC.
	(as needed for good QA/QC) / S^-, Sensitive Inorganics	Minimal Air Contact, Preservation	
	Trace Metals for Mobile Substance Load+++ / Major Cations and Anions		
Storage Transport		Minimal Loss of Sample Integrity Prior to Analysis	Observe maximum sample holding or storage periods recommended by the Agency. Documentation of actual holding periods should be carefully performed.

*   Denotes samples that should be filtered to determine dissolved constituents. Filtration should be accomplished preferably with in-line filters and pump pressure or by N_2 pressure methods. Samples for dissolved gases or volatile organics should not be filtered. In instances where well development procedures do not allow for turbidity free samples and may bias analytical results, split samples should be spiked with standards before filtration. Both spiked samples and regular samples should be analyzed to determine recoveries from both types of handling.
** Denotes analytical determinations that should be made in the field.
+++ See Puls and Barcelona (1989).

Figure 10.25 Generalized flow diagram of groundwater sampling protocol. (From USEPA, 1993a.)

ment. Then a pressurized spray may be used to apply a 50/50 mixture of acetone and distilled water. Finally, the mixture is rinsed by directing a high-pressure hot water wash over the equipment and allowing the equipment to air dry.

All equipment that contacts groundwater (water-level indicator, bailers, funnels, etc.) should be washed in an Alconox or similar solution and then rinsed by distilled/deionized water. A mixture of acetone and distilled water is applied after rinsing. Finally, the

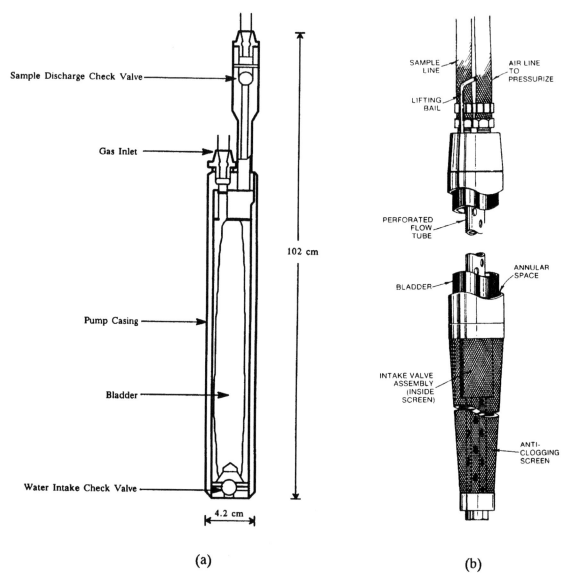

Figure 10.26 *Bladder pump:* (a) *schematic;* (b) *operational unit. (From USEPA, 1993a.)*

mixture is rinsed with distilled/deionized water. Alternatively, dedicated pumps or bailers at each monitoring well may be utilized to prevent any cross-contamination problems.

Fragile equipment such as the pH meter, conductivity meter, and thermometer should be rinsed liberally in distilled water between sample measurements.

QA/QC Samples. The field program must include QA/QC procedures to be used in the field for the sampling of soil and groundwater. This program includes collecting field blanks and trip blanks. *Field blanks* are defined as samples that are obtained by running analyte-free deionized water through sample collection equipment (bailer, pump, auger, spatula, etc.) after de-

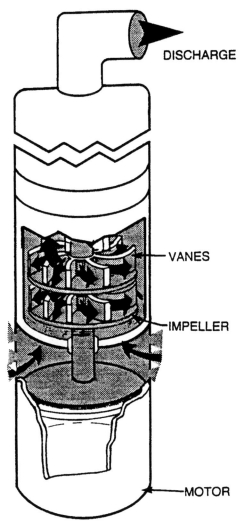

DISCHARGE

VANES

IMPELLER

MOTOR

Figure 10.27 Submersible centrifugal pump. (From USEPA, 1993a.)

after their preparation should the sample containers be opened before they reach the laboratory.

The collection frequency of QA/QC samples is specified for each sample medium. The frequencies are based on the estimated number of samples to be collected, as shown in the work plan. The actual number of QA/QC samples may vary depending on the actual number of samples collected. The following guidelines should be used to determine the number of QA/QC samples to be collected. For soils, field blanks should be submitted at the rate of one field blank for every 20 investigative samples; trip blanks are not required for soils. For groundwater, field blanks should be submitted at the rate of one field blank for every 20 investigative samples. Trip blanks should be included for groundwater at a frequency of one trip blank per every 25 investigative samples.

Chain-of-Custody Procedures. Samples intended for analysis should always be shipped in sealed coolers or containers. Each cooler or container should be affixed with a numbered chain-of-custody seal, and a properly completed chain-of-custody form must accompany the samples. The chain-of-custody form must at least include the following information: the well number or sample location number, name and signature of each person who has transported or received the samples, name and signature of the person(s) who collected the samples, an inventory of all sample bottles and containers in the shipping container (cooler), time and date the sample kit was opened, time and date the samples were collected, and a complete record of the time and date of all changes of sample possession. Most contract laboratories provide their own chain-of-custody forms with the sample bottles and/or sample kits. These forms may be used if they provide the minimum information listed above.

Chemical Analysis Methods. Laboratory testing on the soil and groundwater samples should be performed in accordance with the USEPA test procedures given in SW-846 or applicable ASTM test methods (USEPA, 1986; ASTM, 2003). A summary of laboratory test methods to analyze soil and groundwater samples, including the holding times, is given in Table 10.1. An

contamination. They are then placed in appropriate sample containers for analysis. Field blanks are used to determine the sufficiency of decontamination procedures. Soil field blanks may also be called *rinsate samples. Trip blanks* are prepared prior to the sampling event in the actual sample containers and are kept with the investigative samples throughout the sampling event. Trip blanks are then packaged for shipment with the other samples and sent for analysis. At no time

Table 10.1 **Summary of laboratory test methods for analysis of soil and groundwater samples**

Soil Analytical Methods

Reference Method	Parameter	Technique[a]	Container Type, Number, Volume	Preservation / Storage Requirements	Maximum Holding Time from Collection[b]	Maximum Holding Time (Analysis)
EPA 8240	Volatile organics	GC/MS	2 40-mL VOA vials	Refrigerated at 4°C	NA	14 days[c]
EPA 8270	Semivolatile extractable organics	GC/MS	1 250-mL glass jar	Refrigerated at 4°C	14 days	40 days
EPA 8010	Halogenated volatile organics	GC/HSD	2 40-mL VOA vials	Refrigerated at 4°C	NA	14 days[c]
EPA 8020	Aromatic volatile organics	GC/PID	1 500-mL glass jar	Refrigerated at 4°C	NA	14 days[c]
EPA 8040	Phenols	GC/FID	1 500-mL glass jar	Refrigerated at 4°C	NA	40 days
EPA 8080	Organochlorine pesticides and PCBs	GC/ECD	1 500-mL glass jar	Refrigerated at 4°C	14 days	40 days
EPA 8100 or 8310	Polynuclear aromatic hydrocarbons	GC/FID HPLC	1 500-mL glass jar	Refrigerated at 4°C	14 days	40 days
EPA 8120	Chlorinated hydrocarbons	GC/ECD	1 500-mL glass jar	Refrigerated at 4°C	14 days	
EPA 6010	Metals: Cd, Cr, Pb, Mn, Ba, Si, Fe, Al, Sb, Be, Co, Cu, Mo, Ni, Ag, Tl, V, Zn	ICP, emission spectroscopy	1 250-mL glass jar	Refrigerated at 4°C	NA	6 months
EPA 7060	Arsenic	Furnace AA	1 250-mL glass jar	Refrigerated at 4°C	NA	6 months
EPA 7740	Selenium	Furnace AA			NA	6 months
EPA 7471	Mercury	Cold vapor AA			NA	28 days
EPA 7421	Lead	Furnace AA			NA	6 months
ASTM 3237	Organic lead	AA after extraction	1 500-mL glass jar	Refrigerated at 4°C	14 days	14 days[c]
EPA 9070/9071	Oil and grease	Gravimetric	1 500-mL glass jar	Refrigerated at 4°C	NA	28 days
SM 209F	% Solids	Gravimetric	1 250-mL glass jar	Refrigerated at 4°C	NA	Analyze immediately
EPA 9045	pH	Electrometric		Refrigerated at 4°C	NA	Analyze immediately
EPA 1010	Flashpoint	Pensky–Marten closed cup	1 250-mL glass jar	Refrigerated at 4°C	NA	Analyze as soon as possible[c]
EPA reactivity	Cyanide/sulfide	Acidification, colorimetry/titration	1 250-mL glass jar	Refrigerated at 4°C	NA	Analyze as soon as possible[c]
EPA 340.1/340.2	Fluoride	Bellack distillation/SIE	1 500-mL glass jar	Refrigerated at 4°C	NA	28 days

Groundwater Analytical Methods

EPA Method	Parameter	Method	Container	Preservation		
EPA 8240, EPA 624	Volatile organics	GC/MS	2 40-mL VOA vials	Refrigerated at 4°C; HCl, pH < 2	NA	14 days[c,d]
EPA 8270, EPA 625	Semivolatile extractable organics	GC/MS	1 1-L glass bottle; TFE-lined cap	Refrigerated at 4°C	7 days[c]	40 days
EPA 8280 or EPA 613	Polychlorinated dibenzodioxins (PCDDs) and polychlorinated dibenzofurans (PCDFs)	GC/MS	1 1-L amber glass bottle; TFE-lined cap 1 1-L amber glass bottle; TFE-lined cap	Refrigerated at 4°C Refrigerated at 4°C	30 days[c] 7 days[c]	40 days 40 days
EPA 8010, EPA 601	Halogenated volatile organics	GC/HSD	2 40-mL VOA vials	Refrigerated at 4°C	NA	14 days[c]
EPA 8020, EPA 602	Aromatic volatile organics	GC/PID	2 40-mL VOA vials	Refrigerated at 4°C; HCl, pH < 2	NA	14 days[c,d]
EPA 8040, EP 604	Phenols	GC/FID	1 1-L glass bottle; TFE-lined cap	Refrigerated at 4°C	7 days[c]	40 days
EPA 8080, EPA 608	Organochlorine pesticides and PCBs	GC/ECD	1 1-L amber glass bottle; TFE-lined cap	Refrigerated at 4°C	7 days[c]	40 days
EPA 680	PCBs	GC/MS	1 1-L amber glass bottle; TFE-lined cap	Refrigerated at 4°C	7 days[c]	40 days
EPA 8120 (modified)	Chlorinated hydrocarbons	GC/ECD	1 1-L amber glass bottle; TFE-lined cap	Refrigerated at 4°C	7 days[c]	40 days
EPA 8140	Organophosphate pesticides	GC/FPD	1 1-L amber glass bottle; TFE-lined cap	Refrigerated at 4°C	7 days[c]	40 days
EPA 8150	Chlorinated herbicides	GC/ECD	1 1-L amber glass bottle; TFE-lined cap	Refrigerated at 4°C	7 days[c]	40 days
EPA 8310	Polunuclear aromatic hydrocarbons	HPLC	1 1-L amber glass bottle; TFE-lined cap	Refrigerated at 4°C	7 days[c]	40 days
EPA 6010, EPA 200.7	Metals	ICPES	1 1-L glass bottle	HNO_3 at pH < 2	NA	6 months
EPA 7060, EPA 206.2	Arsenic	Furnace AA	1 500-mL plastic bottle	HNO_3 at pH < 2	NA	6 months
EPA 7740, EPA 270.2	Selenium	Furnace AA	1 500-mL plastic bottle	HNO_3 at pH < 2	NA	6 months
EPA 7470, EPA 245.1	Mercury	Cold vapor AA	1 500-mL plastic bottle	HNO_3 at pH < 2	NA	28 days
EPA 7421, EPA 239.2	Lead	Furnace AA	1 500-mL plastic bottle	HNO_3 at pH < 2	NA	6 months

Table 10.1 (*continued*)

Reference Method	Parameter	Technique[a]	Container Type, Number, Volume	Preservation/Storage Requirements	Maximum Holding Time from Collection[b]	Maximum Holding Time (Analysis)
EPA 7196	Chromium(VI)	Colorimetric	1 500-mL plastic bottle	Refrigerated at 4°C	NA	24 h[c]
ASTM 3237	Organic lead	AA after extraction	1 500-mL plastic bottle	Refrigerated at 4°C	14 days	30 days
EPA 9030, EPA 376.1	Sulfide	Titrimetric	1 250-mL glass jar or plastic bottle	Refrigerated at 4°C; add 2 mL of zinc acetate and NaOH to pH > 9	NA	7 days[c]
EPA 9012, EPA 353.3	Cyanide (CN)	Colorimetric	1 500-mL plastic bottle	Refrigerated at 4°C; NaOH to pH > 12	NA	14 days[c]
EPA 340.2	Fluoride	Selective ion electrode	1 500-mL plastic bottle	NA	NA	28 days
EPA 300.0	Chloride, nitrate, sulfate	IC	1 500-mL plastic bottle	NA	NA	28 days
EPA 353.1	Nitrate/nitrite	Colorimetric	1 500-mL plastic bottle	Refrigerated at 4°C; H_2SO_4 to pH > 12	NA	14 days[c]
EPA 9066, EPA 420.2	Total phenolics	Colorimetric	1 500-mL plastic bottle	Refrigerated at 4°C; H_2SO_4 to pH > 12	NA	28 days
EPA 415.1	Total organic carbon	Oxidation/NDIR	1 500-mL plastic bottle	Refrigerated at 4°C; H_2SO_4 to pH > 12	NA	28 days
EPA 418.1	Petroleum hydrocarbons	IR	1 1-L glass bottle	Refrigerated at 4°C; H_2SO_4 to pH > 12	NA	28 days
EPA 450.1	Total organic halogens	Dohrman DX-20	1 1-mL amber glass bottle	Refrigerated at 4°C; H_2SO_4 to pH > 12	NA	14 days[c]
EPA 9040, EPA 150.1	pH	Electrometric	1 500-mL plastic bottle	Refrigerated at 4°C	NA	72 h[c]
EPA 9050, EPA 120.1	Conductance	Wheatstone bridge	1 500-mL plastic bottle	Refrigerated at 4°C	NA	28 days
EPA 410.1	Chemical oxygen demand	Titrimetric	1 500-mL plastic bottle	Refrigerated at 4°C	NA	28 days
EPA 413.2	Oil and grease	IR	1 1-L glass bottle	Refrigerated at 4°C; H_2SO_4 to pH > 12	NA	28 days
EPA 160.1	Total dissolved solids	Gravimetric	1 500-mL plastic bottle	Refrigerated at 4°C	NA	7 days[c]
EPA 160.2	Total suspended solids	Gravimetric	1 500-mL plastic bottle	Refrigerated at 4°C	NA	7 days[c]
EPA 310.1	Alkalinity	Titrimetric	1 500-mL plastic bottle	Refrigerated at 4°C	NA	14 days[c]

Source: Bedient et al. (1994), from *Laboratory Test Notes*, Radian Analytical Services, Vol. 1, No. 2, August 1998.

[a] HSD, halogen-specific detector; SM, standard methods.

[b] NA, not applicable.

[c] Samples must be shipped within 24 h of collection.

[d] +, Preserve for aromatics only; 7 days of no HCl.

alternative to laboratory testing, some tests may be performed in the field using portable or mobile equipment. For instance, portable gas chromatography is commonly used to evaluate on-site composition of organic compounds. Other specialized test kits may be used to determine the specific chemical concentrations present or may possess an indicator parameter that shows the evidence of the contaminant presence (USEPA, 1997).

10.5.4   Data Analysis and Evaluation

The geologic, hydrologic, and chemical data should be assembled in proper form and then used to define the extent of contamination.

Geologic Data. Detailed logs should be prepared for each boring/sampling location. Logs should document the location, drilling and sampling methods, descriptions of the soil layers encountered, and all available laboratory and field test results. A typical boring log is shown in Figure 10.28. Boring logs may be used to draw soil profiles along selected directions to determine stratigraphical changes as shown in Figure 10.29. For each geologic unit, contour maps of either the top or bottom elevation of the unit encompassing the project site may be prepared; an example of such a map is given in Figure 10.30. In addition, contour maps of the thickness of selected geologic units, also known as isopach maps, may be prepared; an example map is shown in Figure 10.31. These data help to delineate the various geologic units encountered at the site.

Hydrogeologic Data. A summary of the piezometer and monitoring well locations as well as their as-built diagrams should be prepared. The in-situ hydraulic conductivity test results should be analyzed properly to calculate the hydraulic properties of different geologic units. Based on these hydraulic properties, the geologic units should be defined as aquifers, aquitards, or acquicludes. Based on the water level measurements, aquifers may be further delineated as confined or unconfined aquifers. Water-level measurements taken at the same time in each geologic formation should be plotted to develop potentiometric surface or water-level maps, as shown in Figure 10.32. These

maps will define the direction of horizontal flow. Using these maps, horizontal hydraulic gradients may also be calculated. The water levels measured in nested or multiple wells at the same location in different formations may be used to calculate vertical gradients between the units as well. Generally, the rate of groundwater flow is calculated using *Darcy's Law:* The rate of flow is equal to hydraulic conductivity, multiplied by the hydraulic gradient, and divided by the effective porosity. The effective porosity may be estimated from soil test results or assumed based on the typical values reported in the literature. More details are provided in Chapter 8.

Chemical Data. For analysis of chemical data, the validity of the data, based on the QA/QC procedures followed, must first be assessed. The chemical data for soils and groundwater are used to define the background chemical concentration levels and then to define the boundary of the contaminated area. Statistical analyses may be then performed to establish the background concentrations (USEPA, 1989). A review of applicable standards for the site soils and groundwater should also be determined. Finally, results from the chemical data analysis should be plotted on profiles and in contour form on the site plan to delineate the contaminated zones in three dimensions, as shown in Figure 10.33.

Report Preparation. The overall geologic, hydrologic, and contaminant data are reviewed to recognize the consistent patterns. The regional geologic data are useful to help determine if the site geologic features and the data on regional sources or sinks indicate consistency with the site hydrogeology. A project report should be prepared at the completion of the site characterization process documenting the investigation procedures and results. This report should present the S&H plans and QA/QC plans.

10.6   EXPEDITED OR ACCELERATED SITE CHARACTERIZATION

The conventional detailed characterization as presented in the foregoing sections may be time consuming and

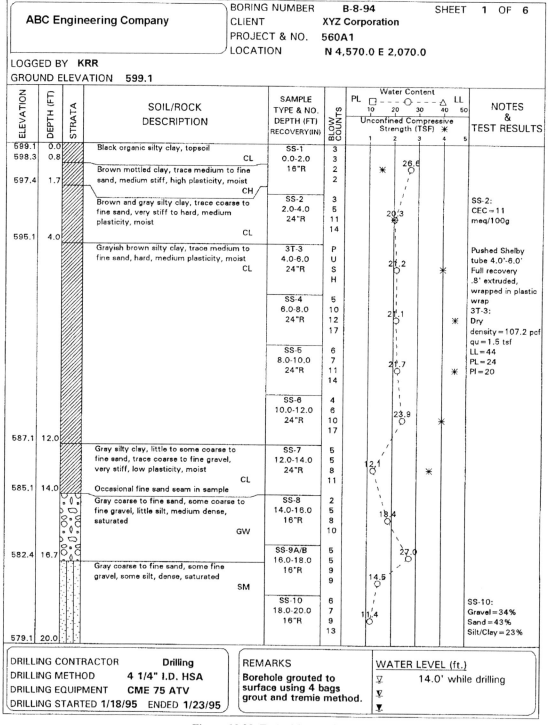

Figure 10.28 Typical boring log.

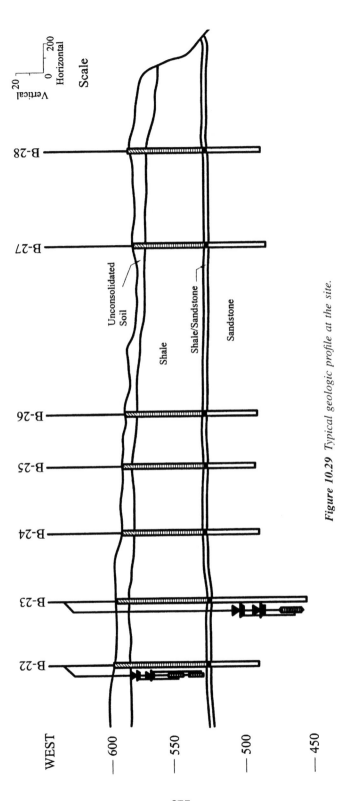

Figure 10.29 Typical geologic profile at the site.

275

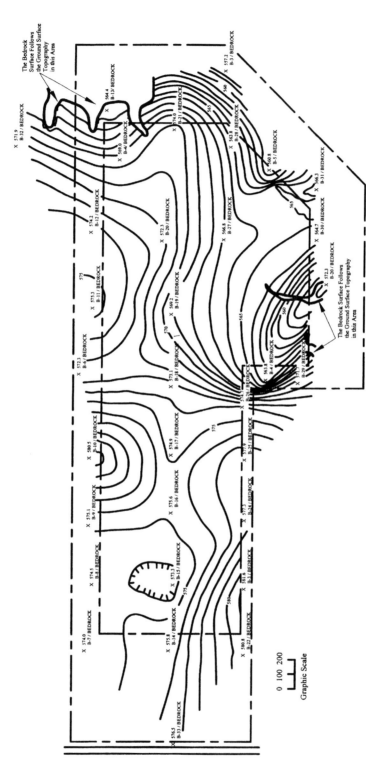

Figure 10.30 Typical elevation contour map showing the top elevation of a selected rock layer at the site.

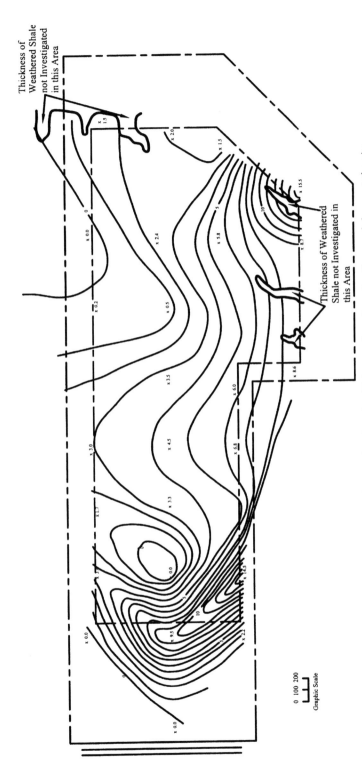

Figure 10.31 Typical isopach map showing the thickness of a selected soil or rock layer at the site.

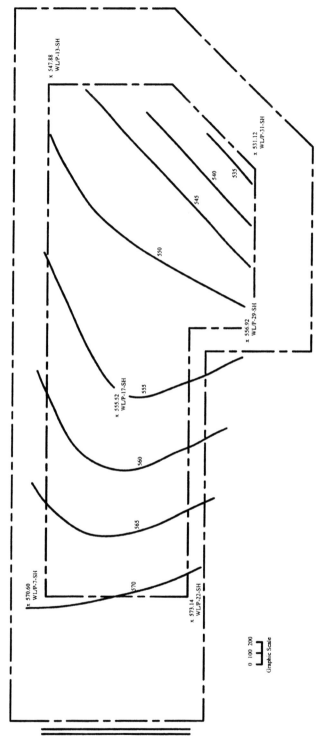

Figure 10.32 Typical potentiometric surface contour map for a selected confined aquifer at the site.

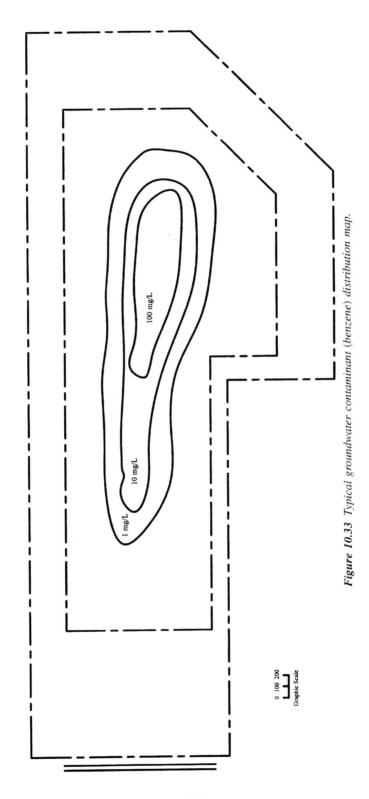

Figure 10.33 *Typical groundwater contaminant (benzene) distribution map.*

100 mg/L

10 mg/L

1 mg/L

0 100 200
Graphic Scale

Table 10.2 **Process comparison of conventional and expedited site characterization**

Process Component	Conventional[a]	Expedited
1. Project duration	Longer because of multiple mobilizations and intervals between where data are compiled and analyzed.	Shorter because of fewer coordinated field mobilizations, analysis and archiving of most data in the field, and improved regulatory and community acceptance.
2. Project leadership	Project leader typically in office; junior staff in field.	Project leader in field with experienced multidisciplinary team.
3. Use of prior data	Reviewed, but often not carefully evaluated or interpreted.	Carefully compiled, evaluated for quality, analyzed and interpreted as part of developing preliminary conceptual model for the site.
4. Technical approach	Different disciplines tend to work independently at different times for field data collection and interpretation.	All phases of field investigation—data collection, compilation, analysis, and interpretation—integrated in the field.
5. Field investigation methods	Measurements usually not corroborated by complementary methods. Emphasis on installation of monitoring wells.	Use of multiple, complementary methods. Emphasis on noninvasive and minimally invasive investigation methods.
6. Work plan	• Field characterization and sampling plan defined before full mobilization and generally not modified during the course of a mobilization (may be modified if DQO process used). • QA/QC plan for chemical sampling and analysis but usually not for other characterization activities. • Data management plan usually not a formal part of the work plan. • Community relations plan not always part of the work plan (except for CERCLA program).	• Dynamic field technical program uses data as they come in to guide type and location of field measurements and samples for analysis. • QA/QC plan for all aspects of field data collection and handling; strong emphasis on individual team member responsibility for QA/QC. • Data management plan formal part of work plan. • Community relations plan considered important part of work plan.
7. Number of mobilizations	Multiple mobilizations, often carried out by different groups with little communication, interspersed with office analysis of data for a given investigation phase.	Normally, one full mobilization for each investigation phase (two total), carried out under direct control and participation of a single core technical team.
8. Data results and analysis	Data results and interpretation usually interpreted in office weeks to months after field work. Computers not normally used in the field for data management and analysis.	Data obtained, interpreted, and archived in the field (hours to days) as part of dynamic technical program. Computers used in field for data management as an aid in analyzing data.

Source: ASTM (2003).

[a] Certain elements included in the "expedited" column may be incorporated into a given "conventional" site investigation, but the characteristics described in this column can be considered typical of most site characterization activities during the 1980s and early 1990s.

expensive for larger-scale projects. As an alternative, expedited site characterization (ESC) procedures, ASTM PS85, have been developed (ASTM, 2003). A comparison of conventional and expedited site characterization procedures is shown in Table 10.2. The ESC procedure focuses on collecting only the information required to meet clearly defined objectives and ensuring that characterization ceases as soon as the objectives are met. A multidisciplinary technical team with expertise in geologic, hydrologic, and chemical systems is present at the site to implement a "dynamic" work plan, which provides the flexibility and responsibility to select the type and location of measurement to optimize data collection activities. The ESC process includes (1) intensive compilation, quality evaluation, and independent interpretation of prior data to develop a preliminary conceptual site model; and (2) use of multiple complementary site geologic and hydrologic investigation methods, rapid data collection and interpretation, rigorous quality control of data collection, and daily analysis and interpretation of the data for on-site decision making. The data are then used to perform exposure and risk evaluation (described in detail in Chapter 11) and to choose a course of action (no action, monitoring, or interim/final remedial action).

10.7 SUMMARY

Site characterization is the most critical task in any remediation project. A phased approach is recommended in order to obtain the most pertinent data in a timely and efficient manner. A phased approach is also recommended to prevent high costs incurred in obtaining unnecessary data. A multidisciplinary team is generally charged with the responsibility of collecting and evaluating the data. A detailed work plan for each phase of investigation should be prepared addressing health, safety, QA/QC plans, and applicable regulations. To characterize the sites, the geology, hydrogeology, and contaminant conditions must be determined. A variety of techniques and equipment are available for this purpose. The data collected are then used to form a basis for defining the extent and potential risks of the contamination present.

QUESTIONS/PROBLEMS

10.1. Explain the primary purpose of site characterization.

10.2. How does the site characterization for contaminated sites differ from standard geotechnical site characterization?

10.3. Compare and contrast different geophysical methods used for site characterization.

10.4. List advantages and disadvantages of drive methods for site characterization.

10.5. Briefly describe various soil and rock drilling and sampling methods. Identify the method(s) that will be most appropriate for drilling and sampling in cohesionless soils below the water table.

10.6. What methods are available to obtain hydrogeologic data? Identify two important hydraulic properties required to understand groundwater flow behavior; define these two properties.

10.7. What are the advantages and disadvantages of slug tests?

10.8. Describe methods available to obtain chemical data. Compare and contrast the methodology with that generally followed in conventional site characterization procedures.

10.9. What are general QA/QC requirements for sampling of groundwater at contaminated sites?

10.10. Discuss innovative technologies that can expedite contaminated site characterization.

REFERENCES

ASTM (American Society for Testing and Materials), *Annual Book of ASTM Standards,* ASTM, West Conshohocken, PA, 2003.

Bedient, P. B., Rifai, H. S., and Newell, C. J., *Ground Water Contamination: Transport and Remediation,* 1st ed., Prentice Hall, Upper Saddle River, NJ, 1994.

Bouwer, H., The Bouwer and Rice slug test: an update, *Ground Water,* Vol. 27, No. 3, pp. 304–309, 1989.

Bouwer, H., and Rice, R. C., A slug test for determining hydraulic conductivity of unconfined aquifers with completely or partially penetrating wells, *Water Resour. Res.,* Vol. 12, pp. 423–428, 1976.

Cooper, H. H., Bredehoeft, J. D., and Papadopulos, I. S., Response to a finite diameter well to an instantaneous charge of water, *Water Resour. Res.,* Vol. 3, pp. 263–269, 1967.

Dawson, K. J., and Isotok, J. D., *Aquifer Testing: Design and Analysis of Aquifer Tests,* Lewis Publishers, Chelsea, MI, 1991.

Haeni, F. P., Application of seismic-refraction techniques to hydrologic studies, Chap. D2 in *Techniques of Water-Resources Investigations of the United States Geological Survey,* U.S. Department of the Interior, U.S. Government Printing Office, Washington, DC, 1988.

Hvorslev, M. J., *Time Log and Soil Permeability in Groundwater Observations,* Bulletin 36, U.S. Army Corps of Engineers Waterways Experimental Station, Vicksburg, MS, 1951.

ITRC (Interstate Technology and Regulatory Cooperation), *Technology Overview: Dense Non-aqueous Phase Liquids (DNAPLs): Review of Emerging Characterization and Remediation Technologies,* ITRC, Washington, DC, 2000.

Kaback, D., *Technology Status Report: A Catalogue of the Horizontal Environmental Wells in the United States,* TS-02-01, Groundwater Remediation Technologies Analysis Center, Pittsburgh, PA, 2002.

Kram, M., Lorenzana, D., Michaelsen, J., and Lory, E., *Performance Comparison: Direct-Push Wells versus Drilled Wells,* TR-2120-ENV, Naval Facilities Engineering Command, Washington, DC, 2001.

Lutenegger, A. J., and DeGroot, D. J., Field techniques for sampling and measurement of soil gas constituents at contaminated sites, in *Remediation Engineering of Contaminated Sites,* Wise, D. L. et al. (eds.), Marcel Dekker, New York, 2000.

McCaulou, D. R., Jewett, D. G., and Huling, S. G., *Nonaqueous Phase Liquids Compatibility with Materials Used in Well Construction, Sampling, and Remediation,* EPA/540/S-95/503, USEPA, Washington, DC, 1995.

Mines, B., and Stauffer, T. B., Using the BAT probe, *Mil. Eng.,* Vol. 84, No. 548, pp. 20–22, 1992.

Osborne, P. S., *Suggested Operating Procedures for Aquifer Pumping Tests,* EPA/540/S-93/503, USEPA, Washington, DC, 1993.

Puls, R. W., and Barcelona, M. J., *Ground Water Sampling for Metal Analyses,* EPA/540/4-89/001, USEPA, Washington, DC, 1989.

———, *Low-Flow (Minimal Drawdown) Ground-Water Sampling Procedures,* EPA/540/S-95/504, USEPA, Washington, DC, 1996.

Thornhill, J. T., *Accuracy of Depth to Water Measurements,* EPA/540/4-89/002, USEPA, Washington, DC, 1989.

USACE (U.S. Army Corps of Engineers), *Geophysical Exploration,* EM 1110-1-1802, U.S. Department of the Army, Washington, DC, 1979.

———, *Engineering and Design: Horizontal Directional Drilling for Environmental Applications,* ETL-110-1-178, U.S. Department of the Army, Washington, DC, 1996.

USBR (U.S. Bureau of Reclamation), *Groundwater Manual- A Water Resources Technical Publication,* 2nd ed., Bureau of Reclamation, U.S. Department of the Interior, Denver, CO, 1981.

USEPA (U.S. Environmental Protection Agency), *Standard Operating Safety Guides,* Office of Emergency and Remedial Response, USEPA, Washington, DC, 1984.

———, *Test Methods for Evaluating Solid Waste,* SW-846, Vols. 1A, 1B, 1C, *Laboratory Manual,* Vol. 2, *Field Manual,* USEPA, Washington, DC, 1986.

————, *Statistical Analysis of Groundwater Monitoring Data at RCRA Facilities: Interim Final Guidance,* EPA/530/SW-89/ 026, USEPA, Washington, DC, 1989.

————, *Seminar Publication: Site Characterization for Subsurface Remediation,* EPA/625/4-91/026, Office of Research and Development, USEPA, Washington, DC, 1991a.

————, *Conducting Remedial Investigations/Feasibility Studies for CERCLA Municipal Landfill Sites,* EPA/540/P-91/001, Office of Emergency Remedial Response, USEPA, Washington, DC, 1991b.

————, *Subsurface Characterization and Monitoring Techniques: A Desk Reference Guide,* Vol. I, *Solids and Ground Water,* App. A and B, EPA/625/R-93/003a, Office of Research and Development, USEPA, Washington, DC, 1993a.

————, *Subsurface Characterization and Monitoring Techniques: A Desk Reference Guide,* Vol. II, *The Vadose Zone, Field Screening and Analytical Methods,* App. C and D, EPA/625/R-93/003b, Office of Research and Development, USEPA, Washington, DC, 1993b.

————, *Field Analytical and Site Characterization Technologies: Summary of Applications,* EPA/542/R-97/011, Office of Solid Waste and Emergency Response, USEPA, Washington, DC, 1997.

————, *Geophysical Techniques to Locate DNAPLs: Profiles of Federally Funded Projects,* EPA/542/R-98/020, Office of Solid Waste and Emergency Response, USEPA, Washington, DC, 1998.

————, *A Resource for MGP Site Characterization and Remediation, Expedited Site Characterization and Source Remediation at Former Manufactured Gas Plant Sites,* EPA/542/R-00/005, Office of Solid Waste and Emergency Response, USEPA, Washington, DC, 2000a.

————, *Innovations in Site Characterization: Geophysical Investigation at Hazardous Waste Sites,* EPA/542/R-00/003, Office of Solid Waste and Emergency Response, USEPA, Washington, DC, 2000b.

Walton, W. C., *Groundwater Pumping Tests: Design and Analysis,* Lewis Publishers, Chelsea, MI, 1989.

Yeskis, D., and Zavala, B., *Ground-Water Sampling Guidelines for Superfund and RCRA Project Managers,* EPA/542/S-02/ 001, Office of Solid Waste and Emergency Response, USEPA, Washington, DC, 2002.

11

RISK ASSESSMENT AND REMEDIAL STRATEGY

11.1 INTRODUCTION

After a thorough site characterization, a risk assessment is performed if contamination at the site is confirmed. A risk assessment, also known as an *impact assessment*, is a systematic evaluation of the potential risk posed by the detected contamination to human health and the neighboring ecosystems under present and future conditions. If the risk assessment determines that the risk is unacceptable, a remedial strategy must be developed to address the problem. The results from the risk assessment will provide the rational remedial goals, if necessary, to reduce the potential risks. In this chapter, the most commonly used risk assessment procedures are described first. Then the remedial strategy and available remedial options are presented.

11.2 RISK ASSESSMENT PROCEDURES

In conjunction with CERCLA (or Superfund) feasibility studies and RCRA corrective measure studies, the USEPA developed a risk assessment procedure to address risk associated with major contaminated sites (USEPA, 1989, 1991a,b, 1997, 2001; USDOE, 1995; Stern and Feinberg, 1996). This procedure is general and comprehensive and may therefore be used to assess risk associated with any contaminated site. The various steps involved in the USEPA procedure are explained in Section 11.3.

The American Society for Testing and Materials (ASTM) developed a standard guide, E 1739, the *Guide for Risk-Based Corrective Action* (RBCA), to specifically address sites contaminated by leaking pe-

troleum tanks (ASTM, 1995). The RBCA procedure is explained in Section 11.4. Various federal and state agencies modified the USEPA procedure or the RBCA procedure to assess risk associated with petroleum and other types of contamination. More details on these procedures are provided in Section 11.5.

11.3 USEPA PROCEDURE

Site characterization data must be available to initiate the risk assessment process. This procedure, also known as *baseline risk assessment,* quantifies potential risk to human health. In addition to this assessment, ecological risk assessment must be performed to determine the potential risk to the living organisms, wildlife, and so on. The ecological risk assessment may be quite complex, depending on the site-specific conditions; therefore, it is not detailed in this section. More details on the ecological risk assessment may be found in USEPA (1997, 2001). The baseline risk assessment procedure consists of four steps: (1) data collection and evaluation, (2) exposure assessment, (3) toxicity assessment, and (4) risk characterization. These four steps are explained below.

11.3.1 Data Collection and Evaluation

The data collection and evaluation consists of analysis of site characterization data to identify potential chemicals of concern. Of specific interest are (1) the contaminant identities, (2) the contaminant concentrations in the key sources and media of interest, (3) the characteristics of the sources of contamination, and (4) the characteristics of the environmental setting that may affect the fate, transport, and persistence of the contaminants. The chemical data are sorted by medium (soil, water, and air) and then evaluated. The evaluation should include analysis of the sampling and analytical methods employed and the quality of the data with respect to quantification limits, QA/QC measures, and so on. After this evaluation, a tentative list of compounds, with corresponding data, is chosen for further assessment.

11.3.2 Exposure Assessment

The exposure assessment consists of estimating the type and magnitude of human exposure to the selected chemicals of potential concern. *Exposure* is defined as the contact of a chemical or biological agent with the outer boundary of an organism (USEPA, 1992). By measuring or estimating the amount of an agent available at the visible exterior of a person (i.e., the skin, mouth, nostrils) during a specified time period, the magnitude of exposure is determined. Exposure assessment is the determination or estimation (qualitative or quantitative) of the route, frequency, duration, and magnitude of exposure. Exposure assessments may consider both current and future exposures. The exposure assessment process consists of the following three tasks: (1) characterization of exposure setting, (2) identification of exposure pathways, and (3) quantification of exposure concentrations.

In task 1, *exposure setting* is characterized to include the physical features of the site. More specifically, the physical features include climate and weather conditions, vegetation, geology, hydrogeology, and surface water identification. In addition, population location relative to the site as well as human activity patterns are characterized as part of exposure setting.

After characterizing exposure setting, the second task is to identify various *exposure pathways*. For this process, the types and location of chemicals present at the site, the sources and release mechanisms of these chemicals, the likely environmental transport and fate of these chemicals, and the location and activities of

populations that may be exposed are identified. For each exposure pathway, exposure points and routes of exposure are distinguished.

In the final task, *exposure concentrations* of chemicals in contact and the frequency and duration of the contact are quantified. Chemical fate and transport models (as explained in Chapter 8) are used for estimating both current and future exposure concentrations. All different exposure pathways that have been identified are analyzed. Using these exposure concentrations, chemical intakes are expressed in terms of the mass of the contaminant in contact with the body, and per unit body weight per unit time (e.g., mg/kg · day). After estimation, the chemical intakes are organized by grouping all applicable exposure pathways for each exposed population. Sources of uncertainty in the exposure estimates must be evaluated and clearly stated.

11.3.3 Toxicity Assessment

The toxicity assessment begins with the identification of potential adverse health effects of a contaminant (e.g., cancer, birth defects). This step is also known as *hazard identification*. Following this, the data on dose–response relationship of the contaminant are obtained. This relationship defines the dose of the contaminant or the amount of exposure received by people or animals, and the incidence of adverse health effects in the exposed population. The toxicity assessment (hazard identification) process involves the review of toxicological data derived mainly from epidemiological studies. Although the toxicology data for chemicals may be obtained from various sources, the primary source is the USEPA's integrated risk information system (IRIS) (USEPA, 1999). Chemicals may be grouped as either noncarcinogens or carcinogens. For noncarcinogens, the *reference dose* (RfD) is used to express the daily exposure levels. The RfD is an estimate (with uncertainty spanning perhaps an order of magnitude) of a daily oral exposure to the human population (including sensitive subgroups) that is likely to be without an appreciable risk of deleterious effects during a lifetime. It can be derived from a No Observed Adverse Effect Level (NOAEL), Lowest Observed Adverse Effect Level (LOAEL), or benchmark dose, with uncertainty factors generally applied to reflect limitations of the data used. RfD values are generally used in EPA's noncancer health assessments.

For carcinogens, the *slope factor* (SF) is used to express the potential risk level associated with their exposure. The SF is an upper bound, approximating a 95% confidence limit, on the increased cancer risk from a lifetime exposure to an agent. This estimate, usually expressed in units of proportion (of a population) affected per mg/kg · day, is generally reserved for use in the low-dose region of the dose–response relationship, that is, for exposures corresponding to risks less than 1 in 100. Table 11.1 summarizes the reported toxicology data (RfD or SF values) for selected chemicals that are commonly found at contaminated sites. Examples 11.1 and 11.2 explain the use of the data presented in this table.

11.3.4   Risk Characterization

The risk characterization combines the exposure and toxicity assessments into quantitative and qualitative expressions of risk. This is accomplished by performing the following tasks:

- *Task 1: Organize results of exposure and toxicity assessments.* The exposure and toxicity information is organized for each exposure pathway. Exposure routes are properly matched for the toxicity and exposure values (oral to oral, inhalation to inhalation).
- *Task 2: Quantify pathway risk.* For individual noncarcinogens, the noncancer *hazard quotient* (HQ) is calculated for noncancerous health effects. Mathematically,

$$HQ = \frac{E}{RfD} \qquad (11.1)$$

where E is the exposure level (or chemical intake) obtained from exposure assessment, and RfD is the reference dose obtained from the toxicity assessment for the same exposure pathway as E (Table 11.1). If the HQ is less than unity, less concern exists for adverse health effects.

For individual carcinogens, risk is calculated using

$$risk = (CDI)(SF) \qquad (11.2)$$

where risk is a unitless probability of a person developing a cancer, CDI the chronic daily intake averaged over 70 years (mg/kg · yr) obtained from the exposure assessment, and SF the slope factor (mg/kg · day)$^{-1}$ obtained from the toxicity assessment (Table 11.1).

When multiple chemicals or chemical mixtures are present, cumulative risk is calculated. For non-carcinogen mixtures, individual HQ values are summed for each exposure pathway; this sum is also known as the *hazard index* (HI). For carcinogens, total carcinogenic risk is the sum of the carcinogenic risk for each substance of concern.

- *Task 3: Combine risks across exposure pathways.* The next task is to sum the risks across exposure pathways to calculate the total risk. For this summation, reasonable exposure pathway combinations are identified. In addition, the likelihood of the same person facing exposure by more than one path is examined.
- *Task 4: Assess and present uncertainty.* Uncertainty associated with the selection of chemicals of concern and their concentrations, toxicity values, site-related variables, and assumptions in exposure assessment should be determined and clearly presented.

Overall, the final result of the baseline risk assessment includes estimated cancer and noncancer risks for all exposed pathways and land uses analyzed and for all chemicals of concern. If the noncancer risk expressed by HI is greater than 1, adverse health effects of the chemicals are inferred. Generally, USEPA considers the cancer risk greater than 10^{-6} (one in 1 million) unacceptable. However, allowable health risks should be based carefully on the site-specific conditions. Examples 11.1 and 11.2 show the application of the USEPA risk assessment procedure for noncarcinogenic and carcinogenic chemicals, respectively.

Table 11.1 **Toxicological data of selected chemicals**[a]

Chemical Name	Oral RfD (mg/kg · day)	Inhalation RfD (mg/kg · day)	Oral SF (mg/kg · day)$^{-1}$	Inhalation SF (mg/kg · day)$^{-1}$
Acenaphthene	6.00 E-02	No data	No data	No data
Acetone	1.00 E-01	No data	No data	No data
Alachlor	1.00 E-02	No data	Pending	Pending
Aldicarb	2.00 E-04	No data	No data	No data
Aldrin	3.00 E-05	No data	1.70E+01	1.70E+01
Anthracene	3.00 E-01	No data	No data	No data
Atrazine	5.00 E-03	No data	Pending	Pending
Benzene	Pending	Pending	2.90E-02	2.90E-02
Bromodichloromethane	2.00E-02	No data	6.20E-02	No data
Bromoform	2.00E-02	Pending	7.90E-03	7.70E-07
Butanol	1.00E-01	No data	No data	No data
Butyl benzyl phthalate	2.00E-01	No data	No data	No data
Carbofuran	5.00E-03	No data	No data	No data
Carbon disulfide	1.00E-01	Pending	No data	No data
Carbon tetrachloride	7.00E-04	No data	1.30E-01	1.30E-01
Chlordane	6.00E-05	Pending	1.30E+00	1.30E+00
4-Chloroaniline (p-Chloroaniline)	4.00E-03	No data	No data	No data
Chlorobenzene (monochlorobenzene)	2.00E-02	Pending	No data	No data
Chlorodibromomethane (dibromochloromethane)	2.00E-02	No data	8.40E-02	No data
Chloroform	1.00E-02	Pending	6.10E-03	8.10E-02
Dalapon	3.00E-02	No data	No data	No data
DDD	No data	No data	2.40E-01	No data
DDE	No data	No data	3.40E-01	No data
DDT	5.00E-04	No data	3.40E-01	3.40E-01
Dibenzo[a,h]anthracene	No data	No data	No data	No data
1,2-Dibromo-3-chloropropane	No data	5.72E-05	Pending	Pending
1,2-Dibromoethane (ethylene dibromide)	No data	Pending	8.50E+01	7.70E-01
1,2-Dichlorobenzene (o-Dichlorobenzene)	9.00E-02	No data	No data	No data
3,3'-Dichlorobenzidine	No data	Message	4.50E-01	No data
1,2-Dichloroethane	No data	No data	9.10E-02	9.01E-02
1,1-Dichloroethylene	9.00E-03	Pending	6.00E-01	1.80E-01
trans-1,2-Dichloroethylene	2.00E-02	No data	No data	No data
1,2-Dichloropropane	No data	1.14E-03	No data	No data
1,3-Dichloropropene (1,3-dichloropropylene, cis + trans)	3.00E-04	5.72E-03	No data	No data
Dieldrin	5.00E-05	No data	1.60E+01	1.60E+01
Diethyl phthalate	8.00E-01	No data	No data	No data
2,4-Dimethylphenol	2.00E-02	No data	No data	No data
2,4-Dinitrotoluene	2.00E-03	No data	No data	No data
Endosulfan	5.00E-05	No data	No data	No data
Endothall	2.00E-02	No data	Pending	Pending
Endrin	3.00E-04	No data	No data	No data
Ethylbenzene	1.00E-01	2.86E-01	No data	No data
Fluoranthene	4.00E-02	No data	No data	No data
Fluorene	4.00E-02	No data	No data	No data
Heptachlor	5.00E-04	No data	4.50E+00	4.50E+00
Heptachlor epoxide	1.30E-05	No data	9.10E+00	9.10E+00
Hexachlorobenzene	8.00E-04	No data	1.60E+00	1.60E+00

Table 11.1 (*continued*)

Chemical Name	Oral RfD (mg/kg · day)	Inhalation RfD (mg/kg · day)	Oral SF (mg/kg · day)$^{-1}$	Inhalation SF (mg/kg · day)$^{-1}$
α-HCH (α-BHC)	No data	No data	6.30E+00	6.30+00
γ-HCH (lindane)	3.00E-04	Pending	Pending	Pending
Hexachlorocyclopentadiene	7.00E-03	No data	No data	No data
Hexachloroethane	1.00E-03	Pending	1.40E-02	1.40E-02
Isophorone	2.00E-01	No data	4.10E-03	No data
Methoxychlor	5.00E-03	No data	No data	No data
Methyl bromide (bromomethane)	1.40E-03	1.43E-03	No data	No data
Methyl *tert*-butyl ether	Pending	1.43E-01	No data	No data
Methylene chloride (dichloromethane)	6.00E-02	Pending	7.50E-03	1.65E-03
2-Methylphenol (*o*-cresol)	5.00E-02	No data	No data	No data
Nitrobenzene	5.00E-04 [I]	Pending	No data	No data
N-Nitrosodiphenylamine	No data	No data	4.90E-03	No data
N-Nitrosodi-*n*-propylamine	No data	No data	7.00E+00	No data
Phenol	6.00E-01	Message	No data	No data
Picloram	7.00E-02	No data	Pending	Pending
Polychlorinated biphenyls (PCBs)	No data	No data	7.70E+00	No data
Pyrene	3.00E-02	No data	No data	No data
Styrene	2.00E-01	2.86E-01	Pending	Pending
Tetrachloroethylene (perchloroethylene)	1.00E-02	No data	Pending	Pending
Toluene	2.00E-01	1.40E+00	No data	No data
Toxaphene	No data	No data	1.10E+00	1.10E+00
1,2,4-Trichlorobenzene	5.00E-03	No data	No data	No data
1,1,2-Trichloroethane	4.00E-03	Pending	5.70E-02	5.70E-02
Vinyl acetate	No data	5.72E-02	Pending	Pending
Xylenes (total)	2.00E+00	Pending	No data	No data

Ionizable Organics

Chemical Name	Oral RfD (mg/kg · day)	Inhalation RfD (mg/kg · day)	Oral SF (mg/kg · day)$^{-1}$	Inhalation SF (mg/kg · day)$^{-1}$
Benzoic acid	4.00E+00	No data	No data	No data
2-Chlorophenol	5.00E-03	No data	No data	No data
2,4-Dichlorophenol	3.00E-03	No data	No data	No data
2,4-Dinitrophenol	2.00E-03	Message	No data	No data
Dinoseb	1.00E-03	No data	No data	No data
Pentachlorophenol	3.00E-02	Pending	1.20E-01	No data
2,4,5-Trichlorophenol	1.00E-01	Message	Pending	Pending
2,4,6 Trichlorophenol	No data	Message	1.10E-02	1.10E-02

Inorganics

Chemical Name	Oral RfD (mg/kg · day)	Inhalation RfD (mg/kg · day)	Oral SF (mg/kg · day)$^{-1}$	Inhalation SF (mg/kg · day)$^{-1}$
Antimony	4.00E-04	No data	No data	No data
Arsenic	3.00E-04	No data	1.75E+00	5.00E+01
Barium	7.00E-02	Pending	No data	No data
Beryllium	5.00E-03	No data	4.30E+00	8.40E+00
Boron	9.00E-02	No data	Pending	Pending
Cadmium	1.00E-03	Pending	No data	6.10E+00
Chromium, ion, trivalent	1.00E+00	Pending	Pending	Pending
Chromium, ion, hexavalent	5.00E-03	Pending	No data	4.10E+01
Cyanide (amenable)	2.00E-02	No data	No data	No data
Fluoride	6.00E-02	No data	No data	No data

Table 11.1 (*continued*)

Chemical Name	Oral RfD (mg/kg · day)	Inhalation RfD (mg/kg · day)	Oral SF (mg/kg · day)$^{-1}$	Inhalation SF (mg/kg · day)$^{-1}$
Manganese	1.00E-02	1.14E-04	No data	No data
Nickel	No data	No data	No data	1.68E+00
Nitrate as N	1.60E+00	No data	No data	No data
Selenium	5.00E-03	No data	No data	No data
Silver	5.00E-03	No data	No data	No data
Thallium	8.50E-05	No data	No data	No data
Vanadium	9.00E-03	No data	Message	Message

Source: USEPA (1999).

[a] Data reported as of October 2002.

EXAMPLE 11.1

At a site, two noncarcinogenic contaminants are found. The exposure level of each contaminant over the exposure period (30 years) is calculated. The reference dose (RfD) for each contaminant derived for a similar exposure period is also obtained. All these data are summarized below.

Contaminant	Exposure Level, E (mg/kg · day)	Reference Dose, RfD (mg/kg · day)
Acetone	0.01	0.1
Dalapon	0.02	0.03

Calculate the risk posed by the site contamination.

Solution:

For noncarcinogens, risk is expressed as a hazard quotient (HQ), given by equation (11.1) as follows:

$$HQ = \frac{E}{RfD}$$

If HQ > 1, there is a concern for potential noncarcinogen health effects.

$$HQ_{(acetone)} = \frac{0.01}{0.1} = 0.1$$

$$HQ_{(dalapon)} = \frac{0.02}{0.03} = 0.7$$

Individually, HQ < 1. But consider contaminants that target the same organ (e.g., liver, kidney) and calculate the hazard index (HI). In this particular case, both acetone and dolapan could target kidney. Therefore, HI = 0.1 + 0.7 = 0.8 (target organ kidney). HI < 1 ⇒ no adverse health effects on the target organ.

EXAMPLE 11.2

At a site, benzene and toxaphene are found. Based on the exposure assessment, chronic daily intake (CDI) of each contaminant over a lifetime (70 years) is reported. The slope factor (SF) that defines the relationship between dose and response for each contaminant is also given. All these data are summarized below.

Contaminant	CDI (mg/kg · day)	SF (mg/kg · day)$^{-1}$
Benzene	0.005	0.03
Toxaphene	0.004	1.1

Calculate the risk posed by the site contamination.

Solution:

For carcinogenic contaminants, risk is estimated as the probability of a person developing cancer over a lifetime as a result of the exposure to the contaminant and is given by equation (11.2), which is as follows:

$$risk = (CDI)(SF)$$

$$risk_{(benzene)} = (0.005)(0.03) = 0.00015$$

$$risk_{(toxaphene)} = (0.004)(1.1) = 0.0044$$

$$total\ risk = 0.00015 + 0.0044 = 0.00455$$

Generally, risk $> 1 \times 10^{-6}$ is unacceptable. The risk posed by the site contamination is greater than the acceptable level; therefore, the risk is unacceptable and remedial action is required.

11.4 ASTM PROCEDURE

The ASTM Standard E 1739 provides a general procedure for performing a risk assessment to determine risk-based corrective action (RBCA) for petroleum release sites (ASTM, 1995). The approach integrates risk and exposure assessment practices with site assessment activities and remedial measure selection, ensuring that the chosen action is protective of human health and the environment. The RBCA process utilizes a tiered approach in which corrective action activities are tailored to site-specific conditions and risks. The different tasks involved in RBCA analysis are shown in Figure 11.1.

Similar to the USEPA risk assessment procedure, the first actions required before risk evaluation are performance of site characterization and site classification. General site characterization procedures were de-scribed in Chapter 10. The site characterization is required to identify all contaminants present at the site, while the site classification is performed to organize the contaminants present by the urgency of initial response. A site may be classified as one of the following four types, depending on the threat posed to human health, safety, or sensitive environmental receptors:

- *Class 1:* immediate threat
- *Class 2:* short-term (0 to 2 years) threat
- *Class 3:* long-term (>2 years) threat
- *Class 4:* no demonstrable long-term threat

The initial response actions may include notifying appropriate authorities, property owners, and potentially affected parties as well as evaluating the need to (a) evacuate occupants; (b) install a vapor barrier (e.g.,

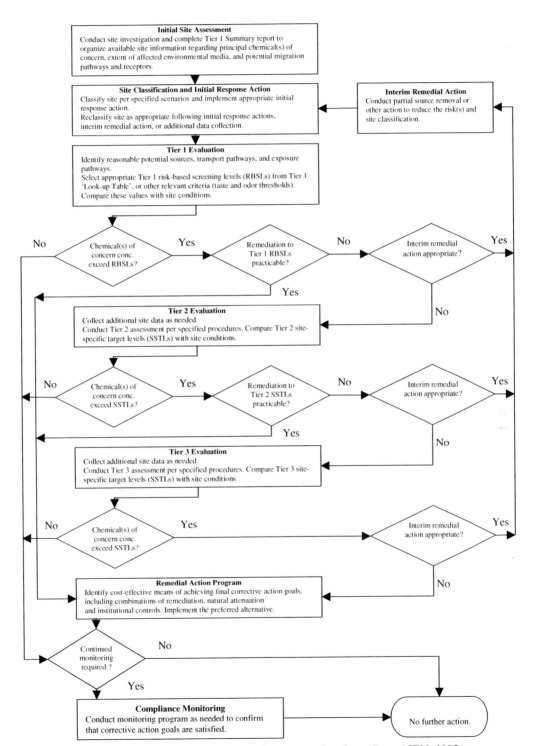

Figure 11.1 *Risk-based corrective action process flowchart (From ASTM, 1995.)*

capping, foams); (c) remove soils, cover soils, or restrict access; (d) monitor groundwater and evaluate the need for an alternative water supply; and (e) other measures, depending on the site-specific conditions. After assessing the contaminants of concern, the tiered evaluation begins. The three tiers (tier 1, tier 2, and tier 3) involved in RBCA, along with examples, are described below.

11.4.1   Tier 1 Evaluation

The tier 1 evaluation begins with a site assessment that usually includes a review of historical records of site activities and past releases; identification of chemicals of concern; location of the major sources of chemicals of concern and their maximum concentrations in soil and groundwater; location of humans and the environmental receptors that could be affected (points of exposure); and identification of potential significant transport and exposure pathways. The tier 1 assessment may also include determination of current or potential future use of the site and surrounding land, groundwater, surface water, and sensitive habitats; determination of regional hydrogeologic and geologic characteristics; and a qualitative evaluation of impacts to environmental receptors.

In tier 1, a "look-up" table of risk-based screening levels (RBSLs) is used. If a look-up table of RBSLs is not available, it should be developed. The look-up table is a tabulation of concentrations of chemicals for potential exposure pathways, media (soil, water, air), a range of incremental carcinogenic risk levels, and hazard quotients equal to unity, and potential exposure scenarios for each chemical of concern. In tier 1, the points of exposure and points of compliance are assumed to be located within close proximity to the source areas or the area where the highest concentration of the chemicals of concern have been identified. Table 11.2 shows an example RBSL look-up table given in the Standard E 1739, which is based on certain specific assumptions. The user should refer to the ASTM Standard E 1739 for details on these assumptions. All of these assumptions should be reviewed prior to using any of the RBSL values given in Table 11.2 for any site.

Tier 1 evaluation essentially involves comparing the concentrations of the site contaminants with tier 1 RBSLs. To complete the tier comparison, potential exposure scenerios for the site are selected. Next, based on the affected media identified, the primary sources, secondary sources, transport mechanisms, and exposure pathways are determined. Then receptors based on current and anticipated future use should be selected, taking into consideration land use restrictions and surrounding land use. If contaminant concentrations are below the RBSLs for the most restrictive route, the contaminants are not of concern, and no further action is necessary. If the contaminant concentrations exceed the RBSLs for the most restrictive exposure route, a decision is made whether (1) further tier evaluation is warranted, (2) implementation of interim remedial action is required, or (3) RBSLs may be applied as remedial target levels. Examples 11.3 and 11.4 show the typical procedures followed for tier 1 evaluation.

11.4.2   Tier 2 Evaluation

Tier 2 provides an option to determine the site-specific point(s) of compliance and corresponding site-specific target levels (SSTLs) for chemicals of concern applicable at the points(s) of compliance and source area(s). For tier 2, the incremental effort over tier 1 assessment is minimal in acquiring additional site data. Particularly, information on the soil and groundwater parameters, such as the soil type and hydraulic conductivity, is generally obtained as a part of tier 2 evaluation. SSTLs are calculated based on the measured and predicted attenuation of the chemical(s) of concern away from the source area(s) using relatively simplistic mathematical models (Table 11.3). In circumstances where the site-specific data are similar among several sites, a table of tier 2 SSTL values may be created. Table 11.4 shows an example of SSTLs, for soil and groundwater, calculated based on certain specific assumptions (ASTM, 1995), and the user should review these assumptions prior to using them for any site. The concentrations of chemicals of concern should then be compared with the SSTLs at the determined points of compliance or source areas. As in the tier 1 evaluation,

Table 11.2 Example of a tier I risk-based screening level look-up table[a]

Exposure Pathway	Receptor Scenario	Target Level	Benzene	Ethyl benzene	Toluene	Xylene	Naphthalene	Benzo[a]pyrene
Air								
Indoor air screening levels for inhalation exposure (µg/m³)	Residential	Cancer risk = 1E-06	3.92E-01					1.86E-03
		Cancer risk = 1E-04	3.92E+01					1.86E-01
		Chronic HQ = 1		1.39E+03	5.56E+02	9.73E+03	1.95E+01	
	Commercial/industrial	Cancer risk = 1E-06	4.93E-01					2.35E-03
		Cancer risk = 1E-04	4.93E+01					2.35E-01
		Chronic HQ = 1		1.46E+03	5.84E+02	1.02E+04	2.04E+01	
Outdoor air screening levels for inhalation exposure (µg/m³)	Residential	Cancer risk = 1E-06	2.94E-01					1.40E-03
		Cancer risk = 1E-04	2.94E+01					1.40E-01
		Chronic HQ = 1		1.04E+03	4.17E+02	7.03E+03	1.46E+01	
	Commercial/industrial	Cancer risk = 1E-06	4.93E-01					2.35E-03
		Cancer risk = 1E-04	4.93E+01					2.35E-01
		Chronic HQ = 1		1.46E+03	5.84E+02	1.02E+02	2.04E+01	
OSHA TWA PEL (µg/m³)			3.20E+03	4.35E+05	7.53E+05	4.35E+06	5.00E+04	2.00E+02
Mean odor detection threshold (µg/m³)			1.95E+05		6.00E+03	8.70E+04	2.00E+02	
National indoor background concentration range (µg/m³)			3.26E+00 to 2.15E+01	2.20E+00 to 9.70E+00	9.60E-01 to 2.91E+01	4.86E+00 to 4.76E+01		
Soil								
Soil volatilization to outdoor air, mg/kg	Residential	Cancer risk = 1E-06	2.72E-01					RES
		Cancer risk = 1E-04	2.73E+01					RES
		Chronic HQ = 1		RES	RES	RES	RES	
	Commercial/industrial	Cancer risk = 1E-06	4.57E-01					RES
		Cancer risk = 1E-04	4.57E+01					RES
		Chronic HQ = 1		RES	RES	RES	RES	
Soil-vapor intrusion from soil to buildings, mg/kg	Residential	Cancer risk = 1E-06	5.37E-03					RES
		Cancer risk = 1E-04	5.37E-01					RES
		Chronic HQ = 1		4.27E+02	2.06E+01	RES	4.07E+01	
	Commercial/industrial	Cancer risk = 1E-06	1.69E-02					RES
		Cancer risk = 1E-04	1.69E+00					RES
		Chronic HQ = 1		1.10E+03	5.45E+01	RES	1.07E+02	
Surficial soil [0 to 3 ft (0 to 0.9 m)] ingestion/dermal/inhalation (mg/kg)	Residential	Cancer risk = 1E-06	5.82E+00					1.30E-01
		Cancer risk = 1E-04	5.82E+02					1.30E+01
		Chronic HQ = 1		7.83E+03	1.33E+04	RES	9.77E+02	
	Commercial/industrial	Cancer risk = 1E-06	1.00E+01					3.04E-01
		Cancer risk = 1E-04	1.00E+03					3.04E+01
		Chronic HQ = 1		1.15E+04	1.87E+04	1.45E+06	1.50E+03	
Soil-leachate to protect groundwater ingestion target level, mg/kg	MCLs		2.93E-02	1.10E+02	1.77E+01	2.08E+05	N/A	9.42E+00
	Residential	Cancer risk = 1E-06	1.72E-02					5.50E-01
		Cancer risk = 1E-04	1.72E+00					RES
		Chronic HQ = 1		5.75E+02	1.29E+02	3.05E+02	2.29E+01	
	Commercial/industrial	Cancer risk = 1E-06	5.78E-02					1.85E+00
		Cancer risk = 1E-04	5.78E+00					RES
		Chronic HQ = 1		1.61E+03	3.61E+02	RES	6.42E+01	

Table 11.2 (*continued*)

Groundwater

Exposure Pathway	Receptor Scenario	Target Level	Benzene	Ethyl benzene	Toluene	Xylene	Naphthalene	Benzo [a]pyrene
Groundwater volatilization to outdoor air (mg/L)	Residential	Cancer risk = 1E-06	1.10E+01					>S
		Cancer risk = 1E-04	1.10e+03					>S
		Chronic HQ = 1		>S	>S	>S		
	Commercial/ industrial	Cancer risk = 1E-06	1.84E+01					>S
		Cancer risk = 1E-04	>S					>S
		Chronic HQ = 1		>S	>S	>S	>S	
Groundwater ingestion, (mg/L)	MCLs		5.00E-03	7.00E-01	1.00E+00	1.00E+01	N/A	2.00E-04
	Residential	Cancer risk = 1E-06	2.94E-03					1.17E-05
		Cancer risk = 1E-04	2.94E-01					1.17E-03
		Chronic HQ = 1		>S	7.30E+00	7.30E+01	1.46E-01	
	Commercial/ industrial	Cancer risk = 1E-06	9.87E-03					3.92E-05
		Cancer risk = 1E-04	9.87E-01					>S
		Chronic HQ = 1		3.65E+00	2.04E+01	>S	4.09E-01	
Groundwater-vapor intrusion from ground water to buildings (mg/L)	Residential	Cancer risk = 1E-06	2.38E-02					>S
		Cancer risk = 1E-04	2.38E+00					>S
		Chronic HQ = 1		1.02E+01	3.28E+01	>S	4.74E+00	
	Commercial/ industrial	Cancer risk = 1E-06	7.39E-02					>S
		Cancer risk = 1E-04	7.39E+00					>S
		Chronic HQ = 1		>S	8.50E+01	>S	1.23E+01	

Source: ASTM (1995).

[a]RES, selected risk level is not exceeded for pure compound present at any concentration; >S, selected risk level is not exceeded for all possible dissolved levels (≤ pure component solubility). This table is presented only as an example set of tier I RBSLs; the user should review all assumptions prior to using any values.

EXAMPLE 11.3

A release from the underground storage tank at a service station is discovered during a real estate investment assessment. It is known that there are petroleum-affected surficial soils in the area of the tank fill ports; however, the extent to which the soils are is not known. In the past, both gasoline and diesel have been sold at the facility. The new owner plans to continue operating the service station facility. Based on the previous knowledge that gasoline and diesel have been dispensed at this facility, chemical analyses of soil and groundwater are limited to benzene, toluene, ethyl benzene, xylenes, and naphthalene. Site assessment results indicate that the extent of petroleum-affected soils is limited to the vicinity of the fill ports for the tanks. Evidence suggests that the soils are affected only due to the spills and overfills associated with the storage tank. The concrete driveway is highly fractured. The site is underlain by layers of fine to silty sands. The groundwater, at a depth of 13 ft below the ground surface, is not affected. Maximum depth at which hydrocarbons are detected is 10 ft. Maximum detected soil concentrations are as follows:

Compound	Depth (ft)	Concentration (mg/kg)
Benzene	8	10
Ethyl benzene	4	4
Toluene	6.5	5.5
Xylenes	3.5	38
Naphthalene	2	17

A survey indicates that two domestic water wells are located near the source area. Perform RBCA tier 1 evaluation and determine if corrective action is necessary.

Solution:

A step-by-step procedure for tier 1 evaluation of the site is presented below.

Site classification and initial response action: This site is classified as class 3, as it is a long-term threat to the human health and environmental resources. The appropriate initial response is to evaluate the need for a groundwater monitoring program. But the decision is deferred until the completion of the tier 1 evaluation.

Development of the tier 1 look-up tables of risk-based screening level: Assumptions used to derive example tier 1 RBSL look-up Table 11.2 are presumed valid for this site.

Exposure pathway evaluation: The exposure pathways of this site are (1) inhalation of the ambient vapors by on-site workers, and (2) the leaching to groundwater, groundwater transport to the downgradient drinking water well, and ingestion of groundwater. A comparison of RBSLs for both pathways of concern indicates that RBSLs associated with the leaching pathway are the more critical of the two.

Comparison of site conditions with tier 1 RBSLs: Based on the look-up table (Table 11.2), concentrations of benzene and toluene are exceeded (e.g., benzene concentration in soil is 10 mg/kg, while the acceptable limit from Table 11.2 is 0.0578 mg/kg).

Evaluation of the tier 1 results: The shallow aquifer is not yet affected. Quick removal of the source will eliminate the need for the groundwater monitoring. Limited excavation of the soils to meet the tier 1 criteria could be performed quickly and inexpensively when the tanks are removed, relative to the cost of proceeding to the tier 2 analysis. If the area is renovated with a new concrete pavement after the excavation to prevent future leaching, no further action will be required.

EXAMPLE 11.4

A 5000-gallon release of super unleaded gasoline occurs from a single-walled tank after repeated manual gauging with a gauge stick. Soils are sandy at this site and groundwater is shallow. Also, free product is observed in a nearby monitoring well within 24 hours. The site is located next to an apartment building that has a basement where coin-operated washers and dryers are located for use by the tenants. An initial site assessment is conducted rapidly and is focused toward identifying if immediate hazardous conditions exist. It is known from the geological assessments that the first encountered groundwater is not potable. Groundwater monitoring wells in the area are inspected periodically for the appearance of the floating product. Strong gasoline odors are found in the basement of the apartment. Perform RBCA tier 1 evaluation.

Solution:

A step-by-step procedure for tier 1 evaluation of the site is presented below:

Site classification and initial response action: The site is classified as class 2, based on the observed vapor concentrations, size of release, and geological concentrations. The initial response implemented is as follows:

- Periodic monitoring of the basement begins to ensure that levels do not increase to the point where evacuation is necessary. All the tenants are informed about the activities at the site.
- A free product recovery system is installed to prevent further migration of the mobile liquid gasoline.
- A subsurface vapor extraction system is installed to prevent vapor intrusion into the building (refer to Chapter 13 for more information on the vapor extraction system).

Development of tier 1 look-up table risk-based screening level (RBSL) selection: Assumptions used to derive example tier 1 RBSL look-up Table 11.2 are valid for this site. Target soil and groundwater concentrations are determined based on 1×10^{-4} chronic inhalation risk for benzene and hazard quotients of unity for all other compounds.

Exposure pathway evaluation: There is a very low potential for groundwater use. Because of the close proximity of the apartment building, the vapor intrusion is a concern.

Comparison of the site conditions with tier 1 RBSLs: All parties agree that currently the RBSLs are likely to be exceeded.

Evaluation of tier 1 results: The owner decides to implement an interim action based on tier 1 RBSLs but reserves the right to propose a tier 2 evaluation in the future.

Tier 1 remedial action evaluation: A vapor extraction system is proposed to remediate the source soils while the operation of hydraulic control system is continued. Placement of a piezometer in the aquifer monitors the groundwater and periodic monitoring of the basement is also done. When hydrocarbon removal rates decline, a soil and groundwater assessment plan will be instituted to collect data to support a tier 2 evaluation.

---

a decision is made whether (1) further tier evaluation is warranted, (2) implementation of interim remedial action is warranted, or (3) tier 2 SSTLs may be applied as remediation target levels. Examples 11.5 and 11.6 show the typical RBCA tier 2 evaluation procedures.

11.4.3 Tier 3 Evaluation

In tier 3 evaluation, SSTLs for the source areas(s) and the point(s) of compliance are developed on the basis of more sophisticated statistical and contaminant fate

Table 11.3(a) **Screening-level transport models**

Description	Mathematical Approximation
Dissolved phase transport Maximum transport rate $U_{d,max}$ (cm/day) of dissolved plume	$U_{d,max} = \dfrac{K_S i}{\theta_S R_c}$
Minimum time $t_{d,min}$ (days) for leading edge of dissolved plume to travel distance, L (cm)	$t_{d,min} = \dfrac{L}{U_{d,max}}$
Steady-state attenuation $[(g/cm^3 H_2O)/(g/cm^3 H_2O)]$ along the centerline $(x, y = 0, z = 0)$ of a dissolved plume	$\dfrac{C(x)}{C_{source}} = \exp\left[\dfrac{X}{2\alpha_x}\left(1 - \sqrt{1 + \dfrac{4\lambda\alpha_x}{U}}\right)\right] \text{erf} \dfrac{S_w}{4\sqrt{\alpha_y X}}\, \text{erf}\left(\dfrac{Sd}{4\sqrt{\alpha_z X}}\right)$ where: $U = F_S i/\theta_S$
Immiscible phase transport Maximum depth D_{max} (cm) of immiscible phase penetration	$D_{max} = \dfrac{V_{spill}}{\theta_R \pi R_{spill}^2}$

Equilibrium Partitioning
  Vapor concentration:
   $C_{v,eq}$ (g/cm^3 vapor)
  Maximum vapor concentration above dissolved hydrocarbons
	$C_{v,eq} = H C_{w,eq}$

Maximum vapor concentration when immiscible hydrocarbon is present	$C_{v,eq} = \dfrac{X_i P_v^i M_w}{RT}$
Maximum vapor concentration in soil pores (no immiscible phase present)	$C_{v,eq} = \dfrac{H C_{soil} \rho_S}{\theta_W + K_S \rho_S + H\theta_v}$

Dissolved concentration
  $C_{w,eq}$ (g/cm^3 H$_2$O)
  Maximum dissolved concentration when immiscible hydrocarbon is present
	$C_{w,eq} = x_i S_i$

Maximum dissolved concentration in soil pores (no immiscible phase present)	$C_{w,eq} = \dfrac{C_{soil} \rho_S}{\theta_W + K_S \rho_S + H\theta_V}$

Equilibrium partitioning:
  Soil concentration [g/cm^{3}H$_2$O)
  Soil concentration (C_{soil} (g/g soil) at which immiscible hydrocarbon phase forms in soil matrix
	$C_{soil} = \dfrac{S_i}{\rho_S}(\theta_w + K_S \rho_S + H\theta_V)$

Vapor-phase transport Effective porous-media diffusion coefficient D^{eff} (cm^2/day) for combined vapor and solute transport, expressed as a vapor-phase diffusion coefficient (no immiscible hydrocarbon present outside the source area)	$D^{eff} = \dfrac{\theta_V^{3.33}}{\theta_T} D^{air} + \dfrac{1}{H}\dfrac{\theta_W^{3.33}}{\theta_T} D^W$
Porous media "retardation" factor R_v (no immiscible hydrocarbon present outside the source area)	$R_V = \dfrac{\theta_W}{H} + \dfrac{K_S \rho_S}{H} + \theta_V$
Maximum convective transport rate $U_{v,max}$ (cm/day) of vapors	$U_{v,max} = \dfrac{1}{R_V}\dfrac{K_V}{\mu_V}\nabla P$
Minimum time $t_{v,min}$ (days) for vapors to travel a distance L (cm) from source area by convection	$t_{c,min} = \dfrac{L}{U_{V,max}}$
Minimum time $t_{v,min}$ (days) for vapor to travel a distance L (cm) from source area by diffusion	$t_{d,min} = \dfrac{L^2}{D^{eff}/R_V}$

Table 11.3(a) (*continued*)

Description	Mathematical Approximation

Vapor emissions from subsurface vapor sources to open surfaces

Maximum diffusion diffusive vapor flux F_{max} (g/cm^2-day) from subsurface vapor source located a distance d (cm) below ground surface (steady-state, constant source)

$$F_{max} = D^{eff} \frac{C_{V,eq}}{d}$$

Maximum time-averaged diffusive vapor flux $<F_{max}>$ (g/cm^2-day) from subsurface soils over period from time = 0 to time = t, single-component immiscible phase present

$$<F_{max}> = \frac{\rho_S C_{soil}}{t} \left(\sqrt{d^2 + \frac{2 C_{v,eq} D^{eff}}{\rho_s C_{soil}}} - d \right)$$

Maximum combined convective and diffusive vapor flux F_{max} (g/cm^2-day) from subsurface vapor source located a distance d (cm) below ground surface

$$F_{max} = R_V U_{V,max} C_{V,eq} - \frac{R_V U_{V,max} C_{V,eq}}{\left[1 - \exp\left(\frac{R_V U_{V,max} d}{D^{eff}} \right) \right]}$$

Vapor emissions from surface soils to open spaces

Maximum time-averaged diffusive vapor flux $<f_{max}>$ (g/cm^2-day) from surface soils over period from time = 0 to time = t, single-component immiscible phase present

$$<F_{max}> = \rho_s C_{soil} \sqrt{\frac{2 C_{v,eq} D^{eff}}{\rho_s C_{soil} t}}$$

Maximum time-averaged diffusive vapor flux $<f_{max}>$ (g/cm^2-day) from surface soils over period from time = 0 to time = t, no immiscible phase present

$$<F_{max}> = 2 \rho_s C_{soil} \sqrt{\frac{D^{eff}}{\pi R_v t}}$$

Maximum time-averaged diffusive vapor flux $<f_{max}>$ (g/cm^2-day) from surface soils over period from time = 0 to time = t, volatile components from relatively nonvolatile immiscible phase (e.g., benzene from gasoline)

$$<F_{max}> = \frac{2 D^{eff} (x_i P_i^v M_{w,1}/RT)}{\sqrt{\pi \alpha t}}$$

where

$$\alpha = \frac{D^{eff}}{\theta_v + \rho_s RT (C_{soil}/M_{w,T})/P_i^v}$$

Vapor emissions to enclosed spaces

Maximum vapor emission rate E_{max} (g/cm^2-day) to enclosed spaces from subsurface vapor sources located a distance d (cm) away from the enclosed spaces

$$E_{max} = Q_B C_{v,eq} \left(\frac{D^{eff} A_B}{Q_B d} \right) \exp\left(\frac{Q_{soil} L_{crack}}{D^{crack} A_{crack}} \right) \Bigg/$$
$$\left\{ \exp\left(\frac{Q_{soil} L_{crack}}{D^{crack} A_{crack}} \right) + \left(\frac{D_{eff} A_B}{Q_{soil}} \right) \left[\exp\left(\frac{Q_{soil} L_{crack}}{D^{crack} A_{crack}} \right) - 1 \right] \right\}$$

Hydrocarbon vapor dispersion

Ambient hydrocarbon vapor concentration resulting from area vapor source $C_{outdoor}$ (g/cm^3)

$$C_{outdoor} = \frac{FL}{U_W \delta}$$

Enclosed space vapor concentration C_{indoor} (g/cm^3)

$$C_{indoor} = \frac{E_{max}}{V_B E_B}$$

Leachate transport impact on groundwater

Groundwater source area concentration C_{source} (g/cm^3 H$_2$O) resulting from leaching through vadose zone hydrocarbon-impacted soils

$$C_{source} = C_{w,eq} \frac{q_i W}{K_s i M + q_i W}$$

Groundwater source area concentration C_{source} (g/cm^3 H$_2$O) resulting from hydrocarbon-impacted soils in direct contact with groundwater

$$C_{source} = C_{w,eq}$$

Source: ASTM (1995).

Table 11.3(b) **Screening-level transport model parameters**

Symbol	Parameter	Units
$C(x)$	Dissolved hydrocarbon concentration along centerline ($x, y = 0, z = 0$) of dissolved plume	g/cm³ H₂O
C_{source}	Dissolved hydrocarbon concentration in dissolved plume source area (g/cm³ H₂O)	g/cm³ H₂O
i	Groundwater gradient	cm/cm
K_s	Saturated hydraulic conductivity	cm/day
k_s	Sorption coefficient	(g/g soil)/(g/cm³ H₂O)
L	Distance downgradient	cm
R_c	Retardation factor $= 1 + k_s\rho_s/\theta_s$	—
S_w	Source width (perpendicular to flow in the horizontal plane)	cm
S_d	Source width (perpendicular to flow in the vertical plane)	cm
U	Specific discharge	cm/day
$U_{d,max}$	Maximum transport rate of dissolved plume	cm/day
x	Distance along centerline from downgradient edge of dissolved plume source zone	cm
y	Depth below water table	cm
z	Lateral distance away from dissolved plume centerline	cm
α_x	Longitudinal dispersivity $\approx 0.01x$	cm
α_y	Transverse dispersivity $\approx \alpha_x/3$	cm
α_z	Vertical dispersivity $\approx \alpha_x/20$	cm
λ	First-order degradation constant	d⁻¹
θ_s	Volumetric water content of saturated zone	cm³ H₂O/cm³ soil
ρ_s	Soil bulk density	g soil/cm³ soil
$t_{d,min}$	Minimum convective travel time of dissolved hydrocarbons to distance L	day
$erf(\eta)$	Error function evaluated for value η	—
C_{soil}	Total soil hydrocarbon concentration	g/g soil
$C_{v,eq}$	Equilibrium vapor concentration	g/cm³ vapor
$C_{w,eq}$	Equilibrium dissolved concentration	g/cm³ H₂O
D_{max}	Maximum depth of immiscible phase penetration	cm
H	Henry's law constant	(g/g soil)/(g/cm³ H₂O)
M_w	Molecular weight	g/mol
P_v^i	Vapor pressure of compound i	atm
R	Gas constant $= 82$	cm³ atm/mol · K
R_{spill}	Radial extent of hydrocarbon impact	cm
S_i	Pure compound solubility	g/cm³ H₂O
T	Absolute temperature	K
V_{spill}	Volume of hydrocarbon released	cm³
X_i	Mole fraction of compound i	—
θ_R	Volumetric residual content of hydrocarbon under drainage condition	cm³ hydrocarbon/cm³ soil
θ_w	Volumetric content of soil pore water	cm³ H₂O/cm³ soil
θ_v	Volumetric content of soil vapor	cm³ vapor/cm³ soil
C_{soil}	Concentration at which immiscible phase forms in soil	g/g soil
D_{air}	Pure compound diffusion coefficient in air	cm²/day
D^{eff}	Effective diffusion coefficient for combined vapor and solute transport, expressed as a vapor-phase diffusion coefficient (no immiscible hydrocarbon present outside the source area)	cm²/day
D^w	Pure compound diffusion coefficient in water	cm²/day
k_v	Permeability to vapor flow	cm²
L	Distance	cm
R_v	Porous media retardation factor (no immiscible hydrocarbon present outside the source area)	—

Table 11.3(b) *(continued)*

Symbol	Parameter	Units
$U_{v,max}$	Maximum convective transport rate of vapors	cm/day
∇P	Vapor-phase pressure gradient	$g/cm^2 \cdot s^2$
θ_T	Total volumetric content of pore space in soil matrix	cm^3/cm^3 soil
μ_v	Vapor viscosity	$g/cm \cdot s$
$t_{c,min}$	Minimum time for vapors to travel to distance L (cm) by convection	days
$t_{d,min}$	Minimum time for vapors to travel to distance L (cm) by diffusion	days
C_{soil}	Total soil hydrocarbon concentration	g/g soil
d	Distance below ground surface to top of hydrocarbon vapor source	cm
D^{crack}	Effective diffusion coefficient through foundation cracks	cm^2/day
L_{crack}	Thickness of foundation/wall	cm
$M_{w,i}$	Molecular weight of i	g/mol
$M_{w,t}$	Average molecular weight of the hydrocarbon mixture	g/mol
Q_B	Volumetric flow rate of air within enclosed space	cm^3/s
Q_{soil}	Volumetric infiltration flow rate of soil gas into enclosed space	cm^3/s
t	Averaging time	s
E_B	Enclosed space air exchange rate	day^{-1}
E_{max}	Vapor emission rate into enclosed space	g/day
F	Vapor flux	$g/cm^2 \cdot day$
M	Groundwater mixing zone thickness	cm
q_i	Water infiltration rate	cm/day
U_w	Wind speed	cm/day
V_B	Volume enclosed space	cm^3
W	Width of affected soil zone	cm
δ	Height of breathing zone	cm

Source: ASTM (1995).

and transport modeling, using site-specific input parameters for both direct and indirect exposure scenerios. This level of evaluation commonly involves collection of significant additional site information and completion of more extensive modeling efforts than was required for either a tier 1 or tier 2 evaluation. The SSTLs are compared with the concentrations of chemicals at the site at the same compliance locations. If the site concentrations are higher than the SSTLs, remedial action must be undertaken to reduce concentrations to SSTLs.

After completion of the RBCA activities, a RBCA report is prepared and submitted to the regulatory agency. This report usually includes a site description, summaries of the ownership and use of the site, past releases, current and completed activities, regional and site-specific hydrogeologic conditions, beneficial use, and a discussion of the remedial action. Summaries of the tier evaluation, analytical data and RBSLs and

SSTLs used, and a summary of the ecological assessment are also included in the report.

11.5 OTHER RISK ASSESSMENT METHODS

Various state agencies have adopted the ASTM standard guide with modifications to provide flexible, site-specific risk assessment approaches for cleanup of UST sites as well as non-UST sites. The tiered evaluations are favored by many states because they take into account the contaminants of concern, exposure routes, and end use of the site. Sites located in a state should contact the local environmental agency about the specific methodology accepted in that state.

Among the states, the risk assessment procedure developed by the Illinois Environmental Protection Agency (IEPA) is one of the most comprehensive pro-

Table 11.4 Example of tier 2 site-specific target levels (SSTLs) for soil and groundwater

Exposure Pathway	Receptor Scenario	Distance to Source [ft (m)]	SSTLs at Source Sandy Soil, Natural Biodegradation Carcinogenic Risk = 1×10^{-5}, HQ = 1				SSTLs at Source Clay Soil, No Natural Biodegradation, Carcinogenic Risk = 1×10^{-5}, HQ = 1			
			Benzene	Ethylbenzene	Toluene	Xylene	Benzene	Ethylbenzene	Toluene	Xylene
Soil										
Soil vapor intrusion from soil to buildings (mg/kg)	Residential	10 (3)	0.052	18	11	450	1.7	570	300	9500
		25 (7.6)	0.47	160	160	1.7[a]	65	11[a]	10[a]	RES[b]
		100 (30)	3.1[a]	RES	RES	RES	RES	RES	RES	RES
	Commercial/ industrial	10(3)	0.13	39	24	980	4.3	1200	650	2.0[a]
		26 (7.6)	1.2	340	340	3.6[a]	950	24[a]	22.5[a]	RES
		100 (30)	8.0[a]	RES	RES	RES	RES	RES	RES	RES
Surficial soil ingestion and dermal (mg/kg)	Residential		22	5100	5400	280	22	5100	5400	280
	Commercial/ industrial		120	9600	1.7[a]	1500	117	9800	1.7[a]	1500
Soil leachate to protect groundwater ingestion target level (mg/kg)	Residential	0 (0)	0.17	47	130	2200	0.17	47	130	2200
		100 (30)	0.32	88	250	4200	0.20	130	760	RES
		500 (152)	4.0	1200	6300	RES	RES	RES	RES	RES
	Commercial/ industrial	0 (0)	0.58	130	350	6200	0.58	130	350	6200
		100 (30)	1.1	250	670	1.2[a]	0.70	380	2100	RES
		500 (152)	13	3300	1.75[a]	RES	RES	RES	RES	RES
Groundwater										
Groundwater ingestion (mg/L)	Residential	0	0.029	3.6	7.3	73	0.029	3.6	7.3	73
		100	0.054	6.8	14	140	0.035	10	43	>S[c]
		500	0.68	90	350	>S	>S	>S	>S	>S
	Commercial/ residential	0	0.099	10	20	200	0.099	10	20	200
		100	0.185	19	38	>S	0.12	29	120	>S
		500	2.3	250	>S	>S	>S	>S	>S	>S
Groundwater vapor intrusion from groundwater to buildings (mg/L)	Residential	10	0.11	32	17	510	5.0	>S	>S	200
		25	0.72	210	160	>S	1200	>S	>S	>S
		100	>S	>S	>S	>S	>S	>S	>S	>S
	Commercial/ industrial	10	0.28	70	36	>S	13	>S	>S	>S
		25	1.9	>S	350	>S	>S	>S	>S	>S
		100	>S	>S	>S	>S	>S	>S	>S	>S

Source: ASTM (1995).

[a] Weight percent.

[b] RES, selected risk level is not exceeded for pure compound present at any concentration.

[c] >S, selected risk level is not exceeded for all possible dissolved levels.

EXAMPLE 11.5

Petroleum-affected soils are discovered in the vicinity of a gasoline station. In the past, both gasoline and diesel were sold at this facility, which has been the operating facility for more than 20 years. Site assessment is conducted at the site. Chemical analysis of soil and groundwater are limited to benzene, toluene, ethyl benzene, xylene, and naphthalene. The extent of petroleum-affected soil is confined to the vicinity of the tanks and the dispensers. Evidence suggests that the soils are affected due only to the spills and overfills associated with the storage tank. The asphalt driveway is intact and there are no cracks. Another service station is located hydraulically downgradient, diagonally across the intersection. The site is underlain by silty sands with a few thin discontinuous clay layers. Groundwater, which is first encountered at 32 ft below the ground surface, is affected. The groundwater flow gradient is very low. The aquifer is considered to be a potential drinking water supply. A survey indicates that no detectable levels of hydrocarbon vapors are found in the southern border of the property or in soils surrounding the service station kiosk. Maximum concentrations detected in the soil and groundwater are as follows:

Compound	Soil (mg/kg)	Groundwater (mg/L)
Benzene	20	2
Ethyl benzene	4	0.5
Toluene	120	5
Xylene	100	5
Naphthalene	2	0.05

A survey indicates that no domestic water wells are located within $\frac{1}{2}$ mile of the site. Perform RBCA tier 1 and tier 2 evaluations.

Solution:

Site classification and responsive action: The site is classified as class 3, as it is long-term threat to the human health and environment. The appropriate initial response is to evaluate a need for a groundwater monitoring program.

Development of a tier 1 look-up table of risk-based screening level (RBSL) selection: Assumptions used to derive tier 1 RBSL look-up Table 11.2 are presumed valid for this site.

Exposure pathway evaluation: Based on the current and projected future use, there are no potential pathways at this site. However, being concerned about the uncontrolled use of this aquifer in future, the regulatory agency requests the owner to evaluate the groundwater transport to the residential drinking water ingestion pathway.

Comparison of site conditions with tier 1 RBSLs: Based on the data and the look-up Table 11.2, tier 1 soil and groundwater concentrations are exceeded for benzene.

Evaluation of tier 1 results: The responsible party decides to proceed to tier 2 evaluation for benzene and the pathway of concern rather than devising a corrective action plan to meet the tier 1 standards due to the following factors:

- The groundwater movement is very slow, and the dissolved plume appears to be stable.
- Excavation of soils to meet tier 1 criteria would be expensive.
- A tier 2 evaluation for this site is estimated to require minimal additional data and is anticipated to result in equally protective and less costly corrective action.

Tier 2 evaluation: The owner collects the additional groundwater monitoring data and verifies that:

- No mobile free-phase product is present.
- The dissolved plume is stable and the concentrations appear to decrease with time.
- Extent of the dissolved plume is limited to 50 ft of the property boundaries.
- Dissolved oxygen concentrations are higher, indicating some level of aerobic degradation.
- Simple groundwater transport modeling indicates that observations are consistent with the expectations of the site conditions.

Remedial action evaluation: The owner negotiates the corrective action plan based on the following:

- Restrictions are enacted to prevent the use of the groundwater within that zone until the dissolved levels decrease below drinking water MCLs.
- Deed restrictions are enacted to ensure that the site land use will not change significantly.
- Continued sampling of the groundwater monitoring well should be done yearly.
- The corrective action plans have to be revised if the concentrations exceed the tier 1 RBSLs in the future.
- Closure will be granted if the dissolved conditions remain stable or decrease for the next two years. (Natural attenuation of contaminants is expected to occur at the site. For more information on natural attenuation of contaminants, refer to Chapter 14.)

EXAMPLE 11.6

Petroleum-affected groundwater is discovered in monitoring wells at a former service station. The underground tanks and the piping were removed and the site is now occupied by an auto repair shop. Site assessment showed that the site contaminants are benzene, toluene, ethyl benzene, and xylenes. The maximum concentrations of these contaminants in soils are summarized below. The area of the hydrocarbon-affected soil is approximately 18,000 ft^2 and the depth of the soil impact is less than 5 ft. The plume is off-site. The site is underlain by clay. Hydrocarbon-affected perched groundwater is encountered 1 to 3 ft below grade. Maximum detected concentrations in this groundwater are shown below.

Compound	Soil (mg/kg)	Groundwater (mg/L)
Benzene	39	1.8
Ethyl benzene	12	0.5
Toluene	15	4
Xylene	140	9

Groundwater velocity is 0.008 ft/day based on slug tests and groundwater elevation survey and an assumed soil porosity of 50%. Survey indicates that the nearest groundwater well is greater than 1 mile away and the nearest surface water body is 0.5 mile. The distance to the nearest sensitive habitat is greater than 1 mile. The nearest home is 1000 ft away. The commercial building is 25 ft away. Perform RBCA tier 1 and tier 2 evaluations.

Solution:

Site classification: The site is classified as class 4, with no demonstrable long-term threat to the human health. As the hydrocarbon-affected soils are covered by asphalt and concrete in good condition and cannot be contacted, only potable perched water with no existing local use is affected, and there is no potential for explosive levels or concentrations that could cause the acute effects in the nearby buildings.

Development of the tier 1 look-up table of risk-based screening level (RBSL): The assumptions used to derive the example tier 1 RBSL look-up Table 11.2 are valid for this site.

Exposure pathway evaluation: The complete pathways are groundwater and soil volatilization to enclosed spaces and to ambient air, and direct exposure to the affected soil or groundwater by construction workers. A comparison of RBSLs for these pathways of concern indicates that RBSLs associated with the soil volatilization to an enclosed space are the most critical RBSLs.

Comparison of site conditions with tier 1: Based on the look-up Table 11.2, the concentrations of benzene in soil and groundwater and the concentration of toluene in the groundwater are exceeded.

Evaluation of the tier 1 results: The responsible party decided to proceed to the tier 2 evaluation for the pathways of concern rather than develop a corrective action plan, for the following reasons:

- Shallow perched water is affected and the plume is moving very slowly in tight clay.
- Excavation of the soils to meet the tier 1 objectives is expensive.
- Treatment technologies such as pump and treat or vapor extraction would be ineffective.
- A tier 2 evaluation would need no additional data and is expected to be an equally protective but less costly corrective action plan.

Development of the tier 2 table of site-specific targets: The tier 2 table is similar to the tier 1 look-up table with the exception that SSTLs for the pathways of concern are presented as functions of both distance from the source to the receptor and the soil type. Assumptions used to derive the example tier 2 SSTL table (Table 11.4) are presumed valid for this site.

Comparison of the site conditions with tier 2 table SSTLs: Based on the SSTLs in Table 11.4, all the concentrations in the soil and groundwater are within limits.

Tier 2 remedial action evaluation: Annual compliance monitoring of groundwater at downgradient monitoring wells will be performed to demonstrate the decreasing concentrations. The corrective action plan should be reevaluated if the levels exceed in the future. Closure will be granted if the dissolved concentrations remain stable or decrease for the next two years.

---

cedures developed to date (2002). This risk assessment procedure, known as the Tiered Approach to Corrective Action Objectives (TACO), has been adopted to determine whether detected site contamination poses environmental hazards (IEPA, 2002). If the site contamination is determined to pose a risk, the TACO procedure, like the ASTM procedure, also determines the remedial target levels in both soil and groundwater.

According to the IEPA (35 Illinois Administrative Code 620), groundwater is classified as:

- *Class I:* potable resource groundwater
- *Class II:* general resource groundwater
- *Class III:* special resource groundwater
- *Class IV:* other groundwater

However, TACO addresses the class I and class II groundwater. The Illinois procedure is applicable for contaminated sites ranging from leaking underground storage tank sites to RCRA corrective action sites.

TACO uses a three-tiered approach to develop soil and groundwater remediation objectives, depending on certain site characteristics as shown in Figure 11.2(*a*) and (*b*), respectively. Consequently, the first step in the TACO approach is to conduct a site characterization that includes characterization of the following parameters: the contaminant sources, the present and future exposure routes, the extent of the contamination, and the significant physical features of the site and vicinity. Based on these data, one must evaluate which tier or combination of tiers should be used to develop the proper cleanup objectives. To develop cleanup objectives, the following factors must be addressed: exposure routes, receptors, contaminants of concern, land use, and groundwater classification.

A tier 1 evaluation compares the concentrations of contaminants detected at a site to baseline contaminant cleanup objectives. To complete tier 1 evaluation, one must know the extent and concentrations of the contaminants of concern, the groundwater class, the land use classification at the site, and the soil pH, if applicable. The cleanup objectives can be based on either residential property or industrial/commercial property. If the cleanup objectives are developed based on industrial/commercial use, institutional controls (legal mechanisms for imposing restrictions on the land use) are required. The following three exposure routes must be evaluated in tier 1 evaluation: ingestion, inhalation, and migration to groundwater. Table 11.5 presents the tier 1 soil remedial objectives for residential properties. Table 11.6 presents the tier 1 groundwater remediation objectives for the groundwater component of the groundwater ingestion route. The user should review the assumptions made in the development of these remedial objectives prior to their use. Illinois TACO also provides tier 1 tables of soil remediation objectives for industrial/commercial properties as well as pH-specific soil remediation objectives for inorganics and ionizing organics for the soil component of the groundwater ingestion route (for class I and class II groundwater), and these can be found in IEPA (2002). If the concentrations of the contaminants of concern detected at the site are below the tier 1 values for any of the exposure routes, no further evaluation of that route is necessary. If the chemical that has been detected on the site is not

listed in tier 1 values, one must either request a site-specific cleanup objective from the IEPA or propose his or her own site-specific cleanup objectives under tier 3 guidelines.

A tier 2 evaluation is required only if it has been found that the chemicals detected under tier 1 evaluation for one of the three exposure routes exceed the tier 1 cleanup objective concentration for that chemical. Approaches to tier 2 evaluation for soil and groundwater are shown in Figure 11.3(*a*) and (*b*). The tier 2 cleanup objectives are developed through the use of analytical models, as presented in Tables 11.7(a) and 11.7(b), which allow site-specific data to be input. If one wants to use site-specific information outside the tier 2 framework, the project must be evaluated under tier 3. The following three categories of equations are used when performing a tier 2 soil evaluation: (1) the combined routes of ingestion of soil, inhalation of vapors and particulates, and chemical contact, (2) the ambient vapor inhalation route from subsurface soils, and (3) the migration to groundwater route. When calculating the soil cleanup objectives for industrial/commercial land uses, any calculation must be performed using both the industrial/commercial exposure default values and the construction worker exposure values. The more stringent soil cleanup objectives derived from these calculation must be used for further tier 2 evaluations.

When performing a tier 2 groundwater evaluation, any contamination must either be remediated to the tier 1 cleanup objectives or a tier 2 or tier 3 evaluation can be performed. It is permissible to develop cleanup objectives that exceed the tier 1 cleanup objectives, but certain actions must be performed. These include, but are not limited to, determining the maximum tier 2 groundwater cleanup value, identifying the extent of plume migration, showing that corrective actions have been taken to remove any free product and to control the source of contamination to the groundwater, and demonstrating that no existing potable water supply wells will be affected by any remaining groundwater contamination. This can be done if it is determined that the concentration levels of the chemicals detected do not pose a significant risk to human health or to the environment, and that it would be unnecessary to spend

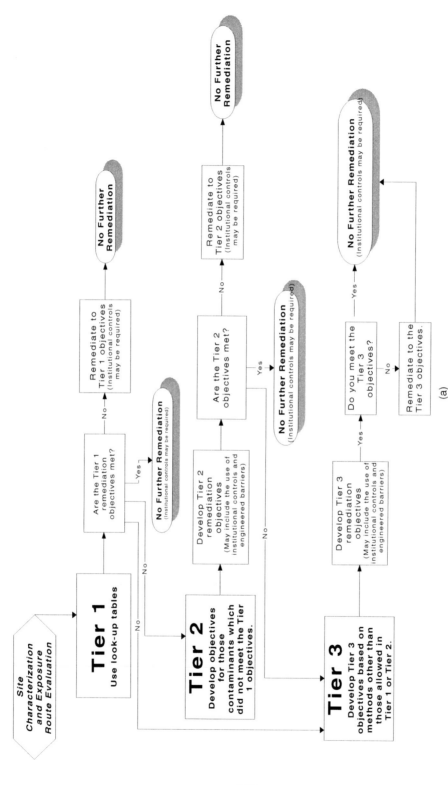

Figure 11.2 (a) *Developing soil remediation objectives under the tiered approach;* (b) *Developing groundwater remediation objectives under the tiered approach.* (*From IEPA, 2002.*)

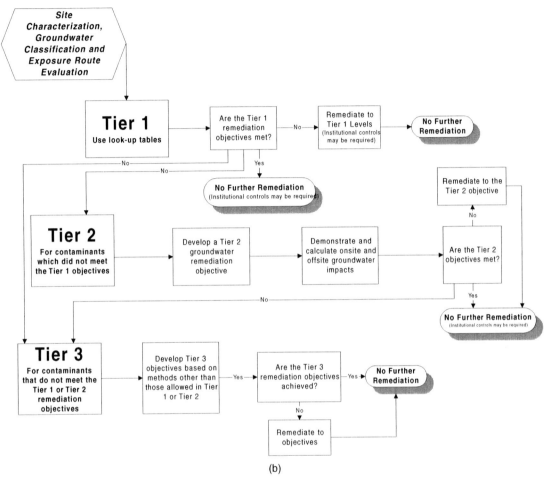

(b)

Figure 11.2 (*continued*)

exorbitant amounts of money in order to adhere to tier 1 goals.

A tier 3 evaluation is conducted when the frameworks of both tier 1 and tier 2 evaluations are inadequate for the site-specific conditions. However, even though tier 1 and tier 2 evaluations are not required when performing a tier 3 evaluation, the data from them can assist in developing an adequate tier 3 evaluation. When conducting a tier 3 evaluation, one must submit his or her proposal to the IEPA for approval as to whether the interpretations and conclusions reached are supported by the information gathered. In many instances, technical information may demonstrate that there is no actual or potential impact of contaminants of concern to receptors from a particular exposure route. Examples 11.7 and 11.8 show typical applications of Illinois's TACO for contaminated sites.

11.6 REMEDIAL STRATEGY

No remedial action is required if the risk assessment results indicate that there is no significant risk to human health or the environment. Generally, a total can-

Table 11.5 **Tier 1 soil remediation objectives**[a] **for residential properties**[†]

CAS No.	Chemical Name	Exposure Route-Specific Values for Soils		Soil Component of the Groundwater Ingestion Exposure Route Values		ADL (mg/kg)
		Ingestion (mg/kg)	Inhalation (mg/kg)	Class I (mg/kg)	Class II (mg/kg)	
83-32-9	Acenaphthene	4,700[b]	—[c]	570[b]	2,900	*
67-64-1	Acetone	7,800[b]	100,000[d]	16[b]	16	*
15972-60-8	Alachlor[o]	8[e]	—[c]	0.04	0.2	NA
116-06-3	Aldicarb[o]	78[b]	—[c]	0.013	0.07	NA
309-00-2	Aldrin	0.04[e]	3[e]	0.5[e]	2.5	0.94
120-12-7	Anthracene	23,000[b]	—[c]	12,000[b]	59,000	NA
1912-24-9	Atrazine[o]	2700[b]	—[c]	0.066	0.33	*
71-43-2	Benzene	12[e]	0.8[e]	0.03	0.17	*
56-55-3	Benzo[a]anthracene	0.9[e]	—[c]	2	8	*
205-99-2	Benzo[b]fluoranthene	0.9[e]	—[c]	5	25	*
207-08-9	Benzo[k]fluoranthene	9[e]	—[c]	49	250	*
50-32-8	Benzo[a]pyrene	0.09[e,f]	—[c]	8	82	*
111-44-4	Bis(2-chloroethyl)ether	0.6[e]	0.2[e,f]	0.0004[e,f]	0.0004	0.66
117-81-7	Bis(2-ethylhexyl)phthalate	46[e]	31,000[d]	3,600	31,000[d]	*
75-27-4	Bromodichloromethane (Dichlorobromomethane)	10[e]	3,000[d]	0.6	0.6	*
75-25-2	Bromoform	81[e]	53[e]	0.8	0.8	*
71-36-3	Butanol	7,800[b]	10,000[d]	17[b]	17	NA
85-68-7	Butyl benzyl phthalate	16,000[b]	930[d]	930[d]	930[d]	*
86-74-8	Carbazole	32[e]	—[c]	0.6[e]	2.8	NA
1563-66-2	Carbofuran[o]	390[b]	—[c]	0.22	1.1	NA
75-15-0	Carbon disulfide	7,800[b]	720[d]	32[b]	160	*
56-23-5	Carbon tetrachloride	5[e]	0.3[e]	0.07	0.33	*
57-74-9	Chlordane	1.8[e]	72[e]	10	48	*
106-47-8	4-Chloroaniline (p-chloroaniline)	310[b]	—[c]	0.7[b]	0.7	*
108-90-7	Chlorobenzene (monochlorobenzene)	1,600[b]	130[b]	1	6.5	*
124-48-1	Chlorodibromomethane (dibromochloromethane)	1,600[b]	1,300[d]	0.4	0.4	*
67-66-3	Chloroform	100[e]	0.3[e]	0.6	2.9	*
218-01-9	Chrysene	88[e]	—[c]	160	800	*
94-75-7	2,4-D[o]	780[b]	—[c]	1.5	7.7	*
75-99-0	Dalapon[o]	2,300[b]	—[c]	0.85	8.5	*
72-54-8	DDD	3[e]	—[c]	16[e]	80	*
72-55-9	DDE	2[e]	—[c]	54[e]	270	*
50-29-3	DDT	2[e]	—[g]	32[e]	160	*
53-70-3	Dibenzo[a,h]anthracene	0.09[e,f]	—[c]	2	7.6	*
96-12-8	1,2-Dibromo-3-chloropropane	0.46[e]	11[b]	0.002	0.002	*

CAS	Chemical					
106-93-4	1,2-Dibromoethane (ethylene dibromide)	0.0075^e	0.17^e	0.0004	0.004	0.005
84-74-2	Di-n-butyl phthalate	7,800^b	2,300^d	2,300^d	2,300^d	*
95-50-1	1,2-Dichlorobenzene (o-dichlorobenzene)	7,000^b	560^d	17	43	*
106-46-7	1,4-Dichlorobenzene (p-dichlorobenzene)	—^c	11,000^b	2	11	*
91-94-1	3,3'-Dichlorobenzidine	1^e	—^c	0.007^{e,f}	0.033	1.3
75-34-3	1,1-Dichloroethane	7,800^b	1,300^b	23^b	110	*
107-06-2	1,2-Dichloroethane (ethylene dichloride)	7^e	0.4^e	0.02	0.1	*
75-35-4	1,1-Dichloroethylene	700^b	1,500^d	0.06	0.3	*
156-59-2	cis-1,2-Dichloroethylene	780^b	1,200^d	0.4	1.1	*
156-60-5	trans-1,2-Dichloroethylene	1,600^b	3,100^d	0.7	3.4	*
78-87-5	1,2-Dichloropropane	9^e	15^b	0.03	0.15	*
542-75-6	1,3-Dichloropropene (1,3-dichloropropylene, cis + trans)	6.4^e	1.1^e	0.004^e	0.02	0.005
60-57-1	Dieldrin^n	0.04^e	1^e	0.004^e	0.02	0.603
84-66-2	Diethyl phthalate	63,000^b	2,000^d	470^b	470	*
105-67-9	2,4-Dimethylphenol	1,600^b	—^c	9^b	9	*
121-14-2	2,4-Dinitrotoluene	0.9^e	—^c	0.0008^{e,f}	0.0008	0.250
606-20-2	2,6-Dinitrotoluene	0.9^e	—^c	0.0007^{e,f}	0.0007	0.260
117-84-0	Di-n-octyl phthalate	1,600^b	10,000^d	10,000^d	10,000^d	*
115-29-7	Endosulfan^o	470^b	—^c	18^b	90	*
145-73-3	Endothall^o	1,600^b	—^c	0.4	0.4	*
72-20-8	Endrin	23^b	—^c	1	5	*
100-41-4	Ethylbenzene	7,800^b	400^d	13	19	*
206-44-0	Fluoranthene	3,100^b	—^c	4,300^b	21,000	*
86-73-7	Fluorene	3,100^b	—^c	560^b	2,800	*
76-44-8	Heptachlor	0.1^e	0.1^e	23	110	0.871
1024-57-3	Heptachlor epoxide	0.07^e	5^e	0.7	3.3	1.005
118-74-1	Hexachlorobenzene	0.4^e	1^e	2	11	*
319-84-6	α-HCH (α-BHC)	0.1^e	0.8^e	0.0005^{e,f}	0.003	0.0074
58-89-9	γ-HCH (Lindane)^n	0.5^e	—^c	0.009	0.047	*
77-47-4	Hexachlorocyclopentadiene	550^b	10^b	400	2,200^d	*
67-72-1	Hexachloroethane	78^b	—^c	0.5^b	2.6	*
193-39-5	Indeno[1,2,3-c,d]pyrene	0.9^e	—^c	14	69	*
78-59-1	Isophorone	15,600^b	4,600^d	8^b	8	*
72-43-5	Methoxychlor^o	390^b	10^b	160	780	*
74-83-9	Methyl bromide (bromomethane)	110^b	10^b	0.2^b	1.2	*
1634-04-4	Methyl tertiary butyl ether	780^b	8,800^d	0.32	0.32	*
75-09-2	Methylene chloride (dichloromethane)	85^e	13^e	0.02^e	0.2	*
95-48-7	2-Methylphenol (o-cresol)	3,900^b	—^c	15^b	15	*
91-20-3	Naphthalene	1,600^b	170^b	12^b	18	*

Table 11.5 (continued)

CAS No.	Chemical Name	Exposure Route-Specific Values for Soils		Soil Component of the Groundwater Ingestion Exposure Route Values		
		Ingestion (mg/kg)	Inhalation (mg/kg)	Class I (mg/kg)	Class II (mg/kg)	ADL (mg/kg)
98-95-3	Nitrobenzene	39[b]	92[b]	0.1[b,f]	0.1	0.26
86-30-6	N-Nitrosodiphenylamine	130[e]	—[c]	1[e]	5.6	*
621-64-7	N-Nitrosodi-n-propylamine	0.09[e,f]	—[c]	0.00005[e,f]	0.00005	0.0018
108-95-2	Phenol	47,000[b]	—[c]	100[b]	100	*
1918-02-1	Picloram[o]	5,500[b]	—[c]	2	20	NA
1336-36-3	Polychlorinated biphenyls (PCBs)[n]	1[h]	—[c,h]	—[h]	—[h]	*
129-00-0	Pyrene	2,300[b]	—[c]	4,200[b]	21,000	NA
122-34-9	Simazine[o]	390[b]	—[c]	0.04	0.37	*
100-42-5	Styrene	16,000[b]	1,500[d]	4	18	*
127-18-4	Tetrachloroethylene (perchloroethylene)	12[e]	11[e]	0.06	0.3	*
108-88-3	Toluene	16,000[b]	650[d]	12	29	*
8001-35-2	Toxaphene[n]	0.6[e]	89[e]	31	150	*
120-82-1	1,2,4-Trichlorobenzene	780[b]	3,200[b]	5	53	*
71-55-6	1,1,1-Trichloroethane	—[c]	1,200[d]	2	9.6	*
79-00-5	1,1,2-Trichloroethane	310[b]	1,800[d]	0.02	0.3	*
79-01-6	Trichloroethylene	58[e]	5[e]	0.06	0.3	*
108-05-4	Vinyl acetate	78,000[b]	1,000[b]	170[b]	170	*
75-01-4	Vinyl chloride	0.46[e]	0.28[e]	0.01[f]	0.07	*
108-38-3	m-Xylene	160,000[b]	420[d]	210	210	*
95-47-6	o-Xylene	160,000[b]	410[d]	190	190	*
106-42-3	p-Xylene	160,000[b]	460[d]	200	200	*
1330-20-7	Xylenes (total)	160,000[b]	320[d]	150	150	*
Ionizable Organics						
65-85-0	Benzoic acid	310,000[b]	—[c]	400[b,j]	400[j]	*
95-57-8	2-Chlorophenol	390[b]	53,000[d]	4[b,j]	4[j]	*
120-83-2	2,4-Dichlorophenol	230[b]	—[c]	1[b,j]	1[j]	*
51-28-5	2,4-Dinitrophenol	160[b]	—[c]	0.2[b,f]	0.2	3.3
88-85-7	Dinoseb[o]	78[b]	—[c]	0.34[b,j]	3.4[j]	*
87-86-5	Pentachlorophenol	3[e,j]	—[c]	0.03[f,j]	0.14[j]	*
93-72-1	2,4,5-TP (silvex)	630[b]	—[c]	11[j]	55[j]	*
95-95-4	2,4,5-Trichlorophenol	7,800[b]	—[c]	270[b,j]	1,400[j]	*
88-06-2	2,4,6 Trichlorophenol	58[e]	200[e]	0.2[e,f,j]	0.77[j]	0.66

Inorganics

CAS	Name					
7440-36-0	Antimony	31[b]	—[c]	0.006[m]	0.024[m]	*
7440-38-2	Arsenic[i,n]	t	750[e]	0.05[m]	0.2[m]	*
7440-39-3	Barium	5,500[b]	690,000[b]	2.0[m]	2.0[m]	*
7440-41-7	Beryllium	160[b]	1,300[e]	0.004[m]	0.5[m]	*
7440-42-8	Boron	7,000[b]	—[g]	2.0[m]	2.0[m]	*
7440-43-9	Cadmium[i,n]	78[b,r]	1,800[e]	0.005[m]	0.05[m]	*
16887-00-6	Chloride	—[c]	—[c]	200[m]	200[m]	
7440-47-3	Chromium, total	230[b]	270[e]	0.1[m]	1.0[m]	*
16065-83-1	Chromium, ion, trivalent	120,000[b]	—[c]	—[g]	—[g]	*
18540-29-9	Chromium, ion, hexavalent	230[b]	270[e]	—	—	*
7440-48-4	Cobalt	4,700[b]	—[c]	1.0[m]	1.0[m]	*
7440-50-8	Copper[n]	2,900[b]	—[c]	0.65[m]	0.65[m]	*
57-12-5	Cyanide (amenable)	1,600[b]	—[c]	0.2[a,m]	0.6[a,m]	*
7782-41-4	Fluoride	4,700[b]	—[c]	4.0[m]	4.0[m]	*
15438-31-0	Iron	—[c]	—[c]	5.0[m]	5.0[m]	*
7439-92-1	Lead	400[k]	—[c]	0.0075[m]	0.1[m]	*
7439-96-5	Manganese	3,700[b]	69,000[b]	0.15[m]	10.0[m]	*
7439-97-6	Mercury[i,n,s]	23[b]	10[b]	0.002[m]	0.01[m]	*
7440-02-0	Nickel[l]	1,600[b]	13,000[e]	0.1[m]	2.0[m]	*
14797-55-8	Nitrate as N[p]	130,000[b]	—[c]	10.0[q]	100[q]	*
7782-49-2	Selenium[i,n]	390[b]	—[c]	0.05[m]	0.05[m]	*
7440-22-4	Silver	390[b]	—[c]	0.05[m]	—[g]	*
14808-79-8	Sulfate	—[c]	—[c]	400[m]	400[m]	*
7440-28-0	Thallium	6.3[b,u]	—[c]	0.002[m]	0.02[m]	*
7440-62-2	Vanadium	550[b]	—[c]	0.049[m]	0.1[m]	*
7440-66-6	Zinc[l]	23,000[b]	—[c]	5.0[m]	10[m]	*

311

Table 11.5 (*continued*)

Source: IEPA (2002); amended at 26 Ill. Reg. 2683, effective February 5, 2002.

†indicates that the ADL is less than or equal to the specified remediation objective; NA means not available; no PQL or EQL available in USEPA analytical methods.

[a] Soil remediation objectives based on human health criteria only.

[b] Calculated values correspond to a target hazard quotient of 1.

[c] No toxicity criteria available for the route of exposure.

[d] Soil saturation concentration (C_{sat}) = the concentration at which the absorptive limits of the soil particles, the solubility limits of the available soil moisture, and saturation of soil pore air have been reached. Above the soil saturation concentration, the assumptions regarding vapor transport to air and/or dissolved-phase transport to groundwater (for chemicals that are liquid at ambient soil temperatures) have been violated, and alternative modeling approaches are required.

[e] Calculated values correspond to a cancer risk level of 1 in 1,000,000.

[f] Level is at or below contract laboratory program required quantitation limit for regular analytical services (RAS).

[g] Chemical-specific properties are such that this route is not of concern at any soil contaminant concentration.

[h] 40 CFR 761 contains applicability requirements and methodologies for the development of PCB remediation objectives. Requests for approval of a tier 3 evaluation must address the applicability of 40 CFR 761.

[i] Soil remediation objective for pH of 6.8. If soil pH is other than 6.8, refer to Appendix B, Tables C and D of this part.

[j] Ingestion soil remediation objective adjusted by a factor of 0.5 to account for dermal route.

[k] A preliminary remediation goal of 400 mg/kg has been set for lead based on *Revised Interim Soil Lead Guidance for CERCLA Sites and RCRA Corrective Action Facilities*, OSWER Directive 9355.4-12.

[l] Potential for soil–plant–human exposure.

[m] The person conducting the remediation has the option to use: (1) TCLP or SPLP test results to compare with the remediation objectives listed in this table; or (2) the total amount of contaminant in the soil sample results to compare with pH specific remediation objectives listed in Appendix B, Table C or D of this part. (See Section 742.510.) If the person conducting the remediation wishes to calculate soil remediation objectives based on background concentrations, this should be done in accordance with Subpart D of this part.

[n] The agency reserves the right to evaluate the potential for remaining contaminant concentrations to pose significant threats to crops, livestock, or wildlife.

[o] For agrichemical facilities, remediation objectives for surficial soils which are based on field application rates may be more appropriate for currently registered pesticides. Consult the agency for further information.

[p] For agrichemical facilities, soil remediation objectives based on site-specific background concentrations of nitrate as N may be more appropriate. Such determinations shall be conducted in accordance with the procedures set forth in Subparts D and I of this part.

[q] The TCLP extraction must be done using water at pH 7.0.

[r] Value based on dietary reference dose.

[s] Value for ingestion based on reference dose for mercuric chloride (CAS No. 7487-94-7); value for inhalation based on reference concentration for elemental mercury (CAS No. 7439-97-6).

[t] For the ingestion route for arsenic, see 742:Appendix A, Table G.

[u] Value based on reference dose for thallium sulfate (CAS No. 7446-18-6).

Table 11.6. **Tier 1 groundwater remediation objectives for the groundwater component of the groundwater ingestion route**

CAS No.	Chemical Name	Groundwater Remediation Objective	
		Class I (mg/L)	Class II (mg/L)
83-32-9	Acenaphthene	0.42	2.1
67-64-1	Acetone	0.7	0.7
15972-60-8	Alachlor	0.002^c	0.01^c
116-06-3	Aldicarb	0.003^c	0.015^c
309-00-2	Aldrin	0.014^a	0.07
120-12-7	Anthracene	2.1	10.5
1912-24-9	Atrazine	0.003^c	0.015^c
71-43-2	Benzene	0.005^c	0.025^c
56-55-3	Benzo[a]anthracene	0.00013^a	0.00065
205-99-2	Benzo[b]fluoranthene	0.00018^a	0.0009
207-08-9	Benzo[k]fluoranthene	0.00017^a	0.00085
50-32-8	Benzo[a]pyrene	0.0002^{a,c}	0.002^c
111-44-4	Bis(2-chloroethyl)ether	0.01^a	0.01
117-81-7	Bis(2-ethylhexyl)phthalate [di(2-ethylhexyl)phthalate]	0.006^c	0.06^c
75-27-4	Bromodichloromethane (dichlorobromomethane)	0.0002^a	0.0002
75-25-2	Bromoform	0.001^a	0.001
71-36-3	Butanol	0.7	0.7
85-68-7	Butyl benzyl phthalate	1.4	7.0
86-74-8	Carbazole	—	—
1563-66-2	Carbofuran	0.04^c	0.2^c
75-15-0	Carbon disulfide	0.7	3.5
56-23-5	Carbon tetrachloride	0.005^c	0.025^c
57-74-9	Chlordane	0.002^c	0.01^c
108-90-7	Chlorobenzene (monochlorobenzene)	0.1^c	0.5^c
124-48-1	Chlorodibromomethane (dibromochloromethane)	0.14	0.14
67-66-3	Chloroform	0.0002^a	0.001
218-01-9	Chrysene	0.0015^a	0.0075
94-75-7	2,4-D	0.07^c	0.35^c
75-99-0	Dalapon	0.2^c	2.0^c
72-54-8	DDD	0.014^a	0.07
72-55-9	DDE	0.01^a	0.05
50-29-3	DDT	0.006^a	0.03
53-70-3	Dibenzo[a,h]anthracene	0.0003^a	0.0015
96-12-8	1,2-Dibromo-3-chloropropane	0.0002^c	0.0002^c
106-93-4	1,2-Dibromoethane (ethylene dibromide)	0.00005^c	0.0005^c
84-74-2	Di-n-butyl phthalate	0.7	3.5
95-50-1	1,2-Dichlorobenzene (o-dichlorobenzene)	0.6^c	1.5^c
106-46-7	1,4-Dichlorobenzene (p-dichlorobenzene)	0.075^c	0.375^c
91-94-1	3,3'-Dichlorobenzidine	0.02^a	0.1
75-34-3	1,1-Dichloroethane	0.7	3.5
107-06-2	1,2-Dichloroethane (ethylene dichloride)	0.005^c	0.025^c
75-35-4	1,1-Dichloroethylene^b	0.007^c	0.035^c
156-59-2	cis-1,2-Dichloroethylene	0.07^c	0.2^c
156-60-5	trans-1,2-Dichloroethylene	0.1^c	0.5^c
78-87-5	1,2-Dichloropropane	0.005^c	0.025^c
542-75-6	1,3-Dichloropropene (1,3-dichloropropylene, cis + trans)	0.001^a	0.005

Table 11.6. (*continued*)

CAS No.	Chemical Name	Groundwater Remediation Objective	
		Class I (mg/L)	Class II (mg/L)
60-57-1	Dieldrin	0.009[a]	0.045
84-66-2	Diethyl phthalate	5.6	5.6
121-14-2	2,4-Dinitrotoluene[a]	0.00002[a]	0.00002
606-20-2	2,6-Dinitrotoluene[a]	0.00031[a]	0.00031
88-85-7	Dinoseb	0.007[c]	0.07[c]
117-84-0	Di-*n*-octyl phthalate	0.14	0.7
115-29-7	Endosulfan	0.042	0.21
145-73-3	Endothall	0.1[c]	0.1[c]
72-20-8	Endrin	0.002[c]	0.01[c]
100-41-4	Ethylbenzene	0.7[c]	1.0[c]
206-44-0	Fluoranthene	0.28	1.4
86-73-7	Fluorene	0.28	1.4
76-44-8	Heptachlor	0.0004[c]	0.002[c]
1024-57-3	Heptachlor epoxide	0.0002[c]	0.001[c]
118-74-1	Hexachlorobenzene	0.00006[a]	0.0003
319-84-6	α-HCH (α-BHC)	0.00011[a]	0.00055
58-89-9	γ-HCH (Lindane)	0.0002[c]	0.001[c]
77-47-4	Hexachlorocyclopentadiene	0.05[c]	0.5[c]
67-72-1	Hexachloroethane	0.007	0.035
193-39-5	Indeno[1,2,3-*c,d*]pyrene	0.00043[a]	0.00215
78-59-1	Isophorone	1.4	1.4
72-43-5	Methoxychlor	0.04[c]	0.2[c]
74-83-9	Methyl bromide (bromomethane)	0.0098	0.049
1634-04-4	Methyl tertiary butyl ether	0.07	0.07
75-09-2	Methylene chloride (dichloromethane)	0.005[c]	0.05[c]
91-20-3	Naphthalene	0.14	0.22
98-95-3	Nitrobenzene[b]	0.0035	0.0035
86-30-6	*N*-Nitrosodiphenylamine	0.0032[a]	0.016
621-64-7	*N*-Nitrosodi-*n*-propylamine	0.0018[a]	0.0018
87-86-5	Pentachlorophenol	0.001[c]	0.005[c]
108-95-2	Phenol	0.1[c]	0.1[c]
1918-02-1	Picloram	0.5[c]	5.0[c]
1336-36-3	Polychlorinated biphenyls (PCBs)	0.0005[c]	0.0025[c]
129-00-0	Pyrene	0.21	1.05
122-34-9	Simazine	0.004[c]	0.04[c]
100-42-5	Styrene	0.1[c]	0.5[c]
93-72-1	2,4,5-TP (silvex)	0.05[c]	0.25[c]
127-18-4	Tetrachloroethylene (perchloroethylene)	0.005[c]	0.025[c]
108-88-3	Toluene	1.0[c]	2.5[c]
8001-35-2	Toxaphene	0.003[c]	0.015[c]
120-82-1	1,2,4-Trichlorobenzene	0.07[c]	0.7[c]
71-55-6	1,1,1-Trichloroethane[b]	0.2[c]	1.0[c]
79-00-5	1,1,2-Trichloroethane	0.005[c]	0.05[c]
79-01-6	Trichloroethylene	0.005[c]	0.025[c]
108-05-4	Vinyl acetate	7.0	7.0
75-01-4	Vinyl chloride	0.002[c]	0.01[c]
1330-20-7	Xylenes (total)	10.0[c]	10.0[c]

Table 11.6. (*continued*)

		Groundwater Remediation Objective	
CAS No.	Chemical Name	Class I (mg/L)	Class II (mg/L)
Ionizable Organics			
65-85-0	Benzoic acid	28	28
106-47-8	4-Chloroaniline (*p*-chloroaniline)	0.028	0.028
95-57-8	2-Chlorophenol	0.035	0.175
120-83-2	2,4-Dichlorophenol	0.021	0.021
105-67-9	2,4-Dimethylphenol	0.14	0.14
51-28-5	2,4-Dinitrophenol	0.014	0.014
95-48-7	2-Methylphenol (*o*-cresol)	0.35	0.35
95-95-4	2,4,5-Trichlorophenol	0.7	3.5
88-06-2	2,4,6 Trichlorophenol	0.01[a]	0.05
Inorganics			
7440-36-0	Antimony	0.006[c]	0.024[c]
7440-38-2	Arsenic	0.05[c]	0.2[c]
7440-39-3	Barium	2.0[c]	2.0[c]
7440-41-7	Beryllium	0.004[c]	0.5[c]
7440-42-8	Boron	2.0[c]	2.0[c]
7440-43-9	Cadmium	0.005[c]	0.05[c]
16887-00-6	Chloride	200[c]	200[c]
7440-47-3	Chromium, total	0.1[c]	1.0[c]
18540-29-9	Chromium, ion, hexavalent	—	—
7440-48-4	Cobalt	1.0[c]	1.0[c]
7440-50-8	Copper	0.65[c]	0.65[c]
57-12-5	Cyanide	0.2[c]	0.6[c]
7782-41-4	Fluoride	4.0[c]	4.0[c]
15438-31-0	Iron	5.0[c]	5.0[c]
7439-92-1	Lead	0.0075[c]	0.1[c]
7439-96-5	Manganese	0.15[c]	10.0[c]
7439-97-6	Mercury	0.002[c]	0.01[c]
7440-02-0	Nickel	0.1[c]	2.0[c]
14797-55-8	Nitrate as N	10.0[c]	100[c]
7782-49-2	Selenium	0.05[c]	0.05[c]
7440-22-4	Silver	0.05[c]	—
14808-79-8	Sulfate	400[c]	400[c]
7440-28-0	Thallium	0.002[c]	0.02[c]
7440-62-2	Vanadium[b]	0.049	0.1
7440-66-6	Zinc	5.0[c]	10[c]

Source: IEPA (2000); amended at 26 Ill. Reg. 2683, effective February 5, 2002.

[a] The groundwater remediation objective is equal to the ADL for carcinogens according to the procedures specified in 35 Ill. Adm. Code 620.

[b] Oral reference dose and/or reference concentration under review by USEPA. Listed values subject to change.

[c] Value listed is also the groundwater quality standard for this chemical pursuant to 35 Ill. Adm. Code 620.410 for class I groundwater or 35 Ill. Adm. Code 620.420 for Class II groundwater.

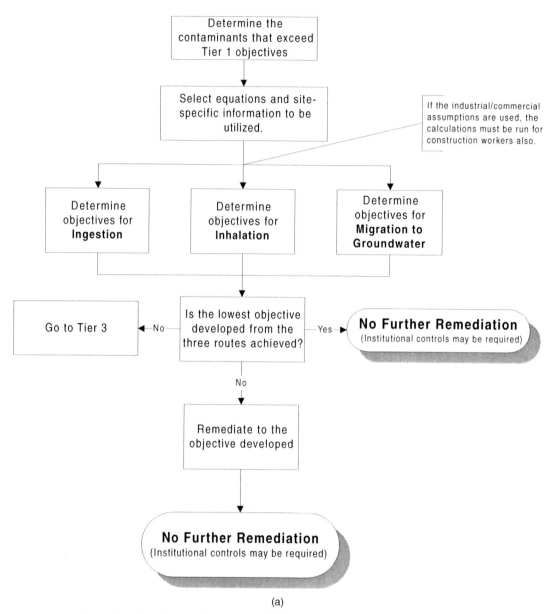

Figure 11.3 *Tier 2 evaluation for* (a) *Soil and* (b) *groundwater. (From IEPA, 2002.)*

(b)

Figure 11.3 (*continued*)

Table 11.7(a) **Tier 2 SSL equations**

	Description	Equation	
Equations for soil ingestion exposure route	Remediation objectives for noncarcinogenic contaminants (mg/kg)	$\dfrac{THQ \cdot BW \cdot AT \cdot 365\ days/yr}{(1/RfD_o) \cdot 10^{-6}\ kg/mg \cdot EF \cdot ED \cdot IR_{soil}}$	S1
	Remediation objectives for carcinogenic contaminants, residential (mg/kg)	$\dfrac{TR \cdot AT_c \cdot 365\ days/yr}{SF_o \cdot 10^{-6}\ kg/mg \cdot EF \cdot IF_{soil-adj}}$	S2
	Remediation objectives for carcinogenic contaminants, industrial/commercial, construction worker (mg/kg)	$\dfrac{TR \cdot BW \cdot AT_c \cdot 365\ days/yr}{SF_o \cdot 10^{-6}\ kg/mg \cdot EF \cdot ED \cdot IR_{soil}}$	S3
Equations for inhalation exposure route (organic contaminants and mercury)	Remediation objectives for noncarcinogenic contaminants, residential, industrial/commercial (mg/kg)	$\dfrac{THQ \cdot AT \cdot 365\ days/yr}{EF \cdot ED \cdot (1/RfC \cdot 1/VF)}$	S4
	Remediation objectives for noncarcinogenic contaminants, construction worker (mg/kg)	$\dfrac{THQ \cdot AT \cdot 365\ \frac{d}{yr}}{EF \cdot ED \cdot \left(\dfrac{1}{RfC} \cdot \dfrac{1}{VF'}\right)}$	S5
	Remediation objectives for carcinogenic contaminants, residential, Industrial/commercial (mg/kg)	$\dfrac{TR \cdot AT_c \cdot 365\ days/yr}{URF \cdot 1000\ \mu g/mg \cdot EF \cdot ED \cdot 1/VF}$	S6
	Remediation objectives for carcinogenic contaminants, construction worker (mg/kg)	$\dfrac{TR \cdot AT_c \cdot 365\ days/yr}{URF \cdot 1000\ \mu g/mg \cdot EF \cdot ED \cdot 1/VF'}$	S7
	Equation for derivation of the volatilization factor, residential, industrial/commercial, VF (m³/kg)	$VF = \dfrac{Q}{C} \cdot \dfrac{(3.14 D_A T)^{1/2}}{2\rho_b D_A} \cdot 10^{-4}\ m^2/cm^2$	S8
	Equation for derivation of the volatilization factor, construction worker, VF' (m³/kg)	$VF' = \dfrac{VF}{10}$	S9
	Equation for derivation of apparent diffusivity, D_A (cm²/s)	$D_A = \dfrac{(\theta_a^{3.33} D_i H') + (\theta_w^{3.33} D_w)}{\eta^2(\rho_b K_d) + \theta_w + (\theta_a H')}$	S10
Equations for inhalation exposure route (fugitive dusts)	Remediation objectives for noncarcinogenic contaminants, residential, industrial/commercial (mg/kg)	$\dfrac{THQ \cdot AT \cdot 365\ days/yr}{EF \cdot ED \cdot (1/RfC \cdot 1/PEF)}$	S11
	Remediation objectives for noncarcinogenic contaminants, construction worker (mg/kg)	$\dfrac{THQ \cdot AT_c \cdot 365\ days/yr}{EF \cdot ED \cdot (1/RfC \cdot 1/PEF')}$	S12
	Remediation objectives for carcinogenic contaminants, residential, industrial/commercial (mg/kg)	$\dfrac{TR \cdot AT_c \cdot 365\ days/yr}{URF \cdot 1000\ \mu g/mg \cdot EF \cdot ED \cdot 1/PEF}$	S13
	Remediation objectives for carcinogenic contaminants, construction worker (mg/kg)	$\dfrac{TR \cdot AT_c \cdot 365\ days/yr}{URF \cdot 1000\ \mu g/mg \cdot EF \cdot ED \cdot 1/PEF'}$	S14
	Equation for derivation of particulate emission factor, PEF (m³/kg)	$PEF = \dfrac{Q}{C} \cdot \dfrac{3600\ s/h}{0.036(1 - V)(U_m/U_t)^3 F(x)}$	S15
	Equation for derivation of particulate emission factor, PEF', construction worker (m³/kg)	$PEF' = \dfrac{PEF}{10}$	S16

Note: PEF must be the industrial/commercial value.

318

Equations for the soil component of the groundwater ingestion exposure route	Remediation objective (mg/kg)	$C_w \left(K_d + \dfrac{\theta_w + \theta_a H'}{\rho_b} \right)$ *Note*: This equation can only be used to model contaminant migration not in the water-bearing unit.	S17
	Target soil leachate concentration, C_w (mg/L)	$C_w = DF \cdot GW_{obj}$	S18
	Soil-water partition coefficient, K_d (cm³/g)	$K_d = K_{oc} f_{oc}$	S19
	Water-filled soil porosity, θ_w (L_{water}/L_{soil})	$\theta_w = \eta \left(\dfrac{I}{K_s} \right)^{1/(2b+3)}$	S20
	Air-filled soil porosity, θ_a (L_{air}/L_{soil})	$\theta_a = \eta - \theta_w$	S21
	Dilution factor, DF (unitless)	$DF = 1 + \dfrac{Kid}{IL}$	S22
	Groundwater remediation objective for carcinogenic contaminants, GW_{obj} (mg/L)	$\dfrac{TR \cdot BW \cdot AT_c \cdot 365 \text{ days/yr}}{SF_o \cdot IR_w \cdot EF \cdot ED}$	S23
	Total soil porosity, η (L_{pore}/L_{soil})	$\eta = 1 - \dfrac{\rho_b}{\rho_s}$	S24
	Equation for estimation of mixing zone depth, d (m)	$d = (0.0112L^2)^{0.5} + d_a \left[1 - \exp\left(\dfrac{-LI}{Kid_a} \right) \right]$	S25
Mass-limit equations for inhalation exposure route and soil component of the groundwater ingestion exposure route	Mass-limit volatilization factor for the inhalation exposure route, residential/ industrial/ commercial, VF (m³/kg)	$VF_{M-L} = \dfrac{Q}{C} \dfrac{T_{M-L}}{\rho_b d_s} \dfrac{(3.15 \cdot 10^7 \text{ s/yr})}{10^6 \text{ cm}^3/\text{m}^3}$ *Note*: This equation may be used when vertical thickness of contamination is known or can be estimated reliably.	S26
	Mass-limit volatilization factor for inhalation exposure route, construction worker, VF' (m³/kg)	$VF'_{M-L} = \dfrac{VF_{M-L}}{10}$	S27
	Mass-limit remediation objective for soil component of the groundwater ingestion exposure route (mg/kg)	$\dfrac{C_w I_{M-L} ED_{M-L}}{\rho_b d_s}$ *Note*: This equation may be used when vertical thickness is known or can be estimated reliably.	S28
	Equation for derivation of the soil saturation limit, C_{sat}	$C_{sat} = \dfrac{S}{\rho_b} \cdot (K_d \rho_b + \theta_w + H' \theta_a)$	S29

Source: IEPA (2002).

Table 11.7(b) **Explanation of tier 2 SSL parameters**

Symbol	Parameter	Units	Source[a]	Parameter Value(s)
AT	Averaging time for noncarcinogens in ingestion equation	yr		Residential = 6 Industrial/commercial = 25 Construction worker = 0.115
AT	Averaging time for noncarcinogens in inhalation equation	yr		Residential = 30 Industrial/commercial = 25 Construction worker = 0.115
AT_c	Averaging time for carcinogens	yr		70
BW	Body weight	kg	SSL	Residential = 15, noncarcinogens; 70, carcinogens Industrial/commercial = 70 Construction worker = 70
C_{sat}	Soil saturation concentration	mg/kg	Appendix A, Table A or equation S29 in Appendix C, Table A	Chemical-specific or calculated value
C_w	Target soil leachate concentration	mg/L	Equation S18 in Appendix C, Table A	Groundwater standard, health advisory concentration, or calculated value
d	Mixing zone depth	m	SSL or equation S25 in Appendix C, Table A	2 m or calculated value
d_a	Aquifer thickness	m	Field measurement	Site-specific
d_s	Depth of source (vertical thickness of contamination)	m	Field measurement or estimation	Site-specific
D_A	Apparent diffusivity	cm²/s	Equation S10 in Appendix C, Table A	Calculated value
D_i	Diffusivity in air	cm²/s	Appendix C, Table E	Chemical-specific
D_w	Diffusivity in water	cm²/s	Appendix C, Table E	Chemical-specific
DF	Dilution factor	unitless	Equation S22 in Appendix C, Table A	20 or calculated value
ED	Exposure duration for ingestion of carcinogens	yr		Industrial/commercial = 25 Construction worker = 1
ED	Exposure duration for inhalation of carcinogens	yr		Residential = 30 Industrial/commercial = 25 Construction worker = 1
ED	Exposure duration for ingestion of noncarcinogens	yr		Residential = 6 Industrial/commercial = 25 Construction worker = 1
ED	Exposure duration for inhalation of noncarcinogens	yr		Residential = 30 Industrial/commercial = 25 Construction worker = 1

Symbol	Description	Units	Source	Value
ED	Exposure duration for the direct ingestion of groundwater	yr		Residential = 30 Industrial/commercial = 25 Construction worker = 1
ED_{M-L}	Exposure duration for migration to groundwater mass-limit equation S28	yr	SSL	70
EF	Exposure frequency	d/yr		Residential = 350 Industrial/commercial = 250 Construction worker = 30
$F(x)$	Function dependent on U_m/U_t	unitless	SSL	0.194
f_{oc}	Organic carbon content of soil	g/g	SSL or field measurement (see Appendix C, Table F)	Surface soil = 0.006 Subsurface soil = 0.002, or Site-specific
GW_{obj}	Groundwater remediation objective	mg/L	Appendix B, Table E, 35 IAC 620, Subpart F, or equation S23 in Appendix C, Table A	Chemical-specific or calculated
H'	Henry's law constant	unitless	Appendix C, Table E	Chemical-specific
i	Hydraulic gradient	m/m	Field measurement (see Appendix C, Table F)	Site-specific
I	Infiltration rate	m/yr	SSL	0.3
I_{M-L}	Infiltration rate for migration to groundwater mass-limit equation S28	m/yr	SSL	0.18
$IF_{soil\text{-}adj}$ (residential)	Age-adjusted soil ingestion factor for carcinogens	mg · yr/(kg · day)	SSL	114
IR_{soil}	Soil ingestion rate	mg/day		Residential = 200 Industrial/commercial = 50 Construction worker = 480
IR_W	Daily water Ingestion rate	L/day		Residential = 2 Industrial/commercial = 1
K	Aquifer hydraulic conductivity	m/yr	Field measurement (see Appendix C, Table F)	Site-specific
K_d (non-ionizing organics)	Soil-water partition coefficient	cm³/g or L/kg	Equation S19 in Appendix C, Table A	Calculated value
K_d (ionizing organics)	Soil-water partition coefficient	cm³/g or L/kg	Equation S19 in Appendix C, Table A	Chemical and pH-specific (see Appendix C, Table I)
K_d (Inorganics)	Soil-water partition coefficient	cm³/g or L/kg	Appendix C, Table J	Chemical and pH-specific
K_{oc}	Organic carbon partition coefficient	cm³/g or L/kg	Appendix C, Table E or Appendix C, Table I	Chemical-specific
K_s	Saturated hydraulic conductivity	m/yr	Appendix C, Table K Appendix C, Illustration C	Site-specific

Table 11.7(b) (*continued*)

Symbol	Parameter	Units	Source[a]	Parameter Value(s)
L	Source length parallel to groundwater flow	m	Field measurement	Site-specific
PEF	Particulate emission factor	m³/kg	SSL or equation S15 in Appendix C, Table A	Residential = 1.32×10^9 or site-specific; Industrial/commercial = 1.24×10^9 or site-specific
PEF′	Particulate emission factor adjusted for agitation (construction worker)	m³/kg	Equation S16 in Appendix C, Table A using PEF (industrial/commercial)	1.24×10^8 or site-specific
Q/C (used in VF equations)	Inverse of the mean concentration at the center of a square source	(g/m²·s)/(kg/m³)	Appendix C, Table H	Residential = 68.81; Industrial/commercial = 85.81; Construction worker = 85.81
Q/C (used in PEF equations)	Inverse of the mean concentration at the center of a square source	(g/m²·s)/(kg/m³)	SSL or Appendix C, Table H	Residential = 90.80; Industrial/commercial = 85.81; Construction worker = 85.81
RfC	Inhalation reference concentration	mg/m³	IEPA (IRIS/HEAST[a])	Toxicological-specific (*Note:* For construction workers, use subchronic reference concentrations)
RfD_0	Oral reference dose	mg/kg·day	IEPA (IRIS/HEAST[a])	Toxicological-Specific (*Note:* for construction worker, use subchronic reference doses)
S	Solubility in water	mg/L	Appendix C, Table E	Chemical-specific
SF_0	Oral slope factor	mg/(kg·day)⁻¹	IEPA (IRIS/HEAST[a])	Toxicological-specific
T	Exposure interval	s		Residential = 9.5×10^8; Industrial/commercial = 7.9×10^8; Construction worker = 3.6×10^6
T_{M-L}	Exposure interval for mass-limit volatilization factor equation S26	yr	SSL	30
THQ	Target Hazard Quotient	unitless	SSL	1
TR	Target cancer risk	unitless	SSL	Residential = 10^{-6} at the point of human exposure; Industrial/commercial = 10^{-6} at the point of human exposure; Construction worker = 10^{-6} at the point of human exposure
U_m	Mean annual windspeed	m/s	SSL	4.69
URF	Inhalation unit risk factor	(µg/m³)⁻¹	IEPA (IRIS/HEAST[a])	Toxicological-specific
U_t	Equivalent threshold value of windspeed at 7 m	m/s	SSL	11.32
V	Fraction of vegetative cover	unitless	SSL or field measurement	0.5 or site-specific

Symbol	Description	Units	Source	Value
VF	Volatilization factor	m^3/kg	Equation S8 in Appendix C, Table A	Calculated value
VF'	Volatilization factor adjusted for agitation	m^3/kg	Equation S9 in Appendix C, Table A	Calculated value
$VF_{M\text{-}L}$	Mass-limit volatilization factor	m^3/kg	Equation S26 in Appendix C, Table A	Calculated value
$VF'_{M\text{-}L}$	Mass-limit volatilization factor adjusted for agitation	m^3/kg	Equation S27 in Appendix C, Table A	Calculated value
η	Total soil porosity	L_{pore}/L_{soil}	SSL or equation S24 in Appendix C, Table A	0.43, or Gravel = 0.25 Sand = 0.32 Silt = 0.40 Clay = 0.36, or calculated value
θ_a	Air-filled soil porosity	L_{air}/L_{soil}	SSL or equation S21 in Appendix C, Table A	Surface soil (top 1 m) = 0.28 Subsurface soil (below 1 m) = 0.13, or Gravel = 0.05 Sand = 0.14 Silt = 0.24 Clay = 0.19, or Calculated value
θ_w	Water-filled soil porosity	L_{water}/L_{soil}	SSL or equation S20 in Appendix C, Table A	Surface soil (top 1 m) = 0.15 Subsurface soil (below 1 m) = 0.30, or Gravel = 0.20 Sand = 0.18 Silt = 0.16 Clay = 0.17, or Calculated value
ρ_b	Dry soil bulk density	kg/L or g/cm^3	SSL or field measurement (see Appendix C, Table F)	1.5, or Gravel = 2.0 Sand = 1.8 Silt = 1.6 Clay = 1.7, or Site-specific
ρ_s	Soil particle density	g/cm^3	SSL or field measurement (see Appendix C, Table F)	2.65, or site-specific
ρ_w	Water density	g/cm^3	SSL	1
$1/(2b+3)$	Exponential in equation S20	unitless	Appendix C, Table K Appendix C, Illustration C	Site-specific

[a]Appendices and tables referenced here can be found in 35 Illinois Administrative Code 620.

Source: IEPA (20020.

EXAMPLE 11.7

At a leaking underground storage tank (LUST) site, the following contaminants are found in the soil and the groundwater. The site is located in a residential area and the groundwater is the sole drinking source of drinking water for that area. Perform an Illinois TACO tier 1 evaluation to determine if remediation is required, and if so, what the remedial objectives are.

Contaminant	Soil Concentration (mg/kg)	Groundwater Contamination (mg/L)
Benzene	8	0.2
Toluene	20	0.7
Ethyl benzene	200	10
Xylene (total)	500	20
Chrysene	250	50
Benzo[a]pyrene	0.1	0.2

Solution:

Using IL-TACO tier 1 look-up Tables 11.5, soil remediation objectives for residential properties are:

	Exposure Route (mg/kg)		
Contaminant	Ingestion	Inhalation	Migration to Groundwater Class I
Benzene	12	0.8	0.03
Toluene	16,000	650	12
Ethyl benzene	7,800	400	13
Xylene	160,000	320	150
Chrysene	88	—	160
Benzo[a]pyrene	0.09	—	8

Based on the above, the critical exposure routes and the corresponding contaminant concentrations are:

Contaminant	Critical Exposure Routes	Concentrations (mg/kg)
Benzene	Migration to groundwater	0.03
Toluene	Migration to groundwater	12
Ethyl benzene	Migration to groundwater	13
Xylene	Migration to groundwater	150
Chrysene	Ingestion	88
Benzo[a]pyrene	Ingestion	0.09

On comparing the contaminant concentrations above with the concentrations in the soil, we can conclude that all contaminant concentrations exceed tier 1 levels. Therefore, remediation is required to reduce the contaminant concentrations lower than the tier 1 remedial objectives.

Using IL-TACO tier 1 look-up Table 11.6, the class I groundwater remediation objectives are:

Contaminant	Groundwater (Class I) Remedial Objectives (mg/L)
Benzene	0.005
Toluene	1
Ethyl benzene	0.7
Xylene (total)	10
Chrysene	0.0015
Benzo[a]pyrene	0.0002

On comparing the contaminant concentrations above with the concentrations in the site groundwater, we can conclude that all the contaminant concentrations exceed tier 1 groundwater remedial objectives. Therefore, remediation is needed to meet the groundwater remediation objectives. This shows that both the site soil and the groundwater require remediation. (If desired, tier 2 evaluation may be performed to determine cleanup objectives based on the site-specific conditions.)

EXAMPLE 11.8

At a LUST site located in a residential area, contaminants found in the soil are given below. The groundwater at that site is classified as class I. Perform Illinois TACO tier 1 and tier 2 evaluations to determine if remediation is needed, and if so, what the remedial objectives are.

Contaminant	Concentration in Soil (mg/kg)
Benzene	0.6
Toluene	7
Ethyl benzene	10
Chrysene	50

Solution:

Tier 1 evaluation: According to the Illinois TACO Tier 1 look-up Table 11.5, soil remediation objectives are:

Contaminant	Exposure Route (mg/kg)		
	Ingestion	Inhalation	Migration to Groundwater Class I
Benzene	12	0.8	0.03
Toluene	100	650	12
Ethyl Benzene	7800	400	13
Chrysene	88	—	160

Based on the tier 1 levels, critical exposure routes and the corresponding contaminant concentrations are:

Contaminant	Critical Exposure Routes	Concentration in Soil (mg/kg)
Benzene	Migration to groundwater	0.03
Toluene	Migration to groundwater	12
Ethyl benzene	Migration to groundwater	13
Chrysene	Ingestion	88

On comparing the concentrations of the contaminants above with the concentrations in the soil, it can be concluded that all the contaminants except benzene have lower concentrations, and hence remediation is not required for the remaining contaminants except benzene.

Tier 2 evaluation: We can now calculate the tier 2 objectives for benzene (migration to groundwater exposure route). IEPA recommends default values for all the parameters except the groundwater objective (GW_{obj}) and the organic carbon fraction in the soil (f_{oc}). Select $GW_{obj} = 0.005$ mg/L (as per IEPA, 2002). f_{oc} is the organic carbon fraction in the soil, and it depends on the soil that we are dealing with. Let us assume that f_{oc} for this procedure as 0.05g/g. Calculate soil leachate concentration using

$$C_w = DF \times GW_{obj} = 20 \times 0.005 = 0.1 \text{ mg/L}$$

where DF is the dilution factor; IEPA recommends a value of 20. Calculate the soil water partition coefficient:

$$K_d = K_{oc} \times f_{oc} \quad (K_{oc} = 58.9 \text{ L/kg for benzene})$$

$$= 58.9 \times 0.05 = 2.945$$

The remedial objective for benzene is determined using equation S17 in Table 11.7:

$$\text{Remedial objectives for benzene} = C_w \left(K_d + \frac{\theta_1 + \theta_R \times H'}{\rho_b} \right)$$

$$= 0.1 \left(2.945 + \frac{0.3 + 0.13 \times 0.22}{1.5} \right)$$

$$= 0.316 \text{ mg/kg}$$

On comparing this value with the benzene concentration in the site soil (0.6 mg/kg) it can be concluded that remediation is needed to meet the remedial objective. (A remedial action can be selected and implemented to achieve the remedial objectives, or tier 3 evaluation can be made with site-specific data to define the remedial objective.)

cer risk of less than 10^{-6}, which implies one in 1 million probability, is considered to be acceptable. In some cases, though, monitoring of the site contaminants may be required to verify that these results are valid. Corrective action is required when the risk posed by the site contamination is deemed unacceptable. It is most beneficial to compose a remedial strategy when the need for corrective action is confirmed. A remedial strategy is essential to ensure that the intended remedial method complies with all technical, economic, and regulatory considerations.

Different remedial options are shown in Figure 11.4. The first option, *natural attenuation,* also known as *monitored natural attenuation* (MNA), is used when there is a potential for the contaminants, mainly organic chemicals, to be degraded by the naturally occurring microorganisms. Many aspects considered in

bioremediation also require consideration in such studies. More details on MNA and bioremediation are given in Chapters 13 and 14. The second remedial option involves containing the waste so that it will not spread and isolating the waste to prevent human exposure. This can be achieved by excavating the contaminated media and disposing of it in either a landfill if it is a solid (in a physical sense) waste or in a surface impoundment if it is a liquid waste. The details on landfills and surface impoundments are presented in Chapters 15 through 22. Instead of landfills and surface impoundments, one may opt to leave the contaminated media in the subsurface and construct in-situ barriers to enclose the contaminated zone. Passive in-situ barriers (horizontal barriers, vertical barriers, and caps) and active in-situ barriers (pumping and subsurface drains) can be used. For a given site, it may be required

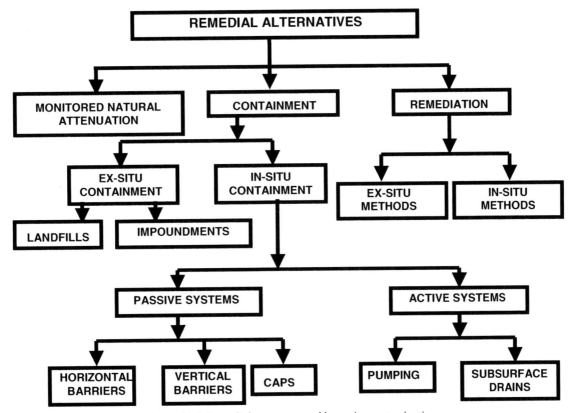

Figure 11.4 *Remedial strategy to address site contamination.*

to use one or a combination of these barriers to contain the waste effectively. The details on in-situ barriers are provided in Chapter 12.

Decontamination or remedial methods aim at extracting, immobilizing, destroying, or detoxifying the contaminants in soils and groundwater. Remediation methods are generally divided into two categories: in-situ and ex-situ. *In-situ methods* treat contaminated soils and/or groundwater in place so that the contaminated soils are not excavated and the groundwater is not extracted. In-situ methods offer many advantages over ex-situ methods, such as lower costs, less disruption to the site, and reduced worker exposure to hazardous materials. Successful implementation of in-situ systems, however, requires adequate understanding of the subsurface conditions.

Conversely, *ex-situ methods* treat excavated contaminated soils and/or extracted groundwater. Aboveground treatment is completed either on-site or off-site, depending on the site-specific conditions. Urban site conditions such as close neighboring buildings or narrow streets often limit the use of on-site treatment facilities for excavated soil and groundwater. Therefore, off-site treatment, which requires transportation of the contaminated materials to a treatment facility, is often a more likely option at a compact urban site. If desired, the treated soil and groundwater may also be returned to the site. The advantage of ex-situ methods is that the contaminated soil and groundwater are subjected to treatment methods without due consideration to subsurface variable conditions. Greater control over remediation operation allows implementation of a wide variety of remediation methods.

Both in-situ and ex-situ remediation methods are based on physicochemical, biological, or thermal manipulations of contaminants in soil and groundwater. Several remediation methods are available and reported in the literature. The different methods to remediate contaminated soils and groundwater are described in Chapters 13 and 14.

QUESTIONS/PROBLEMS

11.1. What is the difference between human risk and ecological risk?

11.2. What are major uncertainties in the USEPA risk assessment procedure?

11.3. What are the major limitations of the ASTM RBCA procedure?

11.4. Explain the assumptions made in the development of example RBSL look-up Table 11.2.

11.5. Explain the assumptions made in the development of example SSTL Table 11.4.

11.6. Redo Example 11.7 assuming that the site is located in industrial/commercial setting with class II groundwater.

11.7. Soil samples in a borehole had the following benzene concentrations:

Depth (ft)	Concentrations (mg/kg)
0.5	ND
2.5	0.1
4.5	0.2
6.5	0.15
8.5	0.1
10.5	ND

where ND indicates "not detected" (detection limit = 0.04 mg/kg). Using Illinois TACO tier 1, determine if remediation is required, and if so, what the remedial objectives are.

11.8. An owner has conducted a site characterization in preparation for a property transfer in the Chicago area. Soil samples were analyzed for volatiles, semivolatiles, and metals. The volatile and semivolatile results were all below the acceptable detection limits. The following metals were reported and the soil pH was found to be 7.3. Groundwater at the site is classified as class II.

Lead (mg/kg)	Cadmium (mg/kg)	Lead (mg/kg)	Cadmium (mg/kg)
52	17	28	1341
43	42	74	243
117	177	90	100
63	16	225	19
311	77	56	90

Based on the Illinois TACO, determine residential soil clean up objectives.

11.9. Contact the environmental agency of the state that you live in and find out if the state has adopted a specific risk assessment procedure to use for contaminated sites. Explain the procedure, if available.

11.10. At an industrial site, soil is contaminated with heavy metals shown below. Groundwater is classified as class I. The pH of the subsurface soil is 7.5. Perform an Illinois TACO tier 1 evaluation to determine if remediation is required, and if so, what the remedial objectives would be.

Contaminant	Concentration in Soil (mg/kg)
Barium	1500
Chromium	1000
Lead	500

REFERENCES

ASTM (American Society for Testing and Materials), *Standard Guide for Risk-Based Corrective Action Applied at Petroleum Release Sites*, E1739, ASTM, West Conshohocken, PA, 1995.

IEPA (Illinois Environmental Protection Agency), 35 Illinois Administrative Code, Part 742, *Tiered Approach to Corrective Action Objectives (TACO)*, IEPA, Springfield, IL, 2002.

Stern, P. C., and Fineberg, H. V. (eds.), *Understanding Risk: Informing Decisions in a Democratic Society*, National Academy of Sciences, National Academy Press, Washington, DC, 1996.

USDOE (U.S. Department of Energy), *CERCLA Baseline Risk Assessment: Reference Manual for Toxicity and Exposure Assessment and Risk Characterization*, Office of Environmental Policy and Assistance, USDOE, Washington, DC, 1995.

USEPA (U.S. Environmental Protection Agency), *Risk Assessment Guidance for Superfund*, Vol. I, *Human Health Evaluation Manual*, Part A, EPA/540/1-89/002, Office of Emergency and Remedial Response, USEPA, Washington, DC, 1989.

———, *Standard Default Exposure Factors: Supplemental Guidance to Human Health Evaluation Manual*, OSWER-9285.6-03, Office of Solid Waste and Emergency Response, USEPA, Washington, DC, 1991a.

———, *Role of the Baseline Risk Assessment in Superfund Remedy Selection Decisions*, OSWER-9355.0-30, Office of Solid Waste and Emergency Response, USEPA, Washington, DC, 1991b.

———, *Understanding Superfund Risk Assessment*, OSWER-9285-7-06FS, Office of Solid Waste and Emergency Response, USEPA, Washington, DC, 1992.

————, *Ecological Risk Assessment Guidance for Superfund: Process for Designing and Conducting Ecological Risk Assessments (Interim Final)*, EPA/R-97/006, Office of Solid Waste and Emergency Response, USEPA Washington, DC, 1997.

————, Integrated Risk Information System (IRIS) database, *www.epa.gov/iris*, 1999.

————, *Risk Assessment Guidance for Superfund*, Vol. III, Part A, *Process for Conducting Probabilistic Risk Assessment*, EPA/540/R-02/002, Office of Emergency and Remedial Response, USEPA, Washington, DC, 2001.

12

IN-SITU WASTE CONTAINMENT

12.1 INTRODUCTION

For sites where the subsurface soils and groundwater have been contaminated with toxic chemicals due to improper past waste disposal practices, short-term in-situ containment may be required as an interim remedial action to prevent the spreading of contaminants, thereby reducing the risk to public health and the environment. Subsequently in the future, long-term in-situ or ex-situ technologies may be applied to clean up the contamination. In some situations, particularly at large waste disposal sites such as abandoned landfills and mine tailing disposal sites, in-situ containment may be used as a permanent remedial action because other types of cleanup technologies are cost-prohibitive and/or impractical.

In-situ waste containment methods can be grouped in two categories: passive systems and active systems. As displayed in Figure 12.1, *passive systems* involve installation of physical low-permeable barriers all around the waste to encompass the entire contaminated zone in order to effectively reduce the potential for spreading of contaminants. These systems can be used in both vadose and saturated zones of the subsurface, and they include one or a combination of the following components:

- Vertical barriers to limit lateral spread of the contaminant
- Bottom barriers to limit downward migration of leachates
- Surface cap or cover to minimize infiltration of surface water and contaminants and to control the possible release of volatile components from the waste

Active systems are generally used in the saturated zone and involve installation of either pumping wells or subsurface drains to manipulate the hydraulic gradients, thereby containing the groundwater contamination. In some situations, a combination of passive and active systems may be used to effectively contain the waste or groundwater plume.

In this chapter we describe various in-situ waste containment systems. First, passive waste containment systems that include vertical barriers, horizontal barriers, and surface caps are described. Then active waste containment systems that include groundwater pumping and subsurface drains are explained.

12.2 VERTICAL BARRIERS

12.2.1 Description

Vertical barriers, also known as *vertical cutoff barriers, vertical cutoff walls,* or *barrier walls,* function in the subsurface to contain contaminants and/or to redirect groundwater flow by blocking lateral flows. Vertical barriers can be installed in several configurations. A vertical barrier could extend to the top of a natural or human-made bottom barrier, if used, as shown in Figure 12.1. Usually, vertical barriers are embedded or keyed into a low-permeability formation (aquitard; Figure 12.2). If the pollutants are predominantly floating at the water table (e.g., LNAPLs), a hanging wall may be used (Figure 12.3). A hanging wall may extend beneath the water table to encompass the contamination plume, and interior pumping may be required to maintain an inward gradient.

Horizontal configuration of a vertical barriers can be circumferential, upgradient, or downgradient, as

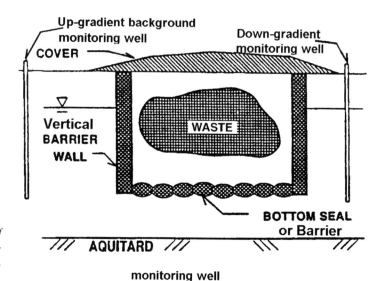

Figure 12.1 *General configuration of waste containment using passive containment systems. (Modified from Mitchell, 1994.)*

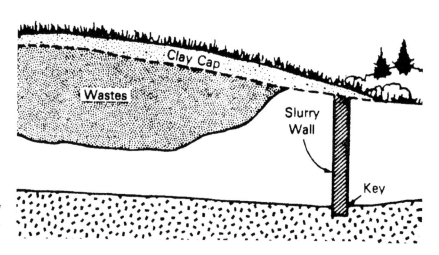

Figure 12.2 *Keyed-in slurry barrier. (From USEPA, 1984a.)*

shown in Figures 12.4, 12.5 and 12.6, respectively. With circumferential configuration, the vertical barrier completely surrounds the waste site. This particular layout is preferred if groundwater flow conditions are uncertain or unknown. However, upgradient and downgradient configurations are also commonly used. An upgradient configuration is used to prevent groundwater flow through the contaminant and consequent contaminant spreading. A vertical barrier with downgradient configuration, in conjunction with groundwater extraction wells, is used to allow groundwater flowing through the contaminated zone to flush contaminants from the site. Table 12.1 provides a summary of various barrier configurations.

Totally impervious barriers are impossible to construct. However, the barriers can be built to meet the low-hydraulic-conductivity requirements of the governing regulatory agencies. They can be made resistant to chemical transport by advection (flow, due to hydraulic gradient, of dissolved and suspended materials

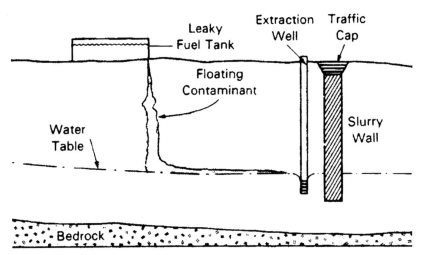

Figure 12.3 *Hanging slurry barrier. (From USEPA, 1984a.)*

within the groundwater) as well as diffusion (chemical flow under a chemical concentration gradient). Low hydraulic conductivity causes a low flow rate of water through the material, causing low advective transport of the contaminants. However, chemical diffusion of contaminants becomes dominant in such materials, particularly if hydraulic conductivity is less than approximately 1×10^{-6} cm/s. More details on the contaminant transport are given in Chapter 8.

In the selection of a suitable type of vertical barrier, one should consider wall function and configuration, coupled with site and subsurface conditions. The following are the most common type of vertical barriers: (1) compacted clay barriers, (2) slurry trench barriers, (3) grouted barriers, (4) mixed-in-place barriers, and (5) steel sheet pile barriers. These vertical barriers are discussed briefly in the following sections. Several excellent publications are available on vertical barriers (e.g., USEPA, 1984a; Johnson et al., 1985; Paul et al., 1992; Sharma and Lewis, 1994; Xanthakos, 1994; Rumer and Ryan, 1995; Rumer and Mitchell, 1996) and should be consulted for further details.

12.2.2 Compacted Clay Barriers

Compacted clay barriers are formed by excavation of a trench which is then backfilled with clay soil com-

pacted in thin layers to a high density (Figure 12.7). Clay sources should be characterized in detail similar to clay liners for landfills, as described in Chapter 17. The selected compaction criteria (moisture content and compaction effort) should provide the lowest permeability. Clay barriers are suitable when there is a shallow depth to the aquitard layer, a low flow rate of water into the open trench, and the ground can be excavated without collapse of the sidewalls. In addition, an adequate supply of suitable clay must be readily available. An advantage of these barriers is that the construction can be monitored and controlled as clay is compacted into the trench; however, compacted clay barriers are difficult to construct to depths of more than a few feet, depending on soil and water conditions.

12.2.3 Slurry Trench Barriers

Slurry trench barriers, also known as *slurry trench cutoff walls* or *slurry walls,* are the most commonly used vertical barriers for waste containment at hazardous waste sites (USEPA, 1992; Evans, 1993; USACE, 1996). They are constructed by excavating a narrow vertical trench, typically 2 to 4 ft wide. As excavation proceeds, the trench is filled with a slurry that stabilizes the walls of the trench, thereby preventing collapse. The slurry penetrates the surrounding permeable

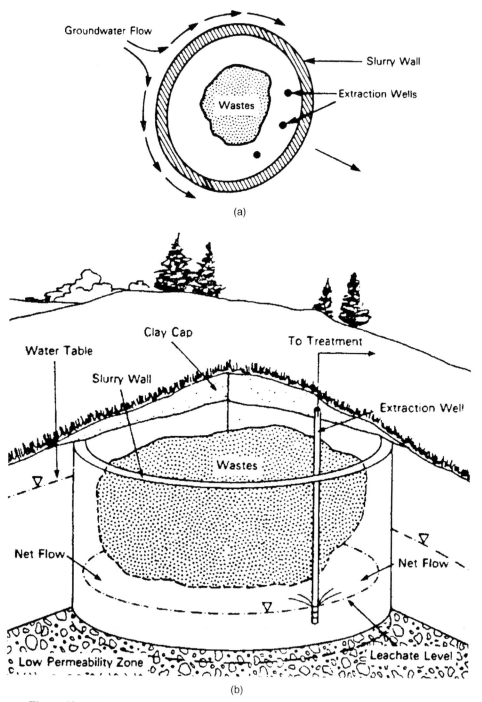

Figure 12.4 *Circumferential barrier:* (a) *plan;* (b) *cross section.* (*From USEPA, 1984a.*)

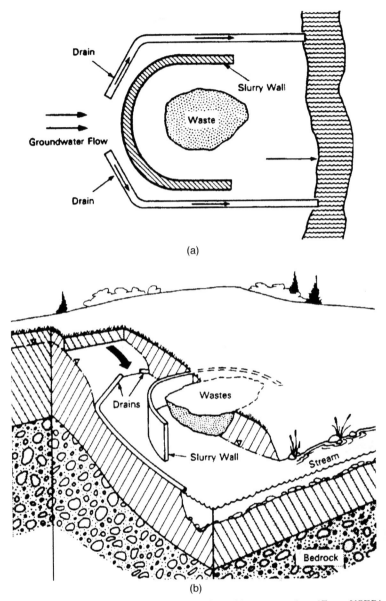

Figure 12.5 *Upgradient barrier with drains:* (a) *plan;* (b) *cross section.* (*From USEPA, 1984a.*)

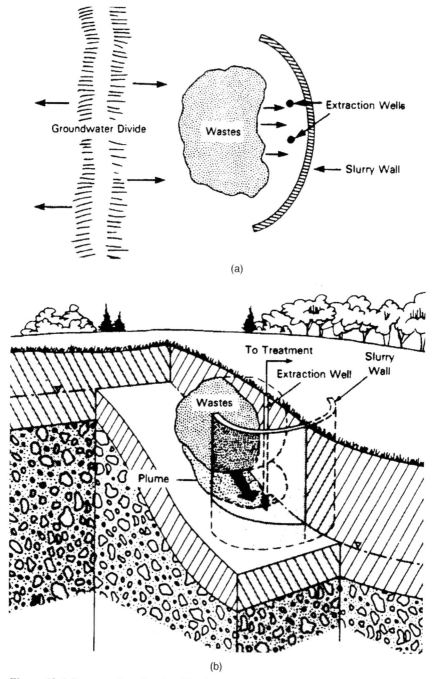

Figure 12.6 *Downgradient barrier:* (a) *plan;* (b) *cross section. (From USEPA, 1984a.)*

Table 12.1 **Summary of slurry wall configurations**

Vertical Configuration	Horizontal Configuration		
	Circumferential	*Upgradient*	*Downgradient*
Keyed-in	Most common but expensive Most complete containment Vastly reduced leachate generation	Not common Used to divert groundwater around site in steep-gradient situations Can reduce leachate generation Compatibility not critical	Used to capture miscible or sinking contaminants for treatment Inflow not restricted; may raise water table Compatibility very important
Hanging	Used for floating contaminants moving in more than one direction (such as on a groundwater divide)	Very rare May temporarily lower water table behind it Can stagnate leachate but not halt flow	Used to capture floating contaminants for treatment Inflow not restricted; may raise water table Compatibility very important

Source: USEPA (1984a).

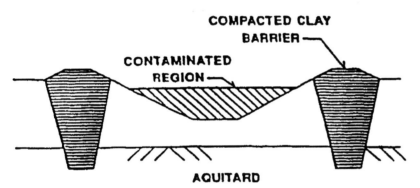

Figure 12.7 Compacted clay barrier. (From Mitchell and van Court, 1992.)

soils, creating a filter cake on the trench walls that seals the soil formations, prevents slurry loss, and contributes to the low permeability of the completed barrier. This narrow trench is then backfilled with a slurry mixture. Depending on the backfill, various slurry barriers can be created.

The design of slurry trench barriers should consider site conditions, barrier requirements to meet design criteria, and general construction requirements. Predesign investigations should include a thorough evaluation of site conditions, including the (1) site geology and hydrogeology, (2) nature and extent of contamination, and (3) geotechnical properties of subsurface materials. Soil borings should be drilled along the potential alignment of the barrier, and samples should be collected

for geotechnical information and contaminant analyses. Groundwater modeling may also be necessary. Table 12.2 lists the data needed for design of slurry trench barriers, along with its purpose and potential sources.

Site-specific conditions, including hydrogeology, chemical compatibility, and permeability, are considered in selecting the width and depth of slurry trench. Slurry trenches can typically be excavated to depths of up to 50 ft using backhoes. Deeper slurry trenches can be excavated using a crane-mounted dragline or clamshell bucket. If it is necessary to install a slurry barrier into bedrock, drilling or blasting may be required to excavate the rock. Special blasting techniques will be required to maintain the integrity of the bedrock. Var-

Table 12.2 **Data requirements for slurry cutoff trench/wall**

Data Description	Purpose(s)	Source(s)/Method(s)
Site accessibility	Select wall type	Site inspection
Topography	Soil–bentonite walls require large land area with relatively flat topography	USGS topography map, site inspection, site-specific topographic/contour maps, water-level maps
Depth to continuous impermeable strata or competent bedrock	Selection of keyed-in or hanging wall	Borings, geophysical survey, bedrock and surficial geology maps
Heterogenity of subsurface formation	Selection of wall type; excavated material may not be appropriate to mix	Surficial geology maps, test pits, soil borings, geophysical survey
Vertical and horizontal hydraulic conductivity of confining layer	Determine suitability of layer as a key	Slug tests, laboratory tests
Excavated soil type	Suitability for use as trench backfill material	Gradation analyses, permeability tests
Degree of bedrock fracturing	Evaluate potential for contaminants to migrate underneath the key	Rock cores, boring logs, geology maps
Groundwater depth; rate and direction of flow	Establish potential for installation of hanging wall with design of inward gradient for pump-and-treat scheme	Existing hydrogeologic maps, boring logs, observation wells, piezometers
Hydraulic conductivity of contaminated soil	Evaluate effectiveness of slurry wall pump-and-treat systems	Pumping tests, slug tests
Soil chemistry	Cement and bentonite can be modified to accommodate chemistry	Soil sampling and analysis
Chemistry of waste and groundwater	Compatibility testing of cement or bentonite and wall material with contaminated groundwater and soil	Groundwater sampling and chemical analysis, filtrate loss, free swell, and permeability testing

Source: (USEPA, 1991).

ious types of slurry trench excavation equipment are compared in Table 12.3.

The slurry, generally a mixture of bentonite and water, or cement, bentonite, and water, is selected to ensure trench stability during excavation. The slurry performs two separate functions. First, it maintains the stability of trench walls by imparting hydrostatic pressure greater than that of groundwater on either side. Second, the slurry deposits a dense, very low permeability bentonite filter cake on the walls of the trench. As shown in Figure 12.8, the depth of the slurry in the trench is maintained at a level above the existing water table so that a net outward hydrostatic pressure acts against the trench walls. The slurry should have a minimum unit weight of 1025 kg/m^3.

The permanent barrier is then constructed by backfilling the trench with a low-permeability backfill, dis-

placing the slurry. For proper displacement of slurry by backfill material, the unit weight of the backfill material should be 240 kg/m^3 (15 lb/ft^3) greater than that of the slurry. Backfilling of the trench is often accomplished with the same equipment as that used to excavate the trench. A bulldozer can be used to mix the soil with the slurry alongside the trench as well as to backfill the portion of the trench. Care must be taken to ensure that no pockets of slurry are trapped during backfilling, as these can greatly reduce the barrier's effectiveness.

Many design issues need to be considered for slurry trench barriers. The slurry composition must be selected to ensure the stability of the trench. Sharma and Lewis (1994) and Xanthakos (1994) reported a general method of slurry trench stability analysis on the basis of limit theory. Figure 12.9 shows the active wedge

Table 12.3 **Excavation equipment used for slurry trench construction**

Type	Trench Width (ft)	Trench Depth (ft)	Comments
Standard backhoe	1–5	50	Most rapid and least costly excavation method
Modified backhoe	2–5	80	Uses an extended dipper stick, modified engine, and counterweighted frame; is also rapid and relatively low cost
Clamshell	1–5	>150	Attached to a Kelly bar or crane; needs ≥ 18-ton crane; can be mechanical or hydraulic
Dragline	4–10	>120	Used primarily for wide, deep SB trenches
Rotary drill, percussion drill, or large chisel	—	—	Used to break up boulders and to key into hard rock aquicludes; can slow construction and result in irregular trench walls

Source: (USEPA, 1984a).

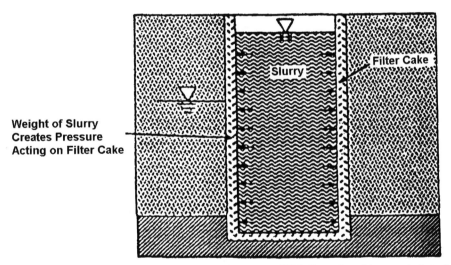

Figure 12.8 Slurry trench stability. (From Daniel and Koerner, 1995. Reproduced by permission of ASCE.)

ABC and forces acting on it that must satisfy equilibrium condition to ensure trench stability. A uniform surcharge pressure (q) is assumed to act at the ground surface. The critical depth (H_{cr}) to which a trench can be excavated in clay ($\phi = 0$) of unit weight γ and cohesion c, filled with slurry of unit weight γ_f, is given by

$$H_{cr} = \frac{4c - 2q}{\gamma - \gamma_f} \quad (12.1)$$

The critical depth calculation above assumes that there is no slurry loss to the ground through the soil–slurry interface and there is a very low permeability barrier at the interface so that the slurry can exert full hydro-

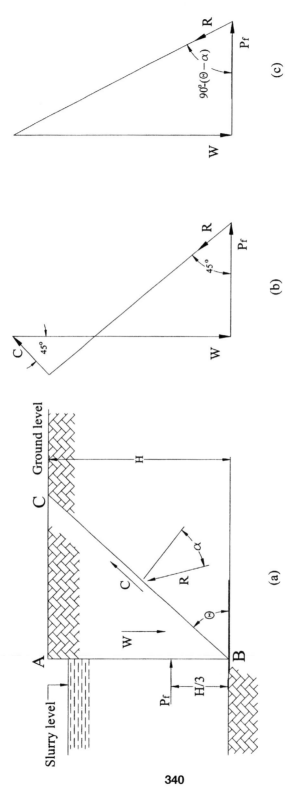

Figure 12.9 *Stability of a slurry trench:* (a) *section through trench;* (b) *force polygon for purely cohesive soil;* (c) *force triangle for purely cohesionless sand.* *(From Xanthakos, 1992.)*

static thrust. It also assumes that the excavation is temporary and remains open for a short period of time. In case of trenches in dry cohesionless sand ($c = 0$), the factor of safety (FS) against sliding can be given by

$$FS = \frac{2\sqrt{\gamma\gamma_f}\tan\phi}{\gamma - \gamma_f} \qquad (12.2)$$

For saturated sand with groundwater near the ground surface, expression (12.2) is modified as

$$FS = \frac{2\sqrt{\gamma'\gamma_f'}\tan\phi'}{\gamma' - \gamma_f'} \qquad (12.3)$$

where $\gamma' = \gamma - \gamma_w$ and $\gamma_f' = \gamma_f - \gamma_w$. If there is a variation in slurry and groundwater levels as shown in Figure 12.10, the following expression can be obtained:

$$\gamma_f = \frac{\gamma(1 - m^2)K_a + \gamma'm^2K_a + \gamma_w m^2}{n} \qquad (12.4)$$

where

$$K_a = \tan^2\left(45 - \frac{\phi'}{2}\right) = \frac{\gamma_f - \gamma_w}{\gamma'}$$

and m and n are as defined in Figure 12.10.

The selection of backfill that ensures low permeability is the most important design parameter. The overall (or effective) permeability of slurry trench barriers depends on the backfill as well as the filter cake formed by the slurry. From Darcy's law and the equation of continuity, the horizontal flow rate (Q) through the barrier can be given by the following expression (D'Appolonia, 1980):

$$Q = Ki = K\frac{h}{2t_c + t_s} = K_c\frac{h_c}{2t_c} = K_s\frac{h_s}{t_s} \qquad (12.5)$$

where K is the hydraulic conductivity, t the thickness, and h the head loss $= h_c + h_s$ (the subscript c is for filter cake and s is for soil backfill). Equation (12.5) can be simplified with the knowledge that t_s is significantly larger than t_c, as follows:

$$K = \frac{t_s}{t_s/K_s + 2(t_c/K_c)} \qquad (12.6)$$

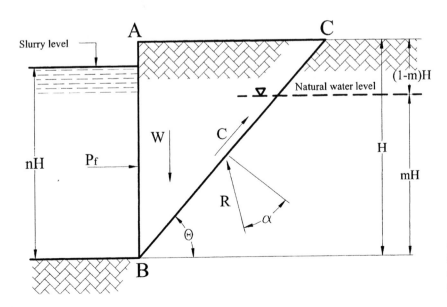

Figure 12.10 Stability of a trench for arbitrary slurry and natural water level. (From Xanthakos, 1992.)

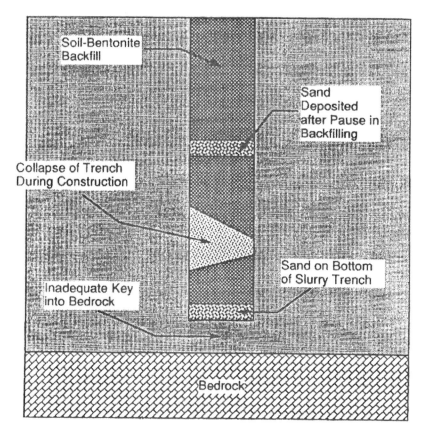

Figure 12.11 Problems caused improper construction of slurry trench. (From Daniel and Ko-erner, 1995. Reproduced by the permission of ASCE.)

Equation (12.6) assumes that the filter cake remains intact. This assumption is valid if the backfill material is of the proper consistency and does not have large particles that may scrape off the sidewalls. If the backfill contains 20% or more fines, water flow does not cause any filter cake loss due to seepage forces.

In addition to permeability, the chemical compatibility of barrier materials with site contaminants needs to be evaluated. Contaminant transport models may be used to evaluate barrier performance. One-dimensional contaminant transport models may be used for simple situations. Because of the low permeability of the barrier walls, advective transport is insignificant compared to diffusive flux when contaminant transport through the wall is considered. More details on transport modeling are given in Chapter 8.

A strict construction quality assurance and construction quality control (CQA/CQC) program is necessary to avoid potential problems such as those depicted in Figure 12.11. These problems include nonhomogeneous mixing of backfill, entrapment of slurry in backfill, segregation of particles in backfill, cave-in of trench sidewalls, and inadequate key-in to underlying low-permeable material or a horizontal barrier. In addition, slurry trench barriers may require postconstruction monitoring to ensure that the barrier is containing the waste effectively. This may include monitoring the ground movement at the barrier location, downgradient groundwater quality, and permeability of the barrier.

There are four types of slurry trench barriers, depending on the material used to backfill the trench: (1)

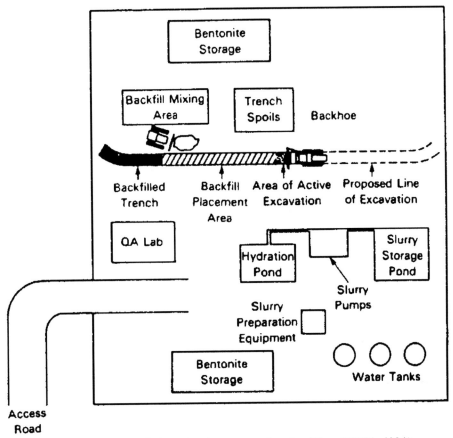

Figure 12.12 Typical slurry barrier construction site. (*From USEPA, 1984.*)

soil–bentonite barrier, (2) cement–bentonite barrier, (3) plastic concrete barrier, and (4) composite barrier. These slurry trench barriers are described next.

Soil–Bentonite Barriers. Soil–bentonite (SB) barriers are the most commonly used barriers at hazardous waste sites. The backfill used for SB walls is composed of well-graded soil, water, and bentonite. The bentonite content of the backfill can range from 1 to 5%. The hydraulic conductivity of this backfill generally ranges from 1×10^{-7} to 1×10^{-8} cm/s. SB barrier installation requires adequate space at the site for bentonite storage, slurry preparation equipment, water storage tanks, hydration pond, circulating pumps and slurry

storage pond, a trench spoils area where excavated soils are placed adjacent to the trench, and a backfill mixing area where bentonite is mixed with the backfill soils. A typical SB barrier construction site is shown in Figures 12.12 and 12.13. Surface grades should be less than 1%; otherwise, steeper slopes will cause slurry to flow and will result in lower slurry levels within the upslope portion of the trench, reducing trench stability. Site grades may need adjustment before construction of the trench begins.

Figure 12.13 shows a cross section of a SB slurry trench, including excavation and backfill operations. The trench is excavated using a backhoe, dragline, or clamshell, depending on depth requirements (Table

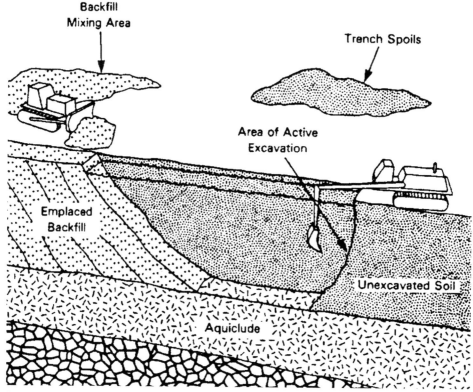

Figure 12.13 Cross section of slurry trench showing excavation and backfilling operations. (*From USEPA, 1984a.*)

12.3). During excavation, a bentonite–water slurry consisting of 4 to 6% bentonite by weight suspension in water is placed in the trench to support the sides of the trench. The slurry overcomes the active earth pressures in the soil adjacent to the trench and forms a filter cake along the trench walls, which stabilizes the trench. The bentonite slurry should be fully hydrated before being placed in the trench, typically for 12 to 24 hours. Temporary ponds or tanks are therefore necessary for hydration and storage of the slurry (see Figure 12.12). The slurry level in the trench is maintained at or near the top of the trench and above the surrounding groundwater table.

The slurry mix design should establish weight, viscosity, and filtrate loss estimates for the slurry. The weight of the slurry should be sufficient to overcome active earth pressures in order to maintain an open trench, but the slurry must also be light enough to be displaced by the soil–bentonite backfill. The quality of the water used to hydrate the bentonite should be determined. Compatibility testing of the water and bentonite may be required, although potable water supplies are usually acceptable.

Soil excavated from a trench may be used for trench backfill unless physical or contaminant characteristics render it unsuitable. The backfill design should specify criteria for unit weight, slump, gradation, and permeability. The backfill should consist of a well-graded mixture of coarse and fine materials. The relationship between the soil type and benotonite content on the permeability of soil–bentonite backfill is shown in Figure 12.14. For maximum permeability reduction, the

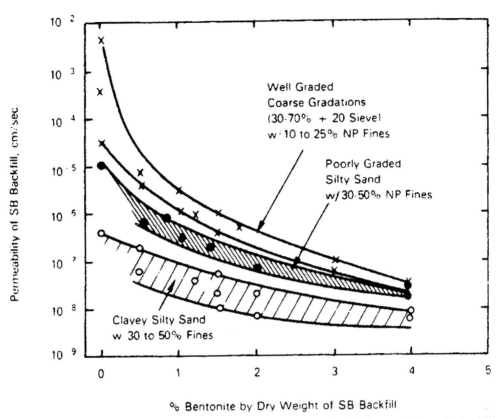

Figure 12.14 *Relationship between permeability and quantity of bentonite added to soil–bentonite backfill. (From USEPA, 1984a.)*

soil–bentonite mixture used for backfilling should contain 20 to 25% fines [soil particles that will pass a 0.075-mm (200-mesh) sieve]. To ensure long-term permeability reduction, as much as 40 to 45% fines may be required. In the event that the on-site soils are too coarse, imported fines or additional bentonite must be added. Plastic fines reduce permeability more than nonplastic fines, and the permeability of SB backfill is reduced with an increase in fines, as shown in Figure 12.15. The compatibility of contaminated groundwater and soil–bentonite backfill should be evaluated. Some contaminants, including some solvents and salts, have been shown to reverse the swelling characteristics of bentonite, which would result in a higher permeability

value. Table 12.4 shows the effects of leaching of various contaminants on soil–bentonite permeability. Bentonite–water slurry is mixed into the soil to form a mixture of soil–bentonite backfill with a slump of approximately 4 to 6 in. The unit weight of backfill should be at least 15 lb/ft^3 greater than the slurry's unit weight to ensure that the slurry will be displaced during backfill placement. Typically, the unit weight of the backfill ranges from 1442 to 1682 kg/m^3 (90 to 105 lb/ft^3). The backfill is then placed in the trench in such a way that it flows down a shallow slope (see Figure 12.13). The backfill should not be free-dropped into the trench. Appropriate placement of soil–bentonite backfill in the trench is necessary to dis-

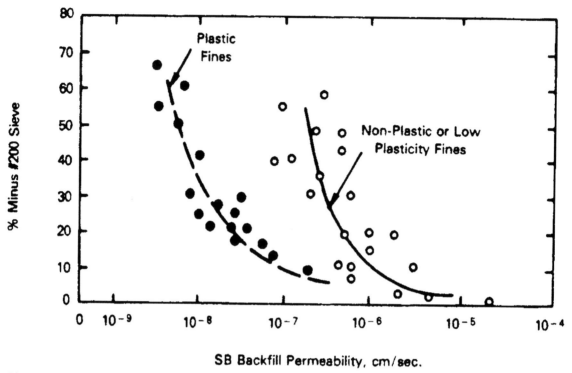

Figure 12.15 *Effect of plastic and nonplastic fines content on soil–bentonite backfill permeability. (From USEPA, 1984a.)*

place the bentonite–water slurry without entrapping lenses of slurry within the backfill.

Construction quality assurance and quality control are very critical for proper performance of SB barriers. This should entail:

- Sampling and testing of the *slurry* for unit weight, viscosity, and filtrate loss to ensure that these pa- rameters meet the design requirements. This should be done on slurry before as well as after placement in the trench. A mud balance, Marsh funnel, and filter press can be used to make field measurements of unit weight, viscosity, and fil- trate loss, respectively.

- Inspection of *trench geometry* width, depth, key penetration, verticality, continuity, stability, and bottom cleaning. The most critical factor is key penetration, which is checked by measuring the depth to the key stratum with a rigid probe.

- Sampling and testing of *backfill* for unit weight, slump, gradation, and permeability to ensure that it meets the design requirements.

- Careful *handling of contaminated backfill* or slurry and prevention of exposure of site workers or other receptors.

Cement–Bentonite   Barriers. Cement–bentonite (CB) barriers involve excavating a trench under a head of slurry composed of water, cement, and bentonite. The bentonite–water slurry is prepared and allowed to fully hydrate before portland cement is added. Once the cement has been added, the CB slurry is pumped to the trench. The CB slurry is left to harden in place, forming a hydraulic barrier. Generally, the CB slurries contain 4 to 7% bentonite, 8 to 25% cement, and 65 to 88% water (USEPA, 1984a).

A comparison of SB and CB barriers is shown in Table 12.5. The permeability of CB barriers ranges

Table 12.4 **Soil–bentonite permeability increases due to leaching with various pollutants**

Pollutant	Permeability Increase[a]
Ca^{2+} or Mg^{2+} at 1000 ppm	N
Ca^{2+} or Mg^{2+} at 10,000 ppm	M
NH_4NO_3 at 10,000 ppm	M
Acid (pH > 1)	N
Strong acid (pH < 1)	M/H*
Base (pH < 11)	N/M
Strong base (pH > 11)	M/H*
HCl (1%)	N
H_2SO_4 (1%)	N
HCl (5%)	M/H*
NaOH (1%)	M
$Ca(OH)_2$ (1%)	M
NaOH (5%)	M/H*
Benzene	N
Phenol solution	N
Seawater	N/M
Brine (SC = 1.2)	M
Acid mine drainage ($FeSO_4$, pH 3)	N
Lignin (in Ca^{2+} solution)	N
Organic residues from pesticide manufacture	N
Alcohol	M/H

Source: (USEPA, 1991).

[a]N, no significant effect, permeability increase by about a factor of 2 or less at steady state; M, moderate effect, permeability increase by factor of 2 to 5 at steady state; H, permeability increase by factor of 5 to 10; *, significant dissolution likely.

from 10^{-5} to 10^{-6} cm/s. The relatively high permeability compared to SB barriers is the result of the portland cement reducing the swelling properties of the bentonite. Because of their relatively high permeability, CB barriers are not used for waste containment applications, which often require permeabilities less than 10^{-7} cm/s.

Alternative cement mixes have been used to lower permeability and improve chemical compatibility. For example, ground granulated blast furnace slag mixed with a portland cement ratio of 3:1 or 4:1 has displayed permeabilities of 10^{-7} to 10^{-8} cm/s (USEPA, 1984b). Bentonite substitutes have also been used. One such substitute is attapulgite, a clay mineral that is more resistant to chemical degradation than bentonite. The use of such additives, however, can increase the overall cost of a barrier significantly.

CB barriers have higher shear strengths than SB barriers. The hardened trench of a CB barrier will exhibit the consistency of stiff clay. Therefore, CB barriers can be used where higher strengths are needed. CB barriers can be constructed with steeper surface grades than can SB barriers. Grade steps can be accomplished easily because the CB slurry hardens daily. An advantage of CB barriers is that backfilling of the trench is not required, minimizing construction defects. Construction of CB barriers does not require as large a working area as construction of SB barriers because backfill mixing areas are not required.

Materials excavated from the trench of a CB barrier will require disposal. If the materials are contaminated, handling and disposal of the materials may represent a significant cost factor and may require additional design considerations. If applicable, use of these materials as site fill may be feasible.

CQA and CQC is an important part of CB barrier construction. The hydrated bentonite–water slurry mixture should be tested for unit weight, viscosity, and filtrate loss before the addition of cement because the cement will begin to cure and these properties will change. The slurry should also be tested for the same parameters after the introduction of cement, and the test results should be compared to the design requirements. CB slurry test cylinders should be prepared in the field, allowed to cure (typically for 28 days in a 100% humidity environment), and tested in the laboratory for shear strength and permeability. Because the CB slurry hardens in the trench, during each day of construction a clean contact should be established with the previous day's hardened slurry. Therefore, at the beginning of each construction day, the end of the CB trench should be excavated to ensure a clean contact with the new CB slurry to be added to the trench.

Plastic Concrete Barriers. Plastic concrete (PC) barriers are constructed by excavating a trench under a head of bentonite–water slurry similar to that of SB barriers and are backfilled with a lean concrete mix of water, cement, aggregate, and bentonite. Plastic concrete barriers are usually constructed in panels, as shown in Figure 12.16. The backfill is placed by tremie

Table 12.5 **Comparison of SB and CB methods of slurry wall construction**

Soil–Bentonite (SB) Method	Cement–Bentonite (CB) Method
Low installation cost	High installation cost (30% higher than SB walls)
Wide range of chemical compatibilities (i.e., greater resistance to most pollutants)	Narrower range of chemical compatibilities—more susceptible to chemical attack
Lower permeability (1×10^{-9} to 1×10^{-10} m/s)	Higher permeability (1×10^{-8} m/s)
Backfill can be specially blended to meet specific design requirements	Not dependent on availability or quality of soil available for backfill
Requires a large work area for backfill mixing—when excavated material can be mixed with backfill, minimal spoil problems are created	No backfill mixing required; therefore, better for excavating in areas with difficult access or inadequate room for backfill mixing
Trenching must be done continuously in one direction	Allows for construction of trench in sections providing more flexibility to meet site constraints
Least strength, highest compressibility	Greater strength, less elasticity
Level work site required because of flowability of slurry backfill	Better for sites where topography is extreme and grading of site impractical

Wall Specifications

Width: 0.5 to 1.8 m	Width: 0.3 to 1.0 m
Slump value = 0.1 to 0.15 m; must stand on a slope of 5:1 to 10:1	Unconfined compressive strength = 1.5×10^4 to 2.8×10^4 kg/m^2

Slurry Specifications

4 to 7% bentonite by weight	4 to 7% bentonite by weight
93 to 97% water by weight	65 to 88% water by weight
Makeup water low in total dissolved solids	Makeup water low in total dissolved solids
Slurry pH 7.5 to 12.0	Slurry pH = 12.0 to 13.0
Density ≈ 1050 kg/m^3	Density ≈ 1090 kg/m^3
Marsh funnel viscosity = 35 to 45; apparent viscosity = 15 to 20 Centi Poise (CP)	Marsh funnel viscosity = 30 to 50 seconds

Backfill Specifications

2 to 4% bentonite by weight of the total mixture	6% bentonite by weight
≈ 25 to 35% water by weight	≈ 55 to 70% water by weight
≈ 10 to 40% fines by weight	≈ 30 to 40% solids by weight
	≈ 18% cement by weight
Density ≈ 1680 to 1920 kg/m^3	Density ≈ 1300 kg/m^3 max.

Source: Keith and Murray (1994). Courtesy of the Society of Mining, Metallurgy, & Exploration (SME), *www.smenet.org.*

pipe. The hydraulic conductivity of plastic concrete barriers ranges from 10^{-6} to 10^{-8} cm/s.

Design considerations for PC barriers are similar to those for SB and CB barriers. Because PC barriers are constructed under a bentonite–water slurry, the trench and slurry design criteria are similar to SB barriers. As with CB barriers, PC barriers (1) require disposal of excavated material, (2) can be constructed over steeper

grades, and (3) have higher strengths that allow their use in situations requiring structural loading.

PC barriers are considerably stronger than SB and CB barriers. Therefore, PC barriers should be considered for applications where additional strength is desirable. The limited data available indicate that PC barriers are more resistant to organic contamination than CB barriers. PC barriers are more expensive than SB

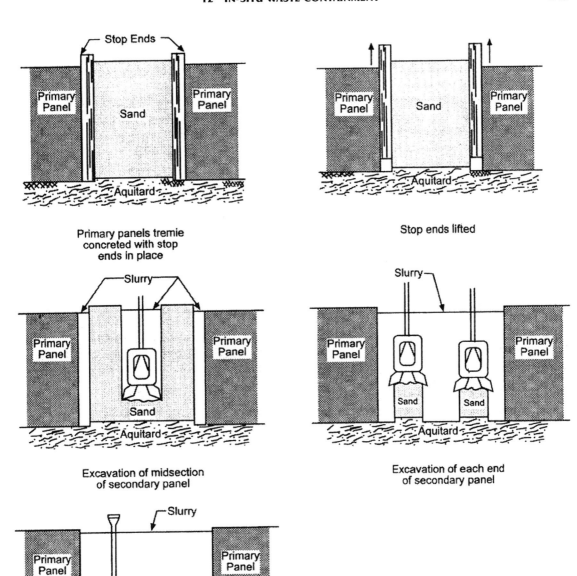

Figure 12.16 *Panel construction method for plastic concrete barrier. (From USEPA, 1998.)*

and CB barriers. The higher costs are the results of (1) disposal costs associated with excavated materials; (2) panel construction, which is more time consuming than continuous trenching; and (3) the added cost of the aggregate used.

CQA and CQC considerations for PC barriers are similar to those for SB and CB barriers. Slurry, trench construction, and backfill parameters should be maintained in accordance with design requirements.

Composite Barriers. Composite barriers are constructed using a combination of materials. Figures 12.17 and 12.18 show one such composite barrier type in which a geomembrane is inserted within a slurry trench. Such composite barriers improve the performance of traditional soil–bentonite barriers by decreasing the permeability of the barrier by as much as four to five orders of magnitude and improving the chemical resistance of the barrier.

Proper installation of a geomembrane within a slurry trench is critical. Care should be taken to install geomembranes without punctures or tears and to develop a continuous seal between adjacent membranes. Commonly, a geomembrane is mounted on an installation frame and the frame is lowered into a slurry-filled trench (Figure 12.19). Interlocking geomembranes (Figure 12.20) have been developed to improve joint seals. Interlocking joints on either side of the geomembrane are sealed with a hydrophilic gasket or by the slurry. The installation frame is then removed. Alternatively, the bottom of the geomembrane is weighted so that the liner sinks into the trench or hardened geomembrane panels are driven into the ground using a pile driver.

12.2.4 Grouted Barriers

Grouted barriers, also known as *grouted curtains,* are formed by injecting a grout into the subsurface. Pressure grouting and jet grouting are two common methods of injection grouting, in which grout mixture is

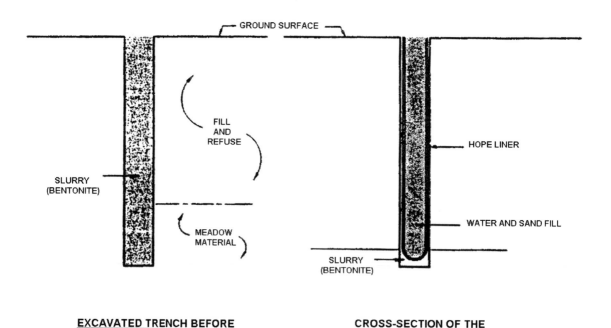

**EXCAVATED TRENCH BEFORE
PLACEMENT OF THE LINER**

**CROSS-SECTION OF THE
TRENCH WITH LINER**

Figure 12.17 *Composite barrier: trench barrier with geomembrane. (From Cavalli, 1992. Copyright ASTM International. Reprinted with permission.)*

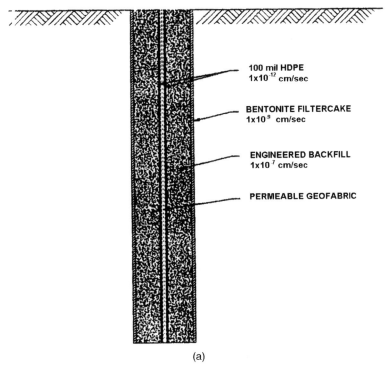

100 mil HDPE
1x10⁻¹² cm/sec

BENTONITE FILTERCAKE
1x10⁻⁹ cm/sec

ENGINEERED BACKFILL
1x10⁻⁷ cm/sec

PERMEABLE GEOFABRIC

(a)

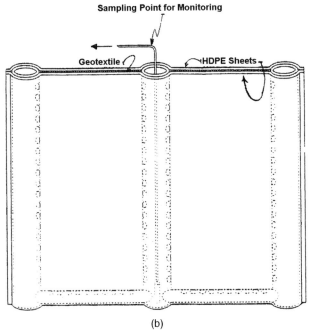

Sampling Point for Monitoring

Geotextile

HDPE Sheets

(b)

Figure 12.18 *Composite barrier incorporating geomembrane. (From Cavalli, 1992. Copyright ASTM International. Reprinted with permission.)*

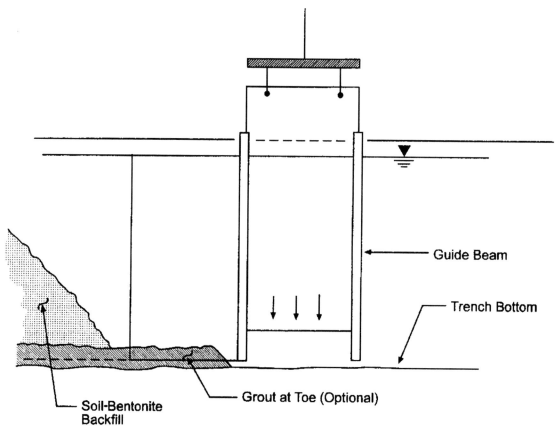

Figure 12.19 *Installation method for a geosynthetic composite barrier. (From USEPA, 1998.)*

injected from a hole into void spaces and fissures in the soil or rock. The holes are spaced so that the treated zones overlap to form a continuous barrier. Often, two or three rows of grout holes are necessary to achieve a continuous low-permeability barrier, as displayed in Figure 12.21. The spacing of the injection holes is site-specific and is determined by the penetration radius of the grout out from the holes. Ideally, grout injected in adjacent holes should touch, as shown in Figure 12.22. Injection of grout may also be accomplished by the vibrating beam method, in which grout is injected through a special H-pile into the space created by the driven pile when the pile is removed (described below). All of the grouted barriers are keyed into the underlying low-permeability soil strata or competent bedrock.

The choice of grout depends on soil permeability, soil grain size, soil and groundwater chemistry, compatibility between the grout and the contaminants present, and rate of groundwater flow (USEPA, 1984a). In general, grouts can be divided into two main categories: particulate grouts (or suspension grouts) and chemical grouts. Particulate grouts include slurries of bentonite, cement, or both and water. Chemical grouts generally contain a chemical base, a catalyst, and water or another solvent. Common chemical grouts include sodium silicate, acrylate, and urethane. Table 12.6 summarizes the significant characteristics of various

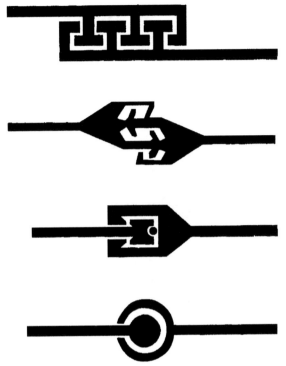

Figure 12.20 *Interlocks for geosynthetic composite barriers. (From USEPA, 1998.)*

types of grouts, and Table 12.7 presents the compatibility of selected grouts for various types of contaminants. Figure 12.23 shows the suitability of various types of grouts for different types of soils. Particulate grouts have higher viscosities than chemical grouts and are therefore used for sands and gravels or to fill wide-open fractures, whereas chemical grouts are generally used in fine sands and silts. A combination of particulate and chemical grouts can also be used; larger voids can first be filled with a particulate grout, followed by application of a chemical grout to fill the remaining smaller voids.

A detailed site characterization is required to determine the suitability of using grouted barriers. Particularly, information regarding soil profile, soil properties that control groutability (e.g., grain size analysis, permeability, porosity), and water loss during rock coring

will be required. Data on groundwater chemistry and the type of contaminants present at a site are required to evaluate the compatibility of the grout with the contaminants.

Because of their high cost, grouted barriers are not used at hazardous waste sites. Slurry barriers are less costly and have lower permeability than that of grouted barriers. However, grouted barriers may be best suited to sealing off fractured rocks. Grouted barriers can be constructed by permeation (pressure) grouting, jet grouting, or by using a vibrating beam method, all described briefly below.

Permeation (Pressure) Grouting Method. In permeation (pressure) grouting, soil voids are filled with a sealing grout. To achieve low permeability, the soil voids must be filled completely, and the lateral extent of grout penetration must be controlled. The design of a permeation grouted barrier must consider soil permeability, grout viscosity, and soil and grout particle size. In general, permeation grouting with chemical grout is suitable in soils with permeability greater than 10^{-3} cm/s, while particulate grouts can be used when soil permeability is greater than 10^{-1} cm/s.

Permeation grouting can be accomplished in two ways: by the point injection or tube-a-manchette (also known as sleeve pipe injection) method. In the *point injection method,* a grout casing is driven to full depth, and grout is injected through the end as the casing is withdrawn. The injection points are typically arranged in a triple line of primary and secondary grout holes. A predetermined amount of grout is pumped into the primary holes. After the grout in the primary holes has had time to set or gel, the secondary holes are injected. The secondary grout holes fill in any gaps left by the primary grout injection. Thus, a continuous barrier is formed, as shown in Figure 12.22. Primary holes are typically spaced at intervals of 3 to 5 ft.

In the *tube-a-manchette method,* (Figure 12.24), a sleeve pipe containing small holes at 1-ft intervals is placed in a grout hole. The small holes are covered by rubber sleeves (manchettes) that act as one-way valves, allowing grout to be forced into the formation. A double packer is placed in the sleeve pipe in such a way

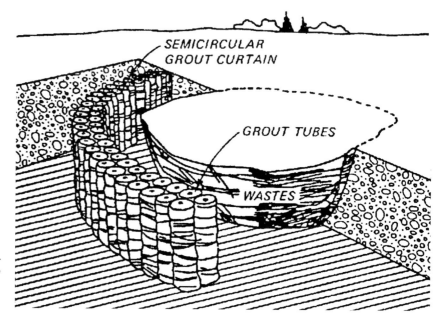

Figure 12.21 Semicircular grouted barrier around waste site. (From USACE, 1994.)

that it straddles the manchettes and grout is injected under pressure. If the required containment permeability is not achieved, the tube-a-manchette method allows for regrouting at the same location. This method also allows different grout types to be used at the same location (e.g., a cement grout to fill larger voids and a chemical grout to fill smaller voids).

The design of permeation grouted barrier must include a thorough evaluation of the grouting pressure to be used. Excessive pressure can cause hydrofracturing, causing grout to enter into the hydrofractures with the natural soil voids unfilled. If this happens, the barrier would not meet the design permeability requirement.

Jet Grouting Method. Jet grouting involves the use of a combination of grout, air, and water delivered by a small jet or jets in the drill rod at very high pressures, often reaching 5000 to 6000 psi. After advancing the drill rod to the desired depth, it is lifted and rotated as the jetted grout cuts away the soil and creates a large cylindrical hole, as shown in Figure 12.25. Portland cement or cement–bentonite grouts are generally used when jet grouting is employed. Cement grout mixes with the soil to form a soil–cement mixture column in

the ground. The excess water and soil are forced to the surface around the drill rod. A horizontally continuous barrier can be created by successive installation of jet grouted columns.

Three types of jet grouting—single rod, double rod, and triple rod—(all referring to the number of passageways in the drill rod)—have been developed. In single-rod jetting, only grout is pumped through the rod and injected into the soil. After the soil–grout column is mixed, excess soil and water are displaced to the surface. Single-rod columns can be up to 1.2 m wide in granular soils and 0.8 m wide in cohesive soils. In double-rod jetting, air is injected into the soil through the same jet as the grout. The injected air keeps the jet stream clear by holding back groundwater and the soil cut by the jet and helps lift the cut soil to the surface. Double-rod columns can be up to twice as large as single-rod columns. However, the high air content of a double-rod column may result in higher permeability. In triple-rod jetting, air and water are injected through a cutting jet that cuts and lifts the soil, resulting in removal of most of the native soil. Grout is injected through a second jet, filling the column as the drill rod is lifted. The resulting column is therefore

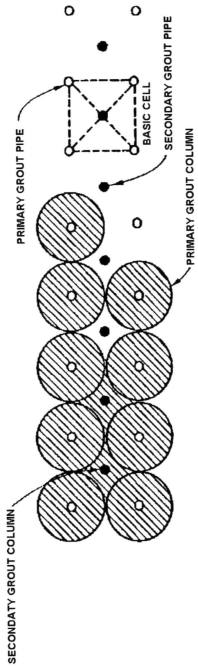

Figure 12.22 Grout pipe layout for grout curtains. (From USACE, 1994.)

PRIMARY GROUT PIPE

SECONDARY GROUT PIPE

BASIC CELL

PRIMARY GROUT COLUMN

SECONDATY GROUT COLUMN

355

Table 12.6 **Significant characteristics of types of grout**

Type	Characteristic
Portland cement or particulate grouts	Appropriate for higher-permeability (larger-grained) soils Least expensive of all grouts when used properly; most widely used in grouting across the United States (90% of all grouting)
Sodium silicate	Most widely used chemical grout At concentrations of 10 to 70% gives viscosity of 1.5 to 50 cP Resistant to deterioration by freezing or thawing Can reduce permeabilities in sands from 10^{-2} to 10^{-8} cm/s Can be used in soils with up to 20% silt and clay at relatively low injection rates; portland cement can be used to enhance water cutoff
Acrylamide	Should be used with caution because of toxicity First organic polymer grout developed May be used in combination with other grouts, such as silicates, bitumens, clay, or cement Can be used in finer soils than most grouts because low viscosities are possible (1 cP) Excellent gel time control, due to constant viscosity from time of catalysis to set/gel time Unconfined compressive strengths of 344 to 1378 kPa (50 to 200 psi) in stabilized soils Gels are permanent below the water table or in soils approaching 100% humidity Vulnerable to freeze–thaw and wet–dry cycles, particularly where dry periods predominate and will fail mechanically; due to ease of handling (low viscosity), enables more efficient installation and is often cost-competitive with other grouts
Phenlolic (phenoplasts)	Rarely used, due to high cost Should be used with caution in areas exposed to drinking water supplies, because of toxicity Low viscosity Can shrink (with impaired integrity) if excess (chemically unbound) water remains after setting; unconfined compressive strength of 344 to 1378 kPa (50 to 200 psi) in stabilized soils
Urethane	Set through multistep polymerization Reaction sequence may be halted temporarily Additives can control gellation and foaming Range in viscosity from 20 to 200 cP Setting time varies from minutes to hours Prepolymer is flammable
Urea–formaldehyde	Rarely used due to high cost; will gel with an acid or neutral salt Gel time control is good Low viscosity Considered permanent (good stability) Solution toxic and corrosive Relatively inert and insoluble
Epoxy	In use since 1960 Useful in subaqueous applications Viscosity variable (molecular weight dependent) In general, setting time difficult to regulate Good durability Resistant to acids, alkalis, and organic chemicals
Polyester	Useful only for specific applications Viscosity 250 to several thousand centipoise Setting time hours to days Hydrolyzes in alkaline media Shrinks during curing Components are toxic and require special handling
Lignosulfonate	Rarely used due to high toxicity Lignin can cause skin problems and hexavalent chromium is highly toxic (both are contained in these materials) Cannot be used in conjunction with portland cement; pH values conflict; ease of handling Loses integrity over time in moist soils Initial soil strengths of 344 to 1378 kPa (50 to 200 lb/in^2)

Source: (USACE, 1994).

Table 12.7 **Interaction between grouts and specific chemical classes**[a]

Grout Type Chemical Group	Portland Cement Type I	Portland Cement Types II and V	Bentonite	Cement–Bentonite	Silicate	Acrylamide
Organic Compounds						
Alcohols and glycols	?d	?d	?d	?d	?	?d
Aldehydes and ketones	?	?	?d	?	?	?a
Aliphatic and aromatic hydrocarbons	2a	2?	?d	?	?	?a
Amides and amines	?	?	?	?	?	?
Chlorinated hydrocarbons	2d	2d	?	?	?	?a
Ethers and epoxides	?	?	?	?	?	?a
Heterocyclics	?	?	?d	?	?	?a
Nitriles	?	?	?	?	?	?
Organic acids and acid chlorides	1d	1d	?d	?a	?	2a
Organometallics	?	?	?	?	?a	?
Phenols	1d	?	?d	?	?	?
Organic esters	?	?	?	1a	?	?
Inorganic Compounds						
Heavy metals, salts, and complexes	2c	2a	?d	2c	?a	2?
Inorganic acids	1d	1a	?c>	?c	?	2c
Inorganic bases	1a	1a+	?c>	?d	?	3d
Inorganic salts	2d	2a	2d	?d*	1a	3d

Source: (USEPA, 1991).

[a]Effect on set time: 1, no significant effect; 2, increase in set time (lengthen or prevent from setting); 3, decrease in set time. Effect on durability: a, no significant effect; b, increase durability; c, decrease durability (destructive action begins with a short period); d, decrease durability (destructive action occurs over a long period); *, except sulfates, which are ?c; +, except KOH and NaOH, which are 1d; >, modified bentonite is d; ?, data unavailable.

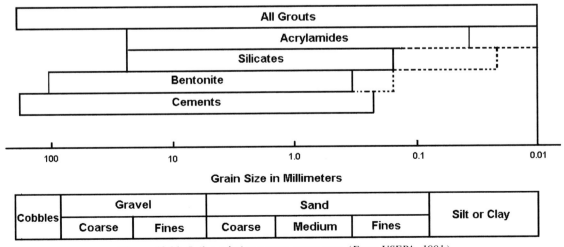

Figure 12.23 Soil gradation versus grout type. (From USEPA, 1991.)

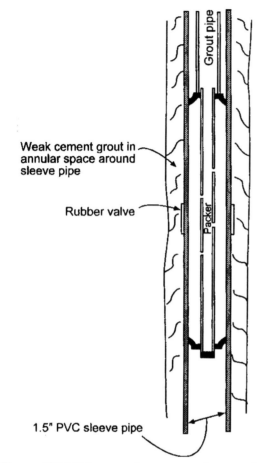

Figure 12.24 *Tube-a-manchette permeation grouting.* *(From USEPA, 1998.)*

almost entirely grout. Triple-rod columns can be up to 3 m wide in granular soils and 1.5 m wide in cohesive soils.

Jet grouting can be used in soils ranging from gravels to heavy clays. Jet grouted barriers have been built to depths greater than 200 ft, although below about 100 ft the verticality and thus the continuity of jet grouted barriers are difficult to control or confirm. A disadvantage of the jet grouting process is that the only exit for the displaced soil cuttings and water is the drill hole opening at the ground surface. Excess pressures can build up, and hydrofracture may occur if this opening

is blocked. Another disadvantage is the potentially large volume of spoil produced. The spoil from a waste site usually has to be cleaned and redeposited on site and can greatly increase the cost.

Vibrating Beam Method. A vibrating beam barrier is a grouted barrier that is suitable for shallow depths. Construction of a vibrating beam barrier consists of driving a modified H-pile into the ground with a vibratory hammer; the H-pile is modified to inject grout through nozzles at the bottom end of the pile. During pile driving, a small amount of grout may be injected through the nozzles to provide lubrication. Grout is injected through the nozzles to fill the void as the pile is withdrawn. This procedure is displayed in Figure 12.26. The driving and filling process is repeated in an overlapping pattern to form a continuous barrier for containing waste or contamination.

Cement bentonite grouts are commonly used for vibrating beam barriers, although bituminous (asphalt-based) grouts have also been used. Vibrating beam barriers are only 2 to 3 in. thick and therefore have a high potential to hydrofracture. The permeability of a vibrating beam barrier depends on the grout used. A permeability of 10^{-5} to 10^{-6} cm/s may be expected where cement–bentonite grout is used.

The principal advantage of a vibrating beam barrier is that handling or disposal of excavated material is not required. The primary disadvantage of vibrating beam barriers is that the H-piles may deflect from vertical, making the continuity of the barrier at depth uncertain. The bottom of a vibrating beam barrier cannot be inspected to confirm verticality or key penetration. Also, preaugering may be necessary to ensure ease of penetration into tighter soils.

12.2.5 Mixed-in-Place Barriers

Mixed-in-place barriers, also known as *deep soil mixed barriers* or *soil mixed walls,* are constructed by in-situ mixing of soil and a slurry. Specially designed equipment consisting of three auger mixing shafts is used to inject and mix a water–bentonite or cement–bentonite slurry into the soil as the augers are advanced. This

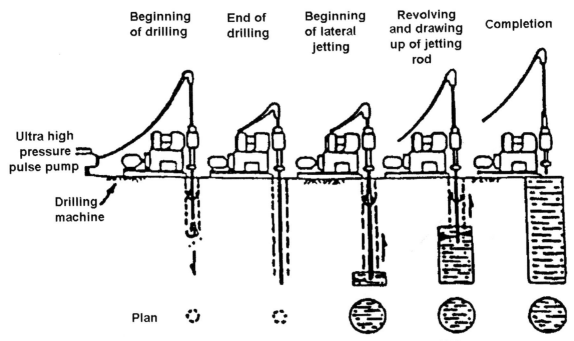

Figure 12.25 *Jet grouting method. (From Mitchell and van Court, 1992.)*

results in a column of thoroughly mixed soil. A continuous wall typically from 0.5 to 0.9 m wide is obtained by overlapping penetration with the installation sequence, as indicated by stroke numbers 1, 2, 3, 4, and 5 in Figure 12.27.

Mixed-in-place barriers are relatively wider than other types of barriers (e.g., vibrating beam barrier) and can achieve permeabilities of 10^{-7} cm/s. Because potentially contaminated soils are not excavated, these barriers reduce health risks and safety issues and eliminate costs associated with handling and disposal of contaminated soils. One major disadvantage is that the bottom of mixed-in-place barriers cannot be inspected to confirm key penetration.

12.2.6 Steel Sheet Pile Barriers

Steel sheet pilings have been used for a wide variety of civil engineering applications, including control of groundwater flow into excavations at construction sites. Steel sheet pile barriers, also known as *sheet pile cutoff walls* or simply *sheet pile barriers,* may be used to contain subsurface contamination and to divert groundwater flow around a contaminated area. They are constructed by driving individual sections of interlocking steel sheets into the ground using single double-acting impact or vibratory pile drivers to form a thin impermeable barrier to groundwater flow, as shown in Figure 12.28. Each sheet pile section interlocks with an adjacent section by means of a ball-and-socket union. Sheet piles with various configurations and fittings can be used, depending on the site-specific conditions, including soil type and driving depth.

To serve as an effective barrier, sheet piles should extend into a low-permeability soil stratum or to bedrock. Driving sheet piles into low-permeability soils (e.g., clays, silty clays) is relatively easy with standard pile-driving practices. However, driving sheet piles into a rock unit is difficult. Pile testing and borings to an impermeable soil or rock layer can be used to determine the effectiveness of the barrier and piling inter-

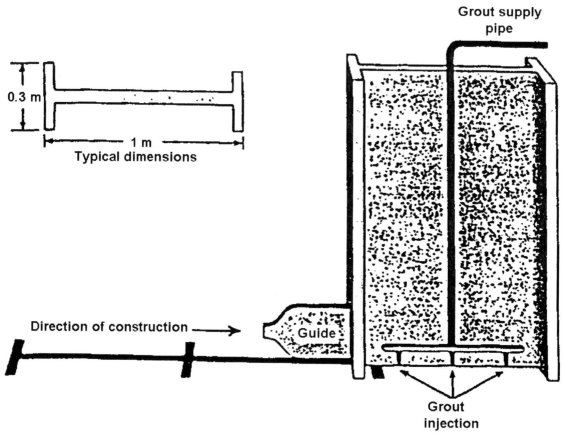

Figure 12.26 *Vibrating beam barrier. (From Rumer and Ryan, 1995. This material is used by permission of John Wiley & Sons, Inc.)*

lock damage. Sheet piles may be driven to a maximum depth of 50 to 100 ft.

The major problem with sheet pile barriers is leakage at the interlocks. Because of this leakage problem, steel sheet piles are not commonly used for waste containment. To address this problem, some new types of piles have recently become available for use in waste containment applications. Interlocking joints can be fit with geomembranes. The cavity in the joint is filled during placement with a material that swells on contact with water to ensure an adequate seal. Also, to reduce seepage through the interlocks, steel sheet piling has been used in conjunction with cement–bentonite. A

composite wall constructed in this manner is expensive but results in a very low permeable barrier. Compatibility of steel sheet piles with the site contaminants is also an important design consideration. They are not suitable for groundwater containing high concentrations of salts or acids unless they can be coated with a coal tar expoxy or cathodic protection can be used.

The advantages of sheet piles include their easy availability, strength, and rapid placement; however, these advantages are offset by the vibrations caused by driving, leakage through the interlocks, and high cost of the piles. Sheet piles are not suitable for very dense soils or soils with boulders present because the sheet

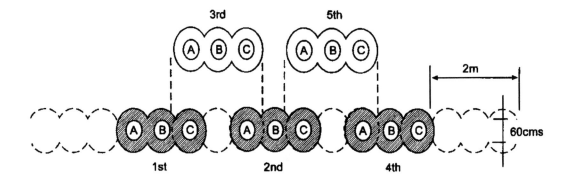

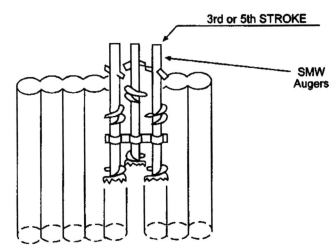

Figure 12.27 Installation sequence for a deep soil mixed barrier. (From USEPA, 1998.)

piles will be damaged during installation. Sheet pile barriers could be effective as a short-term measure to enhance recovery or containment.

12.3 BOTTOM BARRIERS

12.3.1 Description

Bottom barriers are used when no naturally occurring low-hydraulic-conductivity stratum exists at reasonable depth beneath a waste site (see Figure 12.1). This construction can be accomplished in several ways, such as by using grouting techniques or employing a combination of tunneling, installation of geomembranes, and

grout or slurry mix. Other less proven approaches include ground freezing to stabilize soils and the creation of electrochemical barriers to contaminant transport that can be installed across compacted clay liners. The common methods that are used to construct a bottom barrier below buried waste are described in this section.

12.3.2 Permeation Grouting

In coarse-grained soils, permeation grouting (explained in Section 12.2.4) can be used to build a layer of overlapping grout bulbs beneath the contaminated site, as

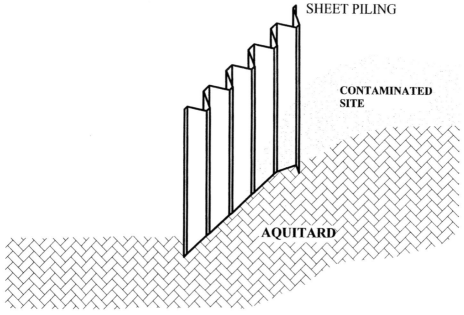

Figure 12.28 Sheet pile barrier. (From USEPA, 1998.)

shown in Figure 12.29. The grout is injected at pressures that will displace water in the void spaces and maintain the soil structure. The pressures should not be so high as to cause hydraulic fracturing. The thickness of the grouted barrier must be great enough to ensure adequate watertightness. The tube-a-manchette method (described in Section 12.2.4) provides the best control over the injection of grout within a desired depth range. Field tests are performed to determine the thickness required. The hydraulic conductivity of the grouted soil will vary from 10^{-6} cm/s with portland cement grout to 10^{-10} cm/s with acrylate grout or microsilica grout.

12.3.3   Jet Grouting

The jet grouting technique, described in section 12.2.4, is also used to form short columns or disks that will overlap each other to form a seal beneath the contaminated region, as shown in Figure 12.30. Similar to

permeation grouting, jet grouting can be used to create both bottom and vertical barriers. Jet grouting, however, provides better barrier characteristics than permeation grouting, for the following reasons:

- Good control of grout column dimensions
- Good control of continuity of grout placement
- Relative independence from heterogeneities in the soil
- Ability to use portland cement grout, which has a high resistance to chemicals and is inexpensive (portland cement grout cannot be injected in many soils but can be used with jet grouting in almost all types of soils).

A typical jet-grouted bottom barrier made with portland cement and bentonite may have a hydraulic conductivity as low as 10^{-7} cm/s.

If drilling through a contaminated zone or waste for the installation of a bottom barrier is not allowed, slant drilling for permeation or jet grouting may be consid-

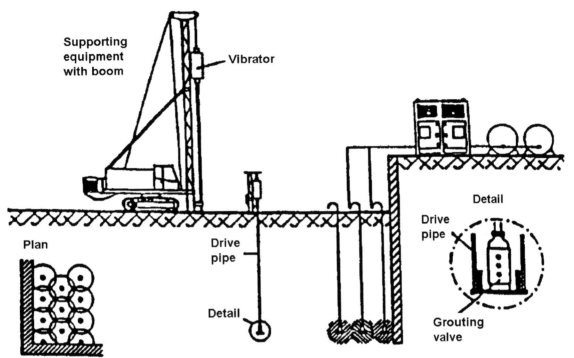

Figure 12.29 *Permeation grouting to create bottom barrier. (From Mitchell and van Court, 1992.)*

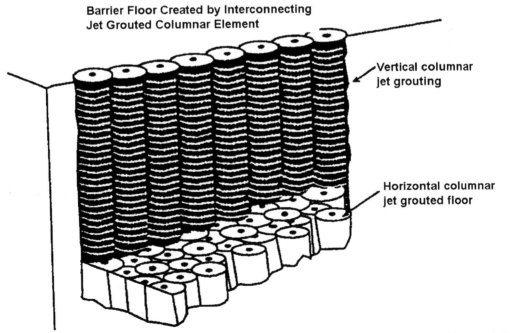

Figure 12.30 *Jet grouting to creat bottom barrier. (From Mitchell and Rumer, 1997. Reproduced by permission of ASCE.)*

ered. However, this method will be applicable only if the waste width of the contaminated zone or waste is small. An example of such a bottom barrier in conjunction with the vertical barriers is depicted in Figure 12.31.

12.3.4 Bottom Barriers by Directional Drilling and Grouting

Directional drilling allows the drilling of a parabola-shaped path from one point on the ground surface to another point on the ground surface. This may be useful for reaching beneath a waste pile or contaminated zone without going through it, as shown in Figure 12.32. The minimum radius of curvature that directional drilling can achieve is on the order of 15 to 30

m. Directional drilling is used in conjunction with either permeation grouting or jet grouting to create bottom barriers.

12.3.5 Bottom Barriers Using Hydrofracturing and Block Displacement Method

Hydrofracturing is the intentional fracturing of a soil formation by water or air pumped into the ground under high pressure. In the block displacement method (Figure 12.33), a series of boreholes are injected simultaneously with fluid (water or air) under high pressure in order to lift a block of waste and surrounding soil and to place a low-permeability bottom barrier beneath the block. This method requires drilling several boreholes around the perimeter of the waste as well as

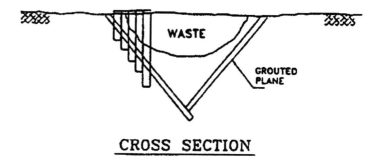

CROSS SECTION

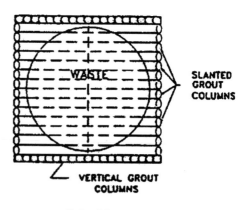

PLAN

Figure 12.31 Bottom barrier formed by slanted grout columns. (From Rumer and Ryan, 1995. This material is used by permission of John Wiley & Sons, Inc.)

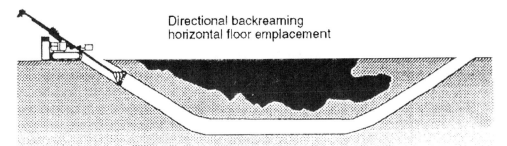

Figure 12.32 *Creation of bottom barrier using directional drilling technique. (From Mitchell and Rumer, 1997. Reproduced by permission of ASCE.)*

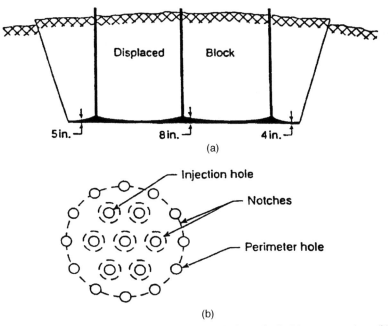

Figure 12.33 *Creation of bottom barrier using displacement block method:* (a) *cross section;* (b) *location of drill holes. (From Rumer and Ryan, 1995. This material is used by permission of John Wiley & Sons, Inc.)*

through the waste. Bentonite grout is pumped at high pressure into each interior hole simultaneously so that horizontal fractures are developed. The bentonite grout fills the fractures, forcing them to open wider and expand laterally. The thickening and expansion of the fractures force the block to displace upward. The bentonite grout also keeps the fractures from closing. The grout-filled fractures act as a bottom barrier.

12.3.6 Sheet Pile Combined with Injection Grouting

A buried waste may also be contained utilizing a combination of steel sheet piles and permeation grouting. The steel piles are driven at angles toward a common point or line located below the buried waste, forming a cone or wedge, similar to slant drilling, as shown in

Figure 12.31. Once the sheet piles are in place, the soil mass between the sheet pile wedge and the waste is filled with grout using permeation grouting through multiple injection holes. The grout fills the voids in the soil and the joints between the individual sheet piles, preventing further leachate migration.

12.4 SURFACE CAPS OR COVERS

12.4.1 Description

Surface caps, also called *covers* or *surface barriers,* are similar to landfill final cover systems, described in Chapter 19. Caps are constructed over the buried wastes similar to landfill covers to prevent infiltration of precipitation, thereby minimizing the generation of leachate (see Figure 12.1). Caps also prevent transfer of contaminants to the atmosphere, reduce erosion, and improve aesthetics (USEPA, 1993; USACE, 1995). The type of cap depends primarily on the nature of the waste, site conditions, and the likelihood of extended maintenance of the cap. Caps may range from a one-layer system of vegetated soil to a complex multilayer system of soils and geosynthetics, as described in Section 12.4.2.

Caps have been used to control wastes containing a variety of contaminants, including volatiles, semivolatiles, metals, radioactive materials, corrosives, oxidizers, and reducers. Caps are used by themselves or in conjunction with other waste containment systems such as vertical barriers and/or remediation technologies such as groundwater pump and treat system and in-situ waste treatment (e.g., in-situ bioremediation). Caps may be used as an interim or final remedial measure. An interim cap can be used to minimize the generation of leachate until a better remedial action is selected or while treatment is being applied. Caps may be used as the final remedial action at sites where waste masses are too large to excavate and remove and/or remedial cleanup action is too difficult to implement because of potential hazards and/or unrealistic costs. Abandoned landfills, waste pits, and tailing piles are examples of sites where caps are often used for final remediation.

12.4.2 Configuration and Materials

The general configuration of a multilayer cap includes (from top to bottom) (1) a surface or erosion layer, (2) a protection layer, (3) a drainage layer, (4) a barrier layer, (5) a gas collection layer, and (6) a foundation layer, as shown in Figure 12.34. Table 12.8 summarizes the functions of these layers and materials that can be used for their construction.

The function of a *surface layer* is to separate the underlying components of the cap from the ground surface and to minimize temperature and precipitation extremes in underlying layers. Topsoil, geosynthetic erosion control material overlying topsoil, cobbles, and paving materials have all been used for the surface layer in a cap. Topsoil is the most commonly used material, although excessive erosion can be a problem. For this reason, as a temporary measure, a geosynthetic erosion control material can be placed over the topsoil to limit the problem.

A *protection layer* may serve several functions: (1) to store water that has infiltrated the cap until the water is later removed by evaporation or transportation, (2) to separate the waste physically from animals or roots, (3) to minimize the possibility of human intrusion, and (4) to protect underlying layers in the cap from excessive wetting/drying, which could cause cracking of some materials. Soil is the most commonly used material for the protection layer. In many cases, the surface and protection layers are combined to form a single topsoil layer.

A *drainage layer* serves the following functions: (1) to reduce the head of water on the underlying barrier layer, which minimizes percolation of water through the cap; (2) to drain the surface water infiltrated from the overlying protection and surface layers, which increases the water storage capacity and helps to minimize erosion of these layers; and (3) to reduce pore water pressures in the cap materials, which improves slope stability. The materials used for drainage layers include sand, gravel, geonet with a geotextile filter, and geocomposite drainage material. Geotextile may be used as a filter between the vegetative/protective layer and the drainage layer. These materials are discussed in Chapter 17.

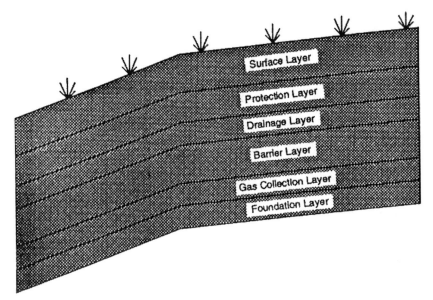

Figure 12.34 Cap cross section. (*From Mitchell and Rumer, 1997. Reproduced by permission of ASCE.*)

A *barrier layer* minimizes percolation of water through the cap by impeding infiltration through it and by promoting storage or drainage of water in the overlying layers. The barrier also restricts upward movement of any gases or volatile constituents that might be emitted by the waste. Soil barrier layers usually consist of clay that is compacted to a hydraulic conductivity of 1×10^{-7} cm/s or less or as required by local regulations or site constraints. Compacted soil barriers are generally installed in 6-in. (or smaller) lifts to achieve a thickness up to 2 ft or more. Composite barriers use both soil and geomembranes. Geomembranes used in barrier layers are manufactured in large rolls and are available in varying thicknesses (20 to 140 mils), widths (15 to 100 ft), and lengths (180 to 840 ft). Geomembranes are much less permeable than clays. A geotextile may be used over the geomembrane as a protective layer to prevent mechanical damage to the geomembrane (e.g., tear, puncture, overstressing). Composite barriers have proven to be most effective in decreasing infiltration.

Satisfactory long-term performance of compacted clay liners cannot always be assured, as desiccation, freezing and thawing, or excessive settlement of the waste below may cause cracking. Because of these problems with compacted clay liners, geosynthetic clay liners (GCLs) have been developed for use as a barrier layer material (more details on GCLs are given in Chapter 17). These are composed of a thin layer of bentonite between two geosynthetic materials. On wetting, the bentonite expands to create a low-permeability, resealable barrier that is "self-healing." The geosynthetic clay barrier material is produced in rolls, but unlike geomembranes, it does not require seaming. Other barrier materials, such as soil–bentonite, fly ash–bentonite–soil mixtures, superabsorbent geotextiles, sprayed-on geomembranes, and soil–particle binders, have been developed for quick and easy installation, better quality control, and cost savings.

The lowermost layer in the cap is the *gas collection layer,* also known as the *vent layer.* This layer transmits gas to collection points for venting or flaring or cogeneration. This layer also serves as the *foundation layer* and should be strong enough to support the overlying layers and construction equipment. The gas collection layer is usually made up of granular materials and piping.

Not all of these layers are needed at all sites. The selection of cap materials and cap design depends on

Table 12.8 **Cap components**

Layer	Primary Functions	Potential Materials
Surface layer	Separate underlying layers from ground surface	Topsoil (vegetated)
	Resist erosion	Geosynthetic
	Reduce temperature and moisture extremes in underlying layers	Paving material
Protection layer	Store infiltration water before removal	Soil
	Separate waste from humans, animals, and vegetation	Cobbles
	Protect underlying layers from wetting and drying	Recycled or reused waste (e.g., fly ash, bottom ash, paper mill sludge)
	Protect underlying layers from freezing and thawing	
Drainage layer	Reduce water head on barrier layer	Sand or gravel
	Reduce uplift water pressure on overlying layers which are saturated after rain	Geonet or geocomposite
		Recycled or reused waste
Hydraulic barrier layer	Impede water percolation through cap	Compacted clay
	Restrict outward movement of gases from waste	Geomembrane
		Geosynthetic clay liner
		Recycled or reused waste
		Asphalt
		Sand or gravel capillary barrier
Gas collection layer	Collect and remove gases	Sand or gravel
		Geonet or geocomposite
		Geotextile
		Recycled or reused waste
Foundation layer	Foundation for the cap, especially during construction	Sand or gravel
		Soil
		Recycled or reused waste
		Select waste

Source: Mitchell and Rumer (1997). Reproduced by permission of ASCE.

site-specific factors such as local availability and costs of cover materials, desired function of the cover, the nature of the wastes being covered, local climate (rainfall), site topography, hydrogeology, and projected future use of the site. Among many cap systems used in practice, Figure 12.35(*a*) and (*b*) show two examples of cap systems that have been used at waste sites.

12.4.3 Design

The design of caps is similar to final covers for landfills as described in Chapter 19. The major design factors that influence the effectiveness of a cap include (1) determination of global waste stability, (2) underlying waste settlement analysis, (3) slope stability analysis of the cap system, (4) drainage analysis, (5) leach-

ate management analysis, and (6) gas management analysis. Determination of global waste stability involves evaluating whether a waste mass will remain stable under all potential loading conditions. Waste mass stability is analyzed under several different loading conditions, including cap loads, seismic stresses, and construction loading. Settlement analysis evaluates the potential for the foundation and waste materials to consolidate under the loading conditions of a cap. Long-term settlement analysis should be conducted, and a monitoring plan should be developed to measure settlement.

The stability of the cap system itself should be analyzed to determine its potential for failure. Analyses should address (1) interface stability for critical interfaces within a cap system (e.g., clay liner and

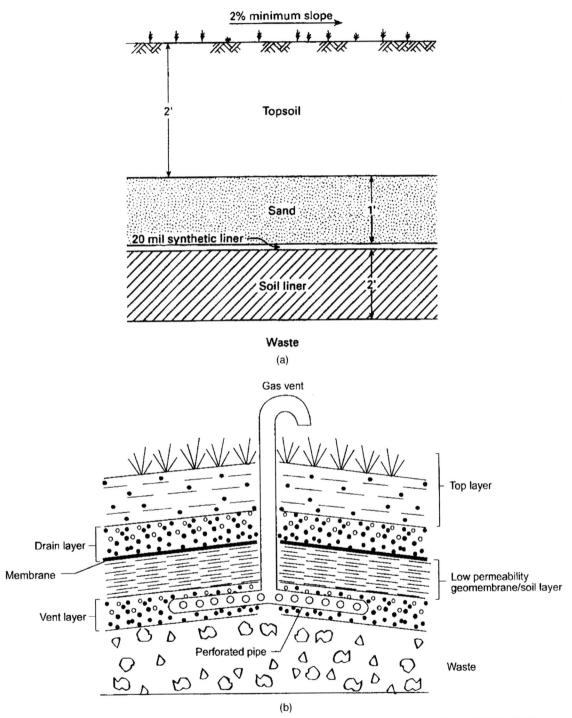

2% minimum slope

2'

Topsoil

Sand 1'

20 mil synthetic liner

Soil liner 2'

Waste

(a)

Gas vent

Top layer

Drain layer

Membrane

Low permeability
geomembrane/soil layer

Vent layer

Perforated pipe

Waste

(b)

Figure 12.35 *Examples of cover systems:* (a) *without gas vents;* (b) *with gas vents.* (*From USEPA, 1998.*)

geomembrane, GCL and geomembrane, geomembrane and geotextile), (2) cap soil tension above a geosynthetic-lined slope, (3) various stresses within cap components, and (4) the impact of differential settlement on geosynthetic materials.

A drainage analysis is necessary to prevent buildup of hydraulic head on the low-permeability cap layer and stability of the overlying layers due to seepage pressures. Drainage analysis should address (1) drainage capacity, (2) geotextile filtration, (3) runoff control, and (4) erosion control. Gas management analysis consists of an evaluation of the proposed gas collection well design and placement and usually includes a pilot test to determine the gas composition and generation rate. If a passive venting system is required, the locations of lateral and vertical vents are determined on the basis of waste characteristics. Other design considerations that affect the performance of the cap include frost penetration of the cover soils and the puncture vulnerability of geosynthetic materials. Site-specific conditions may necessitate consideration of these and/or other design factors for evaluation of caps.

12.4.4   Construction

Caps are usually constructed to enhance runoff. The base layer, which may be a gas collection layer, overlies the waste mass. The low-permeability layer, such as the clay component of the barrier layer, is constructed over this base layer. The clay is spread and compacted in lifts a few inches thick until the desired thickness is achieved. Each lift is scarified (roughed up) following compaction to remove any trace of a smooth surface, which might exhibit binding between it and the next-higher lift. The top lift is compacted and rolled smooth so that the geomembrane can be laid on it with direct and uniform contact.

A geomembrane should be laid without wrinkles or tension. Its seams should be welded fully and continuously, and the geomembrane should be installed before the underlying clay surface can desiccate and crack. If vent pipes are present, they should be carefully attached to the geomembrane to prevent tearing

as a result of subsidence. Punctures and tears should be avoided during geomembrane handling and installation. In addition, the effects of air temperature and seasonal variations on the geomembrane should be taken into account; stiffness and brittleness are associated with low air temperatures. If installed while the air temperature is high, the geomembrane will expand and can then shrink to the point where seams may be overstressed.

A geotextile may be laid on the surface of the geomembrane to protect it from damage by overlying materials, particularly if coarse and sharp granular materials are to be used in the overlying drainage layer. The drainage layer is designed to carry away water that percolates down to the barrier layer. The drainage layer may be either a granular soil with high permeability or a geosynthetic drainage grid or geonet sandwiched between two porous geotextile layers. Another geotextile may be put on the top of the drainage layer to prevent clogging of the drainage layer by soil from above. Fill soil and topsoil are then applied, and the topsoil is seeded with grass or other vegetation.

Construction quality assurance/construction quality control (CQA/CQC), including testing, is estimated to increase the cap installation cost and completion time by 10 to 15% (although this may vary from project to project) but is generally acknowledged as improving the quality and hence the performance of the cap constructed. Soils are tested to determine their grain size, Atterberg limits, hydraulic conductivity, and compaction characteristics. To ensure high seam quality, geomembrane test strip seams are subjected to strength (shear and peel) testing to simulate stress from equipment, personnel, or climatic changes. Installation procedures should specify that seams run up and down slopes rather than across them, to reduce seam stress. Chapter 17 provides further test details.

Construction equipment required during cap installation includes bulldozers, graders, various rollers, and vibratory compactors. Additional equipment is needed for moving, placing, and seaming geosynthetic materials. Storage areas are needed for the materials used in the cap. If site soils are not adequate for use in cap construction, low-permeability soils have to be trucked in. Adequate water supplies are needed to ensure that soils used in construction maintain their optimum soil density.

CQA/CQC processes for soil liners are intended to ensure that (1) soil liner materials are suitable, (2) soil liner materials are properly placed and compacted, and (3) the completed liner is properly protected from damage due to adverse weather conditions. Clay prequalification testing is accomplished through periodic soil classification, generation of compaction curves, and remolded permeability tests. A test pad is generally required to demonstrate that the materials and methods proposed will result in construction of a liner with the required large-scale, in-situ hydraulic conductivity. Construction testing of the soil liner should include generation of compaction curves as well as testing for density, moisture content, undisturbed hydraulic conductivity, Atterberg limits, and grain size.

Manufacturing quality control testing should be performed on geomembrane barriers to ensure that the geomembrane complies with NSF 54 or other applicable standards and displays the desired melt index, resin index, and resistance to environmental stress cracks. Field testing should be conducted periodically to evaluate geomembrane thickness, tensile strength, elongation, and resistance to puncture and tears. CQA inspection should be performed for every roll of geomembrane at the site. Seam testing should be performed using nondestructive (vacuum box or air pressure) and destructive (peel and shear testing) methods. Chapter 17 provides further details on testing requirements.

12.4.5 Performance and Economic Considerations

A well-designed and constructed cap system should usually perform satisfactorily and should require minimal maintenance. However, the impact of differential settlement and the effect of surface water erosion should be monitored quarterly or semiannually. Additional monitoring of leachate quality, quantity, and leak detection, water infiltration, gas quality and quantity,

and groundwater will ensure that the cap system is performing as intended.

The cost of a 0.5- to 1-acre cap can vary from $500,000 for a one-layer system to several million dollars for a multilayer cap (USEPA, 1998). The cost is highly dependent on the local availability of soils suitable for construction and the requirements for monitoring, leachate collection, and gas collection.

12.5 GROUNDWATER PUMPING SYSTEMS

12.5.1 Description

Groundwater pumping systems are active waste containment systems used to manipulate and manage groundwater for the purpose of removing, diverting, and containing a contaminated plume or for adjusting groundwater levels to prevent plume movement. Figure 12.36(a) depicts the plan view of the use of a line of extraction wells to halt the advance of the leading edge of a contaminant plume and thereby prevent contamination of a drinking water supply. Figure 12.36(b) presents the cross-sectional view of the system. Use of extraction wells alone is best suited for situations where contaminants are miscible and move readily with water, the hydraulic gradient is steep, the hydraulic conductivity is high, and quick removal is not necessary. Extraction wells are frequently used in combination with barriers, commonly with vertical barriers, to prevent groundwater from overtopping the barrier and to minimize contact of the leachate with the barrier to prevent barrier degradation due to such conditions as the imposition of a high hydraulic head across the barrier wall, as shown in Figures 12.4 and 12.6.

A combination of extraction and injection wells is frequently used in containment or removal where the hydraulic gradient is relatively flat and hydraulic conductivities are only moderate. The function of the injection well is to direct contaminants to the extraction wells, which then remove the contaminants. This method has been used with some success for plumes that are not miscible with water. Figure 12.37 illustrates an extraction–injection well system for removal. One problem with such an arrangement of wells is that

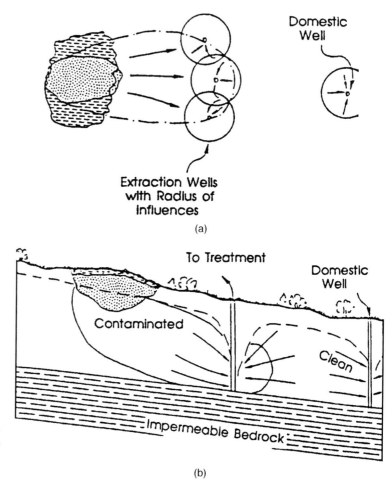

Figure 12.36 Pumping system with a line of extraction wells for waste containment: (a) *plan;* (b) *cross section. (From USEPA, 1985.)*

dead spots (i.e., areas where water movement is very slow or nonexistent) can occur when these configurations are used. The size of the dead spot is related directly to the amount of overlap between adjacent radii of influence; the greater the overlaps, the smaller the dead spots will be. Another problem is that injection wells can suffer from many operational problems, including air locks and the need for frequent maintenance and well rehabilitation.

Extraction or injection wells can also be used to adjust groundwater levels, although this application is not widely used. In this approach, plume development can be controlled at sites where the water table intercepts disposed wastes by lowering the water table with extraction wells. For this pumping technique to be effective, infiltration into the waste pile must be eliminated and liquid wastes must be removed completely. If these conditions are not met, the potential exists for development of a plume of contaminants. The major drawback to using well systems for lowering water tables is the continued costs associated with maintenance of the system.

Groundwater barriers can be created using injection wells to change both the direction of a plume and the

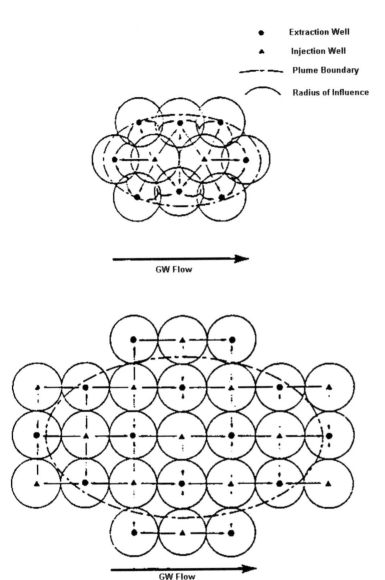

● Extraction Well

▲ Injection Well

— · — · — Plume Boundary

⌒ Radius of Influence

GW Flow

GW Flow

Figure 12.37 Extraction and injection well patterns for plume removal. (*From USEPA, 1985b.*)

speed of plume migration. Figure 12.38 shows an example of plume diversion using a line of injection wells to protect domestic water resources. By creating an area with a higher hydraulic head, the plume can be forced to change direction. This technique may be desirable when short-term diversions are needed or when diversion will provide the plume with sufficient time to degrade naturally, so that containment and removal are not required.

Pumping is most effective in aquifers with high hydraulic conductivity. It has been used with some effectiveness in aquifers with moderate hydraulic conductivity and where pollutant movement is occurring along fractured or jointed bedrock. In fractured bed-

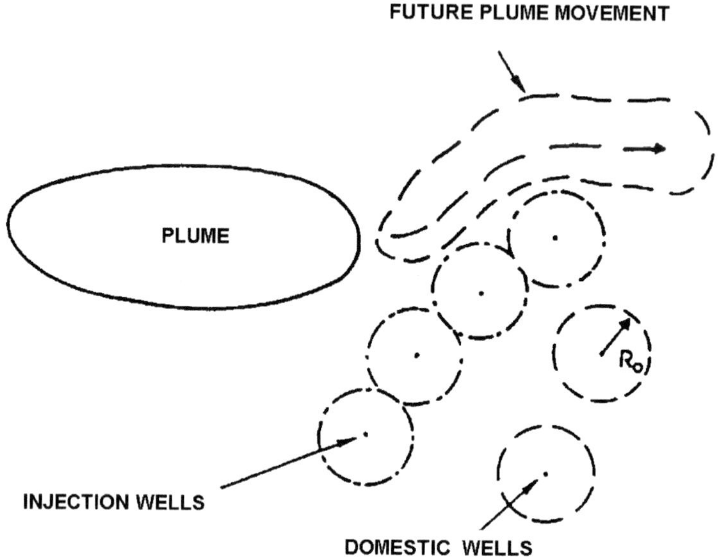

Figure 12.38 Plume diversion using injection wells. (*From USEPA, 1985b.*)

rock, the fracture patterns must be traced in detail to ensure proper well placement.

If contaminated groundwater is pumped, need for its treatment before disposal must be considered. The type of treatment system depends on the contaminants present. Contaminated waters can be treated on- or off-site. If the water must be reinjected, on-site treatment may be necessary. More details on contaminated water treatment systems are given in Chapter 14.

12.5.2 Design

A detailed subsurface characterization must first be made to determine hydrogeologic conditions and the nature and extent of the groundwater contamination. A potentiometric surface map (i.e., a map depicting contours of equal head) and a geologic cross section of the site should be developed. The plume area, depth, flow rate, and direction must be determined. Pump tests should include determination of transmissivity and storage coefficients, pumping rate, and radii of influence of test wells. The presence of perched water tables or other anomalies must also be assessed. Table 12.9 summarizes the types of data needed to determine

if a pumping system is suitable for the site and to design the pumping well system.

The basis of plume containment by pumping depends on incorporating the plume within the radius of influence of an extraction well. The design of a well system is based on the theory of flow toward a well. Well theory differs for confined and unconfined aquifers. Definitions of unconfined and confined aquifers as well as the fundamentals of flow toward a well are provided in Chapter 7. When a well is pumped, the water table is lowered in a cone-shaped manner. This is known as *drawdown*. The amount of drawdown depends on the pumping rate, permeability and thickness of aquifer, groundwater recharge, presence of boundaries, and duration of pumping. For plume containment, a combination of extraction and injection wells may be used. Injection well theory is identical to extraction well theory except that cones of depression and drawdowns are inverted to above the water table or piezometric surface (cone of impression).

For floating contaminants, wells partially penetrating aquifers can be used. When a well does not fully penetrate an aquifer, the water close to the well must move along curved lines to reach the well. Evaluation

Table 12.9 **Data requirements for pumping and subsurface drains**[a]

Data Description	Purpose(s)[b]	Source(s)/Method(s)
Depth to aquifer/water table	Select appropriate extraction type and design P and SD	Hydrogeologic maps, observation wells, boring logs, piezometers
Types, thicknesses, and extents of saturated and unsaturated subsurface materials	Design P and SD	Hydrogeologic maps, surficial geology maps/reports, boring logs, geophysical surveys
Hydraulic conductivities and storativities of subsurface materials	Design P and SD	Pumping tests, slug tests, laboratory permeability tests
Containmant concentrations and areal extent	Locate select depth of wells or drains	Water quality data
Piezometric surface map, groundwater flow rates, and vertical/horizontal gradients	Locate and design wells or drains	Water-level data
Seasonal changes in groundwater elevation	Select depth of wells or drains	Long-term water-level monitoring
NAPL density/viscosity/solubility	Predict vertical distribution of contamination, design wells or drains	Literature
Groundwater/surface water relationship	Design wells or drains	Seepage measurements, stream gauging
Location screen/open interval depths and pumping rates of wells influenced by site	Determine impacts/interference	Well inventory, pumping records
Precipitation/ recharge	Design wells or drains	NOAA reports

Source: (USEPA, 1991).

[a] Applies to hydraulic barrier technology (i.e., well systems).

[b] P, pumping; SD, subsurface drains.

[c] National Oceanic and Atmospheric Administration.

of the effects of partial penetration can be complicated. However, the effect of partial penetration is negligible on the flow pattern and drawdown beyond a radial distance larger than 0.5 to 2 times the saturated thickness, depending on the well penetration (Todd, 1980).

The design of a well system includes (1) design of wells and (2) design and selection of well components (e.g., screens, casing, and pumps). These are described in detail in the following sections.

Well Design. Well system design includes a determination of the number of wells needed, the patterns and spacing of the wells, the design of the individual wells, the pumping cycles and rates needed, and the method of handling discharges. Determining the radius of influence for a well in a given aquifer is critical in remedial action design because it can be used for determining well spacing, pumping rates, pumping cy-

cles, and screen lengths. The radius of influence of a well increases as pumping continues until equilibrium conditions are reached (i.e., when aquifer recharge equals the pumping rate or the discharge rate). The designer has to decide whether equilibrium or nonequilibrium pumping will be used for design purposes because this affects the extent of the radii of influence. Equilibrium and nonequilibrium conditions of pumping wells were reviewed in Chapter 7. Equilibrium pumping is commonly used in designs; however, nonequilibrium pumping may be more realistic for aquifers with low hydraulic conductivity, for nonmiscible plumes, and for sites with groundwater barriers or limited recharge conditions.

The most accurate method for estimating the radius of influence is by pumping test analysis. Pumping tests can identify recharge boundaries, barrier boundaries, and slow storage release conditions. The pumping test should be performed until equilibrium conditions are reached. Typical test durations for a confined aquifer are about 24 hours, whereas it may be several days for an unconfined aquifer. Once equilibrium conditions have been reached, the radius of influence for equilibrium or nonequilibrium conditions can be estimated using the equations in Table 12.10 or graphical methods explained in Chapter 7.

When pumping test data are lacking or incomplete, rough approximations of the radius of influence can be obtained using equations presented in Table 12.10 and using the values of transmissivity or hydraulic conductivity, pumping times, and coefficient of storage. Coefficients of storage values typically range from 0.01 to 0.35 for unconfined aquifers and from 0.00001 to 0.001 for confined aquifers. Because these estimates are approximate and do not take recharge into account, it is advisable that the value of R_0 be adjusted downward so that there will be greater overlap of the cones of depression and therefore a lower probability that contaminants escape between the wells.

Well Spacing. Determination of the proper spacing of wells to capture a groundwater plume completely is arguably the most important item in system design. Field practitioners have had a long-standing rule of thumb for establishing well spacing such that adjacent cones of depression should overlap (i.e., radii of influence should overlap). This method is reasonably accurate for aquifers that have low natural flow velocities but will not be valid for aquifers with high natural flow velocities. For the latter cases (and preferably for all cases), capture zone analysis on the basis of velocity distribution should be used to determine well spacing and ensure capture of the plume (Javandel and Tsang, 1986).

Capture zone analysis is based on a determination of the *stagnation point* (i.e., downgradient location of a pumping well where the discharged water is exactly the same as natural flow to the well). The stagnation point is related directly to the pumping rate of the well (i.e., the higher the pumping rate, the farther downgradient the stagnation point) and inversely related to the natural flow (the higher the natural flow rate, the closer the stagnation point is to the well). Only for the extremely rare case of zero natural flow are the areal boundaries of the capture zone identical to the calculated cone of depression. This means that even though the cones of depression of two pumping wells intersect, they may not be capturing the plume completely unless their capture zones intersect. More details on capture zone analysis are provided in Chapter 14 and in USEPA (1985b), and Javandel and Tsang (1986).

Pumping Rates. For a confined aquifer, capacity and therefore pumping rate are directly proportional to the drawdown. Increasing the pumping rate will not affect the radius of influence but will affect the amount of pumping time required. Therefore, pumping rates can be selected to suit the situation. In situations where the contaminated plume floats, drawdowns and pumping rates will probably be small. Large drawdowns and high pumping rates are desirable where contaminants are dispersed throughout the aquifer, quick removal is desired, and natural groundwater rates are large.

For an unconfined aquifer it has been found that maximum efficiency for well operation occurs when the drawdown is at about 67% of the maximum theoretical drawdown; pumping rates should be adjusted

Table 12.10 **Radius of influence equations**[a]

Condition	Confined Aquifer	Unconfined or Water Table Aquifer
Equilibrium, exact	$\ln R_0 = [T(H - h_w)/229Q] + \ln r_w$	$\ln R_0 = [K(H^2 - h_w^2)/458Q + \ln r_w]$
Nonequilibrium, exact	Drawdown vs. log distance plots or Theis method	
Nonequilibrium, approximate	$R_0 = r_w + (Tt/4790S)^{0.5}$	$R_0 = 3(H - h_w)(0.47K)^{0.5}$

Source: (USEPA, 1985b).

[a] R_0 = radius of influence (ft); K, hydraulic conductivity (gal/day-ft²); H, total head (ft); h_w, head in well (ft); Q, pumping rate (gal/min); r_w, well radius (ft); T, transmissivity (gal/day-ft); t, time (min); S, storage coefficient (dimensionless).

accordingly. This is because part of the formation within the cone of depression is actually unwatered during pumping and the specific yield decreases with increased drawdown. Optimum operating conditions are achieved when the product of specific yield and capacity are greatest, which occurs at about 67% of maximum drawdown.

System Integration. Once the well spacing, pumping rate, and drawdown have been determined, the system can be designed as a unit. At this point, a decision must be made on the pattern and type (i.e., injection or extraction) of wells to be installed. Numerous patterns of extraction wells or injection wells or both are available. The choice is typically based on whether the design is for containment or removal, the time available for recovery, and the amount of dewatering that is allowable. Patterns that combine extraction and injection wells allow for more rapid contaminant removal without greatly affecting groundwater levels. These patterns are also advantageous because the treated water can be reinjected after extraction.

After a well pattern is chosen, the number of wells needed to control the plume must be determined. This is based on the estimated well spacing and the drawdown required. In spacing wells for plume control, it is necessary to have the well capture zones intersect each other so that contaminants will not flow between wells and escape, as described previously. The number of wells needed can be determined by plotting the chosen pattern of wells with their required spacing on the potentiometric surface map of the site. After this is done, the drawdowns are determined. This will result in a new potentiometric surface map of the site that can be used to identify dead spots where contaminants can escape.

12.5.3   Construction

Extraction and injection wells are installed using conventional drilling techniques similar to that of monitoring wells, as explained in Chapter 10. The wells must have sufficient diameter to accommodate a submersible pump and to accommodate expected flow.

12.5.4   Performance Assessment: A Case History

Pumping has been used at several sites to contain groundwater plumes. However, under difficult geologic and contaminant conditions, pumping alone proved to be ineffective to contain contamination completely. Therefore, barrier systems are often combined with pumping to contain contamination completely. The Fairchild Camera and Instrument Corporation site in South San Jose, California, is an example of such a situation. Over 43,000 gal of organic chemicals was lost into the soil, contaminating four aquifers beneath the site. Groundwater in the shallow aquifer was initially contaminated with the chemicals. Subsequently, contaminants migrated into the lower aquifers through interconnecting sand beds, existing wells, and slow seepage through separating strata. The first action taken for plume containment was the installation of an extraction well to create a zone of influence to draw the plume back. The initial pumping rate was 500 gal/min, but was increased to 1500 gal/min when the lower rate failed to contain the plume spread. To control the spread of the plume further, three rows of extraction wells were installed near the source of the contaminant. The combined pumping dewatered two aquifers but left the contaminant behind in the unsaturated soil. Subsequently, contaminants continued to migrate to lower aquifers. Finally, the source was contained with a slurry wall 350 ft long, 70 to 140 ft deep, and 3 ft wide.

12.6   SUBSURFACE DRAINS

12.6.1   Description

Subsurface drains are an alternative to pumping wells for the containment of contaminated groundwater. They consist of a buried conduit to convey and collect groundwater by gravity flow. Subsurface drains function basically like an infinite line of extraction wells and create a continuous zone of influence in which groundwater within this zone flows toward the drain. Figure 12.39 shows a subsurface drainage system. The major components of a subsurface drainage system are:

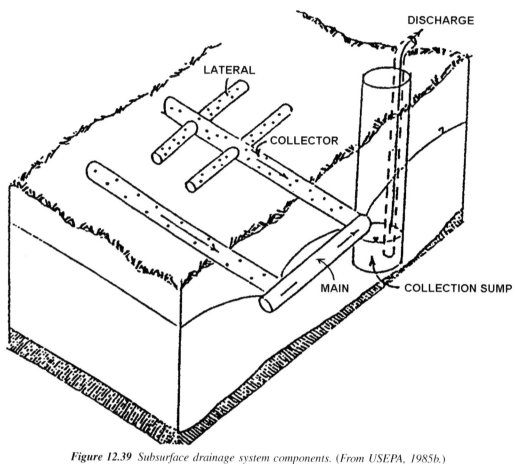

Figure 12.39 *Subsurface drainage system components. (From USEPA, 1985b.)*

- *Drain pipe:* for conveying flow to a storage tank or wet well
- *Envelope:* for conveying flow from the aquifer to the drain pipe
- *Filter:* for preventing fine particles from clogging the system
- *Backfill:* to bring the drain to grade and prevent ponding
- *Manholes or wet wells:* to collect flow and pump the discharge to a treatment plant

Subsurface drains are commonly used to intercept a plume hydraulically downgradient from its source, as shown in Figure 12.40(*a*). Sometimes, these drains are used together with a barrier wall as shown in Figure 12.40(*b*), often called a *drain/barrier wall*. A drain/barrier wall is used to prevent infiltration of large amounts of clean water from nearby streams as depicted in Figure 12.40(*a*) and (*b*). Subsurface drains can also be placed around the circumference of a waste site in order to lower the groundwater table or to contain a plume, as shown in Figure 12.41. A circumferential subsurface drain may be part of a total containment system, which consists of a barrier wall and a cap in addition to the subsurface drain, as shown in Figure 12.42.

Subsurface drains are best suited for sites where the groundwater table is relatively shallow and the contam-

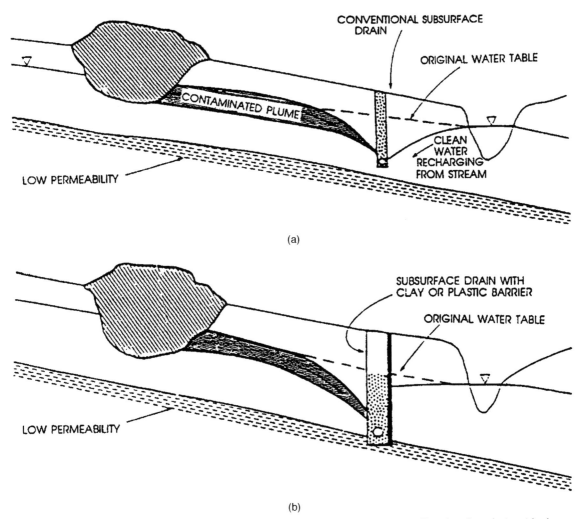

Figure 12.40 Plume containment using: (a) *conventional subsurface drain systems;* (b) *subsurface drain with clay or geomembrane barrier.* (*From USEPA, 1985b.*)

ination is near the water table either because (1) it is an LNAPL, (2) it is confined to a thin upper aquifer by underlying strata of low hydraulic conductivity, or (3) it is prevented from migrating downward due to upward vertical gradients or there has not been sufficient time for vertical migration. Drains are better in relatively low permeability soils because groundwater pumping in such soils would perform poorly. The areal configuration of a plume may also create an advantage

for the use of drains rather than pumping. Unlike pumping, operation and maintenance costs associated with subsurface drains are low.

12.6.2 Design

Detailed site characterization data are needed prior to the design of subsurface drain systems. Table 12.9

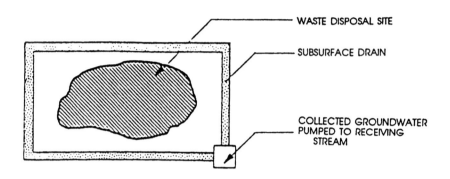

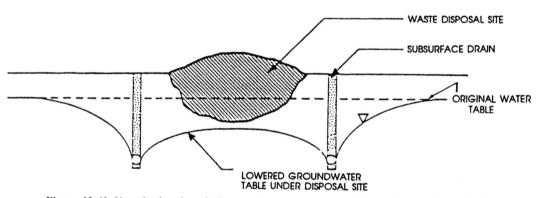

Figure 12.41 Use of subsurface drainage to lower groundwater levels. (*From USEPA, 1985b.*)

shows specific data that are needed for the design of drain systems. Design of a subsurface drainage system includes the following elements:

- Selection of location and spacing of drains to achieve the necessary groundwater levels
- Selection of pipe diameter and gradient to ensure adequate flow and structural stability
- Selection of envelope and filter materials to prevent clogging
- Design of a pumping station (sump and pump system, manhole placement), to ensure effective collection and removal

Location and Spacing of Drains. Subsurface drains are installed perpendicular to groundwater flow and are used to intercept groundwater from an upgradient source. The locations of drains depend on site conditions. Site-specific potentiometric surface (or water table) maps, hydraulic conductivity data, plume boundary limits, and geologic cross sections are used to select the drain line location. Additional borings are then drilled along this line, and the alignment is shifted if necessary to obtain proper interception of the contaminant plume. In stratified soils having greatly differing hydraulic conductivities, the drain should be installed resting on a layer of low hydraulic conductivity.

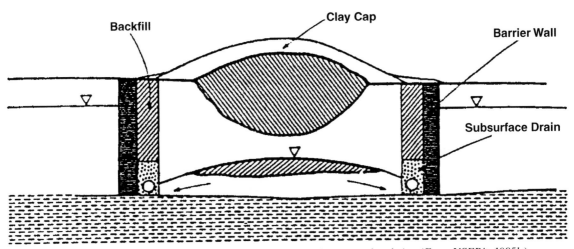

Figure 12.42 *Use of subsurface drainage in a completely encapsulated site. (From USEPA, 1985b.)*

If the trench is cut through an impervious stratum, there is a danger that a significant percentage of the leachate moving laterally will bridge over the drain and continue downgradient. Similarly, if soil layers or pockets with high hydraulic conductivity underlie the drain, the groundwater plume may flow beneath the drain.

To decide where to position a drain, reasonable estimates of the upgradient and downgradient influence of the drain should be known. The theoretical upgradient influence can be expressed by the equation (Donnan, 1946)

$$D_u = 1.33m_sI \qquad (12.7)$$

where D_u is the effective distance of drawdown upgradient (ft), m_s the saturated thickness of the waterbearing strata not affected by drainage (ft), and I the hydraulic gradient (dimensionless).

The depth to which the water table is lowered downgradient of the interceptor is proportional to the depth of the drain. Theoretically, a true interceptor drain lowers the water table downgradient to a depth equal to the depth of the drain. The distance downgradient to which the drain is effective in lowering the water table is infinite provided that recharge is not occurring from a downgradient source. This is never the case, since infiltration from precipitation always re-

charges groundwater. The theoretical determination of the downgradient influence can be obtained from the equation (Figure 12.43)

$$D_d = \frac{KI}{q}(d_e - h_d - D_2) \qquad (12.8)$$

where D_d is the downgradient influence (ft), K the hydraulic conductivity (ft/day), I the hydraulic gradient (dimensionless), q the drainage coefficient (ft/day), d_e the depth of the drain (ft), h_d the desired depth of drawdown (ft), and D_2 the distance from ground surface to water table prior to drainage at the distance D_d downgradient from the drain (ft). In this equation, D_d and D_2 are interdependent variables. To solve this, D_2 is estimated, then a trial computation is made. If the actual value of D_2 at distance D_d is appreciably different, a second calculation is necessary. Where I is uniform throughout the area, D_2 can be considered equal to D_1 (i.e., the distance from the ground surface to the water table measured at the drain). If a second interceptor is needed to lower the water table to the desired depth, it would be located D_d feet downgradient from the first.

Depth and Spacing of Parallel Drains. Basic equilibrium equations are used for estimating the spacing of drains under two conditions: (1) drains resting on

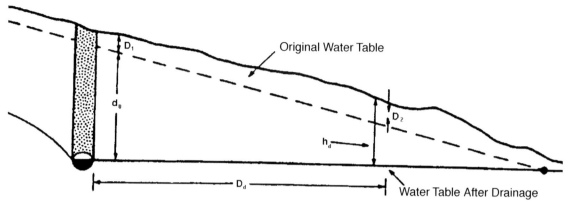

Figure 12.43 *Flow to drain resting on a low-permeability barrier.* (*From USEPA, 1985b.*)

an impermeable layer, and (2) drains above an impermeable layer. The equations assume that steady-state conditions exist, recharge distribution and leachate generation over the area between the drains are uniform, and the soil is homogeneous. To represent more realistic conditions, numerical computer models should be used.

Drains are often used where the depth to a low-permeability barrier is relatively shallow and the drains can be laid just above the barrier. In developing and using drain spacing formulas for this case, an underlying soil layer is considered to be impermeable if the hydraulic conductivity is less than one-tenth that of the soil layer above. Flow to drains resting on a low-permeability layer is shown in Figure 12.44. This flow is represented by (Donnan, 1946)

$$L = \sqrt{\frac{8KDH + 4KH^2}{q}} \qquad (12.9)$$

where L is the drain spacing (ft), K the hydraulic conductivity of the drained material (ft/day), D the distance between the water level in the drain line and the impermeable barrier (ft), H the water table height above the drain levels at the midpoint between two drains (ft), and q the leachate generation rate (ft/day). For a pipe drain resting on an impermeable barrier, the parameter D approximately equals the radius of the pipe and hence is very small in comparison to H (the

water table height above the drain). Therefore, equation (12.9) can be simplified to

$$L = \sqrt{\frac{4KH^2}{q}} = 2H\sqrt{\frac{k}{q}} \qquad (12.10)$$

As shown in Figure 12.44, when two parallel drains are installed, each exerts an influence (L or the drain spacing), which in theory will intersect with each other midway between the drain lines. The influence (L) is the distance from the drain to a point where the drawdown can be considered insignificant and is commonly referred to as the *zone of influence*. Drain spacing (L) and hydraulic head level (H) in the equations above are interdependent design variables which are a function of the leachate generation rate (q) and hydraulic conductivity (K) of the drained material. Assuming constant q and K, the closer two drains are spaced, the more their drawdown curves will overlap and the lower the hydraulic head levels between the drains will be. n many instances, it may not be possible to install drains to the depth of a low-permeability barrier because the cost of installing drains to this level is prohibitively high or because the plume does not extend to the depth of the barrier. In these instances, flow is not described adequately using the equations above.

Figure 12.45 shows the flow to a drain not resting on a low-permeability layer. The flow lines are not parallel and approximately horizontal, as shown in Fig-

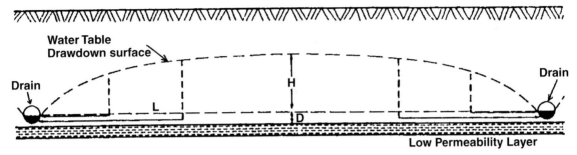

Figure 12.44 *Downgradient influence of an interceptor drain. (From USEPA, 1985b.)*

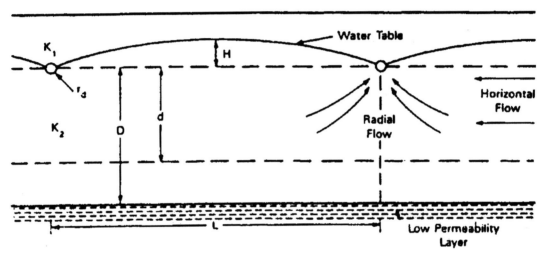

Figure 12.45 *Flow to drain not resting on a low-permeability barrier. (From USEPA, 1985b.)*

ure 12.42; rather, they converge toward the drains. The convergence, or *radial flow* as it is commonly called, causes a more than proportional loss of hydraulic head because the flow velocity in the vicinity of the drain is larger than elsewhere in the flow region. The effect is that elevation of the water table and drain spacing would be larger than would be predicted using the equation (12.10).

Hooghoudt (1940a and b) developed a modified drain spacing formula that accounts for radial flow and head loss. His method accounts for head loss by using an equivalent depth, *d*, to replace *D* in equation (12.9). The equation can be used to describe the conditions shown in Figure 12.45, that is, flow to drains that are

installed at the interface of two layered soils with hydraulic conductivities of K_1 and K_2. For this condition, *L* is given by

$$L = \frac{8K_2 dH + 4K_1 H_2}{q} \qquad (12.11)$$

where *d* is the equivalent depth (ft), K_1 the hydraulic conductivity of the layer above the drain (ft/day), and K_2 the hydraulic conductivity of the layer below the drain (ft/day). In equation (12.11), both drain spacing *L* and equivalent depth *d* are unknowns. The value of *d* is typically calculated from a specified value for *L*, so that equation (12.11) cannot be solved explicitly in terms of L. The use of this equation as a drain formula involves either a trial-and-error procedure of selecting d and L until both sides of the equation are equal or the use of formulas or nomographs that have been developed specifically for equivalent depth and drain spacing (Wesseling, 1964; Van der Molen and Wesseling, 1991).

For saturated thicknessses (D) greater than about 30 ft, the equivalent depth can be calculated from drain spacing using the equation (USEPA, 1985a)

$$d = 0.57L + 0.845 \qquad (12.12)$$

Again, it should be noted that the Hooghoudt equation, although widely used in drain spacing, is accurate only when the level of the drain corresponds with the interface between the two soil levels. When drains are being placed so that the interface lies either above or below the drains, drain spacing may not be predicted accurately by this formula. If hydrogeologic conditions are complex, numerical groundwater flow models should be used for more accurate prediction of drain spacing.

Pipe Diameter and Gradient. Pipe diameter and gradient are the design parameters used to ensure that water which arrives at a drainline can be conveyed without a buildup of pressure. The formula used for hydraulic design of pipes is based on the following Manning formula for pipes:

$$Q = \frac{1}{N} AR^{0.67} I^{0.5} \qquad (12.13)$$

where Q is the design discharge (ft^3/s), N the roughness coefficient, A the drainage area (ft^2), R the hydraulic radius (ft) equal to the wetted cross-sectional areas A_w divided by the wetted perimeter (one-fourth the diameter for full flowing pipes), and I the hydraulic gradient.

In designing subsurface drainage systems, a gradient is chosen that is large enough to result in a flow velocity that prevents siltation (>1.4 ft/s) but will not

cause turbulence (critical velocity) (SCS, 1972). Critical velocities for various soil types are summarized in Table 12.11 (SCS, 1972). Table 12.12 gives the gradients for different sizes of drains which result in the critical velocity for drains with roughness coefficients of 0.011, 0.013, and 0.015. The roughness coefficient is a function of the hydraulic resistance of the drain material.

Design discharge, Q, is equal to the sum of the individual discharges that impinge on a drain. Estimates of the total discharge can be obtained using the water balance method. This method provides an estimate of the amount of percolation that will recharge the water table between the lines of the drain. Once the percolation rate has been calculated, discharge can be obtained by multiplying the percolation route by the drainage area:

$$Q = qA \qquad (12.14)$$

where Q is the design discharge (ft^3/s), q the leachate generation or infiltration rate (ft/s), and A the drainage area (ft^2).

The diameter of the drain pipe is a function of design discharge, hydraulic gradient, and the roughness coefficient. With this information, the appropriate drain diameter can be determined based on the Manning velocity equation or from nomographs prepared using the Manning formula. Figure 12.46 is a nomograph for estimating pipe diameter for pipe with an N value of 0.015. A pipe diameter one size larger than that determined to be necessary is generally recommended.

It is critical that the drain pipes selected be stable under the loadings expected from overlying soils and other construction equipment at the ground surface.

Table 12.11 Critical velocity in different soil types

Soil Type	Velocity (ft/s)
Sand and sandy loam	3.5
Silt and silt loam	5.0
Silty clay loam	6.0
Clay and clay loam	7.0
Coarse sand and gravel	9.0

Source: SCS (1972).

Table 12.12 **Drain grades (feet per 100 feet) for selected critical velocities**[a]

Drain Size (in.)	Velocity (ft/s)					
	1.4	3.5	5.0	6.0	7.0	9.0
For Drains with N = 0.011[a]						
Clay Tile, Concrete Tile, and Concrete Pipe (with Good Alignment)						
4	0.28	1.8	3.6	5.1	7.0	11.5
5	0.21	1.3	2.7	3.9	5.3	8.7
6	0.17	1.0	2.1	3.1	4.1	6.9
8	0.11	0.7	1.4	2.1	2.8	4.6
10	0.08	0.5	1.1	1.5	2.1	3.5
12	0.07	0.4	0.8	1.2	1.6	2.7
For Drains with N = 0.013						
Clay Tile, Concrete Tile, and Concrete Pipe (with Fair Alignment)						
4	0.41	2.5	5.2	7.5	10.2	16.8
5	0.31	1.9	3.9	5.6	7.7	12.7
6	0.24	1.5	3.1	4.4	6.0	10.2
8	0.17	1.0	2.1	3.0	4.1	6.8
10	0.12	0.8	1.6	2.2	3.0	5.0
12	0.09	0.6	1.2	1.8	2.4	3.9
For Drains with N = 0.015						
Corrugated Plastic Pipe						
4	0.53	3.3	6.8	9.8	13.3	21.9
5	0.40	2.5	5.1	7.3	9.9	16.6
6	0.32	2.0	4.0	5.8	7.9	13.2
8	0.21	1.3	2.7	3.9	5.3	8.8
10	0.16	1.0	2.0	2.9	4.0	6.6
12	0.13	0.8	1.6	2.3	3.1	5.1

Source: SCS (1972).

[a]*N* is the roughness coefficient, must be obtained from the pipe manufacturer.

They should not buckle, crush, or undergo large deflections. The procedures for structural stability analysis are the same as those used for the leachate collection pipes in landfills and are described in detail in Chapter 18.

Filter and Envelopes. The primary function of a filter is to prevent soil particles from entering and clogging a drain. Filters should always be used where soils have a high percentage of fines. The function of an envelope is to improve water flow and reduce flow velocity into drains by providing a material that is more permeable than the surrounding soil. Envelopes may also be used to provide suitable bedding for a drain and to stabilize the soil material on which the drain is being placed. Envelopes are required for most applications. Although filters and envelopes have distinctly different functions, well-graded sands and gravels can be used to meet the requirements of both a filter and an envelope. Geotextiles are also widely used as filters. They are generally made of polypropylene, polyethylene, polyster, or polyvinyl chloride. Filter fabric should be selected based on its compatability with the leachate.

The general procedure for designing a gravel filter is to (1) make a mechanical analysis of both the soil and the proposed envelope material, (2) compare the two particle distribution curves, and (3) decide by

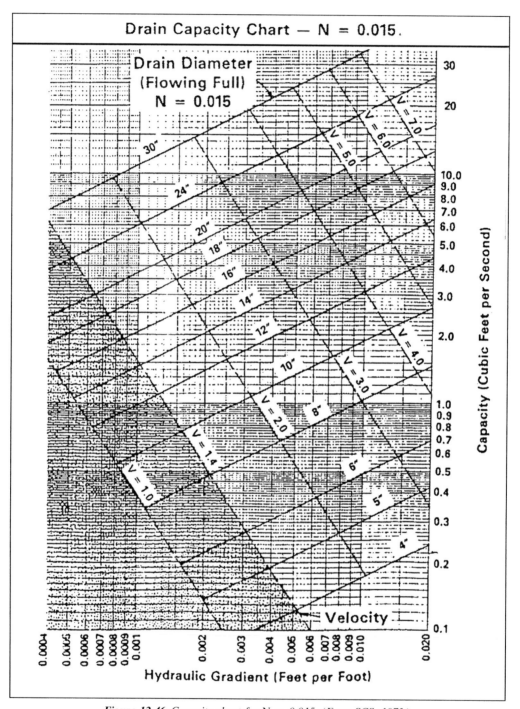

Figure 12.46 *Capacity chart for* N = 0.015. (*From SCS, 1972.*)

some set of criteria whether the envelope is satisfactory. SCS (1972) and others developed various criteria that set size limits for a filter material based on the size of the base material. A filter is considered satisfactory if it allows some of the fine soil particles to pass through so as not to plug the filter but retains larger particles that would deposit in the drain. For synthetic materials, the suitability of a filter can be determined from the ratio of particle size distribution to the pore size of the fabric.

The first requirement of sand and gravel envelopes is that the envelope have a hydraulic conductivity higher than that of the base material. SCS (1972) generally recommends that all of the envelope material should pass a 1.5-in. sieve, 90% should pass a 0.75-in. sieve, and not more than 10% should pass a No. 60 sieve (0.01 in.). This minimum limitation is the same for filter materials. However, the gradation of the envelope is not important since it is not designed to act as a filter (SCS, 1972). Recommendations for gravel envelope thickness have been made by various agencies. The Bureau of Reclamation recommends a minimum thickness of 10 cm around the pipe, and the Soil Conservation Service recommends a minimum of 8 cm for agricultural drains.

Manholes. Manholes are used in subsurface drainage systems to serve as junction boxes between drains; as silt and sand traps; as observation wells; and as access points for pipe location, inspection, and maintenance. Manholes should be located at junction points, changes in alignment or grade, and other designated points. There are no set criteria for manhole spacing. A manhole should extend from 12 to 24 in. above the ground surface for ease of location. The base of the manhole should be a minimum of 18 in. below the lowest pipe to provide a trap for sediments. Manholes are typically designed to have a drop in elevation between the inlet and outlet pipes, to compensate for head losses in the manhole (USBR, 1978). A typical manhole design is shown in Figure 12.47.

Sump and Pumping System. Contaminated groundwater is collected by gravity flow in a drainage sump from which it is pumped to treatment (Figure 12.48).

The major steps in designing the sump and pumping system include the following (USBR, 1978):

1. Determine the maximum inflow (Q_p) to the sump. Maximum inflow is based on total discharge (Q). An extra 20% allowance is usually made for flows in excess of design discharge.

2. Determine the amount of storage required. The cycling operation of the pump determines the amount of storage required. Maximum storage occurs when the inflow rate is one-half the discharge of the pump. Therefore, storage volume (S_v) is equal to one-half the cycling time (six for a 12-minute cycle pump) times Q_p.

3. Determine the pumping rate. The pumping rate Q_m (ft^3/min) is determined from the following equation:

$$Q_m = \frac{S_v + Q_p t_p}{t_p} \qquad (12.15)$$

where Q_m is the pumping rate (ft^3/min), S_v the storage volume (ft^3), Q_p the maximum inflow (ft^3/min), and t_p the running time of the pump (min).

4. Determine the start, stop, and discharge levels. In general, the maximum water level for starting the pump should be about the top of the pipe drain discharging to the sump. The minimum elevation should be about 2 to 4 ft above the base of the sump.

5. Determine the size of the sump. The volume required for storage plus the criterion that the minimum water level should be 2 to 4 ft above the bottom of the sump determine the size of the sump.

6. Select the pump. Selecting a pump for a particular application requires that the total head capacity be determined. Manufacturers' performance curves can then be used to determine pump efficiency and necessary horsepower. Centrifugal or diaphragm pumps are generally used.

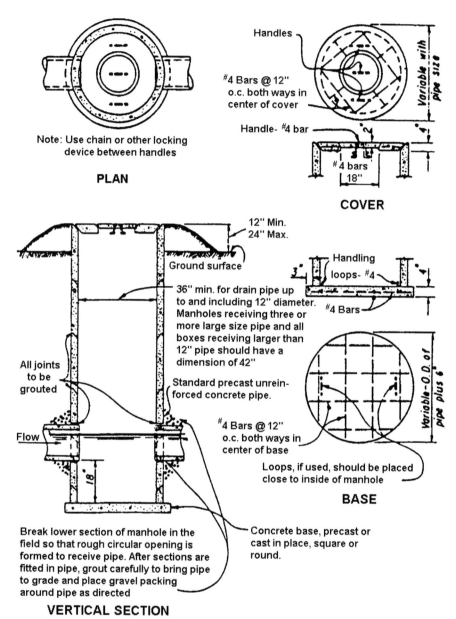

Figure 12.47 *Typical manhole design for a closed drain. (From USBR, 1978.)*

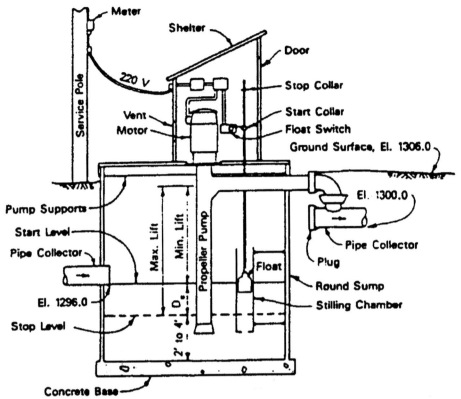

Figure 12.48 *Typical design of an automatic drainage pumping plant. (From USBR, 1978.)*

12.6.3 Construction

The major activities associated with construction of subsurface drains are (1) trench excavation, (2) installation of the drains, (3) placement of envelope and filter materials, and (4) backfilling.

Trench Excavation. Trench excavation is usually accomplished using either trenching machines or backhoes. Cranes, clamshells, and draglines are also used for deep excavation. The factors that influence the rate of trenching include (1) soil moisture; (2) soil characteristics such as hardness, stickiness, and stones; and (3) depth and width of trench. Generally, continuous trenching in suitable materials is much faster than trenching via backhoe. Trenchers can be equipped with back-end modifications to provide shoring, to install a

geotextile envelope, to lay tile of flexible piping, to place the piping, and to backfill with gravel or excavated soil. Backhoes can excavate earth and fragmented rock up to one-half of the bucket diameter to a depth of 40 to 60 ft. A crane and clamshell can be used for deeper excavation or when access excludes use of a backhoe. Use of draglines is generally limited to the removal of loose rock and earth.

Excavation of a trench through material containing numerous large boulders or hard rock layers results in considerable construction delays and increases the cost of construction substantially. Typically, these materials must be fractured to facilitate their removal. The most commonly used method for fragmenting rock in hazardous waste site work involves the use of rotary or percussion drills; backhoe-mounted pneumatically driven impact tools, and tractor-mounted mechanical rippers. Blasting, although commonly used in the construction industry for rock fragmentation, is not recommended for hazardous waste site work.

Proper grade control in subsurface drain ensures against ponding of water and provides for a nonsilting velocity in the drainage pipe. Proper grade control can be accomplished using automatic laser or visual grade-control systems. *Laser systems* are adaptable to a wide range of earth-moving equipment, including trenchers and backhoes. In *visual grade control,* grade stakes of equal length are driven to the design subgrade at selected points along the trench line. A line drawn through the top of the grade stakes would be parallel to the design grade of the trench. Targets are driven next to the grade stakes and are adjusted to a fixed distance above the elevation of the grade stakes. The selection of this distance depends on the depth of the trench and the line of sight between the machine operator and a reference sighting rod on the machine. When the trenching machine is cutting on grade, the target will align with the reference sighting rod.

Proper installation of drains (i.e., maintenance of grade, placement, and alignment of pipes) generally requires dewatering to achieve a dry environment. Three basic options are available for dewatering: open pumping, predrainage using wellpoints or a well system, and groundwater cutoff. These techniques may be used separately or in combination. Open pumping involves construction of a sump hole or pit at the lowest point of excavation so that water can flow toward and collect in the pit. Wellpoints and deep wells can be used to lower the water table near a trench excavation. Wellpoints are one of the most widely used and most versatile dewatering technologies. Groundwater cutoff barriers such as steel sheet piling, concrete, or a bentonite slurry may also be used together with wells and wellpoints to reduce the size of the predrainage system required.

Trench excavations generally require the use of wall stabilization methods to prevent cave-ins during installation of drain pipes. With shallow trenches in stable soils, the need for shoring can be eliminated by cutting the trench with sloped walls so that a stable angle is attained (usually a 1.5 H:1V slope). Shoring, which involves supporting the trench wall with wooden or steel structures, is the most commonly used method

of wall stabilization. Shoring methods for supporting shallow trenches involve the use of slipshields (constructed on-site by welding I-beams between two parallel pieces of sheet steel) and adjustable aluminum bracing. For trenches that are deeper than about 10 ft, steel sheet piling or steel H-piles with horizontal wooden beams between them can be driven and braced to support trench walls.

Installation of Drains. All subsurface drains must be laid on a stable bed with the desired grade. Trenches that have inadvertently been overexcavated should be refilled with dry soil and brought to grade with envelope material. Well-graded gravel is then laid in an even layer several inches thick to provide bedding for the pipes. Pipe installation generally begins at the lowest trench excavation and proceeds upgrade. During installation, water may be removed by allowing it to flow through tubing installed previously (USBR, 1978). The envelope material can then be laid on top of the stabilized floor.

To take advantage of the characteristics of flexible tubing, equipment capable of automatic pipe installation should be used. Trenching machines can be modified to include a hopper for bedding/envelope material with chutes to deliver the material, a rack for a roll of tubing and/or filter fabric that is designed and located to minimize stretching, and a conveyor for automatic backfill (SCS, 1972). Rigid pipes cannot be installed automatically, and long lengths of pipe are either hand carried or lowered by crane into the trench. When extending drainage systems under roadways, structures, root zones, or areas not requiring drainage, unperforated pipe should be specified.

Placement of Envelope and Filters. Gravel envelopes are installed around a pipe drain to increase flow into the drain and reduce the buildup of sediments in the drain line. They may be placed by hand, backhoe, or by a hopper cart or truck. When using continuous trencher drain installation machines, gravel filling may be ongoing along with other operations.

Filter fabrics are sometimes installed around a gravel envelope to prevent fines from clogging the envelope and drain pipe. When constructing a drain using a fabric filter wrapping, the fabric is installed first, followed by the bedding, the pipe, and the envelope prior to backfilling with soil. Fabric filters can be installed manually or by machine.

Backfilling. After the gravel envelope has been installed, the trench must be backfilled to the original grade. Prior to backfilling, the drain should be inspected for proper elevation below the ground surface; proper grade and alignment, broken pipe, and thickness of the gravel envelope. The inspector should ensure that pipe drains and manholes are free of deposits of mud, sand and gravel, or other foreign matter, and are in good working condition.

Almost any type of excavation equipment can be used to backfill trenches, including backhoes, bulldozers, scrapers, and combination backhoe/front-end loaders. During backfilling, care should be taken to ensure that the drain is not disturbed either vertically or horizontally. About 1 ft of fill should be placed carefully over the envelope before starting the general backfill operation (USBR, 1978).

Geotextile fabric may be used on the top of the envelope to prevent siltation of the envelope from the backfill materials. To prevent settling of the backfill after construction, periodic compaction of soil lifts is also required. This may be accomplished using air tamping or a vibrating or sheepsfoot compactor.

12.6.4 Performance and Economic Considerations

After installation of the subsurface drain is complete, the drain should be tested for obstructions. For a short drainage systems, this can be done visually by shining a high-powered flashlight through a drain from one manhole and observing the beam in another. TV camera inspections may be used for large-diameter drains. Mechanical methods can be used both to remove obstructions and to test for obstructions. Manholes and silt traps should be checked frequently for sediment buildup for the first year or two of operation.

Piezometers may be installed in the various parts of a drainage system to identify operational problems with the filter, envelope, pipe, or other components of the system. Piezometers can measure the loss of head

through a medium and thus can identify obstructions to flow, such as a clogged envelope or filter. Monitoring wells can also be installed downgradient of the drainage system. Detection of contaminants would indicate a malfunction or failure of the system.

Malfunction of subsurface drains can be attributed to chemical clogging due to buildup of calcium carbonate precipitates and iron and manganese deposits, clogging due to biological slimes, or a variety of physical mechanisms, such as formation of sinkholes or blowouts due to pipe breakage, and root penetration. Clogged drain pipes can be corrected using high-pressure water jetting equipment, mechanical scrapers, or brushes. In some cases, chemicals may be needed to remove difficult deposits. When there is a structural problem, such as drain breakage or improper drain spacing causing a sinkhole, the drain must be dug up and the condition corrected. Malfunctioning perforated pipe drains located near root systems should be dug up and replaced with nonperforated pipe.

Subsurface drains are most useful in preliminary containment applications for controlling contaminant migration while a final treatment design is developed and implemented. They also provide a measure of long-term protection against residual contaminants following conclusion of treatment and site closure. Subsurface drain systems have been used successfully for groundwater plume containment at several large shallow sites. For example, at the Sylvester hazardous waste site in Nashua, New Hampshire, a subsurface drain system was installed as a method to contain spread of the leachate plume until a remedial cleanup action could be implemented. The system was operated for one year, until a vertical barrier wall and cap were constructed over the 20-acre site (USEPA, 1989a).

Costs for installation and operation of subsurface drains can be divided into four categories: (1) installation costs, (2) material costs, (3) engineering supervision, and (4) operation and maintenance. Installation costs depend primarily on the depth of excavation, stability of soils, extent of rock fragmentation required, and groundwater flow rates. The major material costs include pipes, gravel, manholes, and pumps and other accessories for the drainage sump. Engineering and supervision involves such activities as staking the drain line, checking for grade control and alignment; and checking pipe specification and pipe quality. Engineering and supervision costs are usually about 5 to 19% of the total; however, these costs can be considerably higher, depending on geologic and hydrogeologic conditions. Capital costs associated with installation of subsurface drains are typically much higher than those associated with pumping systems. This is particularly true where substantial rock excavation is required and deep drains requiring extensive shoring are needed. These factors may result in exclusion of drains as cost-effective remedial action. However, operation and maintenance costs associated with drains are generally lower than with pumping, provided that the system is properly designed and maintained. Lower operation and maintenance costs become significant, particularly where contaminant plume removal is needed over a long period of time. Based on the past typical field installations, the total cost of the subsurface drains varied from $6 to $86 per square foot, depending on the complexity of the site conditions.

12.7   SUMMARY

Different passive and active in-situ containment techniques may be employed to contain the waste effectively. The most common passive in-situ containment methods include isolating the waste (or contaminated zone) with low-permeability barriers. All or a combination of vertical barriers, bottom barriers, and surface caps are used to achieve this. Because hydraulic conductivities of the barrier materials are very low (10^{-6} to 10^{-12} cm/s), hydraulic flow is drastically reduced. Molecular diffusion may cause chemical flow across the barriers due to concentration gradients, and it should be assessed. Chemical compatibility between the materials used in these barriers and caps and the wastes is very important. Site-specific compatibility tests should be conducted to determine the suitability of the materials used in the containment system. Long-term performance of the barriers is difficult to assess; therefore, long-term groundwater monitoring programs

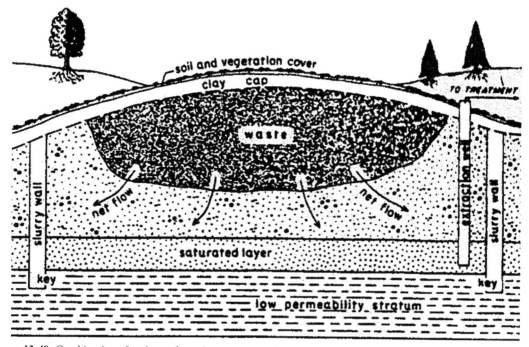

Figure 12.49 *Combination of active and passive waste containment systems. (From Keith and Murray, 1994. Courtesy of the Society of Mining, Metallurgy, & Exploration* (SME), *www.smenet.org.)*

Table 12.13 **Advantages and disadvantages of various in-situ waste containment systems**

Containment Type	Advantages	Disadvantages
Slurry barriers	Long-term economical method of groundwater control No maintenance required over long term Materials inexpensive and available Technology well proven Capital cost of construction of vertical barriers varies from $5 to $15 per square foot; O&M cost is usually small and depends on the monitoring requirements	Groundwater or waste leachate may be incompatible with slurry material Lack of near-surface impermeable layer; large boulders or underground caverns may make installation difficult or impractical Difficult to construct with over 10% slopes
Grout barriers	Can inject grout through relatively small diameter drill holes at unlimited depths Size of the grouted column is a function of pore space volume and volume of grout injected Grout can incorporate and/or penetrate porous materials in the vicinity of the injection well such as boulders or voids Various setting times and low viscosities are also advantages	Limited by the permeability of the soil or rock Uncertainty of complete cutoff Particulate grouts can only be used in the most permeable units Grout barriers are not amenable to inspection; quality control is a difficult issue since small voids or breaks can greatly lessen the effectiveness of a grout curtain
Composite barriers	Geomembrane provides an effective barrier Good compatibility with a wide range of contaminants	Sealing is a difficult process that requires material handling and manipulation not afforded by trench emplacement Keying the membrane adequately to the impervious layer is also difficult Geomembrane installation is dependent on moisture and weather conditions Expensive
Steel sheet pile barriers	Sheet pilings require no excavation; thus, the construction is relatively economical In most cases, no maintenance is required	Lack of effective seal between pilings Corrosion of piles can be a severe problem Many sites contain mineral acids that react readily with iron; reactions can produce hydrogen gas that may diffuse from the soil and create a fire or explosion hazard at the surface
Pumping	System may be less costly than construction of an impermeable barrier High degree of design flexibility Moderate to high operational flexibility, which will allow the system to meet increased or decreased pumping demands as site conditions change Conventional readily available technology Wells can be drilled to significant depth at most locations which access by the required drilling equipment is available	Plume volume and characteristics will vary with time, climatic conditions, and changes in the site resulting in costly and frequent monitoring System failures could lead to contamination spreading O&M costs are higher than for barriers Capture zones not significant compared to the drains

Table 12.13 (*continued*)

Containment Type	Advantages	Disadvantages
Subsurface drains	Operation costs are relatively cheap since flow to underdrains is by gravity Provides a means of collecting leachate without the use of impervious liners Considerable flexibility is available for design of underdrains; spacing can be altered to some extent by adjusting depth or modifying envelope material Systems fairly reliable, provided that there is continuous monitoring They have a relatively simple construction They are relatively inexpensive to install	Not well suited to poorly permeable soils In many instances it may not be feasible to situate underdrains beneath the site System requires continuous and careful monitoring to ensure adequate leachate collection Open systems require safety precautions to prevent fires and explosions They are not useful for sites where contamination is deep They may interfere with other operations at a facility Potential clogging of drains Excavation of large volume of contaminated soil requires special handling and disposal Dewatering of potentially contaminated water may be required during trench excavation, which also needs special handling and disposal

are needed to ensure that the contaminants are not migrating through the barriers.

The most commonly used active in-situ containment methods include groundwater pumping and subsurface drains. Active in-situ waste containment systems are generally less costly to construct than passive containment systems. Such systems offer a high degree of flexibility in their design; new wells/drains can be installed at minimal expense and pumping/injection rates can be varied to meet the objectives of the system. Environmental impact at the surface is negligible. On the other hand, operation and maintenance costs of active containment systems are commonly very high and monitoring of the system may also be expensive.

Integrated containment systems must be properly evaluated. A combination of passive and active systems can be used to ensure effective containment of contaminated zone or waste. When used in combination, the general approach is to use passive barrier systems to minimize the quantity of groundwater that must be addressed by active systems. Pumping can divert groundwater and reduce the contact of contaminants with barrier materials. Pumping can reduce the hydrostatic

pressure exerted against the barrier. Hydraulic gradients across the barrier can be reduced, thus reducing the seepage through the barrier.

The most common integrated containment system is the use of a circumferential slurry barrier, keyed into an underlying aquiclude, combined with an interior pumping system to maintain an inward hydraulic gradient. Barriers are also used with capping to confine a waste area fully and to prevent clean water from leaching through the wastes. An appropriate interface between the SB barrier and the cap should be designed. Figure 12.49 shows an example of how vertical barriers, cap, and pumping may be combined. Vertical barriers and subsurface drains can be used together. Subsurface drains can intercept water before it reaches the disposal area, thus controlling leachate generation. The interception of groundwater in the vicinity of a slurry wall also serve to relieve water pressure on the barrier itself.

Each containment method has several advantages and disadvantages, summarized in Table 12.13. Each method must be assessed carefully for its applicability at a given site based on the site-specific hydrogeologic and contaminant conditions.

QUESTIONS/PROBLEMS

12.1. Cite examples where in-situ containment is preferred over in-situ remediation.

12.2. A slurry trench barrier is proposed in loose saturated sand with an angle of internal friction of 28° and a unit weight of 115 lb/ft^3. Calculate the minimum unit weight of slurry required to ensure a stable trench.

12.3. A vertical slurry trench needs to be made in dry silty sand with an angle of internal friction of 30° and a unit weight of 100 lb/ft^3. Determine the minimum unit weight of the slurry required to ensure trench stability.

12.4. A 30-ft-deep slurry trench barrier is proposed in loose sand with an angle of internal friction of 28° and a unit weight of 115 lb/ft^3. Groundwater is encountered 10 ft below the ground surface. Slurry is proposed to be maintained 5 ft or less below the ground surface. Calculate the minimum unit weight of the slurry required to ensure a stable trench.

12.5. A vertical slurry trench needs to be made to contain the groundwater plume at a site. The site soils are predominantly silty sand with an internal angle of friction of 30° and a unit weight of 110 lb/ft^3. Groundwater at the site is encountered 5 ft below the ground surface. Determine the unit weight of the slurry and the level where it should be maintained to ensure trench stability.

12.6. A vertical slurry barrier is 200 ft long, 50 ft deep, and 3 feet thick. The hydraulic gradient across the barrier is 5, and the barrier's permeability is 1×10^{-7} cm/s. Calculate the leakage rate through the barrier.

12.7. For Problem 12.6, what gradient should be maintained to limit the leakage rate to 50 gal/day.

12.8. A vertical slurry barrier is proposed at a site to contain the groundwater plume. Determine the hydraulic conductivity and thickness of the barrier to maintain a water-level difference across the barrier of 10 ft.

12.9. Explain why CB barriers have low permeability compared to SB barriers.

12.10. Postconstruction monitoring of a CB barrier revealed that the barrier permeability is increasing with time due to interaction with the site contamination. What remedial actions can be taken to provide effective containment of the groundwater plume?

12.11. List advantages and disadvantages of active in-situ waste containment systems.

12.12. The following data are given for a contaminated site: hydraulic conductivity = 1×10^{-3} cm/s, water table at the ground surface, depth to impermeable layer = 20 ft, hydraulic gradient = 0.002, and average rainfall rate = 10 mm/day. To minimize the leachate generation, the water table needs to be deeper than 5 ft from the ground surface. A subsurface drainage system is proposed to achieve this. Determine the drain spacing. Show a layout of the subsurface drainage system.

12.13. A subsurface drainage system is proposed to drop the water table from the ground surface to a depth of 5 ft. The hydraulic conductivity of the aquifer is 1×10^{-4} cm/s and the depth to an impermeable layer is 20 ft. If the depth of the drains is to be 15 ft, calculate the drain spacing required.

12.14. Toxic wastes were disposed of in an unconfined aquifer at a site. The saturated thickness of the aquifer is 30 ft and the aquifer has a hydraulic conductivity of 1×10^{-2} cm/s and a specific yield of 0.3. The water table is required to be lowered by 10 ft to minimize leachate generation. Design a pumping system to accomplish this.

REFERENCES

Cavalli, N. J., Composite barrier slurry wall, pp. 78–85 in *Slurry Walls: Design, Construction and Quality Control*, Paul, D. B., and Davidson, R. R, and Cavalli, N. J. (eds.), ASTM STP 1129, ASTM, West Conshohocken, PA, 1992.

Daniel, D. E., and Koerner, R. M., Vertical cutoff walls, pp. 279–297 in *Waste Containment Facilities: Guidance for Construction, Quality Assurance and Quality Control of Liner and Cover Systems,* ASCE Press, New York, 1995.

D'Appolonia, D. J., Soil–bentonite slurry trench cutoffs, *J. Geotech. Eng.,* Vol. 106, No. GT4, pp. 399–417, 1980.

Donnan, W. W., Spacing and depth criteria for tile drains in irrigated land, thesis, Iowa State College, Ames, IA, 1946.

Evans, J. C., Vertical cutoff walls, in *Geotechnical Practice for Waste Disposal,* Chapman & Hall, New York, 1993.

Hooghoudt, S. B., Bijra qen tot de kennis van eenqe natuurkundige grootheden van den goo nd. 7: Alqemeene beachonwing van het problem van de detail ontwatering ende infiltrate doon middle van parallel loopende drains, greppels, slooten en kanalen, *Verst. Lands. Ond.,* Vol. 46, pp. 515–707, 1940a.

———, *Hooghoudt's Theory of Drainage* (Voor Cultuurtechnik en Waterhuishouding), translated in English from the original publication of Hooghoudt by the Institut Voor Cultuurtechnik en Waterhuishouding, 1940b.

Javandel, I., and Tsang, C. F., Capture-zone type curves: a tool for aquifer cleanup, *Ground Water,* Vol. 24, No. 5, pp. 616–625, 1986.

Johnson, A. I., Frobel, R. K., Cavalli, N. J., and Pettersson, C. B. (eds.), *Hydraulic Barriers in Soil and Rock,* STP 874, ASTM, West Conshohocken, PA, 1985.

Keith, K. S., and Murray, H. H., Clay liners and barriers, in *Industrial Minerals and Rocks,* 6th ed., Carr, D. D., (ed.), Society of Mining, Metallurgy, and Exploration, Littleton, CO, 1994.

Mitchell, J. K., Physical barriers for waste containment, pp. 951–962 in *Proceedings of the First International Congress on Environmental Geotechnics,* Edmonton, Alberta, Canada, 1994.

Mitchell, J. K., and Rumer, R. R., Waste containment barriers: evaluation of the technology, pp. 68–82 in *Proceedings of the Conference on In-Situ Remediation of the Geoenvironment,* Minneapolis, MN, 1997.

Mitchell, J. K., and van Court, W. A., *Contaminant Immobilization and Containment: Barriers, Walls and Covers,* UCB/GT/92-09, University of California, Berkeley, CA, 1992.

Paul, D. B., Davidson, R. R., and Cavalii, N. J. (eds.), *Slurry Walls: Design, Construction, and Quality Control,* STP 1129, ASTM, West Conshohocken, PA, 1992.

Rumer, R. R., and Mitchell, J. K. (eds.), Assessment of barrier containment technologies: a comprehensive treatment for environmental remediation applications, in *Proceedings of the International Containment Technology Workshop,* 1996.

Rumer, R. R., and Ryan, M. E. (eds.), *Barrier Containment Technologies for Environmental Remediation Applications,* Wiley, New York, 1995.

SCS (Soil Conservation Service), *Maintaining Subsurface Drains,* No. 557, U.S. Department of Agriculture, Washington, DC, 1972.

Sharma, H. D., and Lewis, S. P., *Waste Containment Systems, Waste Stabilization, and Landfills: Design and Evaluation,* Wiley, New York, 1994.

Todd, D. K., *Groundwater Hydrology,* Wiley, New York, 1980.

USACE (United States Army Corps of Engineers), *Technical Guidelines for Hazardous and Toxic Waste Treatment and Cleanup Activities,* EM 1110-1-502, U.S. Department of the Army, Washington, DC, 1994.

———, *Checklist for Hazardous Waste Landfill Cover Design,* ETL 1110-1-162, U.S. Department of the Army, Washington, DC, 1995.

———, *Checklist for Design of Vertical Barrier Walls for Hazardous Waste Sites,* ETL 1110-1-163, U.S. Department of the Army, Washington, DC, 1996.

USBR (U.S. Bureau of Reclamation), *Drainage Manual,* U.S. Department of the Interior, Washington, DC, 1978.

USEPA (U.S. Environmental Protection Agency), *Slurry Trench Construction for Pollution Migration Control,* EPA540/2-84-001, USEPA, Cincinnati, OH, 1984a.

———, *Compatibility of Grouts with Hazardous Wastes,* EPA600/2-84-001, USEPA, Washington, DC, 1984b.

———, *Handbook for Remedial Action at Waste Disposal Sites,* EPA/625/6-85/006, USEPA, Washington, DC, 1985a.

————, *Leachate Plume Management*, EPA/540/2-85/004, USEPA, Washington, DC, 1985b.

————, *Corrective Action: Technologies and Applications*, EPA/625/4-89/020, USEPA, Washington, DC, 1989a.

————, *Technical Guidance Document: Final Covers on Hazardous Waste Landfills and Surface Impoundments*, EPA/530-SW-89-047, USEPA, Washington, DC, 1989b.

————, *Stabilization Technologies for RCRA Corrective Actions*, EPA/625/6-91/026, USEPA, Washington, DC, 1991.

————, *Engineering Bulletin: Slurry Walls*, EPA/540/S-92/008, USEPA, Washington, DC, 1992.

————, *Engineering Bulletin: Landfill Covers*, EPA/540/S-93/500, USEPA, Washington, DC, 1993.

————, *Evaluation of Subsurface Engineered Barriers at Waste Sites*, EPA/542/R-98/005, USEPA, Washington, DC, 1998.

Van der Molen, W. H., and Wesseling, J., A solution in closed form and a series solution to replace the tables for the thickness of the equivalent layer in Hooghoudt's drain spacing formula, *Agric. Water Manag.*, Vol. 19, pp. 1–16, 1991.

Wesseling, J., A comparison of the steady state drain spacing formulas of Hooghoudt and Kirkham in connection with design practice, *J. Hydrol.*, Vol. 2, pp. 25–32, 1964.

Xanthakos, P. P., *Slurry Walls as Structural Systems*, 2nd ed., McGraw-Hill, New York, 1994.

13

SOIL REMEDIATION
TECHNOLOGIES

13.1 INTRODUCTION

Excavation of contaminated soil followed by transportation to and disposal in landfills is the most common practice of remediating these soils. The contaminated soil is excavated using conventional excavating equipment, transported by authorized haulers, and disposed of in a permitted landfill. Prior to disposal, the soil may require pretreatment to reduce concentrations below the land disposal restrictions stipulated by the regulations. This disposal approach is relatively simple, fast, and cost-effective for small volumes under any soil and contaminant conditions. In addition, regulatory approval and permits are relatively easy to obtain for this type of disposal. However, when the contaminated soil quantity is large and the contamination is deep, excavation and disposal in landfills may be very expensive and impractical. Moreover, many new environmental regulations require soil treatment rather than disposal in landfills. These factors have led to the development of numerous soil remediation technologies.

The soil remediation technologies can be implemented either (1) in-situ or (2) ex-situ. In-situ remediation methods treat the contaminated soils in place; thus, no excavation of the contaminated soils is required. In-situ remediation methods are preferred over ex-situ methods because they cause fewer disturbances to the site as well as less contaminant exposure to personnel and members of the public near the site. In addition, in-situ methods are less expensive than ex-situ methods.

All ex-situ methods involve excavation of soil from the site. The treatment can take place on-site, or the soils can be transported to another location for treat-ment. Site constraints may place practical limits on the potential for successful application of ex-situ remediation methods because the contaminated soil must be accessible for excavation. A shallow water table, buildings, overhead power lines, or underground utilities may limit the potential for excavating all of the soil requiring remediation. Space requirements may also limit the on-site use of ex-situ technologies. Adequate space is required for treatment equipment and for stockpiles of excavated soil awaiting treatment and cleaned soil awaiting final disposal.

The most popular soil remediation technologies are (1) soil vapor extraction, (2) soil washing, (3) stabilization/solidification, (4) electrokinetic remediation, (5) thermal desorption, (6) vitrification, (7) bioremediation, and (8) phytoremediation. These techniques, with some modifications, can be used for both in-situ and ex-situ remediation methods. All of these methods are based on manipulation of physical, chemical, electrical, thermal, or biological processes and aim to extract, immobilize, or detoxify the contaminants.

In this chapter we describe the following features for each of the foregoing remediation technologies: (1) the process, including applicability, advantages, and disadvantages; (2) the fundamental processes involved; (3) design and implementation; (4) predictive modeling; (5) modified and complementing technologies; (6) economic and regulatory considerations; and (7) case studies. We also discuss soil fracturing, which is not itself a remediation technology, but is often employed in combination with a remediation technology to enhance remediation in low-permeability soils.

13.2 SOIL VAPOR EXTRACTION

13.2.1 Technology Description

Soil vapor extraction (SVE) is a technique for removing volatile organic compounds (VOCs) and motor fuels from contaminated soils. This technology is known in the industry by various names, including *vacuum extraction, soil venting, aeration, in-situ volatilization, and enhanced volatilization.* Figure 13.1 shows a schematic of the SVE implementation in the field. It involves applying a vacuum to the contaminated soil through extraction wells, which creates a negative pressure gradient that causes movement of vapors toward these wells. The contaminant-laden vapors extracted from the wells are then treated aboveground using standard air treatment techniques such as carbon filters or combustion (USEPA, 1990a, 1991b, 1996b; USACE, 1995).

SVE is applicable when contaminants present in subsurface are volatile. This technology has been proven effective in reducing concentrations of volatile organic compounds (VOCs) and certain semivolatile organic compounds (SVOCs) found in petroleum and chlorinated products (USEPA, 1993c). In general, as a simplified guideline, a compound is considered a good candidate for remediation by SVE if it has (1) a vapor pressure value greater than 0.5 mmHg and (2) a Henry's law constant greater than 0.01 (USEPA, 1996c). The technology is applicable where soils are relatively homogeneous and highly permeable.

SVE technology has the following advantages:

- Needs easily available equipment and can be installed easily
- Causes minimum disturbance to the site
- Needs short treatment time, usually ranging from six months to two years under optimal conditions
- Is economical compared to other remediation technologies
- Is effective for treating both dissolved and free-phase (free-product) contaminants
- Is very easy to couple with other remediation technologies

The technology has the following disadvantages:

- Contaminant removal is often found very high in the beginning, and then lingering contaminant is found to exist for prolonged times.
- It is ineffective in low-permeable soils, stratified soils, and high-humic-content soils.
- Air emission treatment systems and permits are required, increasing the project cost.
- It is applicable for unsaturated soils only.

13.2.2 Fundamental Processes

Different processes occur during soil vapor extraction, as depicted in Figure 13.2. The main processes include volatilization, diffusion, advection, and desorption. *Volatilization* is the dominant process (USEPA, 1991b; USACE, 1995). The volatilization of the contaminants occurs as the soil-air (i.e., air in the soil matrix) laden with the vapor form of the contaminant is removed. As the soil-air carrying the contaminant in the vapor phase in the vadose zone gets removed due to the vapor extraction, the remaining soil-air will be less saturated with volatile substance. This, in turn, changes the state of equilibrium that existed prior to soil vapor removal. As the soil-air gets less saturated with contaminants, it paves the way for the dissolved and free-phase contaminants to partition from the saturated zone (portion below the groundwater table) and get into the unsaturated vadose zone. Then the soil-air will again be laden with VOCs. The vapor extraction process then removes it as the SVE operation is continued.

Advection is one of the processes that contribute to the decontamination process. As the vacuum is applied to the vadose zone, it initiates airflow toward the vapor extraction wells. As this air moves to the extraction wells, new air will be introduced into the contaminated soil zones. This circulation of air will occur during the entire vapor removal operation. This continuous movement of air within the subsurface resulting from the application of vacuum is called *advection.* Advection causes contaminant vapor removal as the SVE operation continues.

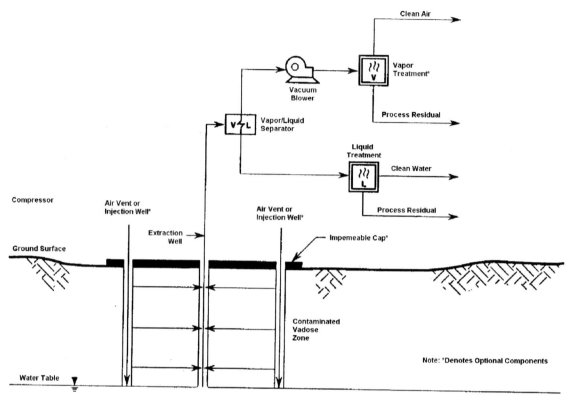

Figure 13.1 *Schematic of SVE implementation in the field. (From USACE,1995.)*

Diffusion of contaminant is an important consideration in SVE when either a free-liquid contaminant phase is floating on the groundwater table or contaminants are trapped within an impermeable soil lens surrounded by permeable material. In these situations, the SVE induces gas flow around, but not through, these contaminated zones, and diffusion processes dictate volatile contaminant removal. Diffusion of contaminants occurs perpendicular to the gas flow, thereby creating a concentration boundary layer. The total contaminant removal rate from the contaminated zone will depend on the vapor velocity induced by the SVE application, the free product thickness, and the vapor flow path, and will be limited by vapor-phase diffusion, liquid-phase diffusion, or a combination thereof. Overall, the efficiency of applying SVE to a heterogeneous contaminated soil will never be as high as an appli-cation to a homogeneous soil. In the case of free-floating phase on the groundwater table, it may be more efficient to remove the free phase by pumping and removing the residual by SVE.

Another major contaminant transport process that occurs during SVE operation is *desorption* of contaminant from the surface of the subsurface soil particles. This is accomplished as the *soil-air* is removed due to the applied vacuum. As the soil-air is sucked out from the subsurface it passes through the void spaces of the soil matrix. When it passes at high speed on a straight/meandering path, it will have a stripping power of dissolved/free-phase contaminant sorbed onto the soil particles. In addition, depending on the microscale vapor movement, it will also have a spiral movement around the soil particles, increasing the stripping capabilities as the vapor removal proceeds.

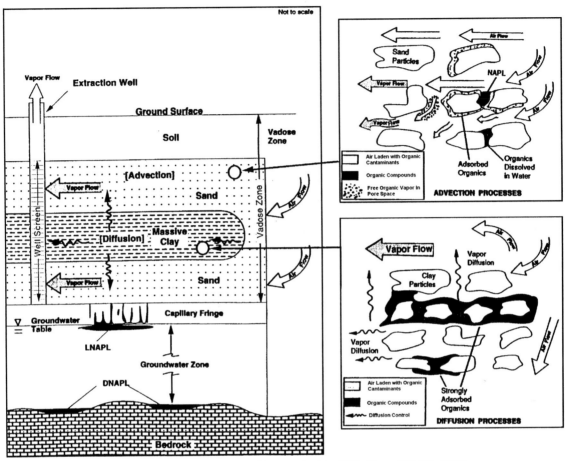

Figure 13.2 *Fundamental processes involved in SVE. (From USACE, 1995.)*

13.2.3 System Design and Implementation

General Design Approach. Figure 13.3 shows various tasks involved in the design and implementation of soil vapor extraction technology at actual contaminated sites. Initially, the site conditions should be evaluated to determine if the site soil and contaminant conditions are favorable to SVE application. SVE can also be integrated with bioventing (BV) if microorganisms are present that can detoxify the contaminants. More details on bioventing are given in Sections 13.2.5 and 13.8. If the site conditions are favorable for SVE, bench- and pilot-scale testing can be conducted to determine the airflow pathways, zone of influence, and cleanup levels that can be achieved by the implementation of SVE. The pilot testing involves the installation of wells, a vapor treatment system, and blower and monitoring devices to perform the actual cleanup operation on a much smaller scale. Then the result obtained from this pilot testing together with the mathematical model results, if available, can be used as a fundamental basis to start the actual design of the soil vapor extraction system.

The pilot-scale test results and mathematical modeling results are then used to design the full-scale system. The full-scale system design will include selection of numbers and spacing of extraction wells and ex-

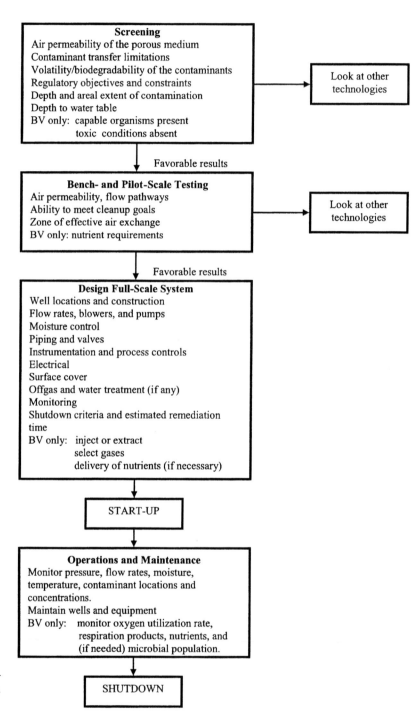

Figure 13.3 *General design approach to SVE and bioventing. (From USACE, 1995.)*

traction rates. The wells and treatment and monitoring systems are then installed and operated. The monitoring program will help document the contaminant concentrations after the system is in operation. If the concentrations are below the actual cleanup targets, remediation is halted but postmonitoring will be continued.

Equipment. Equipment selection can be made after performing pilot-scale testing. The pilot testing is done to evaluate extraction rates, areas of influence, and efficiencies. The methods employed are more or less similar to the procedure employed for testing hydraulic conductivity. The decision tree approach (Figure 13.4) can be used to determine if pilot-scale testing is required. The equipment needed for pilot-scale testing is the same as that required for full-scale implementation of SVE and includes the following:

- *Extraction wells.* Vertical extraction wells are designed and constructed similar to groundwater monitoring wells. Figure 13.5(*a*) shows the typical configuration of SVE wells. Where traffic is expected, flush-mounted wells are used, as shown in Figure 13.5(*b*), with the entire piping network located below the ground surface.
- *Manifold piping.* There is a need to install a pipe layout throughout for the entire project both above and below the surface. In cases where there is a need to use PVC or other plastic piping aboveground, the piping selected should be resistant to degradation from sunlight.
- *Blower or vacuum pump.* SVE requires a blower or vacuum pump typically driven by an electric motor. In the design, compliance with the specifications for type and size of blower, motor rating and horsepower, or vacuum/pressure range and airflow rate must be assured.
- *Condensate tank.* The condensate tank must meet specifications particularly for capacity, pressure/vacuum rating, construction, and level sensors. Usually, these tanks are used for air–water separation as the extraction proceeds.
- *Vapor treatment system.* An off-gas treatment system must be employed to remove VOCs prior to

discharge to the atmosphere. Table 13.1 summarizes the most common vapor treatment technologies. These methods include granular activated carbon (GAC) adsorption, thermal oxidation (incineration and packed-bed thermal processor), and catalytic oxidation.

Two basic performance test methods are typically used in any SVE pilot testing (USACE, 1995): (1) stepped-rate tests for estimating extraction rates, and (2) constant-rate tests for evaluating extraction areas of influence and efficiencies. *Stepped-rate tests* are conducted to evaluate the vapor recovery rates obtainable at various applied vacuums. The stepped-rate test data are used to develop the system curve: the air yield from the well versus the applied wellhead vacuum. This information is then used in designing the extraction rates, determining optimum recovery rates, and specifying blowers for a full-scale SVE system. *Constant-rate performance tests* are usually conducted following the stepped-rate tests to evaluate the actual area of influence and efficiency. Constant-rate performance tests are usually conducted under steady-state conditions to ensure that an empirical and representative area of influence is obtained. Constant-rate performance tests can take several hours to several days to complete.

To size the blower for the stepped-rate test, the steady-state flow equation for a vertical extraction well can be used to estimate the required vacuum to obtain a target flow rate:

$$P_{wt} = \frac{1}{2} \left\{ \frac{Q_T \mu_a \ln(R_w/R_I)}{L k_a} + \left[\left(\frac{Q_T \mu_a \ln(R_w/R_I)}{L k_a} \right)^2 + 4P_A^2 \right]^{\frac{1}{2}} \right\} \quad (13.1)$$

where P_{wt} is the target absolute pressure at test vent, Q_T the target flow rate, μ_a the viscosity of air, R_w the radius of the test vent, R_I the radius of the pressure influence for the test vent, L the effective vent length, k_a the estimated air permeability, and P_A the absolute atmospheric pressure. Once the pressure has been determined, the target flow rate (Q_T) can be determined. The targeted flow should be high enough to remove

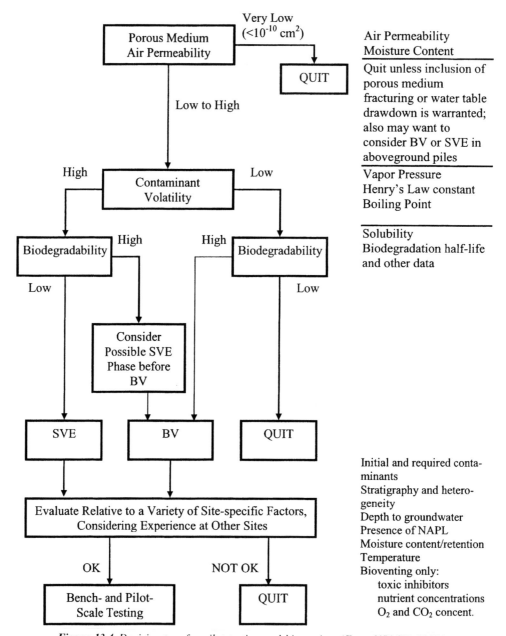

Figure 13.4 *Decision tree for pilot testing and bioventing. (From USACE, 1995.)*

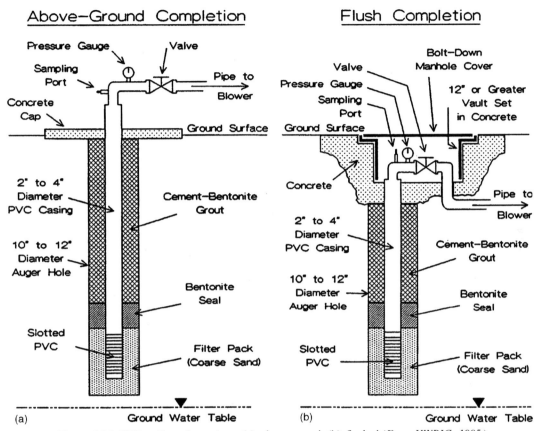

Figure 13.5 SVE well configurations: (a) aboveground; (b) flushed (From HWRIC, 1995.)

from the contaminated zone the number of soil pore volumes required by the final SVE design (USACE, 1995). Therefore, the targeted flow rate can be estimated by

$$Q_T = \frac{n\ \pi R_E^2 b n_a}{8.64 \times 10^4 \text{ s/day}} \qquad (13.2)$$

where n is the soil pore volumes per day, Re the extent of the zone of effective air exchange of the test vent (cm), b the unsaturated zone thickness (cm), and n_a the effective (air-filled) soil porosity (dimensionless).

The test blower should be capable of applying the required vacuum to the well and producing the target

flow rate at that vacuum. Depending on the test equipment layout and piping configuration, it may be prudent to factor in head losses in the test equipment itself. Usually, as much as 80 to 90% of vacuum can be lost in test equipment piping and through the extraction well. Consequently, a larger blower may be required to achieve the desired flow rates and vacuums at the extraction well.

The number and spacing of extraction wells are specified based on the radius of influence as determined from pilot-scale testing. The radius of influence (ROI) of each extraction well is the area in which the pressure change can contain the contaminant vapors. Alternatively, the number of wells may be calculated

Table 13.1 **Various technologies to control air emissions**

Control Technology	Advantages	Disadvantages
Carbon adsorption	Effective for gas streams with variable flow rates. Effective for gas streams with variable VOC content. Effective for gas streams with low VOC content.	Bed fires may occur if oxygenated material is present and bed temperature rises due to heat of adsorption. Spent carbon must be either regenerated or discarded. Filters/mist eliminators may be needed for liquids. Not effective for low-molecular-weight compounds. Less effective for high-humidity gas streams.
Thermal oxidation (incineration)	Widely demonstrated technology. Effective for a variety of VOCs.	Potential generation of PICs, and acid gases. Supplemental fuel required. Only effective for combustibles.
Catalytic oxidation (incineration)	Requires operating lower temperatures than thermal oxidizers. No disposal concerns.	Catalyst easily fouled or degraded. High temperature may cause burnout. Low heat content of waste gas stream will require extra fuel. Not effective for many chlorinated solvents.
Condensers	Effective for very high VOC concentrations. Good for pretreatment of dilute streams prior to other controls.	Performance somewhat sensitive to process conditions (flow rate, temperature). High utility costs.
Internal combustion engine	Compact units can provide usable power. Well-developed technology.	Limited capacity (<1000 ft^3/min) Supplemental fuel may be expensive. Easily fouled or corroded.
Membranes	Reduces waste stream volume. Concentrates VOCs.	Pretreatment only. Limited data available.
Operational control	Improved removal efficiency for minimal cost.	Requires knowledgeable operator.
Soil filters/biotreatment	Low cost. Simplicity. May degrade semi- or nonvolatile organics. May trap some metals. Not susceptible to variations.	Not effective on all VOCs. May require large surface area, biologically sensitive to temperature and humidity. High pressure drop.
Wet absorbers	Good for high-temperature gas streams. Simple to operate. Effective for wide variety of VOCs.	Low removal efficiency for low VOC concentrations. Effluent may pose disposal problems. Susceptible to concentration, flow rate, and temperature changes, less efficient at low flow rates.
UV	High removal efficiencies. No solvent or wastewater generated.	Complex system. Limited data available.

Source: USEPA (1992e).

theoretically. Johnson and Ettinger (1994) recommend the following equation to compute the required number of extraction wells:

$$N = \frac{\alpha M}{Q t_r} \qquad (13.3)$$

where N is the number of wells, α the minimum volume of air per unit contaminant mass required to achieve remediation under ideal flow conditions, M the mass of the contamination to be removed, Q the estimated flow rate to a single well, and t_r the desired remediation time. For single-component systems under ideal flow conditions, $1/\alpha$ is equal to the saturated vapor concentration. Similarly, USEPA (1994e) recommends the following equation to calculate the number of wells:

$$N = \frac{nV}{Q t_p} \qquad (13.4)$$

where n is the soil porosity (dimensionless), V the volume of soil to be treated, Q the air flow rate, and t_p the desired pore volume exchange time. The time required for pore volume exchange (t_p) can be related to the total time required for the site remediation (t_r) by the number of pore volume exchanges required to achieve the remediation goal. The number of pore volumes required for the cleanup generally ranges between 200 to 400, but it may be as high as 2000 to 5000 for some contaminated sites (USACE, 1995).

Operational Parameters. From an operational point of view, selection of vacuum pressure and airflow rates are crucial. For preliminary design purposes, the arrangement of wells and projected pumping rates can be estimated from air permeabilities derived from soil descriptions and laboratory testing using the following equation (Johnson et al., 1990):

$$\frac{Q}{H} = \frac{\pi k P_w}{\mu} \frac{1 - [P_{atm}/P_w]^2}{\ln(R_w/R_l)} \qquad (13.5)$$

where Q is the flow rate (ft^3/min), H the screen length (feet), k the soil permeability (ft^2 or darcy), P_{atm} the atmospheric pressure (atm), P_w the well pressure (atm), R_w the well radius (feet), and R_l the radius of influence (feet). The permeability (k) units can be converted into hydraulic conductivity (K) units: for example, 1 ft^2 = 9.11×10^7 cm/s and 1 darcy = 9.66×10^{-4} cm/s. [Refer to Freeze and Cherry (1979) for other such conversions.] Based on equation (13.5), the relationship between the extraction rates predicted for a range of soil permeabilities and applied vacuums, assuming a well radius of 2 in. and a 40-ft radius of influence during pumping can be presented as shown in Figure 13.6. Similar results for other parameters can be obtained from the equation (13.5). For the final design, it is important to consider overall system pneumatics prior to selecting the vacuum pressure and airflow rate. A systematic approach for estimating the relationship between flow rate and vacuum level for a system consists of the following (USACE, 1995):

1. Develop a relationship for vacuum level versus airflow in the subsurface.

2. Calculate the friction loss for the system components and piping for a range of flow rate.

3. Develop a "system" curve by adding the frictional losses calculated in steps 1 and 2.

4. Research and select a blower and determine the blower curve.

5. Predict the flow rate and vacuum level from the simulations (graphical) solution of the blower curve and system curve.

6. Balance the flows at each well, if necessary, and recalculate the vacuum level.

The *vacuum level and airflow* in the subsurface can be determined with relatively fair accuracy from site modeling or hand calculations based on pilot- or bench-scale studies. This allows the designer a means to predict the flow rate of air removed from the subsurface as a function of the vacuum level applied.

Calculation of *head loss* through the system components for a range of flow rates is fairly routine and not at all unique to soil vapor extraction. The most

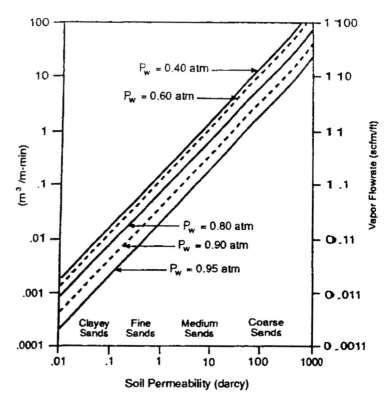

Figure 13.6 *Predicted extraction rates of a typical SVE system. (From Johnson et al., 1990.)*

common method of predicting friction losses in straight pipes is to use the Darcy–Weisback equation:

$$h_f = \frac{fL}{d} \frac{v^2}{2g} \qquad (13.6)$$

where: h_f is the head loss, f the friction factor, L the length of pipe, d the diameter of the pipe, v the average pipe velocity, and g the gravitational constant.

The friction losses from the subsurface, the straight pipe lengths, and the valves and fittings are added together to obtain the total friction loss at a given vacuum level. This calculation is repeated for several flow rates to establish a *system curve*. Then the blower curve is superimposed on the system curve. A specific blower would be selected based on mechanical, electrical, and pneumatic considerations. The pneumatic considerations are of primary concern. The predicted

flow rate and vacuum level obviously occur at the intersection of the two curves, representing the simultaneous solution of two equations. More details on this may be found in USACE (1995).

Monitoring. To determine the effectiveness of an SVE remediation system, monitoring probes are installed adjacent to extraction wells. The monitoring probes can be used to determine the vacuum and soil-air concentrations at any one point (USEPA, 1994e; USACE, 1995). To determine the vacuum at a monitoring probe, the probe is sealed with a threaded removable cap or septum to maintain a vacuum within the probe. A vacuum gauge or manometer may be tightly threaded through the top of the probe to provide a continuous reading, or a pressure transducer may be employed to provide more sensitive readings of the applied vacuum. Soil gas contaminant concentration may be measured with the probe by connecting a small vacuum to the probe through a valve, and pumping the soil-air to gas chromotography (GC) with a flame ionization detector (FID) or photo-ionization detector (PID).

Typically, monitoring probes are constructed with minimal screened intervals so as to characterize parameters at distinct depths. The probes may be installed in clusters with multiple intervals screened to evaluate the variation in parameters with depth. For shallow installations of approximately 7.5 m or less, hydraulically driven probes may be used.

For a single extraction well, this may entail the installation of at least two monitoring point clusters, which form a 90° angle with the extraction point, or three monitoring point clusters at 120° radials from the extraction point. Within each cluster, at least two different depths can be monitored individually, and more than one cluster can be situated along a given radial. For larger sites with many extraction points, the ratio of monitoring point clusters to extraction points can be reduced to between 1 and 2, as careful location of the monitoring points can supply data for more than one extraction point. As the size of the site and the number of extraction wells increases, it is usually not necessary to provide two monitoring points for each extraction well, although a ratio of at least 1 is recommended.

13.2.4 Predictive Modeling

Numerous models have been developed for the assessment of soil vapor extraction systems. A summary of selected models that have commonly been applied to soil vapor extraction is given in Table 13.2. One of the widely accepted models is called Hyperventilate (USEPA, 1993f). This is a user-friendly software program that can help designers to predict the performance of SVE systems. However, it will not help completely to design the vapor extraction system or provide information on how many days it should be operated. But it guides the user through a structured thought process to identify and characterize the required site-specific data, decide if SVE is appropriate at the site, evaluate the air permeability test results, calculate the minimum number of extraction wells needed, and show how the results at the site may differ from an ideal case.

13.2.5 Modified or Complementary Technologies

Instead of vertical wells, *trenches* or *horizontal wells* may be used for SVEs, as shown in Figure 13.7 (USEPA, 1990a; USACE, 1995). Trenches are useful when the contamination is limited to shallow depths. Horizontal wells are useful when treating the contamination under buildings or the areas that are not easily accessible for installing vertical wells or trenches.

Although less common, SVE can be implemented as an *ex-situ process,* as shown in Figure 13.8 (USEPA, 1990a). In this process, soil is excavated and placed over a network of aboveground piping to which a vacuum is applied to volatilize the contaminants. An off-gas system is used to collect and treat air emissions. Advantages of ex-situ SVE over in-situ SVE is that the excavation process creates an increased number of air passageways, shallow groundwater no longer limits the process, leachate collection is possible, and treatment is more uniform and easily monitored. Disadvantages of ex-situ SVE are that air emissions may occur during excavation and material handling, possibly requiring treatment, and a large amount of space is required.

An in-situ soil vapor extraction system may be designed to remediate soils in the vadose zone as well as groundwater; such remediation systems are known as *dual-phase extraction systems* (USEPA, 1996b). In such systems the wells are screened to encompass the contaminated soils in the vadose zone as well as the contaminated groundwater. When a vacuum is applied, contaminated vapors and groundwater are extracted. The extracted vapors and groundwater then proceed through a separator, where water and vapor are separated and sent to different treatment systems.

Soil vapor extraction can be coupled with some other remediation technologies, such as pump and treat, air sparging, bioventing, thermal desorption, and

Table 13.2 **Summary of selected models**

Model Name	Model Type and Use	Developer and Availability	Computer Requirements	Input Parameters	Output Parameters	Ease of Use
AIRFLOW/ SVE	Two-dimensional finite element radial symmetric airflow	Waterloo Hydrologic software available to public	IBM PC 386/486 with minimum of 4-Mb RAM, EGA or VGA display, and a math coprocessor. A mouse is recommended	Permeability, initial pressures, gas characteristics, temperature	Soil pressure distribution; total systems flow	Easy
AIRTEST, AIR2D, AIR3D	Two- and three-dimensional analytical radial-symmetric airflow	A L Baehr. USGS, C. J. Joss, Drexel University Water Resources Division, Mountain View Office Park, 810 Bear Tavern Road, Suite 206, West Trenton, NJ 06628	IBM PC/AT compatible, 512-kb RAM	Permeability test data, initial pressures, flow rates	Permeability, pressure distribution, and flow	Easy
CSUGAS	Three-dimensional finite difference vapor flow	J. Warner, Colorado State University, Civil Engineering Department, (303)481-8381 Available to public.	IBM PC/AT-compatible, DOS 5.1 or higher, 640-kB RAM, graphics monitor, math coprocessor, at least 595K available low memory	Permeability, porosity, initial pressures, topography	Soil pressure distribution; total systems flow	Moderate
Hyper-ventilate	Screening	Developed by USEPA and Shell Oil Company, Chi-Yaun Fan (project officer), 609/292-3131 Available through NTIS, 703/487-4650	Apple Macintosh, 2-Mb RAM HyperCard 2.0 or newer; or IBM 386 or IBM compatible PC, 4-Mb RAM, VGA through the use of Spinnaker PLUS in Microsoft Windows version 3.0 or higher	Permeability, porosity, initial pressures, topography, boiling-point data on spill components, and desired remediation time	Estimates of flow rates; removal rates; residual concentrations, no. of wells required	Easy
MAGNAS	Two- and three-dimensional finite element transport of water, NAPL, and air through porous media; can simulate the flow of air as a fully active phase	HydroGeologic, Inc. 1166 Hemadon Parkway, Suite 900, Hemadon, VA 22079, 703/478-6186	IBM PC/AT-compatible, code documentation and user's manual is available: written in Fortran 77	Heterogeneous and anisotropic media properties, capillary pressures, and permeability	Breakthrough curves of concentration vs. time, flow, and transport mass balances	Difficult

Source: USACE (1995).

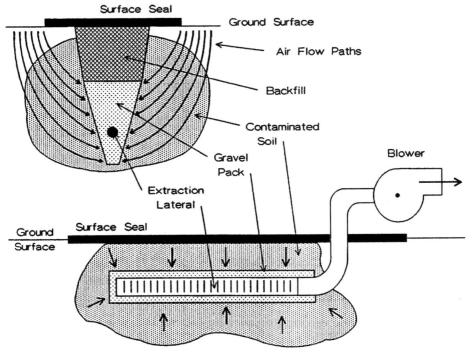

Figure 13.7 *Use of trenches and horizontal wells in SVE. (From HWRIC, 1995.)*

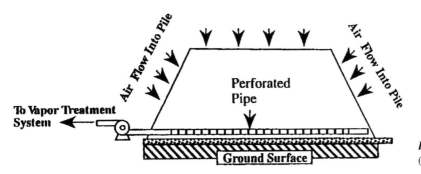

Figure 13.8 *Ex-situ SVE system.*
(From HWRIC, 1995.)

soil fracturing (USEPA, 1995g). When SVE is used with the *pump and treat technique,* the soil vapor extraction handles contaminants with high volatility to be removed from the unsaturated zone, while the pump and treat system removes the dissolved, and to a certain limited extent, sorbed contamination.

In coupling this system with *air sparging,* the air injected into the subsurface will start volatilizing the free phase and dissolved VOCs and strip them into the unsaturated (vadose) zone. Then the SVE will remove the volatilized VOCs in the soil-gas of the unsaturated zone. The vapor removed from VOCs will be treated before releasing it to the atmosphere.

Soil vapor extraction can also be modified or altered into *bioventing.* This can be achieved by injection of nutrition and microbes before the vapor extraction commences. While the vapor extraction is operating, the biodegradation of the contaminant will occur simultaneously, due to the microbes introduced within the subsurface.

The performance of soil vapor extraction is often enhanced with *thermal heating.* Heated air or steam is injected into the contaminated soils through the injection wells. The heated air or steam helps to loosen some less volatile compounds from the soil. The increased temperature also causes higher biodegradation rates. Thermal heating of the soil can also be accomplished by various methods, including *radiofrequency heating,* in order to better volatilize the contaminants and then recover the vapors using the SVE system.

Soil fracturing can enhance the performance of SVE by creating new or enlarging existing fractures in the subsurface. These fractures increase the soil permeability; increase the effective radius of recovery or injection wells, and increase potential contact area with contaminated soils.

13.2.6   Economic and Regulatory Considerations

A number of factors could affect the estimated cost of SVE treatment. Among them are the type and concentration of contaminant, the extent of contamination, groundwater depth, soil moisture, air permeability of the soil, site geology, geophysical site conditions, site accessibility, required support facilities and availability of utilities, and treatment goals. It is important to characterize the site thoroughly and properly before implementing this technology to ensure that treatment is focused on contaminated areas. The overall cost of in-situ SVE ranges from $10 to $150 per ton, and the overall cost of ex-situ SVE is under $100 per ton, including the cost of excavation but excluding treatment of off-gases and collected groundwater (FRTR, 1997).

Although SVE is considered to be a well-established technology, research is still needed on different mass transfer/transformation processes that occur during SVE, and developing and validating

mathematical models for predicting the performance of SVE systems in the field. This technology is readily acceptable to the regulating agencies. Approvals and permits may be required to establish cleanup standards, air emissions, wastewater discharges, and monitoring requirements. A monitoring program is designed such that any spreading of the contaminants in the subsurface during the SVE process can be detected.

13.2.7   Case Studies

Several case studies have been published that document successful use of SVE for the remediation of soils contaminated with volatile organic contaminants (FRTR, 1997, 1998). Two of these case studies are presented in this section.

Case Study 1: Holloman Air Force Base Sites 2/5, New Mexico. During the 1960s and 1970s, several releases of JP-4 jet fuel and diesel fuel occurred from several aboveground storage tanks at Holloman Air Force Base, New Mexico. Releases included chronic leaks and a 30,000-gal spill that occurred in 1978. The site soils consist of clean to silty sand deposits interbedded with silt and clay lenses. The groundwater at the site is located 10 ft below the ground surface. Soil borings showed that an 80 ft wide × 200 ft long area was contaminated, and the contamination extended to a depth of 16 ft below the ground surface. Soil samples collected and tested revealed that the site soils were contaminated with benzene, toluene, xylene, and ethylbenzene in concentrations of 48, 210, 500, and 180 mg/kg, respectively. The total petroleum hydrocarbons (TPHs) in the soil was found to be 17,500 mg/kg. It was determined that groundwater remediation was not required based on the quality of the groundwater and the lack of floating free-phase hydrocarbons at the site.

In 1994 and 1995, a SVE system was constructed at the site. The system included 22 extraction wells, a 2-hp blower, and a knockout tank to separate vapor and liquid phases in the extraction stream. Various extraction well configurations were used. The system was tested for vacuum pressures, exhaust volume, and air emissions. Operation of the system was such that air emissions were below the limits set by the regulating agency for allowable air emissions for organic compounds; therefore, an off-gas treatment system was not required.

As of October 1998, 44,000 lb of total petroleum hydrocarbons had been removed from the soil at the site. Soil testing showed that soil TPH concentrations had been reduced below the regulatory guidelines of 1000 mg/kg. Sampling also indicated that benzene concentrations were below the regulatory limit of 25 mg/kg. The total cost of this remediation was $610,000.

Case Study 2: Seymour Recycling Corporation Superfund Site, Seymour, Indiana. From 1970 to early 1980, the Seymour Recycling Corporation processed, stored, and incinerated chemical wastes at the Seymour site. Because of improper waste management practices, the subsurface was contaminated. The soils at the site consist of sands and silts with groundwater located 6 to 8 ft below the ground surface. Soils and groundwater were found to be contaminated with a wide range of volatile organic compounds. The contaminated soil zone was approximately 12 acres with an average depth of 10 ft. The concentration of VOCs in the soils ranged from 10 to 1000 mg/kg. To remediate this site, a SVE system was selected as the prime soil remediation technology and a groundwater pump and treat system was selected as the groundwater remediation technology.

The SVE system was constructed between July and October 1990. The system consisted of 19 horizontal vapor extraction wells, 11 horizontal air inlet wells (passive), a vacuum blower, a moisture separator, and an activated carbon adsorption system. The horizontal vapor extraction piping was installed 30 in. below the grade. Wells were spaced about 50 ft apart and a multimedia cap (consisting from top to bottom, a 24-in. vegetative cover, geotextile fabric, 12-in.-thick drainage layer, 60-mil geomembrane, and a 2-ft-thick clay layer) was constructed above the wells. The cap was used to prevent short circuiting of airflow in the subsurface. The vacuum blower used in this system was

a 3-hp belt-driven model designed to deliver 40 ft³/min at 27 in. H_2O.

The SVE system began operating in June 1992 at an average flow rate of 104 ft³/min. Monitoring of the remedial progress was conducted by gas sampling. The system extracted a total of 29,166 lb of VOCs (of an estimated 200,000 lb) over a four-and-one-half-year period. The mass of VOCs extracted per year by the SVE system decreased by more than 90% over the four-year period. The total mass of VOCs removed approached an asymptotic value. As of 1997, 430 pore volumes and about 30,000 lb of VOCs had been extracted by the SVE system. Overall, the performance of the system was poor. This may be due to installation of horizontal wells in a very shallow vadose zone. The capital cost for the SVE system was $1.2 million, and the operating and maintenance (O&M) costs are not available.

13.3 SOIL WASHING

13.3.1 Technology Description

Soil washing technology is used to separate contaminants from excavated soils and to reduce the volume of soil requiring final treatment or disposal. The technology relies on the fact that contaminants tend to be associated preferentially with organic matter and fine-grained soil particles (i.e., silt and clay). Volume reduction is achieved by cleaning the coarse-grained soil fraction and leaving the contaminants in the fine-grained fraction and washing fluids (USEPA, 1990b, 1996d). The cleaned coarse fraction can then be returned to the excavation. The concentrated contaminants in the fine fraction and wash fluids are treated using any of the other remediation methods described in this chapter, such as thermal desorption and bioremediation or disposed of in a permitted landfill.

A variety of soil washing processes have been developed that employ different equipment configurations and may be aimed at specific contaminants. Figure 13.9 depicts a typical soil washing process. Following excavation, the soil is screened to separate coarse debris (larger than about 2 in.) such as rocks

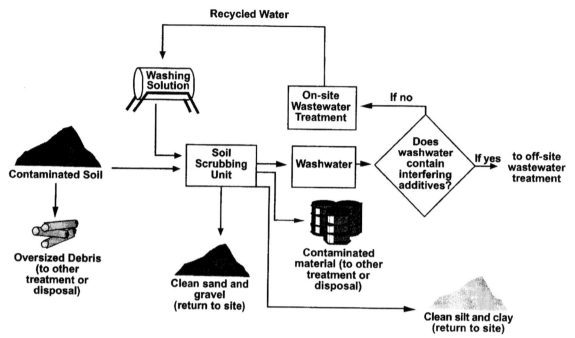

Figure 13.9 *Typical soil washing process. (From USEPA, 1996d.)*

and roots. The remaining soil may be fluidized, or made pumpable, with the addition of water. In the scrubbing unit, a water-based washing solution is used to separate soluble contaminants and fine particles from coarser soil materials. Surficial contamination is removed from the coarse fractions by solution and by the energetic abrasive scouring action maintained in the unit. The scrubbing action also disintegrates soil clumps, freeing contaminated fine particles from the larger grains. Dissolution of contaminants can be enhanced with chemical additives; for example, acidic wash solutions may be used to solubilize lead or other metals derived from waste oil. Surfactants or foaming agents may also be added. Many vendors, providing soil washing services, consider their extraction agents proprietary (Raghavan et al., 1988).

Following washing, the soil slurry undergoes a separation step in which water, cleaned coarse material, and contaminated fines are segregated. Suspended fines may be flocculated and separated by gravity means or

may be removed in a vacuum filter press. The clean soil is then returned to the excavation. Typically, scrubbing solutions are at least partially recycled. Nonrecycled wash solutions are treated using conventional wastewater treatment technologies. The residuals from wastewater treatment (e.g., spent exchange resins, spent carbon, or biological treatment sludges) may then be combined with the contaminated fines and sent for treatment or disposal. Volatiles can also be collected from the soil handling and washing units and treated. Air emissions can be controlled throughout the process. Typical air treatment processes include activated carbon adsorption or catalytic or thermal oxidation.

Soil washing can be effective for treating soils contaminated with a variety of organic and inorganic contaminants (Semer and Reddy, 1996). Developmental studies indicated good to excellent applicability of the process for removal of VOCs and metals from sandy and gravelly soils. Soil washing is less likely to be effective with silt or clay soils. In general, the soil types for which this can be applied effectively are those with relatively high hydraulic conductivities. If the contaminants adsorb to the soil strongly, soil washing may not be effective. At present, soil washing is being included as a pretreatment step in the remedy of several Superfund sites (Table 13.3). At these sites, it is being applied to soils contaminated with heavy metals, radinuclides, cyanides, polycyclic aromatic compounds, pesticides, and PCBs. Typically, on-site treatment of soils using soil washing is not cost-effective unless the site contains at least 5000 tons of contaminated soil.

The soil washing technology has the following advantages:

• This process significantly reduces the volume of contaminated soil in that the contaminants are concentrated in a relatively small mass of material. Ideally, the soil washing process would lead to a volume reduction of about 90% (which means that only 10% of the original volume would require further treatment).

• Closed system unaffected by external conditions. This system permits control of the conditions (such as the pH level and temperature) under which the soil particles are treated.

• Allows the contaminated soil to be excavated and treated on-site. Clean soils can be backfilled at the same site.

• Potential to remove both organic and inorganic contaminants.

• High throughput: about 70 tons of soil per hour.

• Relatively few permits are required to employ this method.

This technology has the following disadvantages:

• Ineffective for soils containing 30 to 50% of silt, clay, or organic matter. Contaminants tend to bind readily, chemically or physically, to silt, clay, and organic material. Silt, clay, and organic material, in turn, bind physically to sand and gravel. When the soil contains a large amount of clay and organic material, the contaminants attach more easily to the soil and are, therefore more difficult to remove than when a small amount of clay and organic material is present.

• Relatively expensive, due to additional costs associated with treating wastewater and air emissions.

• Complex waste mixtures (such as metals with organics) make formulating washing fluid difficult.

• Small volumes of contaminated sludge and washwater remain at the end of the treatment, which require further treatment or disposal.

• Exposure of the contaminants to the public is possible as a result of soil excavation and handling.

Table 13.3 Selected Superfund sites where soil washing was used

Name of Site	Medium	Contaminants
Myers property, NJ	Soil, sediment	Metals
Vineland Chemical, NJ	Soil	Metals
GE Wiring Devices, PR	Soil, sludge	Metals
Cabot Carbon/Koppers, FL	Soil	Semivolatile organic compounds, polyaromatic hydrocarbons (PAHs), metals
Whitehouse waste oil pits	Soil, sludge	Volatile organic compounds, polychlorinated biphenyls, PAHs, metals
Cape Fear Wood Preserving, NC	Soil	Metals, creosote, Pentachlorophenol (PCP)
Moss American, WI	Soil	PAHs, metals

Source: USEPA (1996d).

- A large space is required based on the design of the soil washing system, system throughput rate, and site logistics (including staging for untreated and treated soils).

13.3.2 Fundamental Processes

The soil washing remediation method is controlled by physicochemical processes such as desorption, complexation, dissolution (or solubilization), and oxidation–reduction. When a washing solution is introduced into the soil, it desorbs the contaminants from soil particle surfaces. The washing solution may affect acid–base reactions, leading to pH changes, and consequently, causing the contaminants to solubilize. In some cases, the washing solution may form complexes with the contaminants and these complexes may be soluble. It is also possible that when washing solutions are introduced, oxidation–reduction reactions may occur and the change in oxidation state of the contaminants may lead to desorption or solubilization of the contaminants. At the end of soil washing, the majority of the contamination is transferred into the washing solution; however, some contamination may remain in the soil. This can be represented mathematically as

$$C_{si}M_s = C_{sf}M_s + V_lC_l \qquad (13.7)$$

where C_{si} is the initial contaminant concentration in the soil (mg/kg), M_s the total dry mass of the soil (kg), C_{sf} the final contaminant concentration the soil (mg/kg), V_l the total volume of washing solution (L), and C_l the contaminant concentration in the washing solution (mg/L). If the washing is continued for adequate time, equilibrium conditions will exist and the contaminant concentration in washing solution (C_l) and in the soil (C_{sf}) can be related by the distribution coefficient of the contaminant between the soil and the washing solution (K_d) as $C_{sf} = K_dC_l$. Substituting this in equation (13.7) and rearranging the terms, we get soil washing removal efficiency as

$$\text{removal efficiency} = \frac{C_{sf}}{C_{si}} \times 100(\%)$$

$$= \frac{1}{1 + V_l/M_sK_d} \times 100(\%) \qquad (13.8)$$

13.3.3 System Design and Implementation

General Design Approach. Figure 13.9 illustrates a typical flowchart of the soil washing process. Oversized soils and debris are separated by screening. Washing these materials with water may be enough to achieve adequate contaminant removal. The finer soil is then fed in to the washing apparatus (or soil scrubbing unit). Extraction agents and water (i.e., washing solution) are then added to the soil. After mixing, the remediated soil is separated from the extracting solution. The remediated soil may be returned to the site or disposed of off-site. The finer soil residue containing concentrated contaminants is either treated further or disposed of in a permitted landfill. The wash water may be treated on-site and reused for washing solution.

Equipment. Soils can be excavated with backhoes, front loaders, continuous excavators, or other equipment. The soil is separated or screened by use of a hooper and vibrating grizzly. The grizzly passes material onto another mechanical screening unit, which consists of a double-decked coarse vibrating screen with staking conveyors, to remove process oversize (greater than 2-in. material) from the fall-through. The fall-through (smaller than 2 in. material) is then subjected to the washing process. The washing technique is either the manual procedure or a mechanical device. For the manual procedure the equipment needed consists of metal pans sufficient in size to hold the material sample and large metal spoons for mixing the sample with the wash solution. The mechanical device, such as a Plogg washer, is commonly known as the *washing apparatus*. Large-scale soil washing units can process over 100 yd³ of soil per day.

Operational Parameters and Monitoring. Soil separation and screening should generally be conducted in an environment where the off-gases and dusts are all captured inside a hood or cover. When stored prior to treatment, all excavated soil should be covered securely with plastic liners. Monitoring and evaluation should be performed to prevent any leaks or spills during the remediation process.

During excavation and material-handling activities, meteorological conditions should be considered and evaluated carefully to minimize cross-media transfer. High winds should be avoided. Weather monitoring will help with respect to fugitive dust emission.

Treated soils are sampled and tested for their contaminant levels to verify that the remedial standards have been met. The sampling frequency depends on the following factors: homogeneous process stream, low contaminant concentration in process stream relative to treatment goal, percent of fines and organic matter in the soil, and site-specific data indicating the failure rate for treated batches. Mass-balance calculations are generally included as part of the remedial performance assessment process, as explained in Section 13.3.2.

13.3.4 Predictive Modeling

Specific mathematical models are not available to design or assess soil washing systems. However, geochemical models as described in Chapter 6 are often used in the selection of the flushing solution for any given soil and contaminant conditions.

13.3.5 Modified or Complementary Technologies

Other ex-situ technologies, such as *solvent extraction* and *chemical dechlorination,* are similar to soil washing. Solvent extraction technology uses organic solvents rather than aqueous wash solutions to remove organic contaminants from the soil (USEPA, 1990c, 1991d, 1992b). Typical solvents employed include liquefied gas (propane and/or butane) and triethylamine. Chemical dechlorination is used to remove chlorine from chlorinated organic chemicals (USEPA, 1991d). The soil and dechlorination reagents are mixed in a reactor. The dechlorination reagents may be either an alkaline metal hydroxide or alkaline polyethylene glycol (APEG) reagent. The most common metal hydroxide used in the process is potassium hydroxide (KOH). In conjunction with polyethylene glycol (PEG), KOH forms a polymeric alkoxide referred to as KPEG. Sodium hydroxide (NaOH) and tetraethylene (TEG) have also been used. The soil–reagent mixture is heated between 100 and 180°C for 1 to 5 h, depending on the type, quantity, and concentration of contaminants. Dechlorination reduces the toxicity of chemicals such as solvents, pesticides, and PCBs. It may also make it easier to further treat, biodegrade, recycle, or dispose of these contaminants. Dechlorination is not applicable to nonchlorinated volatile organic compounds.

Soil washing can also be implemented as an in-situ process, called *in-situ soil flushing* (USEPA, 1991e). In-situ flushing often refers to groundwater remediation (as described in Chapter 14); however, it can also be implemented in treating soils in the vadose zone, as shown in Figure 13.10. Similar to soil washing, soil flushing uses water or water containing additive to enhance contaminant solubility. It is accomplished by passing the extraction fluid through in-place soils using injection wells or infiltration systems such as spray irrigation, flooding of the ground surface, or the use of infiltration galleries. After application, the flushing fluid percolates downward through the contaminated soil. Contaminants are mobilized through dissolution, formation of emulsions, or by chemical reaction with the flushing fluid additives. The resulting leachate continues to percolate downward until reaching the water table. At this level, the leachate mixes with the groundwater and flows downgradient to the withdrawal (extraction) point. Extraction of the contaminated groundwater and flushing fluids is conducted via conventional extraction wells or recovery trenches. Recovered groundwater and flushing fluids must meet appropriate discharge standards prior to recycle or release to local, publicly owned wastewater treatment works.

A related technology, *chemical oxidation,* involves chemically converting hazardous contaminants to nonhazardous or less toxic compounds (USEPA, 1991c). In this method, contaminated soil is mixed with an oxidizing agent such as ozone, hydrogen peroxide, hypochlorites, chlorine, or chlorine dioxide. This can be accomplished under in-situ or ex-situ conditions. Redox reactions chemically convert hazardous contaminants into nonhazardous or less toxic compounds. This method can be applied for contaminants such as inorganics, nonhalogenated VOCs and SVOCs, fuel hydrocarbons, and pesticides. The advantage of this method is that secondary waste streams are not generated. However, care should be taken to ensure that oxidation reactions do not form intermediate contaminants,

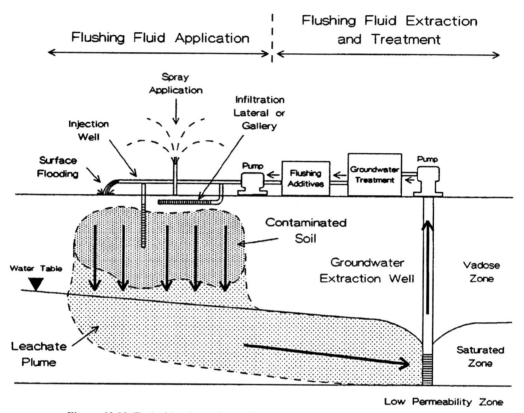

Figure 13.10 Typical in-situ soil washing in vadose zone. *(From USEPA, 1991e.)*

which may be more toxic than the original ones. The cost of oxidizing agents will be excessive if high concentrations of the contaminants are required to be treated.

13.3.6 Economic and Regulatory Considerations

The average cost for use of this technology, including the excavation, can range from $50 to $200 per ton (USEPA, 1991d; FRTR, 1995b; ITRC, 1997a). From a regulatory perspective, permits are required for wastewater treatment and air emission treatment systems. The health and safety of the workers involved in the remediation has to be ensured through an approved plan.

13.3.7 Case Studies

Table 13.3 lists examples of Superfund sites where soil washing has been selected for remediation (FRTR, 1995b). The King of Prussia (KOP) Technical Corporation Superfund site in Winslow Township, New Jersey, is an example case study for the remediation process of soil washing. This site, shown in Figure 13.11(*a*), is a rectangular 10-acre parcel bordered by dense pine forest to the northwest, northeast, and southeast. A drainage swale in the site is dammed by two fire roads. The site runoff flows toward the Great Egg Harbor River. The swale has been designed as wetlands. The site is generally sandy and barren with sparse patches of tall seed grass. This area site was once a waste processing facility.

The processing facility operated from January 1971 to April 1974. The waste management practices that

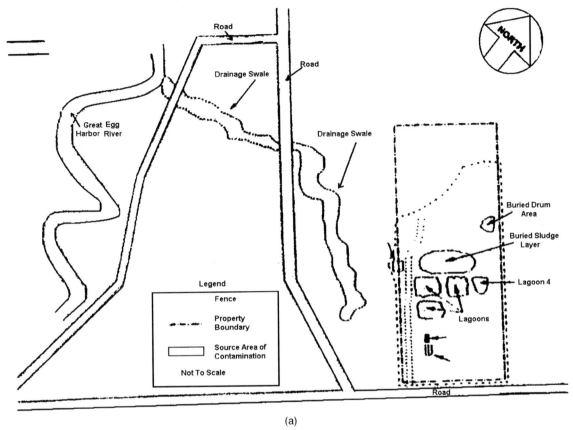

(a)

Figure 13.11 *Soil washing at King of Prussia Technical Corporation Superfund site: (a) site map; (b) soil washing system (From FRTR, 1995a.)*

contributed to the contamination were surface impoundment (lagoons) and unauthorized dumping of trash and hazardous materials. During its operation, it is estimated that at least 15 million gallons of acids and alkaline wastes was processed at this site. A treatability test of soil washing using the soil from the KOP site was conducted in January 1992; the results from the treatability test indicated that the soil at KOP had an acceptable level of sand content and could be treated effectively by soil washing. Samples of surface soil (less than 2 ft deep), subsurface soil (2 feet to 10 ft), and sediment were collected during the investigations to characterize the soil next to the lagoons, the sediments in the swale, and the sludge in the lagoons and adjacent areas. The sampling results indicated that

beryllium, chromium, copper, nickel, and zinc are the primary contaminants in these areas. The highest concentration of surface contamination was located in the sediment at the bottom of the swale. Average soil concentrations were measured as 660 mg/kg for chromium, 860 kg/mg for copper, and 330 mg/kg for nickel.

The soil washing unit used to remediate the contaminated soil and sludge at the KOP site is shown in Figure 13.11(*b*). It consists of four components: screening, separation, froth floatation, and sludge management. This unit has a rated system throughput of 25 tons/h. The screening stage separates the gross oversize fraction (concrete, tree stumps, and branches) from the pile of material to be treated by means of a

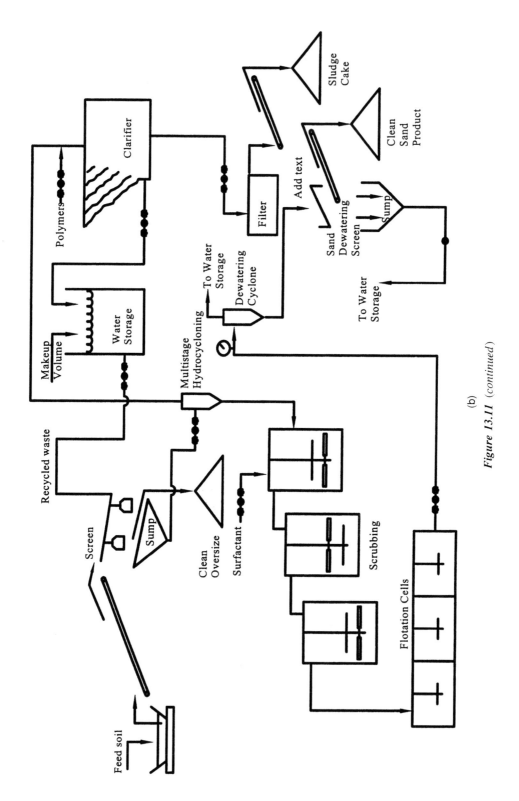

Figure 13.11 (continued)

hopper and vibrating grizzly. The material that passes through the grizzly is directed to another mechanical screening unit, which consists of a double-decked, coarse vibrating screen with stacking conveyors, to remove the process oversize (greater than 2 in.) from the fall-through.

The separation stage consists of screening the soil–water slurry material being separated from the coarse- and fine-grained material through the use of multistage hydrocyclones. The use of multiple cyclones achieves a separation efficiency of >99% of the sands and fines. The froth floatation stage consists of removing the contaminants from the coarse-grained material by means of air flotation treatment units. The air floatation tank is equipped with mechanical aerators. The coarse-grained material was pumped into the tank, where a surfactant was added. The surfactant reduced the surface tension between the contaminant and the sand. The contaminants "float" into a froth and are removed from the surface. The sludge management stage consists of treating the overflow from the hydrocyclones.

The total time frame from the dates of operations to the decontamination and disassembly of the soil washing unit lasted approximately 22 months. The soil washing process was able to clean the materials to meet cleanup goals for 11 metals. For example, chromium levels went from 8000 mg/kg to 480 mg/kg. The total cost of $7.7 million was expended on the soil washing remediation at KOP, including all off-site disposal costs.

13.4 STABILIZATION AND SOLIDIFICATION

13.4.1 Technology Description

The *stabilization and solidification* (S/S) *process*, also referred to as *immobilization, fixation,* or *encapsulation,* uses additives or processes to chemically bind and immobilize contaminants or to microencapsulate the contaminants in a matrix that physically prevents mobility (USEPA, 1986, 1993g; Sharma and Lewis, 1994). Stabilization typically refers to a chemical process that actually converts the contaminants into a less

soluble, mobile, or toxic form. *Solidification* generally refers to a physical process where a semisolid material or sludge is treated to render it more solid. Thus, the S/S process refers to either chemically binding or physically trapping the contaminants in soils. This technology neither removes the contaminants from soils, such as soil washing, nor degrades the contaminants, such as bioremediation; rather, it eliminates or impedes the mobility of contaminants. In some cases, the S/S process is used as a pretreatment to reduce soluble contaminant concentrations in the soils to below the regulatory limits (e.g., land disposal restriction limits) in order to dispose of them in a landfill.

The S/S process can be implemented under ex-situ or in-situ conditions. An *ex-situ S/S process* involves the following steps: (1) excavation of the contaminated soil, (2) mixing of a reagent with the soil, (3) curing of the mixed product, and (4) backfilling or landfilling of the treated soil. An *in-situ S/S process* involves the injection and/or mixing of stabilizing agents into subsurface soils to immobilize the contaminants, to prevent them leaching into groundwater.

The S/S process is applicable to soils contaminated with metals, radionuclides, and other inorganics as well as nonvolatile and semivolatile organic compounds. Soils contaminated solely with volatile organic compounds are not considered to be appropriate for the S/S process because they may be volatilized and released during mixing and curing operations. The S/S process is applicable to all types of soils (clays, silts, or sands).

This technology has the following advantages:

- Low cost due to the use of widely available and relatively inexpensive additives and reagents
- Applicable to a wide variety of contaminants, including organic compounds and heavy metals; metals and organics may be treated in one step
- Applicable to different types of soils
- Uses readily available equipment and is simple
- High throughput rates compared to other technologies

The technology, however, has the following disadvantages:

- Contaminants are not destroyed or removed.
- The volume of treated soil may be increased significantly with the addition of reagents (in some cases this increase could be double the original volume).
- Emissions of VOCs and particulates may occur during mixing procedures, requiring extensive emission controls.
- Delivery of reagent to the subsurface and achieving uniform mixing for in-situ treatment may be difficult.
- In-situ solidification may hinder future site use.
- Long-term efficiency of the process may be uncertain.

13.4.2 Fundamental Processes

Stabilization/solidification processes depend on the type of stabilization reagent used. Depending on the type of reagent used, S/S processes can be grouped into the following types: (1) cement-based S/S, (2) pozzolanic S/S, (3) thermoplastic S/S, and (4) organic polymerization S/S (USEPA, 1989). The first two are inorganic S/S processes, which use cements and pozzolans, and the latter two are organic S/S processes, which use thermoplastic binders and organic polymerization.

Cement-based S/S is a process in which contaminated soils are mixed with portland cement. Water is added to the mixture if the water content in the soil is low. This causes hydration reactions necessary for bonding the cement. The contaminants are incorporated into the cement matrix and undergo physico-chemical changes that reduce their mobility. Because of the high pH value of the cement, hydroxides of metals are formed, which are much less soluble than other ionic species of the metals. Small amounts of fly ash, sodium silicate, and bentonite are often added to the cement to enhance processing. The stabilized soil will be a cohesive solid, depending on the amount of reagent added. This type of S/S process is applicable for metals, PCBs, oils, and other organic compounds.

Pozzolanic S/S involves siliceous and aluminosilicate materials, which do not display cementing action alone, but form cementitious substances when combined with lime or cement and water. The primary immobilization mechanism is the physical entrapment of the contaminant in the pozzolan matrix. Pozzolans that are commonly used are fly ash, pumice, lime kiln dusts, and blast furnace slag. These pozzalans contain significant amounts of silicates. The stabilized soil can vary from soft fine-grained material to a hard cohesive material similar in appearance to cement. Pozzolanic reactions are generally slower than cement reactions. This type of S/S process is applicable for metals, waste acids, and creosote.

Thermoplastic S/S is a microencapsulation process in which the contaminants in the soil do not react chemically with the encapsulating material. Microencapsulation process covers contaminants with a relatively impermeable layer. A thermoplastic material such as asphalt (bitumen) or polyethylene is used to bind the contaminants into a stabilized/solidified mass. The asphalt binder may be heated before it is mixed with the dry contaminated soil, or the asphalt may be applied as a cold mix. This type of S/S process is applicable for metals, organics, and radionuclides.

Organic polymerization S/S relies on polymer formation to immobilize the contaminants in the soil. Urea–formaldehyde is the most commonly used organic polymer for this purpose. This type of S/S is applicable primarily for special wastes such as radionuclides, but is also applicable for metal and organic contaminants.

13.4.3 System Design and Implementation

General Design Approach. Figure 13.12 shows a systematic approach to the selection and implementation of an S/S process at a contaminated site. The first and foremost step is to look at the major contaminants present in the soil and if these contaminants require pretreatment, to make it acceptable for an S/S process. The presence of large amounts of oil and grease and other organics in the soil make the S/S process inefficient or ineffective. If the contaminants are acceptable for S/S treatment, a detailed characterization of the contaminated soil through analysis of major physical and chemical properties is required. Physical properties

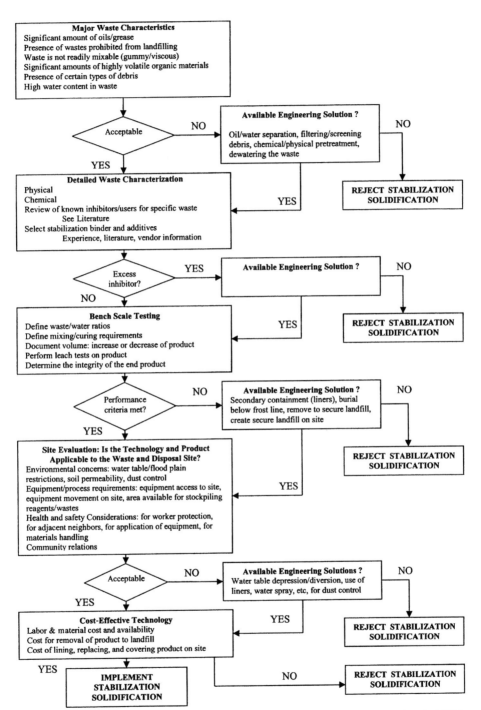

Figure 13.12 *General design approach to stabilization/solidification. (From USEPA, 1989.)*

such as volume, water content, particle size distribution, and pH have an impact on S/S treatment. Bench-scale testing is conducted to select the type and proportion of stabilizing agents needed for effective immobilization of the contaminants. Different contaminated soil-to-reagent ratios are tested to determine the optimal ratio for the final treatment. Bench-scale tests also define mixing requirements, curing time, and quality control parameters. The applicability of the S/S process as determined from bench-scale testing is evaluated for use under site-specific conditions. Factors such as groundwater table location, area of the site, and site drainage are considered. Based on all these aspects, a determination must also be made as to whether the S/S process identified can be implemented economically. This includes the cost of the reagents, equipment, and labor. Finally, the technically and economically feasible S/S treatment selected is implemented at the site.

Equipment and Implementation. Under ex-situ conditions, contaminated soil is excavated, screened to remove oversized material, and then homogenized to provide uniform mixing before being fed into a mixer such as a pug mill. A schematic of the process is shown in Figure 13.13. The contaminated soil is mixed with water, if required, and then mixed with stabilizing reagents. After it is mixed throughly, the treated soil is discharged from the mixer. Treated soil is a solidified mass with significant compressive strength, high stability, and a rigid texture similar to that of concrete. The treated soil is finally backfilled into the same excavation or disposed of in a permitted landfill.

Traditional earth-moving equipment such as tracked backhoes, draglines, bulldozers, and front-end loaders can be used to excavate the contaminated soils. Dump trucks can be used to transport the excavated soils to the storage area or treatment area. The storage area, if required, must be prepared to avoid contamination of the surrounding environment. Bottom liners and drainage systems are designed to achieve this. Air emission controls, if required, are also installed. Stabilizing reagents may also require on-site storage, and bins and hoppers are often used for this purpose. Mixing of contaminated soils and stabilizing reagents is achieved in mixing equipment such as pug mills, ribbon blenders,

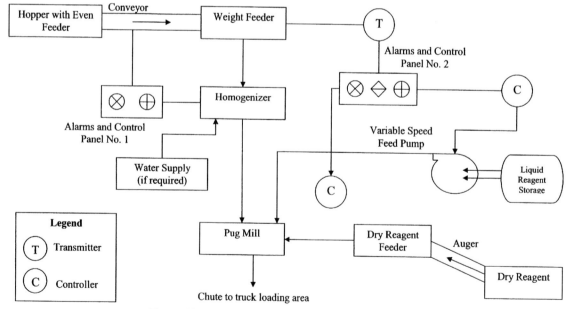

Figure 13.13 *Ex-situ S/S process. (From USEPA, 1986.)*

extruders, and screw conveyers. Stabilized soil can be transported by dump trucks. If the stabilized soil needs to be backfilled, it should be done immediately before the stabilized soil sets up and hardens. Generally, the stabilized soil is placed in 8- to 10-in. lifts and compacted with graders or bulldozers.

The in-situ S/S process uses a combination of an auger and caisson. The stabilizing reagents are fed into the auger and then into the contaminated soil through a hollow stem, as shown in Figure 13.14. Inside the caisson, the auger mixes the reagent with the soil by a lifting and turning action. In this way, a large-diameter (6 ft or greater) "plug" of the contaminated soil is mixed in place. After thorough mixing, the auger is removed and the setting slurry is left in place. The auger is advanced to overlap the last plug slightly

and the process is repeated until the contaminated area is covered, as shown in Figure 13.15. Figure 13.16 shows the overall operations involved in an in-situ S/S process.

Monitoring. Physical and chemical testing is required for the characterization of soils prior to and after treatment (USEPA, 1989). Table 13.4 shows various types of tests and the procedure and purpose of each test for physical characterization of contaminated or treated soils. These tests include index property tests, which provide data that are used to relate general physical characteristics of the contaminated soil to process operational parameters. Density tests are used to determine weight-to-volume relationships of the soils. Permeability tests measure the relative ease with which

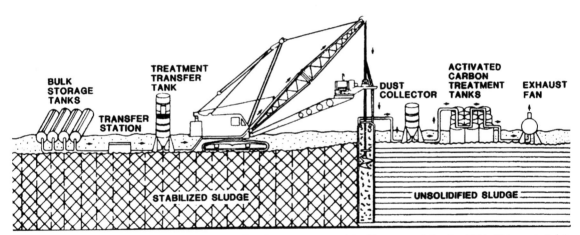

Figure 13.14 In-situ S/S process. (From Geo-Con Inc. as presented in Sharma and Lewis, 1994. This material is used by permission of John Wiley & Sons, Inc.)

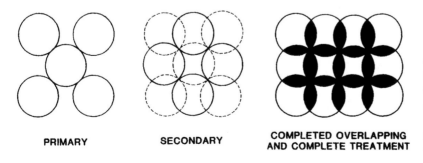

PRIMARY   SECONDARY   COMPLETED OVERLAPPING AND COMPLETE TREATMENT

Figure 13.15 Overlapping to cover the entire contaminated area. (From Geo-Con Inc. as presented in Sharma and Lewis, 1994. This material is used by permission of John Wiley & Sons, Inc.)

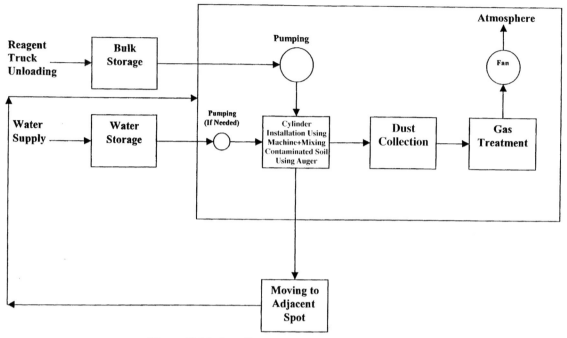

Figure 13.16 Overall operations of in-situ S/S process.

water will pass through the soil, strength tests provide a means for judging the effectiveness of the S/S process under mechanical stresses, and durability tests determine how well the contaminated soil withstands repeated wetting and drying or freezing and thawing cycles.

Table 13.5 shows general parameters and test methods for chemical characterization of untreated and treated soils. As stated earlier, the total concentrations of the chemical constituents in untreated and treated soils will be the same. However, the S/S process makes the chemical constituents immobile and less leachable. A comparison of the total concentrations in untreated soil with leachate concentrations of the treated soil determines the efficiency of the S/S process to stabilize the contaminants within the soil. Numerous leaching procedures have been developed to test wastes and stabilized/solidified soils. These include:

- Toxicity characteristic leaching procedure (TCLP)
- Extraction procedure toxicity test (EP Tox)
- California waste extraction test (Cal WET)
- Multiple extraction procedure (MEP)
- Monofilled waste extraction procedure (MWEP)
- Equilibrium leach test (ELT)
- Acidic neutralization capacity (ANC)
- Sequential extraction test (SET)
- Sequential chemical extraction (SCE)

A description of these tests can be found in USEPA (1989). A comparison of the extraction tests is shown in Table 13.6. Obviously, these tests vary significantly, and the results from these tests must be evaluated with an understanding of the differences in the experimental conditions. The TCLP is used by the USEPA as the criterion for defining hazardous and nonhazardous wastes and as the basis for the promulgation of best

Table 13.4 **Physical characterization of treated and untreated soils**

Test procedure	Reference	Purpose
Index property tests		
Particle size analysis	ASTM D422	To determine the particle size distribution of a material
Atterberg limits		
Liquid limit	ASTM D4318	To define the physical characteristics of a material as
Plastic limit	ASTM D4318	a function of its water control
Plasticity index	ASTM D4318	
Moisture content	ASTM D2216	To determine the percentage of free water in a material
Suspended solids	USEPA Method 208C	To determine the amount of solids that do not settle from a column of liquids
Paint filter test	USEPA Method 9095-SW846	To determine the presence of free liquids in a representative sample of bulk or noncontainerized waste
Density testing		
Bulk density		
Drive cylinder method	ASTM D2937	To determine the in-place density of soils or soil-like material
Sand-cone method	ASTM D1556	To determine the in-place density of soils or soil-like materials
Nuclear methods	ASTM D2922	To determine the in-place density of soils or soil-like materials
Stabilized waste		To determine the density of a monolithic stabilized waste
Compaction testing		
Moisture density relations	ASTM D558	To determine the relation between moisture content
of soil–cement mixtures	ASTM D1557	and density of a material
Permeability testing		
Falling-head permeability	USEPA Method 9100-SW846	To measure the rate at which water will pass through a soil-like material
Constant head	USEPA Method 9100-SW846	To measure the rate at which water will pass through a soil-like material

Source: USEPA (1989).

demonstrated available technologies (BDAT) treatment standards under the land disposal restriction program. The TCLP is generally used to evaluate leachate concentrations of the stabilized soils.

13.4.4 Predictive Modeling

Specific models to predict the performance of an S/S process are not available. However, the geochemical models discussed in Chapter 6 are useful in evaluating the effects of various stabilizing reagents on immobilization of the contaminants.

13.4.5 Modified or Complementary Technologies

When an S/S process is applied to contaminated in-situ soil, the process is known as *in-situ immobilization* or *in-situ fixation*. This process includes injection of chemicals or reagents into the subsurface by means of

Table 13.5 **Chemical characterization of treated and untreated soils**

Parameter	Test Method	Applicability to Untreated and Stabilized/Solidified Wastes
pH	EPA Method SW-9045	Leachability of hazardous constituents (e.g., metals) may be governed by the pH of the solid
Oxidation/reduction potential (E_H)	ASTM D1498	Changes in Eh after treatment can change the leachability of many elements
Major oxides	ASTM C114	Mineralogy of the stabilized/solidified waste may aid in interpretation of leach test results
Total organic carbon (TOC)	Combustion Method	Used to approximate the nonpurgeable organic carbon in wastes and treated solids
Oil and grease	EPA Method 413.2	May be used to compare the leachable oil and grease from the treated and untreated wastes
Elemental analysis	EPA Method SW-846	Used to determine the fraction of metals leached to the total metals content of the untreated and stabilized/solidified wastes
Volatile organic compounds (VOCs)	EPA Method SW-846 (Methods 5030, and 8240)	Used to compare VOC concentrations in stabilized/solidified wastes and untreated wastes with the VOC concentrations in TCLP extracts to determine relative teachability of the treated and untreated wastes
Base, neutral, and acid compounds	EPA Method SW-846 (Methods 3540, 3520, and 8270)	Used to compare BNA concentrations of leachates with respective concentrations in treated and untreated wastes to determine relative leachability of the treated and untreated wastes
Polychlorinated biphenyls (PCBs)	EPA Method SW-846 (Methods 3540, 3520, 680, and 8080)	Same as for VOC with respect to PCB leachability from treated and untreated wastes
Ion	Standard Method 429	Used to determine leachate ionic species concentrations
Heat of hydration	ASTM C186	Measurement of temperature changes during current mixing will allow prediction of VOC emissions in the field
Alkalinity	Titrometry	Alkalinity changes in leachates may be used to determine changes in stabilized/solidified waste form

Source: USEPA (1989).

wells. The reagents will interact with and transform metal or radionuclide contaminants into an immobile (precipitate) form. Although cost-effective, immobilization is ineffective in heterogeneous or low-permeability soils, due to the difficulty in delivering the stabilizing reagents into such soils.

Vitrification is also a form of an S/S process in which heat melts and converts contaminated soils into glass or other crystalline products. This technology is described in detail in Section 13.7.

Soil mixing is combined with *vapor extraction, hot air injection,* or *hydrogen peroxide injection* to remediate organic compounds effectively and is used under certain subsurface conditions. Soil mixing techniques used in the S/S technology may also be used for add-ing nutrients and microorganisms for the purposes of *bioremediation.*

13.4.6 Economic and Regulatory Considerations

The cost of an S/S process varies widely according to materials or reagents used and their availability, project size, and the chemical nature of the contaminants. The in-situ S/S process cost ranges from $40 to $60 per cubic yard for shallow applications and $150 to $250 per cubic yard for deeper applications. The cost of an ex-situ S/S process is approximately $100 per ton, including excavation (USEPA, 1997a).

Table 13.6 **Comparison of various extraction tests**

Test Method	Leaching Medium	Liquid / Solid Ratio	Maximum Particle Size	Number of Extractions	Time of Extraction
TCLP	Acetic acid	20 : 1	9.5 mm	1	18 h
EP Tox	0.04 M acetic acid (pH − 5.0)	16 : 1	9.5 mm	1	24 h
CaL WET	0.2 M sodium citrate (pH − 5.0)	10 : 1	2.0 mm	1	48 h
MEP	Same as EP tox, then with synthetic acid rain (sulfuric acid/nitric acid in 60 : 40 wt.% mixture)	20 : 1	9.5 mm	9 (or more)	24 h per extraction
MWEP	Distilled/deizonized water or other for specific situation	10 : 1 per extraction	9.5 mm or monolith	4	18 h per extraction
Equilibrium leach test	Distilled water	4 : 1	150 μm	1	7 days
Acid neutralization capacity	HNO$_3$ solutions of increasing strength	3 : 1	150 μm	1	48 h per extraction
Sequential extraction tests	0.04 M acetic add	50 : 1	9.5 mm	15	24 h per extraction
Sequential chemical extraction	Five leaching solutions, increasing in acidity	Varies from 16 : 1 to 40 : 1	150 μm	5	Varies from 2 to 24 h

Source: USEPA (1989).

S/S technology is used widely; however, research into finding new reagent additives and improving equipment for effective and efficient implementation of this technology is still needed. Regulatory involvement may include obtaining permits for any air emissions and any underground injection activities. Safety of workers from potential exposure to volatile organic contaminants and drilling operations must always be considered. Also, the noise levels during remedial operations may be high and may be of concern to the surrounding public.

13.4.7   Case Studies

The two case studies presented below show the typical design and performance of S/S systems (USEPA, 1997a).

Case Study 1: Department of Energy Portsmouth Gaseous Diffusion Plant, Ohio. The site is the Department of Energy (DOE) Portsmouth Gaseous Diffusion Plant (PGDP), near Piketon, Ohio. A field demonstration was conducted at the X-231B unit in June 1992. The X-231B unit was used from 1976 to 1983 as a land disposal site for waste oils and solvents. Soils beneath the unit were contaminated with VOCs, such as TCE at approximately 100 ppm, and low levels of radioactive substances (^{235}U and ^{99}Tc). The shallow groundwater (12 to 14 ft deep) was also contaminated, and some contaminants were above the drinking water standards. Approximately 78% of the VOCs were located in the upper 12 ft. Geologically, the site contains low-permeability sediments, composed of silt and clay deposits with hydraulic conductivities of less than 10^{-6} cm/s. Figure 13.17(a) and (b) show the site map and site subsurface stratigraphy. The total area to be treated was approximately 0.8 acre.

The contaminated soil is mixed as a cement grout and is injected under pressure to solidify and immobilize the contaminated soil in a concretelike form as shown in Figure 13.17(c). A mechanical system was employed to mix unsaturated or saturated contaminated soils while simultaneously injecting treatment or stabilizing agents. The main system components include

the following: a crane-mounted soil mixing auger, a treatment agent delivery system, a treatment agent supply, an off-gas collection, and a treatment system. The mixing system comprised a track-mounted crane with a hollow Kelly bar attached to a drilling tool, known as the MacTool, consisting of one or two 3- to 5-feet-long horizontal blades attached to a hollow vertical shaft, yielding an effective mixing diameter of 6 to 10 ft. Depths of 40 ft can be achieved with this equipment. Treatment agents were injected through a hollow, vertical shaft and out into the soil through 0.25- or 0.5-in. diameter orifices in the backside of the soil mixing blades. Treatment is achieved in butted or overlapped soil columns. The soil treatment rate was about 45 yd^3/h. The ground surface above the mixed region was covered by a 14-ft diameter shroud under a low vacuum to contain air emissions and direct them to an off-gas treatment process. The off-gas treatment system consisted of activated carbon filters followed by a HEPA filter. The removal of VOCs was enhanced by moving the mixing auger up and down from 2 to 15 ft below the ground surface.

Because grout was applied before soil mixing was initiated and because the grout application rate was rapid, little volatilization of VOCs is believed to have occurred as mixing proceeded. The TCLP concentrations for regulated constituents were either not detected or were below the USEPA regulatory limits. Physical tests revealed that the treated soil had a decrease in density and in increase in strength and hydraulic conductivity. The decrease in density was attributed to the initial high density of the clay-rich soil, which is reduced as a result of mixing. Also, these effects are due to the entrapment/entrainment of air in the grout during mixing. The increase in hydraulic conductivity was attributed to disruption of the dense clay deposit as a result of mixing and increased porosity within the grout–soil mixture, due to incomplete filling of the pores with grout. The cost of this remediation was about $90 per cubic yard.

Case Study 2: Imperial Oil Company/Champion Chemical Company Superfund Site, Morganville, New Jersey. Ex-situ solidification and stabilization was applied in December 1988 at the Imperial Oil

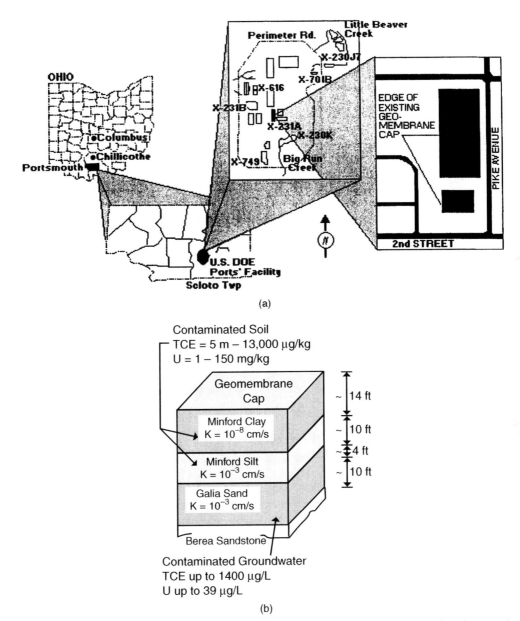

Figure 13.17 *S/S remediation at Portsmouth gaseous diffusion plant site:* (a) *site map;* (b) *subsurface stratigraphy;* (c) *in-situ mixing system. (From USEPA, 1997a.)*

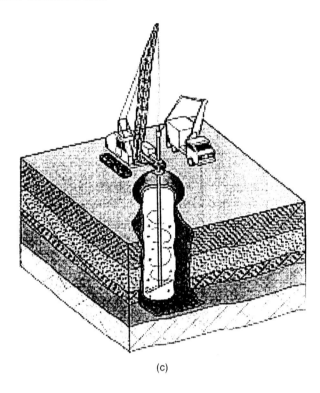

Figure 13.17 (*continued*)   (c)

Company/Champion Chemical Company Superfund site in Morganville, New Jersey. This site formerly contained both chemical processing and oil reclamation facilities. Soils, filter cakes, and oily wastes from an old storage tank were treated during the demonstration. These wastes were contaminated with petroleum hydrocarbons, PCBs, other organic compounds, and heavy metals. Two 10- by 20-ft areas were treated, one to a depth of 18 ft and the other to the depth of 14 ft.

The remedial system consisted of an auger, mixer, pozzalan tank, truck, and dumpster, as shown in Figure 13.18. The process solidified both solid and liquid wastes with high organic content (up to 17%) as well as oil and grease. Extract and leachate analysis showed that the heavy metals in the treated waste were immobilized. Organic compounds were also not detected in the treated waste. Physical tests on the solidified waste showed an increase in density, strength, and durability and a decrease in hydraulic conductivity. The solidified waste increased in volume by an average of 22%. The cost of the remediation was approximately $110 per ton.

13.5 ELECTROKINETIC REMEDIATION

13.5.1 Technology Description

Electrokinetic remediation, also known as *electrokinetics, electromigration, electrorestoration, electroremediation,* and *electroosmosis,* removes the contaminants from soils by applying an electric potential. This technology involves applying an electric potential across contaminated soil through a pair of electrodes, located at the anode and the cathode. Due to a variety of processes, contaminants are transported toward the electrodes. The contaminant-laden liquids are then removed from the electrodes. Figure 13.19 illustrates how the electrokinetic system is implemented under in-situ conditions (Reddy and Shirani, 1997). The system consists of a minimum of two electrodes buried underground and connected to a power supply. The electrodes are located a certain distance apart and are encased by reservoirs or wells. The electrodes are called anodes or cathodes, the *anode* being the positively charged component and the *cathode* the negatively charged component. In simple terms the anode attracts

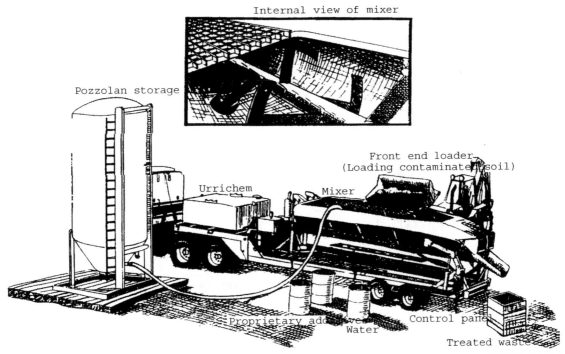

Figure 13.18 *Equipment used for ex-situ S/S process at Imperial Oil Company Superfund site.* (*From USEPA, 1997a.*)

contaminants that have a negative charge, and the cathode attracts positively charged contaminants. In remediating unsaturated soils, water is injected into electrode wells or reservoirs. Removal of contaminants can be achieved by pumping the contaminated water in the reservoirs or wells or by electroplating, precipitation, or coprecipitation at the electrodes. Removal of contaminants from the soils can be optimized by using enhancement solutions, such as weak acids, surfactants, and complexing agents, at the reservoirs.

In the field, electrodes are arranged according to the desired effect. The electrodes may be placed in a two long rows of anodes and cathodes if the contamination is narrow and long. If the contamination is rather large, alternating rows of anodes and cathodes may be needed. Electrodes can be arranged to prevent only the migration of contaminants or their removal.

This technique can be used to remediate soils with high clay or humic content. It can also be used in heterogeneous soils. It can also be used on both saturated

or unsaturated soils. Electrokinetics can be used to treat a wide range of contaminants, such as heavy metals, radionuclides, and organic contaminants (Reddy et al., 1999).

This technology has several advantages:

- Applicable to low-permeability soils and heterogeneous soils
- Applicable for a wide range of contaminants; metal contamination can be moved because of its charge, while noncharged contaminants can be moved with the induced flow
- Flexible to use as both in-situ and ex-situ technology
- Less expensive than other remediation techniques
- Tailored to site-specific contamination

The disadvantages include:

- Electrolysis reactions near the electrodes may change the soil pH significantly from anode to the

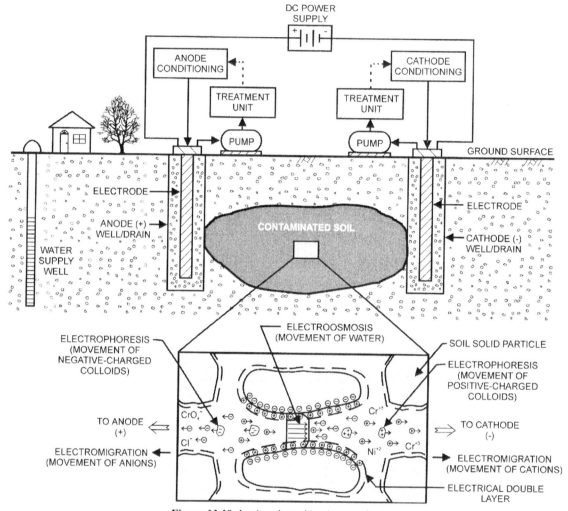

Figure 13.19 In-situ electrokinetic remediation.

cathode, leading to complex geochemical inter-actions (Chinthamreddy and Reddy, 1999; Reddy and Chinthamreddy, 1999).

- Buried metal objects may short-circuit the current path, changing the voltage gradient, which will affect the extraction rate.
- Acidic conditions and electrolytic decay can corrode materials used in the anode.
- There are stagnant zones between wells where the rate of migration is slow.

- VOCs will probably be stripped from the soil and will therefore increase the soil vapor concentration.

13.5.2 Fundamental Processes

Applying an electrical potential to soil induces the transport of water and the contaminants. The main transport mechanisms include electroosmosis, electromigration, electrophoresis, and diffusion. These transport mechanisms are affected by physical–chemical processes: electrolysis, adsorption–desorption, precipitation–dissolution, and oxidation-reduction.

Electroosmosis. *Electroosmosis* is the advective transport or movement of water or moisture under an electric field. Typically, the surfaces of fine-grained soil particles are negatively charged. Cations in the pore water will align themselves along the negatively charged surfaces. Water molecules then align themselves around the excess cations. When there are no excess cations, water molecules will orient themselves around the negatively charged surface of the soil, forming a boundary layer. The water molecules closest to the soil surface are held tighter, due to electrical attraction, and they are free to move in the double layer.

During electrokinetic remediation on the application of an electrical field, water molecules that are positively charged will move toward the cathode. The soil's zeta potential will affect the movement of water molecules in the double layer. The *zeta potential* is defined as the potential between the stationary and movable parts of the double layer surrounding the soil particles. The zeta potential is typically negative for clays and silts that are saturated and range from -10 to -100 mV (Probstein and Hicks, 1993). When the zeta potential is negative, electroosmotic flow will move toward the cathode. Electroosmotic flow can be reversed to the anode if the zeta potential becomes positive, which can occur if the contaminant concentration is high. The transport of contaminants induced by electroosmotic flow depends on the pore water viscosity, ion concentration, temperature, dielectric constant, and the mobility of ions (Mitchell, 1993). The quantity of water moved per unit time (q_e) by electroosmotic flow can be quantified by the improved Helmholtz equation,

$$q_e = \frac{ED\zeta R^2}{4\eta L} \quad (13.9)$$

where E the electric potential, D the dielectric constant, ζ the zeta potential, η the viscosity, and L the length of the specimen. When compared with hydraulic gra-dients, the flow due to electric gradients is a few orders of magnitude higher in fine-grained soils (Mitchell, 1993). This is also supported by Acar and Alshawab-keh (1993), who concluded that electroosmosis is effective only in fine-grained soils with micrometer-sized pores.

Electromigration. *Electromigration* is the movement of ions or charged species toward their respective electrodes. Anions move toward the anode, and cations toward the cathode. Electromigration occurs 10 times faster than advective transport caused by electroosmosis (Acar and Alshawabkeh, 1993). During electrokinetic remediation, electromigration may be the dominant transport mechanism for ions and charged species. The electromigration rate is a function of ionic mobility, valence of the species, and total electrolyte concentration (Acar and Alshawabkeh, 1993). The velocity of ions in solution (v_{xi}) due to electromigration is quantified by the equation (Lindgren et al., 1993)

$$v_{xi} = \frac{Iv_i}{A}\frac{P_w}{\tau\theta} \quad (13.10)$$

where I is the applied current (A), v_i the ion velocity, A the total cross-sectional area (m²), P_w the pore water resistivity ($\Omega \cdot$ cm), τ the tortuosity, θ the volumetric moisture content (cm³/cm³).

Electrophoresis. *Electrophoresis* is the movement of charged colloids to their respective electrodes, similar to electromigration. However, this process is ineffective if the soil is tightly packed, thus restricting movement of the colloids.

Diffusion. *Diffusion* is the spreading of contaminants due to a concentration gradient. Fick's first law describes the diffusion of any charged species that are under chemical potential. Diffusion depends on the porosity and tortuosity of the porous medium and the molar concentration of the species. The diffusion rate (U_{diff}) equation is

$$U_{\text{diff}} = \frac{vRT}{c} \nabla c \qquad (13.11)$$

where v is the mobility, R the gas constant, T the temperature, ∇c the concentration gradient, and c the concentration of contaminant in moles. During electrokinetic remediation, diffusion is slow compared to the electromigration rate. The diffusion rate is typically one to two orders of magnitude slower than the electromigration rate. Therefore, this does not have a significant role in transporting the contaminants in the electrokinetic remediation process.

Electrolysis. When an electric field is applied, *electrolysis reactions* occur at the anode and cathode. Near the anode, hydrogen ions (H^+) and oxygen gas (O_2) are generated. Reactions at the cathode generate hydroxyl ions (OH^-) and hydrogen gas (H_2). These reactions can be written as follows:

Anode $2H_2O \Rightarrow 4H^+ + O_2 + 4e^-$

Cathode $4H_2O + 4e^- \Rightarrow 4OH^- + 2H_2$

The rate of production of H^+ and OH^- ions depends on the current applied. Secondary reactions are also possible, such as the reduction of H^+ to H_2, or the reduction of a metal to a lower valence state.

Oxygen and hydrogen gases are generated due to electrolysis, which can alter the redox conditions of the porewater. An acid front is generated due to the formation of H^+ ions at the anode, and a base front is generated at the cathode due to the formation of OH^- ions. The pH at the anode drops to approximately 2, while at the cathode it rises to 12, depending on the current applied. H^+ ions tend to move toward the cathode and the OH^- ions toward the anode. The movement of these ions occurs due to electromigration as well as diffusion caused by concentration gradients. The mobility of H^+ ions is greater than OH^- ions, because of a smaller ionic radius. This causes the acid front migration to occur twice as fast as the base front migration. The extent of migration of the acid front depends on the acid buffering capacity (capacity to neutralize the acidity) of the soil. In a low-buffering-capacity soil, such as kaolinite, the acid front migration

can occur easily. However, in a high-buffering-capacity soil such as glacial till, the H^+ ions are consumed to neutralize buffering constituents in the soil (Reddy et al., 1997).

The acid front will desorb and dissolve typical cations (Ni and Cd) from the soil surface, or if precipitated, will increase cation removal. If the contaminants are anionic [Cr(VI)], the acid front increases the adsorption and hinders contaminant removal. The pH changes induced by migration of the acid and base fronts will affect the zeta potential of the soil, therefore affecting the electroosmotic flow (Eykholt and Daniel, 1994).

Adsorption–Desorption. *Adsorption* is the partitioning of the solute contaminants from the pore fluid to the soil surface. Soil surfaces are typically negative in charge; however, this depends on the pH of the soil. The pH at which soil has net zero charge is called the *point of zero charge* (PZC). If the pH of the pore water is below the PZC, adsorption of anions will be significant. When the pH is above the PZC, adsorption of cations will be significant. Adsorption depends on the type of contaminant (anionic or cationic), soil type, surface area of the soil, surface charge on the soil, concentration of cationic species, presence of organic matter and carbonates in the soil, and pore fluid characteristics.

Desorption, the opposite of adsorption, is the partitioning of contaminants from the soil surface to the pore fluid. If the pH of the pore water is below the PZC, the desorption of cations will be significant. When the pH is above the PZC, sorption of cations will occur. When applying an electric field in low-buffering soils, low-pH conditions are present at the anode due to electrolysis reactions, and high-pH conditions are present at the cathode. Cationic adsorption and anionic desorption occur at the cathode, while anionic adsorption and cationic desorption occur at the anode.

Precipitation–Dissolution. During application of an electric field, the pH of the soil changes. Precipitation–dissolution reactions are pH dependent. *Precipitation* occurs when the amount of ions equals or exceeds the solubility product of that solid. *Disso-*

lution is the opposite of precipitation. Because of the pH changes experienced during the electrokinetic process, contaminants may be precipitated or dissolved depending on the contaminant's location. When contaminants precipitate, their removal by electrokinetic remediation is hindered; however, when the contaminants are dissolved, they can be removed easily.

Oxidation–Reduction. *Redox conditions* are also altered during the electrokinetic process (Chinthamreddy and Reddy, 1999). Electrons that are removed at the anode cause oxidation, and the electrons that are pushed in at the cathode cause reduction. Near the cathode, metal cations reduce and then precipitate. Gases produced during electrolysis can also cause oxidation–reduction if not allowed to escape the anode and cathode. Some metals may be present in different valence states, depending on the redox conditions. The valence state controls the solubility of the metal and may have an effect on removal of the metal.

13.5.3 System Design and Implementation

General Design Approach. A flowchart detailing the design procedure is shown in Figure 13.20. It shows different steps involved in the process of designing and implementing this technology. First, bench-scale experiments must be performed to determine the appropriate parameters and for the selection of equipment. The voltage needed and distance between electrodes must be determined from the electromigration flow rate equation (13.10) for metal contamination. However, for organic contamination when flushing is more important, the electroosmosis equation should be used. A flow rate is assumed, and then a low voltage can be chosen. After choosing the voltage, the equation will be used to find the appropriate distance between electrodes. The voltage should be feasible, taking into account the cost of electricity. As the voltage decreases, the spacing between electrodes need to be closer. A balance in cost must be found between the two.

After determining the voltage and spacing, additional bench-scale experiments can be performed. The purpose of the tests is to find the best-suited enhancement solution. As discussed earlier, the soil experiences pH changes that can increase or hinder the removal of contaminants. Enhancement solutions known as *surfactants* and *chelators* are added to the anode and cathode, depending on the desired effect. Some enhancment solutions that have been used are acetic, humic, and gallic acids.

Another consideration before installing a field system is to select material for the electrodes and reservoirs or wells. Graphite electrodes are common. However, the casing around the electrode must be considered as well. The reservoirs or wells may be exposed to corrosive contaminated fluid along with an extremely low or high pH.

Now the field system can be designed and installed. This requires drilling wells for the placement of electrodes. Pumps and treatment units must also be designed. Treatment units can be used to clean the removed water or simply to adjust the pH for proper disposal.

Equipment. Electrokinetic remediation requires the following equipment: (1) electrodes, (2) power supply. (3) wells constructed of appropriate material, (4) pumps, and (5) treatment units. Electrodes are typically constructed of carbon, graphite, or platinum because they are inert and will not add more contamination to the site. Power supplies must be able to deliver currents of $1 A/ft^2$ of cross-sectional area between electrodes, and be able to produce a voltage potential of 10 to 30 V/ft between electrodes. Generally, 50 to 250 kWh is needed to treat a cubic yard of saturated soil. Typical values from field and pilot tests indicate a range of 100 to 600 V and 15 to 50 A.

Although standard well construction may be followed to build electrode wells, ceramic wells may also be used to prevent corrosion problems. Standard pumps may be used for the purpose of recycling solutions and removing contaminated water from wells. Treatment units may be needed to adjust the pH in electrode wells as well as to treat effluent.

Operational Parameters. For electrokinetic remediation to be cost-effective, the lowest voltage possible to achieve cleanup objectives should be chosen in the design. Field and pilot studies have used voltages that range between 100 and 600 V and 15 and 50 A.

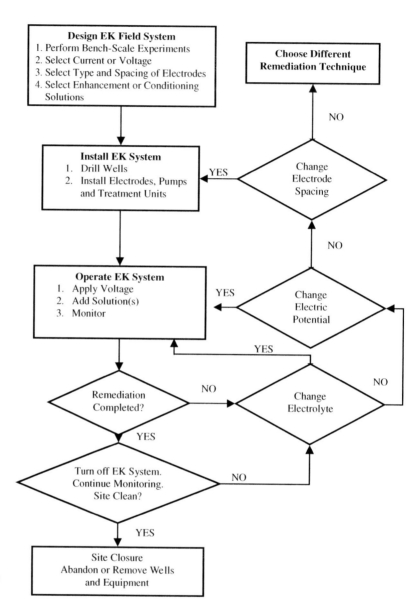

Figure 13.20 *General approach to electrokinetic remediation.*

Another consideration is the spacing between electrodes. The closer the electrodes are, the more wells that must be constructed. The system is not cost-effective if a large number of wells are needed. Typical values for distances between electrodes range from 2 to 7 ft.

Since electrokinetic remediation has demonstrated some difficulties with premature precipitation of metals at the cathode, enhancement solutions should be used. Typical enhancement solutions include the following: acetic acid, ethylenediamine, citrate and carbonate salts, sodium chloride, citric acid, and EDTA. The concentration of enhancement solutions should be optimized to reduce unnecessary costs.

Monitoring. After the system is installed, voltage is applied and enhancement solutions are added; the reservoirs should then be monitored to determine the removal efficiency. When the system no longer removes contaminants and a good mass balance has been determined, the remediation may be complete. After analyzing the amount of contaminant removed, if the amount is not sufficient, adjustments to the system must be made. The first option should be to allow the system to run for a longer duration and then reanalyze the results. If this is not effective, the enhancement solutions may need to be reanalyzed and changed. Finally, if this does not work, the electrode spacing may need to be decreased. The final step would be to remove the equipment and abandon or remove the wells.

13.5.4 Predictive Modeling

Modeling is needed to determine if electrokinetic remediation will perform as expected. If an accurate model can be created, electrokinetic remediation can be compared to other technologies that have been modeled for the same site, without performing any pilot tests, thus reducing the costs of remediation. Table 13.7 summarizes selected models, and their purposes, assumptions, and limitations.

13.5.5 Modified or Complementary Technologies

The same principles involved in in-situ electrokinetics may be used to design aboveground electrokinetic reactors. Figure 13.21 shows a schematic of one such system (Reddy et al., 1997). The excavated soil is placed in the reactor, and an electric potential is applied. The contaminants collected at the electrodes are then removed. Electrokinetically enhanced bioremediation uses the processes of electrokinetics to deliver nutrients, water, extra microorganisms, and heat (GeoKinetics, 1999b). These are necessary for the or-

ganisms to survive or to perform more efficiently at degrading the contaminants.

Electro-heated dual-phase extraction (EH/DPE) uses electrokinetics to remove some contamination with the water that is pumped out, but uses the heat generated by the electrokinetic process to volatilize the contaminant as well. This technology should be combined with a soil vapor and groundwater extraction system (GeoKinetics, 1999a). Electrokinetics can also be combined with soil flushing in a soil that has a low hydraulic conductivity (Reddy, 2001). The processes from electrokinetics introduces the flushing solutions to the soil at a much faster rate than if hydraulic gradients were used.

13.5.6 Economic and Regulatory Considerations

Cost for this technology depends on site-specific conditions, such as the chemical and hydraulic properties of the soils. Pilot-scale studies estimate energy consumption for heavy metal remediation at 500 kWh/m^3 or more for electrodes spaced 1 to 1.5 m. The cost for this amount of energy is approximately $25 per cubic meter. Estimated prices from four companies range from $20 to $225 per cubic yard. Certain factors can affect the price, such as initial and target contaminant concentrations, conductivity of pore water, soil characteristics and moisture content, the amount of waste generated, residual waste, site preparation, and the cost of electricity (ITRC, 1997b).

The electrokinetic processes affect the soil's pH, which can cause precipitation of metal contamination near the cathode. Research has been conducted with acetic acid to help eliminate the premature precipitation; however, more research must be conducted. Predictive modeling for migration rates needs to be developed, and more field studies must be completed to validate the results of the models (ITRC, 1997b).

Regulatory concerns of this technology include:

- *The amount of time for cleanup.* Electrokinetic remediation has low target levels, and the efficiency decreases with time. The technology

Table 13.7 **Selected models of electrokinetic remediation**

Reference	Type of Model	Purpose of Model	Assumptions	Comments
Corapcioglu (1991)	Numerical	To predict the outcome under a variety of conditions	Uses mass balance equations, conservation of charge; the solid matrix is modeled as an elastic solid with a one-dimensional field	Based on electrochemicoosmotic flow
Alshawabkeh and Acar (1992)	Numerical	To predict the removal of ionic contaminants	Uses fluid flux, mass transport, charge flux, conservation of mass, charge, and energy as equations	Relatively good comparison with experiments; however, slight difference in result near cathode
Thornton and Shapiro (1993)	Numerical	To predict cost based on power consumption, number of electrodes, chromate concentrations, and voltage	Saturated soil, zero groundwater velocity	Limited by the assumptions and simplifications
Lindgren et al. (1993)	Numerical	To predict the removal of ionic contaminants	Electroneutrality	Doesn't consider hydraulic flows and electroosmosis; unrealistic boundary conditions; doesn't include acid–base chemistry
Jacobs et al. (1994)	Numerical	To predict the removal of zinc using electrokinetics	Electroosmosis negligible; used equation for electroneutrality in place of transport equations for either H^+ or OH^-	Simulated successfully
Choi and Lui (1995)	Numerical	To predict the removal of cadmium contamination	Heavy metals; water is in excess; dissociation–association of water in H^+ and OH^- is rapid; electroosmois insignificant when compared to electromigration	Not discussed

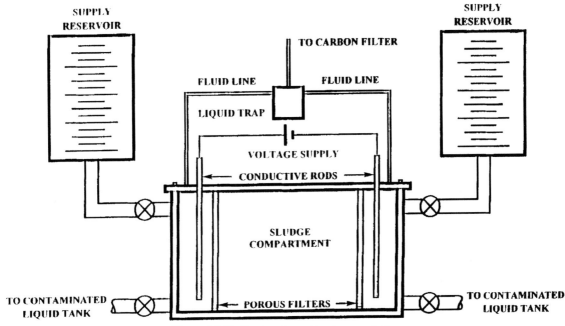

Figure 13.21 *Ex-situ electrokinetic remediation.*

should be used when long-term remediation is an option

- *Impacts to the site and surrounding environment because of electrokinetics.* The soil may not be able to sustain plant growth because of chemical, physiological, and biological impacts to the soil.

- *Prolonged effects to soil properties.* Acid and basic fronts may remain in the soil after remediation is complete.

- *Soil near the cathodes must typically be removed.* Because of premature precipitation, soil near the cathodes may need to be excavated.

13.5.7 Case Studies

Very few number of full-scale implementations of this technology have been reported. This technology is relatively new, and many investigations in the United States have reported bench- or pilot-scale testing (USEPA, 1995f). Two projects that deal with performance of electrokinetics under field conditions are described in this section.

Case Study 1: Remediation of Inorganic Contaminants. Lageman et al. (1989) reported cleanup of a former timber-impregnation plant site using electrokinetics. At this site, Supervolmansalt D ($Na_2HAsO_4 \cdot 7H_2O$) was used to impregnate the timber. The plant was destroyed in a fire and was not rebuilt; when the property was sold, a "statement of unpolluted soil" was needed. During this investigation arsenic contamination from improper storage or disposal was found. The contamination of heavy clay was 2 m deep, and 10 m × 10 m in size in one area, and 1 m × 10 m × 5 m for a total volume of polluted soil equal to 250 m^3. Arsenic concentrations for the entire area were 110 ppm on average.

The design called for two rows of four cathodes installed at a depth of 1.5 m for one row and 0.5 m for the other row. The four cathodes were spaced 3 m apart, and the two rows were placed 25 m apart, except for one cathode spaced only 12.5 m. A plan view of this site is shown in Figure 13.22. In between the two rows of cathodes 36 anodes were installed in two rows of 14 and one row of 8. The anodes installed were at a depth of 1 to 2 m and 1.5 m apart. The operational

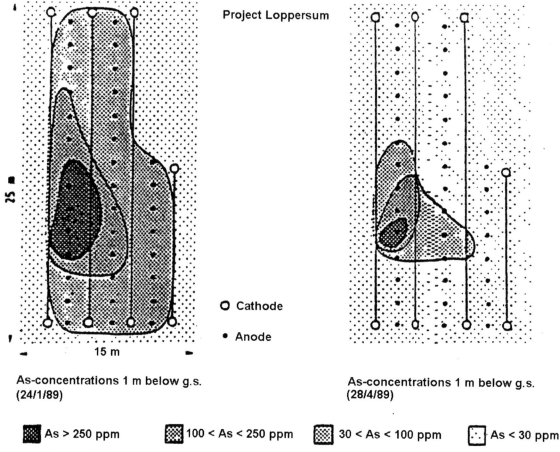

Figure 13.22 *Electrokinetic remediation of metal-contaminated site.* (*From Lageman et al., 1989.*)

time was originally calculated as 50 days, with an energy supply of 44 kW. At the time of application, the soils resistivity was $10 \ \Omega \cdot m$, and the temperature was 7°C. After three to four weeks the temperature had risen to 50°C, and the resistivity dropped by $5 \ \Omega \cdot m$. During this same period the potential dropped from 40 V/m to 20 V/m.

Arsenic concentrations decreased from 110 ppm to 30 ppm. However, in one area, As concentrations remained high. The system was turned off and the soil was excavated. The soil excavated contained many metal objects. Remedial costs for this site were not reported. This site demonstrated that electrokinetics was effective at removing the arsenic contamination

but that the presence of metal objects decreases the efficiency of this process.

Case Study 2: Remediation of Organic Contaminants. Ho et al. (1999a,b) reported the use of electrokinetics at the U.S. Department of Energy Paducah Gaseous Diffusion Plant (PGDP) Cylinder Drop Test Area (SWMU 91), Paducah, Kentucky. At this site electrokinetics was coupled with in-situ treatment zones that were directly installed in the ground. This test site was chosen because of its low-permeability soil and single chlorinated contaminant, TCE. This field test was constructed in November and December 1994 and began operation in January 1995. The study was completed in May 1995. The purpose of this field experiment was to demonstrate the simplest Lasagna configuration, taking into consideration the operating issues and electrokinetic effects.

This site consisted of a 4-ft layer of gravel and clay over 40 ft of sandy clay loam with interbedded sand layers. The hydraulic conductivity of the sandy clay loam was about 1×10^{-7} cm/s. The test area was 15 ft × 10 ft × 15 ft, and the average concentration was 83.2 ppm. Steel panels were used as electrodes, and granular activated carbon was installed near the electrodes to absorb the TCE. Two electrodes were set 10 ft apart and four treatment zones were installed between the electrodes. The treatment zones were constructed with granular activated carbon. The electrodes were constructed with eight steel panels, installed side by side with a $\frac{1}{2}$ in. gap between panels. Each panel

was $\frac{1}{4}$ in. thick, 18 in. wide, and 16 ft long. Fluid was circulated with overflow tubes and pumps. A shed above the soil was constructed to contain the off-gas and then filtered to remove the TCE. Figure 13.23 shows the site layout.

This test was run with a constant current of 40 A. The initial power was 138 V and after one month it decreased to 105 V. The electroosmotic flow rate was 4 to 5 L/h, and the temperature of the soil increased from 15°C to 25 to 30°C. The voltage gradient ranged from 0.35 to 0.45 V/cm. The system moved three pore volumes of water and had a removal efficiency of 98%, with the levels reduced to an average concentration of 1.2 ppm. The TCE levels in the carbon treatment zones were very high; they were then replaced and only a small amount of carbon was uncovered. From these results it was decided that the test should be stopped.

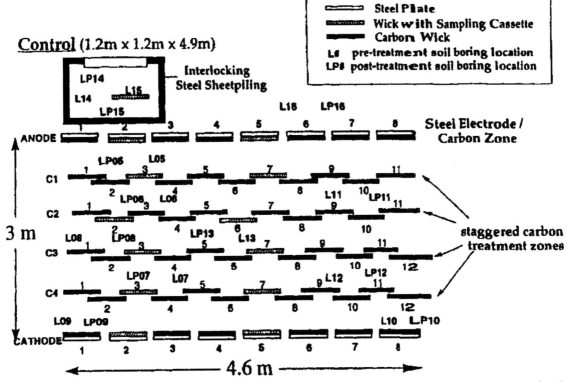

Figure 13.23 Plan view of small-scale electrokinetic system layout for the remediation of TCE-contaminated soil. (Reprinted with permission from Ho et al., 1999a. Copyright 1999 American Chemical Society.)

The results also showed a good contrast between the TCE levels in the treated zone and just outside the treatment zone. The cost for complete remediation of this site was $1310 per cubic yard. It was learned that electroosmosis flow can flush TCE from clay with the passage of two to three pore volumes. This may have been possible because the soil at this site did not absorb much TCE, thus allowing easy movement with the electroosmotic flow.

A large-scale field test was constructed after performing the small field test, in August 1996 until March 1997. The purpose of this field test was to degrade the TCE in-situ. This site consisted of a 4-ft layer of gravel and clay over 40 ft of sandy clay loam with interbedded sand layers. The hydraulic conductivity of the sandy clay loam was about 1×10^{-7} cm/s. The test area was 21 ft × 30 ft × 45 ft deep. The water table is present at about 30 to 40 ft below ground level.

The electrodes were constructed of a 1.5-in.-thick layer of 50/50 by volume iron filings and Loresco coke to a depth of 45 ft. The iron filings were used to promote iron corrosion instead of electrolysis reactions to minimize the H^+ formation. The electrodes were spaced 21 ft apart. The treatment zones were iron filings in a mix with wet kaolin clay. Three treatment zones were installed between the electrodes. Figure 13.24 shows the site layout.

This test was run with a constant voltage of 150 V until high concentrations of TCE was indicated in the wells; at this time the voltage was increased to 200 V. The temperature then increased and the voltage was dropped to 120 V. The voltage gradient was 0.25 V/cm and moved the water with an average speed of 0.43 cm/day. The system had an overall removal efficiency of 99.7%. TCE was believed to exist in free phase at the site, and electrokinetics was able to remove the TCE in the free phase. There were also no significant

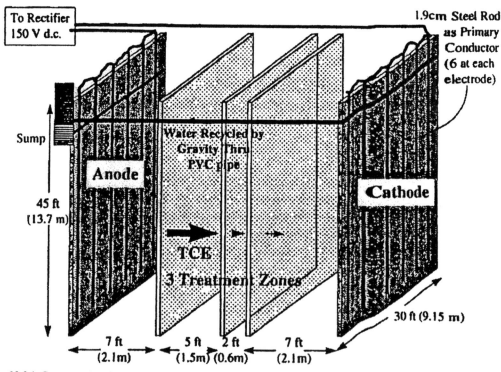

Figure 13.24 Cross-sectional view of large-scale electrokinetic system for the remediation of TCE-contaminated soil (Reprinted with permission from Ho et al., 1999b. Copyright 1999 American Chemical Society.)

TCE emissions. The complete remediation cost of this site is estimated to be $80 per cubic yard for two years of treatment time and an area of $\frac{1}{4}$ acre and 45 ft deep.

13.6 THERMAL DESORPTION

13.6.1 Technology Description

Thermal desorption is a technology that treats contaminated soil by heating soils to temperatures between 200 and 1000°F. This causes contaminants with low boiling points to vaporize and thus separate from the soil. The vapors are collected by a vacuum system and transported to a treatment center (USEPA, 1994d, 1995e, 1998a). One might think that thermal desorption is a form of incineration. However, this is not the case, as the heat used in thermal desorption does not destroy the contaminants. Instead, thermal desorption physically separates the contaminants from the soil. Subsequently, the vapors taken from the treatment are

either condensed for disposal (as in higher-temperature incinerators), or they are reused. Figure 13.25 shows a schematic of the thermal desorption process.

Thermal desorption can be an ex-situ or an in-situ application. In an ex-situ application the soil is excavated and brought to a facility to be processed. The facility can be located at the site or the soil can be transported to another location. In an in-situ application, the process is done completely in place. Thermal blankets are placed on the soil surface to treat any shallow contamination. Also, thermal wells are placed in the ground to treat the deeper contamination. Heat from the wells is transferred via radiation and thermal conduction to the contaminated soil. Extraction wells are added to remove the soil vapors that are produced as a part of the process.

Thermal desorption is effective in removing volatile and semivolatile organics from soils contaminated with oil refining wastes, coal tar wastes, wood-treating wastes, creosotes, hydrocarbons, chlorinated solvents,

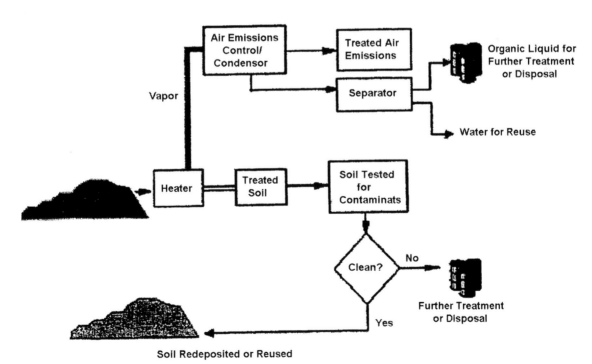

Figure 13.25 Schematic of thermal desorption process. (*From USEPA, 1994d.*)

fuels, PCBs, mixed wastes, synthetic rubber processing waste, pesticides, and paint wastes (FRTR, 1997). The degree of effectiveness against the contaminants will vary depending on the method of thermal desorption that is used (i.e., high or low temperature). The physical properties of soils to be treated by thermal desorption play a major role. The key characteristics considered are soil plasticity, heat capacity, particle size, and bulk density. For instance, soils with high sand and gravel contents are easier to treat than those with cohesive silts and clays. When clays and silts are heated, they emit dust, which can disrupt the air emission machinery (used to treat vaporized contaminants). Furthermore, the moisture content is critical in the process. Soils that contain a moisture content of 15% or lower are acceptable. However, an increase in moisture content affects cost. Moisture in the material acts as a heat sink because it must be evaporated from the soil along with organic contaminants. With this process, more fuel is needed to vaporize all the contaminants with the water in the soil; thus, the costs increase as moisture increases. When soils do contain a high moisture content, usually they must be "de-watered" or mixed with dry materials before they undergo the thermal desorption process. Moreover, the presence of heavy metals in the soil may hamper the process. Generally, most metals are not affected by thermal desorption. Therefore, the soil must be tested for the presence of metals. Soils contaminated with heavy metals will not be a good choice for thermal desorption, as they do not separate from the soil easily. Also, soils that are tightly packed prevent the heat from making contact with all the contaminants, and they do not serve as appropriate candidates for the thermal desorption process.

Advantages of thermal desorption are:

- Readily available equipment for on-site or off-site treatment.
- Treated soil can be redeposited on-site.
- Very rapid treatment time; most systems are capable of over 25 tons/h throughput.
- Can be easily combined with other technologies, such as groundwater extraction.
- Cost competitive for large volumes of soils (greater than 1000 yd^3).

- Very effective in treating volatile organic components.
- Lower temperatures used require less fuel than other methods.

Disadvantages of thermal desorption are:

- Dewatering may be necessary to achieve acceptable soil moisture content levels.
- There are specific particle size and materials handling requirements that can affect applicability or cost at specific sites.
- Clay and silty soils increase reaction time.
- On-site treatment will require sufficient land for the equipment and the treated soil.
- Highly abrasive feed potentially can damage the processor unit.
- Ineffective in treating heavy metals.

13.6.2 Fundamental Processes

Initially, in the thermal desorption process, heat must be transferred to the soil particles to vaporize the contaminants from these particles. Subsequently, the vaporized contaminants are transferred from the soil particles to the gas phase. Figure 13.26 depicts this phenomenon. Three major principles control this phenomenon: volatilization; adsorption–desorption; and diffusion.

Generally, organics possess lower melting and boiling points. When the soil comes into contact with the heat, the organics in the free phase will be easily volatilized. As volatilization and temperature increase, the contaminants start to lose their "hold" from the soil particle surface. From here, the contaminants can easily transfer to the vapor phase. Subsequently, as the temperature starts to increase, the organics will be removed from the liquid phase and then from the adsorbed phase (i.e., volatilization increases).

The adsorption properties of the soil affect the contaminants and "hold" on to the soil surface. Adsorption decreases as the temperature increases. Desorption refers to the removal of the contaminants from the soil particles. Most organic contaminants are more easily adsorbed than they are desorbed, and thus, more en-

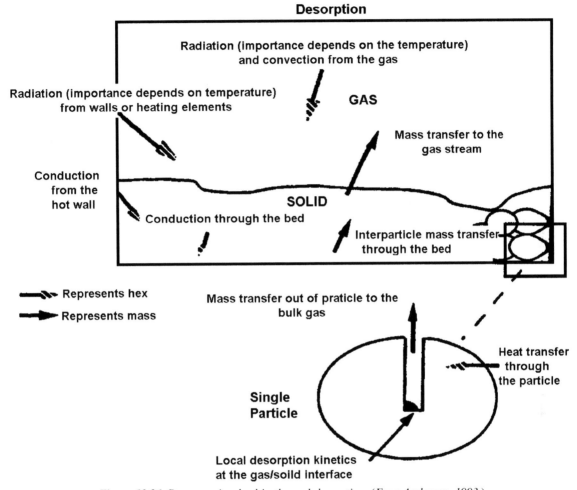

Figure 13.26 *Processes involved in thermal desorption. (From Anderson, 1993.)*

ergy and time are needed for desorption. The effective rate of desorption from the soil is also controlled by diffusion, which in turn is controlled by the type of contaminants in the soil.

These processes are all system specific and must be taken into consideration on a site-by-site basis. The following factors affect the process significantly:

- *Temperature.* Increase in temperature greatly decreases concentrations.
- *Soil matrix.* Coarse particles such as sands will desorb contaminants easier than fine-grained clays and silts.

- *Contaminant.* Some contaminants will bind stronger to soils than others.
- *Moisture content.* Increase in moisture reduces the capacity of the contamination to adsorb on soils with high mineral contents (i.e., silts and clays).

13.6.3 System Design and Implementation

Thermal desorption can be designed and implemented as an ex-situ or in-situ process as detailed in this section.

Ex-Situ Thermal Desorption. Thermal desorption is categorized into two groups:

1. *High-temperature thermal desorption.* The soil is heated to a temperature of 600 to 1000°F. It can produce a final contaminant concentration level below 5 mg/kg for the target contaminants identified (FRTR, 1997). When soil undergoes high temperature desorption, it may lose many of its natural soil properties.

2. *Low-temperature thermal desorption.* The soil is heated to a temperature of 200 to 600°F. Organic components are not damaged; thus, the treated soil can retain support for biological activity. Soil that is treated with low-temperature desorption retains its physical properties.

The two methods of thermal desorption most commonly used within the categories of high- and low-temperature thermal desorption described above are classified as follows:

- *Direct fired.* Fire is applied to the surface of the contaminated soil.
- *Indirect fired.* The soil does not come into direct contact with the fire. Instead, a metal cylinder is heated and the soil is heated indirectly by the metal cylinder.

Each method of thermal desorption consists of following three components:

- *Pretreatment and material handling system.* This system determines where the excavated soil should be placed and conditions the material that is to be placed into the desorber unit. Furthermore, a method is developed to deliver the soil to the desorber unit.
- *Desorption unit.* The unit heats the contaminated soil to a sufficient temperature for a period of time required to dry and vaporize the contaminants. Different units are available depending on the situation. For instance, rotary dryers, the most commonly used units, are horizontal cylinders that can be indirect or direct fired. Thermal screw units, screw conveyors, or hollow augers are used to transport the medium through an enclosed trough.

- *Posttreatment system.* The vapors from the desorption unit are processed to remove residual particulates. Subsequently, the treated soil is tested to measure the effectiveness of the treatment.

When determining the system design, an important factor to consider is the direction of flow of the combustion gases relative to the desorber. The flow configuration of the desorber can either be cocurrent or countercurrent. The flow configuration of the desorber will affect the arrangement and size of the components used in the treatment process. In a cocurrent system, gases exiting the desorber are relatively hot. From this, the system must be designed to handle these hot gases. Figure 13.27(*a*) is a flowchart of a typical cocurrent system. Alternatively, if a countercurrent flow configuration is used, the gases leaving the desorber are cool enough to flow from the cyclone to the baghouse. Figure 13.27(*b*) is a flowchart of a typical countercurrent system. The components of both these types of systems include:

- *Feeders.* Conveyors and augers are used to bring the contaminated soil to the desorber unit. Material larger than 2 in. in diameter is crushed or removed. Crushed material is recycled back into the feed to be processed. Also, if the moisture content is high, dewatering takes place.

- *Desorber unit.* The function of the unit is to heat the soil to a sufficient temperature and maintain it for a period of time to desorb the contaminants from the soil. Two types of desorption units exist:

1. For *direct fire,* combustion gases provide heat to affect the contaminated soil. The heat range is from 7 to 100 MM Btu/h. "As the rule of thumb, a heat input of 25,000 Btu/h is the maximum required for each cubic foot of internal kiln (desorber) volume" (Anderson, 1993). Length-to-diameter ratios vary from 2:1 to 10:1. Rotation speeds range from .25 to 10 rev/min. To increase or decrease the time the soil is in the unit, the horizontal inclination and the rotational speed are adjusted.

2. For *indirect fire,* propane or natural gas is used to heat the metal cylinder. The heat is trans-

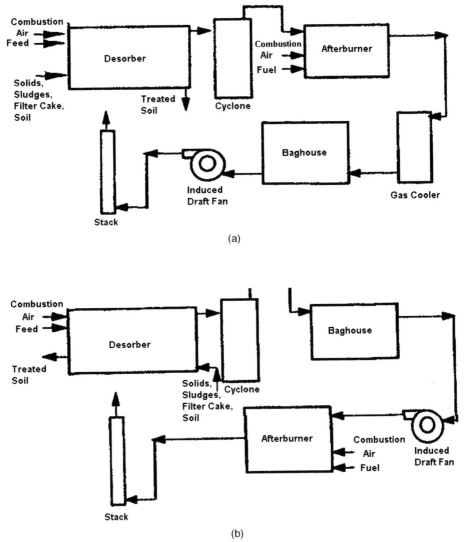

Figure 13.27 *Flow configuration of desorber:* (a) *cocurrent system;* (b) *countercurrent system.* (*From Anderson, 1993.*)

ferred via conduction. The units are less than 2.4 m (8 ft) in diameter and have heated lengths less than 14 m (45 ft). Rotation speeds can be as high as 2.5 rev/min. Inclination angles vary from 1 in. to 2 in. downward. Feed rates depend on the contaminant levels that are required. Typical feed rates range from 1.3 kg/s (5 tons/h) to 2 kg/s (8 tons/h). When moisture is high, the feed rate must be limited.

- *Cyclone.* The cyclone is designed to remove the largest of the entrained particles from the gas stream. Cyclone separators are most efficient in removing larger particles (larger than 15 μm). There are dry and wet cyclone separators, but in thermal desorption systems only dry cyclones are used. The dry cyclone separator is a true inertial separator. Particles entrained in the gas stream enter the cyclone, are directed into a vortex flow pattern, collect on the wall of the separator because of inertial effects, and eventually drop to the receiver part of the unit (Anderson, 1993).

- *Baghouses.* The gas passes through a series of permeable bags, where the particulate matter is collected. They collect the finer entrained particles with cloth filters. Baghouses remove particles of less than 10 μm and are very efficient in removing particles of less than 1 μm. The particles collected must be removed to avoid high pressure drops.

- *Afterburner.* Treatment may or may not require the use of afterburners. The gas may be collected and brought to an off-gas treatment site. This depends on the regulations at the site. Afterburners provide high temperatures to destroy organic compounds that have been desorbed from the soil. Afterburners can be used before or after the baghouse. If the baghouse is located after the afterburner, a gas cooling system is used to cool the soil down.

- *Venturi scrubbers.* Venturi scrubbers are used to remove sulfur dioxide and hydrogen chloride. They can also remove particles larger than 5 μm in the gas stream. The gas passes through a venturi throat, reaching velocities of 60 to 180 m/s. Typically, 8 to 45 L of water per 28 standard cubic meter of the gas is required in the throat section.

- *Wet scrubbers.* Acid neutralization may be required to prevent corrosive attack on the steel and other materials in the system. The scrubbers are used for acid neutralization. They are designed to use an alkali regent to acid gas stoichiometric ratio of slightly over 1.

- *Carbon adsorption unit.* Carbon absorption filters have been used to treat off-gases. If a carbon adsorption unit is used, the design might not require an afterburner.

The following factors must be taken into consideration when designing a desorber system.

- *Temperature.* The temperature controls the desorption rate. With higher temperatures the de-

sorption rate increases; thus the soil retention time is shortened.

- *Solid residence time.* The complexities of the contaminated soils require tests to determine the length of time for which the high temperature must be maintained in the soil. The soil residence time can be controlled by adjusting the rotation rate and angle of inclination of the desorber. The residence time of the soil in the kiln in minutes (*t*) is:

$$t = \frac{0.19 L_T}{(\text{rpm}) D S} \qquad (13.12)$$

where L_T is the length of the kiln in meters, rpm the revolutions per minute, D the inside diameter in meters, and S the slope of the kiln in m/m. The residence time also depends on the rate of desorption of the contaminant and the desired final contaminant concentration in the soil. For desorption, a first-order type of reaction is assumed and the relationship between the initial and final contaminant concentrations in the soil are related by

$$\frac{C_{sf}}{C_{si}} = e^{-kt} \qquad (13.13)$$

where C_{sf} is the final contaminant concentration in the soil (mg/kg), C_{si} the initial contaminant concentration in the soil (mg/kg), k the desorption rate constant (min^{-1}), and t the residence time (min).

- *Solid particle size distribution.* Characteristics of the soil can affect the amount of contaminant that is adsorbed on the soil. For example, soils made up of smaller particles have a greater surface area on which the contaminants can absorb. Also, when dealing with clay and silt type soils, "dusting" will increase.

- *Desorber rotational speed.* Soil residence time and the degree of mixing are directly related to the rotational speed of the drum.

- *Moisture content.* Energy needed to evaporate moisture in the soil increases as the moisture content increases. Furthermore, high moisture content may cause operating problems.

- *Mixing conditions.* Mixing conditions are important in transferring heat and venting the desorbed contaminants.

Monitoring is required to make sure that heat levels within the desorber are within the design range. If the heat goes beyond the designed range, this can cause damage to the system. Moreover, the soil that is cleaned must be monitored. The soil must be tested to determine if the thermal desorption process performed adequately.

In-Situ Thermal Desorption. In-situ thermal desorption applies heat to the soil with the use of blankets and wells. The equipment used is as follows:

- *Thermal blankets.* A typical in-situ system uses thermal blankets placed on the soil surface to treat shallow contamination. These blankets are about 8 ft × 20 ft in size.
- *Thermal wells.* Thermal wells handle the deeper contamination. Each well contains a heater as well as a vacuum. The vacuum is used to collect the vapor produced, which is then sent to an off-gas treatment center. These wells are placed such that the entire contaminated area is affected by heating. Instead of wells, heating can be applied to the soil directly by placing a network of electrodes.

Two different methods are commonly employed for soil heating: (1) powerline frequency heating (PLH) and (2) radio-frequency heating (RFH). Soil may be heated using *powerline frequency heating,* which is characterized by resistive heat generation by a 60-Hz alternating current. The current is carried through the soil via the conductive path of the residual soil water. The soil heating vaporizes the soil water. When the temperature nears 100°C, the resistive heating energy input becomes restricted, due to the increased soil resistance and effectiveness decreases. Powerline frequency (PLF) heating can be used at almost any depth. *Radio-frequency heating* uses high-energy radiowaves (2 to 20 MHz) to heat the soil by dielectric

heating. The RFH energy is transmitted through the soils without using residual soil water as the conductive path. Energy deposition is a function of the frequency applied and the dielectric features of the soil medium. Frequency selection is based on trade-offs of wave penetration depth and the dielectric constant of the soil profile; lower frequencies penetrate further. By adjusting the transmitter frequency to match the impedance of the soil, soil heating can continue up to 250°C or greater. The RFH technique heats a discrete volume of soil encompassed by the electrodes embedded in soil. When energy is applied to the electrode array, heating begins at the top center. It proceeds vertically downward and laterally outward through the soil volume.

In RFH, three rows of electrodes are placed through the contaminated soil in a triplate array configuration. The center-row electrodes are connected to the exciter (energy input) source. The two exterior rows are the ground/guard electrodes, which contain the input energy to the treated zone. Surface hardware connecting the electrodes is installed. A schematic of an RF in-situ soil heating system is shown in Figure 13.28. A horizontal electrode system is effective for a shallow site, and a vertical system is efficient in deep contaminated sites.

If the system is used along with SVE, several vacuum vapor extraction wells are used as part of the exciter array, as shown in Figure 13.29. A vacuum blower and an off-gas treatment system remove the heated soil contaminants. In-situ soil heating enhances SVE in two ways: (1) contaminant vapor pressures are increased by heating, and (2) the soil permeability is increased by drying.

13.6.4 Modified or Complementary Technologies

Vitrification is an immobilization technology to treat media contaminated with organic and inorganic contaminants. This technology uses electrical current to heat the soil. When all the soil becomes molten, the electrodes are discontinued. The molten soil is allowed to cool and becomes dense and hard glassy material.

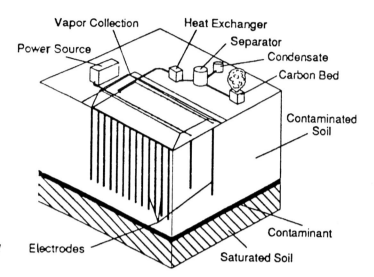

Figure 13.28 Radio-frequency in-situ soil
heating system. (*From HWRIC, 1995.*)

This remediation technology is described in detail in Section 13.7.

Incineration is a high-temperature remediation technology, which aims at first volatilizing the contaminants and then destroying them by combustion. The technology employs temperatures in the range 1600 to 2200°C. Recall as a comparison that low-temperature thermal desorption, applicable only for volatile organic compounds, is designed solely to remove contaminants from the soil. Incineration, on the other hand, can treat a variety of contaminants and achieves total destruction of contaminants. Incinerators have higher capital costs, though, and are more expensive to operate than low-temperature systems. Factors contributing to the high cost of incineration include:

- The high removal efficiencies required, which necessitate high operating temperatures and may result in the use of enhanced combustion techniques or afterburners
- The higher operating temperatures, which require more rugged equipment and use of more energy
- The more sophisticated stack gas cleaning equipment necessary for treatment of the combustion gases

Three types of incinerators are commonly used: rotary kilns, infrared units, and circulating fluidized-bed incinerators. Rotary kilns are inclined rotating cylinders lined with a refractory material. The waste or soil to be treated is fed into the high end of the cylinder. As the cylinder rotates, the soils tumble down its length through the combustion zone, where organic material is oxidized. Flue gases and inert ash leave the cylinder at the low end. The gas is routed to gas cleaning equipment, and the ash is quenched and ejected. Infrared incinerators, also known as *infrared conveyor furnaces,* use silicone carbide elements to produce thermal radiation. The medium to be treated passes through the unit on a conveyor belt. Contaminants are then volatilized from the soil. The off-gases are treated in a secondary thermal treatment unit. Circulating fluidized-bed incinerators consist of a refractory-lined vessel containing an inert sandlike material. Combustion air is forced through the bed material to suspend it. Lower combustion temperatures are used because the agitation of the fluidized bed aids complete combustion of the contaminants.

In-situ steam extraction treatment is also a thermal method that can be considered a form of thermal soil treatment (USEPA, 1991g). In this method, steam is injected into the soil through injection wells. A vacuum applied to the extraction wells recovers the contaminated gases and liquids, which are treated above ground. This method can also be used to treat contam-

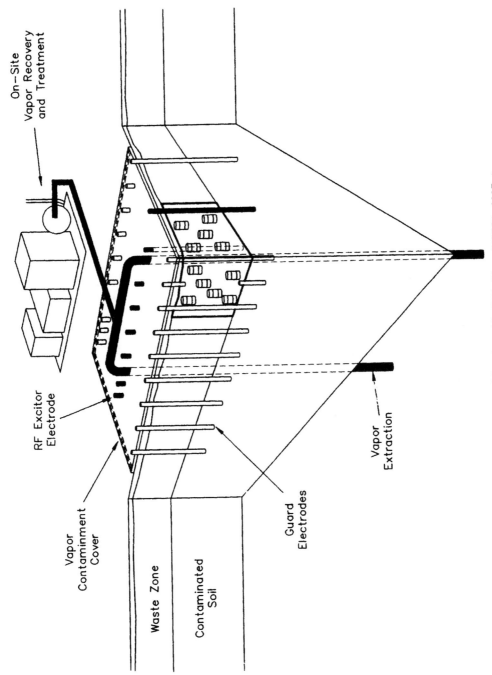

Figure 13.29 In-situ soil heating with SVE. (From USEPA, 1997g.)

On–Site Vapor Recovery and Treatment

RF Excitor Electrode

Vapor Containment Cover

Waste Zone

Contaminated Soil

Guard Electrodes

Vapor Extraction

inated groundwater and is explained further in Chapter 14.

13.6.5 Economic and Regulatory Considerations

Planning and preparation costs are those involved in obtaining the permits to operate the system. The cost of applying for air permits can range from a few thousand to many hundreds of thousands of dollars. Mobilization costs are incurred in the transport and setup of the desorption system. The cost depends on the size of the system, condition of the area where the system will be in use, and the location of the system before it is transported. To estimate the cost of mobilization, the location of the site and nearest utilities and availability must be accounted for. Excavation costs will vary with the type of contamination and soil that is being dealt with. Furthermore, if the soil is to be transported to a posttreatment center, these costs must be considered as well. Variable operating costs include fuel, process water, electrical power, and chemicals used at the site. The variable costs depend on the type of thermal desorption system used and the quantity of the material being treated.

Several issues need to be resolved with regulatory agencies in dealing with remediation technology. Air emissions from excavation and operation of the thermal desorption system have the potential of affecting the air. Excavation of contaminated soil for thermal desorption exposes material to the atmosphere. Highly volatile contaminants can evaporate into the air, presenting dangerous hazards to workers and people living around the excavation area. To minimize fugitive emissions, foams, water sprays, or portable enclosures may be used to prevent the release of harmful substances. Air monitoring systems may be used to protect workers.

Sometime the excavated material must be stored, and care must be taken to store it so as to prevent the release of hazardous materials or objectionable odors from the area. Physical enclosures with independent dust/vapor control or covers can be used to minimize air impacts. Equipment used in the desorption system may develop leaks over time. Leaks must be repaired immediately to contain spills. Wastewater from desorption units must be treated and tested before being discharged.

The transfer and handling of soils pose difficulties. Treated soil exiting the desorber may need to be cooled before being transported to its final destination. Since the treated media from the desorption unit must be tested, and analyses can take up to several days to complete, treated media may require a temporary storage area. Dusting can be a problem if the treated soil is left in an open area. Also, rainwater runoff from ash piles of treated solids can also present problems. Treated material awaiting fixation before final disposal may leach potentially hazardous compounds when it comes into contact with the rainwater (Anderson, 1993).

One of the most common complaints from people living or working near the site is noise. The equipment is equipped with warning sirens to warn when there are problems. Large motors or fans running also cause noise at the site. Noise abatement steps might be necessary to control this nuisance.

13.6.6 Case Studies

Case Study 1: Former Mare Island Naval Shipyard Site. The Mare Island electrical shop is located in the center of Mare Island Naval shipyard. The facility was used as an electrical workshop from 1955 to 1994. Solvents (including methyl ethyl and ketone) were frequently used. The facility had a cleaning area where their equipment, such as motors and transformers, would be washed before repairing them. From 1955 to 1978, transformers washed in the cleaning room contained polychlorinated biphenol (PCB) oil. Transformer washing procedures included draining the oil and pressure washing the interior of the transformers with stream and degreasing solvents and detergents (NFESC, 1999). Sludge and liquid waste would collect in a drain pipeline in the corner of the room. The sludge was removed from the drain pipe periodically. In 1981, the sludge was discovered to contain PCBs, and further investigations revealed PCBs in the cleaning room. Also, oily liquids containing some of the compounds used in the cleaning room were discovered in the soil. Organic compounds discovered in the soil included PCBs, volatile organic compounds, and pesticides. The target compounds in the area were primarily PCBs which had concentrations as high as 2200 mg/kg.

The in-situ thermal desorption system was composed of two heating elements: thermal blankets and thermal wells. Figure 13.30 shows the system setup of the thermal blankets and the thermal wells. The thermal wells were drilled vertically or horizontally to treat deep or hard-to-reach contaminated areas. The well assembly consisted of five components: (1) stainless steel well casing, (2) subsurface heating element, (3) vacuum barrier of shimstock, (4) layer of insulated material, and (5) impermeable sheet.

The heaters initiate a thermal front that moves laterally through the soil by thermal conduction. As the soil is heated, organic compounds and water vapor are desorbed and evaporated from the soil matrix. Negative

pressure is induced throughout the treatment zone by a pressure blower, while an impermeable liner and insulation minimize fugitive emissions and heat loss. The soil vapor is drawn via the vacuum blower and treated in a trailer-mounted air pollution control system (NFESC, 1999).

The thermal blanket system used the same vapor extraction and air pollution control system. The thermal blankets treated the contaminated soil surface, up to 18 in. The components for the blanket system were (1) surface heating element, (2) insulated mat, and (3) impermeable sheet.

The targeted temperatures ranged from 1600 to 1800°F for PCB destruction. Depending on the boiling point of the contaminants of concern and the moisture content, the heating time ranged from 24 to 36 h. Operations were set up on September 1997 and stopped on December 1997. After treatment, the soil samples were analyzed for PCBs using EPA Method 8081. All posttreatment samples had no detectable PCB concentrations at a quantitation limit of 10 μg/kg. Actual construction and operation costs for the project were not available.

Case Study 2: Mckin Company Superfund Site. The Mckin Superfund site is located in Gray, Maine, about 15 miles north of Portland. The Mckin company operated a tank cleaning and waste removal business at the site. Here the company collected, stored, disposed of, and transferred petroleum and industrial wastes. The site included 22 aboveground storage tanks, an asphalt-lined lagoon, and an incinerator. Between the years 1970 and 1973 about 100,000 to 200,000 gal of liquid waste was processed. As a result of these operations, soil at this site was contaminated with volatile organic compounds and heavy oils. The soil was contaminated with trichloroethene, tetrachloroethene and 1,1,1-trichloroethane. Concentrations of these contaminants were as high as 3310 mg/kg.

Low-temperature thermal desorption was used at the site. Over 8000 m³ of soil was excavated. The system consisted of:

- *Feed system.* Excavated soil and debris were separated by screening the soil with a coarse grate.

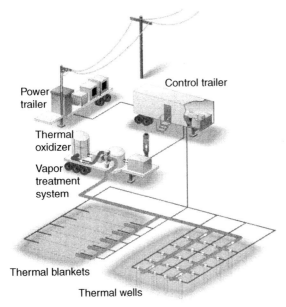

Figure 13.30 In-situ thermal desorption system used at former Mare Island Naval Shipyard site contaminated with PCBs. (From FRTR, 1995b.)

Power trailer

Control trailer

Thermal oxidizer

Vapor treatment system

Thermal blankets

Thermal wells

- *Direct-fired desorber.* A cylindrical drum 7 ft in diameter and 28 ft in length. Speeds reached 6 rpm.
- *HEPA filter.* Used to remove smoky particulates.
- *Baghouse.* Consisted of six banks of fine-mesh fabric filters. Used to remove more particulates.
- *Venturi scrubber.* The countercurrent flow scrubber was a 10-ft tower with a 6-ft diameter. This was used to remove water-soluble chemicals and to remove most of the remaining particulates.
- *Carbon filter system.* Scrubber exhaust was treated here, consisting of 15 tons of activated carbon.

The soil was treated in a continuous operation with a soil retention time of 6 to 8 minutes. During the process, the soil was heated to a temperature of 300°F. The treated soil was solidified and disposed of on-site. Operations lasted from July 1986 to April 1987. Soil samples were analyzed using SW-346 analytical methods. Also, 10% of the samples were analyzed off-site for confirmatory purposes. Within 10 months, thermal desorption reduced the concentrations of TCE from 3310 mg/kg to less than 0.1 mg/kg and other volatile and semivolatile organic contaminants from 320 mg/kg to levels less than 1 mg/kg. The total cost of this remediation was $2,900,000. Over 80% of the costs were for treatment of the soil. Other costs included salaries and wages, rental, supplies, subcontracts, and professional services.

13.7 VITRIFICATION

13.7.1 Technology Description

Vitrification involves the use of heat to melt and convert the contaminated soil into a stable glass or crystalline product. When the contaminated soil is melted, thermally stable inorganic contaminants are surrounded by the molten soil. As the molten soil cools, it forms a solidified mass of waste glass that incorporates these inorganic contaminants. These contaminants are incorporated into the waste glass either

through chemical bonding or through encapsulation (USEPA, 1992a). Also during the vitrification process, the high temperatures necessary to melt the contaminated soil will cause organic contaminants either to be destroyed via pyrolysis or removed as off-gases (USEPA, 1995b).

Vitrification can be applied as an in-situ process, a staged in-situ process, or as an ex-situ process. An in-situ vitrification (ISV) system uses a group of four graphite electrodes arranged in a square array. These four electrodes are inserted into the contaminated soil and an electric current is applied to them. Figure 13.31 presents a schematic diagram of a typical staged ISV process. A staged in-situ vitrification process involves excavating the contaminated soil and consolidating it into an on-site trench. The contaminated soil can then be vitrified in the trench using an ISV system. Ex-situ vitrification processes involve excavating the contaminated soil and transporting it to an adequately equipped facility. The contaminated soil is then fed into a furnace, which is used to heat and vitrify the contaminated soil.

Vitrification is applicable to soils contaminated with mixed contaminants that include radionuclides, metals and other inorganics, and organics. The process can be applied to all types of soils; however, soil moisture content and permeability can limit the applicability of vitrification. While high moisture content in the soil will not preclude the use of vitrification due to any technological limitations of the process, it may limit its applicability due to increased energy costs, resulting from the increased energy used to disperse the water during the vitrification process (USEPA, 1992a). The permeability of the soil must be low enough so that it prohibits water from recharging the vitrification zone faster than the vitrification process can dry and melt the soil. Soils with a permeability of less than 10^{-4} cm/s may require additional steps to remove moisture content prior to attempting vitrification (USEPA, 1992a).

Vitrification has the following advantages:

- Waste glass product has stable chemical and physical properties and excellent weathering properties.

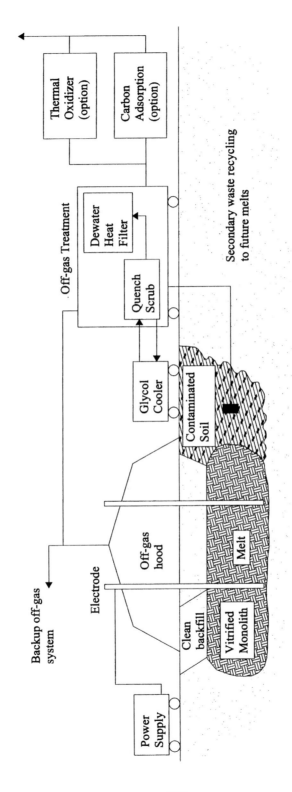

Figure 13.31 Schematics (a) ISV system; (b) staged ISV system. (From USEPA, 1992a.)

(a)

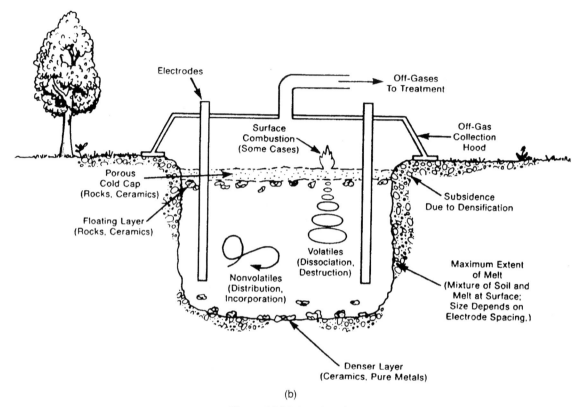

(b)

Figure 13.31 (*continued*)

- Excellent long-term durability of the waste glass product, possibly lasting for millions of years.
- Ability to process a variety of waste types, including contaminated soils, sediments, mine tailings, sludges, and radioactive wastes.
- Ability to process soils with mixed contaminants.
- Significant volume reduction resulting from the vitrification process (25 to 50% for most natural soils) (USEPA, 1997e).
- Potential reuse of the waste glass.
- Cost-effective for difficult sites with mixed contaminants or stringent cleanup standards (USEPA, 1992a, 1995b).
- Good public acceptance of ISV, due to safety benefits of the ISV process and the stable product.

- ISV requires no excavation, transport, or reburial of contaminated soil, improving worker safety and reducing costs.

Vitrification has the following disadvantages:

- Soil water content and water recharge can limit ISV applicability. Wet soil requires extra energy input to dry the soil, if the water recharge rate in the site exceeds the drying and melting rate of the soil, ISV applicability can be limited.
- Inability to process soils that do not provide sufficient electrical conductivity in their molten state.
- Limited processing depth, the maximum processing depth attained using ISV is approximately 22 ft (USEPA, 1995b).

- Ineffective to process contaminated soils containing more than 10% total organic content (USEPA, 1995b).

- Risk of an electrical short caused by the presence of metals settling to the bottom of the vitrification zone.

- Not proven to be safe for processing sites with sealed containers of vaporizable materials such as organic liquids.

- Not applicable to sites containing underground utility lines or other underground facilities.

- Energy-intensive nature of vitrification, high energy costs can potentially be prohibitive (USEPA, 1992a, 1995b).

13.7.2 Fundamental Processes

Vitrification systems employ a variety of fundamental processes to remediate various types of contaminants while limiting the escape of these contaminants into the environment. These processes include chemical incorporation or physical encapsulation of contaminants into the waste glass product, destruction of contaminants via pyrolysis or combustion, and volatilization of contaminants followed by their removal in the off-gas treatment system (USEPA, 1992a).

Both chemical and physical processes are used to remediate metals, nonmetallic inorganics, and radioactive inorganic contaminants. These contaminants are generally treated via immobilization in the waste glass product. Chemical immobilization is achieved when contaminants are chemically bonded into the waste glass product. This is achieved primarily for metals and other inorganic contaminants, such as asbestos (which has a high silicate content) that are capable of forming covalent bonds with oxygen and replacing silicon in the glass matrix structure (USEPA, 1992a). Certain other inorganic contaminants do not bond covalently with oxygen but are capable of bonding ionically with the oxygen atoms that are present in the glass network structure. This is another form of chemical incorporation into the waste glass. However, when this type of

incorporation occurs, the structure of the glass network is altered and the properties of the waste glass (e.g., leachability or physical durability) can be affected (USEPA, 1992a).

Physical immobilization of contaminants is achieved through encapsulation of the contaminant in the waste glass product. This is the primary method for the remediation of heavy metal contaminants and the nonmetallic inorganic contaminants. During the molten phase of the vitrification process, contaminants are surrounded by the vitrified material, and when the molten material cools, the contaminants are encapsulated into the waste glass product. This limits the possibility of the contaminants escape into the environment. The ability of the waste glass to retain a metal will depend on such things as the operational parameters of the vitrification process and the solubility of the contaminant in the waste glass (USEPA, 1992a, 1995b).

Thermal and chemical processes are the primary means of remediating organic contaminants and volatile inorganic contaminants when using a vitrification process. These processes include pyrolysis, combustion, and volatilization. Pyrolysis, which is the thermal destruction of a substance in the absence of oxygen, is one way in which vitrification remediates organic contaminants. During pyrolysis induced by the high temperatures attained during the vitrification process, some organic contaminants are thermally decomposed before ever reaching the surface of the melt zone (USEPA, 1995b).

Organic contaminants that are vaporized but not thermally decomposed in the melt are remediated either through combustion or are removed in the off-gas treatment system. After being vaporized, organic contaminants can migrate toward the surface of the melt. Some of the contaminants that reach the surface of the melt will be destroyed by combustion upon contacting the air contained above the melt surface and below the off-gas collection hood. The products of these combustion processes will be collected and, if necessary, treated in the off-gas treatment system. Organic contaminants that are not destroyed via oxidation upon contacting the air at the surface of the melt must be collected into the off-gas treatment system and either removed or thermally destroyed. Metals, such as mercury, after being volatilized during the vitrification process must also be treated via removal in the off-gas treatment system (USEPA, 1995b).

13.7.3   System Design and Implementation

General Design Approach.   A flowchart for a typical vitrification design is shown in Figure 13.32. The design process involves three primary steps: treatability and bench-scale testing, engineering-scale testing, and pilot- and large-scale testing. The primary purposes of treatability and bench-scale testing are to determine the applicability of vitrification to the soil type at the contaminated site and to determine the characteristics of the off-gas generated during vitrification processing (USEPA, 1992a). During treatability studies, melt parameters such as temperature and electrical power input should also be addressed. The vitrified product can also be tested to determine the resulting concentrations of contaminants and for its leaching characteristics (USEPA, 1992a).

Engineering-scale tests, which are conducted on a smaller scale than pilot- and large-scale tests, can be used to determine vitrification process limits. Finally, pilot- and large-scale testing can be done to verify that the vitrification system performs as predicted in the treatability/bench-scale testing. Possible contaminant migration to surrounding uncontaminated soil during vitrification should also be addressed during each phase of the design process.

Equipment.   The major components of a vitrification system include the electrodes and the associated electrode feed system, the electrical power system, the off-gas collection hood, and the off-gas treatment system. This equipment must be selected so that it meets the operational requirements of the vitrification process, which involves high voltages (up to 4000 V), high currents (up to 4000 A), and high temperatures (up to 3300°F) (USEPA, 1992a, 1995b).

At the start of the vitrification process, particularly ISV, a square array of four electrodes is inserted into the soil with an electrode feed system (EFS) attached. The EFS is designed to grip the electrodes, but also to allow them to sink down into the molten soil as the vitrification process proceeds (USEPA, 1992a). The electrodes remain in the molten soil during the vitrification process, with the EFS attached above the melt surface, so that if problems are encountered during the vitrification process, the EFS can prevent downward movement of the electrodes (USEPA, 1992a). Because the electrodes remain in the molten soil during the vitrification process, they must be able to withstand the corrosive effects of the molten soil as the vitrification process proceeds (USEPA, 1992a). Additionally, the electrodes must maintain sufficient mechanical strength at high temperatures, so that, if necessary, their progress into the melt can be halted by the EFS system during the process. The conductivity of the electrodes must also be sufficiently high to allow high currents to pass through them. These requirements have resulted in the use of threaded graphite electrodes, typically with an outside diameter of 12 in. (USEPA, 1995b, 1997d).

The spacing of the electrodes in the square array will determine the width of the melt zone. Melt width will be approximately $1\frac{1}{2}$ times the electrode spacing (USEPA, 1992a). For example, if the electrodes are spaced 12 ft apart, the melt will grow to an approximate width of 18 ft. The resulting melt zone will be shaped like a cube with slightly rounded corners (USEPA, 1995b).

The primary components of the electrical system are the power supply, a power substation (or power conditioning station), and power cables. The electrical system can either draw its power from the local utility grid or from a diesel generator that is transported to the site. For a typical vitrification system, the power system is required to produce a maximum in the range 3.5 to 4 MW of three-phase ac power (USEPA, 1997d). According to the USEPA (1992a), "since the conductivity of molten glass is ionic, an alternating current (AC) must be used to avoid the risk of electrolysis, annodization of electrodes, and the depletion of charge carriers." During the process, the applied voltage must be varied to allow the system to continue operating at

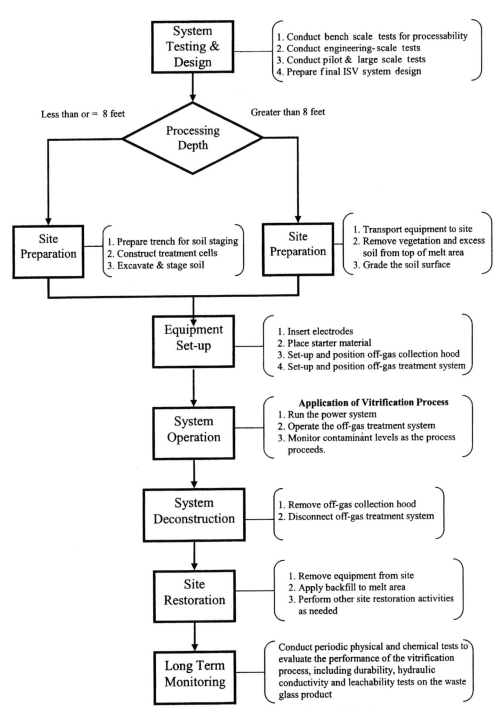

Figure 13.32 *General design approach to vitrification.*

an acceptable power level. Therefore, an electrical substation (or power conditioning station) is used to allow the appropriate variations in the applied voltage, within the range 400 to 4000 V, and to convert the power to two-phase (USEPA, 1997d). Power cables, for connection of the substation to the electrodes, must also be selected. These need to have a sufficient electrical conductivity to sustain the high levels of current [up to 4000 A (USEPA, 1992a)] that will be attained as the vitrification process proceeds.

The off-gas collection hood needs to be sized to ensure that all the gases that migrate to the top of the melt are collected under the hood and directed to the off-gas treatment system. This requires that the off-gas collection hood not only cover the entire melt area but also extend beyond the outer boundary of the melt area. In a typical ISV system, an octagonal collection hood with a maximum width of 60 ft between flat edges was used to cover a 26 ft × 26 ft square melt area (USEPA, 1995b, 1997c). Additionally, the off-gas collection hood should be constructed of steel.

The off-gas collection system for a vitrification system includes the following components in series: a high-efficiency particulate air (HEPA) filter, two scrubbers in series, a condenser, a heater, and two additional HEPA filters in series (USEPA, 1992a). The first HEPA filter provides an initial filtering of the off-gas, prior to its entering the system, in an effort to reduce the production of secondary wastes. The two scrubbers, placed in series, cool the off-gas and remove particulates. A condenser or demister is then used to remove the water vapor from the off-gas. After the water vapor has been removed, a heater is used to heat the remaining gases back above the dewpoint, before they are filtered by two additional HEPA filters in series (USEPA, 1992a). A thermal oxidizer can also be added at the end of the off-gas treatment system to polish the gas being emitted from the system and to eliminate any offensive odors (USEPA, 1995b, 1997c). This system should be operated at negative pressure to help ensure that contaminants do no escape into the atmosphere (USEPA, 1992a).

Soil properties at the contaminated site also need to be considered when designing the vitrification system. Properties such as soil moisture, density, and compo-

sition will affect the configuration of the ISV system, including the spacing and diameter of the electrodes, and the depth to which the electrodes are inserted into the ground (Koegler and Kindle, 1991; USEPA, 1992a). The presence of combustible materials and other buried wastes, in the treatment zone, must also be considered. The presence of combustibles increases the production rate of off-gases as they are burned. This production rate must not be high enough to defeat the off-gas system's capacity to operate at a negative pressure (USEPA, 1992a).

It should also be noted that when selecting the equipment for the process, the depth of the contamination should be considered. For sites where the processing depth is less than 8 ft, the most efficient and cost-effective method for performing vitrification is as a staged process (USEPA, 1997b). Performing vitrification as a staged process involves excavating the contaminated soil and placing it into a trench, with melt cells constructed in the trench. Vitrification can then be applied to each melt cell in succession until all of the contaminated soil has been treated (USEPA, 1995b).

Operational Parameters. The primary operational parameter for a vitrification system is the electrical input to the melt zone (USEPA, 1995b). A vitrification system heats soil via Joule heating, in which an electrical current flows through the contaminated soil. As the soil resists the current, power is lost and transferred to the soil as heat energy (USEPA, 1992a). Joule's law of electrical heating, which states that

$$P = I^2R \qquad (13.14)$$

where P is the dissipated power (watts), I the current through the material (amperes), and R the resistance of the material (ohms) (USEPA, 1992a), can be applied to a vitrification system to predict the power dissipated.

The vitrification process is started when the power source is activated and an electric potential is applied in two phases, via the substation, to the graphite electrodes. Most soils typically have high resistance. Therefore, initially, the applied voltage is high (up to 4000 V per phase), and as predicted by Ohm's law, the resulting current in each phase is low (100 to 200 A range) (USEPA, 1995b). As the vitrification process proceeds, the soil melts and the molten soil becomes more conductive/less resistant. Therefore, the voltage and current must be adjusted to allow the system to maintain operation at an acceptable power level. Power must also be maintained at a level that is sufficient to overcome the heat that is lost at the surface of the melt and dissipated into the surrounding soil. Voltage is decreased (to lows of approximately 400 V) to adjust for the lowered resistance, and this results in an increase in current (as high as 4000 A) (USEPA, 1992a). As mentioned above, the dissipated power can be predicted using Joule's law, and if the heat loss at the edges and the surface of the melt are also considered, this can provide an estimate of the energy being transferred to the melt (Koegler and Kindle, 1991).

During vitrification system operations, contaminants that migrate to the surface of the melt are either destroyed by combustion upon contacting oxygen at the surface of the melt, or are collected with the resulting gases in the off-gas collection hood and directed into the off-gas treatment system. The gases are then treated via the treatment train, which includes HEPA filters, scrubbers, a condenser, a heater, and a thermal oxidizer (if needed) (USEPA, 1992a).

Once the desired treatment depth is reached, the power to the electrodes is shut down. The maximum depth attainable using a large-scale ISV system is approximately 22 ft (USEPA, 1995b). After the power to the electrodes has been discontinued, the melt is allowed to cool naturally. The off-gas collection hood can typically be removed within 24 h of the power shutdown (USEPA, 1995b). Once the hood has been removed, the graphite electrodes are cut at the melt surface and allowed to become part of the vitrified mass. During the vitrification process, there is a significant volume reduction of the treated soil, typically 20 to 50% (USEPA, 1997d). This requires that a backfill of clean soil be applied to the vitrified area once the equipment has been removed.

Monitoring. During the vitrification process, the system should be monitored to assure that it is performing properly. Upon completion of the vitrification process,

preliminary tests should be conducted to determine whether or not the vitrification process has met the treatment goals. Additionally, a plan for long-term monitoring of the site should be created and implemented.

While the vitrification system is operating, the process needs to be monitored to assure that the system is functioning properly. This includes (1) monitoring the progress of the melt zone to ensure that all of the contaminated soil is being treated, and (2) monitoring the waste stream being produced by the off-gas treatment system to assure that it meets cleanup requirements before it is released back into the environment.

The waste streams generated by the off-gas treatment system that need to be monitored during processing are the scrubber liquor and the stack gas. The scrubber liquor is the waste water generated during off-gas treatment. Samples of the scrubber liquor should be collected during and after treatment to determine whether the concentrations of contaminants in the discharge are at acceptable levels. The stack gas is the gas that remains after treatment and is released into the atmosphere. The stack gas can be monitored continuously for the presence of carbon monoxide and hydrocarbons (USEPA, 1995b). It should also be sampled periodically to check for the presence of metals or other contaminants known to be present in the soil being treated (USEPA, 1995b).

Upon completion of the vitrification process, the melt area is allowed to cool sufficiently so that the surface of the melt solidifies. This may occur within 24 h after the system has been shut down (USEPA, 1995b). Once the surface of the vitrified soil has solidified, samples can be taken from just below the melt surface (USEPA, 1995b). Despite the fact that the samples taken from near the surface of the melt are not a representative sample of soil conditions within the melt area, they can be tested for the presence of the contaminants of concern at the site and used to give a preliminary indication of the effectiveness of the vitrification process. In addition to testing for the presence of contaminants, the toxicity characteristic leaching procedure (TCLP) should be conducted on the waste product to determine its leachability (USEPA, 1995b). Other tests that can be performed to determine the characteristics of the vitrified product include density tests, hydraulic conductivity tests, strength tests, and durability tests (USEPA, 1992a).

13.7.4   Predictive Modeling

Very few models exist to predict performance of vitrification process. Koegler and Kindle (1991) developed a mathematical model to provide a tool that predicts vitrification time, depth, width, and electrical consumption, based on input parameters, including electrode configuration, soil properties, and molten glass properties. The model assumes that electrical energy is converted into heat in the molten glass via Joule heating, heat loss occurs at the boundaries and surface of the melt, and the localized rate at which the vitrified zone progresses is proportional to the heat flux from the melt to the soil at that point. Confirmation of the model via comparison testing with actual ISV test results is limited, due to lack of usable existing field data. The heat-flux correlation can be improved by better simulation of the vitrification shape.

Dragun (1991) developed a model to determine whether the net migration of organic contaminants would be toward the melt or into the surrounding uncontaminated soil. The model assumes that five soil zones form during vitrification and remain in a quasi-equilibrium state: a melt zone, a pyrolysis zone, a heat-affected zone, a transient zone, and an ambient soil zone. Each of these zones affects the contaminant migration mechanisms occurring during vitrification. Applicability of this model is limited by the assumptions regarding the five soil zones and by the lack of data available for confirmation testing.

13.7.5   Modified or Complementary Technologies

Vitrification can be applied either in-situ or ex-situ. *Ex-situ vitrification* essentially involves excavating the contaminated soil, transporting it off-site, and feeding it into a melter or furnace. Ex-situ vitrification processes can be divided into two primary categories, those using electrical heating processes and those using

thermal heating processes. Although several electrical heating processes have been used for ex-situ vitrification, Joule heating, based on the same principles used for ISV systems, is also commonly used in ex-situ vitrification systems. In these systems, the contaminated soil, along with additional sand or other glass-forming materials intended to facilitate the formation of the waste glass, is fed into an electrical resistance furnace (USEPA, 1997f). This furnace is equipped with electrodes that are immersed in the material to be vitrified. As in ISV, an electric potential is applied to the electrodes and the vitrification process proceeds. Also, as in ISV, a collection hood with an off-gas collection system is used. This type of system is typically cooled via water that is circulated in the hollow walls of the furnace (USEPA, 1997f). These systems frequently employ the use of a *cold top* (USEPA, 1997f). This involves the addition of soil to the top of the melt during processing, thereby forming a cold top, which restricts the escape of contaminants from the top of the melt and encourages their incorporation into the vitrified product (USEPA, 1992a).

Ex-situ vitrification processes using thermal heating are those process that involve the burning of the waste and/or some type of fuel to provide the heat for vitrification (USEPA, 1992a). These systems typically involve the use of a rotary kiln, which is a cylindrical shell lined with fire brick, mounted at an incline, that rotates as the wastes pass through the combustion zone in the kiln (USEPA, 1992a). According to the USEPA, a rotary kiln operated in the slagging mode may produce a vitrified product (USEPA, 1992a). The rotary kiln must be operated at a temperature high enough to melt the waste feed if vitrification is to occur (USEPA, 1992a). This type of ex-situ process also requires an air pollution control system for the waste gas that is generated and may require the addition of glass-forming constituents to the contaminated soil that is fed into the rotary kiln (USEPA, 1992a, 1997e).

Vitrification is commonly grouped into the category of stabilization/solidification (S/S) processes, despite the fact that it does not neatly fit into this category. While the S/S processes are peripherally related to vitrification processes in that each process seeks to remediate contaminants by immobilizing them into a solid waste product, the methods used to achieve this are very different. These S/S processes do not involve the heating methods used in vitrification systems, nor do they involve the high temperatures that are necessary in vitrification processes. Further, these stabilization/solidification processes do not thermally destroy organic contaminants as vitrification does. Vitrification is generally considered a stand-alone remediation process that can be applied to sites with mixed contamination, without the need for the use of other subsequent remediation technologies.

13.7.6 Economic and Regulatory Considerations

Vitrification can be cost-effective at sites with difficult or mixed contamination. Several factors determine the overall cost of implementing an ISV system. The primary factors affecting ISV operating costs are the cost of electricity at the site, depth and rate of treatment, and soil moisture content (USEPA, 1994f). The volume of soil to be processed will affect the total cost of the ISV system, but as the volume of soil to be treated increases, the cost per unit volume actually decreases. The operations cost for a typical ISV system is in the $375 to $425 per ton range (USEPA, 1997d). However, the total costs depends on other factors such as treatability/pilot testing, site preparation, mobilization, and site restoration and long-term monitoring (USEPA, 1997d).

The primary regulatory concerns related to an ISV system are the emissions of the off-gas treatment system during ISV operations, the wastewater generated during off-gas treatment, and the stability and reduced toxicity of the resulting vitrified product. Federal environmental regulations that are particularly pertinent to ISV systems include the Comprehensive Environmental Response, Compensation and Liability Act (CERCLA), the Clean Air Act (CAA), the Clean Water Act (CWA), and the Safe Drinking Water Act (SDWA) (USEPA, 1995b). Other federal regulations may be found to be relevant depending on the nature and extent of the contamination. Therefore, each site needs to be considered on a case-by-case basis when doing a comprehensive evaluation of the applicable or relevant and appropriate regulations. Further, individual states may have environmental regulations that are more strict than the applicable federal regulations.

13.7.7 Case Studies

In this section, two case studies are considered to assess the performance of vitrification as a remedial technology. In both of these cases, vitrification was used as a large-scale commercial remediation project (USEPA, 1997b).

Case Study 1: Parsons Chemical/ETM Enterprises Superfund Site. The Parsons site in Grand Ledge, Michigan, is the former location of Parsons Chemical Works, Inc. The site was occupied by Parsons for approximately 34 years, from April 1945 until 1979 (USEPA, 1997c). Activities at Parsons included the mixing, manufacturing, and packaging of agricultural chemicals, including pesticides, herbicides, and mercury-based compounds (USEPA, 1995b). These activities resulted in contamination of the surrounding soils and sediments.

The depth of the resulting contamination was limited to a maximum of 5 ft, with a total of approximately 3000 yd^3 (5400 tons) of soil being contaminated (USEPA, 1995b). The soil at the site was predominantly a silty clay soil (USEPA, 1997c). The types of contaminants included pesticides, such as DDT and dieldrin, and heavy metal contaminants such as mercury, lead, and arsenic (USEPA, 1997d). The concentrations of these contaminants ranged from 2200 to 23,100 µg/kg.

The contaminated soil at the Parsons site was treated using a staged ISV process. This involved excavating a treatment trench and then constructing melt cells in that trench. The contaminated soil was then excavated and consolidated into the melt cells in the treatment trench. The contaminated soil was then treated, in the treatment trench, via a series of applications of ISV. A total of eight melts were conducted to remediate the 5400 tons of contaminated soil (USEPA, 1997c). Each melt cell was 16 ft deep and

26 ft × 26 ft square (USEPA, 1997c). The vitrification process used threaded graphite electrodes with an outside diameter of 12 in., supplied in 6-ft sections, and spaced up to 18 ft apart in a square array (USEPA, 1995b). These were supported using an electrode feed system, which allowed the electrodes to sink into the melt as the vitrification process proceeded. Electrical power was obtained from the local utility grid and transformed using a power conditioning station. The steel off-gas collection hood used was octagonal with a diameter of 60 ft. Total power applied to the electrodes during the ISV process was approximately 3.5 MW. Once a steady-state processing condition was attained, the applied voltage to each of the two pairs of electrodes averaged 600 V and the current for each phase averaged 2900 A (USEPA, 1997c). A treatment rate of 4 to 6 tons of soil per hour was attained, with the vitrified zone growing downward at a rate of 1 to 2 in./h (USEPA, 1995b).

These melts were conducted over approximately a one-year period, with the duration of each melt ranging from 10 to 19.5 days and the electrical power consumed for each melt ranging from 559,200 to 1,100,000 kWh (USEPA, 1995a). Approximately one year of cooling time was required before the melt could be sampled to confirm the results of initial sampling, which had indicated contaminant concentrations below acceptable levels (USEPA, 1995a).

The application of ISV at the Parsons site succeeded in reducing contamination to below acceptable levels. Monitoring of off-gas emissions during the application of the ISV process revealed that emissions of contaminants in the stack gas were well below applicable or relevant and appropriate regulations. Confirmation corings were performed approximately one year after the ISV process was completed. These revealed that mercury and pesticide concentrations were reduced below detection limits in the resulting vitrified material (USEPA, 1997c). Further, the ISV process succeeded in reducing the volume of contaminated soil by 30% (USEPA, 1997c).

The total cost for the ISV treatment at the Parsons site was $1,763,000, with $800,000 going directly to vitrification operations. This results in vitrification costs of $270 per cubic yard (USEPA, 1997c). It should be noted that pretreatment costs were particularly high in this case, due to the need to excavate and stage the contaminated soil prior to treatment.

Case Study 2: Wasatch Chemical Company (Lot 6) Superfund Site. The Wasatch site in Salt Lake City, Utah, was, at one time, an active chemical production, storage, and distribution facility. Pesticides, herbicides, and various other chemical products were manufactured or packaged at the site. Much of the facility's waste was directed into an on-site concrete evaporation pond, and some was discharged directly into the ground (USEPA, 1991f). These activities resulted in the contamination of the surrounding soil and groundwater.

A total of 3587 yd³ (5600 tons) of contaminated soil and 650 gal of liquid waste was consolidated in the concrete evaporation pond and targeted for remediation using ISV (USEPA, 1991f). The desired treatment depth in the evaporation pond was 7 to 8 ft (USEPA, 1995a, 1996a). The mixed contaminated site soil was combined with sludge heel, from an evaporation process, and various debris, including wooden timbers, clay pipe, sample containers, scrap metal, smashed 55-gal drums, plastic sheeting, and protective clothing (USEPA, 1997d). Contaminants at the site included dioxins/furans, pentachlorophenol, pesticides, volatile organic compounds (VOCs), and semivolatile organic compounds (SVOCs) (USEPA, 1997d). The concentrations of these contaminants were as high as 272,900 μg/kg.

The remediation process applied at the Wasatch site involved the combined application of ISV, a pump and treat system, and bioremediation. The bioremediation was performed on an area of soil contaminated with hydrocarbons, and the pump and treat system was applied to remediate groundwater that was contaminated primarily with VOCs and pentachlorophenol (USEPA, 1999a). The ISV system was used to treat contaminated soil and mixed waste that had been consolidated in the concrete evaporation pond. This ISV used essentially the same equipment (graphite electrodes, 60-ft-diameter off-gas collection hood, electrode feed system, and power conditioning station) as that used during remediation of the Parsons site. A total of 37 adjacent melts were conducted to remediate the 5600 tons of contaminated materials (USEPA, 1997d). The operational controls of the process (power applied to the electrodes, applied voltage and current per phase) also were comparable to the operational controls used at the Parsons site. However, the desired treatment depth at the Parsons site was 16 ft, approximately twice as deep as the desired treatment depth at the Wasatch site. Also, as noted previously, the medium being treated at the Wasatch site was a mixed soil with sludge and all types of debris in it, as opposed to the predominantly silty clay soil that was treated at the Parsons site. Both of these factors affect the application of the ISV process.

The application of ISV at the Wasatch site was successful in reducing contamination to acceptable levels (USEPA, 1997d). Molten dip samples of the vitrified zone were taken during ISV processing, and these samples showed no detectable organic contamination remaining in the molten soil. Off-gases were also monitored during the process and were found to have levels of dioxins/furans below detection levels (USEPA, 1997d). Additionally, samples were taken from the soil surrounding the concrete evaporation pond before and after application of the ISV process. This confirmed that no contamination had migrated to the surrounding soil during the ISV process (USEPA, 1997d).

The initial total cost for the ISV application at the Wasatch site was estimated to be approximately $3,300,000. However, due to problems experienced during application of the process and the lack of experience with the use of ISV technology at the time of the application, the total cost was reestimated to be $6,370,000 (USEPA, 1995a,1996a).

13.8 BIOREMEDIATION

13.8.1 Technology Description

Bioremediation is a process in which microorganisms degrade organic contaminants or immobile inorganic contaminants. Under favorable conditions, microorganisms can degrade organic contaminants completely into nontoxic by-products such as carbon dioxide and water or organic acids and methane (USEPA, 1991a).

In the natural attenuation process, microorganisms occurring in the soil (yeast, fungi, or bacteria) degrade the contaminants for their survival. However, depending on the type of contaminant and its toxicity levels, specific microbes may be introduced into the soil to be remediated. In addition, for microbial survival and growth, supplies of oxygen, moisture, and nutrients may be needed. The process of bioremediation refers to enhancement of the natural process by adding microorganisms to the soil, referred to as *bioaugmentation,* and/or supplying oxygen, moisture, and nutrients required for microbial survival and growth to the soil, referred to as *biostimulation.* Bioremediation is also called *enhanced bioremediation* or *engineered bioremediation* in the published literature (USEPA, 1991a, 1992c, 1994a).

Bioremediation in the presence of air or oxygen is called *aerobic bioremediation* and in the absence of air or oxygen is known as *anaerobic bioremediation.* Under aerobic conditions, organic contaminants are converted to carbon dioxide and water. Under anaerobic conditions, organic contaminants are converted to methane, limited amounts of carbon dioxide, and traces of hydrogen. Aerobic bioremediation is quicker than anaerobic bioremediation; therefore, it is often preferred. Bioremediation can be accomplished under in-situ conditions, called *in-situ bioremediation,* or under ex-situ conditions, called *ex-situ bioremediation* (USEPA, 1988a,b; Thomas and Ward, 1989; Cauwenberghe and Roote, 1998; Cookson, 1995).

At most sites, microorganisms naturally exist that are capable of degrading the contaminants. However, environmental conditions are not conducive for these microorganisms to degrade the contaminants. Bioremediation basically involves supplying oxygen, moisture, and nutrients to the contaminated soil zone so that the naturally existing microorganisms are activated to degrade the contaminants. For degradation to occur, it has to be ensured that oxygen, moisture, and nutrient concentrations are maintained in sufficient amounts. The monitoring can be done by maintaining monitoring wells and also by measuring the concentrations of carbon dioxide and oxygen. The increase in biological activity will be marked by the decrease in oxygen concentration and an increase in carbon dioxide concentration.

Bioremediation is commonly used for the treatment of soils contaminated with organic compounds. Petroleum hydrocarbons can easily be treated using bioremediation. Other organic compounds, such as PAHs and PCBs, are resistant to degradation, due to their high levels of toxicity to microbial population. Specific microbes that can resist these toxicity levels are often required. Bioremediation cannot degrade inorganic contaminants such as heavy metals, but it can be used to change the valence states of these metals, thus converting them into immobile form. For example, microbes can convert mobile hexavalent chromium into immobile trivalent chromium. Biomediation can be used in any soil type with adequate moisture content, although it is difficult to supply oxygen and nutrients into low-permeability soils. It should be noted that very high concentrations of the contaminants may be toxic to microorganisms and thus may not be treated by bioremediation. Therefore, a feasibility investigation is needed to determine if biodegradation is a fesible option for the site-specific soil and contaminant conditions (USEPA, 1985; Aggarwal et al., 1990).

Bioremediation has the following advantages:

- It may result in complete degradation of organic compounds to nontoxic by-products.
- There are minimum mechanical equipment requirements.
- It can be implemented as an in-situ or ex-situ process. In-situ bioremediation is safer since it does not require excavation of contaminated soils. Also, it does not disturb the natural surroundings of the site.

- Cost is low compared to other remediation technologies.

Bioremediation has the following disadvantages:

- There is a potential for partial degradation to equally toxic, more highly mobile by-products.
- The process is highly sensitive to toxins and environmental conditions.
- Extensive monitoring is required to determine biodegradation rates.
- It may be difficult to control volatile organic compounds during an ex-situ bioremediation process.
- It generally requires a longer treatment time than that of other remediation technologies.

13.8.2 Fundamental Processes

Bioremediation is a common technology for the treatment of organic compounds; however, the use of this technology for the treatment of inorganics such as heavy metals is still new (Means and Hinchee, 1994). Therefore, the fundamental processes involved in biodegradation of organic contaminants will be focused in this section. In simple terms, the microorganisms swallow the contaminants present and metabolize them into compounds of less molecular weight and toxicity. Figure 13.33 depicts how the microbes degrade contaminants (USEPA, 1991a). Unfortunately, the degrading

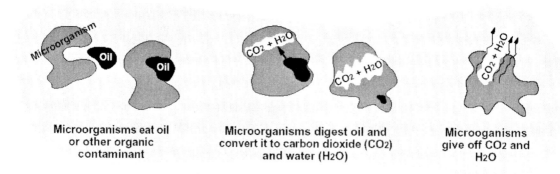

Microorganisms eat oil or other organic contaminant

Microorganisms digest oil and convert it to carbon dioxide (CO2) and water (H2O)

Microoganisms give off CO2 and H2O

Figure 13.33 *Degradation of contaminants by microbes.* (*From USEPA, 1991b.*)

(metabolic) processes are very complicated and depend on the types of microorganisms and contaminants present in the soil. If the microorganisms are native to the site, they are known as *indigenous microorganisms*. To simulate the growth of these indigenous microorganisms, soil conditions such as temperature, pH, and oxygen and nutrient content may need to be adjusted. If the microorganisms needed to degrade the contaminants are not present in the soil, microorganisms are taken from other locations, cultured in the laboratory, and then added to the contaminated soil; these are called *exogenous microorganisms*. Soil conditions may also need to be adjusted to make these microorganisms grow. Most bioremediation applications make use of indigenous microorganisms to make the process more effective and less expensive.

Enzymes are responsible for the degradation of organic carbon, which is used by the bacterial cell to produce both building blocks of life and energy. The degradation of any organic molecule, including contaminants, require the production and efficient utilization of enzymes. In most cases, degradation is a complex oxidation–reduction reaction. The electrons or reducing equivalents (hydrogen or electron-transferring molecules) produced must be transferred to a terminal electron acceptor (TEA). During the transfer process, energy is produced that is utilized by the cell. Depending on TEA, bacteria are grouped into three categories: (1) aerobic bacteria, which can only utilize molecular oxygen as a TEA; (2) facultative aerobes/anaerobes, which can utilize oxygen or when oxygen is low or nonexistent, may switch to nitrate, manganese oxides, or iron oxides as electron acceptors; and (3) anaerobes, which cannot utilize oxygen as an electron acceptor and for which oxygen is toxic, and they utilize sulfate or carbon dioxide as electron acceptors. Contaminant degradation may occur in one of three modes: aerobic, cometabolic, and anaerobic. A detailed discussion on these degradation modes can be found in references such as Alexander (1994) and Cookson (1995).

Bioremediation processes may be directed toward accomplishing (1) destruction of organic contaminants, (2) oxidation of organic chemicals whereby the organic chemicals are broken down into smaller constituents, or (3) dehalogenation of organic chemicals by cleaving chlorine atoms or other halogens from a compound.

Microorganisms need appropriate environmental conditions to survive and grow. These conditions include pH, temperature, oxygen, nutrients, and toxicity (Thomas and Ward, 1989; Cookson, 1995) Typically, the bioremediation is most efficient at a pH near 7. However, bioremediation can be achieved between pH values of 5.5 and 8.5. Most bioremediation systems operate over the temperature range 15 to 45°C. Microorganisms need a certain amount of oxygen not only to survive but also to mediate their reactions. Generally, oxygen concentration greater than 2 mg/L is required for aerobic microorganisms and less than 2 mg/L is required for anaerobic microorganisms. Microorganisms need nutrients for their growth. The major nutrients needed are identified with the generalized biomass formula ($C_{60}H_{82}O_{23}N_{12}P$) and include carbon, hydrogen, oxygen, nitrogen, and phosphorus. The actual quantity of these nutrients depend on the biochemical oxygen demand (BOD) of the contaminated soil. Generally, the C/N/P ratio (by weight) required is 120 : 10 : 1. Other nutrients such as sodium, potassium, ammonium, calcium, magnesium, iron, chloride, and sulfer are needed in minor quantities, in the concentration range 1 to 100 mg/L. In addition, traces (less than 1 mg/L) of nutrients such as manganese, cobalt, nickel, vanadium, boron, copper, zinc, various organics (vitamins), and molybdenum are needed. One must be careful that toxic substances do not exist that will produce adverse conditions for bioremediation. High concentration of any contaminant can be toxic to microbes. Some contaminants even at low concentrations may be toxic to microbes. Generally, toxicity concerns are addressed by dilution or acclimated microbes. It is also desirable to maintain the soil moisture level between 40 and 80% of field capacity.

13.8.3 System Design and Implementation

General Design Approach. Figure 13.34 shows the general approach to the application of bioremediation. This approach can be followed for both in-situ and ex-

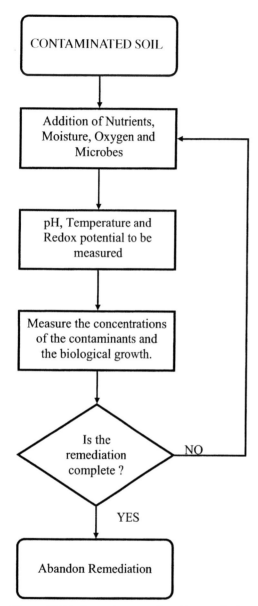

Figure 13.34 General approach to bioremediation.

situ conditions. In-situ bioremediation allows treatment of a large volume of soil at once and it is mostly effective at sites with sandy soils. In-situ bioremediation techniques can vary depending on the method of supplying oxygen to the organisms that degrade the contaminants. Two in-situ methods of oxygen delivery are bioventing and injection of hydrogen peroxide or oxygen-releasing compound (ORC).

- *Bioventing.* Bioventing systems deliver air from the atmosphere into the soil above the water table through injection wells placed in the ground where the contamination exists (Figure 13.35). An air blower may be used to push in or pull out air into or from the soil through injection wells. Air flows through the soil, and the oxygen present in the air is used by the microorganisms. Nutrients may be pumped into the soil through the injection wells. Nitrogen and phosphorus may be added to increase the growth rate of the microorganisms (Hinchee, 1994; USEPA, 1992b, 1993b).

- *Injection of hydrogen peroxide or oxygen releasing compound.* This process delivers oxygen to stimulate the activity of naturally occurring microorganisms by circulating hydrogen peroxide (in liquid form) or ORC through contaminated soils to speed up the bioremediation of organic contaminants. ORC is a patented formulation of magnesium peroxide which when moist releases oxygen slowly (Oencrantz et al., 1995). Since it involves putting a chemical (hydrogen peroxide or ORC) into the ground (which may eventually seep into the groundwater), this process is used only at sites where groundwater is already contaminated. A system of pipes or a sprinkler system is typically used to deliver hydrogen peroxide to shallow contaminated soils. Injection wells are used for deeper contaminated soils.

Ex-situ bioremediation involves excavation of the contaminated soil and treating in a treatment plant located on the site or away from the site. This approach can be faster, easier to control, and used to treat a wider range of contaminants and soil types than in-situ approach. Ex-situ bioremediation can be implemented

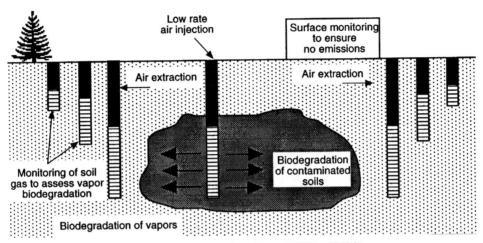

Figure 13.35 *Bioventing. (From USEPA, 1992d.)*

as slurry-phase or solid-phase bioremediation (USEPA, 1988a, 1994b).

In *slurry-phase bioremediation,* the contaminated soil is mixed with water to create a slurry. The slurry is aerated, and the contaminants are biodegraded aerobically. The treatment can take place on-site, or the soils can be removed and transported to a remote location for treatment (USEPA, 1990d). The process generally takes place in a tank or vessel (a bioreactor), but can also take place in a lagoon [Figure 13.36(*a*)]. Figure 13.36(*b*) presents a schematic of the process. Contaminated soil is excavated and then screened to remove large particles and debris. A specific volume of soil is mixed with water, nutrients, and microorganisms. The resulting slurry pH may be adjusted, if necessary. The slurry is treated in the bioreactor until the desired level of treatment is achieved. Aeration is provided by compressors and spargers. Mixing is accomplished either by aeration alone or by aeration combined with mechanical mixers. During treatment, the oxygen and nutrient content, pH, and temperature of the slurry are adjusted and maintained at levels suitable for aerobic microbial growth. Natural soil microbial populations may be used if suitable strains and numbers are present in the soil. More typically, microorganisms are added to ensure timely and effective treatment. The microorganisms can be seeded initially on

startup or supplemented continuously throughout the treatment period for each batch of soil treated. When the desired level of treatment has been achieved, the unit is emptied. The treated soil is then dewatered and backfilled in excavations. The wastewater is treated and disposed or recycled, and a second volume of soil is treated.

In *solid-phase bioremediation,* soil is treated in aboveground treatment areas equipped with collection systems to prevent any contaminant from escaping the treatment. Moisture, heat, nutrients, or oxygen are controlled to enhance bioremediation for application of this treatment. Solid-phase systems are relatively simple to operate and maintain, require a large amount of space, and cleanups require more time to complete than in slurry-phase processes. There are three different ways of implementing solid-phase bioremediation: contained solid-phase bioremediation, composting, and land farming (USEPA, 1988a, 1993a, 1988a; Cookson, 1995).

1. In *contained solid-phase bioremediation,* the excavated soils are not slurried with water; the contaminated soils are simply blended to achieve a homogeneous texture. Occasionally, textural or bulk amendments, nutrients, moisture, pH adjustment, and microbes are added. The soil is

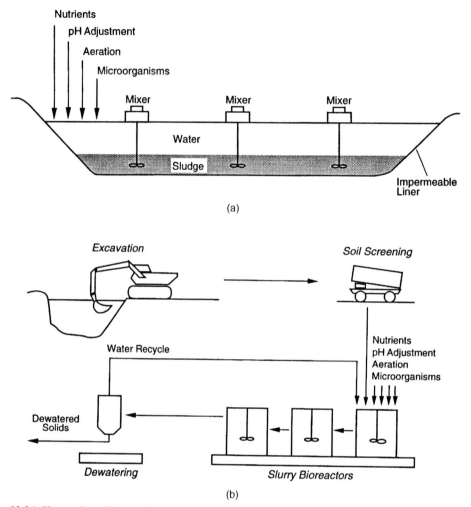

Figure 13.36 *Slurry-phase bioremediation:* (a) *in lagoons;* (b) *aboveground reactors.* (*From USEPA, 1993a.*)

then placed in an enclosed building, vault, tank, or vessel (Figure 13.37). The temperature and moisture conditions are controlled to maintain good growing conditions for the microbial population. In addition, since the soil mass is enclosed, rainfall and runoff are eliminated, and VOC emissions can be controlled. Mechanisms for managing/controlling flammable or explosive atmospheres and special equipment for blending and aeration of the soil may be required.

2. *Composting,* if carried out in an enclosed vessel, is similar to contained solid-phase bioremediation, but it does not employ added microorganisms. Structurally firm material may be added to the contaminated material to improve its handling characteristics, and the mixture may be stirred or mixed periodically to promote aeration and aerobic degradation. If necessary, moisture may also be added. Usually, composting is conducted outdoors rather than in an enclosed space. The two basic types of unenclosed composting

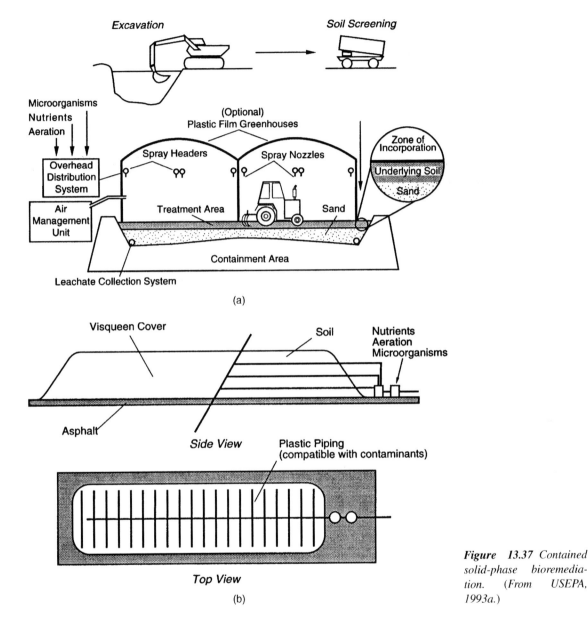

Figure 13.37 Contained solid-phase bioremediation. (From USEPA, 1993a.)

are open and static windrow systems (Figure 13.38). In *open windrow systems*, the compost is stacked in elongated piles. Aeration is accomplished by tearing down and rebuilding the piles. In *static windrow systems*, the piles are aerated by a forced-air system. Composting is com-

monly a less controlled process than other forms of bioremediation (with the possible exception of land farming). The waste is not protected from variations in natural environmental conditions such as rainfall and temperature fluctuations.

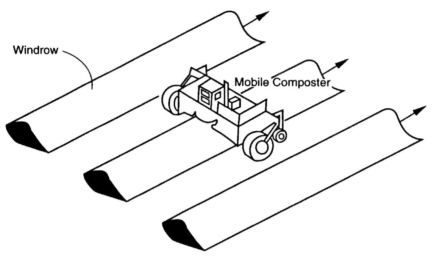

Figure 13.38 Composting: open and static windrow systems. (*From USEPA, 1993a.*)

3. The *land farming* process involves spreading the contaminated soil in fields or limited treatment beds. The soil is spread in thin lifts up to $\frac{1}{2}$ in. thick. Conventional construction and/or farm equipment may be used to spread the soil. The soil is tilled periodically, thereby providing oxygen. Microorganisms, nutrients, and moisture may also be added. Clay or plastic liners may be installed in the field prior to placement of the contaminated soil, which act to retard or prevent migration of contaminants into underlying and adjacent clean soils, groundwater, and surface water. Treatment is achieved through biodegradation, in combination with aeration and possibly photooxidation in sunlight. These processes are most active in warm, moist, sunny conditions. Treatment is greatly diminished or even completely arrested during winter months when temperatures are cold and snow covers the ground.

Equipment. The equipment required for in-situ bioremediation is basically those that involve the injection of nutrients into the soil, which may consist of spraying or sprinkling equipment, injection wells, or extraction wells (Thomas and Ward, 1989; USEPA, 1992c).

They need to be in sufficient number to ensure that the required amount of moisture and nutrients reach the contaminated zone. In the case of ex-situ bioremediation, a bioreactor with adequate volume to hold the contaminated soil is needed, or adequate space is needed to spread the contaminated soil (USEPA, 1990d, 1993a; Cookson, 1995). The volume of bioreactor is estimated roughly using

$$\frac{dC}{dt} = \frac{r_0}{K_{sd}X_s + 1} \tag{13.15}$$

$$\frac{dM}{dt} = Vr_0 \tag{13.16}$$

where V is the volume of the reactor, r_0 the rate of contaminant biodegradation, C the mass concentration of soluble contaminant, M the mass of contaminant to be put in the reactor, t the time in days, K_{sd} the soil distribution coefficient, and X_s the mass concentration of solids (contaminated soils).

Monitoring equipment is also needed to check contaminant spreading and to verify contaminant degradation. Equipment is necessary for the measurement of concentrations of microbes, nutrients, carbon dioxide, and oxygen. Excavating equipment is also necessary for excavating the contaminated soil from the site for the ex-situ treatment.

Operational Parameters. Besides the microbes, the following parameters are monitored and maintained at certain levels: nutrients, oxygen supply, temperature, pH, and moisture content. The addition of organic nutrients is often required to maintain the microbial ecology. Usually, initial laboratory tests are conducted to find the amount of carbon, nitrogen, and phosphorus required for the type and concentrations of contaminants at that particular site. But an approximate estimate of nitrogen, phosphorus, and oxygen uptake rates can be made using (Cookson, 1995)

$$r_N = \frac{0.06r_0}{1 + 0.05\tau} \qquad (13.17)$$

$$r_p = \frac{0.06r_0}{1 + 0.05\tau} \qquad (13.18)$$

$$r_{oxygen} = 0.06r_0 \left(1 - \frac{0.06}{1 + 0.05\tau}\right) \qquad (13.19)$$

where r_N is the rate of nitrogen uptake, r_p the rate of phosphorus uptake, r_{oxygen} the rate of oxygen uptake rate, r_0 the rate of biodegradation, and τ the solids residence time.

The oxygen supply is maintained either by pumping in or extracting out air through the contaminated zone through injection wells. Generally, the concentration of oxygen is maintained between 2 and 20 mg/L, depending on the type and concentration of the contaminants and on the soil type. The temperature of the bioreactor needs to be controlled if remediation is carried out in extreme weather conditions. Generally, the temperature is maintained between 25 and 40°C to ensure microbial activity. Since many soils are acidic in nature, pH adjustment is often needed. This is generally done by adding calcium or calcium/magnesium-containing compounds to the soil. The optimum moisture content required is generally obtained from pilot-scale tests done before the actual remediation is started and spraying or sprinkling equipment or injection wells are installed accordingly.

13.8.4 Predictive Modeling

Models for predicting the contaminant transport and degradation are given in Table 13.8. BIOPLUMEIII is commonly used to assess bioremediation performance at contaminated sites.

13.8.5 Modified or Complementary Technologies

Different technologies use bioremediation as a modification to that technology or as complementary technology. Soil vapor extraction is combined with bioremediation in that the extracted organic vapors are passed through the soil zones, where microbial activity is high, or passed through biofilters or biostrippers, which contain microbial population in chambers and contaminated vapor extracted is passed through the biofilters to facilitate degradation. Soil washing and bioremediation may be combined in that the soil may first be washed with some reagent and then the effluent and the silty and clayey materials are mixed to form a slurry which is then treated in a bioreactor for bioremediation. This process will reduce the volume of soil to be treated and in turn reduce the total costs involved in remediation.

13.8.6 Economic and Regulatory Considerations

This technology is found to be economical compared to the other technologies, especially when the contaminants involved are organic compounds and the soils have sufficient hydraulic conductivity to allow nutrient and oxygen flow through it. The costs involved in the treatment vary from as low as $30 per cubic yard of soil to as high as $750 per cubic yard (USEPA, 1985).

From a regulatory perspective, the air emissions have to be within the limits specified in the regulations, the seepage of the contaminants into the groundwater during remediation must be controlled, and care has to be taken that the volatile organic compounds do not enter into the atmosphere while excavating the soil for ex-situ remediation. Bioremediation has been the sub-

Table 13.8 **Selected models used in bioremediation studies**

Name	Type of Model	About the Model	Comments
BioChlor	Three-dimensional mathematical model	Screening model that simulates remediation by natural attenuation of dissolved solvent at chlorinated solvent release sites and can also be used for modeling contaminant transport with biodegradation as a first-order reaction	Simple to use since it uses an Excel spreadsheet to display results and is easy to understand and run
Bioscreen	Three-dimensional analytical model	Contaminant transport model for dissolved-phase hydrocarbons under the influence of oxygen, nitrate, sulfate, or iron and methane-limited biodegradation	Simple and easy-to-use model
Bioplume-III	Two-dimensional analytical model	Contaminant transport model under the influence of oxygen, nitrate, sulfate, or iron and methanogenic biodegradation.	Has a graphical user interface for preprocessing and output viewing

ject of intense research for the past few years to address issues such as:

- Evaluation of suitable microbes, nutritional requirements, lag times, and degradation rates (in the field) for various types of contaminants
- Optimization of environmental conditions and stimulation of favorable growth conditions under site-specific variations
- In-situ methods for an efficient monitoring process
- Mass balance of electron donors and acceptors within a given system
- Impact on aquifer permeability due to enhanced bioremediation
- Enhancing bioremediation of soils of low permeability
- Understanding bioaugmentation in soils of low permeability
- Understanding bioaugmentation, including which organics degrade specific types of contaminants
- Effect of hydraulic conductivity on microbial activity
- Techniques to minimize well fouling

13.8.7 Case Studies

The efficiency for cleanup of the remediation technology used at a particular site is very important while considering the costs involved in the remediation and effect of contamination on the environment. Several case studies have been reported where bioremediation has been used to remediate site contamination (FRTR, 1995a). To have a better idea about the performance, two full-scale remediation studies are presented below.

Case Study 1: Brown Wood Preserving Superfund Site. For approximately 30 years, the Brown Wood Preserving site in Live Oak, Florida was used for the pressure treatment of lumber products with creosote. Occasionally, pentachlorophenol was also used for this purpose. Lumber was treated in two cylinders and the wastewater from the treatment was dumped into a lagoon. Consequently, the lagoon and the soils (which included clayey soils to fine sand) around it were contaminated with high levels of organics, which consisted of PAHs found in creosote. These contaminants included benzo[a]anththracene, benzo[a]pyrene, benzo-[b]flouranthene, chrysene, dibenzo[a,h]anthracene, and indeno(1,2,3-cd)pyrene. The concentrations of these PAHs ranged from 100 to 208 mg/kg. Consequently, a study was conducted and land treatment technology was selected to clean up the contaminated soils that were stockpiled during the interim removal activities. A cleanup goal of 100 mg/kg of summation concentrations of six PAHs was set.

Construction of the land treatment area included installation of a clay liner, berm, run-on swales, and a subsurface drainage system. A retention pond was set up for runoff control. A portable irrigation system was used for maintaining optimum moisture during remediation. The first lift was inocculated with PAH-degrading microorganisms and the lifts cultivated every two weeks. The soil moisture content was maintained at 10%. Land treatment of the contaminated soils was performed from January 1989 to July 1990. Approximately 8100 yd^3 of soil was treated in three lifts. The concentration of PAHs was found to be varying between 23 and 92 mg/kg after treatment. The cleanup goal was achieved within a period of 18 months. The total treatment cost for this site was $565,400, or approximately $70 per cubic yard of soil, which is low compared to other treatment technologies.

Case Study 2: French Limited Superfund Site. The French Limited Superfund site in Crosby, Texas, was a former industrial waste disposal facility where an estimated 70 million gallons of petrochemical wastes was disposed in an unlined lagoon between 1966 and 1971. The primary contaminants at the site included benzo[a]pyrene, vinyle chloride, and benzene. The contaminant concentrations ranged from 400 mg/kg to 5000 mg/kg. Initial studies were conducted and slurry-phase bioremediation was selected as the treatment technology for the cleanup of the contaminated soils.

The treatment was done in two cells, each having a capacity of 17 million gallons. An innovative technology called the Mixflo system was used for aeration (to maintain a dissolved oxygen concentration of 2 mg/ L), which minimized the air emissions during the treatment. Approximately 300,000 tons of soil and sludge was treated during the cleanup operation. The cleanup took approximately 11 months and the concentrations after the treatment ranged from 7 to 43 mg/kg. The total cost was approximately $49,000,000, which included costs for treatment, pilot studies, technology development, and backfill of the lagoon.

13.9 PHYTOREMEDIATION

13.9.1 Technology Description

Phytoremediation involves removal, stabilization, or degradation of contaminants in soils by plants. The various remediation mechanisms are in the root zone or in the plant itself and are reflected by process-specific terminology (Figure 13.39). Plants are in contact with the contaminated soil in the root zone. Pollutants must pass through root membranes before they are absorbed by the plant, called *rhizofiltration*. Contaminant fate is determined by the plant's capacity to break down the absorbed organic chemicals by plant metabolic processes, called *phytodegradation,* or incorporate inorganic chemicals in plant tissue, called *phytoaccumulation.* Phytodegradation continues outside the plant through the release of root exudates (soluble organic matter, nutrients) and enzymes, which stimulate bacterial and fungal degradation of organic contaminants, called *rhizodegradation* (Schnoor, 1997; ITRC, 1999; USEPA, 1999b,c, 2000).

Contaminants enter the plant through a mechanism called *phytoextraction.* The large amount of water evapotranspirated by some plant species during summer accounts for a natural pump and treat method, which includes translocation of contaminants to stems, shoots, and leaves of the plant. Organic contaminants are degraded by plant-metabolic processes, and inorganic contaminants accumulate in plant tissue (Figure 13.39). The aboveground portions of the plant are then harvested and burned or composted. The removal of water from the soil creates a groundwater depression that limits advection and thus entrains contaminant-plume migration. Groundwater capture together with sorption of contaminants to roots and soil organic mat-

ter provides effective immobilization of the contaminant. This is called *phytostabilization.* Plants also minimize the release of toxic windblown dust, an important pathway for human exposure (Schnoor, 1997; USEPA, 2000).

Phytoremediation is best applied at sites with shallow (<10 ft) contamination. It is well suited for use at very large field sites and at sites with low (below regulatory action levels) contaminant concentrations where only "polishing treatment" is required (Schnoor, 1997). Because the remediation depth is the rooting depth of plants, fast-growing trees such as willow and poplar are most efficient to treat bulk soil amounts. However, the appropriate plant species for a contaminated site depends on soil characteristics and the contaminant type(s) and concentrations, which may be toxic to some species but provide essential nutrition to other species (ITRC, 1999).

A decision-tree flowchart of the applicability of phytoremediation is presented in Figure 13.40 (ITRC, 1999). Sorption of the contaminant to soil initially determines the applicability of phytoremediation. If the pollutant cannot be made available for extraction, phytoremediation is not an efficient cleanup technology. For organics, the partitioning coefficient provides a measure of the availability of the contaminant. Other tiers are geotechnical and agronomical engineering requirements, and modeling predictions, which may demonstrate that phytoremediation of a site is inappropriate cost- or time-wise. If phytoextraction is unsuccessful, phytostabilization (i.e., groundwater entrainment or increased sorption in the root zone), may still provide a remediation alternative. Whether phytoremediation is a suitable remediation technique for a certain region eventually depends on the costs of harvesting and destroying the contaminated plants, which may be substantial if the target area is large.

Target contaminants of phytoremediation include explosives, crude oil and oil products, pesticides (atrazine, cyanazine, alachlor), heavy metals and radioactive nuclides (cesium-137, strontium-90, uranium), and polycyclic aromatic hydrocarbons (PAHs). Plant uptake of metals varies for each metal group: cadmium, nickel, zinc, arsenic, selenium, and copper are easily absorbed, but hydrophobic metals such as lead, chromium, and uranium are not. Addition of chelates im-

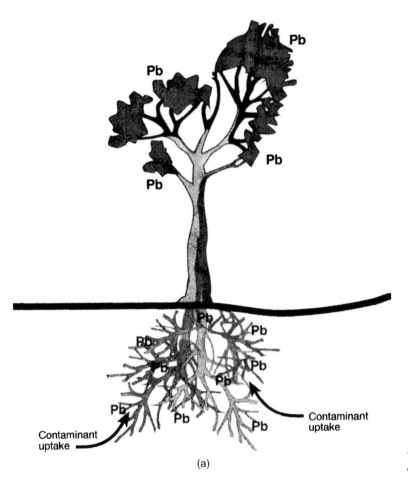

(a)

Figure   *13.39 Phytoremediation processes. (From ITRC, 1999.)*

proves the mobility of lead but increases the risk of leaching. However, complex formation and sorption of lead, chromium, and uranium to soils and root mass immobilizes these metals (Schnoor, 1997; USEPA, 1998b, 2000; Best, 1999). The technology is applicable for any soil that supports vegetation.

Phytoremediation offers the following advantages:

- It uses plants, therefore, is relatively less expensive.
- It is in-situ technology.
- Phytoremediation is a safe and passive technique, "driven" by solar energy and pleasant to the eye, hence likely to be accepted by the public.

Phytoremediation has the following disadvantages:

- Relatively shallow cleaning depths: <3 ft for grasses; <10 ft for shrubs, more (<20 ft) for deep-rooting trees.
- The phytoextraction process is slow because it requires several growing seasons (three to five years) to achieve cleanup standards.
- Knowledge to optimize remediation using plants is still experimental (USEPA, 1999c). Phytoremediation is site specific, the choice of plants is critical, and the cleaning strategy requires detailed site characterization to maximize plant growth and contaminant uptake.

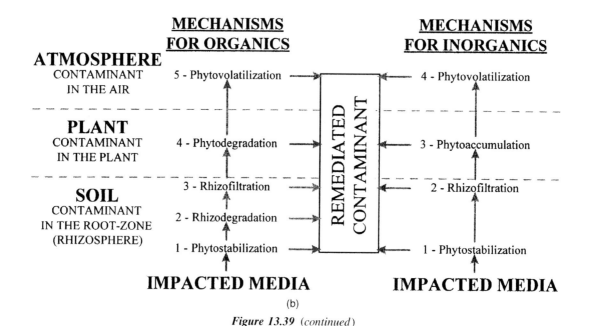

Figure 13.39 *(continued)*

- There is potential contamination of the food chain.
- Phytoremediation relocates contaminants to the plants, creating residual waste that must be disposed of.

13.9.2 Fundamental Processes

Contaminant transport and fate in plants is traced by labeling organic pollutants with radioactive carbon isotopes (^{14}C) as summarized in Figure 13.41. After contaminants have entered the plant, plant metabolism begins to break down the substance. Compounds are transformed by poplar trees to trichloroethanol, trichloroacetic acid, and dichloroacetic acid metabolites, sorbed to soil and roots, and mineralized to $^{14}CO_2$ (Newman et al., 1997; Best, 1999). About 15% of pesticides are incorporated into the biomass as bound residue (Schnoor, 1997). Pollutants released to the atmosphere by evapotranspiration react with atmospheric hydroxyl radicals, yielding half-lives of hours to days (Figure 13.41).

The efficiency of phytoremediation is directly related to physical–chemical properties of the contaminant, such as molecular weight, solubility in water, vapor pressure, and sorption. *Hydrophobicity* determines the sorption of organic chemicals to soil organic matter and mineral surfaces. Sorption is a function of the organic carbon fraction of the soil (f_{oc}) and the partitioning capacity of the contaminant, expressed by the octanol–water coefficient (K_{ow}). This coefficient relates to the partitioning of mass between groundwater and mineral or organic solids, and is found by shaking the organic compound with a mixture of n-octanol and water (see Chapter 8 for details). The K_{ow} is the concentration dissolved into octanol divided by the concentration dissolved into water. With highly hydrophobic contaminants or a high soil organic content, contaminants may be absorbed to soil irreversibly. Low-hydrophobic (log $K_{ow} < 1$) pollutants will be removed by leaching and groundwater advection and pass through plant membranes without plant uptake. Hence, phytoremediation is effective for soils contaminated with moderately hydrophobic (log $K_{ow} = 1$ to 3.5) organic contaminants such as benzene, toluene, ethyl-

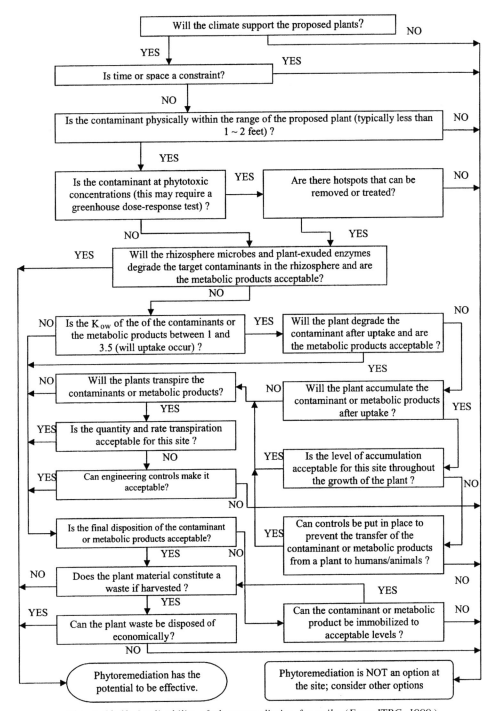

Figure 13.40 *Applicability of phytoremediation for soils.* (*From ITRC, 1999.*)

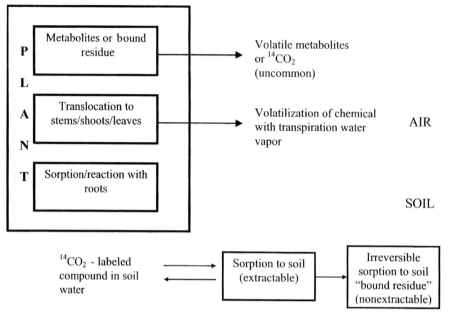

Figure 13.41 *Fate and transport of organic chemicals in phytoremediation. (From Schnoor, 1997.)*

benzene, xylenes, chlorinated solvents, PAHs, nitrotoluene ammunition wastes, excess nutrients (nitrate, ammonium, phosphate), and heavy metals (Schnoor, 1997; USEPA, 2000). Clay minerals are negatively charged and absorb metals under conditions of neutral pH. Thus, metal immobilization can be achieved by adding lime to soils. Alternatively, desorption of inorganic contaminants to enable phytoextraction may be stimulated by manipulation of the soil pH, controlled redox reactions, or addition of chelates.

13.9.3 System Design and Implementation

Figure 13.42 shows the different tasks involved in the design and implementation of phytoremediation. The first task is the selection of plant species and planting density. Plant selection and soil characterization are critical in phytoremediation, because of the long cleanup time involved. Following contaminant identification and quantification, plant selection is dependent on site-specific information. Plant species are subjected to laboratory tests, exposing them to toxic levels similar to those encountered in the polluted soils. Soils

must have sufficient water-holding capacity, to sustain plants and to provide a medium for contaminant solution. Failure to identify the appropriate plant species and soil amendments will result in suboptimal performance.

To optimize contaminant uptake, plant species are selected with high growth rates and biomass production (>3 tons dry matter per acre per year). Plant growth and root formation is stimulated by irrigation and addition of nutrients. Grasses, rye, sunflower, and mustard plants are often selected for their great surface root mass. Wetland conditions may be created to dissolve contaminants and increase evapotranspiration using weeds (e.g., Best, 1999). Phytoextraction of inorganic pollutants (metals) requires hyperaccumulation species [i.e., plants that accumulate 10 times the contaminant level in soil (>1000 mg/kg)].

If groundwater flow needs to be intercepted, species are selected with high evapotranspiration and deep roots. Suitable trees are tolerant of the groundwater environment (phreatophytic), such as poplar, cottonwood, and willow trees (USEPA, 1999c). Slowing groundwater flow decreases advection of the contam-

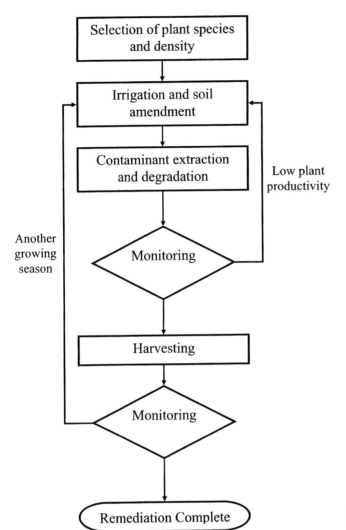

Figure 13.42 *General design approach to phytoremediation.*

inant plume and stimulates retardation by sorption. Groundwater entrainment is achieved by planting trees in multiple rows perpendicular to the plume's direction of movement, with monitoring wells throughout the treatment site (Chappell, 1998).

The second task is the irrigation and soil amendment. Depending on the plant species selected, subsequent soil amendments may be necessary. These include irrigation, addition of nutrients, adjustment of the pH to release contaminants and of the Eh to stimulate plant-assisted biodegradation. When pollution consists

of heavy metals and radioactive isotopes, chelates may have to be added to maintain contaminant concentrations in the soil solution (Huang et al., 1997; Dushenkov et al., 1999). In this situation, predictive modeling should assume zero-order rather than first-order kinetics (refer to Chapter 3 for a reaction kinetics background) because the contaminant will be available for extraction at a constant rate.

The third task is the monitoring. Monitoring of soil contaminant levels (e.g., soil sampling) will indicate the successfulness of the phytoremediation technique.

If plant productivity or contaminant extraction/breakdown remains below the expected levels, soil amendments may need to be reassessed or adjusted to stimulate plant activity.

The fourth task is the harvesting. If phytoextraction proceeds as expected, plants should be harvested toward the end of the growing season, or when saturation levels have been reached (Figure 13.42). After each growing season, soil monitoring will indicate if there is a need for continued phytoremediation. If contaminant levels remain above regulation standards, another growing season is required (Figure 13.42).

Phytoremediation monitoring includes periodic sampling of groundwater, soils, soil gas, plant tissue, and evaporation gas. Plant monitoring is done by growth measurements. Soil water should be monitored, especially during summer drought, to ensure constant water supply. Site maintenance also involves prevention of deer and insect damage. Plants may need to be sprayed with toxins that are specific to various insects. Bars of soap were hung from poplar trees to deter deer (Chappell, 1998).

13.9.4   Predictive Modeling

Table 13.9 summarizes selected applications of predictive modeling. Understanding the processes leading to phytoremediation of soils contaminated with different contaminants is still an active research area. Improvements to the existing models and development of new models is expected to continue.

13.9.5   Modified or Complementary Technologies

Phytoremediation is a complementing technology to remediate low-level (remnant) contamination to EPA regulation standards. A related technology is in-situ bioremediation and the same methods that enhance bioremediation (e.g., bioventing and control of aerobic conditions) apply to stimulate rhizodegradation. Phytoextraction furthermore provides natural, sunlight-driven pump and treat groundwater remediation.

13.9.6   Economic and Regulatory Considerations

The relatively simple procedures and the use of plants make phytoremediation cost-efficient and thus competitive with other remediation technologies, which often require exclusive equipment. Phytoremediation cost can range from $10 to $35 per ton of soil (Schnoor, 1997). Residual waste disposal costs may vary and depend on the toxicity of harvested plants, especially following phytoextraction of metals. Organic pollutants are biodegraded and plants may pose only a low level of toxicity. Despite the long cleanup time involved, phytoremediation of soils with low-level contamination is cost-efficient.

Regulatory issues concern cleanup time and residual levels (Schnoor, 1997). Experiments have demonstrated that release of toxic products to the atmosphere is minimal and is well below air quality standards. Metabolic products are commonly nontoxic, but by

Table 13.9 **Modeling of phytoremediation**

References	Model Used	Study	Assumptions	Limitations
Paterson et al. (1994)	Numerical	Contaminant fate in root, stem, and foliage	Three separate plant compartments	Experimental setting, days to months time scale
Trapp and Matthies (1995)	Analytical	Chemicals—mass balance in aerial plant parts	Exponential growth; nonionic pollutants	Generic, mathematical approach
Nedunuri et al. (1997)	Numerical	Simulate rhizosphere heavy metal immobilization	First-order kinetics groundwater uptake	Hypothetical soil column; partitioning
Schnoor (1997)	Numerical	Contaminant uptake rate	First-order kinetics	Fractional uptake

uptake of the contaminants the plants become toxic wastes themselves, and contamination of the food chain is therefore a major concern.

13.9.7 Case Studies

The effectiveness of either groundwater entrainment with poplar trees or extraction of contaminants using various accumulator plant species is demonstrated in small-scale plots. Table 13.10 summarizes several of these studies. Two of these studies are explained briefly in this section.

Case Study 1: Phytoextraction of Radinuclides and Heavy Metals at Chernobyl, Ukraine. The meltdown of Chernobyl's nuclear fission plant in April 1986 caused dispersal of ^{137}Cs between the Ukraine and Sweden. The highest amounts of radiation were measured in the region north of Chernobyl, where fallout percolated into sandy topsoils with rainwater. Twenty desorption-stimulating surfactants were tested by Dushenkov et al. (1999) with the purpose to increase radiocesium bioavalability. In a similar study, Huang et al. (1997) tested five types of chelates to achieve Pb desorption. The effectiveness of each chelate was established from the Pb levels in shoots of corn (*Zea mays*) and pea (*Pisum sativum*), reflecting Pb availability in soil water. Lead concentrations in the plants increased from 500 to 10,000 mg/kg (Huang et al., 1997).

Phytoextraction of radiocesium was effective during the first three weeks of accumulation, when radioactivity was lessened by 0.3 to 21%, depending on the chelate added (Dushenkov et al., 1999). After those three weeks, cesium uptake leveled, requiring plant harvesting. Plant species tested for ^{137}Cs accumulation were *Amaranthus* sp., Indian mustard, corn, peas, artichoke, and sunflower, the latter being hyperaccumulators. Phytoextraction produces a considerable amount of residual wastes in the form of toxic plants. However, incineration of the plant material reduces the volume of this waste to <10%. Heavy metals can be isolated by further processing of this waste.

Case Study 2: Phytotransformation of TNT. Removal of explosives (trinitrotoluene, TNT) from contaminated soil and groundwater was tested at the Milan Army Ammunition Plants in Tennessee (Best, 1999). To facilitate aquatic species and contaminant dissolution, wetlands are created. The aquatic plant parrot feather (*Myriophyllum aquaticum*) was used in a study demonstrating oxidative transformation of TNT (Best, 1999; Bhadra et al., 1999). Oxidative metabolism is well documented as a phytoremediation mechanism of herbicides and is equally efficient for explosives (Bhadra et al., 1999). Explosives removal by plants is far higher under sunlight-stimulated field conditions than under laboratory conditions (Best, 1999). TNT disappeared completely from groundwater incubated with plants in approximately two weeks (Best, 1999).

13.10 SOIL FRACTURING

13.10.1 Technology Description

Soil fracturing is not a remediation technology by itself, but it is an enhancement technology that is designed to increase the efficiency of other in-situ technologies in low-permeable geologic formations. Soil fracturing basically extends and enlarges fissures (cracks) that are present in the soil and also introduces new fractures. When fracturing has been completed, the cracks increase the permeability of the formation, leading to high vapor or liquid extraction during the remedial operations (Murdoch et al., 1991; Schuring et al., 1993; USEPA, 1994c).

There are two different methods of soil fracturing: pneumatic fracturing and hydraulic fracturing (USEPA, 1995d). Pneumatic fracturing involves injecting highly pressurized air beneath the surface to develop cracks in the soil [Figure 13.43(*a*)]. Hydraulic fracturing involves injecting a fluid, which contains water or slurry of water, sand, and a thick gel, polymers, or other compounds, under high pressures to create fractures [Figure 13.43(*b*)]. Generally, hydraulic fracturing produces larger fractures and it can be performed at greater depths than pneumatic fracturing.

Table 13.10 **Selected studies on phytoremediation**

Location	Application	Plants	Contaminants	Performance
Chernobyl, Ukraine	Rhizofiltration demonstration pond near nuclear disaster	Sunflowers, *Helianthus annuus*	137Cs, 90Sr	90% Reduction in 2 weeks; roots concentrated 8000-fold
Ashtabula, OH	Rhizofiltration demonstration DOE energy wastes	Sunflowers, *H. annuus*	U	95% removal in 24 hours from 350 ppb to < 5 ppb
Trenton, NJ	Phytoextraction demonstration 200 ft × 300 ft plot brownfield location	Indian mustard, *Brassica juncea*	Pb	Pb cleaned up to below action level in one-season SITE[a] program
Rocky Flats, CO	Rhizofiltration from landfill leachate	Sunflowers and mustard	U and nitrate	Just beginning SITE program
Dearing, KS	Phytostabilization demonstration 1-acre test plot abandoned smelter, barren land	Poplars, *Populus* spp.	Pb, Zn, Cd cones. > 20,000 ppm for Pb and Zn	50% survival after 3 years; site was successfully revegetated
Whitewood Cr., SD	Phytostabilization demonstration 1-acre test plot mine wastes	Poplars, *Populus* spp.	As, Cd	95% of trees died; inclement weather, deer browse, toxicity caused die-off
Pennsylvania	Phytoextraction pilot mine wastes	*Thlaspl caerulescens*	Zn, Cd	Uptake is rapid but difficult to decontaminate soil
San Francisco, CA	Phytovolatilization refinery wastes and agricultural soils	*Brassica* sp.	Se	Selenium is partly taken up and volatilized, but difficult to decontaminate soil
J-field site, Aberdeen, MD	Phytotransformation groundwater capture on 1-acre plot	Hybrid poplars, *Populus* spp.	TCE, PCA (1,1,2,2-tetrachloroethane)	Only in second-year demonstration project
Carswell AFB, Ft. Worth, TX	Phytotransformation groundwater capture on 4-acre plot	Hybrid poplars, *Populus* spp.	TCE	Only in second-year SITE project

Source: Schoor (1997).

[a]SITE—Superfund Innovative Technology Evaluation

Pneumatic Fracturing

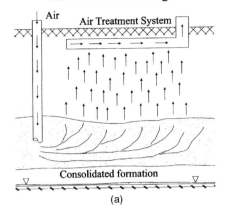

(a)

Hydrofracturing

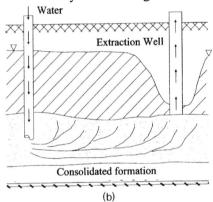

(b)

Figure 13.43 Soil fracturing types: (a) pneumatic fracturing; (b) hydraulic fracturing.

Fracturing is used primarily to fracture low-permeability geologic materials, such as fine-grained soils and overconsolidated sediments, including silts and clays, as well as bedrock. Fracturing can enhance the remediation of any chemical contaminants usually treated by the specific technology with which fracturing is combined. Often, fracturing is combined with soil vapor extraction, in-situ bioremediation, in-situ thermal treatment, in-situ vitrification, and in-situ electrokinetics (Frank et al., 1994; Davis-Hoover et al., 1995; Murdoch et al., 1995).

Advantages of the soil fracturing technology are:

- It can increase the effectiveness of other traditional technologies in low-permeability soils and bedrock.
- It decreases the remediation time.
- Overall cost of the remediation decreases.
- Proppants injected during fracturing can be made part of the remediation process, such as heating zones in vitrification, nutrients in bioremediation, or electrodes in electrokinetics.

Disadvantages of this technology are:

- Water-filled fractures increase the volume of liquid to be extracted and later treated, which increases the cost of remediation.
- If not controlled properly, fractures may aid the spreading of the contaminants into clean areas of the subsurface.
- Fracturing may induce the ground surface settlement and may endanger the stability of structures nearby.

13.10.2 Fundamental Processes

Pneumatic fracturing injects highly pressurized air beneath the surface to develop cracks in low-permeability and overconsolidated sediments. The injection of pressurized air into soil, sediments, or bedrock causes existing fractures to extend and creates a secondary network of conductive subsurface fissures and channels to create additional subsurface airflow (Figure 13.44). This network of fractures increases the exposed surface area of the contaminated soil as well as its permeability to liquids and vapors. The pore gas exchange rate, often a limiting factor during vapor extraction, can be increased significantly as a result of pneumatic fracturing, thereby allowing accelerated removal of contaminants. These new passageways increase the effectiveness of in-situ processes, including soil vapor extraction (SVE), and enhance extraction efficiencies by increasing contact between contaminants adsorbed

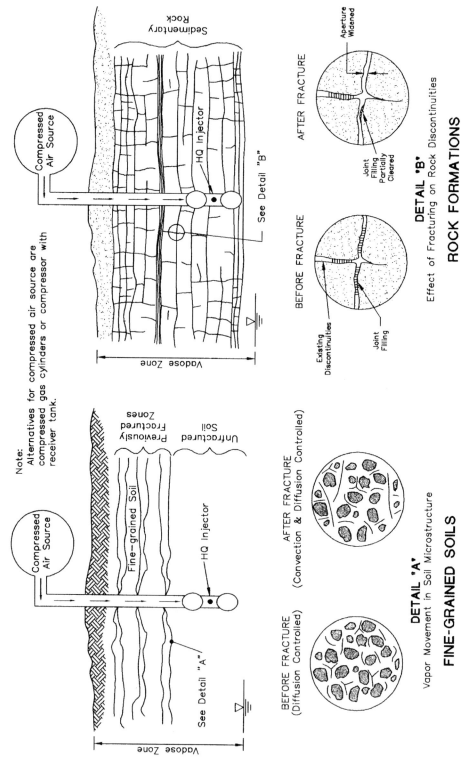

FINE-GRAINED SOILS

DETAIL "A"
Vapor Movement in Soil Microstructure

ROCK FORMATIONS

DETAIL "B"
Effect of Fracturing on Rock Discontinuities

Figure 13.44 Creation of cracks during pneumatic fracturing. (From USEPA, 1995g.)

488

onto soil particles and the extraction system (Schuring et al., 1993; USEPA, 1993d,e, 1995c).

Hydraulic fracturing involves injecting a fluid that contains water or slurry of water, sand, and a thick gel, polymers, or other compounds to maintain open fractures in the subsurface soils and increase soil permeability. This is done by creating distinct, subsurface fractures that may be filled with sand or other granular material. The fractures are created through the use of fluid pressure to dilate a well borehole and open adjacent cracks. Once fluid pressure exceeds a critical value, a fracture begins to propagate. The fracture continues to grow until injection ceases or pressure dissipates, the fracture intersects a barrier, a permeable channel or the ground surface, or the injected fluid leaks out through the boundary walls of the formation being fractured. The hydraulic fracturing process is repeated at varying depths (typically, 5 to 30 ft), creating a "stack" of sand-filled fractures. Figure 13.45 depicts a conceptual stack (Murdoch et al., 1995; USEPA, 1995d).

The pressure required to initiate a fracture in a borehole depends on several factors, such as confining stresses, toughness of the geologic formation, initial rate of injection, size of existing fractures, and the presence of pores or defects in the borehole wall. In pneumatic fracturing, the fracture initiation pressure can be calculated by $P_i = Cd'Z + t_a + P_o$, where Z is the overburden depth, P_o the hydraulic pressure, t_a the tensile strength, C a coefficient (ranging from 2 to 2.5), and d' the effective unit weight of the formation. For example, at a depth of 20 ft for bedrock, the fracture initiation pressure would be 200 psi. In general, the injection pressure increases with increasing depth, injection rate, and fluid viscosity (hydraulic fracturing).

Fractures may remain open naturally, or they may be held open by permeable materials, known as *proppants* (typically, sand) injected during fracture propagation. Since the resulting fracture interval is designed to be more permeable than the adjacent geologic formation (by filling it with highly permeable materials— air, sand, etc.), extraction of contaminants can be achieved at a higher rate. This holds for both pneumatic and hydraulic fracturing.

The effectiveness of pneumatic and hydraulic fracturing depends largely on the size and shape of the fracture with respect to the borehole. Propagation would continue indefinitely if the fracture were created in indefinitely impermeable material. However, in real-time situations, several factors limit the size and shape of fractures. The volume of injected fluids and the rate in which they are injected are the primary controllable variables that affect the size of the fracture.

The volume of injected fluid determines the size of the resulting fracture. Maximum dimensions of fractures created by gases or liquids are limited by the tendency of the fracture to climb and intersect the ground surface or by the loss of fluid through the fracture walls. The maximum horizontal dimension of a fracture also increases with increasing depth and decreasing permeability of the soil formation. At a depth range of about 5 to 16 ft in overconsolidated silty clay, the typical maximum dimension of a fracture is about three to four times its depth (USEPA, 1995d). The shape of a fracture created by pneumatic and hydraulic fracturing depends largely on the fracturing technique, including the type of liquid used during injection, the rate and pressure of injection, the configuration of the borehole, and the site conditions (USEPA, 1995d). Critical site conditions that affect the shape of the fracture include loading at the ground surface (buildings, equipment, etc.), the permeability and heterogeneity of the geologic formation, the presence of subsurface borings, other wells within vicinity, and backfilled excavations.

In soil fracturing, higher-permeability zones are created for enhancement of advective flow through the contaminated zone. Furthermore, pathways for diffusion-controlled migration of the contaminants are also created. The creation of advective flow channels and shortened diffusive pathways result in an enhanced mass removal rate during extraction. Soil fracturing can expand the applicability of other in-situ remedial technologies. These technologies include in-situ biodegradation (enhancing delivery of oxygen and nutrients), in-situ electrokinetics (enhancing fluid flow into fractured zones), and in-situ vitrification (creating heating zones by injecting graphite into the fractures) (Murdoch et al., 1993; USEPA, 1994c, 1995d).

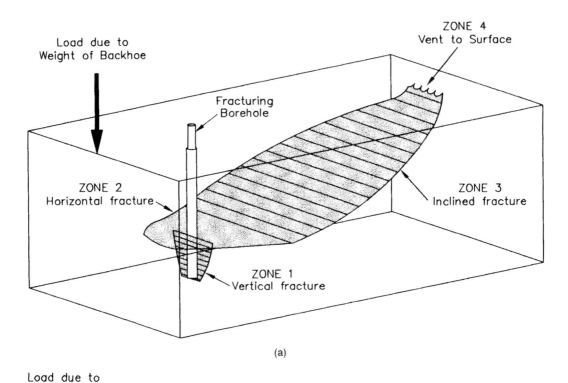

(a)

(b)

Figure 13.45 Creation of a stack of sand-filled fractures. (From USEPA, 1997g.)

13.10.3 System Design and Implementation

General Design Approach. Figure 13.46 summarizes the steps necessary for system design and implementation. At first, the specific proppant solution needs to be determined (lab-oriented work). This is needed when hydraulic fracturing is used. Then the geotechnical evaluation takes place to determine all necessary properties, such as water content, liquid and plastic limits, grain size, and unconfined compressive strength. These properties are needed to establish the fluid pressures and might help establish the pressures needed to induce fractures. This will lead to a more economical design and will help in selection of the process.

The next step is to conduct a pilot test. This is the most crucial step for this technology because it will yield very useful information. This information will include clues as to whether the technology will work for the site-specific conditions, how much will it cost, and so on. After the test area is selected (this is based on information compiled previously), test wells are drilled, and the system is applied on a small scale. After evaluation of the pilot test data has been completed, the full-scale design is implemented. For this, a fracture location plan needs to be generated. Then the wells are installed and full-scale operation takes place. After this, monitoring and progress testing take place to help ascertain the degree of effectiveness of the system design. The final step is adjustment of the system design in order to optimize performance.

Selection of Equipment. The equipment used to create fractures consists of both an aboveground system that must be capable of injecting the desired fracture medium at the required pressures and rates and a belowground system that must be capable of isolating the zone where injection will take place. The type of medium to be injected largely determines the specifications of the aboveground equipment.

The pneumatic fracturing requires the following equipment: a high-pressure air source (typically, a pressurized gas cylinder), a pressure regulator, and a receiver tank with an inline flow meter and pressure gauge. Typically, a packer system is used to isolate (usually) 2-ft intervals so that short (20-s) bursts of compressed air is injected by the use of a proprietary nozzle (Figure 13.47). The hydraulic fracturing requires a pump, a slurry mixer, a lance and a notch tool, and a 6- or 8-in. hollow stem auger.

Both hydraulic and pneumatic fracturing can use straddle packers that allow spacing of fractures approximately every 1.5 ft along an open borehole. Straddle packers are appropriate in rock and some unlithified sediment. An alternative to the use of straddle packers during hydraulic fracturing of unlithified sediments is the driving of a casing with an inner pointed rod to the specified depth (Figure 13.48). After the rod is removed, a high-pressure pump injects a water jet to cut a notch in the sediments at the bottom of the borehole. The notch reduces the pressure required to start propagation and ensures that the fracture starts in a horizontal plane at the bottom of the casing. A fracture can be created at the bottom of the casing by injecting air, a liquid, or slurry. After the fracture has been created, the rod is reinserted and driven to greater depths to create another fracture, or the casing can be left in place for access to the fracture during recovery.

In selecting the pumps, the usual choice is a variable-speed pump with additional air tanks to eliminate problems arising from different pressure requirements. For slurry injection it is common to use variable-pressure pumps and mixers to compensate for indifferences in the soil formations.

Operational Parameters. In hydraulic fracturing, to create a fracture by injecting a liquid into soil at 20 gal/min and at a 6-ft depth requires approximately 8 to 10 lb/in^2 of pressure. This required pressure increases by 1 lb/in^2 for each additional foot of depth. In comparison, the pressure required to create a fracture by injecting air, with injection rates of 700 to 1000 ft^3/min, is about 70 to 150 lb/in^2 (USEPA, 1995d). Usually, the pressure during propagation decreases. However, the specifics of the pressure depend on a variety of factors, such as a slight increase in the concentration of sand in the slurry during injection.

Some of the injected air or liquid may leak through the walls of the fractures into the pores of surrounding soil or rock. The rate of such leakoff controls the size of fractures. For example, injecting gas at 800 to 1800

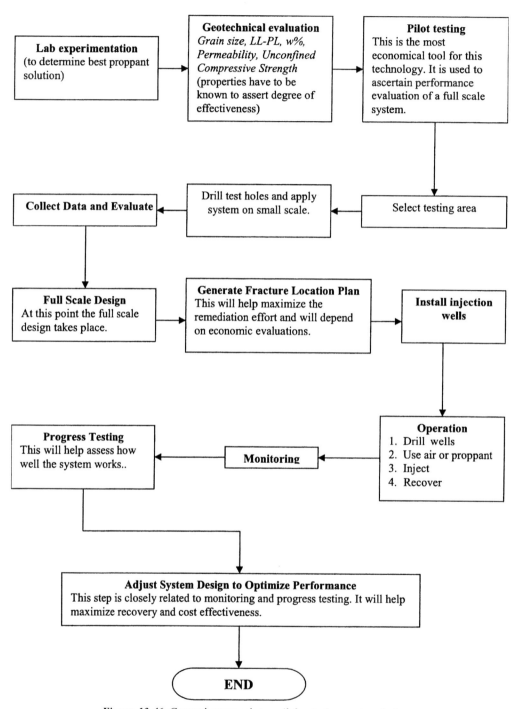

Figure 13.46 General approach to soil fracturing system design.

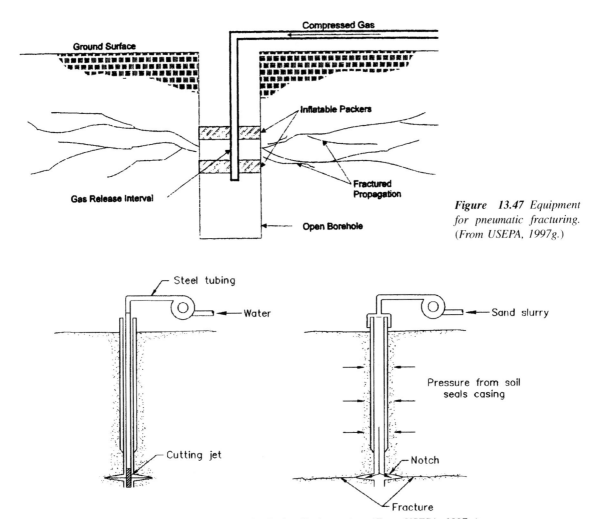

Figure 13.47 Equipment for pneumatic fracturing. (From USEPA, 1997g.)

Figure 13.48 Equipment for hydraulic fracturing. (From USEPA, 1997g.)

ft³/min into sandstone for approximately 20 s typically results in fractures approximately 20 to 70 ft in maximum dimension (USEPA, 1995d). A longer injection period does not greatly affect the dimensions of the fracture. However, increasing the rate of injection will generally increase the size of the fracture. So, in pneumatic fracturing, the rate of injection is a critical design variable that affects the size of the fracture.

Operation. The operation of the equipment to induce fractures is straightforward. It requires inexpensive equipment and minimal labor. The procedure of op-

eration is simple and includes drilling of a well, fitting a rod with packers attached or a hollow-stem auger with 1.5-ft spacers, connecting the hollow rod (or auger) to the air tank (or slurry pump), and turning the power on.

Monitoring. The most commonly used method of monitoring the location on a formation of fractures in the subsurface is by measuring the vertical displacement of the ground surface. Net displacements can be determined by surveying a field of measuring staffs with finely graduated scales, before and after fractur-

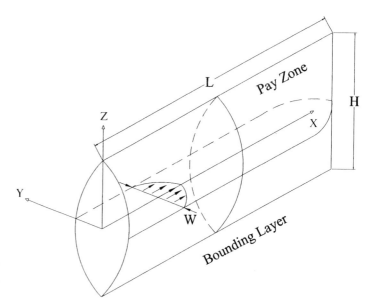

Figure 13.49 Modeling of hydraulic frac-turing. (*From Bai and Abousleiman, 1997, by courtesy of Marcel Dekker, Inc.*)

ing. This is a very inexpensive method that can provide reliable data on final displacements. As a more expensive alternative, tilt sensors that detect changes in electrical resistivity can measure extremely gentle slopes of the ground of the ground surface in real time while fracturing is in process. This technique can provide such information as strain, displacements, and tilt.

13.10.4 Predictive Modeling

Mathematical models of fracturing in soils are based on several simplified assumptions. Nevertheless, these models provide useful information on the effects of system variables on fracture initiation and propagation. Poroelastic models are useful to delineate the impacts of pore pressure buildup, energy dissipation, and elastic deformation of the hydraulic fracture geometry. One such model is developed by Bai and Abousleiman (1997). This model assumes that:

- A vertical hydraulic fracture is embedded within a horizontal permeable layer and confined vertically by impermeable layer and overburden.

- Because of fluid injection, the fracture subjected to high pressure propagates only in the lateral direction.

- One-dimensional fluid flow in a homogeneous infinite porous medium under isothermal conditions was used for calculating the pore pressure.

- For instantaneous fluid injection, constant fluid pressure and temperature at the perforation point and zero pressure and temperature perturbation at the reservoir boundary was assumed.

The formula for pressure for a reservoir with infinite boundary and pressure in the fracture in real space and the fracture configuration assumed are shown in Figure 13.49. In determining the parameters associated with this model, it is important to consider that a hydraulic fracture tends to propagate if it is subjected to an excess load as a result of such induced gradients as stresses, pressures, and temperatures. Because of their relative importance, the analysis only emphasized the influence of the latter two parameters. For this, the fluid injection was induced until a valve was turned off. During backflow, the inlet valve is opened at a controlled setting allowing the fluid in the cavity to flow back to the well and then back to the surface. It was determined that the amount of flow was not the same as that of inflow because of fluid infiltration into the formation.

By using a pressure-related leakoff, the pressure profiles were established for two cases: in the first of which the leakoff was constant, and in the second, the leakoff was pressure related. The result implied that the pressure-related leakoff could improve the physical representation and computation accuracy of the test significantly. It was noticed that there is an effect in fracturing widths with constant leakoff at different thermal influences. In effect, depending on the thermal properties of the formation, temperature variations might exert a significant influence on the changes in pore pressure and fracturing geometry, such as width and length.

13.10.5 Modified or Complementary Technologies

Hydraulic and pneumatic fracturing are not stand-alone remediation techniques. After a network of fractures has been created, the contaminants are ready to be removed by other technologies. Some of the technologies used in conjunction with soil fracturing include soil vapor extraction, in-situ bioremediation, in-situ heating, and in-situ electrokinetics (USEPA, 1994a, 1995d).

13.10.6 Economic and Regulatory Considerations

The cost associated with this technology can vary and it is greatly dependent on the number of fractures needed and the type of soil to be fractured. According to the USEPA estimates, the cost per single fracture can range from $950 to $1425. Other costs include about $1000 for site preparation, $5000 for permits, $1000 for equipment rental, $2000 for labor (highly dependent on experience), $700 for monitoring, and $400 for demobilization. Generally, the costs for hydraulic and pneumatic fracturing are comparable (USEPA, 1994a, 1995d).

Some of the areas of concern to a regulatory authority would include gels or liquids that might alter the chemistry of the subsurface, biodegradable and nontoxic injectants, lack of control over fracture generation, and health and safety issues (high pressures). Also, the environmental impact of this technology could be an area of concern, especially where local authorities are involved to ensure a minimal or nonexistent environmental impact on their communities. However, a first concern would be the state or local requirements, including permits and disposal of cuttings (the soil that the drill brings to the surface via the auger).

13.10.7 Case Studies

Selected sites where soil fracturing was applied to enhance remediation are shown in Table 13.11 (FRTR, 1997). Two of these sites are described briefly in this section.

Case Study 1: Pneumatic Fracturing at Industrial Site, Hillsborough, New Jersey. This case was conducted under the SITE demonstration program in the summer of 1992. Fracture wells were drilled in the contaminated vadose zone of a siltstone formation and were then left as open boreholes. The pneumatic fracturing process was applied to isolated 2-ft intervals of the formation. Short bursts (less than 20 s) of air were injected into the formation at successive depth intervals of the fracture well to create an intensely fractured unsaturated zone. Each injection extended and enlarged existing fissures in the formation and created new fissures, primarily in the horizontal direction. Following fracturing, contaminated vapors were extracted from the fracture well utilizing a vacuum. Contaminants encountered at the site included VOCs and SVOCs, including TCE, PCE, and benzene.

The fracturing process was observed to increase extracted airflow by more than 600% relative to that achieved in the site formation prior to the application of pneumatic fracturing. Even higher airflow rate increases (19,000%) were observed when one or more of the monitoring wells were opened to serve as a pas-

Table 13.11 **Selected sites where soil fracturing was used**

Site Name	Contaminant/Geology[a]	Mass Recovery Factor Increase	Radius of Influence Improvement
Oak Brook, IL	VOCs in silty clay	7–14	30 times
Dayton, OH	VOCs in silty clay	25–40	Not reported (NR)
Bristol, TN	DNAPLs/fractured bedrock	2.8–6.2	30 times
Regina, Saskatchewan	VOCs in silty clay	NR	25 times
Calgary, Alberta	VOCs in silty clay	10	NR
Linemaster, CT	Solvents in till	4–6	NR
Manufacturing facility, NY	TCE, PCE, BTEX, VOCs	8	NR
Hillsborough, NJ	VOCs, TCE, PCE	6	35 times
Military facility, OK	No. 2 fuel oil	5–7	15
Beaumont, TX	DNAPLs in silty clay	50	~25 times

[a] VOCs, volatile organic compounds; DNAPLs, dense non-aqueous-phase liquids.

sive air inlet to enter the formation. The effective radius of influence was observed to increase from 380 ft^2 to at least 1254 ft^2, an increase of over threefold. Pressure data, collected at perimeter monitoring wells, and surface heave measurements indicate that fracture propagation extended well past the farthest monitoring wells (at 20 ft) to at least 35 ft.

While TCE concentrations in the airstream remained approximately constant at roughly 50 parts per million, the increased airflow rate resulted in an increase in TCE mass removal of 675%. When wells were opened to passive air inlet, the increase in TCE mass removal was 2300% following the application of pneumatic fracturing. Additional chemical analysis of the extracted air during postfracture testing showed high concentrations of organic compounds that had only been detected in trace amounts prior to application of pneumatic fracturing. This confirmed that the pneumatic fracturing process had effectively accessed pockets of previously trapped VOCs. The cost for full-scale remediation was estimated at $140 per pound of TCE removed based on the demonstration and information provided by the developer.

Case Study 2: Hydraulic Fracturing at Superfund Sites, Oak Brook, Illinois and Dayton, Ohio.

Hydraulic fracturing was demonstrated under EPA's Superfund Innovative Technology Evaluation (SITE) program in July 1991 at sites in Oak Brook, Illinois and Dayton, Ohio. Both sites contained low-

permeability soils ($<10^{-7}$ cm/s) that were contaminated with volatile organic compounds (VOCs). Fracturing was accomplished to a depth of 15 ft below ground surface. In Illinois, contaminants removed by soil vapor extraction were increased by 7 to 14 times; the area of influence was 30 times greater after fracturing. In Ohio, flow of water into the fractured well was increased 25 to 40 times; the bioremediation rate was increased by approximately 75%.

13.11 SUMMARY

The remediation of contaminated soils is one of the most expensive and time-consuming tasks. There are numerous conventional and innovative methods available to remediate the contaminated soils. These remediation methods can be grouped as either in-situ or ex-situ methods. Furthermore, each group of remediation methods can be classified into physical, chemical, thermal, or biological methods. In in-situ methods, soils are decontaminated in-place. In ex-situ methods, soils are excavated and treated with either on- or off-site resources. In-situ treatment methods are preferred due to low costs and less exposure of contamination to the personnel employing the operation.

Physical treatment separates contaminants from the soil and transfers them into either air or water, where they are then collected in concentrated form. *Soil washing, soil flushing,* and *soil vapor extraction* come under this category. Soil washing and soil flushing work on the same principle in that a suitable extractant is used to take the contaminant out of soil. The methods are named differently because one is an ex-situ method (soil washing) and the other is an in-situ method (soil flushing). Soil vapor extraction can be applied to soils contaminated with volatile organics; volatile organics present in the gaseous phase can be extracted by applying a vacuum. Pneumatic fracturing and bioventing are two techniques that can be used with soil vapor extraction to increase the effectiveness of the process.

Chemical and electrical processes remove or chemically transform contaminants into substances that are less toxic and less mobile. *Solidification/stabilization* and *electrokinetics* are examples of such processes. Electrokinetics is well suited for low-permeable soils where other remediation methods are ineffective or very expensive.

Thermal processes involve heating the soil and volatilizing and separating contaminants from it. Low *thermal desorption* and high thermal desorption separate the contaminants from soils without destroying them. *Incineration,* on the other hand, separates as well as destroys contaminants. The above-mentioned thermal techniques are suitable for volatile and semivolatile organics. *Vitrification,* another thermal process,

heats soils to very high temperatures to melt the soils and its contaminants, forming a glasslike material. Vitrification is suitable for all types of contaminants.

Biological treatment can be applied to treat organics, which can be degraded by aerobic and anaerobic microbes. This method can be applied to soils with relatively high hydraulic permeability. For soils with low hydraulic permeability, biological treatment can be used along with an enhancement technique. Pneumatic fracturing is one such enhancement technique that creates new fractures and enlarges existing ones. Pneumatic fracturing may be used to ensure easy injection of nutrients. Bioventing is another enhancement technique. Bioventing supplies oxygen to the microbes, thus increasing the efficiency of the process. Depending on the type of application, in-situ or ex-situ, biological treatment may be named in-situ biological treatment, slurry-phase bioremediation, contained solid-phase remediation, or land farming.

Application of any of the foregoing techniques involves careful site characterization and contaminant characterization. A suitable method should then be employed based on the type of soil, type of contaminants, and target percent reduction in contaminants. A combination of different technologies (often known as a *treatment train*) may be needed for effective remediation of the site contamination.

QUESTIONS/PROBLEMS

13.1. Draw a flowchart showing in-situ and ex-situ soil remediation technologies. Subdivide these technologies according to physical, chemical, electrical, thermal, or biological methods.

13.2. What are the advantages and disadvantages of ex-situ and in-situ remediation approaches?

13.3. List technologies that aim to remove the contaminants from soils.

13.4. List technologies that aim to immobilize the contaminants in soils.

13.5. List technologies that aim to degrade the contaminants in soils.

13.6. List technologies that are applicable to remediate organic contamination in soils.

13.7. List technologies that are applicable to remediate metal-contaminated soils.

13.8. List technologies that are applicable to remediate soils contaminated with radioactive contamination.

13.9. List technologies that can used to remediate low-permeability soils such as clays.

13.10. A SVE system is used to remediate a TCE-contaminated site. The contaminated zone is approximately 5000 yd^3, with an average TCE concentration of 800 mg/kg. The contaminated soil is mainly clayey sand with a water content of 15% and a moist unit weight of 120 lb/ft^3. The SVE system was operated under a airflow rate of 200 ft^3/min for 30 days. The average effluent vapor concentration measured was 100 mg/L. Calculate the removal efficiency of the SVE system.

13.11. Two sites exist where the remediation of soil in the vadose zone is required. One site is contaminated with TCE and the other with phenenthrene. Both sites contain sandy soil with similar permeabilities. Is SVE suitable for remediating these two sites? Justify your answer.

13.12. What causes removal of residual contamination by SVE to be inefficient?

13.13. At a LUST site, a 500 ft × 200 ft area was contaminated due to leakage of gasoline. Soil was contaminated to a depth of 20 ft. Site soil is predominantly silty sand. Design a preliminary SVE system for the site.

13.14. A site was contaminated due to leakage of 1000 gal of TCE from an underground storage tank. The contaminated soil zone begins at 15 ft below the ground surface and is approximately 100 ft in diameter and 20 ft thick. The soil in the contaminated zone is predominantly silty sand with a water content of 15%, porosity of 40%, and total unit weight of 110 lb/ft^3. Groundwater at the site was located at 40 ft depth. A SVE system is proposed to remediate the site. Determine the following: **(a)** Calculate the distribution of TCE different phases, assuming that the K_d value of soil and TCE is 0.025 L/kg. **(b)** If pilot-scale testing showed an average effluent concentration of 100 g/m^3 under an extraction rate of 200 ft^3/min, calculate the time required for 90% removal of the TCE.

13.15. Explain why prewashing of coarser soil fraction is performed during the soil washing process.

13.16. Why is soil washing not effective for treating soils with a high fines content?

13.17. A soil is contaminated with 300 mg/kg of PCE. Soil washing is proposed with a solids concentration of 0.1 kg/L. Calculate the soil washing removal efficiency if **(a)** water is used as the washing solution (with K_d = 100 L/kg), and **(b)** a solvent is used as the washing solution (with K_d = 10 L/kg).

13.18. Explain the differences in stabilization processes with the use of portland cement versus lime as the stabilizing agent for a contaminated soil.

13.19. What are the difficulties associated with in-situ stabilization?

13.20. At a contaminated site, subsurface soils were contaminated with lead and zinc. The soils consist predominantly of silty sand. The site contamination extends from the ground surface to a depth of 5 ft. The groundwater at the site is located at 10 ft below the ground surface. Design a preliminary S/S process for remediating this site. Justify all your assumptions.

13.21. A contaminated sandy soil was stabilized using lime. The moisture content, porosity, and wet unit weight of the soil were 30%, 40%, and 110 lb/ft^3, respectively. Estimate the change in the volume of the soil as a result of this stabilization process.

13.22. A soil column 6.2 cm in diameter and 19.1 cm in length was subjected to 25 V dc. If the electroosmotic permeability of the soil is 1×10^{-5} cm/s, calculate the electroosmotic flow volume in 24 h.

13.23. What are the difficult aspects of in-situ electrokinetic remediation?

13.24. What are the differences between thermal desorption and vitrification?

13.25. A soil contaminated with jet fuel is proposed to be treated with ex-situ thermal desorption. The initial total concentration of the organic contaminants in the soil is 10,000 mg/kg, and the required cleanup level is 100 mg/kg. If the rate of desorption in the desorber is 0.2 min^{-1}, how long should the soil be treated?

13.26. Explain the different methods of in-situ thermal desorption.

13.27. What are the advantages and disadvantages of different in-situ heating (thermal) methods?

13.28. In-situ vitrification was used at a site where 100 ft × 200 ft × 20 ft of soil was contaminated with mixed contamination (combination of organics, metals, and radionuclides). The soil had an initial porosity of 40%. The contaminated site consists of silty sand with a porosity of 40%. Calculate the expected surface settlement as a result of this remediation. (*Hint:* The porosity of the soil after vitrification is zero.)

13.29. Explain the advantages and disadvantages of various ex-situ bioremediation methods.

13.30. At a contaminated site, 500 yd³ of sandy soil contaminated with gasoline is to be treated using ex-situ bioremediation. Determine the moisture, nutrients, and oxygen requirements for the bioremediation to be effective. State all of your assumptions.

13.31. Explain the differences between soil vapor extraction and bioventing.

13.32. What are the limitations of in-situ bioremediation?

13.33. What are the various decontamination processes involved in phytoremediation?

13.34. As a result of fracturing, the soil permeability doubled at a contaminated site. Estimate the increase in extraction rates of a SVE system as a result of this permeability increase.

13.35. A former industrial site was used for manufacturing and chemical storage over 25 years. During that time, waste and chemical spills from various chemical handling, storage, and transfer activities contaminated the site. Preliminary site characterization data indicate that the contaminants of concern are trichloroethane, benzene, 1,2-dichlorobenzene, and styrene. Soil concentrations of all these contaminants are above 1000 ppm. Previous soil borings had shown that most of the contamination was located 20 to 30 ft below the ground surface. The zone of contamination covers 3 acres. Groundwater occurs at 50 ft below the ground surface, 20 ft above the bedrock surface; it is not contaminated. The soils at the site are sandy clay and fairly homogeneous. Based on this information, which remediation method do you recommend for the site? Why?

13.36. A former agricultural distributorship contained 10,000 yd³ of pesticide-contaminated silty sand, having combined concentrations of 100 ppm for 2,4-dichlorophenoxyacetic acid and 4-chloro-2-methyl-phenoxyacetic acid. The regulatory cleanup requirement for the soil is 10 ppm. Which method(s) are potentially applicable to remediate the soil? Explain.

REFERENCES

Acar, Y. B., and Alshawabkeh, A. N., Principles of electrokinetic remediation, *Environ. Sci. Technol.*, Vol. 27, No. 13, pp. 2638–2647, 1993.

Aggarwal, P. K., Means, J. L., Hinchee, R. E., Headington, G. L., and Gavaskar, A. R., *Methods to Select Chemicals for In-Situ Biodegradation of Fuel Hydrocarbons*, Air Force Engineering and Services Center, Tyndall AFB, FL, 1990.

Alexander, M., *Biodegradation and Bioremediation*, Academic Press, San Diego, CA, 1994.

Alshawabkeh, A. N., and Acar, Y. B., Removal of contaminants from soils by electrokinetics: a theoretical treatise, *J. Environ. Sci. Health*, Vol. A27, No. 7, pp. 1835–1861, 1992.

Anderson, W. C., *Innovative Site Remediation Technology*, Vol. 6; *Thermal Desorption*, American Academy of Environmental Engineers, Annapolis, MD, 1993.

Bai, M., and Abousleiman, Y., Modeling thermal effects on hydraulic fracturing, *In Situ J.*, Vol. 21, No. 2, pp. 161–186, 1997.

Best, E. H., *Phytoremediation of Explosives in Groundwater Using Constructed Wetlands,* U.S. Army Engineer Research and Development Center, Environmental Laboratory, Vicksberg, MS, 1999.

Bhadra, R., Spanggord, D. G., Hughes, J. B., and Shanks, J. V., Characterization of oxidation products of TNT metabolism in aquatic phytoremediation systems of *Myriophulum aquaticum, Environ. Sci. Technol.* Vol. 33, pp. 3354–3361, 1999.

Cauwenberghe, L. V., and Roote, D. S., *In Situ Bioremediation,* TO-98-01, Groundwater Remediation Technologies Analysis Center, Pittsburgh, PA, 1998.

Chappell, J., *Phytoremediation of TCE in Groundwater Using Populus,* Technology Innovation Office, USEPA, Washington, DC, 1998.

Chinthamreddy, S., and Reddy, K. R., Oxidation and mobility of trivalent chromium in manganese enriched clays during electrokinetic remediation, *J. Soil Contam.,* Vol. 8, No. 2, pp. 197–216, 1999.

Choi, Y. S., and Lui, R., A mathematical model for the electrokinetic remediation of contaminated soil, *J. Hazard. Mater.,* Vol. 44, pp. 61–75, 1995.

Cookson, J. T., *Bioremediation Engineering Design and Application,* McGraw-Hill, New York, 1995.

Corapcioglu, M. Y., Formulation of electro-chemico-osmotic processes in soils, *Transp. Porous Media,* Vol. 6, pp. 435–444, 1991.

Davis-Hoover, W. J., Roulier, M., Bryndzia, T., Herrmann, J., Vane, L., Murdoch, L. C., and Vesper, S. J., *Hydraulic Fractures as Anaerobic and Aerobic Biological Treatment Zones,* EPA/600/R-95/012, USEPA, Washington, DC, 1995.

Dragun, J., Geochemistry and soil chemistry reactions occurring during in situ vitrification, *J. Hazard. Mater.,* Vol. 26, pp. 343–364, 1991.

Dushenkov, S., Mikheev, A., Prokhnevsky, A., Ruchko, M., and Sorochinsky, B., Phytoremediation of radiocesium-contaminated soil in the vicinity of Chernobyl, Ukraine, *Environ. Sci. Technol.,* Vol. 33, pp. 469–475, 1999.

Eykholt, G. R. and Daniel, D. E., Impact of system chemistry on electroosmosis in contaminated soil, *J. Geotech. Eng.,* Vol. 120, No. 5, pp. 797–815, 1994.

FRTR (Federal Remediation Technologies Roundtable), *Remediation Case Studies: Bioremediation,* EPA/542/R-95/002, USEPA, Washington, DC, 1995a.

———, *Remediation Case Studies: Thermal Desorption, Soil Washing, and In Situ Vitrification,* EPA/542/R-95/005, USEPA, Washington, DC, 1995b.

———, *Remediation Case Studies: Soil Vapor Extraction and Other In Situ Technologies,* EPA/542/R-97/009, USEPA, Washington, DC, 1997.

———, *Remediation Case Studies: In Situ Soil Treatment Technologies (Soil Vapor Extraction, Thermal Processes),* EPA/542/R-98/012, USEPA, Washington, DC, 1998.

Frank, U., Skovronek, H. S., Liskowitz, J. J., and Schuring, J. R., *Site Demonstration of Pneumatic Fracturing and Hot Gas Injection,* EPA/600/R-94/011, USEPA, Washington, DC, 1994.

Freeze, R. A., and Cherry, J. A., *Groundwater,* Prentice Hall, Upper Saddle River, NJ, 1979.

GeoKinetics, Fuel oils, DNAPL's, solvents: EH/DPE, *www.geokinetics.com/giievac.htm,* 1999a.

———, PCB's and explosives, electrokinetically enhanced bio-remediation, *www.geokinetics.com/giiebio.htm,* 1999b.

HWRIC (Hazardous Waste Research and Information Center), *LUST Remediation Technologies,* HWRIC, Champaign, IL, 1995.

Hinchee, R. E., *Bioventing of Petroleum Hydrocarbons* in *the Handbook of Bioremediation,* Lewis Publishers, Boca Raton, pp. 39–59, 1994

Ho, S., Athmer, C., Sheridan, P., Hughes, B., Orth, R., Mckenzie, D., Brodsky, P. Shapiro, A., Thornton, R., Salvo, J., Schultz, D., Landis, R., Landis, R., Grifith, R., and Shoemaker, S., The lasanga technology for in situ soil remediation: 1. Small field test, *Environ. Sci. Technol.,* Vol. 33, pp. 1086–1091, 1999a.

Ho, S., Athmer, C., Sheridan, P., Hughes, B., Orth, R., Mckenzie, D., Brodsky, P., Shapiro, A., Sivavec, T., Salvo, J., Schultz, D., Landis, R., Landis, R., Grifith, R., and Shoemaker, S., The lasanga technology for in situ soil remediation: 2. Large field test, *Environ. Sci. Technol.,* Vol. 33, pp. 1092–1099, 1999b.

Huang, J. W., Chen, J., Berti, W. R., and Cunningham, S. C, Phytoremediation of lead-contaminated soils: role of synthetic chelates in lead phytoextraction, *Environ. Sci. Technol.* Vol. 31, pp. 800–812, 1997.

ITRC (Interstate Technology and Regulatory Cooperation) *Technical and Regulatory Guidelines for Soil Washing,* ITRC, Washington, DC, 1997a.

——, Metals in Soils Work Team, Emerging Technologies Project, *Emerging Technologies for the Remediation of Metals in Soils: Electrokinetics,* ITRC, Washington, DC, Dec. 1997b.

——, *Phytoremediation Decision Tree,* ITRC, Washington, DC, 1999.

Jacobs, R. A., Sengun, M. Z., Hicks, R. E., and Probstein, R. F., Model and experiments on soil remediation by electric fields, *J. Environ. Sci. Health,* Vol. A29, No. 9, pp. 1933–1955, 1994.

Johnson, P. C., and Ettinger, R. A., Considerations for the design of in situ vapor extraction systems: radius of influence vs. zone of remediation, *Ground Water Monitor. Rev.,* summer issue, pp. 123–128, 1994.

Johnson, P. C., Stanley, C. C., Kemblowski, M. W., Byers, D. L., and Colhart, J. D., A practical approach to the design, operation, and monitoring of in situ soil venting systems, *Ground Water Monitor. Rev.,* Vol. 10, No. 2, 1990.

Koegler, S. S., and Kindle, C. H., Modeling of the in situ vitrification process, *Am. Ceram. Soc. Bull.,* Vol. 70, pp. 832–835, 1991.

Lageman, R., Pool, W., and Seffinga, G. A., Theory and practice of electro-reclamation, presented at the forum on innovative hazardous waste treatment technologies, Atlanta, GA, June 1989.

Lindgren, E. R., Rao, R. R., and Finlayson, B. A., Numerical simulation of electrokinetic phenomena. pp. 46–82 in *Emerging Technologies in Hazardous Waste Management IV,* ACS Symposium Series, Tedder, D. W., and Pohlan, F. G. (eds.), Atlanta, GA, Sept. 1993.

Means, J. L., and Hinchee, R. E., *Emerging Technology for Bioremediation of Metals,* Lewis Publishers, Boca Raton, FL, 1994.

Mitchell, J. K., *Fundamentals of Soil Behavior,* 2nd ed., Wiley, New York, 1993.

Murdoch, L. C., Losonsky, G., Cluxton, P., Patterson, B., and Klich, I., *Feasibility of Hydraulic Fracturing of Soil to Improve Remedial Actions,* EPA/600/2-92/012, USEPA, Washington, DC, 1991.

Murdoch, L. C., Kemper, M., Wolf, A., Spencer, E., and Cluxton, P., *Hydraulic and Impulse Fracturing to Enhance Remediation,* EPA/600/R-93/040, USEPA, Washington, DC, 1993.

Murdoch, L. C., Chen, J., Cluxton, P., Kemper, M., Anno, J., and Smith, D., *Hydraulic Fractures as Subsurface Electrodes: Early Work on the Lasagna Process,* EPA/600/R-95/012, USEPA, Washington, DC, 1995.

Nedunuri, K. V., Govindaraju, R. S., and Erickson, L. E., *Modeling and Simulation of Heavy Metal Transport in Rhizosphere Soil: Influence of Active Biomass,* Departments of Civil and Chemical Engineering, Kansas State University, Manhattan, KS, 1997.

Newman, L. A., Strand, S. E., Choe, N., Duffy, J., Ekuan, G., Ruszaj, M., Shurtleff, B. B., Wilmoth, J., Heilman, P., Gordon, M. P., Uptake and biotransformation of trichloroethylene by hybrid poplars, *Environ. Sci. Technol.,* Vol. 31, No. 4, pp. 1062–1067, 1997.

Naval Facilities Engineering Service Center, *Cost and Performance Summary Report: In Situ Thermal Desorption at the Former Mare Island Naval Shipyard,* NFESC, Port Hueneme, CA 1999.

Odencrantz, J. E., Johnson, J. G., and Koenigsberg, S. S., Enhanced intrinsic bioremediation of hydrocarbons using an oxygen releasing compound: *Remediation Journal,* Vol. 6, No. 4, pp. 99–114, 1995.

Paterson, S., Mackay, D., and McFarlane, C., A model of organic chemical uptake by plants from soil and the atmosphere, *Environ. Sci. Technol.,* Vol. 28, pp. 2259–2266, 1994.

Probstein, R. F., and Hicks, R. E., Removal of contaminants from soils by electric fields, *Science* Vol. 260, pp. 498–503, Apr., 1993.

Raghavan, R., Dietz, D. H., and Coles, E., *Cleaning Excavated Soil Using Extraction Agents: A State-of-the-Art Review,* EPA 600/2-89/034, USEPA, Washington, DC, 1988.

Reddy, K. R., Integrated electrokinetic remediation of contaminated soils and groundwater, presented at the Symposium on Environmental Science and Technology Research, University of Illinois at Chicago, 2001.

Reddy, K. R., and Chinthamreddy, S., Electrokinetic remediation of heavy metal contaminanted soils under reducing environments, *Waste Manag.*, Vol. 26, No. 1, pp. 269–282, 1999.

Reddy, K. R., and Shirani A. B., Electreokinetic remediation of metal contaminated glacial tills, *Geotech. Geol. Eng. J.*, Vol. 15, No. 1, pp. 3–29, 1997.

Reddy, K. R., Parupudi, U. S., Devulapalli, S. N., and Xu, C. Y., Effects of soil composition on the removal of chromium by electrokinetics, *J. Hazard. Mater.*, Vol. 55, pp. 135–158, 1997.

Reddy, K. R., Donahue, M. J., Saichek, R. E., and Saasoka, R., Preliminary assessment of electrokinetic remediation of soil and sludge contaminated with mixed waste, *J. Air Waste Manag. Assoc.*, Vol. 49, pp. 174–181, 1999.

Schnoor, J. K., *Phytoremediation*, TE-98-01, Groundwater Remediation Technology Analysis Center, Pittsburgh, PA, 1997.

Schuring, J. R., Chan, P. C., Liskowitz, J. J., Fitzgerald, C. D., and Frank, U., *Pneumatic Fracturing of Low Permeability Formations*, EPA/600/R-93/040, USEPA, Washington, DC, 1993.

Semer, R., and Reddy, K. R., Evaluation of soil washing process to remove mixed contaminants from a sandy loam, *J. Hazard. Mater.*, Vol. 45, No. 1, pp. 45–57, 1996.

Sharma, H. D., and Lewis, S. P., *Waste Containment Systems, Waste Stabilization, and Landfills: Design and Evaluation*, Wiley, New York, 1994.

Thomas, J. M., and Ward, C. H., In situ bioremediation of organic contaminants in the subsurface, *Environ. Sci. Technol.*, Vol. 23, No. 7, pp. 760–766, 1989.

Thornton, R. F., and Shapiro, A. P., *Modeling and Economic Analysis of In-Situ Remediation of Cr(VI) Contaminated Soil by Electromigration*, in *Proc. Emerging Technologies in Hazardous Waste Management*, Atlanta, GA, 1993.

Trapp, S., and Matthies, M., Generic one-compartment model for uptake of organic chemicals by foliar vegetation, *Environ. Sci. Technol.*, Vol. 29, pp. 2333–2338, 1995.

USACE (U.S. Army Corps of Engineers), *Soil Vapor Extraction and Bioventing*, EM 1110-1-4001, U.S. Department of the Army, Washington, DC, 1995.

USEPA (U.S. Environmental Protection Agency), *EPA Guide for Identifying Cleanup Alternatives at Hazardous Waste Sites and Spills: Biological Treatment*, EPA/600/3-83/063, USEPA, Washington, DC, 1985.

———, *Handbook for Stabilization/Solidification of Hazardous Wastes*, EPA/540/2-86/001, Office of Research and Development, USEPA, Cincinnati, OH, 1986.

———, *Bioremediation of Contaminated Surface Soils*, EPA/600/2-89/073, USEPA, Washington, DC, 1988a.

———, *Determination of Aerobic Degradation Potential for Hazardous Organic Constituents in Soil: Interim Protocol*, Risk Reduction Engineering Laboratory, USEPA, Cincinnati, OH, 1988b.

———, *Stabilization/Solidification of CERCLA and RCRA Wastes: Physical Tests, Chemical Testing Procedures, Technology Screening, and Field Activities*, EPA/625/6-89/022, Office of Research and Development, USEPA, Washington, DC, 1989.

———, *State of Technology Review: Soil Vapor Extraction System Technology*, EPA/600/2-89/024, Hazardous Waste Engineering Research Laboratory, USEPA, Cincinnati, OH, 1990a.

———, *Engineering Bulletin: Soil Washing Treatment*, EPA/540/2-90/017, USEPA, Washington, DC, 1990b.

———, *Engineering Bulletin: Solvent Extraction Treatment*, EPA/540/2-90/013, USEPA, Washington, DC, 1990c.

———, *Engineering Bulletin: Slurry Bioremediation*, EPA/68/C8/0062, Office of Emergency and Remedial Response, USEPA, Washington, DC, 1990d.

———, *Understanding Bioremediation: A Guidebook for Citizens*, EPA/540/2-91/002, Office of Research and Development, USEPA, Washington, DC, 1991a.

———, *Guide for Conducting Treatability Studies under CERCLA: Soil Vapor Extraction, Interim Guidance*, USEPA, Washington, DC, 1991b.

——, *Engineering Bulletin: Chemical Oxidation Treatment*, EPA/530/2-91/025, USEPA, Washington, DC, 1991c.

——, *Innovative Treatment Technologies: Overview and Guide to Information Sources*, EPA/540/9-91/002, USEPA, Washington, DC, 1991d.

——, *Engineering Bulletin: In Situ Soil Flushing*, EPA/540/2-91/021, USEPA, Washington, DC, 1991e.

——, *Record of Decision Abstract, Wasatch Chemical Co. (Lot 6).*, EPA/ROD/R08-91/048, USEPA, Washington, DC, Mar. 1991f.

——, *Engineering Bulletin: In Situ Steam Extraction Treatment*, EPA/540/2-91/005, USEPA, Washington, DC, 1991g.

——, *Handbook: Vitrification Technologies for Treatment of Hazardous and Radioactive Wastes*, Office of Research and Development, EPA/625/R-92/002, USEPA, Washington DC, May 1992a.

——, *A Citizen's Guide to Solvent Extraction*, EPA/542/F-92/004, USEPA, Washington, DC, 1992b.

——, *Bioremediation of Hazardous Wastes*, EPA/600/R-92/126, Office of Research and Development, USEPA, Ada, OK, 1992c.

——, *A Citizen's Guide to Bioventing*, EPA/542/F-92/008, USEPA, Washington, DC, 1992d.

——, *Control of Air Emissions from Superfund Sites*, EPA/625/R-92/012, Office of Research and Development, USEPA, Washington, DC, 1992e.

——, *Bioremediation Using the Land Treatment Concept*, EPA/600/R-93/164, USEPA, Washington, DC, 1993a.

——, *Guide for Conducting Treatability Studies Under CERCLA, Biodegradation Remedy Selection, Interim Guidance.* EPA/540/R-93/519a, USEPA, Washington, DC, 1993b.

——, *Presumptive Remedies: Site Characterization and Technology Selection for CERCLA Sites with Volatile Organic Compounds in Soils*, USEPA, Washington, DC, Sept. 1993c.

——, *Accutech Pneumatic Fracturing Extraction and Hot Gas Injection, Phase I, Applications Analysis*, EPA/540/AR-93/509, USEPA, Washington, DC, 1993d.

——, *Pneumatic Fracturing Increases VOC Extractor Rate*, EPA/542/N-93/010, USEPA, Washington, DC, 1993e.

——, *Decision Support Software for Soil Vapor Technology Application: HyperVentilate*, EPA/600/R-93/028, USEPA, Washington, DC, 1993f.

——, *Engineering Bulletin: Solidification/Stabilization of Organics and Inorganics*, EPA/540/S-92/015, Office of Research and Development, USEPA, Cincinnati, OH, 1993g.

——, *Demonstration Bulletin: Augmented In-Situ Subsurface Bioremediation Process*, EPA/540/MR-93/527, RREL, USEPA, Washington, DC, 1994a.

——, *Demonstration Bulletin: Ex-situ Anaerobic Bioremediation System*, EPA/540/MR-94/508, RREL, USEPA, Washington, DC, 1994b.

——, *Alternative Methods for Fluid Delivery and Recovery*, EPA/625/R-94/003, USEPA, Washington, DC, 1994c.

——, *Engineering Bulletin: Thermal Desorption Treatment*, USEPA, Washington, DC, Feb. 1994d.

——, *How to Evaluate Alternative Cleanup Technologies for Underground Storage Tank Sites: A Guide for Corrective Action Plan Reviewers*, EPA/510/B-94/003, USEPA, Washington, DC, 1994e.

——, *Geosafe Corporation In Situ Vitrification Technology*, EPA/540/R-94/520a, Office of Research and Development, USEPA, Washington DC, Nov. 1994f.

——, *Emerging Technology Bulletin: Waste Vitrification through Electric Melting*, EPA/540/F-95/503, USEPA, Washington, DC, Mar. 1995a.

——, *Geosafe Corporation In Situ Vitrification: Innovative Technology Evaluation Report*, EPA/540/R-94/520, Risk Reduction Engineering Laboratory, Office of Research and Development, USEPA, Washington DC, Mar. 1995b.

——, *In Situ Remediation Technology Status Report: Hydraulic and Pneumatic Fracturing*, EPA/542/K-94/005, USEPA, Washington, DC, 1995c.

——, *Soil Vapor Extraction (SVE) Enhancement Technology Resource Guide*, EPA/542/B-95/003, USEPA, Washington, DC, 1995d.

———, *Thermal Desorption Implementation Issues: Engineering Forum Issue Paper*, USEPA, Washington, DC, 1995e.

———, *In Situ Remediation Technology: Electrokinetics*, EPA/542/K-94/007, USEPA, Washington, DC, Apr. 1995f.

———, *Soil Vapor Extraction Enhancement Technology Resource Guide: Air Sparging, Bioventing, Fracturing, Thermal Enhancements*, EPA/542/B-95/003, USEPA, Washington, DC, 1995g.

———, *Review of Barriers to Superfund Site Cleanup: Wasatch Chemical Case Study*, Office of the Inspector General, No. 6400016, USEPA, Washington, DC, Oct. 1995 to Mar. 1996a.

———, *A Citizen's Guide to Soil Vapor Extraction and Air Sparging*, EPA/542/F-96/008, USEPA, Washington, DC, 1996b.

———, *Engineering Forum Issue Paper: Soil Vapor Extraction Implementation Experiences*, EPA/540/F-95/030, USEPA, Washington, DC, 1996c.

———, *A Citizen's Guide to Soil Washing*, EPA/542/F-96/002, USEPA, Washington, DC, 1996d.

———, *Innovative Site Remediation Technology: Design and Application, Stabilization/Solidification*, Vol. 4, EPA/542/B-97/007, Office of Solid Waste and Emergency Response, USEPA, Washington, DC, 1997a.

———, *Recent Developments for In Situ Treatment of Metal Contaminated Soils*, Work Assignment 011059, Contract 68-W5-0055, USEPA, Washington, DC, Mar. 1997b.

———, Chemical/ETM Enterprises Superfund Site, Grand Ledge, Michigan, in *Remediation Case Studies: Bioremediation and Vitrification*, Vol. 5, prepared by member agencies of the Federal Remediation Technologies Roundtable, EPA/542/R-92/002, USEPA, Washington, DC, July 1997c.

———, In situ vitrification, U.S. Department of Energy, Hanford Site, Richland, Washington; Oak Ridge National Laboratory WAG 7, Oak Ridge, Tennessee; and various commercial sites, in *Remediation Case Studies: Bioremediation and Vitrification*, Vol. 5, prepared by member agencies of the Federal Remediation Technologies Roundtable, EPA/542/R-92/002, USEPA, Washington, DC, 1997d.

———, *Emerging Technology Summary: Vitrification of Soils Contaminated by Hazardous and/or Radioactive Wastes*, EPA/540/S-97/501, USEPA, Washington, DC, Aug. 1997e.

———, *Demonstration Bulletin: Cold Top Ex-Situ Vitrification Process*, EPA/540/MR-97/506, USEPA, Washington, DC, Aug. 1997f.

———, *Analysis of Selected Enhancements for Soil Vapor Extraction*, EPA/542/R-97/007, Office of Solid Waste and Emergency Responses, USEPA, Washington, DC, 1997g.

———, *Cost and Performance Summary Report: In Situ Thermal Desorption at the Missouri Electric Works Superfund Site Cape Girardeau, Missouri*, USEPA, Washington, DC, 1998a.

———, *Technology Fact Sheet: A Citizen's Guide to Phytoremediation*, EPA 542-F-98-011, Office of Solid Waste and Emergency Response, Technology Innovation Office, USEPA, Washington, DC, 1998b.

———, *Wasatch Chemical Company (Lot 6) Fact Sheet*, EPA ID No. UTD000716399, USEPA, Washington, DC, Nov. 1999a.

———, *Phytoremediation Resource Guide*, Office of Solid Waste and Emergency Response, Technology Innovation Office, USEPA, Washington, DC, 1999b.

———, Phytoremediation of petroleum in soil and groundwater, *www.epa.gov/ordntrnt/ORD/NRMTL/rcb/phytopet.htm*, 1999c.

———, *Introduction to Phtoremediation*, EPA/600/R-99/107, Office of Research and Development, USEPA, Washington, DC, 2000.

14

GROUNDWATER REMEDIATION TECHNOLOGIES

14.1 INTRODUCTION

Groundwater is the source of about 40% of the water used for drinking water in the United States. It provides drinking water for the more than 97% of the rural population who do not have access to public water supply systems. Between 30 and 40% of the water used for agriculture comes from groundwater. Groundwater is also critical because it supplies roughly 40% of the average annual flow in U.S. surface water bodies, which provide the balance of drinking water to those areas that do not rely on groundwater as their primary source for drinking water. Because groundwater is such a valuable resource, it must be protected from any type of contamination that will have adverse effects on human health and the environment (USEPA, 1995b; USEPA, 2000b).

There are different technologies to remediate contaminated groundwater. These technologies aim to remove, immobilize, or degrade/destroy the contaminants in the groundwater and are based on physical–chemical, thermal, or biological processes. The most common groundwater remediation technologies are (1) pump and treat, (2) in-situ flushing, (3) permeable reactive barriers, (4) in-situ air sparging, (5) monitored natural attenuation, and (6) bioremediation. In this chapter we present the following for each of these remediation methods: (1) description, applicability, advantages, and disadvantages; (2) fundamental processes; (3) design methodology, including system components, operational parameters, and monitoring requirements; (4) predictive modeling; (5) modified and complementary technologies; and (6) case studies.

14.2 PUMP AND TREAT

14.2.1 Technology Description

Pump and treat is the most common technology used for groundwater remediation. The treatment system involves pumping groundwater to the surface, removing the contaminants, and recharging treated water into the ground or discharging it either to surface water or to a municipal sewage plant. Figure 14.1 shows typical pump and treat systems with recharge (or injection) wells. When groundwater has been pumped to the surface, it can be treated to reduce contaminants to very low levels with established wastewater treatment technologies. However, pumping contaminated water from an aquifer does not guarantee that all contaminants have been removed from the subsurface. Contaminant removal is limited by the behavior of contaminants in the subsurface (i.e., a function of contaminant characteristics), site geology and hydrogeology, and extraction system design (USEPA, 1996a).

Pump and treat systems can be designed to meet two objectives: (1) *containment:* to prevent contamination from spreading, and (2) *restoration:* to remove contaminant mass. In pump and treat systems that are designed for containment, extraction rate is generally established as the minimum rate sufficient to prevent enlargement of the contaminated zone. More details on this type of containment system are given in Chapter 12. In systems designed for restoration, the pumping rate is generally larger than for containment, so that clean water will flush through the contaminated zone at an expedited rate.

The applicability of pump and treat systems depends on the site's contamination and hydrogeologic

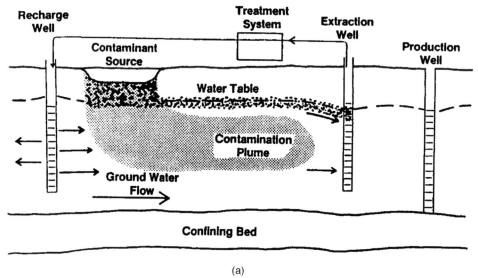

(a)

Figure 14.1 (a) *Schematic of pump and treat system;* (b) *well configurations for pump and treat systems.* (*From USEPA, 1996a.*)

conditions. Assessment of a site's contamination would require information on the nature and extent of contamination. The distribution of the contaminants in different phases (sorbed and aqueous phases for inorganic contaminants, and sorbed, NAPL, aqueous, and gaseous phases for organic liquids) should also be assessed. The subsurface soils and hydrogeology should be characterized in detail. This includes determining particle size distribution, sorption characteristics, and hydraulic conductivity. All of these data are required for evaluating the extent to which tailing and rebound (explained below) may present problems at a site. Generally, residual contamination in soil pores cannot be removed by groundwater pumping because of high sorption of the contaminant to soil matrix. Groundwater pumping is not applicable to contaminants with high residual saturation, contaminants with high sorption capabilities, low permeability (less than 1.0×10^{-5} cm/s) soils, and heterogeneous soils (USEPA, 1996a).

Pump and treat remediation offers the following advantages:

- Effective for source zone removal, where free-phase contamination is present
- Requires simple equipment

However, disadvantages of pump and treat remediation are:

- Lingering residual contamination due to tailing and/or rebound
- Long time period required to achieve remediation
- Biofouling of extraction wells and associated treatment stream that can severely affect system performance
- High cost of treating large quantities of wastewater
- High operation and maintenance costs

14.2.2 Fundamental Processes

The basic operating principle of a pump and treat system is the locating of wells and then pumping at rates

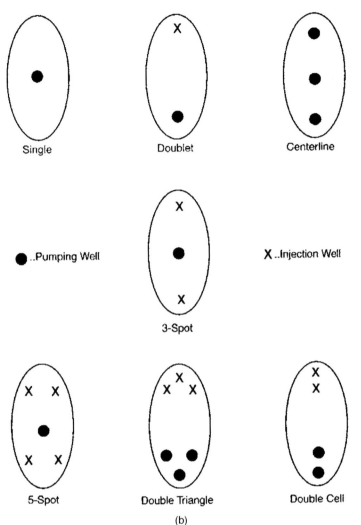

Figure 14.1 (*continued*)

that cause all water in a contaminant plume to enter the well instead of traveling through the subsurface. The extent of contaminant removal depends on the contaminant chemistry, hydrogeologic conditions, and system design. Monitoring the performance of pump and treat systems at different sites for over a decade indicates that these systems cannot reduce and maintain dissolved contaminant concentrations below cleanup levels in a reasonable time frame (Keely, 1989;

USEPA, 1996a). This is attributed to tailing and rebound phenomena. *Tailing* refers to the progressively slower rate of decline in dissolved contaminant concentration with continued operation of a pump and treat system (Figure 14.2). *Rebound* refers to the rapid increase in contaminant concentration that can occur after pumping has been stopped. The tailing effect increases significantly the time that pump and treat systems must be operated to accomplish groundwater res-

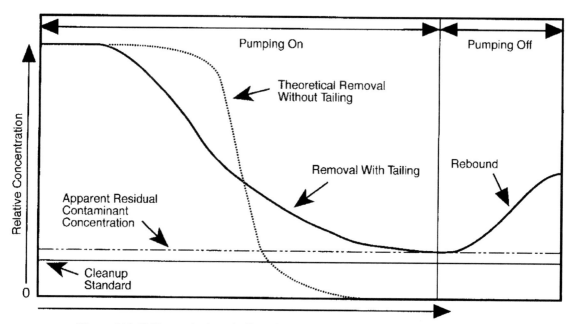

Figure 14.2 Tailing and rebound effects during pump and treat. (From USEPA, 1996a.)

toration goals. When tailing occurs, it often results in rapid initial decline in the rate of contaminant concentrations, followed by a more gradual decline that eventually stabilizes at an apparent residual concentration level the above the cleanup standard (Figure 14.2). Rebound is a problem when a pump and treat system attains cleanup standard but concentrations subsequently increase to a level that exceeds the standard. The degree to which tailing and rebound complicate remediation efforts at a site is a function of physical and chemical characteristics of contaminants, subsurface soils, and groundwater.

The major factors and processes that contribute to tailing and rebound are (1) the presence of non-aqueous-phase liquids (NAPLs), (2) contaminant desorption, (3) contaminant precipitation–dissolution, (4) matrix diffusion, and (5) groundwater velocity variation (USEPA, 1996a). Light NAPLs such as benzene and dense NAPLs such as TCE are relatively insoluble in water, but they are soluble enough to cause concentrations in groundwater to exceed MCLs (maximum contaminant levels). Therefore, the presence of NAPLs

will continue to contaminate groundwater that makes sufficient contact to dissolve small amounts from NAPL surfaces [Figure 14.3(*a*)]. When groundwater is moving slowly, contaminant concentrations can approach the solubility limit for NAPL [Figure 14.3(*c*)]. Pump and treat systems increase groundwater velocity, causing an initial decrease in concentration. The decline in concentration will later tail off until the NAPLs' rate of dissolution is in equilibrium with the velocity of pumped groundwater. If pumping stops, groundwater velocity slows and concentrations can rebound, rapidly at first, then gradually reaching equilibrium concentration [Figure 14.3(*c*)], unless pumping is restarted.

Movement of many organic and inorganic contaminants in groundwater is retarded by sorption processes that cause some of dissolved contaminant to attach to solid surfaces. The amount of contaminant sorbed is a function of concentration (with sorption increasing as concentrations increase) and the sorption capacity of subsurface materials. Sorbed contaminants tend to concentrate on organic matter and clay-sized mineral ox-

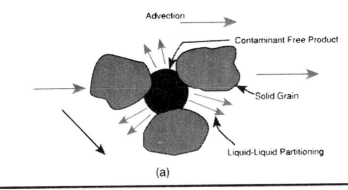

(a)

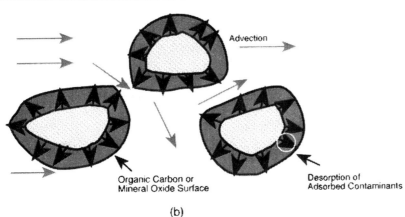

(b)

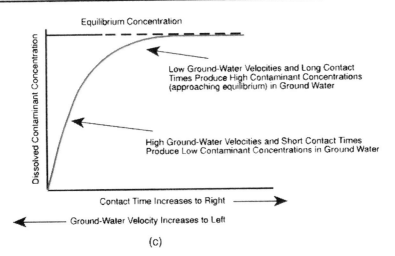

(c)

Figure 14.3 Removal of NAPLs during pumping: (a) dissolution of residual and pooled free product; (b) desorption and solubilization; (c) dissolution kinetics. (From USEPA, 1996a.)

ide surfaces [Figure 14.3(b)]. Sorption is a reversible process; as dissolved contaminant concentrations are reduced by pump and treat system operation, contaminants sorbed to subsurface media can desorb from the matrix into groundwater. Contaminant concentrations resulting from sorption and desorption show a relationship to groundwater velocity and contact time similar to that of NAPLs [Figure 14.3(c)], causing tailing of contaminant concentrations during pumping as well as rebound after pumping stops.

As with sorption–desorption reactions, precipitation–dissolution reactions are reversible. Large quantities of inorganic contaminants such as chromate in $BaCrO_4$ may be found with crystalline or amorphous precipitates in the subsurface. In this situation, if pumping stops before the solid phase is depleted, rebound can occur.

As contaminants move through relatively permeable pathways in heterogeneous media, concentration gradients cause diffusion of contaminant mass into less permeable media. Matrix diffusion occurs with dissolved contaminants that are not strongly sorbed (inorganic anions and some organic chemicals). During a pump and treat operation, dissolved contaminant concentrations in relatively permeable zones are reduced by advective flushing, causing a reversal in initial concentration gradient and slow diffusion of contaminants from low- to high-permeability media. The significance of matrix diffusion increases as the length of time between contamination and cleanup increases. In a heterogeneous aquifer, matrix diffusion contributions to tailing and rebound can be expected as long as contaminants have been diffusing into less-permeable materials.

Tailing and rebound also result from variable travel times associated with a different flow path taken by contaminants to an extraction well. Groundwater at the edge of a capture zone created by a pumping well travels a greater distance under a lower hydraulic gradient than does groundwater closer to the center of the capture zone. In addition, contaminant-to-well travel time varies as a function of hydraulic conductivity in heterogeneous aquifers. The extent of tailing and rebound depends on the hydrogeologic condition and subsur-

face distribution of contaminants. The rate of mass transfer and sorption characteristics of contaminants are the most important parameters affecting cleanup time.

14.2.3 System Design and Implementation

General Design Approach. The procedure for evaluating the feasibility of the pump and treat approach at a contaminated groundwater site and designing an appropriate system consists of the following steps: (1) contaminant characterization, (2) hydrogeologic characterization, (3) extraction well network, and (4) extracted groundwater treatment and disposal (USEPA, 1996a; Cohen et al., 1997).

Contaminant Characterization. Contaminant concentrations and areal extent determine the seriousness of the problem. The existence of NAPLs determines the applicability and effectiveness of the process. Soil and water quality sampling can be performed to determine contaminant distribution. Contaminant/soil properties such as density, aqueous solubility, octanol–water/carbon partitioning coefficient, soil organic carbon content, and sorption parameters are used to determine mobility properties and applicability of the pump and treat approach. As a guidance, published literature and laboratory tests are often used to determine these properties.

Hydrogeologic Characterization. Hydraulic conductivities and storativities of subsurface materials are needed to determine the feasibility of extracting groundwater and applicability of the pump and treat approach. Pumping tests, slug tests, and laboratory permeability tests can be used to determine hydraulic properties of subsurface materials. *Types, thickness,* and *extent of saturated* and *unsaturated* subsurface materials are used to develop conceptual design and applicability for implementation. Hydrologic maps, surficial geology maps/reports, boring logs, and geophysics can be used as sources for this information. *Depth of the aquifer/water table* is used to select the appropriate extraction system type and consideration for implementation. Hydrogeological maps, observation wells, boring logs, and piezometers can be used to obtain these data. *Groundwater flow direction* and vertical/horizontal *gradients* are used to determine proper well locations/spacing considerations for implementation. Water-level data and potentiometric maps for the site can be used to obtain these data. Seasonal changes in groundwater elevation and long-term water-level monitoring are used to locate wells, screened intervals, and considerations for implementation.

Extraction Well Network. This includes the selection of locations, screen/open interval depth, pumping rates, and groundwater treatment system. The selection of well locations, spacing, and pumping rates are often determined by using mathematical groundwater flow models. Several mathematical models have been developed and applied to compute the capture zone, groundwater path lines, and related travel times to extraction wells. For simple hydrogeologic conditions (homogeneous isotropic aquifers), analytical equations solved manually or graphically may be adequate. For complex sites, numerical models based on finite difference or finite element methods may be required (see Section 14.2.4 as well as Chapters 7 and 8 for more information on commonly used mathematical models). These models provide insight to flow patterns generated by alternative pump and treat approaches and to selection of monitoring points and frequency. Contaminant transport models may be used to evaluate the effectiveness of different well patterns for restoring contaminated groundwater. Because these contaminant transport models do not account for contaminant–aquifer solid interactions, tailing and rebound effects are not considered in these models; therefore, these models should be used with caution. In addition, optimization-programming methods may be used to optimize the number of wells and pumping rates. Optimization involves defining an objective function, such as minimizing the sum of pumping rates from a number of wells. A set of restrictions or constraints specify various conditions, such as maximum pumping rates

and minimum hydraulic heads at individual wells that must be satisfied by an optimal solution alternative. Common modeling techniques used for the design of pump and treat systems are presented in Section 14.2.4.

Extracted Groundwater Treatment and Disposal. Selection of the treatment method depends on the type and concentrations of the contaminants present in the extracted groundwater. Pilot-scale tests are often conducted to study the effect of varying system parameters, such as flow rate, on the treatment efficiency selected. Air stripping and granular activated carbon methods are commonly used for organic contaminants, while precipitation and ion exchange methods are commonly used for inorganic contaminants. More details on these are given in the next section. The treated water may be reinjected into the ground to enhance hydraulic control and flushing of the contamination zone. The reinjection may be accomplished by using wells, trenches, drains, or surface applications. The reinjection systems are designed with careful consideration of site conditions to ensure that the environmental problem is not exacerbated. Regulatory approvals/permits are often required for implementation of reinjection systems. Other options are to dispose of the treated groundwater in surface water bodies or into public sewer or stormwater systems.

Equipment. The selection of equipment used for pump and treat systems depends on site-specific conditions and remediation goals (USEPA, 1996a; Cohen et al., 1997). Wells, pumps, and extracted groundwater treatment are the major components of any pump and treatment system. Other peripheral equipment needed is summarized in Table 14.1.

Extraction wells capture and remove contaminated groundwater. The construction of these wells is similar to the construction of monitoring wells. The well diameter must be large enough to accommodate the pump. The casing size should be such that uphole velocity during pumping is less than 5 ft/s, to prevent excessive head losses. Screens and filter packs are appropriately sized to the native media. Well screens and

Table 14.1 **Peripheral equipment needed in pump and treat systems**

Equipment	Description
Piping	Conveys pumped fluids to treatment system and/or point of discharge. Piping materials will dictate if the system may be installed above or below grade with or without secondary containment measures. Piping materials (i.e., steel, HDPE, PVC, etc.) are selected based on chemical compatibility and strength factors.
Flowmeters	Measure flow rates at a given time and/or cumulative throughput in a pipe. Typically installed at each well, at major piping junctions, and after major treatment units. Some designs allow for the instrument to act as an on–off switch or flow regulator. Many different types are available.
Valves	The primary use of valves (i.e., gate, ball, check, butterfly) is to control flow in pipes and to connections in the pipe manifold. Valves may be operated manually or actuated by electrical or magnetic mechanisms. Check valves are used to prevent backflow into the well after pumping has ceased and siphoning from tanks or treatment units. Other uses for valves include sample ports, pressure relief, and air vents.
Level switches and Sensors	Float, optical, ultrasonic, and conductivity switches/sensors are used to determine the level of fluids in well or tank. Used to actuate or terminate pumping to warn operators of rising or falling fluid levels in wells and tanks.
Pressure switches	Used to shut off pumps after detecting a drop in discharge pressure caused by a loss in suction pressure.
Pressure and vacuum indicators	Used to measure the pressure in pipes, across-pipe connections, and in sealed tanks and vessels.
Control panels	Device that provides centralized, global control of pump and treat system and monitors and displays system status. Control panels are typically designed for specific applications.
Remote monitoring, data acquisition, and telemetry devices	Provides interactive monitoring and control of unattended pump and treat systems. Allows for real-time data acquisition. Alerts operators to system failures and provides an interface for remote reprogramming of operations. Remote monitoring devices should also be accessible from the control panel.
Pull and junction boxes	Above- and/or below-grade installations that allow access to connections in the piping manifold, electric wiring, and system controls. Strategic placement provides flexibility for the system expansion.
Pitless adaptor unit	Allows for the transfer of extracted groundwater from the well to buried piping outside the well casing.
Well cover	Available with padlock hasps with and without a connection to the electrical conduit for submersible pumps.

Source: Cohen et al. (1997).

casings made of black low-carbon steel, stainless steel, or PVC are used. Well screen diameter is selected to provide sufficient open area so that the velocity of water entering the screen is less than 0.1 ft/s, to minimize friction head losses and corrosion.

Many different types of pumps are available, but standard electric submersible pumps are most com-

monly used. Final selection of the pipe depends on the desired pumping rate, well yield, and the total hydraulic head lift. Performance curves and data provided by the pump manufacturers are used for the selection of pumps. Pneumatic pumps may be used at sites where providing electrical service is a problem, combustible vapors are present, or excessive drawdown has the potential to damage electric submersible pumps. Pump intake is not placed in the well screen, in order to prevent corrosion, sand pumping, or dewatering of the screen.

A water treatment system is needed to clean up the contaminated water that is extracted. Air stripping, granular activated carbon, chemical/ultraviolet oxidation, and aerobic biological reactors are commonly used to treat groundwater contaminated with organic contaminants. Chemical precipitation, ion exchange–adsorption, and electrochemical methods are commonly used to treat groundwater contaminated with inorganic contaminants such as heavy metals. A combination of these technologies is used to treat groundwater contaminated with both organic and inorganic contaminants. The most commonly used groundwater treatment technologies that are commercially available as packages are described in Table 14.2.

Operational Parameters. Pumping rate is an important operational parameter of pump and treat systems and should establish design objectives. For example, the pumping rate should maintain specified pore volume flushing rates. Pump replacement and well development may be performed to increase the efficiency of operation of the pump and treat system. Pumping operations can be made more efficient in the following ways (USEPA, 1996a):

- *Adaptive pumping:* involves designing the well field such that extraction and injection can be varied to reduce zones of stagnation. Some extraction wells can be shut off periodically while the others are turned on, and pumping rates are varied to ensure that contaminant plumes are remediated at the fastest rate possible.
- *Pulsed or intermittent pumping:* has the potential to increase the ratio of contaminant mass removed to groundwater volume where mass transfer limitations restrict dissolved contaminant concentrations (USEPA, 1996a; Sullivan, 1996). Figure 14.4 indicates the concept of pulsed pumping. During the resting phase of pulse pumping, contaminant concentrations increase due to diffusion, desorption, and dissolution in slower-moving

groundwater [Figure 14.3(b) and (c)]. When pumping is restarted, groundwater with a higher concentration of contaminants is removed, increasing mass removal during pumping. Special care must be taken to ensure that contamination does not spread during pump rest periods.

Monitoring. Monitoring involves measurement of contaminant concentrations in water during pumping and observation wells to determine the rate and effectiveness of mass removal. The overall system performance is monitored by measuring hydraulic heads and gradients, groundwater flow directions and rates, pumping rates, pumped water and treatment system effluent quality, and contaminant distributions in groundwater and porous media. These data are evaluated to interpret capture zones, flushing rates, and contaminant transport and removal, and to improve system performance (Cohen et al., 1994). Periodic monitoring of the pumping rate, drawdown, and efficiency may be used to determine maintenance requirements of wells and pumps.

The remediation progress can be assessed by comparing rates of contaminant mass removal to estimates of dissolved and/or total contaminant mass-in-place. If rate of contaminant mass extracted approximates rate of dissolved mass-in-place reduction, contaminants removed by pumping are derived primarily from the dissolved phase. If the mass removal rate exceeds the rate of dissolved-phase reduction, it indicates the removal of contaminants in the free phase. The time needed to remove dissolved mass-in-place may be projected by extrapolating the trend of the mass removal rate curve or cumulative mass-removed curve. However, the effects of tailing and rebound should be considered in determining the cleanup time. If the mass removal trend indicates greater cleanup duration than estimated originally, system modification may be necessary.

14.2.4 Predictive Modeling

Many mathematical models have been developed or applied to optimize pump and treat system designs. These models compute capture zones, groundwater

Table 14.2 **Extracted groundwater treatment technologies**

Method	Process Description	Package Plant Components and Sizes (Dimensions Are for Overall Plant Envelope)	Advantages and Limitations
Air stripping Widely used to remove volatile contaminants from groundwater	Volatile contaminants are transferred from water phase to gas phase by passing air or steam through water in a tall packed tower, shallow tray tower or stripping lagoon. The airstream containing volatile contaminants may require treatment (e.g., with vapor-phase carbon). Stripping with steam may be cost-effective for water containing a mix of relatively nonvolatile and volatile compounds, particularly at industrial facilities where steam is readily available.	Package plants include tall packed tower or compact low-profile diffuser tray units, feed pump, air blower, and effluent pump. Flow meters for influent and airflow are required. An influent throttle valve and a blower damper are required to adjust the air/water ratio. Acid or chlorine is used to wash the tower packing (e.g., of Fe precipitates). Heights are for packet tower units. 1–10 gal/min, 4 × 4 × 20 ft, 2 hp 10–50 gal/min, 6 × 8 × 25 ft, 5 hp 50–100 gal/min, 7 × 10 × 30 ft, 8 hp 100–400 gal/min, 8 × 12 × 40 ft, 20 hp	Effective for VOCs. Equipment is relatively simple. Startup and shutdown can be accomplished quickly. Modular design is well suited for contaminant P&T. Package systems are widely available. Dissolved Fe and Mn can be precipitated and foul the packed media, resulting in head loss and reduced system effectiveness. Pretreatment (oxidation, precipitation, sedimentation) of effluent may be required. Biological fouling may also occur (requiring cleaning via chlorination or biocide). Sensitive to pH, temperature, and flow rate. May be cost-prohibitive at temperatures below freezing (may need to heat). May need GAC polishing of water effluent and treatment of airstream.
Granular activated carbon (GAC) adsorption Widely used to remove metals, volatile and semivolatile organics, pesticides, PCBs, etc., from groundwater and leachate	Aqueous contaminants are sorbed to GAC or synthetic resin packed in vessels in parallel or series. Used sorbent is regenerated or replaced. Extent of adsorption depends on strength of molecular attraction, molecular weight of contaminants, type and characteristic of adsorbent, pH, and surface area.	Package system includes one to three pressure valves on a skid, interconnecting piping, a feed pump, optionally a back wash pump, pressure gauges, differential pressure gauges, influent flow meter, backwash flow meter, and a control panel. Spent adsorbers are disconnected and sent to regeneration centers or landfills. 1–10 gal/min, 12 × 8 × 8 ft, 2 hp 10–50 gal/min, 14 × 8 × 8 ft, 7 hp 50–100 gal/min, 20 × 10 × 8 ft, 10 hp 100–200 gal/min, 20 × 20 × 8 ft, 20 hp	Effective for low-solubility organics. Useful for a wide range of contaminants over a broad concentration range. Not affected adversely by toxics. High O&M costs. Intolerant of suspended solids (will clog). Pretreatment required for oil and grease greater than 10 mg/L. Synthetic resins intolerant of strong oxidizing agents.

Chemical precipitation, flocculation, sedimentation Widely used to remove metals from contaminated groundwater and landfill leachate

Metals are precipitated to insoluble metal hydroxides, sulfides, carbonates, or other salts by the addition of a chemical (e.g., to raise pH), oxidation, or change in water temperature. Flocculent aids may be added to hasten sedimentation.

Package plants include a rapid mix tank, flocculation chamber, and a settling tank. Inclined plate gravity separation or circular clarifiers are used for settling. Typical equipment includes a rapid mixer, flocculator and drive, feed pump, sludge pump, acid and caustic soda pumps for pH control, and a polymer pump.
1–10 gal/min, 8 × 4 × 9 ft, 3 hp
10–50 gal/min, 10 × 4 × 13 ft, 5 hp
50–100 gal/min 11 × 6 × 14 ft, 7 hp

Useful for many contaminated ground water streams, particularly as a pretreatment step. Effectiveness limited by presence of complexing agents in water. Precipitate sludge may be a hazardous waste.

Source: Cohen et al. (1997).

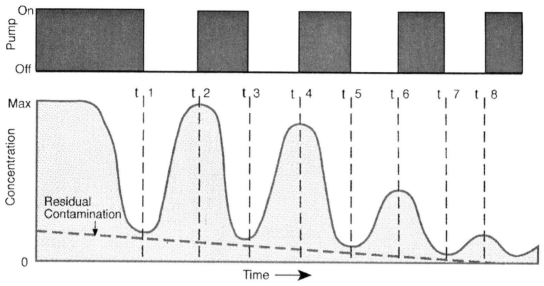

Figure 14.4 *Pulsed pumping. (From USEPA, 1996a.)*

path lines, flushing rates, and associated travel times to extraction wells. These models provide capture zones generated by alternative pump and treat schemes and aid in the selection of monitoring well spacing and pumping rates. The capture zone of an extraction well refers to that portion of the subsurface containing groundwater that will ultimately discharge to the well. Fundamentals of groundwater flow and contaminant transport modeling have been provided in Chapters 7 and 8. Models used specifically for pump and treat system design are summarized in Table 14.3.

Type curves are often used to define capture zones and in the selection of the arrangement or pattern and spacing of wells. Type curves are available for different well patterns, such as the one shown in Figure 14.5. Use of these type curves requires the following procedure (Javandel and Tsang, 1986):

1. A map of the contaminant plume is constructed at the same scale as the type curves. The edge or perimeter of the plume should be clearly indicated, together with the direction of regional groundwater flow.

2. Superimpose the type curve for one well on the plume, keeping the *x* axis parallel to the direc-

tion of regional groundwater flow and along the midline of the plume so that approximately equal proportions of the plume lie on each side of the *x* axis. The pumped well on the type curve will be at the downstream end of the plume. The type curve is adjusted so that the plume is enclosed by a single Q/BU curve.

3. The single-well pumping rate (Q) is calculated using the known values of aquifer thickness (B) and the Darcy velocity for regional flow (U), along with the value of Q/BU indicated on the type curve (TCV) with the equation $Q = B \times U \times \text{TCV}$.

4. If the pumping rate is feasible, one well with pumping rate Q is required for cleanup. If the required production is not feasible due to a lack of available drawdown, it will be necessary to continue adding wells.

5. Repeat steps 2, 3, and 4 using the two-, three-, or four-well type curves in that order until a single-well pumping rate is calculated that the aquifer can support. Then calculate the optimum spacing between wells using the following simple rules for centerline spacing (L):

Table 14.3 **Mathematical models used in pump and treat system design**

Method	Example	Description
Aquifer tests and pilot testing	See Chapter 10	Controlled and monitored pilot tests are conducted to assist pump and treat (P&T) design. Suggested operating procedures for aquifer tests and analytical methods are described in Chapters 7 and 10 and many others. Test results should be used to improve P&T design modeling, where applicable.
Graphical capture zone-type curves	(Javandel and Tsang, 1986)	A simple graphical method can be used to determine minimum pumping rates and well spacings needed to maintain capture using one, two, or three pumping wells along a line perpendicular to the regional direction of groundwater flow in a confined aquifer.
Semianalytical groundwater flow and pathline models	WHPA (Blandford et al., 1993) WHAM (Strack et al.,1994; Haitjema et al., 1994)	These models superposition analytical functions to simulate simple or complex aquifer conditions, including wells, line sources, line sinks, recharge, and regional flow (Strack,1989). Advantages include flexibility, ease-to-use speed, accuracy, and no model grid. Generally limited to analysis of two-dimensional flow problems
Numerical models of groundwater flow	MODFLOW (McDonald and Harbaugh, 1988)	Finite-difference (FD) and finite element (FE) groundwater flow models have been developed to simulate two-dimensional areal or cross-sectional quasi-or fully, steady or transient flow in anisotropic heterogeneous-layered aquifer systems. These models can handle a variety of complex conditions allowing analysis of simple and complex groundwater flow problems, including P&T design analysis. Various pre and post processors are available. In general, more complex and detailed site characterization data are required for simulation of complex problems.
Path line and particle tracking post processors	MODPATH (Pollock, 1994) GPTRAC (Blandford et al., 1993)	These programs use particle tracking to calculate path lines, capture zones, and travel times based on groundwater flow model output. Programs vary in assumptions and complexity of site conditions that may be used to simulate (e.g., two- or three-dimensional flow, heterogeneity, anisotropy).
Numerical models of groundwater flow and contaminant transport	MT3D (Zheng, 1992) MOC (Konikow and Bredehoeft, 1989)	These models can be used to evaluate aquifer restorations issues such as changes in contaminant transport mass distribution with time due to P&T operation.
Optimization models	MODMAN (Greenwald, 1993)	Optimization programs designed to link with groundwater flow models yield answers to such questions as: (1) Where should pumping and injection wells be located? and (2) at what rate should water be extracted or injected at each well? The optimal solution maximizes or minimizes a user-defined objective function and satisfies all user-defined constraints. A typical objective may be to maximize the total pumping rate from all wells, while constraints might include upper and lower limits on heads, gradients, or pumping rates. A variety of objectives and constraints are available to users, allowing many P&T issues to be considered.

Source: Cohen et al. (1997).

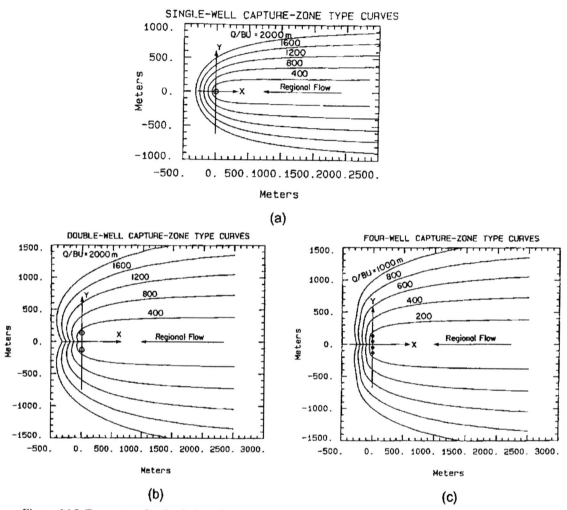

Figure 14.5 *Type curves for the design of pump and treat system design. (From Javandel and Tsang, 1986.)*

For two-well system: $L = \dfrac{Q}{\pi BU}$ (14.1*a*)

For three-well system: $L = \dfrac{1.26Q}{\pi BU}$ (14.1*b*)

For four-well system: $L = \dfrac{1.2Q}{\pi BU}$ (14.1*c*)

To model complex field conditions, semianalytical or numerical models are used. Although easy to use, semianalytical solutions are generally used for two-dimensional flow problems. Numerical models are commonly used to simulate groundwater flow in complex three-dimensional hydrogeologic conditions. Among the numerical models, MODFLOW is the most popular groundwater flow model (Uhlman and Portman, 1991). More details on the MODFLOW model are given in Chapter 7. The module WELL in MODFLOW allows the user to specify withdrawal or injection from the area modeled. Wells are assumed to be placed at the center of grid blocks and to fully penetrate the layer for which they are specified. The output of these models is then input to particle-tracking models (e.g., PATH3D) to assess path lines and to define capture zones. These models may also be combined with linear programming methods to optimize pump and treat design by specifying an objective function subject to various constraints.

Contaminant transport models may be used to predict the contaminant removal rates during pump and treat remediation. However, these models do not consider the effects of contaminant tailing and rebound, and the results should be used with caution. Analytical solutions may be used for simple cases, but numerical models may be needed for complex site conditions (Javandel et al., 1984; Fetter, 1999).

The selection of a particular model depends on the complexity of the site, data available, and familiarity of the analyst with different codes. In general, the simplest tool applicable to site conditions and desired degree of uncertainty should be used in design. However, conditions at many sites will be sufficiently complex that screening-level characterizations and design tools will result in significant uncertainty. Regardless of the design tools that are used, capture zone analysis should be conducted, and well locations and pumping rates optimized.

14.2.5 Modified or Complementary Technologies

Many types of variations and alternatives to pump and treat methods exist, as described below (Palmer and Fish, 1991; USEPA, 1996a).

Interceptor Trenches. After vertical wells, trenches are the most widely used method for controlling subsurface fluids and recovering contaminants. They function similarly to horizontal wells, but can also have a significant vertical component that allows access to permeable layers in interbedded sediments. For shallower applications, trenches can be installed at relatively low cost. Figure 14.6 shows a typical intercepting trench system. As the contaminant plume flows into the trench, floating pumps extract and deposit the contaminant in containers for treatment. This process

is used primarily for gasoline contamination and free-phase LNAPLs, as extraction is simpler for contaminants on or near the surface of the groundwater flow. Large contaminant exposure to the atmosphere is a major limitation on this method. Therefore, interceptor trenches are not commonly employed.

Horizontal and Inclined Wells. Directional drilling technology uses specialized bits to curve bores in a controlled arc. Directional drilling methods can create wellbores with almost any trajectory. Wells that curve to a horizontal orientation are especially suited to environmental applications (Figure 14.7).

Induced Fractures. EPA research has shown that petroleum engineering technology used to induce fractures for increased productivity of oil wells can also improve the performance of environmental wells. Induced fractures are used mainly where low-permeability aquifer materials create problems for recovery of contaminants.

Physical Enhancements. Air sparging, known as insitu aeration, is similar to soil vapor extraction except that air is injected into the saturated zone rather than the vadose zone [shown in Figure 14.22 (Section 14.5)]. Air sparging can remove a substantial amount of volatile aromatic and chlorinated hydrocarbons in a variety of geologic conditions, but significant questions remain about the ability of this technology to achieve health-based standards throughout the saturated zone. Thermal enhancement, such as steam and hot-water flooding, increase the mobility of volatile and semivolatile contaminants. Use of induced fractures is also considered to be a physical enhancement to pump and treat systems.

Chemical Enhancements. Chemically enhanced pump and treat systems require the use of injection wells to deliver reactive agents to the contaminant plume and extraction wells to remove reactive agents and contaminants. The two major types of chemical enhancements are:

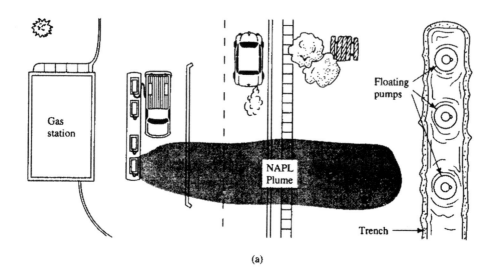

(a)

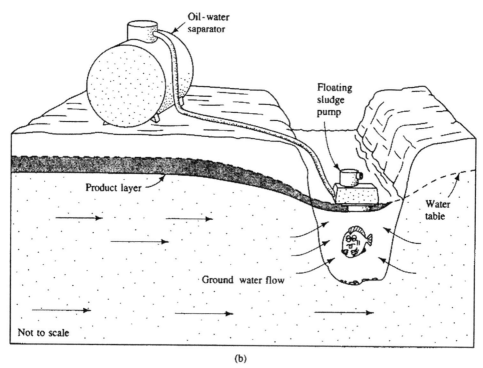

(b)

Figure 14.6 *Use of interceptor trenches in pump and treat systems: (a) location of an interceptor trench used to capture a floating plume of light nonaqueous phase liquid; (b) cross section of trench and floating pump to capture the floating product and depress the water table. (From Fetter, 1999. Reprinted by permission of Pearson Education, Inc., Upper Saddle River, NJ.)*

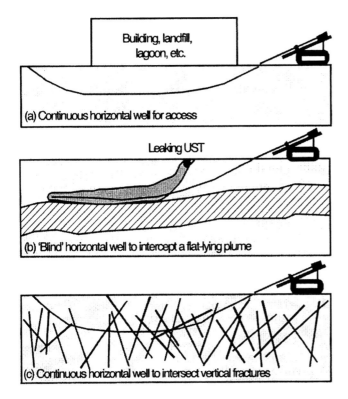

Figure 14.7 *Use of horizontal wells in pump and treat systems. (From Cohen et al., 1997.)*

1. Soil flushing enhances recovery of contaminants with low water solubility, free and residual NAPLs, and sorbed contaminants (Figure 14.8). Two types of chemical agents can be used: (1) cosolvents, when mixed with water, increase the solubility of some organic compounds, and (2) surfactants, which may cause contaminants to desorb and may increase NAPL mobility by lowering interfacial tension between NAPLs and water, increasing solubility. Soil flushing is one of most promising technologies for dealing with separate-phase DNAPLs in the subsurface. This technology is discussed in Section 14.3.

2. In-situ chemical treatment involves reactive agents that oxidize or reduce contaminants, converting them to nontoxic forms or immobilizing them to minimize contaminant migration (Figure 14.12). This technology is still in the early stages of development.

Biological Enhancements. Biological enhancements to pump and treat systems stimulate subsurface microorganisms, primarily bacteria, to degrade contaminants to harmless mineral and products, such as carbon dioxide and water. In-situ bioremediation of some types of hydrocarbons, by addition of oxygen and nutrients to groundwater, is an established technology. Other biodegradable substances, such as phenol, cresols, acetone, and cellulosic wastes, are also amenable to aerobic in-situ bioremediation. Key elements are delivery of oxygen and nutrients by use of an injection well or an infiltration gallery (discussed in Section 14.7). A limitation of in-situ bioremediation is that minimum contaminant concentrations required to maintain microbial populations may exceed health-based cleanup standards, particularly where heavier hydrocarbons are involved.

In-situ bioremediation of chlorinated solvents has not been well demonstrated because metabolic pro-

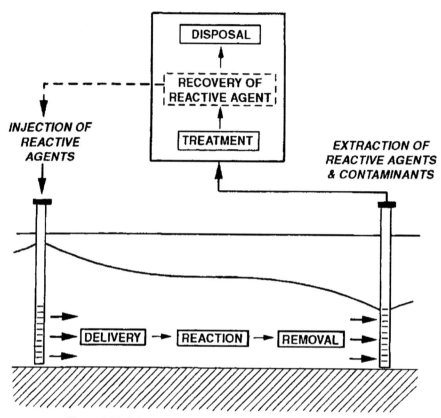

Figure 14.8 *Schematic of in-situ flushing. (From USEPA, 1996b.)*

cesses for their degradation are more complex than those for hydrocarbon degradation. Methanotrophs are able to degrade some chlorinated solvents under aerobic conditions if methane is supplied as an energy source. Also, the ability of anaerobic bacteria to degrade a variety of chlorinated solvents is documented in the literature. Two major obstacles to use of anaerobic processes for in-situ bioremediation are: (1) hazardous intermediate degradation products can accumulate in the system, and (2) undesirable water quality changes (such as dissolution of iron and manganese) can occur in the system.

14.2.6   Economic and Regulatory Considerations

The total cost of pump and treat systems depends on the cost of equipment, installation, permitting, material

handling, chemical and/or biological enhancements, wastewater treatment, and operation and maintenance. Typical costs associated with groundwater pump and treatment systems are provided in Section 14.2.7. In general, pump and treat systems are costly to construct and expensive to operate and maintain for the long periods of time commonly required for site cleanup.

Groundwater treatment systems involve permits associated with discharge of treated water and air emissions. Discharge into surface water is regulated and permitted under the National Pollutant Discharge Elimination System (NPDES) under the Clean Water Act (CWA). The permit specifies the treatment system to be used, allowable discharge flows and contaminant concentrations, water quality parameters, and monitoring requirements. Reinjection of treated water into the subsurface is regulated and permitted through the underground injection control program under the Safe

Drinking Water Act. Discharge of treated groundwater to a POTW does not require an NPDES permit. However, the treatment system must meet all pretreatment standards specified under the CWA, along with any other standards specified by the POTW.

Air emissions from groundwater treatment systems are not an issue, with air strippers as the most common exception. The need for an air permit depends on the nature and quantity of the air emissions and the specific regulatory requirements of the permitting authority. Federal permitting requirements are specific in 40 CFR 70. These regulations are geared toward larger industrial sources. The federal regulations also provide for delegation of the permitting authority to state and local agencies, and this delegation has taken place for virtually all of the states. In practice, the need for an air permit can vary widely depending on the location of the system, and must be worked on a case-by-case basis with the appropriate local, state, and other regulatory authorities.

14.2.7 Case Studies

Several case studies documenting the performance of pump and treat systems to remediate groundwater contaminated with different types of organic and inorganic chemicals have been reported (FRTR, 1988a,b). Two of these case studies are presented in this section.

Case Study 1: Baird and McGuire Superfund Site, Holbrook, Massachusetts. Baird and McGuire, Inc. conducted chemical mixing operations at the site from 1912 to 1983. The wastes generated from these operations were stored in surface impoundments and also discharged into a septic system and wetlands. The infiltration of the contaminants from these wastes created a contamination plume in the groundwater. The contamination included a wide range of compounds, including BTEX, PAHs, pesticides/herbicides, and metals, and the concentration of these chemicals were as high as 10,000 μg/L. The areal extent of the plume was estimated to be more than 700,000 ft^2 in area and 70 ft in thickness, with approximately 111 million gallons of groundwater contaminated. The site contains stratified material consisting of silty sands, sand, and

silt underlain by stratified material consisting of fine to coarse sand. The sand layer is underlain by an unstratified glacial till that is underlain by fractured bedrock. Groundwater at the site was encountered at 10 to 15 ft below the ground surface. The groundwater flow velocity within the upper two layers is estimated to range between 50 and 500 ft/yr.

A pump and treat system was selected to remediate the contaminated groundwater. The design consisted of six wells. Four wells were screened in the first two stratified materials and two were screened in both the till and bedrock. The wells were located in the part of the plume where the highest contamination was detected. Groundwater was extracted at an average total pumping rate of 60 gal/min. Extracted groundwater was treated using an elaborate treatment plant consisting of (1) equalization and removal of free-floating product, (2) chemical treatment (with ferric chloride and lime in one stage, and phosphoric and sulfuric acids and ammonium sulfate in a second stage), (3) flocculation–clarification, (4) aeration, (5) pressure filtration, and (6) carbon adsorption. Treated groundwater was reinjected through infiltration basins.

During the first two years of operation, the pump and treat system reduced average VOC and SVOC concentrations by 16% and 48%, respectively. Approximately 2100 lb of organic contaminants have been removed from the groundwater. However, contaminant concentrations in some individual wells did not decline and concentrations have not decreased to cleanup goals. The cleanup goals were established based on the maximum contaminant levels (MCLs) as defined by the primary drinking water standards and the state of Massachusetts's drinking water criteria. Because of the spreading of the contaminant, plume containment and enhancement of pump and treat system design were recommended. The actual cost for this treatment was $22,726,000, which corresponds to $284 per 1000 gal of groundwater extracted and $10,822 per pound of contaminant removed. This case study clearly reveals the difficulties in designing pump and treat system and achieving the cleanup goals.

Case Study 2: Former Firestone Facility Superfund Site, Salina, California. The former Firestone facility was operated as a tire manufacturing plant from

1963 to 1980. Accidental release of chemicals at the plant caused subsurface contamination. Groundwater was contaminated with volatile organic compounds and the plume extended 2.5 miles downgradient. The major contaminants in groundwater included benzene, 1,1-dichloroethane (1,1-DCA), 1,1,1-trichloroethylene (1,1,1-TCA), TCE, PCE, toluene, and xylene. 1,1-DCE was selected during the design process as the index compound for the remedial action, and it was detected at maximum concentration of 120 μg/L. Groundwater at the site occurred in three interconnected aquifers, designated as shallow, intermediate, and deep aquifers. The shallow aquifer consisted of permeable sands and gravels underlain by a thin, discontinuous clay. This aquifer extends from ground surface to a depth of 90 ft. The intermediate aquifer, approximately 10 to 45 ft in thickness, consisted of alluvial channels of sands underlain by a discontinuous clay layer. The deep aquifer consisted of sands and gravels with discontinuous clay aquitards at different depths. Groundwater was located near the ground surface at this site. The hydraulic conductivity of the aquifers ranged from 100 to 1200 ft/day. The average velocity of groundwater in the aquifers was 2 to 4 ft/day.

The pump and treat system consisted of 25 extraction wells located on- and off-site and screened in three aquifers. A computer model was used for the design of the system. Groundwater was extracted at an average total pumping rate of 480 gal/min. The groundwater extracted was treated with oil–water separation, air stripping, and carbon adsorption, and discharged to surface water under a NPDES permit.

From 1986 to 1992, 496 lb of total VOCs was removed from the groundwater. 1,1-DCE concentrations decreased from 120 μg/L to 4.8 μg/L in 1994 and 6 μg/L in 1995. The cleanup levels for this site were selected based on chemical-specific MCLs and health-based restrictions. The concentration reductions met the cleanup levels. The actual cost of treatment was $12,884,813, which corresponds to $7 per 1000 gal of groundwater extracted and $26,000 per pound of contaminant removed. The remediation goal at this site was accomplished in seven years of treatment. The extraction system was adjusted during the operations to maximize removal of contaminants from the ground-

water. High permeability of the aquifers at the site contributed to better remedial performance.

14.3   IN-SITU FLUSHING

14.3.1   Technology Description

In-situ flushing remediation involves using an aqueous solution to purge contaminants from the subsurface media. The flushing solution consists of plain water or a carefully developed solution (surfactant/cosolvent) that will optimize contaminant desorption and solubilization. As shown in Figure 14.8, the flushing solution is pumped into groundwater via injection wells. The solution then flows downgradient through the region of contamination where it desorbs, solubilizes, and/or flushes the contaminants from the soil and/or groundwater. After the contaminants have been solubilized, the solution is pumped out via extraction wells located farther downgradient. At the surface, the contaminated solution is treated using typical wastewater treatment methods, and then recycled by pumping it back to the injection wells (USEPA, 1996b; Roote, 1997).

In-situ flushing is applicable for many types of contaminants, both organic and inorganic, that exist in soils and/or groundwater. Table 14.4 shows the parameters and characteristics of the site and the contaminants that will be vital to the success or failure of the in-situ flushing process (Roote, 1997). The hydraulic conductivity of the contaminated media is of primary concern since the contaminants and the flushing solution must interact efficiently. Therefore, regions of high permeability greater than 1.0×10^{-3} cm/s are considered optimal, and regions of low permeability less than 1.0×10^{-5} cm/s are poor candidates for soil flushing (USEPA, 1991; Roote, 1997). Similarly, soils with low carbon contents (less than 1% by weight) are good candidates for soil flushing, whereas soils with high carbon contents (greater than 10% by weight) are generally difficult. Furthermore, if the depth to the contamination is large or the soil profile has lenses or layers of low hydraulic conductivity and/or highly organic soils, in-situ flushing may not be appropriate or economical. It is of paramount importance that the solution is capable of sufficient interaction with the

Table 14.4 **Applicability of in-situ flushing**

Critical Success Factor	Likelihood of Success[a]			Basis	Data Needs
	Less Likely	Marginal	More Likely		
Critical Success Factor Site Related					
Dominant contaminant phase equilibrium partitioning coefficient[b]	Vapor	Liquid	Dissolved	Contaminant preference to partition to extractant is desirable	Equilibrium partitioning coefficient of contaminant between soil and flushing solution
Hydraulic conductivity[b]	Low (<10^{-5} cm/s)	Medium (10^{-5}–10^{-3} cm/s)	High (>10^{-3} cm/s)	Good conductivity allows efficient delivery of flushing fluid	Geologic characterization (hydraulic conductivity ranges)
Soil surface area[b]	High (>1 m²/kg)	Medium (0.1–1 m²/kg)	Small (<0.1 m²/kg)	High surface area increases sorption on soil	Specific surface area of soil
Carbon content	High (>10% wt)	Medium (1–10% wt)	Small (<1% wt)	Flushing typically more effective with lower soil organic content	Soil total organic carbon (TOC)
Soil pH and buffering capacity[b]	NS	NS	NS	May affect flushing additives and construction material choice	Soil pH, buffering capacity
Cation exchange capacity (CEC) and clay content[b]	High (NS)	Medium (NS)	Low (NS)	Increased binding of metals, sorption, and inhibited contaminant removal	Soil CEC, composition, texture
Fractures in rock	Present	—	Absent	Secondary permeability characteristic renders flushing fluid contact more difficult	Geologic characterization
Critical Success Factor Contaminant Related					
Water solubility[b]	Low (<100 mg/L)	Medium (100–1000 mg/L)	High (>1000 mg/L)	Soluble compounds can be removed by flushing	Contaminant solubility
Soil sorption	High (>10,000 L/kg)	Medium (100–10,000 L/kg)	Low (<100 L/kg)	Higher capacity of contaminant to sorb to soil, decreasing flushing efficiency	Soil sorption constant
Vapor pressure	High (>100 mmHg)	Medium (10–100 mmHg)	Low (<10 mm Hg)	Volatile compounds tend to partition to vapor phase	Contaminant vapor pressure at operating temperature
Liquid viscosity	High (>20 cP)	Medium (2–20 cP)	Low (<2 cP)	Fluid flows through the soil more readily at lower viscosity	Fluid viscosity at operating temperature
Liquid density	Low (<1 g/cm³)	Medium (1–2 g/cm³)	High (>2 g/cm³)	Dense insoluble organic fluids can be displaced and collected via flushing	Contaminant density at operating temperature
Octanol/water partition coefficient[b]	NS	NS	10–1000 (dimensionless)	More hydrophilic compounds are amenable to removal by water-based flushing fluids	Octanol/water partition coefficient

Source: Roote (1997).

[a]NS, no action level specified.

[b]Higher-priority factors.

contaminated media, and it must be able to transport or mobilize the contaminants. The chemical properties of the contamination are also critical. As seen in Table 14.4, water solubility, soil sorption, vapor pressure, viscosity, density, and octanol–water partition coefficient are important parameters that should be characterized to evaluate the applicability of in-situ flushing.

In-situ flushing has the following advantages:

- It is an in-situ technology, so it causes less exposure of the contaminants to cleanup personnel and the environment.

- It is usually simple and economical to implement, and operation is also easy compared to other technologies.

- It is applicable for a wide variety of contaminants, both organic and inorganic.

- It is applicable to both saturated and unsaturated zones.

- It may be used with many other remediation technologies.

In-situ flushing has the following disadvantages:

- It may be a slow process when heterogeneities such as soil layers or lenses of less permeable (less than 1.0×10^{-5} cm/s) or organic materials are located within the soil horizon.

- Since the contaminants are solubilized into the solution, they may be transported beyond the extraction well and unintentional spreading of the contamination can occur.

- Remediation times may be long and the effectiveness of the process depends largely on solution, contaminant, soil or groundwater interaction.

- Remediation depends strongly on the ability of the solution to desorb and solubilize the contaminant.

- The process may be costly, with contamination located at large depths or with expensive solutions and long remediation times.

14.3.2 Fundamental Processes

Although in-situ flushing is a relatively simple remediation process to implement and operate, many phys-ical, chemical, and biological processes are involved. Transport mechanisms such as advection, dispersion, and/or diffusion may all be significant in spreading the contamination away from the initial source. The primary mechanism responsible for the remediation flushing process, however, is advection, or movement of the contaminant along with the flow of solution. Chemical properties such as sorption and interfacial tension between the contaminant and soil, as well as the solubility, density, and viscosity of the contaminant, are crucial to soil flushing. Literature reviews and laboratory studies should be performed to determine partition coefficients such as soil–water (K_d) and octanol–water (K_{ow}), and during solution screening these should be optimized to partition the contaminant into the solution. It is also important to assess the sorption of the flushing solution, because in some instances it may exacerbate the situation by increasing contaminant sorption to soil solids. A full and complete understanding of the hydrogeology of the site from a macro- and microscale perspective is crucial. Microscale characteristics of the soil geochemistry, such as mineralogy, cation exchange and buffering capacity, carbon content, pH, redox, electrical conductivity, and point of zero charge, and the chemical properties of the groundwater, such as pH, redox, electrical conductivity, salinity, and carbon and colloidal content, all need to be assessed. Additionally, macroscale parameters such as the stratigraphy, hydraulic conductivity and the tortuosity of the flow paths through pore space, as well as the amount of soil surface area, will all greatly affect the flushing process. Biological processes should also be considered because in-situ flushing can be used to enhance biological degradation, and some solutions may degrade rapidly in subsurface environments.

The primary objective of in-situ flushing is to use a carefully selected solution so that advection will carry or push the contaminants toward the extraction well. Thus, dissolution–solubilization and/or mobilization are key fundamental processes affecting contaminant transport and fate. The solutions chosen for enhancing dissolution–solubilization depend primarily on the characteristics of the contaminants that are present. Since inorganic contaminants and organic contaminants have quite different properties, they need solutions that possess different properties. Organic contaminants such as dense or light non-aqueous phase liquids (NAPLs) often require solutions such as surfactants or cosolvents to lower contaminant–solution interfacial tension (Jafvert, 1996). Conversely, inorganic contaminants need complexing or chelating agents, or acidic solutions, in order to increase dissolution–solubilization. Most important, it must be remembered that for both types of contaminants, the flushing solution must meet regulatory requirements for toxicity and cause minimal or negligible impact to human health and the environment.

To comprehend the basics of how surfactants (i.e., surface-active agents) lower interfacial tension, it is necessary to examine the process at the molecular level. Surfactant molecules, or *monomers,* are composed of two separate groups, or *moieties.* Each monomer has a polar hydrophilic head moiety that often includes an anion or cation, and a hydrophobic tail moiety that usually consists of a long hydrocarbon chain (CH2M HILL, 1997). The water-loving and water-hating nature of the molecule attracts it to interfacial regions where nonpolar and polar-phase liquids exist. At these interfaces the head of the molecule resides in the polar phase and the tail in the nonpolar phase. At a special surfactant concentration called the *critical micelle concentration* (CMC), the monomers begin to accumulate and form aggregates called *micelles.* When a micelle forms in an aqueous solution, all the monomer tails are directed toward the center of the aggregate, which creates a nonpolar region within the polar solution. These nonpolar regions are locations where non-aqueous phase liquid or contaminants may reside, thus increasing contaminant solubility.

Cosolvents (in combination with water) usually consist of alcohols or liquids that are soluble in water and NAPLs. Alcohols such as methanol, ethanol, and propanol are commonly used as cosolvents. When alcohols are used in water, they reduce the polarity of the hydrogen bond and allow greater solubilization of non-aqueous phase liquids. When larger amounts of cosolvent are used, the cosolvent may partition into both the NAPL and water phases and can result in reduction of the NAPL–water interfacial tension to zero, facilitating mobilization of NAPL. Combinations of a cosolvent and a surfactant are also used to increase the efficiency of the flushing to lessen surfactant loss due to sorption and manipulate the viscosity of the surfactant solution. Once the contaminants are dissolved/solubilized and/or are being mobilized by surfactants and cosolvents, the controlling fundamental process is commonly advective transport.

14.3.3 System Design and Implementation

General Design Approach. Figure 14.9 depicts a flowchart that provides a general plan and procedure that may be followed once the site has been characterized and in-situ flushing has been selected as the remediation technology (Roote, 1997). There are basically only four major steps to accomplish. The first one is to conduct a comprehensive laboratory analysis to decide on the safest, most effective and economical flushing solution, as well as its concentration. Batch and soil column testing using actual field soil and contaminants are essential laboratory procedures. The next step is to decide on the physical layout and design of the system, such as the number of wells, injection, extraction, and monitoring, the depths of the wells, and the pumping rates that will be used. Numerical simulations and computer modeling will be helpful in determining many of the factors in the second step. The third major step is then to get regulatory approval for the project and conduct a small field demonstration. Field and laboratory conditions are often quite different, and the field demonstration allows the process to be evaluated under actual conditions so that corrections and adjustments can be performed to optimize system efficiency. Generally, field demonstrations are conducted in small confined regions, so contaminant migration will be prevented if miscalculations were made or problems occur. The final major step is to gain regulatory approval and proceed with full-scale implementation. The completed process should be monitored and evaluated constantly so that corrections and adjustments can be made to optimize system performance.

Equipment. The components used in in-situ flushing remediation consist of injection and extraction equipment, such as drilling equipment, pumps, and pipes, mixing equipment for specialized solutions such as

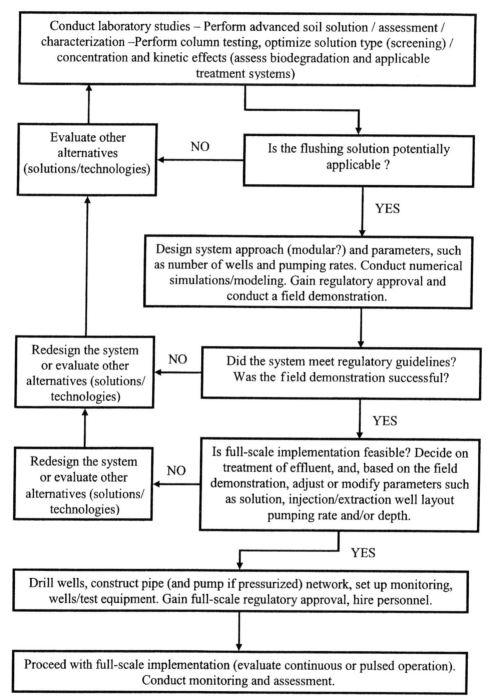

Figure 14.9 General design approach to in-situ flushing. (*From CH2M HILL, 1997.*)

surfactants/cosolvents, and wastewater treatment equipment if it is to be treated on site. Testing analysis instruments are also necessary to monitor contaminant removal continually. The modes of the injection and extraction systems are either by gravity or by pressure. Gravitational systems rely on the hydraulic gradient for movement of the solution through the subsurface. Gravitational collection is attained through the use of wells. Pressurized systems commonly deliver the solution by using pumps combined with vertical wells, and effluent recovery occurs through the use of extraction wells (Roote, 1997). Physical barriers such as sheet piling, or slurry walls may also be used to contain the region of contamination or direct the flow to areas of concern.

Table 14.5 shows different types of flushing agents that may be used depending on the type of contaminants to remediate. The flushing solution selected depends on the properties of the contaminant and the site-specific soils. Other essential considerations concern the environmental conditions, such as temperature or precipitation, and the disposal method or treatment of the effluent. The effluent treatment methods are similar to those used for treating the extracted groundwater in pump-and-treat remediation, as described in Section 14.2.3. Specific effluent treatment methods containing surfactants and cosolvents as well as recovery–reconcentration processes for surfactants are summarized in Table 14.6.

Operational Parameters. The type of solution, mode of operation (gravitational or pressurized), amount and depth of injection–extraction wells, pumping rates, and

Table 14.5 **Types of flushing agents**

Flushing Agents	Contaminants Targeted
Clean water	High-solubility organics; soluble inorganic salts
Surfactants	Low-solubility organics; petroleum products
Water/surfactants	Medium-solubility organics
Cosolvents	Hydrophobic contaminants
Acids	Basic organic contaminants, metals
Bases	Phenolics, metals
Reductants/oxidants	Metals

process duration are all dependent on the site and contaminant characteristics. Table 14.7 shows typical surfactants and cosolvents that are used and some respective properties. The number of injection–extraction wells ranges from one to hundreds (Roote, 1998). Since in-situ flushing is similar to pump and treat, an estimate of the annual amount of groundwater that could be treated was obtained from an overview of 28 different groundwater remediation sites using primarily pump and treat technology. A range of 1.7 to 554 million gallons per year was treated at these groundwater sites (USEPA, 1999b), so it is expected that in-situ flushing could produce similar results.

The remediation process may be accomplished in different phases of flushing whereby the phases apply different solutions and/or pumping rates. For instance, the region may be flushed with plain water first, then with a surfactant, and then a postflush with water is provided to remove any residual surfactant. The delivery of the plain water is often faster at flow rates from 30 to 320 gal/min, depending on the permeability of the aquifer, while surfactants and cosolvents are delivered at slower flow rates such as between 6 and 32 gal/ (CH2M HILL, 1997). Chemicals are usually delivered in a concentrated form, diluted on site, and then stored in large 10,000-gal tanks.

Monitoring. Typically, monitoring wells are used and periodic sampling is done to assess the flow rate, contaminant removal rate, and process efficiency. The type and frequency of monitoring are highly dependent on the site characteristics, such as the type of contaminant or soil present. The method used to determine the contaminant concentrations are based on the type of contaminant. Advancements in instrumentation and water quality analysis have made it possible to detect contaminant levels to the parts-per-trillion level (Fetter, 1999).

14.3.4 Predictive Modeling

Basically, any numerical simulation that can model the mass transfer of a contaminant can be applied to in-situ flushing. The most comprehensive numerical model appears to be UTCHEM, developed by Brown

Table 14.6 **Effluent treatment and surfactant recovery and reconcentration methods**

(a) Decontamination Processes for Surfactant Solutions

Contaminant Removal Process	Separation Process	Potential Contaminants	Main Advantages	Primary Concerns
Air stripping	Effluent contacts air stream, contaminants partition to airstream	Volatile	Low cost, effective, commercially available	May require anti-foaming compound; addition of water treatment chemicals
Liquid/liquid extraction	Effluent contacts liquid extractant, contaminant partitions to extractant	Volatile, semivolatile, inorganics	Wide applicability, no foaming, minimal fouling	Surfactant may partition to extractant, and viceversa
Pervaporation	Consists of nonporous hydrophobic membrane with gas purge/vacuum on the other side; contamintants partition into membrane and evaporate into purge vacuum	Volatile	No foaming, contaminants collected in condensed form, can recover alcohols from surfactants	Precipitates may cause plugging, membrane leaks may lead to foaming
Precipitation	Properties of effluent altered to achieve precipitation of surfactant	Volatile, semivolatile, inorganics	Generally inexpensive, wide range of contaminants, easy reuse of separate phase surfactant	Contaminants may partition into surfactant, surfactant may not flocculate well

(b) Recovery/Reconcentration Processes for Surfactants

Surfactant Recovery/ Reconcentration Process	Separation Process	Main Advantages	Primary Concerns
Micellar enhanced ultrafiltration (MEUF)	Surfactant passes through membrane, micelles are retained while water, monomers, salts/alcohols pass through	Low cost, commercially available, efficiently recovers surfactant micelles	Not effective if particular surfactant's CMC is high or influent surfactant concentration is low
Nanofiltration (NF)	Surfactant passes through nanofiltration membrane, monomer and micelles retained	Recovers monomers and micelles, commercially available	Higher pressure required (vs. MEUF), more susceptible to fouling
Foam fractionation	Foam is generated by sparging air through surfactant solution, then separated and resulting water is allowed to coalesce, creating high concentration surfactant solution	Low cost, efficient recovery of surfactant monomers	If influent surfactant concentration is high (in term of number of CMCs), many fractionation stages are necessary

(c) Decontamination Processes for Cosolvent Solutions

Contaminant Removal Process	Separation Process	Potential Contaminants	Main Advantages	Primary Concerns
Distillation	Cosolvent solution exposed to sequences of vapor liquid equilibrium stages with temperature profile in packed or tray-type column	Volatile, semivolatile, inorganics	Commercially available with existing design equations, may achieve both contaminant and water removal, cost-effective	Energy intensive, formation of water alcohol azeotropes; fate of contaminant in distillation column (may stay with cosolvent)
Liquid/liquid extraction	As with surfactants effluent contacts liquid extractant, contaminants partition to extractant	Volatile, semivolatile, inorganics	Applicable for all types of contaminants, great deal of research in this area	Solvent may partition to extractant, extractant may dissolve in cosolvent
Pervaporation	Same as used in surfactant system; contaminants partition into membrane, evaporate into purge/vacuum	Volatile	Contaminants collected in condensed form	Requires larger membrane area due to reduced activity of contaminant in cosolvent solution

Source: Strbak (2000).

531

Table 14.7 **Properties of surfactants and cosolvents**

Surfactant	CMC (mg/L)
Witconol 2722	13
Triton X-100	130
Triton X-114	110
Triton X-405	620
Brij 35	74
Sodium docecyl sulfate	2100
Synperonic NP4	23.7
Marlophen 86	32.5
Synperonic NP9	48.9
Marlophen 810	55.4
1:1 Blend Rexophos 25/97, Witconol NP-100	2000

Cosolvent	Density (g/cm³)	Viscosity (cP)	Aqueous Solubility at 25°C
Methanol	0.791	0.597	Miscible
Ethanol	0.789	1.2	Miscible
1-Propanol	0.804	2.25	Miscible
2-Propanol	0.785	2.5	Miscible

Source: CH2M HILL (1997).

et al. (1994). Table 14.8 shows some additional models that have been used. UTCHEM is a three-dimensional finite difference program that can simulate a multiphase, multicomponent system and model fluid flow and mass transport. Up to 18 transport components, such as surfactant, contaminant, and water can be modeled, and the system may include up to four phases (Freeze et al., 1995). UTCHEM uses equilibrium phase behavior and composition to compute the viscosity, density, and interfacial tensions of the particular phase. The model also accounts for surfactant adsorption, capillary pressure, and relative permeability (CH2M HILL, 1997).

Another recent simulation model developed specifically for cosolvent flushing was developed by Augustijn et al. (1994). It is a one-dimensional transport model (analytical) that describes contaminant removal when solvents are used. The model incorporates relationships between hydrophobic organic contaminants and cosolvents and the sorption, solubility, and nonequilibrium effects that occur. The cosolvent was assumed to be nonsorbed and the contaminants are assumed to be in homogeneous media under one-

dimensional flow conditions. The model is presented to examine cosolvent effects, and heterogeneous flow is not considered. The initial conditions between the solution and the sorbed contaminant are assumed to be at equilibrium, and the contaminant is assumed to be distributed equally throughout the media.

14.3.5 Modified or Complementary Technologies

In-situ flushing can also be applied to treat vadose zone soil as discussed in Chapter 13. For this, the flushing solution is sprayed on the ground surface, which then infiltrates the contaminated region by gravitational force, as shown in Figure 14.10. The in-situ flushing process itself is basically an enhancement of the pump-and-treat technique with an emphasis on using the mechanism of flushing to accomplish contaminant removal. The in-situ flushing process is also commonly modified or enhanced by using other innovative remediation techniques. Bioremediation may complement the in-situ flushing process because nutrients can be injected into the flushing solution and then pumped into the ground to nourish and increase the population of existing microorganisms. Soil fracturing may also be used with in-situ flushing in fine-grained soils to increase solution, soil and contaminant interactions. Recently electrokinetics and soil washing have been combined to treat heterogeneous soils or low permeable soils contaminated with a wide range of contaminants (Reddy and Saichek, 2002).

Remediation by *in-situ steam injection* is also based on the principles of soil flushing. In this, steam is injected into the subsurface through injection wells to stimulate volatilization of the contaminants (Figure 14.11). The high temperature of the steam alters the phase of the contaminants into recoverable forms (a vapor phase, a separate liquid phase, and a dissolved phase). Steam also physically displaces the various phases of the contaminants and aids in recovery through extraction wells for further treatment aboveground.

In-situ chemical oxidation/reduction or *in-situ chemical treatment* may be considered as modified in-situ flushing technology. In this, treatment is accom-

Table 14.8 **Mathematical models used in in-situ flushing system design**

Reference	Type of Model	Purpose	Assumptions	Limitations
Freeze, et al. (1995)	Numerical	Multiphase, multi-component, three-dimensional modeling applicable for heterogeneous soil layers	Phase equilibrium, but authors state that a nonequilibrium approach is available	Relies on calibration and accurate parameters
Augustijn et al. (1994)	Analytical	Simulate cosolvent enhanced remediation using one-dimensional nonequilibrium	Homogeneous media, initial conditions are at equilibrium, and contaminant is equally distributed	Only one-dimensional and for homogeneous media; relies on calibration and accurate parameters
Abriloa et al. (1993)	Numerical	Simulates surfactant-enhanced solubilization of NAPL with nonequilibrium	Stationary residual NAPL phase	Only one-dimensional; relies on calibration and accurate parameters
Mason and Kueper (1996)	Numerical	Simulates surfactant-enhanced solubilization of pooled NAPL with nonequilibrium	Stationary residual NAPL phase	Only one-dimensional; relies on calibration and accurate parameters
Reitsma and Kueper (1996)	Numerical	Simulates alcohol flooding to remove DNAPL from below water table	Stationary residual NAPL phase	Only one-dimensional; relies on calibration and accurate parameters
Gao et al. (1996)	Numerical	Three-dimensional simulation of the alkaline–surfactant–polymer process	Appropriate constitutive relationships are necessary	Relies on calibration and accurate parameters

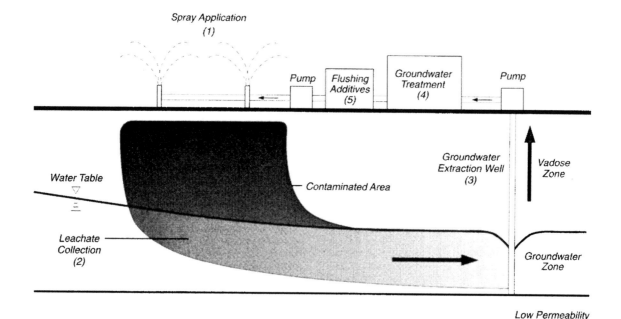

Figure 14.10 *Schematic of in-situ flushing in vadose zone.* (*From USEPA, 1996b.*)

plished by reaction, such as oxidation or reduction. It involves the injection of reagents to stimulate chemical reactions to detoxify the contaminants in the groundwater (Figure 14.12). For example, hydrogen peroxide (H_2O_2) is injected with ferrous sulfate (Fenton's reagent) to produce hydroxyl radicals. The hydroxyl radicals then oxidize the organic compounds into carbon dioxide, water, and chloride ions (in the case of chlorinated hydrocarbons). Potassium permanganate ($KMnO_4$) or sodium permanganate ($NaMnO_4$) by itself or in combination with hydrogen peroxide may be used to remediate VOCs, SVOCs, and PCBs. Oxidation may also be achieved by introducing oxygen releasing compounds (ORCs). Other in-situ oxidants include ozone and chlorine, which are injected into groundwater as a gas. Chemical reduction may also be used to remediate groundwater contamination. For example, Cr(VI) may be reduced to Cr(III) by using reducing agents such as Fe(II), Fe(0), and calcium polysulfide.

In-situ immobilization or *fixation* is similar to chemical treatment and includes injection of chemicals or reagents which interact with and transform metal or radionuclide contaminants into an immobile (precipitate) form (Figure 14.13).

14.3.6   Economic and Regulatory Considerations

The cost of in-situ flushing has been estimated roughly at about $75 to $210 per cubic yard of contaminated soil, including operation and maintenance costs (Roote, 1997). However, as with almost any remediation technique, the costs may increase due to site-specific factors. Factors that exacerbate costs for in-situ flushing include the type of contaminants and contaminant concentrations that are present and the existence of heterogeneities such as low-permeability and/or organically rich soils. Certain compounds, such as hydrophobic organics, may require expensive surfactant/cosolvents, more flushing or higher concentrations, and low-permeability and/or organically rich regions may require additional flushing because of the reduced solution, soil, and contaminant interaction.

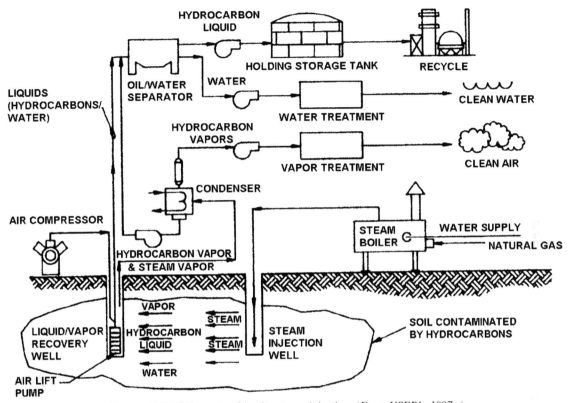

Figure 14.11 *Schematic of in-situ steam injection. (From USEPA, 1997c.)*

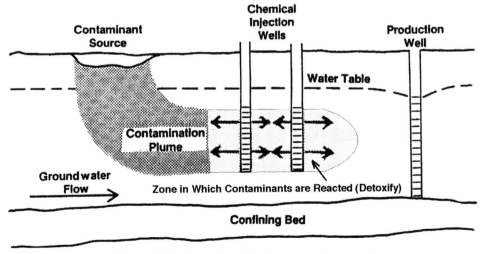

Figure 14.12 *Schematic of in-situ chemical treatment.*

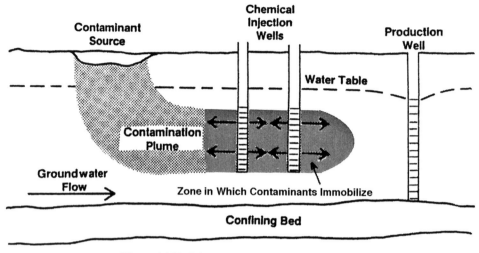

Figure 14.13 Schematic of in-situ immobilization.

Primary concerns for regulatory approval of the in-situ flushing process include the toxicity of the surfactant or cosolvent, any deleterious effects the flushing solution could produce, such as transfer or spread of the contamination to the soil or groundwater, and adequacy of flow control and monitoring. Basically, the regulators need to make sure that the flushing process will not exacerbate the problem (USEPA, 1995a). Even before pilot testing is performed, contact and interaction with regulators should be established. Corrective action proposals may take months or more to review and gain authorization. A major part of the review is to determine all the constituents of the flushing solution and make sure that they will not be hazardous to human health or the environment. Furthermore, although the solution may be safe, it is also necessary to demonstrate that a substantial portion of it will be recovered, and any residual solution, with or without contamination, must not be hazardous or affect the environment. The expected effectiveness of the flushing solution and the amount of time to remediate also need to be reported. It is essential to prove that the goal will be accomplished, whether it is to use the surfactant or cosolvent to reduce interfacial tension and desorb/solubilize or mobilize the contaminant, decrease possible migration using the solution to increase contaminant sorption, or use the flushing solution to promote bio-

remediation. Finally, a plan for the implementation of the monitoring program must be presented. The plan should also consider that the flushing process is contained and surrounding aquifers are not affected.

14.3.7 Case Studies

In-situ flushing is a developing technology that has had limited full-scale applications. An increasing number of investigative studies have been performed to evaluate in-situ flushing as a remediation alternative. These studies include laboratory, pilot, and field demonstrations, and full-scale or commercial projects (Roote, 1998). To gain a better comprehension of the in-situ flushing process, a discussion of two completed field demonstration projects are presented. Table 14.9 shows some important data concerning the two sites.

Case Study 1: Canadian Forces Base (CFB), Borden near Alliston, Ontario, Canada. The demonstration site consisted of a $3 \times 3 \times 3$-m^3 cell with a pore volume of about 9100 L. Sheet piling was installed and driven into an underlying glacial till clay layer to isolate the area and prevent contaminant migration. The region was then contaminated by the controlled release of approximately 231 L of tetrachloro-

Table 14.9 **Field demonstration project information**

Parameter	Depth Below Ground Surface	Initial Value	Final Value	Pore Volumes	Comments
Case Study 1: CFB Borden					
PCE residual saturation[a]	0–1 m	10%	<1%	14.4	Sand, no visible difference at 2-cm scale
PCE pool height	2.5–3 m	50 cm	2–3 cm	14.4	Perched on clay aquitard
PCE pool saturation		20%	3%	14.4	
Case Study 2: Christi, Texas					
CTET concentration	10–12 ft	>2000[b] mg/L	>2000 mg/L	3	Clay
			>2000 mg/L	12.5	
CTET concentration	12–14 ft	77–2956 mg/L	>2000 mg/L	3	Clayey sand (10–30% clay)
			>10 mg/L	12.5	
CTET concentration	16–24 ft	574–2674 mg/L	<10 mg/L	3	Sand with 1–5% clay
			<10 mg/L	12.5	

[a] Saturation is the ratio of the volume of PCE in the sample to the pore volume of the sample.
[b] Maximum concentration.

ethylene (PCE), and the PCE was left to equilibrate for about two months (Fountain, 1992). The groundwater was fresh and the aquifer soil was unconsolidated, well-sorted, clean beach sand or gravel with less than 1% clay and less than one-tenth of 1% of organic carbon (Roote, 1998). The hydraulic conductivity varied within the cell, but it was generally greater than 10^{-3} cm/s. Originally the cell was 4 m thick, but the top 1 m of sand was removed to study PCE migration. Bentonite was placed over the remaining 3 m, and the top 1-m layer was replaced with clean sand. Five injection wells and five extraction wells (5 cm of PVC), parallel and opposite each other on each side of the cell, were drilled into different depths of the aquifer. The wells were screened throughout the contaminated zone. Multilevel monitoring wells were also installed. A blend of 1% nonylphenol ethoxylate (Witconol NP-100) and 1% phosphate ester of nonylphenol ethoxylate (Rexophos 25-97) surfactant solution was injected, and the cell was kept saturated with surfactant during the demonstration.

The initial solubility of PCE in water was about 150 ppm, and the surfactant produced a PCE solubility of about 12,000 ppm in the laboratory. Field sampling points yielded values near 11,500 ppm and correlated well with laboratory values. Water flushing was used first and a hydraulic gradient of 0.094 was maintained. Air stripping was used to remove the PCE from the effluent. Periodically during the water flushing, smaller-diameter stainless steel tubing was inserted to the bottom of the extraction wells and free-phase PCE was pumped out. The free phase recovered during the initial water flushing amounted to about 60 L, and the PCE removed during the excavation of the upper layer of sand was estimated at 52 L. When the volume of PCE recovery from the water flushing decreased to insignificant amounts, the process switched to using surfactant. After approximately 16 pore volumes of surfactant, 79 L of PCE was removed. Pumping rates varied from about 2000 to 400 L/day. The test duration was approximately 15 months.

PCE concentrations plummeted from an initial 4000 ppm after adding surfactant to 200 ppm after flushing with 10 pore volumes (Fountain et al., 1995). Within the top meter of the aquifer, PCE decreased from 10% to 1% after 14 pore volumes, and 2.5 to 3.0 m below the ground, the PCE decreased from 20% to 3% after 14 pore volumes. Overall, nearly 80% (139/179) of the PCE was removed and the residual 20% was determined to be located in low-hydraulic-conductivity zones or was volatilized. The data from the monitoring wells indicated that the perched lenses of PCE did not mobilize vertically, and a conservative tracer indicated that sorption of the surfactant was very low. After 20 pore volumes, PCE concentrations were still high in locations adjacent to the aquitard at the bottom of the aquifer. This is where the sand was more infiltrated with silt or clay particles and indicates the effects of heterogeneity and tailing due to low-hydraulic-conductivity layers. No estimates were supplied for full-scale costs because this project was strictly for research/demonstration purposes.

Case Study 2: Dupont Chlorocarbon Manufacturing Plant near Corpus Christi, Texas.

The contaminant at this site was a preexisting plume of carbon tetrachloride (CTET), and it was contained in an area approximately $25 \times 35 \times 12$ ft^3. The zone of contamination was within a 12-ft-high clayey-sand lens that had variable smectite clay content (1 to 15%) and low (0.025 to 0.031%) organic carbon content. The sand and gravel was unconsolidated, well-sorted, and interbedded with clay. The location of the clayey–sand lens was 10 to 24 ft below the ground surface within a regional clay unit. The contaminated lens itself was divided into different soil regions, as shown in Table 14.9. At the 24-ft depth, a clay aquitard existed, so vertical migration would be limited. The groundwater was highly saline (12,000 ppm total dissolved solids), and the hydraulic conductivity of the clayey sand was greater than about 10^{-3} cm/s. The initial concentration of CTET was greater than 1000 ppm both in soil cores and in the water samples.

The project employed six injection wells, one central extraction well, and several monitoring wells. In the first phase of the experiment, a 1% low-toxicity biodegradable surfactant solution (Witco 2722, Witco Corp.) was injected, extracted, air stripped, and then recycled back to the injection wells. In the second phase, a new extraction well was installed because the first one became unusable due to sanding (Roote, 1998). During the second phase, excessive sorption of the surfactant occurred and the surface tanks and delivery wells experienced biofouling. Due to the trouble encountered in the second phase, a different surfactant

(Tergitol 15-S012, Union Carbide Corp.) was used in phase three, and in phase four, a smaller area was tested because of reduced flow rates resulting from a lowering of the regional groundwater table. No estimates were available for the initial mass of CTET, but initial concentrations fell from over 1000 ppm to 790 ppm during the first phase, and then reached levels of 219 ppm by the fourth phase after the injection of 12.5 pore volumes. Final concentrations within the lens are shown in Table 14.9. There were no estimates of project cost, but it was concluded that progress was greatly improved over pump and treat technology. Data from the monitoring wells indicated that zones of different hydraulic conductivity lead to decreasing steps in concentration as the contaminant is cleaned from higher- to lower-hydraulic-conductivity regions.

14.4 PERMEABLE REACTIVE BARRIERS

14.4.1 Technology Description

The permeable reactive barrier (PRB), also called a *treatment wall, passive treatment wall,* or *permeable barrier,* is a treatment technology designed to degrade or immobilize contaminants contained in groundwater as it flows through the barrier (Figure 14.14). The barrier is put in place by constructing a trench across the flow path of contaminated groundwater and filling it with a reactive medium. As the contaminated groundwater passes through the PRB, the contaminants are either immobilized or transformed into nontoxic substances (USEPA, 1995c; 1996c; 1998c; Vidic and Pohland, 1996; USACE, 1997b). Therefore, a PRB is not a barrier to the water, but it is a barrier to the contaminants.

Typical reactant media contained in the barriers include reactant media designed for degrading volatile organics, chelators for immobilizing metals, or nutrients and oxygen to facilitate bioremediation (Bowman, 1999). The media are often mixed with a porous material such as sand to enhance groundwater flow through the barrier. A permeable barrier may be installed as a continuous reactive barrier or as a funnel-and-gate system (Figure 14.15). A continuous reactive barrier consists of a reactive cell containing the per-

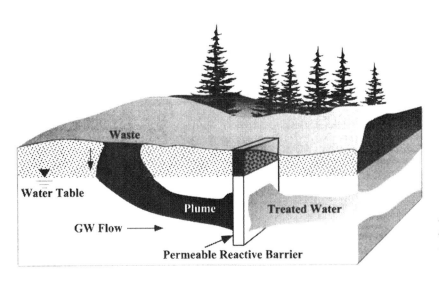

Figure 14.14 *Permeable reactive barrier for groundwater treatment. (From USEPA, 1998c.)*

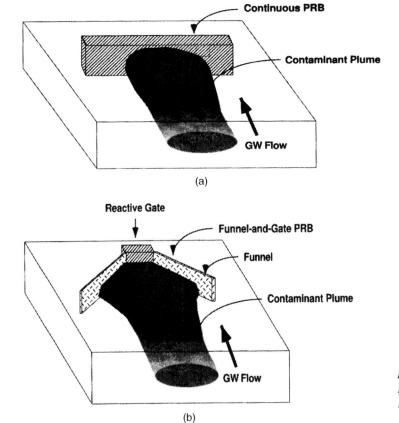

Figure 14.15 *Basic configurations of PRBs: (a) Continuous PRB and (b) Funnel-and-Gate PRB. (From USEPA, 1998c.)*

meable reactive medium. A funnel-and-gate system has an impermeable section, called the *funnel*, that directs the captured groundwater flow toward the permeable section, called the *gate*. The funnel walls may be aligned in a straight line with the gate, or other geometric arrangements of funnel-and-gate systems can be used, depending on the site conditions (Figure 14.16). This funnel-and-gate configuration allows better control over reactive cell placement and plume capture. At sites where the groundwater flow is very heterogeneous, a funnel-and-gate system can allow the reactive

cell to be placed in the more permeable portions of the aquifer. At sites where the contaminant distribution is very nonuniform, a funnel-and-gate system can better homogenize the concentrations of contaminants entering the reactive cell. A system with multiple gates can also be used to ensure sufficient residence times at sites with a relatively wide plume and high groundwater velocity (Figure 14.17). Figure 14.17(*a*) shows an example of a funnel-and-gate system with two gates emplaced with caissons, while Figure 14.17(*b*) shows an example of a funnel-and-gate system with two reactive

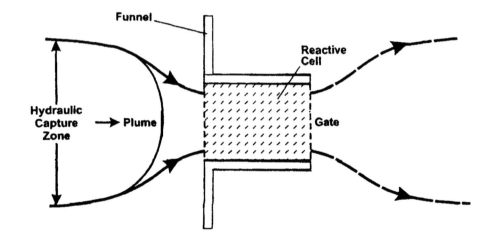

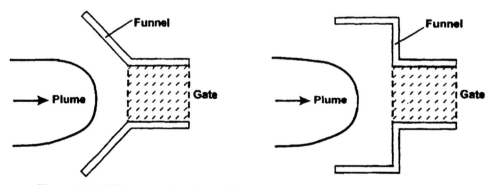

Figure 14.16 *Different configurations of funnel-and-gate system. (From USACE, 1997b.)*

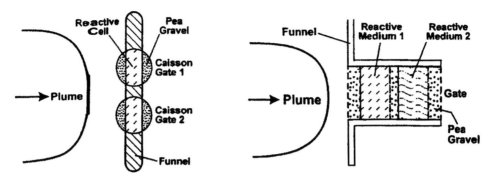

(a) Multiple Caisson Gates   (b) Serial Reactive Medium

Figure 14.17 PRB configurations with multiple reactive cells. (*From USACE, 1997.*)

media emplaced in series within the gate. PRBs are installed as permanent, semipermanent, or replaceable units across the flow path of a contaminant plume.

PRBs can treat a variety of groundwater contaminants, including organic and inorganic contaminants (Table 14.10). Reactive media must be tailored for the specific contaminants present in the groundwater to be treated. Examples of reactive media and their application are also listed in Table 14.10. Organic contaminants such as DCE, TCE, PCE, BTEX, nitrobenzene, DCA, TCA, PCBs, and PAHs can be treated using barriers that either degrade or sorb the contaminants. Re-

Table 14.10 **Contaminants treated by different reactive media in PRBs**

Contaminant	Barrier Type	Reactive Media
Organics	Degradation	Zero-valent iron
DCE, TCE, PCE		Iron(II) porphyrins
BTEX		Resting-state microorganisms
Nitrobenzene		Oxygen-releasing compound
DCA, TCA		Dithionite
PCBs, PAHs	Sorption	Zeolite
		Surfactant-modified silicates
		Organobentonites
		Activated carbon
Inorganics	Sorption	Peat
Heavy metals (Ni, Pb, Cd, Cr, V, Hg)		Ferric oxyhydroxide
Radioactive isotopes (U, Ra, Sr, Cs, Tc)		Bentonite
Nitrate		Zeolites and odified zeolites
		Chistosan beads
	Precipitation	Hydroxyapatite
		Zero-valent iron
		Dithionite
	Degradation	Lime or limestone
		Sawdust

Source: Vidic and Pohland (1996).

active media for degradation treatment of organics includes zero-valent iron, iron(II) porphyrin, resting-state microorganisms, and dithionite. Media used in sorption includes zeolite, organobentonites, and activated carbon. Barriers that sorb, precipitate, or degrade the contaminants can treat inorganics, including heavy metals (e.g., Cr, Ni, Pb, Cd, Zn, and As), radioactive isotopes (e.g., U, Tc, Se, and Co), nitrates, sulfates, and phosphates. Reactive media for sorption of inorganics includes peat, bentonite, and zeolites. Media used for precipitation includes zero-valent iron, dithionite, and limestone. Sawdust can be used as a degradation medium for nitrates. An important characteristic of inorganic contaminants is that these contaminants undergo redox reactions in the presence of reactive media and can form solid precipitates with common groundwater constituents, such as carbonate, sulfide, and hydroxide. The inorganic chemicals that do not participate in redox reactions should have high affinity to sorb onto the reactive medium. The ideal site for a treatment wall is one having permeable soil, contamination no deeper than about 50 ft below ground, and relatively high groundwater flow (Vidic and Pohland, 1996).

The main advantages of this system are:

- No pumping or aboveground treatment is required; the barrier acts passively after installation.
- There are no aboveground installed structures, so the affected property can possibly be put to productive use while it is being cleaned up.
- PRBs can be modified to treat several different types of contaminants.
- The reactive medium is often used up very slowly and has the potential to treat contaminated plumes passively for several years or decades.
- There are very low operating costs other than site monitoring.
- There are no disposal costs or requirements for successfully treated wastes.

There are several factors that may limit the applicability and effectiveness of the PRB; these include:

- Lengthy treatment time relative to other active remediation methods (e.g., in-situ flushing, air sparging)

- Potential for losing reactivity of the media, requiring replacement of the reactive medium
- Potential for decrease in reactive media permeability due to biological clogging and/or chemical precipitation
- Potential for plume bypassing the PRB as a result of seasonal changes in flow regime
- Currently limited to shallow depths
- Longevity of PRB performance is unknown

14.4.2 Fundamental Processes

The fundamental processes involved in a PRB depend on the reactive media used. In addition to degradation or immobilization of the groundwater contaminants, other factors are considered in the selection of reactive media (USACE, 1997b). These factors include: (1) The reactive media should not cause adverse reactions and produce toxic by-products; this requires a thorough understanding of the properties of reactive media as well as the reactive media–contaminant interactions; (2) for longevity, the reactive media should not be readily soluble or depleted in reactivity; (3) the reactive agent should easily be available at low to moderate cost; (4) the reactive media should not alter the hydraulic conductivity of the barrier during the course of the treatment; and (5) it should be safe to personnel handling it during emplacement.

The selection of reactive media depends mainly on the type of contaminant to be treated (Table 14.10). Treatment by reactive media is achieved through sorption, precipitation, and degradation processes. A *sorption* process causes the removal of contaminants from the groundwater by physically sorbing the contaminants onto the surfaces of reactive media. Examples of these adsorbents are zeolites and activated carbon. *Precipitation* reactions cause the groundwater contaminants to form a precipitate. The insoluble products fall out within the barrier. Precipitation reactions are often used to treat groundwater contaminated with metals. For example, limestone is used to neutralize the acid from batteries, causing the lead to precipitate out. Chromium(VI) can be treated in the same way by converting it into Cr(III) precipitates. A *degradation* process involves reactive media such as iron fillings to break down contaminants into harmless products. Degradation may also be accomplished if reactive media contain a mixture of nutrients and oxygen sources to stimulate the activity of microorganisms in the groundwater to reduce the contaminants to nontoxic products.

Much of the research conducted on PRBs deals with the use of zero-valent metals, particularly metallic or granular iron, F(0), as reactive media for the treatment of chlorinated organic compounds such as TCE and PCE in groundwater (USEPA, 1999a). Zero-valent iron is favored because scrap iron is relatively inexpensive (about $300 to 400 per/ton) and easy to obtain in large quantities. Commercially available iron fillings with grain sizes ranging from 0.25 to 2 mm, a bulk density of 2.6 g/cm^3, a specific surface area of 1 m^2/g, and a hydraulic conductivity of 0.05 cm/s are often used as reactive media. As the zero-valent iron in the reactive cell corrodes, the resulting electron activity is believed to reduce the chlorinated compounds to potentially nontoxic products. The iron granules are dissolved in this process, but the metal is used up very slowly. If some oxygen is present in the groundwater as it enters the reactive iron cell, the iron is oxidized and hydroxyl ions are generated. This reaction proceeds quickly, as evidenced by the fact that both the dissolved oxygen and the redox potential drop quickly as the groundwater enters the iron cell. The importance of this reaction is that oxygen can quickly corrode the first few inches of iron in the reactive cell. Under highly oxygenated conditions, the iron may precipitate out as ferric oxyhydroxide (FeOOH) or ferric hydroxide [Fe(OH)$_3$], in which case the permeability could potentially become considerably lower in the first few inches of the reactive cell at the influent end (Vidic and Pohland, 1996). Therefore, the aerobicity of the groundwater can potentially be detrimental to the technology. However, contaminated groundwater at many sites is not highly oxygenated. Also, engineering controls can possibly be used to reduce or eliminate oxygenation from the groundwater before it enters the reactive cell.

Once oxygen has been depleted, the reducing conditions created lead to a host of other reactions (USEPA, 1999a). Chlorinated organic compounds such as TCE are in an oxidized state because of the presence of chlorine. Iron, a strong reducing agent, reacts with the chlorinated organic compounds through electron transfers, in which ethene and chloride are the primary products. In one study it was found that ethene and ethane (in the ratio 2:1) constitute over 80% of the original equivalent TCE mass (Vidic and Pohland, 1996). Partially dechlorinated by-products, such as *cis*-1,2-dichloroethene (*c*-DCE), *trans*-1,2-dichloroethene (*t*-DCE), 1,1-dichlorothene, and vinyl chloride (VC) of the degradation reaction were found to constitute only 3% of the original TCE mass. Additional by-products included hydrocarbons (C$_1$ to C$_4$) such as methane, propene, propane, 1-butene, and butane.

Reduction and precipitation of inorganic anions such as Cr(VI), Se(VI), As(III), As(V), and Tc(VII) can also be accomplished by using zero-valent iron. The mechanisms of Cr(VI) reduction have been investigated extensively. The overall reactions for the reduction of Cr(VI) by Fe(0) and the subsequent precipitation of Cr(III) and Fe(III) oxyhydroxides are as follows:

$$CrO_4^{2-} + Fe^0 + 8H^+ = Fe^{3+} + Cr^{3+} + 4H_2O$$

$$(1 - x)Fe^{3+} + xCr^{3+} + 2H_2O = Fe_{1-x}Cr_xOOH(s) + 3H^+$$

USEPA (1999a) describes other processes and mechanisms to treat contaminants in PRBs. These include biologically mediated precipitation of anions; adsorption and precipitation of inorganic anions by using selected reactive media; reduction of inorganic cations, which include metals and radionuclides such as Cd, Co, Cu, Mn, Ni, Pb, Zn, and U(VI), into precipitates using elemental iron; and biologically mediated reduction and precipitation of cations. In addition to zero-valent iron, reactive media used for the precipitation of inorganics include sodium dithionite (NaS$_2$O$_4$), polysulfide or bisulfide compounds, and limestone (CaCO$_3$), as well as organic materials (e.g., leaves and wood chips) that contain microbes capable of reducing metals into precipitates. Surface-modified zeoloites are also used to sorb inorganic contaminants, thus making them immobile (Ott, 2000).

14.4.3 System Design and Implementation

General Design Approach. The general design approach of PRB consists of the following steps: (1) site characterization, (2) reactive media selection, (3) treatability testing, (4) PRB design using computer modeling, (5) emplacement of PRB, and (6) performance monitoring. A preliminary assessment, as detailed in Figure 14.18, should be made to determine the suitability of a site for PRB application. If the preliminary assessment shows that the site is suitable, a detailed site characterization should be conducted to determine aquifer characteristics and groundwater characteristics. The aquifer characteristics include groundwater depth, depth to aquitard, aquifer thickness and continuity, groundwater velocity, lateral and vertical gradients, site stratigraphy/heterogeneities, hydraulic conductivities of different layers, and porosity. Groundwater characteristics include the dimensions of the plume and composition of groundwater within the plume, including organic and inorganic contaminant constituents. Other parameters, such as pH, redox potential, and dissolved oxygen of the groundwater, are also determined.

Once the site characterization has been completed, potential reactive media are assessed. The reactive media should be selected based on the following considerations: reactivity, hydraulic performance, stability, environmentally compatible byproducts, and availability and price. The reactivity of the media should be such that the contaminant is degraded or immobilized within an acceptable residence time. The higher the reaction rate for the degradation or immobilization of the contaminant by the reactive media, the better the media. Hydraulic performance indicates the ease with which contaminated groundwater can flow through reactive media. Usually, higher reaction rates are associated with media consisting of smaller particle size and higher total surface area, whereas higher hydraulic conductivity is associated with larger particle size.

Therefore, reactive media must be fairly stable, meaning that they should retain their reactivity and hydraulic conductivity over time. The by-products generated by the degradation or immobilization reactions should not be environmentally active or toxic. Any reactive media selected should also be readily available in large quantities at a reasonable price.

Once potential reactive media have been selected, a treatability study is conducted to generate contaminant- and site-specific design data. This study includes conducting batch tests and column tests. Details on testing procedures are given in USACE (1997b). Batch tests are conducted to screen the reactive media. The column tests are conducted using the reactive medium selected and groundwater obtained from the site and under representative groundwater flow velocities. In this test, the column is filled with reactive media and groundwater obtained from the site is supplied to the influent end of the column at a constant velocity using a pump. The schematic of a typical column apparatus is shown in Figure 14.19. Contaminant concentrations are measured periodically at the inlet, outlet, and sampling ports along the column. For each test column at each velocity, contaminant concentrations are plotted as a function of distance along the column. The flow rate is used to calculate the residence time at each sampling position relative to the influent for each profile. The concentration change in the influent is expressed by a first-order decay equation:

$$C_t = C_0 e^{-kt} \qquad (14.2)$$

where C_t is the contaminant concentration in solution at time t, C_0 the initial contaminant concentration of the influent solution, k the first-order rate constant, and t the time. Equation (14.2) can be rewritten as

$$\ln \frac{C_t}{C_0} = -kt \qquad (14.3)$$

The slope of $\log(C_t/C_0)$ versus t gives the k value. The time at which the initial concentration declines by one-half ($C_t/C_0 = 0.5$) is the half-life ($t_{1/2}$). $t_{1/2}$ and k are related by

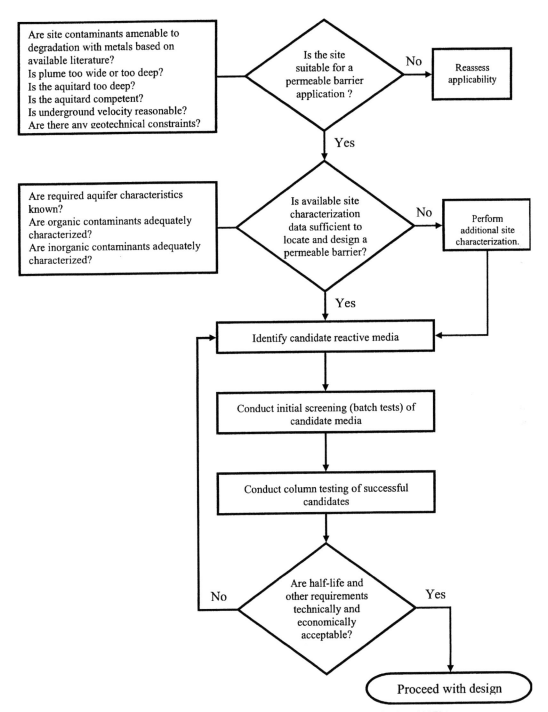

Figure 14.18 *General approach to PRB design. (From USACE, 1997b.)*

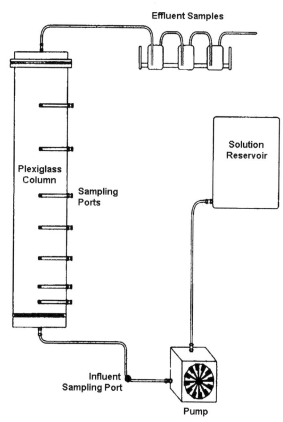

Figure 14.19 *Schematic of column setup used in bench-scale tests. (From USEPA, 1998c.)*

$$t_{1/2} = \frac{-0.693}{k} \qquad (14.4)$$

Half-lives of contaminants with the use of elemental iron reactive media are shown in Table 14.11. Knowing the half-life of a contaminant, k can be calculated using

$$k = \frac{-0.693}{t_{1/2}} \qquad (14.5)$$

The residence time (t_{res}) required for PRB is then calculated using

$$t_{res} = \frac{1}{k} \ln \frac{C_i}{C_e} \qquad (14.6)$$

where C_i is the influent concentration and C_e is the desired outflow concentration. However, this estimate does not account for the residence time required to degrade by-products. For example, the contaminant may be TCE, but the degradation of TCE results in by-products such as 1,2-DCE and VC. The design residence time should be the longest time required to bring all of the by-products to its treatment goals, as depicted in Figure 14.20. If the site conditions are too complex and cannot be simulated in column tests, a field pilot test may be desirable. The geometry of the barrier can then be determined based on the maximum residence required. The width of the barrier (W) is calculated using

$$W = vt_{res} \qquad (14.7)$$

where v is the expected groundwater flow velocity through the PRB.

The site characterization and treatability data are then used to design the PRB. Generally, groundwater flow models (such as MODFLOW) and geochemical models (such as MINTEQA2, PHREEQC, and EQ3/EQ6) are utilized to optimize the location of the barrier, barrier configuration (continuous reactive barrier or funnel-and-gate system), and barrier dimensions based on the predicted long-term performance of the PRB. This includes the estimation of hydraulic capture zones as well as potential for flow changes resulting from precipitate formation. This modeling also helps in designing an appropriate monitoring program. Simple calculations can be made to estimate the weight of reactive material needed per unit cross section of the plume on the basis of laboratory kinetic data and basic knowledge of the plume and the remediation goals. Such calculations involving chlorinated organics using metallic iron as reactive media may be found in USEPA (1999a).

The next task in the design process is the selection of an emplacement technique. Various emplacement

Table 14.11 **Half-lives of organic compounds normalized to 1 m² of iron surface per milliliter of solution**

Organic Compounds	Pure Iron $t_{1/2}$ (hr)	Commercial Iron $t_{1/2}$ (hr)
Methanes		
Carbon tetrachloride	0.02, 0.003, 0.023	0.31–0.85
Chloroform	1.49, 0.73	4.8
Bromoform	0.041	
Ethanes		
Hexachloroethane	0.013	
1,1,2,2-Tetrachloroethane	0.053	
1,1,1,2-Tetrachloroethane	0.049	
1,1,1-Trichloroethane	0.065, 1.4	1.7–4.1
1,1-Dichloroethane		
Ethenes		
Tetrachloroethene	0.28, 5.2	2.1–10.8, 3.2
Trichloroethene	0.67, 7.3–9.7, 0.68	1.1–4.6, 2.4, 2.8
1,1-Dichloroethene	5.5, 2.8	37.4, 15.2
trans-1,2-Dichloroethene	6.4	4.9, 6.9, 7.6
cis-1,2-Dichloroethene	19.7	10.8–33.9, 47.6
Vinyl chloride	12.6	10.8–12.3, 4.7
Other		
1,1,2-Trichlorotriflouroethane (Freon 113)	1.02	
1,2,3-Trichloropropane		24.0
1,2-Dichloropropane		4.5
1,3-Dichloropropane		2.2
1,2- Dibromo-3-chloropropane		0.72
1,2-Dibromoethane		1.5–6.5
n-Nitrosodimethylamine (NDMA)	1.83	
Nitrobenzene	0.008	
No apparent degradation		
Dichloromethane, 1,4-dichlorobenzene, 1,2-dichloroethane, chloromethane		

Source: USACE (1997b).

techniques for PRB installation are described below. Once the emplacement of the barrier is complete, the performance monitoring of the PRB is selected. Finally, the economic feasibility of the PRB remediation is assessed. This includes estimation of capital costs as well as operating and maintenance costs. Comparison of these costs with other potential technologies will provide fiscal benefits of using PRBs.

Equipment. Equipment is needed to emplace PRBs, and the type of equipment depends on the type of emplacement method. No postconstruction equipment is needed except for the standard equipment needed for periodic monitoring purposes. The emplacement meth-

ods used for PRBs are (1) conventional excavation, (2) trenching machines, (3) tremie tube/mandrel, (4) deep soil mixing, (5) high-pressure jetting, and (6) vertical fracturing and reactant sand fracturing (USEPA, 1999a). A comparison of these different emplacement methods is shown in Table 14.12.

Conventional excavation involves using standard equipment such as backhoes, excavators, and cranes to create trenches. Trenches up to about 35 ft may be made with excavators and backhoes, but for greater depths, cranes fitted with clamshells are used. Very shallow trenches may remain open before backfilling; however, deeper depths require use of trench boxes or slurries to keep the trenches open. Trench boxes may

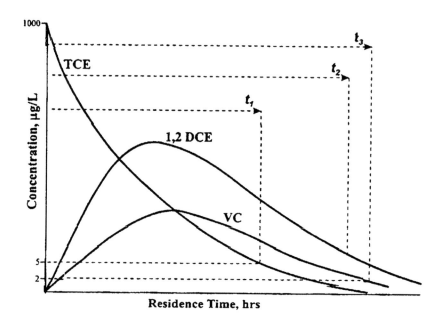

Figure 14.20 Estimation of residence time of organic compounds. (From USACE, 1997b.)

[t_1 is the required residence time] [not to scale]

be used if excavations extend only to shallow depths up to 30 ft. Slurry is used for deeper trenches. Biodegradable polymers are used for the continuous PRBs and for gate portions of funnel-and-gate systems. Bentonite slurry may be used in funnel sections of funnel-and-gate PRBs. Trenches are backfilled with materials containing reactive media.

Trenching machines commonly used for installing underground utilities are used to create trenches up to 30 ft in depth. In this method, the trench is held open by a trench box attached to a chain saw cutting belt mechanism. The trench box is pulled along after the cutter chain. A hopper attached to the top of the trench box can be filled with reactive material that feeds the material into the trench.

The *tremie tube/mandrel* installation method involves driving a hollow rectangular tube with an expendable drive shoe with hydrostatic force or a vibratory hammer. The tube can then be filled with dry granular material or slurry containing the reactive media. The tube is then extracted, leaving the drive shoe and added materials in the ground. The process is re-

peated along the length of the PRB with sufficient overlapping.

Deep soil mixing uses large augers to mix soil insitu while injecting a reactive agent. Large amounts of reactive medium may be needed in such an installation method. High-pressure jetting may be used to inject reactive media. This involves using jetting nozzles in a drill string through which the reactive medium in slurry form is injected under high pressure. The drill string is extracted at the desired rate, creating a large-diameter permeable reactive zone.

Vertical hydraulic fracturing involves boring holes to create fractures first and then pumping reactive media in slurry form into these fractures, creating a continuous wall of reactive media. Alternatively, reactant sand fracturing involves high-pressure fracturing with sand propant and then injecting foamed reactive media.

Operational Parameters and Monitoring. In many of the emplacement methods, reactive medium is injected in a slurry form. This slurry is prepared by mixing the reactive medium and biodegradable viscosify-

Table 14.12 **Comparison of different PRB emplacement techniques**

Emplacement Techniques	Maximum Depth (ft)	Vendor Quoted Cost	Comments
Impermeable Barrier Techniques			
Soil-bentonite slurry wall			
By standard backhoe excavation	30	$2–8/ft^2	Requires a large working area to allow for mixing of backfill. Generates some trench spoil. Relatively inexpensive when a backhoe is used.
By modified backhoe excavation	80	$2–8/ft^2 $6–15/ft^2	
By clamshell excavation	150		
Cement-bentonite slurry wall			
By standard backhoe excavation	30	$4–20/ft^2	Generates large quantities of trench spoil. More expensive than other slurry walls.
By modified backhoe excavation	80	$4–20/ft^2	
By clamshell excavation	200	$16–50/ft^2	
Composite slurry wall	100+	NA	Multiple barrier wall.
HDPE geomembrane barrier[a]	40–50	$35/ft^2	Permeability less than 1×10^{-7} cm/s
Steel sheet piles	60	$17–65/ft^2	No spoils produced.
Sealable-joint piles	60	$15–25/ft^2	Groutable joints.
Permeable or Impermeable Barrier Techniques			
Caisson-based emplacement	45+	NA	Does not require personal entry into excavation; relatively inexpensive.
Mandrel-based emplacement	190	$7/ft^2	Relatively inexpensive and fast production rate. Multiple void spaces constitute a reactive cell.
Continuous trenching	35–40	$5–12/ft^2	High production rate; high mobilization cost.
Jetting	200	$40–200/ft^2	Ability to install barrier around existing buried utilities.
Deep soil mixing	150	$80–200/yd^3	May not be cost effective for permeable barriers. Columns are 3 to 5 ft in diameter.
Hydraulic fracturing	80–120	$2300 per fracture	Can be emplaced at deep sites. Fractures are only up to 3 in. thick.
Jetting saw beam	50	$3–4/ft^2	Used for impermeable barriers.
Vibratory beam	100	$7/ft^2	Driven beam is only 6 in. wide.

Source: USACS (1997b).

[a] HDPE, high-density polyethylene.

ing agents such as guar gum (a natural food thickener). The reactions between the gel ingredients and the reactive media should be assessed carefully. After the PRB system is installed, its integrity is often checked using geophysical methods or tracer tests.

Once installation of the PRB is complete, the system becomes operational. Since it is a passive system, no active operations are required at the site. However, the system must be monitored to evaluate adequate capture and treatment of the plume and ensure accept-able downgradient water quality, evaluate how well the barrier meets design objectives, such as residence time in the reactive cell, and evaluate the longevity of the barrier.

A monitoring system consisting of a network of monitoring wells is installed. Placement of monitoring wells should be designed to ensure that breakthrough or bypass of the contaminants from the barrier is detected. Monitoring wells are located upstream and downstream as well as within the barrier itself. Possible monitoring well configurations are shown in Figure 14.21. Actual placement depends on the site-specific characteristics.

The following parameters are important to monitor and analyze at PRB sites:

- Contaminant concentration and distribution
- Presence of possible by-products and reaction intermediates
- Groundwater velocity and pressure levels
- Permeability assessment of the reactive barrier
- Groundwater quality parameters, such as pH, redox potential, and alkalinity

- Dissolved gas concentrations, such as oxygen, hydrogen, and carbon dioxide

Typical field and laboratory PRB monitoring parameters are listed in Table 14.13. The frequency of monitoring of the parameters depends on the site conditions. Tables 14.14 and 14.15 show suggested PRB monitoring frequency for chlorinated solvent and inorganic contamination, respectively. This monitoring will ensure that the plume is being captured and treated, and can also determine if the treatment system is adversely affecting the groundwater in any way. System breakthrough of contaminant will indicate the need

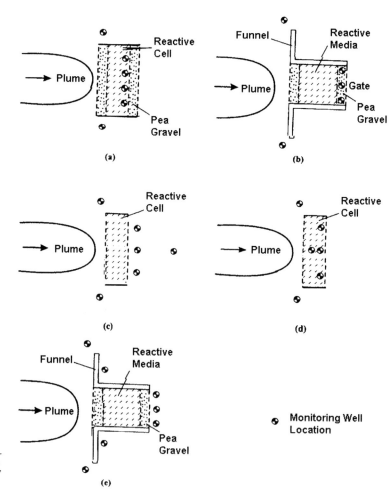

Figure 14.21 *Monitoring well configurations for PRBs. (From USACE, 1997b.)*

Table 14.13 **PRB field and laboratory monitoring parameters**

Analyte or Parameter	Analytical Method	Sample Volume[a]	Sample Container	Preservation	Sample Holding Time
Field Parameters					
Water level	In-hole probe	None	None	None	None
pH	In-hole probe or flow-through cell	None	None	None	None
Groundwater temperature	In-hole probe	None	None	None	None
Redox potential	Flow-through cell	None	None	None	None
Dissolved oxygen	Flow-through cell[b]	None	None	None	None
Specific conductance	Field instrument	None	None	None	None
Turbidity	Field instrument	None	None	None	None
Salinity	Field instrument	None	None	None	None
Organic Analysis					
Volatile organic compounds (VOCs)[c]	USEPA SW846, Method 8240	40 mL	Glass VOA vial	4°C, pH < 2 No pH adjustment	14 days 7 days
	USEPA SW846, Method 8260 a or b	40 mL	Glass VOA vial	4°C, pH 2 No pH adjustment	14 days 7 days
	40 CFR, Part 136, Method 624	40 mL	Glass VOA vial	4°C, pH < 2 No pH adjustment	14 day 7 days
Inorganic Analysis					
Metals[d]: K, Na, Ca, Mg, Fe, Al, Mn, Ba, V, Cr^{+3}, Ni	40 CFR, Part 136, Method 200.7	100 mL	Polyethylene	4°C, pH < 2 HNO$_3$	180 days
Metals: Cr^{6+}	40 CFR, Part 136, HACH method	200 mL	Glass, plastic	4°C	24 h
Anions: SO$_4$, Cl, Br, F	40 CFR, Part 136, Method 300.0	100 mL	Polyethylene	4°C	28 days
NO$_3$	40 CFR, Part 136, Method 300.0	100 mL	Polyethylene	4°C	48 hours
Alkalinity	40 CFR, Part 136, Method 310.1	100 mL	Polyethylene	4°C	14 days

Table 14.13 (*continued*)

Analyte or Parameter	Analytical Method	Sample Volume[a]	Sample Container	Preservation	Sample Holding Time
Other					
TDS	40 CFR, Part 136, Method 160.2	100 mL	Glass, plastic	4°C	7 days
TSS	40 CFR, Part 136, Method 160.1	100 mL	Glass, plastic	4°C	7 days
TOC	40 CFR, Part 136, Method 415.1	40 mL	Glass	4°C, pH < 2 (H_2SO_4)	28 days
DOC	40 CFR, Part 136, Method 415.1	40 mL	Glass	4°C, pH < 2 (H_2SO_4)	28 days
Radionuclides					
Field screening	HPGe gamma spectroscopy FIDLER	None	None	None	None
Gross α/Gross β activities (screening)	Gas proportional counting	125 mL[e]	Polyethylene[e]	pH < 2 (HNO_3)[e]	N/A[e]
Specific isotopes (Am, Cs, Pu, Tc, U)	Alpha spectroscopy Gamma spectroscopy	4 L[e]	Polyethylene[e]	pH < 2 (HNO_3)[e]	6 months[e]

Source: USEPA (2000a).

[a] See Section 7.4 of USEPA (2000a) for variances in sample volumes.

[b] If < 1.0 mg/L use photometric field kit for analysis.

[c] GC methods may be substituted once identity of compounds and breakdown products is verified.

[d] Other metals analytes which are Characteristic of the media should be included.

[e] General guidelines, the parameter is a laboratory-specific parameter.

Table 14.14 PRB Monitoring frequency for chlorinated solvent contamination

Parameter	Frequency
First Quarter After Installation	
Field parameters	Monthly
Organic analytes	Monthly
Inorganic analytes	Monthly
Groundwater levels	Weekly (until equilibrium is reached)
Initial Monitoring Program (1–2 years)	
Field parameters	Quarterly
Organic analytes	Quarterly
Inorganic analytes	Quarterly
Groundwater levels	Monthly, then to be determined
Long-Term Monitoring	
Field parameters	
Organic analytes	Quarterly (may be reduced
Inorganic analytes	based on operational
Groundwater levels	stability)
Postclosure Monitoring	
Inorganic parameters (Fe and other leachable constituents)	To be determined when data collected during operation

Source: USEPA (2000a).

Table 14.15 PRB monitoring frequency for inorganics and radionuclide contamination

Parameter	Frequency
First Quarter After Installation	
Field parameters	Monthly
Inorganic analytes	Monthly
Inorganic contaminants	Monthly
Radionuclides	Monthly
Groundwater levels	Weekly (until equilibrium is reached)
Initial Monitoring Program (1–2 years)	
Field parameters	Quarterly
Inorganic analytes	Quarterly
Inorganic contaminants	Quarterly
Radionuclides	Quarterly
Groundwater levels	Monthly, then to be determined
Long-Term Monitoring	
Field parameters	
Inorganic analytes	Quarterly (may be reduced
Inorganic contaminants	based on operational
Radionuclides	stability)
Groundwater levels	
Postclosure Monitoring	
Inorganic contaminants of concern	To be determined based on
Leachable constituents from reactive media	the closure method and data collected during
Radionuclides of concern	operation of the PRB

Source: USEPA (2000a).

for replacement of the reactive media. Changes in groundwater velocity could indicate the formation of precipitates or biofouling of the reactive media. This problem can be remedied by flushing of the media or replacement. Regulations on the closure of PRBs do not exist. Organic contaminants are degraded into non-toxic compounds; however, metals and radionuclides are sorbed and precipitated within the reactive media. Under long-term conditions, the reactivity of the reactive media may be exhausted and a potential for desorption and/or dissolution of the contaminants exist, which may result in recontamination of the groundwater. Many such longevity issues, including options for the removal and disposal of reactive media, need to be addressed.

14.4.4 Predictive Modeling

Mathematical models are used to optimize the location and configuration of permeable barriers as well as to evaluate long-term performance of the barrier under different scenarios. Both groundwater and geochemical models are used in the evaluation of permeable barrier design and performance. Groundwater models such as MODFLOW (combined with particle tracking codes such as MODPATH) are used to determine hydraulic capture zone width and residence time. Capture zone width refers to the width of the zone of groundwater that will pass through the reactive cell or gate rather than pass around the ends of the barrier or beneath it. Residence time refers to the amount of time that contaminated groundwater is in contact with the reactive medium within the gate. Modeling results are used to determine the optimal configuration of the barrier that can yield adequate capture zone width and residence time in order to treat the entire contaminated groundwater plume effectively for any given site conditions.

Geochemical modeling is used to predict chemical reactions between groundwater and reactive media. This type of modeling is mainly of concern for the dissolved inorganic constituents. Most geochemical models, such as MINTEQA2, assume equilibrium reactions, and such an assumption may not be valid for barriers. Models such as EQ3/EQ6 incorporate reaction rate constants and may be appropriate for barrier design. Often, very limited data are available for geochemical modeling; therefore, quantitative prediction of reaction products is difficult. Model results are commonly used to determine potential reactions associated with different reactive media and conservatively incorporate these effects in the design. In addition, models such as NETPATH are used to interpret net geochemical mass-balance reactions between initial and final waters along a hydrologic flow path. The NETPATH program can help determine the total number of material changeouts in the cell and the rates of inorganic chemical reactions in combination with measured flow velocities.

14.4.5 Modified or Complementary Technologies

PRB technology can be used in combination with other technologies. For example, soil fracturing within the contaminated zone can enhance groundwater flow into the permeable barriers for efficient remediation. If the reactive medium involves microbes, it essentially becomes an engineered bioremediation method (Stavnes, 1999). An attempt to integrate electrokinetics and PRBs has recently been initiated (Reddy, 2001).

14.4.6 Economic and Regulatory Considerations

The economical benefit of permeable barriers has been one of the greatest promoters of the widespread interest

in this technology. A passive technology that requires almost no annual energy or labor input (other than for site monitoring) has obvious cost advantages over a conventional pump and treat system. A cost-benefit approach that includes both tangible and intangible costs and benefits should be used in evaluating the economic feasibility of a permeable barrier at a contaminated site (USACE, 1997b). Cost ranges for the use of PRBs are approximately $250 to $800 per liter per minute (Reddy et al., 1999).

The capital cost of installing a permeable barrier includes the cost of the reactive media, installation of the barrier itself, technology licensing costs, disposal costs of the spent media, and restoration of the ground surface. Granular iron is the most inexpensive medium. Operating and maintenance costs include compliance and performance monitoring costs and periodic maintenance costs. Maintenance costs could include flushing of the reactive media or replacement of the media.

Monitoring to ensure that contaminant concentration requirements are meeting cleanup goals is typically determined by the state regulations where the site is located (USEPA, 1997b). Normal compliance monitoring parameters include determining contaminant levels after treatment, potential degradation products, and general water quality parameters. The USEPA recognizes this technology as being a feasible remediation technology for subsurface contamination. The Interstate Technology and Regulatory Cooperation Workgroup (Permeable Barrier Wall Subgroup) is currently providing regulatory guidance on the installation of PRBs (ITRC, 1999a).

14.4.7 Case Studies

Several studies have been reported where PRBs are used to treat contaminated groundwater (Feltcorn and Breeden, 1997; USEPA, 1999a; Ott, 2000). Two studies are discussed in this section to show the expected performance and typical application of the PRB remediation technique to contaminated groundwater.

Case Study 1: Pilot-Scale Testing of a Surfactant-Modified PRB. In 1998, a pilot-scale demonstration PRB was used to remediate a contained simulated aquifer at the Graduate Institute of Science and Technology in Oregon. The aquifer was constructed at the OGIST Large Experimental Aquifer Program (LEAP) site with a plume containing 22 mg/L chromate and 2 mg/L perchloroethylene contaminant levels.

The reactive medium used in the PRB was a modified zeolite. The zeolite, which is a naturally occurring aluminosilicate with an open, cagelike structure, has very high internal and external cation exchange capacities. Replacing the base cations on the external exchange sites with a cationic surfactant modifies the zeolite. The surfactant forms a bilayer on the surface, resulting in a new positive charge on the zeolite, thereby increasing the organic carbon content to about 5% by weight. Sorption of oxyanions occurs via ion exchange to this new surface; sorption of cations occurs via ion exchange and surface complexation to remaining zeolite surface sites; sorption of organic compounds occurs via partitioning into the new organic stationary phase. The modified zeolite has the ability to sorb the cations, oxyanions, and organic contaminants simultaneously (USEPA, 1999a).

For this demonstration project, the aquifer was created by filling the LEAP site with sand having a hydraulic conductivity of 56.7 ft/day. To simulate emplacement in front of an advancing plume in a shallow unconfined aquifer, the barrier was placed in the center of the aquifer approximately 3 ft above its base. Contaminants were introduced through upper injection wells in order to simulate a shallow plume in the upper half of the aquifer to provide a three-dimensional test of a plume captured by the PRB.

Performance of the barrier system was monitored over a two-month period in which the contaminated plume was injected into the aquifer. Weekly sampling from several sample points in the aquifer and from within the barrier was conducted. Test results indicated that the barrier performed within design specifications, with retardation factors for chromate and PCE on the order of 50. The entire plume was captured by the PRB. Based on these experiments, researchers recommended a minimum of 100-fold permeability contrast between the PRB and the aquifer material (i.e., PRB to be 100-fold more permeable than the aquifer). Costs for construction of the barrier system were approxi-

mately $100,000, including $75,000 for design and $25,000 for installation.

Case Study 2: U.S. Coast Guard Support Center, Elizabeth City, North Carolina. The USCG facility support center included an electroplating shop, which operated for more than 30 years, until 1984 (FRTR, 1997; USEPA, 1997a). In December 1988, a release was discovered during demolition of the former plating shop. Soil excavated beneath the floor of the shop was found to contain high levels of chromium. Subsequent investigation showed substantial groundwater contamination by chromium and chlorinated solvents. A full-scale PRB was constructed as part of an interim corrective measure associated with a voluntary RCRA facility investigation where the electroplating shop was identified as a solid waste management unit under the facility's RCRA Part B permit.

The barrier consisted of 450 tons of granular zerovalent iron placed into an underlying low-conductivity layer at a depth of approximately 22 ft below ground surface. The required residence time in the treatment zone has been estimated as 21 h, based on a highest concentration scenario. The average velocity through the wall was reported as 0.2 to 0.4 ft/day. Analytical data from the first year of full-scale operation showed that the cleanup goal for Cr(VI) had been met but the goal for TCE had not.

Cleanup goals for the site were based on primary drinking water standards: TCE (5 μg/L) and Cr(VI) (0.1 μg/L). Cr(VI) concentrations were below cleanup goals in all downgradient monitoring wells. However, TCE concentrations were above the cleanup goal in four of the six downgradient wells. The reason for the elevated TCE concentrations in some of the downgradient wells had not been identified.

Estimated costs for the PRB were $585,000, which corresponded to $225 per 1000 gal of groundwater treated. By using a PRB rather than the typical pump and treat method, nearly $4 million was saved in construction and long-term maintenance costs.

14.5 IN-SITU AIR SPARGING

14.5.1 Technology Description

Air sparging is a relatively new technique that has enjoyed success in the treatment of saturated soils and groundwater that have been contaminated with volatile organic compounds (VOCs). It is being used extensively to remediate sites where underground storage tanks containing petroleum products have leaked and contaminated the surrounding soils and groundwater (Loden, 1992; Miller, 1996; USEPA, 1996d; Leeson et al., 1999; Reddy and Adams, 2001). In a typical field system, as depicted in Figure 14.22, compressed air is delivered to a manifold system, which in turn delivers the air to an array of air injection wells. The wells inject the air into the subsurface below the lowest known point of contamination. Due to buoyancy, the air will begin to rise toward the surface, and through a variety of mechanisms, the contaminants are partitioned into the vapor phase. As the contaminant-laden air rises toward the surface, it will eventually reach the vadose zone. At this point, the contaminated air is collected with a soil vapor extraction (SVE) system. Air sparging also increases oxygen content in groundwater, which may enhance the aerobic degradation of the contaminants if an indigenous microbial population is present within the contaminated zone.

Ideal conditions where the air sparging is applicable are summarized in Table 14.16 (USEPA, 1997c). Air sparging is applicable to volatile organic compounds (VOCs). Common VOCs that are the target of air sparging are petroleum products, including gasoline and its BTEX (benzene, toluene, ethylbenzene, and the xylenes) constituents, and chlorinated solvents, including trichloroethylene (TCE) and tetrachloroethylene (PCE). A compound is capable of being remediated through air sparging if it is deemed strippable; that is, having a Henry's constant greater than 10^{-5} atm · m^3/mol and vapor pressure greater than 1 mmHg. Reddy and Adams (2001) have reported that air sparging can be applied successfully to remediate dissolved as well as free-phase NAPLs in source zones.

The patterns of airflow resulting from air injection are controlled by the subsurface stratigraphy. The permeability of soil within each layer must be assessed to assure that effective airflow will result; soil permeability should be at least 10^{-3} cm/s for adequate airflow to occur. Because airflow may be affected by the presence of adjacent layers of differing permeability, the thickness and properties of each layer need to be determined. Additionally, lenses or inclusions of soil with properties different from those of the surrounding soil may affect airflow and overall remedial performance and must be delineated. When applying air sparging, the depth to the water table should be at least 5 ft, but the most successful applications are made when the depth to the water table is at least 10 ft. Additionally, the groundwater flow velocity and direction must be assessed to assure that the air sparging system is designed to prevent downgradient contaminant migration away from the treatment zone.

The advantages of air sparging technology are:

- Being an in-situ method, it causes minimal site disruption and reduces worker exposure to contaminants.
- It is applicable to treat contamination in groundwater and capillary fringe.
- It does not require removal, storage, or discharge consideration for groundwater.
- Equipment needed is simple and easy to install and operate.
- It requires a short treatment time (one to three years)
- Overall cost is significantly lower than conventional remediation methods such as pump and treat.

The disadvantages of air sparging are:

- Contamination in low permeability and stratified soils poses a significant technical challenge to air sparging remediation efforts.
- Confined aquifers cannot be treated by this remediation technique.
- Airflow dynamics and contaminant removal or degradation processes are not well understood.
- If not designed properly, it could cause spreading of the contaminants into clean areas.
- It requires detailed data and pilot testing prior to its application.

14.5.2 Fundamental Processes

Air sparging is based on the principles of airflow dynamics as well as contaminant transport, transfer, and

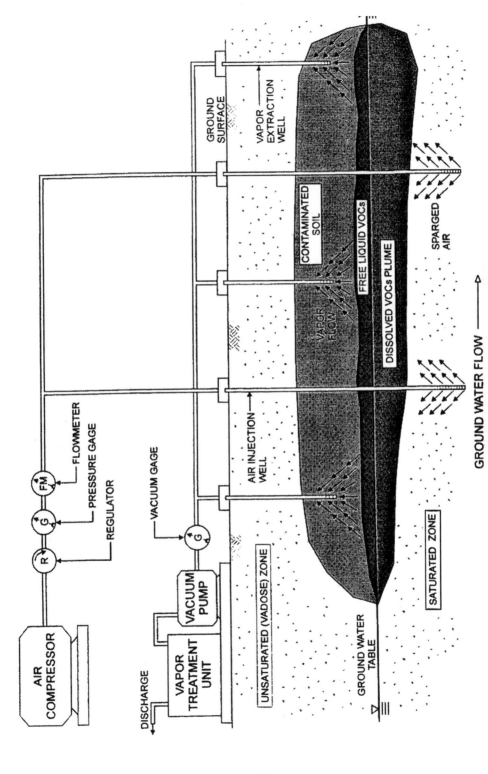

Figure 14.22 Schematic of in-situ air sparging system. (From Reddy and Adams, 2001.)

Table 14.16 **Factors affecting applicability of air sparging**

Factor	Parameter	Desired Range or Conditions[a]
Contaminant	Volatility	High ($K_H > 1 \times 10^{-5}$ atm · m^3/mol)
	Solubility	Low (<20,000 mg/L)
	Biodegradblity	High (BOD$_5$ > 0.01 mg/L)
	Presence of free product	None or thin layer
Geology	Soil type	Coarse-grained soils
	Heterogeneity	No impervious layers above sparge interval; permeability increases toward grade if layering present
	Permeability in the saturated zone	>1 × 10^{-5} cm^2 if horizontal:vertical is <2:1; >1 × 10^{-4} cm^2 if horizontal:vertical is <3:1
	Hydraulic conductivity	>1 × 10^{-3} cm/s
	Depth to groundwater	>5 ft
	Aquifer type	Unconfined
	Saturated thickness	5 to 30 ft

Source: USEPA (1997c).

[a] K_H, Henry's law coefficient; BOD, biological oxygen demand.

transformation processes. These principles are presented briefly in this section.

Airflow Dynamics. When injected into the subsurface, air will enter into pores with lower capillary pressure (larger pore sizes) and produce a fingering pattern of preferential pathways, due to low-viscosity air invading higher-viscosity groundwater. Injected air moves through aquifer materials in the form of either bubbles or microchannels. In coarser soils such as fine gravels, airflow has been observed to be in the bubble form. In finer soils such as sands, the airflow has been observed to be in microchannel form. The density of bubbles or microchannels is found to depend on the injected airflow rate. Soil heterogeneities are found to affect the airflow patterns and the zone of influence significantly.

Contaminant Transport, Transfer, and Transformation Processes. Air sparging owes its effectiveness to a variety of mechanisms that occur during this subsurface air movement. The mechanisms can be classified into three categories: mass transfer mechanisms, mass transformation mechanisms, and mass transport mechanisms (Semer and Reddy, 1998). Mass transfer mechanisms include volatilization, dissolution, and adsorption–desorption. Biodegradation is the mass

transformation that occurs during air sparging. The three mass transport mechanisms consist of advection, dispersion, and diffusion. Air sparging is a dynamic process; during different stages of a remedial program, different mechanisms will control remedial rates and efficiency. The contributions of each of the mechanisms will also vary from site to site and will ultimately determine the effectiveness of an air sparging remedial program.

Volatilization has been shown to be the dominant removal mechanism occurring during the use of air sparging. It is defined as the partitioning of a contaminant from the aqueous or nonaqueous phase into the vapor phase. During air sparging, the injected air provides both a medium and a transport network to allow for such partitioning to occur. Additionally, the injected air disturbs the equilibrium that exists between the liquid and vapor phases. By pulling vapor phase away from its liquid source, the resulting nonequilibrium will force additional partitioning into the vapor phase. Generally speaking, the higher the vapor pressure of a contaminant, the easier it is to remove through volatilization. However, for a given contaminant, volatilization is two orders of magnitude lower in soil than when the contaminant resides in a liquid.

Volatilization is the dominant removal mechanism in the early stages of air sparging, but its importance is diminished with increasing remedial time. When air injection begins, contaminant near a channel will be easily volatilized. If a high-density channel network exists within the subsurface, this will account for a high percentage of removal and a rapid reduction in initial contaminant levels. Eventually, as the contaminant in the vicinity of channels is removed, removal rates decrease as contamination is forced to diffuse toward a channel. Thus, volatilization and overall removal rates drop when the concentration gradient drops at the air–water interface.

Dissolution plays an important role, as it assists in the volatilization and ultimate removal of the contaminant. While volatilization controls during the early stages of an air sparging program, contaminant dissolution from the NAPL phase to the aqueous phase is a controlling factor at later times. Solubility limits vary greatly for different contaminants, but even in the case of compounds with low solubility, maximum contaminant levels for drinking water are often at least two orders of magnitude lower than solubility limits. Yet the solubility limit of a contaminant can often indicate whether or not a specific compound may be removed in a timely fashion.

There is a growing body of evidence that dissolution is a nonequilibrium process. It is suggested that nonequilibrium descriptions are needed for high pore water velocities, hydrophobic solutes, high dispersivities, as well as small spills, large blobs, low residual saturations, and heterogeneous aquifers. After air sparging has ended and the subsurface is allowed to reach equilibrium once again, additional dissolution of residual contaminants may take place, serving to increase groundwater concentrations. This effect is known as *rebound*. Bass and Brown (1997) reported the following equation for rebound:

$$\text{rebound} = \frac{\log (C_r/C_f)}{\log (C_0/C_f)} \tag{14.8}$$

where C_r is the dissolved concentration during postmonitoring, C_f the final dissolved concentration; and C_0 the initial dissolved concentration. If the rebound is less than 0.2, permanent reductions in contaminant concentrations have been achieved.

For air sparging to be successful, *desorption* must occur if organic or clayey soils are present at a given site. Although organic matter is usually confined to the upper layers of a soil stratum, clay may be present at all depths. If desorption takes place at an acceptable rate, air sparging will be successful at remediating contaminated saturated soils. If it does not occur, saturated soils will remain contaminated and will act as a source to further contaminate the surrounding groundwater.

Biodegradation can often account for a substantial portion of the remedial process by acting on the less volatile, more strongly adsorbed contaminants during the later stages of remediation and when dissolved concentrations have decreased. Optimal subsurface conditions are necessary for biodegradation to occur successfully. While suitable microbial populations or nutrient levels may be a limiting factor to the rate of biodegradation, the type of electron acceptor is often the controlling factor. The most efficient electron acceptor for biodegradation is oxygen.

When sufficient oxygen is lacking and anoxic conditions exist within the subsurface, biodegradation becomes less efficient. During anoxic conditions, other subsurface electron acceptors are utilized. Following oxygen as the preferred subsurface electron acceptor, in descending order of efficiency, possible electron acceptors include nitrate, ferric iron, sulfate, and carbon dioxide. Despite preferences in anoxic conditions, field tests have shown that sulfate reduction is the primary electron-accepting process. By adding oxygen to the subsurface through air sparging to create oxic conditions, more efficient, aerobic degradation is allowed to occur. In some cases, oxygen is needed for any sufficient biodegradation to occur. When injecting air into the subsurface, oxygen levels can reach 8 to 10 mg/L, while the injection of pure oxygen can increase levels to 40 mg/L. The injection of hydrogen peroxide can increase subsurface oxygen levels to 500 mg/L.

Advective transport of vapors during air sparging depends on the pressure gradient. The migration due to advection depends on soil permeability, which in turn depends on soil grain size and grain size distribution, the type and structure of the soil, the soil porosity, and the water content of the soil. Dispersion, the mixing or spreading of the contaminant, in either the groundwater or soil gas is due to differences in microscale flow velocities. Once again, soil permeability affects dispersion as pore size, path length, and pore friction create these flow-velocity differences.

Advection and *dispersion* act to disrupt the saturated zone by creating turbulence. The turbulence created by air injection serves to disrupt equilibrium. This action can help to introduce water into dead-end pores previously occupied by NAPL ganglia as well as to create groundwater flow or circulation. Such groundwater movement will aid dissolution by transporting dissolved contaminant away from NAPL sources, creating concentration gradients that will force additional dissolution. This type of behavior will also aid in desorption, as concentration gradients will be established near the surface of soil particles with sorbed contaminants. Caution should be used, however, when inducing subsurface turbulence. Although this turbulence allows for desirable effects, too much movement may force unwanted migration into areas previously free of contamination. Additionally, some mechanisms, especially adsorption–desorption, are reversible, and thus advection–dispersion can act to trap contaminant in dead-end pores as well as to force additional contaminant adsorption. Therefore, subsurface airflow should be monitored carefully to help minimize any negative effects of advection–dispersion.

Diffusion is the primary mechanism for the removal of NAPL trapped in dead-end pores; diffusion must occur to allow for movement from such trapped sites into regions where other processes may occur for removal. It is also a means of transporting contaminant from source locations to air channels in order for volatilization or biodegradation to take place. Diffusion of gas within soil is important, as it will help determine any type of air or vapor movement. Additionally, trapped gas may also be removed through diffusion.

14.5.3 System Design and Implementation

General Design Approach. Figure 14.23 is a flowchart summarizing the design and implementation of in-situ air sparging (IAS) systems. Prior to the design

of an air sparging system, a detailed site characterization must be performed to determine the site hydrogeologic conditions as well as contaminant conditions. If the contaminants and hydrogeologic conditions present at the site are conducive to using air sparging, a pilot test is conducted. A conceptual model must be developed to ensure that the pilot testing is conducted in the proper location. The conceptual model includes the location of the equipment for the air sparging and the depth of the injection wells. The primary objective of pilot testing is to discover quickly the effectiveness of air sparging and to establish its feasibility.

Equipment used for pilot tests is similar to that of full-scale systems. It consists of the injection well, injection blower or pump, and ancillary equipment such as flow control valves. Table 14.17 shows different monitoring methods that can be used during pilot tests. Pilot tests provide the data regarding the approximate extent of zone of influence (ZOI), optimal injection rates and pressures, and off-gas handling.

The collected data are evaluated, and when the air sparging feasibility has been established, the determination is made based on the results to proceed with full-scale design installation and operation of the air sparging system. The pilot test results are utilized to determine system parameters, such as the number, spacing, and depth of air injection wells. Flow rates and pressures need to be determined, as well as the details of the soil vapor extraction system. In addition to pilot test results, mathematical modeling may be performed to optimize the system variables.

Once the system is designed and implemented, the performance of the system needs to be monitored to assure adequate performance. If the expected efficiency is not reached, changes need to be made and the performance reevaluated. This iterative design process continues until the air sparging system is performing to expectations. Typical design parameters for in-situ air sparging systems are summarized in Table 14.18.

Equipment. An air compressor is used to force air into the air injection wells. The compressor must be capable of injecting air at a proper flow rate under suitable air pressure. The number of wells that are fed by a given compressor also dictates the size of the air compressor. Typically, reciprocating or rotary screw

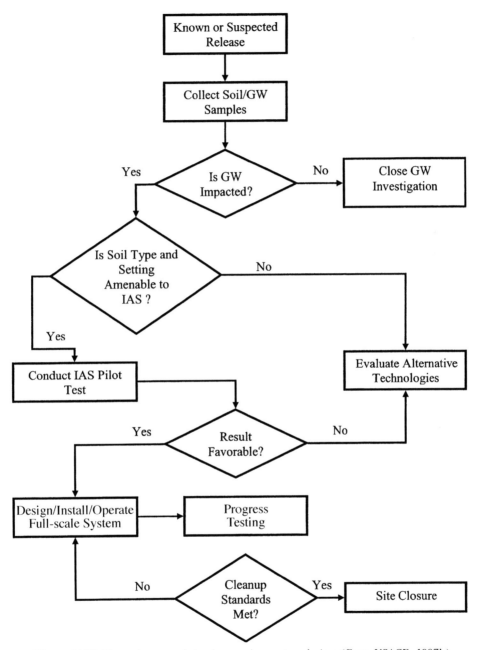

Figure 14.23 *General approach to air sparging system design. (From USACE, 1997b.)*

Table 14.17 **Pilot test monitoring methods**

Method	Applicable Installation(s)	Analytical Equipment	Results
Injection pressure and airflow	Ports in wellhead or manifold	Pressure gauge, anemometer or pilot tube, datalogger	Apparent IAS air-entry pressure, well capacity, system requirements
Neutron thermalization	Access tube consisting of bottom-capped 5-cm (2-in.) Schedule 40 carbon steel pipe	Neutron probe with source, and counter/detector	Vertical profile of saturation, ZOI
Electrical resistance tomography	Electrode array attached to parallel PVC pipes, 1.5–7.5 m (5–25 ft) apart	Power supply, current and volt meters, analyzer	Saturation within plane of electrodes, ZOI
Time-domain reflectometry	Steel waveguide pushed into bottom of soil boring	Electrical pulse generator/detector	Saturation in proximity of wave guide
Tracer gas	Monitoring wells, soil gas monitoring point, SVE wellhead	Tracer gas detector	ZOI, airflow velocities, percent capture
Dissolved Oxygen (DO)	Galvanic "implants," monitoring wells	DO meter, flow cell, datalogger, in-situ ampoules	Dissolved gas ZOI
Pressure (unsaturated zone)	Monitoring wells, soil gas monitoring points	Differential pressure gauge	Airflow ZOI within unsaturated zone
Pressure (below water table)	Monitoring wells, soil gas monitoring points	Differential pressure gauge	Steady-state air flow ZOI
Hydrocarbon off-gas concentrations	SVE wellhead, soil gas monitoring points	FID, PID; vapor sampling equipment	Evidence that IAS is or is not causing significant increase in volatilization
Groundwater elevation	Monitoring wells	Pressure transducer/datalogger	Groundwater mounding; optimal pulse interval

Source: USACE (1997b).

Table 14.18 **Design parameters for air sparging systems**

Parameter	Typical Range
Well diameter	2.5 to 10 cm (1 to 4 in.)
Well screen length	15 to 300 cm (0.5 to 10 ft)
Depth of top of well screen below water table	1.5 to 6 m (5 to 20 ft)
Air sparging flow rate	0.04 to 1.1 m^3/min (1.3 to 40 ft^3/min)
Air sparging injection overpressure[a]	2 to 120 kPa (0.3 to 18 lb/in^2 gauge)
IAS ZOI	1.5 to 7.5 m (5 to 25 ft)

Source: USEPA (1997c).

[a] Overpressure is injection pressure in excess of hydrostatic pressure, P_n.

compressors are used during field application. The air leaves the compressor at a flow rate and pressure that is controlled through the use of regulators. It passes into a manifold system used for delivery to the injection wells. Because the air leaving the compressor may be elevated in temperature, rubber hose or metal pipes are often used in the manifold system. Valves are included in the manifold system to assist in the delivery of the desired amount of air to each well.

The injection well usually consists of 1- to 4-in.-diameter PVC pipe, although stainless steel may be used if steam or heated air is to be injected. Typical injection well construction details are shown in Figure 14.24. One- to 2-in.-diameter PVC pipe is often used because it is cheaper to install. A borehole for well placement is often drilled, but the use of direct push

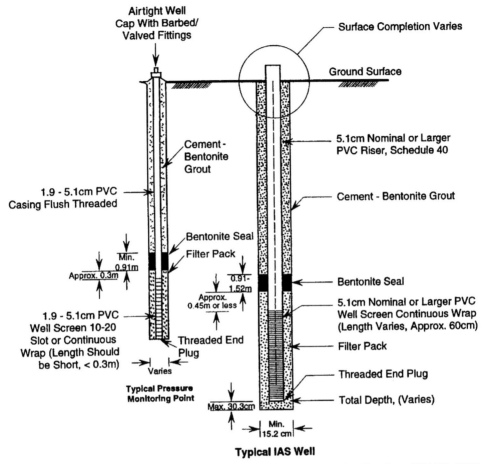

Figure 14.24 *Schematic of air sparging/monitoring well construction details.* (*From USACE, 1997b.*)

methods for well placement may provide significant cost savings. The well screen is placed below the lowest known point of contamination to assure that the entire plume is addressed. Typically, injection points during field applications are located 10 ft below the lowest point of contamination. Wells are effective in delivering air to a depth of 150 ft, but if depths are below 40 ft, a series of nested wells should be used. The well screen is usually 1 to 3 ft long. The screens help minimize any effects of clogging as well as provide for even air distribution. The well is backfilled with 6 in. to 2 ft of sand or gravel to act as a filter and

is then sealed with a bentonite seal to prevent air from short-circuiting to the surface. The remaining annulus is grouted to the surface to assist in providing a seal.

The spacing of injection wells is a very important design parameter. Well spacing should be such that the entire contaminated zone is affected by the treatment process, and it is decided based on the radius of influence as determined from the pilot tests.

Once air is injected into the subsurface, it will migrate toward the ground surface due to the effect of buoyancy. Eventually, the contaminated injected air will enter the vadose zone. At this point, a soil vapor extraction system may be implemented. Soil vapor extraction applies a vacuum to the vadose zone to assist in the collection of vapors. The vacuum will help vapor collection, as well as assist in preventing unwanted off-site migration. Equipment used in a soil vapor extraction system is similar to that used in an air sparging system, with the exception of a vacuum pump in place of an air compressor. Additionally, if the vapors being collected by the soil vapor extraction system are moist, a dewatering system must be employed. The extraction flow rate is very important to control air and contaminant vapor migration. Simply stated, the extraction rates must be higher than the injection rates to prevent soil pressure buildup. The ratio of extraction to injection commonly used during field application is between 4:1 and 5:1. Typical extraction rates are between 100 and 1000 standard cubic feet per minute.

Operational Parameters. The first choice that must be made is the type of gas to be injected into the subsurface. The choice of gas type is especially important in assuring the delivery of oxygen to the subsurface in order to stimulate biodegradation. Air is the gas most commonly used in remedial programs. Air is low in cost and is not explosive or flammable. When air is injected into the subsurface, dissolved oxygen levels may be raised from typical concentrations of about 2 mg/L to levels between 9 and 10 mg/L, thereby providing suitable oxygen levels for biodegradation. If higher levels of dissolved oxygen are desired, pure oxygen or air enriched with oxygen may be used. When pure oxygen is injected into the subsurface, dissolved oxygen levels may reach 40 mg/L.

Another injection alternative is the use of ozone for injection, a method that is being adopted increasingly for the treatment of both contaminated soils and groundwater. The injection of ozone allows subsurface ozonation to occur. Ozonation is an advanced chemical oxidation process that may be used to break complex organic contaminants into less persistent or less toxic molecules. The ozone accomplishes this by assisting in the degradation of double-bonded organics.

Steam may be used in place of air or oxygen for injection into the subsurface. The injection of steam increases the temperature of the subsurface, helping to improve the removal efficiency of semivolatile organic compounds as well as stimulating microbial activity in colder soils and climates. During the use of steam injection, contaminant mass transport occurs by advection in all three phases and by multicomponent diffusion in the gas phase. Additionally, the use of steam allows for the possibility of NAPL recovery.

Once the type of gas to be injected is determined, the flow rate and mode of injection must be determined. Flow rates of 2 to 16.5 ft^3/min are typical for field application. When soil vapor extraction systems are used at a site, injection flow rates between 4 and 10 ft^3/min are used. When injecting air, either continuous or pulsed airflow may be used. Pulsed flow involves using a cycle that consists of injection followed by a time of cessation. After the period of cessation has elapsed, air is once again injected and the cycle is repeated. Pulsing helps improve removal efficiency by inducing mixing. Frequent pulsing is needed to optimize oxygen distribution within the subsurface.

When performing air sparging, air must be injected under proper pressure. If air is injected under too low a pressure it will be unable to enter the subsurface. If air is injected under too high a pressure, the contaminant plume may be forced to migrate into previously uncontaminated areas. Additionally, soil heaving or fracturing may result within the subsurface if the injection pressure is too high. The necessary pressure needed for air to enter the subsurface is the sum of hydrostatic pressure at the injection point, friction due to exiting from the well, and the capillary pressure due to the interface of two fluids within the porous media. One pound per square inch of pressure is needed to overcome every 2.3 ft of static water above the injection point. The capillary or air entry pressure that must be overcome may range between a few centimeters of pressure head in coarse, sandy soils to several meters in low-permeability clay.

When soil vapor extraction is used, the contaminated vapors that have been collected must be addressed. The effluent gases are passed through a diffuser stack and released into the air. This is applicable only when local regulations allow and compliance with air emission standards can be assured. When vapor treatment is needed, a popular option is the use of carbon adsorption. The vapors are passed through an activated carbon matrix, effectively removing contaminated vapors from the effluent gas. Either virgin or recharged carbon may be used during application. Thermal incineration is another option. Temperatures between 1000 and 1400°F are used to destroy the contaminants, providing contaminant destruction rates between 95 and 100%. Care must be taken not to exceed the explosive limit; this can be avoided by diluting and mixing the vapors with fresh air. Lower temperatures may be used with the use of catalytic oxidation. Because a catalyst is employed, temperatures may be lowered to between 600 and 800°F. Such a process allows for contaminant destruction efficiencies of up to 85%. Additionally, if off-vapors provide proper conditions, internal combustion engines may be used to provide power while treating the vapors.

Monitoring. Several parameters are measured such as:

- Pressure in the aquifer, which must be maintained to assure the supply of air
- Dissolved oxygen and carbon dioxide levels before remediation and after air sparging, compared to assess the progress of remediation
- The contamination of extracted air during remediation
- The airflow and distribution of the air channel density, the distribution of sparge air around the sparge well, and the degree to which it is reaching the treatment zone
- The density and population of microbes to assure that bioremediation is accommodated to a desirable level
- The operation of compressors to assure continuous operation and that the unit will produce sufficient pressure to deliver the required airflow in each well
- Evaluation of water table elevation changes to assure proper placement of screens and compressor and pressure selection
- Extraction rates and contaminant concentration in the air and groundwater in order to compare the

data during the remediation to provide data on the effectiveness of air sparging

Detailed information on the methods to measure the parameters above are given in USACE (1997a). Remediation by air sparging is relatively quick, usually under two years. Remedial efficiency can be calculated using

$$\text{remedial efficiency} = \left(1 - \frac{C_f}{C_0}\right) \times 100 \ (\%)$$

(14.9)

where C_f is the final contaminant concentration after the termination of air sparging and C_0 is the initial contaminant concentration at the start of air sparging. As explained earlier, rebound effects (diffusion and dissolution) may increase the contaminant concentrations in groundwater. To check these rebound effects, groundwater sampling and testing is conducted six to 12 months after shutdown of the air sparging system. Using these residual concentrations (C_r), rebound can be calculated using equation (14.8). A rebound value of less than 0.2 represents permanent reduction, while a value greater than 0.5 represents significant rebound (Bass and Brown, 1997).

14.5.4 Predictive Modeling

Methods for predicting air sparging performance will eliminate a great deal of the guesswork involved in system component design and layout. Several researchers have developed mathematical models to simulate air sparging performance (Reddy et al., 1995). Simple models may be used to study specific mass transfer, transformation, or transport processes, or more rigorous models may be required to study macroscale performance. Unfortunately, because air sparging creates dynamic, nonequilibrium processes, models based on simplifying assumptions may provide misleading results. Additionally, many models have yet to be validated with controlled laboratory test data and field data.

14.5.5 Modified or Complementary Technologies

Even though traditional air sparging (vertical well, injection of air) has proven to be successful in a wide variety of field applications, alternative strategies exist that can make air sparging applicable to other conditions. One such strategy is the use of horizontal wells for air injection (Figures 14.25 and 14.26). Horizontal wells may be implemented under a wide range of field applications, but they are especially useful in treating long, shallow spills, such as those that may occur under leaking aboveground or in-ground pipelines. There are four proven methods for horizontal well placement: directional drilling, trenching, boring, and backhoe excavation. These placements and subsequent improvements have made the use of horizontal wells an attractive option with many benefits over the use of vertical wells . One horizontal well can replace several vertical wells, leading to a reduction in operating expense that can help offset the additional placement costs. Because of the continuous nature of a horizontal well, the injected air is able to interact with the subsurface contaminant over a larger surface area, accelerating removal. The continuous nature of a horizontal well also allows for easier interception of a migrating contaminant plume. Additionally, the flexibility of directional drilling allows for the placement of injection wells in locations that may be difficult or impossible with the use of vertical wells, especially in cases of surface or subsurface obstructions caused by roads, buildings, or utilities.

In-situ air sparging can be implemented in a different manner, known as in-well aeration or vacuum vapor extraction. In this, air is injected into a well, lifting contaminated groundwater in the well and allowing additional groundwater to flow into the well (Figure 14.27). Once inside the well, some of the VOCs in the contaminated groundwater are transferred from the water to air bubbles, which rise and are collected at the top of the well by vapor extraction.

Dual-phase treatment is similar to air sparging in that it also combines soil and groundwater treatment of VOC contamination. Simultaneous treatment is achieved by removing both contaminated water and soil gases from a common extraction well under vacuum conditions, thereby reducing time and cost requirements (Figure 14.28).

Sparge trenches or curtains may be used instead of horizontal and/or vertical wells. Trenches and curtains are especially useful when the contaminant plume is migrating in a region of high groundwater flow or where the prevention of off-site migration is of the utmost concern (Figure 14.29). After digging, the trench is shored up with sheet piling and may be left open or backfilled with soil. The backfill must be of equal or higher permeability in order to prevent the flowing groundwater from circumventing the treatment zone. Additionally, cutoff wells are often used to assure that the groundwater is flowing in the proper direction. When designing a trench, larger dimensions actually do not lead to greater efficiency because it only leads to increased volumes of water that need to be treated; the parameters that have the greatest effect on efficiency include the Henry's constant(s) of the contaminant(s) and the gas sparge rate (Pankow et al., 1993).

An alternative to the use of combined air sparging and soil vapor extraction system is known as *biosparging*. During the implementation of biosparging, lower injection flow rates are used to prevent the buildup of excess soil gas pressure. The injected air increases the concentration of dissolved oxygen within the groundwater, stimulating aerobic biodegradation of subsurface by the native microbial population. To prevent the buildup of pressure, flow rates of approximately 1 ft^3/min are used. The feasibility of biosparging is dependent on subsurface conditions; in addition to the introduction of dissolved oxygen to serve as an electron acceptor, a suitable microbial population as well as essential microbial nutrients must be present in sufficient quantity. When the subsurface conditions are favorable, the contamination may be degraded and mineralized into harmless by-products such as carbon dioxide and water, depending on the specific metabolic reaction occurring. Because a soil vapor extraction system is not required, biosparging is cost-effective.

Fracturing of low-permeable soils and bedrock within the contaminated zone can enhance air sparging performance. In addition, air sparging can be comple-

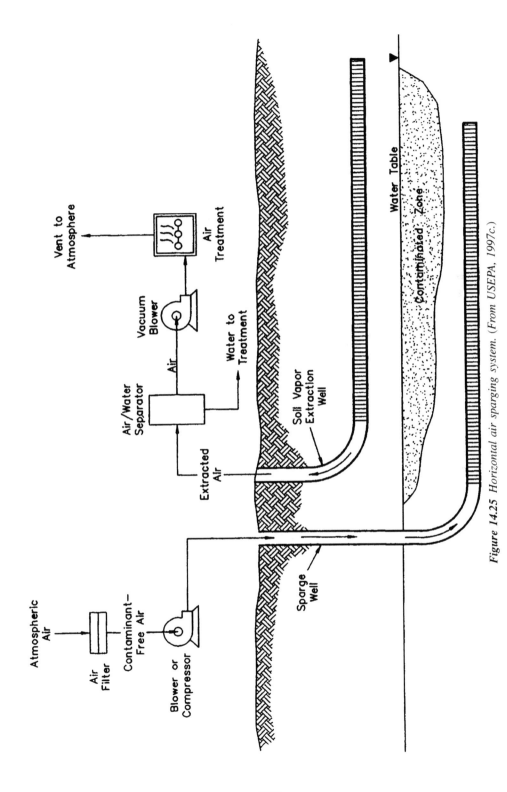

Figure 14.25 *Horizontal air sparging system. (From USEPA, 1997c.)*

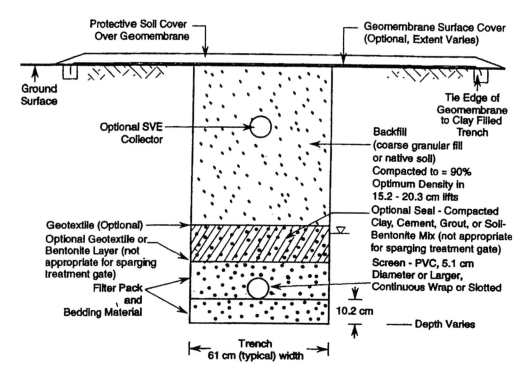

NOT TO SCALE

Figure 14.26 Horizontal air sparging well construction details. (From USACE, 1997a.)

mented with other technologies, such as pump and treat, thermal desorption, natural attenuation, and bioremediation. These technology combinations can lead to effective removal or degradation of the contaminants.

14.5.6 Economic and Regulatory Considerations

There are number of factors that affect the cost of air sparging such as type, concentration of contaminants, and extent of contamination, type of soils, site geology and geographic location, and remediation goals. One important consideration when evaluating the cost of air sparging as compared to other methods is the time required to complete remediation. Due to rapid remediation at air-sparged sites, these sites are generally remediated more rapidly than is typical with other

methods, limiting the actual cost of remediation, especially the cost of monitoring, and saving money by returning contaminated sites to productive use more quickly. The cost of air sparging is estimated to be $10 to $120 per cubic yard (USEPA, 1997c). Operation and maintenance costs are minimal.

From a regulatory perspective, many states require contractors to obtain construction-related permits and approvals from local governmental agencies to install air sparging systems. Permits may be required for all emissions associated with air sparging systems. Disposal or treatment of contaminated soils in conjunction with the construction, installation, or modification of air sparging systems also requires permits or approvals.

14.5.7 Case Studies

A number of case studies have been conducted to assess the effectiveness of air sparging. A summary of

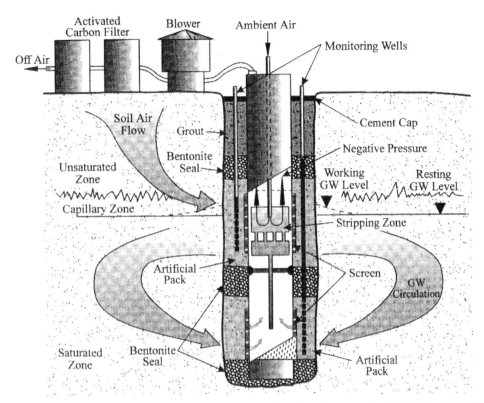

Figure 14.27 *In-well aeration system.* (*Reprinted with permission from Nyer et al., 1996.* (*Copyright CRC Press, Boca Raton, Florida.*)

published information on air sparging sites is shown in Table 14.19. Two sites where air sparging is used as the remediation technology are described briefly in this section.

Case Study 1: U.S. Department of Energy Savannah River Integrated Demonstration Site. Between 1954 and 1985, the USDOE dumped processed wastewater containing chlorinated solvents into a settling basin and stream in Aikin, South Carolina. The basin was not lined. Subsequently, high levels of trichloroethene (TCE) and tetrachloroethene (PCE) were detected in the nearby soils and groundwater. The USDOE used air sparging to remediate part of the site.

The soils on site consisted of permeable sand laced with clayey sediments. The concentration of TCE ranged from 0.67 to 0.629 mg/kg. The concentrations

of PCE ranged from 0.44 to 1.05 mg/kg. The ground water concentrations of TCE and PCE ranged from 10 to 1031 µg/L and 3 to 124 µg/L, respectively. The groundwater table is 120 ft below grade. The process used involved the horizontal injection well inserted 175 ft below grade, 55 ft below the water table, with a screened length of 310 ft. The horizontal extraction well was inserted 80 ft below grade in the vadose zone. It had a screened length of 205 ft. Initially, the vacuum pressure was 240 ft³/min and the injection pressure of the injection well was 200 ft³/min. A number of different types of injection were used in this demonstration. They included continuous injection of methane, pulsed injection of methane, and pulsed injection of methane plus continuous injection of trethyl phosphate to supply nitrogen for enhanced biodegradation. The monitoring and system control at this demonstration site were automated almost completely.

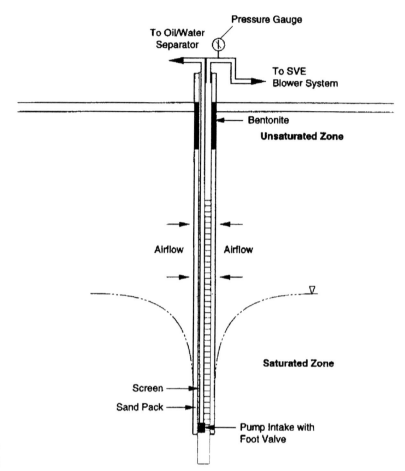

Figure 14.28 *Dual-phase extraction system. (From USACE, 1997a.)*

Over the course of this demonstration, about 13 months from February 1992 to April 1993, almost 17,000 lb of VOCs was degraded or removed. Of that sum, about 12,100 lb was removed by vacuum extraction. The remaining 4900 lb was mineralized or left to degrade by bioremediation. When compared to air sparging alone, mass-balance calculations suggest that 41% more VOCs was destroyed or removed using methane and nutrient injection. Increases in microbial populations, coupled with a decrease in TCE levels, indicate that biostimulation was greatest when pulsed methane injection was used.

This demonstration showed considerable success. TCE and PCE levels in the groundwater decreased by up to 95%, ultimately reaching undetectable levels (less than 2 $\mu g/L$ in some wells). This level was well below the drinking water standard of 5 $\mu g/L$. During the same period, soil gas TCE, and PCE declined by more than 99%. In most areas, the total sediment concentrations of TCE and PCE declined from 0.100 mg/kg to undetectable levels. After the 13 month duration of the test, the site was considered 80 to 90% clean.

Case Study 2: U.S. Department of Energy Mound Facility at Miamisburg, Ohio. In this example, air sparging in conjunction with high vacuum extraction has been used to expedite a cleanup of this site as an

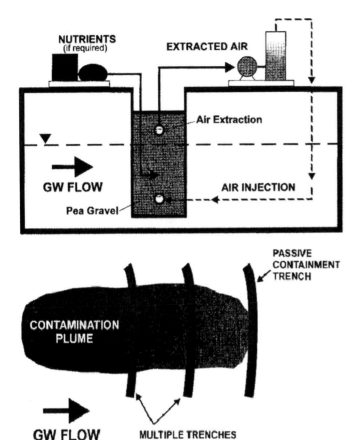

Figure 14.29 *Air sparging trench/curtain system. (Reprinted with permission from Nyer et al., 1996. Copyright CRC Press, Boca Raton, Florida.)*

addition to existing pump and treat remediation techniques. Pump and treat technology was effective in the hydraulic containment of the contamination to the aquifer, but that system's ability to clean the aquifer proved to be limited. Consequently, an air sparging system coupled with a high-vacuum extraction system was installed to remove dissolved-phase contaminants in the groundwater and to remove source contaminants in the vadose zone.

The groundwater contaminants include TCE, PCE, and 1,2-*cis*-dicloroethene (DCE), with the highest concentration being 200, 300, and 600 ppb, respectively. In addition vinyl chloride was present with a maximum concentration of 50 ppb. Finally, and unexpectedly, a high concentration of toluene, up to 220 ppm, were

encountered early in the operation of the system. The Mound facility was formerly a nuclear weapons component, research, and manufacturing facility. The southwestern portion of this site is located directly above the section of a sole-source aquifer providing drinking water to a large population.

This aquifer is made up of highly permeable silty sand and gravel and is quite shallow, in places only 20 ft below grade. Only a small portion of aquifer, approximately 2 acres in size, has been contaminated with the chlorinated solvents referred to above. Remediation of this site was complicated by the fact that this site is located partially below a landfill. To further complicate remediation, the soils above the aquifer include mixture lenses of glacial silts and sand mixed

Table 14.19 **Summary of published information on air sparging sites**[a]

Site	Soil Type	Contaminants[a]	Cleanup Time[b] (months)	Initial Groundwater Concentration (mg/L)	Final Groundwater Concentration (mg/L)
Isleta	Alluvial sands, silts, clays	Leaded gasoline	2	MW-1, -3, -5 BTEX 4, 18, 25	MW-1, -3, -5 BTEX 0.25, 8, 5
Conservancy	Silty sand, interfacing clay layer	Gasoline	5	Benzene 3–6	59% average benzene reduction after 5 months
Buddy Beene	Clay	Gasoline	2	NR	8.5% reduction/month
Bernalillo	NR	Gasoline	17	NR	BTEX and MTBE <5.5
Los Chavez	Clay	Gasoline	9	NR	40% benzene, xylene reduction, 60% toluene reduction, 30% ethyl benzene reduction
Arenal	NR	Gasoline	10	Benzene >30	Benzene <5
BF1	NR	Fuel	12	Benzene 22,000–32,000	Benzene 29–50
Bloomfield	NR	Fuel	48	NR	BTEX below cleanup standards
Firehouse	NR	Fuel	30	Benzene 400–600	Benzene 0.5–4
Dry cleaning facility	Coarse sand, natural clay barrier	PCE, TCE, DCE, TPH	4	Total VOCs 41	Total VOCs 0.897
Savannah River	Sands, thin clay lenses	TCE, PCE	13	TCE 10–1031 PCE 3–124	TCE <5, PCE <5
Former industrial facility	Sands	Toluene	23	NR	NR
Electro Voice	NR	VOC	12	NM[c]	NM[c]
Berlin	Sand, silty lenses, aquitard—clay	cis-1,2-DCE, TCE, PCE	24	cis-1,2-DCE > 2	cis-1,2-DCE >0.440
Bielefeld, Nordrhein Westfalen	Fill, sand, silt, aquitard—siltstone	PCE, TCE, TCA	11	PCE 27; TCE 4.3; TCA 0.7	Total VOCs 1.207
Munich, Bavaria	Fill, sand, gravel aquitard—clayey silt	PCE, TCE, TCA	4	PCE 2.2; TCE 0.4; TCA 0.15	PCE 0.539; TCE 0.012; TCA 0.002
Nordrhein, Westfalen	Clayey silt, sand, aquitard—siltstone	Halogenated hydrocarbons	4, 6	Location A: THH1.5–4.5 Location B: THH 10–12	Location A: THH 0.010; location B: THH 0.200
Bergischesland	Fractured limestone	Halogenated hydrocarbons	15	THH 80	THH 0.4
Pluderhausen, Baden, Wurternburg	Fill, silt gravel, aquitard—clay	TCE	2	1.20	0.23
Mannhelm-Kaesfertal	Sand	PCE, chlorinated hydrocarbons	NR	NR	NR
Gasoline service station	Sand and silt	Gasoline	24	Total BTEX 6–24	Total BTEX 0.38–7.6

Savannah River	Sand, silt, and clay	TCE, PCE	3	TCE 0.5–1.81; PCE 0.085–0.184	TCE 0.010–1.031; PCE 0.003–0.124
Gasoline service station	Fine coarse sand, gravel	Gasoline	3	Total BTEX 21	Total BTEX <1
Solvent spill	Quaternary sand and gravel	TCE, PCE	3	Total VOCs 33	Total VOCs 0.27
Solvent leak at degreasing facility	Fill, sandy and clayey silts	TCE	2	0.200–12	<0.010–0.023
Chemical manufacturer	Sandy gravel aquitard—clay	Halogenated hydrocarbons	9	THH 1.9–5.417	THH 0.185–0.320
Truck distribution facility	Sands	Gasoline and diesel fuel	Ongoing	Total BTEX 30	NR
Irvine	Clays, sandy silts, clayey sands and silts, gravel	Gasoline	9	NR	Below cleanup standards
New Paris	Sand with some gravel, clay layers	PCE, TCE	18	PCE 250	PCE 9

Source: USEPA (1997c).

[a] BTEX, benzene, toluene, ethylbenzene, and xylenes; DCE, dichloroethene; EPA, U.S. Environmental Protection Agency; MTBE, methyl *tert*-butyl ether; MW, monitoring well; NM, not measured; NR, not reported; PCE, tetrachloroethene; TCA, trichloroethane; TCE, trichloroethene; THH, total halogenated hydrocarbons; TPH, total petroleum hydrocarbons; VOC, volatile organic compounds.

[b] Cleanup times represent the time interval between initial and final groundwater concentration reported in the table. Actual remediation time may be longer.

[c] Demonstration assessed remediation capabilities for the vadose zone soil only.

with fill material. All these soils contained the same contaminants found in the groundwater. The concentration of these contaminants ranged around 1 ppb, but reached as high as 300 ppm. The soil in the entire area has been disturbed by past activities. Consequently, the distribution of both contaminants and soil type is inconsistent.

Pilot tests conducted in 1996 removed approximately 13 lb of VOC over a four-day period. This pilot test used an extraction well that was not screened within the till unit. It operated at a steady airflow rate at 80 ft^3/min with 13 in. Hg vacuum at the well bore. In a second test a screen extraction well located within a low-permeability till resulted in a very low mass removal. This test operated at a steady flow around 4 ft^3 with 24 in. Hg vacuum at the well bore. The radius of influence of air sparging in the gravel and permeable sand was approximately 20 ft at a flow rate of 20 ft^3/min. These pilot tests were used to design the wells. This was complicated by the uneven distribution of the contaminants and the uneven permeability of the overlying soils. In areas where little till was located, it was necessary to use a relatively long length screen. In contrast to areas of contaminated glacial till it was necessary that the SVE well be screened only within the low-permeability till, and the screen did not extend into an area of higher permeability. During the first six weeks of operation, results have shown that the system has removed 450 lb of contaminants.

14.6 MONITORED NATURAL ATTENUATION

14.6.1 Technology Description

Monitored natural attenuation (MNA) is the use of natural attenuation processes within the context of a carefully controlled and monitored site cleanup approach that will reduce contaminant concentrations, within a reasonable time frame, to levels that are protective of human health and the environment (USEPA, 1999c). MNA is also referred to as *intrinsic bioremediation, intrinsic remediation, natural attenuation, natural assimilation, natural restoration, natural recovery,* or *passive remediation.* MNA involves physical, chemi-

cal, or biological processes that act without human intervention to reduce the mass, toxicity, mobility, volume, or concentration of contaminants. Specifically, these processes include biodegradation, dispersion, dilution, sorption, volatilization, and chemical or biological stabilization or destruction of contaminants. These processes reduce the potential risk posed by the site contamination in three different ways: (1) transformation of contaminants into less toxic form, (2) reduction of contaminant concentrations, and (3) reduction of contaminant mobility and bioavailability through immobilization. MNA is appropriate if it can be demonstrated that remedial objectives within a reasonable time frame can be achieved. A reasonable time frame depends on the site-specific conditions (e.g., groundwater use at the site and public acceptance), but it is generally decided based on time comparable to that required for other remedial options.

MNA is considered at a contaminated site if (1) natural attenuation processes are observed or strongly expected to be occurring; (2) there are no human or ecological receptors that are likely to be affected or potential receptors in the vicinity of the plume are, or can be, protected; (3) it is protective of human health and the environment; (4) a continuing source that cannot cost-effectively be removed or contained and will require a long-term remedial effort; (5) alternative remediation technologies are not cost-effective or technically impractical; and (6) alternative remedial technologies pose significant added risk by transferring contaminants to other environmental media, spreading contamination or disrupting adjacent ecosystems (ITRC, 1999b).

Although MNA is primarily applicable for the degradation of organic contaminants (e.g., chlorinated solvents and BTEX), it can also be used for immobilization of inorganic contaminants, including heavy metals (lead, chromium, and cadmium) and radionuclides (cesium, uranium, and uranium) (USEPA, 1999c). It is applicable for sites where the contamination plume in groundwater is stable. In addition, MNA is applicable only if microorganisms capable of degrading or immobilizing the contaminants and the environmental conditions that favor the growth of the microorganisms exist at the site. MNA is typically used in conjunction with other remediation measures, such as source control or as a follow-up to such measures.

MNA has the following advantages (USEPA, 1999c):

- It does not generate remediation wastes, which eliminates the potential for cross-contamination. Being an in-situ method, it also reduces the risk of human exposure to contaminated soils and/or groundwater.
- It causes less disruption to the site. It only requires the installation of groundwater monitoring wells and piezometers.
- It causes the destruction of organic contaminants.
- It can be applied to all or part of given site.
- It can be used in conjunction with or as a follow-up to other remediation measures.
- It costs less compared to other remediation methods.

MNA has the following disadvantages (USEPA, 1999c):

- A longer time frame may be required to achieve remediation objectives.
- Site characterization may be more complex and costly. This is due to the need to identify the microbial activity already occurring on the site and the need to quantify the potential nutrient sources for further microbial degradation of the original contaminant.
- The toxicity of transformation products may exceed that of the parent compound. The toxicity of the daughter products, from microbial degradation, may be more toxic than the original parent compound.
- Long-term monitoring is required.
- Institutional controls may be required to ensure long-term productiveness.
- There is a potential for contaminant migration.
- Because of the longer time frames involved in MNA, the hydrogeologic and geochemical conditions, which originally made the site conducive to MNA, are likely to change over time. This

could result in the renewed mobility of previously stabilized contaminants. This is particularly true of those compounds that have been immobilized due to sorption.

- It requires extensive educational efforts to gain public support, convincing the public that this is not a "do-nothing approach" and is the most feasible option in terms of remediation techniques.

14.6.2 Fundamental Processes

The attenuation of contaminants during MNA is attributed to various physical, chemical, or biological processes (USEPA, 1998b; Azadpour-Keeley et al., 1999; NRC, 2000). The processes can be grouped in two categories: nondestructive processes and destructive processes. Nondestructive processes result in the reduction of the contaminant concentrations but not of the total contaminant mass. These processes include advection, mechanical dispersion, diffusion, sorption, dilution, and volatilization. Destructive processes are those that result in the degradation of the contaminant and are primarily the result of biodegradation.

Nondestructive Processes. Advection is a nondestructive attenuation process and is the most important process driving contaminants through soil. It is the transport of contaminants by the bulk movement of groundwater. This process is usually referred to as *plug flow*. The result is a *slug* of the contaminant. Following the passage of the slug, the concentration of the contaminant is again equal to that of the background concentration.

Hydrodynamic dispersion is the sum of molecular and mechanical dispersion. Except at extremely low groundwater velocities, mechanical dispersion is the dominant process of the two. Mechanical dispersion is the process in which the contaminant plume spreads in both the longitudinal and transverse directions of plume migration. Mechanical dispersion occurs as the result of variations in the speed of groundwater flow at the local level versus the overall speed of groundwater flow. These variations are the result of changing unit geometry, discontinuous units, contrasting lithologies, nonuniform permeabilities, and changes in directional permeabilities. The three physical properties that affect mechanical dispersion at the microscopic level are pore size, tortuosity, and friction in the pore throat. The larger the pore size, the slower groundwater will travel through the individual pore spaces. When the pore size is smaller than average, the groundwater is "forced" to travel faster through the pore spaces. This variation in velocities results in mixing of the contaminant with the groundwater. Tortuosity is the number of twists and turns that groundwater has to make while moving through the pore spaces. The less tortuous the path, the quicker the groundwater will travel through the pore spaces. Similar to a complex maze, the groundwater that has to travel a more tortuous path will take longer to "find its way through" the pore spaces. This variation in speed results in further mixing. The final mixing is the result of friction in the pore throat. Next to the individual soil particles, the friction is high and results in a slower-than-average groundwater velocity. In the open spaces between the individual soil particles, the friction is low, and this results in a higher-than-average groundwater velocity.

Sorption is the process in which the dissolved contaminants adhere to the individual soil particles as the contaminated groundwater passes through the soil. The mechanisms that cause sorption include London–van der Waals forces, Coulomb forces, hydrogen bonding, ligand exchange, covalent bonding between soil and contaminant (chemisorption), dipole–dipole forces, induced dipole forces, and hydrophobic forces. The mathematics behind predicting sorption amounts is based on several different models, including the Langmuir sorption model and the Freundlich sorption model. Sorption also results in one of the possible disadvantages of MNA. Following sorption of the individual contaminants, the concentrations will decline, but over time the characteristics of the groundwater may change (alkalinity and temperature), resulting in the desorption of contaminants, thereby increasing the concentrations of contaminants above acceptable levels at a later date.

Volatilization will generally occur only in the capillary fringe region of the groundwater table. The fac-tors that affect volatilization are contaminant concentration, the change in contaminant concentration with depth, the Henry's law constant and diffusion coefficient of the contaminant, and mass transport and the temperature of the groundwater. Except for volatile LNAPLs, volatilization is not taken into consideration when calculating the effects of natural attenuation.

Destructive Processes. The destructive natural attenuation processes result in the actual reduction in mass and concentration of contaminants. Biological degradation is the most important and dominant destructive process. It is due to the action of a number of microorganisms, including simple prokaryotic bacteria, cyanobacteria, eukaryotic algae, fungi, and bacteria. Microorganisms obtain their energy for growth from physiologically coupling oxidation and reduction reactions and using the energy that becomes available from these reactions. Under aerobic conditions, the microorganisms couple the oxidation of organic compounds with the reduction of oxygen. Under anaerobic conditions, the microorganisms can use compounds other than oxygen as electron acceptors such as nitrate, iron(III), sulfate, carbon dioxide, and many chlorinated hydrocarbons.

To understand biological processes involved in MNA, one first needs to understand oxidation–reduction reactions. The combination of the two is known as redox reactions, and the redox potential is the potential capacity of a soil to drive oxidation–reduction reactions. The redox potential of a soil is a site characterization parameter that needs to be measured for deciding if MNA is a potentially workable remediation technique. In a redox reaction, one compound donates an electron (oxidation) and another compound accepts (reduction) an electron. The flow of electrons from donors to acceptors is capable of doing work, and microorganisms use the work done by flowing electrons for growth and sustaining life functions. The biodegradation of hydrocarbons uses the hydrocarbons (contaminants) as the electron donor. The release of the energy drives further growth of microorganisms and thus the degradation of contaminants. The destruction of the contaminants will continue as long as there is an available food supply (contaminant as electron donor) and electron acceptors.

Microorganisms reproduce through the energy gained from oxidation–reduction reactions. From these reactions, cellular components such as membranes, proteins, deoxyribonucleic acid, and cell walls are created. The building blocks for these components come from the surrounding environment. Natural attenuation can occur when these building blocks happen to be contaminants such as chlorinated hydrocarbons. The chemical reactions are made possible by enzymes.

Aerobic and anaerobic respiration are the major pathways through which microorganisms reproduce and sustain themselves. In aerobic respiration, the electron donor (contaminant) is oxidized to produce an oxidized donor product. These oxidized donor products are generally referred to daughter products. In this case that daughter product is CO_2. This can potentially cause one of the disadvantages of MNA in that sometimes the daughter products may be more toxic than the original contaminant of concern at the site.

Microorganisms can use a number of other electron acceptors other than oxygen. These include Fe(III), sulfate, CO_2, and chlorinated solvents. The processes involved include methanogenesis, sulfate reduction, denitrification, reductive dechlorination, and iron(III) reduction. The electron donors can vary widely, but in general they are natural organic carbon or some type of organic contaminant. In general, the simpler the organic contaminant in a structure, the more susceptible the contaminant is to microbial degradation. The organic contaminants that resist biodegradation usually have complex molecular structures, low water solubility, an inability to support microbial growth, or they themselves may be toxic to the microorganisms.

The natural attenuation of metals in groundwater occurs through the following pathways: ion exchange and adsorption, oxidation or reduction reactions, precipitation and dissolution of solids, acid–base reactions, and complex formation. The concentration of metals in solution is most simply described by the distribution coefficient (K_d). The K_d value is almost directly dependent on the pH of the groundwater. (K_d was discussed in Chapter 8). The retardation of metals

also has an effect on concentration and is related to K_d by the retardation factor. Ion exchange and sorption affect the concentration, with the relative order of importance being lead > copper > zinc > cadmium > nickel.

14.6.3 System Design and Implementation

General Design Approach. MNA is generally evaluated using a *lines of evidence approach* (ITRC, 1999b). This approach forms the basis for all current protocols and guidance documents. The suggested lines of evidence are (1) documented reduction of contaminant mass at the site, (2) the presence and distribution of geochemical and biological indicators of natural attenuation, and (3) direct microbiological evidence. The first line of evidence is documented by reviewing historical trends in contaminant concentrations and distribution in conjunction with site geology and hydrogeology to show that a reduction in the total mass of contaminants is occurring at the site. The second line of evidence is documented by examining changes in the concentrations and distributions of geochemical and biochemical indicator parameters that have been shown to be related to specific natural attenuation processes. The third line of evidence is documented through laboratory microcosm studies, used to confirm specific natural attenuation processes and/or estimate site-specific biodegardation rates that cannot be demonstrated conclusively with field data alone. The need to collect the third line of evidence is evaluated on a case-by-case basis and is generally required only when field data supporting the first two lines of evidence are insufficient to adequately argue the feasibility of the natural attenuation approach.

The types of data that are required to support the three lines of evidence depend on the site conditions and the nature and extent of attenuation processes that are occurring. A conceptual model for a given site should be developed to determine what types of data are needed. The data needed may be collected in tiers (I, II, and III) as outlined in Table 14.20. Table 14.20 provides various geologic, hydrogeologic, chemical,

Table 14.20 **Data requirements for evaluation and implementation of natural attenuation**

Parameter	Data Type	Ideal Use, Value, Status, and Comments	Method	Data Collection Tier I	II	III	a
Geological							
Area geology	Topography/soil type/surface water/climate	Provides inferences about natural groundwater flow systems, identifies recharge/discharge areas, infiltration rates, evaluation of types of geological deposits in the area that may act as aquifers or aquitards	Consult published geological/soil/topographic maps, air photo interpretation, field geological mapping	×	×		
Hydrogeological							
Subsurface geology	Lithology/stratigraphy/structure	Identify water-bearing units, thickness, confined/unconfined aquifers, effect on groundwater flow and direction (anisotropy)	Use published hydrogeologic survey/maps	×	×	×	
			Review soil boring/well installation logs	×	×	×	
			Conduct surface or sub-surface geophysics			×	
Velocity	Hydraulic conductivity (K)/permeability (k)	Measure of the saturated hydraulic conductivity of the geological matrix (K) times the gradient gives the specific discharge (v); if site is very layered or complex, measure the vertical/horizontal K.	Estimate range based on geology	×	×	×	
			Conduct: pump, slug, or tracer tests		×	×	
			Estimates with grain size analysis			×	
			Permeability test			×	
			Downhole flow meter/dilution test				*
Gradient (h)		Measure of potential of the fluid to move (hydraulic gradient)	Water table and piezometric surface measurements	×	×	×	
Porosity (n)		Measure of the soil pore space; dividing the specific discharge by porosity gives the average linear groundwater velocity	Estimate range based on geology			×	
			Measure bulk and particle mass density			×	
Direction	Flow field	Estimate direction of groundwater flow	Water and piezometric contour maps	×	×	×	
			Downhole flow meter				*
Dispersion/sorption	Foc	Fraction of organic carbon (foc): used to estimate the retardation of chemical migration relative to the average linear groundwater velocity	Estimate or measure foc in soil samples, estimate from published values, or compare migration of reactive and nonreactive (tracer) chemicals in the groundwater	×	×	×	
	Dispersion	Longitudinal and horizontal dispersion (mixing) spreads out the chemical along the groundwater flow path	Estimate based on distribution of chemicals or use tracer tests	×	×	×	

578

Chemistry

	Parameter	Description	Method			
Organic chemistry	VOC	Identify parent solvents and degradation products; assess their distribution; certain specific isomers/degradation products provide direct evidence of biodegradation (e.g., *cis*-1,2- DCE), while others are formed due to abiotic degradation processes (e.g., formation of 1,1-DCE from 1,1,1-TCA); In addition, aromatic hydrocarbons (BTEX) and ketones can support biodegradation of cVOC	USEPA Method 8240	X	X	X
	Semi VOC	Selected semi-VOCs (e.g., phenol, cresols, alcohols) may support biodegradation of cVOCs	USEPA Methods 8270, 8015M			X
	Volatile fatty acids	Organic chemicals such as acetic acid can provide insight into the types of microbial activity that is occurring and can also serve as electron	Standard analytical methods or published modified methods using ion chromatography			X
	Methane, ethene, ethane, propane, propene	Provide evidence of complete dechlorination of chlorinated methanes, ethenes, and ethanes; methane also indicates activity of methanogenic bacteria; isotope analysis of methane can also be used to determine its origin	Modified analytical methods, GC-FID		X	X
Area geology	TOC/BOD/COD /TPH	Potential availability of general growth substrates	USEPA Methods 415.1, 405.1		X	X
	Alkalinity	Increased levels indicative of carbon dioxide production (mineralization of organic compounds)	USEPA Method 310.1		X	X

Table 14.20 (*continued*)

Parameter	Data Type	Ideal Use, Value, Status, and Comments	Method	I	II	III	a
Inorganic/physical	Ammonia	Nutrient; evidence of dissimilatory nitrate reduction; serves as aerobic co-metabolite	USEPA Method 350.2		×	×	
	Chloride	Provides evidence of dechlorination, possible use in mass balancing, may serve as conservative tracer; road salts may interfere with chloride data interpretation	USEPA Method 300.0	×	×		
	Calcium/potassium	Used with other inorganic parameters to assess the charge-balance error and accuracy of the chemical analysis	USEPA Method 6010			×	
	Conductivity	Used to help assess the representativeness of water samples and assess well development after installation (sand pack development)	Electrode measurement in the field; standard electrode	×	×		
	Dissolved oxygen (DO)	Indicator of aerobic environments; electron acceptor	Use flow-through apparatus to collect representative DO measurements by electrode	×	×		
	Hydrogen	Concentrations in anaerobic environments can be co-related with types of anaerobic activities (i.e., methanogenesis, sulfate and iron reduction); therefore, this parameter is an excellent indicator of the redox environment; hydrogen may be the limiting factor for completing dechlorination of cVOCs	Field measurement; flow-through cell equipped with bubble chamber; as groundwater flows past chamber, hydrogen gas will partition into headspace; headspace sampled with gastight syringe and analyzed in the field using GC; equipment for analysis not yet widely available; relationship to dechlorination activity is still unclear and subject to further R&D				*
	Iron	Nutrient; ferrous (soluble reduced form) indicates activity of iron reducing bacteria; ferric (oxidized) is used as an electron acceptor	USEPA Method 6010A		×	×	
	Manganese	Nutrient; indicator of iron and manganese reducing conditions.	USEPA Method 6010			×	
	Nitrate	Used as an electron acceptor by denitrifying bacteria, or is converted to ammonia for assimilation	USEPA Method 300.0		×	×	
	Nitrite	Produced from nitrate under anaerobic conditions	USEPA Method 300.0		×	×	

Inorganic/physical	pH	Measurement of suitability of environment to support wide range of microbial species; activity tends to be reduced outside pH range of 5–9, and anaerobic microorganisms are typically more sensitive to pH extremes; pH is also used to help assess the representativeness of the water sample taken during purging of wells	pH measurements can change rapidly in carbonate systems and during degassing of groundwater; therefore, pH measurements must be measured immediately after sample collection or continuously through a flow-though cell	×	×	×

Microbiology

Biomass	Microorganisms per unit soil or groundwater	Microbial population density between impacted and nonimpacted/treated areas can be compared to assess whether microbial population responsible for degradation observed; the value of biomass measurements is still being explored for cVOC biodegradation	Three general techniques available: culturing (plate counts, BioLog, MPN enumerations); direct counts (microscopy); and indirect measurement of cellular components (ATP, phospholipid fatty acids)	*
	Bioremediation rate and extent	Demonstrate that indigenous micro organisms are capable of performing the transformations predicted; determine nutrient requirements and limitation; measure degradation rates and extent	Varied; shake flasks, batch, column, bioreactor designs	*
	Species/general functional group	The presence of certain microbial species of functional groups (e.g., methanogenic bacteria) that have been correlated with cVOC biodegradation can be assessed; research is being conducted to identify patterns of microbial composition that are predictive of successful cVOC biodegradation	Three general techniques available: culturing and direct counts, indirect measurement of cellular components, and molecular techniques (16S RNA, DNA probes, RFLP)	*

Source: ITRC (1999b)

*indicates that parameter is optional depending on site complexity.

and microbiological parameters and their measurement methods. Collection of all parameters may not be required for all sites.

A systematic approach may be followed to evaluate whether natural attenuation is occurring. This would consist of identifying and collecting additional data that support the three lines of evidence of natural attenuation and integrating natural attenuation into the long-term site remediation/management strategy. Figure 14.30 shows a general approach to evaluate and implement MNA at sites. This approach involves the following steps: (1) review of available site data, (2) review/develop the site conceptual model, (3) screen the data for evidence of natural attenuation and develop hypothesis to explain the attenuation processes, (4) identify additional data requirements, (5) collect additional data, (6) refine the site conceptual model, (7) interpret the data and test/refine conceptual model, (8) conduct an exposure pathway analysis, and (9) if accepted, integrate natural attenuation into the long-term site management strategy. More details on each of these steps may be found in ITRC (1999b).

Equipment. No equipment is required other than installing monitoring wells. Standard equipment such as pumps and samplers is needed for collecting groundwater samples.

Operational Parameters. MNA is a passive remediation technology; therefore; no operational controls are involved.

Monitoring. A monitoring program is crucial to evaluating remedy effectiveness and to ensure protection of human health and the environment. The monitoring program is designed to accomplish the following (USEPA, 1999c):

- Demonstrate that natural attenuation is occurring according to expectations.
- Detect changes in environmental conditions (hydrogeologic, geochemical, microbiological, or other changes) that may reduce the efficiency of any of the natural attenuation processes.

- Identify any potentially toxic and/or mobile transformation products.
- Verify that the plume is not expanding (either downgradient, laterally, or vertically).
- Verify that there is no unacceptable impact on downgradient receptors.
- Detect new releases of contaminants to the environment that could affect the effectiveness of the natural attenuation remedy.
- Demonstrate the efficiency of institutional controls that were put in place to protect potential receptors.
- Verify the attainment of remedial objectives.

The monitoring program includes sampling of monitoring wells installed within the plume to determine contaminant concentrations as well as sampling of monitoring wells installed outside the contaminated zone to determine background concentrations. The data from monitoring wells within the plume are compared to background wells. The number and location of monitoring wells are determined based on the plume geometry, site complexity, source strength, groundwater flow, and distance to receptors (ASTM, 1998). Specifically, site-specific factors such as depth to water table, hydraulic conductivity and gradient, direction of groundwater flow, storage coefficient or specific yield, vertical and horizontal hydraulic conductivity distribution, direction of plume movement, and the effects of any human-made or natural influences (e.g., underground utilities and ravines) are considered carefully in selection of the monitoring well locations.

There are three types of monitoring: (1) site characterization to describe disposition of contamination and forecast its future behavior, (2) validation monitoring to determine whether the prediction of site characterization is accurate, and (3) long-term monitoring to ensure that the behavior of the contaminant plume does not change (USEPA, 1998b). After the initial site characterization and development of a conceptual model for the site, validation monitoring is required to verify if the forecast of the conceptual model is adequate. The frequency of validation monitoring depends on the natural variability in contaminant concentra-

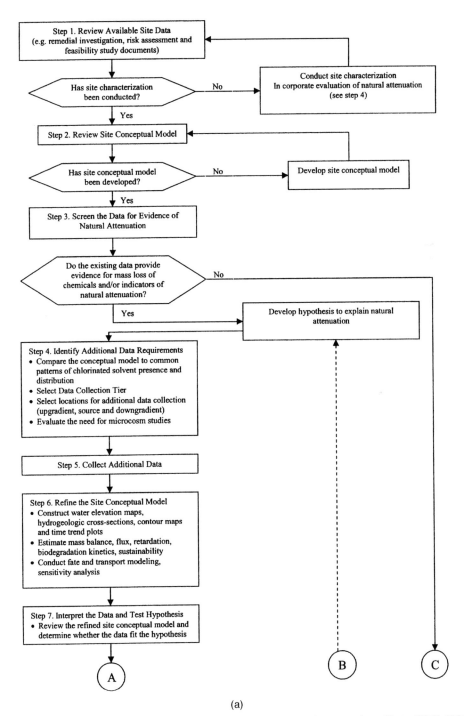

(a)

Figure 14.30 *General approach to evaluate and implement natural attenuation.* (*From ITRC, 1999b.*)

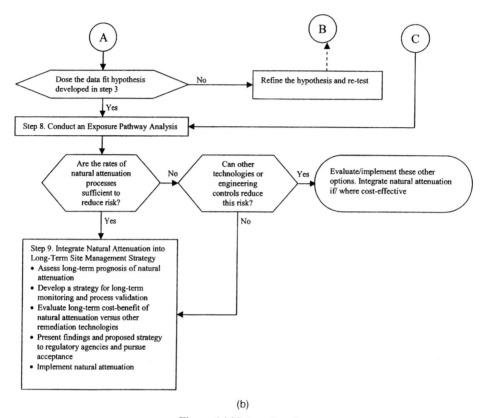

(b)

Figure 14.30 (*continued*)

tions, the distance and time of travel from the source to the location where the acceptance criteria are applied, and the reduction in contaminant concentration required to meet the acceptance criteria. If validation monitoring documents that natural attenuation will meet the acceptance criteria, a program of long-term monitoring should be implemented. The interval of sampling should be related to the expected time of travel of the contaminant along the flow path from one monitoring well to the next.

Figure 14.31 shows a typical monitoring well network, and Table 14.21 lists various elements of the monitoring program. An adequate number of samples is required if statistical analysis of the data is to be performed. Monitoring is continued until remediation objectives have been achieved, and longer if necessary to verify that the site no longer poses a threat to human health or the environment. A contingency remedy is required if MNA fails to perform as anticipated. A contingency remedy can include any of the remediation technologies described in this chapter, including engineered bioremediation.

14.6.4 Predictive Modeling

Two types of modeling are required for MNA. The first one involves conceptual modeling to provide a general understanding of the site conditions based on the site characterization data. This conceptual modeling also allows for determination if additional monitoring wells are needed to monitor the progress of natural attenuation. Once a conceptual model is developed, validation monitoring (as explained earlier) is required to validate the conceptual model. When validation monitoring shows that the conceptual model can reasonably

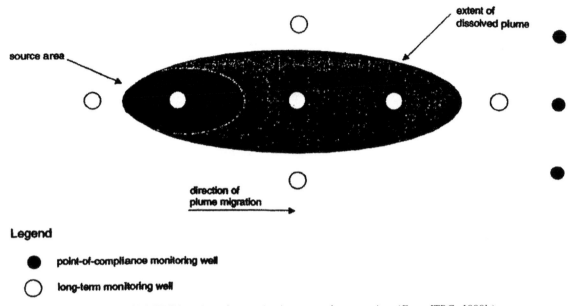

Figure 14.31 Well locations for monitoring natural attenuation. (*From ITRC, 1999b.*)

represent the natural attenuation processes, mathematical models are used to predict long-term performance. Long-term monitoring is performed to validate the model results.

The mathematical models selected to predict the long-term performance of MNA should incorporate all of the physical, chemical, and biological processes considered in the conceptual model. Contaminant fate and transport models such as BIOPLUME can be used to model MNA (Rifai et al., 1989). Screening models such as BIOSCREEN and BIOCHLOR are often used to assess the MNA (Newell et al., 1996, 1998). Calibration of these models based on the site-specific data is performed first. Then the model is used to predict future extent and concentrations of the contaminants in the groundwater.

14.6.5 Modified or Complementary Technologies

Source removal is required prior to using MNA. For the source removal, remediation methods such as pump and treat and air sparging may be used. If the monitoring results show that MNA is not adequate to treat the contaminant plume, other remediation technologies described in this chapter should be evaluated.

14.6.6 Economic and Regulatory Considerations

Monitoring costs are the only costs involved in MNA; therefore, it is cost-effective compared to any other remediation technique. USEPA issued a directive on the use of MNA at Superfund, RCRA corrective action, and underground storage tank sites (USEPA, 1999c). The directive's aim is to promote consistency in the manner in which MNA remedies are proposed, evaluated, and approved. USEPA does not consider MNA to be a presumptive or default remedy—it is merely one option that should be considered with other applicable remedies. Thus, during the process of selecting a site remedy, MNA may be evaluated and compared with other remediation technologies and chosen upon the attainment of remedial goals within a time frame that is comparable to the other remedial techniques. EPA considers source control and long-term monitor-

Table 14.21 **Elements of natural attenuation monitoring plan**

Monitoring Well	Location[a]	Purpose	Parameters	Sampling Frequency
Upgradient	Upgradient of the source, preferably along the groundwater flow path	Purpose is to be most representative location for background conditions	VOCs, all organic, inorganic, and field parameters that are sampled in downgradient wells	Every event as dictated by the regulatory agency involved; frequency may decline over time[b]
Source	Closest well to source area along groundwater flow path	Purpose is to be representative of source term; installation of new well is usually not warranted	VOCs, all organic, inorganic, and field parameters that are relevant to site pattern[c]; measure NAPL if present	As above
Downgradient 1	Downgradient of source area, about one-third of the way along the groundwater flow path	Primary purpose is to monitor VOCs and other parameters as they migrate from source area	VOCs, all organic, inorganic, and field parameters that are relevant to site pattern	As above
Downgradient 2	Downgradient of source area, about two-thirds of the way along the groundwater flow path	Primary purpose is to monitor change (decline) in VOCs and other parameters along flow path	VOCs, all organic, inorganic, and field parameters that are relevant to site pattern	As above
Downgradient 3	Downgradient of source area, slightly beyond the end of the plume	Primary purpose is to document downgradient limit of VOCs and change in redox and other parameters along the flow path	VOCs, all organic, inorganic, and field parameters that are relevant to site pattern	As above
Transgradient 1	Transgradient of the plume, about halfway along the groundwater flow path	Primary purpose is to document lateral movement (stability) of VOC plume	VOCs, all organic, inorganic, and field parameters that are relevant to site pattern	As above
Transgradient 2	Transgradient of the other side plume, about halfway along the groundwater flow path	Primary purpose is to document lateral movement (stability) of VOC plume	VOCs, all organic, inorganic, and field parameters that are relevant to site pattern	As above
Compliance (1–3)	Two or three wells perpendicular to the groundwater flow path and slightly upgradient of the compliance boundary	Primary purpose is regulatory; documenting that the plume is stable and has not crossed the compliance boundary	VOCs alone should be sufficient; any organic or inorganic parameters still present at high concentrations at downgradient 3 suggested	As above

Source: ITRC (1999b).

[a] All wells must be screened in the hydrogeological unit that contains the plume.

[b] Frequency may be lower for sites with low hydraulic conductivity; frequency may also decline for some organic or inorganic parameters over time; frequency for field parameters and VOC will probably remain constant.

[c] Field parameters would include DO and redox; inorganic parameters would include chloride and relevant electron acceptor indicators such as sulfate or iron; organic parameters would include methane, VFAs and any specific or general electron donor indicators (hydrocarbons, ketones, alcohols, TOC, etc.).

ing to be the fundamental components of any MNA remedy.

Several protocols and policies have been developed to assist in the use of MNA at contaminated sites by the USEPA, ASTM, and state regulating agencies (e.g., USEPA, 1998a,b; 1999c; ASTM, 1998). These protocols and policies are very helpful in planning, implementing, and interpreting MNA investigations.

14.6.7 Case Studies

Several case studies have been reported where MNA has been used successfully to remediate contaminated sites. Selected case studies are summarized in Table 14.22.

14.7 BIOREMEDIATION

14.7.1 Technology Description

Bioremediation, also referred to as *in-situ bioremediation, biorestoration, bioreclamation,* or *engineered bioremediation,* is a technology to restore contaminated groundwater (Thomas and Ward, 1989; Sims et al., 1992; USEPA, 1993). Unlike MNA, which occurs naturally, bioremediation requires human intervention to create conditions that stimulate the growth of microorganisms to degrade or immobilize contaminants. The principles of bioremediation of groundwater are the same as bioremediation of soils, as discussed in Chapter 13. Bioremediation of soils can be accomplished under ex-situ or in-situ conditions; however, bioremediation of groundwater is commonly accomplished under in-situ conditions. Bioremediation of contaminated groundwater is achieved by stimulation of indigenous microorganisms to degrade or immobilize the contaminants present in the groundwater. The microorganisms are stimulated by injection of nutrients, additional oxygen sources, or other electron acceptors; the process is known as *biostimulation.* In addition to stimulating indigenous microbial populations, microorganisms with specific metabolic capabilities may be introduced into the aquifer; this process is known as *bioaugmentation.*

A typical bioremediation system consists of pumping and injection wells. Groundwater is recovered, enriched with biomass and/or nutrients, and reintroduced through injection wells as shown in Figure 14.32. Biomass and/or nutrients may also be introduced to the aquifer through the use of infiltration galleries as shown in Figure 14.33. Infiltration galleries allow movement of the injection solution through the unsaturated zone, resulting in potential treatment of source materials that may be trapped in the pore spaces of the unsaturated zone.

Bioremediation is ideally applicable for homogeneous and permeable ($k > 10^{-4}$ cm/s) aquifers with low groundwater gradient. If aquifer hydraulic conductivity is very low, it is difficult to add nutrients to or to remove by-products from. It is used at sites where a contaminant is originating from a single source and as a follow-up action to complete free-product removal. Bioremediation is commonly used for organic contaminants, but its use for metals is also considered in a limited number of studies. The complexity of the molecular structure is directly related to biodegradation. A simple benzene ring is readily degradable, but complex long-chain compounds with multiple rings or rings with substituted chlorines are degraded more slowly. A scoring system that considers the site and contaminant characteristics, shown in Table 14.23, may be used to indicate if bioremediation is feasible at a site.

Bioremediation has the following advantages:

- It transforms the toxic contaminants into harmless compounds.
- It can be used to treat contaminants that are sorbed to aquifer materials or trapped in pore spaces.
- Being an in-situ method, it causes less disturbance to the site and low contaminant exposure to the public.
- Time required to treat can be faster than technologies that involve withdrawal and treatment processes (e.g., pump and treat).
- It costs less than other remediation technologies.
- There is a larger areal zone of treatment than for other remediation technologies.

Table 14.22 **Selected natural attenuation case studies**

Site	Location, Facility Type, Date	Geology	Predominant Redox	Parent / Daughter Chemicals	Electron Donors	Microbial Process	Studies to Date
1	Toronto, Ontario Chemical transfer facility 1989–present	Low-K silt till Shallow, <30-ft bgs	Anaerobic	PCE, ethene	Methanol, acetate	Methanogenesis Acetogenesis Sulfate reduction	NA investigation Laboratory study Eight in-situ microcosms NA remedy
2	Sacramento, CA Industrial facility 1994–present	Unconsolidated alluvium Silty sand, gravel Deep, 70-ft bgs	Generally aerobic Anaerobic source	TCE, ethene, VC, CO₂, TCA ethane, CF, DCM	Septage	Methanogenesis Aerobic oxidation Cometabolic oxidation	NA investigation (groundwater) laboratory study NA investigation (unsaturated zone) Site-wide intrinsic review (>50 sites)
3	Auburn, NY Industrial facility 1992–present	Overburden fractured bedrock	Anaerobic	TCE, ethene	Acetone, methanol	Methanogenesis Acetogenesis Sulfate reduction Iron reduction	Intrinsic biodegradation investigation conceptual design
4	Portland, OR Chemical transfer facility 1995–present	Sand and fill Shallow, <30-ft bgs Clay confining layer	Anaerobic	PCE/TCE ethene, DCM, toluene, xylene	DCM (acetate), alcohols, TEX	Methanogenesis Acetogenesis Iron reduction Sulfate reduction	NA investigation (groundwater) Laboratory study of NA and enhanced (proposed)
5	Kitchener, Ontario Industrial facility 1995–present	Silty sand Shallow, 30-ft bgs Clay confining layer	Aerobic/anaerobic source	TCE ethene	Acetone, DCM, TX	Methanogenesis Acetogenesis Sulfate reduction Iron reduction	NA investigation (groundwater) Laboratory study of NA and enhanced Pilot test (ongoing)
6	Farmington, NH Landfill 1995–present	Landfill Silty sand to 65-ft bgs Bedrock	Aerobic/anaerobic source	TCE, DCE, VC, trace ethene DCM, TEX, ketones	TX, DCM, ketones	Acetogenesis Methanogenesis Cometabolic oxidation	NA investigation (groundwater) laboratory study
7	St. Joseph, MI Industrial facility 1991–present	Fine and medium sand 65 to 95-ft bgs	Background aerobic Anaerobic plume	TCE VC, ethene	Unidentified TOC	Methanogenesis Sulfate reduction	NA investigation (groundwater) Laboratory tests of enhanced (cometabolic)
8	Plattsburg, NY Air force base 1995–present	Fine and medium sand 0 to 90 ft bgs	Background aerobic Anaerobic plume	TCE, VC, ethene	BTEX, jet fuel	Methanogenesis Sulfate reduction Iron reduction	NA investigation (groundwater)

No.	Site	Geology/Material	Conditions	Chlorinated compounds	Other contaminants	Redox processes	Study type
9	New Jersey Picatinny Arsenal 1992?–1999?	Fine to course sand, Discontinuous silt/clay To 50 to 70 ft bgs	Background aerobic Anaerobic plume	TCE, VC, ethene	Unidentified TOC	Methanogenesis Sulfate reduction Iron reduction	NA investigation (groundwater)
10	Dover, DE Dover Air Force Base 1995–present	Fine to course sand, Some silt To 30 to 60-ft bgs	Background aerobic Anaerobic source	TCE, VC, ethene	BTEX, jet fuel, unidentified TOC	Methanogenesis Cometabolic oxidation Aerobic oxidation	NA Investigation (groundwater) Lab study of cometabolic bioventing Cometabolic bioventing pilot test Enhanced aerobic pilot test
11	Alaska Eielson Air Force Base 1992–present	Coarse sand and gravel To 180 to 300-ft bgs	Background aerobic Anaerobic plume	TCE, VC, ethene	BTEX, jet fuel	Methanogenesis Sulfate reduction Iron reduction	NA investigation (groundwater)
12	Oscoda, MI Wirtsmith Air Force Base 1994–present	Medium to fine sand Coarse sand and gravel To 60 to 90-ft bgs	Background aerobic Anaerobic plume	PCE/TCE DCE, trace VC	BTEX, jet fuel	Methanogenesis Sulfate reduction Iron reduction	NA investigation (groundwater)
13	Richmond, CA Chemical plant 1996–present	Esturial deposits of clay, silts and sands To 130-ft bgs	Background aerobic Anaerobic source	PCE/TCE VC and ethene	Unidentified TOC	Some unidentified Sulfate reduction	NA investigation (groundwater)
14	Niagara Falls, NY Landfill ~1994	Overburden Fractured bedrock	Background aerobic Anaerobic plume	PCE/TCE VC TCA, DCA, CA, CT, CF, DCM, CM, ethene and ethane	Landfill leachate, other chemicals	Methanogenesis Sulfate reduction	NA investigation (groundwater)
15	Niagara Falls, NY Chemical plant ~1994	Overburden Fractured bedrock	Background aerobic Anaerobic plume	PCE/TCE VC, TCS, DCA, DCM	DCM, others	Methanogenesis Sulfate reduction	NA investigation (groundwater)
16	Hawkesbury, Ontario Carpet manufacturing plant ~1992	Till, reworked sand Silts over unweathered Sandy silt and fractured bedrock	Background aerobic Anaerobic source	PCE/TCE VC, TCA DCA and CA, ethene and ethane DCM	DCM, methanol, naphtha	Methanogenesis Sulfate reduction Iron reduction Acetogenesis	NA investigation (groundwater)
17	Gulf Coast Chemical Plant ~1995	Peat, clay, and silt layers	Aerobic and anaerobic	1,2-DCA, 2-chloroethanol, ethanol, ethene, ethane	1,2-DCA, unidentified TOC	Methanogenesis Sulfate reduction Iron reduction	NA investigation (groundwater) Laboratory study
18	Netherlands VC production plant ~1995	4 m of sand fill over natural material	Aerobic and anaerobic	1,2-DCA, VC, ethene, ethane	1,2-DCA, unidentified TOC	Methanogenesis Sulfate reduction	NA investigation (groundwater)

Table 14.22 (*continued*)

Site	Location, Facility Type, Date	Geology	Predominant Redox	Parent/ Daughter Chemicals	Electron Donors	Microbial Process	Studies to Date
19	Louisiana VC production plant ~1995	Very fine sand to 20-ft clay aquitard	Background aerobic anaerobic plume	VC release, ethene	VC, unidentified TOC	Methanogenesis Sulfate reduction Iron reduction Aerobic oxidation	NA investigation (groundwater)
20	Cecil County, MD Landfill 1995–1996	Sand and fill over fractured saprolitic bedrock	Background aerobic anaerobic plume	VC release	VC	Aerobic oxidation Anaerobic oxidation	NA investigation (groundwater)
21	Pinellas, FL DOE facility 1995–present	Marine deposits Fine sand, some silt, clay <30-ft bgs	Anaerobic	TCE, VC (suspect ethene), DCM CM	BTEX, ketones, DCM	Unidentified	Pilot test of enhanced
22	Canoga Park, CA Industrial facility 1996–present	Shallow overburden over fractured bedrock	Generally aerobic Anaerobic source	TCE VC, TCA DCA, CT, CF, DCM, 1,2-DCA	Benzene, DCM	Methanogenesis Nitrate reduction Aerobic degradation	NA investigation (groundwater)
23	Sacramento, CA Industrial facility 1996–present	Unconsolidated alluvium Silty sand, gravel Deep, 70-ft bgs	Generally aerobic Anaerobic source	TCE VC and ethene, CF, DCM, 1,2-DCA ethene, 1,2-DCA CO_2	Alcohols, acids, ketones, BTEX	Methanogenesis Nitrate reduction Sulfate reduction Aerobic degradation	NA investigation (groundwater)
24	Ogden, Utah Hill Air Force Base 1995–present	Interbedded clay, silt, sand, gravel 15 to 110-ft bgs	Generally aerobic	PCE/TCE, DCE	None	Unidentified (abiotic processes and possibly natural phytoremediation)	NA investigation (groundwater and vadose)

Source: ITRC (1999b).

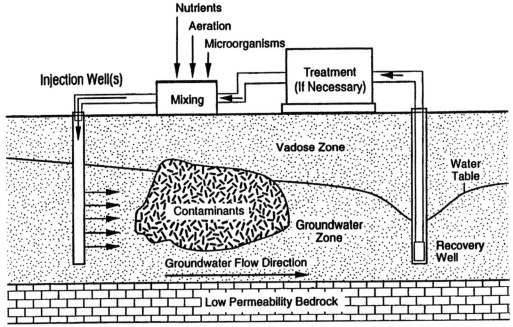

Figure 14.32 *In-situ bioremediation of groundwater using extraction–injection system.* (*From USEPA, 1993.*)

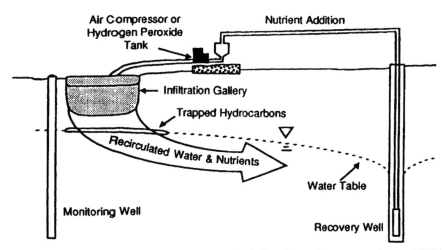

Figure 14.33 *In-situ bioremediation of groundwater using infiltration gallery system.* (*From USEPA, 1996a.*)

Table 14.23 **Screening criteria for in-situ bioremediation**

Parameters	Score
1. Contaminant characteristics	
A. Structure	
i. Simple hydrocarbon C_1 to C_{15}	0
ii. C_{12}–C_{20}	−1
iii. $> C_{20}$	−2
iv. Alcohols, phenols, amines	0
v. Acids, esters, amides	0
vi. Ethers, monochlorinated, nitro	−1
vii. Multichlorinated	−2
B. Sources	
i. Well-defined point sources	+1
ii. Underdefined multiple sources	−1
2. Hydrogeology	
A. Aquifer permeability (cm/s)	
i. $> 10^{-3}$	0
ii. 10^{-3} to 10^{-4}	−1
iii. 10^{-4} to 10^{-5} or less	−2
B. Aquifer thickness (ft)	
i. > 20	+1
ii. 5–20	0
iii. >5	−1
C. Depth to aquifer (ft)	
i. > 20	+1
ii. 5–20	0
iii. <5	−1
D. Homogeneity	
i. Uniform	+1
ii. Nonuniform	−1
3. Soil and groundwater chemistry	
A. Groundwater pH	
i. >10	−2
ii. 8–10	−1
iii. 6.5–8	0
iv. 4.5–6.5	−1
v. < 4.5	−2
B. Groundwater chemistry	
i. High NH_4 and Cl	−0.5
ii. Heavy metals (As, Cd, Hg)	−0.5

Interpreting the total score

0 or greater	Site appears suitable
−1 to −2	Possible areas of concern
−2 to −4	Areas of significant concern or other options advantageous
less than −4	Success is unlikely

Source: HWRIC (1995).

Bioremediation has the following disadvantages:

- Many organic and inorganic contaminants are resistant to degradation.
- Partial degradation of the contaminants may result in toxic by-products.
- Heavy metals and higher concentrations of organic contaminants may inhibit activity of indigenous microorganisms.
- Low degradation rates cause longer treatment times relative to active remedial technologies (e.g., air sparging).
- Clogging of injection wells from microbial growth results from addition of nutrients and oxygen.
- It is difficult to implement in low permeability or heterogeneous aquifers that do not allow the transport of adequate supplies of nutrients and oxygen to the microbes in the contaminated zone.
- It difficult to control and/or maintain ideal environmental conditions (e.g., temperature, pH).
- Extensive monitoring and maintenance is required.

14.7.2 Fundamental Processes

The fundamental processes involved in bioremediation of groundwater are the same as the ones described for bioremediation of soils. Bioremediation of organic contaminants is more common than bioremediation of metal contaminants. Bioremediation of organic compounds is accomplished by the complete mineralization of contaminants into carbon dioxide, water, inorganic salts, and cell mass in the presence of oxygen. For hydrocarbons, this can be represented as

$$\text{hydrocarbon} + O_2^- \rightarrow CO_2 + H_2O$$
$$+ \text{ cell material } + \text{ energy}$$

Under anaerobic metabolisms (absence of oxygen), the organic contaminants are converted into methane, carbon dioxide, and cell mass. A variety of aerobic and anaerobic microbes are capable of degrading the organic contaminants, but anaerobic degradation is generally slower and less complete.

The requirements for biodegradation are (1) microorganisms acclimated to the contaminant and environment, (2) nutrients, and (3) favorable environmental conditions. Bacteria are the predominant microorganisms found in the subsurface. Sometimes, genetically engineered strains are used to inoculate soils and/or supplement the natural population; however, the naturally occurring bacteria, already adapted to the environment, tend to perform better than the nonindigenous bacteria.

The microbes use a wide range of compounds as carbon and energy sources, including hydrocarbons, chlorinated solvents, and PAHs. The rate and extent of biotransformation of organic compounds at a site depends on the geochemical and hydraulic properties of the subsurface. Microbial growth continues as long as mineral nutrients, substrates for growth, or suitable electron acceptors are available. In aquifers, the microbial activity may be limited because of low solubility of oxygen in water and inadequate distribution of nutrients and electron acceptors in the contaminated zone.

Bioremediation aims to increase biological activity by supplying required oxygen (in case of aerobic metabolism), nutrients, and microbes, if necessary. Nutrients necessary for microbial growth include inorganic phosphates, nitrogen in the form of ammonia, and micronutrients (e.g., potassium, iron, sulfur, magnesium, calcium, and sodium). Sources of oxygen include air, hydrogen peroxide, or oxygen-releasing compounds (ORCs). A pumping and injection well system (shown in Figure 14.32) or infiltration gallery system (shown in Figure 14.33) is used to introduce the nutrients, oxygen, and microbes into the contaminated zone of the aquifer.

The rate at which contaminants biodegrade depends not only on the oxygen supply but also on the rate of bioavailability. The rate of contaminant dissolution or the rate at which contaminants diffuse into a flow channel controls the rate of remediation. This phenomenon is most prevalent if the contamination exists as residual NAPL phase or has migrated into low-permeability soils. The remediation in such situations is limited by the diffusion rates of contaminants, leading to longer treatment time.

Although aerobic biodegradation is common, anaerobic biodegradation can be accomplished in oxygen-poor environments. For anaerobic biodegradation, nitrate is commonly used as an electron acceptor instead of oxygen because of its high solubility in water, low cost, and support of the same general biochemical pathways as oxygen-based degradation. Researchers have also been looking at the role of sulfate-reducing, iron-reducing, and methanogenic biodegradation processes.

14.7.3 System Design and Implementation

General Design Approach. The general design approach to bioremediation involves the following tasks: (1) performance of a thorough site investigation, (2) performance of treatability studies, (3) design and implementation of the bioremediation system, and (4) performance evaluation through a monitoring program. A detailed site investigation is performed to determine aquifer, contaminant, and biological characteristics of the contaminated zone. Aquifer characteristics provide information on the suitability of the specific environment for biodegradation processes as well as information required for the design of hydraulic system design and operation. Important aquifer characteristics include the composition, heterogeneity of aquifer material, hydraulic conductivity, specific yield, groundwater flow and direction, bulk density, and porosity. All of these can be determined using the standard subsurface exploration methods discussed in Chapter 10. Contaminant characteristics provide information on determining if the contaminants present are biodegradable. The source of contamination as well as the geometry of contaminant plume in the aquifer are determined by using conventional monitoring wells as explained in Chapter 10. Biological characterization provides information on the presence of a viable population of microorganisms that are capable of degrading the contaminants present at the site. Representative samples of aquifer materials are obtained and tested for biomass and metabolic activities by different methods, such as direct light and epifluorescence microscopy, viable counts (e.g., plate counts, most probable number counts, and enrichment culture procedures), and biochemical indicators of metabolic activity such as ATP, GTP, phospholipid, and muramic acid (Ghiorse and Wilson, 1988).

A treatability study is performed to determine if bioremediation is possible at the site. Microcosms, ranging from simple batch incubation systems to large and complex flow through devices, are generally used in the treatability studies. These studies provide an estimate of the rate and extent of biodegradation. These studies also determine nutrient and electron acceptor requirements for the specific site microorganisms.

The results of site characterization and treatability study are then used to design and implement field systems. Two types of systems are commonly used for supplying nutrients, oxygen and electron acceptors, as shown in Figures 14.32 and 14.33. The first system is a pumping-injection system that includes a pair of injection-pumping wells, a line of downgradient pumping wells, or a pattern of injection-pumping wells around the boundary of a plume. The pumping-injecting system allows better control of the groundwater flow regime. Microbial nutrients are mixed on site and circulated through the contaminated zone using injection and pumping wells. A preliminary pilot-scale evaluation will help in determining the most appropriate spacing of injection and extraction wells. Instead of the pumping-injection well system, infiltration galleries as shown in Figure 14.33 may be used. Injection solution flows through the unsaturated soils and then enters into the saturated zone. This is beneficial if some residual contamination exists in the unsaturated zone that needed to be removed or remediated.

A comprehensive monitoring system involving a network of monitoring wells is used to monitor the progress of bioremediation. Contaminant concentrations as well as indicator parameters such as dissolved oxygen, nutrient levels, and carbon dioxide are measured for this purpose.

Equipment. Equipment needed for bioremediation is relatively simple. Injection and extraction wells are needed. Nutrient preparation equipment is needed. Peripheral equipment such as pumps, flow meters, and pressure regulators are also needed.

Operational Parameters. Operational factors to consider are the delivery of nutrients and electron acceptor, the point of the delivery within the aquifer, and the mode of application. Nutrients are added first, followed by the oxygen source. Simultaneous injection of both may result in excessive microbial growth around the injection point.

Monitoring. A network of monitoring wells are placed within the contaminant plume. Frequent sampling is performed to determine the distribution of dissolved oxygen, nutrient levels, carbon dioxide, and contaminant concentrations. Microbial activity can be assessed based on the dissolved oxygen, nutrients, and carbon dioxide levels. Measurement of intermediate products may also be used to determine if bioremediation is occurring. Frequency of sampling depends on the site-specific conditions such as time required for the water to flow from injection location to extraction location, hydraulic gradient, and changes in dissolved oxygen and electron acceptor.

14.7.4 Predictive Modeling

Groundwater flow models are used for the design of pumping and injection well system. Groundwater flow models such as MODFLOW are used to predict water-level responses to injection and extraction for specified parameters such as boundary conditions, aquifer geometry, hydraulic properties, well locations, injection rates, and extraction rates. Based on the site characterization and treatability study results, biodegradation rates of the contaminants may be predicted using models such as BIOPLUME (Rifai et al., 1989). However, caution must be exercised in using these models because of lack of adequate hydrogeologic and geochemical data.

14.7.5 Modified or Complementary Technologies

Biosparging is a form of in-situ bioremediation. As explained earlier, air is injected into the contaminated groundwater to increase the oxygen concentration and enhance the rate of biological degradation of organic contaminants by naturally occurring microorganisms. Enhanced aerobic bioremediation can be achieved by adding oxygen and/or oxygen sources such as hydrogen peroxide or ozone to groundwater. In addition, oxygen-releasing compounds (ORCs) are also used to release oxygen slowly into the contaminated groundwater. ORC is a patented formulation of magnesium peroxide (MgO_2), which when moist, releases oxygen slowly. Alternatively, enhanced anaerobic bioremediation of organic compounds can be accomplished by using hydrogen-releasing compounds (HRCs).

Bioremediation systems may employ horizontal wells instead of vertical wells to deliver bioremediation agents into the contaminated zone. Also, fracturing of low-permeable soils and bedrock within the contaminated zone may be used to allow injection and distribution of bioremediation agents throughout the contaminated zone. The use of electrokinetics to deliver nutrients and terminal electron acceptors in low-permeability formations has also been suggested by (Reddy, 2001).

14.7.6 Economic and Regulatory Considerations

Bioremediation is more cost-effective than are other remediation technologies. The major cost involved in bioremediation is the supply of air and nutrients, which usually involves pumping the contaminated groundwater out of the aquifer, seeding with acclimated microorganisms and/or nutrients, and recycling it through the system until satisfactory remediation is achieved. The average cost of bioremediation is $35 to $200 per 1000 gallons of groundwater treatment, depending on the site-specific conditions (Cauwenberghe and Roote, 1998).

Bioremediation involves: (1) withdrawal of contaminated groundwater, (2) addition of amendments to the groundwater, and (3) injection of the contaminated groundwater back into the contaminant plume without substantially reducing the concentration of contaminants in the injected fluid. All of these activities may require regulatory approvals or permits. According to the RCRA, withdrawn groundwater may be considered a contaminated medium [40 CFR 261.33(b)], withdrawal may constitute active management of hazardous waste triggering land disposal restrictions, contaminated medium is to be treated as hazardous waste until it no longer contains the listed contaminants or TCLP concentrations are lowered below the levels and can be treated as nonhazardous, and injection of hazardous waste into a usable aquifer constitutes land disposal. Efforts are being made to eliminate regulatory barriers in order to design and implement effective in-situ bioremediation systems (ITRC, 1998).

14.7.7 Case Studies

Several case studies involving the use of bioremediation for the cleanup of contaminated groundwater have been reported (Cauwenberghe and Roote, 1998; USEPA, 1999b). Two of these case studies are presented in this section.

Case Study 1: French Limited. Superfund Site, Crosby, Texas. From 1966 to 1971, the French Limited site was permitted to accept about 80 million gallons of industrial waste material for disposal in a 7-acre lagoon created from an open sand pit. In 1981, a flood caused the dike surrounding the waste lagoon to breach. In 1982, the dike was repaired and most of the waste in sludge form was pumped back into the lagoon. Field investigation revealed that site soils and groundwater were contaminated. The sludge within the lagoons and the contaminated soils were treated using slurry-phase bioremediation. Sheet pile wall containment and pump and treat were used to remediate the source zone, while in-situ bioremediation was used to remediate the contaminated groundwater plume.

Contaminants of concern in groundwater were benzene, vinyl chloride, and 1,2-dichloroethane (1,2-DCA) with maximum concentrations of 19,000, 8200, and 92,0000 μg/L, respectively. The groundwater plume was 500 ft long and 1500 ft wide, with depth ranging from 15 to 20 ft. The site consisted of inter-bedded layers of sands, silts, and clays extending to a depth of 55 ft. The hydraulic conductivity of the soils ranged from 0.283 to 2.835 ft/day. Groundwater was encountered at a depth of 10 to 12 ft below the ground surface.

A phased groundwater remediation strategy was de-veloped for the site. Source control was achieved by installing sheet pile walls around the lagoon. Pump and treat and bioremediation were used to actively reme-diate the groundwater contamination. An in-situ bio-remediation sequence of flushing, nitrifying conditions, and finally, aerobic conditions was designed to stimu-late different types of microorganisms. The design cre-ated cometabolic biodegradation processes to biode-grade a wide variety of chlorinated and nonchlorinated constituents throughout the plume. First, clean water only was injected for 30 days. Second, the nitrate and diammonium phsophate was mixed with clean water and injected for 90 days. Finally, the oxygen was mixed with clean water and injected for 44 months. Nitrate concentration in the groundwater was not al-lowed to exceed the drinking water standard of 10 mg/L, and the oxygen concentration in the injected water was maintained between 35 and 40 mg/L. After four years of treatment, the concentrations of the contami-nants decreased from 640 to 2 μg/L for benzene, 917 to 1 μg/L for 1,2-DCA, and 420 to 1 μg/L for vinyl chloride. Subsequently, natural attenuation has been used to further reduce remaining concentrations of the contaminants. The total cost of this project, including pump and treat and bioremediation, was $33,700,000, which corresponds to $110 per 1000 gallons of groundwater extracted.

Case Study 2: Balfour Road Site, Brentwood, California. At Balfour Road, pipeline leaks were discov-ered in a gasoline supply pipeline in 1990. This caused groundwater contamination at the site. The site consists of a shallow unconfined aquifer that consisted of a sandy and clayey mixture of 18 ft, hydraulic conduc-tivity of 0.001 cm/s, and hydraulic gradient of 0.0009.

The depth to groundwater at the site ranged from 15 to 22 ft.

From 1990 to 1995, groundwater was extracted at the site through an excavation trench and treated. Once the majority of the free product was recovered, the trench system was no longer a cost-effective solution to treat the groundwater plume. Benzene, total BTEX, and TPH were detected at concentrations in the groundwater plume ranging from 0.43 to 10 mg/L. Enhanced bioremediation of the groundwater using ORC was selected and implemented in 1995 to reme-diate the groundwater plume. ORC is a proprietary for-mulation based on magnesium peroxide and is availa-ble from Regenesis Bioremediation Products, Inc. When it comes in contact with groundwater, ORC slowly releases oxygen to the groundwater and is con-verted to a magnesium hydroxide by-product. The in-creased levels of dissolved oxygen in groundwater in-creases the rate of natural bioremediation. The levels of dissolved oxygen (DO) increase depends on the dos-age of ORC, biological activity, and amount of time that has elapsed since ORC was applied. Enhanced bioremediation using ORC was selected at this site be-cause it was expected to reduce the mass of contami-nants in the aquifer by more than 50% in only six months, and it required a smaller capital investment and lower operating expenses.

Filter socks containing ORC and an inert carrier matrix (silica sand) were applied to the groundwater through a system of 10 wells (with well screens located from 10 to 33 ft below the ground surface) located in two rows along the downgradient of the source areas. Approximately 200 lb of ORC was used. The moni-toring data showed that DO levels increased to as high as 3.4 mg/L and benzene concentrations reduced by more than 50% in six months. The cleanup goal of a federal MCL level for benzene of 0.005 mg/L, how-ever, was not achieved at the site. The total cost of this remediation project was approximately $50,000.

14.8   SUMMARY

Groundwater is a valuable source of drinking water. It is also used extensively for agricultural and industrial applications. Remediation of contaminated groundwater is critical to protect human health and the environment. Numerous technologies exist for the remediation of contaminated groundwater. These technologies include pump and treat, in-situ flushing, permeable reactive barriers, air sparging, monitored natural attenu-ation, and bioremediation. Many other innovative technologies have been developed by modifying the conventional remedial technologies. Remediation technology for a particular site is selected based on the site-specific hydrogeologic and contaminant conditions, desired cleanup levels, remedial time, and cost.

QUESTIONS/PROBLEMS

14.1. List groundwater remediation technologies that are based on physical–chemical processes.

14.2. List groundwater remediation technologies that are based on thermal processes.

14.3. List groundwater remediation technologies that are based on biological processes.

14.4. Which methods are not applicable to remediate source zones in groundwater? Why?

14.5. Explain why intermittent pumping is more effective than continuous pumping for groundwater remediation.

14.6. What are the major limitations of pump and treat remediation?

14.7. What is the difference between the *zone of influence* and *capture zone* of a pumping well?

14.8. Estimate capture zone for a well pumping at 100 gal/min in an unconfined aquifer with saturated thickness of 30 ft, hydraulic conductivity of 10^{-3} cm/s, and hydraulic gradient of 0.05.

14.9. An extraction well is proposed to remediate the contaminant plume in an aquifer. The aquifer is 20 ft thick and consists of sand with hydraulic conductivity of 10^{-2} cm/s. The groundwater gradient is 0.01. **(a)** Estimate the capture zone of the pumping well. **(b)** Determine the optimal spacing and capture zone if two wells are to be used with the same pumping rate.

14.10. What is the difference between in-situ flushing and soil washing?

14.11. What are the different flushing agents used in in-situ flushing? Explain their applicability for groundwater remediation.

14.12. What are the major challenges in designing permeable reactive barriers?

14.13. An aquifer is contaminated with chlorinated organics, with the concentrations of TCE, cDCE, and VC equal to 10000, 900, and 101 μg/L, respectively. The aquifer has a hydraulic conductivity of 55 ft/day, a porosity of 35%, and a hydraulic gradient of 0.003. PRBs are proposed for remediation. The cleanup goals are 5, 70, and 2 μg/L for TCE, cDCE, and VC, respectively. Column studies using groundwater from the site produced the half-lives of TCE, cDCE, and VC to be 1.9, 3.5, and 6.4 h, respectively. Estimate the PRB thickness required.

14.14. What is the difference between air sparging and biosparging?

14.15. Why are air sparging pilot tests needed?

14.16. Explain relative applicability of air sparging for remediating the following types of contaminants in groundwater: **(a)** benzene; **(b)** TCE; **(c)** phenanthrene; **(d)** MTBE.

14.17. Under what conditions is pulsed air injection beneficial?

14.18. Explain air distribution using vertical wells versus horizontal wells for air injection.

14.19. What causes the rebound of contaminant concentrations after the termination of air sparging? What are its implications?

14.20. Calculate the minimum air entry pressure needed for each foot below the water table.

14.21. Calculate the maximum injection pressure that can safely be used during air sparging at a site with a soil specific gravity of 2.7, a soil porosity of 30%, a water table depth of 18 ft, and a sparging system screen interval of 30 to 35 ft. (*Hint:* Injection pressure must not exceed the weight of the soil column above the screen.)

14.22. What are the differences between the monitored natural attenuation and bioremediation?

14.23. What are the differences between aerobic bioremediation and anaerobic bioremediation?

14.24. Explain why indigenous microorganisms are preferred over genetically engineered microorganisms for bioremediation.

14.25. Explain the effects of pH, redox potential, moisture content, temperature, and solubility of the contaminant on bioremediation.

14.26. Explain the significance of electron acceptors for bioremediation.

14.27. What types of nutrients are needed for bioremediation? Why?

14.28. Describe methods of oxygen delivery for enhanced bioremediation of groundwater.

14.29. A dissolved gasoline plume exists in groundwater at a site. The contaminant and aquifer characteristics are as follows: an average gasoline concentration of 50 mg/L, an aquifer porosity of 0.3, an aquifer organic content of 2%, a subsurface temperature of 20°C, a bulk density of aquifer material of 1.8 g/cm^3, and a DO concentration in the aquifer of 4.0 mg/L. Calculate the additional oxygen necessary in the aquifer for bioremediation.

14.30. Calculate the amount of oxygen that 500 mg/L hydrogen peroxide can provide for groundwater bioremediation.

14.31. A dissolved hydrocarbon groundwater plume exists at a site. The site characteristics are an average hydrocarbon concentration of 100 mg/L, an aquifer porosity of 0.35, an organic content of 1%, a subsurface temperature of 20°C, and a bulk density of aquifer material of 1.8 g/cm^3. Bioremediation is proposed to treat this contamination. Assuming that no nutrients are available, calculate the amount of nutrients needed to support bioremediation.

REFERENCES

Abriola, L. M., Dekker, T. J., and Pennel, K. D., Surfactant enhanced solubilization of eesidual dodecane in soil columns: 2. Mathematical modeling, *Environ. Sci. Technol.*, Vol. 27, No. 12, 1993.

ASTM (American Society for Testing and Materials), *Standard Guide for Remediation of Groundwater by Natural Attenuation at Petroleum Release Sites,* E-1943-98, ASTM West Conshohocken, PA, 1998.

Augustijn, D. C. M., Jessup, R. E., Suresh, P., Rao, C., and Wood, A. L., Remediation of contaminated soils by solvent flushing, *J. Environ. Eng.*, Vol. 120, No. 1, Jan./Feb. 1994.

Azadpour-Keeley, A., Russell, H. H., and Sewell, G. W., *Microbial Processes Affecting Monitored Natural Attenuation of Contaminants in the Subsurface,* EPA/540/S-99/001, USEPA, Washington, DC, 1999.

Bass, D. H., and Brown, R. A., Performance of air sparging systems: a review of case studies, pp. 117–122 in *Proceedings of the 4th International Symposium on In-Situ and On-Site Bioremediation,* Vol. 1, Batelle Press, Columbus, OH, 1997.

Blandford, T. N., Huyakorn, P. S., and Wu, Y., *WHPA-A Modular Semi-analytical Model for the Delineation of Wellhead Protection Areas*, Office of Drinking Water and Ground Water, USEPA, Washington, DC, 1993.

Bowman, R., Pilot-scale testing of a surfactant-modified zeolite PRG. *Groundwater Currents*, Mar. 1999.

Brown, C. L., Pope, G. A., Abriola, L. M., and Sepehrnoori, K., Simulation of surfactant enhanced aquifer remediation, *Water Resour. Res.*, Vol. 30, No. 11, 1994.

Cauwenberghe, L. V., and Roote, D. S., *In Situ Bioremediation*, TO-98-01, Groundwater Remediation Technologies Analysis Center, Pittsburgh, PA, 1998.

CH2M HILL, *Technology Practices Manual for Surfactants and Cosolvents*, 1997.

CFR, (*Code of Federal Regulations*), Title 40, Vol. 15, Parts 136–149, National primary drinking water regulations, U.S. Government Printing Office, Washington, DC, 1999.

Cohen, R. M., Vincent, A. H., Mercer, J. W., Faust, C. R., and Spalding, C. P., *Methods for Monitoring Pump-and-Treat Performance*, EPA/600/R-94/123, R. S. Kerr Environmental Research Laboratory, USEPA, Ada, OK, 1994.

Cohen, R. M., Mercer, J. W., Greenwald, R. M., and Beljin, M. S., *Design Guidelines for Conventional Pump-and-Treat Systems*, R.S. Kerr Environmental Research Laboratory, USEPA, Ada, OK, 1997.

Feltcorn, E., and Breeden, R., *Groundwater Currents*, Dec. 1997. Reactive barriers for uranium removal.

Fetter, C. W., *Contaminant Hydrogeology*, Prentice Hall, Upper Saddle River, NJ, 1999.

Fountain, J. C., Field tests of surfactant flooding, Chap. 15 in *Transport and Remediation of Subsurface Contaminants, Colloidal, Interfacial, and Surfactant Phenomena*, ACS Symposium Series 491, Sabatini, D. A., and Knox, R. C. (eds.), American Chemical Society, Washington, DC, 1992.

Fountain, J. C., Waddell-Sheets, C., Lagowski, A., Taylor, C., Frazier, D., and Byrne, M., Enhanced removal of dense nonaqueous-phase liquids using surfactants, Chap. 13 in *Surfactant-Enhanced Subsurface Remediation: Emerging Technologies*, ACS Symposium Series 594, Sabatini, D. A., Knox, R. C., and Harwell J. H., (eds.), American Chemical Society, Washington, DC, 1995.

Freeze, G. A., Fountain, J. C., Pope, G. A., and Jackson, R. F., Modeling the surfactant-enhanced remediation of perchlorethylene at the Borden test site using the UTCHEM compositional simulator, Chap. 14 in *Surfactant-Enhanced Subsurface Remediation:* Emerging Technologies, ACS Symposium Series 594, Sabatini, D. A., Knox, R. C., and Harwell J.H., (eds.), American Chemical Society, Washington, DC, 1995.

FRTR (Federal Remediation Technologies Roundtable), *In Situ Permeable Reactive Barrier for Treatment of Contaminated Groundwater at the U.S. Coast Guard Support Center, Elizabeth City, North Carolina*, USEPA, Washington, DC, July 1997.

———, *Remediation Case Studies: Groundwater Pump and Treat (Chlorinated Solvents)*, Vol. 9, EPA/542/R-98/013, USEPA, Washington, DC, 1998a.

———, *Remediation Case Studies: Groundwater Pump and Treat (Nonchlorinated Contaminants)*, Vol. 10, EPA/542/R-98/014, USEPA, Washington, DC, 1998b.

Gao, S., Li, H., Yang, Z., Pitts, M. J., Surkalo, H., and Wyatt, K., Alkaline-surfactant-polymer pilot performance of the West Central Saertu, Daqing Oil Field, *SPE Reservoir Eng.*, Vol. 11, No. 3, 1996.

Ghiorse, W. C., and Wilson, J. T., Microbial ecology of the terrestrial subsurface, *Adv. Appl. Microbiol*, Vol. 33, pp. 107–172, 1988.

Greenwald, R. M., *MODMAN: An Optimization Module for MODFLOW, Version 2.1, Documentation and User's Guide*, GeoTrans, Inc., Sterling, VA, 1993.

Haitjema, H. M., Wittman, J., Kelson, V., and Bauch, N., *WhAEM: Program Documentation for the Wellhead Analytical Element Model*, EPA/600/R-94/210, R. S. Kerr Environmental Research Laboratory, USEPA, Ada, OK, 1994.

HWRIC (Hazardous Waste Research and Information Center), *LUST Remediation Technologies*, Part III, *Options for Groundwater Corrective Action*, HWRIC, Champaign, IL, 1995.

ITRC (Interstate Technology and Regulatory Cooperation), *Technical and Regulatory Requirements for Enhanced In Situ Bioremediation of Chlorinated Solvents in Groundwater*, ITRC, Washington, DC, 1998.

———, *Regulatory Guidance for Permeable Reactive Barriers Designed to Remediate Inorganic and Radionuclide Contamination*, ITRC, Washington, DC, 1999a.

———, *Natural Attenuation of Chlorinated Solvents in Groundwater: Principles and Practices*, ITRC, Washington, DC, 1999b.

Jafvert, C. T., *Technology Evaluation Report; Surfactants/Cosolvents*, Groundwater Remediation Technologies Analysis Center, Pittsburgh, PA, Dec. 1996.

Javandel, I., and Tsang, C. F., Capture-zone type curves: a tool for aquifer cleanup, *Ground Water*, Vol. 24; pp. 616–625, 1986.

Javandel, I., Doughty, C., and Tsang, C. F., *Groundwater Transport: Handbook of Mathematical Models*, Water Resources Monograph 10, American Geophysical Union, Washington, DC, 1984.

Keely, J. F., *Performance Evaluations of Pump-and-Treat Remediations*, EPA/540/4-89/005, R. S. Kerr Environmental Research Laboratory, USEPA, Ada, OK, 1989.

Konikow, L. F., and Bredehoeft, J. D., *Computer Model of Two-Dimensional Solute Transport and Dispersion in Ground Water*, U.S. Geological Survey, Reston, VA, 1989.

Leeson, A., Johnson, P. C., Johnson, R. L., Hinchee, R. E., and McWhorter, D. B., *Air Sparging Design Paradigm (Draft)*, Battelle, Columbus, OH, 1999.

Loden, M. E., *A Technology Assessment of Soil Vapor Extraction and Air Sparging*, EPA/600/R-92/173, Office of Research and Development, USEPA, Cincinnati, OH, 1992.

Mason, A. R., and Kueper, B. H., Numerical simulation of surfactant flooding to remove pooled DNAPL from porous media, *Environ. Sci. Technol.*, Vol. 30, No. 11, 1996.

McDonald, M. G., and Harbaugh, A. W., *A Modular Three-Dimensional Finite-Difference Groundwater Flow Model*, USGS Techniques of Water-Resources Investigations, Book 6, Chap. A1, U.S. Geological Survey, Reston, VA, 1988.

Miller, R. R., *Air Sparging*, TO-96-04, Groundwater Remediation Technologies Analysis Center, Pittsburgh, PA, 1996.

Newell, C. J., McLeod, R. K., and Gonzales, J. R. "*BIOSCREEN: Natural Attenuation Decision Support System, User's Manual, Version 1.3*, EPA/600/R-96/087, USEPA, Washington, DC, 1996.

Newell, C. J., Smith, A. P., Aziz, C. E., Khan, T. A., Gozales, J. R., and Hass, P. E., BIOCHLOR: a planning-level natural attenuation model and database for solvent sites, in *Proceedings of the Symposium of Natural Attenuation of Chlorinated and Recalcitrant Compounds*, Battelle, Columbus, OH, 1998.

NRC (National Research Council), *Natural Attenuation for Groundwater Remediation*, National Academy Press, Washington, DC, 2000.

Nyer, E. K., Kidd, D. F., Palmer, P. L., Crossman, T. L., Fam, S., Johns, F. J., II, Boettcher, G., and Suthersan, S. S., *In Situ Treatment Technology*, CRC Lewis Publishers, Boca Raton, FL, 1996.

Ott, N., *Permeable Reactive Barriers for Inorganics*, Technology Innovation Office, USEPA, Washington, DC, 2000.

Palmer, C. D., and Fish, W., *Chemical Enhancements to Pump-and-Treat Remediation*, EPA/540/S-92/001, R. S. Kerr Environmental Research Laboratory, USEPA, Ada, OK, 1991.

Pankow, J. F., Johnson, R. L., and Cherry, J. A., Air sparging in gate wells in cutoff wells and trenches for control of plumes of volatile organic compounds (VOCs), *Ground Water*, Vol. 31, No. 4, pp. 654–663, 1993.

Pollock, D. W., *User's Guide for MODPATH, MODPATH-PLOT, Version 3: A Particle Tracking Post-processing Package for MODFLOW*, USGS Open-File Report 94-464, U.S. Geological Survey, Reston, VA, 1994.

Reddy, K. R., Integrated electrokinetic remediation of contaminated soils and groundwater, in *Proceedings of the Symposium on Environmental Science and Technology*, University of Illinois at Chicago, 2001.

Reddy, K. R., and Adams, J. A., Cleanup of chemical spills using air sparging, Chap. 14 in *Handbook of Chemical Spill Technologies*, Fingas, M. (ed.), Mc-Graw-Hill, New York, 2001.

Reddy, K. R., and Saichek, R. E., Electrokinetic removal of phenanthrene from kaolin using different surfactants and cosolvents, pp. 138–161 in *Evaluation and Remediation of Low Permeability and Dual Porosity Environments*, ASTM STP 1415, Sara M. N., and Everett, L. G. (eds)., ASTM, West Conshohocken, PA, 2002.

Reddy, K. R., Kosgi, S., and Zhou, J., A review of In situ air sparging for the remediation of VOC-contaminated saturated soils and groundwater, *Hazard. Waste and Hazard. Mater.,* Vol. 12, No. 2, pp. 97–118, 1995

Reddy, K. R., Adams, J. A., and Richardson, C., Potential technologies for remediation of brownfields, *Pract. Per. Hazard. Toxic Radioact. Waste Manag.,* Vol. 3, No. 2, pp. 61–68, 1999.

Reitsma, S., and Kueper, B. H., Compositional modeling study of alcohol flooding for recovery of DNAPL with varied injection concentration and slug size, in *Proceedings of NAPLs in the Subsurface Environment: Assessment and Remediation,* ASCE, Reston, VA, 1996.

Rifai, H. S., Bedient, P. B., and Wilson, J. T., *BIOPLUME: Model for Contaminant Transport Affected by Oxygen Limited Biodegradation,* EPA/600/M-89/019, USEPA, Washington, DC, 1989.

Roote, D. S., *Technology Overview Report: In-Situ Flushing,* Groundwater Remediation Technologies Analysis Center, Pittsburgh, PA, June 1997.

———, *Technology Status Report: In-Situ Flushing,* Ground-water Remediation Technologies Analysis Center, Pittsburgh, PA, Nov. 1998.

Semer, R., and Reddy, K. R., Mechanisms controlling toluene removal from saturated soils during in situ air sparging, *J. Hazard. Mater.,* Vol. 57, pp. 209–230, 1998.

Sims, J. L., Suflita, J. M., and Russell, H. H., *In-Situ Bioremediation of Contaminated Ground Water,* EPA/540/S-92/002, R. S. Kerr Environmental Research Laboratory, USEPA, Ada, OK, 1992.

Stavnes, S., Bioremediation barrier emplaced through hydraulic fracturing, *Groundwater Currents,* Mar. 1999.

Strack, O. D. L., Anderson, E. I., Bakker, M., Olsen, W. C., Panda, J. C., Pennings, R. W., and Steward, D. R., *CZAEM User's Guide: Modeling Capture Zones of Ground-Water Using Analytic Elements,* EPA/600/R-94/174, Office of Research and Development, USEPA, Cincinnati, OH, 1994.

Strbak, L., *In Situ Flushing with Surfactants and Cosolvents,* Technology Innovation Office, USEPA, Washington, DC, 2000.

Sullivan, R. A., Pump and treat and wait, *Civ. Eng.,* Vol. 66, pp. 8A–12A, 1996.

Thomas, J. M., and Ward, C. H., In situ biorestoration of organic contaminants in the subsurface, *Environ. Sci. Technol.,* Vol. 23, pp. 760–766, 1989.

Uhlman, K., and Portman, M. E., Predicting the efficiency of a groundwater remediation extraction program using MODFLOW, in *Proceeding of the 34th Annual Meeting of the Association of Engineering Geologists,* Chicago, 1991.

USACE, (U.S. Army Corps of Engineers), *In-Situ Air Sparging,* EM 1110-1-4005, U.S. Department of the Army, Washington, DC, 1997a.

———, *Design Guidance for Application of Permeable Barriers to Remediate Dissolved Chlorinated Solvents.* CEMP DG 1110-345-117, U.S. Department of the Army, Washington, DC, Feb. 1997b.

USEPA (U.S. Environmental Protection Agency), *Engineering Bulletin: In Situ Soil Flushing,"* EPA/540/2-91/021, USEPA, Washington, DC, 1991.

———, *Guide for Conducting Treatability Studies under CERCLA, Biodegradation Remedy Selection, Interim Guidance,* EPA/540/R-93/519a, USEPA, Washington, DC, 1993.

———, *Surfactant Injection for Ground Water Remediation: State Regulators' Perspectives and Experiences,* EPA/542/R-95/011, Technology Innovation Office, USEPA, Washington, DC, 1995a.

———, *National Water Quality Inventory: 1994 Report to Congress,* Office of Water, USEPA, Washington, DC, 1995b.

———, Assessment of barrier containment technologies, presented at the International Containment Technology Workshop, Baltimore, MD, Aug. 29–31, 1995c.

———, *Pump-and-Treat Ground-Water Remediation: A Guide for Decision Makers and Practitioners,* EPA/625/R-95/005, Office of Research and Development, USEPA, Washington, DC, 1996a.

———, *Technology Fact Sheet: A Citizen's Guide to In-situ Soil Flushing,* EPA/542/F-96/006, USPEA, Washington, DC, Sept. 1996b.

———, *A Citizen's Guide to Treatment Walls.* EPA 542-F-96-016, USEPA, Washington, DC, Sept. 1996c.

———, *A Citizen's Guide to Soil Vapor Extraction and Air Sparging,* EPA 542-F-96-008, USEPA, Washington, DC, 1996d

———, *Permeable Reactive Subsurface Barriers of the Inception and Remediation of Chlorinated Hydrocarbon and Chromium (VI) Plumes in Ground Water,* EPA 600-F-97-008, USEPA, Washington, DC, 1997a.

———, *Recent Developments for In-Situ Treatment of Metal Contaminated Soils,* Work Assignment 011059, prepared by PRC Environmental Management, Inc., Office of Solid Waste and Emergency Response, Technology Innovation Office, USEPA, Washington, DC, 1997b.

———, *Analysis of Selected Enhancement for Soil Vapor Extraction,* EPA/542/R-97/007, USEPA, Washington, DC, 1997c.

———, *Monitored Natural Attenuation for Ground Water,* EPA/625/K-98/001, USEPA, Washington, DC, 1998a.

———, *Technical Protocol for Evaluating Natural Attenuation of Chlorinated Solvents in Groundwater,* EPA/600/R-98/128, USEPA Washington, DC, 1998b.

———, *Permeable Reactive Barrier Technologies for Contaminant Remediation,* EPA/600/R-98/125, Office of Research and Development, USEPA, Washington, DC, 1998c.

———, *Field Applications of In Situ Remediation Technologies: Permeable Reactive Barriers,* EPA/542/R-99/002, USEPA, Washington, DC, Apr. 1999a.

———, *Ground Water Cleanup Overview of Operating Experience at 28 Sites,* EPA/542/R-99/006, Office of Solid Waste and Emergency Response, USEPA, Washington, DC, Sept. 1999b.

———, *Use of Monitored Natural Attenuation at Superfund, RCRA Corrective Action, and Underground Storage Tank Sites,* OSWER Directive 9200.4-17P, Office of Solid Waste and Emergency Response, USEPA, Washington, DC, 1999c.

———, *In Situ Permeable Reactive Barriers: Application and Deployment, Training Manual,* EPA/542/B-00/001, USEPA, Washington, DC, 2000a.

———, *Handbook of Groundwater Policies for RCRA Corrective Action,* EPA/530/D-00/001, USEPA, Washington, DC, 2000b.

Vidic, R. D., and Pohland, F. G., *Technology Evaluation Report: Treatment Walls.* Groundwater Remediation Technologies Analysis Center, Pittsburgh, PA, 1996.

Zheng, C., *MT3D: A Modular Three-Dimensional Transport Model for Simulation of Advection, Dispersion, and Chemical Reactions of Contaminants in Groundwater Systems,* S. S. Papadopulos and Associates, Bethesda, MD, 1992.

PART III
LANDFILLS AND SURFACE IMPOUNDMENTS

15

SOURCES AND CHARACTERISTICS OF WASTES

15.1 INTRODUCTION

What is waste? This is an essential question that must be addressed before permanently disposing of any waste material. Technically speaking, waste is a by-product of life and civilization: the material that remains after a useful component has been removed or consumed. From an economic sense, waste is a material involved in life or technology whose value today is less than the cost of its utilization. However, from a regulatory perspective, a waste is anything that is being discarded or can no longer be used for its original purpose. Something may be a waste if it no longer has a dollar value or if it is used carelessly, spilled, burned, buried, or poured down the drain. The term *solid waste,* as used in the regulations, can be misleading because it includes not only solid materials but also liquids and gases. If one wants to reuse something, it may not be a waste.

This chapter presents different sources of wastes, classifications of wastes based on composition, typical chemical and physical characteristics of wastes, major concerns regarding wastes, and strategies to manage wastes, including the option of disposal in engineered waste disposal facilities.

15.2 SOURCES OF WASTES

The major sources of wastes are (1) dredging and irrigation; (2) mining and quarrying; (3) farming and ranching; (4) residential, commercial, and institutional; (5) industrial; and (6) nuclear power and nuclear defense. These sources are described below.

15.2.1 Dredging and Irrigation

Dredging involves the removal of soil or sediments from waterways, harbors, or irrigation canals in order to keep them navigable or flowing properly. The amount of dredged soil and sediments may be considerable, and they may also contain residual hazardous materials discharged from industrial or farming activities.

15.2.2 Mining and Quarrying

Mining is necessary for coal, in some cases oil, such as oil sand and oil shale projects, a wide variety of metals, and nonmetallic minerals. The wastes generally consist of by-products of mining activities. These wastes, often called *mine tailings,* are usually composed of silts, fine sands, or other aggregate materials. Quarrying for rocks, aggregates, and sands is also an important activity, because these materials are used constantly in industry and construction for building stone, cement products, roofing materials, and so on. Mining and quarrying wastes may pose significant problems to the environment due to the large quantities of wastes produced and, in some cases, their hazardous nature. The weight of waste generated from mine tailings in the United States has been estimated as 3 billion tons per year (Alexander, 1993).

15.2.3 Farming and Ranching

Farming, ranching, or dairy activities produce wastes that are composed largely of spoiled food, manure, excess crop waste, and waste from chemical or pesticide applications. The weight of agricultural wastes generated in the United States has been estimated at 2 billion tons per year (Alexander, 1993), but some of this waste is returned to the land as part of good agricultural practice.

considerable care be used in evaluating these data, since they are highly subjective, and it should not be taken as precise information. For example, the amount and type of construction and demolition waste depends significantly on the region of the country and on the health of the local, state, or national economy (Tchobanoglous et al., 1993).

15.2.4 Residential, Commercial, and Institutional

Municipal residences, commercial or institutional businesses, construction and demolition activities, municipal services, and treatment plants sites or waste incinerators are all major contributors to the waste stream. Table 15.1 shows the wastes generated from these various sources. The data concerning waste sources, composition, and rates of generation are basic to the design and management of wastes, but it is important to note that the definitions of terms and classifications vary widely in the literature. Therefore, it is essential that

15.2.5 Industrial

Industries perform services and produce a wide variety of products and materials. Some of the major industries that are sources of industrial wastes include industrial construction and demolition, fabrication, light and heavy manufacturing, refineries, chemical plants, and nonnuclear power plants. Large industrial plants may have their own recycling programs and own and operate their own landfills, so the actual amount of industrial waste discarded is difficult to ascertain. Alexander (1993) estimated the weight of manufacturing

Table 15.1 **Sources of wastes within a community**

Source	Typical Facilities, Activities, or Locations Where Wastes Are Generated	Types of Wastes
Residential	Single- and multifamily detached dwellings; low-, medium-, and high-rise apartments; etc.	Food wastes, paper, cardboard, plastics, textiles, leather, yard wastes, wood, glass, tin cans, aluminum, other metals, ashes, street leaves, special wastes (including consumer electronics, batteries, oil, and tires), household hazardous wastes, etc.
Commercial	Stores, restaurants, markets, office buildings, hotels, motels, print shops, service stations, repair shops, etc.	Paper, cardboard, plastics, wood, food waste, glass, metals, special wastes (including electronics, batteries, oil, and tires), hazardous wastes, etc.
Institutional	Schools, hospitals, prisons, governmental centers	As above, commercial
Construction and demolition	New construction sites, road repair/renovation sites, razing of buildings, broken pavement	Wood, steel, concrete, dirt, etc.
Municipal services (excluding treatment facilities)	Street cleaning, landscaping, catch basin cleaning, parks and beaches, other recreation areas	Rubbish, street sweepings, landscape and tree trimmings, catch basin debris, and general wastes from parks, beaches, and recreational areas
Treatment plant sites; municipal incinerators	Water, wastewater, and industrial treatment processes; etc.	Plant wastes principally composed of residual sludges

Source: Tchobanoglous et al. (1993).

waste and sludges disposed of yearly in the United States as approximately 165 million tons.

15.2.6 Nuclear Power and Nuclear Defense

The civilian nuclear power industry generates electrical energy for millions of people, governmental defense facilities produce weapons, and nuclear research projects in a wide variety of fields are advancing technology. Even though the amount of nuclear waste generated by these organizations is small compared to the energy produced, the radionuclides in this type of waste may be dangerous. The toxicity due to the exposure of radioactive material is still being studied, and concerns over long-range health and environmental effects make nuclear waste extraordinarily difficult to dispose of safely. Dolan and Scariano (1990) estimated that government nuclear defense operations have accumulated over 306,000 m^3 of high-level wastes (HLWs), and another 4000 m^3 is from civilian operations. Low-level wastes (LLWs) have accumulated at much higher rates, and they estimate that nuclear defense operations are responsible for 2 million cubic meters, with another 1 million cubic meters coming from civilian operations.

15.3 CLASSIFICATION OF WASTES

Wastes are classified into the following types based on their composition: solid waste, hazardous waste, radioactive waste, and infectious (medical) waste. Since even small amounts of hazardous wastes, radioactive wastes, or infectious (medical) wastes may adversely affect human health or the environment, these wastes often require special handling procedures and disposal facilities. The following is a general overview of the composition and sources of these classes of wastes.

15.3.1 Solid Waste

The USEPA defines *solid waste* (SW) as "any garbage, refuse, sludge from a waste treatment plant, water supply treatment plant, or air pollution control facility and other discarded material, including solid, liquid, semi-

solid, or contained gaseous material resulting from industrial, commercial, mining, and agricultural operations, and from community activities" (40 CFR257.2). The USEPA further divides the solid waste into the following categories: (a) household solid waste, (b) commercial solid waste, and (c) industrial solid waste. *Household solid waste* is any solid waste derived from households, including residences, hotels and motels, campgrounds, picnic grounds, and so on. *Commercial solid waste* refers to all types of solid waste generated by stores, offices, restaurants, warehouses, and other nonmanufacturing activities. The household waste and commercial solid waste together are often known as *municipal solid waste* (MSW). Table 15.1 summarizes the main sources of MSW. According to the USEPA (1994), the United States generated approximately 162 million tons of MSW in 1993, and the quantity of waste generated has been increasing yearly. Table 15.2 shows typical composition of MSW and includes estimations of various components. The data include only wastes typically disposed in landfills and does not include recycled wastes, construction or demolition wastes, industrial process wastes, transportation equipment, or municipal sludges. For example, scrap tires are often stockpiled because they do not easily biodegrade and they occupy large volume in landfills. It is estimated that 230 to 240 million rubber tires are discarded annually (Tchobanoglous et al., 1993). It should also be noted that the yard waste category depends significantly on local ordinances of the community.

Industrial solid waste is solid waste generated by manufacturing or industrial processes that is not hazardous waste as defined in Section 15.3.2. Such industrial solid waste, also commonly referred to as *industrial nonhazardous waste*, includes nonhazardous wastes resulting from the following manufacturing processes: electric power generation, fertilizer/agricultural chemicals, food and related products and by-products, iron and steel manufacturing, leather and leather products, plastics and resins manufacturing, and water treatment, among others. The sources and types of industrial wastes are grouped according to their Standard Industrial Classification (SIC) code. Table 15.3 lists some of these wastes. Depending on the waste composition, it may be classified as nonhazardous, and if

Table 15.2 **Generation and composition of municipal solid waste**[a]

Materials	1960	1970	1980	1990	1995	1998	1999	2000
Thousands of Tons								
Paper and paperboard	24,910	37,540	43,420	52,500	48,970	49,800	52,180	47,370
Glass	6,620	12,580	14,380	10,470	9,690	9,700	9,910	9,830
Metals								
Ferrous	10,250	12,210	12,250	10,410	7,510	8,070	8,760	8,880
Aluminium	340	790	1,420	1,800	2,030	2,190	2,240	2,300
Other nonferrous	180	350	620	370	450	430	450	460
Total metals	10,770	13,350	14,290	12,580	9,990	10,690	11,450	11,640
Plastics	390	2,900	6,810	16,760	17,910	21,170	22,800	23,370
Rubber and leather	1,510	2,720	4,070	5,420	5,490	6,000	5,430	5,590
Textiles	1,710	1,980	2,370	5,150	6,500	7,480	7,830	8,110
Wood	3,030	3,720	7,010	12,080	9,990	11,600	11,870	12,220
Other[b]	70	470	2,020	2,510	2,900	3,040	3,140	3,170
Total materials in products	49,010	75,260	94,370	117,470	111,440	119,480	124,610	121,300
Other Wastes								
Food scraps	12,200	12,800	13,000	20,800	21,170	24,330	24,610	25,220
Yard trimmings	20,000	23,200	27,500	30,800	20,690	15,170	13,560	11,960
Miscellaneous inorganic wastes	1,300	1,780	2,250	2,900	3,150	3,290	3,380	3,500
Total other wastes	33,500	37,780	42,750	54,500	45,010	42,790	41,550	40,680
Total MSW discarded (weight)	82,510	113,040	137,120	171,970	156,450	612,270	166,160	161,980
Percent of Total Discards								
Paper and paperboard	30.2	33.20	31.70	30.50	31.30	30.70	31.40	29.20
Glass	8.00	11.10	10.50	6.10	6.20	6.00	6.00	6.10
Metals								
Ferrous	12.40	10.80	8.90	6.10	4.80	5.00	5.30	5.50
Aluminium	0.40	0.70	1.00	1.00	1.30	1.30	1.30	1.40
Other nonferrous	0.20	0.30	0.50	0.20	0.30	0.30	0.30	0.30
Total metals	13.10	11.80	10.40	7.30	6.40	6.60	6.90	7.20
Plastics	0.50	2.60	5.00	9.70	11.40	13.00	13.70	14.40
Rubber and leather	1.80	2.40	3.00	3.20	3.50	3.70	3.30	3.50
Textiles	2.10	1.80	1.70	3.00	4.20	4.60	4.70	5.00
Wood	3.70	3.30	5.10	7.00	6.40	7.10	7.10	7.50
Other[b]	0.10	0.40	1.50	1.50	1.90	1.90	1.90	2.00
Total materials in products	59.40	66.60	68.80	68.30	71.20	73.60	75.00	74.90
Other wastes								
Food scraps	14.80	11.30	9.50	12.10	13.50	15.00	14.80	15.60
Yard trimmings	24.20	20.50	20.10	17.90	13.20	9.30	8.20	7.40
Miscellaneous inorganic wastes	1.60	1.60	1.60	1.70	2.00	2.00	2.00	2.20
Total other wastes	40.60	33.40	31.20	31.70	28.80	26.40	25.00	25.10
Total MSW discarded (%)	100.00	100.00	100.00	100.00	100.00	100.00	100.00	100.00

Source: USEPA (2002), from Franklin Associates, Ltd.

[a]Discards after materials and compost recovery. Does not include construction and demolition debris, industrial process wastes, or certain other wastes. Details may not add to totals due to rounding.

[b]Includes elctrolysis in batteries and fluff pulp, feces, and urine in disposable diapers.

Table 15.3 **Sources and types of industrial wastes**

Code	SIC Group Classification	Waste-Generating Process	Expected Specific Waste
19	Ordinance and accessories	Manufacturing and assembling	Metals. plastic, rubber, wood, cloth, chemical residues
20	Food and kindred products	Processing, packaging, shipping	Food wastes
22	Textile mill products	Weaving, dyeing, shipping	Cloth and fiber residues
23	Apparel	Cutting, sewing, pressing	Cloth, fibers, metals, plastics
24	Lumber and wood products	Sawmills, millwork, misc. wood products	Scrap wood, sawdust, and sometimes glues, sealers, paints, solvents
25a and b	Furniture wood and metal	Manufacture of household furniture	Same as code 24 plus cloth, padding, metals, plastics, adhesives, rubber, glass
26	Paper and allied products	Paper manufacturing	Paper and fiber residues, coatings, chemicals, inks, glues, fasteners
27	Printing and publishing	Newspaper publishing, lithography, printing, bookbinding	Paper, newsprint, cardboard, metals, chemicals, cloth, inks glues
28	Chemical and related products	Manufacture and preparation of inorganic chemicals (soaps, paints, drugs, etc.)	Organic and inorganic chemicals, metals, plastics, rubber, glass, oils, paints, solvents, pigments
29	Petroleum refining and related industries	Manufacture of paving and roofing products	Asphalt and tars, felts, asbestos, paper, cloth. Fiber
30	Rubber and misc. plastic products	Manufacture of fabricated rubber and plastic products	Scrap rubber and plastics, curing compounds, dyes
31	Leather and leather products	Leather tanning and finishing, manufacture of leather belting and packing	Scrap leather, thread, dyes, oils, processing and curing compounds
32	Stone, clay, and glass products	Manufacture of glass, concrete, gypsum, and plaster products, forming and processing of stone and stone products, abrasives, asbestos, etc.	Glass, cement, clay, ceramics, gypsum, asbestos, stone, paper, abrasives
33	Primary metal industries	Melting, casting, forging, etc.	Ferrous and nonferrous metals, scrap, slag, sand, bonding agents
34	Fabricated metal products	Manufacture of metal cans, hand tools, plumbing fixtures, farm machinery, etc.	Metals, ceramics, sand, slag, scale, coatings, solvents
35	Machinery (except electrical)	Manufacture of equipment for construction, mining, machine tools, trailers, conveyers, elevators, etc.	Slag, sand, metal scrap, wood, plastics, resins, rubber, cloth, petroleum products, paints
36	Electrical	Manufacture of electrical equipment, appliances, and communication apparatus, etc.	Metal scrap, glass, exotic metals, rubber, plastics, resins, fibers, cloth residues
37	Transportation equipment	Manufacture of motor vehicles, truck and bus bodies, parts and accessories, aircraft and parts, ships and boats, etc.	Metal scrap, glass, fiber, wood, rubber, plastics, cloth, paints, solvents, petroleum products
38	Professional, scientific instruments	Manufacture of engineering, laboratory, and research instruments	Metals, plastics, resins, glass, wood, rubber, fibers
39	Misc. manufacturing	Manufacture of jewelry, silverware, toys, sporting gear, buttons, etc.	Metals, glass, plastics, resins, leather, rubber, paints, etc.

Source: Tchobanoglous et al., (1993).

so, it can be reused as beneficial material or disposed of in landfills as MSW or special waste. However, if the waste is characterized as hazardous waste, as discussed in the next section, it must be classified and disposed of differently.

Construction and demolition waste is also considered as a solid waste. This type of waste includes wood, metal, concrete, bricks, glass, plastics, plumbing, and so on. These are the materials used in construction of buildings or roadways. The quality and components of such wastes can vary significantly. Such wastes are disposed of as MSW or special waste.

15.3.2 Hazardous Waste

Hazardous waste includes a variety of chemical compounds and materials, and it is defined in specific terms in the Resource Conservation and Recovery Act (RCRA). Subtitle C of the act deals with hazardous waste management. U.S. Congress defined the term *hazardous waste* in Section 1004(5) of RCRA as a "solid waste, or combination of solid wastes, which because of its quantity, concentration, or physical, chemical, or infectious characteristics may: (1) cause, or significantly contribute to an increase in mortality or an increase in serious irreversible, or incapacitating reversible, illness or (2) pose a substantial present or potential hazard to human health or the environment when improperly treated, stored, transported, or disposed of, or otherwise managed."

The definition of a hazardous waste in specific terms has been defined by the USEPA. As given in 40 CFR 261, a solid waste is hazardous if it is not *excluded* from regulation and it meets the following conditions:

1. It exhibits any characteristics of a hazardous waste listed in Table 15.4. If the TCLP concentration of any of the contaminants listed in Table 15.5 is exceeded, the waste is classified as hazardous waste.

2. It has been named as a hazardous waste and listed as such in the regulations.

3. It is a mixture containing a listed hazardous waste and a nonhazardous solid waste.

4. It is a waste derived from the treatment, storage, or disposal of a listed hazardous waste.

The RCRA also stipulates that the responsibility for determining if a particular waste is hazardous falls on the generators. If the waste is neither excluded nor listed, the generators must test their waste or have a sufficient knowledge of it to determine whether the waste exhibits any hazardous characteristic. Excluded wastes are generally common wastes that do not pose a significant threat to human health or the environment, or they are wastes managed under other regulatory programs, such as Subtitle D of the RCRA (state or regional solid waste plans), the Clean Water Act, or the Safe Drinking Water Act. In 40 CFR Part 261.4(b), a few exempted wastes included are oil and gas production wastes, mining wastes, wastes from the combustion of coal or other fossil fuels, cement kiln dust wastes, household wastes, solid wastes that are returned to the soil as fertilizer, various chromium-containing wastes, and discarded arsenically treated wood.

Hazardous wastes are also listed by the EPA and characterized under the following three headings based on their sources:

1. *Nonspecific source wastes* (40 CFR 261.31). These include generic wastes produced by manufacturing and industrial processes. Examples are halogenated solvents used in degreasing, wastewater treatment sludge from electroplating processes, and dioxin wastes.

2. *Specific source wastes* (40 CFR 261.32). These are wastes from specifically identified industries, such as wood preserving, petroleum refining, and organic chemical manufacturing. Examples include various types of sludges, wastewaters, such as from pigment production, and many types of residues.

3. *Commercial chemical product source wastes* [40 CFR 261.33(e) and (f)]. These are wastes from commercial chemical products or manufacturing.

Table 15.4 **USEPA characteristics of hazardous wastes**

Characteristic	40 CRF Subpart	Considerations
Ignitability	261.21	(1) Liquids, except aqueous solutions containing less than 24% alcohol, with flashpoints of less than 140°F (60°C)
		(2) A nonliquid capable under normal conditions of spontaneous and sustained combustion
		(3) An ignitable compressed gas per the Department of Transportation (DOT) regulations or
		(4) An oxidizer per DOT regulations
Corrosivitiy	261.22	(1) An aqueous material with a pH less than or equal to 2 or greater than or equal to 12.5, *or*
		(2) A liquid that corrodes steel at a rate greater than 0.25 in./yr at a temperature of 130°F (55°C)
Reactivity	261.23	(1) A solid waste that is normally unstable and reacts violently without detonating
		(2) Reacts violently with water
		(3) Forms an explosive mixture with water
		(4) Generates toxic gases, vapors, or fumes when mixed with water
		(5) Contains cyanide or sulfide and generates toxic gases, vapors, or fumes at a pH between 2 and 12.5
		(6) Is capable of detonation if heated under confinement or subjected to a strong initiating source
		(7) Is capable of detonation at standard temperature and pressure, *or*
		(8) Listed by the DOT as a class A or B explosive
Toxicity characteristic as defined by the TCLP leaching procedure test	261.24	The following steps are required in the TCLP test:
		(1) If the waste is liquid (i.e., contains less than 0.5% solids), after it is filtered the waste itself is considered the extract (simulated leachate).
		(2) If the waste contains greater than 0.5% solid material, the solid phase is separated from the liquid phase, if any. If required, the particle size of the solid phase is reduced until it passes through a 9.5-mm sieve.
		(3) For analysis other than for volatiles, the solid phase is then placed in an acidic solution and rotated at 30 rev/min for 18 h. The pH of the solution is approximately 5, unless the solid is more basic, in which case a solution with a pH of approximately 3 is used. After extraction (rotation), solids are filtered from the liquid extract and discarded.
		(4) For volatiles analysis a solution of pH 5 is used, and a zero headspace extraction vessel is used for liquid–solid separation, agitation, and filtration.
		(5) Liquid extracted from the solid–acid mixture is combined with any original liquid separated from the solid material and is analyzed for the presence of specified contaminants.
		(6) If any of the contaminants in the extract meets or exceeds any of the maximum concentration levels allowed for the specified contaminants as given in Table 15.5, the waste is classified as a TC hazardous waste.

Source: Tchobanoglous et al. (1993).

Table 15.5 **Maximum concentration of contaminants for the toxicity characteristic**

Contaminant	Regulatory level (mg/L)
Arsenic	5
Barium	100
Benzene	0.5
Cadmium	1
Carbon tetrachloride	0.5
Chlordane	0.03
Chlorobenzene	100
Chloroform	6
Chromium	5
m-Cresol	200
o-Cresol	200
p-Cresol	200
Cresol (total)	200
2,4-D	10
1,4-Dichlorobenzene	7.5
1,2-Dichloroethane	0.5
1,1-Dichloroethylene	0.7
2,4-Dinitrotoluene	0.13 (detection limit)
Endrin	0.02
Heptachlor and its epoxide	0.008
Hexachlorobenzene	0.13 (detection limit)
Hexachlorobutadiene	0.5
Hexachloroethane	3
Lead	5
Lindane	0.4
Mercury	0.2
Methoxychlor	10
Methyl ethyl ketone	200
Nitrobenzene	2
Pentachlorophenol	100
Pyridine	5 (detection limit)
Selenium	1
Silver	5
Tetrachloroethylene	0.7
Toxaphene	0.5
Trichloroethylene	0.5
2,4,5-Trichlorophenol	400
2,4,6-Trichlorophenol	2
2,4,5-TP (silvex)	1
Vinyl chloride	0.2

Source: 40 CFR 261.

Examples include chemicals such as chloroform, creosote, various acids, and pesticides.

15.3.3   Radioactive Waste

Radioactive wastes are characterized according to the following four categories: (1) high-level waste (HLW), (2) transuranic waste (TRU), (3) low-level waste (LLW), and (4) mill tailings. Various radioactive wastes decay at different rates, but health and environmental dangers due to radiation may persist for hundreds or thousands of years.

HLW. HLW is typically liquid or solid waste that results from government defense-related activities or from nuclear power plants and spent fuel assemblies. These wastes are extremely dangerous, due to their heavy concentrations of radionuclides, and humans must not come into contact with them. HLW is determined by the Nuclear Regulatory Commission (NRC) to require permanent isolation. The Department of Energy (DOE) has the responsibility for establishing disposal sites, and the NRC has the responsibility for establishing disposal guidelines that meet EPA standards (Schumacher, 1988). High-level wastes lose about 50% of their radioactivity after about three months in storage and about 80% after a year, but they still may require another 10,000 years before they drop to a safe level (Dolan and Scariano, 1990).

TRU. TRU results primarily from the reprocessing of spent nuclear fuels and from the making of nuclear weapons for defense projects. They are characterized by a moderately penetrating radiation and a decay time of approximately 20 years to reach safe levels. After a ban of reprocessing in 1977, most of this waste stopped being generated. TRU continues to be rare even though the ban was lifted in 1981, because reprocessing of nuclear fuel is expensive, and the plutonium extracted may be used to manufacture nuclear weapons. The U.S. defense reactors still generate TRU by reprocessing their spent nuclear fuels, and they use the uranium and plutonium in nuclear weaponry (Dolan and Scariano, 1990).

LLW. LLW wastes include the remainder of the radioactive waste materials. They constitute over 80% of the volume of all the nuclear wastes but only about 2% of the total radioactivity. Sources of LLW include all of the previously cited sources of HLW and TRU, plus hospitals, industrial plants, universities, and commercial laboratories. LLW is much less dangerous than HLW, and NRC regulations allow some very low level wastes to be released to the environment. LLW may also be stored or buried until the isotopes decay to levels low enough that it can be disposed of as normal waste. LLW disposal is managed by the states, but requirements for operation and disposal are established by the EPA and NRC. The Occupational Safety and Health Administration (OSHA) is the agency in charge of setting the standards for workers that are exposed to radioactive materials.

Mill Tailings. Mill tailings are basically residues from the mining and extraction of uranium from its ore. There are more than 200 million tons of radioactive mill tailings in the United States, and all of it is stored in sparsely populated areas of such western states as Arizona, New Mexico, Utah, and Wyoming (Dolan and Scariano, 1990). These wastes emit low-level radiation, and much of it is buried to reduce dangerous emissions.

15.3.4 Infectious (Medical) Waste

The major governmental agencies concerned with medical waste include the EPA, OSHA, the Center for Disease Control (CDC) of the U.S. Department of Health and Human Services, and the Agency for Toxic Substances and Disease Registry (ATSDR) of the Public Health Service, U.S. Department of Health and Human Services. In 1988, when medical wastes washed up on beaches along the east coast, Congress passed the Medical Waste Tracking Act (MWTA) to evaluate management issues and potential risks related to medical waste disposal. The seven types of wastes listed under MWTA include:

1. Microbiological wastes (cultures and stocks of infectious wastes and associated biologicals that can cause disease in humans)
2. Human blood and blood products (including serum, plasma, and other blood components)
3. Pathological wastes of human origin (including tissues, organs, and body parts removed during surgery or autopsy)
4. Contaminated animal wastes (i.e., animal carcasses, body parts, and bedding exposed to infectious agents during medical research, pharmaceutical testing, or production of biologicals)
5. Isolation wastes (wastes associated with animals or humans known to be infected with highly communicable diseases)
6. Contaminated sharps (includes hypodermic needles, scalpels, broken glass)
7. Uncontaminated sharps

All medical wastes represent a small fraction of the total waste stream, and it is estimated that it is a maximum of about 2%. It is important to understand whether or not infectious medical wastes are much worse than typical MSW wastes that also contain pathogens. Pathogens in MSW may be contributed from sanitary napkins, disposable diapers, tissues, and so on; however, medical wastes contain much higher concentrations of pathogens.

The current trend for disposal of medical wastes is through incineration, because, as with most wastes, it greatly reduces the volume, and it assures the destruction and sterilization of infectious pathogens. Disadvantages of incineration include the potential air pollution risks from dioxins or the disposal of hazardous ash wastes. New options for disposal of medical (infectious) wastes are still being explored as well as some other technologies, including irradiation, microwaving, autoclaving, and mechanical or chemical disinfection (OTA, 1990).

15.4 WASTE CHARACTERIZATION

The variation of waste characteristics within the United States is quite significant, due to the wide range of industries and the various living styles of people in different localities. The waste produced by small towns with farming and agricultural industries will not resemble the waste generated from a large metropolitan area with many manufacturing industries. Despite this variability, chemical and physical characterization of wastes is necessary: (1) to determine the type of waste, (2) to identify possible waste reduction and recycling alternatives, (3) to determine whether or not the waste can be landfilled or remediated, (4) to determine the probable leachate constituents (necessary for determining liner compatibility, treatment plant design, and groundwater monitoring program design), and (5) to determine the physical properties of the waste for design and operation purposes.

15.4.1 Chemical Characterization

The chemical characterization includes determination of chemical composition of the waste itself, the leachate that it generates, or both, as explained in this section.

Bulk Waste. The goal of bulk waste characterization is to determine the chemical composition of the waste itself. This is accomplished generally by grinding the waste, if needed, and stirring it in various acidic or basic solutions to leach the constituents out of the waste material. Once the constituents are in solution, different standard chemical analysis procedures are used to determine the type and amount of different chemicals. It is important to recognize the fact that in many cases, determination of the bulk composition may not be necessary. However, composition of the leachate that the bulk waste generates is of environmental concern, as explained below.

Leachate. Leachate is defined as the liquid that has percolated, due to precipitation or to surface water or groundwater flow, through waste and has extracted dissolved or suspended materials. Leachate may be extremely hazardous and corrosive, so it is very important to prevent it from entering the environment untreated. Chemical analysis of leachate samples col-

lected from existing waste disposal facilities will provide the type and amount of different chemical constituents. If this is not possible, different extraction procedures are used to generate simulated leachate. In addition to the TCLP test, outlined in Table 15.4, there is a water leach test, standard leach test, extraction procedure toxicity test, a synthetic precipitation leachate procedure, and others. Typical compositions of leachates from different types of wastes are summarized in Tables 15.6(*a*) and 15.6(*b*).

15.4.2 Physical Characterization: Engineering Properties of Wastes

In this section we present typical physical characteristics of wastes, particularly the engineering properties of wastes, such as porosity, moisture content, field capacity, wilting point, unit weight, hydraulic conductivity, shear strength, and compressibility. The properties presented in this section are primarily for municipal landfill wastes. Some information on the mineral wastes, such as fly ash, compacted copper slag, and flue-gas desulfurization sludge, has also been presented for preliminary engineering evaluation. More information on the properties of other types of wastes is provided in Chapter 23. The engineering properties presented here are merely as a guide, and wherever possible, site-specific information should be collected for project design.

Particle Size and Size Distribution. The particle size, size distribution, and shape characteristics of various components within the waste may be needed in evaluating biological transformation and potential reuse applications. When the wastes possess components in sizes similar to typical soil particles, standard sieve and hydrometer analysis can be performed to determine the particle size distribution. However, wastes such as MSW contain large components and are not amenable to be tested using the standard soil particle size analysis procedures. In such cases, the size of the components of the waste is defined by measuring the length, width, and height as well as the mass of individual components in a randomly selected represen-

Table 15.6(a) **Composition of leachates from different wastes**[a]

Parameter	Papermill Sludge (mg/L except as indicated)	Municipal Incinerator Ash (mg/L except as indicated)	Municipal Waste (mg/L except as indicated)
Aluminum	0.008–18	2.3–88.8	ND–85.0
Ammonia-nitrogen	< 0.1	—	ND–1200
Antimony	—	—	ND–3.19
Arsenic	0.029	0.005–0.218	ND–70.2
Barium	ND	0.055–2.48	ND–12.5
Benzaldehyde	—	ND–0.008	—
Beryllium	—	—	ND–0.36
Biphenyl	—	ND–0.051	—
BOD	36–10,000	—	ND–195,000
Boron	—	0.42–3.2	0.87–13.0
Bromide	ND	—	—
Cadmium	0.006–0.02	< 0.001–0.3	ND–0.4
Calcium	5.5–2400	21–3200	3.0–2500
Chloride	1–1200	32.6–305	2.0–11,375
Chromium	0–0.15	< 0.002–1.53	ND–5.6
Cobalt	0.005–0.014	0.007–0.04	—
COD	4–43,000	—	6.6–99,000
Color	1,315–38,300 color units	—	—
Copper	< 0.01–0.21	< 0.005–24	ND–9.0
Cyanide	0.017	—	ND–6.0
Dimethyl propane diol	—	ND–0.120	—
Dioxins total	—	0.06–543 (ng/L)	—
2,3,7,8-TCDD	—	0.025–1.6	—
Ethyl hexyl phthalate	—	ND–0.08	—
Fluoride	—	0.1–3.39	—
Furans, total	—	0.04–280	—
Hardness	682–6600	49–742	0.1–225,000
Hexa tiepane	—	ND–0.082	—
Iron	< 0.1–950	< 0.01–121	ND–4000
Kjeldahl nitrogen	34.5–385	—	2.0–3320
Lead	0.037–0.1	< 0.0005–2.92	ND–14.2
Magnesium	3.8–6000	0.006–41	4.0–780
Manganese	0.1–200	0.103–22.4	ND–400
Mercury	< 0.01–7 (μg/L)	< 0.00005–0.008	ND–3.0
Molybdenum	—	< 0.03	0.01–1.43
Nickel	< 0.005–0.024	< 0.005–0.412	ND–7.5
Nitrate	< 0.1–15	—	—
Nitrate-nitrogen	—	0.011–0.59	ND–250
Nitrite	< 0.01–0.018	—	—
Nitrite-nitrogen	—	—	ND–1.46
PCBs	—	< 1.0 (ng/μL)	—
pH	5.4–9.0 units	8.47–9.94 units	3.7–8.9 units
Phenols	0.0011–4.5	—	—
Phosphate	0.11–0.58	0.16–0.43	—
Phosphorus	0.65	—	ND–234
Potassium	140	3.66–4,300	ND–3,200
Selenium	75	0.0025–0.037	ND–1.85

Table 15.6(a) (*continued*)

Parameter	Papermill Sludge (mg/L except as indicated)	Municipal Incinerator Ash (mg/L except as indicated)	Municipal Waste (mg/L except as indicated)
Silicon	< 3	—	—
Silver	—	< 0.001–0.07	ND–1.96
Sodium	9–4,500	11.5–7,300	12.0–6,010
Specific conductance	70–14,370 (μmho/cm)	253–1,874 (μmho/cm)	480–72,500 (μmho/cm)
Strontium	—	0.07–1.03	—
Sulfate	0.9–550	105–4,900	ND–1,850
Sulfide	ND	—	—
Sulfite	4–64	—	—
Sulfonylbissulfur	—	ND–0.011	—
Tannin-lig	13–90	—	—
Thallium	—	—	ND–0.78
Thiolane	—	ND–0.400	—
Tin	< 0.1	0.005–0.013	ND–0.16
Titanium	0.04	—	—
Total alkalinity	174–5,500	60.9–243	ND–15,050
Total fixed solids	144–266	—	—
Total organic carbon (TOC)	1350	—	ND–40,000
Total dissolved solids (TDS)	289–9,810	—	584–55,000
Total suspended solids (TSS)	80–320	—	2–140,900
Total volatile solids (TVS)	211–483	—	—
Turbidity	NR turb. units	—	40–500 Jackson units
Vanadium	< 0.01	—	—
Zinc	< 0.018–0.03	0.002–0.32	ND–731

Source: Bagchi (1994).

[a] ND, not detectable; NR, not reported.

Table 15.6(b) **Composition of leachates from various wastes**[a]

Parameter	Construction/Demolition Waste (mg/L except as indicated)	Coal Burner Fly Ash (mg/L except as indicated)	Iron Foundry Waste (mg/L except as indicated)
Aluminum	—	0.85–1.7	—
Ammonia-nitrogen	30–184	—	—
Antimony	—	< 0.02	—
Arsenic	0.017–0.075	0.135–0.41	0.113
Barium	1.5–8.0	< 0.1	< 0.2
Bicarbonate	2090–7950	—	—
BOD	100–320	—	—
Boron	1.4–3.9	1.8–2.3	—
Cadmium	0.02–0.03	0.01	< 0.01
Calcium	148–578	60–22	—
Carbonate	0	—	—
Chloride	125–240	< 1.0–1.7	139
Chromium	0.1–0.25	0.03–0.29	< 0.05

Table 15.6(b) (*continued*)

Parameter	Construction/Demolition Waste (mg/L except as indicated)	Coal Burner Fly Ash (mg/L except as indicated)	Iron Foundry Waste (mg/L except as indicated)
Chromium (hexavalent)	0.18–4.92	—	—
Cobalt	—	< 0.01–0.02	—
COD	3080–11,200	6.0–16	150
Color	—	—	—
Copper	0.14–0.49	< 0.01–0.04	—
Cyanide	< 0.10	—	< 0.05
Fluoride	< 0.1–0.4	—	1.60
Germanium	—	< 3.0	—
Hardness (as $CaCO_3$)	597–1,516	216–596	—
Iron	29–172	0.07–0.24	< 0.03
Iron (filtered)	0.24–11.0	—	—
Lead	0.22–2.13	< 0.01–0.04	< 0.005
Magnesium	92–192	1.1–4.3	—
Manganese	1.0–4.9	0.01–0.05	< 0.01
Mercury	< 0.002–0.009	< 0.009	< 0.0002
Molybdenum	—	0.29–3.8	—
Nickel	0	0.03–0.06	< 0.04
Nitrate	4–13	—	—
Oil and grease	18–47	—	—
pH	6.5–7.3 units	7.83–9.05 units	12.3 units
Phenolphthalein alkalinity (as $CaCO_3$)	0	—	—
Phenols	0.7–2.99	—	0.118
Phosphorus	2.5–3.89	0.04–0.08	—
Potassium	118–618	20–29	—
Rubidium	—	< 0.09	—
Selenium	< 0.001	0.05–0.18	0.030
Silica	—	5.1–51.0	—
Silver	< 0.01–0.03	—	—
Sodium	256–1290	9.0–50.0	—
Specific conductance	2920–6850 (μmho/cm)	409–1213 (μmho/cm)	—
Strontium	—	< 0.04–0.99	—
Sulfate	< 40	—	5.1
Sulfur	—	53.3–222	—
Tin	—	< 1.0	—
Titanium	—	< 0.5	—
Total alkalinity (as $CaCO_3$)	1710–6520	37–50	—
Total organic carbon (TOC)	76–1080	—	—
Total dissolved solids (TDS)	2412–4270	390–1240	1990
Total suspended solids (TSS)	1000–43,000	—	—
Uranium	—	< 0.005	—
Vanadium	—	0.26–0.92	—
Zinc	1.7–8.63	0.02–0.04	< 0.01

Source: Bagchi (1994).

[a] BOD, biochemical oxygen demand; COD, chemical oxygen demand.

tative sample. The particle size distribution is then described by plotting the percent weight versus maximum, minimum, or average size of components. The shape of the components, if desired, is generally described as cylindrical, spherical, or plate line in shape. The sizes of components in MSW typically range from 1 to 20 in.

Porosity. The *porosity* (n) of waste is defined as the ratio of volume of voids to total volume of waste. In geotechnical engineering practice, the *void ratio* (e) is often used instead of porosity and is defined as the ratio of volume of voids to the volume of solids of the waste. As explained in Chapter 5, the porosity and void ratio of wastes can related by

$$n = \frac{e}{1 + e} \tag{15.1}$$

$$e = \frac{n}{1 - n} \tag{15.2}$$

For municipal refuse, porosity ranges from 0.40 to 0.62, depending on the composition and compaction of the waste. For comparison, a typical compacted soil liner material will have a porosity of about 0.40. Typical porosity for compacted electric plant fly ash is 0.541 and for bottom ash is 0.578; the value for compacted fine copper slag is 0.375.

Moisture Content. The *moisture content* of wastes is expressed in one of three ways: (1) dry gravimetric moisture content, (2) wet gravimetric moisture content, and (3) volumetric moisture content. The *dry gravimetric moisture content* (w) is the same as the moisture content used in geotechnical engineering practice and is defined as the ratio of the mass of water to the mass of dry solids of the waste. The *wet moisture content* of waste (w_w) is defined as the ratio of mass of water to the total mass of the waste. The *volumetric moisture content* of waste (θ) is defined as the ratio of volume of water to the total volume of the waste. The gravimetric and volumetric moisture contents can be related as follows (Zornberg et al., 1999):

$$\theta = \frac{\gamma_d}{\gamma_w} w \tag{15.3}$$

$$\theta = \frac{\gamma_t}{\gamma_w} \frac{w}{1 + w} \tag{15.4}$$

$$\theta = (1 - n)G_s w \tag{15.5}$$

where γ_d is the bulk dry unit weight of the waste, γ_w the unit weight of water, γ_t the total (wet) unit weight of the waste, and G_s the specific gravity of the solids in the waste.

In the literature, moisture contents of wastes are reported using the definitions above, and one must pay attention when making a direct comparison of the moisture contents from different studies. For most MSW, the wet gravimetric moisture content varies from 15 to 40%, depending on humidity and weather conditions, waste composition, and season of the year (Tchobanoglous et al., 1993). Zornberg et al. (1999) reported the variation of dry gravimetric moisture content of MSW with depth in a landfill and found an average gravimetric moisture content of MSW of approximately 28%, and it did not show a significant increasing trend with depth. However, the same results expressed in terms of volumetric moisture content showed an increasing trend of moisture with depth. In literature, the volumetric moisture content of the MSW is reported to vary from 0.05 to 0.30. The moisture content of other types of wastes vary depending on source, composition, and subsequent handling procedures.

Field Capacity. The *field capacity* of waste (θ_{FC}) is defined as the volumetric water content after a prolonged period of gravity drainage. It can also be defined as volumetric water content at a capillary pressure of 0.33 bar. There is only limited information on the field capacity of wastes. For MSW, McBean et al. (1995) reported a field capacity value of 0.55, while Blight et al. (1992) reported field capacity values under semiarid areas in South Africa to range as high as 1.25 to 1.5. Fungaroli and Steiner (1979) and Zornberg et al. (1999) found that field capacity depends on the unit weight of the waste (see below for a definition of the unit weight of waste) and provided the following relationship:

$$\theta_{FC} = 21.7 \ln \gamma_t - 5.4 \qquad (15.6)$$

where θ_{FC} is expressed as a percentage and γ_t is the total (wet) unit weight of the waste expressed in kN/m^3. For a 13.5-kN/m^3 unit weight of MSW, the field capacity value was found to be around 0.51. With regard to other types of wastes, field capacity for compacted electric plant fly ash is typically 0.187, for electric plant bottom ash is 0.266, and for compacted fine copper slag is 0.055. For comparison, the field capacity of compacted clay liners is approximately 0.356.

Wilting Point. The *wilting point* of waste is defined as the lowest volumetric water content that can be achieved by plant transpiration. The wilting point of MSW is reported to range from 0.084 to 0.17 (Sharma and Lewis, 1994; Zornberg et al., 1999). The typical values of wilting point for compacted coal-burning electric plant fly ash, bottom ash, and compacted fine copper slag are 0.0471, 0.0649, and 0.020, respectively. This compares with a typical value of 0.29 for a compacted liner in soil.

Unit Weight. The *unit weight* of waste (γ) is defined as the ratio of total weight to the total volume of the waste. The unit weight of wastes depends primarily on their composition. However, the in-place unit weight of MSW depends on factors such as the degree of variation in the waste constituents, state of decomposition, degree of relative compaction, total thickness of the landfill, and thickness of daily cover. Household refuse is usually heavier than industrial waste. Also, as would be expected, a landfill with large percentages of rubble will have a high unit weight. A summary of various municipal waste landfill units indicates that depending on compaction, method of density measurements, and whether the waste is shredded or baled, unit weights may range between 18 and 67 lb/ft^3 (Oweis and Khera, 1990). Sharma et al. (1990) reviewed unit weight data from various sources, as well as performing their own measurements, and reported that the unit weight for municipal refuse may range from 20 to 84 lb/ft^3 (Table 15.7). Fassett et al. (1994) indicate that the unit weight range is from 18.5 to 99.6 lb/ft^3 and recommended that for good compaction procedures the total unit weight of the upper 30 ft of waste would be 65 $\pm$ 10 lb/ft^3. Kavazanjian et al. (1995) considered the initial and average unit weights reported by landfill operators as well as the values of unit weights and waste com-

Table 15.7 **Refuse fill average unit weights**

Source	Refuse Placement Conditions	Unit Weight kg/m^3	Unit Weight lb/ft^3
U.S. Dept. Navy (1983)	Sanitary landfill		
	Not shredded		
	Poor compaction	320	20
	Good compaction	641	40
	Best compaction	961	60
	Shredded	881	55
Sowers (1968)	Sanitary refuse: depending on compaction effort	481–961	30–60
NSWMA (1985)	Municipal refuse		
	In a landfill	705–769	44–49
	After degradation and settlement	1009–1121	63–70
Landva and Clark (1986)	Refuse landfill (refuse to soil cover ratio varied from about 2:1 to 10:1)	913–1346	57–84
EMCON Associates (1989)	For 6:1 refuse to daily cover soil	737	46

Source: Sharma et al. (1990).

pressibility reported in the literature and developed a unit weight profile as shown in Figure 15.1. If site-specific information on unit weight for a municipal waste is not available, an estimate of unit weight of MSW can be obtained from Figure 15.1 for an old degraded and settled landfill or for a very well compacted new municipal solid waste landfill for the purpose of an engineering evaluation.

Based on information obtained from Shoemaker (1972), Sowers (1973), Collins (1977), Tchobanoglous et al. (1977), Bromwell (1978), Oweis and Khera (1986), Peirce et al. (1986), and Sargunan et al. (1986), Oweis and Khera (1990) summarized the values for unit weights of various wastes. This summary indicates that incinerator residue ash weighs between 41 and 52 lb/ft³, poorly burned residue weighs 46 lb/ft³, and well-burned residue weighs 81 lb/ft³. The data summarized by Oweis and Khera (1990) also indicate that 40- to 50-ft-deep hazardous waste dry dust and soil

may weigh between 30 and 110 lb/ft³, and 75-ft-deep waste with soil cover at a hazardous waste landfill may weigh 101 lb/ft³. Similarly, a unit weight of 90 lb/ft³ was cited for 30- to 40-ft-deep landfill, with 90 to 95% of its landfill waste in metal drums. Kiln dust, sludge tar, creosote, and soil in about 62-ft-deep (average thickness) waste have been reported to weigh 75 lb/ft³ (Sharma and Lewis, 1994).

The information above indicates that waste unit weights vary significantly and that site-specific data should be collected for use in an engineering evaluation. The data cited above may, however, be used in preliminary design.

Saturated Hydraulic Conductivity. The hydraulic conductivity of wastes determines the transport rate of leachates and other constituents within the waste. Wastes such as MSW are highly heterogeneous and anisotropic, and accurate determination of properties

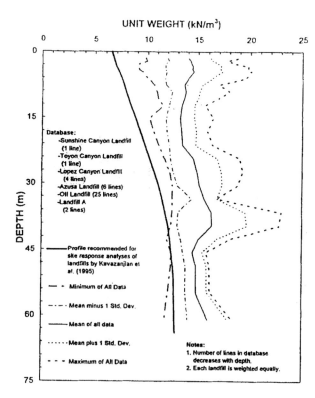

Figure 15.1 *Unit weight profile of MSW in a landfill. (From Kavazanjian, 1999.)*

of such wastes is difficult. Saturated hydraulic conductivity for municipal waste is typically 1×10^{-3} cm/s. However, field permeabilities measured by Landva and Clark (1990) in municipal waste pits ranged from 1×10^{-3} to 4×10^{-1} cm/s. Typical hydraulic conductivities for compacted clay liners range from 1×10^{-7} to 1×10^{-8} cm/s. Typical saturated hydraulic conductivities for compacted coal-burning electric plant fly ash, bottom ash, and compacted fine copper slag are 5×10^{-5}, 4×10^{-5}, and, 4×10^{-2} cm/s, respectively (Aziz et al., 1992).

Shear Strength. The shear strength of wastes is an important engineering consideration because it determines potential beneficial reuse of the waste as well as it is required for stability analysis of the landfill in which the waste is to be disposed. Due to variability in municipal refuse, reliance only on conventional laboratory results is not prudent. Various laboratory and field tests on MSW have been reported in the literature. Landva and Clark (1990) and Singh and Murphy (1990) evaluated the results presented by various investigators.

Based on the critical valuation of the published and unpublished shear strength data for municipal landfill waste, Singh and Murphy (1990) presented a cohesion (c) and friction angle (ϕ) relationship. The data obtained from vane shear tests, SPT tests, and back-calculated c–ϕ values from field load tests, failures, and performance records were used to develop this relationship. Kavazanjian et al. (1995) performed a systematic assessment of the shear strength data on MSW obtained from laboratory and in-situ testing supplemented by the strengths obtained from back-analysis of existing stable solid waste landfill slopes. Based on these data, a bilinear strength envelope (Figure 15.2) was presented as a lower-bound envelope for the drained shear strength of MSW. Based on this, the shear strength parameters of MSW are $\phi = 0$ and $c = 24$ kPa (500 lb/ft^2) at normal stress below 30 kPa (625 psf) and $\phi = 33°$ and $c = 0$ at higher normal stresses.

In general, shear strength characteristics of municipal waste is not yet well understood and need further evaluation. No useful information is yet available on the stress–strain behavior of MSW. Until further information is obtained, the bilinear strength envelope shown in Figure 15.2 may be used as a guide in the landfill design.

Oweis and Khera (1990) reviewed the results of strength tests performed by various investigators on mineral wastes. Evaluations indicated that addition of a small amount of lime or cement to fly ash may result in a strength increase. Table 15.8 summarizes strength

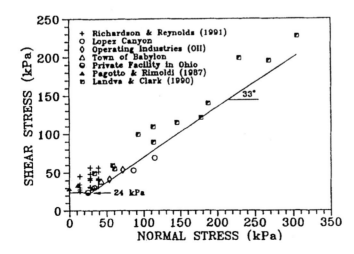

Figure 15.2 *Shear strength of MSW. (From Kavazanjian, 1999.)*

Table 15.8 **Strength properties of mineral wastes**

Description	Cohesion (kPa)	Friction Angle (deg)	Undrained Compressive Strength (kPa)	Water Content (%)
Coal refuse				
Undrained condition	10–40	10–28		
Effective stress	0–40	25–43		
Fly ash, Arizona 7-day				
Unit wt, 12.6 kN/m^3			223	
Unit wt, 13.4 kN/m^3			331	
Unit wt, 13.8 kN/m^3			587	
Fly ash (silica 46%, aluminum 34%, calcium 7%)				
Slurry samples		37		
Compacted, undrained		41		
Effective stress		37		
Compacted, drained				
West Virginia fly ash				
Shelby tube samples (consolidated, undrained)		34		
Shelby tube samples (consolidated, drained)		37.5		
West Virginia bottom ash		38–43		
Flue-gas desulfurization (FGD) sludge				
Consolidated, drained test	0	41.5		
Compacted	0–40	10–40		
Red mud (bauxite residue)				
Unleached			63	52
Leached			0	49
Mud:sand (5:1)			38	41

Source: Oweis and Khera (1990).

properties of mineral wastes, and Table 15.9 presents strength variations of fly ash and mixtures of fly ash over time.

Compressibility. Compressibility of waste materials is an important consideration in settlement analysis. Although compressibility of MSW is irregular due to its heterogeneous nature, it generally exhibits many characteristics similar to those of cohesive organic soils, especially peat (Sharma and Lewis, 1994). Significant settlement occurs during waste placement and shortly after waste placement is completed. This is followed by substantial additional settlements occurring at a slower rate over extended periods of time. The initial settlements are classified as primary settlement, with long-term settlement being classified as secondary

compression (Sowers, 1973). The primary settlement in each distinct layer can be calculated using

$$S = H \frac{C_c}{1 + e_0} \log \frac{P_0 + \Delta P}{P_0} \qquad (15.7)$$

where S is the primary compression occurring in the layer under consideration, H the initial thickness of the waste layer under consideration, C_c the primary compression index, e_0 the initial void ratio of the layer, P_0 the existing overburden pressure acting at the midlevel of the layer, and ΔP the increment of overburden pressure at the midlevel of the layer. The primary settlement calculated using equation (15.7) occurs due mainly to mechanical compression, with little or no pore pressure buildup. The primary compression index

Table 15.9 **Strength variation of fly ash and mixtures of fly ash**

Material	q_u (kPa) 7 days	28 days
Michigan	0.170×10^3	0.210×10^3
New Jersey	0.310×10^3	0.425×10^3
UK	0.550×10^3	0.660×10^3
West Virginia		
Ash A with 3% lime	19.6×10^3	31.6×10^3
Ash A with 3% cement	7.4×10^3	12.8×10^3
Ash B with 3% lime	2.8×10^3	4.0×10^3
Ash B with 3% cement	3.2×10^3	8.0×10^3
Lignite fly ash (fly ash 10%, sand 90%)	2.6×10^3	4.6×10^3
Lime/fly ash/FGD sludge		
6.3:43.7:50.0	3.3×10^3	5.5×10^3
2.5:55.8:41.7	3.0×10^3	4.0×10^3
0:41.2:58.8	0.3×10^3	0.4×10^3
Fly ash/lime/cement/ dredge waste		
4:0 :0 :96		8
0:4:0:96		10.5
0:0:4:96		38
10:0:0:90		10
0:10:0:90		19
0:0:10:90		78

Source: Oweis and Khera (1990).

C_c is assumed to be proportional to the initial void ratio of the layer and is calculated by multiplying e_0 by a value between 0.15 and 0.55, depending on the organic content of the waste. A value of 0.15 would be used for waste with a low organic content, while a value of 0.55 would be used for a high organic content. For older landfills (10 to 15 years) that are subjected to external loads, $C_c/(1 + e_0)$ values range from 0.1 to 0.4 (U.S. Department of the Navy, 1983).

The secondary compression is estimated using

$$S_s = HC_\alpha \log \frac{t_2}{t_1} \qquad (15.8)$$

where S_s is the secondary compression occurring in layer under consideration, H the initial thickness of the waste layer under consideration, C_α the secondary compression index, t_1 the starting time for the long-term period under consideration, and t_2 the ending time for the long-term period under consideration. Values of C_α for MSW have been estimated to be similar to those of peat. NAVFAC DM 7.3 (U.S. Navy, 1983) recommends C_α ranging from 0.02 to 0.07 for landfills between 10 and 15 years old. Oweis and Khera (1990) recommend C_α values between 0.01 and 0.04. Secondary compression values are calculated for each layer under consideration for the given period. The total long-term settlement occurring during the period under consideration is calculated by summing the settlement in each layer. Sharma (2000) presents C_α values for refuse under two loading conditions: (1) when subjected to external loads, and (2) under self-weight. These values are presented in Section 19.4.6.

It should be recognized that the actual settlement behavior of a landfill is complex and can be influenced by various factors, such as movement of smaller particles into larger voids, chemical reactions, biodegradation of organics within the landfill, dissolving of soluble substances by percolating water, creep, and changes in deformation properties with time (Sowers, 1968; Huitric, 1981; U.S. Navy, 1983; Fassett et al., 1994). Generally, it has been observed that a new landfill may settle up to 50% of its thickness or depth, while a closed landfill may settle between 15 and 20% of its total thickness after 10 to 15 years. Since landfill settlements vary significantly depending on specific waste types and placement methods, a settlement monitoring program should be established to estimate project-specific settlements.

Dynamic Properties. The dynamic properties of wastes are needed to determine the stability of the wastes in various applications as well as to analyze the stability of landfills in which the wastes are disposed under seismic conditions. The dynamic properties include dynamic shear modulus (G_s), dynamic shear damping (D), and Poisson's ratio (v). The maximum dynamic shear modulus (G_{max}) is often calculated directly from the shear wave velocity and total unit weight. The shear wave velocity is determined in the field using various geophysical techniques, such as cross-hole and downhole soundings, and surface wave measurements. Sharma et al. (1990) measured the in-situ shear wave velocity of a MSW landfill. Based on this and other in-situ measurements and published data, Kavazanjian (1999) presented a shear wave velocity profile of MSW landfill (Figure 15.3). Based on the data obtained from a MSW landfill, Kavazanjian (1999) also presented modulus reduction and damping curves (Figure 15.4). Based on the shear wave velocity and compressional wave velocity measurements in the same borehole at a MSW landfill, Kavazanzian (1999) presented a Poisson's ratio profile through MSW as shown in Figure 15.5. These dynamic properties may be used in the preliminary design of MSW landfills in seismic areas, but a site-specific investigation on the dynamic properties of the wastes is recommended for the project design.

15.5 ENVIRONMENTAL CONCERNS WITH WASTES

As described earlier, there are many sources of wastes and the wastes vary widely in composition. Some of the wastes may be hazardous, radioactive, or infec-

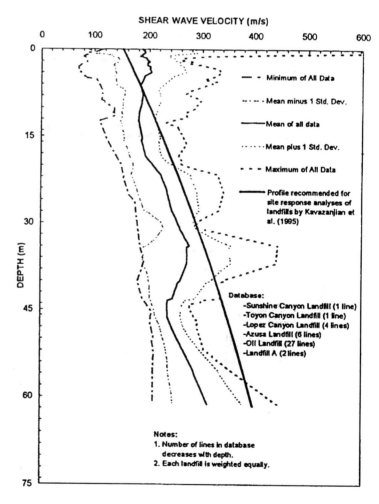

Figure 15.3 Shear wave velocity profile of MSW in a landfill. (*From Kavazanjian, 1999.*)

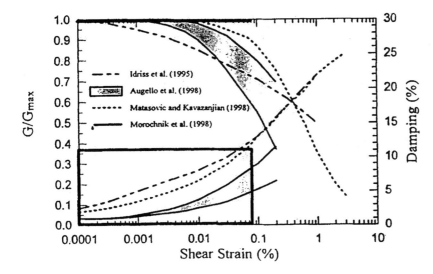

Figure 15.4 Modulus reduction and damping of MSW. (*From Kavazanjian, 1999.*)

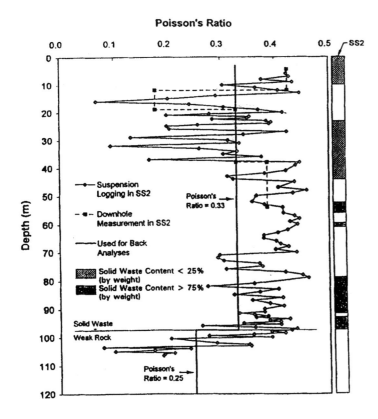

Figure 15.5 Poisson's ratio profile of MSW in a landfill. (*From Kavazanjian, 1999.*)

tious. It is important to manage and dispose of these wastes prudently because of the concerns described below.

15.5.1 Growing Quantities

The quantity of wastes generated and disposed of annually is large and growing. Table 15.10 shows the quantities of different wastes generated annually in the United States. From 1970 to 1988, while the population increased 1.0% the MSW discards increased 1.8% (Alexander, 1993). According to the World Wildlife Fund (1989), in less than three decades, the amount of MSW more than doubled, from 88 million tons to 180 million tons between 1960 and 1988. The waste generated also increased by nearly 20 million tons between 1985 and 1988. This represents an 11% increase in only three years. In addition, many landfills have opted to close rather than meet new, stricter regulations, and it is estimated that in a very short period of time, hundreds more existing landfills are expected to reach their authorized capacity.

Overall, concerning the growing quantities of waste, it is important to note that (1) U.S. waste per capita is definitely increasing, and (2) U.S. landfill capacity is declining, due primarily to (a) older landfills reaching their authorized capacity, (b) strengthened environmental regulations, and (c) difficulty in siting new landfills, due to public opposition. The public opposition is based on poorly operated facilities and potential health and environmental risks. The acronym NIMBY ("not in my backyard") reflects the public's view toward landfills and wastewater treatment facilities.

15.5.2 Improper Handling and Disposal

Prior to the passage of environmental regulations, wastes were disposed of without due consideration of potential effects on the public health and environment (Figure 15.6). That practice led to numerous contaminated sites where soils and groundwater have been contaminated and pose a risk to public safety. There are more than 1400 sites listed under the Superfund program's national priority list that require immediate cleanup. The USEPA has identified about 2500 additional contaminated sites that will eventually require remediation. This list does not include the estimated 15 to 20% of 5 to 7 million underground storage tanks that are leaking. In addition, the U.S. Department of Energy is responsible for the cleanup of more than 3500 inactive sites where complex contaminants exist. It is estimated that more than 4700 facilities in the United States currently treat, store, or dispose of hazardous wastes. Of these, about 3700 facilities that house approximately 64,000 solid waste management units (SWMUs) may require corrective action.

There are many choices for the disposal of newly generated wastes, and as the costs for disposal in new, more stringently operated landfills or incinerators has increased, if waste generators opt for less expensive alternative methods, contamination of the water, groundwater, air, or land may result, leading to health problems and environmental degradation.

15.5.3 Toxic Chemicals

A wide variety of chemicals are found in municipal wastewaters, MSW, and especially in industrial and hazardous waste streams. Typical chemicals and ranges of concentrations found in leachates from a variety of wastes are shown in Table 15.6. Chemicals of significant environmental concern are listed in Table 15.11. Although these leachate sources contain toxic chemicals, the concentrations and variety of toxic chemicals are quite small compared to hazardous waste sites.

Table 15.10 **Quantities of wastes generated**

Waste Type	Quantity
Municipal solid waste	196 million tons/yr
Industrial nonhazardous waste	7.6 billion tons/yr
Hazardous waste	380 million tons/yr
Radioactive waste	
Low level	24,000 m³/yr
High level	350,000 m³ (total as of year 2000 and increasing every year)
Medical waste	465,000 tons/yr

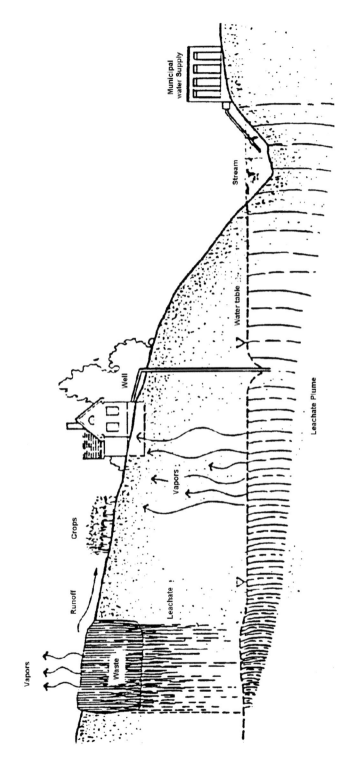

Figure 15.6 Problems associated with improper disposal of wastes.

Table 15.11 **Typical hazardous waste compounds produced by commercial, industrial, and agricultural activities that are typically found in MSW and their health effects**

Name	Formula or Symbol	Concern
Nonmetals		
Arsenic	As	Carcinogen and mutagen. Long-term: can sometimes cause fatigue and loss of energy; dermatitis.
Selenium	Se	Long-term: red staining of fingers, teeth, and hair; general weakness; depression; irritation of nose and mouth.
Metals		
Barium	Ma	Flammable at room temperature in powder form. Long-term: increased blood pressure and nerve block.
Cadmium	Cd	Flammable in powder form. Toxic by inhalation of dust or fume. A carcinogen. Soluble compounds are highly toxic. Long-term: concentrates in the liver, kidneys, pancreas, and thyroid; hypertension suspected effect.
Chromium	Cr	Hexavalent compounds are carcinogens and corrosive on tissue. Long-term: skin sensitization and kidney damage.
Lead	Pb	Toxic by ingestion or inhalation of dust or fumes. Long-term: brain, nervous system, and kidney damage; birth defects.
Mercury	Hg	Highly toxic by skin absorption and inhalation of fume or vapor. Long-term: toxic to central nervous system; may cause birth defects.
Silver	Ag	Toxic metal. Long-term: permanent gray discoloration of skin, eyes, and mucous membranes.
Organic Compounds		
Benzene	$C_6 H_6$	A carcinogen. Highly toxic. Flammable.
Ethylbenzene	$C_6 H_5 C_2 H_5$	Toxic by ingestion, inhalation, and skin.
Toluene	$C_6 H_5 CH_3$	Flammable; dangerous fire risk. Toxic by
Halogenated Compounds		
Chlorobenzene	$C_6 H_5 Cl$	Moderate fire risk. Toxic by inhalation and
Cloroethene	$CH_2 CHCl$	Extremely toxic and hazardous by all avenues.
Dichloromethane	$CH_2 Cl_2$	Toxic. A carcinogen, narcotic.
Tetrachloroethane	$C Cl_2 C Cl_2$	Irritant to eyes and skin.
Pesticides		
Endrin	$C_{12} H_8 OCl_6$	Toxic by inhalation and skin absorption.
Lindane	$C_6 H_6 Cl_6$	Toxic by inhalation, ingestion, and skin.

Source: Tchobanoglous et al., (1993).

In 1979, the estimated number of hazardous waste sites in the United States was approximately 30,000 to 50,000, and at least 2000 of those sites posed an imminent hazard to the environment. The most hazardous sites should be assessed on a case-by-case basis, because the conditions and site characteristics widely vary, as do the types and concentrations of contaminating chemicals. An analysis of 115 priority sites divided the hazardous wastes into five types and listed the percentage of sites that contained each type of waste, as shown in Table 15.12 Explosives and radioactive wastes are located primarily at Department of Energy (DOE) sites, because weapons were built on many of their sites. In 1993, the DOE spent $5 billion

Table 15.12 **Types of hazardous wastes and percentages at 115 priority sites**

Waste Type	Sites
Pesticides and highly toxic organics	65%
Other organics	52%
Inorganics	56%
Radioactive substances	4%
Explosives and flammables	9%

Source: Greenberg and Anderson (1984).

on cleaning up contaminated sites, one-third of their annual budget (Bredehoeft, 1994). Organic contaminants are found largely at oil refineries or petroleum storage sites, and inorganic and pesticide contamination usually is the result of a variety of industrial activities.

The events at the Crystal Chemical Plant located near Houston, Texas represent an example of an environmental emergency action. The plant produced herbicides and used one of the most toxic forms of arsenic. Poor maintenance, spills, and heavy rainfall combined to contaminate the entire site with arsenic. The Texas Department of Water Resources issued a "no discharge order," and the company constructed a dike around the plant to contain runoff. After more heavy rains, company managers said that it was beyond their capability to contain the wastes, and the USEPA initiated emergency action. A total of 815,000 gal of liquid containing an average arsenic concentration of 20,000 mg/L was removed from the site. Other dangerous chemicals were also removed from the site, such as methyl chloride, sulfur dioxide, sulfuric acid, and other arsenic compounds (HMCRI, 1984).

15.5.4 Health Effects

Since there are so many various types of toxic chemicals found in waste streams, the number of possible human reactions and diseases produced from exposure to them is enormous. Hazardous wastes have been linked to many types of cancer, chronic illnesses, and abnormal reproductive outcomes, such as birth defects, low birthweights, and spontaneous abortions. Table

15.11 shows the health effects of toxic chemicals found in MSW. Since there are often so many chemicals and chemical reactions within hazardous wastes, it is often difficult to characterize their health effects. Generally, studies have been performed with major toxic chemicals found at hazardous waste sites with epidemiological or animal tests to determine their toxic effects (Greenberg and Anderson, 1984).

The effects of radioactive materials are classified as somatic or genetic. The somatic effects may be immediate or occur over a long period of time. Immediate effects from large doses often produce nausea and vomiting and may be followed by severe blood changes, hemorrhage, infection, and death. Delayed effects include leukemia and many types of cancer, including bone, lung, and breast cancer. Genetic effects involve gene mutations or chromosome abnormalities that have harmful effects, such as a slight decrease in life expectancy and an increase in the susceptibility to disease, infertility, or even the death of humans in the embryonic stages of life. Occupational dosage limits have been recommended by the National Council on Radiation Protection (Schumacher, 1988).

15.5.5 Effects on Ecosystems

The chemicals found in wastes can have profound effects on entire ecosystems. Contaminants may enter the food chain through plants or microbiological organisms, and higher, more evolved animals bioaccumulate the wastes by ingesting them. As the contaminants move farther up the food chain, the concentrations of the chemicals increase. Then, as these important species begin to die off, the ecosystem becomes unbalanced, and disastrous consequences may occur. Contaminants may also change the chemistry of waters and destroy aquatic life and underwater ecosystems that higher life forms depend on. Examples include the near extinction of the bald eagle due to DDT pesticide ingestion from eating fish, and the declining numbers of oysters, crabs, and fish in the Chesapeake Bay due to excessive quantities of fertilizers, toxic chemicals, farm manure wastes, and power plant emissions (Pollack, 1996).

15.6   WASTE MANAGEMENT STRATEGIES

In general, waste problems involve first defining waste, identifying its major sources, classifying it according to composition, and presenting methods to determine chemical and physical waste characterization. Then, major concerns with wastes and broad objectives or goals are established. Finally, different plans or strategies for the management of wastes are evaluated and the best available, most feasible, and economical technology is selected. The preferred hierarchy of waste management practice is pollution prevention, waste minimization, recycling, incineration, and landfilling. These different waste management practices are described briefly in this section.

15.6.1   Pollution Prevention

Pollution prevention is a basic goal of all waste management strategies. To protect and preserve the environment for future generations, waste management professionals will have to make pollution prevention a top priority. As stated earlier, pollutants or contaminants that invade the air, water, groundwater, and land may be responsible for destroying ecosystems and creating havoc with the environment. Using substitute products that are less hazardous or harmful, employing modern leakage detection systems for material storage, and using chemical neutralization or dewatering techniques to reduce reactivity are just some of the methods available to prevent environmental pollution.

15.6.2   Waste Minimization

Waste minimization or source reduction is an extremely important concept for professionals in the waste industry. By working constantly with waste, professionals become aware and informed about trends or products that may be causing problems in the waste stream, and halting these problems and informing the public about them before they become major catastrophes is of major importance. In industry, waste can be reduced by reusing materials, using less hazardous, substitute materials, or by modifying components of design and processing (Wilson, 1981).

15.6.3   Recycling

Recycling is another key to waste minimization. By reusing products or materials or composting garden and yard wastes, people will save time, energy, trees, and land space for the future. The predominantly recycled materials include paper, plastics, glass, aluminum, steel, and wooden pallets. Many industries also recover products and clean solvents for reuse. Examples are copper and nickel recovery from metal finishing processes; the recovery of oils, fats, and plasticizers by solvent extraction from filter media such as activated carbon and clays; and acid recovery by spray roasting, ion exchange, or crystallization (Wilson, 1981).

15.6.4   Incineration

New incinerators are cleaner, more flexible, and efficient; have been used to convert waste to energy, and are excellent at reducing the volume of waste. Despite all these advantages, they are viewed negatively because they still emit some air pollution, they may create new chemical compounds, and the ash produced is usually highly toxic. The EPA is developing new regulations to carefully monitor air quality from incinerators under the Clean Air Act. Incineration is often used for hazardous wastes such as chlorinated hydrocarbons, oils, solvents, medical wastes, and pesticides.

15.6.5   Landfilling

Landfilling is the primary method of disposing waste in the United States. The amount of MSW has been increasing and the number of remaining landfills has been decreasing. New regulations concerning proper waste disposal and the use of high-tech liner systems to control contamination due to leachate have caused a substantial increase in the costs of landfill disposal. Public opposition to landfills is based largely on

1. Liner System
2. Leachate Collection System
3. Final Cover System
4. Groundwater Monitoring System
5. Gas Collection System (Not Shown)

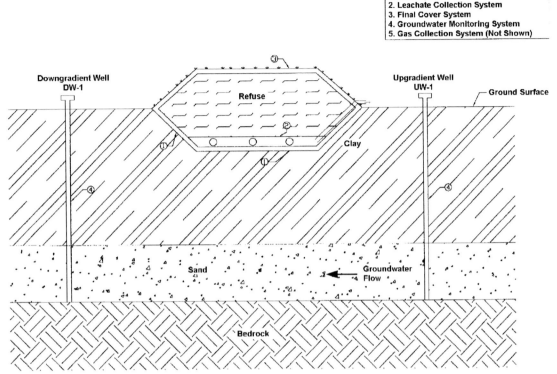

Figure 15.7 *Components of a typical MSW landfill.*

NIMBY and the concept of the old vermin-infested dumps that contaminated groundwater, smelled horrendous, and lowered nearby home values.

15.7 ENGINEERED WASTE DISPOSAL FACILITIES

Although landfilling is the last priority, a well-developed waste management strategy must incorporate landfilling as an essential element to dispose of wastes. Landfilling represents disposal of all types of wastes in engineered facilities. There are basically three types of engineered waste disposal facilities: (1) municipal solid waste (MSW) landfills, (2) hazardous waste landfills, and (3) surface impoundments. MSW landfills contain general (less toxic) wastes from sources such as private homes, institutions, schools,

and businesses without hazardous wastes. Since these landfills are the most common facility type, they are the primary focus of this book.

Unlike in the past, landfills are engineered to protect public health and the environment. Figure 15.7 shows the typical MSW landfill components, including a liner system, a leachate collection and removal system, a final cover system, gas management, and a groundwater monitoring system. Each of these components is designed based on the sound engineering principles detailed in Chapters 16 through 21.

The hazardous waste landfills are disposal facilities for the more toxic chemicals and dangerous by-products. These landfills must be extremely well designed to reduce any chance of the escape of hazardous compounds into the environment. The surface impoundments are facilities that deal with liquid waste disposal. Many of the design procedures used for

MSW landfills are also applicable to hazardous waste landfills and surface impoundments and are discussed in Chapters 16 through 22.

15.8 SUMMARY

Throughout history, humans have had the difficult and complex task of safely disposing of wastes. As populations and technologies advance, this task has become even more difficult and complex. Many wastes, such as high-level radioactive wastes, will stay dangerous for thousands of years, and even MSW often produces a dangerous leachate that could devastate an entire ecosystem if allowed to migrate into the environment. To protect human health and the environment, environmental professionals must solve many problems, such as reducing the sources of wastes and safely disposing of wastes. It is, therefore, extremely important to know the sources, classifications, chemical compositions, and physical characteristics of wastes, and to understand the strategies for managing them.

QUESTIONS/PROBLEMS

15.1. List sources of wastes.

15.2. Explain different types of wastes.

15.3. How do you classify the wastes generated from the following sources: **(a)** hospitals, **(b)** nuclear power plants, **(c)** electroplating plant, **(d)** schools, and **(e)** restaurants?

15.4. What is solid waste?

15.5. If a waste material can be recycled and reused, would you call this material a waste? Why?

15.6. What is municipal solid waste (MSW)?

15.7. How is hazardous waste defined by USEPA regulations?

15.8. What is the typical municipal solid waste composition?

15.9. Observe the waste generated at your home or apartment. Based on this, estimate the types and quantities of wastes that are recycled and disposed of. Estimate the composition of the waste that is disposed of and compare it with typical MSW composition.

15.10. List the types of hazardous chemicals that may get into a municipal solid waste stream.

15.11. What are the physical properties of MSW? List typical values.

15.12. Define *field capacity* and *wilting point*. What happens to water in excess of field capacity?

15.13. Explain why leachate composition is more important than the bulk chemical composition of wastes.

15.14. Explain methods to determine leachate composition.

15.15. What are consequences of mismanagement of wastes?

15.16. Explain the components of a comprehensive waste management plan.

15.17. Explain why pollution prevention is given utmost priority in any waste management strategy.

15.18. What are the advantages and disadvantages of landfilling compared to incineration?

15.19. What is a landfill?

15.20. What are the major components of an engineered landfill?

15.21. What is the major difference between a MSW landfill and a hazardous waste landfill?

15.22. Why can't we dispose of liquid wastes in landfills?

15.23. What is an impoundment?

REFERENCES

Alexander, J. H., *In Defense of Garbage,* Praeger Publishers, Westport, CT, 1993.

Aziz, N., Schroeder, P. R., and Lloyd, C. M., *The Hydrologic Evaluation of Landfill Performance (HELP) Model,* Hazardous Waste Engineering Research Laboratory, USEPA, Cincinnati, OH, 1992.

Bagchi, A., *Design, Construction, and Monitoring of Landfills,* Wiley, New York, 1994.

Blight, G. E., Ball, J. M., and Blight, J. J., Moisture and suction in sanitary landfills in semiarid areas, *J. Environ. Eng.,* Vol. 118, No. 6, pp. 865–877, 1992.

Bredehoeft, J. D., Hazardous waste remediation: a 21st century problem, *Groundwater Monitor. Remed.,* pp. 95–100, winter 1994.

Bromwell, L. G., Properties, behavior and treatment of waste fills, presented at the Seminar on Improving Poor Soil Conditions, Metals Section, ASCE, New York, Oct. 1978.

CFR (*Code of Federal Regulations*), 40 CFR, U.S. Government Printing Office, Washington, DC, 2002.

Collins, R. J., Highway construction of incinerator residue, pp. 246–266 in *Proceedings of the Conference on Geotechnical Practice for Disposal of Solid Waste Materials,* ASCE, Ann Arbor, MI, 1977.

Dolan, E. F., and Scariano, M. M. *Nuclear Waste: The 10,000-Year Challenge,* Franklin Watts, New York, 1990.

EMCON Associates, personal communications between L. Burch and H. Sharma, 1989.

Fassett, J. B., Leonards, G. A., and Repetto, P. C., Geotechnical properties of municipal solid wastes and their use in landfill design, in *Proceedings of Waste Tech '94,* Landfill Technology, Charleston, SC, 1994.

Fungaroli, A. A., and Steiner, R. L., *Investigation of Sanitary Landfill Behavior,* Vol. 1, *Final Report,* EPA/600/2-79/053a, USEPA, Cincinnati, OH, 1979.

Greenberg, M. R., and Anderson, R. F., *Hazardous Waste Sites: The Credibility Gap,* Center for Urban Policy Research, New Brunswick, NJ, 1984.

HMCRI (Hazardous Materials Control Research Institute), National Conference and Exhibition on Hazardous Wastes and Environmental Emergencies, Silver Spring, MD, 1984.

Huitric, R., Sanitary landfill settlement rates, Los Angeles County Sanitation District, paper presented at Technische Universitat, Berlin, 1981.

Kavazanjian, E., Seismic design of solid waste containment facilities, in *Proceedings of the 8th Canadian Conference on Earthquake Engineering,* Vancouver, British Columbia, Canada, 1999.

Kavazanjian, E., Matasovic, N., Bonaparte, R., and Schmertmann, G. R., Evaluation of MSW properties for seismic analysis, pp. 1126–1141 in *Geoenvironment 2000,* ASCE Geotechnical Special Publication 46, Vol. 2, 1995.

Landva, A. O., and Clark, J. I., Geotechnics of wastefill, pp. 86–103 in *Geotechnics of Waste Fills: Theory and Practice,* ASTM STP 1070, Landva, A., and Knowles, G. D. (eds.), ASTM, West Conshohocken, PA, 1990.

McBean, E. A., Rovers, F. A., and Farquhar, G. J., *Solid Waste Landfill Engineering and Design,* Prentice Hall, Upper Saddle River, NJ, 1995.

NSWMA (National Solid Waste Management Association), *Basic Data: Solid Waste Amounts, Composition and Management Systems,* Technical Bulletin 85-6, NSWMA, Washington, DC, 1985, p. 8.

OTA (Office of Technology Assessment), *Facing America's Trash,* OTA-O-424, OTA, Washington, DC, Oct. 1989.

———, *Finding the Rx for Managing Medical Wastes,* OTA-O-459, OTA, Washington, DC, Sept. 1990.

Oweis, I. S., and Khera, R. P., Criteria for geotechnical construction on sanitary landfills, pp. 205–222 in *Proceedings of the International Symposium on Environmental Geotechnique,* Vol. 1, Fang, H. Y. (ed.), 1986.

———, *Geotechnology of Waste Management,* Butterworth, Sevenoaks, Kent, England, 1990.

Peirce, J., Sallfors, G., and Murray, L., Overburden pressures exerted on clay liners, *Environ. Eng.,* Vol. 112, No. 2, p. 284, 1986.

Pollack, S., Holding the world at bay, *Sierra,* Vol. 81, No. 3, May/June 1996.

Sargunan, A., Malli Karjun, N., and Ranapratap, K., Geotechnical properties of refuse fills of Madras, India, pp. 197–204 in *Proceedings of the International Symposium on Environmental Geotechniques,* Vol. 1, Fang, H. Y. (ed.), 1986.

Schumacher, A., *A Guide to Hazardous Materials Management,* Quorum Books, New York, 1988.

Sharma, H. D., Solid waste landfills: settlements and post-closure perspectives, pp. 447–455 in *Proceedings of the ASCE National Conference on Environmental and Pipeline Engineering,* July 2000.

Sharma, H. D., and Lewis, S. P. *Waste Containment Systems, Waste Stabilization, and Landfills: Design and Evaluation,* Wiley, New York, 1994.

Sharma, H. D., Dukes, M. T., and Olsen, D. M., Field measurements of dynamic moduli and Poisson's ratios of refuse and underlying soils at a landfill site, in *Geotechnics of Waste Fills: Theory and Practice,* ASTM STP 1070, Landva, A., and Knowles, G. D. (eds.), ASTM, West Conshohocken, PA, 1990, pp. 57–70.

Shoemaker, N. B., Construction techniques for sanitary landfill, *Waste Age,* pp. 24–25, 42–44, Mar./Apr. 1972.

Singh, S., and Murphy, B. J., Evaluation of the stability of sanitary landfills, pp. 240–258 in *Geotechnics of Waste Fills: Theory and Practice,* ASTM STP 1070, Landva, A., and Knowles, G. D. (eds.), ASTM, West Conshohocken, PA, 1990.

Sowers, G. F., Foundation problems in sanitary landfill, *J. Sanit. Eng. Div. ASCE,* Vol. 94, No. SA1, pp. 103–116, 1968.

———, Settlement of waste disposal fills, pp. 207–210 in *Proceedings of the 8th International CSMFE,* Moscow, USSR National Society for Soil Mechanics and Foundation Engineering, 1973.

Tschobanoglous, G., Theissen, H., and Eliassen, R., *Solid Wastes,* McGraw-Hill, New York, 1977.

Tchobanoglous, G., Theisen, H., and Vigil, S. A., *Integrated Solid Waste Management,* McGraw-Hill, New York, 1993.

USEPA (U.S. Environmental Protection Agency), *Characterization of Municipal Solid Waste in the United States: 1994 Update,* EPA/530/R–94/042, USEPA, Washington, DC, Nov. 1994.

———, *Municipal Solid Waste in the United States: 2000 Facts and Figures,* EPA/530/R-02/001, USEPA, Washington, DC, June 2002.

U.S. Navy, *Soil Dynamics, Deep Stabilization, and Special Geotechnical Construction,* NAVFAC DM 7.3M, Naval Facilities Engineering Command, Alexandria, VA, Apr. 1983, pp. 7.3–7.9.

Wilson, D. C., *Waste Management, Planning Evaluation, Technologies,* Oxford University Press, New York, 1981.

World Wildlife Fund and Conservation Foundation, *Getting at the Source,* WWF Publications, Baltimore, MD, 1989.

Zornberg, J. G., Jernigan, B. L., Sanglerat, T. R., and Cooley, B. H., Retention of free liquids in landfills undergoing vertical expansion, *J. Geotech. Geoenviron. Eng.,* Vol. 125, No. 7, pp. 583–594, 1999.

16

LANDFILL REGULATIONS, SITING, AND CONFIGURATIONS

16.1 INTRODUCTION

Finding a suitable site for a landfill is a very difficult task because of negative public opinion. The poor reputation of landfills was ascertained before government regulations were implemented. Unregulated landfills were often vector-infested, smelly, filthy, contaminating eyesores. The squalid conditions at unsanitary landfills were what brought about increased government regulations and greatly improved conditions. However, their bad reputation precedes them, and most communities have perceptions that landfills are notoriously bad and dangerous. Consequently, most residents are adamantly opposed to the construction of a landfill in the neighborhood. This syndrome is known as the "not in my backyard" (NIMBY) mentality. Thus, finding a location to construct a landfill is a daunting task. Federal, state, or local regulations also have location criteria for landfills that must be taken into account when citing a landfill.

The need for new landfills cannot be overemphasized. A large amount of municipal solid waste (MSW) requires disposal in landfills. In addition, waste quantities are increasing rapidly. Many existing landfills have been closed due to new regulatory requirements or have reached their capacity, and many more landfills are expected to be closed within the next 15 years. The siting of new landfills has become difficult because of very limited suitable sites. This means that if the waste volumes generated remain as expected, there will be a need for more landfill space for waste disposal.

Before the search for a suitable waste disposal site or the design of a waste facility is begun, one must obtain a thorough knowledge of the applicable federal,

state, and local government regulations. This knowledge is essential because the site selection, design, and operation of landfills and impoundments are regulatory driven. The regulations have been developed with the aim of protecting the environment, and they stipulate the minimum requirements that must be met by a waste disposal facility at its designated location. All laws and regulations promulgated by federal, state, and local governments must be complied with when designing waste disposal facilities. This chapter presents information on (1) regulations applicable to landfills, (2) landfill siting methodology, and (3) general information on landfill configurations.

16.2 FEDERAL REGULATIONS

As explained in Chapter 2, the Congress enacts laws or acts that are meant to accomplish certain goals, such as keeping the air and water safe and clean. It is then the job of the USEPA to give details, specifics, and guidelines for achieving goals. The most relevant federal laws and regulations dealing with wastes and waste disposal are the (1) Resource Conservation and Recovery Act (RCRA), (2) Clean Water Act, (3) Clean Air Act, and (4) Comprehensive Environmental Response, Compensation and Liability Act (CERCLA) or Superfund law.

16.2.1 Resource Conservation and Recovery Act

The RCRA was passed in 1970 and amended in 1980 and 1984 (HSWA Hazardous and Solid Waste Amendments). The definition of solid waste and classification of solid waste into municipal solid waste (MSW) and hazardous waste according to RCRA are provided in Chapter 15. A solid waste is considered hazardous if (a) it is listed, or (b) it has been characterized as ignitable, corrosive, reactive, or TCLP toxic (Refer to Section 15.3.2 for more details.) It is also important to know the following three rules:

1. *Mixture rule.* A mixture of a listed waste and a solid waste is considered a listed waste.

2. *Derived-from rule.* Any waste generated from the treatment, storage, or disposal of a liquid waste remains listed.

3. *Contained-in rule.* Any soil, groundwater, or other material containing a hazardous waste must be managed as a hazardous waste.

The listed hazardous wastes are:

- *Hazardous wastes from nonspecific sources:* includes 28 different wastes, also known as the F list, stipulated in Title 40, *Code of Federal Regulations,* Section 261.31 (40 CFR 261.31). Wastes from nonspecific sources include such wastes as spent halogenated solvents used in degreasing, wastewater treatment sludge from the electroplating process, and dioxin wastes. Certain wastes are excluded from nonspecific sources (refer to 40 CFR 261, Appendix IX).

- *Hazardous wastes from specific sources:* includes 111 different wastes, also known as the K list, as stipulated in 40 CFR 261.32. Wastes from specific sources include wastes from wood preserving, petroleum refining, organic chemical manufacturing including sludges, still bottom, wastewaters, and spent catalysts. Certain wastes are excluded from specific sources (refer to 40 CFR 261, Appendix IX).

- *Discarded commercial chemical products:* includes all off-specification species, containers, and spill residues thereof, as stipulated in 40 CFR 261.33.

- *Acutely hazardous wastes that are fatal in low doses:* includes 107 wastes, also known as the P list, as stipulated in 40 CFR 261.33e.

- *Toxic wastes that are carcinogenic, mutagenic, or tertogenic:* includes 248 wastes, also known as the U list, as stipulated in 40 CFR 261.33f.

Under the RCRA, the USEPA developed regulations on the proper disposal of solid waste. The USEPA classified landfills into two categories: municipal solid waste (MSW) landfills and hazardous waste landfills. The MSW landfills contain only nonliquid wastes, and regulations governing these landfills are known as RCRA Subtitle D regulations (40 CFR 257 and 258). The hazardous solid waste landfills contain nonliquid hazardous wastes and must conform to RCRA Subtitle C regulations (40 CFR 264 and 265). The main features of these regulations are outlined in the following sections.

RCRA Subtitle D Regulations. RCRA Subtitle D regulations cover municipal solid waste landfills (MSWLFs) and the regulations are covered in 40 CFR 257 and 258. 40 CFR 257 lists regulations concerning criteria for the classification of solid waste disposal facilities and practices. 40 CFR 258 covers other criteria for municipal solid waste landfills and includes the topics described below.

Location Restrictions. There are location restrictions for landfills in relation to airports, floodplains, wetlands, fault areas, seismic impact zones, and unstable areas. These location restrictions are detailed in Section 16.4.3.

Operating Criteria. Operating requirements stipulate that municipal solid waste landfills cannot accept hazardous wastes and that their operations must have the following features:

- Provide daily cover.
- Control on-site disease vectors.
- Provide routine methane monitoring.
- Eliminate most open burning.

- Control public access.
- Construct run-on and runoff controls.
- Control discharges to surface water.
- Cease disposal of most liquid wastes.
- Keep records that demonstrate compliance.

Design Criteria. There are two choices for landfill liners as depicted in Figure 16.1. These are:

(a) *Composite liner with a leachate collection system.* The liner must have a flexible membrane liner (FML) and at least 2 ft of compacted soil beneath it with a hydraulic conductivity value less than or equal to 1×10^{-7} cm/s. The re-

quirement for leachate depth over the liner is that it should be less than 30 cm (1 ft). This is discussed further in Chapter 17.

(b) *Performance-based liner approved by the director of an approved state.* The liner must be designed to ensure that the contaminant concentration values at the point of compliance remain below the values listed in Table 16.1.

Groundwater Monitoring and Corrective Action. The uppermost aquifer must be monitored for contamination. The monitoring program includes the installation of a monitoring system and the establishment of a sampling and analysis program. The next step is detection

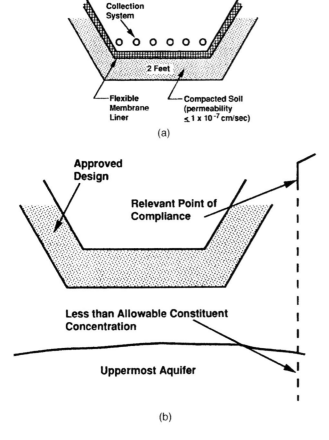

Figure 16.1 New municipal solid waste landfill units design criteria. (From Federal Register, 1991.)

Table 16.1 **Maximum concentration of constituents in the uppermost aquifer at the relevant point of compliance**

Chemical	MCL (mg/L)
Arsenic	0.05
Barium	1.0
Benzene	0.005
Cadmium	0.01
Carbon tetrachloride	0.005
Chromium (hexavalent)	0.05
1,4-Dichlorobenzene	0.075
1,2-Dichloroethane	0.005
1,1-Dichloroethylene	0.007
2,4-Dichlorophenooxyacetic acid	0.1
Endrin	0.0002
Fluoride	4
Lead	0.05
Lindane	0.004
Mercury	0.002
Methoxychlor	0.1
Nitrate	10
Selenium	0.01
Silver	0.05
Toxaphene	0.005
Trichloroethylene	0.005
1,1,1-Trichloromethane	0.2
2,4,5-Trichlorophenoxyacetic acid	0.01
Vinyl chloride	0.002

Source: 40 CRF 258.40.

monitoring for 40CFR 258, Appendix I constituents. If the monitoring indicates a statistically significant increase in the constituents in the uppermost aquifer, an assessment monitoring must be performed. If a statistical increase in 40CFR, Appendix II constituents is observed, corrective action must be taken. More details on groundwater monitoring are provided in Chapter 21.

Closure. Closure of a landfill involves the installation of a final cover system with a minimum of two layers. The infiltration layer, the lower layer, must consist of material that is at least 18 in. in depth with a hydraulic conductivity value less than or equal to that of the liner system or natural subsoils, or 1×10^{-5} cm/s. The upper layer is an erosion layer that must be at least 6 in. in depth, composed of material that is capable of sustaining vegetation. Figure 16.2 shows such a cover system. Further information on a final cover liner system is presented in Chapter 19.

A written closure plan must also be prepared that documents the following four important points:

1. Final cover design and installation
2. Largest area requiring final cover at any time during the active service life
3. Maximum on-site inventory of waste
4. Closure schedule

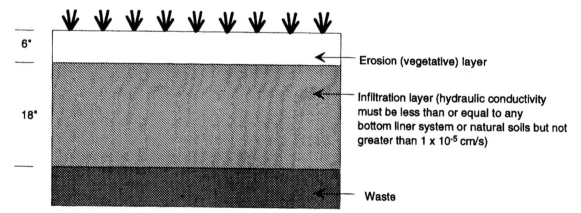

Figure 16.2 *Final cover system for a MSW landfill. (From USEPA, 1994.)*

Additionally, the closure activities must start within 30 days of the final receipt of waste, and they must be completed within 180 days following startup of the closure activities.

Postclosure. Postclosure is for a minimum of 30 years and consists of a documented maintenance program. The program must include actions for the safety and quality of the landfill. These actions must include the following:

1. Maintenance of the integrity and thickness of the final cover
2. Maintenance of the operation and control of the leachate collection system
3. Groundwater monitoring
4. Maintenance of the gas monitoring system

Additionally, a written postclosure plan must be submitted that documents the type and frequency of the monitoring activities, the contact persons in charge, and planned uses of the property after closure.

Financial Assurance Criteria. State and federal government entities are exempted from financial assurance, but all other entities must provide a detailed estimate of their financial assurance. The detailed estimate must include the current dollars and annual updated estimates for modifications and/or adjustments due to inflation.

RCRA Subtitle C Regulations. The RCRA Subtitle C regulations stipulate the minimum standards for proper management and disposal of hazardous wastes and are given in 40 CFR 264 and 265. These regulations set the standards for owners and operators of hazardous waste treatment, storage, and disposal facilities. 40 CFR 264 outlines the following main criteria:

- Location standards
- Preparedness and prevention
- Contingency plan and emergency procedures
- Manifest system, record keeping, and reporting
- Release from solid waste management units
- Closure and postclosure
- Financial requirements
- Use and management of containers
- Tank systems
- Surface impoundments
- Waste piles
- Landfills
- Incinerators
- Miscelleneous units

Details of these issues are presented in 40 CFR 264; location standards and some design criteria specifically for the hazardous waste landfills are presented here.

Location Standards. Based on seismic considerations, hazardous waste treatment, storage, and disposal facilities must not be located within 200 ft of a Holocene fault. A hazardous waste management facility located within a 100-year floodplain must be designed, constructed, operated, and maintained to prevent washout of any hazardous waste by a 100-year flood. Placement of any noncontainerized or bulk liquid hazardous waste salt dome formation, salt bed formation, underground mines, and caves is prohibited. These requirements may have some exceptions, as outlined in 40 CFR 264.18.

Groundwater Protection Standards. Facilities must be designed to ensure that hazardous constituents specified under 40 CFR 264.93 entering the groundwater from a regulated unit do not exceed the concentration limits under 40 CFR 264.94 in the uppermost aquifer underlying the waste management area beyond the point of compliance, as specified in 40 CFR 264.95 (Table 16.2).

Design Criteria. A landfill used for the disposal of hazardous waste must have a leachate collection system; a top FML; a leachate collection, detection, and removal system; and a composite liner system (Figure 16.3). A landfill undergoing closure must be covered with a final cover that minimizes long-term migration of liquids through the closed landfill. In addition, it

Table 16.2 **Maximum concentration of constituents for groundwater protection**

Constituent	Maximum Concentration (mg/L)
Arsenic	0.05
Barium	1.0
Cadmium	0.01
Chromium	0.05
Endrin (1,2,3,4,10.10-hexachloro-1,7-epoxy-1,4,4a,5,6,7,8,9a-octahydro-1,4-endo, endo-5,8-dimethanonaphthalene)	0.0002
Lead	0.05
Lindane (1,2,3,4,5,6-hexachlorocyclohexane, γ isomer)	0.004
Mercury	0.002
Methoxychlor (1,1,1-trichloro-2,2-bis-p-methoxyphenylethane)	0.1
Selenium	0.01
Silver	0.05
2,4-D (2,4-dichlorophenoxyacetic acid)	0.1
2,4,5-TP (silvex) (2,4,5-trichlorophenoxypropionic acid)	0.01
Toxaphene ($C_{10}H_{10}C_{l6}$, technical chlorinated camphene, 67–69% chlorine)	0.005

Source: 40 CFR 264.94.

must function with minimum maintenance, promote drainage, and minimize erosion of the cover, accommodate settling, and have a permeability value less than or equal to that of any bottom liner system or natural subsoil present (40 CFR 264.310). Figure 16.4 shows a multilayer cover system for a closed hazardous waste landfill. Further details on liner system, leachate collection system, and cover system are provided in Chapters 17 through 19.

16.2.2 Clean Water Act

The Clean Water Act must be conformed in the design and operation of municipal solid waste and hazardous waste landfills. It includes regulations concerning wetlands and covers the National Pollutant Discharge Elimination System (NPDES) program. The NPDES program covers the following three situations: (1) discharges from stormwater management systems, (2) leachate treatment plants, and (3) stormwater discharges during construction.

16.2.3 Clean Air Act

The Clean Air Act must be conformed to in the design and operation of municipal solid waste and hazardous wastes landfills. The Clean Air Act regulates gas emissions, particularly those of methane and nonmethane organic compounds (NMOCs).

16.2.4 Comprehensive Environmental Response, Compensation and Liability Act

CERCLA regulates municipal solid waste and hazardous waste landfills. However, CERCLA is applicable only when the landfills have been abandoned and need immediate fixing due to a danger to public health and the environment.

16.3 STATE AND LOCAL REGULATIONS

It is important to realize that federal regulations are the minimum requirements that must be met by state and local governments. State and local regulations are, therefore, extensions or in addition to federal regulations. State regulations covering municipal solid waste and hazardous waste landfills vary from state to state. However, every state must comply with federal regulations as the minimum requirement, but they may also impose stricter provisions. The responsible state agency must be contacted to obtain the applicable regulations. For example, MSW landfill regulations are covered in Illinois's Title 35: Environmental Protection, Subtitle G, Waste Disposal; in California's CCR Title 27, and in New York's 6NYCRR Part 360, Solid Waste Management Facility.

Local regulations covering municipal solid waste and hazardous wastes landfills vary significantly from

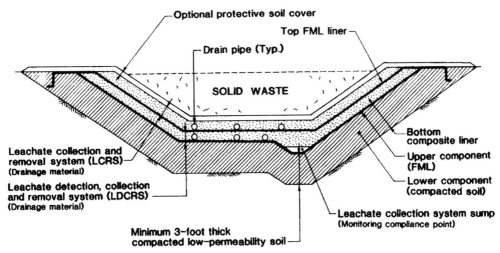

Figure 16.3 Base liner system for a hazardous waste landfill. (*From USEPA, 1989.*)

location to location. Local regulations may include both county requirements and city or village requirements. Typically, they include the setback distance from the following:

- Surface water bodies
- Sources of drinking water
- Public roadways
- Structures such as homes, schools, hospitals, and nursing homes
- Cultural resources such as historical parks, monuments, structures, and museums
- Recreational parks
- Designated wild and scenic rivers
- Coal seams or limestone outcrops
- Gas or oil wells

Local restriction or regulations may also include other considerations, such as the following:

- Zoning
- Buffers
- Deed restrictions
- Special protection areas
- Traffic/access roads
- Height restrictions

- Groundwater separation
- Maximum/minimum final cover slopes

16.4   SITING METHODOLOGY

Prior to initiation of the site selection process, one must first decide on the type of waste to be disposed of: hazardous or nonhazardous. If it is hazardous waste, decide if it will be liquid or solid waste. Note that MSW landfills can accept only nonliquid nonhazardous waste, which is the focus of the procedure outlined below.

16.4.1   Calculating Landfill Acreage

Based on the service area of a landfill, quantities of wastes generated are estimated. All feasible recycling options must be taken into account in calculating the wastes that require disposal in landfill. The design life of a landfill must be decided. Based on the quantities of waste and the design life of the landfill, an estimate of the required volume, or air space, can be calculated. After the required air space is estimated, an approximation of the necessary waste depth will lead to a projection of the site acreage or size of the landfill footprint. The other landfill requirements, such as buffer

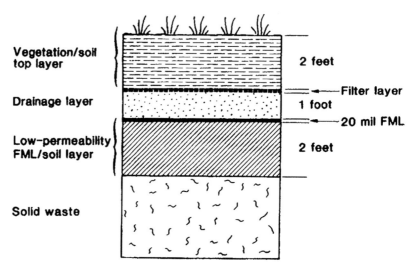

Figure 16.4 *Final cover system for a hazardous waste landfill. (From Landreth, 1990.)*

zones, buildings, access roads, and entrance facilities, should be included to obtain the approximate size or minimum limits for the general area.

An estimate of the volume of waste in a landfill may be made using

$$V = \frac{R}{D(1 - P/100)}\,CV \qquad (16.1)$$

where V is the volume of compacted solid waste plus cover soil (cubic yards per person per year), R the solid waste generated per person on an annual basis [i.e., weight/year (consider recycling options)], D the uncompacted density of solid waste, P the percent volume reduction per unit volume (i.e., $\Delta V/V$) achieved from solid waste compaction, and CV the soil cover factor, which is equal to [1 + (soil cover thickness/ height of landfill)]. The required area of the landfill may be calculated using

$$A = \frac{27VN}{d(43,560)} \qquad (16.2)$$

where A is the landfill area needed (acres/year), N the population, d the landfill height in feet, and 43,560 the

unit conversion factor from square feet to acres. An additional area used to buffer the waste limits should be added.

16.4.2   Review of the Literature

A comprehensive literature review must be performed to identify potential sites. Information is gathered concerning the conditions existing near the site that may affect future landfill construction and operation. There are a number of categories into which information can be compiled and reviewed, such as the following:

1. General information, which may include information on or from;
   a. County maps
   b. Soil surveys
   c. Residential developments
   d. Utilities
   e. Road maps and road conditions
   f. Topographic maps
   g. Soil maps
   h. Land use maps
   i. Transportation maps
   j. Geologic maps and/or reports

2. Floodplain maps or flood insurance rating maps
3. Wetlands and national wetlands inventory maps
4. Nature preserves and forest preserves
5. Aerial photographs
6. Public water wells
7. Real estate costs

16.4.3   Regulatory Location Restrictions

There are a number of location restrictions for MSW landfills, so the regulations should be reviewed carefully to assure that the site meets the minimum requirements for federal, state, and local governments. A general review of various regulations was presented in Sections 16.2 and 16.3. Next, we review briefly landfill regulations applicable to siting requirements.

Federal Regulations. Subtitle D regulations in 40 CFR, Section 258.10-16 are a number of federal restrictions for MSW landfill locations. These restrictions include siting landfills near or in airports, floodplains, wetlands, fault areas, seismic impact zones, and unstable areas. These siting restrictions are outlined below.

1. Landfills must be greater than 5000 ft from a piston-type aircraft airport and greater than 10,000 ft from a turbojet aircraft airport. This restriction is primarily to deter birds from inhabiting close to the airports.
2. Landfills should not be located in floodplains.[1] If a landfill is located in a 100-year floodplain, it cannot restrict the flow of the 100-year flood, reduce the temporary storage capacity of the floodplain, or result in washout of MSW.

3. Landfills must not be located in wetlands[2] unless another practical alternative not involving wetlands does not exist. Wetlands generally include swamps, marshes, bogs, or similar areas. Additionally, construction and operation of landfills must not violate applicable state water quality standards or toxic effluent standards of the Clean Water Act.
4. Landfills must not be located within 200 ft of a fault[3] area that has experienced displacement within the Holocene Epoch (the last 10,000 years).
5. Landfills cannot be located in seismic impact zones[4] unless all containment structures (e.g., liners, leachate collection system) are designed to resist the maximum horizontal acceleration and the site will remain stable. Seismic impact zones in the United states are shown in Figure 16.5.
6. Landfills must not be located on unstable-site subgrades unless engineering measures are taken to ensure the integrity of the landfill structural components. These unstable-site subgrades may include poor foundation conditions (e.g., highly compressible soil layers), sites susceptible to mass movements (e.g., landslides), and karst terrain that may have hidden sinkholes.

State Regulations. State regulations vary from state to state concerning location restrictions. Many of these regulations contain stricter siting requirements than the

---

[1] *Floodplains* are defined as lowland and other flat areas adjacent to inland or coastal waters that are inundated during a 100-year flood.

[2] Under 40 CFR 232.2, *wetlands* are defined as those areas that are inundated or saturated by surface water or groundwater at a frequency and duration sufficient to support, and that under normal circumstances do support, a prevalence of vegetation typically adapted for life in saturated soil conditions.

[3] A *fault* is a fracture or zone of fracture in geologic material along which strata on one side have been displaced with respect to strata on the other side.

[4] *Seismic impact zones* are defined as regions having a 10% or greater probability that maximum horizontal acceleration at the site caused by an earthquake will exceed $0.1g$ in 250 years (g is acceleration due to gravity).

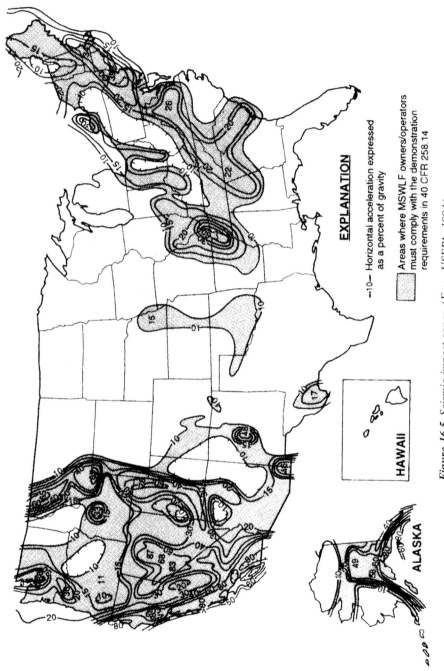

EXPLANATION

~-10— Horizontal acceleration expressed as a percent of gravity

Areas where MSWLF owners/operators must comply with the demonstration requirements in 40 CFR 258.14

HAWAII

ALASKA

Figure 16.5 Seismic impact zones. (From USEPA, 1994.)

federal requirements, such as restrictions on development in critical watershed areas, wellhead protection areas, sole-source aquifers, minimum buffer zones, or agricultural lands. For example, some restrictions stipulated by the State of Illinois in accordance with Title 35 IAC, Section 811.302, are as follows:

1. Landfills are required to have a low-permeability soil layer that is at least 50 ft in thickness and a hydraulic conductivity value of less than 1×10^{-7} cm/s above the upper sole-source aquifer.

2. Landfills must be greater than 500 ft from any township, or county road, or highway.

3. Landfills require a screen greater than 8 ft in height around the perimeter.

4. Landfills must be greater than 500 ft from an occupied dwelling, school, or hospital.

Local Restrictions. Local restrictions will also vary from location to location, so it is important to check with county and local municipalities to review their requirements. Often, public hearings are held to publicize and seek the opinion of the population affected.

16.4.4 Map Overlay Procedure to Identify Potential Sites

To assess possible site locations, it is advisable to generate a number of maps that are to the same scale and overlay them on top of one another to observe open spaces where a landfill may be sited. Figure 16.6 depicts this overlay procedure and shows two potential sites, site A and site B, for landfilling. These maps should include:

- Map for airports and runways
- Map for occupied dwellings, schools, and hospitals
- Map for nature/forest preserves
- Map for waterwells
- Map for floodplains
- Map for wetlands
- Other applicable maps

Based on this overlay procedure, potential sites for siting the landfill are determined.

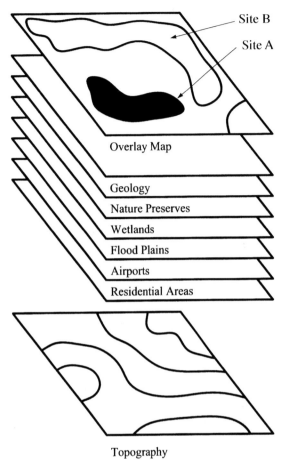

Figure 16.6 *Map overlay procedure to identify potential sites.*

16.4.5 Identification of Specific Sites Based on Local Features

Make a list of the potential sites and evaluate the merits of each potential site according to the following considerations:

- Available area
- Site access
- Traffic impact
- Landowners
- Distance from centroid of area (hauling distance)
- Power and sewer services
- Land costs

16.4.6 Preliminary Site Exploration to Determine Site Hydrogeology

Assess the site according to the procedures of phase 1 evaluation, and proceed to phases 2 and 3 if necessary. These procedures are the same as those described in Chapter 10. Assuming that the site passes the environmental assessment, proceed to a site investigation.

1. Perform a limited site investigation that consists of the following: (a) soil borings, (b) test pits, (c) soil and groundwater testing and sampling, (d) field and laboratory testing (hydraulic conductivity, etc.), (e) geophysical testing, and (f) monitoring wells. The procedures for conducting these tasks are explained in Chapter 10.

2. Compile a hydrogeological report. The components of this report are described in detail in Chapter 10.

16.4.7 Conceptual Design of a Landfill

A preliminary design report should be prepared that addresses the following important topics concerning landfill design:

1. Design the landfill geometry and configuration.
2. Design the liner system.
3. Design the leachate collection and removal system.
4. Design the final cover system.
5. Design the surface drainage system.
6. Design the monitoring programs for groundwater and other site conditions.
7. Design the gas collection system.
8. Decide on the final use of the landfill property.
9. Develop plans for corrective response action.

16.5 SITE PERMIT APPLICATION

After completion of the site selection process, the formal approval of local and state agencies is required. The first objective is to obtain a development permit. This will include submission of all documentation,

public hearings, and addressing all other issues dealing with the landfill. It is a very difficult and lengthy process to obtain a development permit, so once that has been accomplished, a major obstacle has been removed.

After the development permit has been approved, a detailed design employing actual materials is prepared. Construction plans and specifications are prepared, and bids from selected contractors are reviewed. After the contractors are selected, construction begins and stringent safety and health and quality assurance and quality control plans are put into effect to document the construction fully. Once the landfill is constructed, the owner must obtain an operating permit, and the landfill may then become operational.

16.6 LANDFILL CONFIGURATIONS

There are four basic forms of landfill configurations, as shown in Figure 16.7.

1. *Area fill.* Area fill requires no excavation and is generally used in flat areas when the groundwater elevation is very shallow and close to the ground surface elevation. (Figure 16.7a)

2. *Above- and below ground fill.* Above- and below ground fill involves large cells that are excavated one at a time and filled. Finally, areas between them are covered. This landfill type is suited for relatively flat areas where the water table is deep. (Figure 16.7b)

3. *Valley fill and canyon fill.* Valley and canyon fills are used in locations that have mountainous topography. The natural ground surface consists of a valley and/or canyon scenario so the waste can be confined and placed inside the valley or canyon walls with very low or no excavation costs. (Figure 16.7c)

4. *Trench fill.* Trench fill is similar to the above- and belowground fills, except that the cells are narrow and parallel. These are used for small waste streams. (Figure 16.7d)

Sharma and Lewis (1994) present these and other landfill configurations and discuss their stability implications. The above- and belowground fill configuration

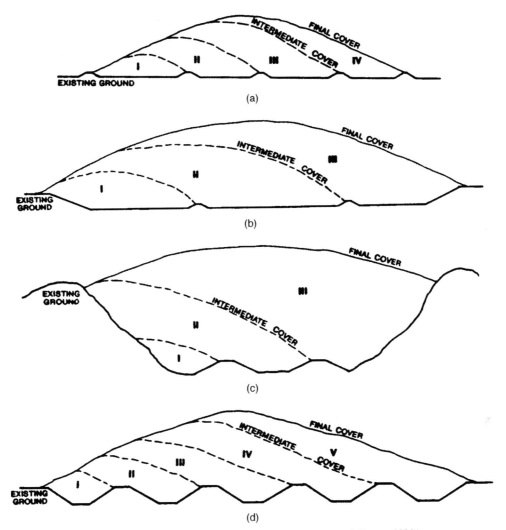

Figure 16.7 *Landfill configurations. (From Repetto and Foster, 1993.)*

is very common and a detailed procedure to determine its geometrical configuration is explained below.

16.6.1 Footprint and Cell Layout

The footprint area is the maximum area available at the site after allocating the area required for other facilities (e.g., access roads, buffer zones, office building, entrance facilities). The waste area is divided into cells as shown in Figure 16.8 so that (1) cells can be constructed one at a time, thereby requiring less initial capital investment, (2) leachate generated can be minimized, and (3) excessive stockpiling of excavated soils can be avoided. The typical cell size depends on the waste stream and the desired life of the landfill, but it often ranges from 2 to 8 acres, depending on the life required for each cell, which may vary from one to three years.

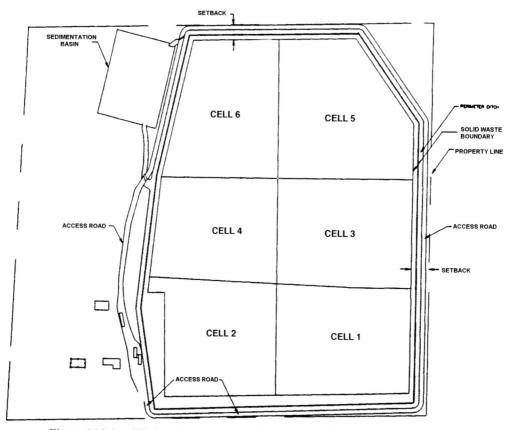

Figure 16.8 Landfill footprint area and cell layout. (*From Repetto and Foster, 1993.*)

16.6.2 Base Grading

The subbase grading defines the elevation of the deepest level of excavation at different areas of the footprint. The subbase grading and the final cover grading (discussed later) determine the air space or volume available for waste disposal in the landfill. To maximize the air space, one wants excavations to be as deep as possible. However, many factors must be considered in deciding the base grades, as explained below.

Depth to the Water Table or Uppermost Aquifer. Regulations stipulate a minimum separation between the subbase and the water table. If no regulatory restrictions exist, subgrades can be below the groundwater; however, the excavation costs may increase substantially if the excavation goes below the water table,

due to dewatering costs. When the water table or potentiometric surface is lower than the base of the landfill, an outward gradient exists, as depicted in Figure 16.9; whereas if the water level or potentiometric surface is higher than the landfill base, inward gradient conditions exist, as depicted in Figure 16.10.

Depth to Bedrock. Excavation costs increase dramatically if a large amount of rock blasting is required. Therefore, where possible, the landfill base is kept very close to the elevation where bedrock is encountered.

Stability of the Foundation. Three hazards should also be assessed for foundation stability evaluation: (1) previously mined areas, (2) sinkholes, and (3) uplift. Uplift may occur if a confined aquifer exists at the site,

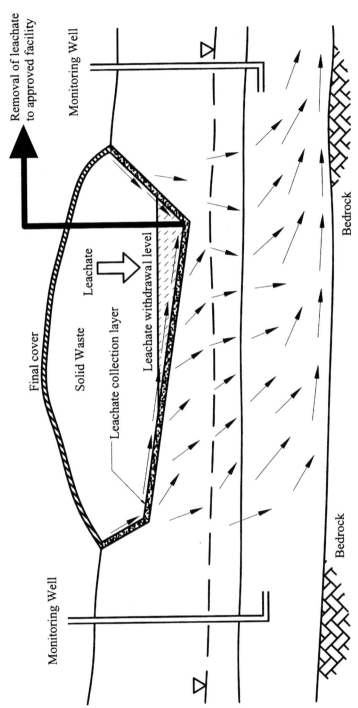

Figure 16.9 Outward gradient design concept with leachate collection system.

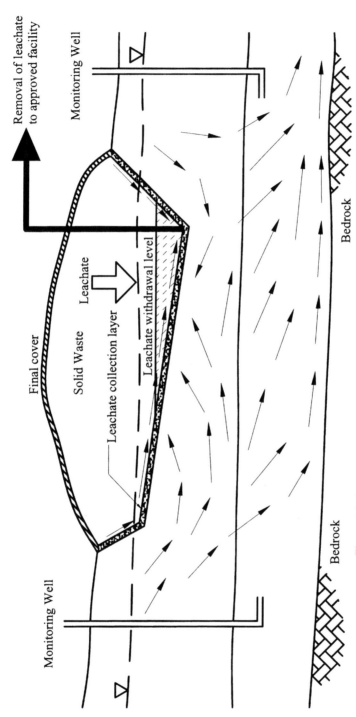

Figure 16.10 *Inward gradient design concept with leachate collection system.*

as shown in Figure 16.10. To prevent uplift failure, adequate soil layer must overlay the confining aquifer. The uplift condition is very critical during construction and depends on the potentiometric level in the confined aquifer. If the potentiometric surface is lower than the base elevation of the landfill, uplift failure will not occur. However, if the potentiomentric surface is higher than the landfill base, uplift may occur. As shown in Figure 16.11, the following equation should be used to determine the factor of safety:

$$FS = \frac{\gamma_t h}{\gamma_w h_w} \qquad (16.3)$$

where h is the separation distance between the top of the aquifer and the subbase of the landfill, h_w the distance from the top of the aquifer to the height to which the water would rise in a piezometer, γ_t the total unit weight of soil, and γ_w the unit weight of water. The factor of safety should be much greater than 1 for design purposes.

Site Topography and Stability of Slopes. The site topography and stability of slopes are extremely important in selecting base grades. The stability of any excavation slopes as well as landfill mass slopes must be examined to prevent possible slope failures. A detailed explanation of slope stability analysis methods is generally covered in standard geotechnical engineering textbooks (e.g., Terzaghi and Peck, 1967; Lambe and Whitman, 1969; Das, 1998). Sharma and Lewis (1994) provide information on static and seismic slope stability methods applicable to landfills.

Limit equilibrium methods are commonly used for slope stability analysis. The common aspect of all types of limit equilibrium analyses is that the results may be represented in terms of a factor of safety. The factor of safety is commonly defined as the ratio of the summation of resisting forces and/or moments to the summation of driving forces and/or moments. Assuming that the exact values are known for the resisting and driving forces, a factor of safety of 1 indicates imminent slippage, and a factor of safety below 1 indicates unstable conditions. The slope is theoretically stable if the factor of safety is greater than 1. There are three common types of limit equilibrium methods: infinite slope method, wedge method, and the method of slices. The *infinite slope method* is one-directional and movement is parallel to the slope, as shown in Figure 16.12(*a*). Such a situation may arise when a relatively thin soil veneer is placed on a slope. The *wedge method* is used when the anticipated geometry of the failure mass is relatively simple and can be divided into wedge-shaped sections [Figure 16.12(*b*)]. The infinite slope method is commonly used to analyze the stability of landfill cover slopes and is described in Chapter 19. The wedge method is used to analyze the stability of a side-slope liner and drainage system, as detailed in Chapter 18. The *method of slices* can be used for circular or wedge/block-type slip surfaces, as shown in Figure 16.12(*c*) and (*d*), and is commonly used to analyze the excavation slopes and landfill mass. In this method, the potential sliding mass is divided into a series of vertical slices and the equilibrium of each individual slide evaluated. Figure 16.13 presents a general failure surface and the forces that exist on an individual slide. The forces on each slide within the potential slide mass are summed to obtain a factor of safety for the mass. Several potential slide masses must be evaluated to determine the slide mass with the minimum factor of safety. Different methods of slices are developed, depending on the assumptions made regarding the normal or interslice force distribution (Sharma and Lewis, 1994). Among these methods, the modified Bishop method is commonly used for circular failure surface analysis, whereas the Janbu and Spencer methods are used for generalized failure surface analysis. Several computer programs have been developed to perform these methods of slices, the most commonly used programs being PCSTABL, UTEXAS3, XSTABL, and SLOPE/W.

Figure 16.14 illustrates some of the potential slip surfaces that may occur in landfill excavation slopes. All landfill configurations, with the exception of the aboveground landfill, have some potential for excavation slope instability. Typically, a minimum factor of safety of 1.3 is acceptable for temporary excavation slopes, and a minimum factor of safety of 1.5 is acceptable for permanent excavation slopes.

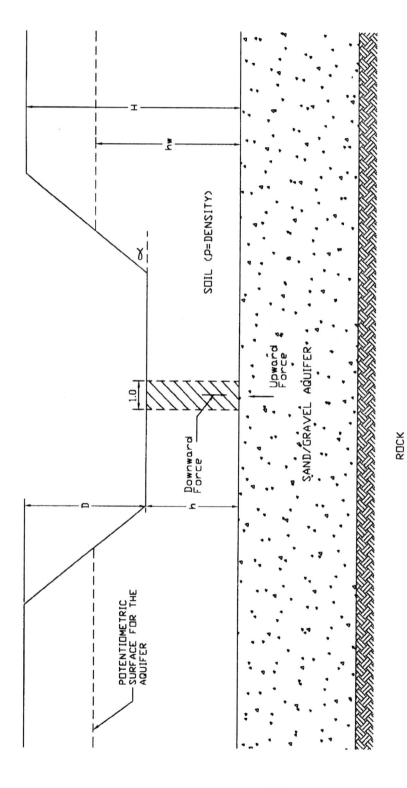

Downward Force (DF) = Wt of Soil = Pg(h×1×1)
Upward Force(UF)= Pwg(hw×1×1)

$$FS = \frac{DF}{UF} = \frac{\rho g h}{\rho_w g h} = \frac{\rho h}{\rho_w h_w}$$

Figure 16.11 Uplift analysis.

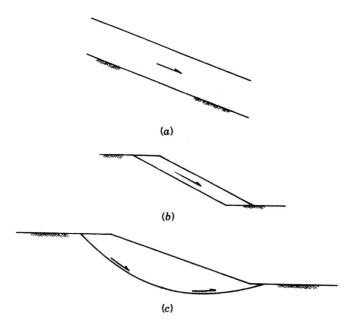

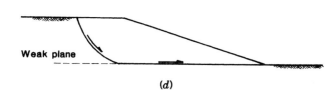

Weak plane

(d)

Figure 16.12 Types of slip surface: (a) *infinite-slope slip;* (b) *wedge-shaped slip surface;* (c) *circular slip surface;* (d) *composite slip surface.* (*From Sharma and Lewis, 1994. This material is used by permission of John Wiley & Sons, Inc.*).

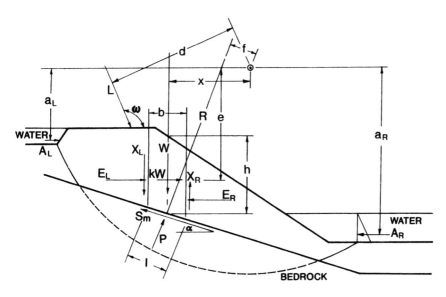

Figure 16.13 Method of slices force diagram. (*From Fredlund and Krahn, 1977.*)

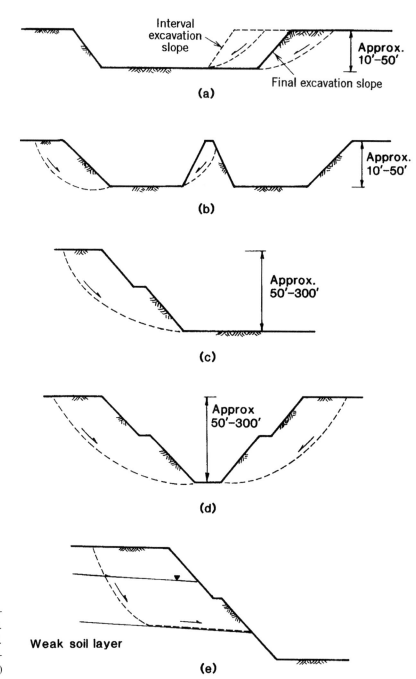

Figure 16.14 *Potential slip surfaces in excavation slopes. (From Sharma and Lewis, 1994. This material is used by permission of John Wiley & Sons, Inc.)*

Refuse fill stability analyses are important in assessing the stability of a landfill in its final configuration. However, since landfills are typically constructed in phased modules, the critical stability concern may occur during interim refuse fill conditions, especially where lining systems utilizing geosynthetics are employed. It is important, therefore, to assess both final and intermediate refuse slopes. Typical potential slip surfaces through refuse generally occur in one of the three ways shown in Figure 16.15: (1) through refuse alone, (2) along the liner system, and (3) a composite surface through refuse and along the liner. The refuse

properties needed for this stability analysis are shear strength and unit weight (refer to Chapter 15 for these properties). Liner material and liner interface strength properties are also needed for the analysis, and these properties are presented in Chapter 17.

In addition to the static analyses noted above, seismic analyses of excavation slopes and landfill mass slopes will be required if the landfill is located in a seismic impact zone. When an earthquake occurs, seismic accelerations are imposed on the mass that generate inertial forces. To account for seismic effects, a horizontal inertial force is applied at the center of grav-

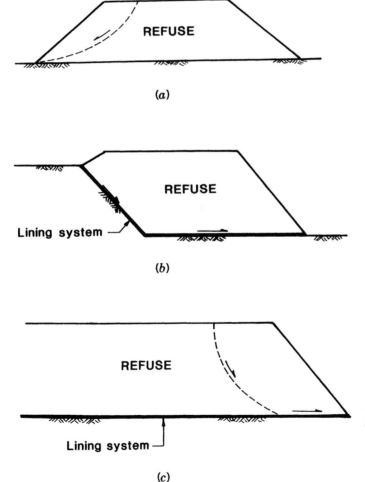

Figure 16.15 Potential slip surface in refuse fills: (a) failure through refuse alone; (b) failure along lining system; (c) composite surface through refuse and lining system. (From Sharma and Lewis, 1994. This material is used by permission of John Wiley & Sons, Inc.)

ity of the mass, and its effect on the factor of safety is determined. This method is known as the *pseudostatic approach.* Another approach to the evaluation of the seismic stability of slopes is to estimate cumulative seismic displacement based on the Makdisi and Seed (1977) method, which is a modified Newmark (1965) analysis that allows for the nonrigid dynamic response of sliding soil or refuse mass. Sharma and Lewis (1994), Bray et al. (1995), and Kavazanjian (1999) provide details on this analysis methodology. The dynamic properties needed for the seismic analysis include dynamic modulus and damping ratio of sliding mass, and these properties for the waste are presented in Chapter 15.

Soil Balance. During excavation it is economical to maintain a soil balance (cut and fill) so that the volume of excavated soil is approximately equal to the volume of fill needed for daily cover, backfill on subgrade, or embankment fill.

16.6.3   Final Cover Grading

Preparation of a preliminary final cover grading plan is necessary in order to estimate the air space. The final cover design is covered in detail in Chapter 19. The final cover grading must ensure stable slopes, meet the height restrictions, if any, and control runoff. In addition, erosion control measures must be outlined.

16.6.4   Other Facilities and Site Development Master Plan

Other facilities within the landfill permitted area may include the following:

- Access roads
- Barrow areas
- Buffer zones

- Fences and gates
- Leachate holding tanks
- Office buildings
- Scale and scale house
- Sedimentation basins
- Stockpiling areas
- Truck loading or washing facilities

A site master plan should be prepared to depict the locations of the property boundary, the optimized landfill area, cell development phases, and optimized locations of the facilities.

16.7   SUMMARY

Laws and regulations on waste disposal aim at protecting public health and the environment. The laws and regulations at the federal, state, and local levels may exist and must be determined based on the location of the waste disposal facility. The USEPA has developed extensive regulations on classifying the waste and disposal facilities. State and local governments may impose additional requirements on the site location and design and operation of waste disposal facilities. One must be careful in meeting all of the regulatory requirements when a landfill siting evaluation is performed.

The selection of the landfill geometry is the basic step in landfill design. The geometry should be such that it maximizes the air space available for waste disposal. The subbase grading and side slopes should be selected to ensure stability. The final grading is also selected based on the stability considerations. Any regulatory requirements applicable for site-specific conditions must be satisfied. The waste boundary must be strategically located in relation to all the ancillary facilities at the landfill site.

QUESTIONS/PROBLEMS

16.1.   What is the main objective of regulations on waste disposal?

16.2.   What federal regulations affect waste disposal?

16.3. What is the difference between Subtitle D and Subtitle C regulations?

16.4. What types of landfill issues are addressed under CERCLA, or Superfund?

16.5. How would state and local regulations differ from federal regulations in dealing with disposal of wastes?

16.6. Contact your state environmental agency and obtain a copy of the state's regulations on landfills. Compare these regulations with those of the federal (USEPA) regulations.

16.7. Contact your county or city office and obtain a copy of the requirements for landfills. Compare these regulations with the state regulations.

16.8. Why do we need more new landfills?

16.9. Why is it very difficult to find a suitable site for a new landfill?

16.10. Outline various tasks performed in selecting a site for a landfill.

16.11. Outline various tasks involved in obtaining a regulatory approval of a new landfill.

16.12. Contact your town or city office and get annual estimates of (*a*) solid waste generation, (*b*) recycling rates, and (*c*) population growth in your community. Based on this information, estimate the volume of landfill required accommodating the wastes that will be generated in next 10 years. If the average depth of the landfill is 30 ft, what is the required landfill area?

16.13. List federal regulatory restrictions on landfill siting.

16.14. What are the additional landfill siting restrictions in the state, county, and city in which you live?

16.15. Explain the advantages of a map overlay procedure in identifying potential landfill sites.

16.16. What information is obtained from a limited site exploration program during the landfill siting process?

16.17. How do the public participate in the landfill siting process?

16.18. What is a development permit? Who issues this?

16.19. What is an operating permit? Who issues this?

16.20. Explain various landfill configurations. What factors control the selection of a particular configuration at a landfill site?

16.21. Draw a typical above- and belowground landfill configuration.

16.22. What factors need to be considered in deciding the landfill footprint area?

16.23. Why is the landfill footprint area is divided into cells?

16.24. List factors controlling the base grading of a landfill.

16.25. What is uplift failure?

16.26. What are the various slope stability analysis methods?

16.27. Why is computer-aided slope stability analysis needed in selecting landfill side slopes?

16.28. Calculate the maximum excavation depth and side slopes for a proposed landfill at a site consisting of a 50-ft-thick silty clay layer (unit weight = 120 lb/ft^3, undrained shear strength: c = 2000 lb/ft^2, ϕ = 0; drained shear strength: c' = 0 and ϕ' = 32°) underlain by a confined sand aquifer with a potentiometric surface located at 20 ft from the top of the sand aquifer layer.

16.29. What factors control the selection of a final cover configuration?

REFERENCES

Bray, J. D., Augello, A. J., Leonards, G. A., Repetto, P. C., and Byrne, R. J., Seismic stability procedures for solid waste landfills, *J. Geotech. Eng.* Vol. 121, No. 2, pp. 139–151, 1995.

CFR, (*Code of Federal Regulations*), 40 CFR, U.S. Government Printing Office, Washington, DC, 1992.

Das, B. M., *Principles of Geotechnical Engineering,* PWS Publishing, Boston, 1998.

Federal Register, Rules and regulations, Vol. 58, No. 196, Oct. 9, 1991.

Fredlund, D. G., and Krahn, J., Comparison of slope stability methods, *Can. Geotech. J.,* Vol. 14, p. 429, 1977.

Kavazanjian, E., Seismic design of solid waste containment facilities, in *Proceedings of the 8th Canadian Conference on Earthquake Engineering,* Vancouver, British Columbia, Canada, June 1999.

Lambe, T. W., and Whitman, R. V., *Soil Mechanics,* Wiley, New York, 1969.

Landreth, R. E., Landfill containment systems regulations, pp. 1–13 in *Waste Containment Systems: Construction, Regulation, and Performance,* Geotechnical Special Publication 26, Bonaparte, R. (ed.), ASCE, Reston, VA, 1990.

Makdisi, F. I., and Seed, H. B., *A Simplified Procedure for Estimating Earthquake-Induced Deformations in Dams and Embankments,* UCB/EERC-77/19, University of California, Berkeley, CA, 1977.

Newmark, N. M., Effects of earthquakes on dams and embankments, *Geotechnique,* Vol. 5, No. 2, 1965.

Repetto, P. C., and Foster, V. E., Basic considerations for the design of landfills, pp. 1–34 in *Proceedings of the First Annual Great Lakes Geotechnical/Geoenvironmental Conference,* Toledo, OH, 1993.

Sharma, H. D., and Lewis, S. P., *Waste Containment Systems, Waste Stabilization, and Landfills: Design and Evaluation,* Wiley, 1994.

Terzaghi, K., and Peck, R. B., *Soil Mechanics in Engineering Practice,* Wiley, New York, 1967.

USEPA (U.S. Enviornmental Protection Agency), *Requirements for Hazardous Waste Landfill Design, Construction and Closure,* EPA/625/4-89/022, USEPA, Washington, DC, 1989.

———, *Design, Operation, and Closure of Municipal Solid Waste Landfills,* EPA/625/R-94/008, Office of Research and Development, USEPA, Washington, DC, 1994.

17

WASTE CONTAINMENT LINER SYSTEMS

17.1 INTRODUCTION

According to the Resource Conservation and Recovery Act (RCRA) Subtitle D program (40 CFR 258), the new municipal solid waste landfill (MSWLF) units and lateral expansions must be constructed with a composite liner and leachate collection system. The composite liner must consist of an upper component of a flexible membrane liner (FML) [minimum 30-mil FML; at least 60-mil high-density polyethylene (HDPE)] and a lower component of a minimum 2-ft-thick compacted soil material with a hydraulic conductivity (k) value less than or equal to 1×10^{-7} cm/s. The regulations also stipulate that the final cover system for all MSWLFs must have an erosion layer consisting of a minimum 6 in. of a vegetative layer, overlaying an infiltration layer consisting of a minimum of 18 in. of earthen material with k less than or equal to 1×10^{-5} cm/s or less than or equal to the k of the base liner system. An alternative liner system that achieves the same reduction in infiltration as the specified liner system may be approved. Figure 17.1 presents various components of a MSWLF bottom (base and side slope) and final cover liner systems.

According to the RCRA Subtitle C program (40 CFR 264), a hazardous solid waste landfill must have two or more liners and a leachate collection and removal system between such liners. Figure 17.2 presents various components of a hazardous solid waste liner system.

The minimum technology requirements specified in the Hazardous Solid Waste Amendments (HSWA) of 1984 for new hazardous waste surface impoundments require a double liner with a leak detection and col-

lection layer placed between the liners (USEPA, 1991). The system must have a top liner (e.g., a geomembrane) and a composite bottom liner (e.g., a geomembrane over compacted soil). The top liner and the upper component of composite liner must be designed and constructed of materials to prevent the migration of hazardous constituents into such a liner during the active life and postclosure care period. The lower component must be designed and constructed of materials to minimize the migration of hazardous constituents. This requirement of minimizing the migration of hazardous constituents is specified so that if a breach in the upper liner component (geomembrane) were to occur, the lower component will act as a second line of defense. As shown in Figure 17.3 for hazardous waste surface impoundment, the lower component of the composite liner must be constructed of at least 3 ft (91 cm) of compacted soil material with a hydraulic conductivity (k) of no more than 1×10^{-7} cm/s (40 CFR 264.221).

USEPA's minimum technology requirements also recommend that the upper or primary geomembrane be at least 30 mil (0.76 mm) thick where covered by a protective soil and/or geotextile layer. This thickness should be at least 40 mils (1.14 mm) if the geomembrane is uncovered or exposed directly to the elements. Some geomembranes may require greater thickness to prevent failure or to accommodate unique seaming requirements. For HDPE liners (discussed in Section 17.3), a thickness of 60 to 100 mils (1.52 to 2.54 mm) is generally recommended.

A leak detection, collection, and removal system (LDCRS) is placed between the two liners (i.e., below

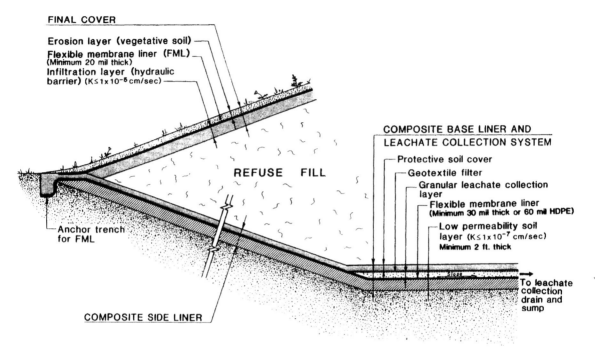

Figure 17.1 Subtitle D liner and final cover for MSWLF.

the primary geomembrane and above the secondary geomembrane). As the name indicates, the purpose of this system is to detect, collect, and remove leaks of hazardous constituents at the earliest practicable time during the active life and postclosure care period. The system should be constructed with a bottom slope of 1% or more. Also, as exhibited in Figure 17.3 for surface impoundment, if constructed with granular material it should have a thickness of 12 in. (30.5 cm) or more and a hydraulic conductivity of 1 cm/s or more. If the system is constructed of synthetic drainage material, its transmissivity should be 3×10^{-3} m²/s or more (USEPA, 1991). Granular or synthetic drainage materials must be resistant to chemical constituents of the waste and should be designed and operated to minimize clogging during the active life and postclosure care period. The system must have sumps and pumps of sufficient size to collect and remove liquids.

The double-liner requirements discussed above may be waived by the regional administrator for any monofill that contains only hazardous wastes from foundry furnace emission controls or metal casting molding sand. It should also be ensured that such monofill wastes do not contain constituents that would render the wastes hazardous for reasons other than the EP toxicity characteristics.

As shown is Figure 17.4, many owners and operators of surface impoundments like to use two composite (geomembrane/compacted soil layer) liners separated by a leak detection/collection layer. In many cases a separating geotextile is placed between the upper compacted soil and the drainage layer to prevent the migration of soil particles into the drainage layer. This layer also facilitates compaction of the upper soil layer. This double-composite liner should provide further assurance against the escape of waste chemical constituents from the impoundment.

For surface impoundments that contain nonhazardous wastes, single composite liners are generally acceptable. Actually, in many situations these liners may have 2-ft (61-cm)-thick compacted soil with $k \leq 1 \times 10^{-6}$ cm/s underlying the synthetic liner. As shown in

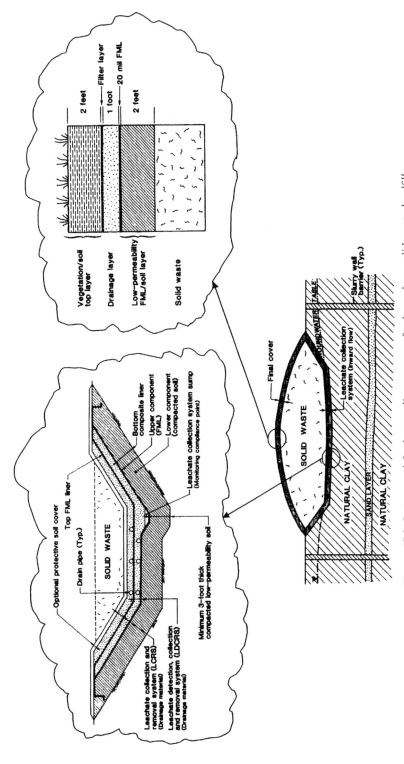

2 feet — Filter layer
1 foot — 20 mil FML
2 feet

Vegetation/soil top layer
Drainage layer
Low-permeability FML/soil layer
Solid waste

Bottom composite liner
Upper component (FML)
Lower component (compacted soil)

Optional protective soil cover
Drain pipe (Typ.)
Top FML liner

SOLID WASTE

Leachate collection system sump
(Monitoring compliance point)

Leachate collection and removal system (LCRS)
(Drainage material)

Leachate detection, collection and removal system (LDCRS)
(Drainage material)

Minimum 3-foot thick compacted low-permeability soil

Final cover

GROUND WATER TABLE

Slurry wall barrier (Typ.)

SOLID WASTE

Leachate collection system (Inward flow)

NATURAL CLAY

SAND LAYER

NATURAL CLAY

Figure 17.2 Bottom and final cover liner systems for hazardous solid waste landfills.

661

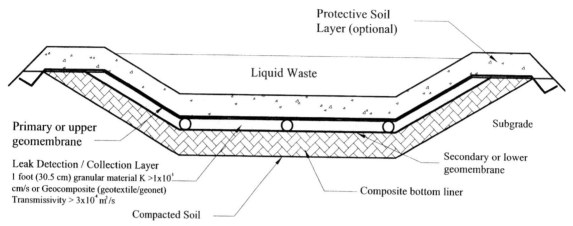

Figure 17.3 *Surface impoundment cross section for hazardous waste: Double liner with composite bottom liner (as required by 40 CRF 264).*

Figures 17.1 through 17.4, liners consist of various components made of natural and synthetic materials. These consist of low-permeability materials such as natural clays, soil-admixed materials, synthetic materials (e.g., high-density polyethylene) and geosynthetic clay liners, and drainage materials. Drainage materials could be naturally obtained sands and gravels, or synthetic materials such as geotextiles, geonets, and geocomposites.

This chapter presents the material properties of these liner component materials and the test methods used to determine these properties. Wherever applicable, American Society for Testing and Materials (ASTM) test methods are referenced because these methods are widely used and accepted in engineering practice in North America.

17.2 LOW-PERMEABILITY SOIL LINERS

As discussed in Section 17.1, low-permeability soil material is used in liner systems for both landfills and for surface impoundments. Typically, compacted clays and soil admixed (soil mixed with clays such as bentonite) are examples of low-permeability liner materials. Since hydraulic conductivity is an important parameter for low-permeability liner material selection,

in this section we discuss factors affecting hydraulic conductivity (k), methods of measuring k, and a brief presentation of other factors that need to be considered in design and construction aspects of low-permeability liner materials.

17.2.1 Hydraulic Conductivity

According to an empirical relationship (called *Darcy's law*) established by Darcy (1856), the coefficient of permeability[1] (k) for flow through a porous medium can be presented in the following form:

$$k = \frac{q}{iA} \qquad (17.1)$$

where q is the flow rate through an area A under the driving force (called the *hydraulic gradient*) i. The value of k is dependent on the properties of a porous medium and the flowing fluid (Olson and Daniel, 1981; Mitchell, 1993). Considering the unit weight (γ) and viscosity (μ) of the permeating fluid, an additional

---

[1] The terms *coefficient of permeability* and *hydraulic conductivity* are used interchangeably herein.

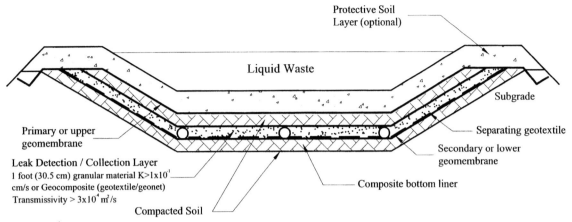

Figure 17.4 *Surface impoundment cross section for hazardous waste: Double composite liner.*

term, intrinsic permeability (K) is used in the literature and can be given by

$$K = \frac{k\mu}{\gamma} \qquad (17.2a)$$

$$k = \frac{K\gamma}{\mu} \qquad (17.2b)$$

Expressions for k of porous media like soil that consider properties of both the porous medium and permeating fluid have been proposed by Kozeny and improved by Carman (in the Kozeny–Carman equation). Based on these, Taylor (1948) developed a general expression as presented by Lambe and Whitman (1969):

$$k = CD_s^2 \frac{\gamma}{\mu} \frac{e^3}{(1 + e)} S^3 \qquad (17.3)$$

where C is the shape factor, D_s the effective particle diameter, γ the unit weight of fluid, μ the viscosity of fluid, e the void ratio, and S the degree of saturation. Equation (17.3) describes the behavior of coarse-grained (cohesionless) soils reasonably well, but serious discrepancies are found if it is used to explain the permeability of fine-grained cohesive (clays) soils (Mitchell, 1993).

In the following paragraphs we present the effects of permeant and soil characteristics (both physical and chemical) on hydraulic conductivity of soils, followed by typical k values for various natural and compacted soils.

Effects of Permeating Liquid. As indicated by equation (17.2b), k will generally decrease when the permeating fluid has a lower unit weight (γ) and/or a higher viscosity (μ). Also, it should be noted that since temperature variations can change viscosity, k will also be (indirectly) affected by the temperature. The effect of factors other than γ and μ on k is presented in Figure 17.5. This is specifically true for clayey soils because characteristics such as dielectric constant, ion exchange, and cation concentration need to be considered in evaluating the influence of the permeant on the k value of clays. Equation (17.2) introduces the term *intrinsic* (also called *specific* or *absolute*) *permeability*, K; using k to represent permeability, the influences of γ and μ have been eliminated in the presentation. As shown in Figure 17.5, when saturated K is plotted against $e^3/(1 + e)$, the K is influenced significantly by the nature of the permeant. In these experiments the kaolinite soil samples were initially molded in the same fluid as that used as the permeant. However, as shown in Figure 17.6, when water was used during

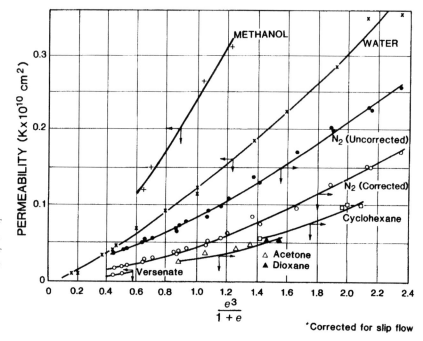

Figure 17.5 Permeability of kaolinite to various fluids to various fluids as a function of $e^3/(1 + e)$, where e is the void ratio. (Reprinted with permission from Michaels and Lin, 1954. Copyright 1954 American Chemical Society.)

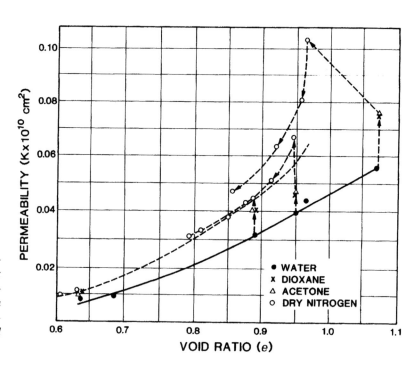

Figure 17.6 Effects of desolvation on kaolinite permeability; initial permeant, water. (Reprinted with permission from Michaels and Lin, 1954. Copyright 1954 American Chemical Society.)

sample remolding (preparing) followed by first performing a permeability test with water which was then displaced with each successive permeant, the permeability differences for different permeants were smaller. In both cases, however, it was evident that in fine-grained soil (clays), the hydraulic conductivity is not only influenced by γ and μ but also by the chemical constituents of the permeant.

Effects of Soil Properties. In the following we discuss how various soil properties may affect the hydraulic conductivity of a soil:

- *Void ratio.* As shown in Figure 17.7 and as can be interpreted from equation (17.3), the hydraulic conductivity of a soil will decrease as the void ratio (*e*) decreases, and vice versa.
- *Particle size.* Equation (17.3) also shows that *k* varies with the square of the particle size (D_x^2). This means that the smaller the particle size, the lower the permeability. As shown in Figure 17.8, experience also exhibits that the quantity of percent fines in soils (i.e., percent passing a No. 200 sieve) in coarse-grained soils can have a significant influence on the *k* of soils. This figure also indicates that types of fine particles also influence *k* (i.e., clay fines reduce *k* more that silt fines or coarse silt fines).
- *Soil minerals.* As can be seen from Figure 17.9, for the same void ratio, montmorillonite has significantly lower *k* than kaolinite, indicating that clay mineral types influence the permeability. Also, for the same void ratio, calcium montmorillonite will have higher permeability than potassium montmorillonite.
- *Soil fabric.* Tests on compacted fine-grained soils indicate that the most flocculated soil structure will exhibit the highest permeability, and the most dispersed structure will show the lowest. Generally, a clay compacted at or dry of optimum moisture content (OMC) has a flocculated structure, while a clay compacted wet of OMC has a dispersed structure. This means that for a compacted clay, *k* at or on the dry side of OMC will be higher than *k* on the wet side of OMC.

- *Degree of saturation.* As presented in equation (17.3), hydraulic conductivity (*k*) varies with the cube of degree of saturation (S^3); this means that *k* increases with *S*. Figure 17.10 shows such behavior for compacted clay; similar trends are exhibited by various sands (Lambe and Whitman, 1969).

Effects of Chemicals. The thickness of the diffused double layer of fine-grained soils (clays) can influence its fabric (i.e., flocculated or dispersed structure), resulting in permeability changes. When fluids containing chemicals permeate a clay liner the chemicals may influence the thickness of the diffused double layer, due to various chemical parameters, such as the dielectric constant, cation valence, electrolyte concentration, and so on. These chemical parameters of the permeating fluids can therefore change the permeability of the clay material.

Based on the results of laboratory testing and field monitoring of chemical influence on *k* of a clayey till placed below a municipal landfill, King et al. (1993) report that for a period of over four years, *k* values remain lower than the specified design criterion of 1 $\times$ 10^{-8} cm/s for the clay liner when permeated by *municipal landfill leachate*. Storey and Peirce (1989) investigated the effects of alcohol–water mixtures on clay and found that for concentrated aqueous concentrations of methanol (more than 80%) did increase the *k* of clay. However, concentrations lower than 80% did not change *k* even after 14 months of testing. Bowders and Daniel (1987) present permeability test results *when methanol, acetic acid, heptane, trichloroethylene,* and *water* are separately permeated through compacted specimens of kaolinite and illite–chlorite clays. As discussed by Sharma and Lewis (1994) in detail, the following conclusions can be drawn from the results of Bowders and Daniel (1987) tests:

- Permeation of *methanol* through clay did not increase *k* value until its concentrations were more than 80%. The reason for increased *k* may be due to the lower dielectric constant of methanol (a value of 32.6) as compared with water (a value of 80.1) resulting in shrinkage of double layer

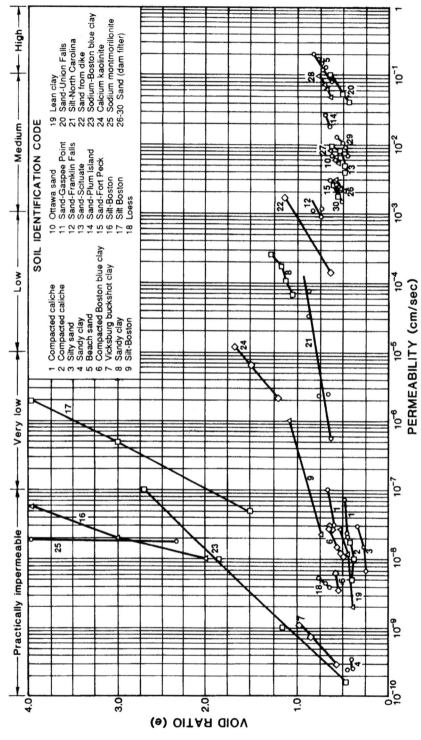

Figure 17.7 Permeability test data. (From Lambe and Whitman, 1969. This material is used by permission of John Wiley & Sons, Inc.)

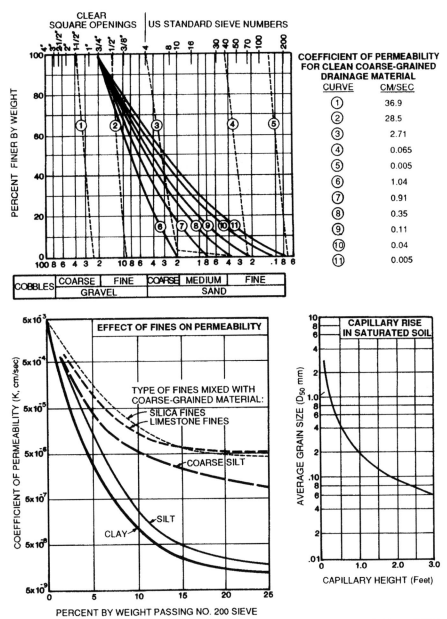

Figure 17.8 Permeability and capillarity of drainage materials. *(From U.S. Navy, 1982.)*

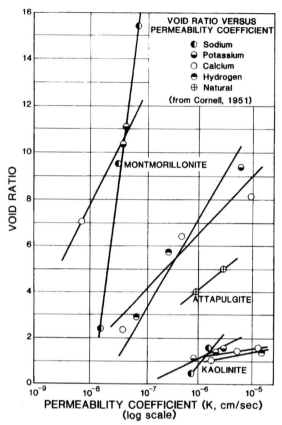

Figure 17.9 *Void ratio versus permeability. (From Lambe and Whitman, 1969. This material is used by permission of John Wiley & Sons, Inc.)*

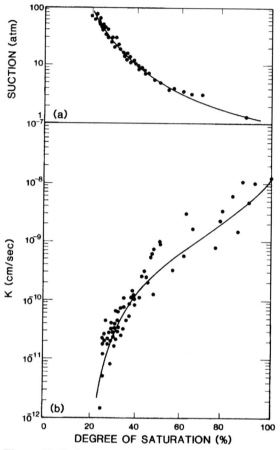

Figure 17.10 *Suction and hydraulic conductivity versus degree of saturation for compacted fire clay. (From Olson and Danial, 1981.)*

causing the development of macropores and cracks for faster flow of fluids resulting in higher permeability.

- Permeation of *diluted acid* reduced the k value; this decrease in k value can be attributed to dissolution and precipitation that would potentially clog the clay soil pores. It is, however, possible that these precipitates would probably redissolve on a long-term basis if sufficient volumes of acids were allowed to permeate. This could result in an increase in permeability over time; this needs further evaluation, however.

- Permeation of *pure heptane* and *pure TCE* increased k significantly. However, the permeability

was not changed when heptane and TCE were at their solubility [2] limits.

- Results of the tests also supported the hypothesis that if an organic liquid does not alter the Atterberg limits (compared with water), the permeating organic liquid will not change the permeability of compacted clay. If the organic liquid does change

[2] Like melting point and boiling point, solubility is a physical property characteristic of a particular substance. The amount of solute per unit of solvent needed to form a saturated solution, at a given temperature, is called *solute solubility*.

Atterberg limits, however, the permeating liquid may or may not influence the hydraulic conductivity.

Based on a literature review of the test data related to chemical effects on the k value of clay, Mitchell and Madsen (1987) drew the following general conclusions:

- Permeation with pure hydrocarbons may affect k, however, hydrocarbons to concentrations at or below their solubility limit do not influence k.
- Acetone solutions at concentrations less than 75% have little influence on hydraulic conductivity. Similarly, alcohol with concentrations of less than 80% have little effect on k.

As presented in *compatibility testing* (Section 17.2.3), at the final design phase, site-specific chemicals must be permeated through the clay to evaluate and quantify the effects of chemicals on the k value of the project-specific clay.

Typical Hydraulic Conductivity Values for Compacted Soils.

As presented in Table 17.1, the k values for most clays, silts, and other clayey mixtures when compacted at maximum dry density and optimum moisture content (ASTM D 1557) is less than 1×10^{-6} cm/s. These values can vary significantly, depending on grain size distribution, clay mineralogy, percent compaction, and placement moisture content. Permeability values presented in Table 17.1 are for guidance purposes only and may be used at feasibility design phase. For comparison purposes, k values for sand and gravel are also presented in the table. For final design, it is recommended that the permeability testing be performed by permeating site-specific construction water and chemical fluids through the project-specific soils.

17.2.2 Laboratory Hydraulic Conductivity Testing

Evaluation of hydraulic conductivity of both remolded (compacted) and relatively undisturbed soils can in most cases be performed in laboratory. The main ad-

vantages of laboratory tests are that the tests can be economically (when compared with field tests) conducted by varying parameters, such as soil types, moisture content, density, and permeants. The three types of permeameters (permeability cells) commonly available for laboratory permeability tests are compaction mold (cell), consolidation cell, and flexible wall (triaxial cell) permeameters. The compaction mold permeameter is used for compacted samples while the consolidation cell permeameter is for relatively undisturbed soil samples. The flexible wall permeameter is suitable for testing both the compacted and the relatively undisturbed soils. In addition, this permeameter (flexible wall) can control many testing variables, such as confining pressures, drainage conditions, and degree of saturation. Also, unlike in compaction mold, the sidewall leakage problems can be controlled in these cells. In flexible wall permeameters the hydraulic conductivity can be performed by two test methods: (1) the constant-head test method and (2) the falling-head test method.

Constant-Head Test Method.

As the name indicates, in this method the hydraulic head (h) is maintained constant and the quantity of water (Q) discharged is measured during a time (t) through a soil sample of area (A) and length (L). The permeability (k) is then determined by using the following expression:

$$k = \frac{QL}{tAh} \qquad (17.4)$$

The constant-head test method is generally used for coarse-grained soils because for fine-grained soils the k is very small (less than or equal to 1×10^{-5} cm/s). Therefore, to have a reasonable testing time the conventional equipment needs to be modified to apply high heads if this test method is to be used for fine-grained soils.

The laboratory permeability test method commonly used for coarse-grained soil (with $k > 1 \times 10^{-3}$ cm/s) is specified by ASTM test method D 2434. This procedure is limited to granular soils containing not more than 10% soil passing the No. 200 sieve (0.075

Table 17.1 **Typical hydraulic conductivity of compacted soils**

Group Symbol	Soil Type	Typical Coefficient of Permeability[a] (cm/s)
GW	Well-graded, clean gravels, gravel–sand mixture	2×10^{-2}
GP	Poorly graded, clean gravels, gravel–sand mixture	5×10^{-2}
GM	Silty gravels, poorly graded, gravel–sand–silt	$5 \times >10^{-7}$
GC	Clayey gravels, poorly graded, gravel–sand–clay	$5 \times >10^{-8}$
SW	Well-graded, clean sands, gravelly sands	$5 \times >10^{-4}$
SP	Poorly graded, clean sands, sand–gravel mixture	$5 \times >10^{-4}$
SM	Silty sands, poorly graded, sand–silt mixture	$2 \times >10^{-5}$
SM – SC	Sand–silt clay mixed with slightly plastic fines	$1 \times >10^{-6}$
SC	Clayey sands, poorly graded, sand–silt mixture	$2 \times >10^{-7}$
ML	Inorganic silts and clayey silts	$5 \times >10^{-6}$
ML – CL	Mixture of inorganic silt and clay	$2 \times >10^{-7}$
CL	Inorganic clays of low to medium plasticity	$5 \times >10^{-8}$
OL	Organic silts and silt–clays, low plasticity	—
MH	Inorganic clayey silts, elastic silts	$2 \times >10^{-7}$
CH	Inorganic clays of high plasticity	$5 \times >10^{-8}$
OH	Organic clays and silty clays	—

Source: U.S. Navy (1982).

[a] All permeabilities are for conditions of modified Proctor maximum density (ASTM D 1557); > indicates that a typical property is greater than the value shown; — indicates insufficient data available for an estimate.

mm). The test method is presented here briefly; for detail, the original ASTM standard should be referenced. Figure 17.11 shows the test apparatus. The test is performed by measuring the quantity of flow (Q) over time (t) through the specimen of known length (L) and area (A) by maintaining a constant head (h). Permeability (k) is then determined by using equation (17.4).

Falling-Head Test Method. In this method, heads h_1 at time t_1 and h_2 at time t_2 are measured in a buret of area a when the fluid flows through a soil specimen of length L and area A. The permeability (k) is then determined using the formula

$$k = \frac{aL}{At} \ln \frac{h_1}{h_2} \qquad (17.5)$$

The small flows through the test specimen can be measured with greater accuracy by using a buret of small area of cross section a. Also, the hydraulic head for flow can be increased by superimposing an air pressure in the inlet system; therefore, the testing time for

low-permeability soils can be reduced. This method is thus more applicable for low-permeability (fine-grained) soils.

Flexible Wall Permeameter. The commonly used test method for measurement of hydraulic conductivity of saturated porous soils (materials) with $k \leq 1 \times 10^{-3}$ cm/s is by using flexible wall permeameter. The ASTM designated test method for such soils is D 5084. This method is discussed briefly here; for further details the original ASTM standard (ASTM, 2003) should be referenced. This test method provides saturated k for undisturbed or compacted low-permeability soils. The saturation for low-permeability soils is usually achieved by applying backpressure to the soil specimen. The required back pressure to achieve a final degree of saturation (between 97 and 100%) for various initial degrees of saturation, as suggested by Lowe and Johnson (1960), is presented in Figure 17.12.

It is recommended that during permeation, when possible, the hydraulic gradient (i) should be similar to that expected to occur under actual site conditions. Smaller i can lead to very long testing time and larger

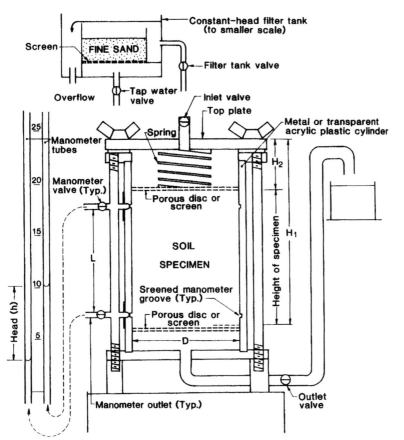

Figure 17.11 Constant-head per-meameter. (*From ASTM, 2003.*)

i can result into larger seepage pressures. The larger seepage pressures may wash material from the specimen, carry fines from specimen to the effluent end of the specimen, and thus plug the system or consolidate the specimen. For a soft and compressible specimen, an $i < 10$ value may be required to minimize consolidation and reduction in k. ASTM recommends an i value of 5 for k between 1×10^{-4} and 1×10^{-5} cm/s, 10 for k between 1×10^{-6} and 1×10^{-7} cm/s, and 30 for k less than 1×10^{-7} cm/s.

Figure 17.13 exhibits the schematic of the test setup. The ASTM D 5084 describes three different test procedures: constant head (method A), falling head (methods B and C), and constant rate of flow (method D). The permeability can be calculated by using equation (17.4) in test methods A and D. For test method

B, where the falling-head test is performed by keeping the tailwater pressure, the k can be calculated by using equation (17.5). For test method C, where the test is performed as a falling-head test by increasing tailwater pressure, k can be calculated as:

$$k = \frac{aL}{-2At} \ln \frac{h_1}{h_2} \qquad (17.6)$$

All terms are as defined for equation (17.5).

Issues with Laboratory Permeability Testing. Laboratory permeability tests are widely performed because various test parameters can be varied during testing. Thus, the laboratory testing has practical and ec-

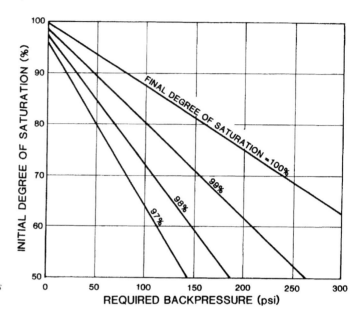

Figure 17.12 *Backpressure to attain various degree of saturation. (From ASTM, 2003.)*

onomical advantages. However, laboratory tests have some limitations. The two main problems with these tests are that (1) errors, such as nonrepresentative samples, simulation of flow directions, and so on, may be introduced during laboratory testing, and (2) small laboratory specimens may not represent actual site conditions. Table 17.2 summarizes these limitations; most projects require that field permeability testing be performed to confirm the results of laboratory tests.

17.2.3 Laboratory Compatibility Testing

Laboratory compatibility testing is performed to evaluate the effect of contained chemical fluids or permeating leachate on hydraulic conductivity of the liner material. A flexible wall permeameter, as discussed in Section 7.2.2, can also be used in conducting compatibility testing. The test method is similar to the ASTM D 5084 method except that as shown in Figure 17.14, bladder accumulators are inserted between the cell and the inlet and outlet ports. Prior to testing, the sample is first saturated with water, followed by permeating with one pore volume of water. Two to four pore volumes of leachate or the chemical fluid is then perme-

ated through the sample. Permeability values are determined for water and the leachate; results are plotted as shown in Figure 17.15. The test results presented in this figure exhibit that for a specific silty clay when permeated with organic chemicals, obtained from a hazardous waste management site, for all practical purposes the hydraulic conductivity either deceased or remained constant as these chemical fluids were permeated through the site soils. Further information on compatibility testing is provided by Sharma et al. (1991) and USEPA (1986b).

17.2.4 Field Hydraulic Conductivity Testing

In field permeability testing, larger and more representative field samples are permeated by water, allowing flow through features such as macropores and fissures. These tests are performed on a test fill pad, generally constructed as a part of the preconstruction quality assurance program. The field permeability tests can be classified into the following four categories (Daniel, 1989): (1) borehole tests, (2) porous probe permeater tests, (3) infiltrometer tests, and (4) lysimeter tests. In this section, only general principles of these tests are

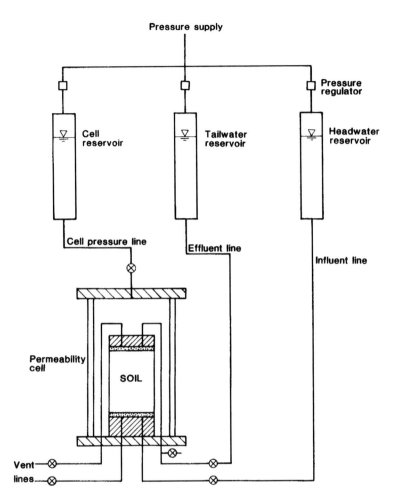

Pressure supply

Pressure regulator

Cell reservoir

Tailwater reservoir

Headwater reservoir

Cell pressure line

Effluent line

Influent line

Permeability cell

SOIL

Vent lines

Figure 17.13 Permeameter cell. (From ASTM, 2003.)

presented; detailed information on these tests are provided in various references, such as USEPA (1986a); Daniel (1989); Fernuik and Haug (1990); Sai and Anderson (1990, 1991); and ASTM (2003). These tests are also discussed by Sharma and Lewis (1994).

Borehole Tests. The two commonly used borehole tests are the Boutwell permeameter and the constant-head permeameter. As shown in Figure 17.16(*a*) and (*b*), the Boutwell permeameter is installed by placing a casing in a predrilled hole. After casing installation the annular space between the casing and the hole is sealed with grout. Following this, a falling-head per-

meability test is performed; this is stage I of the test. Hydraulic conductivity (k_1) for stage I is calculated at various time periods until k_1 reaches a steady state by using Hvorslev (1949) equation, as given by (Daniel, 1989)

$$k_1 = \frac{\pi d^2}{11D(t_2 - t_1)} \ln \frac{H_1}{H_2} \qquad (17.7)$$

Stage II of the test is performed by first disassembling the permeameter, deepening the hole (either by auger or pushing a tube), removing the smear by a wire brush, and reassembling the permeameter. A falling-

Table 17.2 **Sources of errors in tests performed in laboratory fine-grained soils**

Source of Error	Cause or Effect of Error	Method of Minimizing Effect of Error	Influence on Hydraulic Conductivity (k), Measurement
Nonrepresentative samples	Variability of deposit, tendency to select homogeneous zones	Thorough field investigaion, development of accurate geological model	± One order of magnitude or more
Sample preparation	Voids, stress relief, fissures, smear zones, variation in molding water content	Careful trimming, apply stress to close fissures, use as large a specimen as possible, control moisture density of compacted soils	± One order of magnitude or more
Sample placement in cell	Dispersion of fines, irregular cell wall contact, leading to piping during test	Careful placement and tamping in cell, removal of pebbles in trimmed sample, use cell constructed of transparent material (e.g., Lucite), apply sufficient cell pressure in triaxial tests	± One order of magnitude or more
Flow direction	Sample generally oriented vertically, whereas horizontal flow may dominate in site	Orient laboratory specimen to model dominant flow direction in site	K_H/K_V may range from 1 to 10 higher
Alterations in clay chemistry (permeant)	Leaching sample with distilled water, permeant with different chemical characteristics than in field	Where possible, use permeant with same chemistry as field, recycle fluids, use tap water rather than distilled water	± Two orders of magnitude in extreme cases
Growth of microorganisms	Clogging of flow channels by organic matter which grows in sample during prolonged tests	Use disinfectant in permeant if microorganisms would not grow naturally in field situation studied	± 8 to 50 times k for prolonged tests (measured values generally too low)
Air in sample	Attempted saturation of compacted sample	Apply backpressure to dissolve air into water, conduct test under a reasonably large pressure gradient (see "Exessive hydraulic gradients")	± 2 to 5 times k (measured values generally too low)
Equipment	Meniscus problems in capillary tubes, leaks, evaporation of permeant, volume changes of cell	Measure pressure drops instead of heads in capillary tubes; tests should be performed in well-constructed laboratory equipment, and a dummy cell should be set up to measure evaporation and volume changes due to temperature fluctuations	0.3 to 3 times k, depending on equipment problems
Excessive hydraulic gradients	Used to reduce testing time, increase seepage quantity; may consolidate sample, cause piping	Use gradients as close as possible to field conditions; use falling-head test	± 1 to 5 times k; may reduce k due to particle migration; limited influence for carefully run tests

Table 17.2 (*continued*)

Source of Error	Cause or Effect of Error	Method of Minimizing Effect of Error	Influence on Hydraulic Conductivity, (k), Measurement
Volume change of specimen due to stress change	Large gradients increase effective stresses in lower half of sample and may cause consolidation; upper half experiences reduced effective stress (unless loaded) and may swell	Measure inflow and outflow, plot flow rate, q, versus time $t^{-1/2}$ and continue test until linear relationship is observed (steady-state seepage condition)	Small, provided that accurate measurements of sample swelling/compression are taken (generally 1 to 20 times too high)
Temperature	Viscosity changes of permeant	Perform tests at temperature relevant to field situation modeled	Viscosity decreases 3% per °C rise; minor influence on k (generally ±0.5 to 1.5 times k)

Source: Alberta Environment (1985).

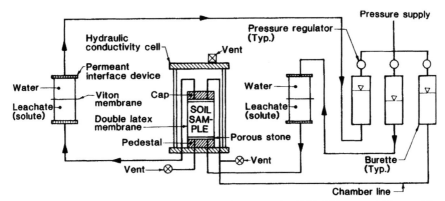

Figure 17.14 *Apparatus for compatibility testing. (From Sharma et al., 1991. This material is used by permission of John Wiley & Sons, Inc.)*

head test is again performed until permeability at this stage (k_2) becomes relatively constant. This is given by

$$k_2 = \frac{\alpha}{\beta} \ln \frac{H_1}{H_2} \qquad (17.8)$$

where

$$\alpha = d^2 \left\{ \ln \left[\frac{L}{D} + \left(1 + \frac{L^2}{D^2} \right)^{0.5} \right] \right\} \quad \text{and}$$

$$\beta = 8D \frac{L}{D} (t_2 - t_1) \left[1 - 0.562 \exp\left(-1.57 \frac{L}{D} \right) \right]$$

Knowing k_1, k_2, L, and D, m (which accounts for anisotropy) can be obtained from Figure 17.16(*c*). The values of horizontal permeability ($k_h = m k_1$) and ver-

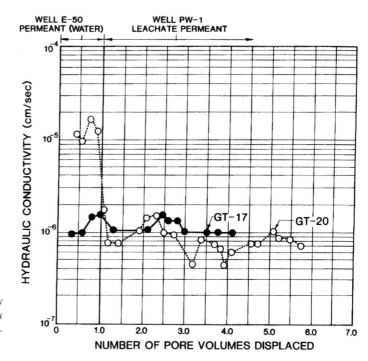

Figure 17.15 *Leachate–soil compatibility test results. (From Sharma et al., 1991. This material is used by permission of John Wiley & Sons, Inc.)*

tical permeability $[k_v = (1/m)\, k_1]$ can then be calculated (Daniel, 1989). It should be recognized that only a small amount of soil is tested by this method. This means that flow through flaws in a soil liner and soil macropores may not be represented; consequently, the resulting k values may not show the actual field conditions.

As shown in Figure 17.17, a constant-head permeameter measures the rate of flow (q) needed to maintain a constant head (h) in a standpipe. Daniel (1989) comments that a comparison of these tests with the laboratory results shows that field tests, by this method, may be about 10 times larger than laboratory tests when the k value is calculated from the following expression.

$$k = \frac{Cq}{2\pi H^2 + \pi r^2 C + 2\pi H/\alpha} \qquad (17.9)$$

where C can be estimated from Figure 17.18 and $\alpha = 0.01$ cm^{-1} for compacted clay.

Porous Probe Permeameter Tests. As shown in Figure 17.19, these permeameters typically consist of a cone-shaped porous probe that is either pushed or driven into the soil. A commercially available porous probe, the BAT permeameter, is reported to measure permeability for soils with k less than or equal to 1×10^{-7} cm/s. As for Boutwell permeameter, this method only tests a small amount of soil; therefore has the disadvantage of nonrepresentative of the overall soil liner.

Infiltrometer Tests. The five types of infiltrometers are the (1) open single-ring infiltrometer, (2) open double-ring infiltrometer, (3) sealed single-ring infiltrometer, (4) sealed double-ring infiltrometer, and (5) air-entry permeameter. Figure 17.20 shows the first four of these infiltrometers. All these five infiltrometer tests are discussed briefly below.

Open Single-Ring Infiltrometer (OSRI). As shown conceptually in Figure 17.20, it consists of using an open ring of area A through which the quantity of flow

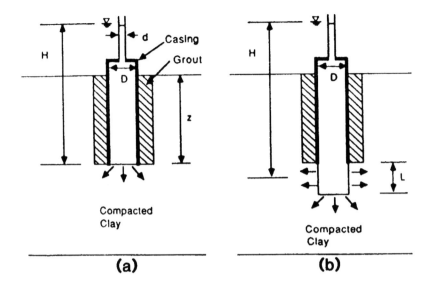

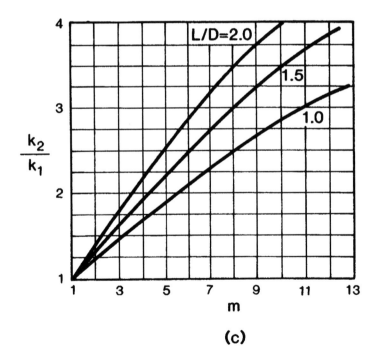

Figure 17.16 *Schematic diagram of two-stage in-situ hydraulic conductivity test with Boutwell permeameter for case in which potentiometric level is below base of permeameter:* (a) *stage I;* (b) *stage II;* (c) *curves of* K_2/K_1 *versus* m *required to satisfy equation (17.8) for* L/D = 1.0, 1.5, *and 2.0. (From Daniel, 1989. Reproduced by permission of ASCE.)*

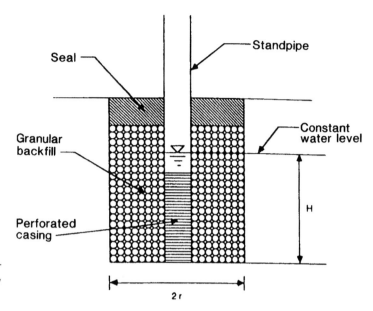

Figure 17.17 *Borehole test with water level. (From Daniel, 1989. Reproduced by permission of ASCE.)*

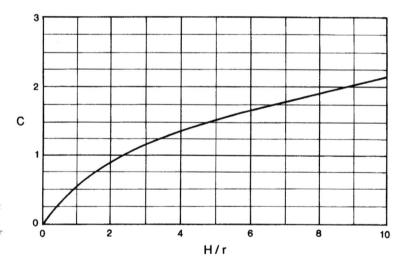

Figure 17.18 *Factor C [equation (17.9)] versus H/r. (From Daniel, 1989. Reproduced by permission of ASCE.)*

(*V*) over time (*t*) is recorded. The infiltration rate (*I*) is then calculated by using $I = V/At$. If the depth of ponded water is *H* and the depth of wetting front (L_f) is measured either by tensionmeter or by measuring moisture content in a soil sample obtained by driving a tube after the test into the ground, the hydraulic gradient $i = (H + L_f)/L_f$. The hydraulic conductivity (*k*) can then be estimated by using the relationship $k = I/i$. The main problems with this method are (1) the difficulty in measuring *k* lower than 1×10^{-6} to 1×10^{-7} cm/s, due to accuracy problems; (2) problems with accounting for evaporation losses; and (3) interpreting vertical *k*, especially when the wetting front passes below the ring and water spreads laterally.

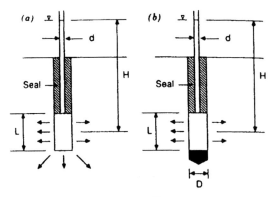

Constant Head: $k = \dfrac{q}{FH}$

Falling Head: $k = \dfrac{\pi d^2/4}{F(t_2 - t_1)}\ \ln\left(\dfrac{H_1}{H_2}\right)$

Case A: $F = \dfrac{2\pi L}{\ln\left(\dfrac{L}{D} + \sqrt{1 + (L/D)^2}\right)}$

Case B: $F = \dfrac{2\pi L}{\ln\left(\dfrac{L}{D} + \sqrt{1 + (L/D)^2}\right)} - 2.8\,D$

Figure 17.19 Hydraulic conductivity from porous probe tests: (a) case A, probe with permeable test; (b) case B, probe with impermeable base. (From Daniel, 1989. Reproduced by permission of ASCE.)

Open Double-Ring Infiltrometer (ODRI). As shown in Figure 17.20, it consists of two rings that are filled with water. The method is described in ASTM D 3385. As discussed above for OSRI, by this method k is also equal to I/i, where all terms are described earlier. The method has the advantage over OSRI because in this method the direction of vertical flow can be maintained by maintaining the same water levels in the inner and outer rings. The test results, however, may be unreliable if $k > 1 \times 10^{-2}$ cm/s or $k < 1 \times 10^{-6}$.

Sealed Single-Ring Infiltrometer (SSRI). This test also requires measuring V, t, H, and L_f and then estimating k as described in OSRI. Overall, it is similar to OSRI except that the ring is sealed to minimize evaporation losses. Other than that, it has problems similar to those with OSRI results. The apparatus for the infiltrometer is shown in Figure 17.21.

Sealed Double-Ring Infiltrometer (SDRI). As shown in Figure 17.22, the apparatus has two rings and the water pressure in the inner (sealed) and outer (open) rings is kept the same so that vertical flow below the inner ring can be maintained. The quantity of flow (V) over time (t) is measured by connecting a flexible bag filled with a known weight of water to a port in the inner ring. Hydraulic conductivity (k) is then calculated by using $k = I/i$. All terms have been defined above.

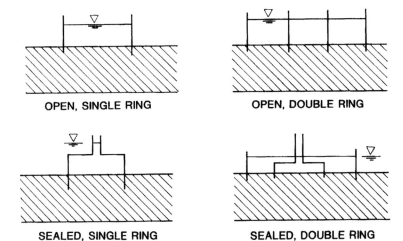

OPEN, SINGLE RING   OPEN, DOUBLE RING

SEALED, SINGLE RING   SEALED, DOUBLE RING

Figure 17.20 Types of infiltrometers. (From Daniel, 1989. Reproduced by permission of ASCE.)

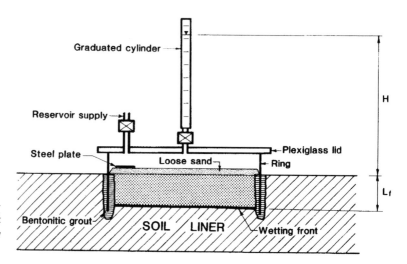

Figure 17.21 Sealed single-ring in-filtrometer (SSRI). (From Fernuik and Haug, 1990, Reproduced by permission of ASCE.)

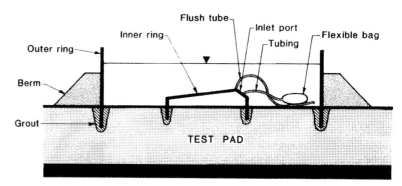

Figure 17.22 *Schematic of a double-ring infiltrometer with a sealed inner ring. (From ASTM, 2003.)*

The test method is described in ASTM D 5093 and is suitable for fine-grained clayey or low-permeability soils.

Air-Entry Permeameter (AEP). The apparatus for AEP is shown in Figure 17.23. In this test, the k is first determined similar to the SSRI test, as discussed earlier, then the valve to the flow-measuring device is closed; this seals the AEP. As the underlying partially saturated soil tends to suck water downward due to suction pressure, a negative head develops that can be measured by a vacuum gauge. The hydraulic conductivity of the saturated soils (k) then be determined from the following relationship:

$$k = \frac{VL_f}{At(H + L_f + \Psi_f)} \qquad (17.10)$$

where

$$\Psi_f = \frac{1}{2}\left(\frac{U_w}{\gamma_w} + L_f + G\right) \qquad (17.10a)$$

U_w is the minimum water pressure (a negative value) measured with a pressure gauge located a distance G above the ground surface, the term Ψ_f is expressed by equation (17.10a) and γ_w is the unit weight of water. All other terms are as defined earlier.

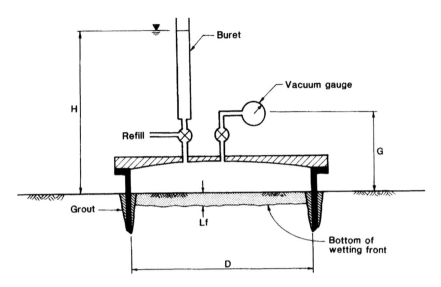

Figure 17.23 Air-entry permeameter. (*From Daniel, 1989. Reproduced by permission of ASCE.*)

Lysimeter Tests. These tests are performed by placing underdrains beneath a clay liner to collect percolating liquids. k can then be estimated by using the quantity of liquid (V) percolating during time period (t) through a liner section. Although the tests have been used successfully for estimating k, they are time consuming and require an elaborate installation of liner, sumps, and pumps; therefore are not in common use. Regulations generally require that in addition to performing laboratory permeability tests, the hydraulic conductivity for a waste containment system be confirmed by an appropriate field test method. Table 17.3 summarizes the advantages and disadvantages of various field test methods. An engineer should evaluate the need for a specific project and the validity of an appropriate field test before selecting one or a combination of field permeability tests.

17.2.5 Low-Permeability Compacted Clay Liners

Low-permeability compacted liners may either be constructed from naturally occurring cohesive soils or a bentonite–soil blend (admixed) material. In this section we present the compaction and permeability issues for compacted cohesive soil liners; a soil–bentonite blend alternative is described in Section 17.2.6.

Compaction and Permeability Requirements. The factors that can primarily influence permeability of compacted clays are (1) clay mineralogy, (2) compaction moisture content, (3) method of compaction, and (4) degree of compaction (or relative compaction). These factors influence pore size distribution, particle orientation, and structure of a compacted clay, resulting in strength and permeability variations. In addition, freeze–thaw considerations may also affect the strength and permeability of a compacted clay. These issues are presented in the following paragraphs.

As shown in Figure 17.24(*a*) and suggested by Lambe (1958), from a micropore point of view, clay mineral particles compacted on the dry side of optimal moisture content (OMC) have a flocculated structure and when compacted on the wet side of OMC have a dispersed structure. Flocculated structure results in higher pore voids exhibiting higher k. Also, since the soil is drier (i.e., has a moisture content lower than OMC), it has higher strength. In comparison, soils compacted at the wet side of OMC have a lower void space, therefore a lower k. Since the moisture content

Table 17.3 **Advantages and disadvantages of various methods of field permeability testing**

Type of Test	Device	Advantages	Disadvantages
Borehole	Boutwell permeameter	Low equipment cost (<$200 per unit) Easy to install. Hydraulic conductivity is measured in vertical and horizontal direction. Can measure low hydraulic conductivity (down to about 10^{-9} cm/s). Can be used at great depths and on slopes.	Volume of soil tested is small. Unsaturated nature of soil not properly taken into account. Testing times are somewhat long (typically, several days to several weeks for hydraulic conductivities $<10^{-7}$ cm/s).
	Constant-head permeameter	Low equipment cost ($1000 per unit). Easy to install. Unsaturated nature of soil taken into account relatively rigorously. Relatively short testing times (a few hours to several days). The hydraulic conductivity that is measured is primarily the horizontal value (which is an advantage if this is the desired value). Can be used at great depths.	Volume of soil tested is small. The hydraulic conductivity that is measured is primarily the horizontal value. (In some applications, the value in the vertical direction is desired.) The device is not well suited to measuring very low hydraulic conductivities (less than 10^{-7} cm/s).
Porous probe	BAT permeameter	Easy to install. Short testing times (usually a few minutes to a few hours). Probe can also be used to measure pore water pressures. Can measure low hydraulic conductivity (down to about 10^{-10} cm/s). The hydraulic conductivity that is measured is primarily the horizontal value (which is an advantage if this is the desired value). Can be used at large depths.	High equipment cost (>$6000). Volume of soil tested is very small. Soil smeared across probe during installation may lead to underestimation of hydraulic conductivity. The hydraulic conductivity that is measured is primarily the horizontal value. (In some applications the value in the vertical direction is desired.) The unsaturated nature of the soil is not properly taken into account.
	Open, double-ring infiltrometer	Low equipment cost (<$1000). Hydraulic conductivity in the vertical direction is determined. Minimal lateral spreading of water that infiltrates from inner ring.	Low hydraulic conductivity (10^{-7} cm/s) is difficult to measure accurately. Must eliminate or make a correction for evaporation. Testing times are somewhat long (usually, several days to several months for hydraulic conductivities $<10^{-7}$ cm/s). Must estimate wetting-front suction head. Cannot be used on steep slopes unless a flat bench is cut.

Table 17.3 (*continued*)

Type of Test	Device	Advantages	Disadvantages
	Closed, single-ring infiltrometer	Low equipment cost (<$1000). Hydraulic conductivity in the vertical direction is measured. Can measure low hydraulic conductivity (down to 10^{-8} to 10^{-9} cm/s).	Volume of soils tested is somewhat small because diameter of ring is <1 m. Need to correct for lateral spreading of water if wetting front penetrates below the base of the ring. Testing times are long (usually several weeks to several months). Must estimate wetting-front suction head. Very difficult to use on steeply sloping ground.
Porous probe	Sealed, double-ring infiltrometer	Moderate equipment cost (<$2500). Hydraulic conductivity in the vertical direction is determined. Can measure low hydraulic conductivity (down to about 10^{-8} cm/s). Minimal lateral spreading of water that infiltrates from inner ring. Relatively large volume of soil is permeated.	Testing times are relatively long (usually several weeks to several months). Must estimate wetting-front suction head. Cannot be used on slopes unless a flat bench is cut.
	Air-entry permeameter	Relatively short testing times (a few hours to a few days). Modest equipment cost (<$3000). Hydraulic conductivity in the vertical direction is measured. Can measure low hydraulic conductivity (down to 10^{-8} to 10^{-9} cm/s). Wetting-front suction head is estimated in second stage of test.	A relatively small volume of soil is permeated because the wetting front usually does not penetrate more than a few centimeters into compacted clay. Cannot be used on slopes unless flat bench is cut. Several important assumptions are required.
Underdrain	Lysimeter pan	Low cost. The hydraulic conductivity in the vertical direction is measured. Large volumes of soil can be tested. Few experimental ambiguities. No disturbance of soil.	Must install underdrain before the liner is constructed. Relatively long testing times (usually, several weeks to several months for hydraulic conductivities less than 10^{-7} cm/s). Must collect and measure seepage from underdrain, which usually necessitates a sump and a pump.

Source: Daniel (1989). Reproduced by permission of ASCE.

of these soil samples will be higher than that of OMC, they will exhibit lower strength (than compacted either below OMC or at OMC). Thus, according to the micropore concept, the orientation or arrangement of individual particles, affected by the molding water content, influences k.

The *macropore concept,* suggested by Olsen (1962) and referred to as *clod theory,* looks into an arrangement of aggregates. According to this concept, as shown in Figure 17.24(*b*), when cohesive soils are compacted on the wet side of OMC, the soft and wet clods of soil get remolded, resulting in smaller inter-

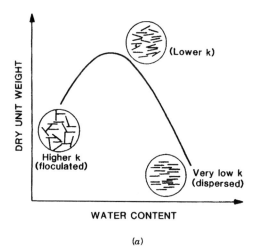

WATER CONTENT

(a)

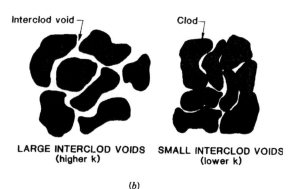

(b)

Figure 17.24 Pore spaces, moisture, and hydraulic conductivity: (a) arrangement of soil particles suggested by
Lambe (1958) and comments added; (b) flow of water
through relatively large pores between clods of soil proposed by Olsen (1962). (From Benson and Daniel, 1990.
Reproduced by permission of ASCE.)

clod pore spaces, causing lower k values than if compacted on the dry side of OMC (which has higher voids
and higher k).

In general, it can be concluded that based on the
micropore concept, a flocculated structure (soils compacted on the dry side of OMC) yields a higher k value
than a dispersed structure (soils compacted on the wet
side of OMC). Similar conclusions on permeability are
made based on macropore concept or clod theory.

Garcia-Bengochea et al. (1979) also suggested that k
is influenced by changes in macropore distribution represented by the second moment of pore size distribution; this is exhibited in Figure 17.25(a).

The effect of clay mineralogy on k is shown in Figure 17.25(b). This figure also exhibits the fact that Na-
saturated soils will have lower k values than those of
Ca-saturated soils, because lower-valence Na will yield
a thicker double layer, resulting in a dispersed structure, whereas the higher-valence Ca will have a flocculated structure.

Figure 17.25(c) shows that in general k decreases
as molding moisture content increases. Test results also
show that beyond a certain level of moisture content
increase additional moisture will not decrease k and
may even slightly increase permeability. A similar pattern was exhibited by two different methods of compaction [Figure 17.25(d)]. Results indicated that
kneading compaction was more efficient than static
compaction.

Figure 17.26 shows that in general, for a given compaction method, k decreases with (1) increase in compactive effort (as indicated by various Proctor methods), (2) increase in molding water content, and (3)
increase in density at the same molding water content.
The traditional method for compaction specifications
for a clay liner meeting a certain permeability value
would require the approach shown in Figure 17.26.
According to this approach, a soil liner typically would
achieve a maximum required k within remolded water
contents of w_1 and w_2 with a minimum dry density of
γ_d [or a relative compaction (RC) = γ_d/γ_{dmax}] for a
specified method of Proctor compaction. This would
result in specifying an acceptable zone, as shown in
Figure 17.27. This approach can be extended to include other liner requirements, such as soil shear
strength, liner interface strengths, shrink–swell considerations, and other design factors that may need inclusion in the specification. This is explained in Figures
17.28 and 17.29. Figures 17.28(a) shows a water
content–dry unit weight zone for acceptable k value,
and Figure 17.28(b) shows a zone to account for other
design factors. To satisfy design needs for all factors,
these zones would need to be superimposed and an

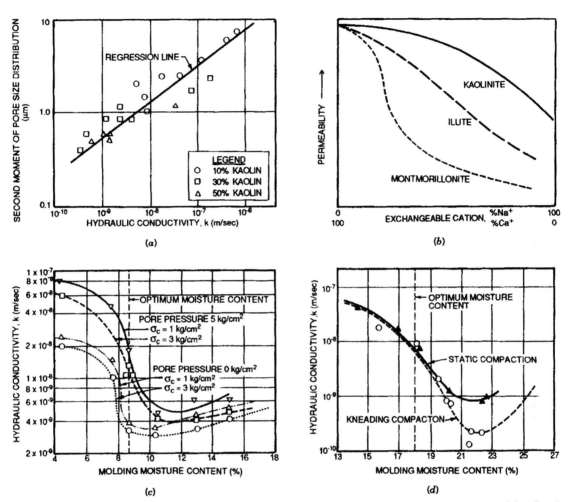

Figure 17.25 *Factors affecting compacted clay permeability: (a) pore size distribution as suggested by Garcia-Bengochea et al. (1979); (b) clay mineralogy and exchangeable cations as suggested by Yong and Warkentin (1975); (c) moisture content during compaction as suggested by Mitchell at al. (1965); (d) methods of compaction as suggested by Mitchell et al. (1965). (From Folkes, 1982. Reproduced by permission of National Research Council, Canada.)*

overall acceptable zone would need to be determined. As an example, Daniel and Wu (1993) present a case for compacted clay liners and covers for an arid site in West Texas. The liner had to meet (1) $k \leq 1 \times 10^{-7}$ cm/s, (2) unconfined compressive strength ≥ 30 lb/in^2, and (3) volumetric shrinkage upon drying $\leq 4\%$ (this was required for low shrinkage and cracking potential on drying). This is shown in Figure 17.29(b).

Daniel and Wu's (1993) study concluded that for this site the specified clay would meet the foregoing requirements if RC varied between 96 and 98% and placement water content was within 2% of OMC.

Clay Liner Material Requirement: General Guidelines. The primary purpose of a clay liner is to provide a low-permeability (typically $k \leq 1 \times 10^{-7}$ cm/

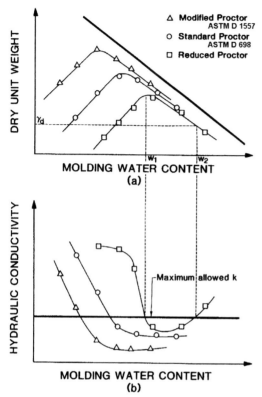

Figure 17.26 *Recommended design procedure:* (a) *determine compaction curves with modified, standard, and reduced proctor compactive effort;* (b) *determine hydraulic conductivity of compacted specimen.* (*From Daniel and Benson 1990. Reproduced by permission of ASCE.*)

s) material layer for a waste containment unit liner system. For an overall system evaluation, such as the slope stability requirement, other liner properties, such as undrained strength of the compacted clay and the interface strength between a synthetic liner component and the soil component, also become important. However, these property requirements may be accommodated by adjusting the waste containment unit's geometry (e.g., height and slope).

Benson et al. (1999) present the analyses of a database consisting of 85 full-scale compacted clay liner's hydraulic conductivity tests. The results indicate that these liners can be constructed with a broad variety of clayey soils, with primary emphasis on compaction on the wet side of the optimal moisture content. A

naturally occurring soil material that has maximum particle size between 25 and 50 mm, percent gravel less than about 30%, fine content (percent passing sieve number 200) greater than about 40%, and plasticity index greater than about 15%, when compacted properly (moisture and density range to be determined by laboratory testing), should generally meet the permeability criteria ($k \leq 1 \times 10^{-7}$ cm/s). These criteria should, however, be used only as a guideline, and detailed testing on project specific soils should be performed prior to selecting a borrow source for a natural soil liner material.

17.2.6 Low-Permeability Soil-Admixed Liners

At sites where the required amount of clay is not available in sufficient quantities, the available on-site higher-permeability soil can be mixed with the available on-site or imported clay. Alternatively, on-site higher-permeability soil should be mixed with commercially imported bentonite. In all cases, an appropriate laboratory and field-testing program should be initiated to determined appropriate proportions of the mix. In the author's experience, about 5 to 10% (by dry weight) sodium bentonite mixed with well-graded soils may achieve $K \leq 1 \times 10^{-7}$ cm/s. For poorly graded (uniformly graded) soils, about 10 to 15% bentonite may be required to achieve similar permeabilities.

Various case histories are available that indicate that low-permeability soil-admixed liners have been designed and constructed successfully. For example, Knitter et al. (1993) cite a case in which loess was admixed with 5% bentonite to achieve $k \leq 1 \times 10^{-7}$ cm/s. Similar permeability was obtained in a case where a mixture of gravel and sand (with about 27% fines) was mixed with 6% sodium bentonite (Sharma and Kozicki, 1988). In another case low k ($\leq 1 \times 10^{-7}$ cm/s) was achieved when a silty sand was mixed with 5% sodium bentonite (Sharma and Hullings, 1993). Further information on soil-admixed material can be found in USEPA (1988a and 1989). Sivapullaiah et al. (2000) present methods for predicting hydraulic conductivity of bentonite–sand mixtures. Authors show that a reasonable correlation can be established be-

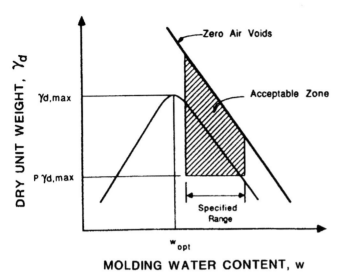

Figure 17.27 Traditional method for specification of acceptable water content and dry unit weight for compacted clay liners. (From Daniel and Benson 1990. Reproduced by permission of ASCE.)

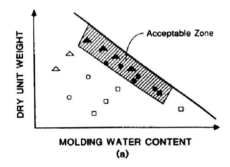

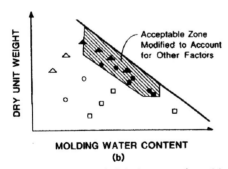

Figure 17.28 Recommended design procedure: (a) replot compaction curves using solid symbols for compacted specimen with hydraulic conductivies maximum allowable value; (b) modify acceptable zone based on other consideration such as shear strength or local construction practice. (From Daniel and Benson, 1990. Reproduced by permission of ASCE.)

tween k, liquid limit, and the void ratio of the mixture. This may need site-specific verification.

Dennis and Turner (1998) present a laboratory study where the hydraulic conductivity of a compacted silty sand was reduced from about 10^{-6} cm/s to about 10^{-8} cm/s by treating the soil. The soil is treated by permeating it with a nutrient solution and creating a biofilm. The durability of this biofilm was tested by permeation with saline, acidic, and basic solutions. This, however, needs field confirmation.

17.3 GEOMEMBRANE LINERS

According to ASTM D 4439, geomembrane liners are very low permeability (also loosely called *impermeable*) membrane liners or barriers used to control fluid migration in engineered projects. The three major types of geomembranes that are in current use are:

1. *Thermoplastic polymers.* This means that the polymer can be melted, and on cooling it nearly reverts to its original structure. Examples of such geomembranes are polyvinyl chloride (PVC), polyethylene (PE) [which are of various densities, e.g., very low density (VLDPE), linear low density (LLDPE), medium density (MDPE), and high density (HDPE)], chlorinated polyethylene

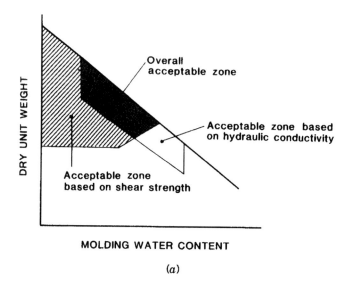

Figure 17.29 Dry unit weight versus molding water content and acceptable zone: (a) use of hydraulic conductivity and shear strength data to define single acceptable zone; (b) acceptable zone based on low hydraulic conductivity, low desiccation-induced shrinkage, and high unconfined compressive strength. [(a) From Daniel and Benson, 1990. Reproduced by permission of ASCE (b) From Daniel and Wu, 1993. Reproduced by permission of ASCE.]

(CPE), chlorosulfonated polyethylene (CSPE), and ethylene interpolymer alloy (EIA).

2. *Thermoset polymers.* These polymers can be processed only once because further processing results in degradation of the material. Examples of these are butyl or isoprene–isobutylene (IIR),

ethylene propylene (EPDM), polychloroprene (neoprene), ethylene propylene terpolymer (EPT), and ethylene vinyl acetate (EVA).

3. *Combination.* As the name suggests, their examples are PVC–nitrile rubber, PE–EPDM, and PVC–ethyl vinyl acetate.

In waste containment applications, thermoplastic (PE) geomembranes are commonly used. In the various examples of thermoplastic geomembranes, mentioned above, additives (which include antioxidant, processing aids, and lubricants), fillers, and/or plasticizers are used with the base resin, which gives the geomembrane its name: for example, HDPE and VLDPE; each have 96 to 97% resin, 0.5 to 1% additive, 2 to 2.5% carbon black and filler, and no plasticizer. In comparison, PVC has 45 to 50% resin, 3 to 5% additive, 10 to 15% carbon black and filler, and 35 to 40% plasticizer. This indicates that each geomembrane type, mentioned here, is actually a compound that a geomembrane manufacturer makes into a final sheet (Koerner, 1993, 1994, 1996, 1998).

Geomembranes are generally manufactured in one of three ways: (1) In the *extrusion process* (used for HDPE and VLDPE), a continuous screw extruder takes the formulated compound from a hopper, melts it under vacuum, and then delivers it through rollers. This delivery can be done in either of two forms; one is in a wide flat-sheet final product form, and the other way is in the form of a circular disk. In the latter case, the circular disk is first carried upward in the form of a large cylinder, then brought over a nip bar. This is then cut, unfolded, and finally rolled into a sheet form; this is called the *blown film process* (Koerner, 1993). These sheets are transported directly to the project site. (2) In the *calendering process* (used for PVC and CPE) the compound is blended in a mixer and then fed into the calender. The calender consists of counterrotating sets of rollers forming the individual plys of the geomembrane. (3) In the *spread coating process* the compound is screeded over a fabric-reinforcing substrate. It can be placed either on one or both sides of the fabric. This is the least widely used manufacturing process.

The two commonly used geomembrane types in waste containment systems are polyethylene (PE) and polyvinyl chloride (PVC). Although PVC have been recommended and used in many waste containment applications because of economics and ease of installation (solvent weld for PVC versus heat welding for PE), PE are most commonly used as base and cover liners in landfills. For surface impoundment liners, PE geomembranes are also widely used, primarily because of their high chemical resistance and durability properties.

17.3.1 Material Properties and Tests

The material properties of geomembranes can be divided into (1) physical properties, (2) mechanical properties, (3) chemical properties, (4) biological properties, and (5) thermal properties (Koerner, 1993). These properties and test methods are presented briefly here. Koerner (1998), Sharma and Lewis (1994), and various ASTM standards provide further details on these test methods. These properties have been grouped here into (1) physical, (2) mechanical, and (3) endurance properties, as discussed below.

Physical Properties. The physical properties include measurement of membrane thickness, density, melt flow index (MI), water vapor transmission (WVT), and solvent vapor transmission (SVI).

Thickness. The thickness of a geomembrane is measured in accordance with ASTM D 5199, the standard test method for measuring nominal thickness of geotextile and geomembranes. The test is performed by using a thickness gauge (micrometer) under a specified pressure of 20 kPa (2.9 lb/in^2) for 5 s. For a textured geomembrane the measurement is done along the smooth edge strips. This test method measures the nominal thickness, not necessarily minimum thickness.

Density and Mass per Unit Area. The density or specific gravity of a geomembrane can be determined by either ASTM D 792 or by ASTM D 1505 (which is used for materials with density less than 1). Since all commercially available HDPE have density less than 1, ASTM D 1505 is a more accurate method of density determination. Mass per unit area (or simply weight) of a representative geomembrane specimen (oz/yd^2 or g/m^2) is generally determined in accordance with the ASTM D 1910 test method.

Melt Flow Index (MI). The melt index test determines the fluidity of the molten geomembrane resin. It is measured by heating the polymer in a furnace until it melts and then pushing it through an orifice under a constant load for 10 min. The test method used is ASTM D 1238, standard test method for flow rates of thermoplastics by extrusion plastometer. All things being equal, a higher MI indicates lower density, and vice versa. Impurities and inconsistencies in the resin will be reflected in MI. Typical MI values for PE range from 0.2 to 1.0 g per 10 min (Sharma and Lewis, 1994).

Water Vapor Transmission (WVT). Geomembranes have very low permeability; a typical indirectly interpreted k range based on WVT tests is 0.5×10^{-10} to 0.5×10^{-13} cm/s (Koerner, 1994). Also, Koerner (1998) shows that based on WVT tests on a 0.75-mm-thick PVC geomembrane, one can estimate its equivalent $k = 0.5 \times 10^{-11}$ cm/s. Testing on thicker geomembranes, particularly HDPE, whose WVT values are very low, will either require a very long testing time (which may have permeant or water evaporation problems) or a very high hydraulic gradient requirement, which may cause material specimen failure. Therefore, conventional geotechnical permeability determination methods are not applicable. The indirect method for geomembrane k is therefore by using WVT test. The applicable test method is ASTM E 96. In this test the geomembrane specimen is sealed over an aluminum cup with water in it. With water in the cup having a 10% relative humidity inside and lower relative humidity outside, a weight loss over time is measured. WVT are then calculated and equivalent k interpreted. Typical WVT for 30-mil PVC is 1.8 g/m$^2 \cdot$ day, and 30-mil HDPE is 0.017 g/m$^2 \cdot$ day. By comparison, for a WVT value of 3.08 g/m$^2 \cdot$ day, Koerner (1994) estimates an equivalent k of 0.78×10^{-11} cm/s.

Solvent Vapor Transmission (SVT). A waste containment unit has liquids other than water. The geomembrane used may need to contain fluids and need to be tested for such fluids. Examples for such fluids would be landfill leachate, methane in landfill cover, and hydrocarbon vapor containment in waste covers. In such situations the molecular size and attraction of liquid vis-à-vis the polymeric liner material might result in very different vapor transmission values than when the tests are done using water. The test is similar to E96 except that the site-specific solvent is used within the cup (Koerner, 1994). Matrecon (1988) provides SVT data for various solvents and geomembranes. For example, SVT for methyl alcohol for 32-mil HDPE is 0.16 g/m$^2 \cdot$ day and for 30-mil LDPE is 0.74 g/m$^2 \cdot$ day; for xylene these two values are 21.6 and 116 g/m$^2 \cdot$ day, respectively (Matrecon, 1988).

Mechanical Properties. The mechanical properties include measurements of tensile behavior, shear and peal mode of seam behavior, tear, impact and puncture resistances, interface shear strength between geomembrane and other materials, and environmental stress cracking behavior. Koerner (1998) provides a detailed table that provides information on some of the recommended mechanical test methods for geomembrane and geomembrane seams. In summary, some of these tests are:

Geomembrane Type	Test Type	ASTM Standard
PVC		
Membranes	Tensile	D 882
Seams	Shear	D 3083
Seams	Peal	D 413
VLDP		
Membranes	Tensile	D 638
Seams	Shear	D 4437
Seams	Peal	D 4437
HDPE		
Membranes	Tensile	D 638
Seams	Shear	D 4437
Seams	Peal	D 4437

Tensile Tests. There are many types of tensile tests. The three types commonly used are:

1. Tensile tests performed on small narrow-width samples and used routinely for quality control and quality assurance of geomembrane. These tests are therefore called *index tests*. The commonly used ASTM methods for these tests are

D 638, D 882, and D 751. Specimen shapes for D 638 tests are dumbbell, for D 882 are strip, and for D 751 are grab; corresponding specimen widths are 0.25, 1, and 4 in. and lengths are 4.5, 6, and 6 in., respectively. The major issue with these types of tests are their contraction within the central part of a specimen representing one-dimensional behavior, thus not representative of field conditions. A second type of test that would represent a plane strain condition is therefore recommended.

2. *Uniform, wide width,* or *strip tests* may be performed on 8-in. (200-mm)-wide strips in accordance with ASTM D 4885. These tests take a long time (e.g., at a strain rate of 1.00 mm/min it takes about 3 h for 200% strain and about 17 h for 1000% strain), therefore are not used for quality control. However, they are useful as performance and are design-oriented tests. In certain types of field situations, such as landfill covers, the waste settlement causes out-of-plane movements of the cover geomembrane. None of the tests mentioned above would simulate such conditions and requires a third type of test.

3. Multiaxial or axisymmetric tensile tests are performed to simulate geomembrane's tensile behavior under biaxial or out-of-plane stress or deformation conditions. Although at the present time there is no ASTM test procedure for multiaxial tests, several methods have been reported in the literature (Koerner et al., 1990; Hoekstra, 1991; Norbert, 1993); the Geosynthetic Research Institute (GRI) test method GM4 is one such test. Most test methods use a circular test specimen bolted to a circular chamber with pressure exerted on the specimen by air or water. The stress–strain curve for the specimen can be obtained by measuring pressure and volume changes of the deformed specimen.

Figure 17.30(*a*), (*b*), and (*c*) show stress–strain plots for some geomembranes under index tensile tests, wide-width specimen tensile tests, and axisymmetric tensile tests, respectively. These results indicate different quantitative (stress–strain) values for different test types. It is therefore important to perform appropriate tests to simulate field conditions, as discussed above. A typical range of tensile strength values, in the machine direction, reported by manufacturers for 60-mil HDPE are from 225 to 245 lb/in., for 80-mil HDPE are from 280 to 325 lb/in., for 40-mil VLDPE are from 125 to 140 lb/in., and for 40-mil PVC are from 90 to 95 lb/in. These tests were performed in accordance with ASTM D 638 for PE and ASTM D 882 for PVC (Sharma and Lewis, 1994).

Tear Resistance. Tear resistance of geomembrane can become important when it may have a potential for tear, due to high winds or handling stresses during panel placement. There are no specific required tear resistance values for geomembranes; obviously, the higher the tear resistance, the better it is. Therefore, tear resistance tests can be considered as index tests. There are many ASTM methods for the measurement of tear resistance of geomembranes (e.g., ASTM D 2263, D 1004, D 751, D 1424). For thermoplastic geomembranes, manufacturers have most commonly reported tear resistance using ASTM D 1004. For example, tear resistance by this method for 60-mil HDPE is from 40 to 45 lb, for 80-mil HDPE is from 50 to 60 lb, for 40-mil VLDPE is from 15 to 20 lb, and for 40-mil PVC is from 10 to 12 lb (Sharma and Lewis, 1994). As the geomembrane gets thicker, its tear resistance increases and installation tearing may not be an issue.

ASTM 1004 test method uses a template to form a 90° notch in a test specimen as shown in Figure 17.31. Both ends of the test specimen are then gripped in a tension machine, and a tensile load is applied until the specimen tears, as indicated by extension of the notch across the specimen. The tear resistance is the maximum load during the test.

Impact Resistance. This test provides an assessment of geomembranes resistance to impact from falling objects. These impacts may either tear the geomembrane and cause leaks or weaken the membrane, which may initiate leaks at a later stage. Various ASTM methods are available for such assessments. For example, ASTM D 1709 uses a free-falling dart, ASTM D 3029

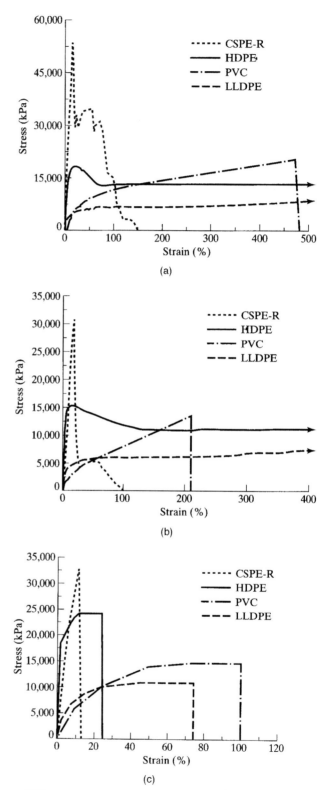

Figure 17.30 Stress–strain results for geomembranes from various tests. (From Koerner, © 1998. Reprinted by permission of Pearson Education, Inc., Upper Saddle River, NJ.)

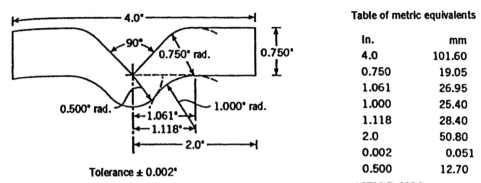

Figure 17.31 *Specimen dimensions for tear resistance test (ASTM D 1004).*

uses a falling weight, and ASTM D 746, D 1822, and D 3998 use pendulum-type impacts for testing. In general, impact tests indicate that thicker geomembranes have higher impact resistance, scrim reinforcement does not provide any significant advantage over non-reinforced geomembrane, and geotextiles underlying and/or overlying the geomembrane have a significantly higher impact resistance. Also, test results indicate that when geotextile is placed on both sides of the geomembrane, its impact resistance is improved significantly over the situation when geotextile was placed on only one side of the geomembrane (Koerner, 1994; Koerner et al., 1986).

Puncture Resistance. Evaluation of puncture resistance of geomembrane is important because when placed on or below rough surfaces, stones or other sharp objects may puncture the membrane, causing leaks. Detection and repair for leaks may be very difficult and costly. A representative test method should simulate actual subgrade conditions (e.g., density, grain sizes, shapes) and the loading conditions that geomembrane may be subjected to in the system. Various test methods are available in literature that evaluate the puncture resistance of geomembranes (Hulling and Koerner, 1991; Narejo et al., 1993). GRI Test Method GM3, the Federal Test Method Standard 101C, Method 2031, ASTM D 4833 and D 5494). All these methods are basically index tests.

ASTM D 4833 is a standard test method for the index puncture resistance of geotextiles, geomem-

branes, and related products. In this method a geomembrane sample is clamped over a 1.75-in. (45-mm)-diameter ring. A $\frac{5}{16}$-in. (8-mm)-diameter rod with a 45° chamfer is then punched through the membrane at a 12-in. (300-mm)/min speed. The maximum load measured on the test machine is the puncture resistance. ASTM D 5494 is a standard test method for determination of pyramid puncture resistance of unprotected and protected geomembranes.

Both ASTM methods mentioned above are index tests and do indicate the comparative puncture resistances of geomembranes under different conditions. For example, reinforced geomembranes have higher puncture resistance than nonreinforced geomembranes. Geomembrane resistance to puncture increases with thickness [e.g., tests indicate that 80-mil HDPE has a puncture resistance of about 200 lb (900 N), in comparison to 60-mil HDPE, which has a puncture resistance of about 155 lb (700 N)]. The 80-mil HDPE's puncture resistance increased from 200 lb (900 N) to about 255 lb (1150 N) when a 12-oz/yd² (400-g/m²) needle-punched nonwoven geotextile was placed on top (in front) of the geomembrane. This resistance increased to about 355 lb (1600 N) when the geotextile was placed on both sides of the geomembrane (Koerner, 1994; Koerner et al., 1986).

Although there is a need to develop a test that would represent field condition, the index test results discussed above indicate the importance of using thick, reinforced geomembrane protected by geotextile against potential leaks caused by membrane puncture.

Environmental Stress Cracking. Environmental stress cracking (ESC) in a geomembrane can be defined as a rupture or crack (internal or external) caused by an applied tensile stress that is less than its (short-term) tensile strength. Although ESC can occur in PE, PVC, or polyester, it is more likely to occur in higher-density (hence crystallinity) PEs. ESC usually occurs at seams where scratches may be caused by grinding. Also, most ESC have been reported in surface impoundments, where membranes are exposed to atmosphere, and tensile stresses are mobilized due to temperature changes.

ESC resistance of a geomembrane can be increased by blending and manufacturing processes so as to have appropriate polymeric properties, such as molecular weight, orientation, and distribution. Two ASTM test procedures D 1693 and D 5397, are available to measure ESC resistance; of these two test methods, ASTM D 5397 [standard test method for evaluation of stress crack resistance of polyolefin geomembranes using notched constant tensile load (NCTL)] is appropriate to measure the ESC resistance of HDPE. This test method uses a dumbbell-shaped notched test specimen to a constant tensile load at various percentages of their yield stress. The testing is done in the presence of a surface wetting agent (usually, Igepal) and at an elevated temperature of 122°F (50°C). Test results are plotted as a percentage yield stress versus failure time; Figure 17.32 presents three possible types of curves from the NCTL tests. In this figure (T_t) shows the transition time and (σ_t) is the transition stress, which indicates the onset of the brittle region and signifies the transition of material into a slow crack growth regime (Hsuan et al., 1993a,b). Typical T_t values range from 10 to 5000 h; this is indicated in Figure 17.33, where percent yield stress and failure time (transition time, T_t) are plotted from NCTL test for 18 virgin geomembranes and seven field-retrieved geomembranes. The current recommendation for T_t for an acceptable HDPE geomembrane is 100 h (Hsuan et al., 1993a).

Endurance Properties. There are various phenomena, such as ultraviolet, radioactive, biological, chemical, thermal, and oxidation degradations, that may cause polymeric bond breaking within the polymer

structure of a geomembrane. These may influence geomembrane's long-term performance. As discussed in the following paragraphs, each situation may affect specific behavioral change in the geomembrane. The general trend is to render the geomembrane brittle in its stress–strain behavior over time, and therefore, as discussed previously, physical and mechanical properties could be used to assess the impact of polymeric degradation (Koerner, 1994).

Ultraviolet Resistance. The ultraviolet (UV; from sunlight) degradation of geosynthetics is caused by the shorter wavelength energy from ultraviolet (the UV-B wavelength between 315 and 280 nm), resulting in severe polymer damage. This damage can be minimized by adding carbon black or pigments to geomembrane (actually, to other geosynthetics as well, such as geotextiles). This blocks or screens the UV and provides UV resistance to the geomembrane. For permanently exposed geomembranes (e.g., used in surface impoundments and floating covers for reservoirs), manufacturers' warranties should be obtained. Other options are covering the exposed geomembranes with cover soil or providing sacrificial geotextile that is periodically replaced. If carbon black content in HDPE is about 2 to 3% and in PVC is about 10 to 15%, it is generally considered that the geomembrane has acceptable UV resistance. ASTM D 1603 provides a standard test method for carbon black content, and ASTM D 3015 provides a standard practice for microscopical examination of pigment distribution in plastic compounds.

Radioactive Resistance. Very little information (other than internal reports by various organizations) is available on the effects of radioactive material on geomembranes. It is generally recognized that high-level radioactive waste will cause polymer degradation of geomembrane. However, low-level radioactive waste can be disposed of in an HDPE-lined containment system (Kane and Widmayer, 1989).

Chemical Resistance. Geomembranes in waste containment systems are used in a liner system so that contaminants are contained within the system. There-

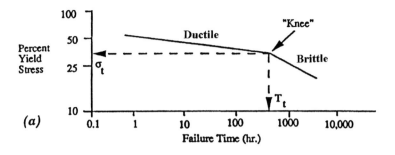

(a)

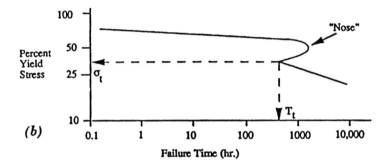

(b)

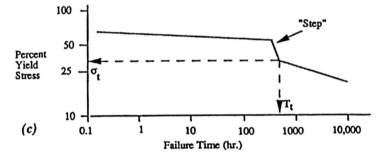

(c)

Figure 17.32 Typical response curves resulting from complete NCTL test: (a) *bilinear (or knee) response curve;* (b) *overshoot (or nose) response curve; and* (c) *trilinear (or step) response curve. (From Hsuan, et al., 1993a. Reproduced by permission of ASCE.)*

fore, it is important that geomembranes' chemical resistance to the contained chemicals be evaluated. Table 17.4 presents general guidelines for chemical resistance of various geomembranes to chemicals in a single-component material. This does not provide information on synergistic effects. Therefore, tests on site-specific leachate should be performed (Rigo and Cazzuffi, 1991). This table can be used as a guideline. However, there are other problems in using it, such as for gemembranes that are modified at the seams, seams with material different from the geomembranes, test conditions different from those used for this table (i.e.,

extreme temperatures), and geomembrane made of a composite of different materials. It is therefore important to perform chemical resistance tests.

ASTM D 5322 (standard practice for immersion procedures for evaluating the chemical resistance of geosynthetics to liquids) and EPA test method 9090 (compatibility tests for wastes and membrane liners) are commonly used to determine the chemical resistance of geomembranes. The test method, in general, consists of first determining physical and mechanical properties of the geomembrane and then incubating in an immersion tank in a site-specific chemical at room

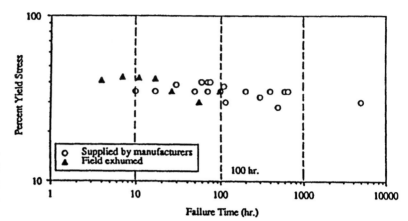

Figure 17.33 Transition points of virgin geomembranes and field geomembranes (From Hsuan 1993a. Reproduced by permission of ASCE.)

temperature (22°C) and at elevated temperatures (50°C) for different time periods (e.g., 30, 60, 90, and 120 days). At the end of each time period samples are taken out and their changes in properties are evaluated. The incubation is done in a closed tank to prevent loss of volatiles in the chemicals (Figure 17.34). Based on the evaluations by Little (1985), O'Toole (1986), and Koerner (1994), it can be recommended that PVC, LLDPE, VLDPE, and CSPE will be resistant to specific chemicals if the tests indicate change in weight less than 10% change in tensile strength less than 20%, and change in elongation at break less than 30%. For HDPE to be resistant, the change in weight should be less than 2%, change in yield strength should be less than 20%, change in yield elongation should be less than 30%, change in tear strength should be less than 20%, and change in puncture strength should be less than 20%.

Thermal Resistance. Hot and cold temperatures may affect the physical and mechanical properties of geo-membranes. ASTM D 794 and D 1870 are recommended to evaluate the effect of heat on plastics. Cold temperatures generally result in decreasing flexibility and making seams more difficult; this would create installation problems.

Oxidation Resistance. Oxygen combines with the free radical (created on a carbon atom in polyethylene chain) "to form a hydroperoxy radical, which is passed

around within the molecular structure" (Koerner, 1994). This results in "chain scission" and degradation if a geomembrane is exposed or covered with unsaturated soils for a long period of time. If oxygen is removed from the geomembrane surface, oxidation degradation should not occur. Thus membranes covered with waste and liquids should not experience this problem. Usually, manufacturers use proprietary antioxidation additives and stabilizers to make geomembranes oxidation resistant. Hsuan and Koerner (1998) present the results of an ongoing study to evaluate the antioxidant depletion lifetime in HDPE geomembranes. The results from this study published in 1998 indicate that the lifetime of antioxidants at 20°C is more than about 200 years.

Biological Resistance. Geomembranes are used in the field under conditions that raise questions about their resistance to borrowing animals, fungi, and bacteria. Detailed test and evaluations for biological resistance are not available. The general thinking is that high-molecular-weight polymer geomembranes may have good biological resistance. Further research is needed in this area.

17.3.2 Geomembrane Seams and Their Properties

Geomembranes are manufactured in various dimensions. However, in order to install several acres of geo-

Table 17.4 General chemical resistance guidelines of commonly used geomembranes[a]

Chemical	Butyl Rubber 100 (°F)	Butyl Rubber 158 (°F)	Chlorinated Polyethylene 100 (°F)	Chlorinated Polyethylene 158 (°F)	Chlorosulfonated Polyethylene 100 (°F)	Chlorosulfonated Polyethylene 158 (°F)	Elasticized Polyolefin 100 (°F)	Elasticized Polyolefin 158 (°F)	Epichlorohydrin Rubber 100 (°F)	Epichlorohydrin Rubber 158 (°F)	Ethylene Propylene Diene Monometer 100 (°F)	Ethylene Propylene Diene Monometer 158 (°F)	Polychloroprene (Neoprene) 100 (°F)	Polychloroprene (Neoprene) 158 (°F)	Polyethylene 100 (°F)	Polyethylene 158 (°F)	Polyvinyl Chloride 100 (°F)	Polyvinyl Chloride 158 (°F)
General																		
Aliphatic Hydrocarbons			X	X			X		X	X			X	X	X	X		
Aromatic Hydrocarbons							X		X	X			X	X	X	X		
Chlorinated solvents	X	X					X		X	X	X		X		X	X		
Oxygenated Solvents	X	X					X		X		X	X	X	X	X	X		
Crude Petroleum Solvents			X	X			X		X	X			X	X	X	X		
Alcohols	X	X	X	X	X		X		X	X	X	X	X	X	X	X	X	X
Acids																		
Organic	X	X	X	X	X		X		X		X	X	X	X	X	X	X	X
Inorganic	X	X	X	X	X		X		X		X	X	X	X	X	X	X	X
Bases																		
Organic	X	X	X	X	X		X		X	X	X	X	X	X	X	X	X	X
Inorganic	X	X	X	X	X		X		X	X	X	X	X	X	X	X	X	X
Heavy metals	X	X	X	X	X		X	X	X	X	X	X	X	X	X	X	X	X
Salts	X	X	X	X	X		X	X	X	X	X	X	X	X	X	X	X	X

Source: Vandervoort.

[a] X, Generally good resistance. The definition of suitable performance may vary significantly depending on the application and desired life expectancy. Seemingly similar materials may perform quite differently in the same exposure. This comparison is provided for general information purposes only and is not intended as a guide for selecting specific materials for specific applications.

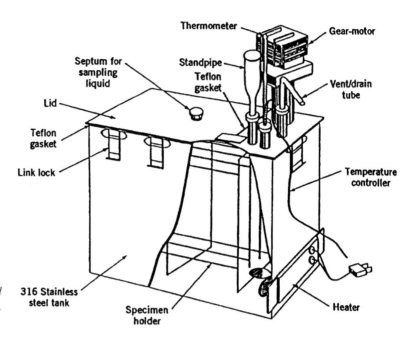

Figure 17.34 EPA test method 9090 immersion tank. (*USEPA, 1988b.*)

membranes at the site, field seams are required. PVC geomembranes are seamed using heat or solvent welds, while PE geomembranes are typically heat welded. Table 17.5 presents field-seaming techniques for HDPE geomembranes; of these the two most commonly used seaming methods for PE geomembranes are:

1. *Fillet extrusion weld.* This is formed by extruding a bead of molten parent material (as shown in Table 17.5) by a hand-operated extrusion welder. Since this is labor intensive, it is generally used in small curved areas and for small repair work.

2. *Double hot-wedge weld.* This is formed "by melting the surface of two PE geomembrane panel edges, and then running a hot metal wedge between the two surfaces" (Sharma and Lewis, 1994). A homogeneous bond is then formed by applying pressure with a roller, as shown in Figure 17.35; a typical cross section through a double wedge weld is shown in Figure 17.36.

Seam strengths can be measured by (1) destructive testing and (2) nondestructive testing. Two types of

destructive (peel and shear tests) *testing* are common. They are described in ASTM D 4437 (standard practice for determining the integrity of field seams used in joining flexible polymeric sheet geomembranes) and are performed using a tensile testing machine. Figure 17.37 is a schematic of specimens in shear and peel tests. *Nondestructive tests* do not determine seam strengths but are used to detect if any holes (leaks) exist in the seams. The two commonly nondestructive tests are vacuum and pressure tests. In a vacuum test a soapy solution is placed over a seam, a vacuum box is placed over the seam, and a vacuum of approximately 5 psi is applied. If a stream of soap bubbles is detected, it indicates a leak and must be repaired. The *pressure test method,* performed only on double-wedge weld seams, consists of sealing both ends of an unobstructed double-wedge weld and then applying approximately 30 psi air pressure for approximately 5 minutes. A reduction in air pressure greater than 2 psi indicates a leak in the seam and must be repaired.

The generally required seam strengths, based on shear tests, for HDPE are typically 90% of the parent geomembrane yield strength. For PVC 80% of the

Table 17.5 **Field-seaming techniques for HDPE geomembranes**

Method	Seam Configuration	Typical Rate	Comments
Fillet extrusion		100 ft/h	Upper and lower sheets must be ground. Upper sheet must be beveled. Height and location are hand-controlled. Can be rod or pellet fed. Extrudate must use same polymer compound. Air heater can preheat sheet. Used routinely for difficult details.
Flat extrusion		50 ft/h	Good on long flat surfaces. Highly automated machine. Difficult for side slopes. Cannot be used for close details. Extrudate must use same polymer compound. Air heater can preheat sheet.
Hot air		50 ft/h	Good to tack sheets together. Handheld and automated devices. Air temperature fluctuates greatly. No extrudate added.
Hot wedge		300 ft/h	Single and double tracks available. Double track patented. Built-in nondestructive test. Cannot be used for close details. Highly automated machine. No extrudate added. Controlled pressure for squeeze-out.
Ultrasonic		300 ft/h	New technique for geomembranes. Sparse experience in the field. Capable of full automation.
Electric welding		Unknown	New technique for geomembranes. Still in development stage. Extrudate must use same polymer compound. Wires provide possibility of doing spark test.

Source: Koerner and Bove (1989).

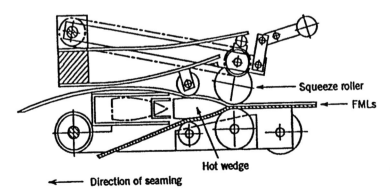

Squeeze roller
FMLs
Hot wedge
Direction of seaming

Figure 17.35 Typical hot wedge welding apparatus. (From USEPA, 1988b.)

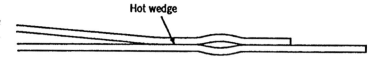

Figure 17.36 *Typical cross section through double-wedge weld. (From Cadwallader and Burkinshaw, 1991.)*

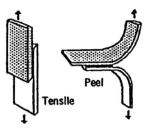

Figure 17.37 *Schematic of specimens used in peel and shear tests. (From Rollin et al., 1991.)*

parent geomembrane at break strength is acceptable (Sharma and Lewis, 1994). The recommended seam strengths, based on peel tests, are that peel strength should be greater than 60% of the parent geomembrane yield strength and less than 10% separation at the end of the test (Giroud and Peggs, 1990; Carlson et al., 1993).

17.4   GEOTEXTILES

Geotextile is defined as a permeable geosynthetic comprised solely of textiles (ASTM D 4439). The majority of geotextiles are made from polyester or polypropylene polymers; some geotextiles are composed of polyethylene or polyamide. The various types of fibers used in the manufacture of geotextiles are (1) monofilament fibers, created by extruding molten polymer through an apparatus containing several small-diameter holes, called *spinnaret,* and then stretched and cooled; (2) staple fibers, also manufactured by extruding through spinnaret, with strings twisted together and cut into 1- to 4-in. lengths; and (3) silt-film fibers, manufactured by extruding a continuous sheet of polymer and cutting it into fibers.

The fibers, discussed above, can then be made into either woven or nonwoven geotextiles. Woven geotextiles are produced using traditional weaving methods,

while nonwoven geotextiles are made by needle punching or melt bonding or resin bonding the fibers together.

In waste containment facilities, geotextiles are typically used for the following functions:

1. Woven geotextiles are generally used for reinforcement (i.e., to increase total system tensile strength).
2. Nonwoven geotextiles are generally used for:
   a. *Separation* between two dissimilar materials (e.g., coarse-grained aggregate and fine-grained soils)
   b. *Filtration* when fluid flows between two materials that have significantly different particle sizes (e.g., between granular leachate collection layer and the mixed-grained operation layer)
   c. *Drainage* that allows for adequate liquid flow within the plane of geotextile
   d. *Cushion* to protect geosysnthetic memebrane from overlying large angular rocks (e.g., between HDPE and leachate collection layer in landfills)

17.4.1   Material Properties and Tests

The geotextile material properties can be generally categorized into (1) physical properties, (2) mechanical properties, (3) hydraulic properties, (4) endurance properties, and (5) degradation properties (Koerner, 1994, 1998). These are presented briefly here; for further details, reference should be made to Sharma and Lewis (1994), Koerner (1998), and various ASTM standards, as noted in the following paragraphs.

Physical Properties. The physical properties, sometimes called *index properties,* are for the final manufactured geotextile product. These include properties, such as specific gravity, thickness, mass per unit area, and stiffness. Among these the most commonly specified properties in practice are the thickness and mass per unit area.

Thickness. The thickness is measured according to ASTM D 5199 (standard method for measuring nominal thickness of geotextile and geomembrane), which specifies that the thickness is the measurement between the upper and lower fabric surfaces measured under a pressure of 42 lb/ft² (2 kPa). Thickness is not used directly in the design but is related to material properties, especially for strength and permeability, as discussed in the following paragraphs.

Mass per Unit Area. This property as measured in accordance with ASTM D 5261 and is determined by weighing test specimens from various locations of the fabric sample of known length, width, and thickness under zero tension and is measured within an accuracy of 0.01 g. The mass is specified in oz/yd² or g/m², with typical values ranging between 4 and 20 oz/yd².

Although there is no direct correlation between mass per unit area and various mechanical properties of the fabric, it is qualitatively considered that heavier geotextiles have better mechanical properties.

Mechanical Properties. The various mechanical properties of geotextile discussed here are uniaxial tensile strength, multiaxial tensile or burst strength, tear strength, and puncture resistance.

Uniaxial Tensile Strength. Tensile strength is probably the most important property of a geotextile. As shown in Figure 17.38, the test is performed by holding the ends of specimen with clamps, applying the tensile load and measuring the elongation; typical load per unit width–percent elongation plot is shown in Figure 17.39. From this curve, values such as maximum tensile stress (also called *geotextile strength*) and strain at failure given as elongation and modulus of elasticity are obtained. Figure 17.40 shows typical tensile responses of various geotextiles manufactured by different processes.

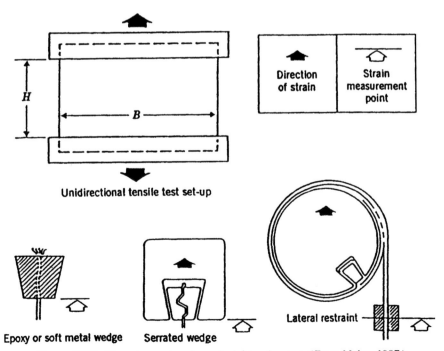

Figure 17.38 *Clamping system for uniaxial tension test.* (*From Myles, 1987.*)

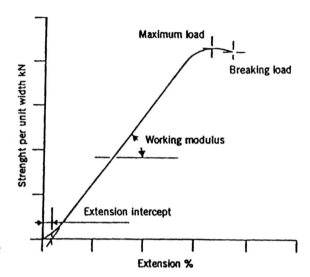

Figure 17.39 Strength per unit width versus extension curve for uniaxial tension test. (From Myles, 1987.)

Various factors that affect the tensile strength of geotextile include specimen size, aspect ratio (width-to-length ratio), clamping condition, fabric type, and strain rates. Because of the influence of these factors on tensile strength, ASTM standards D 4632 (test method for breaking load and elongation of geotextiles or grab method), D 751 (narrow strip test), and D 4595 (tensile properties by the wide-width strip method) have been developed. These methods are presented in Figure 17.41, while the grab tensile test (ASTM D 4632) will continue to be used as an index test by manufacturers and users as a quality indicator; it, however, has the problem of providing artificially high value. Wide-width tests (ASTM D 4595) are therefore generally used for design because as mentioned, narrow specimens tend to give artificially high strength values, due to severe Poisson's ratio effect.

The tests mentioned above are performed without the use of soil adjacent to them. This is unlike the actual field application, where geotextile will by supported or confined by soils. It is generally recognized that although rarely performed, confined tests yield higher strength than tests without soil confinement (Chang et al., 1993).

Multiaxial Tensile or Burst Strength. Burst tests are used to evaluate the effect of three-dimensional (out-of-plane) loading, such as wheel loads, landfill settlements, or in a puncture situation, on geotextiles. These field conditions may stress the geotextile until failure. Two test methods are available to simulate these conditions:

1. *ASTM D 3786* (test method for hydraulic bursting strength of knitted goods and nonwoven fabrics—diaphragm bursting strength tester method). In this method a 5-in.-diameter geotextile is tested, by applying pressure through an expandable diaphragm, until bursting of the geotextile occurs. This method is discussed further by Sharma and Lewis (1994).

2. *Diaphragm test.* In this a large rectangular test specimen is deformed by an underlying rubber membrane and the pressure versus strain response is monitored. This is a difficult test to set up. Generally, ASTM test results are reported by manufacturers and have limited value other than as a quality control test (Sharma and Lewis, 1994).

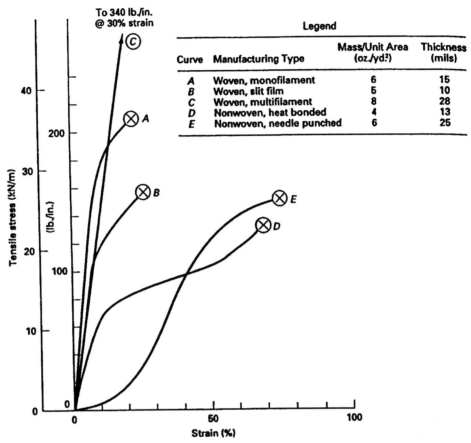

Figure 17.40 *Tensile test response of various goetextiles manufactured by different processes. All are polypropylene fabrics; test specimens were initially 8 in. (200 mm) wide by 4 in. (100 mm) high. (From Koerner, © 1998. Reprinted by permission of Pearson Education, Inc., Upper Saddle River, NJ.)*

Tear Strength. Geotextiles may be subject to tearing stresses during installation. The tear strength tests are therefore intended to measure the ability of a geotextile to resist tear once it is initiated. The three commonly used tear tests are (1) the tongue tear test (ASTM D 751), (2) the Elmendorf tear test (ASTM D 1424), and (3) the trapezoidal tear test (ASTM D 4533). The most commonly used test is the trapezoidal tear test. In this test an isosceles trapezoid is marked on a rectangular specimen and a 0.6-in. (15-mm) cut is made to start the test. The specimen is tested in a tensile testing machine by pulling at a constant rate, forcing

a tear. The trapezoidal tear strength is the maximum recorded force causing tearing of the specimen.

Puncture Resistance. In situations where a geotextile is placed in contact with granular angular materials, such as a drainage material, its resistance to puncture needs to be evaluated. The commonly used tests to measure this resistance are (1) ASTM D 4833 (test method for index puncture resistance of geotextiles, geomembranes, and related products) and (2) the CBR puncture method. Both these methods are not representative of field conditions since a flat-ended probe is

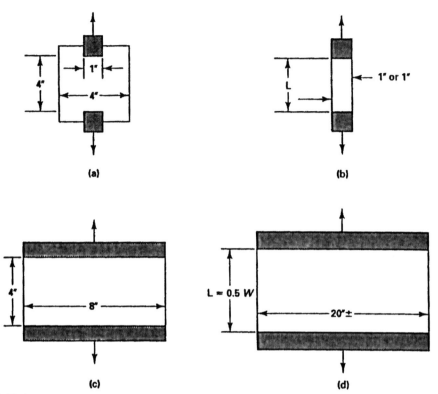

Figure 17.41 *Various tensile test specimen sizes used to obtain fabric strength properties:* (a) *ASTM D 4632 grab;* (b) *ASTM D 751 narrow strip;* (c) *ASTM D 4595 wide width;* (d) *very wide width.* (*From Koerner,* © *1998. Reprinted by permission of Pearson Education, Inc., Upper Saddle River, NJ.*)

used instead of a sharp angular shape. Also, these tests are performed without the use of supporting soil, as is the case in the field. These tests have limited design value and are used as index tests and can provide comparative puncture resistance values of several geotextiles.

Hydraulic Properties. When geotextiles are used for filtration and drainage, their properties, such as apparent opening size (a measure of void sizes), permittivity (cross-plane permeability) and transmissivity (in-plane permeability) need to be evaluated.

Apparent Opening Size (AOS). AOS or O_{95} is defined: "for a geotextile, a property which indicates the approximate largest particle that would effectively pass

through the geotextile" ASTM (2003). In this method, according to ASTM D 4751, the procedure consists of (1) placing a geotextile specimen in a sieve frame, (2) placing sized glass beads on the geotextile surface, (3) shaking the geotextile and frame so that beads pass through the specimen, and (4) repeating the procedure on the same specimen with glass beads of various sizes. The AOS is then assigned to the bead size for which 5% of beads (by weight) pass the geotextile specimen. AOS is synonymous terminology with equivalent opening size (EOS). ASTM D 4751 provides further test details.

AOS values are useful in geotextile applications when it is used as a medium to retain soil particles between it and the adjacent soil. This property becomes important in designing geotextile applications for filtration and drainage. In Table 17.6, a typical range of AOS values for selected nonwoven geotextiles reported by manufacturers is presented.

Permittivity (Ψ). It is the measure of the cross-plane (or perpendicular) flow of water through a geotextile that is used for filtration. If can be defined as follows:

$$\Psi = \frac{k_n}{t} \qquad (17.11)$$

where k_n is the cross-plane hydraulic conductivity of a geotextile having thickness t. Table 17.6 gives a typical range of permittivity values for some nonwoven geotextiles.

ASTM D 4491 (test methods for water permeability of geotextile by permittivity) provides procedures for determining water permeability of geotextiles in terms of permittivity under standard testing conditions. The following two test methods are described in this procedure:

1. *Constant-head test.* In this the quantity of water flow (Q) is measured versus time (t) when a head of 2 in. (50 mm) is maintained on the geotextile. The permittivity (Ψ) is then expressed as follows:

$$\Psi = \frac{QR_t}{hAt} \qquad (17.12)$$

where $R_t = U_t/U_{20^{\circ}C}$ is the temperature correction factor (U_t the viscosity of water at temperature t and U_{20} the viscosity of water at 20°C), A the cross-sectional area of test specimen, and h the head of water on the specimen.

2. *Falling-head test.* In this method a column of water is allowed to flow through the geotextile of cross-sectional area (A) and reading of water head change from initial head (h_0) to final head (h_1) during a time period (t) is measured in a standpipe of cross-sectional area (a). The permittivity (Ψ) can be expressed by the relationship

$$\Psi = \frac{aR_t}{At} \ln \frac{h_0}{h_1} \qquad (17.13)$$

Figure 17.42 shows the permeability apparatus.

If the permeating fluid is other than water (e.g., leachate), the following relationship can be used to account for fluid density (ρ_f) and fluid viscosity (μ_f) (Koerner, 1994).

$$\Psi_f = \Psi_w \frac{\rho_f \mu_w}{\rho_w \mu_f} \qquad (17.14)$$

The subscript w is for water and f is for fluid. The permittivity of geotextile above is measured with compressing normal loads. Test method ASTM D 5493 can be used to evaluate the effect of normal load. Based

Table 17.6 Typical range of selected nonwoven geotextile properties[a]

| | Textile Weight (oz/yd²) | | | | |
Material Property	4	6	8	12	16
AOS (U.S. Standard Sieve)	50–140	70–140	70–140	80–200	100–200
Puncture resistance (lb)	40–70	25–100	95–145	90–210	160–300
Mullen burst strength (psi)	140–260	210–350	360–450	250–700	500–900
Trapezoidal tear (lb)	35–60	53–90	75–110	90–145	110–220
Grab tensile (lb)	90–130	100–225	200–225	265–390	340–500
Wide-width tensile (lb/in.)	50–62	70–83	90–98	130–184	150–206
Permittivity (s⁻¹)	0.7–2.3	0.1–2.0	0.9–1.9	0.05–1.2	0.5–1.1

[a] Values represent the typical range of values reported by manufacturers, based on values published in the *Geotechnical Fabrics Report*, 1992. These values are updated yearly. Readers are encouraged to use the most recent report.

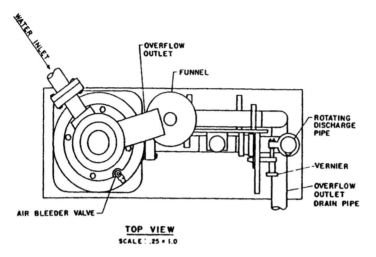

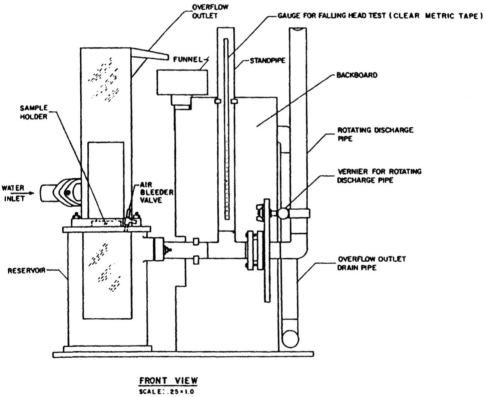

Figure 17.42 Constant- and falling-head permeability apparatus (ASTM D 4491.)

on tests, it can generally be concluded that Ψ will change only slightly under normal loads for woven geotextiles but may decrease for nonwoven geotextile, depending on the normal load. Specific tests should be performed if this is a design concern. The permittivity tests above do not consider the field situation where the presence of overlying soil may influence Ψ due to clogging. For measuring the clogging potential of a soil–geotextile system, a gradient ratio test in accordance with ASTM D 5101 may be performed.

The gradient ratio test (ASTM D 5101) is "applicable for determining the soil–geotextile system permeability and clogging behavior under unidirectional flow condition." As shown in Figures 17.43 and 17.44, the test setup consists of a cylindrical permeameter with a geotextile and soil. The water is passed through the system by applying various differential heads. The *gradient ratio* is defined as the ratio of the hydraulic gradient through a soil–geotextile system to the hydraulic gradient through the soil alone. ASTM D 5101 provides further details on the test and the applicability and acceptability criteria.

Transmissivity (In-Plane Permeability). Transmissivity (θ) is a measure of the flow of water within the plane of geotextile and is given by the expression

$$\theta = k_p t = \frac{QL}{WH} \qquad (17.15)$$

where k_p is the in-plane hydraulic conductivity, t the thickness of the geotextile, Q the average quantity of fluid discharged per unit time, L the length of specimen, W the width of specimen, and H or Δh the difference in total head across the specimen.

ASTM D 4716 [method for constant-head hydraulic transmissivity (in-plane flow) of geotextiles and geotextile-related products] is recommended to measure θ, and Figure 17.45 shows the testing device. Typical values for θ range between 1.2×10^{-8} and 3×10^{-8} m^2/s for woven geotextile and are about 2×10^{-6} m^2/s for nonwoven (needle punched) geotextile and about 3×10^{-9} m^2/s for nonwoven (heat-bonded) geotextile

(Gary and Raymond, 1983). As shown qualitatively in Figure 17.46, θ depends on confining normal stresses and hydraulic gradient. Therefore, tests should be performed to represent field conditions. Sharma and Lewis (1994) discuss this issue further.

Endurance Properties. These properties attempt to evaluate long-term performance of geotextiles. However, due to the lack of a confirmation test on actual long-term evaluation of samples recovered from field installation (especially from waste containment structures), the results generally provide qualitative trends. For example, to minimize installation damage, "no geotextile less than 8 oz/yd^2 should be used unless special precautions are taken to avoid installation damage" Koerner (1994). Koerner also presents various endurance tests, such as creep tests, abrasion tests, and clogging tests. Since these are generally long-term tests and need further correlation with field performance, these tests are generally not used in practice at this time.

Degradation Properties. Long-term properties of geotextiles are often questioned due to various degradation processes (e.g., temperature degradation, oxidation degradation, hydrolysis degradation, chemical degradation, biological degradation, ultraviolet degradation) along with general polymeric aging. Evaluation of all these issues is time consuming and expensive. Koerner (1994, 1998) discusses these factors in detail. We discuss here only the ultraviolet (sunlight) degradation because it is an important cause of degradation to all geosynthetic polymers.

Sunlight Ultraviolet Resistance. As discussed in Section 17.3.1 for geomembranes, UV-B from solar (rays) energy causes polymer damage, and therefore when exposed to sunlight, long-term properties of geotextiles are often threatened under sunlight exposure. ASTM D 4355 (test method for deterioration of geotextiles from exposure to ultraviolet light and water (xenon-arc-type apparatus)) is widely used for geotextile evaluation. Most manufacturers will provide information on their geotextiles' ultraviolet resistance. Koerner

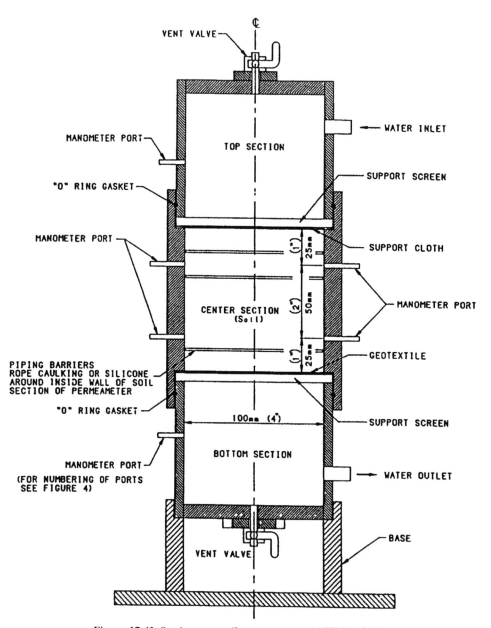

Figure 17.43 *Section-geotextile permeameter (ASTM D 5101).*

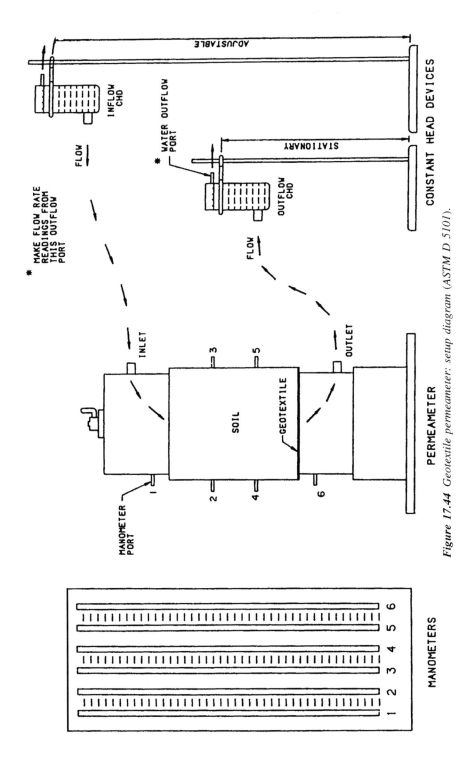

Figure 17.44 Geotextile permeameter: setup diagram (ASTM D 5101).

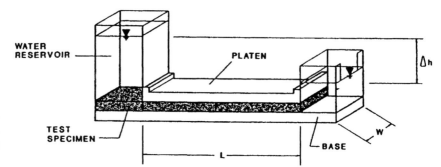

Figure 17.45 Constant-head hydraulic transmissivity testing devise (ASTM D 4716).

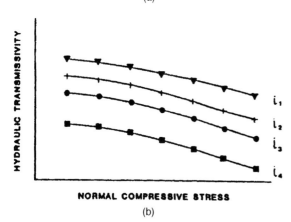

Figure 17.46 Relationship between hydraulic transmissivity and hydraulic gradient and normal stress (ASTM D 4716).

(1994) and Sharma and Lewis (1994) review various test results available in the literature and generally recommend the following:

• Geotextiles must be protected form long-term ultraviolet exposure. Also, instead of providing a thicker geotextile to allow for some sacrificial geotextile, it is preferable to specify a removable protective layer.

• Protective plastic covering provided on geotextile rolls when shipped from the plant should not be unrolled or removed until the material is ready for use.

17.4.2 Geotextile Seams and Their Properties

Geotextile seaming is often done by overlapping the panels. The geotextile panel overlaps generally range from a minimum of 10 in. to over 24 in., depending on the compressibility of the underlying material and the required geotextile strength to resist imposed loads that is transferred by geotextile-to-geotextile friction across the panel overlaps. This overlap can be replaced economically by field sewing of the geotextile panels.

Figure 17.47 illustrates various types of sewn seams for joining geotextiles. The various choices for thread type are Kevlar, nylon, polyester, and polypropylene. It is important to match the thread and geotextile composition so that their moduli are similar. The flat or prayer-type seams are weakest, and butterfly seams are strongest. Also, a stitch density of two to four stitches per inch is typical. The "401" two-thread chainstitch

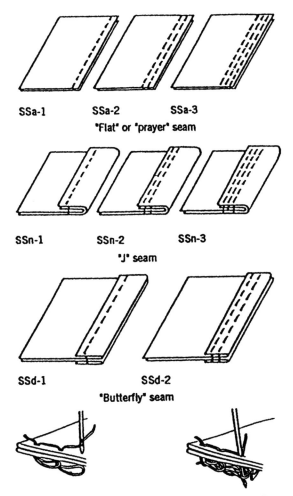

Figure 17.47 *Various types of sewn seams.* (*From Diaz, 1985.*)

is recommended. It is generally recommended that the seam strength be at least about 70% of the fabric strength. The strength of sewn geotextile seams can be evaluated in accordance with the *grab tensile test.* (ASTM D 1683, failure in sewn seams of woven fabrics) and *wide-width tensile test* (ASTM D 4884, standard test method for seam strength of sewn geotextiles). Since the ASTM D 4884 test was developed specifically for geotextiles, it is recommended that this method be used to evaluate the field seams. The test method consists of gripping an 8-in.-wide sewn seam

in a tensile testing machine and applying the tensile force perpendicular to the seam at a prescribed rate until rupture of seam or fabric; the seam strength (in tension) is reported in pounds per inch width.

17.5 GEOSYNTHETIC CLAY LINERS

Geosynthetic clay liners (GCLs) are factory manufactured rolls of hydraulic barriers that typically consist of bentonite placed between geotextiles or bonded to a geomembrane by an adhesive. Bentonite, a naturally occurring clay mineral, provides low permeability, and geosynthetics provide containment unit(s) for GCLs. Figure 17.48 shows cross-sections of available GCLs. As shown in Figure 17.48(*a*) adhesive-bounded bentonite clay may be placed between two geotextiles and is available by the trade name Claymax (produced by CETCO). To add midplane shear strength to the adhesive-bounded clay, stitch bonding is provided and is available in Claymax and NaBento (Huesker Inc.); this is shown in Figure 17.48(*b*). Figure 17.48(*c*) shows that the added midplane shear strength can also be achieved by "locking of fibers that penetrate through the thickness of needle-punched GCLs." These are available as Bentofix and Bentomat. A product called GundSeal uses "a geomembrane as a carrier material for a layer of adhesively bonded bentonite" (Koerner, 1996). This is shown in Figure 17.48(*d*).

17.5.1 Technical Equivalency Issues of GCLs and CCLs

Table 17.7 compares traditional hydraulic barrier compacted clay liners (CCLs) and GCLs. The summary presented in the table indicates that in addition to meeting hydraulic conductivity requirements, GCLs have many advantages over CCLs. Due to their newness, GCLs, however, have a shorter history in waste containment applications. Sharma et al. (1997) present successful case histories where GCLs have been used as containment liners on the base, side slopes, and cover of landfills. In the authors' experience, GCLs are now (2003) being used as liners for various waste containment systems.

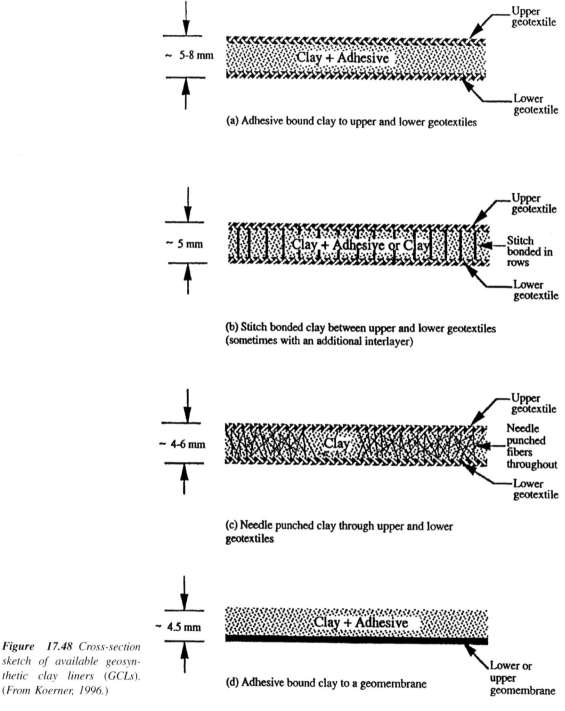

Figure 17.48 Cross-section sketch of available geosynthetic clay liners (GCLs). (From Koerner, 1996.)

Table 17.7 **Some selected differences between GCLs and CCLs**

Characteristic	Geosynthetic Clay Liner (GCL)	Compacted Clay Liner (CCL)
Materials	Bentonite, adhesives, geotextiles, and geomembranes	Native soils or blends of native soils and bentonite
Thickness	Typically, 7 to 10 mm	Typically, 300 to 900 mm
Permeability	$\leq (1$ to $5) \times 10^{-11}$ m/s	$\leq 1 \times 10^{-9}$ m/s (U.S.); $\leq 1 \times 10^{-10}$ m/s (Germany)
Speed of construction	Rapid, simple installation	Slow, complicated construction
Need for MQC and MQA	Factory-manufactured material requiring constant inspection	Naturally found materials or mineral layers generally requiring little inspection
Status of CQC and CQA	Relatively simple, straightforward, common-sense procedures	Complex procedures requiring highly skilled and knowledgeable people
Field desiccation sensitivity	GCLs cannot desiccate during construction unless prematurely hydrated	CCLs are nearly saturated; can desiccate during construction
Availability of materials	Materials are readily shipped to any site	Varies widely from readily available to not available at all
Experience	Limited due to newness and nonfamiliarity	Has been used for many decades with great confidence as a liner material

Source: Koerner and Daniel (1995).
MQC = manufacturing quality control
MQA = manufacturing quality assurance
CQC = construction quality control
CQA = construction quality assurance

Table 17.8 summarizes various technical equivalency issues that need to be addressed for GCLs. Koerner and Daniel (1995) discuss these issues in detail. They suggest that many equivalency issues depend on the particular GCL product and the unique site-specific conditions. Koerner and Daniel (1995) cite that an important site-specific issue will be slope stability. This will require extensive material testing as presented by Sharma et al. (1997) and is discussed further in subsequent sections on interface shear strength testing. Overall, with proper engineering, GCLs can be shown to provide equivalent or better performance than CCLs, as liners for most waste containment units.

17.5.2 Material Properties and Tests

These are many tests available for bentonite and the geosynthetics components used in GCL. However, the most relevant properties and tests for GCLs can be classified into two categories: (1) manufacturing quality control (MQC) tests and (2) engineering design tests. MQC testing first focuses on testing of individual GCL components (e.g., bentonite, geotextile, and geomembrane), as applicable, and then the testing of composite GCL testing is considered. For a design engineer the tests of most importance are hydraulic conductivity and direct shear tests. However, site-specific conditions may require additional tests. Readers are encouraged to refer to Koerner (1996) for other standard test methods and details.

Permeability or Hydraulic Properties of GCL. The hydraulic conductivity (k) of GCL can be determined by ASTM D 5084 and GRI-GCL-2 methods. Tests show that hydraulic conductivity of a GCL depends on the compressive stress applied to a specimen. At low stresses (less than 20 kPa) the typical k is 1×10^{-9} to 1×10^{-8} cm/s, while at high stresses (greater than 100 kPa), k is in the range of 1×10^{-10} to 1×10^{-9} cm/s (Daniel, 1996). Daniel (1996) also presented the

Table 17.8 **Technical equivalency categories and specific issues to be addressed**

Category	Criterion for Evaluation	Possibly relevant for:	
		Liners	Covers
Hydraulic issues	Steady flux of water	X	X
	Steady solute flux	X	
	Chemical adsorption capacity	X	
	Breakout time:		
	(1) Water	X	X
	(2) Solute	X	
	Horiz. flow in seams or lifts	X	X
	Horiz. flow beneath geomembranes	X	X
	Generation of consolidation water	X	X
	Permeability to gases		X
Physical/mechanical issues	Freeze–thaw behavior	X[a]	X
	Wet–dry behavior		X
	Total settlement response	X[b]	X
	Differential settlement response	X[b]	X
	Slope stability considerations	X	X
	Vulnerability to erosion	X	X
	Bearing capacity (sqeezing)	X	X
Construction issues	Puncture resistance and resealing	X	X
	Subgrade condition considerations	X	X
	Ease of placement or construction	X	X
	Speed of construction	X	X
	Availability of materials	X	X
	Requirements for water	X	X
	Air pollution concerns	X	X
	Weather constraints	X	X
	Quality assurance considerations	X	X

Source: Koerner and Daniel, 1995.

[a] Relevant only until liner is covered sufficiently to prevent freezing.
[b] Settlement of liners usually of concern only in certain circumstances (e.g., vertical or lateral expansions).

results of a round-robin k tests and concluded industry's ability to consistently measure k with minimal variability.

Effect of Chemicals on k. It is well known that high concentrations of chemicals, such as strong acids, bases, and hydrocarbons, can affect adversely the value k of GCLs. However, GCLs maintain their hydraulic conductivity when they are exposed to leachates from solid-waste landfills. In situations where a concern may be expressed about a specific chemical regarding its influence on k of GCL, compatibility testing should be performed for the site-specific project.

Hydraulic Equivalency. If water with a head or depth (h) is ponded on top of a low-permeability layer with permeability (k) and thickness (t), the steady downward flux or velocity of water (v) is given by the relation

$$v = ki = k\,\frac{h + t}{t} \qquad (17.16)$$

where i is the hydraulic gradient and k is the hydraulic conductivity of the low-permeability layer. For CCL it will represent k for compacted clay, and for GCL it will represent k for the bentonite component.

Koerner and Daniel (1995) evaluated the hydraulic equivalency of GCL and CCL by equating the fluxes (i.e., $v_{GCL} = v_{CCL}$) and arrived at the following relationship for GCL and CCL equivalent performance in terms of steady flux of water:

$$(k_{GCL})_{required} = k_{CCL} \frac{t_{GCL}}{t_{CCL}} \frac{h + t_{CCL}}{h + t_{GCL}} \quad (17.17)$$

Based on the relationship above, one can calculate that for a 2-ft (600-mm)-thick CCL with a value $k \leq 1 \times 10^{-7}$ cm/s, the equivalent GCL, which is about 7 mm thick (after hydration), should have a k value of 3.4×10^{-9} cm/s. Most commercially available bentonites can meet this hydraulic conductivity requirement.

Effects of Miscellaneous Factors. Various other factors that may influence hydraulic properties of GCL are overlaps, wet–dry cycles, freeze–thaw cycles, differential settlements, vulnerability to puncture, permeability to gas and migration of bentonite toward the area under the wrinkle in HDPE geomembrane. Generally, dry powdered bentonite is added in the overlap to provide a proper seal. With proper CQC/CQA monitoring it is generally accepted that such overlaps will self-seal (LaGatta, 1992). Hydrated GCL will crack on drying, but the desiccation cracks "shut swell" on rehydrating (Kim and Daniel, 1992; Boardman and Daniel, 1996). Thus, wet–dry cycles do not seam to adversely impact the hydraulic properties of GCLs. One should, however, be careful about the presence of natural pore waters primarily containing divalent cations. For example, Lin and Benson (2000) show that the k values of GCLs exposed to wet–dry cycles with water having a higher concentration of divalent (CaCl$_2$) will eventually increase. Hydraulic conductivity of GCLs does not change when subjected to freeze–thaw cycles (Hewitt and Daniel, 1997). LaGatta (1992) and Boardman (1993) performed tests to evaluate the effects of differential settlements on the k values of GCLs. Test results indicate that GCLs can withstand large distortions (Δ/L up to 0.5) and large tensile strains (up to 10 to 15%) without undergoing a significant increase in k

(Koerner and Daniel, 1995), where Δ is settlement over a length L. In comparison, normal compacted clay will crack if tensile strains exceed about 0.85%. Generally, GCLs will self-heal if punctured by small objects such as nails or small rocks. However, because of their small thickness, GCLs are more vulnerable to puncture than CCLs, especially due to accidental puncture caused by construction equipment. In situations where gas permeability is an issue, such as in cover of solid waste landfills, and GCLs are used without a synthetic liner, the design should include features that will cause GCLs to hydrate to inhibit gas permeation. Otherwise, dry GCL, without synthetic membranes, will be permeable to gases. Finally, the issue of bentonite squeezing, and thus reducing the effective thickness of GCL when GCL is overlain by polyethylene liner, needs further evaluation. At this time, data indicate that there is some migration of wet bentonite toward the wrinkles, thus effectively increasing the permeability of GCLs (Anderson, 1996). This needs further evaluation, however.

17.6 GEONETS AND GEOCOMPOSITE DRAINS

ASTM D 4439 defines a *geonet* as "a geosynthetic consisting of integrally connected parallel sets of ribs overlying similar sets at various angles for planar drainage of liquids or gas." There are various types of geonets available in the market. Figure 17.49 shows a typical geonet configuration. Although geonets may be made from polypropylene, polystyrene, and other materials, the most commonly used material for geonets is polyethylene (PE).

17.6.1 Material Properties and Tests for Geonets

There is a number of physical, mechanical, endurance, and environmental properties of geonets that are important for their long-term performance in field applications. The two properties that are important for drainage are the thickness and in-plane hydraulic prop-

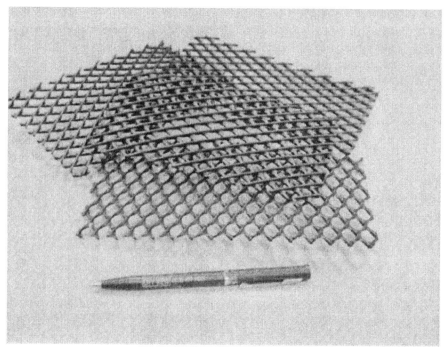

Figure 17.49 *Typical geonet photograph.*

erty or transmissitivity. Readers are encouraged to refer to Koerner (1998) for other geonet properties.

Thickness. The applicable test methods for thickness measurement are ASTM D 1777 (method for measuring thickness of textile materials) and ASTM D 5199 (method for measuring nominal thickness of geotextiles and geomembranes). It is recommended that geonet thickness be measured under a normal pressure of approximately 3 psi (20 kPa). Although thickness itself may not be a significant property of a geonet; intuitively, the thicker a geonet, the larger its flow capacity.

Transmissivity. Transmissivity is defined as the volumetric flow rate per unit width of specimen per unit flow gradient in a direction parallel to the plane of the specimen. The transmissivity test is performed in accordance with ASTM D 4716 (method for constant-head hydraulic transmissivity, in-plane flow, of geotextile and geotextile-related products). The flow rates depend on normal stress and hydraulic gradient. In real field conditions, geonet drains are sandwiched between geotextiles or geotextile and geosynthetic membrane. Table 17.9 shows comparative data on flow rate behavior of a 6.3-mm-thick geonet under various hydraulic gradients (*i*) for two cases: (1) when geonet has 1.5-mm HDPE geomembranes on both sides, and (2) when geonet is sandwiched between a 540-g/m² non-woven needle-punched geotextile with clay above and a 1.5-mm HDPE geomembrane below. The surrounding material intrusion into geonet would reduce the flow rate, keeping normal stress and hydraulic gradient constant. Koerner (1998) and Sharma and Lewis (1994) present further information on hydraulic properties of geonet.

17.6.2 Geocomposite Drains

A geocomposite drain will generally use a geonet for drainage and geotextile for separation and filtration.

Table 17.9 **Flow rate comparisons for two cases of geonet sandwiches**[a]

Normal		Hydraulic Gradient (i)					
Stress (kPa)	Cross Section	0.03	0.06	0.12	0.25	0.50	1.00
50	HDPE (both sides)	0.005	0.011	0.019	0.032	0.055	0.095
	GT/clay (one side)	0.004	0.008	0.013	0.022	0.038	0.059
	Difference	0.001	0.003	0.006	0.010	0.017	0.036
	Reduction	20%	27%	31%	31%	31%	38%
250	HDPE (both sides)	0.005	0.011	0.017	0.028	0.048	0.077
	GT/clay (one side)	0.004	0.008	0.012	0.019	0.032	0.052
	Difference	0.001	0.003	0.005	0.009	0.016	0.025
	Reduction	20%	27%	29%	32%	33%	32%
500	HDPE (both sides)	0.005	0.010	0.017	0.027	0.047	0.072
	GT/clay (one side)	0.004	0.007	0.012	0.018	0.028	0.043
	Difference	0.001	0.003	0.005	0.009	0.019	0.029
	Reduction	20%	30%	29%	30%	40%	40%
950	HDPE (both sides)	0.005	0.008	0.013	0.021	0.035	0.054
	GT/clay (one side)	0.004	0.007	0.011	0.017	0.024	0.035
	Difference	0.001	0.001	0.002	0.004	0.011	0.019
	Reduction	20%	12%	15%	19%	31%	35%

Source: Koerner (1998). Reprinted by permission of Pearson Education, Inc., Upper Saddle River, New Jersey.

[a] The two cases are when a 6.3-mm-thick geonet is sandwiched between (1) two 1.5-mm-thick geomembranes, and (2) a nonwoven geotextile with clay above and 1.5-mm-thick HDPE geomembrane below; The flow rate is in $m^3/min \cdot m$.

Thus, a geocomposite drain may typically have a geonet sandwiched between two geotextiles or a geotextile on one side and a geomembrane on the other side. At the present time, geotextile is typically heat (or thermally) bonded with the geonet for added strength. Material properties and tests for individual components (e.g., geonet and geotextile) as discussed in previous sections, will also apply for geocomposite drain components.

17.7 GEOGRIDS

Geogrids are made of geosynthetic materials (typically, polyethylene, polypropylene, or polyester) and have large (typically, 0.5 to 4 in.) open spaces (or apertures) between ribs (Figure 17.50). They can be made by several methods. Knitted and punched drawn sheets are commonly available. Geogrids that have the highest strength in the longitudinal direction (uniaxial geogrids) have punched circular holes in the geomembrane sheet. When the sheet is drawn and stretched over rollers, an elliptical shape is created. Geogrids that have

strength in both the longitudinal and transverse directions (biaxial geogrids) have punched square holes in a geomembrane sheet which when drawn over rollers longitudinally and stretched transversely retain nearly square or rectangular apertures (Sharma and Lewis, 1994; Koerner, 1998). Geogrids are used as reinforcement. For example, in waste management applications they can be used to support a lining system over a weak subgrade, in "piggyback" or vertical expansion of landfills, and to support final covers in steep landfill closure slopes.

17.7.1 Material Properties and Tests

Geogrids have several physical, mechanical, endurance, and degradation properties. Koerner (1998) discusses these properties in detail. In this section, tensile strength is discussed because this is the most applicable property of geogrid when used in waste containment applications. The tensile strength can be determined using ASTM D 4595 (method for tensile properties of geotextiles by the wide-width strip

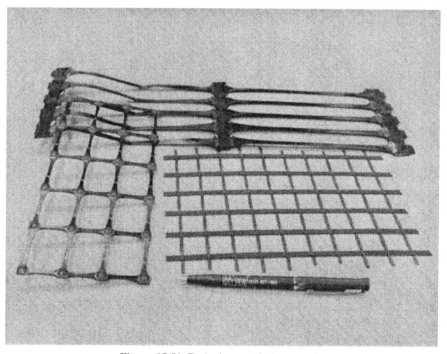

Figure 17.50 *Typical geogrid photograph.*

method). For uniaxial geogrids the test is done in the machine direction. For biaxial geogrids the testing is performed in both machine and cross-machine directions. The tensile strength can vary over a wide range depending on the product. Product manufacturers can provide information on the allowable tensile strengths of their products. The specifier's guide in the *Geotechnical Fabrics Report* can be useful for obtaining such information. When a product is selected, properties specific to a certain design must be confirmed by testing.

17.8   INTERFACE STRENGTH AMONG VARIOUS LINER COMPONENTS

As shown in Figures 17.1 through 17.4, one needs to consider the strengths (internal or interface strengths) between various liner component interfaces. These interfaces generally consist of the following:

- Low-permeability soil and synthetic membrane
- Suitable subgrade and GCL or other geosynthetic materials
- Internal strength of GCL
- Geosynthetics and geosynthetics

These interface and internal strengths are important for overall stability evaluation of a containment system. Research on strength value estimation has been performed by many investigators (Sharma and Hullings, 1993; Stark and Peoppel, 1994; Reddy et al., 1996; Stark and Eid, 1996; Stark et al., 1996; Sharma and Vargas, 1997; Sharma et al. 1997, 1998). Sharma and Lewis (1994) also cite many other references and present further discussions on interface strength tests.

Although there are various interface strength test procedures, such as torsional ring, tilt table, and Texas double-ring, it is generally agreed that direct shear testing conducted in accordance with ASTM D 5321 and ASTM D 6243 is an appropriate test method for developing interface strengths for waste containment design applications.

17.8.1 Direct Shear Test Methods: ASTM D 5321 and ASTM D 6243

These test methods cover a procedure for determining the shear stress-displacement relations and then interpret the shear strength of the following:

- A soil against a geosynthetic
- A geosynthetic against a different or similar type of geosynthetic
- A soil and geosynthetic in any combination
- An interval strength of any type of GCL

The test is performed by placing the geosynthetic and one or more contact surfaces into a direct shear box. The recommended shear device generally consists of two square or rectangular containers, which, according to ASTM D 5321 and ASTM D 6243, should have a minimum dimension that is the greater of the following:

- 300 mm (12 in.)
- 15 times D_{85} of the coarser soil used in the testing, or
- A minimum of five times the maximum opening size (in plan) of the geosynthetic tested

The depth of each container should be either 50 mm (2 in.) or six times the maximum particle size of the coarser soil tested, whichever is greater. Finally, the minimum specimen width to thickness ratio is 2:1 (ASTM D 5321). The minimum container dimensions mentioned above may be changed if by comparative testing with smaller containers it can be shown that data generated by smaller devices contain no scale or edge effect bias for the materials to be tested.

The test is performed by assembling various test components and fabricating soil specimen(s), if they are a part of the testing, at the test conditions (moisture, density, hydration, shear rate, and consolidation conditions) specified by the engineer. A constant normal compressive stress is applied to the specimen followed by application of a tangential (shear) force so that the upper and lower sections of the box move in relation to one another.[3] The shear force is recorded as a function of horizontal displacement of the moving section of the shear box. The tests are repeated for different normal loads (which are then converted to shear stresses) that represent field conditions.

Figure 17.51 shows two typical shear stress versus displacement plots for the interfaces discussed above. The plots generally show peaks and postpeaks. In situations where either large displacements (postpeak) are encountered in the field or various interfaces, such as liners and waste, form a part of potential failure surface strain compatibility requires the use of postpeak strengths. Therefore, for such situations, the engineering practice is to plot postpeak shear stresses and normal stresses to obtain a strength envelope.

17.8.2 Interface Shear Strength: Shear Strength Envelopes

Unlike the ideal Mohr–Coulomb strength envelope (which represents a linear relationship between the shear stress and normal stress), the strength envelopes at peak and postpeak displacements for liner interfaces are generally curvilinear over a range of normal stresses. Figure 17.52 shows such envelopes for a low-permeability compacted clay and textured HDPE geomembrane. Figure 17.53 shows strength envelopes for textured HDPE and nonwoven geotextile-based reinforced GCL. In this figure the applicable interface shear strength failure envelope is along ABC. These envelopes may be considered linear over a narrow range of normal stresses but are nonlinear or multilinear (two or more lines) over a larger range of normal stresses.

A general mistake in interpreting shear strength data is to assume linear shear stress (τ) behavior for interface strengths over a wide range of normal

---

[3] For soil–geosynthetic interface friction, the shear is applied at a rate of 1 mm/min (0.04 in./min). For geosynthetic/geosynthetic interface friction, the shear is applied at a rate of 5 mm/min (0.2 in./min).

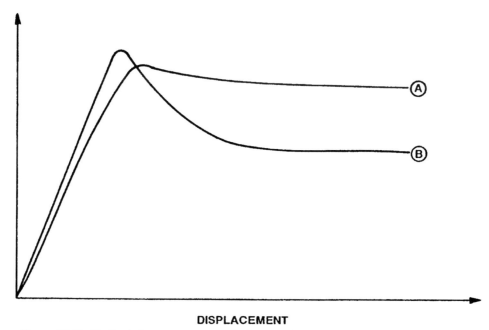

DISPLACEMENT

Figure 17.51 *Idealized shear stress versus displacement behavior.* (*From Sharma et al., 1997.*)

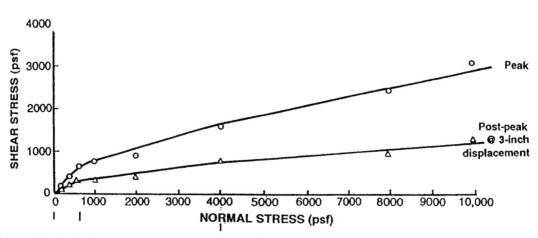

Figure 17.52 *Typical shear stress versus normal stress for low-permeability clays and textured HDPE.* (*From Sharma et al., 1997.*)

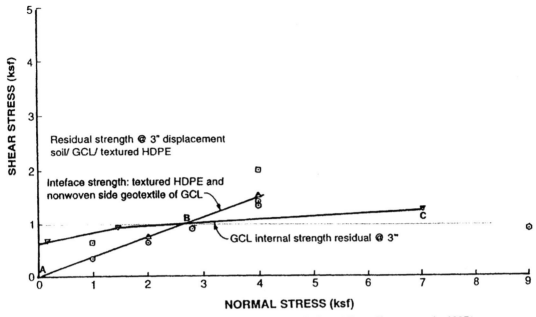

Figure 17.53 GCL internal and interface strength data. (From Sharma et al., 1997.)

stresses. Also, typical interface friction values for various interfaces are generally cited in the literature for various materials irrespective of normal stress range. One should, however, recognize that the interface strength is affected by several factors, such as various material types in the system (e.g., textured versus smooth HDPE geomembrane, woven versus nonwoven geotextile, reinforced versus unreinforced GCLs, and degree of GCL hydration), the level of normal stress (e.g., 400 lb/ft^2 versus 5000 lb/ft^2), and the amount of deformation (peak versus postpeak) allowed in the test. In addition, drainage and consolidation conditions may also influence these strengths significantly. Therefore, it is recommended that project-specific conditions, representing site-specific materials, normal stresses, and drainage and hydration conditions, must be considered when interface strength tests are planned. Finally, interface strengths must be described in terms of tested normal stresses. Therefore, it is generally preferable to specify a strength envelope instead of specifying the conventional adhesion strength parameter (c_a) or friction strength parameter (δ).

17.9   MANUFACTURING AND CONSTRUCTION QUALITY ASSURANCE

According to USEPA (1993), even the best designs and regulatory requirements will not necessarily translate to facilities that are protective of human health and the environment unless the following are met:

1. The geosynthetic materials' manufacturing quality assurance (MQA) and manufacturing quality control (MQC) are performed appropriately.
2. The construction quality assurance (CQA) and construction quality control (CQC) for the facility are also performed appropriately.

In other words, in addition to performing a good design and meeting regulatory compliance, the waste containment facility also needs to be constructed properly. This requires performing MQA/CQA and MQC/CQC as defined below. *MQA* refers to measures taken by the MQA organization to determine if the manufacturer is in compliance with the product certification and project plan. *MQC* refers to measures taken by the requirements for materials and workmanship as stated in certification documents. *CQC* refers to measures taken by the installer or contractor to determine compliance with the requirements for materials and workmanship as stated in project construction documents. *CQA* refers to measures taken by the CQA organization to assess if the installer or contractor is in compliance with the project plans and specifications (USEPA, 1993). It should be recognized that MQC and CQC are performed independent of MQA and CQA. Figure 17.54 presents various elements of MQA/CQA inspection activities. Guidelines for CQA and CQC programs and recommendations for testing frequency are available in various documents (Gorgon et al., 1984; USEPA, 1993). Although local and project-specific variations do exist, recommendations provided in these documents provide useful guidance for MQA/CQA and MCQ/CQC programs.

17.10 ESTIMATION OF LEAKAGE THROUGH LINER SYSTEMS

For a waste containment system it is important to have an estimation of the amount of liquid that could leak through the liner system. In the following sections we discuss the leakage mechanism through the liner systems and the quantitative values estimated for a composite liner system. This discussion is based primarily on information provided by Giroud and Bonaparte (1989a,b) and applications summarized by Schroeder et al. (1994).

17.10.1 Leakage Mechanisms

Leakage through soil liners (q_s) can be estimated based on the equation

$$q_s = k \frac{\Delta h}{L} A \qquad (17.18)$$

where k is the coefficient of permeability, $\Delta h/L$ the head loss per unit length, and A the cross-sectional area through which flow occurs.

Leakage through geomembranes liners occurs by (1) permeation of fluids (liquids and vapors) due to differential pressures, and (2) permeation through defects, such as pinholes, defective seams, and puncture or tear in geomembrane. Evaluations indicate that permeation through defects is significantly higher than permeation through intact geomembrane. Therefore, the following discussion will be limited to leakage through liner defects only. Giroud and Bonaparte (1989a) recommend the following regarding geomembrane defects:

* For sizing lining system and LCRS components, use a large hole (0.16 in^2 in area).
* For liner performance calculations, such as estimating the flow in a leak detection and collection layer under typical operating conditions, use a small hole (0.005 in^2 in area).
* With good quality assurance and quality control (QA/QC), use a frequency of defect of one hole per acre.
* With poor QA/QC, use a frequency of defect of 10 holes per acre or greater.

The flow of liquid through a pinhole (Q_p) can be estimated using the relationship

$$Q_p = \frac{\pi \rho g h_w d^4}{128 \eta T_g} \qquad (17.19)$$

where a pinhole is defined as a defect that creates a hole with dimensions less than the thickness (T_g) of the geomembrane, ρ is the density of the liquid, g the acceleration due to gravity, h_w the liquid depth above the geomembrane, d the pinhole diameter, and η is the viscosity of the liquid.

The flow of liquid through a hole (Q_h), an opening that has dimensions equal to or greater than the geomembrane thickness, can be calculated by

$$Q_h = C_B a \sqrt{2ghw} \qquad (17.20)$$

where C_B is a dimensionless coefficient related to the shape of aperture edges; for shape edges it is equal to 0.6 and a is hole area. Relation (17.20) is Bernoulli's

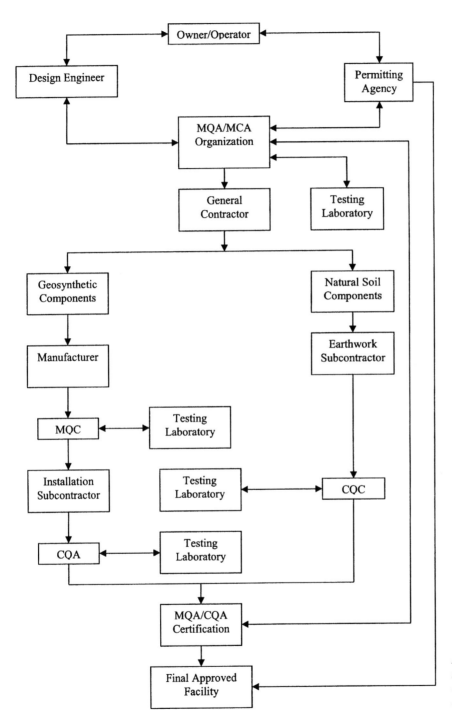

Figure 17.54 *Organization structure of MQA/CQA inspection activities. (From USEPA, 1993.)*

orifice equation and assumes that the materials on either side of the geomembrane that have permeability greater than 0.1 cm/s may be assumed to have infinite permeability.

17.10.2 Leakage through Composite Liners

Composite liners consist of a low-permeability material, such as soil liner overlain by a geomembrane. The flow through a composite liner defect has the following mechanism:

- The flow occurs first by passage of the liquid through a geomembrane hole.
- The liquid then travels along the interface space between the low-permeability soil liner and the geomembrane.
- Finally, the liquid flows downward through the low-permeability soil liner.

If the contact between the geomembrane and the low-permeability soil liner is perfect, interface flow will not occur. However, depending on the QA/QC level, there may be good or poor contact between the two components of the composite liner system. For good construction practice, where the soil is well compacted, the surface is smooth and is free of ruts, clods, and cracks, Giroud and Bonapate (1989b) recommended the following relations for estimating leakage rates (Q) and the radius of wetted area (R) for a composite liner system with a hole in the geomembrane:

$$Q = 0.7a^{0.1}K_s^{0.88}h_w \qquad (17.21)$$

$$R = 0.5a^{0.05}K_s^{-0.06}h_w^{0.5} \qquad (17.22)$$

where a is the area of the hole in the geomembrane, K_s the hydraulic conductivity of low-permeability soil liner, and h_w the liquid depth on top of the geomembrane.

EXAMPLE 17.1

A composite liner system consists of a 2-ft-thick 10^{-7}-cm/s compacted clay liner overlain by a 60-mil HDPE geomembrane liner. Considering a good QA/QC condition, estimate the liner leakage (a) under operating conditions, and (b) for sizing LCRS components. For both cases, consider a 6-in. liquid head on the lining system and assume that the total liner area is 20 acres.

Solution:

(a) Using equation (17.21) and substituting $K_s = 10^{-7}$ cm/s, $a = 0.005$ in^2, and $h_w = 6$ in.:

$$Q = 0.7 \, (0.005)^{0.1} \left(10^{-7} \frac{1}{2.54}\right)^{0.88} (6) \text{ in}^3/\text{s per acre}$$

$$= (7.5 \times 10^{-7}) \, 20 = 1.5 \times 10^{-5} \text{ in}^3/\text{s}$$

(b) The only difference will be using a 0.16-in^2 hole. Substituting in equation (17.21) we obtain the following:

$$Q = 0.7 \, (0.06)^{0.1} \left(10^{-7} \frac{1}{2.54}\right)^{0.88} (6)(20) \text{ in}^3/\text{s}$$

$$= 2.1 \times 10^{-5} \text{ in}^3/\text{s}$$

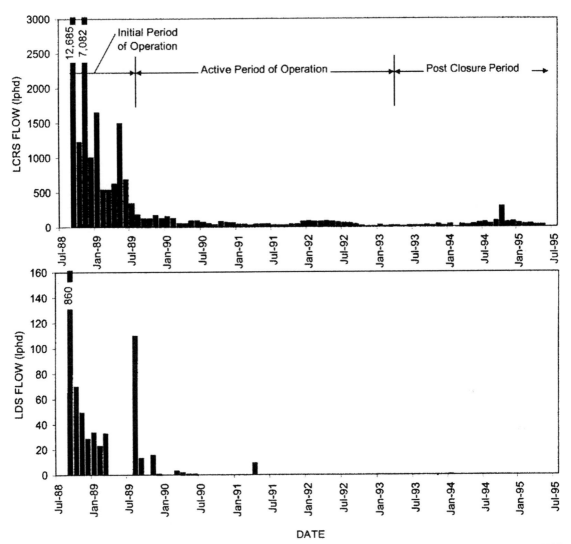

Figure 17.55 *LCRS and LDS flow rates over time at a MSW landfill in Pennsylvania. (From Bonaparte et al., 2002.)*

17.11 PERFORMANCE OF LINERS IN WASTE CONTAINMENT SYSTEMS

A broad-based study, performed under a USEPA Cooperative Agreement, addressed the design, construction, and performance of liners in waste containment systems. The waste containment systems are used in landfills, surface impoundments, waste piles, and in the remediation of contaminated sites. The various liner system components studied were geomembranes (GMs), geotextiles (GTs), geonets (GNs), geosynthetic clay liners (GCLs), and compacted clay liners (CCLs). Based on various laboratory tests, field test plots, field performance evaluations of numerous landfill cells, and analyses of various liner systems, the following can generally be, concluded (Bonaparte et al., 2002):

1. Needle-punched nonwoven GTs can provide adequate protection to GMs against puncture by adjacent granular soils.

2. Analysis of CCLs constructed in the final cover test sections generally showed that CCLs placed without a GM overlain by soil tended to desiccate and lose their low hydraulic conductivity within a few years.

3. Buried HDPE geomembranes have an estimated service life that is measured in terms of at least hundreds of years.

4. Based on the field performance evaluation of liquid management data from double-lined landfill cells, the following can be concluded:

   a. GM liners can achieve true hydraulic efficiencies (E) in the range 90 to 99%; where E = (LCRS flow rate – LDS flow rate)/(LCRS flow rate). Therefore, GMs should not be used alone where a hydraulic efficiency above 90% must reliably be achieved.

   b. GM/GCL, GM/CCL, and GM/GCL/CCL composite liners can achieve true hydraulic efficiencies of 99% to more than 99.9%.

   c. GM/CCL and GM/GCL/CCL composite liners are capable of substantially preventing leachate migration over the entire period of significant leachate generation of typical landfill operation scenarios.

   d. As shown in Figure 17.55, the leachate collection and removal system (LCRS) flow rates were highest at the beginning of the cell operations and decreased as the waste thickness increased and daily and intermediate cover were applied to waste. The corresponding small to negligible flow rates in a leak detection system (LDS) over time confirms that the composite liner systems perform efficiently over time.

17.12 SUMMARY

This chapter presents information on various liner components used in the waste containment liner systems. The liner components discussed are low-permeability soil liners, geomembrane liners, geotextiles, geosynthetic clay liners, geonets and geocomposite drains, and geogrids. Various applicable material properties, such as hydraulic conductivity, transmissivity, durability, strength, and interface shear strength, as they apply to various components, are also presented in this chapter. It has also been emphasized that for a good waste containment design to be effective in protecting the environment, it is very important that manufacturing quality assurance and construction quality assurance are performed as a part of construction. Finally, key results of a broad-based USEPA-sponsored study of the performance of liners in waste containment systems are also summarized here.

QUESTIONS/PROBLEMS

17.1. What is a composite liner system? In accordance with federal regulations, describe and draw sketches for the following liner systems: **(a)** municipal solid waste landfill base and cover liner systems; **(b)** hazardous waste landfill base and cover liner systems; **(c)** hazardous and nonhazardous surface impoundment liner systems.

17.2. List the names of various liner component materials.

17.3. Show from experimental data presented in the literature that hydraulic conductivity is influenced by chemical constituents of the permeant.

17.4. List various laboratory hydraulic conductivity test methods. Describe each method; compare and contrast the methods.

17.5. The intrinsic permeability (K) of a kaolinitic soil in water is 0.04×10^{10} cm^2. Estimate the following soil properties: **(a)** void ratio; **(b)** the K value for acetone permeation (when initially, water was used during sample remolding, followed by performing a permeability test with water, which was then displaced with acetone).

17.6. A coarse-grained soil has grain size as follows: 100% passing $\frac{3}{4}$ in., 70% passing $\frac{3}{8}$ in., 5% passing sieve no. 4, 25% passing sieve no. 10, and 2% passing sieve no. 40. Its hydraulic conductivity is 3.5×10^{-1} cm/s. Estimate its k value for the following conditions: **(a)** if 15% limestone fines are mixed with the soil; **(b)** if 15% silt is mixed with the soil; **(c)** if 15% clay fines are mixed with the soil.

17.7. Explain how chemicals can influence some soil structures, which in turn can change their hydraulic conductivity. By citing references, discuss how the hydraulic conductivity of clays can be influenced with permeation of methanol, acetic acid, and trichloroethylene.

17.8. What are the advantages and disadvantages of laboratory and field hydraulic conductivity tests? Identify the most appropriate laboratory and field hydraulic conductivity tests for a clay liner, and discuss the reasons for your answer.

17.9. Under what circumstances would one use lab and field hydraulic conductivity tests for evaluation of a low-permeability liner? Which lab and which field tests are most appropriate and in common use for clay liner hydraulic conductivity tests, and why?

17.10. Support your answers to parts (a) to (d) with sketches.

  (a) Explain changes in hydraulic conductivity (k) based on micropore and macropore theories.

  (b) How is k influenced by clay mineralogy?

  (c) For the same relative compaction, how will k change with compaction moisture content?

  (d) How do you achieve an acceptable zone for γ_d and moisture content for a maximum specified k?

17.11. With reference to the Boutwell test shown in Figure 17.16, for a 4-in.-internal-diameter casing and 1-in. tube, if $z = 2$ ft, the initial head is 3.2 ft at time 5 days and 2 ft at 8 days, estimate the initial stage I hydraulic conductivity. If a stage II test is done by pushing the casing 6 in. into the underlying clay, estimate the value of the horizontal (K_h) hydraulic conductivity of the compacted clay.

17.12. With reference to a constant-head permeameter, as shown in Figure 17.7, calculate the hydraulic conductivity of the formation if $H = 2$ ft, $r = 7$ in., $\alpha = 0.01$ cm^{-1}, and $q = 3.5$ cm^3/h.

17.13. **(a)** How do you define a geomembrane? What are the major types of geomembranes, and name each type by citing examples.

  (b) Name the geomembrane that is most commonly used in a waste containment system, and why.

  (c) Discuss the method of estimating hydraulic conductivity (k) of an HDPE geomembrane. Why is it difficult to determine the conductivity of an HDPE geomembrane by conventional laboratory permeability tests?

17.14. **(a)** Define *geotextile* and describe the various types of fibers used in the manufacture of geotextiles.

  (b) What are the uses of woven and nonwoven geotextiles?

  (c) What is (are) the most common physical, mechanical, and hydraulic properties of geotextiles, and why?

  (d) What is the purpose of a gradient ratio test?

17.15. Discuss the uses of geonets and geocomposite drains in waste containment systems.

17.16. **(a)** What is a GCL? What are its uses?

  (b) Show that a typical GCL is hydraulically equivalent to 2-ft-thick compacted clay with $k \leq 1 \times 10^{-7}$ cm/s.

17.17. Based on leakage analysis, which of the liner systems shown in Figure P17.17 is the most effective system?

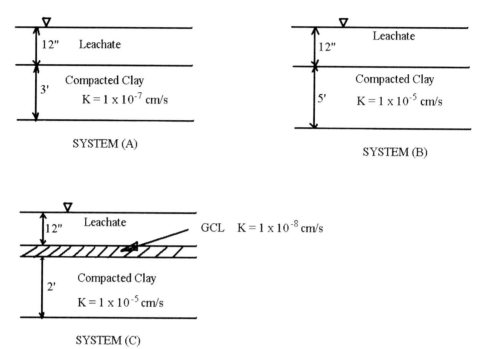

Figure P.17.17

REFERENCES

Alberta Environment, *Design and Construction of Liners for Municipal Wastewater Stabilization Ponds,* prepared by Komex Consultants, Ltd., Alberta Environment Publication, Edmonton, Alberta, Canada, Mar. 1985.

ASTM (American Society for Testing and Materials), *Annual Book of ASTM Standards,* Vol. 04.08, ASTM, West Conshohocken, PA, 2003.

Anderson, J. D., Are geosynthetic clay liners (GCLs) really equivalent to compacted clay liners?, *Geotech. News,* pp. 20–23, June 1996.

Benson, C. H., and Daniel, D. E., Influence of clods of hydraulic conductivity of compacted clays, *J. Geotech. Eng.,* Vol. 116, No. 8, pp. 1231–1248, 1990.

Benson, C. H., Daniel, D. E., and Boutwell, G. P., Field performance of compacted clay liners, *J. Geotech. Geoenviron. Eng.,* Vol. 125, No. 5, pp. 390–403, 1999.

Boardman, B. T., The potential use of geosynthetic clay liners as final covers in arid regions, M.S. thesis, University of Texas, Austin, TX, 1993.

Boardman, B. T., and Daniel, D. E., Hydraulic conductivity of desiccated geosynthetic clay liners, *J. Geotech. Eng.,* Vol. 123, No. 3, pp. 204–208, 1996.

Bonaparte, R., Daniel, D. E., and Koerner, R. M., *Assessment and Recommendations for Improving the Performance of Waste Containment Systems,* EPA Cooperative Agreement CR-821448-01-0, USEPA, Washington, DC, July 2002.

Bowders, J. J., and Daniel, D. E., Hydraulic conductivity of compacted clay to dilute organic chemicals, *J. Geotech. Eng.*, Vol. 113, No. 12, pp. 1432–1448, 1987.

Cadwallader, M. W., and Burkinshaw, J. R., Molecular and rheological changes in polythylene occurring from heat seaming HDPE liners, in *Geosynthetics '91, Conference Proceedings,* IFAI, Atlanta, GA, 1991.

Carlson, D. S., et al., Laboratory evaluation of HDPE geomembrane seams, in *Geosynthetics '93, Conference Proceedings,* IFAI, Vancouver, British Columbia, Canada, Mar. 1993.

Chang, D. T. T., Wey, W. T., and Chen, T. C., Study of geotextile behaviors of tensile strength and pull-out capacity under confined condition, pp. 607–618 in *Geosynthetics '93, Conference Proceedings,* IFAI, Vancouver, British Columbia, Canada, Mar. 1993.

CFR (*Code of Federal Regulations*), 40 CFR, Parts 257, 258, 264, and 265, U.S. Government Printing Office, Washington, DC, 1994.

Cornell, *Final Report on Soil Solidification Research,* Cornell Univ., Ithaca, NY, 1951.

Daniel, D. E., In situ hydraulic conductivity tests for compacted clay, *J. Geotech. Eng.*, Vol. 115, No. 9, pp. 1205–1226, 1989.

———, Geosynthetic clay liners, part two: Hydraulic properties, *Geotech. Fabr. Rep.*, pp. 22–26, June/July 1996.

Daniel, D. E., and Benson, C. H., Water content—density criteria for compacted soil liners, *J. Geotech. Eng.*, Vol. 116, No. 12, pp. 1811–1830, 1990.

Daniel, D. E., and Wu, Y. K., Compacted clay liners and covers for arid sites, *J. Geotech. Eng.*, Vol. 119, No. 2, pp. 223–237, 1993.

Darcy, A., *Les Fontaines publiques de la ville de Dijon,* Dalmont, Paris, 1856.

Dennis, M. L., and Turner, J. P., Hydraulic conductivity of compacted soil treated with biofilm, *J. Geotech. Geoenviron. Eng.*, Vol. 124, No. 2, pp. 120–127, 1998.

Diaz, V., Thread selector for geotextiles, *Geotech. Fabr. Rep.*, Vol. 3, No. 1, pp. 15–19, Jan./Feb., 1985.

Fernuik, N., and Haug, M. M., Evaluation of in situ permeability testing methods, *J. Geotech. Eng.*, Vol. 116, No. 2, pp. 297–311, 1990.

Folkes, D. J., Fifth Canadian Geotechnical Colloquium: Control of Contaminant Migration by the Use of Liners, *Can. Geotech. J.*, Vol. 19, No. 3, pp. 320–344, 1982.

Garcia-Bengochea, I., Lovell, C. W., and Altschaeffl, A. G., Pore distribution and permeability of silty clays, *J. Geotech. Eng. Div. ASCE*, Vol. 105, No. GT7, pp. 839–856, 1979.

Gary, G. S., and Raymond, G. P., The in-plane permeability of geotextiles, *Geotech. Test. J.*, Vol. 6, No. 4, pp. 181–189, 1983.

Giroud, J. P., and Bonaparte, R., Leakage through liners constructed with geomembranes: geomembrane liners, *Geotext. Geomembr.*, Vol. 8, No. 1, 1989a.

———, Leakage through liners constructed with geomembrane liners: composite liners, *Geotext. Geomembr.*, Vol. 8, No. 2, 1989b.

Giroud, J. P., and Peggs, I. D., Geomembrane construction quality assurance, in *Waste Containment Systems: Construction, Regulation, and Performance,* Bonaparte, R. (ed.), ASCE Geotechnial Special Publication 26, ASCE, Reston, VA, 1990.

Gordon, M. E., Huebner, P. M. and Kmet, P., An evaluation of the performance of four clay-lined landfills in Wisconsin, pp. 399–460 in *Proceedings of the 7th Annual Madison Waste Conference,* pp. 399–460, 1984.

Haxo, H. E., and Haxo, P. D., Chemical compatibility of four geosynthetics with two MSW incinerator ash leachates, in *Geosynthetics '93, Conference Proceedings,* IFAI, Vancouver, British Columbia, Canada, Mar. 1993.

Hewitt, R. D., and Daniel, D. E., Hydraulic conductivity of geosynthetic clay liners subjected to freeze–thaw, *J. Geotech. Eng.*, Vol. 123, No. 3, 1997, pp. 305–313.

Hoekstra, S. E., Burst testing, in *Geomembranes Identification and Performance Testing,* Rilern Report 4, Rollin, A. and Rigo, J. M. (eds.), Chapman & Hall, London, 1991.

Hsuan, Y. G., and Koerner, R. M., Antioxidant depletion lifetime in high density polyethylene geomembranes, *J. Geotech. Geoenviron. Eng.*, Vol. 124, No. 6, pp. 532–541, 1998.

Hsuan, Y. G., Koerner, R. M., and Lord, A. E., Stress-cracking resistance of high-density polyethylene geomembranes, *J. Geotech. Eng.*, Vol. 119, No. 11, pp. 1840–1855, 1993a.

———, Notched constant tensile load (NCTL) test for high-density polyethylene geomembranes, *Geotech. Test. J.* Vol. 16, No. 4, pp. 450–457, 1993b.

Hullings, D. E., and Koerner, R. M., Puncture resistance of geomembranes using a truncated cone test," pp. 273–286 in *Geosynthetics '91, Conference Proceedings,* St. Paul, MN, 1991.

Hvorslev, M. J., *Time Lag in the Observation of Groundwater Levels and Pressures,* U.S. Army Corps of Engineers Waterways Experiment Station, Vicksburgh, MS, 1949.

Kane, J. D., and Widmayer, D. A., Considerations for long-term performance of geosynthetics at radioactive waste disposal facilities, pp. 13–27 in *Durability and Aging of Geosynthetics,* Koerner, R. M. (ed.), Elsevier, London, 1989.

Kim, W. H., and Daniel, D. E., Effects of freezing and hydraulic conductivity of a compacted clay, *J. Geotech. Eng.*, Vol. 118, No. 7, pp. 1083–1097, 1992.

King, K. S., Quigley, R. M., Fernandez, F., Reads, D. W., and Bacopoulos, A., Hydraulic conductivity and diffusion monitoring of Keele Valley landfill liner, Maple, Ontario, *Can. Geotech. J.,* Vol. 30, No. 1, pp. 124–134, 1993.

Knitter, C. C., Haskell, K. G., and Peterson, M. L., Use of low plasticity silt for soil liners and covers, pp. 1255–1259 in *Proceedings of the 3rd International Conference on Case Histories in Geotechnical Engineering,* St. Louis, MO, 1993.

Koerner, R. M., Geomembrane liners, pp. 164–185 in *Geotechnical Practice for Waste Disposal,* Daniel, D. (ed.), Chapman & Hall, London, 1993.

———, *Designing with Geosynthetics,* 3rd and 4th eds., Prentice Hall, Uppere Saddle River, NJ, 1994, 1998.

———, Geosynthetic clay liners, Part one: An overview, *Geotech. Fabr. Rep.,* pp. 22–24, May 1996.

Koerner, G. R., and Bove, J. A., Inspection of HDPE geomembrane installations, in *Geosynthetics '89, Conference Proceedings,* IFAI, San Diego, CA, 1989.

Koerner, R. M., and Daniel, D. E., A suggested methodology for assessing the technical equivalency of GCLs and CCLs, pp. 73–97 in *Proceedings of an International Symposium: Geosynthetic Clay Liners,* A. A. Balkema, Rotterdam, The Netherlands, 1995.

Koerner, R. M., Monteleone, M. J., Schmidt, J. R., and Roethe, A. T., Puncture and impact resistance of geosynthetics, pp. 677–682 in *Proceedings of the 3rd International Conference on Geotextiles,* Vienna, Austria, 1986.

Koerner, R. M., Koerner, G. R., and Hwu, B. L., Three dimensional anti-symmetric geomembranes tension test, pp. 170–184 in *Geosynthetic Testing for Waste Applications,* ASTM STP 1081, Koerner R. M., (ed.), ASTM, West Conshohocken, PA, 1990.

LaGatta, M. D., Hydraulic conductivity tests on geosynthetic clay liners subjected to differential settlement, M.S. thesis, University of Texas, Austin, TX, 1992.

Lambe, T. W., The structure of compacted clay, *J. Soil Mech. Found. Eng. Div., SACE,* Vol. 84, No. 2, pp. 1–35, 1958.

Lambe, T. W., and Whitman, R. V., *Soil Mechanics,* Wiley, New York, 1969.

Lin, L.-C., and Benson, C. H., Effect of wet–dry cycling on swelling and hydraulic conductivity of GCLs, *J. Geotech. Geoenviron. Eng.*, Vol. 126, No. 1, pp. 40–49, 2000.

Little, A. D., Inc., *Resistance of Flexible Membrane Liners to Chemicals and Wastes,* PB86-119955, USEPA, Washington, DC, Oct. 1985.

Lowe, J., and Johnson, T. C., Use of back pressure to increase degree of saturation of triaxial test specimen, in *Proceedings of the ASCE Research Conference on Shear Strength of Cohesive Soils,* Boulder, CO, 1960.

Matrecon, Inc., *Lining Waste Containment and Other Impoundment Facilities,* EPA/600/2-88/052, USEPA, Washington, DC, Sept. 1988.

Michaels, A. S., and Lin, C. S., Permeability of kaolinite, *Ind. Eng. Chem.*, Vol. 46, pp. 1239–1246, 1954.

Mitchell, J. K., *Fundamentals of Soil Behavior*, Wiley, New York, 1993.

Mitchell, J. K., and Madsen, F. T., Chemical effects on clay hydraulic conductivity, pp. 87–116 in *Geotechnical Practice for Waste Disposal '87*, Geotechnical Special Publication 13, ASCE, Reston, VA, 1987.

Mitchell, J. K., Hooper, D. R., and Companella, R. G., Permeability of compacted clay, *Journal of the Soil Mechanics and Foundations Division*, ASCE, Vol. 91, No. SM 4, 1965, pp. 41–65.

Myles, B., A review of existing geotextile tension testing methods, *Geotextile Testing and Design Engineering*, ASTM STP 952, ASTM, West Conshohocken, PA, 1987.

Narejo, D. B., Wilson-Fahmy, R., and Koerner, R. M., Geomembrane puncture evaluation and use of geotextile protection layers, pp. 1–16 in *Proceedings of the Penn DOT/ASCE Conference on Geotechnical Engineering*, Hershey, PA, 1993.

Norbert, J., The use of the multi-axial burst test to assess the performance of geomembranes, in *Geosynthetics '93, Conference Proceedings*, IFAI, Vancouver, British Columbia, Canada, Mar. 1993.

Olsen, H. W., Hydraulic flow through saturated clay, *Clay Miner*, Vol. 9, No. 2, pp. 131–161, 1962.

Olson, R. E., and Daniel, D. E., Measurement of hydraulic conductivity of fine-grained soils, pp. 18–64 in *Permeability and Groundwater Contaminant Transport*, ASTM STP 746, Zimmie, T. F. and Riggs, C. O. (eds.), ASTM, West Conshohocken, PA, 1981.

O'Toole, J. L., Design guide, *Modern Plastics Encyclopedia, 1985–1986*, McGraw-Hill, New York, 1986.

Reddy, K. R., Kosgi, S., and Motan, E. S., Interface shear behavior of landfill composite liner system: a finite element analysis, *Geosynth. Int.*, Vol. 3, No. 2, pp. 247–275, 1996.

Rigo, J. M., and Cazzuffi, D. A., Test standards and their classification, in *Geomembranes Identification and Performance Testing*, Rilem Report 4, Rollin, A. and Rigo, J. M. (eds.), Chapman & Hall, London, 1991.

Rollin, A. L., et al., Non-destructive and destructive seam testing, in *Geomembranes Identification and Performance Testing*, Rilem Report 4, Rollin, A. and Rigo, J. M. (eds.), Chapman & Hall, London, 1991.

Sai, J. O., and Anderson, D. C., Field hydraulic conductivity tests for compacted soil liners, *Geotech. Test. J.*, Vol. 13, No. 3, pp. 215–225, 1990.

———, *State-of-the-Art Field Hydraulic Conductivity Testing of Compacted Soils*, EPA/600/S2-91/022, Office of Research and Development, USEPA, Washington, DC, 1991.

Schroeder, P. R., Dozier, T. S., Zappi, P. A., McEnroe, B. M., Sjostrom, J. W., and Peyton, R. L., *The Hydrologic Evaluation of Landfill Performance (HELP) Model: Engineering Documentation for Version 3*, USEPA, Cincinnati, OH, 1994.

Sharma, H. D., and Hullings, D. E., Direct shear testing for HDPE/amended soil composites, in *Geosynthesis '93, Conference Proceedings*, IFAI, Vancouver, British Columbia, Canada, 1993.

Sharma, H. D., and Kozicki, P., The use of synthetic and/or soil–bentonite liners for groundwater protection, pp. 1140–1156 in *Proceedings of the 2nd International Conference on Case Histories in Geotechnical Engineering*, St. Louis, MO, 1988.

Sharma, H. D., and Lewis, S. P., *Waste Containment Systems, Waste Stabilization, and Landfills: Design and Evaluation*, Wiley, 1994.

Sharma, H. D., and Vargas, J. C., Alternative liners: equivalency, testing and design, and construction case histories, pp. 35–60 in *Proceedings of Waste Tech '97*, Tempe, AZ, Feb. 1997.

Sharma, H. D., Olsen, D. M., and Sinderson, L. K., *Contaminant Migration Evaluation at a Hazardous Waste management Facility*, Geotechnical Engineering Congress, Geotechnical Special Publication 27, ASCE, New York, 1991, pp. 1256–1267.

Sharma, H. D., Hullings, D. E., and Greguras, F. R., Interface strength tests and application to landfill design, pp. 913–926 in *Geosynthetics '97, Conference Proceedings*, IFAI, Long Beach, CA, 1997.

———, Soil parameters influencing liner strengths and impacts on landfill stability, pp. 405–410 in *Proceedings of the 6th International Conference on Geosynthetics*, Atlanta, GA, Mar. 1998.

Sivapullaiah, P. V., Sridharen, A., and Stalin, V. K., Hydraulic conductivity of bentonite–sand mixtures, *Can. Geotech. J.*, Vol. 37, No. 2, pp. 406–413, 2000.

Stark, T. D., and Eid, H. T., Shear behavior of reinforced geosynthetic clay liners, *Geosynth. Int.*, Vol. 3, No. 6, pp. 770–786, 1996.

Stark, T. D., and Poeppel, A. R., Landfill liner interface strengths from torsional-ring shear tests, *J. Geotech. Eng.*, Vol. 120, No. 3, pp. 597–614, 1994.

Stark, T. D., Williamson, T. A., and Eid, H. T., HDPE geomembrane/geotextile interface shear strength, *J. Geotech. Eng.*, Vol. 122, No. 3, pp. 197–203, 1996.

Storey, J. M. E., and Peirce, J. J., Influence of changes in methanol concentration on clay particle interaction, *Can. Geotech. J.*, Vol. 26, No. 1, pp. 57–63, 1989.

Taylor, D. W., *Fundamentals of Soil Mechanics*, Wiley, New York, 1948.

USEPA (U.S. Environmental Protection Agency), *Construction Quality Assurance for Hazardous Waste Land Disposal Facilities*, EPA/530/SW-86/031, USEPA, Washington, DC, 1986a.

———, *Saturated Hydraulic Conductivity, Saturated Leachate Conductivity and Intrinsic Permeability*, EPA Method 9100, USEPA, Washington DC, 1986b, pp. 9100-1 to 9100-7.

———, *Guide to Technical Resources for the Design of Land Disposal Facilities*, EPA/625/6-88/018, Risk Reduction Engineering Laboratory, Center for Environmental Research Information, USEPA, Cincinnati, OH, 1988a.

———, *Lining of Waste Containment and Other Impoundment Facilities*, EPA/600/2-88/052, USEPA, Washington, DC, 1988b.

———, *Requirements for Hazardous Waste Landfill Design, Construction, and Closure*, EPA/625/4-89/022, USEPA, Cincinnati, OH, 1989.

———, *Design, Construction, and Operation of Hazardous and Non-hazardous Waste Surface Impoundments*, EPA/530/SW-91/054, Office of Research and Development, USEPA, Washington, DC, 1991.

———, *Quality Assurance and Quality Control for Waste Containment Facilities*, EPA/600/R-93/182, USEPA, Washington, DC, 1993,

U.S. Navy, *Soil Mechanics*, DM 7.1 and 7.2, Naval Facilities Engineering Command. Alexandria, VA, 1982.

Vandervoort, J., *The Use of Extruded Polymers in the Containment of Hazardous Wastes*, Schlegal Lining Technology, The Woodlands, TX.

Yong, R. N., and Warkentin, B. P., Soil Properties and Behavior, *Geotechnical Engineering*, Vol. 5, Elsevier Science Publishing Company, New York, 1975, 449 pages.

18

LEACHATE COLLECTION AND REMOVAL SYSTEMS AND LINER DESIGN

18.1 INTRODUCTION

The purpose of the leachate collection and removal systems (LCRS) is to minimize the leachate accumulation above the liner by collecting leachate and removing it from the landfill. The Subtitle D regulations for a standard composite liner system (consisting of compacted soil overlain by a flexible membrane liner) require a LCRS that must be designed so that the depth of leachate is less than or equal to 12 in. above the liner. In addition, appropriate analyses need to be performed that would exhibit that (1) the LCRS material does not puncture the underlying liner system, and (2) the liner system is anchored appropriately without being overstressed.

A typical LCRS, shown in Figure 18.1, consists of the following components: (1) a permeable drainage layer placed directly on the graded liner or placed on a protective layer; the protective layer (generally, a geotextile) is placed between the liner and the drainage layer to protect the liner from any mechanical damage resulting from contact of the drainage layer material due to construction loads and waste loads; (2) a network of perforated leachate collection pipes; (3) a protective/filter layer above the drainage layer to prevent clogging of the drainage layer as well as to protect the drainage layer and pipes from sharp objects in the waste during construction and operations; and (4) manholes, sumps, and pumps to allow leachate removal from the landfill.

18.2 DESIGN CRITERIA

The following criteria is used for the design of LCRS:

1. It should be capable of collecting and removing the design leachate volume. The design leachate volume may be estimated from hydrologic evaluation of landfill performance (HELP) model analyses using site-specific weather data.

2. It should show by analysis that the maximum depth of leachate over the liner is less than 12 in. Because of this maximum allowable depth of the leachate, the drainage layer thickness is often selected as 12 in. unless a geosynthetic material or a very permeable material is used which guarantees a very low leachate head at any time during or after the waste placement.

3. It should minimize the clogging of pipes and drainage material.

A systematic design approach to design LCRS consists of the following tasks: (1) estimate leachate impingement on LCRS; (2) select drainage layer material and thickness; (3) select leachate collection pipe size and layout; and (4) design sumps, risers, header pipes, and pumps. Details on these tasks are provided in this chapter. In addition, underlying liner puncture resistance to overlying LCRS material and liner anchor trench design guidelines are also presented here.

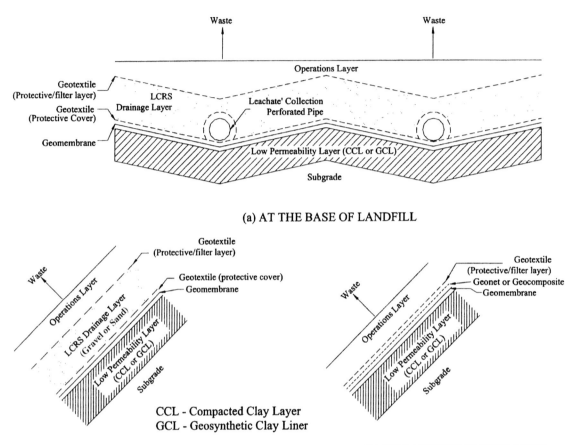

(a) AT THE BASE OF LANDFILL

CCL - Compacted Clay Layer
GCL - Geosynthetic Clay Liner

(b) ON LANDFILL SLOPES WITH
SOIL AS DRAINAGE MATERIAL

(b) ON LANDFILL SLOPES WITH
GEOSYNTHETICS AS DRAINAGE
MATERIAL

Figure 18.1 Typical leachate collection and removal system.

18.3 LEACHATE GENERATION AND MANAGEMENT

Leachate is contaminated liquid that is generated due to contact of the infiltrating water with the waste. Water infiltration into the waste could occur due to precipitation, surface run-on, and groundwater intrusion. The amount of leachate generated depends on the availability of water, which depends on the climate at the site, the landfill surface conditions (including the type of the final cover system), waste conditions, and subgrade conditions.

The major concern with the landfills is the potential for the leachate to migrate into the groundwater. Therefore, the design of the landfill components, as described in earlier chapters, is focused primarily on effectively containing this leachate and also removing it from the bottom of the landfill through LCRS. To design these landfill components, particularly the liner system and the LCRS, the rate of generation and chemical composition of the leachate is needed. The final covers or caps (as described in Chapter 19) are designed to minimize infiltration into the refuse, thereby limiting the leachate generation.

18.3.1 Leachate Quantity Estimation

To estimate the amount of leachate generation, a water balance is performed. As shown in Figure 18.2, the percolation or infiltration is given as follows:

$$\text{infiltration} = P - \text{RO} - \text{ET} \pm \Delta S \quad (18.1)$$

where P is the precipitation, RO the runoff, ET the evapotranspiration, and ΔS the change in underlying material moisture storage. This type of analysis does not account for the leachate generated due to decomposition of the waste.

Many water balance models are available in hydrology, and one that is specifically developed for estimating the leachate generation rate at landfills under the site-specific climatologic conditions is called hydrologic evaluation of landfill performance (HELP) model (Schroeder et al., 1994a). The HELP model was developed by the U.S. Army Corps of Engineers Wa-

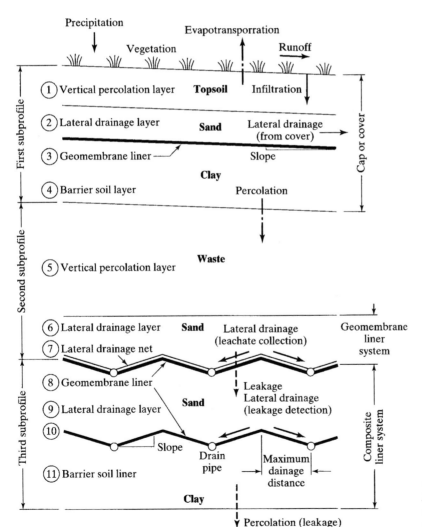

Figure 18.2 Schematic of landfill profile illustrating typical landfill features. (From Schroeder et al., 1994a.)

terways Experiment Station, Vicksburg for the U.S. Environmental Protection Agency. HELP is a quasi-two-dimensional hydrologic model of water movement across, into, through, and out of landfills. The model is deterministic and quasi-two-dimensional with one-dimensional vertical flow and one-dimensional lateral flow above the barrier layer. The model requires climatologic data, material data, and design data and utilizes a solution technique that accounts for the effects of runoff, infiltration, percolation, evapotranspiration, soil moisture storage, and lateral drainage. Landfill systems, including various combinations of vegetation, cover soils, waste cells, special drainage layers, and relatively impermeable barrier soils, as well as synthetic membrane covers and liners, may be modeled. The model, applicable to open, partially closed, and fully closed sites, is a tool for designers to optimize the design of landfill components. Engineering background for HELP is provided by Schroeder et al. (1994b).

Input to HELP Analysis. The specific *climatologic data* needed for HELP include precipitation, temperature data, solar radiation, and evapotranspiration information, such as maximum leaf area and evaporative zone depth. The weather database available for a five-year time period for 102 cities in the United States may be used or the weather data may be generated stochastically for 1 to 100 years for 139 cities in the United States.

The *material data* needed for HELP analysis include the type of layer, thickness, initial water content, and soil texture. The material layers are divided into four types: vertical percolation layer, lateral drainage layers, barrier soil liners, and barrier soil liner with synthetic membrane. All of the layers in the cover system, leachate collection system, and liner system can be modeled by one of these layer types.

The *design data* include layer data such as drainage length and drain slope as well as general site data, which include the Soil Conservation Service (SCS) runoff curve number, landfill surface condition, landfill area, and surface vegetation.

Output of HELP Analysis. HELP analysis simulation can be performed for 2 to 100 years. The results of the analysis include daily, monthly, annual, and long-term average changes in runoff, evapotranspiration, lateral drainage, seepage through liner, head buildup on liner, and soil moisture storage. The HELP does not model the following: inflow through the liner if inward gradient conditions exist, pollutant transport, plant maturation over time (except during growing season), surface water run-on from adjacent watershed, characteristics associated with landfill aging (settlement, cracking, desiccation, etc.), leachate produced by decay and decomposition, and local variations in water or soil properties. Despite these limitations, HELP is used extensively because of its simplicity and reasonable assumptions.

The HELP model is commonly used to simulate two conditions:

1. *An open condition for the active period.* As the worst-case scenario, this condition is generally modeled with a 10- or 20-ft-thick layer of waste and a 6-in. daily cover placed over the liner and LCRS. No runoff from the site is assumed for this condition.

2. *A closed condition for the period of postclosure.* This condition is modeled with the final cover in place, and the water contents of the materials within the landfill are stabilized.

Other intermediate conditions may also be considered, depending on the design engineer's requirements. A 25-year 24-hour event is commonly used for estimating leachate quantities. The HELP model results are used to optimize the design of the liner, LCRS, and final cover. Example 18.1 provides the details of HELP analysis for typical landfill conditions.

As with most models, the HELP model incorporates several assumptions and has some significant limitations. For example, the use of Darcy's law requires the assumption that all materials are saturated. This may be untrue, especially for flow through waste. There is also no way to account for the preferential channeling of leachate through cracks in refuse or cover material,

EXAMPLE 18.1

For a MSW landfill cell, the collection pipes are centered 140 ft apart. The base liner system is sloped at 1.5% in the y direction and 0.54% in the x direction. (a) Calculate the amount of leachate generated and the maximum head on top of the composite liner system over a 30-year period by performing HELP analysis. (b) Using the estimated flows, calculate the leachate collection pipe sizes.

Solution:

(a) We consider three cases:

Case (1): 10 ft of MSW over a lined cell for 30 years based on local climate conditions. The input data are as follows:

- *Geometry:* shown in Figure E18.1A.

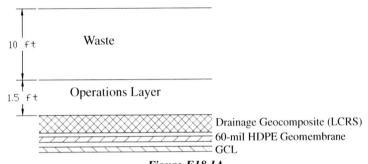

Figure E18.1A

- *Waste:* material no. 18; set the initial moisture to field capacity for a conservative estimate.
- *Operations layer:* material no. 27 (SC), set the initial moisture to field capacity for a conservative estimate.
- *Geocomposite LCRS:* material no. 0 (Note that the material numbers are needed for HELP analysis. A similar analysis can be performed for granular LCRS material underlain by a HDPE geomembrane and compacted clay layer. Appropriated material numbers from HELP will, however, need to be specified.)
- *Hydraulic conductivity* (k): from the manufacturer's specifications, transmissivity (T) = 0.1×10^{-3} m^2/s and thickness (t) $\approx$ 220 mils ($\approx$ 5 mm):

$$k = \frac{T}{t}$$

$$= \frac{0.1 \times 10^3 \text{ m}^2/\text{s} \ (1000 \text{ cm}^2/\text{m}^2)}{220 \text{ mils} \ (\text{in.}/1000 \text{ mils}) \ (2.54 \text{ cm/in.})}$$

$$\approx 2.0 \text{ cm/s}$$

- *Drainage length and slope* (Figure E18.1B):

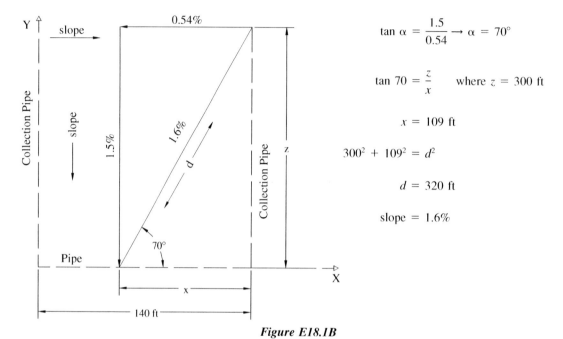

$$\tan \alpha = \frac{1.5}{0.54} \rightarrow \alpha = 70°$$

$$\tan 70 = \frac{z}{x} \quad \text{where } z = 300 \text{ ft}$$

$$x = 109 \text{ ft}$$

$$300^2 + 109^2 = d^2$$

$$d = 320 \text{ ft}$$

$$\text{slope} = 1.6\%$$

Figure E18.1B

- *Geomembrane:* material no. 35, two holes per acre (i.e., two defects per acre), good installation conditions.
- *GCL:* material no. 17, set the initial moisture to field capacity for a conservative estimate.
- *General design data:*

      Slope length = 300 ft
      Slope = 2% (assumed)
      No runoff allowed, bare ground, default SCS runoff curve
      Evaporative depth = 40 in. (assumed)
      Area = 1 acre (to simplify scaling)
      Initial snow water = 0
      Subsurface inflow = 0

- *Weather Data:*

      Precipitation: assume a value of 16.5 in./yr for the site
      Temperature: assume values for the site
      Solar radiation coefficients for the site adjusted for latitude of the site

Case 2: 100 ft of waste and 2 ft of cover over a lined cell for 30 years based on local climate conditions. This case represents the conditions for the portion of the landfill underlying the top deck (3% slope, 275 ft length). All other input is the same as in case 1.

Case 3: 50 ft of waste and 2 ft of cover over a lined cell for 30 years based on local climate conditions. This case represents the conditions for the portion of the landfill underlying the side slopes (29% slope, 230 ft length). All other input is the same as in case 1.

The HELP input and output file for case 1 is shown in Figure E18.1C but the input–output files for cases 2 and 3 are not presented; these can be obtained as for case 1. A summary of results for the three cases is presented below.

Case Description	h_{max} (peak daily) (in.)	Q_{max} (peak daily) (ft^3)	Q_{avg} (year) (ft^3)
Case 1 10 ft waste no runoff 2% slope at 300 ft	0.093	96.3	2356
Case 2 100 ft waste 2 ft int. cover runoff allowed 3% slope at 275 ft	0.064	66.1	130.6
Case 3 50 ft waste 2 ft int. cover runoff allowed 29% slope at 230 ft	0.064	66.1	133.1

Q_{max} = peak daily flow in cubic feet per acre

Q_{avg} = yearly average flow in cubic feet per acre

h_{max} = peak daily maximum head on liner

As shown above, in all cases the maximum head over the liner is less than 1 ft; therefore, the design is acceptable.

```
********************************************************
**                                                    **
**                                                    **
**    HYDROLOGIC EVALUATION OF LANDFILL PERFORMANCE   **
**    HELP MODEL VERSION 3.07  (1 NOVEMBER 1997)      **
**    DEVELOPED BY ENVIRONMENTAL LABORATORY           **
**    USAE WATERWAYS EXPERIMENT STATION               **
**    FOR USEPA RISK REDUCTION ENGINEERING LABORATORY **
**                                                    **
**                                                    **
********************************************************


PRECIPITATION DATA FILE:        C:\HELP3\EXAMPLE\DATA4.D4
TEMPERATURE DATA FILE:          C:\HELP3\EXAMPLE\DATA7.D7
SOLAR RADIATION DATA FILE:      C:\HELP3\EXAMPLE\DATA13.D13
EVAPOTRANSPIRATION DATA:        C:\HELP3\EXAMPLE\DATA11.D11
SOIL AND DESIGN DATA FILE:      C:\HELP3\EXAMPLE\DATA10.D10
OUTPUT DATA FILE:               C:\HELP\EXAMPLE\EX18.OUT


TIME:   ██████        DATE:


TITLE:  EXAMPLE 18.1 CASE 1

********************************************************

NOTE:  INITIAL MOISTURE CONTENT OF THE LAYERS AND SNOW WATER
       WERE SPECIFIED BY THE USER.


                    LAYER   1
                    -------

           TYPE 1 - VERTICAL PERCOLATION LAYER
                    MATERIAL TEXTURE NUMBER 18
THICKNESS                     =    120.00   INCHES
POROSITY                      =     0.6710  VOL/VOL
FIELD CAPACITY                =     0.2920  VOL/VOL
WILTING POINT                 =     0.0770  VOL/VOL
INITIAL SOIL WATER CONTENT    =     0.2920  VOL/VOL
EFFECTIVE SAT. HYD. COND.     =  0.100000005000E-02 CM/SEC


                    LAYER   2
                    -------

           TYPE 1 - VERTICAL PERCOLATION LAYER
                    MATERIAL TEXTURE NUMBER 27
THICKNESS                     =     18.00   INCHES
POROSITY                      =     0.4000  VOL/VOL
FIELD CAPACITY                =     0.3660  VOL/VOL
WILTING POINT                 =     0.2880  VOL/VOL
INITIAL SOIL WATER CONTENT    =     0.3660  VOL/VOL
EFFECTIVE SAT. HYD. COND.     =  0.779999993000E-06 CM/SEC
```

```
                    LAYER   3
                    -------

           TYPE 2 - LATERAL DRAINAGE LAYER
                    MATERIAL TEXTURE NUMBER  0
THICKNESS                     =      0.20   INCHES
POROSITY                      =     0.8500  VOL/VOL
FIELD CAPACITY                =     0.0100  VOL/VOL
WILTING POINT                 =     0.0050  VOL/VOL
INITIAL SOIL WATER CONTENT    =     0.0100  VOL/VOL
EFFECTIVE SAT. HYD. COND.     =  2.00000000000   CM/SEC
SLOPE                         =      1.60   PERCENT
DRAINAGE LENGTH               =    320.0    FEET


                    LAYER   4
                    -------

           TYPE 4 - FLEXIBLE MEMBRANE LINER
                    MATERIAL TEXTURE NUMBER 35
THICKNESS                     =      0.06   INCHES
POROSITY                      =     0.0000  VOL/VOL
FIELD CAPACITY                =     0.0000  VOL/VOL
WILTING POINT                 =     0.0000  VOL/VOL
INITIAL SOIL WATER CONTENT    =     0.0000  VOL/VOL
EFFECTIVE SAT. HYD. COND.     =  0.199999960000E-12 CM/SEC
FML PINHOLE DENSITY           =      2.00   HOLES/ACRE
FML INSTALLATION DEFECTS      =      2.00   HOLES/ACRE
FML PLACEMENT QUALITY         =      3 - GOOD


                    LAYER   5
                    -------

           TYPE 3 - BARRIER SOIL LINER
                    MATERIAL TEXTURE NUMBER 17
THICKNESS                     =      0.25   INCHES
POROSITY                      =     0.7500  VOL/VOL
FIELD CAPACITY                =     0.7470  VOL/VOL
WILTING POINT                 =     0.4000  VOL/VOL
INITIAL SOIL WATER CONTENT    =     0.7500  VOL/VOL
EFFECTIVE SAT. HYD. COND.     =  0.300000003000E-08 CM/SEC


         GENERAL DESIGN AND EVAPORATIVE ZONE DATA

NOTE:  SCS RUNOFF CURVE NUMBER WAS COMPUTED FROM DEFAULT
       SOIL DATA BASE USING SOIL TEXTURE #18 WITH BARE
       GROUND CONDITIONS, A SURFACE SLOPE OF  2.% AND
       A SLOPE LENGTH OF  300. FEET.

SCS RUNOFF CURVE NUMBER               =     79.80      PERCENT
FRACTION OF AREA ALLOWING RUNOFF      =      0.0       PERCENT
AREA PROJECTED ON HORIZONTAL PLANE    =      1.000     ACRES
EVAPORATIVE ZONE DEPTH                =     40.0       INCHES
INITIAL WATER IN EVAPORATIVE ZONE     =     11.680     INCHES
```

Figure E18.1C

740

UPPER LIMIT OF EVAPORATIVE STORAGE   = 26.840   INCHES
LOWER LIMIT OF EVAPORATIVE STORAGE   = 3.080   INCHES
INITIAL SNOW WATER   = 0.000   INCHES
INITIAL WATER IN LAYER MATERIALS   = 41.818   INCHES
TOTAL INITIAL WATER   = 41.818   INCHES
TOTAL SUBSURFACE INFLOW   = 0.00   INCHES/YEAR

EVAPOTRANSPIRATION AND WEATHER DATA

NOTE: EVAPOTRANSPIRATION DATA WAS OBTAINED FROM ▆

        STATION LATITUDE   = 34.49   DEGREES
        MAXIMUM LEAF AREA INDEX   = 0.00
        START OF GROWING SEASON (JULIAN DATE) = 145
        END OF GROWING SEASON (JULIAN DATE)   = 272
        EVAPORATIVE ZONE DEPTH   = 40.0   INCHES
        AVERAGE ANNUAL WIND SPEED   = 7.40   MPH
        AVERAGE 1ST QUARTER RELATIVE HUMIDITY = 58.00   %
        AVERAGE 2ND QUARTER RELATIVE HUMIDITY = 41.00   %
        AVERAGE 3RD QUARTER RELATIVE HUMIDITY = 54.00   %
        AVERAGE 4TH QUARTER RELATIVE HUMIDITY = 58.00   %

NOTE: PRECIPITATION DATA WAS SYNTHETICALLY GENERATED USING ▆
       COEFFICIENTS FOR ▆

           NORMAL MEAN MONTHLY PRECIPITATION (INCHES)

JAN/JUL	FEB/AUG	MAR/SEP	APR/OCT	MAY/NOV	JUN/DEC
1.70	1.58	1.58	0.82	0.49	0.33
2.17	2.58	1.42	1.11	1.20	1.49

NOTE: TEMPERATURE DATA WAS SYNTHETICALLY GENERATED USING ▆
       COEFFICIENTS FOR ▆

          NORMAL MEAN MONTHLY TEMPERATURE (DEGREES FAHRENHEIT)

JAN/JUL	FEB/AUG	MAR/SEP	APR/OCT	MAY/NOV	JUN/DEC
43.00	46.00	49.90	57.10	65.10	74.80
79.60	77.40	72.60	62.60	51.50	44.40

NOTE: SOLAR RADIATION DATA WAS SYNTHETICALLY GENERATED USING ▆
       COEFFICIENTS FOR ▆
       AND STATION LATITUDE = 34.49 DEGREES

****

           ANNUAL TOTALS FOR YEAR   1

	INCHES	CU. FEET	PERCENT
PRECIPITATION	14.78	53651.414	100.00

RUNOFF	0.000	0.000	0.00
EVAPOTRANSPIRATION	17.764	53593.949	99.89
DRAINAGE COLLECTED FROM LAYER 3	2.0161	7318.497	13.64
PERC./LEAKAGE THROUGH LAYER 5	0.000003	0.011	0.00
AVG. HEAD ON TOP OF LAYER 4	0.0097		
CHANGE IN WATER STORAGE	-2.000	-7261.052	-13.53
SOIL WATER AT START OF YEAR	41.817	151797.328	
SOIL WATER AT END OF YEAR	39.815	144536.266	
SNOW WATER AT START OF YEAR	0.000	0.000	0.00
SNOW WATER AT END OF YEAR	0.000	0.000	0.00
ANNUAL WATER BUDGET BALANCE	0.0000	0.008	0.00

****

           ANNUAL TOTALS FOR YEAR   2

	INCHES	CU. FEET	PERCENT
PRECIPITATION	20.35	73870.516	100.00
RUNOFF	0.000	0.000	0.00
EVAPOTRANSPIRATION	20.299	73683.711	99.75
DRAINAGE COLLECTED FROM LAYER 3	0.7662	2731.166	3.76
PERC./LEAKAGE THROUGH LAYER 5	0.000002	0.009	0.00
AVG. HEAD ON TOP OF LAYER 4	0.0037		
CHANGE IN WATER STORAGE	-0.715	-2594.398	-3.51
SOIL WATER AT START OF YEAR	39.817	144536.266	
SOIL WATER AT END OF YEAR	39.102	141941.875	
SNOW WATER AT START OF YEAR	0.000	0.000	0.00
SNOW WATER AT END OF YEAR	0.000	0.000	0.00
ANNUAL WATER BUDGET BALANCE	0.0000	0.031	0.00

****

           ANNUAL TOTALS FOR YEAR   3

	INCHES	CU. FEET	PERCENT
PRECIPITATION	14.04	50965.207	100.00

Figure E18.1C (continued)

RUNOFF	0.000	0.000	0.00
EVAPOTRANSPIRATION	15.512	56307.965	110.48
DRAINAGE COLLECTED FROM LAYER 3	0.3572	1296.529	2.54
PERC./LEAKAGE THROUGH LAYER 5	0.000002	0.008	0.00
AVG. HEAD ON TOP OF LAYER 4	0.0017		
CHANGE IN WATER STORAGE	-1.829	-6639.334	-13.03
SOIL WATER AT START OF YEAR	39.102	141941.875	
SOIL WATER AT END OF YEAR	37.273	135302.531	
SNOW WATER AT START OF YEAR	0.000	0.000	0.00
SNOW WATER AT END OF YEAR	0.000	0.000	0.00
ANNUAL WATER BUDGET BALANCE	0.0000	0.039	0.00

****

AVERAGE ANNUAL TOTALS & (STD. DEVIATIONS) FOR YEARS 1 THROUGH 30

	INCHES		CU. FEET	PERCENT
PRECIPITATION	16.73	(3.127)	60745.6	100.00
RUNOFF	0.000	(0.0000)	0.00	0.000
EVAPOTRANSPIRATION	16.208	(2.8460)	58836.33	96.857
LATERAL DRAINAGE COLLECTED FROM LAYER 3	0.64917	(0.76635)	2356.497	3.87929
PERCOLATION/LEAKAGE THROUGH LAYER 5	0.00000	(0.00000)	0.008	0.00001
AVERAGE HEAD ON TOP OF LAYER 4	0.003	(0.004)		
CHANGE IN WATER STORAGE	-0.123	(1.0061)	-447.21	-0.736

****

PEAK DAILY VALUES FOR YEARS 1 THROUGH 30

	(INCHES)	(CU. FT.)
PRECIPITATION	2.10	7623.000
RUNOFF	0.000	0.0000
DRAINAGE COLLECTED FROM LAYER 3	0.02653	96.31059
PERCOLATION/LEAKAGE THROUGH LAYER 5	0.000000	0.00006

AVERAGE HEAD ON TOP OF LAYER 4	0.047
MAXIMUM HEAD ON TOP OF LAYER 4	0.093
LOCATION OF MAXIMUM HEAD IN LAYER 3 (DISTANCE FROM DRAIN)	3.2 FEET
SNOW WATER	1.31 4756.4434
MAXIMUM VEG. SOIL WATER (VOL/VOL)	0.3292
MINIMUM VEG. SOIL WATER (VOL/VOL)	0.1609

*** Maximum heads are computed using McEnroe's equations. ***

Reference: Maximum Saturated Depth over Landfill Liner by Bruce M. McEnroe, University of Kansas ASCE Journal of Environmental Engineering Vol. 119, No. 2, March 1993, pp. 262-270.

****

****

FINAL WATER STORAGE AT END OF YEAR 30

LAYER	(INCHES)	(VOL/VOL)
1	31.3440	0.2612
2	6.5880	0.3660
3	0.0020	0.0100
4	0.0000	0.0000
5	0.1875	0.7500
SNOW WATER	0.000	

****

****

Figure E18.1C (continued)

(b) Assume the drainage pattern shown in Figure E18.1D.

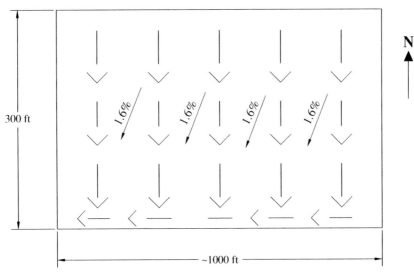

Figure E18.1D

$$\text{Area} \simeq 300 \text{ ft} \times 1000 \text{ ft} \simeq 300,000 \text{ ft}^2 = 6.9 \text{ acres}$$

• Peak flow: header pipe. All flow is toward a header pipe at the south end of the cell. Header pipe flow to the southwestern corner of the cell, where it connects with a pipe that drains to the main sump. Based on the results of HELP for 10 ft of waste (case 1):

$$Q_{\text{total}} = \frac{Q_{\text{max}}}{\text{acre}} (\text{area}) = \frac{96.3 \text{ ft}^3}{\text{acre} \cdot \text{day}} \times 6.9 \text{ acres} = 665 \text{ ft}^3/\text{day}$$

$$= 7.7 \times 10^{-3} \text{ ft}^3/\text{s}$$

• *Pipe sizing according to Manning's equation:*

$$Q \text{ (ft}^3/1) = \frac{1.486}{n} R_h^{2/3} S^{1/2} A$$

$$= \text{pipe flow capacity}$$

$$n = \text{roughness coefficient} = 0.009 \text{ (\emph{Driscopipe Design Manual})}$$

$$R_h = \text{hydraulic radius (ft)} = \frac{A}{P} = \frac{\pi D^2/4}{\pi D} = \frac{D}{4}$$

$$S = \text{pipe slope} = 0.0054$$

$$A = \text{flow area (ft}^2)$$

From the authors' experience and as recommended by Sharma and Lewis (1994), a minimum diameter of pipe should be 6 in., due to the potential for biological clogging of smaller-diameter pipes.

Verify the capacity of 6-in. header pipe ($D = 0.5$ ft):

$$R_h = \frac{0.5 \text{ ft}}{4} = 0.125 \text{ ft}$$

$$A = (0.25)^2 \pi = 0.196 \text{ ft}^2$$

$$Q_{pipe} = \frac{1.486}{0.009}(0.125)^{2/3}(0.0054)^{1/2}(0.196) = 0.59 \text{ ft}^3/\text{s}$$

The factor of safety for 6-in.-diameter pipe:

$$\frac{Q_{pipe}}{Q_{max}} = \frac{0.59 \text{ ft}^3/\text{s}}{7.7 \times 10^{-3} \text{ ft}^3/\text{s}} = 77 \qquad \text{OK}$$

Although the 6-in. pipe has significantly higher capacity than leachate flow, the header pipe should have a minimum 6 in. diameter, as discussed above. Secondary pipes crossing are less critical to drainage of leachate and may be 4 in. in diameter.

---

although the location and depth of the crack would be difficult to define in any model. There is considerable experience, however, in using the HELP model. It has been used by practicing engineers since the first version was released in 1984. Generally, the model has proven to be a valuable tool in predicting leachate generation rates.

It is generally recognized that in arid and semiarid regions, the HELP model overestimates infiltration rates. However, in humid and semihumid climates, HELP provides reasonable estimates for infiltration rates (EPRI, 1981; Sharma et al., 1989). It is the author's experience that the infiltration model UNSAT-H (Fayer, 2000) predicts reasonable infiltrations in arid and semiarid climates and should be used for such sites.

18.3.2 Leachate Collection Pipes

A network of slotted piping is placed within the drainage layer to collect and drain leachate toward collection points, where it can be removed out of the landfill.

A careful selection of the type, size, and spacing of the pipes should be made. The type of pipe selected is based on several factors, including the leachate compatibility, durability, physical properties, ease of installation, and cost. Due to leachate compatibility and perforations required, flexible thermoplastic pipes are generally used. Commonly, slotted high-density polyethylene (HDPE) pipes with inside diameter between 4 and 8 in. are used. The pipe type and size are selected based on hydraulic and structural analysis to ensure that the estimated flow can be accommodated and stability under loads (construction loads and waste loads) can be ensured. The spacing of the pipes should be selected such that leachate head in the drainage layer does not exceed 12 in. In the following sections, analysis to select pipe spacing is presented first. Then hydraulic analysis and structural analysis of leachate collection pipes is explained.

Pipe Spacing. The maximum spacing of the pipes is determined by calculating the maximum depth of leachate that will result due to leachate impingement into the drainage layer. The landfill bases should be graded (longitudinally and transversely) to promote flow toward leachate collection pipes and sumps. LCRSs are therefore more typical of the configuration presented in Figure 8.3. The pipe spacing (L) can be obtained from the following equation (Moore, 1980):

$$L = \frac{2h_{max}}{\sqrt{c}\left[\left(\dfrac{\tan^2 \phi}{c}\right) + 1 - \left(\dfrac{\tan \phi}{c}\right)\sqrt{\tan^2 \phi + c}\right]}$$

(18.2)

where $c = q/k$ and ϕ is the slope angle of liner.

In some cases, a geonet or geocomposite drainage net may be used to replace the granular drainage material in the LCRS. The hydraulic properties of a geonet are expressed in terms of transmissivity rather than permeability. Therefore, for LCRSs containing geonets or geocomposite drainage nets, the hydraulic conductivity of the drainage material used in equation (18.2) should be converted to an equivalent transmissivity. This conversion is derived in equation (18.3) using Darcy's law:

$$q = kiA \qquad (18.3a)$$

$$q = ki(wt) \qquad (18.3b)$$

$$kt = \theta = \frac{q}{iw} \qquad (18.3c)$$

$$k = \frac{\theta}{t} \qquad (18.3d)$$

where q is the flow rate, k the hydraulic conductivity, i the hydraulic gradient, A the cross-sectional area perpendicular to flow, w the flow width, t the thickness of the geosynthetic drainage material, and θ the transmissivity of geosynthetic drainage media. Therefore, use of this conversion assumes implicitly that the geosynthetic drainage media is saturated and that flow is laminar. In some cases, these assumptions may not be valid, and the flow rate per unit width of the geosynthetic drainage media should be used. HELP analysis can be used to estimate pipe spacing for the calculated infiltration and limiting the 12-in. head on the liner system (see Example 18.1).

A part of the leachate flows into the collection pipes and the rest may leak through the liner. To quantify this leachate flow distribution, the model developed by Wong (1977) is commonly used. This model assumes that leachate moves toward a collection point as an instantaneously developed rectilinear slab with a phreatic surface parallel to the liner. Sharma and Lewis (1994) provide further details on this model.

A parametric analysis based on modified Wong's analysis was performed by Kmet et al. (1981) to identify the variables that affect the drainage performance significantly. These results suggest that to have the minimum (negligible) leachate leakage, it is necessary to have base liner permeability at least four orders of magnitude less than the drainage layer, and the liner slope or base grade (α) should be approximately 2%. In addition, the closer the pipe spacing, the higher will be the efficiency of the drainage system.

Hydraulic Analysis. After estimating the spacing of the leachate collection pipes, a hydraulic analysis must be performed to determine the correct pipe diameter,

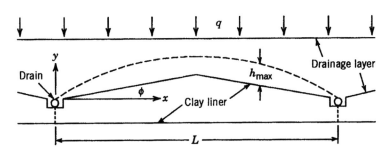

Figure 18.3 *Sloping liner leachate collection system. (From Demetracopoulos et al., 1984. Reproduced by permission of ASCE.)*

slot size, and slot spacing. To check that the pipe capacity will be adequate, Manning's formula is used to calculate the pipe flow capacity (Simon and Korom, 1997):

$$Q = \frac{1.486}{n} A R_h^{2/3} S^{1/2} \qquad (18.4)$$

where A is the cross-sectional area in square inches, R_h the hydraulic radius in inches, n Manning's roughness coefficient, and S the slope in ft/ft. The factor of safety (FS) for flow capacity (Q) can be calculated using

$$\text{FS} = \frac{Q}{Q_L} \qquad (18.5)$$

Q_L, the expected leachate flow rate, is determined by using HELP analysis.

For pipe slots, square orifice theory is used as follows (Simon and Korom, 1997):

$$Q = CA(2gH)^{1/2} \qquad (18.6)$$

where C is about 0.6, A the cross-sectional area, and H is the head above the liner. Based on this, the number of slots required per foot length of pipe (Q per foot) can be calculated. To prevent clogging or the washing of particles from the drainage layer, the following criteria are generally satisfied:

$$\frac{D_{85} \text{ drainage soil}}{\text{diameter or width of slots}} \geq 2 \qquad (18.7)$$

Structural Analysis. High-density polyethylene (HDPE) pipes are currently the most commonly used pipes for LCRS applications, due to their strength characteristics and chemical resistance to many constituents found in leachate. Since HDPE pipe is a flexible pipe with the ability to transfer load to surrounding soils, it also performs well under several hundreds feet of refuse that may be disposed in a landfill. Polyvinyl chloride pipe (PVC) may also be used, but due to the superior chemical resistance of HDPE to leachate,

HDPE has become the industry standard for LCRSs. The analytical discussion presented in this section therefore focuses primarily on HDPE pipe, although some of the theory may also apply to other flexible pipes.

The three pipe parameters typically found to be most critical in the performance analysis of buried flexible pipes are deflection, wall crushing, and wall buckling (Driscopipe, 1988). The vertical pipe *deflection* is calculated using

$$\frac{\delta}{D} = FKP(1.1) \frac{2E}{3(\text{SDR} - 1)^3} + 0.061E' \qquad (18.8)$$

where F is the lag factor (usually equal to 1.0), K the bedding constant = 0.085, P the pressure (be sure to include the drainage layer, the waste, and the final cover), E the elastic modulus in lb/in², E' the soil modulus in lb/in², and SDR the pipe standard dimension ratio (= pipe diameter/pipe thickness). To check the potential for pipe wall *crushing*, calculate

$$P_W = \frac{(\text{SDR} - 1)P}{2} \qquad (18.9)$$

where P_w is the pipe wall stress and P is the pressure in lb/in². Then, compare P_w to the allowable material stress in terms of factor of safety:

$$\text{FS} = \frac{P_{w \text{ allowable}}}{P_{w \text{ calculated}}} \qquad (18.10)$$

Finally, to check the potential for pipe *buckling*, calculate the critical buckling pressure without the surrounding soil, using the equation

$$P_{cr} = \frac{2.32E}{(\text{SDR})^3} \qquad (18.11)$$

With soil surrounding the pipe, the critical buckling pressure is calculated by the following equation:

$$P_b = 1.15(E'P_{cr})^{1/2} \qquad (18.12)$$

Then the factor of safety against pipe buckling is given by

$$FS = \frac{P_{b \text{ allowable}}}{P_{b \text{ calculated}}} \qquad (18.13)$$

For allowable deflection and allowable material strengths, reference should be made to pipe manufacturers' data sheets.

18.3.3 Selection of Drainage Layer Material and Thickness

Three major types of materials are commonly used as the drainage material: (1) granular soils (e.g., sands and gravels), (2) geosynthetics (e.g., geonets and geocomposites), and (3) alternative materials (e.g., recycled materials such as shredded tires and glass cullet). Ideally, these materials should be selected such that (1) the material should be permeable enough to collect and transport leachate and avoid accumulation of leachate head greater than 1 ft, (2) the material should be compatible with the waste, and (3) the material should not damage the liner. Some state regulations stipulate a minimum material thickness of 12 in. for soils and materials alike, and the material must possess a hydraulic conductivity (k) value greater than or equal to 1×10^{-2} cm/s.

In case of soils, the permeability tests are performed using standard test methods (e.g., ASTM D 2434). Special tests may be required if the hydraulic conductivity is greater than 1.0 cm/s. To check chemical compatibility, the soil sample is immersed in a representative leachate for two to three months, and then a grain size analysis is performed. If there is a significant difference in the grain size distribution between the original and tested samples, the effects of gradation changes on the hydraulic conductivity are assessed.

If there are sharp particles in the drainage layer, there is a potential for damage to the liner, especially a geomembrane liner during and after construction. Therefore, to protect the liner, drainage soils containing small (generally smaller than $\frac{3}{8}$-in.), rounded to subrounded particles should be used, or a geotextile

should be used between the geomembrane and the drainage layer (Reddy et al., 1996; Reddy and Saichek, 1998).

The procedure for evaluating the use of alternative materials (such as scrap tires and glass cullet) for a drainage layer is similar to that used for soils. For geosynthetics, the equivalency in the hydraulic conductivity (k) is considered.

Regardless of what material is used in the drainage layer, its stability on slopes must be evaluated. In addition, the clogging potential of drainage material must be assessed and measures taken to prevent it from occurring. The analyses of slope stability and clogging potential are discussed below.

Stability on Slopes. The drainage materials have to be stable; generally, this becomes an issue on the side slopes. As the soils are commonly used as leachate drainage material, methods to analyze the stability of these soil slopes are presented in this section. For alternative materials (such as recycled shredded scrap tires), the same analysis methods may be used. However, when using geosynthetic materials as drainage material (such as geonets, geotextiles, and geocomposites), it is important to consider the geosynthetic properties, such as thickness, tensile strength, and interface shear strength, in determining their stability on the slopes.

Since the thickness of the drainage layer is relatively thin (generally, 12 in.) as compared to the length of the slope, a simple infinite slope stability analysis may be performed . Figure 18.4 shows a schematic of drainage layer overlying a liner system. The factor of safety (FS), with no water table or seepage conditions, can be calculated using

$$FS = \frac{\tan \delta}{\tan \alpha} (c_a = 0) \quad \text{or} \quad \frac{\tan \phi}{\tan \alpha} (c = 0) \quad (18.14)$$

where δ is the interface friction angle between the drainage material and the underlying material (compacted clay or geomembrane or geotextile); ϕ the friction angle for the drainage material; and α the slope angle. If c_a is not equal to zero, FS is given by

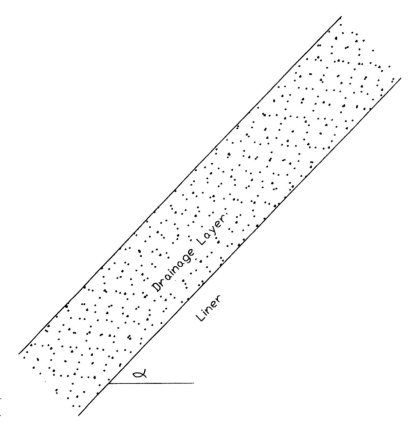

Figure 18.4 *Schematic of drainage layer overlying a liner system.*

$$FS = \frac{c_a}{\gamma H \cos^2\alpha \tan \alpha} + \frac{\tan \delta}{\tan \alpha} \quad (18.15)$$

Several different analysis methods that take into account of the slope geometry as well as loading conditions have been reported to analyze the drainage layers on the slopes. Koerner and Hwu (1991) proposed a method that assumes two discrete zones, as shown in Figure 18.5. A small passive wedge that resists a long, thin active wedge extending the length of the slope is assumed. At the top of the slope, or at an intermediate berm, a tension crack in the drainage soil is considered to occur, thereby separating from the soil above it.

Resisting the tendency for the drainage soil to slide is the adhesion and/or interface friction of the drainage soil to the specific type of underlying material. The values of c_a and δ must be obtained for this interface.

The passive wedge is assumed to move on the underlying drainage soil so that the shear parameters c and ϕ of the drainage material are used. By taking free bodies of the passive and active wedges with the appropriate forces being applied, the following formulation for the factor of safety is derived:

$$FS = \frac{-b \pm \sqrt{b^2 - 4ac}}{2a} \quad (18.16)$$

where

$a = 0.5\gamma LH \sin^2 2\alpha$

$b = -[\gamma LH \cos^2\alpha \tan \delta \sin 2a + c_a L \cos \alpha \sin 2\alpha + \gamma LH \sin^2 \alpha \tan \phi \sin 2\alpha + 2cH \cos \alpha + \gamma H^2 \tan \phi]$

$c = (\gamma LH \cos \alpha \tan \delta + c_a L) \tan \phi \sin \alpha \sin 2\alpha$

$H = $ layer height

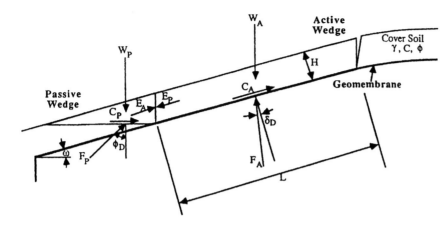

Figure 18.5 Stability analysis taking into account of slope geometry and loading conditions (From Koerner, © 1994. Reprinted by permission of Pearson Education, Inc., Upper Saddle River, NJ.)

L = slope length
α = slope angle
c_a, δ = adhesion and interfacial friction
c, ϕ = shear strength parameters

If the factor of safety becomes unacceptably low for the site-specific conditions, a possible solution to the situation is to add a layer of geogrid or geotextile reinforcement as shown in Figure 18.6. The tensile stresses that are mobilized in the reinforcement are carried to an individual anchor trench. If the reinforcement is a geogrid, it is placed within the drainage soil so that the soil can strike through the apertures and the maximum amount of anchorage against the transverse ribs can be mobilized. When using geotextiles, they can be placed directly on the liner or embedded within the drainage soil so as to mobilize friction in both sur-

faces. Koerner and Hwu (1991) derived the following equation for the tensile stress of the reinforcement layer per unit width to yield a factor of safety of 1. The reinforcement is selected to provide adequate safety; thus an alternative definition of the factor of safety is defined in terms of tensile stress and tensile strength of reinforcement layer:

$$T_{\text{reqd}} = \frac{\gamma LH \sin (\alpha - \delta)}{\cos \delta} - c_a L$$
$$+ \frac{-\cos \phi \left[\dfrac{cH}{\sin \alpha} + \dfrac{\gamma H^2}{\sin \alpha \, (2\alpha)} \tan \phi \right]}{\cos(\phi + \alpha)}$$

$$\text{FS} = \frac{T_{\text{allowable}}}{T_{\text{reqd}}} \tag{18.17}$$

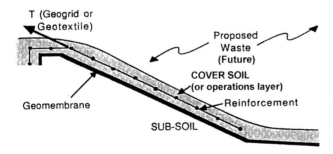

Figure 18.6 Geogrid or geotextile reinforcement in landfill cover (From Koerner, © 1994. Reprinted by permission of Pearson Education, Inc., Upper Saddle River, NJ.)

The analysis method above does not consider the effects of seepage forces and construction equipment loading on the slope stability. Druschel and Underwood (1993) considered these (seepage and construction equipment) loading conditions as shown in Figure 18.7 and derived the required anchorage force:

$$T_{\text{reqd}} = \frac{\gamma_w H_w^2}{2 \tan \alpha} \left(\frac{\tan \phi}{\cos^2 \alpha} + \frac{2D \tan \delta}{\cos \alpha} - \frac{\tan \delta}{\cos \alpha} \right)$$
$$+ W_e \left[0.3 + \frac{\sin (\alpha - \delta)}{\cos \delta} \right]$$
$$- \frac{\gamma H^2 \sin (\alpha - \delta)}{2 \sin \alpha \cos \alpha \cos \delta}$$
$$\left[\frac{\sin \phi \cos \delta}{\cos(\alpha + \phi) \sin(\alpha - \delta)} + 1 \right.$$
$$\left. - \left(\frac{2D \cos \alpha}{H} \right) \right] \quad (18.18)$$

where D is the vertical distance from the base of the landfill to the top of the slope, H_w the height of water within the drainage layer, H the height of the drainage layer, W_e the weight of the construction equipment, and α, δ, ϕ, γ, are as defined previously.

The drainage material on the slopes may not be easily placed in uniform thickness. These materials can unravel and slump very easily, even under static conditions. To alleviate this situation, common practice is to taper the drainage soil, laying it thicker at the bottom and gradually thinning going toward the top, as shown in Figure 18.8. For this configuration, the infinite slope stability analysis cannot be used. Stability analysis based on a wedge concept similar to that used for analyzing zoned earth dams by the Corps of Engineers is used. Shown in Figure 18.8 are the forces acting on the active wedge (the long, tapered section tending to cause failure) and the neutral block (the smaller triangular section tending to resist failure). The individual forces involved and their determination are the following: W_A is the weight of active wedge (area times unit weight), F_A the frictional force of drainage soil acting on the liner (unknown in magnitude, but known in direction, which is an angle δ perpendicular to the liner), δ the angle of shearing resistance of the drainage soil to the liner, E_A the force from the neutral block acting on the active wedge (unknown in magnitude but assumed parallel to the drainage soil slope), E_{NB} the force from the active wedge acting on the neutral block (equal in magnitude but opposite in direction to E_A), W_{NB} the weight of the neutral block (area times unit weight), F_{NB} the frictional force of the soil below the neutral block (unknown in magnitude but known in direction, which is at an angle ϕ to the vertical), and ϕ the angle of shearing resistance of the drainage soil.

The procedure is one of graphic statics, whereby force polygons for both the active wedge and the neutral block are drawn repeatedly until the polygons converge. Each iteration is made using a different factor of safety, which is applied to the δ and ϕ angles.

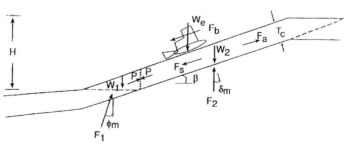

Figure 18.7 *Free-body diagram of side-slope forces.* (*From Druschel and Underwood, 1993.*)

Note: P, F_s, F_a, and F_b are assumed to be parallel to β

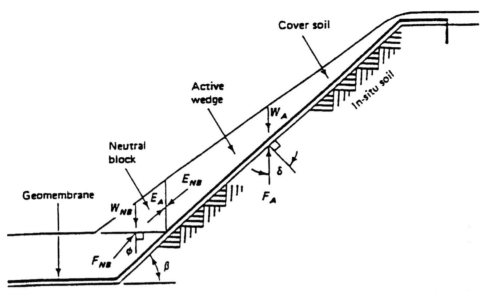

Figure 18.8 *Schematic diagram for forces involved with tapered corner soil on geomembrane-lined slope. (From Koerner, © 1994. Reprinted by permission of Pearson Education, Inc., Upper Saddle River, NJ.)*

Clogging Potential. Leachate drainage media may be clogged with the excessive migration of particles from the overlying waste or protective layer. Such clogging potential must be taken into consideration during the design stage. Geotextile filters are used to prevent such clogging. Geotextile filters function in a manner similar to graded soil filters; that is, they promote the formation of a filter zone consisting of either a filter bridge or a filter cake (Sharma and Lewis, 1994). Clogging of leachate drainage media may also occur due to various chemical, biological, and biochemical mechanisms. For example, formation of insoluble chemical precipitates such as calcium carbonate may cause blockage or cementation of drainage material. Biological growth by the organic and nutrient materials in leachate may cause blockage of pore spaces in drainage media. Although difficult to ascertain chemical and biological clogging, research study conducted by Koerner and Koerner (1990) suggest that the drainage media selected should have porosity large enough to pass the majority of sediment and microorganisms contained in the leachate in a steady-state manner. In ad-

dition, the effective long-term coefficient of permeability of drainage media should be used in the LCRS analysis and design.

18.3.4 Leachate Management

Leachate flows by gravity into the slotted collection pipes and then flows into the slotted header pipes, incorporated within the LCRS. The header pipes will then flow by gravity into the main collection sump, where the leachate will be pumped up into a tanker truck or on-site storage tanks. Generally, the leachate levels will be monitored at leachate collection structures and risers. The volume of leachate removed from each sump is measured.

The pumped leachate at landfills may be managed in three different ways: (1) recycling, (2) on-site treatment (better if large quantities of leachate are generated), and (3) off-site treatment (leachate is transported to a local wastewater treatment plant). The recycling option involves spreading the leachate on the top of the waste, and leachate infiltration through the waste increases the moisture content of the waste and hence accelerates degradation of the waste. Although this approach is not common, experimental studies at different landfills appear to yield positive results, and as a result, this option may find increased use in the future. Chapter 25 provides more details on leachate recycling and bioreactor landfills. Pumping and treating leachate on- or off-site has been the most common practice at landfills. If large quantities of leachate are generated and/or a local wastewater treatment facility is not closeby, it may be feasible to treat on-site. However, if the leachate generated is in small quantities and a local wastewater treatment facility is closeby, it may be economical to transport the leachate periodically to the wastewater treatment facility for treatment. Different methods can be used to treat leachate, and details on these methods may be found in Chapter 14 and in Metcalf & Eddy (2003).

18.4   CONTAINMENT SYSTEM LINER DESIGN

As discussed earlier and presented in Figure 18.1, the key elements of a containment system (from top down) are a LCRS, an HDPE geomembrane, and a low-permeability layer. The basic focus in geomembrane liner design should be on a relatively stress-free material that is resistant to puncture.

18.4.1   Material Stresses

The geomembrane liner within a landfill may be subjected to stresses in two phases: (1) during construction and (2) after liner construction is complete and waste has been placed on top of the liner. *During liner construction,* the exposed geomembrane is subjected to stress due to self-weight, wind, and temperature variations. On the side slopes, additional stresses may be imposed due to loads imposed by overlying operations (or protective soil) layer and the equipment used to construct the operations layer. *After liner construction* has been completed and the waste has been placed,

the side-slope liner will also experience down-drag stresses caused by waste settlement.

Material Stresses during Construction. On the containment system slopes, a lining system will experience stresses due to self-weight, wind, temperature, the operations layer, and construction equipment. The lining system will have a tendency to move downslope unless anchored at the top of the lined slope. The anchoring system at the top will resist the loads and will keep the lining system in place. This will impose stresses on the liner, however. The anchor trench design philosophy will therefore be that the "anchor" resistance should not exceed the allowable stress of the liner material under consideration. In this section we first present methods to estimate the loads imposed by various elements discussed above followed by methods to estimate anchor system resistance. A design example (Example 18.2) will then be presented to show how the anchor system should be sized so that the anchor resistance does not exceed the material allowable stress.

Stresses due to Liner Self-Weight. The liner self-weight on steep slopes may impose excessive tensile stress on a liner component, such as the geomembrane. Generally, this is not a problem on most slopes but should be checked on steep slopes. The tensile stress (σ_T) induced on the liner can be calculated by using the following relationships:

$$\sigma_T = \frac{W \sin \alpha - W \cos \alpha \tan \delta}{1 \times t} \quad (18.19)$$

$$W = S_g \, \gamma_w \, A \, \frac{h}{\sin \alpha} \quad (18.20)$$

where

S_g = specific gravity of geomembrane
γ_w = unit of weight of water
t = geomembrane thickness
h = slope height

W = geomembrane weight

A = cross-sectional area of unit geomembrane = 1 × t

δ = interface friction angle between geomembrane and underlying soil

α = slope angle

Compare the tensile stress (σ_T) with the yield strength σ_y of the geomembrane obtained from laboratory testing. Typical yield strength value for HDPE geomembrane is provided in Chapter 17 . The calculated design ratio $D_R = \sigma_Y/\sigma_T$ should be more than 10 (Koerner and Richardson, 1987).

Stresses due to Temperature Expansion and Contraction. In many geographic regions, the geomembranes, either during installation in landfill cells or during operations in liquid containment ponds, may experience temperature fluctuations between −30 and 40°C. These temperature fluctuations will result in significant contractions and elongations, resulting in tensile strains in geomembranes that are anchored. Adequate slack must therefore be incorporated during geomembrane installation to compensate for such temperature-induced expansion and contraction. The required slack or potential change in length (ΔL) to compensate for temperature fluctuation can be estimated as:

$$\Delta L = \Delta T L \mu \qquad (18.21)$$

where ΔT is the change in temperature, L the original length, and μ the coefficient of thermal expansion for the geomembrane.

Stresses due to Wind. In a majority of cases geomembranes can be damaged during installation due to wind because of improper anchorage of the geomembrane by the contractor. This problem can be resolved by placing sandbags on top of appropriately anchored geomembrane. In situations where the geomembrane is anchored permanently at the edges, the wind uplifts are caused by suction pressures. The wind uplift or suction pressure (q) can be estimated by the following simplified equation (Swanson, 1987):

$$q = 0.002556V^2 \qquad (18.22)$$

where q is in lb/ft² and V is wind speed in miles per hour. V can be obtained from the local weather station or from the Uniform Building Code.

Zornberg and Giroud (1997) provide a wind suction pressure on an anchored geomembrane considering the slope angle, elevation, geomembrane weight, and wind velocity. The relationship is as follows:

$$q = 0.05\lambda V^2\, e^{-(1.252\times10^{-4})z} - 9.81\mu_{GM}\cos\beta \qquad (18.23)$$

where q is in pascal, V is the wind velocity in km/h, Z the height above sea level in meters, and μ_{GM} the geomembrane mass in kg/m²; $\lambda = 0.4$ at the base flat surface of a slope, $\lambda = 0.7$ along the slope (an average value); this value varies between 0.55 at the toe of the slope and 1.0 at the crest of the slope. For unit conversions, standard conversion tables can be used.

Sandbags can be placed over the liner to counter wind uplift pressures. Alternatively, wind vents alone or in combination with sandbags can be utilized for such purposes. Sharma and Lewis (1994) discusses practical implications of wind uplift design in landfill liner construction application.

Stresses due to Operations Soil Layer and Equipment Loads. As shown on Figure 18.9, placement of an operations soil layer on top of a liner, such as a HDPE geomembrane, may induce tensile stress on the liner that is anchored at the top. The tensile force T can be estimated by the following equation:

$$T = W_{soil}\sin\beta + W_{equip}\sin\beta + \text{braking force} - P_p \qquad (18.24)$$

where

T = tensile force along the upper surface of geomembrane

W_{soil} = weight of operations soil layer (lb/ft)

W_{equip} = equipment weight

β = slope angle

Assume no breaking force is applied

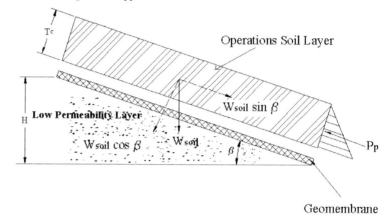

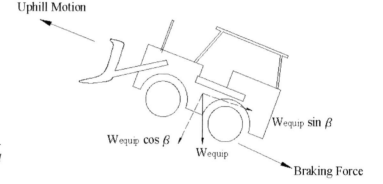

Uphill Motion

Figure 18.9 Stresses on geomembrane due to operations soil layer and placement equipment.

$$P_p = 0.5\, K_p\, \gamma_{soil}\, T_C^2$$
$$K_p = \tan^2(45 + \phi/2)$$
$$\phi = \text{friction strength parameter of soil}$$
$$\gamma_{soil} = \text{unit weight of } \textit{operations soil layer}$$
$$T_C = \text{thickness of } \textit{operations soil layer}$$

Generally, it is recommended that during placement of the *operations soil layer* the equipment should not apply braking force. This will reduce the induced tensile stress on the liner. This requirement should be specified in the specifications.

Liner Anchor Trench Design. As shown in Figure 18.1, a typical landfill liner system may consist of multilayer components. These layers, especially when placed on side slope, may be subject to stresses im-

posed during construction or in exposed applications due to self-weight, temperature variations, and wind. An anchor trench is typically designed to resist stresses either alone or together with other restraints, such as sandbags.

Design Philosophy. The main design consideration of a liner system is to have relatively "stress-free" scenarios. From a material resistance point of view, HDPE geomembrane is a critical element of the composite liner system. This means that stresses induced by various forces, discussed above, should not exceed the allowable tensile strength of various geosynthetic components (e.g., HDPE geomembranes). The implication of this requirement is that the anchor trench capacity or resistance should not exceed the geosynthetic liner allowable stress. In other words, it is preferable for the liner to slip or pull out of the anchor trench rather than to tear (Sharma and Lewis, 1994).

Anchor Trenches. As shown in Figure 18.10, anchor trenches are usually used at the top of a side-slope liner system to hold the installed geosynthetics in place. The selection of one of the following anchor trench types (a) flat runout, (b) V-trench, (c) rectangular, and (d)

narrow trench—depends on one or a combination of the following factors: (1) available space at the anchor trench location, (2) available construction equipment, (3) site access considerations, and (4) required anchor trench holding (resisting) capacity.

Anchor Trench Holding Capacity. The anchor trench capacity for various configurations, shown in Figure 18.10, can be obtained by using the simple mechanics

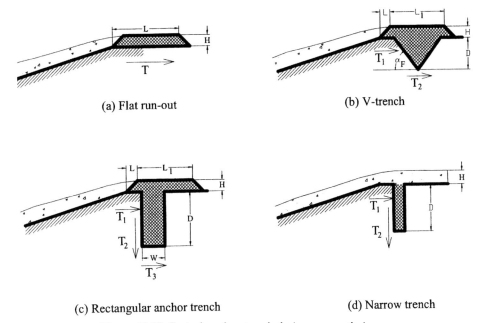

(a) Flat run-out

(b) V-trench

(c) Rectangular anchor trench

(d) Narrow trench

Figure 18.10 Typical anchor trench designs currently in use.

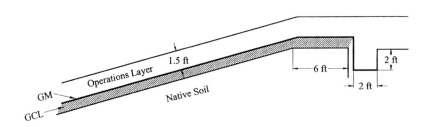

- GCL placed onto 6 ft run-out at access road, but not into trench
- Fill placed across 6 ft run-out and into anchor trench 2 ft wide and 2 ft deep
- Total slope length = 115 ft
- Slope length = 60 ft

Figure 18.11 Side-slope liner anchorage.

of free-body diagrams. Druschel and Underwood (1993), Sharma and Lewis (1994), Hullings and Sansone (1996), and Koerner (1998) discuss this in greater detail. The following discussion covers the basic formulation for anchor trench capacities for various anchor trench types. Example 18.2 presents a design problem and solution for an anchor trench design.

Flat Runout Anchor. As shown in Figure 18.10(*a*), the anchor force (*T*) for a flat runout anchor can be obtained by simple mechanics and is given by

$$T = W \tan \delta_L \qquad (18.25a)$$

$$T = LH\gamma \tan \delta_L \qquad (18.25b)$$

where the dimensions *L* and *H* are shown in Figure 18.10(*a*), *W* is the weight of the overlying soil, γ the unit weight of soil, and δ_L the interface friction angle between the lower side of the geosynthetics and the underlying soil or subgrade.

V-trench Anchor. As shown in Figure 18.10(*b*), the anchor trench force (*T*) can be divided into two components (T_1 and T_2). T_1 is calculated based on equation (18.25), and T_2 is calculated by the following two expressions:

1. If the overlying soil block moves:

$$T_2 = W \cos \alpha \tan \delta_L + W \sin \alpha \qquad (18.26a)$$

$$T_2 = \left(L_1 H + \frac{L_1 D}{2} \right)$$

$$\gamma \left[\cos \alpha \tan \delta_L + \sin \alpha \right] \qquad (18.26b)$$

2. If the overlying soil block does not move:

$$T = W (\tan \delta_L + \tan \delta_U) \qquad (18.27a)$$

$$T_2 = \left(L_1 H + \frac{L_1 D}{2} \right)$$

$$\gamma (\tan \delta_L + \tan \delta_U) \qquad (18.27b)$$

where δ_u is the interface friction angle between the upper side of geosynthetics and the overlying soil. Resistance due to T_2 will be the lower of the values obtained from (18.26*b*) and (18.27*b*). Finally, the total resistance will be given by

$$T = (LH \gamma \tan \delta_L) + T_2 \qquad (18.28)$$

Rectangular Trench Anchor. The anchor trench resistance *T* will be a sum of T_1, T_2, and T_3, as shown in Figure 18.10(*c*) and can be obtained from the following relationship:

$$T = T_1 + T_2 + T_3 \qquad (18.29a)$$

$$= (LH \gamma \tan \delta_L) + [K_0 \gamma H_{av} (\tan \delta_L + \tan \delta_U) D]$$

$$+ [W(H + D) \gamma(\tan \delta_L + \tan \delta_U)] \qquad (19.29b)$$

where K_0 is the coefficient of earth pressure at rest and is typically $= (1 - \sin \phi)$, where ϕ is the frictional strength parameter of soil backfill, and H_{av} is equal to $(H + D/2)$. Other terms are as defined previously.

Narrow Trench Anchor. As shown in the Figure 18.10(*d*), this is a special case of rectangular anchor trench where T_3 is zero. Therefore, equation (18.29*a*) will apply except that the last term will be zero.

Downdrag Stresses. The settlement of the waste overlying the liner system may induce stresses on the liner and the anchor trench. It is difficult to quantify these stresses (Sharma and Lewis, 1994). However, for slopes steeper than 3:1 (horizontal to vertical), these stresses do not appear to pose problems (Von Pein and Prasad, 1990). A design strategy to resolve this problem is to provide a liner system that has a lower interface friction angle above the geomembrane than below the geomembrane (Von Pein and Prasad, 1990; Koerner, 1992). One should, however, check the overall stability if this method is chosen (Sharma and Lewis, 1994).

EXAMPLE 18.2

A landfill containment system has a liner system consisting of GCL placed on a 3H:1V side slope. An HDPE geomembrane is placed over GCL overlain by a 1.5-ft-thick operations soil layer. The GCL is anchored as flat runout and the geomembrane is placed in a rectangular anchor trench as shown in Figure 18.11. (a) Calculate anchor forces on a geomembrane and the GCL assuming that (1) for a geomembrane, $\delta_U = 18°$ and $\delta_L = 19°$, and (2) for GCL $\delta_U = 19°$ and $\delta_L = 34°$. (b) If the operations layer is placed using a D6H dozer, calculate the factor of safety against tear for the geosynthetics. The dozer has an operating weight of 45,400 lb with a constant base width of 10.3 ft. Material data indicate that (1) a single-side textured HDPE geomembrane has a tensile strength (at yield) of 1584 lb/ft, and (2) a nonwoven (8 oz/yd²) geotextile component of GCL has a tensile strength (ultimate wide width) of 960 lb/ft (80 lb/in.).

Solution:

(a)(1) *Geomembrane anchorage (3:1 Slope)*: landfill liner

- *Geosynthetic:* interface friction angles (Figure E18.2A) are upper $\delta_U = 18°$ and lower $\delta_L = 19°$.

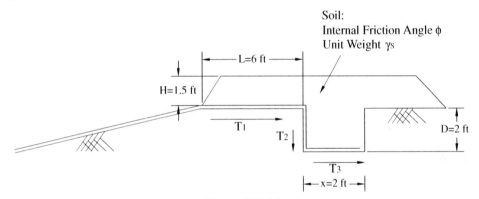

Figure E18.2A

- *Total anchor capacity:* using equation (18.29a) yields

$$T = T_1 + T_2 + T_3$$

where

$$T = L\,H\,\gamma_S \tan \delta_L$$

$$T_2 = \left[(1 - \sin \phi)\gamma_S \left(H + \frac{D}{2} \right) \right] (\tan \delta_L + \tan \delta_U)\, D$$

$$T_3 = X(H + D)\gamma_S (\tan \delta_L + \tan \delta_U)$$

For pullout of the geomembrane due to *operations soil* downdrag, consider cover soil overlying geomembrane overlying GCL. Find the total anchor capacity given by the following dimensions and properties:

Dimensions (ft)		Properties	
H	1.5	ϕ	34°
L	6	δ_U	18°
D	2	δ_L	19°
$W = X$	2	γ_S	115 lb/ft³

$$T_1 = 356.4 \ \text{lb/ft}$$

$$T_2 = 228.7 \ \text{lb/ft}$$

$$T_3 = 726.3 \ \text{lb/ft}$$

$$T = 1311.4 \ \text{lb/ft}$$

(a)(2) *GCL anchorage (3:1 slope)*: landfill liner

· *Geosynthetic:* interface friction angles (Figure E18.2B) are upper $\delta_U = 19°$ and lower $\delta_L = 34°$.

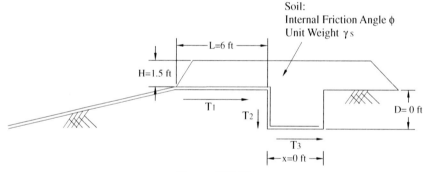

Figure E18.2B

· *Total anchor capacity:* Using equation (18.29a) yields

$$T = T_1 + T_2 + T_3$$

where

$$T_1 = LH\gamma_S \tan \delta_L$$
$$T_2 = \left[(1 - \sin \phi)\gamma_S \left(H + \frac{D}{2}\right)\right](\tan \delta_L + \tan \delta_u)D$$
$$T_3 = X (H + D)\gamma_S (\tan \delta_L + \tan \delta_U)$$

For pullout of the GCL due to operations soil downdrag, consider geomembrane overlying GCL overlying compacted soil. Find the total anchor capacity given by the following dimensions and properties:

Dimensions (ft)		Properties	
H	1.5	ϕ	34°
L	6	δ_U	19°
D	0	δ_L	34°
$W = X$	0	γ_S	115 lb/ft³

$$T_1 = 698.1 \ \text{lb/ft}$$

$$T_2 = 0.0 \ \text{lb/ft}$$

$$T_3 = 0.0 \ \text{lb/ft}$$

$$T = 698.1 \ \text{lb/ft}$$

(b) *Placement of operations layer (3H : 1V slope). Assumptions:* operating layer is placed using a D6H dozer, over a length of 115 ft, measured along the slope (Figure 18.11). No braking force is applied (Figures E18.2C and E18.2D).

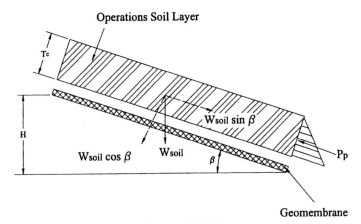

Figure E18.2C

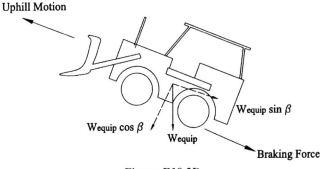

Figure E18.2D

Weight of operating layer soil	$W_{soil} = (115 \text{ ft})(115 \text{ ft})(1.5 \text{ ft}) = 19,837.5 \text{ lb/ft}$
Soil unit weight	$\gamma_{soil} = 115 \text{ lb/ft}^3$
Soil thickness	$T_C = 1.50 \text{ ft}$
Soil shear strength	$\phi = 34°$
Slope angle	$\beta = 18.34°$
Slope height	$H = 36.37 \text{ ft}$
Slope length	$L = 115.00 \text{ ft}$
Equipment type	Cat D6H LGP Series II (operating weight/total width = 45,400 lb/10.3 ft)
Equivalent equipment weight	$W_{equip} = 4407.77 \text{ lb/ft}$
Equipment braking force	0.30
Braking force	1322.33 lb/ft
$W_{soil} \cos \beta$	18,819.50 lb/ft
$W_{soil} \sin \beta$	6273.17 lb/ft
$W_{equip} \cos \beta$	4181.57 lb/ft
$W_{equip} \sin \beta$	1393.86 lb/ft
$K_P = \tan^2 (45 + \phi/2)$	3.54
$P_P = 0.5(K_P \gamma_{soil} T_c^2)$	457.62 lb/ft

- Total force along the upper surface of the geomembrane:

$$F_{upper} = W_{soil} \sin \beta + W_{equip} \sin \beta + \text{braking force} - P_P = 8531.74 \text{ lb/ft}$$

- Total force normal to the geomembrane interface:

$$N = W_{soil} \cos \beta + W_{eqiup} \cos \beta = 23,001.08 \text{ lb/ft}$$

Distribution of forces on the geomembrane (Figure E18.2E)

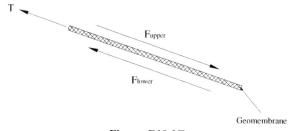

Figure E18.2E

- Friction angle at the bottom of the geomembrane

$$\delta = 19.00° = 0.33 \text{ rad}$$

- Frictional force at the bottom of the geomembrane:

$$F_{lower} = 7919.91 \text{ lb/ft} = N \tan \delta$$

- Force at the upper surface of the geomembrane (calculated previously):

$$F_{upper} = 8531.74 \text{ lb/ft}$$

$$F_{lower} < F_{upper}$$

- Tension in geomembrane:

$$T = F_{upper} - F_{lower} = 611.83 \text{ lb/ft}$$

- Factor of safety against tear, FS_t = tensile strength/tension:

$$FS = \frac{1584}{611.83} = 2.59$$

- Factor of safety against anchor pullout, FS_p = anchor capacity/tension:

$$FS = \frac{1311}{611.83} = 2.14$$

Note: This is a conservative approach where it is assumed that the entire force applied on the top of the geomembrane is acting along its upper surface (i.e., it does not take into account the friction angle at the upper interface, which will limit the transfer of forces to the geomembrane).

Distribution of Forces on the GCL:

- Total force along the upper surface of the GCL = total force along the lower surface of the geomembrane. For the GCL,

$$F_{upper} = 7919.91 \text{ lb/ft}$$

- Friction angle at the bottom of the GCL:

$$\delta = 34.00° = 0.59 \text{ rad}$$

- Frictional force at the bottom of the GCL:

$$F_{lower} = 15,514.42 \text{ lb/ft} = N \tan \delta$$

$$F_{lower} > F_{upper}$$

- Tension in the GCL, $T = F_{upper} - F_{lower}$:

$$T = 0.00 \text{ lb/ft}$$

• Factor of safety against tear, FS_t = tensile strength/tension:

$$FS = \frac{960}{0.00} = NA$$

• Factor of safety against anchor pullout, FS_p = anchor capacity/tension:

$$FS = \frac{698}{0.00} = NA$$

18.4.2 Geomembrane Puncture Resistance

As shown in Figure 18.1, geomembrane, in a composite liner system, may be underlain by a low-permeability compacted soil layer and/or may be overlain by an LCRS granular drainage layer. These underlying or overlying layers may have particles that may cause damage by puncturing (by cutting or penetrating) the geomembrane. As shown in Figure 18.12, preventing geomembrane puncture means preventing the stress concentrations due to puncturing objects (Koerner et al., 1996; Koerner, 1998; Narejo and Corcoran, 2002). The factors that will influence geomembrane puncture are (1) the effective particle size of the potentially puncturing object, (2) the shape of the potentially puncturing object, and (3) the type and amount of stress imposed.

The effective *particle size* is the maximum dimension of the largest soil particle protrusion to which the geomembrane may be exposed. The *shape of the protrusion* is quantitatively specified as (1) angular, (2) subangular and subrounded, and (3) rounded and well rounded. *Stress* on the geomembrane may be imposed by (1) construction equipment, and (2) overlying media, such as waste, liquid, or soil (Narejo and Corcoran, 2002).

Protecting Geomembrane with Geotextile from Puncturing Due to Overlying Static Loads. Based on extensive testing and evaluation performed at the

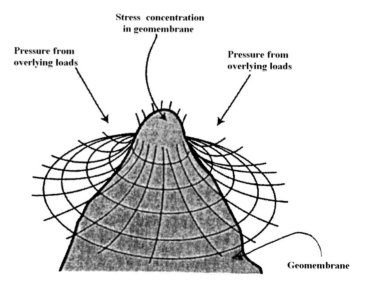

Figure 18.12 *Schematic of stress concentration due to loads and puncturing stone.*

Geosynthetic Institute, Drexel University, an empirical equation has been derived that can be used to design protection of the geomembrane from puncture due to *static loads*. The governing equation is (Narejo and Corcoran, 2002)

$$P_a = \left(450 \frac{M}{H_e^2}\right) \frac{1}{MF_{ps}FS_{cr}FS_{CBD}} \quad (18.30)$$

where P_a is the allowable pressure on the membrane (kPa), M the mass per unit area of nonwoven needle-punched geotextile (g/m^2), H_e the effective protrusion height (mm), MF_{ps} the modification factor for protrusion shape, FS_{cr} the factor of safety for creep, FS_{CBD} the factor of safety for chemical and biological degradation. The empirical equation was established based on a relatively truncated cone height and mass per unit area of protective geotextile used as a protection for 60-mil (1.5-mm)-HDPE geomembrane obtained from hydrostatic truncated cone puncture tests performed according to ASTM D 5514. As discussed by Narejo and Corcoran (2002), modifications to the original empirical equation were made as follows.

Effect of Geomembrane Thickness. In general, the effect of geomembrane thickness (t, mm) is small on failure pressures. However, if an engineer wants to consider the effect of geomembrane thickness, equation (18.30) can be modified as follows:

$$P_{at} = P_a - 1.3 \times 10^5 (1.5 - t) H^{-2.4} \quad (18.31)$$

where P_{at} is the allowable pressure on 1.5-mm-thick HDPE geomembrane (kPa), MF_{pd} the modification factor for packing density (1.0 for isolated stones, 0.5 for packed stones), H the effective protrusion height (mm), and $H_e = H \cdot MF_{pd}$ the effective protrusion height considering packing (mm).

Effect of Geotextile and Geomembrane Creep. Since geomembrane and geotextile are viscoelastic materials, short-term puncture test results need to be modified to account for viscoelastic (creep) behavior. Based on long-term tests, Narejo et al. (1996) proposed a factor

of safety for creep (FS_{cr}). The recommended FS_{cr} values are given in Table 18.1.

Effect of Protrusion Shape and Arrangement. The truncated cone test results need to be modified to account for shape and arrangement of soil, aggregate, or stone. This was done by performing tests on angular, subrounded, and rounded stones of various sizes placed in the same manner as the truncated cones. Modification factors for protrusion shape were then obtained (Narejo et al., 1996):

Angular shape particles	$MF_{ps} = 1.0$
Subangular and rounded particles	$MF_{ps} = 0.5$
Rounded particles	$MF_{ps} = 0.25$

The effect of packing the stones was discussed earlier when MF_{pd} and H_e were introduced.

Effect of Biological and Chemical Degradation. A factor of safety FS_{CBD} has been recommended to account for this degradation. Since biological degradation is not a concern in polypropylene and polyester geotextiles and geomembranes, only chemical degradation is considered here. Koerner (1998) recommends FS_{CBD} as follows: potable water ponds and canal-type inert waste, $FS_{CBD} = 1.0$; landfill leachate, $FS_{CBD} = 1.5$; and containments of brine or diluted acid-type aggressive environment use, $FS_{CBD} = 2.0$.

Finally, a global factor of safety FS against the puncture of geomembrane is recommended as follows:

$$FS = \frac{P_a}{P_r} \quad (18.32)$$

Table 18.1 FS$_{cr}$ values

Geotextile Mass		FS_{cr}[a]		
oz/yd^2	g/m^2	$H_e = 0.5$ in.	$H_e = 1.0$ in.	$H_e = 1.5$ in.
8	270	> 1.5	NR	NR
16	540	1.3	1.5	NR
32	1080	1.1	1.2	1.3

[a]NR not recommended.

EXAMPLE 18.3

Calculate the mass per unit area (M) of a nonwoven needle-punched geotextile required to protect a 60-mil HDPE geomembrane from a puncture when subjected to a 98.5-ft-high municipal solid waste with a unit weight of 76.5 lb/ft^3. Assume that a 3.28-ft (1-m)-thick packed LCRS granular layer with a maximum stone size of 1 in. (25 mm) overlies the geomembrane. The stones are angular in shape and have a unit weight of 127 lb/ft^3.

Solution:

From equation (18.30),

$$P_a = \left(450 \frac{M}{H_e^2}\right) \frac{1}{\mathrm{MF_{ps}FS_{cr}FS_{CBD}}}$$

where $H_e = \mathrm{MF_{pd}} \cdot H = 0.5(25) = 12.5$ mm, $F_{pd} = 0.5$ for packed stones, $\mathrm{MF_{ps}} = 1.0$ for angular particles, $\mathrm{FC_{CBD}} = 1.5$ for municipal solid waste leachate, and

$$P_a = \frac{450M}{(12.5)^2} \frac{1}{1 \cdot \mathrm{FS_{CR}} \cdot 1.5} = \frac{1.92M}{\mathrm{FS_{CR}}}$$

$$P_r = (76.5)(98.5) + (3.28)(127) = 7951.8 \text{ lb/ft}^2 = 380 \text{ kPa (1 kPa} = 20.89 \text{ lb/ft}^2)$$

From equation (18.32),

$$\mathrm{FS} = \frac{P_a}{P_r}$$

$$\mathrm{FS} = \frac{1.92M}{\mathrm{FS_{CR}} (380)} \qquad \text{FS against puncture} = 3$$

$$M = \frac{3(380)}{1.92} \mathrm{FS_{CR}} = 593.8 \mathrm{FS_{CR}}$$

From Table 18.1, for $H = 25$ mm:

- $M = 8$ oz/yd^2 (270 g/m^2) is not recommended.
- If we use $M = 16$ oz/yd^2 (540 g/m^2), $\mathrm{FS_{CR}} = 1.5$; substituting into the equation above does not work (i.e., $540 < 593.8 \times 1.5 = 890$).
- If we use $M = 32$ oz/yd^2 (1080 g/m^2), $\mathrm{FS_{CR}} = 1.2$; substituting into the equation above does work, i.e. left hand side quantity of the equation is larger than right hand side quantity (i.e., $1080 < 593.8 \times 1.2 = 712$. So 32-oz/yd^2 (1080-g/m^2) geotextile can be used.

Therefore, select 32-oz/yd^2 (1080-g/m^2) nonwoven needle-punched geotextile protection for the geomembrane.

where P_a is as defined in equation (18.30) and P_r is the site-specific overburden pressure. To ensure that the yield of the geomembrane over the design life is prevented, Koerner et al. (1996) recommended the following minimum global factor of safety against yield (FS_y); in all cases, the minimum global factor of safety against puncture is 3. For an effective protrusion height of 0.25 in. (6 mm), $FS_y = 3$; for 0.5 in. (12 mm), $FS_y = 4.5$; for 1.0 in. (25 mm), $FS_y = 7.0$; and for 1.5 in. (38 mm), $FS_y = 10.0$.

Protecting Geomembrane from Puncture During Installation. Experience indicates that a geomembrane can be protected from puncture damage during installation from *subgrade material* if the in-situ soil or compacted clay liner is prepared with a smooth drum roller and does not have particles coarser than $\frac{3}{8}$ in. (10 mm) protruding from the surface. To protect the geomembrane from damage, soil particles should be (a) round to subround and (b) should have particles less than or equal to $\frac{3}{8}$ in. (10 mm) in size unless a suitable nonwoven needle-punched geotextile protection layer (as discussed earlier) is placed over the geomembrane (Reddy et al., 1996; Richardson and Johnson, 1998). Based on the results of project-specific testing performed by these investigators and others, it is recommended that during installation the following mass per unit area (M) of nonwoven needle-punched geotextile be placed over the geomembrane (Narejo and Corcoran, 2002): For a maximum stone size ≤ 0.5 in., $M \geq 10$ oz/yd²; ≤ 1.0 in., $M \geq 12$ oz/yd²; ≤ 1.5 in., $M \geq 16$ oz/yd²; and ≤ 2.0 in., $M \geq 32$ oz/yd².

Experience and field testing also demonstrate that a minimum soil cover thickness over geomembrane for the *operation of construction equipment* should be maintained at all times. It is recommended that a minimum lift thickness of 12 in. (30 cm) be maintained for equipment ground pressure of less than 10 lb/in² (70 kPa), 24 in. (60 cm) for equipment ground pressure between 10 and 20 lb/in² (70 to 140 kPa), and 36 in. (90 cm) for equipment ground pressure greater than 20 lb/in² (140 kPa) (Narejo and Corcoran, 2002).

18.5 SUMMARY

The Leachate Collection and Removal System (LCRS) in a landfill, is placed at the bottom of the landfill above the liner system and is used to collect and remove the leachate from a landfill. A typical LCRS consists of a permeable drainage layer, a network of perforated leachate collection pipes, a protective filter layer above the drainage layer, and a system of sump/manhole and pump to allow leachate removal from the landfill. This chapter presents the analytical methods to estimate the quantity of leachate generation, hydraulic and structural design methods for leachate collection pipes, stability analyses for drainage layer when placed on the side-slope, and the clogging potential for the drainage layer. The methods to estimate stresses in the underlying liner system and design methods to protect it against the possible puncture from the LCRS granular material are also discussed. Finally, these analytical methods have been supported by practical numberical examples.

QUESTIONS/PROBLEMS

18.1. Calculate the design ratio due to tensile stress by 60 mil-HDPE geomembrane weight placed on a 30° inclined slope that is 120 ft high. The interface friction angle between HDPE geomembrane and underlying soil is 20°.

18.2. **(a)** For a geomembrane placed on a 3H:1V slope, calculate the wind uplift pressure for a 32-km/hr (20 mph) wind. Assume the mass per unit area of the geomembrane to be 1.41 kg/m² and the height (z) above sea level to be 600 m. **(b)** If 13.9-kg (20-lb) sandbags are placed on top of the geomembrane to resist uplift, estimate the number of sandbags.

18.3. A 50-ft-high 3H:1V side slope is lined with 60-mil single-sided textured HDPE (textured side down against underlying clay and smooth side facing up). Calculate various stresses in the liner and determine the anchor trench capacity assuming that it is 3 ft deep and 2 ft wide. At the base, a 3-ft thickness of soil, consisting of a 1-ft drainage layer and a 2-ft-thick operations layer, is already in place.

18.4. Determine the effective protrusion height (H_e) for a layer of 1.5-in.-maximum-diameter gravel placed on the surface of a geomembrane as a LCRS layer. Based on the result of this H_e value, determine the mass per unit area of nonwoven needle-punched geotextile required to protect a 60-mil (1.5-mm) HDPE geomembrane from construction damage.

18.5. What are the design criteria for a LCRS, and how does one manage the leachate?

18.6. What is the HELP model used for? What are its limitations? For an arid climate, would you recommend using HELP analysis? Discuss and present an alternative model.

18.7. For a 0.3-ft²/s leachate flow rate, estimate a leachate collection pipe size. For a base 400 ft × 600 ft wide where the collection pipes are to be placed leading to a collection sump, show the base grades draining to these pipes. Make appropriate assumptions for material properties and slopes.

18.8. A 12-in.-thick drainage material is placed on top of a composite liner system on a 2H:1V side slope. Discuss the method(s) you would use to evaluate this drainage material's stability on the slope. What measures would you suggest if it is found to be unstable?

18.9. Determine the optimal spacing of the leachate collection pipes shown in Figure P18.9.

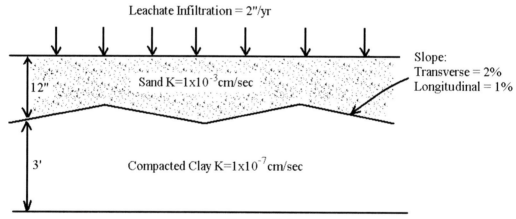

Leachate Infiltration = 2"/yr

Slope:
Transverse = 2%
Longitudinal = 1%

Sand K=1x10⁻³cm/sec — 12"

Compacted Clay K=1x10⁻⁷cm/sec — 3'

Figure P18.9

18.10. The landfill shown in Figure P18.10 is nearly filled. Three of the four liner corrugation sections (cells) have received a final cover, and cell 4 is near the disposal capacity. Each cell is 2.5 acres in size and is served by a leachate collection pipe. The site is located near Chicago, Illinois. The active area is bermed and graded so that none of the runoff is allowed to drain from the surface (it must all be collected and treated as leachate). The vegetative layer and daily cover is a USCS ML material with a porosity of 0.501. Assume that the average waste thickness is applicable to all cells. The construction specification (materials and permeability) for the clay cover is identical to that for the liner. Use the HELP model to estimate the average annual leachate volume and peak daily discharge from a closed section of the landfill (cell 3) and from the active area (cell 4).

   (a) Compute the ratio (active/closed) of average annual leachate volume and peak daily discharge rate for this landfill.

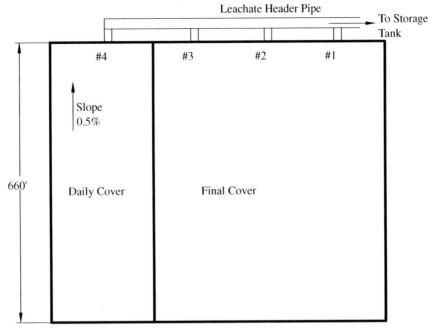

PLAN VIEW
NTS

(a)

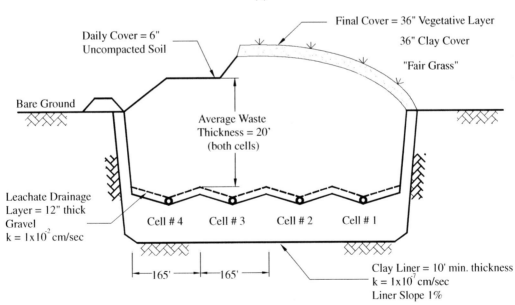

CROSS SECTION
NTS

(b)

Figure P.18.10

(b) For each cell: Of the average annual volume of water passing through the waste, what percentage is collected by the leachate collection system? What percentage passes through the clay liner?

(c) What is the relative distance in leachate passing through the liner from the open versus closed cells?

REFERENCES

Demetracopoulos, A. C., Korfiatis, G. P., Bourodimos, E. L., and Nawy, E. G., Modeling for design of landfill bottom liners, *J. Environ. Eng.*, Vol. 110, No. 6, pp. 1084–1098, 1984.

Driscopipe, *System Design*, Phillips Driscopipe, Richardson, TX, 1988.

Druschel, S. J., and Underwood, E. R., Designing of lining and cover systems sideslopes, in *Geosythetics'93*, IFAI, Vancouver, British Columbia, Canada, 1993.

EPRI (Electric Power Research Institute), *Unsaturated Groundwater Flow Model (UNSAT-1D) Computer Code Manual*, CS-2434-CCM, EPRI, Palo Alto, CA, 1981.

Fayer, M. J., *UNSAT-H Version 3.0: Unsaturated Soil Water and Heat Flow Model—Theory, User Manual, and Examples*, USDOE, Washington, DC, 2000.

Hullings, D. E., and Sansone, L. J., Design concerns and performance of geomembrane anchor trenches, in *GRI Conference Proceedings*, Philadelphia, 1996.

Kmet, P., Quinn, K. J., and Slavic, C., Analysis of design parameters affecting the collection efficiency of clay lined liners, pp. 204–227 in *Proceedings of the 4th Annual Madison Conference of Applied Research and Practice on Municipal and Industrial Waste*, University of Wisconsin, Madison, WI, 1981.

Koerner, R. M., Preservation of the environment via geosynthetic containment, in *Geotextiles, Geomembranes, and Related Products*, Den Hoedt, G. (ed.), A.A. Balkema, Rotterdam, The Netherlands, 1992.

———, *Designing with Geosynthetics*, 3rd and 4th eds., Prentice Hall, Upper Saddle River, NJ, 1994, 1998.

Koerner, R. M., and Hwu, B. L., Stability and tension considerations regarding cover soils on geomembrane lined slopes, *Geotext. Geomembr.*, Vol. 10, No. 4, pp. 335–355, 1991.

Koerner, G. R., and Koerner, R. M., Biological activity and potential remediation involving geotextile landfill leachate filters, in *Geosynthetics Testing for Waste Containment Applications*, ASTM STP 1081, Koerner, R. M. (ed.), ASTM, West Conshohocken, PA, 1990.

Koerner, R. M., and Richardson, G. N., Design of geosynthetic systems for waste disposal, in *Proceedings of the Conference on Geotechnical Practice for Waste Disposal '87*, ASCE, Ann Arbor, MI, 1987.

Koerner, R. M., Wilson-Fahnry, R. F., and Narejo, D., Puncture protection of geomembrane: III. Examples, *Geosynth. Int.*, Vol. 3, No. 5, 1996.

Metcalf & Eddy, Inc., *Wastewater Engineering: Treatment and Reuse*, 4th ed., McGraw-Hill, New York, 2003.

Moore, C. A., *Landfills and Surface Impoundments Evaluation*, EPA/530/SW-869C, USEPA, Cincinnati, OH, 1980.

Narejo, D., and Corcoran, G., *Geomembrane Protection Design Manual*, GSE Lining Technology, Inc., Houston, TX, 2002.

Narejo, D., Koerner, R., and Wilson-Fahnry, R. F., Puncture protection of geomembrane: II. Experimental, *Geosynth. Int.*, Vol. 3, No. 5, 1996.

Reddy, K. R., and Saichek, R. E., Performance of protective cover systems for landfill geomembrane liners under long-term MSW loading, *Geosynth. Int.*, Vol. 5, No. 3, pp. 287–307, 1998.

Reddy, K. R., Bandi, S. R., Rohr, J. J., Finy, M., and Siebken, J., Field evaluation of protective covers for landfill geomembrane liners under construction loading, *Geosynth. Int.*, Vol. 3, No. 5, 1996.

Richardson, G. N., and Johnson, S., Field evaluation of geosynthetic protective cushions: phase 2, *Geotech. Fabr. Rep.*, Oct./Nov., 1998.

Schroeder, P. R., Aziz, N. M., Lloyd, C. M., and Zappi, P. A, *The Hydrologic Evaluation of Landfill Performance (HELP) Model: User's Guide for Version 3,* EPA/600/R-94/168a, Office of Research and Development, USEPA, Washington, DC, Sept., 1994a.

Schroeder, P. R., Dozier, T. S, Zappi, P. A, McEnroe, B. M., Sjostrom, J. W., and Peyton, R. L., *The Hydrologic Evaluation of Landfill Performance (HELP) Model: Engineering Documentation for Version 3,* EPA/600/R-94/168b, Office of Research and Development, USEPA, Washington, DC, Sept., 1994b.

Sharma, H. D., and Lewis, S. P., *Waste Containment Systems, Waste Stabilization, and Landfills: Design and Evaluation,* Wiley, New York, 1994.

Sharma, H. D., Dukes, M. T., and Olsen, D. M., Evaluation of infiltration through final cover systems: analytical methods, in *Waste-Tech '89,* Washington, DC, 1989.

Simon, A. L., and Korom, S. F., *Hydraulics,* 4th ed., Prentice Hall, Upper Saddle River, NJ, 1997.

Swanson, D., Formula helps determine wind load characteristics of fabric, *Int. Fabr. Prod. Rev.,* Apr. 1987, p. 6.

Von Pein, R. T., and Prasad, S., Composite lining system design issues, in *Proceedings of the 4th GRI Seminar on the Topics of Landfill Closure: Geosynthetics, Interface Friction, and New Developments,* GRI, Philadelphia, 1990.

Wong, J., The design of a system for collecting leachate from a lined landfill site, *Water Resour. Res.,* Vol. 13, No. 2, 1977.

Zornberg, J. G., and Giroud, J. P., Uplift of geomembrane by wind: extension of equations, *Geosynth. Int.,* Vol. 4, No. 2, pp. 187–206, 1997.

19

FINAL COVER SYSTEMS

19.1 INTRODUCTION

After the base and side liner system is in place, landfill operations begin. Landfill operations consist of refuse filling, refuse compaction, and the application of daily cover material. After the landfill cell is completely filled with refuse to its permitted final elevation, the final cover system is placed over the area of the completed landfill section (Figure 19.1). As discussed in Chapter 12, final cover systems are also used to cover contaminated sites as a remediation action. In this chapter, the purpose, general design criteria, and minimum regulatory requirements for final cover are explained and a detailed design and analysis of final cover systems is presented.

19.2 PURPOSE AND DESIGN CRITERIA

There are several important reasons for the placement of a final cover system on top of landfills and they are as follows:

- To minimize the infiltration of precipitation into the refuse and thus minimize leachate generation and lower costs for leachate treatment
- To minimize the hydraulic head on the liner system, leading to a low potential for subsurface contamination
- To resist erosion due to wind or runoff
- To control the migration of landfill gases or enhance gas recovery
- To separate the refuse from vectors such as animals, insects, and rodents
- To improve aesthetics

In general, the following criteria are necessary to consider during final cover design: (1) low permeability to limit infiltration and enhance surface drainage; (2) durability to avoid erosion, desiccation, freeze–thaw problems, burrowing animals, and root penetration; (3) flexibility to accommodate large differential settlements and localized subsidence without cracking; and (4) stability on slopes.

To meet the design criteria above, the cover system, in general, could consist of five different layers as depicted in Figure 19.2. Depending on the site-specific conditions, a cover may consist of all or some of these layers. An *erosion layer* could be vegetated top soil, asphalt, or cobbles. A *protective soil* layer in certain cases may be specified between the erosion and the drainage layer. A *drainage layer* could be the natural materials, such as sand and gravel, or synthetic material, such as geocomposite. A typical *barrier layer* or infiltration layer may consist of one or a combination of compacted clay liner (CCL), flexible membrane liner (FML), and geosynthetic clay liner (GCL). A *gas collection layer* may consist of natural materials, such as sand or gravel; synthetic materials, such as geotextile or geocomposites, may also be used as gas collection layer materials.

19.3 REGULATORY MINIMUM REQUIREMENTS

19.3.1 Cover System for MSW Landfills

The federal regulations stipulated in RCRA Subtitle D (40 CFR 258.60) state that the final cover must consist of an *erosion layer* underlain by an *infiltration layer* (USEPA, 1992). The infiltration layer must be com-

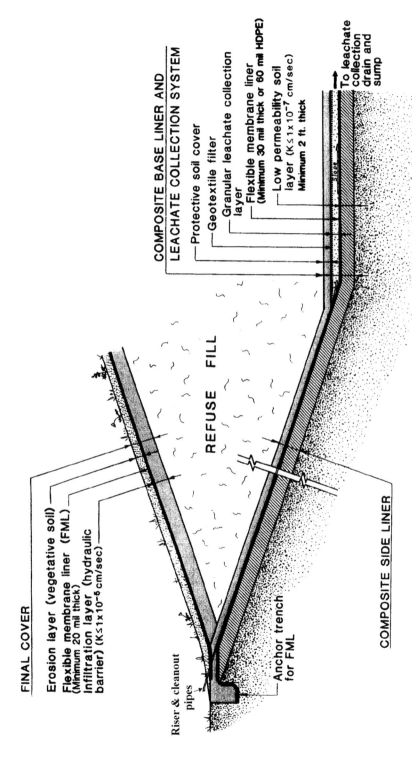

FINAL COVER

Erosion layer (vegetative soil)
Flexible membrane liner (FML)
(Minimum 20 mil thick)
Infiltration layer (hydraulic
barrier) (K≤1×10⁻⁶cm/sec)

Riser & cleanout
pipes

Anchor trench
for FML

COMPOSITE SIDE LINER

REFUSE FILL

**COMPOSITE BASE LINER AND
LEACHATE COLLECTION SYSTEM**

Protective soil cover
Geotextile filter
Granular leachate collection
layer
Flexible membrane liner
(Minimum 30 mil thick or 60 mil HDPE)
Low permeability soil
layer (K≤1×10⁻⁷ cm/sec)
Minimum 2 ft. thick

To leachate
collection
drain and
sump

Figure 19.1 Subtitle D landfill liner and final cover systems. (From Sharma and Lewis, 1994. This material is used by permission of John Wiley & Sons, Inc.)

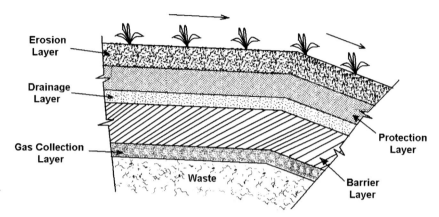

Figure 19.2 Final cover system components.

prised of a minimum of 18 in. of earthen material that has a permeability less than or equal to the permeability of any bottom liner system or natural subsoils present, or a permeability no greater than 1×10^{-5} cm/s, whichever is less. This implies that a composite liner must be used in the cover system if one is used in the base liner. The erosion layer must consist of a minimum of 6 in. of earthen material that is capable of sustaining native plant growth. Figure 19.3 depicts the recommended minimum final cover system. The federal regulations provide flexibility for the states to approve an alternative final cover design that includes an infiltration layer that achieves an equivalent reduction in infiltration and an erosion layer that achieves an equivalent protection from wind and water erosion. This is discussed further in Section 19.3.3.

State regulations vary from state to state. For instance, in Illinois, the regulations stipulated in IAC 811.314 require that the final cover systems must consist of a *low-permeability layer* overlain by a *final protective layer*. The low-permeability layer must consist of any one of the following: (1) a compacted earth layer with a minimum thickness of 3 ft and permeability of 1×10^{-7} cm/s; (2) a geomembrane with performance equal or superior to the compacted earth layer; or (3) any other low-permeability layer that provides equivalent or superior performance to the compacted earth layer or geomembrane. The final protective layer should consist of a minimum 3-ft-thick soil

material capable of supporting vegetation and preventing desiccation, cracking, freezing, or other damage to the low-permeability layer. Figure 19.4 shows a schematic of this cover system.

19.3.2 Cover System for Hazardous Waste Landfills

As shown in Figure 19.5, the minimum requirements for a hazardous waste (Subtitle C) landfill final cover system, as outlined by the USEPA, is from top to bottom:

- A minimum 2-ft-thick vegetated topsoil layer (The layer should be graded at a slope between 3 and 5%.)
- A soil or geosynthetic (e.g., geotextile) filter layer to prevent soil or a root system from clogging the underlying drainage layer
- A drainage layer consisting of either a minimum 1-ft-thick granular material with a minimum hydraulic conductivity of 1×10^{-2} cm/s or an equivalent geosynthetic drainage material, such as a geocomposite
- A minimum 20-mil geomembrane liner
- A minimum 24-in.-thick low-permeability soil layer with a maximum hydraulic conductivity of 1×10^{-7} cm/s

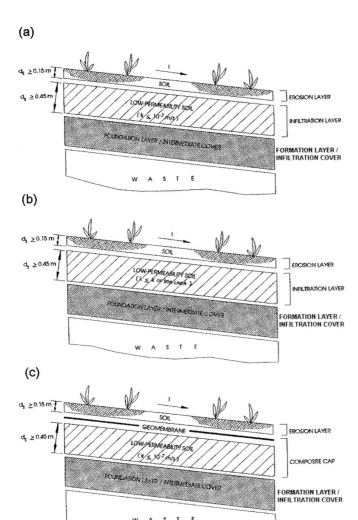

(a)

(b)

(c)

Figure 19.3 USEPA (1992) recommended minimum final cover systems for (a) unlined MSW landfills; (b) MSW landfills underlain by a soil liner; (c) MSW landfills underlain by a geomembrane/soil composite liner. d_s represents the soil layer thickness. (From Othman et al., 1995, Reproduced by the permission of ASCE.)

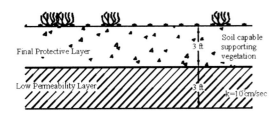

Figure 19.4 Final cover system as required by Illinois administrative code (IAC 811.314).

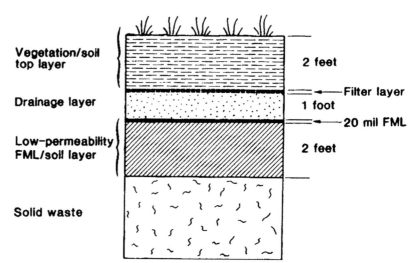

Figure 19.5 *Final cover system as required by RCRA Subtitle C: hazardous waste landfill. (From Landreth, 1990.)*

19.3.3 Alternative Final Cover Systems

As discussed in previous sections, the dominant feature of a *conventional cover* system is the use of one or more barrier layers. For example, Subtitle D, MSW conventional final cover in accordance with Title 258.60 (a)(1), requires a barrier layer with permeability no greater than 1×10^{-5} cm/s. Similarly, Subtitle C, hazardous waste landfill final cover system, requires a clay barrier to have a maximum saturated hydraulic conductivity value no greater than 1×10^{-7} cm/s, in accordance with Title 264.310 (a)(5). In addition to the permeability requirement, the final cover must also have a permeability less than or equal to the bottom layer or natural subsoil.

The Air Force Center for Environmental Excellence landfill survey reveals that construction and maintenance costs for conventional final landfill covers may be reduced significantly if *alternative covers*, which have equivalent or better performance, are used. For example, the data indicate that alternative cover construction cost savings could exceed $200,000 per acre. Similarly, estimates indicate that repairing an alternative cover would cost 60 to 65% less than for a conventional barrier cover (Hauser et al., 2001).

A key feature of an alternative cover is the replacement of barrier layer(s) of conventional cover with other cost-effective equivalently performing layers. For

example, as shown in Figure 19.6(*b*), a *capillary barrier* is formed by two layers (e.g., a layer of fine soil cover and a layer of coarse soil material). The barrier is created by the large change in pore sizes between the layers of fine and coarse materials, resulting in capillary forces that cause the fine soil overlying the coarse soil to hold more water than if there were no change in particle size between the layers. This system can fail if either the desired large change in pore size is missing in spots or too much water accumulates in the fine-particle layer. Ankeny et al. (1997), Gee and Ward (1997), Stormont (1997), Khire et al. (1999), and Hauser et al. (2001) present more information about capillary barrier covers. Figure 19.6(*b*) also shows an *asphalt barrier*, where the compacted clay barrier is replaced with the asphalt. This replacement is most suitable in arid climates, where a clay barrier may fail because of desiccation (Gee and Ward, 1997).

An alternative landfill cover with no barrier layer works on the principle of water-holding properties of soils and that most of the precipitation returns to the atmosphere via evapotranspiration. Such covers are called *evapotranspirative* (ET) covers. ET cover uses two processes to control infiltration of water into the waste. The first process is that of soil stores infiltrating water, and the second process is that of evapotranspiration removing water from the soil water reservoir.

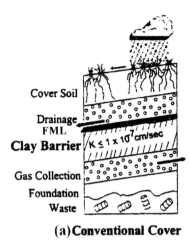

(a) Conventional Cover

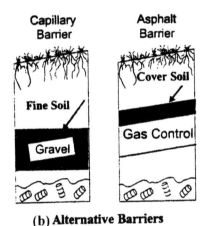

(b) Alternative Barriers

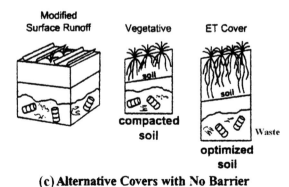

(c) Alternative Covers with No Barrier

Figure 19.6 Schematic representation of comparisons between conventional and alternative final cover systems. (From Hauser et al., 2001.)

Figure 19.6(c) shows an ET cover system. The authors' experience and information available in the literature (Hauser et al., 2001; Kavazanjian, 2001) show that ET cover is less costly to build and maintain than conventional cover. ET covers are most suitable in arid and semiarid regions. However, a designer should carefully evaluate the vegetation type and availability of suitable soil for ET covers. This is because if not evaluated properly, short growing seasons and unsuitable soils could become major disadvantages for the ET cover at some locations. Typical values of saturated hydraulic conductivity for the storage layer of an ET cover are between 1×10^{-4} and 1×10^{-5} cm/s (Kavazanjian, 2001).

19.4 DESIGN PROCEDURE

19.4.1 Layout and Grading

The first and foremost important step in the design of final cover system is to decide on the layout and grading of the cover system. This will depend on waste capacity, waste settlements, regulatory requirements, operational and maintenance concerns, environmental and aesthetic considerations, surface water drainage structures, cover component limitations, and stability concerns.

As shown in Figure 19.7, the cover systems depend on the type of landfill: either aboveground landfill or pit-type landfill. The slope of final covers may be divided into side slopes and top deck as shown in Figure 19.7(a). The side slopes are made as steep as possible, but making sure that construction and maintenance can be achieved easily. Typically, a slope of 2H:1V is used if the cover consists of soil layers only. The slopes may have to be flatter (equal to or flatter than 3H:1V) if geosynthetic components such as geomembrane are used in the cover. The top deck generally consists of an area exceeding 100 ft × 100 ft and sloped at 2 to 5%. The final determination of the slopes is made based on a detailed slope stability analysis as discussed later in the chapter. The refuse settlement can be substantial (10 to 20% of the height of the refuse), and it must be accounted for properly in the final cover grad-

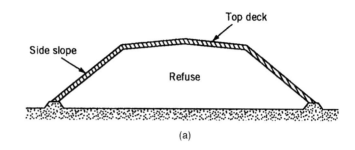

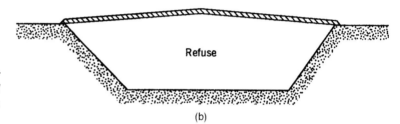

Figure 19.7 Landfill final cover configuration: (a) *aboveground landfill;* (b) *pit-type landfill. (From Sharma and Lewis, 1994.)*

ing so that, after settlement, positive grades are maintained.

19.4.2 Cover Profile and Material Selection

Figures 19.1 through 19.6 show general cross sections of a final cover system. The selection of material and thickness for each layer will depend on the performance assessment for the site-specific conditions described in subsequent sections. A general description of the materials and considerations in the selection of thickness for each layer is provided below.

For the conventional landfill final cover system, the most important layer in the final cover is the barrier layer. This layer may consist of a single layer of compacted soil, geomembrane, or geosynthetic clay liner, or a composite layer consisting either of compacted soil layer overlain by geomembrane or GCL overlain by geomembrane. The purpose of the barrier layer is to minimize infiltration into the refuse. The selection of compacted soil, geomembrane, and GCL for use in the barrier layer will be based on the considerations similar to those in the base and side liner systems. Unlike the base and side liner systems, the barrier layer

materials in the final cover are subjected to low overburden stresses and are not exposed to leachate that consists of various chemicals. However, the barrier layer materials in the final cover must be carefully designed to account for subsidence, desiccation cracking, and freeze–thaw cycling. To ensure this, geomembrane placed on the compacted soil layer should be at least 20 mils thick, or 60 mils thick for HDPE geomembrane. Stress situations such as bridging over subsidence and friction between the geomembrane and other cover components (e.g., compacted clay or drainage material) should be considered in the final selection of the geomembrane.

The cover soil that overlies the barrier layer may consist of two layers of different soil types. The upper layer or surface layer generally consists of topsoil to support vegetation that will allow runoff from major storms while *inhibiting erosion*. In areas where vegetation may be unsuitable, particularly in arid areas, other materials, such as stones and cobbles, may be used to prevent wind-induced erosion. The bottom (or protective) layer should consist of soils that are capable of sustaining nonwoody plants, have adequate water-holding capacity, and are sufficiently deep to allow for expected long-term erosion losses. A medium-textured soil such as a loam may be used in this layer. The cover soil slope should be uniform and at least 3% and should not allow erosion rills and gullies to form. Slopes greater than 5% may promote erosion unless controls are built in to limit erosion. The thickness of the cover soil should be adequate to maintain the vegetation and prevent erosion.

The *drainage layer* is often used between the barrier layer and the cover soil. It is designed to minimize the time the infiltrated water is in contact with the barrier layer and hence to lessen the potential for the water to reach the refuse. Water that filters through the cover soil is intercepted and rapidly moved to an exit drain, such as by gravity flow to a toe drain; this prevents buildup of hydraulic head above the barrier layer, consequently increases the slope stability. The drainage layer can consist of granular soils, geosynthetic drainage materials (e.g., geocomposites), or other highly permeable materials (e.g., shredded scrap tires or glass cullet). If granular soils such as sand and gravel are used, the drainage layer should be at least 1-ft thick with a hydraulic conductivity of 1×10^{-2} cm/s or greater. The soils and the drainage pipes used within these soils should not damage the underlying liner, particularly geomembrane, similar to that of leachate drainage material overlying geomembrane in the base and side liner systems. If geosynthetic materials such as geocomposites are used in the drainage layer, the same physical and hydraulic requirements should be met (e.g., equivalency in hydraulic transmissivity, compatibility with geomembrane, compressibility, conformance to surrounding materials, and resistance to clogging). If alternative highly permeable materials are used in the drainage layer, the equivalency in the performance to that of granular soils should be demonstrated.

Between the drainage layer (particularly when granular materials are used) and the cover soil layer, a *geotextile filter fabric* is used to prevent the drainage layer from clogging by the fines from the cover soil layer. The design of the filter fabric is similar to that used for the design of filter fabric between the leachate drainage layer and the refuse at the bottom of the landfill.

A *gas collection layer* may be needed on a site-specific basis. This layer is located above the refuse and below the barrier layer. Coarse-grained materials, geosynethetic materials, or alternative materials, similar to those used in the drainage layer, may be used as a gas collection layer. Perforated horizontal venting pipes may be used to channel gases to a minimum number of vertical risers located at high points to promote gas ventilation. To prevent clogging, a geotextile filter fabric may be used between the gas collection layer and the barrier layer. As an alternative to horizontal venting pipes in gas collection layers, vertical standpipe gas collectors may be built as the landfill is filled with refuse.

In addition to the strength and hydraulic properties of each material, interface shear strengths between any two contact materials in the cover system should be determined. The interface with the lowest shear strength will dictate the design of slopes of the final cover, as discussed in a later section. Strength and other properties of various liner material components are discussed in Chapter 17.

19.4.3 Infiltration Analysis

One of the main functions of the final cover system for a landfill is to limit the amount of precipitation that infiltrates the waste so that the amount of leachate generated is minimized. Infiltration or hydraulic analyses are therefore performed to estimate infiltration rates through the cover system and into the underlying wastes. At the present time, the practice is to use hydraulic computer models to perform these analyses. The following is a partial list of computer models that are available to perform these analyses for the landfill final cover systems.

- Hydraulic evaluation of landfill performance (HELP) model (Schroeder et al., 1994a,b)
- Water balance analysis program (MBALANCE) model (Kmet, 1982; Scharch, 1985)
- Leaching estimation and chemistry model (LEACHM) (Hutson and Wagenet, 1992)

• Unsaturated soil water and heat flow model (UNSAT-H) (Fayer and Jones, 1990; Fayer, 2000)

As presented in Chapter 18, HELP is the most widely used infiltration analysis model for both final cover and LCRS designs. HELP simulates the hydrologic processes for a landfill by utilizing a quasi-two-dimensional deterministic approach, as discussed by Schroeder et al. (1994a,b). The model is more appropriate for humid and semihumid areas. An infiltration model that will be more appropriate for arid and semiarid regions is one that considers unsaturated flows and relies on the moisture storage capacities of the surficial soils (EPRI, 1984; Sharma et al., 1989). One such model that has recently been extensively used for landfill cover design in arid and semiarid regions is the UNSAT-H model.

UNSAT-H Model. UNSAT-H computer code employs a finite difference approach to model unsaturated flow (Fayer, 2000). The theoretical approach implemented in the computer model uses the following mass-balance equation:

$$I = P - R - (E + T) \pm \Delta S \qquad (19.1)$$

where I is infiltration through the bottom of the cover, P the precipitation, R the runoff or overland flow, E the evaporation, T the transpiration, and ΔS the change in soil-moisture storage. Although UNSAT-H code is specifically designed for unsaturated flow modeling in arid and semiarid climates, it has been field verified under both arid and humid climatic conditions (Khire et al., 1997).

The input parameters and variables in the UNSAT-H model are soil properties, weather data, vegetation data, program control variables (maximum and minimum soil suction and allowable mass balance error, etc.), and initial conditions. The soil properties required for input to the UNSAT-H model consist of saturated hydraulic conductivity and soil-water retention properties. The soil-water retention properties are represented by fitting parameters that describe the shape

of the Van Genuchten soil-water retention function (Mualem, 1976; Van Genuchten, 1980).

The Van Genuchten parameters are used in the Van Genuchten–Mualem unsaturated hydraulic conductivity model as expressed by the following relationships:

$$\theta = \theta_r + \frac{\theta_s - \theta_r}{[1 + (\alpha h)^n]^m} \qquad (19.2a)$$

$$K_h = K_{sat} \frac{\{1 - (\alpha h)^{n-2}[1 + (\alpha h)^n]\}^{-m}}{[1 + (\alpha h)^n]^m} \qquad (19.2b)$$

where

θ = volumetric moisture content
θ_s = saturated volumetric moisutre content
θ_r = residual volumetric moisture content
α, n = Van Genuchetn fitting parameters
$m = 1 - 1/n$
$l = 0.5$
K_h = unsaturated hydraulic conductivity
K_{sat} = saturated hydraulic conductivity
h = suction head

The code solves the water balance equations by utilizing the modified Picard finite difference scheme to solve the Richards equation for moisture movement in unsaturated soils. Simulation results are the rates of transpiration, evaporation, runoff, and infiltration or percolation through the cover system. The following presents an example (19.1) where UNSAT-H model analysis was performed for an evapotranspirative (ET) cover. For comparison, a HELP analysis was also performed for the same cover system and the infiltration results were then compared from both model analyses.

19.4.4 Erosion Assessment

The cover soil should be designed and maintained to minimize its erosion potential. Adequate vegetation should be provided to promote evapotranspiration and surface grades should be such as to allow surface runoff. A surface runoff collection system should be provided using benches, V-ditches, and downdrains. The

EXAMPLE 19.1

For a monofill ET cover system over a landfill near Phoenix, Arizona, obtain the infiltration rate through a 4.5-ft-thick soil cover with a saturated hydraulic conductivity of 1×10^{-4} cm/s using the UNSAT-H model. Compare the infiltration rate with the values estimated using the HELP model.

Solution:

Using weather data from a set of CD-ROMs containing records gathered by the National Climatic Data Center (NCDC), precipitation and temperature data were used from the Tempe ASU weather station since it had a long period of record (48 years, 1953 through 2000). Additional weather data (dew point temperature, solar radiation, wind speed, and cloud cover) were obtained from the Phoenix Sky Harbor International Airport station. *Soil parameters* for the model were taken from data presented by Carsel and Parrish (1988), which contain tables representing mean values of the Van Genuchten parameters for various soil types. The soil selected from this table for soil parameters was for sandy loam (SL), which is similar to silty sand (SM) and has a saturated hydraulic conductivity value of 1×10^{-4} cm/s. Vegetation data, such as the leaf area index (LAI = 1) and a plant coverage of 25% of the ground surface, were assumed. This information for such a geographical location is available in the HELP manual.

Figure 19.8 presents the results of a 30-year UNSAT-H simulation using the parameters described above. The infiltration rate through the cover system was 0 mm/yr. HELP analysis showed that the corresponding infiltration rate was 2 mm/yr. As expected, HELP analysis showed a slightly higher infiltration rate. However, from a practical point of view, both of these analyses showed similar values for the infiltration rates through the cover.

---

erosion is measured by the soil loss rate in tons per acre per year. This is calculated using the *universal soil loss equation* (USLE). The USLE was developed by the U.S. Department of Agriculture and is often used. The USLE is given as

$$A = RKL_sCP \tag{19.3}$$

where A is the average annual soil loss (tons/acre), R the erosivity of rainfall/runoff, K the slope erodibility factor, L_s the slope length, C the cover management factor, and P the practice factor. These factors can be obtained from local offices of the U.S. Soil Conservation Service or published data from a local agency. The cover design should ensure the lowest value of A.

19.4.5 Drainage Layer Capacity

The drainage capacity of the final cover drainage layer, either granular soil or geosynthetic material, should be evaluated in relation to the actual drainage capacity requirement based on the HELP analysis results (Section 19.4.3). Perforated drainage pipes are installed within the drainage layer at a regular spacing. The spacing and slope of the drainage pipes are therefore known and used in calculation of the flow rate in the drainage layer.

Darcy's law is used to calculate the allowable flow rate in the drainage layer:

$$Q_{\text{allowable}} = KiA \tag{19.4}$$

where $Q_{\text{allowable}}$ is the allowable flow rate, K the hydraulic conductivity of drainage material [in case of geosynthetics, K can be calculated by dividing transmissivity (T) by the thickness (t)], i the hydraulic gradient, and A the cross section through which flow occurs (equal to thickness multiplied by the width of the cross section). The actual flow rate (Q_{actual}) that occurs

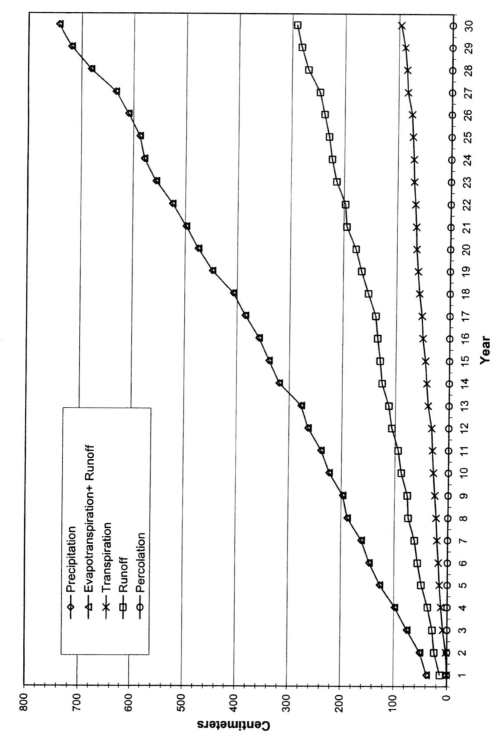

Figure 19.8 Summary of 30-year UNSAT-H simulation—cumulative results for a landfill ET cover located near Phoenix, Arizona.

in the cover drainage layer based on HELP analysis is calculated using

$$Q_{\text{actual}} = ILb \qquad (19.5)$$

where I is the infiltartion rate (either average annual or peak daily) obtained from the HELP analysis results; L the drainage pipe spacing, b the thickness of the flow cross section (generally 1 ft for granular material). An effective drainage system design can be measured by a factor of safety, defined as

$$FS = \frac{Q_{\text{allowable}}}{Q_{\text{actual}}} \qquad (19.6)$$

It is also important to determine the size of the perforated drainage pipes in the drainage layer. For this determination, the following data are used: (1) the slope of pipes, and (2) the infiltration rate from HELP analysis. It is assumed that the flow occurs under gravity and the pipe is flowing full. The maximum expected water infiltration rate is defined by

$$Q_{\text{actual}} = IAt \qquad (19.7)$$

where I is the maximum infiltration rate contributing to the pipe based on HELP analysis, A the total area of the landfill, and t the duration of infiltration. Assuming that $Q_{\text{allowable}} = Q_{\text{actual}}$, Manning's equation is used to calculate the required perforated pipe radius:

$$Q = \frac{1.486}{n} AR^{2/3} S^{1/2} \qquad (19.8)$$

where n is the roughness coefficient, A the area in flow, r the radius of the pipe (to be calculated), R the hydraulic radius (area of flow divided by wetted perimeter, equal to $r/2$ for full-flow conditions), and S the hydraulic gradient.

19.4.6 Cover Geomembrane Analysis

The landfill final cover is subject to settlement, due primarily to the underlying waste compressibility.

Settlement Due to Waste Compression. The final landfill cover system may be subject to settlements resulting from (1) waste compression, (2) localized compression or collapse of objects within the waste mass, and (3) compression of soil layer(s) below the landfill foundations. These settlements may result in (a) slope reversal causing surface drainage problems, and (b) excessive stresses and strains in the final cover system components, causing tear of geosynthetic materials and cracks in the soil components of the final cover system. In this section we discuss waste settlement estimation methods.

The mechanism of refuse settlement is complex and can be attributed to (1) physical and mechanical processes, (2) chemical processes, (3) dissolution processes, and (4) biological decomposition. Refuse settlement estimation methods and parameters needed for such estimation have been addressed by numerous investigators (Sowers, 1973; Dodt et al., 1987; Edil et al., 1990; Sharma et al., 1999; Sharma, 2000). Sowers's method is widely used in practice, due to the familiarity of the consolidation-based approach in the practicing engineering community. A waste settlement estimation approach based on this method is presented below.

Refuse settlement can be separated into primary and secondary settlements. The primary settlement has been reported to occur anywhere between one month and five years after waste filling had ceased. Secondary compression, due to decomposition and creep, may take up to 50 years after completion of the waste fill. Equation (19.9) may be used to estimate secondary refuse settlement. These relationships have been established based on numerous field settlement measurements. Because final covers are placed some time after refuse fill placement and the designer is interested in long-term performance evaluation, only secondary settlement estimation is discussed here.

Long-term secondary refuse settlement can be divided into two categories: (1) settlement under self-weight and (2) settlement under external loads (Sharma, 2000).

- *Settlement under self-weight.* The long-term secondary settlement under refuse self-weight can be expressed by the relationship

$$\Delta H_{(SW)} = C_{\alpha(SW)} H \log \frac{t_2}{t_1} \qquad (19.9)$$

where $\Delta H_{(SW)}$ is the settlement at time t_2 after landfill placement, t_1 the initial period (typically, one to four months) of settlement, H the thickness of the refuse fill, and $C_{\alpha(SW)}$ the coefficient of secondary compression of refuse fill due to self-weight.

- *Settlement under external loads.* The long-term secondary settlement due to external loads can be expressed by the relationship

$$\Delta H_{(EL)} = C_{\alpha(EL)} H \log \frac{t_2}{t_1} \qquad (19.10)$$

where $C_{\alpha(EL)}$ is the coefficient of secondary compression of refuse fill due to external loads. All other terms were defined earlier.

Both $C_{\alpha(EL)}$ and $C_{\alpha(SW)}$ values depend on site-specific environmental conditions and the organic content of the waste fill. Typical values of $C_{\alpha(EL)}$ range between 0.01 and 0.07 and $C_{\alpha(SW)}$ range between 0.1 and 0.4. Higher values indicate higher organic content, higher humidity, and/or a higher degree of decomposition of the waste (Sharma, 2000).

These settlements may result in differential settlements on a final liner cover system, depending on the thickness of the refuse at various locations. Resulting differential settlements may cause excessive tensile strains in clay and/or flexible membrane components of the cover system. Differential settlement-induced tensile strains should be compared with the tensile strains at failure for typical liner materials. For example, according to LaGatta et al. (1997), tensile strains at failure range from 0.1 to 4% for compacted clay

liners, 1 to 10% for geosynthetic clay liners, and 20 to 100% for geomembranes.

For the final cover system consisting of a geomembrane, the following analyses should be performed to evaluate the stability of the geomembrane. The procedures that were used for the geomembrane in a base and side liner system are also used for analyzing the geomembrane integrity in the final cover system. In particular, the stability of the geomembrane should be analyzed for (1) localized subsidence, (2) bending, and (3) tension due to unbalanced shear.

Check for Localized Subsidence. Tensile stresses are induced in the geomembrane due to localized subsidence, which often occurs after closure of a landfill, especially when biological decomposition of refuse is taking place. The magnitude of tensile stress in the geomembrane is a function of the dimensions of the subsidence zone and the cover soil properties. A general scheme of the assumed shape of geomembrane deformation is illustrated in Figure 19.9. The deformed geomembrane is in a spheroid shape, the center point of which is gradually decreasing along the symmetric axis. The geomembrane is assumed to be fixed at the circumference of the subsidence zone. The tensile stress induced in the geomembrane is given by (Koerner, 1993)

$$\sigma = \frac{2DL^2 \gamma_{cs} H}{3t(D^2 + L^2)} \qquad (19.11)$$

where D is the vertical deflection, L the depression radius, γ_{cs} the unit weight of cover soil, H_{cs} the thickness of cover soil, and t the thickness of geomembrane. For the geomembrane, the allowable tensile stress at yield ($\sigma_{allowable}$) is generally known. The factor of safety can then be defined as

$$FS = \frac{\sigma_{allowable}}{\sigma} \qquad (19.12)$$

The FS should be greater than 1.0 to ensure the tensile stress of geomembrane within the working range.

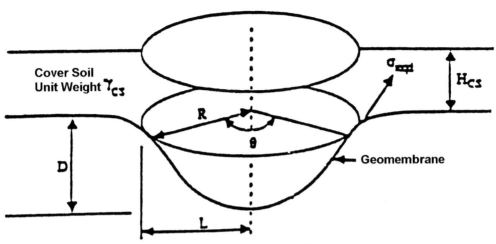

Figure 19.9 *Cover geomembrane analysis for localized subsidence. (Reprinted with permission from Koerner, 1993. Copyright CRC Press, Boca Raton, Florida.)*

Check for Bending. As shown in Figure 19.10, due to sloping, tensile stress may be induced in the geomembrane due to self-weight (before cover soil placement) and is given by

$$\sigma = \frac{w \sin \beta - w \cos \beta \tan \delta}{t} \qquad (19.13)$$

The factor of safety against tensile failure is defined by

$$FS = \frac{\sigma_{\text{yield}}}{\sigma} \qquad (19.14)$$

where $\sigma_{\text{allowable}}$ is the yield stress from a wide-width tensile test.

After the cover soil is placed, the tensile stress induced in the geomembrane may be given by (Koerner, 1993)

$$\sigma = \frac{\gamma_{cs} H_{cs}}{\cos \beta} \frac{x}{t} (\tan \delta_U + \tan \delta_L) \qquad (19.15)$$

where γ_{cs} is the unit weight of cover soil, H_{cs} the thickness of cover soil, β the slope, x the mobilized distance, t the geomembrane thickness, δ_U the upper friction angle, and δ_L the lower friction angle. Again, the factor of safety is defined by

$$FS = \frac{\sigma_{\text{yield}}}{\sigma} \qquad (19.16)$$

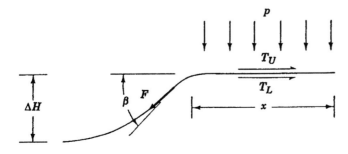

Figure 19.10 *Cover geomembrane analysis for bending. (Reprinted with permission from Koerner, 1993. Copyright CRC Press, Boca Raton, Florida.)*

Check for Tension Due to Unbalanced Shear.
Similar to side liner systems, the geomembrane in the final cover should also be checked for tension due to unbalanced shear. With the conditions shown in Figure 19.11, the tensile stress in the geomembrane is given by (Koerner, 1993)

$$\sigma = \frac{[(C_{aU} - C_{aL}) + \gamma_{cs}H_{cs}\cos\omega(\tan\delta_U - \tan\delta_L)]LW}{t}$$

$$(19.17)$$

where C_{aU} is the adhesion on the upper surface, C_{aL} the adhesion on the lower surface, γ_{cs} the unit weight of the cover soil, H_{cs} the thickness of the cover soil, β the slope, δ_U the upper friction angle, γ_L the lower friction angle, L the length, W the width, and t the thickness of the geomembrane. The factor of safety may again be defined as

$$FS = \frac{\sigma_{yield}}{\sigma} \qquad (19.18)$$

19.4.7 Cover Slope Stability Analysis

The components of the final cover should be stable during and after construction. When geosynthetic materials (geomembrane or geosynthetic drainage layers such as geocomposites) are used, the stability is often governed by the interface strengths of the materials. The slope stability of the final covers should be evaluated for the following conditions: (1) static condition: this condition represents the long-term performance of the final cover, (2) construction condition: this condition is represented by a vehicle operating on the final cover during and after the construction, (3) hydraulic head buildup: impact of head buildup on the final cover, and (4) seismic condition: if the landfill is located in the seismic impact zone, stability analysis should also be performed considering the seismic condition.

Static Slope Stability of Final Cover System.
Figure 19.12(a) shows an example of an infinite-slope slip surface where potential movements are parallel to the slope. Such a situation normally would arise when a relatively thin soil veneer is placed on a slope. As shown in Figure 19.12(b), such situations arise when a final cover system (soils and liners) is placed on a refuse side slope. This will require performing infinite slope stability analyses. These analyses assume that movement of an earth mass occurs parallel to the slope.

Infinite slope stability analyses assume that the forces causing movement are due to the weight of the materials and the resisting forces are due to material strength(s). The factor of safety (FS) against instability is then defined as the ratio of resisting forces and driv-

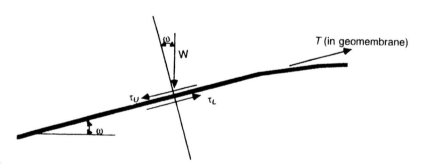

Figure 19.11 Cover geomembrane analysis for tension due to unbalanced shear. (*Reprinted with permission from Koerner, 1993. Copyright CRC Press, Boca Raton, Florida.*)

where

$$\tau_U = c_{aU} + (W\cos\omega)\tan\delta_U$$
$$\tau_L = c_{aL} + (W\cos\omega)\tan\delta_L$$

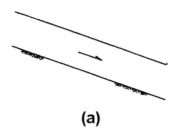

(a)

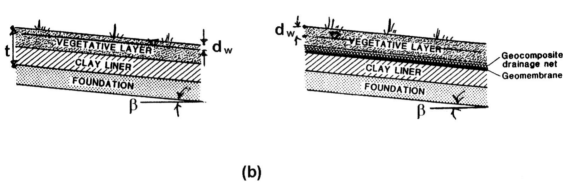

(b)

Figure 19.12 *Final cover slope stability:* (a) *infinite slope slip surface;* (b) *stability evaluation of cover lining system.*

ing forces and is given by the following relationship [for a derivation, see Sharma and Lewis (1994)]:

$$FS = \frac{c' + \gamma_b(t \cos^2\beta \tan \phi')}{\gamma_t(t \sin \beta \cos \beta)} \quad (19.19)$$

where c' and ϕ' are effective strength parameters of soil or the liner interface materials, representing the weakest plane. The soil–liner interface strengths have been presented in Chapter 17. γ_b = buoyant unit weight = $\gamma_t - \gamma_w$, where γ_t is the total unit weight and γ_w is the unit weight of water, t the depth to the failure surface or thickness of the element, and β is the slope inclination.

For *cohesionless soils, $c' = 0$,* the factor of safety can be given by the following relationship:

$$FS = \frac{\gamma_b}{\gamma_t} \frac{\tan \varphi'}{\tan \beta} \quad (19.20)$$

For a special case where the water table is drawn down to the failure surface or liner level, which is the weakest plane, then $\gamma_b = \gamma_t$. FS is then given by the following relationship:

$$FS = \frac{\tan \phi'}{\tan \beta} \quad (19.21)$$

For a case where the water table is at a depth d_w below the ground surface, the factor of safety is given by the following relationship (Matasovic, 1991):

$$FS = \frac{(c'/\gamma_t t \cos^2\beta) + \tan \varphi'[1 - \gamma_w(t - d_w)/\gamma_t t]}{\tan \beta}$$

$$(19.22)$$

All terms have been defined earlier. If $d_w = 0$, the equation is the same as equation (19.19).

Seismic Slope Stability of Final Cover Systems. In accordance with federal regulation 40 CFR258, landfill containment systems, including a final cover system, require evaluation of the impact of seismic accelerations when located in seismic impact zones. A *seismic impact zone* is defined as having 10% or greater probability that the peak horizontal acceleration in lithified earth material (i.e., bedrock) will exceed $0.1g$ within a 250-year period.

Generally, the seismic slope stability evaluation of the final cover system will consist of the following steps.

1. Estimate the peak horizontal acceleration in lithified earth material. The USGS seismic probability maps present peak bedrock horizontal acceleration having a 90% probability of not being exceeded over a 250-year period. Some local state agencies also have studies available for such accelerations.

2. Estimate the peak horizontal accelerations ($\ddot{u}_{max}$) at the top of the landfill. These accelerations can

be estimated by performing one-dimensional site response analysis, such as SHAKE (Schnabel et al., 1972) and Idriss and Sun (1992). Alternatively, a Harder's curve (Figure 19.13), presented by Kavazanjian and Matasovic (1995), can be used to estimate $\ddot{u}_{max}$. As discussed above, in a landfill scenario the final cover system is a thin veneer on top of the landfill mass. Therefore, $\ddot{u}_{max}$ can be approximately equated to K_{max}, which is the maximum average acceleration for a potential sliding mass of the final cover system in an infinite-slope stability case.

3. Calculate K_y, yield acceleration (i.e., an acceleration at which a potential sliding surface would develop a factor of safety of unity). K_y can be calculated by the following relationship (Matasovic, 1991):

$$K_y = \frac{(c'/\gamma_t t \cos^2\beta) + \tan \varphi' \, [1 - \gamma_w(t - d_w)/\gamma_t t] - \tan \beta}{1 + \tan \beta \tan \varphi'}$$

(19.23)

EXAMPLE 19.2

A 2-ft-thick soil liner with $c' = 100$ lb/ft^2, $\varphi' = 15°$, and $\gamma_t = 120$ lb/ft^3 is placed at an inclination of $\beta = 18.4°$ with the horizontal (i.e., a 3:1 slope). If the water table is at the ground surface, calculate the factor of safety of the soil liner assuming that the weakest plane is at 2 ft depth. Discuss if a soil with $\phi' = 30°$ and $c' = 0$ was proposed instead of c', ϕ' soil.

Solution:

Since water is at the ground surface, seepage will be parallel to the slope. The factor of safety can then be calculated by the equation

$$FS = \frac{c' + \gamma_b \, (t \cos^2\beta \tan \phi')}{\gamma_t(t \sin \beta \cos \beta)}$$

$\gamma_b = \gamma_t - \gamma_w = 120 - 62.4 = 57.6$ lb/ft^3; $t = 2$ ft. All other values are given above. Using the equation above, FS = 1.8.

If it was a cohesionless soil, FS = (γ_b/γ_t) (tan φ'/tan β), and using $\phi' = 30°$ for cohesionless soil, FS will be 0.8, which will not be stable. This means that a soil with either a higher ϕ' value and/or with c' and ϕ' will need to be used.

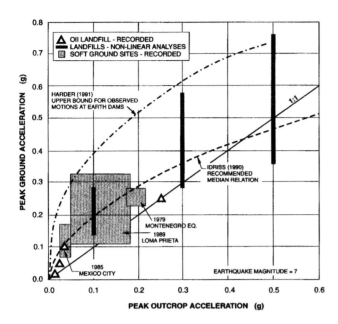

Figure 19.13 *Comparison of peak horizontal bedrock accelerations and peak horizontal accelerations at soft soil sites and landfill sites. (Based on Kavazanjian and Matasovic, 1995.)*

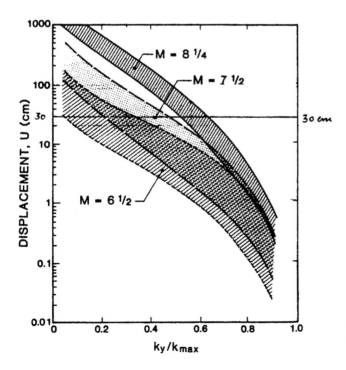

Figure 19.14 *Variation of permanent displacement with yield acceleration. (From Makdisi and Seed, 1997.)*

All terms have been defined previously.

4. Estimate the seismically induced deformation by double integrating the acceleration time history above the K_y values. Alternatively, seismic deformation can be estimated for K_y/K_{max} values using the chart in Figure 19.14 (Makdisi and Seed, 1977).

19.5   SUMMARY

After the landfill operations have been completed and the waste has reached its final permitted elevations in a landfill cell, the final cover is placed over the completed section of the landfill. The primary purposes of constructing a final cover system are to: (1) resist erosion due to wind or water runoff; (2) minimize the infiltration of precipitation into the underlying refuse; and (3) control the migration of landfill gas and/or enhance gas recovery. This means the final cover system, as a minimal, should have an erosion control layer underlain by a low permeability layer over a permeable layer for gas conveyance. Two typical final cover systems, one for Municipal Solid Waste Landfills and the second for Hazardous Waste Landfills are presented in this chapter. In addition, information on Alternative Cover Systems has also been discussed. Finally, analytical methods for design followed by design examples are presented.

QUESTIONS/PROBLEMS

19.1. Why is a final cover system required over a closed landfill? Describe final cover system components for both MSW and hazardous waste landfills.

19.2. Describe an alternative final cover system. Compare and contrast the conventional and various alternative cover systems.

19.3. What is an evapotranspirative cover? Describe the principle that would explain infiltration control by an ET cover system.

19.4. What are the various infiltration models that can be used for cover system analysis? Discuss advantages and disadvantages of HELP and UNSAT-H infiltration models.

19.5. Discuss various design and construction measures that you would recommend to minimize surface erosion of a cover system. When answering this question, include consideration of the USLE.

19.6. A MSW landfill is more than 15 years of age and is closed with a cover system that has clay and HDPE geomembrane. Estimate the tensile strains between two locations in the cover materials after 30 years when these locations are 35 ft apart. Also assume that location 1 has no waste (anchor trench location) and that location 2 has 40 ft of waste. Comment on the impact on clay and HDPE geomembrane integrity due to differential settlement between two locations.

19.7. A 1-ft vegetative soil layer with $c' = 10$ lb/ft^2 and $\phi' = 20°$ and $\gamma_t = 115$ lb/ft^3 is placed on a 3H:1V slope. The soil layer is underlain by a geomembrane over a 1.5-ft-thick clay layer. The interface strength between soil and geomembrane is 22° and that between clay and geomembrane is 18°. Is this slope stable? If not, how would you make it stable? Assume that the water level is at the ground surface.

19.8. For the statically stable cover system in Problem 19.7, estimate the seismic deformation for an earthquake of magnitude 7.5 and peak horizontal acceleration at the rock outcrop of 0.15g. Note that seismic analysis does not consider the existence of a water level.

REFERENCES

Ankeny, M. D., Coons, L. M., Majumdar, N., Kelsey, J., and Miller, M., Performance and cost consideration for landfill caps in semi-arid climates, in *Landfill Capping in the Semi-arid West: Problems, Perspectives, and Solutions,* Reynolds, T. D., and Morris, R. C. (eds.), Environmental Science and Research Foundation, Idaho Falls, ID, 1997.

Carsel, R. F., and Parrish, R. S., Developing joint probability distribution of soil water retention characteristics, *Water Resour. Res.,* Vol. 24, No. 5, pp. 755–769, 1988.

Dodt, M. E., Sweatmen, M. B., and Bergstrom, W. R., Field measurements of landfill surface settlements, in *Geotechnical Practice for Waste Disposal '87,* Woods, R. D. (ed.), ASCE Specialty Conference, Ann Harbor, MI, 1987.

Edil, T. B., Ranguette, V. J., and Wuellner, W. W., Settlement of municipal refuse, pp. 225–239 in *Geotechnical Waste Fills: Theory and Practice,* Landra, A., and Knowles, G. D. (eds.), STP1070, ASTM, West Conshohocken, PA, 1990.

EPRI (Electrical Power Research Institute), *Comparison of Two Groundwater Flow Models UNSATID and HELP,* EPRI CS-3695, Project 1406-1, EPRI, Palo Alto, CA, Oct. 1984.

Fayer, M. J., *UNSAT-H, Version 3.0: Unsaturated Soil Water and Heat Flow Model,* prepared for the U.S. Department of Energy, Pacific Northwestern National Laboratory, Richland, WA, 2000.

Fayer, M. J., and Jones, T. L., *UNSAT-H, Version 2.0: Unsaturated Soil Water and Heat Flow Model,* prepared for the U.S. Department of Energy, Pacific Northwestern National Laboratory, Richland, WA, 1990.

Gee, G. W., and Ward, A. L., Still in quest of perfect cap, in, *Landfill Capping in the Semi-arid West: Problems, Perspectives, and Solutions,* Reynolds, T. D., and Morris, R. C. (eds.), Environmental Science and Research Foundation, Idaho Falls, ID, 1997.

Hauser, V. L., Weand, B. L., and Gill, M. D., *Alternative Landfill Covers,* prepared for the Air Force Center for Environmental Excellence, Technology Transfer Division, Brooks, AFB, TX, July 2001.

Hutson, J. L., and Wagenet, R. J., *Leaching Estimation and Chemistry Model, LEACHM,* New York State College of Agriculture and Life Sciences, Cornell University, Ithaca, NY, 1992.

Idriss, I. M., and Sun, J. I., *User's Manual for SHAKE '91,* Center for Geotechnical Modeling, University of California, Davis, CA, 1992.

Kavazanjian, E., Design and performance of evapotranspirative cover system for arid region landfill, pp. 11–26 in *Proceedings of the Annual Western States Engineering Geology and Geotechnical Engineering Symposium,* University of Nevada, Las Vegas, NV, Mar. 2001.

Kavazanjian, E., and Matasovic, N., Seismic analysis of solid waste landfills, pp. 1066–1080 in *Proceedings of Geoenvironment 2000,* ASCE Specialty Conference, New Orleans, LA, Feb. 1995.

Khire, M. V., Benson, C. H., and Bosscher, P. J., Water balance modeling of earthen final covers, *J. Geotech. Geoenviron. Eng.,* Vol. 123, No. 8, p. 744–754, 1997.

———, Field data from a capillary barrier and model predictions with UNSAT-H, *J. Geotech. Geoenviron. Eng.,* Vol. 125, No. 6, p. 518–527, June 1999.

Kmet, P., *EPA's 1975 Water Balance Method: Its Use and Limitations,* Department of Natural Resources, State of Wisconsin, Madison, WI, Oct. 1982.

Koerner, R. M., Geomembrane liners, pp. 164–186 in *Geotechnical Practice for Waste Disposal,* Daniel, D. E., (ed.), Chapman & Hall, London, 1993.

LaGatta, M. D., Boardman, B. T., Cooley, B. H., and Daniel, D. E., Geosynthetic clay liners subjected to differential settlement, *J. Geotech. Geoenviron. Eng.,* Vol. 123, No. 5, pp. 402–410, 1997.

Landreth, R. E., Landfill containment systems regulations, pp. 1–13 in *Waste Containment Systems: Construction, Regulation, and Performance,* Geotechnical Special Publication 26, Bonaparte, R. (ed.), ASCE, New York, 1990.

Makdisi, F. I., and Seed, H. B., *A Simplified Procedure for Estimating Earthquake-Induced Deformation in Dams and Embankments,* UCB/EERC-77/19, University of California, Berkeley, CA, 1977.

Matasovic, N., Selection of method for seismic slope stability analyses, pp. 1057–1062 in *Proceedings of the 2nd International Conference on Recent Advances in Geotechnical Earthquake Engineering and Soil Dynamics*, Mar. 1991.

Mualem, Y., A new model for predicting the hydraulic conductivity of unsaturated porous media, *Water Resour. Res.*, Vol. 12, pp. 513–522, 1976.

Othman, M. A., Bonaparte, R., Gross, B. A., and Shmertmann, G. R., Design of MSW landfill final cover systems, pp. 218–257 in *Landfill Closures: Environmental Protection and Land Recovery*, Dunn, J., and Sugh, U. P., (eds.), Geotechnical Special Publication 53, ASCE, Reston, VA, 1995.

Scharch, P. E., *Water Balance Analysis Program for the IBM-PC MicroComputer*, Department of Natural Resources, State of Wisconsin, Madison, WI, May 1985.

Schnabel, P., et al., *SHAKE: A Computer Program for Earthquake Response Analysis of Horizontally Layered Sites*, EERC 72-2, Earthquake Engineering Research Center, University of California, Berkeley, CA, 1972.

Schroeder, P. R., Lloyd, C. M., and Zappi, P. A., *The Hydrologic Evaluation of Landfill Performance (HELP) Model: User's Guide for Version 3*, EPA/600/R-94/168a, Office of Research and Development, USEPA, Washington, DC, Sept., 1994a.

Schroeder, P. R., Dozier, T. S., Zappi, P. A., McEnroe, B. M., Sjostrom, J. W., and Peyton, R. L., *The Hydrologic Evaluation of Landfill Performance (HELP) Model: Engineering Documentation for Version 3*, EPA/600/R-94/168b, Office of Research and Development, USEPA, Washington, DC, Sept., 1994b, p. 116.

Sharma, H. D., Solid waste landfills: settlements and post-closure perspectives, pp. 447–455 in *Proceedings of the ASCE National Conference on Environmental and Pipeline Engineering*, Surampalli, R. Y. (ed.), July 2000.

Sharma, H. D., Dukes, M. T., and Olsen, D. M., Evaluation of infiltration through final cover systems: analytical methods, presented at Waste-Tech '89, Washington, DC, 1989.

Sharma, H. D., and Lewis, S. P., *Waste Containment Systems, Waste Stabilization, and Landfills: Design and Evaluation*, Wiley, New York, 1994.

Sharma, H. D., Fowler, W. L., and Cochrane, D. A., Evaluation and remediation of ground cracking associated with refuse settlements, presented at the 7th International Waste Management and Landfill Symposium, Cagliari, Italy, 1999.

Sowers, G. F., Settlement of waste disposal fills, in *Proceedings of the 8th International Conference on Soil Mechanics and Foundation Engineering*, Moscow, 1973.

Stormont, J. C., Incorporating capillary barriers in surface cover systems, in *Landfill Capping in the Semi-arid West: Problems, Perspectives, and Solutions*, Reynolds, T. D., and Morris, R. C. (eds.), Environmental Science and Research Foundation, Idaho Falls, ID, 1997.

USEPA (U.S. Environmental Protection Agency), *Fed. Reg.* Vol. 57, No. 124, pp. 28626–28628, June 26, 1992.

Van Genuchten, M. T., A closed-form equation for predicting the hydraulic conductivity of unsaturated soils, *Soil Sci. Soci. Am. J.*, Vol. 44, pp. 892–898, 1980.

20

GAS GENERATION AND MANAGEMENT

20.1 INTRODUCTION

Gases are generated due to decomposition of municipal solid waste (MSW) in landfills. Composition of the landfill gas (LFG) depends strongly on the composition and age of the waste. Generally, landfill gas consists of approximately 50% methane (CH_4) and 50% carbon dioxide (CO_2). Trace amounts of oxygen and non-methane organic compounds (NMOCs) may also be present. There are several concerns with the landfill gas (USACE, 1995):

- Methane gas is highly combustible, making it a potential hazard in the landfill environment or in structures on adjacent properties.

- Landfill gas is capable of migrating significant distances through soil, thereby increasing the risk of explosion and exposure. Serious accidents resulting in injury, loss of life, and extensive property damage may occur where landfill conditions favor gas migration.

- As landfill gas is produced, the pressure gradient upward may create cracks and disrupt the geomembrane in the landfill cover.

- Methane gas is an asphyxiant to humans and animals in high concentrations.

- Migrating gas may cause adverse effects such as stress to vegetation, by lowering the oxygen content of soil gas available in the root zone.

- Gas generated at landfills and vented to the atmosphere frequently emits nuisance odors, causing annoyance to people residing nearby.

- Emissions of NMOCs in landfill gas may be contributing to the degradation of local air quality.

Vinyl chloride from landfills has been found to be present in substantial concentrations in landfill gases, representing health and safety concerns.

- Methane gas, a "greenhouse gas," contributes to the possibility of global warming of Earth's climate.

- Uncontrolled landfill gas is a loss of potential resources; instead, it can be a satisfactory fuel for a wide variety of applications.

Federal, state, and local regulations require control of landfill gases and require gas management systems as a component of the landfill cover. RCRA regulates landfill gas from MSW landfills under 40 CFR Part 258, which states that owner/operators of MSW landfills must ensure that the concentration of CH_4 gas generated by the facility does not exceed 25% of the lower explosion limit (LEL) for CH_4 in facility structures (excluding gas control or recovery system components) or the LEL at the facility property boundary. The owner/operator must also implement a routine CH_4 monitoring program with at least a quarterly monitoring frequency. Other federal regulations, such as the Clean Air Act, also require landfill facilities to install gas collection and control systems to reduce NMOC emissions. In addition, state regulations may require stringent monitoring and control of landfill gas emissions at landfill sites.

In this chapter we describe the mechanisms of landfill gas generation followed by the physical and chemical characteristics of the landfill gas. Then gas production rates and gas migration processes are presented. Next, gas collection systems that include passive and active collection systems are described. This is followed by a description of landfill gas emission control by flaring or using for energy recovery.

20.2   GAS GENERATION MECHANISMS

Municipal solid waste (MSW) consists of mostly degradable organic materials and generates gases as a result of one or more of three mechanisms: (1) evaporation/volatilization, (2) biological decomposition, and (3) chemical reactions.

Evaporation/volatilization occurs as a result of the change of chemical phase equilibrium that exists within a landfill. Volatile organic compounds in the landfill will vaporize until the equilibrium vapor concentration is reached. This process is accelerated when the waste becomes biologically active, as a result of heat. The rate at which compounds are evolved depends on the physical and chemical properties of compounds. *Henry's law* determines the extent of volatilization of a contaminant dissolved in water, and the Henry's law constant describes the equilibrium partitioning between the vapor and aqueous phases at given temperature and pressure. More details on volatilization and Henry's law are given in Chapter 6.

Biological degradation or decomposition of solid wastes generates methane (CH_4) and carbon dioxide (CO_2) along with traces of other compounds. The bacteria involved in biological decomposition exist in the

waste and soil used in landfill operations. As shown in Figures 20.1 and 20.2, the biological decomposition of solid waste occurs in the following three distinct phases.

- Phase I, *aerobic decomposition,* begins shortly after waste is placed in a landfill and continues until all of the entrained oxygen is depleted from the voids and from within the organic waste. Aerobic bacteria produce a gas of high CO_2 content (approximately 30%) and low CH_4 content (approximately 2 to 5%). Aerobic decomposition may last for as little as six months to as long as 18 months for waste in the bottom lifts of a landfill.
- In phase II, *anaerobic/thermophilic decomposition,* the waste decomposition changes from aerobic to anaerobic and facultative bacteria thrive and decompose waste into simpler molecules, such as hydrogen, ammonia, CO_2, and organic acids.
- In phase III, *anaerobic/methanogenic decomposition,* methanogenic bacteria become more prominent. These methanogens degrade the volatile acids, primarily acetic acid, and use the hydrogen to generate CH_4 (45 to 57%) and CO_2 (40 to 48%).

More details on microbial processes that cause waste degradation and gas generation are given by Barlaz et al. (1989) and Bogner et al. (1996).

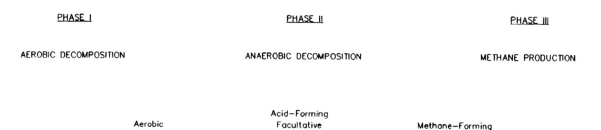

PHASE I   PHASE II   PHASE III

AEROBIC DECOMPOSITION   ANAEROBIC DECOMPOSITION   METHANE PRODUCTION

Figure 20.1 *Three-stage biological decomposition of solid waste.* (*From USACE, 1995.*)

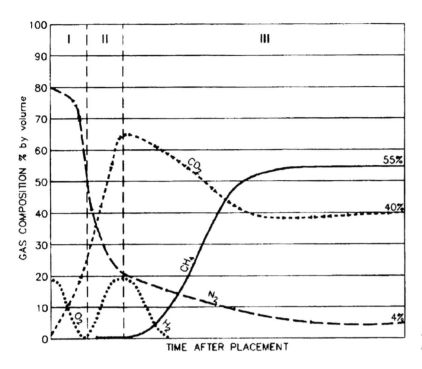

Figure 20.2 *Typical landfill gas evaluation. (From USACE, 1995.)*

Chemical reactions between materials in the waste can produce gases. Such reactions are likely to occur when incompatible materials are co-disposed of in landfills. For example, aliphatic chlorinated solvents are incompatible with aluminum, so if solvent-soaked rags come in contact with aluminum cans in the waste, hydrogen chloride gas is produced. Heat generated by biological decomposition may accelerate chemical reaction rates.

Gas generation in a landfill is affected by the following factors (USACE, 1995): (1) availability of nutrients, (2) temperature, (3) moisture, (4) pH, (5) atmospheric conditions, and (6) age of waste.

Availability of Nutrients. Bacteria in a landfill require various nutrients for growth, primarily carbon, hydrogen, oxygen, nitrogen, and phosphorus (macronutrients), but also small amounts of other elements, such as sodium, potassium, sulfur, calcium, and magnesium (micronutrients). The availability of macronutrients in the landfill mass has an effect on both the volume of water generated from microbial processes

and the composition of the gases generated. Landfills that accept MSW and use daily soil cover have an adequate nutrient supply for most microbiological processes to proceed. If deficient, nutrients can be improved by the addition of sewage sludge, manure, or agricultural wastes.

Temperature. Temperature in the landfill can influence the types of bacteria that are predominant and the level of gas production. The optimum temperature for aerobic decomposition is 54 to 71°C, while the optimum temperature range for anaerobic bacteria is 30 to 41°C. Landfill temperatures are reported to be in the range 29 to 60°C as a result of aerobic decomposition, but may be dropped to 19 to 21°C range as a result of anaerobic activity.

Moisture. Moisture content is the most important parameter affecting waste decomposition and gas production. A high moisture content of the waste (between 50 and 60% by weight) favors maximum methane generation. In general, the moisture content of MSW ranges from 10 to 20% to a high of 30 to 40% with an average of 25% on a wet weight basis. Very low moisture content, such as the case of solid waste in arid regions, may prevent decomposition of waste and thus limit gas production. Leachate circulation in bioreactor landfills allows control of moisture inside the landfill. Typically, when waste achieves 50% moisture, it has reached the field capacity and will tend to leach continuously downward thereafter for additional moisture added to lower layers. In-situ moisture content as high as 70% is possible. At this level, a decrease in the efficiency of a gas collection system can be expected.

pH. The pH of waste in a landfill ranges from 5 to 9 and may change depending on the biological processes. Most landfills have an acidic environment initially, but when the aerobic and acidic anaerobic stages have been completed, the methanogenic processes return the pH to approximately neutral (7 to 8) due to the buffering capacity of the system pH and alkalinity. One concern during the acidic stages of the biological process is that the reduced pH will mobilize metals that may leach out of the landfill or become toxic to the bacteria generating the gas. In some cases, the addition of sewage sludge, manure, or agricultural wastes during waste placement will improve methane gas generation.

Atmospheric Conditions. Atmospheric conditions, particularly temperature, barometric pressure, and precipitation, affect landfill conditions. Atmosphereic temperature and pressure affect the surface layer of waste by reducing the surface concentrations of gas components and creating advection near the surface. Precipitation affects the gas generation significantly by supplying water to the process and by carrying dissolved oxygen into the waste with the water.

Age of Waste. As shown in Figure 20.1, three biological decomposition processes occur, depending on the age of the waste and thus affect the landfill gas production.

20.3   GAS CHARACTERISTICS

The characteristics of landfill gas include both physical and chemical characteristics. The physical characteristics include density, viscosity, temperature, heat value content, and moisture content. The *density* of landfill gas depends on the proportion of gas components present. For example, a mixture of 10% hydrogen and 90% CO_2, such as might be produced in the first stage of anaerobic decomposition, will be heavier than air, while a mixture of 60% CH_4 and 40% CO_2, such as might be produced during the methanogenic phase of decomposition, will be slightly lighter than air. Methane density is 0.714×10^{-4} kg/m^3 and typical composite gas density is 1.07 kg/m^3. *Viscosity* is the property describing resistance to flow due to the existence of internal friction within the fluid. The viscosity values of methane and composite gas are 1.04×10^{-5} N·s/m^2 and 1.15×10^{-5} N·s/m^2, respectively. The *temperature* of landfill gas varies with location, depth, and phase composition. Concentrated mixtures of landfill gas can have a *calorific value* of 500 Btu/ft^3 during the methane generation stage. This value is about half that of natural gas. The amount of *moisture* in the gas depends on the temperature and pressure and can be saturated or undersaturated.

The chemical characteristics of landfill gas depend on the waste type and the stage of decomposition. As discussed in Section 20.2 and shown in Figure 20.2, the major components of the landfill gas are methane, CO_2, NMOC, and water vapor. Methane is the major constituent of landfill gas. Methane is lighter than air and is colorless and odorless. Landfill gas is flammable due to the presence of CH_4 and can be an asphyxiant if present in high concentrations without O_2. CH_4 is explosive at about 5 to 15% by volume in air. Another major constituent of landfill gas is CO_2. CO_2 is heavier than air, colorless, and odorless. CO_2 can be a simple asphyxiant and health hazard if present in high concentrations. Besides CH_4 and CO_2, many minor constituents, collectively known as *nonmethane organic compounds* (NMOCs) are present in landfill gas at low concentrations. Landfill gas created during the decomposition of organic compounds typically includes between 4 and 7% by volume of water vapor.

Site-specific landfill gas composition can be determined by collecting gas samples and analyzing them. Landfill gas samples can be collected using borehole probes, permanent gas monitoring probes, and gas extraction wells. Figure 20.3 shows typical gas probe monitoring details. The samples collected are analyzed for CH_4, CO_2, and NMOCs using standard USEPA methods.

20.4 GAS PRODUCTION RATES

The amount of landfill gas produced is generally a function of the type, extent, and rate of decomposition. The major environmental conditions that affect the type, rate, and extent of biochemical decomposition in a landfill are oxygen availability, moisture, rainfall infiltration, temperature, pH, amount of solid waste, and availability of microbes. Site-specific landfill gas generation rates can be determined by installing gas extraction wells. Extraction wells are installed in a cluster of three or five dispersed locations in the landfill, and a blower extracts the LFG from the wells. Landfill pressure and orifice pressure differentials are measured and the landfill gas production rate is calculated.

The maximum gas yield has been estimated to be 15,000 yd^3 per ton of waste, with an average estimated gas composition by volume of 54% methane and 46% CO_2 and trace amounts of NMOCs. On a wet-weight basis, the theoretical cubic feet of gas generated per pound of solid wastes was determined to be 6.5 for CO_2 and CH_4 and 3.3 for CH_4 alone (USACE, 1994). Gas monitoring at landfills has shown that the gas production rate is proportional to the waste decomposition rate. The higher the moisture content, the higher the rate of gas production will be. Monitoring data from three MSW landfills in California showed gas production to range from 22 to 45 mL/kg of waste per day (USACE, 1994). Landfills in arid climates can generate gas, at lower rates, over a very long period (over 100 years) of time. On the other hand, landfills in a humid climate have been reported to have a gas generation life of 20 years (EMCON, 1980). In bioreactor landfills, where a considerable amount of moisture is added, the landfill gas generation rate will be higher, but over a shorter period of time, 8 to 15 years.

Several mathematical models are reported for estimating the landfill gas generation rates using site-specific input parameters. Of these, the following three relatively simple models are commonly used (USACE, 1995): (1) the Scholl canyon model, (2) the theoretical model, and (3) the regression model.

The *Scholl canyon model* assumes that CH_4 generation is a function of first-order kinetics. This model ignores the first two stages of bacterial activity and is based simply on the observed characteristics of substrata-limited bacterial growth. The parameters of this model are determined empirically by fitting the empirical data to the model to account for variations in the waste moisture content and other landfill conditions. The gas production rate is assumed to be at its peak upon initial placement after a negligible lag time, during which anaerobic conditions are established and decreases exponentially (first-order decay) as the organic content of the waste is consumed. For methane generation rate estimation, average annual waste placement rates are used, and the time measurements are in years. The model equation has the form

$$Q_{CH4} = L_0 R (e^{-kc} - e^{-kt}) \qquad (20.1)$$

where Q_{CH4} is the CH_4 generation rate at time t, L_0 the potential CH_4 generation capacity of the waste, R the average annual acceptance rate of waste, k the CH_4 generation rate constant, c the time since landfill closure ($c = 0$ for active landfills), and t the time since initial waste placement. The model can be refined further by dividing the landfill into smaller submasses to account for the landfill age over time. If a constant annual waste acceptance rate R is assumed, the CH_4 generated from the entire landfill (sum of each submass contribution) is maximum at the time of landfill closure. Lag time due to the establishment of anaerobic conditions could also be incorporated into the model by replacing c with "$c + lag\ time$" and t by "$t + lag\ time$." The lag time before which anaerobic conditions are established may range from 200 days to several years. The refined Scholl canyon equation then takes the form

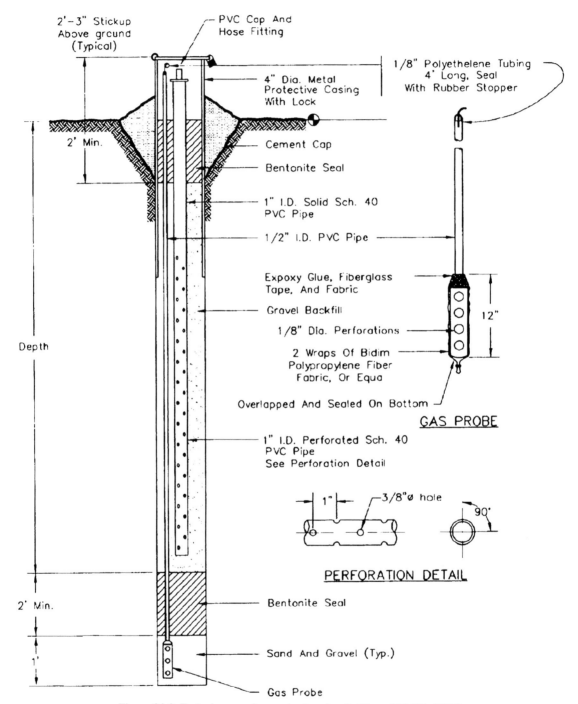

Figure 20.3 *Typical gas probe monitoring detail. (From USACE, 1995.)*

$$Q = 2kL_0 \, \mathrm{Re}^{-k(t-\text{lag})} \qquad (20.2)$$

$$Q_{CH4} = 4.52W \qquad (20.4)$$

where Q is the gas generation rate at time t, L_0 the potential CH_4 generation capacity of the waste, R the average annual acceptance rate of waste, k the CH_4 generation rate constant, t the time since initial waste placement, and "lag" the time to reach anaerobic conditions.

The *theoretical model* determines the theoretical CH_4 generation capacity (L_0) based on stoichiometry using the following empirical formula representing decomposition of waste ($C_aH_bO_cN_dS_e$) :

$$C_aH_bO_cN_dS_e \rightarrow vCH_4 + wCO_2 + xN_2$$
$$+ yNH_3 + zH_2S + \text{humus} \qquad (20.3)$$

The composition of landfill gas during anaerobic conditions is approximately 50% CH_4, 40 to 50% CO_2, and 1 to 10% other gases. The value of L_0 is most directly proportional to the waste's cellulose content. The theoretical CH_4 generation rate increases as the cellulose content of the refuse increases. However, if the landfill conditions are not favorable to methanogenic activity, there would be a reduction in the theoretical value of L_0. The L_0 values have been reported to range from 6 to 270 m^3 CH_4 per metric ton of MSW. The CH_4 generation rate constant, k, estimates how rapidly the CH_4 production rate falls after the waste has been placed. The value of k depends on temperature, moisture content, availability of nutrients, and pH. CH_4 generation increases as the moisture content increases up to a level of 60 to 80%, at which the generation rate does not increase. Values of k obtained from the literature range from 0.003 to 0.21 per year. Once these constants have been estimated, the rate of waste placement and the time in the landfill life cycle determine the estimated gas emission rate.

The *regression model* estimates the CH_4 generation rate based on statistical analysis of the actual data from 21 landfills in the United States. Based on this analysis, the methane generation rate can be estimated by the expression

where Q_{CH4} is the CH_4 flow rate in m^3/min and W is the mass of waste in metric tons. The regression coefficient (r^2) for this correlation was 0.5. The upper and lower 95% confidence limits for the slope in equation (20.4) are 6.52 and 2.52 m^3 CH_4 per ton of waste, respectively.

All of the models above are simple and they do not account accurately for various landfill processes. A generalized mass-balance approach developed by Bogner and Spokas (1993) provides a useful framework for the analysis of landfill gas, particularly methane, at a site. This approach entails examining gas production, gas emission, gas recovery, and gas migration (as discussed in Section 20.5). This approach has been applied to analyze landfill gas emissions at several landfill sites as well as toward optimization of landfill gas recovery strategies (Bogner et al., 1997a,b, 1999; Diot et al., 2001).

20.5 GAS MIGRATION

Gases generated in the landfills may migrate as a result of three processes, as shown in Figure 20.4: (1) molecular effusion, (2) molecular diffusion, and (3) convection.

Molecular effusion occurs at the surface boundary of the landfill with the atmosphere. When the material has been compacted and has not been covered, effusion is the process by which diffused gas releases from the top of the landfill. For dry solids, the principal release mechanism is direct exposure of the waste vapor phase to the ambient atmosphere. *Raoult's law* (Chapter 6) predicts the release rate based on the vapor pressure of the compounds present.

Molecular diffusion occurs in gas systems when a concentration difference exists between two different locations within the gas. The diffusive flow of gas is in the direction in which its concentration decreases. The concentration of a volatile constituent in the landfill gas will almost always be higher than that of the surrounding atmosphere, so the constituent will tend to

EXAMPLE 20.1

The following parameters are given for a landfill:

- *Site characteristics*
  Landfill footprint: 12 acres
  Landfill volume (fill volume) = 872.827 yd^3
  Maximum depth at the center point: 70 ft
  Landfill side slope: 3:1 horizontal:vertical
  Landfill top slope: 5%
  Landfill cover area: 620,000 ft^2

- *Refuse characteristics*
  Ratio of refuse/cover material: 4:1
  Age of refuse: 20 years
  In-place refuse density: 800 lb/yd^3 (at the time of placement; overall density, including cover soil and long-term settlement, will be higher)
  Capping material: 40-mil HDPE
  Refuse void ratio: 4%

- *Gas characteristics*
  Gas constant: 0.08 yr^{-1}
  Gas production potential: 7400 ft^3/ton
  Concentration of methane in gas: 50%
  Diameter of influence/well: 200 ft (USEPA defined)
  Vacuum pressure at wellhead: 10 in water column
  Temperature of landfill gas: 110°F
  Landfill gas viscosity: 2.8 × 10^{-7} lb-s/ft^2
  Landfill gas density: 7.6 × 10^{-2} lb/ft^3

Calculate the gas generation rate after 5 years of closure using the Scholl canyon model.

Solution:

First, estimate the volume of refuse in the landfill:

$$\text{total landfill volume} = 872{,}827 \text{ yd}^3 \text{ (given)}$$

Assuming that a 12-in. intermediate/final cover is constructed across the entire landfill area, we can calculate:

$$\text{volume of intermediate/final cover} = 620{,}000 \text{ ft}^2 \times 1 \text{ ft} = 620{,}000 \text{ ft}^3 = 22{,}962 \text{ yd}^3$$

Assuming that there are six layers of refuse, we can calculate:

total cover material	$= 22{,}962 \times 6 = 137{,}772$ yd^3
volume of refuse	$= 872{,}827 - 137{,}772 = 735{,}055$ yd^3
refuse density	$= 800$ lb/yd^3 (given)
tonnage of refuse	$= 735{,}055 \times 800 = 588{,}044{,}000$ lb $= 294{,}022$ tons
annual refuse disposal rate in landfill	$= 294{,}022/15 = 19{,}600$ tons/yr

Now, estimate the landfill gas generation. Assume that waste composition can be approximated by average municipal waste composition data compiled by the USEPA and that landfill settling is a humid environment establishing conditions affecting biological degradation. Also assume that landfill gas generation is due principally to anaerobic bacteria and can be simulated by first-order kinetics. Use the Scholl canyon model (equation 20.2) to estimate the gas generation rate:

$$Q = 2 \times k \times L \times R \times e^{-k(t-\text{lag})}$$

where: Q is the landfill gas generation rate at time t (ft^3/yr), L the potential gas generation capacity of refuse $= 7400$ ft^3/ton (given), R the annual refuse acceptance rate in landfill $= 19{,}600$ tons/yr, k the gas generation rate, or refuse decay rate $= 0.08$ yr^{-1} (given), t the time since refuse placement $= 5$ years, and lag the time to reach anaerobic conditions $= 2$ years (assumed). Substituting these values yields

$$Q = 1.825 \times 10^7 \text{ ft}^3/\text{yr}$$

---

migrate to a lower concentration area (the ambient air). The rate of diffusion depends primarily on the concentration gradient. Specific compounds exhibit various diffusion coefficients, which are the rate constants for this transport.

Convection is the movement of landfill gas in response to pressure gradients developed within the landfill. Gas will flow from higher- to lower-pressure regions and from the landfill to the atmosphere. Where it occurs, the convective flow of gas will overwhelm the other two release mechanisms in its ability to release materials into the atmosphere. The source of the pressure may be the production of vapors from biodegradation processes, chemical reactions within the landfill, compaction effects, or methane generation at the lower regions of the landfill which drive vapors toward the surface. The rate of gas movement is generally orders of magnitude faster for convection than for diffusion. For a particular gas, convective and diffusive flow may be in opposing directions, resulting in

an overall tendency toward cancellation. However, for most cases of landfill gas recovery, diffusive and convective flows occur in the same direction.

The transport mechanisms above are affected by the following factors: (1) permeability, (2) depth of groundwater, (3) condition within the waste, (4) moisture content, (5) human-made features, and (6) landfill daily cover and cap systems.

The permeability or the coefficient of permeability is a measure of the ease with which a porous medium can transmit landfill gas, water, or other fluid through its media. The permeability distribution to gas has a profound influence on gas flow rates and gas recovery rates. Coarse-grained waste typically exhibits higher permeability than fine-grained waste.

The water table surface tends to act as a no-flow boundary for gas flow within the unsaturated zone. As a result, it is generally used to estimate the thickness of the zone from which a gas can be moved. Gas migration is restricted by the relative insolubility of the

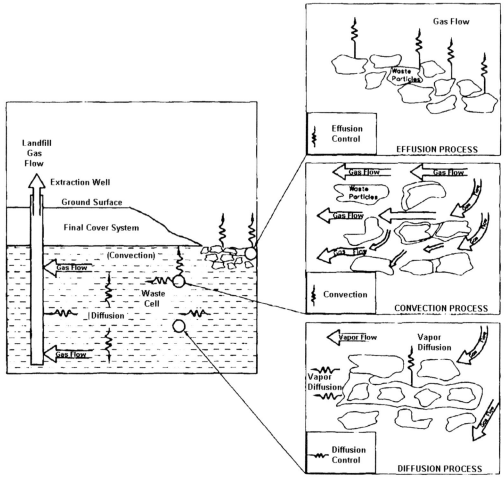

Figure 20.4 Landfill gas migration processes. (From USACE, 1995.)

gas in water. The presence of perched water or water table underneath the landfill base can inhibit the depth of gas migration and increase lateral gas movement.

The conditions within the waste, particularly heterogeneities, porosity, and moisture retention, affect the gas transport. Heterogeneities such as spatial variations, layering, and so on, will affect the gas flow patterns and thus gas recovery rates within the landfill. The waste porosity and moisture retention also affect the air pathway and thus the gas recovery rates.

The subsurface soils, daily cover soil, and the final cover system can influence vertical and lateral migration of landfill gas as depicted in Figure 20.5. Several other human-made features also affect the landfill gas migration. Sewers, drainage culverts, and buried utilities lines running near landfills can provide corridors for gas migration. In addition, breaks in subsurface utility structures such as manholes, vaults, catch basins, or drainage culverts near landfills not only provide corridors for gas migration but also provide areas

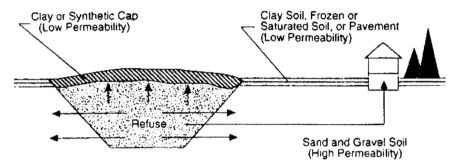

EXTENSIVE LATERAL MIGRATION

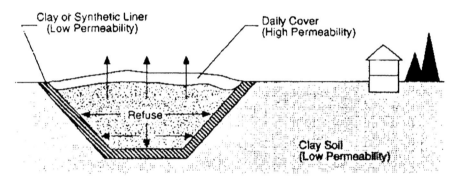

EXTENSIVE VERTICAL MIGRATION

Figure 20.5 *Potential effects of surrounding geology on gas migration. (From USEPA, 1993.)*

for potentially dangerous concentrations of gas to accumulate. Additionally, natural features such as gravel and sand lenses and void spaces, cracks, and fissures resulting from landfill differential settlement can lead to substantial landfill gas migration.

20.6 GAS COLLECTION SYSTEMS

To prevent an explosive hazard as well as adverse effects on human health and the environment due to exposure to landfill gas, control systems are installed to collect and remove gases from a landfill (O'Leary and Tansel, 1986; SWANA, 1992; McBean et al., 1995; Qian et al., 2002). From a regulatory perspective, land-

fill gas should be controlled so that methane concentration at a landfill property line is less than 5% by volume (40 CFR 256.23). In general, the landfill gas control systems can be grouped into two types. A *passive system* functions on the principle that natural pressure gradient and convection mechanisms move the landfill gas. Passive systems provide corridors to intercept lateral gas migration and channel the gas to a collection point or a vent. These systems use barriers to prevent migration past the interceptors and the perimeter of the landfill. *Active systems* move the landfill gas under induced negative pressure (vacuum). The zone of negative pressure created by the applied vacuum induces a pressure gradient toward a collection point that is either a well or horizontal collection pipe.

20.6.1 Passive Gas Collection Systems

Passive gas collection systems are used to collect land-fill gases when gases can freely migrate to a collection point and when little or no lateral migration occurs. These systems control convective gas flow; however, they are less effective in controlling diffusive gas flow. To control lateral gas migration, an impermeable barrier layer needs to be incorporated into the design. Also, these systems should be deeper than the landfill in order to intercept all lateral gas flow. Generally, these systems are tied into an impermeable zone such as a continuous impermeable geologic unit or water table. The passive gas collection systems include well vents, trench vents, or combinations of these.

Well Vents. The well vents can be employed to control lateral and vertical migration of landfill gas. The components and basic configurations of gas vents are shown in Figures 20.6 and 20.7. These are located at points where gas is collecting and building up pressure. The lateral gas migration can be controlled only if the vents are located very close together. Preliminary sampling should be conducted to determine gas collection points for proper vent placement. For optimum effectiveness, gas vents are placed at maximum concentration and/or pressure contours. To ensure proper ventilation, vent depth should extend to the bottom of the waste. Atmospheric gas vents, shown in Figure 20.7(*a*) and (*b*), are generally not recommended for control of lateral gas migration. If these wells are to be used for lateral gas migration control, they must be spaced very close together, generally less than 50 ft apart.

The gas vents are constructed using 4- to 6-in. PVC perforated pipes (Figure 20.6). A surrounding layer of gravel pack is installed to prevent clogging. The pipe vent should be sealed off from the atmosphere with cement or soil grout so that excess air is not introduced into the system and landfill gas is not leaked. Passive wells should generally be located about 30 to 50 ft from the edge of the wastes and typically not more than one well per acre. Additional wells may be needed further within the body of the wastes to intercept their full depth if the site is benched or sloping.

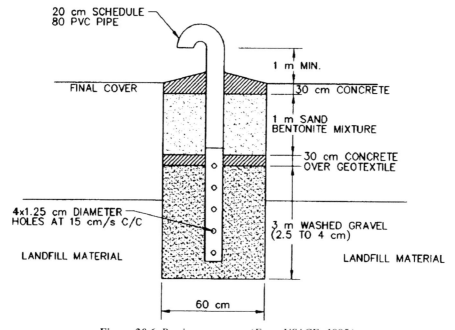

Figure 20.6 *Passive gas vents. (From USACE, 1995.)*

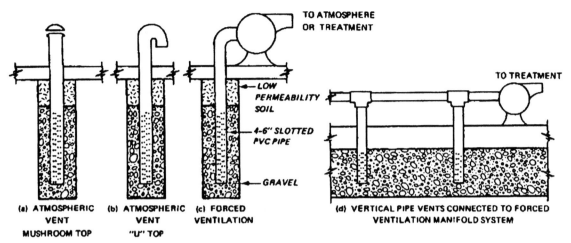

Figure 20.7 *Design configuration of pipe vents. (From USACE, 1994.)*

The gas vents can be incorporated into the final cover systems as shown in Figure 20.8. Geomembranes in landfill cover, if used, should be properly sealed at the gas vent locations, as shown in Figure 20.9.

Trench Vents. Trench vents are employed primarily to control lateral gas migration. The basic configurations of trench vents are shown in Figure 20.10. These systems are suitable to sites where the depth of gas migration is limited by an impervious formation or groundwater. If the trench can be excavated to this depth, trench vents can offer full containment and control of gases. Passive open trenches [Figure 20.10(a)

and (b)] may be used as a permanent control for gas migration; however, their efficiency is expected to be low. An impervious liner can be added to the outside of the trench to increase control efficiency. Open trenches are suitable for sparsely populated areas where they are not likely to be accidentally covered, planted over, or otherwise plugged by outsiders. Passive trench vents may be covered over by clay or other impervious materials and vented to the atmosphere. Such a system ensures adequate ventilation and prevents infiltration of rainfall into the vent. Also, an impermeable clay layer can be used as an effective seal against the escape of landfill gas.

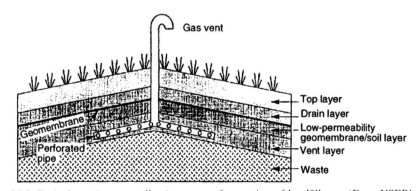

Figure 20.8 *Typical passive gas collection system for venting of landfill gas. (From USEPA, 1994.)*

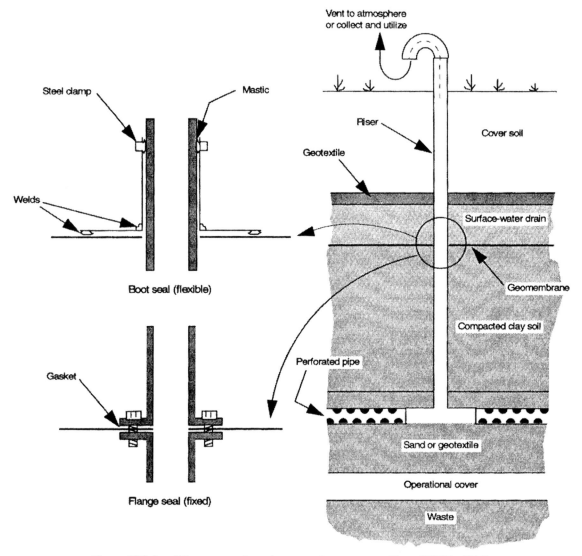

Figure 20.9 *Landfill gas vents through geomembrane covers. (From USEPA, 1994.)*

Open trenches as shown in Figure 20.10(*a*) are subject to infiltration by rainfall runoff and could become clogged by solids. Therefore, these are not located in an area of low relief, and the surrounding ground at the trench locations is sloped to direct runoff away from the trench as shown in Figure 20.10(*a*). The gravel pack in the trench should be permeable enough, relative to the surrounding strata, to transport the gas adequately. As shown in Figure 20.10(*b*), a barrier may be installed on the outside of the trench to prevent bypass of landfill gas. Three types of barriers are commonly used: geosynthetic liners, natural clays, or admixed materials. These are similar to the liner materials used in the base and side-slope liner systems. Closed trenches [Figure 20.10(*c*) and (*d*)] consist of lateral and risers in trench filled with gravel pack. An impermeable clay layer is used for sealing to protect against escape of the landfill gas.

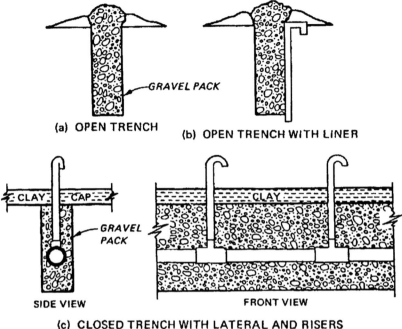

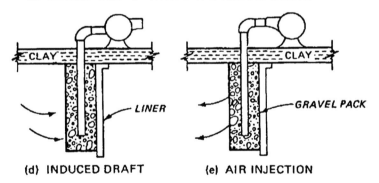

Figure 20.10 *Design configuration of trench vents. (From USACE, 1994.)*

Gas collection trenches can be used where vertical extraction wells are not practical, such as in areas where the waste depth is shallow or where the leachate levels are high. A drawback of trenches is their tendency to draw in air if the seal over each trench is inadequate. Extreme care should be taken in the design of all vent systems to prevent them from being a source of infiltration through the cover. Major advantages of trench systems include ease of construction and relatively uniform withdrawal influence areas. However, these trenches are susceptible to crushing as subsequent lifts of waste are placed and are susceptible to

severing and severe damage as a result of differential settlement of the waste pack. When placed below groundwater levels, the trenches are also subject to flooding. When designing trenches that will be installed below the expected high groundwater or leachate levels, measures should be taken to avoid drawing water into the gas collection system.

The trenches can be vertical or horizontal at or near the base of the landfill. A vertical trench is constructed in much the same manner as a vertical well is constructed. For a new site, horizontal trenches are installed within a landfill cell as each layer of waste is applied, as shown in Figure 20.11. The distance between layers should be no greater than 15 ft. This allows for gas collection as soon as possible after gas generation begins and avoids the need for aboveground piping, which can interfere with landfill maintenance equipment. Additional "legs" of the system are connected to the manifold as the landfill grows in areal size or height.

The horizontal trench pipes may be constructed of perforated PVC, HDPE, or other nonporous material of suitable strength. The material used should also be corrosion-resistant. The trench should be about 3 ft wide, filled with gravel of uniform size, and extend into the waste about 5 ft below the landfill cap layer. Trenches should be located between the waste fill and the gas barrier or side of the site. The side of the trench nearest the property boundary should be sealed with a low-permeability barrier material, such as a geomembrane, to prevent gas migration. The remainder of the trench should be lined with a filter fabric to prevent clogging of the permeable medium.

The gas collection piping enclosed in the trench gravel pack is connected to surface vent pipes of construction similar to that of the collection piping. Vent pipe spacing should be determined from monitoring and site investigation data, but should generally be greater than 50 m apart. Passive vents can be used in combination with horizontal trenches by connecting vents to the pipes with flexible hosing. The flexible hose between the extraction well or trench and the collection header system allows differential movement. Because of its horizontal layout, the collection header system would be expected to settle more than a vertical extraction well. This flexible connection allows more movement than would be possible if the two pipes were rigidly connected. Sampling ports can be installed, allowing monitoring of pressure, gas temperature and concentration, and liquid level.

20.6.2 Active Gas Collection Systems

Active gas collection systems are very effective for controlling migration of gas. An active collection system consists of a mechanical blower or compressor attached to a system of gas extraction wells or collection trenches. A pressure gradient is created in the wells or trenches, thereby forcing the removal of gas from the landfill. The gas is then piped to a flare or other treatment system. The effectiveness of an active landfill gas collection system depends on the design and operation of the system. An effective collection system should be designed and configured so as to handle the maximum landfill gas generation rates, effectively collect

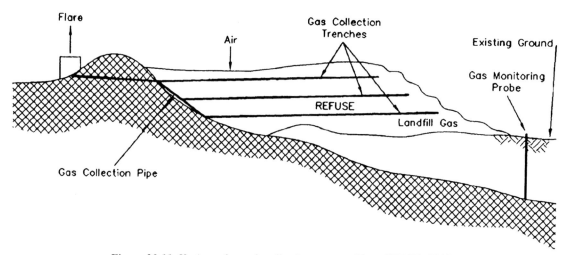

Figure 20.11 *Horizontal trench collection system.* (*From USACE, 1995.*)

landfill gas from all areas of the landfill, and provide the capability to monitor and adjust the operation of individual extraction wells and trenches.

Active gas collection systems can be divided into extraction and pressure systems. Both systems usually incorporate some type of impermeable gas barrier system. Extraction systems usually incorporate a series of gas extraction wells installed within the perimeter of the landfill. Extraction wells are similar to gas monitoring wells, only larger, and construction and materials are the same. A typical gas extraction well design is shown in Figure 20.12. The number and spacing needed for the extraction wells for any particular landfill are site dependent. Often, a pilot system of only a few wells is installed first to determine the radius of influence in the area of the wells. Once the wells are installed, they are connected using gas valving and

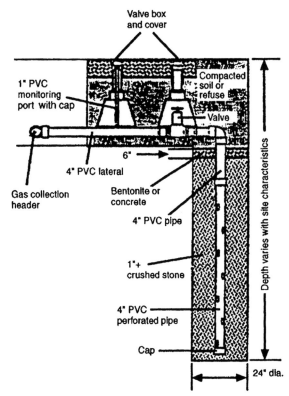

Figure 20.12 *Schematic of gas extraction well. (From USEPA, 1994.)*

condensation traps to a suction system. A centrifugal blower creates a vacuum on the manifold, drawing gas from the wells and causing the gas in the refuse and soil to flow toward each well. Depending on the location, the gas is either exhausted to the atmosphere, flared to prevent malodors, or recovered and treated. A pressure gas control system is sometimes considered when structures are built or already exist on abandoned landfills. The system uses a blower to force air under the building's slab to flush away any gas that has collected and to develop a positive pressure to prevent gas from migrating toward the structure.

Landfill gas migration control can be accomplished effectively by installing forced-ventilation systems in which a vacuum pump or blower is connected to the discharge end of a vent pipe. Such a system is applicable for controlling both vertical and lateral movement of gas in the landfill by installing vents along the perimeter of the site. The gas collected can be vented to the atmosphere, flared, or recovered and treated. Figure 20.13 depicts a series of pipe vents installed in a trench connected to a manifold that leads to a blower and finally, to gas treatment. Such a configuration can be used to prevent emission of gases to the atmosphere across the entire area of the site. Another type of forced ventilation, in a trench for landfill gas control, is air injection; in this method, air injected into the trench by a blower forces the gas back. This system should work well in conjunction with pipe vents installed close to the landfill and inside the circumferences of the trench.

With active systems, the flow rate can be increased or decreased as the gas generation rate increases or decreases. This offers a great deal of flexibility of control in the system. Thus, extraction rates can be reduced with time, and operating costs will decrease. It is expected that gas vents from active systems may get clogged with time and will need to be replaced. Also, it is expected that more maintenance will be required for forced ventilation than for passive atmospheric vent systems. An active collection system has four major components: (1) gas extraction wells (or horizontal trenches), (2) gas-moving equipment, (3) landfill gas treatment units, and (4) condensate removal and disposal units.

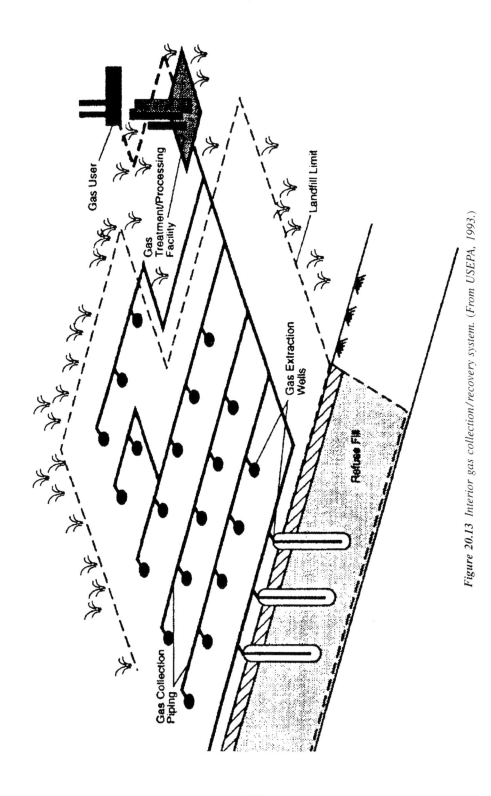

Figure 20.13 *Interior gas collection/recovery system. (From USEPA, 1993.)*

808

Gas Extraction Well Construction. Landfill gas extraction wells are installed around the perimeter and into the center of the landfill. A typical active vertical extraction well configuration is shown in Figure 20.12. The extraction well is generally constructed of PVC, HDPE, or other suitable nonporous material. Pipe diameters vary but generally are no smaller than 2 in. in diameter and no larger than 12 in. in diameter. It is recommended that the bottom three-fourths of the pipe be perforated with $\frac{1}{2}$-in.-diameter holes spaced at 90° every 6 in. Slotted pipe having equivalent perforations is also suitable. Wells are constructed in 12- to 36-in.-diameter boreholes. Upon insertion of the casing into the borehole, the remainder of the well excavation is backfilled with crushed stone. The crushed stone gives the extraction well a larger effective diameter from which gas can be drawn. In unlined landfills, wells are constructed to either the base of the landfill or the water table. However, in a lined landfill, wells are typically constructed to 75% of the landfill's depth, to avoid damaging the liner. The screened interval of a landfill gas extraction well typically extends from the bottom of the well to a point at least 5 ft below the landfill surface. Each wellhead is typically designed with a butterfly or ball valve for regulating the applied pressure to the wellhead.

Spacing and Radius of Influence. The spacing of landfill gas extraction wells is generally determined from the radius of influence of individual wells. This radius is described as the distance from the center of a well to a point away from the well where the steady-state pressure gradient resulting from the blower is 0.1 in. H_2O. Accordingly, any CH_4 generated beyond the radius of influence would not be collected by the extraction wells.

To obtain a representative well spacing for the landfill, several pump tests should be performed so that the waste compaction variability can be taken into consideration. Due to the costs associated with conducting these tests, several theoretical models have been developed to estimate the vacuum radius of influence relationship. Typical negative pressures at the wellhead range from 5 to 15 in. H_2O column. Typically, well spacing ranges from 50 to 300 ft, depending on the

radius of influence for each well. The desired method for determining effective well spacing at a specified landfill is to use field measurement data. Pump tests with monitoring probes at incremental distances from the test well will indicate the influence of a given negative pressure at that location.

Number of Extraction Wells. The factors affecting the number of extraction wells selected are the well radius of influence and spacing, and landfill geometry. Some overlap of influence zone is desirable for the perimeter wells of a system designed for control of gas migration to ensure that effective control is obtained at points between wells along the landfill boundary. The gas extraction rate and radius of influence are dependent on one another, and individual well flow rates can be adjusted after the recovery system is in operation to provide effective migration control and/or efficient CH_4 recovery.

Gas-Moving Equipment. Gas-moving equipment includes pipeline header system and compressors and blowers. A pipeline header system conveys the flow of collected landfill gas from the well or trench system to the blower or compressor facility. A typical header pipe is made of PVC or HDPE and is generally 6 to 24 in. in diameter depending on the flow rate through each section of the pipe. The size and type of blower is a function of the total gas flow rate, total system pressure drop, and vacuum required to induce the pressure gradient.

20.6.3 Comparison of Various Gas Collection Systems

Table 20.1 presents a comparison of various gas collection systems. The efficiency of a passive collection system depends on good containment of the landfill gas to prevent direct emission to the ambient air. Generally, passive collection systems have lower collection efficiencies than active systems, since they rely on natural pressure or concentration gradients to drive gas flow rather than a stronger, mechanically induced pressure gradient. A well-designed passive system, how-

EXAMPLE 20.2

Based on the information given in Example 20.1, determine the number of gas wells required.

Solution:

Given a landfill surface area of 620,000 ft^2 and a diameter of influence of 200 ft (as recommended by USEPA):

$$\text{area of influence} = \frac{\pi D^2}{4} = \frac{3.14 \times (200)^2}{4} = 31,400 \text{ ft}^2$$

$$\text{number of wells required} = \frac{\text{area of landfill}}{\text{area of influence}}$$

$$= \frac{620,000}{31,400} = 19.74$$

$$= 20 \text{ wells}$$

These wells should be located such that the landfill gas from the entire landfill area can be collected.

ever, can be nearly equivalent in collection efficiency to an active system if the landfill design includes geosynthetic liners in the landfill liner and cover. Since passive systems rely on venting, in the event that the vent is blocked by moisture or frost, the gas seeks other escape routes, including moving into surrounding formations. Passive systems are not considered reliable enough to provide an exclusive means of protection. With their concentrated vent gas, passive systems may be considered as an uncontrolled air emissions point source by regulatory agencies. In addition, passive venting systems raise the potential for nuisance odor problems because there is no positive system for odor management. The construction of passive systems is less critical than that of active systems because the collection well is under positive pressure, and air infiltration from the surface is not as great a concern. Additionally, elaborate wellhead assemblies are not required for passive systems since monitoring and adjustment are not usually necessary in these systems. Active systems are usually utilized where a higher degree of system reliability is required than can be accomplished with a passive collection system. Based on

theoretical evaluations, a well-designed active collection system is considered the most effective means of gas collection.

20.7 GAS FLARING AND ENERGY RECOVERY

Landfill gas can be either combusted with no energy recovery, combusted with energy recovery, or released to atmosphere without treatment. The nonenergy recovery techniques use flares and thermal incinerators. The energy recovery techniques include gas turbines and internal combustion engines that generate electricity from the combustion of landfill gas.

20.7.1 Gas Flaring

Flares are used at landfills as the main method of air emission control and as a backup to an energy recovery system. Flaring is an open combustion process in which the oxygen required for combustion is provided by either ambient air or forced air. Figure 20.14 shows

Table 20.1 **Comparison of various collection systems**

Collection System Type	Preferred Applications	Advantages	Disadvantages
Active vertical well collection systems	Landfills employing cell-by-cell landfilling methods Landfills with natural depressions such as canyons	Cheaper or equivalent in costs when compared to horizontal trench systems	Difficult to install and operate on the active face of the landfill (may have to replace wells destroyed by heavy operative equipment)
Horizontal trench collection systems	Landfills employing layer-by-layer landfilling methods	Easy to install since drilling is not required	The bottom trench layer has a higher tendency to collapse and difficult to repair once it collapses Has tendency to flood easily if water table is high Difficult to maintain uniform vacuum along the length (or width) of the landfill Must be installed while the landfill is being constructed; not applicable for constructed landfills
Passive collection systems	Landfills with good containment (side liners and cap)	Cheaper to install and maintain if only a few wells are required Lower in operation and maintenance cost	

Source: USACE (1995).

a typical landfill gas flare system. Landfill gas is conveyed to the flare through the collection header and transfer lines by one or more blowers. A knockout drum is normally used to remove gas condensate. The landfill gas is usually passed through a water seal before going to the flare. This prevents possible flame flashbacks, which can occur when the gas flow rate to the flare is too low and the flame front moves down into the stack.

Two types of flare systems are generally available: open-flame flare and enclosed flare. An open-flame flare represents the first generation of flares. The open-flame flare was used primarily for safe disposal of combustible gas when emission control had not been a requirement. Open-flame flares have also been used

widely in landfill gas combustion. Open-flame flare design and the conditions necessary to achieve 98% reduction of total hydrocarbon are described in 40 CFR 60.18. The advantages of open-flame flares are (1) simple design, since combustion control is not possible; (2) ease of construction; (3) most cost-effective way of safely disposing of landfill gases; and (4) open-flame flares can be located at ground level or elevated. The major disadvantages of all open-flame flares are (1) that they do not have the flexibility to allow temperature control, air control, or sampling of combustion products due to its basic design; and (2) that it is not possible to design a closed-loop system to accurately measure flow rates or emissions from an open-flame flare. This is for two reasons: (1) sample probes placed

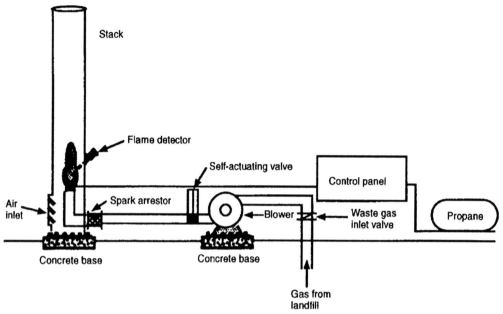

Figure 20.14 *Schematic of a landfill flare system with blower. (From USEPA, 1994.)*

too close to the flame will measure high CO_2 and hydrocarbon levels, and (2) samples taken farther away from the flame are diluted unpredictably by air. Therefore, if emissions sampling and testing are required, an enclosed type of flare will be needed.

Enclosed flares differ from open flares in that both landfill gas and airflows are controlled. Figure 20.15 shows a typical enclosed flare system. While landfill gas is pushed through the flame arrestor and burner tips by a blower, the flare stack pulls or drafts the air through air dampers and around burner tips. The stack acts as a chimney, so its height and diameter are critical in developing sufficient draft and residence time for efficient operation. Enclosed flares are used in landfill gas applications for two reasons: (1) they provide a simple means of hiding all or parts of the flame (i.e., neighbor friendly), and (2) emission monitoring may be mandatory. Depending on air regulations in each state, enclosed flares with an automatic air damper control may be required. Periodic sampling of these flares is conducted to ensure that an emission reduction of 98% is being achieved.

A flare system is used to burn the landfill gas in a controlled environment to destroy harmful constituents and dispose of it safely to the atmosphere. The operating temperature is a function of gas composition and flow rate. The factors that need to be considered in the design of a landfill gas flare are residence time, operating temperature, turbulence, O_2, and flame arrestor. Adequate time must be available for complete combustion. The temperature must be high enough to ignite the gas and allow combustion of the mixture of fuel and O_2. The residence time varies from 0.25 to 2 seconds, and the operating temperature of the combustor is higher than 760°C, which is the temperature at which CH_4 autoignites.

20.7.2 Energy Recovery

In many large MSW landfills, landfill gas is being developed as an energy resource. The following four approaches have been adopted for recovering energy from landfill gas:

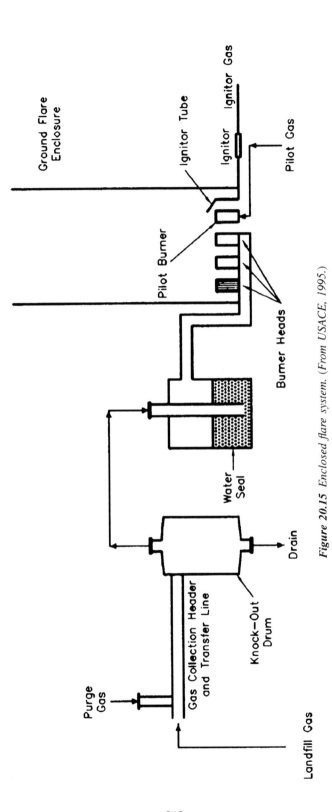

Figure 20.15 Enclosed flare system. (*From USACE, 1995.*)

813

1. Use of landfill gas to fuel gas turbines
2. Generation of electricity by the operation of an internal combustion engine with landfill gas
3. Use of landfill gas directly as a boiler fuel
4. Upgrading the gas quality to pipeline quality for delivery to utility distribution systems

Typically, landfill gas contains approximately 500 Btu per standard cubic foot (4450 kcal/m³) of energy, whereas pipeline-quality gas contains 1000 Btu per standard cubic foot (6900 kcal/m³). The energy content of landfill gas varies widely depending on the performance of the gas collection system and the stage of decomposition within the landfill. Generally, the collection of gas for energy recovery purposes has been limited to large landfills with over 1 million tons of solid waste in place.

Gas turbines aspire ambient air, compress it, and combine it with fuel in the combustor. Figure 20.16(a) illustrates landfill gas to turbine electric generation. The combustor exhaust flows to the power turbine, which burns the fuel to heat it, then expands it in the power turbine to develop shaft horsepower. This shaft power drives the inlet compressor and an electric generator. Two basic types of gas turbines have been used in landfill applications: simple cycle and regenerative cycle. The gas temperatures from the power turbine range from 430 to 600°C. The regenerative cycle gas turbine is essentially a simple cycle gas turbine with an added heat exchanger. Thermal energy is recovered from the hot exhaust gases and used to preheat the compressed air. Since less fuel is required to heat the compressed air to the turbine inlet temperature, the regenerated cycle improves the overall efficiency of the gas turbine. Based on field tests, these turbines are capable of achieving greater than 98% destruction of NMOCs. The applicability of gas turbines depends on the quality of landfill gas generated, the availability of customers, the price of electricity, and environmental issues.

Reciprocating internal combustion engines produce shaft power by confining a combustible mixture in a small volume between the head of a piston and its surrounding cylinder, causing this mixture to burn, and allowing the resulting high-pressure products of combustion gas to push the piston. Power is converted from linear to rotary form by means of a crankshaft. Figure 20.16(b) shows a landfill gas-IC generation plant. The major problem with the use of combustion engines for these applications is selection of the fuel gas compressor. Internal combustion engines are being used for landfill gas control because of their short construction time, ease of installation, and operating capability over a wide range of speeds and loads. IC engines fueled by landfill gas are available in capacities ranging from 500 kW up to well over 3000 kW.

In general, the selection of a recovery technique depends on the gas generation rate, the location of the plant, the availability of markets for the recovered energy, and the environmental impacts.

20.8   SUMMARY

Landfill gases are the result of microbial decomposition of solid waste. Gases produced include methane, carbon dioxide, and lesser amounts of other gases (e.g., hydrogen, volatile organic compounds, hydrogen sulfide). Landfill gas production rates vary spatially within a landfill. Migration of landfill gases occurs due to concentration gradients and pressure gradients. Generally, landfill gas moves through the path of least resistance. Uncontrolled migration of landfill gas can cause explosion hazards, undesirable odors, physical disruption/damage of covers and toxic vapor emissions. Regulations require that landfill gas should be monitored to ensure that methane concentration does not exceed 25% of the lower explosion limit. Generally, gas collections systems that include passive and active gas collection systems are used at landfills. At large landfills, the collected gases are being used for energy recovery. Thus, landfill gas recovery systems can reduce landfill gas odor and migration, can reduce the danger of explosion and fire, and may be used as a source of revenue that may help to reduce the cost of closure.

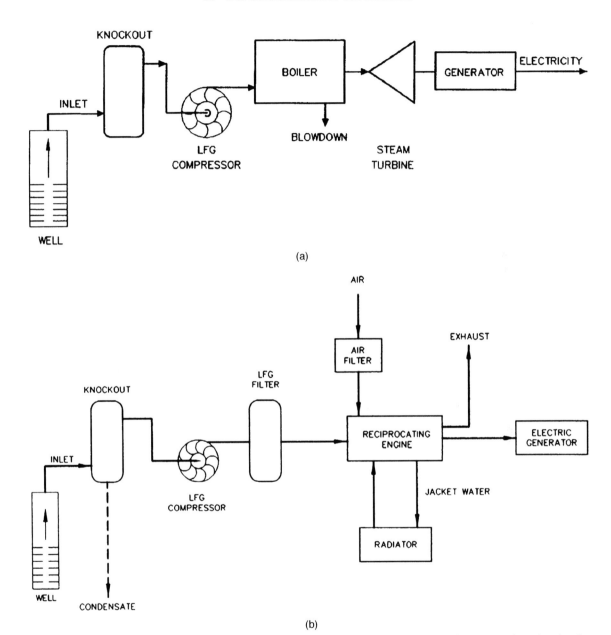

Figure 20.16 (a) *Schematic of LFG to turbine electric generator;* (b) *schematic of LFG to internal combustion for electric generator.* (*From USACE, 1995.*)

QUESTIONS/PROBLEMS

20.1. How are the gases in landfills generated?

20.2. What is the typical composition of landfill gas?

20.3. Why do landfill gases have to be collected and removed?

20.4. Why are landfill gas generation rates higher in the presence of higher moisture content?

20.5. Compare the Scholl canyon and regression models for predicting landfill gas generation rates.

20.6. List advantages and disadvantages of passive and active gas collection systems.

20.7. What are the advantages and disadvantages of LFG for gas turbines for the purpose of energy recovery?

20.8. What are the advantages and disadvantages of LFG for internal combustion engines for the purpose of energy recovery?

20.9. Based on the data given in Example 20.1, determine the gas generation rate after one and 10 years of closure. Discuss the results.

20.10. Based on the data given in Example 20.1, determine the number of extraction wells if the radius of influence is 50, 100, 200, and 300 ft. Explain how the radius of influence can be determined.

REFERENCES

Barlaz, M., Schaefer, D. M., and Ham, R. K., Bacterial Population Development and Chemical Characteristics of Refuse Decomposition in a Simulated Sanitary Landfill, *Applied Environmental Microbiology*, Vol. 55, pp. 55–65, 1989.

Bogner, J., and Spokas, K., Landfill CH_4: rates, fates, and role in global carbon cycle, *Chemosphere*, Vol. 26, No. 1–4, pp. 369–386, 1993.

Bogner, J. E., Sweeney, R. E., Coleman, D., Huitric, R., and Ririe, G. T., Using isotopic and molecular data to model landfill gas processes, *Waste Manag. Res.*, Vol. 14, pp. 367–376, 1996.

Bogner, J. E., Spokas, K. A., and Burton, E. A., Kinetics of methane oxidation in a landfill cover soil: temporal variations, a whole-landfill oxidation experiment, and modeling of net CH_4 emissions, *Environ. Sci. Technol.*, Vol. 31, No. 9, pp. 2504–2514, 1997a.

Bogner, J., Meadows, M., and Czepiel, P., Fluxes of methane between landfills and the atmosphere: natural and engineered controls, *Soil Use Manag.*, Vol. 13, pp. 268–277, 1997b.

CFR (*Code of Federal Regulations*), 40 CFR 256, U.S. Government Printing Office, Washington, DC, 2002.

Diot, M., Bogner, J., Chanton, J., Guerbois, M., Hebe, I., Moreau Le Golvan, Y., Spokas, K., and Tregoures, A., LFG mass balance: a key to optimize LFG recovery, pp. 515–524 in *Proceedings of Sardinia 2001, 8th International Waste Management and Landfill Symposium*, Cagliari, Italy, 2001.

EMCON Associates, *Methane Generation and Recovery from Landfills*, Ann Arbor Science, Ann Arbor, MI, 1980.

McBean, E. A., Rovers, F. A., and Farquhar, G. J., *Solid Waste Landfill Engineering and Design*, Prentice Hall, Upper Saddle River, NJ, 1995.

O'Leary, P., and Tansel, B., Landfill gas movement, control and uses, *Waste Age*, pp. 104–115, 1986.

Qian, X., Koerner, R. M., and Gray, D. H., *Geotechnical Aspects of Landfill Design and Construction*, Prentice Hall, Upper Saddle River, NJ, 2002.

SWANA (Solid Waste Association of North America), *A Compilation of Landfill Gas Field Practices and Procedures*, SWANA, Silver Spring, MD, 1992.

USACE (U.S. Army Corps of Engineers), *Technical Guidelines for Hazardous and Toxic Waste Treatment and Cleanup Activities*, EM 1110-1-502, U.S. Department of the Army, Washington, DC, 1994.

———, *Landfill Off-Gas Collection and Treatment Systems,* ETL 1110-1-160, U.S. Department of the Army, Washington, DC, 1995.

USEPA, (U.S. Environmental Protection Agency), *Solid Waste Disposal Facility Criteria,* EPA/530/R-93/017, Office of Solid Waste and Emergency Response, USEPA, Washington, DC, 1993.

———, *Design, Operation, and Closure of Municipal Solid Waste Landfills,* EPA/625/R-94/008, Office of Research and Development, USEPA, Washington, DC, 1994.

21

GROUNDWATER MONITORING

21.1 INTRODUCTION

Monitoring of groundwater at a landfill is required to evaluate the performance of a properly designed and constructed landfill as well as to discover leaks and begin remediation if groundwater is affected by the landfill. By implementing the groundwater monitoring program, it is possible to identify problems or releases in a timely fashion and take appropriate measures to limit contamination. In this chapter we present regulatory requirements of groundwater monitoring and describe a general framework for designing groundwater monitoring systems.

21.2 REGULATORY REQUIREMENTS

Federal regulations, stipulated in 40 CFR 258.50 to 258.58, establish groundwater monitoring and corrective action requirements for all MSW landfills. These regulations specify requirements for the location, design, and installation of groundwater monitoring systems and set standards for groundwater sampling and analysis. They also specify statistical methods and decision criteria for identifying a significant change in groundwater quality. If a significant change in groundwater quality occurs, regulations require an assessment of the nature and extent of contamination, followed by an evaluation and implementation of remedial measures.

Figure 21.1 shows a flowchart summarizing the federal regulatory requirements for groundwater monitoring and corrective action plan for landfill sites. These requirements consist primarily of a groundwater monitoring system, detection monitoring program, assessment monitoring program, and a corrective action pro-

gram. The detailed information on these requirements is provided in subsequent sections of this chapter; however, the main features of these regulatory requirements are as follows:

- The groundwater monitoring program requires a sufficient number of wells, installed at appropriate locations and depths, to yield groundwater samples from the uppermost aquifer. The uppermost aquifer is defined (in 40 CFR 258.2) as "the geologic formation nearest to the natural ground surface that is an aquifer, as well as lower aquifers that are hydraulically interconnected with this aquifer within the facility property boundary."

- Monitoring wells must be located to obtain the quality of both the upgradient (or background) groundwater as well as the downgradient groundwater passing through the relevant point of compliance. The relevant point of compliance can be either at the waste boundary or at a state-specified distance (no greater than 150 m) from the waste boundary.

- The groundwater monitoring program must include consistent sampling and analysis procedures that are designed to ensure monitoring results that provide an accurate representation of groundwater quality at the background and downgradient wells.

- The number of samples collected to establish groundwater quality data must be consistent with the appropriate statistical procedures.

- Statistical methods that are appropriate to use for evaluating groundwater quality data are stipulated.

- Detection monitoring required for all landfills specifies a list of chemical constituents to monitor.

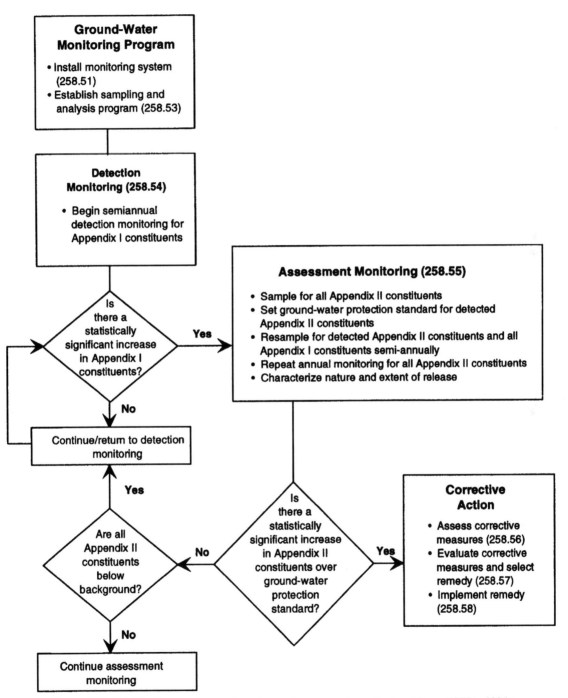

Figure 21.1 *Subtitle D groundwater detection and assessment monitoring.* (*From USEPA, 1994.*)

These chemical constituents, also called the *Appendix I list,* include 15 inorganic chemicals and 47 organic chemical compounds, as shown in Table 21.1.

- Assessment monitoring is required whenever a statistically significant increase over background has been detected for one or more of the Appendix I list items. During this monitoring, the groundwater samples must be tested for an expanded list of chemicals that include a total of 213 chemicals, also known as the *Appendix II list.*
- If any of the constituents listed in Appendix II have been detected at a statistically significant level exceeding the groundwater protection standards, an assessment of corrective measures should be undertaken. The appropriate remedial action is selected and implemented to meet a specified set of conditions to protect human health and the environment.

The federal regulations also provide flexibility for the approved states to suspend groundwater monitoring requirements if it is demonstrated that there is no potential for migration of hazardous constituents from the landfill to the uppermost aquifer during the active life of the landfill and the postclosure care period. The "no potential migration" demonstration must be based on (1) site-specific field-collected data and analysis of physical, chemical, and biological processes affecting contaminant fate and transport, and (2) prediction of maximum contaminant concentrations migrating to the uppermost aquifer to assess impacts on human health and the environment.

States may impose their own, often stringent regulations. Depending on the location of the landfill, the applicable state regulations must be reviewed and complied with. For example, in Illinois, the regulations on groundwater monitoring at landfills are given in Illinois Administrative Code, Sections 811.318 to 811.320. These regulations specify the design, construction, and operation of groundwater monitoring systems, sampling and analysis procedures, assessment methods for potential groundwater impact, remedial action, and groundwater quality standards. In California, these regulations are presented in the California Code of Regulations, Title 27, Environmental Protection Sections 20380 through 20435.

21.3 GROUNDWATER MONITORING SYSTEMS

The objective of a groundwater monitoring system is to intercept groundwater that has been contaminated by leachate from a landfill. Early contaminant detection is important to allow sufficient time for corrective measures to be developed and implemented before sensitive receptors are affected significantly. To accomplish this objective, the monitoring wells should be located to sample groundwater from the uppermost aquifer at the closest practical distance from the waste boundary. Since the monitoring program is intended to operate through the postclosure period, the location, design, and installation of monitoring wells should address both existing conditions as well as expected changes in groundwater flow.

The groundwater monitoring system for a site must incorporate federal, state, and local regulations or requirements. Federal and state agencies developed guidance documents addressing the development and implementation of groundwater monitoring programs (USEPA, 1980, 1986a, 1987). These guidance documents should be followed to develop a comprehensive groundwater monitoring system for given site-specific conditions. In general, the groundwater monitoring system design consists of the following tasks:

- Characterization of site hydrogeology
- Placement of monitoring wells
- Monitoring well design and construction
- Sampling and analysis
- Statistical analysis of monitoring data

These tasks are described briefly in the following sections.

21.3.1 Characterization of Site Hydrogeology

An accurate hydrogeological characterization is essential for an effective groundwater monitoring system.

Table 21.1 **Constituents for detection monitoring**[a]

Common Name[b]	CAS NR[c]	Common Name[b]	CAS NR[c]
Inorganic Constituents		32 *trans*-1,4-Dichloro-2-butene	110-57-6
		33 1,1-Dichloroethane; ethylidene chloride	75-34-3
1 Antimony	(Total)	34 1,2-Dichloroethane; ethylidene dichloride	107-06-2
2 Arsenic	(Total)	35 1,1-Dichloroethylene; 1,1-dichloroethene;	75-35-4
3 Barium	(Total)	vinylidene chloride	
4 Beryllium	(Total)	36 *cis*-1,2-Dichloroethylene; *cis*-1,2-	156-59-2
5 Cadmium	(Total)	Dichloroethene	
6 Chromium	(Total)	37 *trans*-1,2-Dichloroethylene; *trans*-1,2-	158-60-5
7 Cobalt	(Total)	Dichloroethene	
8 Copper	(Total)	38 1,2-Dichloropropane; propylene dichloride	78-87-5
9 Lead	(Total)	39 *cis*-1,2-Dichloropropene	10061-01-5
10 Nickel	(Total)	40 *trans*-1,3-Dichloropropene	1006-02-6
11 Selenium	(Total)	41 Ethylbenzene	100-41-4
12 Silver	(Total)	42 2-Hexanone; methyl butyl ketone	591-78-6
13 Thallium	(Total)	43 Methyl bromide; bromomethane	74-83-9
14 Vanadium	(Total)	44 Methyl chloride; chloromethane	74-87-3
15 Zinc	(Total)	45 Methylene bromide; dibromomethane	74-59-3
		46 Methylene chloride; dichloromethane	75-09-2
Organic Constituents		47 Methyl ethyl ketone, MEK; 2-butanone	78-93-3
		48 Methyl iodine; idomethane	74-88-4
16 Acetone	67-64-1	49 4-Methyl-2-pentanone; methyl isobutyl	108-10-1
17 Acrylonitrile	107-13-1	ketone	
18 Benzene	71-43-2	50 Styrene	100-42-5
19 Bromochloromethane	74-97-5	51 1,1,1,2-Tetrachloroethane	630-20-6
20 Bromodichloromethane	75-27-4	52 1,1,2,2-Tetrachloroethane	79-34-5
21 Bromoform; tribromomethane	75-25-2	53 Tetrachloroethylene; tetrachloroethene;	127-18-4
22 Carbon disulfide	75-15-0	perchloroethylene	
23 Carbon tetrachloride	56-23-5	54 Toluene	108-88-3
24 Chlorobenzene	108-90-7	55 1,1,1-Trichloroethane; methylchloroform	71-55-6
25 Chloroethane, ethyl chloride	75-00-3	56 1,1,2-Trichloroethane	79-00-5
26 Chloroform, trichloromethane	67-66-3	57 Trichloroethylene; trichloroethane	79-01-6
27 Dibromochloromethane;	124-48-1	58 Trichloroflouromethane; CFC-11	75-69-4
chlorodibromomethane		59 1,2,3-Trichloropropane	96-18-4
28 1,2-Dibromo-3-chloropropane; DBCP	96-12-8	60 Vinyl acetate	108-05-4
29 1,2-Dibromoethane; ethyl dibromide; EDB	106-93-4	61 Vinyl chloride	75-01-4
30 *o*-Dichlorobenzene; 1,2-dichlorobenzene	95-50-1	62 Xylene	1330-20-7
31 *o*-Dichlorobenzene; 1,4-dichlorobenzene	106-46-1		

Source: USEPA (1994).

[a] This list contains 47 volatile organics for which possible analytical procedures provided in EPA Report SW-846, Test Methods Evaluating Solid Waste, 3rd ed., November 1986, as revised December 1987, includes Method 8260, and 15 metals for which SW-846 provides either Method 6010 or a method from the 7000 series of method.

[b] Common names are those widely used in government regulations, scientific publications, and commerce; synonyms exist for many chemicals.

[c] Chemical Abstracts Service registry number. Where "Total" is entered, all species in the groundwater that contain this element are included.

Hydrogeologic conditions determine groundwater flow and influence contaminant transport. It is important to thoroughly understand the groundwater flow beneath a site to understand the complexity and to decide on the location of a groundwater monitoring system that will provide representative background and downgradient water measurements.

The goal of a hydrogeological characterization is to acquire site-specific data to enable the development of an appropriate groundwater monitoring program. The site-specific data that should be developed include:

- The lateral and vertical extent of the uppermost aquifer
- The lateral and vertical extent of the upper and lower confining units/layers
- The geology at the waste management unit's site, such as stratigraphy, lithology, and structural setting
- The chemical properties of the uppermost aquifer and its confining layers relative to local groundwater chemistry and wastes managed at the unit
- Groundwater flow, including:
  - The vertical and horizontal directions of groundwater flow in the uppermost aquifer
  - The vertical and horizontal components of the hydraulic gradient in the uppermost and any hydraulically connected aquifer
  - The hydraulic conductivities of the materials that comprise the uppermost aquifer and its confining units/layers
  - The average linear horizontal velocity of groundwater flow in the uppermost aquifer

To perform a hydrogeologic characterization and develop an understanding of site hydrogeology, a number of tasks have to be performed. These tasks include literature review, determination of site geology, and determination of site hydrogeology. These tasks are explained in detail in Chapter 10.

21.3.2 Placement of Monitoring Wells

After the site's hydrogeological conditions have been determined, the number and the lateral and vertical po-

sitions of monitoring wells should be selected. The placement of monitoring wells should consider the following factors:

- Geology of site
- Groundwater flow direction and velocity, including seasonal and temporal fluctuations
- Hydraulic conductivity of water-bearing formations
- Physical–chemical characteristics of contaminants

A groundwater monitoring system requires a minimum of one upgradient or background monitoring well and three downgradient monitoring wells to make statistically meaningful comparisons of groundwater quality. The actual number of upgradient and downgradient wells varies depending on the actual site-specific conditions. A larger number of monitoring wells may be needed at sites with complex hydrogeology. The upgradient or background wells allow the assessment of background quality of the on-site groundwater, and the downgradient wells allow detection of any contaminant plumes from waste disposal area.

Monitoring wells must yield groundwater samples that are representative of both quality of background groundwater and quality of groundwater at a downgradient monitoring point. Therefore, the lateral and vertical placement of wells must be planned carefully. The lateral and vertical placement of monitoring wells is very site-specific. Monitoring wells should be located at the closest practical distance from the boundary of the waste disposal area to detect contaminants before they migrate farther. This will allow early detection of the contamination and allows time to implement appropriate corrective measures and potentially eliminate the need for more extensive corrective action.

Monitoring wells should be placed laterally along the downgradient edge of the waste disposal area to intercept potential contaminant migration pathways. Groundwater flow direction and gradient are two major determining factors in monitoring well placement. The number and spatial distribution of potential contaminant migration pathways and the depths and thickness of stratigraphic horizons that can serve as contaminant migration pathways also must be considered. In homogeneous isotropic hydrogeologic sites, groundwater flow direction and gradient, along with the potential contaminant chemical and physical characteristics, will primarily determine lateral well placement. Computer-aided analyses assist in selecting the monitoring well locations. One such approach is the use of the analytical monitoring efficiency model (MEMO) (Wilson et al., 1992), which assists in the design of monitoring well networks. This model simulates the migration of hypothetical contaminant plumes from a site and quantifies the efficiency of alternative well network designs in detecting the plumes. Monitoring efficiency is defined as the ratio of the area of detection to the total area of the site. For example, a monitoring efficiency of 90% implies that releases occurring over 90% of the

site would be detected by the monitoring wells and releases occurring over 10% of the site would not be detected. An illustration of the application of MEMO is shown in Figure 21.2. A grid of potential chemical source points is defined within the potential source area. At each potential source point, a contaminant plume is generated using an analytical contaminant transport solution. If the plume is intercepted by a monitoring well before it migrates beyond a specified boundary, the source point is considered to be detected. After checking each grid point to determine whether the plume released from the point is detected or not detected, the monitoring efficiency is calculated, and a map showing areas from which chemical releases would not be detected is produced. The contaminant transport used in this model is based on the two-

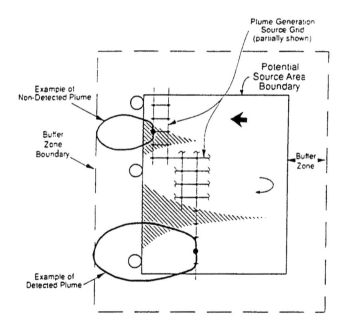

EXPLANATION

○   Monitoring Well

▨   Area from Which a Release Would Not Be Detected

◀   Direction of Groundwater Movement

Figure 21.2 *Selection of monitoring well network using MEMO. (From Wilson et al., 1992.)*

dimensional analytical transport model derived by Domenico and Robbins (1985) and Domenico (1987). This model includes advection, diffusion, dispersion, retardation (sorption), and first-order decay.

In more complex sites where hydrogeology and geology are variable and preferential pathways exist, the well placement determination becomes more complex. Potential migration pathways are influenced by site geology, including varying hydraulic conductivity, fractured zones, and soil chemistry. Human-made features that influence groundwater flow must also be considered. These features may include ditches, filled areas, buried piping, buildings, leachate collection systems, and other adjacent waste disposal areas.

Multiple landfills or landfill cells at the same site may be monitored by using a multiunit groundwater monitoring system as shown in Figure 21.3. The purpose of this system is to reduce the number of monitoring wells that can provide the same information.

Seasonal change in groundwater due to tidal influences, lake or river stage fluctuations, well pumping, or land use pattern changes should also be considered in the determination of well locations. Because of these effects, the groundwater flow direction may change. In some situations, monitoring wells may be placed in a circular pattern to monitor all sides of the waste disposal area. Seasonal fluctuations may cause certain wells to be downgradient only part of the time, but such configurations ensure that releases will be detected.

Similar to lateral placement, vertical well placement in groundwater around a waste disposal area is determined by hydrogeologic conditions as well as the chemical and physical characteristics of the potential contaminants. Potential contaminant migration pathways and the location, size, and geometry of potential contaminant plumes depend on the site-specific geology, hydrogeology, and contaminant characteristics. The monitoring well depths and screen depths should be such that any contaminant release into the subsurface can be detected.

The *chemical and physical characteristics of potential contaminants* from the waste disposal area play a significant role in determining vertical placement of wells. Properties of the contaminants, such as solubil-

ity, density, and partition coefficients, will influence the vertical placement and screen lengths of monitoring wells. A DNAPL (dense nonaqueous phase liquid), for instance, will sink to the bottom of an aquifer and migrate along geologic gradients (rather than hydrogeologic gradients), thus requiring a monitoring well's vertical placement to correspond with the depth of the appropriate geologic feature. LNAPLs (light nonaqueous phase liquids), on the other hand, move along the top of an aquifer and require wells and well screens at the surface of the aquifer. Well screen lengths should be determined to allow for the sampling of the selected zone within an aquifer. To monitor more than one depth at a single location, monitoring wells should be installed in clusters, each with screen in the desired depth. The monitoring well network design should be presented to regulating officials for their approval prior to implementing in the field.

21.3.3 Monitoring Well Design and Construction

The design and construction of monitoring wells will affect the accuracy of samples collected. The monitoring well design must be based on site-specific hydrogeologic data. The formation materials will determine the selection of proper packing and sealant materials, and the stratigraphy will determine the screen length for the interval to be monitored. Construction of monitoring wells should be specified and overseen to ensure that the monitoring well is constructed as designed and will perform as intended. Each well must be tailored to suit the hydrogeological setting, the contaminants to be monitored, and other site-specific factors. Figure 21.4 depicts the components of a typical monitoring well. More details on the monitoring well design, installation, and development are provided in Chapter 10.

21.3.4 Sampling and Analysis

Groundwater samples are collected from each well at an appropriate frequency during the active life of the landfill and its postclosure care period. Monitoring fre-

Single-Unit System

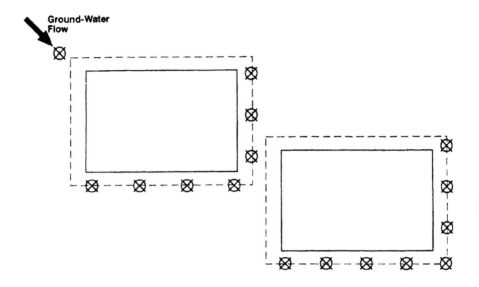

Multi-Unit System

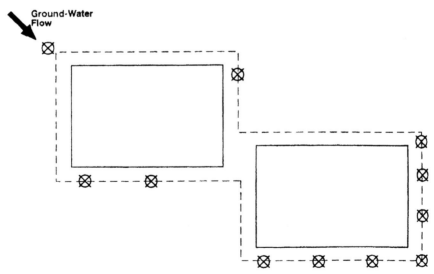

Figure 21.3 *Comparison of single- and multiunit monitoring system. (From USEPA, 1993.)*

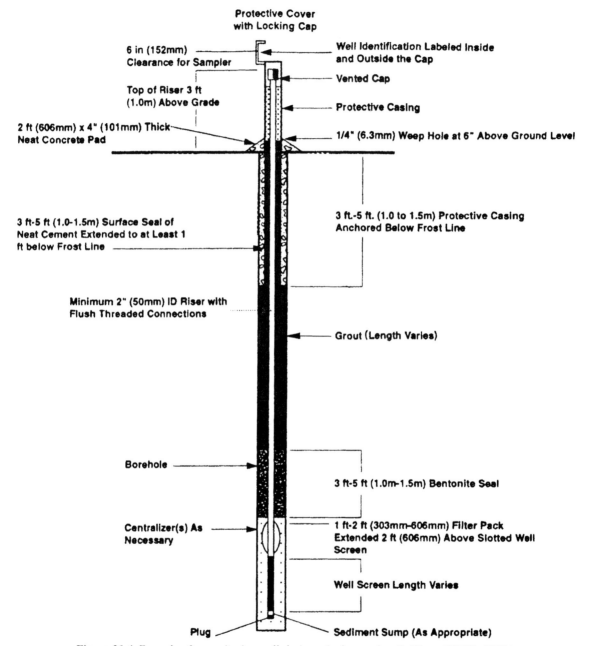

Figure 21.4 *Example of a monitoring well design: single cased well. (From USEPA, 1993.)*

quency should be sufficient to allow detection of groundwater contamination. Background groundwater conditions, such as flow directions and velocities and seasonal groundwater fluctuations, are analyzed to help determine a suitable monitoring frequency for a site. This frequency usually ranges from quarterly to annually.

A list of chemical constituents, based on the specifics of the site, is selected whose levels are monitored to indicate if the site's groundwater quality is changing. Analyzing a large number of groundwater quality parameters in each sampling event can be costly. To minimize expense, only contaminants that can be reasonably expected to migrate to the groundwater and other geochemical indicators of contaminant migration (e.g., pH, conductivity) are selected. These are all called *sampling parameters.* A broad range of parameters should be tested initially to establish background groundwater quality, and these results could then be used to select the sampling parameters. The parameters should provide an early indication of a release from the landfill. Table 21.2 lists potential parameters for a basic groundwater monitoring program. Once contamination is detected, sampling parameters can be expanded for additional constituents to fully characterize the chemical makeup of the release.

Sampling and analytical protocols should be developed and implemented to ensure that the sampling results accurately represent groundwater quality and can be compared over time. A *quality assurance/quality control (QA/QC) plan* should be developed. At a minimum, a sampling protocol should include the following:

- *Data quality objectives.* Objectives of monitoring, list of target contaminants, and level of accuracy requirements for data to be conclusive should be specified.
- *Sampling.* Sampling procedure to ensure sampling quality and avoid cross-contamination of groundwater samples should be specified. The details on groundwater sampling are provided in Chapter 10. Typically, the sampling procedure includes inspecting the well for cracks and/or loose casing,

measuring the water-level elevation, purging stagnant water in the well, collecting a groundwater sample using a bailer or a pump, and field testing for basic parameters such as pH and dissolved oxygen, as shown in Figure 21.5.

- *Sample preservation and handling.* To avoid altering sample quality, groundwater samples should be transferred directly into a contaminant-free container. Different types of containers and preservation methods (such as pH adjustment, chemical addition, and refrigeration), as recommended by the USEPA, are used for analyzing different chemical constituents (see Table 21.3).
- *Sample transport.* To document sample possession from the time of collection to the laboratory, include a chain-of-custody record in every sample shipment. A chain-of-custody record includes the date and time of collection, signatures of those involved in the chain of possession, time and dates of possession, and other notations to trace samples.
- *QA/QC.* To verify the accuracy of field sampling procedures, collect field-quality control samples, such as trip blanks, equipment blanks, and duplicates. These samples are also analyzed for the required monitoring parameters.

The analytical protocols specify the analytical methods that measure accurately the constituents being monitored. Generally, the USEPA-approved analytical methods, also known as SW-846 methods, are generally recommended. Commercial laboratories with qualified lab personnel and good analytical equipment are generally used to conduct the analysis of the samples. A good QA/QC program helps ensure the accuracy of laboratory data.

21.3.5 Statistical Analysis of Monitoring Data

Once groundwater monitoring data have been collected, the data are analyzed to determine if contaminants are released from the landfill into subsurface and migrating in groundwater. Anomalous data may result from sampling uncertainty, laboratory error, or sea-

Table 21.2 **Potential parameters for basic groundwater monitoring**[a]

Category	Specific Parameters
Field-measured parameters	Temperature pH Specific electrical conductance Dissolved oxygen Eh oxidation–reduction potential Turbidity
Leachate indicators	Total organic carbon (TOC-filtered) pH Specific conductance Manganese (Mn) Iron (Fe) Ammonium (NH_4, as N) Chloride Sodium (Na) Biochemical oxygen demand (BOD) Chemical oxygen demand (COD) Volatile organic compounds (VOCs) Total halogenated compounds (TOX) Total petroleum hydrocarbons (TPHs) Total dissolved solids (TDSs)
Additional major water quality parameters	Bicarbonate (HCO_3^-) Boron (Bo) Carbonate Calcium (Ca) Fluoride (F) Magnesium (Mg) Nitrate (as N) Nitrogen (dissolved N_2) Potassium (K) Sulfate (SO_2) Silicon (H_2SiO_4) Strontium (Sr) Total dissolved solids (TDS)
Minor and trace inorganics	Initial background samplings of inorganics for which drinking water standards exist (arsenic, barium, cadmium, chromium, lead, mercury, selenium, silver); ongoing monitoring of any constituents showing background near or above drinking water standards

[a] Potential parameters should be selected based on site-specific circumstances.

sonal changes in natural site conditions. Therefore, statistical procedures are used to determine if statistically significant changes in water quality have occurred or whether the quantified differences could have arisen solely because of one of the above-listed factors.

An appropriate statistical approach will minimize false positives or negatives in terms of potential re-leases. The approach should account for historical data, site hydrogeologic conditions, site operating practices, and seasonal variations. There are numerous statistical approaches, but the methods commonly used for evaluating groundwater monitoring data are detailed in an USEPA guidance document (USEPA, 1993). These methods include:

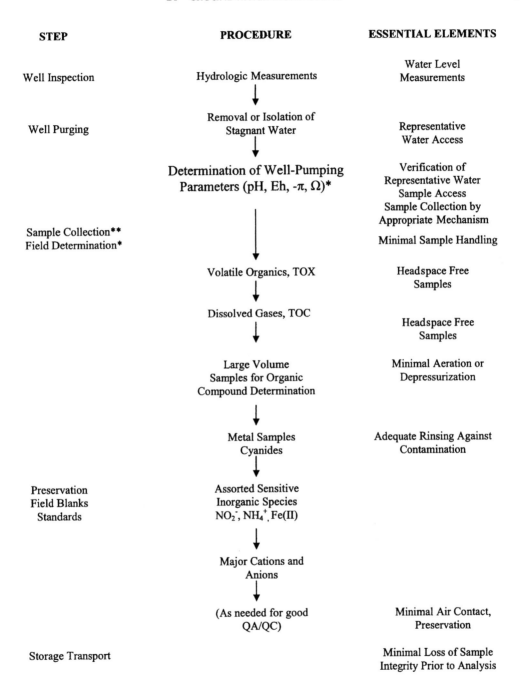

STEP	PROCEDURE	ESSENTIAL ELEMENTS

Well Inspection — Hydrologic Measurements — Water Level Measurements

Well Purging — Removal or Isolation of Stagnant Water — Representative Water Access

Determination of Well-Pumping Parameters (pH, Eh, -π, Ω)* — Verification of Representative Water Sample Access / Sample Collection by Appropriate Mechanism

Sample Collection** / Field Determination* — Minimal Sample Handling

Volatile Organics, TOX — Headspace Free Samples

Dissolved Gases, TOC — Headspace Free Samples

Large Volume Samples for Organic Compound Determination — Minimal Aeration or Depressurization

Metal Samples Cyanides — Adequate Rinsing Against Contamination

Preservation / Field Blanks / Standards — Assorted Sensitive Inorganic Species NO_2^-, NH_4^+, $Fe(II)$

Major Cations and Anions

(As needed for good QA/QC) — Minimal Air Contact, Preservation

Storage Transport — Minimal Loss of Sample Integrity Prior to Analysis

* Denotes analytical determinations which should be made in the field.
**This is a suggested order for sampling, not all parameters are required by Part 258

Figure 21.5 *Generalized flow diagram of groundwater sampling steps. (From USEPA, 1993.)*

Table 21.3 **Sampling and preservation procedures for detection monitoring**[a]

Parameter	Container Recommended[b]	Preservative	Maximum Holding Time	Minimum Volume Required for Analysis
Indicators of Groundwater Contamination				
pH	T, P, G	Field determined	None	25 mL
Spec. conductance	T, P, G	Field determined	None	100 mL
TOC	G, amber, T-lined cap	Cool 4°C, HCl to pH < 2	28 days	4 × 15 mL
TOX	G, amber, T-lined septa or caps	Cool 4°C, add 1 mL of 1.1 *M* sodium sulfite	7 days	4 × 15 mL
Groundwater Quality Characteristics				
Chloride	T, P, G	4°C	28 days	50 mL
Iron, manganese	T, P	Field acidified to pH < 2 with HNO_3	6 months	200 mL
Sodium phenols	G	4°C/H_2SO_4 to pH < 2	28 days	50 mL
Sulfate	T, P, G	Cool, 4°C	28 days	50 mL
EPA Interim Drinking Water Characteristics				
Arsenic, barium, cadmium, chromium	T, P	Total metals Filed acidified to pH < 2 with HNO_3	6 months	1000 mL
Lead, mercury, selenium, silver	Dark bottle	Dissolved metals 1. Field filtration (0.45 μm) 2. Acidify to pH < 2 with HNO_3		
Fluoride	T, P	Cool, 4°C	28 days	300 mL
Nitrate/nitrite	T, P, G	4°C/H_2SO_4 to pH < 2	14 days	1000 mL
Endrin; lindane; methoxychlor; toxaphene; 2,4-D; 2,4,5-TP silvex	T, G	Cool, 4°C	7 days	2000 mL
Radium; gross alpha; gross beta	P, G	Field acidified to pH < 2 with HNO_3	6 months	1 gal
Coliform bacteria	PP, G (sterilized)	Cool, 4°C	6 hours	200 mL
Other Groundwater Characteristics of Interest				
Cyanide	P, G	Cool, 4°C, NaOH to pH > 12 0.6 g ascorbic acid	14 days	500 mL
Oil and grease	G only	4°C/H_2SO_4 to pH < 2	28 days	100 mL
Semivolatile, nonvolatile organics	T, G	Cool, 4°C	14 days	60 mL
Volatiles	G, T-lined	Cool, 4°C	14 days	60 mL

Source: USEPA (1986a).

[a] References: *Test Methods for Evaluating Solid Waste—Physical/Chemical Methods,* SW-846, 2nd ed., 1982; *Methods for Chemical Analysis of Water Wastes,* EPA-600/4-79-020; *Standard Methods for the Examination of Water and Wastewater,* 16th ed., 1985.

[b] Container types: P, plastic (polyethylene); G, glass; T, fluorocarbon resins (PTFE, Teflon, FEP, PFA, etc.); PP, polypropylene.

- *Parametric analysis of variance* (ANOVA). This analysis, as well as rank-based (nonparametric) ANOVA, attempt to determine whether different wells have significantly different average concentrations of constituents.
- *Tolerance intervals.* Tolerance intervals are statistical intervals constructed from data designed to contain a portion of a population, such as 95% of all sample measurements. These intervals can be used to compare data from a downgradient well to data from an upgradient well.
- *Prediction intervals.* These intervals approximate future sample values from a population or distribution with a specific probability. Prediction intervals can be used both for comparison of downgradient wells to upgradient wells (interwell comparison) and for comparison of current well data to previous data for the same well (intrawell comparison).
- *Control charts.* These charts are historical data for comparison purposes and therefore are appropriate only for initially uncontaminated wells.

Detailed information on the statistical analysis methods above can be found in USEPA (1989a, 1992) and other standard references, such as Dixon and Massey (1969) and Gilbert (1987). Recently, different software systems have been developed specifically to perform statistical analysis of monitoring data. Some of these software systems include:

- *Geo-EAS.* This program, Geostatistical Environmental Assessment Software, generates contour maps of interpolated estimates from sample data (USEPA, 1989b). Other functions included in the program are univariate statistics, scatter plots/linear regression, and variogram computation and model fitting.
- *GEOPACK.* This is a user-friendly geostatistical software system (USEPA, 1990). It consists of a package of programs for conducting analyses of the spatial variability of one or more random functions. It includes several statistical analyses options. Basic statistics such as the mean, median,

variance, standard deviation, skew, and maximum and minimum values can be determined for the database selected. The program also allows performing analyses such as linear regression and polynomial regression. Various graphics capabilities are included, such as linear or logarithmic plots, contour plots, and bloc diagrams.
- *GRITSTAT.* This is the groundwater information tracking with statistical analysis capability software distributed by the USEPA (1992).

21.4 DETECTION MONITORING PROGRAM

Figure 21.6 shows detection monitoring program implementation. Once a statistically significant increase in one or more contaminants has been detected, the possibility of contamination caused by other factors unrelated to the landfill must be assessed. Factors unrelated to the landfill that may cause an increase in the detected concentrations are (1) contaminant sources other than the landfill being monitored, (2) natural variations in groundwater quality, (3) analytical errors, (4) statistical errors, and (5) sampling errors. If the increase was caused by a factor unrelated to the landfill, additional measures may not be necessary and the original groundwater monitoring program can be resumed. If, however, these factors have been ruled out, an assessment monitoring program must be implemented.

21.5 ASSESSMENT MONITORING PROGRAM

Figure 21.7 shows implementation of an assessment monitoring program. The purpose of assessment monitoring is to evaluate the rate, extent, and concentrations of contamination. Assessment monitoring typically involves resampling all wells, both upgradient and downgradient, and analyzing the samples for a larger list of parameters than that used during the basic monitoring program. More than one sampling event may be necessary, and the installation of additional

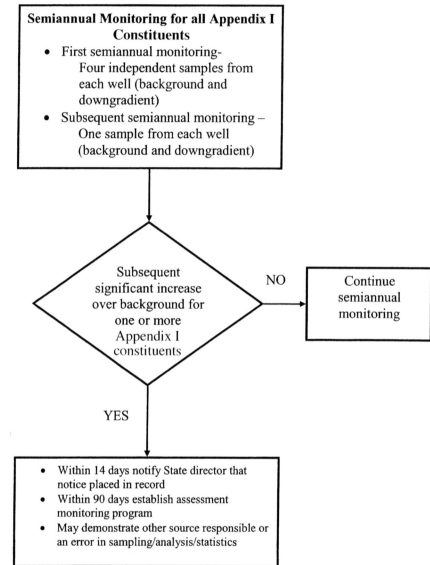

Figure 21.6 Detection monitoring program. (From USEPA, 1993.)

monitoring wells may be needed to adequately determine the extent of any contamination.

21.6 CORRECTIVE ACTION PROGRAM

If assessment monitoring results indicate that there is not a statistically significant increase in the concentra-

tions of one or more of the constituents over the established groundwater protection standards, the original groundwater monitoring program is resumed. If, however, there is a statistically significant increase in any of these constituents, the nature and extent of contamination may be delineated with additional monitoring wells, if necessary. Following this, corrective ac-

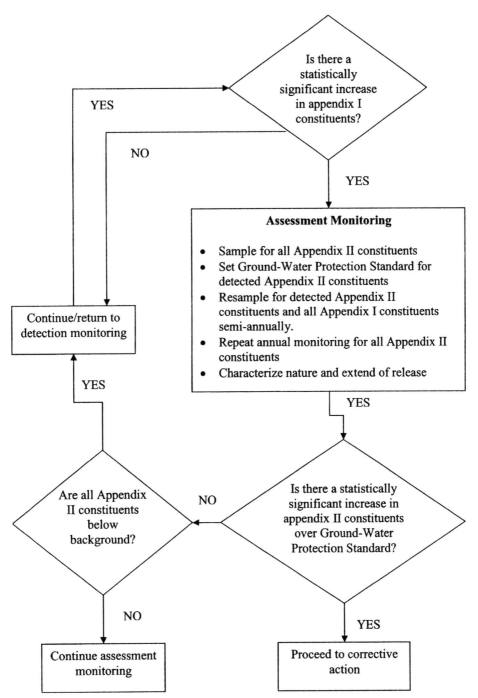

Figure 21.7 Assessment monitoring program. (From USEPA, 1993.)

tion to remedy the site should proceed as described in Chapters 10 to 14.

21.7 SUMMARY

Groundwater monitoring program provides information that the landfill is containing the waste effectively and not affecting the site groundwater. Regulations provide minimum requirements on the groundwater monitoring systems, including the placement of wells, sampling and analysis, and statistical analysis of the monitoring data. Detection monitoring involves monitoring for a set of chemical parameters, and if any of these parameters is increased statistically, assessment monitoring, which involves evaluation of a larger set of chemical parameters, is to be undertaken. If a statistically significant increase of any of the monitored contaminants is confirmed, a corrective program must be implemented. Groundwater monitoring systems should be designed and implemented carefully to determine if the data are accurate and to be able to discover leaks and begin remediation, if required.

QUESTIONS/PROBLEMS

21.1. What is the significance of monitoring the uppermost aquifer?

21.2. Explain site conditions where groundwater monitoring may be not required.

21.3. Explain how background groundwater quality data can be obtained at a site where groundwater flow direction is not well defined.

21.4. Explain the components of a quality assurance/quality control (QA/QC) plan for sampling of groundwater samples.

21.5. Explain the differences between detection monitoring and assessment monitoring.

21.6. Explain the different statistical methods used to analyze groundwater monitoring data.

21.7. The measured lead concentrations in a monitoring well at a landfill site are 0.4, 1.8, 1.8, 2.5, 2.9, 3.8, 5.0, 5.0, 6.0, 9.3, 9.7, 11.0, 11.6, 13.8, and 14.6 mg/L. Calculate the mean, standard deviation, and coefficient of variation. Is the data fit normal distribution?

21.8. Chloride and sulfate concentrations measured in a monitoring well at a landfill site are given below. Determine if there is a statistically significant increase in chloride and sulfate concentrations in the well.

Sampling Event	1Q99	2Q99	3Q99	4Q99	1Q00	2Q00	3Q00	4Q00	1Q01	2Q01	3Q01	4Q01
Chloride (mg/L)	1000	1250	950	1100	1150	1400	800	1000	1150	1300	1100	1100
Sulfate (mg/L)	760	40	70	40	150	270	190	140	760	222	140	210

21.9. Several monitoring wells have been installed at a landfill site and the groundwater quality parameters were measured on a regular basis, as shown in Figure P21.9. Concerns have been expressed that the groundwater at the site has been contaminated by the landfill. Based on Subtitle D regulations, perform a statistical analysis of the monitoring data and determine if "statistically significant groundwater contamination" has occurred at the site.

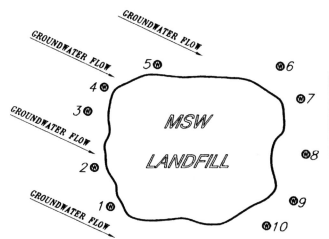

WELL NO.	NORTHING FEET	EASTING FEET	CHLORIDES PPM	TDS PPM	SULFATES PPM
1	100	100	3.1	10.4	28.4
2	325	90	4.8	11.6	31.8
3	575	85	4.8	10.5	33.3
4	650	92	6.1	8.5	30.1
5	700	150	6.5	8.2	33.1
6	700	695	6.5	8.0	28.4
7	650	715	6.5	7.5	46.2
8	360	720	9.0	7.6	53.2
9	135	710	7.2	9.5	57.4
10	70	685	5.2	9.8	54.1

Figure P21.9

21.10. Explain the conditions under which groundwater monitoring may be ceased at a closed landfill site. Will this be allowed in accordance with the RCRA Subtitle D regulations?

REFERENCES

Dixon, W. J., and Massey, Jr., *Introduction to Statistical Analysis,* McGraw-Hill, New York, 1969.

Domenico, P. A., An analytical model for multidimensional transport of a decaying contaminant species, *J. Hydrol.,* Vol. 91, pp. 49–58, 1987.

Domenico, P. A., and Robbins, G. A., A new method of contaminant plume analysis, *Groundwater,* Vol. 23, No. 4, pp. 476–485, 1985.

Gilbert, R. O., *Statistical Methods for Environmental Pollution Monitoring,* Van Nostrand Reinhold, New York, 1987.

USEPA (U.S. Environmental Protection Agency), *Procedures Manual for Groundwater Monitoring at Solid Waste Disposal Facilities,* SW-611, Office of Water and Waste Management, USEPA, Washington, DC, 1980.

———, *RCRA Groundwater Monitoring Technical Enforcement Guidance Document,* OSWER-9950.1, Office of Solid Waste and Emergency Response, USEPA, Washington, DC, 1986a.

———, *Test Methods for Evaluating Solid Waste: Physical/Chemical Methods,* SW-846, Office of Solid Waste and Emergency Response, USEPA, Washington, DC, 1986b.

———, *Groundwater Monitoring Seminar Series: Slide Copies,* CERI-87-8, Office of Research and Development, USEPA, Washington, DC, 1987.

———, *Statistical Analysis of Groundwater Monitoring Data at RCRA Facilities: Interim Final Guidance,* EPA/530-SW-89-026, Office of Solid Waste, USEPA, Washington, DC, 1989a.

———, *Geo-EAS (Geostatistical Environmental Assessment Software) User's Guide,* EPA/600/S4-88/033, Office of Research and Development, USEPA, Washington, DC, 1989b.

———, *Geostatistics for Waste Management: A User's Manual for the GEOPACK (Version 1.0) Geostatistical Software System,* EPA/600/S8-90/004, Office of Research and Development, USEPA, Washington, DC, 1990.

———, *User Documentation of the Groundwater Information Tracking System (GRITS) with Statistical Analysis Capability, GRITSTAT Version 4.2,* EPA/625/11-91/002, Office of Research and Development, USEPA, Washington, DC, 1992.

————, *Solid Waste Disposal Facility Criteria,* EPA/530-R/93/017, Office of Solid Waste and Emergency Response, USEPA, Washington, DC, 1993.

————, *Design, Operation, and Closure of Municipal Solid Waste Landfills,* EPA/625/R-94/008, Office of Research and Development, USEPA, Washington, DC, 1994.

Wilson, C. R., Einberger, C. M., Jackson, R. L., and Mercer, R. B., Design of groundwater monitoring networks using the monitoring efficiency model (MEMO), *Groundwater,* Vol. 30, No. 6, pp. 965–970, 1992.

22

SURFACE IMPOUNDMENTS

22.1 INTRODUCTION

In this chapter we present the regulatory settings as they apply to owners, operators, and designers of surface impoundment[1] facilities that are used to treat, store or dispose hazardous wastes.[2] Other design-related issues, such as cover system and closure issues, are also discussed. In accordance with the regulatory requirements, surface impoundment units must be fitted with liners and leachate collection systems. Description of various liner materials (soil and synthetics) and related design issues are presented in Chapters 17 and 18.

22.2 REGULATORY SETTING

In 1976, the U.S. Congress passed the Resource Conservation and Recovery Act (RCRA). RCRA was amended and authorized by various public laws in 1978, 1980, 1982, 1983, 1984, 1986, and 1988. The guidelines and standards that apply to surface impoundments for treatment, storage, or disposal of hazardous waste are included under 40 CFR 264. Readers are encouraged to read the regulations for detailed information.

40 CFR 264, Subpart K, establishes minimum national standards to define the acceptable management of hazardous waste in surface impoundments. The design, construction, operation, and closure of hazardous waste in surface impoundments are regulated under authority of the Hazardous and Solid Waste Amendments (HSWA) of 1984 to the RCRA of 1976. Surface impoundments that were in existence as of November 19, 1980, became eligible for interim status, and all owners and operators had to notify EPA of their waste management activities. The requirements for owners and operators of hazardous waste treatment, storage, and disposal facilities under interim status are outlined in 40 CFR 265.

40 CFR 264 outlines the following main requirements for *new* surface impoundments used for treatment, storage, and disposal of hazardous waste.

22.2.1 Design and Operating Requirements

A surface impoundment must have a liner which must be designed, constructed, and installed to prevent any waste migration out of the impoundment. The liner is considered as a part of the impoundment. Therefore, contaminant migration to the adjacent subsurface soil or groundwater (or surface water) is to be prevented during the active life of the impoundment; closure period is also included in the active life. Since the liner is considered as a part of the impoundment, waste migration into the liner is allowed. The liner must, therefore, have the following key characteristics:

- The liner material must have sufficient strength and thickness to prevent failure due to pressure gradients.
- The liner material must be chemically compatible with the migrating waste constituents.

---

[1] A *surface impoundment* is a facility that is a natural topographic depression, human-made excavation, or diked area formed primarily of earthen materials. The facility is designed to hold an accumulation of liquid wastes or wastes containing free liquids [e.g., holding, settling, and aeration pits, ponds, and lagoons (40 CFR 257)].

[2] Hazardous wastes are described and defined in Chapter 15.

- The liner material must be protected against climatic conditions, installation, and operation stresses and pressure gradients caused due to settlements, compression, or uplift.

Typical liner system components in accordance with 40 CFR 264.221 are presented in Section 22.3.

If it can be demonstrated by the owner or operator that the operating practices and location characteristics of the surface impoundment facility are such that hazardous waste migration can be prevented into the ground or surface water at any future time, an *alternative design* can be submitted for approval.

22.2.2 Action Leakage Rate

The action leakage rate (ALR)[3] must be approved by the regional administrator (RA) and should have an adequate safety margin to allow for uncertainties in design, construction, and operations. Uncertainties in design may include flow capacity changes over time resulting from siltation and clogging, creep of synthetic components, and effects of leachate on the leak detection system materials.

22.2.3 Response Action Plans

A response action plan (RAP) for a surface impoundment unit must be approved before waste is received at a site. At a minimum, the RAP must describe that if the flow rate into the leak detection system (LDS) exceeds the ALR for any sump, the owner or operator must comply with the following:

- Within seven days of the determination that the flow rate has been exceeded, the RA must be notified in writing.
- Within 14 days of the determination, a written assessment of the amount and source of liquids

and short-term action taken and planned must be submitted to the RA.

- Within 30 days after the notification that the ALR has been exceeded, evaluate and then submit to the RA the location, size, and cause of leak and short- and long-term actions to be taken to mitigate or stop any leaks, and determine if the waste receipt should cease or be curtailed.
- As long as the flow rate exceeds the ALR, the owner or operator must submit to the RA a monthly report summarizing the results of any remedial action planned and taken.

22.2.4 Monitoring and Inspection

During *construction* and *installation* of liners and cover systems, all components must be inspected for uniformity, damage, and imperfections. Similarly, after construction or installation, soil-based liners and covers must be inspected for imperfections that could affect the permeability adversely. Also, it should be ensured that synthetic liners and covers do not have tears, punctures, or blisters. During *operations,* a surface impoundment must be inspected weekly and after storms to detect any evidence of damage and deterioration of any system component. Both during active life and during closure and postclosure periods, the surface impoundment regulations establish a schedule for recording the amount of liquids removed from each LDS sump.

22.2.5 Emergency Repairs: Contingency Plans

A surface impoundment must be removed from service if either dike leaks or the level of liquids in the impoundment change suddenly for unplanned and unknown reasons. In this situation, the owner or operator must immediately stop the addition of waste into the impoundment, contain any surface leak, and take any necessary steps to stop or prevent catastrophic failure. If a leak cannot be stopped, empty the impoundment and notify the RA within seven days after detecting

---

[3] It is the maximum design flow rate that the leak detection system (LDS) can remove without the fluid head on the bottom liner exceeding 1 ft.

the problem. As a part of the permit, a contingency plan is already in the hands of the permitting agency. The owner or operator must specify a procedure for complying with the emergency repairs designated as a part of the contingency plan.

22.2.6 Closure and Postclosure Care

At the closure of the surface impoundment facility, the owner or operator must comply with the following requirements:

- Remove or decontaminate all waste residues, liners and collection systems, and contaminated subsoils and manage them as hazardous waste.
- The other alternative is to eliminate free liquids by removing liquid wastes and solidify the remaining waste. In doing so, the remaining wastes must be stabilized so that they have enough bearing capacity to support the final cover that will be placed over the stabilized and closed impoundment. The final cover system must be designed and constructed so that it minimizes liquid infiltration through the impoundment, promotes surface drainage, minimizes erosion, accommodates settlements, has a permeability less than or equal to the bottom liner, and functions with minimum maintenance.

If some waste residue or contaminated materials are left in place at the final closure, it must comply with all postclosure requirements of hazardous wastes.

22.3 LINER SYSTEMS

The minimum technology requirements specified in HSWA for new hazardous waste surface impoundments require a double liner with a leak detection and collection layer placed between the liners (USEPA, 1991). The system must have a top liner (e.g., a geomembrane) and a composite bottom liner (e.g., a geomembrane over compacted soil). The top liner and the upper component of the composite liner must be designed and constructed of materials to prevent the migration of hazardous constituents into such a liner during the active life and postclosure care period. The lower component must be designed and constructed to *minimize* the migration of hazardous constituents. This requirement of minimizing the hazardous constituents is specified so that if a breach in the upper liner component (geomembrane) were to occur, the lower component will act as a second line of defense. As shown in Figure 22.1, the lower component of the composite liner must be constructed of at least 3 ft (91 cm) of compacted soil material with a hydraulic conductivity (K) value of no more than 1×10^{-7} cm/s (40 CFR 264.221).

USEPA's minimum technology also recommends that the upper or primary geomembrane be at least 30 mils (0.76 mm) thick and be covered by a protective soil and/or geotextile layer. This thickness should be at least 40 mils (1.14 mm) if the geomembrane is uncovered or exposed directly to the elements. Some geomembranes may require greater thickness to prevent failure or to accommodate unique seaming requirements. For HDPE liners, thicknesses of 60 to 100 mils (1.52 to 2.54 mm) are generally recommended.

A leak detection, collection, and removal system is placed between the two liners (i.e., below the primary geomembrane and above the secondary geomembrane). As the name indicates, the purpose of this system is to detect, collect, and remove leaks of hazardous constituents at the earliest practicable time during the active life and postclosure care period. The system should be constructed with a bottom slope of 1% or more. Also, as exhibited in Figure 22.1, if constructed with granular material, it should have a thickness of 12 in. (30.5 cm) or more and a hydraulic conductivity (K) of 1×10^{-1} cm/s or more. If the system is constructed of synthetic drainage material, its transmissivity should be 3×10^{-4} m^2/s, or more. The granular or the synthetic drainage materials must be resistant to chemical constituents of the waste and should be designed and operated to minimize clogging during the active life and postclosure care period.

The double liner requirements discussed above may be waived by the RA for any monofill that contains

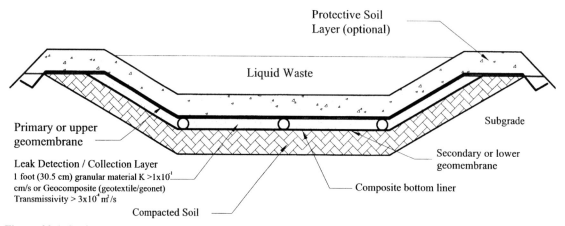

Figure 22.1 *Surface impoundment cross section for hazardous waste: double liner with composite bottom liner (as required by 40 CFR 264). Not to scale.*

only hazardous wastes from foundry furnace emission control or metal casting molding sand. It should also be ensured that such monofill wastes do not contain constituents that would render the waste hazardous for reasons other then the EP toxicity characteristics.

As shown in Figure 22.2, many owners and operators of surface impoundments like to use two composite (geomembrane/compacted soil layer) liners separated by a leak detection/collection layer. In such cases, a separating geotextile is placed between the up-

per compacted soil and the drainage layer to prevent the migration of soil particles into the drainage layer. This double-composite liner should further provide assurance against the escape of waste chemical constituents from the impoundment.

In surface impoundments that contain nonhazardous wastes, single-composite liners are generally acceptable. Actually, in many situations, these liners may have 2-ft (61-cm)-thick compacted soil with hydraulic conductivity (K) less than or equal to 1×10^{-6} cm/s

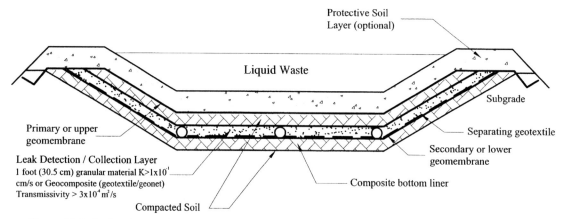

Figure 22.2 *Surface impoundment cross section for hazardous waste: double composite liner. Not to scale.*

underlying the synthetic liner. A leachate collection and removal system below the composite liner system is generally recommended.

22.4   SURFACE IMPOUNDMENT DESIGN

22.4.1   Predesign Activities

In the predesign phase of the surface impoundment design, the following factors should be considered: (1) site topography, (2) site climate, (3) geology and subsurface soil conditions, and (4) surface water and hydrology.

It is important that the most recent (for the year of design) site *topographic map* is available. The features, such as 100-year flood elevation, depressions, and ridges and valleys, should be identifiable on such a map. Ideally, a surface impoundment should require minimal cut and fill operations. This means that it should be located in areas of low relief and away from the 100-year floodplain so as to minimize the height of berms around the impoundment.

Site *climatic* factors, such as precipitation, evaporation, freeze–thaw cycles, and wind direction–circulation play a role in locating, sizing, and designing the surface impoundment facility. Precipitation adds additional unwanted liquids to the impoundment. In addition, siting the impoundment in the 100-year floodplain makes permitting more difficult and may require special protective design features. These features may require erosion protection measures, protective dikes, and inflow–outflow control structures. Evaporation is generally desirable since it reduces the volume of unwanted liquids. Background air quality monitoring prior to construction of an impoundment will help in defending against later accusations of potential emission violations. In northern climates, due to freeze–thaw conditions, the shear strength and permeability of the soil liner and berms may be affected. Such potential impact should be evaluated at the predesign phase. It may require provision of a protective cover over the liner to the depth of maximum frost penetration (USEPA, 1991).

Site *geology* and *subsurface soil* conditions play an important role in siting and designing a surface impoundment. The factors that need to be investigated and evaluated are:

- *Site soil characteristics.* The factors that need to be considered are strength, compressibility, and permeability. Highly porous compressible soils with shallow water table conditions are least desirable.
- *Depth to bedrock and type of bedrock.* It is generally considered that fractured and porous bedrock near the ground surface may not be a desirable site for a surface impoundment. Generally, bedrocks such as limestone may cause solution cavities and channels, and should be avoided. Thick, dense, and undisturbed shale generally make desirable foundations.
- *Special design considerations.* These may be required for seismically active areas and areas prone to landslides.
- *On-site materials.* Availability and suitability of on-site materials for construction of liners, berms, and other protective features, such as erosion control measures, should also be evaluated at this phase of design.

Information on *surface water* and *hydrologic* conditions is very important in the sense that proximity to any source of water is related directly to the risk of contamination from the impoundment. Also, design issues, such as liner uplift due to the existence of shallow groundwater and special requirements to handle the potentially large amounts of construction water, are important factors that need to be considered at the predesign phase. Therefore, the investigations should be carried out to evaluate the following factors:

- Amounts and sources of surface water flows
- Depth to the uppermost saturated zones
- Effects of seasonal and temporal factors on groundwater flow
- Presence of a perched water table, if encountered

22.4.2 Surface Impoundment Configuration

Prior to deciding the surface impoundment configuration (i.e., shape, aerial dimensions, and depth), one needs to determine the type of surface impoundment being designed at a site. As shown in Figure 22.3, a surface impoundment can be classified into the following three types:

1. *Treatment-type surface impoundments.* As shown in Figure 22.3(*a*), such an impoundment is used as a treatment facility. In these facilities, the incoming and outflowing wastes may be either in steady state or in an intermittent state. In addition, precipitation and evaporation factors need to be considered in the liquid balance estimates.

2. *Storage-type surface impoundments.* As shown in Figure 22.3(*b*), storage-type impoundments are used to accommodate some liquid surge or equalization.

3. *Nondischarge-type surface impoundments.* As shown in Figure 2.3(*c*), these impoundments are used for waste evaporation or disposal. These impoundments generally rely on natural evaporation to maintain liquid levels.

The most common and most economical *shape* for a surface impoundment is rectangular. Other shapes increase the cost of earthwork and liner installation (Unger et al., 1985; USEPA, 1983). Generally, a combination of excavation *below grade* surrounded by *above-grade dikes* is the most economical to construct. Entirely above-grade impoundments have the disadvantage of a potential for environmental damage if dikes are breached.

The *depth* of a surface impoundment depends on the waste volumes to be handled. The depth or liquid levels will depend on (1) the rates of waste inflow and outflow, (2) storm surges, (3) precipitation, (4) wind speed, and (5) dike slopes, (Figure 22.4). As shown in this figure, the normal operating level (d_n) takes into account only the inflow and outflow of the waste. The maximum operating level then considers the wettest month in the design storm (e.g., 100-year, 24-hour storm). Other levels are clear from Figure 22.4. The *normal operating depth* (d_n) can be estimated using the equation

$$d_n = \frac{(Q_i - Q_o)t}{S}$$

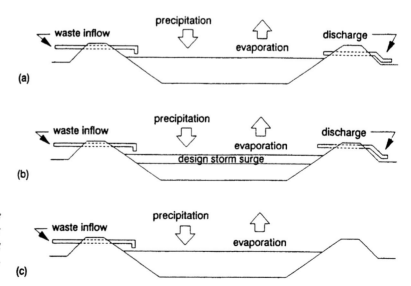

Figure 22.3 Types of surface impoundments: (a) *treatment;* (b) *storage;* (c) *nondischarging (evaporation) type. (From USEPA, 1991.)*

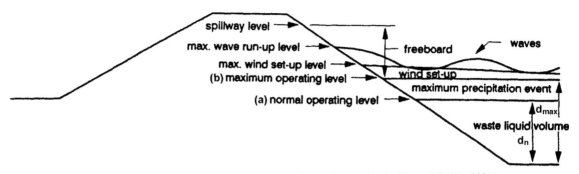

Figure 22.4 *Components making up impoundment design depth. (From USEPA, 1991.)*

where Q_i is the maximum inflow rate during an average detention time t, Q_o the outflow rate during the time period t, and S the liquid surface area. This depth should be adjusted for the cleanout schedule so that the expected depth of sediments are considered in estimating the normal operating depth.

The *maximum operating depth* (d_{max}) is typically estimated using one of the following three methods (USEPA, 1991): (1) hydrologic models, (2) water budget approach, and (3) design storm approach. For further information on these methods, a standard text on hydrology should be consulted. USEPA (1991) also provides information on these methods.

As shown in Figure 22.4, *freeboard* is the distance between the maximum operating level and the liquid level that would result in the release of stored liquid, such as the spillway level. Generally, freeboard accounts for wave run-up and wind setup plus a factor of safety. USEPA (1991) provides further information on freeboard estimation methods.

22.4.3 Component Design

Various component design issues to be considered for surface impoundment consist of the following:

- Foundation condition evaluation
- Side-slope design
- Liner systems
- Geomembrane protective layer
- Leak detection and collection system
- Gas venting layer
- Liquid-level control system
- Surface water management

Foundation Condition Evaluation. The foundation soils below the surface impoundment should be able to support the load imposed by the waste and dikes. Two important issues that need to be considered are (1) the compressibility and hence the settlement, and (2) the shear strength and hence the bearing capacity. *Settlement:* Fine-grained and/or peaty compressible soil will consolidate under additional loads and may result in differential settlements. This may result in the cracking of the dike and the liner system. Similar consequences can also occur if foundation soils are saturated, loose, cohesionless soils and are subject to earthquake loading. Figure 22.5 shows an example of the effect of differential foundation settlement on a surface impoundment dike. *Bearing capacity* is another issue that needs to be evaluated for a surface impoundment dike. Standard geotechnical engineering principles can be used to estimate differential settlement and bearing capacity considerations as identified above.

Side-Slope Design. As shown in Figure 22.6, a surface impoundment sideslope needs to be evaluated for stability of (1) cut slopes, (2) dike slopes, and (3) along the interfaces of double liner components. Cut or excavation and dike slope stability should be evaluated

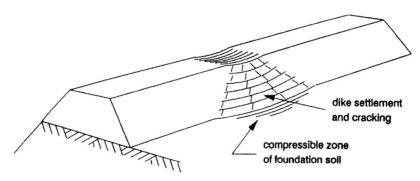

Figure 22.5 *Example of the effect of differential foundation settlement on a surface impoundment dike. (From USEPA, 1991.)*

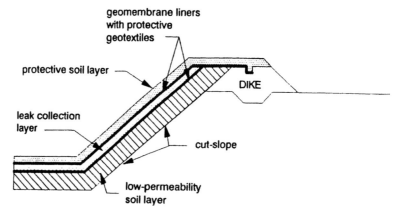

Figure 22.6 *Surface impoundment dike, excavated slope, and liner interface that needs to be considered for stability evaluations. (From USEPA, 1991.)*

under (a) end-of-construction under short-term conditions, (b) steady seepage or long-term conditions, (c) rapid drawdown conditions, and (d) seismic loading conditions. Standard geotechnical slope stability analytical methods such as modified Bishop's method and Spencer's method are generally used for such analyses.

Liner Systems. As shown in Figures 22.1 and 22.2, surface impoundment liner systems generally consist of either (1) a double liner system with a composite bottom liner, or (2) double composite liner systems. These have been discussed in Section 22.2. Chapter 17 presents information on various liner materials, such as HDPE geomembrane, geotextile, geocomposite drainage layer, geosynthetic clay liner (GCL), and compacted clay liner (CCL). Liner stress analyses and anchor trench design requirements are discussed in Chapter 18.

Geomembrane Protective Layers. As shown in Figure 22.7, a protective layer of soil or a soil/geotextile layer may be used on the surface of the liner system to protect the underlying geomembrane from (1) construction and operational loads and (2) weathering damage. USEPA (1983) recommends that a protective soil on a liner be (1) placed at slopes 3H:1V or flatter and (2) at least 18-in. (46-cm) thick before allowing any construction equipment over the liner. Slope stability analysis methods, discussed in Chapter 19, can also be used for protective cover slope stability evaluation.

Leak Detection and Collection System. As shown in Figures 22.1 and 22.2, a leak detection/collection and removal system between the liners of the double-liner system is required. The system should have a drainage layer that should be designed to rapidly de-

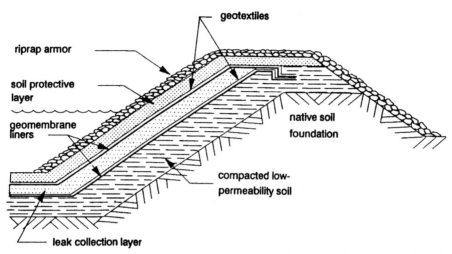

Figure 22.7 *Surface impoundment cross section showing optional protective layers. (From USEPA, 1991.)*

tect, collect, and remove liquid that could migrate through the upper liner system. The material could either be a granular or synthetic drainage material and should have a minimum hydraulic conductivity of 1 cm/s. If granular material is used, it should be round or subround and should not be larger than $\frac{3}{8}$-in. if placed directly on the geomembrane liner (USEPA, 1991). The drainage material should be chemically compatible with the waste contained in the surface impoundment. The system should have a sump, with a depth of at least 12 in. (30.5 cm), below the drainage layer grade. Figure 22.8 shows details of leak detection, collection, and removal system (USEPA, 1991).

Gas Venting Layer. In foundation conditions, where a potential for a source of gas exists, the gas venting layer should be incorporated beneath the bottom liner. The material specified could be either a granular layer or a geosynthetic material. As shown in Figure 22.9, the gas venting layer, if required, should be brought to the surface outside the liner.

Liquid-Level Control. The greatest environmental risk in surface water impoundment operations is unintentional liquid outflow (USEPA, 1991). Therefore, reliable liquid-level controls, such as a spillway (a passive system) and a level-activated pump or valve (an active system), should be designed appropriately.

Surface Water Management. Surface water impoundments should be isolated hydrologically from the runoff contributed from the surrounding terrain. Therefore, diversion structures such as berms and ditches may need to be designed to intercept and redirect the surface water flow from surrounding areas.

22.5 COVER DESIGN

Generally, covers are not provided for surface impoundments that contain waste. However, in many situations, where reduction of air pollution may be an important factor and/or the reservoir needs to be protected from addition of liquids due to precipitation, covers may be required. Cover materials should be resistant to the chemicals stored in the reservoirs and should have ultraviolet and exposed weathering resistance. There are two types of covers in general use: (1) nonremovable covers and (2) removable covers.

22.5.1 Nonremovable Covers

Nonremovable covers are of two types: (1) fixed covers and (2) floating covers.

Fixed Covers. These covers are typically used for smaller [less than about 15 ft (4.6 m) span] structures such as tanks. In this type of cover, the geomembrane

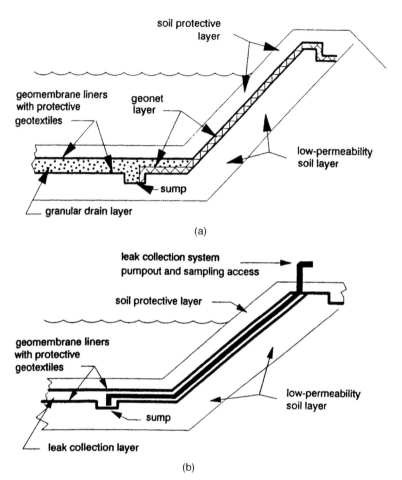

(a)

Figure 22.8 *Leak detection, collection, and removal system: (a) junction of sidewall geonet and bottom granular layers of leak detection, collection, and removal system; (b) example of access for leak collection system and liquid removal. (From USEPA, 1991.)*

(b)

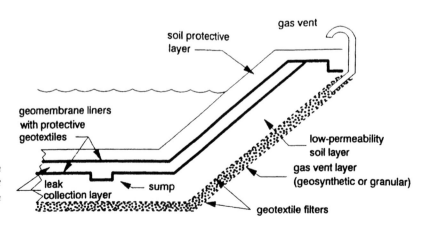

Figure 22.9 *Liner system showing gas-vent layer and exit through top of dike. (From USEPA, 1991.)*

is fixed at the upper edge of the tank and takes a catenary shape toward the center. Koerner (1998) discusses this cover in greater detail.

There are cases where fixed concrete roof structures have been designed and built for large surface impoundments. For example, a fixed concrete roof supported on columns (no walls) had been constructed over a leachate evaporation pond approximately 100×300 ft (30×90 m) at the Sonoma landfill in northern California. The roof was built to prevent rainwater from falling into the impoundment.

Floating Covers. Floating covers are more suited for larger span impoundments than are fixed covers, which are more economical for impoundments with smaller spans (less than 15 ft). Floating covers rest directly on the surface of the stored liquid and move up and down as its surface elevation fluctuates. As shown in Figures 22.10 and 22.11, these covers can either have a peripheral sump or well-defined location(s) of sump to collect surface water. The figures also show floats that are made of lightweight foam or beaded polymer materials. These are usually attached to the underside of the cover. The cover liner materials can be CSPE-R and HDPE geomembranes (Koerner, 1998). The critical liner material properties for design are tensile strength, tear, puncture, and impact resistance. Since stresses caused by wind can exert tear and shear forces, edge anchorage becomes very important. Figure 22.12 shows anchorage details. For further design and construction information on floating covers, readers are encouraged to refer to Gerber (1984) and Koerner (1998).

22.5.2 Removable Covers

In some situations, such as leachate ponds, it may be cost-effective to minimize the collection of rainwater into leachate stored in the surface impoundment. For such situations, the impoundment can be covered with a removable, inflated geomembrane. The main advantage of such an inflated removable geomembrane covers is the ability to take the covers on and off for operations in the rainy season (to prevent rainwater

from falling into the impoundment) and in the summer (for optimal evaporation) (Thiel, 2001). Thiel (2001) also describes a case history for such a cover system. Some of the features of this system are presented below. For further information, the reader is encouraged to refer to the original paper.

Figure 22.13 presents the schematics and cross section of such a cover system. This consists of a geomembrane anchored around the perimeter, a fan to blow in air to inflate the geomembrane, and cables across the impoundment to take the majority of stresses caused by inflation, as well as the forces induced by winds. Figure 22.14 exhibits the design for perimeter anchorage. The geosynthetic material should have high tensile strength, high tear strength, excellent resistance to ultraviolet light, and a low-expansion/contraction coefficient. One such geomembrane that has been used successfully for an inflated and removable cover is reinforced polypropylene (PP-R), which has a tensile strength of almost 200 lb/in., a tear strength of 100 lb (ASTM D 751, tongue shear), and excellent resistance to UV light (Thiel, 2001).

22.6 CLOSURE AND POSTCLOSURE CARE

According to 40 CFR 264.111, owners and operators of a hazardous waste surface impoundment are required to close the facility such that threats to human health and the environment are avoided. For nonhazardous waste impoundments, closure regulations have not yet been promulgated. Under RCRA regulations, two closure options are available:

1. *Clean closure (or removal).* This consists of removal, sampling to verify decontamination, and backfilling.

2. *In-place closure.* This consists of treatment of contaminated media at the site and then placing a final cover over the treated waste. The final cover must meet the minimum technology requirements detailed in USEPA (1989).

Figure 22.15 presents a flowchart that describes the logistics to be followed for each of the two options

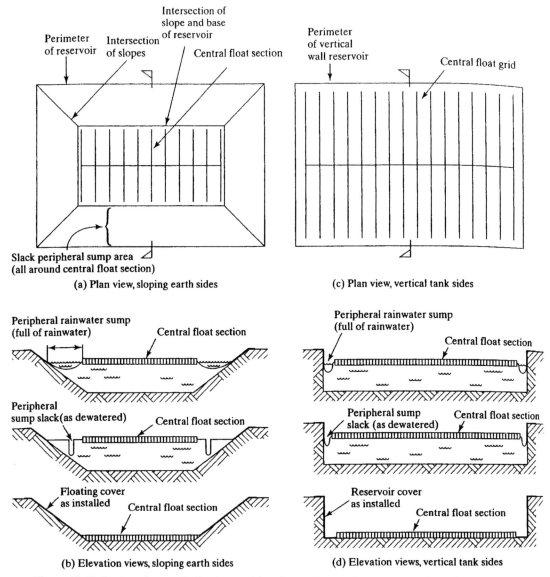

Figure 22.10 *Design of centrally floating peripheral sump impoundment covers. (From Gerber, 1984.)*

above. If the site is not clean closed, *postclosure care* and *monitoring* of the facility are required. Postclosure care period needs to continue for 30 years after final closure in accordance with 40 CFR 264.117. A written postclosure care activities plan must be submitted and must include as a minimum the following elements:

- Frequency of inspections
- A list of components inspected
- Remedial action plan description to repair damage facility components
- Frequency of monitoring and sampling

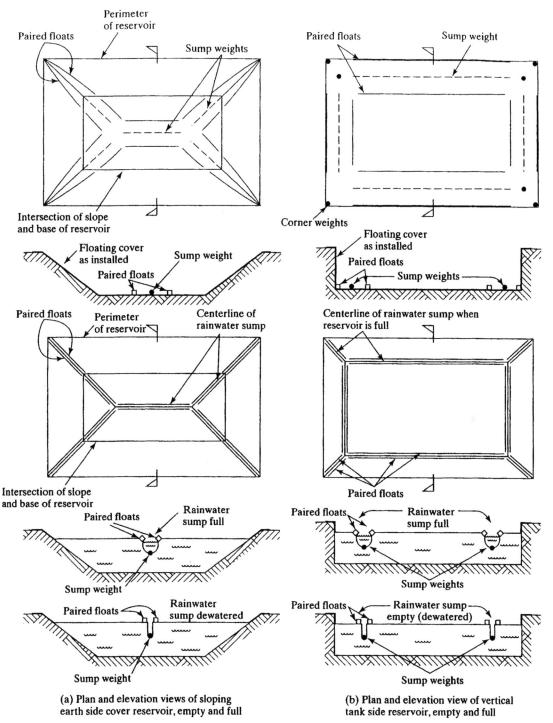

(a) Plan and elevation views of sloping earth side cover reservoir, empty and full

(b) Plan and elevation view of vertical tank side reservoir, empty and full

Figure 22.11 Designs for sump surface impoundment covers. (*From Gerber, 1984.*)

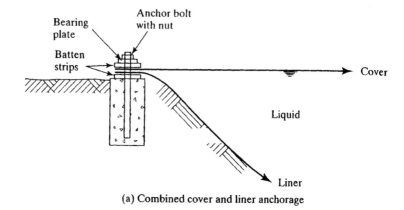

(a) Combined cover and liner anchorage

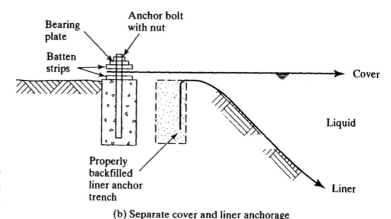

Figure 22.12 Various configurations for anchoring geomembrane floating covers and their edges. (From Koerner, 1998.)

(b) Separate cover and liner anchorage

USEPA (1991) provides further details on closure and postclosure care for hazardous waste surface impoundments.

22.7   SUMMARY

In this chapter we summarize various Resource Conservation and Recovery Act (RCRA) requirements as they apply to surface impoundments. These include design and operating requirements, action leakage rate, response action plan, monitoring and inspection, emergency repair, and closure and postclosure care. RCRA's minimum technology requirements specify double-lined structures with a leak collection and removal layer between the two liners. These liner systems were discussed in this chapter. Following this, various surface impoundment component design issues were presented. Cover types for surface impoundments, if required to meet specific project needs, were also discussed. Finally, closure and postclosure care for surface impoundments were described briefly.

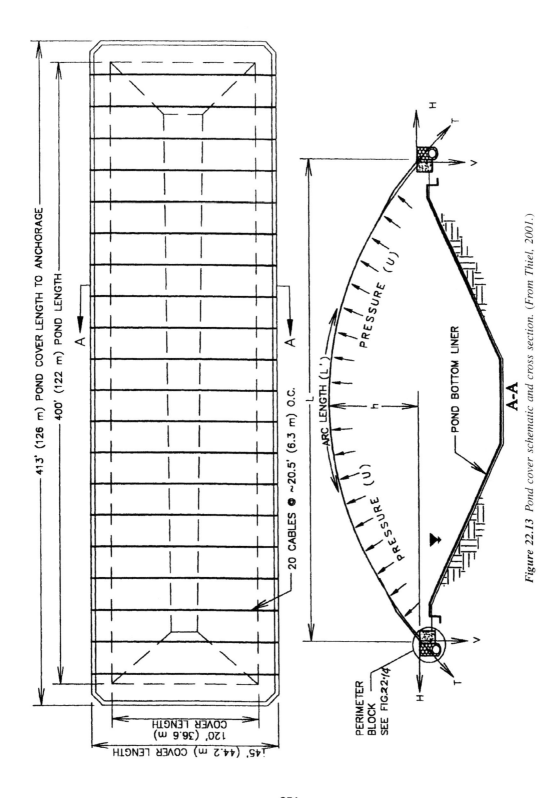

Figure 22.13 *Pond cover schematic and cross section. (From Thiel, 2001.)*

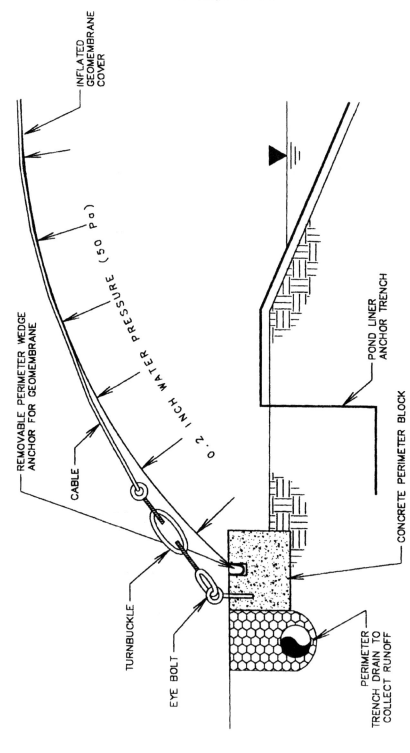

Figure 22.14 Perimeter concrete block detail. (From Thiel, 2001.)

INFLATED GEOMEMBRANE COVER

0.2 INCH WATER PRESSURE (50 Pa)

REMOVABLE PERIMETER WEDGE ANCHOR FOR GEOMEMBRANE

CABLE

TURNBUCKLE

EYE BOLT

CONCRETE PERIMETER BLOCK

POND LINER ANCHOR TRENCH

PERIMETER TRENCH DRAIN TO COLLECT RUNOFF

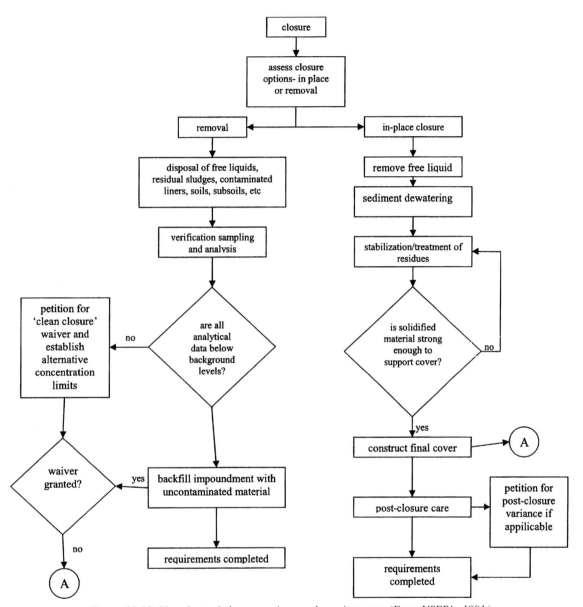

Figure 22.15 *Flowchart of closure options and requirements. (From USEPA, 1991.)*

QUESTIONS/PROBLEMS

22.1. What federal regulation applies to hazardous waste surface impoundments? Describe these regulatory requirements briefly.

22.2. Define a surface impoundment and describe hazardous waste.

22.3. What is a response action plan? Describe the plan briefly.

22.4. Describe the minimum technology requirement for the liner systems of a new hazardous waste surface impoundment.

22.5. Describe various components of a double-composite liner system for a surface impoundment. Also prepare a sketch for this system.

22.6. As a design consultant, you are asked to make a surface impoundment design presentation to a client. What various design steps would you consider for this presentation to your client?

22.7. Describe various surface impoundment types and explain with sketches.

22.8. What are different surface impoundment cover systems? Describe them. What are advantages of floating covers?

22.9. Show by drawing a sketch of the surface impoundment protective cover. How would you evaluate its stability? Explain by relating the factor of safety with geometry and material properties.

22.10. Discuss **(a)** leak detection, collection, and removal systems, and **(b)** a gas venting system when performing a surface impoundment component design. Explain by drawing various components in sketches.

REFERENCES

CFR (*Code of Federal Regulations*), 40 CFR, Parts 257, 264, and 265, U.S. Government Printing Office, Washington, DC, 2002.

Gerber, D. H., Floating reservoir cover designs, pp. 79–84 in *Proceedings of the International Conference on Geomembranes,* IFAI, St. Paul, MN, 1984.

Koerner, R. M., *Designing with Geosynthetics,* Prentice Hall, Upper Saddle River, NJ, 1998, p. 761.

Thiel, R., Unique inflated pond cover using reinforced flexible polypropylene, pp. 337–347 in *Geosynthetic Conference Proceedings,* Portland, ME, 2001.

Unger, M., Sai, J. O., Shiver, R., Jeyapalan, J., Evans, G. B., Jr., and Anderson, D. C., *Comparison of Multiple Small versus Single Large Surface Impoundments* (*Draft Report*), prepared by K.W. Brown and Associates under EPA contract 68-03-1816, USEPA, Washington, DC, 1985.

USEPA (U.S. Environmental Protection Agency), *Lining of Waste Impoundments and Disposal Facilities,* SW-870, Office of Solid Waste and Emergency Response, USEPA, Washington, DC, 1983, p. 448.

———, *Technical Guidance Document: Final Covers on Hazardous Waste Landfills Surface Impoundments,* EPA/1530/SW-89/047, Office of the Solid Waste and Emergency Response, USEPA, Washington, DC, 1989, p. 39.

———, *Technical Resources Document: Design, Construction, and Operation of Hazardous and Non-hazardous Waste Surface Impoundments,* EPA/530/SW-91/054, Office of Research and Development, Washington, DC, June, 1991.

PART IV
EMERGING TECHNOLOGIES

23

BENEFICIAL USE OF WASTE MATERIALS: RECYCLING

23.1 INTRODUCTION

As the world's population grows, wastes generated by the population also grow. It has been perceived that there is a "garbage crisis," meaning that it is not known what to do with all the wastes. To combat this, recycling has become a large part of our lives. Approximately 4.5 billion tons of nonhazardous wastes are produced per year in the United States (Tarricone, 1993). The average U.S. citizen produces over 6 tons of solid waste a year (Hendrickson et al., 1995). These numbers are constantly rising, but the rates at which products are recycled are also changing.

Recycling has been used since the nineteenth century, when scavengers would go through trash and find discarded items to reuse. This was the beginning of an industry that is playing a crucial part in today's environment and business. Many laws have been passed to encourage recycling because of the large amounts of waste that are generated. Forty-one states plus the District of Columbia have set goals of recycling in ranges of up to 70%. Twenty-nine states require municipalities to enact programs to encourage recycling (Hendrickson et al., 1995). To date, the recycling rates for different states range from 5 to 50%. Recycling has been so common now that most people recycle in some way or another. Curbside recycling is available to 30% of the U.S. population (Anon., 1991).

Economically, recycling makes sense only if the markets for the recycled wastes exist. Recycling of aluminum and paper has been the most successful because of their cost-effective reuse as source materials to manufacture products from again. Despite recycling, the volume of waste materials continues to increase. At the same time, disposal of these wastes in landfills has become cost-prohibitive due to high disposal costs associated with the design and operation of the limited number of landfills that must comply with strict environmental regulations. In addition, siting new landfills has become more difficult due to public perception of problems with landfills, especially the "not in my backyard" (NIMBY) syndrome.

The use of waste materials in various civil and environmental engineering applications can provide an attractive way to reduce the wastes to be disposed of, and it may also provide fiscal benefits. In this chapter, various types of waste materials are listed first. Then the following related issues are presented for selected wastes: (1) sources and quantities generated, (2) problem of disposal, (3) potential reuses, (4) physical and chemical properties, (5) environmental concerns, (6) economic considerations, and (7) typical field applications/demonstrations.

23.2 TYPES AND EVALUATION OF WASTE MATERIALS

23.2.1 Types of Waste Materials

The various types of waste materials can be divided according to their source into four categories:

- *Agricultural wastes.* The agricultural wastes include wastes such as animal manure, crop wastes, and lumber and wood wastes. Animal manure constitutes approximately 75% of the agricultural waste produced in the United States. The agricultural wastes are generated at a rate of more than 2 billion tons annually in the United States.

- *Domestic wastes.* The domestic wastes include wastes such as compost, glass and ceramics, incinerator ash, plastics, sewage sludge, scrap tires, and waste paper. Approximately 200 million tons of domestic waste is generated annually in the United States.

- *Industrial wastes.* Industrial wastes include wastes such as blast-furnace (iron) and steel slags, cement and lime kiln dusts, coal ash by-products, construction and demolition debris, foundry wastes, papermill sludge, reclaimed concrete and asphalt, and silica fume. Over 400 million tons of industrial wastes are generated annually in the United States. Of these, approximately 100 million tons are recycled pavement and 70 million tons are coal ash.

- *Mineral wastes.* Mineral wastes include wastes such as coal refuse, mill tailings, phosphogypsum, quarry waste, spent oil shale, and waste rock. About 1.8 billion tons of mineral wastes are generated annually in the United States. Of these, over 1 million tons are waste rocks.

23.2.2 Evaluation Process

A systematic evaluation is needed to decide if a waste material can be used for a beneficial purpose in engineering applications. The evaluation process generally consists of the following six steps:

1. Identification of all relevant engineering, environmental, occupational health and safety, recyclability, and economic issues associated with the proposed waste material and application

2. Establishment of laboratory testing and assessment procedures and criteria that the material should meet prior to acceptance

3. Testing and assessment of the results of the material and application for approval or disapproval using the established procedures and criteria

4. Consideration of the possibility of modifying the material prior to rejecting the material if the material does not meet the established criteria

5. Identification of issues that could impose significant constraints on the implementability of the proposed application

6. Determination of whether a field demonstration is necessary to supplement evaluation and assessment tests and criteria and implement the demonstration, if required

Applicable environmental federal and state regulations must be complied with when proposing to reuse waste materials. The federal regulations include those under RCRA and CERCLA. Many states have regulations and/or guidelines promoting the use of waste materials. In some states, permits, known as *beneficial use permits* or *beneficial use determinations* (BUDs), are needed. Regulations are generally concerned with the environmental suitability of the waste material for the proposed application. The general approach to environmental impact assessment consists of three tasks: identification of potential hazards posed by the use of the material, identification of persons or media (air, water, soils) likely to be affected by the identified hazard, and identification of the magnitude of the potential impact.

In addition to the technical and regulatory issues, cost analysis is needed to decide if the waste material could be used in a cost-effective manner. The costs associated with the material as well as the construction are evaluated for using waste material versus a conventional material.

23.2.3 Common Waste Materials

Of the various types of wastes described in Section 23.2.1, the following specific waste materials have the potential to be useful in civil and environmental engineering applications; these materials are also the major constituents of MSW:

- Fly ash
- Blast furnace slag
- Foundry sand
- Papermill sludge
- Sewage sludge
- Incinerator ash
- Glass
- Plastics
- Scrap tires
- Construction debris
- Wood

It should be noted that waste materials such as fly ash, slag, foundry sand, papermill sludge, sewage sludge, and incinerator ash are known as *process-generated waste materials,* whereas wastes such as glass, plastics, and tires are known as consumer-generated waste materials, and wastes such as construction debris and wood are known as *construction-generated waste materials.* Various aspects of these waste materials are presented in the following sections of this chapter.

23.3 FLY ASH

23.3.1 Sources and Quantities Generated

Fly ash is a fine inorganic by-product of coal combustion from the power-generating industry. Fly ash became readily available from electric power plants in the 1930s. In 1972, approximately 30 million tons of fly ash was produced in the United States (DiGioia and Nuzzo, 1972). In 1994, 774 million tons of coal was needed to meet the demand for electricity, which resulted in 80 million tons of fly ash. Based on projected electricity demands, it is expected that 1008 million tons of coal will be needed annually, which would in turn result in approximately 105 million tons of fly ash produced annually.

23.3.2 Problems with Disposal

With such large quantities of fly ash being produced, the problem of disposal is not trivial. The USEPA issued its final regulatory decision on fly ash on August 2, 1993, which stated that fly ash would not be considered a hazardous waste material under Subtitle C of the Resource Conservation and Recovery Act (RCRA) but instead, classified fly ash under Subtitle D, as solid waste. Thus, coal by-products, such as fly ash, are under state regulations (USDOE, 1994). Until recently, the cost of disposal of fly ash was relatively cheap, at about $5 per ton. However, recently, states have passed stricter "contaminant control" measures for landfills, which have led to an increase in disposal costs, $10 to $20 per ton, the cost of which is borne by the consumer in the form of higher electric costs. The problem

of disposal is not limited to economic concerns but by the sheer fact that available landfill space is already in limited quantity. Approximately 70 to 75% of the fly ash generated is still disposed of in landfills even though much of the fly ash is capable of being used for beneficial purposes.

23.3.3 Potential Reuses

The cementitious properties of fly ash are well documented and derivatives of fly ash can be traced back to Roman times, when it was used as a binding agent in mortar. Fly ash up to about 15% per unit weight of cement required in a concrete mix design can be substituted without a reduction in strength. Therefore, the potential for reuse of this by-product is dramatic (Faber and Babcock, 1987). The Federal Highway Administration recognized back in 1974 that due to the demand and wide use of concrete construction, portland cement would be in short supply. Therefore, the proactive approach to stem the possible shortage of portland cement was to replace 10 to 15% of cement with fly ash as an alternative when feasible (FHWA, 1974). Fly ash used in portland cement concrete must meet the requirements of ASTM C 618, which classified the fly ash as class F fly ash and class C fly ash.

Fly ash has been used as a substitute mineral filler in asphalt paving mixtures for many years. Mineral filler in asphalt paving mixtures consists of particles (less than No. 200 sieve in size) that fill the voids in a paving mix and serve to improve the cohesion of the binder (asphalt cement) and the stability of the mixture. Fly ash meets this gradation requirement and other pertinent physical (nonplastic) and chemical (organic content) requirements of mineral filler specifications.

Fly ash has been used for soil stabilization and subbases. Combining fly ash with lime yields a product that is comparable to bituminous aggregate material (BAM) commonly used in subbases for road construction. Fly ash has also been used as trench backfill, embankment material, berms, and cover material. Fly ash has been used in flowable fill applications as a fine aggregate and as a supplement to cement. Fly ash has been used for several decades as an embankment fill

Table 23.1 **Typical chemical composition of fly ash (percentages by weight)**

Chemical Constituent	Michigan Ash	Eastern U.S. Ash	European Ash
Silicon dioxide (SiO_2)	43.4	30–58	38–58
Aluminum oxide (Al_2O_3)	30.1	7–38	20–40
Iron oxide (Fe_2O_3)	9.8	10–42	6–16
Calcium oxide (CaO)	0.05–0.07	0–13	2–10
Magnesium oxide (MgO)	2.6	0–3	1–3.5
Sulfate (SO_3)	0.8	0.2	0.5–2.5
Unburned carbon	9.1–14.9	0–48	0

Source: Leonards and Bailey (1982).

material. Researchers in Ohio found that in repairing an embankment, fly ash enhanced mix provided a greater stability than soil alone as well as soil and ash mixed together (Anon., 1997). The fly ash was used as a type of weak concrete to prevent washouts and scouring from occurring on the embankment slopes.

Perhaps one of the most promising reuses of fly ash is as structural fill. This is borne out best in some construction projects where large volumes of soil would otherwise be needed as fill. One of the largest uses occurred in Pennsylvania, where 400,000 tons of ash was used as structural fill on the I-279 expressway by the Pennsylvania Department of Transportation (Turgeon, 1988).

23.3.4   Chemical Composition and Engineering Properties

The chemical properties of fly ash depend on the type of coal burned and the techniques used for handling and storage. Table 23.1 compares the typical composition of fly ash from different locations. The principal components of fly ash are silica, alumina, iron oxide, calcium oxide, magnesium oxide, sulfates, and carbon. These chemical constituents come primarily from the different types of clay that are found in various types of coal. The types of clay found in coal belong to one of three groups of clay minerals: smectite, illite, or kaolinite; alumina in fly ash is the resultant from these minerals. The iron oxide in fly ash is from the presence of iron in the coal. The calcium in fly ash results from the calcium carbonates and calcium sulfates commonly found in coal. Similarly, magnesium found in fly ash

results from organic material in some sources of coal. The sulfates found in fly ash result from pyrite and gypsum being present in the coal mineral. The carbon content in fly ash is the result of incomplete combustion of the coal (Leonards and Bailey, 1982).

The engineering properties of fly ash vary somewhat depending on the source and location from where the coal was mined (Leonards and Bailey, 1982; DiGioia and McLaren, 1987). Table 23.2 shows the engineering properties of fly ash from specific sources. Fly ash is predominantly a silt-sized nonplastic material, with 50 to 90% finer than No. 200 (0.075 mm) sieve. When fly ash is well compacted at optimum moisture content, which varies from 20 to 35%, the dry unit weight could range from 70 to 100 lb/ft³. Fly ash derives its shear strength from internal friction that ranges from 26 to 42° with a mean value of 34°. The permeability of well-compacted fly ash ranges from

Table 23.2 **Engineering properties of fly ash (averaged values)**

Property	Value
Percent passing No. 200 sieve	54
Specific gravity	2.3
Optimum moisture content (%)	31
Maximum dry unit weight (lb/ft³)	77.9
Degree of saturation (%)	75
Void ratio	0.96
Air void ratio	0.37
Permeability (cm/s)	5×10^{-7}
Friction angle	34°
Compacted moist unit weight (lb/ft³)	95–100

Source: Leonards and Bailey (1982).

10^{-4} to 10^{-7} cm/s. Because of the higher void ratio and greater permeability, the time rate of consolidation and consolidation settlements are lower than that of typical silty clay soils. California bearing ratio (CBR) values of fly ash are found to range from 6.8 to 15.4, depending on the testing conditions (soaked versus unsoaked). It should be noted that these properties vary, depending on the source of fly ash and must be determined for the specific source available for a project.

23.3.5 Environmental Concerns

One of the environmental concerns pertaining to coal burning is the by-product fly ash. This is why coal-burning plants are equipped with special "scrubbers" to control the emissions of fly ash into the air, thus controlling airborne pollution from fly ash (McKenna et al., 1976). As the demand for electricity is ever increasing, more coal is used by coal-burning electric plants, which results in tons of fly ash a year, thus disposal of fly ash taking up landfill space that is already limited by the other sources of municipal wastes. Therefore, the ASTM E 50 classifies coal by-products as a resource rather than a waste (USDOE, 1994). Perhaps the greatest environmental concern regarding the by-product of coal combustion fly ash is the taboo associated with a by-product material. Both chemical and physical properties of fly ash do vary but only slightly, thus producing a relatively homogeneous material with somewhat consistently defined properties. Furthermore, research has shown that by-products from coal combustion meet the regulatory standards. In 1993 the USEPA "found no justification for the inclusion of coal by-products as a hazardous waste. In effect, environmental concern may be based on a perception of a potential problem (due to ignorance or popular public opinion) and not on actual risk data."

23.3.6 Economic Considerations

As mentioned earlier, the disposal cost of fly ash was relatively cheap, at about $5 per ton. However, recently, due primarily to the increased costs of new disposal facilities with synthetic liners and leachate collection and groundwater monitoring systems, the disposal costs have risen to $15 to $20 per ton (USDOE, 1994). These costs will undoubtedly continue to climb as the available landfill space is used up. The bottom line is that the consumer of the electricity will in the end bear the brunt of these increased overhead costs. The costs of dumping otherwise recyclable material in Europe already carries a high penalty in the form of disposal costs. For example, 1 ton of fly ash disposed at the landfill may cost about $40 per ton. To avoid such high costs of disposal, countries such as the Netherlands and Denmark have close to 100% utilization of coal ash (USDOE, 1994). If disposal costs remain minimal and low pressure from local and state agencies continues, there will continue to be little motivation to utilize by-product fly ash completely.

23.3.7 Field Applications and Demonstrations

As mentioned earlier, fly ash has been used successfully as structural fill, embankment fill, flowable fill, and stabilized base for highway construction projects in a number of locations throughout the United States. At least 21 highway projects have been reported where more than 2 million cubic yards of fly ash has been used as a structural backfill behind retaining walls and bridge abutments or as an embankment fill. Standard construction equipment and procedures were found to be adequate. Monitoring data obtained from these projects did not show any signs of undue settlement or adverse environmental impacts. Fly ash was advantageous to use because it has *relatively low unit weight and high shear strength,* which results in good bearing support and minimal settlement. The ease of moisture control and compaction can reduce construction time and equipment costs. Among the drawbacks, dust control from dry fly ashes during construction may be necessary. In addition, fly ash consists of silt-sized particles that are prone to *erosion,* requiring control measures. Depending on the source, fly ash may have *corrosion potential,* and protection of metal pipes placed within or near the fly ash fill may be needed. Depending on the *sulfate content* of the fly ash, the possibility of sulfate attack on adjacent concrete foundation and walls need to be considered, and protection measures (such as using waterproof membrane or sulfate-resistant cement) may be necessary.

Fly ash has been used as a component in the production of *flowable fill* [also called controlled low-strength material (CLSM)], which is used as self-leveling, self-compacting backfill material in lieu of compacted granular soils. There are two basic types of flowable fill mixes that contain fly ash: high- and low-fly-ash-content mixes. High-fly-ash-content mixes typically contain nearly all fly ash, with a small percentage of portland cement and enough water to make the mix flowable. Low-fly-ash content mixes typically contain a high percentage of fine aggregate or filler material (usually, sand), a low percentage of fly ash and portland cement, and enough water to make the mix flowable. Many projects have used the flowable fill as trench backfill for storm drainage and utility lines on streets and highways. Flowable fill was also used as backfill for abutments and retaining walls, fill for abandoned pipelines and utility vaults, and to fill cavities and settled areas. The advantage of this material is that it can be pumped just about anywhere, and if needed it can be removed relatively easily compared to regular strength concrete (3000 psi). The drawback is its low compressive strength, and depending on the loads applied might not be appropriate for given site characteristics. Currently, it is commonly used to encase buried conduits as well as certain applications of sewer pipes. Primarily, to reinforce and resist crushing of underground pipes, it is also common to use a geotextile material as an integral part of this system as a type of reinforcing mesh.

The use of fly ash as a *soil stabilization medium* is commonly used in today's construction industry. The surface stabilization usually consists of the hardening of surface soils with binding agents such as fly ash, cement, or lime. Water can be added in some applications, and these components are actually sprayed topically onto the surface. Some applications require the use of a soil stabilizer to actually mix the surface soil with the fly ash, lime, or cement and then reapply this stabilized soil to the appropriate grade. Many projects have been completed successfully throughout the United States. The largest single reported project is the Newark International Airport, in Newark, New Jersey, where runways, taxiways, and aprons were built with fly ash-stabilized systems.

23.3.8 Summary

The generation of fly ash will continue to increase due to the ever-increasing demand for electricity. The amount of coal needed to generate this electrical power will obviously continue to grow, and with that, the coal by-product of fly ash will multiply. The diminishing space available in landfills, coupled with the increasing cost of disposal, should lead us toward broadened ways of using fly ash. Fly ash possesses desirable physical and chemical properties to serve as a resource material in several civil and environmental applications. Although further research is needed to define the specific tolerances that exist in both the chemical and physical properties in the varied sources of fly ash, the chemical and physical properties of fly ash remain consistent, provided that the source of the coal used remains the same. The environmental concerns regarding fly ash are minimal and are comprised mostly of the airborne pollutants that are not contained. The incentive to find or implement ways to use fly ash will certainly increase. A combination of governmental incentives for use of fly ash in construction projects and an increase in disposal costs for not reusing recyclable material should initiate increased awareness. The role of engineers in this matter should then be to implement recyclable materials, if applicable, into their design projects to further encourage their use.

23.4   BLAST FURNANCE SLAG

23.4.1   Sources and Quantities Generated

Blast furnace slag is produced during the production of iron and steel (Peacey and Davenport, 1979). Iron blast furnace slag is a nonmetallic by-product produced in the process of making iron. Steel blast furnace slag is a by-product of the steelmaking process. Different types of slag are produced depending on the method and materials used for iron and steel production. It is estimated that approximately 15 million tons of iron slag and 8 million tons of steel slag are produced annually in the United States.

23.4.2 Problems with Disposal

Historically, slag has been disposed of simply by piling it and leaving it uncovered. Slag is alkaline (having a pH above 7.0), and it can contain high amounts of toxic elements, such as chromium and manganese. These and other toxic elements, and compounds are often present due to past mishandling and improper disposal of other chemicals. For example, acidic waste materials were often dumped over the slag piles in an effort to neutralize the waste. If left uncovered, contaminants can be transported via runoff from precipitation that falls on the slag piles. This can lead to direct surface water contamination or to groundwater contamination by leachate infiltration through the piles and underlying soils. For these reasons, slag is often disposed of in landfills, consuming valuable space at an alarming rate given the tremendous quantities of slag generated.

23.4.3 Potential Reuses

Many proven alternative uses exist for iron and steel slag. Some slag never makes it to the slag reuse market, for it is recycled directly to the blast furnace as a residual source of iron or as flux material. It is estimated that up to 80% of the slag produced is recycled directly. In addition, iron extracted from slag during slag processing is often recycled through the furnace.

The majority of iron and steel slag is used as construction material. Accepted uses for blast furnace slag include aggregate in base and surface course in asphalt and as an additive to portland cement in concrete. Research has shown that the material has performed well in these applications. Other accepted but limited uses include embankment and ice control abrasive.

23.4.4 Chemical Composition and Engineering Properties

The chemical composition of slag is usually expressed in terms of simple oxides calculated from elemental analysis determined by x-ray fluorescence. Typical chemical compositions of iron slag and steel slag are summarized in Table 23.3. Both iron and steel slags are mildly alkaline, with a solution pH generally in the range 8 to 10. However, the pH of leachate can exceed 11, a level that can be corrosive to aluminum or galvanized steel pipes placed in direct contact with the slag. Free calcium and magnesium oxides that are not completely consumed in the steel slag are responsible for the expansive nature of steel slags.

Iron slag is angular, roughly cubical, and has textures ranging from rough to smooth surfaces. Steel slag aggregates are highly angular in shape and have rough surface texture. Iron and steel slags can be crushed and screened to meet the specified gradation requirements for a particular application. Table 23.4 summarizes the typical range of physical and engineering properties of slags. These properties are comparable to typical granular soils, making slags an excellent material for a wide variety of fill applications ranging from lightweight to well compacted (e.g., mechanically stabilized earth wall backfill to utility trench backfill).

23.4.5 Environmental Concerns

One of the concerns with steel slag is its potential to expand due to the presence of free lime and magnesium oxides that have not reacted with the silicate structures, and they can hydrate and expand in humid environments. This expansive nature (with volume changes of up to 10% or more attributable to the hydration of calcium and magnesium oxides) could cause

Table 23.3 **Typical chemical composition of iron and steel slags (percentages by weight)**

Chemical Constituent	Iron Slag	Steel Slag
Calcium oxide (CaO)	34–43	40–52
Silicon dioxide (SiO_2)	27–38	10–19
Aluminum oxide (Al_2O_3)	7–12	1–3
Magnesium oxide (MgO)	7–15	5–10
Iron oxide (FeO or Fe_2O_3)	0.2–1.6	70–80
Manganese oxide (MnO)	0.15–0.76	5–8
Sulfur (S)	1.0–1.9	<0.1
Metallic Fe	—	0.5–10

Source: FHWA (1997).

Table 23.4 **Typical properties of iron and steel slags**

Property	Iron Slag	Steel Slag
Specific gravity	2.0–2.5	3.2–3.6
Unit weight (lb/ft³)	70–85	100–120
Absorption (%)	1–6	Up to 3
Los Angeles abrasion (ASTM C 131) (%)	35–45	20–25
Sodium sulfate soundness loss (ASTM C 88) (%)	<12	<12
Angle of internal friction	40–45°	40–50°
Hardness (measured by Moh's scale of mineral hardness)	5–6	6–7
California bearing ratio (CBR) [% top size 19 mm ($\frac{3}{4}$ in.)]	Up to 250	Up to 300

Source: FHWA (1997).

difficulties with products containing steel slag, and is one reason why steel slag aggregates are not suitable for use in portland cement concrete or as compacted fill beneath concrete slabs.

Slag may pose significant environmental threats, depending on its chemical composition and its proximity to surface and/or groundwater. When slags are in extended contact with stagnant or slow-moving water, sulfurous, discolored leachate may result. The leachate can result in sulfur-related odors and typically exhibits a yellow/green tinge, resulting from the presence of free sulfur and sulfur dioxide and a high pH. If slags contain any toxic metals, release of these metals is also of concern.

23.4.6 Economic Considerations

The amount of iron and steel slag consumed each year has increased over the decade. This increase in demand came from the portland cement industry. As the amount of available portland cement experienced some decline, cement companies began exploring the large-scale use of granulated iron slag. Iron slag imports have also increased steadily since 1992. This inclusion of slag in cement mixtures has helped keep the cost of portland cement stable.

Iron and steel slag transport is another significant component of our economy. Eighty-five percent of the product is shipped by truck, 4% by waterway, and 4% by rail, with average marketing distances of 30, 250, and 175 miles, respectively. The balance, 7%, is used at the steel plant where it originates. In 1995, it was estimated that some 21 million tons of iron and steel slags with a market value of $154 million were sold or used.

23.4.7 Field Applications and Demonstrations

Blast furnace iron slag has been used successfully as an aggregate in portland cement concrete and asphalt concrete, granular subbase for roads, and embankment or fill material. Steel slag has been used extensively as a granular base or as an aggregate material in construction applications. Several state departments of transportation (California, Illinois, Indiana, Michigan, New Jersey, New York, and Pennsylvania) have allowed the use of iron slag as granular base or subbase material. Desirable features of using iron slag as granular base material include its ability to stabilize wet, soft underlying soils at early construction stages, good durability to be placed in almost any weather, extremely high stability, and almost complete absence of settlement after compaction. Additionally, the high insulating value of blast furnace iron slag granular bases can be used to minimize frost heaving. Crushed slag with maximum sizes ranging from $\frac{3}{4}$ to $1\frac{1}{2}$ in. has been used.

At least seven states have reported the use of blast furnace iron slag in embankment construction. They include Indiana, Kentucky, Maryland, Michigan, Missouri, New York, and Ohio. Some advantages of using iron slag include their low compacted density, high stability and friction angle, ability to stabilize wet, soft underlying soils at early construction stages, and almost complete absence of settlement after compaction.

Steel slag can be used as aggregate in granular base applications. Good interlock between steel slag aggregate particles provides good load transfer to weaker subgrades. Since volumetric instability is a problem, steel slag is not utilized in confined applications such as backfill behind structures. In addition, the formation of tufalike precipitates (white, powdery precipitates formed by the chemical reaction of atmospheric carbon dioxide and free lime in the steel slag) has resulted in deposits that have clogged subdrains and drain outlets.

23.4.8 Summary

Since the mid-nineteenth century, the industrialized world has had to handle and process slag. It is the by-product of iron and steel, which the world still depends on heavily. Past practices have resulted in the improper disposal and mishandling of these materials. Only recently have the potential reuses and multiple benefits of slag been realized; these include road bases and embankments. As more alternative uses are discovered, market opportunities for slag reuse continue to grow.

23.5 FOUNDRY SAND

23.5.1 Sources and Quantities Generated

Foundry sand is a by-product of the metal casting process. The sand cast system is the most common metal casting process used in the foundry industry. Sand cast molds are generally made of green sand, which consists of high-quality silica sand, about 10% bentonite clay (as the binder), 2 to 5% water, and about 5% sea coal (a carbonaceous mold additive to improve the casting finish). The type of metal being cast determines which additives and what gradation of sand is used. Although the sand is clean prior to use, after casting the sand may contain ferrous (iron and steel), heavy metals, and phenols. The automotive industry and its parts suppliers are the major generators of foundry sand. The annual generation of foundry sand in the United States ranges from 10 to 15 million tons. Typically, about 1 ton of foundry sand is required for each ton of iron or steel casting produced.

23.5.2 Problems with Disposal

Silica sand is the principal basic molding material, and it is relatively inexpensive. However, costs skyrocket when disposal is considered. As an example, an 8000-ton pile of spent sand could cost $680,000 to dispose of properly. Despite high cost, more that 90% of all spent foundry sand is disposed of in landfills. If toxicity characteristic leachate procedure (TCLP) metal (As, Ag, Ba, Cd, Cr, Ni, Zn, Cu, Pb, Hg, Se) concentrations exceed the regulatory limit, the waste is classified as hazardous waste, which requires that it be disposed of in a hazardous waste landfill, increasing the disposal costs significantly.

Traditionally, foundry sands have been landfilled because of the belief that trace metals from the casting process would leach into and contaminate groundwater. However, research has shown that foundry sands can be used in numerous geotechnical applications without significant risk of environmental damage. Again, the motivation for reusing the waste sand is the fact that it is very expensive to dispose of. As time passes, there will be less and less space available for disposal, making the price of disposal go even higher. The foundry industry is in the process of "educating" the state agencies as well as a possible customer about the sand reuses. One of the hardest things to change is the perception of the entire foundry process.

23.5.3 Potential Reuses

In foundry processes, sand from collapsed molds can be reused to make new molds. Most of the spent foundry sand from green sand operations is currently landfilled. Recently, beneficial reuse of foundry sands has been emphasized to preserve valuable landfill space, save the generator the costs associated with disposal, and provide an inexpensive construction material to the user. The potential uses of foundry sand include landfill daily cover, construction fill/roadbase material, flowable fill, cement manufacturing, asphalt, and precast concrete.

23.5.4 Chemical Composition and Engineering Properties

Spent foundry sand consists of silica sand, coated with a thin film of burned carbon, residual binder (bentonite, sea coal, resins), and dust. Table 23.5 shows the typical

Table 23.5 **Typical chemical composition of spent green foundry sand (percentages by weight)**

Chemical Constituent	Value (%)
SiO_2	87.91
Al_2O_3	4.70
Fe_2O_3	0.94
CaO	0.14
MgO	0.30
SO_3	0.09
Na_2O	0.19
K_2O	0.25
TiO_2	0.15
P_2O_5	0.00
Mn_2O_3	0.02
SrO	0.03
LOI (Loss On Ignition)	2.1–12.1
Total	99.87

Source: FHWA (1997).

chemical composition of spent foundry sand. Depending on the binder and type of metal cast, the pH of spent foundry sand varies from 4 to 8. Because of the presence of heavy metals and phenols in foundry sand, potential for leaching of these is a concern.

Table 23.6 summarizes the typical physical and engineering properties of spent green foundry sand. The grain size distribution of spent foundry sand is very uniform, with approximately 85 to 95% of the material between 0.6- and 0.15-mm sieve sizes. Five to 12% of

Table 23.6 **Typical properties of spent green foundry sand**

Property	Value
Specific gravity	2.39–2.55
Unit weight (lb/ft³)	160
Absorption (%)	0.45
Atterberg limits	Nonplastic
Magnesium sulfate soundness loss (ASTM C 88) (%)	6–47
Angle of internal friction	33–40°
Hydraulic conductivity (cm/s)	10^{-3}–10^{-6}
California bearing ratio (CBR) [% top size 19 mm ($\frac{3}{4}$ in.)]	4–20

Source: FHWA (1997).

fines finer than 0.075 mm (No. 200 sieve) can be expected. The particle shape is typically subangular to rounded. The engineering properties of spent foundry sand are comparable to those of conventional sands.

23.5.5 Environmental Concerns

Foundry sand, prior to use, is a uniformly graded material. The spent material, however, often contains metal from the casting and oversized mold and core material containing partially degraded binder. Spent foundry sand may also contain some leachable contaminants, including heavy metals and phenols that are absorbed by the sand during the molding process and casting operations. Phenols are formed through high-temperature thermal decomposition and rearrangement of organic binders during the metal-pouring process. The presence of heavy metals is of greater concern in nonferrous foundry sands generated from nonferrous foundries. Spent foundry sand from brass or bronze foundries, in particular, may contain high concentrations of cadmium, lead, copper, nickel, and zinc.

23.5.6 Economic Considerations

The landfill daily cover criteria, for example, vary from state to state. However, in Pennsylvania, where the study was conducted, the requirements are to produce 25 to 50% runoff, maintain a slope of 3%, and be able to support vegetation. It was suggested that every foundry take advantage of this. As an example, a landfill of 100 acres uses about 1700 tons per day of cover material. Typically, the landfills do not pay for the daily cover; hence cost savings on disposal to the foundry are possible.

23.5.7 Field Applications and Demonstrations

Foundry sand has been used as a substitute for the fine aggregate in asphalt paving mixes. However, foundry sand has been found to be too fine to satisfy some specifications for fine aggregate. Another reported application of foundry sand is using it as a fine aggregate substitute in flowable fill applications. The flowable fill, also called controlled low-strength material (CLSM), is generally composed of a mixture of sand, fly ash, a small amount of cement, water, and admixtures. The mixture ratios can vary depending on the application. Flowable fill can be used as backfill for sewer and utility cuts, underground storage tank bedding, bridge abutments, and other structural applications. It should be noted that many jurisdictions specify the use of fine aggregates conforming to ASTM C 33 in flowable fill, which generally precludes using spent foundry sand unless it is blended with natural sand or other suitable materials.

23.5.8 Summary

The American Foundryman's Society has been around since the 1950s. Together with different state agencies, tests are currently being conducted, which the foundries fund, to try to educate everyone and prove that in fact foundry sand is not hazardous or toxic and may be used in many day-to-day applications. This certainly is a step forward. However, the key to a successful program is accurate characterization of the waste foundry sand and the management perspective that the sand is not a "waste" product but a marketable material.

23.6 PAPERMILL SLUDGE

23.6.1 Sources and Quantities Generated

Papermill sludge is a by-product of the papermaking process and is produced at the papermills themselves. Approximately 2.1 million metric tons of papermill sludge is produced each year (Fuller and Warrick, 1985).

23.6.2 Problems with Disposal

As is the case for other wastes, the most pressing concern in disposing of papermill sludge is the use of landfill space. More and more waste is produced each year,

landfill space availability is dropping, and costs of disposal are increasing. Another problem is the high water content inherent in papermill sludge, which contributes to leachate generation.

23.6.3 Potential Reuses

There are two significant ways of dealing with papermill sludge waste. It can be used as an additive to damaged soils to restore organic materials necessary for plant recovery, or can be used in place of clay liners in landfill covers.

23.6.4 Physical and Chemical Properties

Papermill sludge is typically made up of arsenic, cadmium, chromium, copper, mercury, molybdenum, nickel, lead, selenium, and zinc as trace elements. Major elements are nitrogen, phosphorus, and potassium. Typical physical properties are: water content (150 to 268%), organic content (40 to 60%), specific gravity (1.80 to 1.97), and average initial hydraulic conductivity (1×10^{-7} to 5×10^{-6} cm/s). These ranges cover a variety of sludges from papermills and wastewater treatment plants (Moo-Young and Zimmie, 1995a,b). It is generally a neutral acidic–basic material, with pH in the range 5.4 to 9.0 (Bagchi, 1994).

23.6.5 Environmental Concerns

Because it is composed mostly of organic materials, papermill sludge does not pose a significant environmental threat, other than the fact that it is a waste and must be disposed of or reused. The only environmental concern is the space that it occupies in landfills and leachate generation.

23.6.6 Economic Considerations

As mentioned earlier, the primary economic incentive for finding uses of papermill sludge is reducing disposal costs. Limited landfill space means rising prices, and it is in the paper industry's best interest to consider alternatives to sludge disposal. Papermill sludge can be obtained very cheaply and when used in place of clay liners can save from $20,000 to $50,000 per acre in landfill closure schemes (Moo-Young and Zimmie, 1997).

23.6.7 Field Applications and Demonstrations

Papermill sludge can be used in two different ways. It can be added to soil directly as an organic agent. It can also be used as a substitute for compacted clay in landfill closure operations. Papermill sludge can be applied to soils for abandoned mineland rehabilitation. Revegetation on the soils at mine lands is difficult because of high acidity and low levels of organic matter and plant nutrients. Researchers at Ohio State University showed that applying papermill sludge to abandoned mine soil results in greatly enhanced survival and growth rates of trees, depending on the method of applying the sludge to the soil (Kost et al., 1997). The Department of Agronomy in Baton Rouge, Louisiana, investigated a similar application to mine soil. Paper sludge applied to Bermudagrass, along with fertilizer, adds organic material necessary to promote grass growth and enhances mine soil recovery (Feagley et al., 1994).

Extensive research has shown that papermill sludge can be used effectively in place of compacted clay in landfill cover liners, and tests have revealed that despite high water content, papermill sludge can be compacted effectively to yield low permeabilities suitable for landfill covers (Moo-Young and Zimmie, 1995a). Based on laboratory testing to obtain the properties of a particular papermill sludge, a design procedure has been developed as a basis for using compacted sludge as landfill cover. The procedure was applied to landfill in New York, where a 36-in. layer of papermill sludge was applied as a cover system. Important characteristics governing the design of such a liner are the initial hydraulic conductivity, water content, and organic content of the sludge. Using these initial characteristics, the design procedure involves consolidating the sludge in order to reduce its permeability. Consolidation re-

duces the water content, which in turn makes the sludge less permeable. The consolidation process can usually be completed within one year (Moo-Young and Zimmie, 1997).

23.6.8 Summary

Disposal of papermill sludge is a serious problem facing the paper industry today. A clear alternative in today's world is discovering ways to use papermill sludge for other purposes, thus saving landfill space. Recycling papermill sludge for different uses also saves natural resources from being expended. Research has shown that papermill sludge can be used beneficially for many applications, most prominently as a soil conditioner to restore soil-organic content for plant growth and as substitute material for landfill covers.

23.7 MUNICIPAL SLUDGE

23.7.1 Sources and Quantities Generated

Municipal sludge, also known as *biosolids,* is a by-product of wastewater treatment plants. The sludge is created from coagulants and filter-backwash material settling out of the water suspension. Typically, in older treatment plants, a greater amount of sludge is produced, unlike more modern facilities, which have the ability to recycle the filter backwash to the raw water intake and keep it moving without having to dispose of it. Also, some of the sludge-producing material is very often routed directly to sewage plants. The reason behind the treatment of water and therefore the creation of sludge is to remove pathogens and potentially harmful organic compounds. The usual course of action is to allow the sludge to settle in large lagoons, and then dry and dispose of it accordingly. The entire process, from creation to disposal, can take approximately six months.

Quantities generated can vary depending on the size of the municipality, city, or town. The amount of sludge produced, for example, depends on the wastewater flow. In a study conducted by the USEPA, 10,000 tons per day of sludge was being produced, with a prediction of 13,000 tons per day (USEPA, 1980). A study by the Army Corp of Engineers, based on military facilities, shows a sludge production rate of 0.1 to 0.4 lb per capita (USACE, 1987).

23.7.2 Problems with Disposal

A major problem with disposal is ensuring that the levels of pathogens and organics have been sufficiently reduced so as not to pose a threat to human health. The following methods of treatment of sludge now exist (USACE, 1999): (1) thickening, (2) digestion, (3) conditioning, (4) dewatering and drying, and (5) incineration. Thickening, conditioning, and dewatering methods are commonly used for the removal of water from the sludge prior to disposal. The digestion and incineration involve the specific removal of organics from the sludge material.

The greatest problem with the disposal of sludge is the potential for direct transportation of contaminants and the introduction of those same contaminants to the groundwater. Also, in the case of incineration, the introduction of pathogens to the atmosphere is possible. The problem rests not so much in the final disposal as much as in the treatment before it can be disposed of.

23.7.3 Potential Reuses

The reuses of sludge can involve drying/incineration to create a powder that has potential to be remanufactured into bricks for use in building. Other uses in the civil engineering field include the mixing of portland cement and stabilized ash from sludge incineration. This mixture is then used as a controlled density fill material. It may be possible to use sludge ash in a manner that is similar to fly ash mixed with traditional cement to form new materials.

Sludge is used for land reclamation where strip mining has occurred and also as a land application for agricultural uses (Pietz et al., 1978). The land reclamation involves the placement of sludge in lifts and allowing extensive decomposition in order to recreate topsoil. Once this is completed, forestry or other similar plantings, not for public consumption, could be

placed and the land is beautified. In the case of agriculture, greater care must be taken to ensure that pathogen levels are below risk levels and that the crops grown in such soils do not transport the same to humans. Typically, in agricultural applications, the grains or crops are used for the feeding of livestock. In both cases, the potential for contamination of the groundwater must also be observed to ensure public safety.

23.7.4 Physical and Chemical Properties

Typical composition of municipal sludge is shown in Table 23.7. The chemical concerns when disposal is in a landfill is to ensure that concentrations of pathogens are not so high or volatile as potentially to damage a liner system or leak right through it and contaminate the groundwater supply. Treatment converting sludge into powder (solids) not only improves handling as stated above, it can reduce the pollutant solubility from chemical changes that have occurred during treatment. In the case of land applications, it is desirable to have both nitrogen and phosphorus present in the final product after treatment to assist in the growth of plants/crops.

The physical properties of the sludge depend on the method of treatment that the wastewater discharge has been subjected to during dewatering or incineration. Table 23.8 shows typical physical and engineering properties of municipal sludge. Typically, if the intended disposal is in a landfill, the sludge will probably be in powder form, reducing the volume and improving the appearance, for the purpose of making the final waste a little more appealing to the general public (if seen). If the disposal method is land application, either the powder can be remixed with water or the treated sludge is applied directly.

23.7.5 Environmental Concerns

Regulations governing the monitoring of sludge are covered by Section 405 of the Clean Water Act (CWA). Federal regulations (40 CFR 503.13) specify the ceiling concentrations of chemicals allowed in sludge for land application. The greatest environmental

Table 23.7 **Chemical composition of municipal sludge**

Constituent	Units	Mean[a]	Minimum	Maximum
pH		7.1	5.4	8.2
Electrical conductivity	dS/m	9.70	1.38	21.08
Total solids	%	75.6	59.4	94.3
Total volatile solids	% of total	35.6	18.6	45.4
Total Kjeldahl-N	mg/dry kg	24,476	11,054	40,463
NH_3-N	mg/dry kg	3,339	128	13,937
$NO_2 + NO_3$-N	mg/dry kg	269	36	2,011
Total P	mg/dry kg	19,690	9,970	28,336
K	mg/dry kg	4,802	2,049	6,547
As	mg/dry kg	8	1	14
Cd	mg/dry kg	9	5	38
Cr	mg/dry kg	384	213	625
Cu	mg/dry kg	439	219	609
Hg	mg/dry kg	0.76	0.11	1.28
Mn	mg/dry kg	633	515	769
Mo	mg/dry kg	15	4	21
Ni	mg/dry kg	68	40	01
Pb	mg/dry kg	186	85	283
Se	mg/dry kg	3	<1	7
Zn	mg/dry kg	1,015	505	1,460
Al	mg/dry kg	23,772	15,825	31,090
Ca	mg/dry kg	50,481	40,782	79,213
Fe	mg/dry kg	21,278	17,452	24,130
Mg	mg/dry kg	24,188	18,097	3,021
Na	mg/dry kg	1,156	420	3,021

Source: MWRDGC (2002).

[a] Calculations based on 62 samples.

concern regarding sludge is the potential contamination of groundwater. The sludge may contain heavy metals, especially cadmium in soluble form, creating potential problems. Soluble materials such as nitrogen, which is desired to provide nutrition for plant growth, is also soluble and could potentially contaminate a water source. In cases where sewage sludge is the material being placed, the sludge must be tested thoroughly for the existence of pathogens, in addition to the presence of other organic contaminants (USEPA, 1978).

23.7.6 Economic Considerations

Economic considerations would entail, in the case of land spreading, acquiring equipment capable of spreading the material in an effective manner. Also, transportation of the material requires consideration with respect to travel distance, type of vehicle used for transporting, possible permits required, and so on. In the case of the Fulton County, Illinois, landspreading application, the total cost for the project was $338.92 per metric ton (Pietz et al., 1978), with a capital cost of only $38.13 per metric ton. It could also be stated that the cost could be spread out over a number of projects in the case of equipment and therefore reducing cost even further. Depending on the location and the actual land reuse, such as forestry or crops, the cost of monitoring should also be included and could probably be based on the same criteria as those of landfills.

23.7.7 Field Applications and Demonstrations

Land spreading can be done and could involve two alternatives, depending on the desired outcome: (1) use

Table 23.8 **Physical and engineering properties of municipal sludge**

Property		Range
Water content	Initial (%)	42–75
Grain size distribution	% sand	39–49
	% silt	45–52
	% clay	2–11
Atterberg limits	Liquid limit	71–119
	Plastic limit	59–85
	Plasticity index	17–53
Specific gravity		1.81–2.17
USCS classification		OH
Moisture-density relationship (Proctor test)	Standard Proctor	
	γ_{dmax} (lb/ft^3)	50–68
	OMC (%)	37–70
	Modified Proctor	
	γ_{dmax} (lb/ft^3)	52–72
	OMC (%)	30–64
Shear strength	CU triaxial test (total strength)	
	C (kPa)	0–40
	ϕ (deg)	21.1–29.7
	CU triaxial test (effective strength)	
	C (kPa)	0–50
	ϕ (deg)	32.2–42.0
	UU triaxial test	
	C (kPa)	0–20
	ϕ (deg)	24.6–39.6
	UC strength test	
	UC strength (kPa)	23–126
	Strain at failure	4.0–5.2
Consolidation	Compression index (C_c)	0.26–0.50
	Recompression index (C_r)	0.03–0.10
Bearing strength	Illinois bearing ratio (IBR)	1.6–4.8
	Illinois bearing value (IBV)	2.2–9.4
	Swell (%)	0.93–3.37

Source: GLSEC (2002).

as a reclamation product to replace the soil layer where strip mining has occurred, and (2) use as a fertilizer for crops. A demonstration project involved the spreading of municipal sludge in Fulton County, Illinois (Pietz et al., 1978). This project involved spreading sludge on a 6289-ha area where extensive strip mining had occurred. The goal was to restore the topsoil conditions for the purpose of creating row crop fields. The initial method used for the application of sludge was a sprinkler system, because the sludge had a 95% water content. It was later determined that a disk system, probably similar to a standard plow but with disks in the place of teeth, would be used for application of the sludge. The advantages of a disk system over spray are twofold. First, the disk allows for an increase in the sludge application because of the direct mixing of soil and sludge. Second, the disk allows for a more accurate application of the sludge. This meant that a spray or mist of sludge floating over the area does not exist, which can be of concern to neighboring communities. Also, the disk allows the application to be completed into all the corners of the site in question. The disk system contains a tank holding the sludge and a piping system that delivers the sludge to the disk. The apparatus has the ability to till a row approximately 10 ft wide and 8 in. deep. A second disk device with 24-in. disks and wider spacing was found to be more effective in applications where the natural soil was compacted clay. The overall result of the Fulton County project was deemed successful. Notes were made regarding the labor intensity of the application, and through changing the sludge by densifying, it was thought that the cost would go down by a factor of 2 to 3.

The second instance involved a study by the Madison Metropolitan Sewage District (MMSD) on the transfer of PCBs from sludge-treated soils to crops grown on treated lands. The primary goal of the MMSD was to dispose of 20 to 30-year-old sludge that was contaminated with PCBs. The tests were done on a site that contained a silty loam type soil, similar to the soil type where sludge applications were already taking place. Plot sizes were 20 ft × 60 ft. The sludge was applied using an injection method and then disked to ensure complete mixing of the sludge and soil to a depth of 12 in. Several plots were set up, with each having a different application rate. The concentrations of PCB in the experiment were also varied to help in determination of safe levels of application. This was done to verify the regulation at the time of a sludge concentration limit of 50 ppm PCBs in sludges applied to farmland. It was determined by this study that PCBs containing four or fewer attached chlorine atoms disappear from the soil in 38 months. Isomers containing five or more chlorines do not disappear from the soil and can create a problem (Berthouex and Gan, 1991). The MMSD study may prove otherwise and justify further analysis in the transport mechanisms from crop to livestock to humans. The Fulton County project also concluded that increased metal concentrations can occur in corn grown on the site.

23.7.8   Summary

It would appear that land application is a very efficient way of disposal of the sludge and with good planning can solve or fulfill other needs that are required. The research of using incinerated sludge mixed with ce-ment products for mechanical fills should focus specifically on landfill applications. Use of sludge for building materials could be feasible, but it would probably be better to see it work as a soil substitute first; case studies show that such substitutes do work. However, its effect on human health should be investigated considering food chain impacts.

23.8   INCINERATOR ASH (SEWAGE SLUDGE ASH)

23.8.1   Sources and Quantities Generated

Ashes are generated by incinerating various types of municipal solid waste. The source of ash production that is considered here is the municipal sludge, and the ash generated is also known as *sewage sludge ash*. Ashes are generated during the combustion of dewatered sewage sludge in an incinerator. Municipal sludge incineration in well-operated facilities produce an odorless ash weighing between 30 and 60% of the weight of the original sludge on a dry basis. Municipal wastewater plants in the United States generate about 72 million dry tons of sludge each year. Approximately 2 million tons of this is incinerated, resulting in approximately 70 million tons of ash that might finally be disposed of or reused in an environmentally compatible manner (USEPA, 1991).

23.8.2   Problems with Disposal

Most of the sludge ash generated in the United States is presently disposed of in landfills. Concern with the disposal is the transport of certain trace elements with the leachate, especially arsenic, cadmium, copper, chromium, lead, mercury, selenium, and zinc. The leaching rate is normally very slow, resulting in continuous leaching for hundreds or thousands of years. The ash consists primarily of insoluble silicates, phosphates, sulfate, and refractory metal oxides, some of which may be soluble. A metals analysis of incinerator ash shows a range of metals concentrations (e.g., cadmium 70 to 9000 mg/kg, zinc 900 to 24,000 mg/kg).

A notable deviation is mercury (2 to 9 mg/kg), the low level reflecting the very high percentage transfer to the exhaust gases (USEPA, 1980).

23.8.3 Potential Reuses

Applications that could potentially make use of sewage sludge ash include the use of ash as a raw material in portland cement concrete production, as aggregate in flowable fill for backfilling trenches, as mineral filler in asphalt paving mixes, and as a soil conditioner mixed with lime and sewage sludge.

23.8.4 Physical and Chemical Properties

The chemical composition and properties of sludge ash depend on the wastewater treatment process and the chemicals used in the treatment and sludge handling and incineration process. Table 23.9 shows the typical composition of sludge ash. Prior to incineration, sludge is dewatered, and the dewatering process may include the addition of ferrous salts, lime, organics, and polymers. As a result of this pretreatment, sludge ash may contain higher quantities of ferrous and calcium. The pH of sludge ash varies from 6 to 12, but sludge ash is generally alkaline.

Sewage sludge ash is primarily a silty material with some sand-size particles. Table 23.10 summarizes the

Table 23.9 **Typical chemical composition of sewage sludge ash (percentages by weight)**

Chemical Constituent	Value (%)
SiO_2	14.4–57.7
Al_2O_3	4.6–22.1
Fe_2O_3	2.6–24.4
CaO	8.9–36.9
MgO	0.8–2.2
SO_3	0.01–3.4
Na_2O	0.1–0.7
K_2O	0.07–0.7
P_2O_5	3.9–15.4

Source: FHWA (1997).

Table 23.10 **Typical properties of sewage sludge ash**

Property	Value
Specific gravity	2.44–2.96
Unit weight (lb/ft³)	80–110
Absorption (%)	1.6
Atterberg limits	Nonplastic
Hydraulic conductivity (cm/s)	10^{-4}

Source: FHWA (1997).

typical properties of sludge ash. The specific size range and properties of the sludge ash depend on the type of incineration system and the chemical additives used during the wastewater treatment process.

23.8.5 Environmental Concerns

The major concern considering reusing MSW ashes for civil engineering practices is the release of toxic substance in the air and groundwater. An assessment of the impact of the municipal sludge incineration system on the environment shows it to be a major concern. The greatest concern is associated with air emissions from incinerators because of the potential public health risk associated with the inhalation of airborne gases and particles. Certain emitted contaminants are suspected human carcinogens, and others can exert other acute or chronic effects (USEPA, 1991).The emission from the sludge incinerator must not result in violation of ambient air quality standards and must meet the EPA air pollution emission standards of performance contained in the New Source Performance Standards for Sludge Incinerators (40 CFR 60.15, Appendix IV) (USEPA, 1977). Proper design should assure encapsulation and stabilization of metals and postconstruction monitoring of leachate.

23.8.6 Economical Considerations

Economical impact of the effect of the incineration system on the value of the site itself and of adjacent property must be considered. The system may affect the overall economy in terms of jobs created and land removal from agricultural production. Public ownership of the incineration system property may affect the community tax base. The effects of the system on the supply of the resources and energy must be considered (USEPA, 1980). When dealing with incinerators, fuel is generally the most expensive part of the process. The minimal cost of operation and equipment maintenance is another economic parameter for sludge incineration. Preventive maintenance is the single most important factor in the reduction of operating costs. The following is a breakdown of the cost of each incinerator; the lower the moisture content in the sludge, the cheaper the incinerator will be to operate.

Multiple heart furnace	7000 lb/h 0% moisture $11,000,000
Field bed incinerator	1000 lb/h 0% moisture $900,000
	2900 lb/h 0% moisture $1,600,000
Electric furnace	2400 lb/h 30% moisture $1,3000,000
	2400 lb/h 0% moisture $950,000
Cyclonic incinerator	2000 lb/h 20% moisture $1,000,000

The design cost will be a function of the incinerator cost plus installation, which is normally in the range 4 to 7%. This cost should be doubled to include engineering services during project construction (Maher et al., 1992).

23.8.7   Field Applications and Demonstrations

Research studies indicate that utilization of the ashes as aggregate for portland cement or bituminous concrete in several applications in civil construction, or as aggregate for soil improvement and road construction, is feasible; to a lesser extent, the employment as additive in soil stabilization has also been studied. Maher et al. (1992) mentions several satisfactory in-situ applications, such as construction of an experimental reef

in Long Island Sound with large blocks made of 15% portland cement and 85% ashes by weight; the strength of the submerged blocks increased with time (compressive strength was 1372 lb/in^2 after 358 days of exposition to salt water, which means an increase of 397 lb/in^2 relative to strength before submersion). Twelve projects in the Chicago area used lime and ashes in the construction, in bituminous concrete and in a mixture with lime and asphalt.

23.8.8   Summary

Sludge incineration and disposal or reuse of the resulting ash is an environmentally acceptable method for the disposal of sludge when the environmental assessment shows it to be appropriate. Incineration alone is a volume reduction method rather than an ultimate disposal solution. Volume reduction can be an important consideration, however, where land availability is a problem. After incineration, ash, either dry or in scrubber water, can be reused or disposed of to the land. Ash that is disposed of must be designed to protect groundwater, to minimize dust production, and to ensure protection of surface waters. The basic goal is to provide for municipal sludge management in a cost-effective and environmentally acceptable manner. To accomplish this goal, one alternative method, presented above, involves a sludge incineration unit that produces ash, thus reducing sludge volume and reuse it in an economical way.

23.9   GLASS

23.9.1   Sources and Quantities Generated

Glass is a product of the supercooling of a molted liquid mixture consisting of sand (silicon dioxide) and soda ash (sodium carbonate) to a rigid condition. The source of glass in municipal solid waste comes from households that purchase material placed in glass containers. Once that material has been used, the consumer has no further use for the container and it is disposed of. This type of glass is *post-consumer glass*. *Preconsumer glass* is glass made at the manufacturing plant that does not meet the required specifications. This glass is often recycled within the facility and does not end up going into a landfill. Therefore, the concern is the postconsumer glass, because this is the material that takes up landfill space (Woods, 1994; Miller, 1998).

Approximately 11.5 millions tons of glass containers were generated in 1995. Of the 11.5 million tons of glass containers generated, 3.1 million tons was recovered, which constitutes 27.2%. The remaining almost 8.4 million tons of glass required landfilling (USEPA, 1997).

23.9.2 Problems with Disposal

The use of high compaction, co-collection trucks, and overhandling of bottles collected result in the breakage of glass containers (Woods, 1994). For glass manufacturers to make clear or colored glass, the glass cullet that is purchased from the glass processing facility needs to be of the same color. If the cullet does not meet the standards that the glass manufacturer requires, the entire load can be rejected. For example, using mixed cullet will result in colors of glass that do not meet the manufacturer's standards, so the manufacturer will end up disposing of it.

The cullet used by the manufacturer must also be free of contamination. One of the largest concerns is that broken ceramic dishes can also be mixed in with broken glass in the curbside recycling bin. When pieces of ceramic are mixed in with the cullet, it causes major problems for glass manufacturers. Ceramic has a higher melting point than glass, and when it is put into a glassmaking furnace with the cullet, instead of melting it tends to explode into pieces that will be present in the end product (Malloy, 1997). These two problems, mixed glass colors and contamination, can result in loads of glass cullet being rejected by the glass manufacturer, which would otherwise recycle the material into new containers.

Many of the material recycling facilities (MRFs) in the United States do not have the technology available to easily separate the colors or remove the contaminants. When the MRFs receive a mixed cullet, they can sort the cullet manually, or through waste separation machines if they have them. Separating glass manually poses a safety hazard to workers at the facilities. Also, many existing MRFs do not have the most up-to-date separation equipment. The efficiency of the existing machines in separating the colors and contaminants and the speed with which the cullet is processed may end up costing the MRF more than the end product is worth, resulting in the mixed glass being landfilled (Malloy, 1997). If the glass is landfilled, it is taking up valuable landfill space. Because glass is a dense material, the cost of transportation and disposal of the glass to the landfill is high. Other alternatives of reusing glass need to be studied to find ways of increasing its recycling rate.

23.9.3 Potential Reuses

There are several applications available to reuse glass and glass cullet (CWC, 1993; 1996; 1998). Crushed glass or cullet, if properly sized and processed, can possess characteristics similar to that of a gravel or sand. The cullet can be used in place of aggregate in many construction applications, including backfill, road construction, and retaining wall fill. This application does not require that the glass be sorted out by color, and the contamination in the cullet does not have to be removed completely, allowing 5 to 10% debris allowed in the cullet to be used. This is a possible use of the cullet that is collected by municipalities in areas where glass recycling is done, but a glass container manufacturer is not within a reasonable proximity such that transportation costs to the facility would be too great. This application could also be used if natural aggregate is not readily available and the municipality has a recycling facility that collects glass. The municipality would need to purchase a machine to crush the glass to ensure that the required gradation of the materials is met. Studies have been done in several areas of the United States that have used glass aggregate in place of natural aggregate. Another civil engineering application that uses mixed glass is *glassphalt*, produced by replacing a percentage of the natural aggregate in asphalt with crushed glass. The processed cullet is not limited to replacing gravel in civil engineering projects but can be used to replace fine aggregate as well (Kirby, 1993).

23.9.4 Physical and Chemical Properties

There are three basic types of glass manufactured in the United States: borosilicate, soda-lime, and lead glass. Soda-lime glass makes up 90% of the glass produced commercially (Han, 1993). Table 23.11 shows the chemical composition of these three types of glass. Glass is considered an inert material; however, it is not chemically resistant to hydrofluoric acid and alkalis. Expansive reactions between amorphous silica (glass) and alkalis (such as sodium and potassium found in high concentrations in high-alkali portland cement) could have deleterious effects if glass is used in portland cement concrete structures.

Crushed glass particles are generally angular in shape and can contain some flat and elongated particles. The degree of angularity and quantity of flat and elongated particles depend on the degree of crushing. Smaller particles, resulting from extra crushing, will exhibit somewhat less angularity and reduced quantities of flat and elongated particles. Proper crushing can virtually eliminate sharp edges and the corresponding safety hazards associated with manual handling of the

product. Many studies have exhibited that the engineering properties of glass aggregate are comparable to natural aggregate. Table 23.12 summarizes the typical properties of glass cullet. The properties of glass cullet depend on the gradation of the glass, which in turn depends on the degree of crushing. Higher levels of debris (up to 5%) do not affect the engineering properties negatively.

23.9.5 Environmental Concerns

Although glass is a relatively inert material, problems can exist with the disposal of glass into the landfill or to a glass processing facility, due to its composition. The ingredients to make glass are sand, limestone, soda ash, and recycled glass cullet. In a study conducted by the USEPA, representative samples of glass were collected and analyzed for heavy metal contents. The results indicate that glass can contain a variety of heavy metals, and many of the heavy metal concentrations exist in large concentrations, as shown in Table 23.13. Although the concentrations of many heavy metals are high, it may not be a large concern because many of the materials used to make glass, sand and limestone in particular, are natural soils. The element concentration of many of the heavy metals in natural soils can also be high (refer to Table 23.13). As shown in Table 23.13, the heavy metals that exist in levels above natural soils that can still be a concern are mercury, lead, silver, and zinc. The USEPA study provides analytical results of the total heavy metal content in glass. A toxicity characteristic leaching procedure (TCLP) anal-

Table 23.11 **Typical chemical composition of glass (percentages by weight)**

Chemical Constituent	Soda-Lime	Borosilicate	Lead
SiO_2	70–73	60–80	60–70
Al_2O_3	1.7–2.0	1–4	—
Fe_2O_3	0.06–0.24	—	—
CaO	9.1–9.8	–	1
MgO	1.1–1.7	—	—
Cr_2O_3	0.1	—	—
BaO	0.14–0.18	—	—
Na_2O	13.8–14.4	45	7–10
K_2O	0.55–0.68	—	7
PbO	—	—	15–25
B_2O_3	—	10–25	—

Source: FHWA (1997).

Table 23.12 **Engineering properties of glass cullet**

Property	Value
Specific gravity	1.96–2.52
Optimum moisture content (%)	5.7–7.5
Maximum dry unit weight (lb/ft³)	111–118
Los Angeles abrasion (%)	30–42
California bearing ratio (%)	42–132
Permeability (cm/s)	10^{-1}–10^{-2}
Friction angle	51–53°
Compacted moist unit weight (lb/ft³)	95–100

Source: FHWA (1997).

Table 23.13 **Typical elemental concentrations in glass and natural soils (mg/kg)**

Chemical	Glass	Natural Soil
Aluminum (Al)	6036–13449	10,000–3,000,000
Antimony (Sb)	25.4–154.3	2–20
Arsenic (As)	0.4–9.8	1–50
Barium (Ba)	190.7–784.7	100–3000
Beryllium (Be)	0.01–1.0	0.1–40
Boron (B)	21.5–88.8	2–100
Cadmium (Cd)	0.3–4.8	0.01–0.7
Chromium (Cr)	28–943	1–1000
Copper (Cu)	6–92	2–100
Iron (Fe)	1921–7568	—
Manganese (Mn)	76–256	20–3000
Mercury (Hg)	0.1–0.6	0.01–0.3
Nickel (Ni)	10.1–62.7	5–500
Lead (Pb)	20–109.3	2–200
Selenium (Se)	0.06–0.77	0.1–2
Tin (Sn)	27–166	2–200
Zinc (Zn)	21–1671	10–300

Source: USEPA (1995).

ysis of the glass would need to be done to determine how much of the total metals would actually leach out if the glass were placed in the landfill. These would be the contaminants that would be removed by the leachate collection system but could penetrate to the liner if there are any problems in the leachate collection design. A few studies indicate that the amount of lead that would leach from the glass is below the 5 mg/L USEPA regulatory limit for lead (CWC, 1998).

23.9.6 Economic Considerations

For municipalities dedicated to recycling but not located near a glass manufacturer, it may not be feasible to pay for the cost of transportation to haul the material to the manufacturer. One option these municipalities have is using the cullet as a replacement or supplement to natural aggregate in the field. These municipalities can consider purchasing machines to process the glass into sizes that allow the municipality to use the cullet in engineering applications. This cullet does not need to be color sorted, so any containers, broken or intact, can be put through the processing machine. Also, the

debris levels allowed in the cullet are not as strict as required by the glass manufacturer. The processed cullet can then be used as a replacement for natural aggregate in building roadways or as backfill. Gravel or sand would not need to be purchased by the municipality, and the initial cost of the processing equipment can be offset by the cost savings of not having to purchase the natural aggregates. By using glass cullet, the municipality also does not have to pay the cost of transportation and disposal of the cullet into a landfill.

For municipalities to determine which glass disposal option is most feasible, an economic analysis should be done evaluating the three typical options available: landfilling the glass, selling the glass cullet to a glass bottle manufacturer, or producing glass aggregate. For each region of the United States, typical costs, such as labor, transportation, and disposal, will vary (CWC, 1993). An economic analysis provides a good estimate for the municipality to determine which option is most suitable.

23.9.7 Field Applications and Demonstrations

Some municipalities are currently studying the benefits of using recycled glass in place of construction aggregates. One project was done in 1996 by Hennepin County, Minnesota, where crushed glass was used in place of some natural aggregates in a road base. A pavement test strip was made where 10% and 30% natural class 5 gradation aggregate was replaced with crushed glass. Once the aggregate and glass were mixed, samples were taken to ensure that proper mixing of the material was done by the contractor. Although the weather during placement of the mixed aggregate was very wet at times, the material remained relatively stable. The stability of the material may be due to the fact that the mixed aggregate appeared more permeable than natural aggregate of the same gradation. After the materials were placed as the road base, the test strip was paved over with asphalt. Tests were then done on the pavement to determine how it compared with the control strip that only used natural aggregate (Hennepin County, 1997). The field tests done on the pavement included compaction tests, bituminous tests, and the falling weight deflectometer (FWD) test. The compaction tests indicated that the density of the material had increased from initial placement of the material but not up to the specifications of the standard proctor density of 95%. The bituminous test results were similar to the compaction tests results. The bituminous range of compaction was between 92 and 93%. Normal specifications would be 95% of the Marshall density. Although these two tests indicated lower than specified material values, the FWD test results showed that the pavement moduli had increased more with the mixed aggregate than with the natural aggregate. The increased moduli would indicate that the strength gain in the mixed road base was higher than the natural road base even though the compaction and bituminous tests were lower than required by specifications. From these tests, the county determined that the mixed aggregate road base performed as well as, if not better than, the natural aggregate road base (Hennepin County, 1997). The county was interested to pursue other projects using mixed glass and recommended that other counties and cities pursue demonstration projects that would provide additional information on eventually using mixed aggregates to develop standardized specifications (Hennepin County, 1997).

Another civil engineering application of the mixed glass aggregate is making glassphalt by replacing a portion of the natural aggregate in asphaltic cement (AC) with the crushed glass (CWC, 1994). King County, Washington, is one community that has made a test road using glassphalt to study its performance in the field. In 1992, the county roads division constructed a 1-mile road replacing 5 to 10% of the natural aggregate with mixed cullet. Prior to using the glassphalt in the field, several lab tests were done on the materials, including the Hveem stabilometer, density, and cohesion tests. The stability tests showed that the most stable form of the glassphalt mix was made when 10% of the natural fine aggregate was removed and replaced with mixed glass. The stability value of this mix was 43 compared to other mixes, which ranged from 33 to 37. The other mixes were made by combining the natural aggregate with 5 to 10% of the mixed glass. The density of the trial mixes fell within the required AC class B specifications ranging from

147.3 to 148.1 lb/ft^3 with an air content of 3.4 to 4.2%. These cohesionometer values of the lab mixes all exceeded the minimum values required by the specification (CWC, 1994).

The test pavement was made in September 1992 with the various glassphalt mixes. Additional lab tests were done on the field batches, and field observation of the road was done to determine how the material fared in the field. The lab tests indicated that the mixes met the required AC class B specifications, with some exceptions. Some of the in-place mixes were 1 to 3% below the minimum requirements. The amount of air in the mixes was 1 to 3% over the maximum values specified. For mixes that contained 10% mixed glass, the stability was 1 or 2 points below the minimum required. The field observations of glassphalt roads included checking for corrugation, alligator, longitudinal, longitudinal edge, and transverse cracking. None of the problems above were noted and the road was in excellent condition. The skid resistance of the glassphalt was comparable to the control mix of AC class B section (CWC, 1994). Glassphalt has been used in other communities with more mixed results. In other test surfaces that were done, the initial construction costs using the glass increased. Contractors may charge more for using the glass material, depending on the availability of natural aggregate. Also, if natural aggregate is readily available, the cost of transporting glass aggregate to the construction area may cost more than hauling natural aggregate. On roads with allowable speeds over 40 mph, the roads did not provide traction as well as regular asphalt, and "rutting may occur in the road when studded tires are driven over the glassphalt." More studies may need to be done on roads where the maximum speed limit is less than 40 mph, to determine if the glassphalt performs better or worse than traditional asphalt roads (USEPA, 1992).

23.9.8 Summary

Currently, two different recycling options are available for glass and glass cullet. These options are selling the material to glass manufacturers and processing the cullet into materials that can be used in civil engineering construction projects. The first option, using the cullet to make new glass containers, is a good option if a glass manufacturer is located close to the municipality that collects glass because the manufacturers pay for the cullet used in their operations. The second option, processing the glass cullet into usable construction materials, has been studied and has shown to be a good option for mixed cullet or cullet contaminated with debris. The cullet is used to replace fine or coarse natural aggregate in construction projects from road base and asphalt to landfill cover and leachate collection removal systems. The glass aggregate can replace anywhere from 15% of the natural aggregate for structural application to 100% of the natural aggregate in nonstructural or drainage applications.

Several states currently have specifications that allow the use of glass cullet as a construction aggregate. The states of Washington, Oregon, California, Connecticut, New York, and New Hampshire have specifications that allow glass cullet to replace a portion of the natural aggregate used in construction projects. This allows local municipalities to use the mixed glass cullet, which results in an increased glass recycling rate instead of having to landfill the glass. Studies need to be done in other states that currently have an excess of glass cullet that is unusable and is ending up in landfills. When the recycling rate of glass can be increased, valuable landfill space is saved for materials that presently have no reuse value.

23.10 PLASTICS

23.10.1 Sources and Quantities Generated

Plastics are made from natural resources and we can find them in our daily life without even knowing where they are coming from. For example, with *polypropylene* it is possible to make food containers, disposable trays, and other types of packaging. *Polystyrene* is used in industry to make beverage containers, hangers, plates, and egg cartons. From *polyvinyl chloride* we get pipe and vinyl siding. With polyethylene it is possible to make food packaging (baked goods, meat, etc.) and nonpackaging (industrial liners, shipping sacks, shrink wraps, etc.). The global consumption of plastics

was 135 million tons in 2000 and is expected to rise 3.8% per year. Plastics constitute about 10% of MSW.

23.10.2 Problems with Disposal

There are three principal problems with the disposal of plastics. First, for plastic degradation to occur, a combination of biological, chemical, and photochemical reactions is necessary, and because of that, it takes a very long time for plastics to degrade. Second, when plastics go to the sewage system, it becomes a hazard for marine life. Third, polymers occupy 8% by weight or 20% by volume of landfills in the United States.

23.10.3 Potential Reuses

The recycling of used plastic bottles has been increasing. For example, it increased from 363 million pounds in 1990 to 891 million pounds in 1993, more than 100%. Four modes of plastic recycling are possible: (1) use as generic plastic, (2) use of mixed plastic, (3) regeneration of raw materials, and (4) use of energy recovery.

Polypropylene is used in a wide variety of durable and nondurable products, and just a few of them can be recovered in order to recycle again after their intended purpose. There are various ways to make products derived from plastics: from polypropylene it is possible to get new products formed by extrusion processing (fabric and filaments, film, sheet, others), by injection molding (rigid packaging, consumer products, transportation, appliances, others), by blow molding (consumer and medical), and by compounding and resellers.

Polystyrene is an amorphous thermoplastic produced by either batch suspension or mass-continuous polymerization of styrene monomer. The recycled *polystyrene* could be found in the electronics, horticultural, houseware, and construction industries. From polystyrene and through the following processes, it is possible to get many different products: by solid injection molding (beverage containers, cutlery, hangers, dairy containers, plates, baked goods containers, vending cups, closures, trays), by foam-extruded sheet (poultry trays, hinged-lid containers, egg cartons, cafeteria trays, foam cups), by blow molding (bottles), and by foam molding (cups and containers, loose-fill packaging).

Many people assume that *polyvinyl chloride* cannot be recycled, and that is not true. The largest market for this type of plastics is the construction field, which is about 60% of all vinyl sales, while packaging is the most common use for the other plastic types. In the construction industry, PVC provides many products to be used, such as water distribution pipes and flexible baseboard moldings. Also, it could be used in packaging, for electrical and electronic uses, in transportation, in furniture and furnishings, and in consumer goods.

Polyethylene is also useful, although one of the problems with PET is the high price, so that is one of the reasons why people are leaning toward recycling. From this product one can manufacture items such as trash containers, oil bottles and drainage pans, book binders and dividers, money bags, and consumer products containers as well as in various film and household/industrial/chemical bottle applications.

In the Great Lakes area it is possible to find many companies that are manufacturing many products made from the recycled plastics. For example, the plastic packaging of milk, soda, and detergent bottles are now living a second life as new bottles, bags, or toys. There are many other uses for recycled plastics, such as fencing, electric fence posts for cattle and vine stakes, picnic benches, and playground equipment made from milk containers and bread bags.

23.10.4   Physical and Chemical Properties

The process used to make recycled plastic materials is generally composed of two steps: the first, compounding or preparation of the final compound, and the second, shaping or forming the final product. To do that, one must consider the properties of the plastics (Harper, 1992). Some plastic types have special properties: *polyethylene* is lightweight, flexible, and inert toward water and chemicals; *polypropylene* is lightweight, rigid, and has excellent chemical resistance; *polyvinyl chloride* is stable, useful, versatile in compounding, and can be made rigid or flexible; and *polystyrene* is rigid, shiny, and easily tinted.

23.10.5   Environmental Concerns

The chemicals as well as the environmental resistance of plastics reach all industries and segments of our daily life. Their influence in the composition of things in our surroundings, as in the composition of construction materials, affects the entire world and its economy. The fabrication of processed plastic is subjected to highly corrosive or aggressive chemical environments. The fact is that the designers and engineers have had the challenge of providing good-quality products, safely, and with low maintenance costs.

23.10.6   Economic Considerations

Most of the people agree that even though plastics are products in our daily trash, which is difficult to recycle or get rid of, it is also true that the plastics have many advantages over the conventional materials. These advantages include better quality and lower cost. For example, plastics are becoming more popular than any other packaging material. Over 8 billion pounds of plastics is consumed in the United States per year, and in building and construction the increase in the use of the resin has reached 15 billion pounds. The global plastic consumption has been increasing dramatically. In 1996, more than 95 million metric tons of plastic was consumed. The United States is the largest consumer and producer of plastics in the world. The employment increase in the last few years in the plastic industries reached 3%, compared with the overall manufacturing employment increase of just 0.3%.

23.10.7   Field Applications and Demonstrations

In construction, plastics are taking the place of wood in such matters as remodeling and rehabilitation, because as the price of wood increases, plastics become more attractive for construction. In appliances, the U.S. appliance industry shipped 52.6 million units in 1994 (Anon, 1996). In the car industry, polymeric materials have been used over the past 25 years, their uses appear to be growing, and it is projected to continue increasing. Some of the advantages in this field are to reduce vehicle weight, increase quality and performance, extend warranties, and cut costs (*Modern Plastics Encyclopedia '96*). In the medical field, from flexible tubing for intravenous solutions to semirigid connectors and luers, from film-based blood bags to permeable membranes, and in implantable shunts and valves (*Modern Plastics Encyclopedia '96*), plastics have improved medical services. In the civil engineering field, the most used plastic is polyvinyl chloride, which is used in water pipes and flexible baseboard moldings. On the other hand, the use of plastics is most commonly found in the packaging business.

23.10.8 Summary

Different types of plastics are discarded and many of these can be reused for useful purposes. Various products can be made using recycled plastics. Applications of recycled plastics in civil engineering applications are limited compared to those found in other fields.

23.11 SCRAP TIRES

23.11.1 Sources and Quantities Generated

Scrap tires originate from passenger cars, light and heavy trucks, heavy equipment, aircraft, and off-road vehicles. Approximately 266 million scrap tires are generated each year in the United States; a figure that translates into an average of more than one tire per person per year (STMC, 1996a). At an estimated weight of 3.3 million tons, scrap tires constitute 1.8% of the total quantity of solid wastes generated. Two to 3 billion tires are currently stockpiled all over the country.

23.11.2 Problems with Disposal

Stockpiling of scrap tires has led to at least the following two significant environmental threats:

- *Mosquito hazard.* With their ability to collect and retain rainwater, stockpiled tires form ideal breeding grounds for mosquitos such as the Asian tiger mosquito. These tend to concentrate in the upper fringes of the stockpiles, where the most recent rainwater has collected, and can produce a significant number of offspring. They are also carriers of deadly disease vectors such as encephalitis and dengue fever (STMC, 1996b).

- *Fire hazard.* Stockpiled tires pose a serious fire hazard. Scrap tire fires are difficult to extinguish, due to the fact that 75% void space is present in a tire and tires have greater heat energy than that of coal and can burn for considerable periods of time, releasing heavy black smoke. This has proven to be so serious that due to a series of well-publicized scrap tire stockpile fires (an example is the nine-month Rhinehart tire fire in Winchester, Virginia), legislative action was initiated at the state and national levels (Sikora, 1991). At the beginning of 1991, 44 states had drafted, introduced, and enacted laws to control this scrap tire problem.

Landfilling of scrap tires, even monofills, is not a good disposal alternative. Whole tires are difficult to compact. The California Integrated Waste Management Board (CIWMB, 1996) reports that 75% of the space occupied by a tire is void, which clearly translates into valuable landfill space lost. If shredded to overcome this problem, they may as well be put to other uses, as the cost of shredding ranges from $30 to $65 per 100 tires.

Whole tires also capture explosive methane gas and "float" upward, sometimes shooting to the surface with tremendous force and piercing the landfill cover. Landfilling of scrap tires either whole or shredded is costly, squanders any conceivable resource recovery potential of the scrap tires, may cause fire hazards, may leach specific contaminants into the surrounding soil, and should be adopted only if no other options are available.

23.11.3 Potential Reuses

As neither stockpiling nor landfilling appeared feasible options, other uses were developed for scrap tires.

Scrap tires have been used extensively as tire-derived fuel (TDF). Tires have greater heat energy than coal (37,600 kJ/kg versus 27,200 kJ/kg) (CIWMB, 1996), which is the main advantage in using scrap tires as fuel. The disadvantage is that for burning, whole tires have to be shred to 2- to 6-in. chunks, which increases the costs involved in the process. Another disadvantage is that tire contains steel beads and belts and if the combustion process requires the removal of these (usually, by further fine shredding and the use of powerful magnets), the use of scrap tire becomes very unattractive.

Scrap tires have found a wide range of applications in civil engineering. Some of their applications are described below.

1. *Tire chips in embankments, retaining walls, and bridge abutment backfills.* Shredded tires have a very low density (compacted dry density 38.6 to 40.1 lb/ft^3, compared to the traditional material for such applications, gravel at 125 lb/ft^3), are free draining and absorb only a small quantity of water. These properties have made scrap tires a good construction material in lightweight embankments and retaining walls (Humphrey et al., 1992, 1997; Manion and Humphrey, 1992; Humphrey and Sandford, 1993; Upton and Machan, Edil and Bosscher, 1992; Bosscher et al., 1994, 1993; Drescher and Newcomb, 1994; Cecich et al., 1996). The lightweight offers several advantages, such as permitting construction on weak foundations, better stability against slope failure due to reduced weights, and lower differential settlements. These advantages translate into a reduction in the construction costs of these structures, as the low densities of shredded tires will reduce the thicknesses of the structural components used in their retention, leading to a light design, and expensive foundation treatment procedures to combat differential settlements in weak soils can be avoided.

2. *Whole tire retaining structures.* Whole tires can be used in the construction of low-height retaining structures, typically around 5 ft high. Construction up to 10 ft in height can be accomplished by reinforcing the tires with a geotextile. Typically, one or two rows of tires on a firm bench of indigenous material are used. They may be connected or unconnected and are usually backfilled with shot rock, then seeded and mulched. Such walls may be vertical or with a batter (CALTRANS, 1988).

3. *Frost penetration limitation.* In climates where the ground is subjected to freeze–thaw cycles, it has been found that the gravel pavement surface suffers significantly due to weakening of the granular surface course, rutting due to traffic, and loss of cross drainage. These problems are caused by the earlier melting of ice lenses that are formed in the subsurface. Insulation can be used to prevent the underlying ground from freezing. Tire chips have been found to be about eight times better than gravel for reducing frost penetration .Tire chips have a low thermal conductivity, making them ideal for this application (Humphrey and Eaton, 1993, 1994 and 1995).

4. *Rubberized asphalt concrete.* The use of rubber in the construction of asphalt pavements offers several advantages, such as long pavement life due to decreased thermal cracking, less potholing, decreased deformation, and less reflective cracking. Driving noise can also be reduced, due to the fact that rubber is resilient in nature. Projects in Alaska have shown that the stopping distances could be decreased by the use of rubber in asphalt (Clark et al., 1993). Mixtures with asphalt rubber displayed less stripping than for mixtures without asphalt rubber when tested with an indirect tensile stripping test conducted in the state of Virginia (Maupin, 1994).

5. *Organic compound removal from landfill leachate.* Tire chips have been found to have a relatively high sorption capacity for organic compounds (Park et al., 1995). They are reported to have 1.4 to 5.6% of the sorption capacity of activated carbon on a volumetric basis. Single- and multisolute systems produce the same adsorption capacities. Only 3.5 to 7.9% of the organic compounds were desorbed. It is also recommended by this study that a 30-cm tire chip layer as a primary leachate collection layer could lead to significant levels of adsorption of organic compounds, leading to a decrease in the liquid-phase concentration of the leachate.

6. *Railroad crossings.* Crumb rubber finds an application in railroad crossings. Crumb rubber is rubber that has been cut to a size of $\frac{3}{8}$ in. or less space. Ten to 12 pounds of crumb rubber can be produced from

one passenger tire. The process involves initial shredding to 2.5 to 4 in. and final shredding done by two or three successively narrower blades to obtain the desired size. The sized particles are processed to even smaller sizes by cracking, grinding, or rolling mills. Finally, screens and gravity separators are used to remove the metal, and aspiration equipment is used to remove the fibers. The rubber is molded to form panels that fit between the tracks and can be fastened to the ties. This application typically consumes about 350 lb of rubber per foot of track. Crumb rubber compares well with timber crossings with respect to life span (10 to 20 years, compared to three to four years for timber) (DeGroot and Switzenbaum, 1995).

7. *Highway applications.* Discarded whole tires can be used in a variety of highway maintenance applications, of which the one main application is stabilization of and widening shoulders of roads. Shredded tires may be used as alternative aggregate material (Bressette, 1984).

8. *Landfill applications.* Potential uses of shredded and chipped tires have been studied in the areas of:

   a. *Gas migration control trenches.* A gas migration and control trench is used to control the lateral migration and discharge of landfill gas under active and passive extraction.

   b. *Gas collection and venting layer in closure cap systems.* This layer is incorporated in the landfill cap and is beneath the infiltration layer in landfill caps.

   c. *LCRS drainage layers.* This layer has the purpose of collection of the leachate and conveying it to the LCRS (Hall, 1990).

   d. *Leachate recirculation trenches.* These trenches are located in the waste itself and are used to convey the recirculated leachate back into the landfill in a landfill operated as a wet cell. Conventionally, these layers consist of 6 to 12 in. of granular material, reinforced with geotextiles.

   e. *Drainage layer in covers.* This layer drains the surface infiltration, thus preventing or minimizing the generation of leachate. It has been estimated that 43,500 tires could be consumed per acre of the landfill in the drainage layer application, 450 tires per foot of depth per 100 lineal feet could be consumed in the gas migration control trench application, and 87,000 tires per acre could be consumed in the gas collection and venting layer in closure cap systems.

9. *Retarding VOC movement.* Traditionally, bentonite slurry cut of walls has been used to reduce the migration of groundwater contaminants, particularly VOCs. A study by the University of Wisconsin has shown that tire shreds added to the bentonite in the cutoff slurry walls improves the performance of the walls in retarding the migration of contaminants.

10. *Removal of VOCs in wastewater treatment plants.* Ground tires have been tested successfully to ascertain their sorption capacity for VOCs in wastewater treatment plants (Park and Jhung, 1993). It is reported that the ground tires had a slightly higher sorption capacity than tire chips; VOC removal efficiencies were greater than 95% for polar compounds and greater than 80% for nonpolar compounds.

11. *Use of tires as a noise barrier.* Scrap tires were used in the construction of noise barriers. The cost of these walls is not much different from the cost of conventional walls used for this purpose (Horner 1993).

12. *Miscellaneous uses.* Scrap tires are also used in several other uses, such as stair treads, mats, flooring tiles, sewer rings, guardrails, golf driving mats, and playground covers.

23.11.4 Physical and Chemical Properties

Tire composition varies by manufacturer and type. Automobile tires are made of natural rubber, synthetic rubber elastomers, polymers, and other additives. Steel reinforcing is also provided to improve strength. According to the Rubber Manufacturers Association (RMA), passenger tires are composed of 27% synthetic rubber, 14% natural rubber, 16% fabric (polyester, nylon), 28% carbon black, and 15% steel wire. Tires are designed to withstand the rigors of the environment so that they are durable and safe when used on a vehicle. Even discarded tires maintain their chemical composition, requiring hundreds of years to fully decompose (Marella, 2002).

TCLP studies were carried out in the laboratory for a waste mix of seven shredded old tires (15 to 20 years of age) and seven shredded new tires (5 to 10 years of age) by the Twin City Testing (TCT) Corporation for the Minnesota Pollution Control Agency (TCT, 1990). The test was carried out under four leaching environments; acidic (pH 3.5) , acidic (pH 5), 0.9 % NaCl leach, and basic (pH 8.0). The study concluded that metals leached out in highest concentrations under acidic conditions; polynuclear aromatic hydrocarbons (PAHs) and total petroleum hydrocarbons (TPHs) are leached out in highest concentrations under basic conditions.

Another TCLP study was conducted by the Radian Corporation (Radian, 1989) to ascertain the levels of chemicals that leach from the products of the RMA. The products that were tested were from the manufacturers of tires, roofing products, belts and hoses, molded products, gaskets, and printer rollers. The study concluded primarily that none of the products that were tested exceeded the proposed TCLP regulatory limits. Most of the concentrations were near the method detection limits, and the results of EP toxicity procedures with the proposed TCLP products were comparable. It was concluded that the chemical characteristics of the leachate from scrap tires are fairly innocuous.

The physical properties of scrap tires depend on the size to which they are shredded (Ahmed, 1993; Ahmed and Lowell, 1993). Whole tires will have different properties from shredded tires. Table 23.14 summarizes the typical properties of shredded scrap tires from various sources. It can be seen that the shredded tires possess a unit weight approximately one-third of the unit weight of the soils; thus, it can be used as lightweight fill material. The shear strength is high and hydraulic conductivity is high; however, the high compressibility of tire shreds is a concern in applications such as embankment fill material, where settlement occurs under traffic loading. These properties are used to find various applications for shredded tires.

23.11.5   Environmental Concerns

The principal concern in the use of tire chips either whole or shredded is the quality of the leachate produced when water comes in contact with the tire chips. Table 23.15(*a*) and 23.15(*b*) show the results of eight studies on leaching from new and old scrap tires. These studies reveal that the concentration of contaminants that were leached in most cases satisfied TCLP and EP regulatory levels. Only a few tests show that the concentrations of some specific pollutants were exceeded.

Table 23.14 Physical and engineering properties of shredded scrap tires

Property		Units	Minimum	Maximum	Mean
Unit weight	γ	lb/ft^3	15.3	53	36.3
Hydraulic conductivity	K	cm/s	0.01	59.3	6.8
Shear strength	C	lb/ft^2	0	818	255
	ϕ	deg	14	85	33.7
Compressibility		%	18	65	37.3
Interface shear strength with soils	C_a	lb/ft^2	0	43.8	17.5
	δ	deg	33	39	35.8
Interface shear strength with geotextile	C_a	lb/ft^2	0	0	0
	δ	deg	30	34	32
Interface shear strength with smooth geomembrane	C_a	lb/ft^2	6.5	12	9.8
	δ	deg	15	21	18
Interface shear strength with textured geomembrane	C_a	lb/ft^2	11	21.5	15.2
	δ	deg	30	35	33

Source: Reddy and Marella (2001).

Table 23.15(a) **Studies on leaching from new and scrap tires**

Research	Reference	Type of Tire	Type of Study	Variables	Leachate Analysis	Results
Tire chip evaluation permeability and leachability assessments	Waste Management of Pennsylvania (1989)	Tire chips	Column testing with leachate	Temp. 23°C, 50°C; analysis at 0, 30, 60, 90 days	pH, sulfides, cyanides, EPA tox-metals	No appreciable change in concentration in 90 days.
TCLP (toxicity characteristic leaching procedure) assessment project	Rubber Manufacturers Association (RMA, 1990)	Shredded tires (new, old)	Batch tests	Leaching sol. pH 4.9, pH 2.9 cured, uncured ground, unground	Volatiles, semivolatiles, metal ions	None of the samples exceeded TC_P levels.
Environmental study of the use of shredded waste tires for roadway subgrade support	Minnesota Pollution Control Agency (1990)	Shredded tires (new, old)	Laboratory and field test	pH 3.5, 5, 7, 8	14 metal ions, hydrocarbons	Drinking water standards were exceeded under worst-case conditions, but EP-toxicity limits were not exceeded. Field tests were not reliable.
Evaluation of the potential toxicity of automobile tires in the aquatic environment	Environment Canada National Water Institute (1992)	Whole tires (breakwater, old, new)	Effect of tire leachate on fish and *Daphnia*	Time: 5, 10, 20, 40 days	Toxicity analysis on living organism	Substances contributing to toxicity are water soluble and persistent.
Development of engineering criteria for shredded waste tires in highway applications	Edil and Bosscher (1992)	Shredded tires (old)	Lab (AFS leach and EP-toxicity) test, field samples (lysimeter under embankment fill)	Samples taken 10 times in 2 years in the field	Metal ions, water quality index tests (pH, alkalinity, BOD, COD, etc.)	Slight alkaline conditions but no specific health concern. Field samples were affected from other sources. Laboratory samples leached Ba, Fe, Mn, and Zn, but leachate showed no likelihood of hazardous levels.
Study of waste tire leachability in potential disposal and usage environments	Miller and Chadik (1993)	Shredded tires (old)	Lab (batch) test, field samples (septic tank drain fill field)	Shred size; pH 5.4, 7.0, 8.6; time	Semivolatiles, volatiles, metal ions	Metal ions, VOCs (benzene, toluene, and semivolatiles leached. Biological activity may have affected results.
Water quality testing for Dingley Road tire chip test project	Humphrey and Katz (1995)	Shredded tires (old)	Field samples (monitoring wells near tire chip embankment)	15- and 30-cm-thick tire chip layer above groundwater table	Metal ions, water quality index tests (pH, alkalinity, BOD, COD, etc.)	All substances (except Mn) had concentrations below drinking water standards.
Analysis of leachate from column tests of soil–tire chip and soil	Kim (1995)	Shredded tires (old)	Column tests	Soil–tire chips (830 days); soil (790 days)	Metal ions Zn, Pb, Ba, As, Se	No sample exceeded drinking water limits in terms of metal concentration.

Source: Radian (1989).

Table 23.15(b) **Compounds in tire chip leachates**

Compound	Concentration Limits Based on Tire Chip Mass			Reported Concentrations Based on Tire Chip Mass				
	TCLP[a] (mg/kg)	USEPA's RMCLs[b] (mg/kg)	WI PAL s[b] (mg/kg)	Grefe[b] (1989) (mg/kg)	RMA[c] (1990) (mg/kg)	MPC (1990) (mg/kg)	Miller and Chadik[d] (1993) (mg/kg)	Kim[e] (1995) (mg/kg)
As	25.9	0.06	6.15×10^{-3}	—	—	—	0.02	—
Ba	528.0	2.27	0.25	0.55	0.1	1.08	—	0.37
Cd	5.18	6.15×10^{-3}	1.23×10^{-3}	—	—	0.27	—	—
Cr	25.9	0.15	6.15×10^{-3}	—	0.008	0.51	—	0.019
Pb	25.9	0.025	6.15×10^{-3}	0.075	0.003	0.92	—	0.14
Fe	—	0.37	0.18	1.15	—	1081	—	—
Mn	—	0.06	0.03	1.5	—		—	—
Zn	—	6.15	3.1	3.15	—	50.3	5.02	1.13
Se	5.18	0.055	1.23×10^{-3}	—	—	0.44	—	0.05
Hg	1.03	3.7×10^{-3}	2.5×10^{-3}	—	7.2×10^{-5}	—	—	—
NO$_2$-NO$_3$	—	1.23	2.46	1.85	—	—	—	—
Toluene	333.79	2.46	84.3×10^{-3}	—	0.034	—	0.28	—
Carbon disulfide	332.64	—	—	—	0.012	—	—	—
Phenol	332.64	—	1.23×10^{-3}	—	0.01	—	—	—
Benzene	0.36	6.15×10^{-3}	0.08×10^{-3}	—	—	—	0.63	—

Source: Radian (1989).

[a] TCLPs are converted to mass of compound per kilogram of tire chips assuming that 100 g of tire chips with a specific gravity of 1.22 is used. (Volume of extractor = 600 mL, mass of solid = 100 g of inorganics and 25 g for volatile organics).

[b] RMCLs and WI PALs of 1.22. (1 mg/L = 1.23 mg/kg of tire chips.)

[c] The liquid concentrations are converted to mass of compound per kilogram of tire chips based on the given quantities of tire chips and liquid used in the study.

[d] Concentrations for compounds are taken from approximately 16-h readings, which are usually the highest concentrations achieved during the test. The liquid concentrations are converted to mass of compound per kilogram of tire chips based on the given quantities of tire chips and liquid used in the study.

[e] Concentrations for compounds are the final concentrations reached in 800 days. The liquid concentrations are converted to mass of compound per kilogram of tire chips based on the given quantities of tire chips and liquid used in the study.

However, it cannot be said conclusively that scrap tires do not leach hazardous contaminants into the environment.

The other important concern is the internal heating of tire-shred fills, which were reported to have experienced exothermic reaction at three locations; SR 100 in Ilwaco, Washington (the fill was used to repair a landslide in the SR 100 loop road); Falling Springs road in Garfield County, Washington; and a retaining wall alongside I-70 in Glenwood Canyon, Colorado (Humphrey, 1996). Although none of the incidences were catastrophic, internal heating is a cause of concern. As the number of such tire-shred fills increase, internal heating will have to be addressed firmly, so that there will be no danger of local or regional pollution due to such fills. Some of the potential causes of tire-shred fills can be as follows (Humphrey, 1996):

- Oxidation of the exposed steel wires
- Microbial activity that produces sulfur and organic acids that aid the oxidation process
- Oxidation of rubber, especially of crumb rubber
- Petroleum-consuming bacteria that oxidize the liquid hydrocarbons in an exothermic reaction

The Federal Highway Administration (FHWA) has published certain guidelines for the use of tire chips in fills to minimize the problem of internal heating.

23.11.6 Economic Considerations

The cost of shredding the tires into chips costs between $30 and $65 per ton of tires. Hence, in any application that uses shredded tires, this cost has to be considered. As a lightweight fill material, tire chips are economical, with cost only $9.5 per cubic yard, compared to traditional materials like shale at $45 per cubic yard (Manion and Humphrey, 1992). Cecich et al., 1996 studied the effectiveness of scrap tire shreds as a lightweight retaining wall backfill and found that on average a 60% cost savings can be achieved by using shredded tires. In one of the configurations that was studied, a 30-ft-high, 100-ft-long wall saved $97,500 by the use of scrap tire shreds.

23.11.7 Field Applications and Demonstrations

Several projects have been initiated recently to document the performance of scrap tires in various engineering applications. Some of these are explained below. A study of an experimental shredded tire landfill near Williamsburg, Virginia, showed promising results (Hoppe, 1994). The project used 1.7 million shredded tires mixed with conventional fill material. The study results indicate that settlements within the embankments are larger than usual in an embankment. Stability, however, is not a concern for the embankments, as they have remained for the four years after construction. Infrared scanning revealed no potential sources of an exothermic reaction.

A full-scale 16-ft-high retaining wall test facility was constructed to investigate the properties of tire chips as a backfill material. The at-rest horizontal stress for the project above was 45% less than that of a typical granular fill (Tweedie et al., 1998). Tire chips have also demonstrated the ability to reduce the lateral earth pressure against the abutments of a concrete rigid-frame bridge. The tire chips compressed horizontally as the approach fills were placed; this allowed the fill to strain and mobilize its strength and achieved active pressure conditions, reducing significantly the active pressures against the abutment.

A project was undertaken to ascertain the best type of material for the insulation of a landfill side slope liner. In cold regions the side liners get exposed to frost and can get severely damaged. The materials tested were leachate collection sand, sand covered with waste tire chips, polyurea foam geoinsulation and encapsulated fiberglass geoinsulation. The results of the study indicated that the sand covered with waste tire chips performed best. This is an insulation application in which scrap tires have performed well (Benson et al., 1996).

In cold climates, ice lenses formed in the ground during the winter begin to melt during spring thaw. The lenses that are on the top melt first, and since the ground below is frozen, these meltwaters have nowhere to go but toward the pavement surface on the top as even the side drains are blocked; this saturates and weakens the ground surface and causes rutting of the pavement. To overcome this problem, scrap tires were investigated as a potential solution. At Dingley Road in Richmond, Maine, five test sections, one control section, and two transition sections were constructed. Two sections of the existing road were also used as controls. Two different types of tire chip thicknesses, 6 and 12 in., and three granular soil layers, 12, 18, and 24 in., were used. For the first two winters, 6 in. of tire chip overlain by 12 in. of gravel reduced the frost penetration by 22 to 28%; 12-in. tire chips with 18- or 24-in. gravel reduced the frost penetration depth by 15 to 17% in the 1992–1993 season and by 33 to 37% in the in the 1993–1994 season (Humphrey and Eaton, 1994). These results show that tire chips are effective in controlling the depth of frost penetration.

Another project in Oregon involved the use of tire shreds as lightweight fill used in the construction of an embankment used to repair a landslide. Results indicated that the pavement constructed on top of the tire chips meets the 20-year design life criteria, although the deflection was found to be more than what would be expected from conventional earthen embankments. The cost of the project would have been higher than $26.91 per cubic yard, and other materials would have been suitable, but the project was subsidized in part by the Department of Environmental Quality, as this project involved the beneficial use of tires (Upton and Machan, 1993).

23.11.8   Summary

The problem of scrap tire disposal is considerable. Currently, large quantities of scrap tires are consumed as tire-derived fuel (TDF), but stringent emission control requirement is expected to limit the use of TDF in the future. The civil engineering market is growing and shows promise for engineering applications of tire chips.

23.12   DEMOLITION DEBRIS AND RECYCLED CONCRETE

23.12.1   Sources and Quantities Generated

Demolition debris and recycled concrete are generated due to construction demolition of structures and main-

tenance on roads, bridges, and buildings. These waste materials consist of lumber, bricks, wallboard, gypsum, soil, plastic, paper, glass concrete, concrete blocks, steel, and asphalt. Concrete waste material is about 75% by weight of all construction material. Millions of tons are generated every year (Canada, 1999).

23.12.2   Problems with Disposal

It is not prudent to dispose this type of waste in MSW landfills because of the scarcity of landfill space and the need to dispose of other wastes (Turley, 1999). However, proper disposal must be considered and followed strictly with special construction and demolition landfills. This is easier to accomplish, since siting landfills of this kind is less difficult than siting of MSW landfills. The reason for this is that demolition materials of the type are inert; therefore, the environmental impact is minimal. According to solid waste management legislation in 1989–1991, waste should be reduced by 40% by year 2001 by reuse of material and recycling. Disposing fees increased significantly after stricter regulations were imposed on landfill operations. Contractors were forced to separate material into recyclable and nonrecyclable, as effective separation is a prerequisite to maximizing the benefits of recycling.

23.12.3   Potential Reuses

Recycled concrete can be used as a raw material in cement clinker, as admixtures in cement, or as aggregates in concrete. It can also be used as aggregate base, subbase for new pavements, shoulders, base foundation, or backfill for utility trenches. Demolition materials can be recycled into new materials; scrap lumber can be processed and used in landscaping, boiler fuel, and building products. Metals can be sold in scrap metal yards; cardboard, gypsum drywall, soil, and rubble can be used as aggregate; glass recycles into fiberglass; and asphalt recycles into asphalt paving and pothole repair.

23.12.4 Physical and Chemical Properties

Asphalt, concrete, dirt, brick, and other rubble are for the most part inert material unless contamination from an outside source is introduced at the construction or demolition site. Compressive strength and modulus of elasticity of concrete containing recycled concrete are lower than those of regular concrete. High water/cement ratios are a concern; however, water-reducing agents and higher cement contents seem to correct this problem, yielding higher strengths. Shrinkage of recycled concrete is greater by 10 to 30%. However, further studies are needed to develop a more concise attitude toward recycled concrete and its properties (Canada, 1999). To enhance the quality and durability of the recycled mix, bitumen or cement addition often takes place after the material is placed on site; then liquid is injected into the subbase. This promotes stabilization and durability of the material.

23.12.5 Environmental Concerns

Environmental concerns have pushed the industry to develop techniques for demolition of wastes. They are based on the concept of prevention, reuse, and recycling of waste. Inert waste can be used as fill material or road base. Another alternative is selective demolition. This involves dismantling of selected parts of buildings to be demolished before the wrecking process is initiated in order to use them at a later phase of the project. Other environmental-friendly methods are on-site separation systems using multiple smaller containers in place of a single roll-off or compactor. Crushing, milling, and reuse of secondary stone and concrete materials are often practiced as well.

In 1993, federal and state standards were adopted for solid waste landfills. Accordingly, construction and demolition debris should be separated into recyclable and nonrecyclable materials. Demolition debris and concrete may no longer be landfilled. Inert debris should be recycled, and volume reduction is required before disposing to a landfill.

One of the important issues of environmental concern of recycling operations are the machinery air emission regulations, permit of operations, and restriction of hazardous materials of reuse and recycle. The EPA has established regulations for such material in order to minimize or avoid harm to human health or to the environment (see Chapters 2 and 15).

23.12.6 Economic Considerations

Economic benefits of recycled concrete and demolition debris are numerous. However, the demolition industry is not used to recycling on an organized level. Local governments and other purchasers can save money by using recycled material. Business opportunities can be created. Energy can be saved when recycling is done on site. Conserving diminishing resources of urban aggregates helped meet the state of California's goal of reducing disposal by 50% by year 2000. Primary markets for demolition and construction material are concrete–asphalt subbase, base, and road projects (State of California, 1999).

Furthermore, other economic considerations are to save space (landfill space), transportation costs, and disposal fees, which are $30 to $45 per ton. According to a U.S. Navy report, it is estimated that a 7 to 20% savings on material cost can be obtained if organized recycling is utilized (U.S. Navy, 1999).

23.12.7 Field Applications and Demonstrations

According to California's recycling program, 28% of the waste stream or 11 million tons is generated from demolition and construction development (State of California, 1999). From that, 8.2 million tons is asphalt and concrete "inert wastes," and due to organized recycling, a 51% recycling rate was achieved by the state. In 1995, Los Angeles passed a motion which required that in all city projects, road base including crushed miscellaneous base (CMB) be 100% recycled concrete–asphalt. Concrete brick can be used in various size fractions as gravel substitutes in concrete and mortar. Demonstration of the feasibility of using recycled material was shown at a Nashville, Tennessee, football stadium. The use of recycled concrete and demolition debris used as subbase and fill saved the project $222,661 (Block, 1999).

To take advantage of recycling prior to demolition, the following may be utilized: Promote the use of standard material sizes and reduce packaging. Leftover masonry material can be crushed on site and used for fill, and joist off cuts can be cut up on site and used as stakes or for forming. Salvageable material can be forwarded to businesses that collect and resell material of that nature.

23.12.8   Summary

Construction and demolition wastes have been difficult to deal with. Due to the large volume of the waste and high disposing costs, industry and government have promulgated several alternatives. Due to the rising cost of disposal, contractors were forced to reconsider landfilling and to try to find other options. Separation and reuse, or recycling, have been utilized over the years. Currently, industry is able to recycle almost half of demolition and construction wastes. Building materials and most asphalt and concrete can be recycled. Recycling materials from demolition and renovation projects, such as rebar, asphalt, and concrete, and selling salvaged lumber and fixtures are more economical and can result in a lower bid price for a project. Using recycled materials can result in less expensive projects for local governments, free up space for landfills, and preserve virgin materials for future use.

23.13   WOOD WASTES

23.13.1   Sources and Quantities Generated

Wood and wood fiber actually represent about half of all MSW waste by weight, making it the largest single component in MSW landfills. Wood and wood fiber can be found in the form of discarded paper, wood product waste, and yard waste. A 1994 statistic shows that about 37 million tons of paper and wood materials were recovered for recycling. In addition, other studies showed that other potential quantities of recyclable wood materials exist from three main sources, including MSW, new construction and demolition (C&D), and wood residues from timber-processing facilities. It

was found that about 99% of the material recovered was from paper and only about 1% from recycled solid wood, and about 95% of this was used to produce recycled paper products (Ince, 1995, 1996). Recycled solid wood is increasing, however, as a source for recyclable wood materials, and new innovative products are being manufactured from these materials. Of the various sources and types of wood-based waste, recycled solid wood from C&D projects and wood residues from lumber processing are discussed in this section.

23.13.2   Problems with Disposal

Recycling of wood-based wastes is becoming a necessity because disposal in landfills is no longer the best solution. A shortage of landfill space is making it very difficult to keep up the current rate of dumping, including of wood-based waste, which currently makes up the biggest portion of all waste dumped in MSW landfills. The main problem with disposing of wood waste has to do with the sheer quantities that are generated. It is getting increasingly more convenient to recycle these wastes than to landfill them, due to the shortage of landfill space and the high supply of refuse wood products. Landfill disposal costs are increasing steadily, and new restrictions are also being imposed, making recycling a very appealing option.

23.13.3   Potential Reuses

When broken down into smaller parts (chips, pellets, fibers, etc.), waste wood products can be recycled successfully into various different end products of some economic value. Wood fibers have been used to produce various structural products. *Fiberboard* products can be used for various purposes, including subflooring, light sheathing, and structures for damping noise and insulating. Wood fibers have been mixed with other materials, such as portland cement, to make more *rigid boards,* which could be used as fire barriers, sheet roofing systems, or subflooring for ceramic tiles. Wood fibers can also be used to make fiber strips which fill expansion joints between sections of concrete in roadway and walkway construction. Wood residues have also been mixed with plastic fibers to create a composite material called *plastic lumber*. An advantage of this type of lumber is its water resistance. Plastic lumber can be used in place of treated lumber in the construction of waterfront docks, piers, and decking. Plastic panels can also be produced in a similar fashion and can be used as sheathing and subflooring. Other boards, called *oriented strand boards*, can be manufactured using wood fibers on the outside and a thick inner layer of polystyrene foam. This type of sheathing has proven to have very good strength and is being used extensively in residential construction. One very innovative use of wood fiber in construction was in a permanent type of wall system made up of a composite of reinforced concrete and wood fiber. This system, which contains up to 90% wood fiber, has proven to provide good insulation and strength properties (Stafford-Harris, 1995). Another option for wood particles is to use them as decorative wood chips, mulch, or other ground cover products, including wood-based geotextiles. Bio-based geotextiles from recycled wood can be very useful in civil engineering projects. Biodegradable geotextile provides an environmentally friendly option to control erosion and allows soil to better filter and absorb water. Wood-based sheet mulches can prove useful in civil engineering projects by helping seeds take root and increase the chance of survival. Wood-based geotextiles and sheet mulches could be especially useful in helping provide percolation and vegetative layers in landfill designs (Ince, 95).

23.13.4   Physical and Chemical Properties

Wood chips, mulch, and other ground cover products can be manufactured fairly easily from almost any source of wood debris; the wood just has to be separated from the other waste. Higher-market recycled wood products, such as the structural materials mentioned earlier, must, however, be processed more precisely. Wood waste from mills and lumberyards is usually very clean and easy to process, but wood recovered from construction and demolition projects and from MSW are much more difficult to process. For higher-value materials to be made successfully

from processed wood particles, the material must be clean and uniform. This means that all contaminants must be removed, and different wood types or species must be separated and then chopped or ground into particles of uniform size to fit certain particle geometry requirements (Glenn, 1997). Depending on the exact end product being manufactured from the recycled wood particles, a certain consistency and grade of wood particles may be needed. Chemically treated wood may cause an even greater problem and could be very difficult to process. The main chemical used in treated woods is pentachlorophenol.

23.13.5   Environmental Concerns

The use of recycled wood products not only helps decrease waste volumes dumped in landfills, but also reduces environmental concerns by reducing the number of trees being logged in our forests. Presently, there is great public outcry for an end to cutting trees in old-growth forests and to other wood harvesting on public lands. With the mass cutting of trees to supply the construction industry, ecosystems are altered negatively and the environment suffers. By encouraging the recycling of wood products, forest disruption can be minimized and the environment will benefit greatly. In addition to deforestation, another environmental concern that exists in the manufacture and disposal of wood products is the disposal of chemically or preservative-treated woods. Although the chemical content of preservative-treated woods currently falls within EPA standards, there is definite public concern about disposing of this type of wood in landfills. Studies are currently being done to find possible options for the recycling and reuse of this type of wood, which creates environmental concern when disposed of in landfills (Ince, 1995). Wood products are currently being treated with pentachlorophenol (PCP) to increase durability. This chemical can hinder the process of recycling wood into structural products and is therefore usually not accepted for this type of recycling. PCP-treated wood is even more harmful when used in the production of mulch and geotextiles because of the leachate created as water percolates through them.

PCP-treated wood is generally avoided for recycling purposes; instead, efforts are being made to improve technology in biodegrading the material by chipping it and introducing a fungus that helps break down the material into a fully remediated soil.

23.13.6 Economic Considerations

Waste wood processing into a recyclable material along with actually manufacturing end products from this material can prove to be a successful business. The supply of waste wood is definitely plentiful, as stated earlier, and many new products are currently being manufactured from the processed waste wood. As the technology develops for processing the collected waste wood supply and creating new products from the processed material, operating costs will be reduced and the wood recycling industry should flourish. Waste wood is already very cheap, since it is cheaper to dispose of at recycling plants than at landfills, and it is abundant, so all that is needed for the success of this industry is more companies to develop wood recycling technology.

23.13.7 Field Applications and Demonstrations

Many waste wood processing plants already exist and are being run successfully. One such plant, WPAR in West Paterson, New Jersey, began processing wood in 1985 and has steadily increased production over the years. The company accepts clean loads of wood for $30 to $35 per ton, much less than local refuse transfer fees, about $80 per ton. This company has proven to be successful because they search for many outlets for the wood instead of focusing on just one end user. They also get the wood from a variety of sources, including building contractors, landscapers, municipali-ties, refuse haulers, land clearing operations, and even local residents. The largest market for WPAR's processed wood is decorative mulch, but their premium-grade wood products are sold to higher-value markets, including particle and fiber board manufacturers for new products such as mixed wood/plastic composites. Their local customers also purchase reusable lumber that needs no major processing at all (Steuteville, 1997).

23.13.8 Summary

The potential for additional solid wood recovery for recycling from MSW, new construction and demolition wastes, and timber processing facilities was estimated at 39 million tons annually based on 1993 data presented by Ince (1995). With the current landfill problem, the rapid cutting of trees, and the demand for woody material, this potential for waste wood recovery must be explored further. An increase in waste wood processing and recycling technology is definitely in order to fully utilize today's huge volume of waste wood.

23.14 SUMMARY

Every year, hundreds of millions of tons of municipal solid waste (MSW) is generated in the United States. A large majority of this waste ends up into landfills. With the limited amount of space available in existing landfills and the difficulty in finding new areas to construct new landfills, the best way to make use of what is still available is to find alternatives for disposal of MSW, which then results in landfill space being better utilized. Recycling is one alternative to disposal that has shown a great potential for many materials currently being landfilled.

QUESTIONS/PROBLEMS

23.1. Explain the major difference between the class F and class C fly ash.

23.2. Grain size distribution analysis was performed on sewage sludge ash obtained from four different incinerators, with the results (as percent fines) shown below. Plot the grain size distribution and classify these ashes according to the Unified Soil Classification System (USCS). Discuss how the differences in grain size distribution will potentially affect their beneficial reuse.

Sieve No.	Ash A	Ash B	Ash C	Ash D
4	99	100	100	100
8	99	98	100	100
10	—	—	100	100
20	—	—	100	—
40	99	73	98	—
80	—	—	83	—
100	85	53	—	—
200	66	38	56	47

23.3. Sieve analysis on a glass cullet yielded the following results. Plot the grain size distribution and classify the material according to the USCS. Which soil closely represents this material? Discuss potential uses of this material.

Sieve Size	Percent Finer
1 in.	100
$\frac{1}{2}$ in.	98.7
$\frac{1}{4}$ in.	86.0
$\frac{1}{8}$ in.	32.6
No. 20	6.4
No. 40	3.2
No. 80	1.5
No. 200	0.6

23.4. Identify sources of four waste materials, discuss their environmental concerns, and identify their potential reuses.

REFERENCES

Ahmed, I., *Laboratory Study on the Properties of Rubber Soils*, FHWA/IN/JHRP-93/4, Purdue University, West Lafayette, IN, 1993.

Ahmed, I., and Lowell, C. W., *Rubber Soils as Lightweight Geomaterials*, Transportation Research Record 1442, Transportation Research Board, Washington, DC, 1993.

Anon., Modern Plastics, Vol. 73, McGraw-Hill, NY, 1996.

———, States employ new approaches to waste reduction, recycling, *J. Air and Waste Manag. Assoc.*, Vol. 41, p. 1426, 1991.

———, Coal combustion product is good for highway use, *Civ. Eng. News*, Dec. 1997.

Bagchi, A., *Design, Construction, and Monitoring of Landfills*, Wiley, New York, 1994.

Benson, C. H., Olson, M. A., and Bergstrom, W. R., Temperatures of an insulated landfill, in *Proceedings of the 75th Annual Meeting of the Transportation Research Board*, Washington, DC, 1996.

Berthouex, P. M., and Gan, D. R., Loss of PCBs from municipal-sludge-treated farmland, *J. Environ. Eng.*, Vol. 117, No. 1, pp. 5–24, 1991.

Block, D., Processing C&D debris for markets, *BioCycle,* Oct. 1998; *www.//medusa.prod.oclc.org.,* Mar. 20, 1999.

Bosscher, P. J., Tuncer, E. B., and Eldin, N. N., Construction and performance of a shredded waste embankment, *Transp. Res. Rec.* No. 1345, 1994.

Bressette, T., *Used Tire Material as an Alternative Permeable Aggregate,* FHWA/CA/TL- 84/07, Office of Transportation Laboratory, California Department of Transportation, Sacramento, CA, 1984.

CALTRANS, *The Use of Discarded Tires in Highway Maintenance,* Translab Information Brochure TL/REC/1/88, California Department of Transportation, Sacramento, CA, 1988.

Canada, *Waste and By-products as Concrete Aggregates,* CBD-215, *www.ncr.ca/irc/cdb215,* 1999.

Cecich, V., Gonzales, L., Hoisaeter, A., Williams, J., and Reddy, K., Use of Shredded Tires as Lightweight Backfill Material for Retaining Structures, *Waste Management & Research,* Vol. 14, No. 5, pp. 433–451, 1996.

CIWMB, California Integrated Waste Management Board, *Effects of Waste Tires, Waste Tire Facilities and Waste Tire Projects on the Environment,* Publication 432-96-029, CIWMB, Sacramento, CA, 1996.

Clark, C., Meardon, K., and Russel, D., *Scrap Tire Technology and Markets,* Office of Solid Waste, USEPA, Washington, DC, 1993.

CWC (Clean Washington Center), *Glass Feedstock Evaluation Project: Using Glass as a Construction Aggregate,* GL-93-01, CWC, Seattle, WA, 1993.

———, *King County Glasphalt Demonstration Project,* Fact Sheet, GL-03-09, CWC, Seattle, WA, Dec. 1994.

———, *Recycled Glass in Asphalt,* Fact Sheet, GL-4-02-01, CWC, Seattle, WA, Nov. 1996.

———, *A Toolkit for the Use of Post-consumer Glass as a Construction Aggregate,* GL-97-5, CWC, Seattle, WA, Jan. 1998.

DeGroot, D. J., and Switzenbaum, M. S., *Use of Recycled Materials and Recycled Products in Highway Construction,* Final Report to the Executive Office of Transportation and Construction and the Massachusetts Highway Department, Boston, 1995.

DiGioia, A. M., Jr., and McLaren, R. J., *The Typical Engineering Properties of Fly Ash,* ASCE, Reston, VA, 1987.

DiGioia, A. M., Jr., and Nuzzo, W. L., Fly ash as structural fill, *J. Power Div.,* ASCE, Vol. 98, No. 1, pp. 77–92, 1972.

Drescher, A., and Newcomb, D. E., *Development of Design Guidelines for the Use of Shredded Tires as a Lightweight Fill in Road Subgrade and Retaining Walls,* MN/RC-94/04, Department of Civil and Mineral Engineering, University of Minnesota, Minneapolis, MN, 1994.

Edil, T. B., and Bosscher, P. J., *Development of Engineering Criteria for Shredded or Whole Tires in Highway Applications,* WI 14-92, Department of Civil and Environmental Engineering, University of Wisconsin, Madison, WI, 1992.

Faber, J. H., and Babcock, A. W., 50 years of ash marketing and utilization in the USA, 1936–1986, presented at the CSIR Conference Center, Feb. 2–6, 1987.

Feagley, S. E., Valdez, M. S., and Hudnall, W. H., Bleached, primary papermill sludge effect on Bermudagrass grown on a mine soil, *Soil Sci.,* Vol. 157, No. 6, pp. 389–397, 1994.

FHWA (Federal Highway Administration), *Use of Fly Ash in Portland Cement Concrete and Stabilized Base Construction,* N 5080.4, FHWA, Washington, DC, Jan. 17, 1974.

———, *User Guidelines for Waste and Byproduct Materials in Pavement Construction,* FHWA-RD-97-148, FHWA, Washington, DC, 1997.

Fuller, W. H., and Warrick, A. W., *Soils in Waste Treatment and Utilization,* Vol. I, *Land Treatment,* CRC Press, Boca Raton, FL, 1985.

Glenn, J., Processing woody materials for higher value markets, *Biocycle,* Vol. 38, pp. 30–33, Feb. 1997.

GLSEC (Great Lakes Soil and Environmental Consultants, Inc.), *Geotechnical Characterization of Biosolids,* report submitted to the Metropolitan Water Reclamation District of Greater Chicago, 2002.

Hall, T. J., *Reuse of Shredded Waste Tire Material for Leachate Collection Systems at Municipal Waste Landfills,* report for the Iowa Department of Natural Resources Waste Management and Authority Division, prepared by Shive-Hattey Engineers and Architects Inc., 1990.

Han, C. A., *Waste Products in Highway Construction*, prepared for the Minnesota Local Road Research Board, St. Paul, MN, Apr. 1993.

Harper, C., *Handbook of Plastics, Elastomers, and Composites*, 2nd ed., McGraw-Hill, New York, 1992.

Hendrickson, C., Lave, L., and McMichael, F., Reconsider recycling, *Chemtech*, Vol. 25, pp. 56–60, 1995.

Hennepin County, *Glass Aggregate Test Strip Construction Project*, County Road 47, Hennepin County, MN, Apr. 1997.

Holmes, A. J., In depth recycling, A Cost Effective Alternative to Traditional Pavement Construction, *Munic. Eng.*, Institute of Civil Engineers, Vol. 7, No. 4, pp. 185–192, 1990.

Hoppe, E. J., *Field Study of Shredded-Tire Embankment*, FHWA/VA-94-IR1, Virginia Department of Transportation, Richmond, VA, 1994.

Horner, A. J., Use of recycled tire rubber in the Carsonite noise barrier, in *Proceedings of the Symposium on Recovery and Effective Reuse of Discarded Materials and By-Products for Construction of Highway Facilities*, Denver, CO, 1993.

Humphrey, D. N., *Investigation of Exothermic Reaction in Tire Shred Fills Located on SR 100 in Ilwaco, Washington*, prepared for the Federal Highway Administration, Washington, DC, 1996.

Humphrey, D. N., and Eaton, R. A., Tire chips as subgrade insulation: field trial, in *Proceedings of the Symposium on Recovery and Effective Reuse of Discarded Materials and By-Products for Construction of Highway Facilities*, Denver, CO, 1993.

———, Performance of tire shreds as insulating layer beneath gravel surface roads, in *Proceedings of the 4th International Symposium on Cold Region Development*, Espoo, Finland, 1994.

———, Field performance of tire shreds as subgrade insulation for rural roads in *Proceedings of the 6th International Conference on Low Volume Roads*, Vol. 2, Transportation Research Board, FHWA, Washington, DC, 1995.

Humphrey, D. N., and Sandford, T. C., Tire chips as lightweight subgrade fill and retaining wall backfill, in *Proceedings of the Symposium on Recovery and Effective Reuse of Discarded Materials and By-Products for Construction of Highway Facilities*, Denver, CO, 1993.

Humphrey D. N., Sandford, T. C., Cribbs, M. M., Ghareghat, H., and Manion, W. P., *Tire Chips for Lighweight Backfill for Retaining Walls, Phase 1*, a study for the New England Transportation Consortium, Department of Civil Engineering, University of Orono, ME, 1992.

Humphrey, D. N., Cosgrove, T., Whetten, N. L., and Herbert, R., Tire chips reduce lateral earth pressure against the walls of a rigid frame bridge, presented at the ASCE technical seminar, Renewal Rehabilation and Upgrades in Civil and Environmental Engineering, Maine, 1997.

Ince, P. J., *Recovery of Paper and Wood for Recycling: Actual and Potential*, Forest Products Laboratory, U.S. Forest Service, USDA, Washington, DC, 1995.

———, *Recycling of Wood and Paper Products in the U.S.*, Forest Products Laboratory, U.S. Forest Service, USDA, Washington, DC, 1996.

Kirby, B., Secondary markets for post-consumer glass, *Resour. Recycl.*, June 1993.

Kost, D. A., Boutelle, D. A., Larson, M. M., Smith, W. D., and Vimmerstedt, J. P., Papermill sludge amendments, tree protection, and tree establishment on an abandoned coal minesoil, *J. Environ. Qual.*, Vol. 26, No. 5, pp. 1409–1416, 1997.

Leonards, G. A., and Bailey, B., Pulverized coal ash as structural fill, *J. Geotech. Eng.*, Vol. 108, No. GT4, pp. 517–531, 1982.

Maher, M. H., Bultziger, J. M., and Chae, Y. S., Utilization of municipal solid waste ash as a construction material, p. 25 in *Proceedings of the Mediterranean Conference on Environmental Geotechnology*, Cesme, Turkey, 1992.

Malloy, M. G., Seeing the light in Baltimore, *Waste Age*, Mar. 1997.

Manion, W. P., and Humphrey, D. N., *Use of Tire Shreds as Lightweight and Conventional Embankment Fill*, Technical Paper 91-1, Technical Services Division, Maine Department of Transportation, Augusta, ME, 1992.

Marella, A., Use of shredded scrap tires as drainage material in landfill cover, M.S. thesis, Department of Civil and Materials Engineering, University of Illinois at Chicago, 2002.

Maupin, G. W., Jr., Virginia's experiment with asphalt rubber concrete. *Transp. Res. Rec.*, No. 1339, 1994.

Mckenna, J. D., Ross, J. M., and Foster, M. J., *Continued Assessment of a High Velocity Fabric Filtration Used to Control Fly Ash Emissions,* EPA/600/2-76/182, USEPA, Washington, DC, 1976.

Miller, C., Profiles in garbage: glass containers, *Waste Age,* Sept. 1998.

Moo-Young, H. K., and Zimmie, T. F., Design of landfill covers using paper mill sludges: research transformed into practice, pp. 14–28 in *Implementation of NSF Research,* Colville, J., and Made, M. (eds.), ASCE, Reston, VA, 1995a.

———, Hydraulic conductivity of paper sludges used for landfill covers, pp. 932–946 in *Proceedings of GeoEnvironment 2000,* Geotechnical Special Publication, No. 46, Vol. 2, Acar, Y. B., and Daniel, D. E. (eds.), ASCE, Reston, VA, 1995b.

———, Utilizing a paper sludge barrier layer in a municipal landfill cover in New York, pp. 125–140 in *Testing Soil Mixed with Waste or Recycled Materials,* ASTM STP 1275, Wasermiller, M. A., and Hoddinott, K. B. (eds.), ASTM, West Conshohocken, PA, 1997.

MWRDGC (Metropolitan Water Reclamation District of Greater Chicago), *Contract Documents for Utilization of Biosolids and Biosolids for Landfill Final Cover from LASMA,* Contract 03-948-11, MWRDGC, Chicago, 2002.

Park, J. K., and Jhung, J. K., Removal of volatile organic compounds emitted during waste water treatment by ground tires, presented at the 66th Annual Water Environment Federation Conference, Anaheim, CA, 1993.

Park, J. K., Kim, J. Y. Edil, T. B., and Madsen, C. D., Retardation of organic compound movement by bentonite slurry cut off wall amended with ground tires, in *Proceedings of the 68th Annual Water Environment Federation Conference,* 1995.

Peacey, J. G., and Davenport, W. G., *The Iron Blast Furnace: Theory and Practice,* Pergamon Press, New York, 1979.

Pietz, R. I., Peterson, J. R., Lue-Hing, C., and Halderson, J. L., Rebuilding Topsoil with Municipal Sewage Sludge Solids, in Proc. American Society of Agricultural Engineers (ASAE), Summer Meeting, Logan, UT, 1978.

Radian, *A Report on the RMA TCLP Assessment Project,* prepared for the RMA, Washington, DC, 1989.

Reddy, K. R., and Marella, A., Properties of different size scrap tire shreds: implications on using as drainage material in landfill cover systems, in *Proceedings of the 17th International Conference on Solid Waste Technology and Management,* Philadelphia, Oct. 2001.

Sikora, M., *Scrap Tire News,* Third Annual Legislative Update, Vol. 5, No. 1, 1991.

Stafford-Harris, Inc., Directory of recycled content building materials, in *Harris Directory,* 2nd ed. commercially available electronic data base, Stafford-Harris, Seattle, WA, 1995.

State of California, Construction and demolition program, *www.ciwmb.ca.gov,* Mar. 29, 1999.

STMC (Scrap Tire Management Council), *Scrap Tire Use/Disposal Study, 1996 Update,* STMC, Washington, DC 1996a.

———, *Controlling mosquitos,* STMC, Washington, DC 1996b.

State of California, Construction and demolition program, *www.ciwmb.ca.gov,* Mar. 29, 1999.

Steuteville, R., Large scale wood processing and marketing, *Biocycle,* Vol. 38, pp. 50–53, Jan. 1997.

Tarricone, P., Recycled roads, *Civ. Eng.,* Vol. 63, pp. 46–49, 1993.

Turgeon, R., Fly ash fills a valley, *Civ. Eng. News,* Dec. 1988.

Turley, W., Recycling C&D in the fight for landfill space, *World Wastes,* Nov. 1998; *www.medusa.prod.oclc.org.* Mar. 25, 1999.

Tweedie, J. J., Humphrey, D. N., and Sandford, T. C., *Tire Chips as Lightweight Backfill for Retaining Walls, Phase II,* a study for the New England Transportation Consortium, Department of Civil and Environmental Engineering, University of Maine, Orno, ME, 1998.

Upton, R. J., and Machan, G., Use of shredded tires for lightweight fills, *Transp. Res. Rec.,* No. 1442, 1993.

USACE (U.S. Army Corps of Engineers), *Use of Waste Materials in Pavement Construction,* EM 1110-3-503, U.S. Department of the Army, Washington, DC, 1999.

USACE, *Evaluation Criteria Guide for Water Pollution, Control and Abatement Programs,* TM5-814-8, U.S. Department of Army, Washington, DC, 1987.

USDOE, (U.S. Department of Energy), *Barriers to the Increased Utilization of Coal Combustion / Desulfurization By-products by Governmental and Commercial Sectors,* Office of Fossil Energy, USDOE, Washington, DC, July 1994.

USEPA (U.S. Environmental Protection Agency), *Municipal Sludge Management: Environmental Factors,* Office of Water Program Operations, USEPA, Washington, DC, 1977.

———, *Municipal Sludge Management: Environmental Factors,* Vol. 6, USEPA, Washington, DC, 1978.

———, *Evaluation Sludge Management Systems,* Office of Water Program Operations, USEPA, Washington, DC, 1980.

———, *Development of Risk Assessment Methodology for Municipal Sludge Incineration,* Office of Health and Environmental Assessment, USEPA, Washington, DC, 1991.

———, *Markets for Recycled Glass,* EPA / 530 / SW-90 / 071A, Office of Research and Development, USEPA, Washington, DC, 1992.

———, *Analysis of the Potential Effects of Toxics on Municipal Solid Waste Management Options,* EPA / 600 / R-95 / 047, Office of Resesarch and Development, USEPA, Washington, DC, 1995.

———, *Characterization of Municipal Solid Waste in the United States, 1996 Update,* EPA / 530 / R-97 / 015, Office of Research and Development, USEPA, Washington, DC, 1997.

U.S. Navy, Construction and demolition material recycling, Aug. 1996; *www.enviro.nfecs.navy.mil.,* Mar. 15, 1999.

Woods, R., Automation, *Waste Age,* Jan. 1994.

24

END USES OF CLOSED LANDFILLS

24.1 INTRODUCTION

All landfills will eventually reach their permit capacity, which means that they will not be able to receive any more waste. When this happens, the landfill must be closed and monitored for over a period of time (generally, 30 years). Landfill closure is generally accomplished by constructing the final cover system as described in Chapter 19. The question, then, is what to do with the closed landfill. There are many options for end uses of closed landfills, including park sites, nature centers, recreation centers, golf courses, and commercial buildings. The actual end use might also be a combination of several different uses.

There are several advantages and disadvantages to the end uses of closed landfills. These advantages and disadvantages vary greatly depending on what the specific end use actually is. The major advantage of landfill end use is improving aesthetics. Since landfills are generally much larger than their neighboring structures, they stick out as an eye sore. The end use of a closed landfill usually includes some sort of landscaping plan and design. The planting of trees, grass, bushes, and so on, helps to make the large landfill look better to the eye and gives the surrounding area a more appealing character. Other advantage of end use includes utilization of closed landfills located in urban areas for recreational, commercial, or industrial purposes because the land in urban areas is scarce and expensive. On the other hand, the main disadvantage of end use if planned at the time of initial landfill design is that the proposed profile and grade of a specific end use might limit the end size of the landfill mound, which in turn might limit the waste capacity. Other disadvantages of closed landfill end use include landfill gas emission, uneven settling or subsidence of waste,

and leachate outflows. Landfill gas consists of methane gases that smell bad and cause a flash fire. Landfill settlements are generally high and they are uneven and difficult to predict, leading to difficulty in the design of foundations for structures on landfills. Leachate outflows may occur at lower elevations of the landfill, leading to a costly remediation effort. All of these issues must be addressed because they might limit what actually will be the end use.

Ideally, the issue of end use of a landfill should be discussed during the initial planning and design period of the landfill. The end use selected greatly affects the end shape and slope of the landfill. However, most landfill end-use issues are discussed after the landfill has been in operation for several years or closed, thus making the design of end use a difficult and challenging task.

In this chapter we describe different end uses of closed landfills, followed by a discussion of engineering issues that must be addressed in planning and designing for these end uses. Finally, a few case studies are described briefly where the closed landfills have been used successfully for a beneficial end use.

24.2 VARIOUS END USES OF CLOSED LANDFILLS

There are several end uses of closed landfills. These end uses can be categorized as (1) methane gas energy generation uses, (2) reopening after closed landfill, (3) recreational uses, and (4) commercial and industrial uses. Although less common, residential development has also occurred at some closed landfills. However, the residential development will be limited due to negative public perception and increased liability exposure regarding the landfills. However, the potential to use closed landfills for the above-mentioned uses is great, and these end uses are described briefly below.

24.2.1 Methane Gas Energy Generation Uses

Landfill gas (consisting primarily of methane and carbon dioxide) is a by-product of the anaerobic decomposition of the organic wastes. Methane is colorless, odorless, and lighter than air, and it can sometimes accumulate to an explosive concentration from 5 to 15% by volume in air (WMI, 1990). To control landfill gas, a collection system consisting of vertical extraction wells, gas pipeline, blower units, and possibly a flare to burn off gases safely is used. These collection wells are placed strategically throughout the landfill site. An extraction well is typically an 8-in.-diameter pipe set in a 36-in.-diameter boring (WMI, 1990). More detailed information on landfill gas collection systems is given in Chapter 20. If the gas production is significant, each extraction well is connected to the main gas line, which carries the gas to a station for a beneficial use. Since methane gas has a heating value of 500 Btu per cubic foot, roughly half that of natural gas, landfill operators can either sell the gas or use it for energy. Figure 24.1 illustrates a methane utilization system. After collecting the gas from the wells, the methane is then pumped through a gas scrubber, then a gas compressor, and finally, to an engine generator.

Examples of Methane Gas-Generated Energy. The Settler's Hill landfill in Geneva, Illinois, operated by Waste Management, Inc. (WMI), produces enough gas to generate electricity, up to 4.3 MW, for 7500 homes and businesses in the city of Geneva, which is sold through a local utility. As an example of the longevity of the supply source, this particular landfill gas recovery facility is expected to be operational for at least 20 years beyond complete closure of the landfill (WMI, 1990). WMI has a similar system in the Bradley landfill in Sun Valley, California, which produces 4.5 million cubic feet of gas per day, which is piped to a Los Angeles electrical generating plant (Rabasca, 1996). The community of Hartford, Vermont, transformed its old landfill into an educational landfill where the gas collected is used to heat the buildings as well as selling the gas for electric power (Gruber et al., 1993).

Examples of Methane Gas Venting Instead of Energy Generation. Mostly for economic reasons, some landfills may vent methane gas rather than take advan-

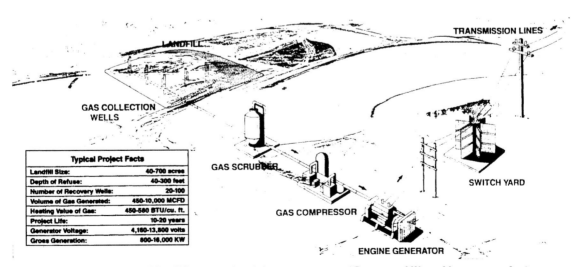

Typical Project Facts	
Landfill Size:	40-700 acres
Depth of Refuse:	40-300 feet
Number of Recovery Wells:	20-100
Volume of Gas Generated:	450-10,000 MCFD
Heating Value of Gas:	450-550 BTU/cu. ft.
Project Life:	10-20 years
Generator Voltage:	4,180-13,800 volts
Gross Generation:	800-16,000 KW

Figure 24.1 *Typical landfill gas-to-electricity power system.* (*Courtesy of Waste Management, Inc.*)

tage of this naturally occurring energy source. Mult-
nomah County, Oregon, built a county maintenance facility on a landfill with an elaborate gas collection
system. As a sanitary landfill, the location made methane gas collection feasible, and a perimeter drain system around the footing was designed as a collection
system for the methane gas generated by decomposing
waste. The drain system is connected to an underfloor
system which is connected, in turn, to a mechanical
exhaust device that sends contaminated air into the atmosphere before it can build up pressure or enter the
building proper (Freece, 1994). Consequently, the
methane gas could have been used to heat the buildings
as well as the gas being sold for electricity production;
the county chose solar heat instead. Similarly, a park
in Fairfield, Connecticut, used PVC pipe and gravel to
displace methane concentrations rather than collecting
and using the methane (Arent, 1989). However, with
tax incentives, a large number of landfills are installing
landfill methane collectors for beneficial uses.

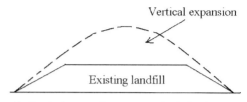

(a) Vertical expansion on top of existing landfill

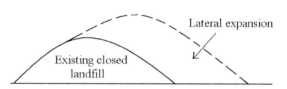

(b) Vertical and lateral expansion of existing
closed landfill

Figure 24.2 *Piggybacking of landfill.*

24.2.2 Reopening after Closure

Once a landfill has reached its capacity, it is closed and
maintained for a certain period of time (generally, 30
years). Today, a new option is to reopen the closed
landfill. The benefit is that a reopened landfill may prevent another landfill from being opened, or at least
until a later time. After closure, a landfill can be reopened in the following two ways: (1) by vertical and/
or lateral expansion (as shown in Figure 24.2), or (2)
landfill mining resulting in additional landfill capacity.

The most common approach to reopening a closed
landfill is "piggybacking," or vertical expansion,
which basically entails putting more waste on top of
an already existing currently capped landfill, as shown
in Figure 24.2. The Blydenburgh landfill in New York
is best known for this method when they reopened an
existing 44.8-acre landfill mound. The project was the
first attempt at putting a new, lined landfill over an
existing one. This expansion provided 1.1 million cubic yards of solid waste disposal capacity. The landfill
also used a new cap when the piggyback portion was

filled. The combination of piggybacking and the cap
saved the agency in charge of the landfill an estimated
$180 million (Hansen and Druback, 1995)

Another approach to reopening of a closed landfill
can be to recover recyclable goods that were disposed
of in the landfill. Before recycling was implemented,
recyclable goods were also disposed of in landfills.
These recyclable products, such as plastics and metals
(e.g., aluminum), don't decompose like organics such
as food waste, so that recyclable material is still intact:
waiting to be recycled. This recycling or reclamation
is basically a process of excavating a landfill using
conventional surface mining technology to recover
metals, glass, plastics, and combustibles. The site can
then be either upgraded into a state-of-the-art landfill,
closed, or redeveloped for some other purpose (Spencer, 1991). The first step, using a bulldozer, is to push
old refuse into a hydraulic excavator. The excavator
operator will sort out large metal items for recycling
and drops the rest into a conveyer to feed a trammel
(Spencer, 1995). Although the method is similar, some

projects have used a series of vibrating screens instead of a trummel to separate recyclables. The first screen separates larger materials such as textiles, scrap metal, tires, and construction waste. The next screen searches for bottles, bottle caps, and so on. In the end, decomposed soil and soil used previously for daily cover is recovered and typically can be reused instead of buying a fresh cover. The remaining material obtained by the screen or the trammel process contains the most potentially recyclable and combustible materials, such as bottles, cans, plastic, paper, and tires. In some landfills, newspapers from years ago were found mostly intact. The recyclable material will, of course, be recycled, but the combustible material will either be put back into the landfill or used as a fuel.

WMI, through recycling a landfill, replaced the new space with a recycling park at the Bradley landfill in California. The facility contains two outdoor manual sorting lines where three streams, commingled containers, residential waste, and commercial waste, are processed. Everything from wood to tires to metal to brick is separated. Less than 5 acres of the 209-acre site is used as a traditional landfill (Rabasca, 1996). Similarly, a landfill in Hartford, Vermont, slated for closure was transformed into an environmental community center through recycling the old landfill. Aside from the everyday operation of recycling, such as sorting material and methane gas collection, this landfill involves the public. The community center includes a recycling building that can handle 30 types of materials, a permanent household hazardous waste facility, an environmental museum and education building, an administration building, a computerized truck scale and scale house, the Good Buy store for reusable goods distribution to the public, yard waste composting, and waste transfer facilities, of which most of the complex is supported on waste.

New USEPA regulations required nearly all unlined landfills to close in 1993. This action was taken as an opportunity for recycling an old landfill, putting a new liner, and reusing the landfill. A landfill without a liner in Bethlehem, Pennsylvania, originally closed in 1991, is currently being recycled. Some of the waste dates back to 1942, of which the separated soil will be used to provide a subgrade for a new liner. Similarly, a site in McDougal, Ontario, is going through the same process of recycling and lining. After completion, the landfill is expected to have an additional capacity of 5 to 10 years (Nelson, 1995). In lining a recycled landfill, the area is typically regraded, and then liner and leachate collection and removal systems are constructed. Refer to Chapters 17 and 18 for detailed information on the liner and leachate collection and removal systems, respectively.

24.2.3 Recreational Uses

Closed landfills have been used extensively to create recreational facilities such as golf courses, nature or recreation parks, animal refuge, tennis courts, toboggan hills and ski runs, parking lots, or similar low-value facilities (Keech, 1996b). Predominantly, many closed landfills have been used to create golf courses successfully (Bier, 1998; Cammarene, 1998; Haughey, 1998; Hurdzan, 1998). For any recreational use to be effective, the major features that need design are capping, trees and landscape, and settlements.

Capping. Capping of the closed landfills only protects the landfill from erosion and rainwater infiltration. If the cap is to fulfill its engineering role in the long run, it needs to be viewed as a self-sustaining, integrated landscape in which plant communities, soil type, and hydrology all interact (Fish, 1993). For instance, when designing a park or animal refuge, which typically contains trees, the depths as well as the slope have to be taken into consideration. The cap is also cut and designed with berms and swales to look natural as well as function. A park with steep slopes needs a liner that resists slipping down steep slopes. Six inches of topsoil for meadow areas is fine, but trees need at least $3\frac{1}{2}$ ft for the wooded vegetation, due to deeper root growth zone. Consequently, the pH of the topsoil may have to be altered so that the cap can sustain life (Cameron, 1995). In some landfills, such as the landfill turned public park in Elmhurst, Illinois, the cap included an additional 8 to 10 ft of fill to achieve desired aesthetic results as well as incorporating geogrid mats 3 ft below the playing surfaces of athletic fields (Paukstis, 1993). Cover soil type is also important, since a site that is expected to sustain trees should have a capping soil rich in humus (Gordon, 1991).

Trees and Landscape. The vegetation on a landfill from trees all the way down to the grass actually has to be picked carefully. Many things change over time in a landfill. Even though the cap acts as a barrier, often it is not totally impermeable, allowing some interaction with the environment. On landfill sites, water, oxygen, nutrients, and physical stability are often inadequate. Drought conditions can be caused by increased runoff due to surface compaction and by higher soil temperature resulting from biological decomposition of trash. The discontinuity of texture between the refuse and cover can also inhibit water being pulled to the surface. Additionally, landfill cover soils may frequently have low nutrient contents and unsatisfactory pH ranges. Therefore, tree species, especially, should be tolerant to a range of poor soil conditions, also taking local climate into consideration (Gordon, 1991). Trees need to be placed in deeper soils since deeper soils retain more moisture and provide a deeper root growth zone (Hansen and Druback, 1995). Poor planting practices can actually be more hazardous to trees than some landfill conditions. As mentioned earlier in capping, it is suggested to have a cap intended for trees with high-humus-content topsoil. Recommended minimum depths range from $\frac{1}{5}$ to 1 m for soils containing humus (Gordon, 1991).

Woody plants and grasses that would develop a root mass close to the surface and have a spreading growth pattern are best suited for slopes. This helps maximize slope stability and prevent erosion and scour (Hansen and Druback, 1995). It is suggested that plants having a shallow, spreading root system and are adaptable to climate extremes, which are common in landfills, are typically better equipped to live in a landfill. As long as methane gas and leachate are controlled, grasses typically grow rather easily on landfill sites. Other plants and flowers typically do fine with the cap, but some herbaceous plants should be planted only after woody plants are well established due to the protection provided by the wood plants (Cameron, 1995). A good

variety of plants brings an increase to the diversity of insect life. This in turn also helps to attract more of the wildlife chain, such as the variety of birds.

Settlement. After closure and as waste continues to decompose, settlement is expected to occur. This happens even though the refuse is compacted daily by the machinery. In predicting settlement, the biggest problem is that the type of waste in the landfill is not necessarily consistent throughout, which means different settlement rates as well as different settlement overall. A closed landfill in Palm Beach County, Florida, is currently being transformed into a park complete with paved and unpaved bike paths. The total site acreage encompasses three landfill cells, including one that is still settling and generating methane. To avoid problems of settlement, this cell will not be part of the park (Anon., 1995). Despite the solution chosen in Florida, there are ways to combat settlement in park areas. Many parks will have athletic playing fields, such as the landfill turned park in Elmhurst, Illinois. Their fields began to warp and adding a thin layer of soil to smooth out the warping did not help over time. After a professional investigation, three solutions were suggested that could apply to other landfills: Use dynamic compaction by dropping a significant weight from a height to compact the soil, fill the entire site with an average of 7 feet of fill, or install a system of stone and geogrid fabric to act as a "bridge" over settled areas. This particular site was filled with 8 to 10 ft of fill, and as stated earlier, geogrid mat was placed 3 ft below playing field surfaces to reinforce the fields, acting to bridge any pockets of settlement that may continue to occur farther down within the landfill (Paukstis, 1993). Overall, an ongoing plan of continuous change can also overcome settlement, initially, keeping the recreational areas simple, such as using nonpermanent and yielding materials. Walking and bike paths can be made of gravel and stone dusts to combat uneven settling effects. In time, when the landfill site becomes more stable, permanent structures such as paved walkways, ballparks and soccer fields, and rest rooms can be added (Arent, 1989). A predicted settlement contour map for the site should be prepared and be used in the design and operation of these developments (Sharma, 2000).

24.2.4 Commercial and Industrial Uses

There are several cases where closed landfills have been used to construct buildings, roads, parking lots, and so on, for commercial and industrial uses (O'Leary and Walsh, 1992). This has become possible because of the long-term monitoring programs that have been implemented at several landfill sites, allowing engineers to have a better understanding of how landfills change with time. These facilities include retail stores, warehouses, office buildings, and manufacturing facilities. The construction of these facilities must consider (Dunn, 1995 and Sharma et al., 2003):

- Large total and differential settlements
- Constructability and health and safety of construction personnel
- Impact of construction on waste containment efficiency
- Impact of landfill environment on embedded facility components (e.g., foundations, pipes)

Long-term settlements due to waste self-weight and external loads (due to regrading and foundation loads) generally result in differential settlements that may result in tilting of a building support system, ponding of water in parking lots, cracking of slabs supported on ground, breakage in utility lines, and downdrag forces

on piles that support heavy building loads. Predicted settlement maps for different time periods should be prepared and used for the design and operation of development facilities. More details on the settlement estimation methods and design considerations for these developments are provided in Section 24.3.

24.3 DESIGN CONSIDERATIONS

There are many engineering issues that come into play when designing for the end uses of a closed landfill. These issues vary in priority depending on the specific end use. Table 24.1 illustrates some engineering issues related to landfill closure and how critical they are depending on the specific end use. The general design process and various design issues involved in the use of closed landfills for commercial and industrial development are presented below.

24.3.1 Design Process

When closed landfills are considered for commercial and industrial use consisting of permanent building structures, a four-step design process should be employed (Keech, 1995) (1) analyze historical documents and geotechnical reports, (2) define site characteristics

Table 24.1 **Engineering issues related to end use of landfills**

	End Use of Landfill[a]								
Problem	Residential	Light Industry	Arable	Grazing	Sports	Recreation	Woodland	Public Open Space	Wildlife
Settlement	1	1	1	3	1	2	4	4	4
Leachate	1	2	1	2	2	3	3	2	4
Gas	1	2	1	2	1	3	2	3	4
Contamination	1	3	1	2	3	3	3	3	3
Litter/hazard	1	2	1	2	1	2	4	2	4
Plant growth	1	4	1	2	2	4	3	3	4
Soil strength	1	4	2	3	1	3	4	2	4
Soil profile	1	4	1	4	2	4	4	4	4
Total score	8	22	9	20	13	24	27	23	31

Source: Crawford and Smith (1985).

[a] 1, major consideration; even small amounts would have serious consequences; 2, important consideration; some, although small, amounts could be tolerated; 3, minor consideration, not likely to have serious consequences; and 4, needs to be checked but only extreme conditions would be important, if at all; total scores: low total = expensive, high total = relatively low cost.

(boundary conditions), (3) design for predicted rates of differential settlement, and (4) define future inspection and maintenance requirements.

The first step in designing for a building in a landfill is to do a document search that entails original landfill grading plans, the sequence of landfill operations, existing topography/utilities and landfill gas collection systems, and previous uses and topography. Settlement monitoring points are then decided upon and it is suggested that intermittent readings should be taken every two to four months for no less than one year, and preferably for two years, to provide an adequate amount of empirical data to make predictions on future rates of settlement (details on settlement estimation methods are provided in Section 24.3.2). A geotechnical report follows to provide predictions of site performance, such as a contour map of depth to the barrier layer and to refuse, and a contour map of anticipated settlement (Keech, 1995).

The second step in the design is identification of site constraints (boundary condition) that lead to differential settlement. Differential settlement is always found on landfill sites due to the variable nature and depth of the waste plus varying depths of fill over the barrier layer. Designs must consider soft- or hard-edge boundary conditions. *Soft-edge boundary conditions* are areas of significant differential settlement occurring gradually over an identifiable horizontal distance. Typ-

ical identifiable causes of these boundary conditions are changes in depth of refuse/topography of landfill bottom, composition and age of landfill refuse, thickness of landfill cover, and previous uses. Figure 24.3 depicts typical examples of soft-edge boundary conditions and their effect on future finished elevations after settlement. *Hard-edge boundary conditions* (artificial boundaries) occur at the edges of the structures supported by deep foundations, where an abrupt change in settlement occurs at the site–structure interface. A hard-edge boundary is most commonly associated with the interaction between the pile-supported structures and site improvements supported by the landfill mass. Causes of vertical shear include edges of supported structures, isolated pile caps and grade beams, and similar vertical elements as depicted in Figure 24.4.

The third step in the design is designing for differential settlement. The settlement estimation methods and the building foundations are presented in Section 24.3.2. In addition to building foundations, finish slopes, site utilities, and pedestrian/utility connections to buildings should be designed to tolerate the differential settlements. It is generally a good design practice to slope pavements and other surface improvements in the direction of increasing settlement to ensure no future reversals. Changes of 1 to 2% in future surface slopes due to settlement are not uncommon in deep

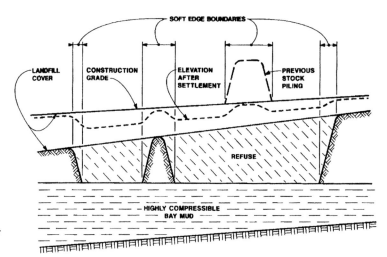

Figure 24.3 Soft-edge boundary conditions. (*From Keech, 1995. This material is reproduced by permission of ASCE.*)

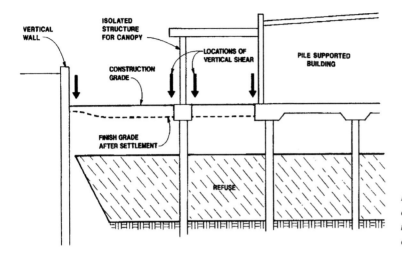

Figure 24.4 Hard-edge boundary conditions. (*From Keech, 1995. This material is reproduced by permission of ASCE.*)

landfills. Roadways, in particular, should use flexible material such as asphaltic concrete, thus avoiding or limiting the use of portland cement concrete and other nonflexible materials. One method of connecting the building to a parking lot is to use a hinge slab system (Figure 24.5) to accommodate vertical dislocations. When possible, utilities should be minimized within the landfill area, and design should provide a positive overflow and leak detection system. Site utilities should also be designed for ultimate slope caused by anticipated settlement from additional fill. Gravity util-

ities need the adequate slope at construction so that after settlement, reversal does not occur across soft-edge boundary conditions. A typical gravity connection (Figure 24.6) is handled by overlapping the building utility line with the site utility (Keech, 1995). Pressure utility systems (Figure 24.6), such as domestic water and fire lines, should be anchored on both the building and site side of the flexible utility connection so that pressure forces at the utility bends do not expand the flexible connection and cause pipe distress above or below the flexible connection.

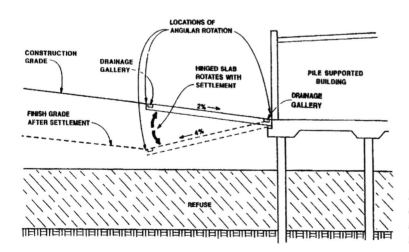

Figure 24.5 Hinged slab at building entrance. (*From Keech, 1995. This material is reproduced by permission of ASCE.*)

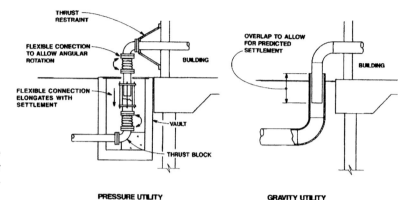

Figure 24.6 Flexible utility connections. (From Keech, 1995. This material is reproduced by permission of ASCE.)

The fourth step of design is defining inspection and maintenance requirements. Although the design is performed to minimize maintenance needs, inspection and maintenance is an ongoing requirement for any facility located on a landfill. The inspection and maintenance include settlement monitoring, pavement condition, hinged slabs, and utility connections.

24.3.2 Design Issues

The major design issues that relate to end use of closed landfills, especially for commercial and industrial uses, are (1) settlement and foundation systems, (2) gas generation, (3) slope stability, and (4) drainage.

Settlement and Foundation Systems. The total settlement of the facilities located on landfills is estimated by summing the waste settlement and the consolidation settlement of soft clays underlying the landfill. The consolidation settlement of clays can be calculated using standard geotechnical procedures; however, this settlement is generally insignificant compared to the waste settlement. Waste settlement is almost always the controlling factor in the selection of foundations for structures that are to be built on closed landfills.

Waste settles under its own weight and as external loads are placed. External loads include daily soil cover, additional waste layers, final cover, and facilities such as buildings and roads. The following processes produce the waste settlement (Sharma, 2000):

- Physical and mechanical processes that include the reorientation of particles, movement of the fine materials into larger voids, and collapse of void spaces
- Chemical processes that include corrosion, combustion, and oxidation
- Dissolution process that consists of dissolving soluble substances by percolating liquids and then forming leachate
- Biological decomposition of organics with time, depending on the humidity and amount of organics present in the waste

Generally, waste settlement behavior is similar to that of organic soils, especially peat. Significant settlement occurs during waste placement and shortly after waste placement (due to physical and mechanical processes). This is followed by substantial additional settlement, occurring at a slower rate over an extended period of time (due to chemical and biological processes). The initial settlement is classified as primary, with the long-term settlement being classified as secondary compression.

Several methods have been reported to estimate the waste settlement (Sowers, 1973; Edil et al., 1990; Sharma and Lewis, 1994; El-Fadel and Khoury, 2000; Sharma, 2000). However, Sower's method (1973), which is based on consolidation theory, is commonly used to estimate the waste settlement in engineering practice. According to this method, the primary settlement in each distinct layer is calculated using the expression

$$S = H \frac{C_c}{1 + e_0} \log \frac{P_0 + \Delta P}{P_0} \qquad (24.1)$$

where S is the primary settlement occurring in the layer under consideration, H the initial thickness of the waste layer under consideration, C_c the primary compression index, e_0 the initial void ratio of the layer, P_0 the existing overburden pressure acting at the midlevel of the layer and ΔP the increment of overburden pressure at the midlevel of the layer. Typical unit weight (needed to calculated overburden pressure) and void ratio values of waste are presented in Chapter 15. For older landfills (10 to 15 years) which are subjected to external loads, $C_c/(1 + e_0)$ values are reported to range from 0.1 to 0.4 (U.S. Navy, 1983). Total primary settlement is estimated by summing the primary settlement of each layer of waste as waste placement proceeds. Based on literature and experience, Sharma (2000) reported that the primary settlement of waste occurs during one to four months of filling. Therefore, all of the primary settlement would occur by the time of landfill closure.

Following the primary settlement, the time-dependent or secondary settlement (ΔH) occurs. ΔH due to self-weight of the waste is given by

$$\Delta H = \Delta H_{(SW)} = C_{\alpha(SW)} H \log \frac{t_2}{t_1} \qquad (24.2)$$

where $\Delta H_{(SW)}$ is settlement at time t_2 after waste placement; t_1 the initial period (typically 1 to 4 months) of settlement, H the thickness of waste fill, and $C_{\alpha(SW)}$ the coefficient of secondary compression due to self-weight. Typically, $C_{\alpha(SW)}$ values range between 0.1 and 0.4 (U.S. Navy, 1983). The ΔH under external loads (especially due to the construction of final cover and facilities such as buildings and roads) is given by

$$\Delta H = \Delta H_{(EL)} = C_{\alpha(EL)} H \log \frac{t_2}{t_1} \qquad (24.3)$$

where $\Delta H_{(EL)}$ is settlement at time t_2 after external load application; t_1 the time for primary settlement, most of which will occur as the load is applied and may continue up to four months after external load application; H the thickness of waste fill; and $C_{\alpha(EL)}$ the coefficient of secondary compression due to external loads. Literature review indicates that older waste fills that have undergone decomposition for some period of time (typically, 10 to 15 years), $C_{\alpha(EL)}$ ranges between 0.01 and 0.07 (Sharma and Lewis, 1994). Both $C_{\alpha(SW)}$ and $C_{\alpha(EL)}$ values depend on site-specific environmental conditions and organic content of the waste fill. The higher C_α value indicates the higher organic content, the higher humidity, and/or the higher degree of decomposition of the waste. Based on the monitored settlements at three different landfills in California, Sharma (2000) calculated value of $C_{\alpha(EL)}$ to be 0.02 and $C_{\alpha(SW)}$ to range from 0.19 to 0.28, and these values fall within the range of values reported by the U.S. Navy (1983) and Landva et al. (2000).

In addition to estimating the waste settlement, if feasible, it is recommended that site-specific measurements of settlement versus time be performed to provide more reliable information on the waste settlement. In addition, a test fill program which essentially involves placing a fill over the waste and monitoring the settlement is useful to provide the site-specific waste settlement data.

Depending on the waste settlement, careful consideration must be taken in deciding on what type of structure or cover is going to be built. In most cases, the secondary settlements can cause distress in a foundation or cause unwanted deformations. To reduce the secondary settlement, one or more of the following site improvement methods could be implemented: (1) surcharging, with settlement monitoring; (2) dynamic compaction; and (3) grouting or fly ash injection. If the use of these methods is not economical or ineffective, structures should be designed with foundation systems that can tolerate expected settlements. Shallow foundations may be used to support lightly loaded structures, but deep foundations, particularly piles, may be used to support larger structures.

Figure 24.7 shows the typical shallow foundation systems that can be used on closed landfills and in-

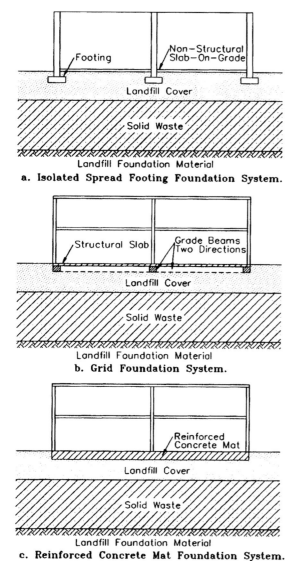

a. **Isolated Spread Footing Foundation System.**

b. **Grid Foundation System.**

c. **Reinforced Concrete Mat Foundation System.**

Figure 24.7 *Shallow foundation types for landfills. (From Dunn, 1995.)*

clude isolated spread footing systems, grid foundation systems, or mat foundation systems. These systems can be used for structures that can tolerate differential settlements. A flexible structure with provisions for jacking and grouting column bases could be used. Figure 24.8 illustrates how adjustable column bases could be used to relevel a light building. Bearing capacity of the foundations should be checked based on the engineered fill that is provided below the foundation to sufficient depth to provide essentially all of the bearing capacity.

Driven piles are the common deep foundation system used to support structures with significant loads. Due to obstructions such as concrete rubble and steel debris present in the landfill, driving may be difficult and result in damage of the piles. Some areas may require excavation and removal or breaking of the obstruction. In evaluating the pile capacity, downdrag loads (negative skin friction loads) generated due to waste settlement should be considered. The downdrag loads could be very high, and as a result, different methods to reduce downdrag loads should be used. Figure 24.9 shows different methods to reduce downdrag loads and include predrilling or spudding of the piles with a steel mandrel, installation of an outer shell or casing (known as double-pile system), or friction-reducing coatings. The lateral resistance of a pile may be affected by loss of contact with surrounding material because of underlying waste settlement or use of a pile installation method. This should be considered in the pile design.

Figure 24.10 is a cross-sectional view of the slab support system that can be used on a compacted landfill material. A subgrade vent system is required to remove landfill gas, and this system consists of a layer of coarse crushed stone separated from the supporting soil by a geofabric (geotextile) and a network of perforated pipe placed at the upper limit of crushed stone. The layer is covered with an impermeable membrane to trap the gas. This membrane may be placed on another layer of geofabric (geotextile) and/or filter sand to prevent punctures. This membrane also serves as the lower bound of a foundation mattress. This mattress consists of a 0.3-m-thick layer of compacted sand and gravel. The membrane should be protected from puncture. This mattress is designed to support the floor slab and allows releveling of the slab by selectively pressure injecting grout beneath the slab. Keyed reinforced slabs may be used to localize differential settlement and limit faulting of the slabs at the joints. Alternatively, floor slabs may be supported on piles to prevent

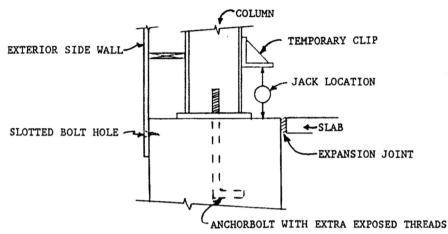

Figure 24.8 Adjustable column base. (*From Gifford et al., 1990. Copyright ASTM International. Reprinted with permission.*)

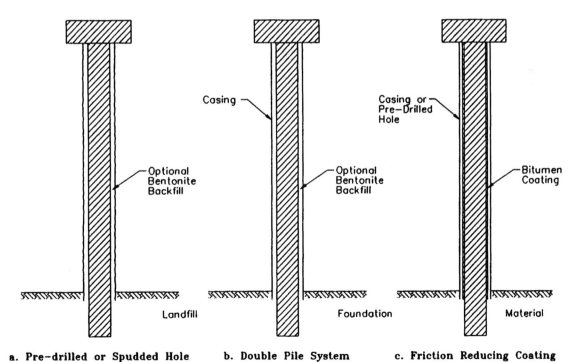

a. Pre-drilled or Spudded Hole   b. Double Pile System   c. Friction Reducing Coating

Figure 24.9 Method of downdrag reduction for piles. (*From Dunn, 1995.*)

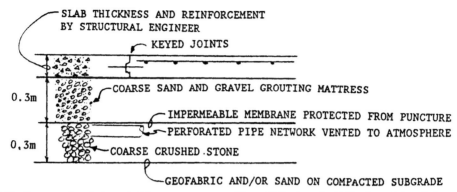

Figure 24.10 Cross-sectional view of slab support system. (*From Gifford et al., 1990. Copyright ASTM International. Reprinted with permission.*)

excessive settlements, but this option may be too expensive to use.

Gas Generation. As the waste decomposes, gas that is approximately 50% methane is generated. This landfill gas can be a fire and explosion hazard as well as a health hazard. To minimize the health, fire, and explosion hazards of landfill gas, a control system is located in the landfill and the areas between the landfill and nearby structures. The landfill gas control system can be an active or passive system. In a passive control system, gravel trenches or perforated pipes installed in gravel trenches are used to allow gas to be collected and removed. In an active control system, landfill gas movement is controlled mechanically. Vertical or horizontal landfill gas collection wells, or both, are installed to collect landfill gas. A blower is used to apply a vacuum to the collection wells. Landfill gas collected is flared or used for beneficial energy use. More details on the gas collection and utilization systems are provided in Chapter 20.

Slope Stability. Slope stability also comes into play because the landfill is usually built as a mound or hill. The slope of the hill is governed by the characteristics of the fill and cover. If these factors aren't taken into account, the slope movement might be an issue; it may erode, or even slide off the waste. This characteristic is also very important when deciding what to build over the fill later on. Addition of a structure or devel-

opment on the landfill will add to the driving force on the slope. Therefore, analyses should be performed by considering existing and additional loads on the fill.

Drainage and Leachate Collection. Waste settlement will change the final surface grades and may affect the drainage and final cover erosion. The differential settlement of waste may result in (1) grade reversals causing surface water ponding, and (2) cracks in the final cover system. Based on the settlements estimated, a postclosure grading and maintenance program should be prepared (Sharma, 2000). The final cover slopes should always allow proper drainage of the stormwater runoff. This runoff could erode the slopes if it is not collected and removed through a proper drainage collection system. The runoff may also contain contaminants or leachates that percolate through the waste. The leachate and surface water runoff with leachate in it needs to be collected and treated before it is allowed to run off the site. Drainage and leachate collection systems should be designed to keep the contaminated rainwater from running off the site and polluting the surrounding groundwater and vegetation.

24.4 CASE STUDIES

Case Study 1: Golf Course. Settler's Hill landfill in Geneva, Illinois, has been transformed into a golf course. This is a 400-acre landfill, where the methane gas is also used to produce power for the city of Geneva (WMI, 2000). Leachate is collected and treated in a wastewater treatment plant. Any settlement that occurs is repaired with a thin layer of soil each spring. Another part of the site is built up and covered to be a ski run. Parts of landfill are also used as a nature area and for horseback riding, as a small lake for paddle boating, and there are picnic areas, two clubhouses, and a restaurant.

Case Study 2: Commercial Building. A Home Depot hardware store was placed on a landfill in Colma, California (Keech, 1996a). The building is supported on deep piles that reach below the landfill's subgrade and will not settle, but the parking and access areas were placed on the fill material and will experience settlement. The store posed unique engineering challenges since the total site measures 7 acres, compared to the normal 9-acre sites for Home Depots elsewhere. The design cantilevers the building out over a 75-ft slope. This was reportedly the deepest landfill, 130 ft, on which any commercial development has ever been built. Engineers had to design for settlements in excess

of 4 ft. To minimize settlement, the site was surcharged with 20 ft of soil on top of key parts of the landfill to preload the waste. The preloading is estimated to have reduced the anticipated settlement to half. Due to customers using shopping carts containing heavy items, the slope of the parking lot must not exceed 3% to avoid runaway carts and possible damage or injury. The parking lot uses a hinge slab connection with an adjacent ramp sloping up into the air 1 ft over its 32-ft length. As the waste settles, the ramp should gradually become level with the building. When the waste settles another foot, it slopes down from the building, at which time engineers can pick up the ramps and raise them 2 ft through hydraulic lifts. The entire process repeats itself until settling ceases.

Case Study 3: Recreational Facility. A closed landfill, which was turned into a functional recreation facility, was the Dyer Boulevard landfill in Palm Beach County, Florida. This landfill was closed and opened as a recreational facility in 1997 (Magnuson, 1999). This facility consisted of baseball and softball fields, three soccer fields, two volleyball courts, four basket-

Figure 24.11 *Recreational use of closed landfill: Dyer Boulevard Landfill, Florida. (From Magnuson, 1999.)*

ball courts, a children's playground, a 3.2-mile mountain bike trail, a 3.4-mile equestrian trail, a model airplane runway, and a 4.2-mile bike and jogging path (Figure 24.11). The development, known as Dyer Park, is special because there was open natural ground between the landfill mounds. This characteristic allowed for the construction of buildings, and park structures on solid ground. A gas collection system was installed in 1990 to collect the excess gas buildups in voids and crevices under the various structures and fields (Magnuson, 1999).

Case Study 4: Industrial Development. An old 640-acre landfill in Snohomish County, Washington, was used for light-industrial development. This landfill, also known as the Cathcart sanitary landfill, was closed. The local county council discussed whether or not to turn a part of the landfill into a school. It would have to be built on the part of the landfill that was not actually filled with waste. The rest of the landfill site would then be given to the local park district so that it could be turned into a recreational facility (Sanders, 1999).

24.5   SUMMARY

All landfills eventually reach their capacity and are subject to closure. When this occurs, the final step in the landfill process begins, which is rehabilitation of the site. This may lead to one of many end uses that are valid for landfills. Overall, several end uses are possible if the end-use alternatives are considered during the planning and design stages of a landfill. Several engineering issues have to be addressed to convert and operate landfills properly for a selected end use. Typical end uses are recreational (e.g., golf courses and parks) and for commercial building.

QUESTIONS/PROBLEMS

24.1. Explain why end use should be considered during the design of a landfill.

24.2. What are advantages and disadvantages of the end use of landfills? Explain them.

24.3. List three major concerns with regard to the end use of landfills.

24.4. Explain soft- and hard-edge boundary conditions and discuss how they affect waste settlement in landfill.

24.5. What are the engineering considerations in using landfills as golf courses?

24.6. What are the engineering considerations for constructing roadways on landfills?

24.7. What are the major engineering challenges in locating structural foundations on landfills?

24.8. Explain the differences in waste settlements of MSW landfills and construction and demolition (C&D) landfills. What will be their implications on end uses?

24.9. Contact your state's environmental agency for information on a closed landfill in the state. Explain the end uses of such landfills, if any.

24.10. If the rate of waste settlements is slow, what methods would you recommend to accelerate waste settlement? Explain advantages and disadvantages of these methods.

REFERENCES

Anon., Palm Beach County, FL, converting landfill to park, *Am. City and County,* p. 30, Feb. 1995.

Arent, K. A., Creation of a park, *Publ. Works,* pp. 58–60, May 1989.

Bier, J., Effects of landfill gas management at the industry Hills Recreation and Conference Center, pp. 50–56 in *Landfill Golf Courses: Environmentally Beneficial Community Assets,* National Golf Foundation, Jupiter, FL, 1998.

Cameron, L. G., Environmental park grows on a landfill. *Publ. Works,* pp. 48–51, June 1995.

Cammarene, M., Case study: Fairwinds Golf Course, pp. 46–48 in *Landfill Golf Courses: Environmentally Beneficial Community Assets,* National Golf Foundation, Jupiter, FL, 1998.

Crawford, J. F., and Smith, P. G., *Landfill Technology,* Butterworth, London, 1985.

Dunn, R. J., Design and construction of foundations compatible with solid wastes, pp. 139–159 in *Landfill Closures: Environmental Protection and Land Recovery,* Geotechnical Special Publication 53, Dunn, R. J., and Singh, U. P., (eds.), ASCE, Reston, VA, 1995.

Edil, T. B., Ranguette, V. J., and Wuellner, W. W., Settlement of municipal refuse, in *Geotechnics of Waste Fills: Theory and Practice,* ASTM STP 1070, Landva, A. and Knowles, G. D. (eds.), ASTM, West Conshohocken, PA, 1990.

El-Fadel, M., and Khoury, R., Modeling settlement in MSW landfills: a critical review, *Crit. Rev. Environ. Sci. Technol.,* Vol. 30, No. 3, pp. 327–361, 2000.

Fish, W., The afterlife of soil wastes: managing a postclosure landfill, *Environ. Sci. Technol.,* pp. 10–11, Sept. 1993.

Freece, L., Building in landfill is energy efficient, *Am. City County,* pp. 44–45, Aug. 1994.

Gifford, G. P., Landva, A. O., and Hoffman, V. G., Geotechnical considerations when planning construction on a landfill, pp. 41–56 in *Geotechnics of Waste Fills: Theory and Practice,* Landva, A., and Knowles, G. D. (eds.), ASTM STP 1070, ASTM, West Conchohocken, PA, 1990.

Gordon, S., Tree planting on landfills, *Biocycle,* p. 80, Oct. 1991.

Gruber, J. S., Greg, K., and Delia, C., Conversion of a dump into an environmental community center, *Publ. Works,* pp. 52–55, Mar. 1993.

Hansen, J. M., and Druback, G., Environmental benefit plan for Blydenburg landfill, *Waste Age,* pp. 241–247, Apr. 1995.

Haughey, R., Golf course development on closed landfills, pp. 32–39 in *Landfill Golf Courses: Environmentally Beneficial Community Assets,* National Golf Foundation, Jupiter, FL, pp. 32–39, 1998.

Hurdzan, M. J., Developing golf courses on sanitary landfills, pp 20–26 in *Landfill Golf Courses: Environmentally Beneficial Community Assets,* National Golf Foundation, Jupiter, FL, 1998.

Keech, M. A., Design of civil infrastructure over landfills, pp. 160–183 in *Landfill Closures: Environmental Protection and Land Recovery,* Geotechnical Special Publication 53, Dunn, R. J., and Singh, U. P. (eds.), pp. 160–183, 1995.

———, Construction of retail building on closed landfill sites, *Publ. Works,* pp. 75–85, May 1996a.

———, Post closure landfill development: is there life after closure? *Waste Age,* pp. 75–82, Apr. 1996b.

Landva, A. O., Valasangkar, A. J., and Pelkey, S. G., Lateral earth pressure at rest and compressibility of municipal solid waste, *Can. Geotech. J.,* Vol. 37, No. 6, pp. 1157–1165, 2000.

Magnuson, A., Landfill closure: end uses, *MSW Manag.,* Vol. 9, No. 5, pp. 82–93, 1999.

Nelson, H., Landfill reclamation projects on the rise. *Biocycle,* pp. 83–84, Mar. 1995.

O'Leary, P., and Walsh, P., Landfill closure and long-term care, *Waste Age,* pp. 87–94, 1992.

Paukstis, S., Landfill transformed into recreation area,. *Am. City County,* p. 30, June 1993.

Rabasca, L., Bringing new life to old landfills, *Waste Age,* pp. 71–76, Jan. 1996.

Sanders, E., Snohomish County to evaluate new uses for landfill site. *Seattle Times,* Aug. 26, 1999; *www.seattletimes.nwsource.com/news/local/html98/cath_19990826.html.*

Sharma, H. D., Solid Waste landfills: settlements and post-closure perspectives, *Environmental and Pipeline Engineering 2000,* ASCE, Surampalli, R. Y. (ed.), ASCE, Reston, VA, 2000, pp. 447–455.

Sharma, H. D., and Lewis, S. P., *Waste Containment Systems, Waste Stabilization, and Landfills: Design and Evaluation,* Wiley, New York, 1994.

Sharma, H. D., Settepani, F. W., and Burns, P. F., *Design and Construction of a Foundation System for an Industrial Building Above a Closed Sanitary Landfill,* Ninth International Waste Management and Landfill Symposium, Sardinia, Cagliari, Italy, 2003.

Sowers, G. F., Settlement of waste disposal fills, in *Proceedings of the 8th International Conference on Soil Mechanics and Foundation Engineering,* Moscow, 1973.

Spencer, R., Mining landfill for recyclables, *Biocycle*, pp. 34–36, Feb. 1991.

———, Looking back at a landfill, *Biocycle*, p. 60, Feb. 1995.

U.S. Navy, *Soil Mechanics*, DM 7.3, Naval Facilities Engineering Command, Alexandria, VA, 1983.

Waste Management, Inc., *Modern Landfill Technology*, WMI, Oak Brook, IL, 1990.

———, *Settler's Hill*, WMI, Oak Brook, IL, 2000.

25

BIOREACTOR LANDFILLS

25.1 INTRODUCTION

Conventional municipal solid waste (MSW) landfill design, described in Chapters 15 through 21, is intended primarily to minimize the production of leachate and landfill gas. By using base and cover containment systems, the waste is kept effectively dry; the system essentially entombs the waste. The full containment and restriction of liquid infiltration retards degradation of waste materials, causing waste to decompose slowly, less completely, and less predictably.

In contrast to the conventional landfill, a bioreactor landfill is designed and operated to enhance microbial processes significantly, to degrade and stabilize the organic constituents of the waste (Pohland, 1994; Mehta et al., 2002; USEPA, 2003). This is accomplished through the control of moisture, temperature, pH, nutrients, and/or other properties within the waste. The bioreactor landfill was initially defined by the Solid Waste Association of North America (SWANA) as "a sanitary landfill operated for the purpose of transforming and stabilizing the readily and moderately decomposable organic waste constituents within five to ten years following closure by purposeful control to enhance microbiological processes. The bioreactor landfill significantly increases the extent of waste decomposition, conversion rates and process effectiveness over what would otherwise occur within the landfill." However, recently, SWANA (2001) defined a *bioreactor landfill* as "any permitted Subtitle D landfill or landfill cell, subject to New Source Performance Standards/Emissions Guidelines, where liquid or air, in addition to leachate and landfill gas condensate, is injected in a controlled fashion into the waste mass in order to accelerate or enhance biostabilization of the waste."

A schematic of a typical bioreactor landfill, including various components, is shown in Figure 25.1. In a bioreactor landfill, liquid is added until the liquid holding capacity (field capacity) of the waste is reached. As the liquid, typically referred to as *leachate,* drains from the waste, it is recirculated back into the landfill to maintain the moisture content throughout the waste. In turn, it helps to distribute nutrients and contaminants throughout the landfill for microbial biodegradation. A distinction should be made between the bioreactor landfill operation and the conventional leachate recirculation performed at some landfills. The bioreactor process requires significant liquid addition to reach and maintain optimal conditions. Leachate alone is usually not available in sufficient quantity to sustain the bioreactor needs. Water or other nontoxic or nonhazardous liquids and semiliquids are suitable amendments to supplement the leachate (depending on climate and regulatory approval). At the closure stage of bioreactor landfills: (1) leachate quantity will be a finite amount, amenable to on-site treatment with limited need for off-site transfer, treatment, and/or disposal; (2) landfill gas generation will be at its declining stage; and (3) long-term environmental risk will be minimized.

This chapter provides an overview of bioreactor landfills, including the types and advantages, relevant regulations, and design and operational issues. Also presented in this chapter are selected case studies where performance of bioreactor landfills has been documented.

25.2 TYPES AND ADVANTAGES OF BIOREACTOR LANDFILLS

Depending on the microbial process that degrades and stabilizes solid waste, bioreactor landfills can be cate-

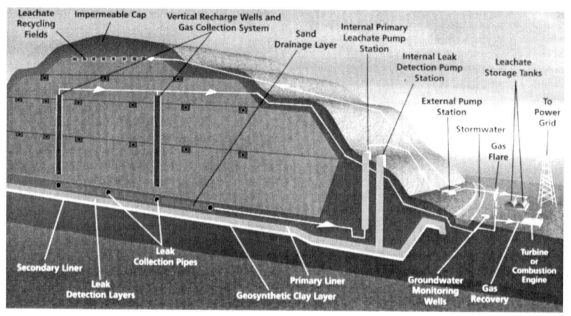

Figure 25.1 Configuration of landfill bioreactor. (From Phaneuf and Vana, 2000.)

gorized into three types (Baker and Eith, 2000): (1) anaerobic bioreactor landfills, (2) aerobic bioreactor landfills, and (3) hybrid or anaerobic–aerobic bioreactor landfills.

Anaerobic bioreactor landfills seek to accelerate waste degradation by anaerobic microorganisms. Anaerobic microorganisms do not use oxygen for cellular respiration. These microbes (includes methanogens) are responsible for the conversion of organic wastes into organic acids and ultimately into methane and carbon dioxide. Oxygen is naturally deficient in most landfills; therefore, the anaerobic conditions exist without intervention. However, the moisture content of solid waste ranges from 10 to 20%, and to optimize the anaerobic degradation process, moisture contents at or near field capacity (typically, 45 to 65%) are required. Under optimal conditions, waste stabilization can be accomplished within six or seven years.

Aerobic bioreactor landfills rely on aerobic microorganisms to accelerate degradation of waste. Aerobic microbes require oxygen for cellular respiration. Aerobic microbes derive their energy from organic waste

in processes that consume oxygen and produce carbon dioxide. Thus, the aerobic process does not produce methane. Because of deficient oxygen conditions in landfills, aerobic activity is promoted by injecting air or oxygen into waste. Air or oxygen can be injected into a landfill using the same vertical and horizontal wells that are used to extract gas or inject liquids. Aerobic microbes grow more quickly than anaerobic microbes, leading to faster degradation rates. Under optimal conditions, waste stabilization can be accomplished in less than two years. Similar to anaerobic bioreactors, moisture content is increased to near field capacity. But because of generation of microbial heat, additional quantities of liquids will be required to develop optimal moisture contents. Aerobic bioreactor landfills are promoted to accelerate waste decomposition and to reduce methane content in landfill gas. However, the potential for landfill fires and additional costs often limit the use of aerobic bioreactor landfills.

Recently, *hybrid bioreactor landfills*, which combine anaerobic and aerobic microorganisms, are being investigated to provide an optimum approach for effective decomposition of wastes. One of the implementation schemes suggested involves treating the uppermost layer of waste aerobically for 30 to 60 days before it is buried by the next layer of waste, at which time it is treated anaerobically. This approach is operationally simple and can be effective because during the initial phase of aerobic treatment, organic wastes are quickly converted into the acid phase and are then effectively treated by anaerobic methanogens.

Table 25.1 shows the advantages and disadvantages of different types of bioreactors. All of bioreactor landfills require specific management activities and operational modifications to enhance microbial decomposition processes. The single most important factor for effective operation is moisture control, and it is the easiest to manipulate in a landfill. It is also the most important because it can influence all the other factors. For example, by assuring moisture accessibility, soluble nutrients also become more available. Other strategies, including waste shredding, pH adjustment, nutrient addition and balance, waste predisposal and postdisposal conditioning, and temperature management, may also serve to optimize the bioreactor process. The successful operation of a bioreactor landfill also requires the development and implementation of focused operational and development plans to ensure that optimal conditions for bioprocesses exist.

Bioreactor landfills offer the following benefits (Pacey et al., 1999; Yolo County, 1999; Sullivan, 2000; Warith, 2002):

- More predictable long-term landfill performance and reduced groundwater contamination risk
- Rapid stabilization of waste to a stage where the ongoing leachate and landfill gas generation is environmentally acceptable without continued collection and treatment activities
- Controllable gas yields during operation with a more economical production profile and reduced waste of gases, which could help reduce the greenhouse gas burden
- Leachate with reduced treatment requirements during operation
- Longer landfill lifetime, resulting from better use of landfill air space and resulting in a reduced need to site new landfills and increased revenue from existing landfills
- Reduced postclosure maintenance requirements and costs
- A reduction in the financial uncertainty associated with potential long-term liabilities
- Increased land value after closure as a result of better waste stabilization and more options for land use, including construction on top of the landfill
- Improved public perception in that any concerns over the long-term risks associated with landfills are reduced and useful products are being produced

Table 25.1 **Comparison of bioreactor landfills**

	Conventional Landfill	Anaerobic Bioreactor	Aerobic Bioreactor
Typical settlement after:			
2 Years	2–5%	10–15%	20–25%
10 Years	15%	20–25%	20–25%
Anticipated waste-stabilization time frame	30–100 years	10–15 years	2–4 years
Methane generation rate	Base case	Two times base case	10–50% base case
Liquid storage capacity utilized in waste mass	None	30–60 gal/yd^3	30–60 gal/yd^3
Liquid evaporation	Negligible	Negligible	50–80%[a]
Average capital cost	Low	Medium	High
Average O&M cost	Low	Medium	High
Average closure/postclosure cost	High	Medium	Low

Source: Campman and Yates (2002).

[a] Liquid evaporation rate is highly dependent on site-specific characteristics.

- The potential for mining of the dredged waste to recover humic material and other recyclables and allow inspection of any lining system

25.3 REGULATORY ISSUES

Like conventional MSW landfills, bioreactor landfills must meet the requirements of the Resource Conservation and Recovery Act (RCRA) Subtitle D, the Clean Water Act, the Clean Air Act, and other federal, state, and local regulations. For both conventional and bioreactor landfills, the liner and leachate collection and removal system should be designed to limit the leachate head on the liner to no more than 12 in. In conventional landfills, the moisture content remains close to that of the entering waste; however, in bioreactor landfills, moisture content is increased due to the addition of liquids. Federal regulations stipulated in 40 CFR 258.28 allow recirculation of leachate in MSW landfills lined with a single-composite liner system that meets 40 CFR 258.40 regulations. Although not explicit, these regulations are also interpreted to allow introduction of water and other amendment additions, such as nutrients to enhance biodegradation processes (Pacey et al., 1999). Thus, the bioreactor landfill can be the same lined landfill in which we are disposing our solid wastes today.

The USEPA issued a memorandum on December 16, 1998 that clarified its Office of Solid Waste's policy on leachate recirculation in MSW landfills. Under current 40 CFR Part 258 requirements, leachate circulation is only allowed in MSW landfills that have a composite liner consisting of a 2-ft-thick compacted clay with permeability equal to or less than 1×10^{-7} cm/ s. Alternative composite liners using geosynthetic clay liners (GCLs) are not currently allowed where leachate will be recirculated to the waste (CFR, 2002).

Several federal policies toward the bioreactor landfills have been initiated. These include the Federal Climate Change Action Plan of 1993 that recommended (1) creation of a joint state/federal coordination program to facilitate siting/permitting, and (2) modification of environmental performance standards and regulatory requirements to remove unnecessary barriers to bioreactor landfills. In addition, federal agencies have initiated and supported several bioreactor landfill field demonstration studies (USEPA, 1995, 2002) through the XL (excellence and leadership) program, which allows for innovative projects to waive those federal requirements that may impede an adequate evaluation of the technology.

A survey conducted by Gou and Guzzone (1997) showed that several states developed specific state regulations on leachate recirculation that closely follow those stipulated under RCRA Subtitle D regulations. Most states allow leachate recirculation at landfills designed with a composite liner system and a leachate collection system to maintain leachate head levels below 1 ft. For example, in Washington State, the Washington Administrative Code 173-351-200(9) specifically permits bioreactor landfills. The pertinent section on operating criteria on liquid restriction states that bulk or noncontainerized liquid waste may not be placed in an MSW landfill unit unless the waste is leachate or gas condensate derived from the MSW landfill unit or water added in a controlled fashion and is necessary for enhancing decomposition of solid waste, as approved during the permitting process of WAC 173-351-700, whether it is a new or existing MSW landfill or lateral expansion.

The New York State Department of Environmental Conservation's Division of Solid and Hazardous Waste Materials developed regulations for new lined landfills to operate as bioreactor landfills and established thresholds for primary liner system leakage (Phaneuf, 2000; Phaneuf and Vana, 2000). The regulations (6 NYCRR Part 360) were developed to provide not only flexibility but also to encourage environmentally sound and resource-conscious landfill management. The following provisions must be complied in leachate recirculation:

- Existing landfills operating under a Part 360 permit that have received department approval to recirculate leachate may continue for the duration of the permit or subsequent permit renewals as long as the landfill meets all of the operating requirements of this part and provided that groundwater monitoring data verify no landfill-induced contamination pursuant to the provisions of Part 703 of this title.

- For all new landfills, or an existing landfill that does not have department approval to recirculate leachate, a double-liner system acceptable to the department is required along with a demonstration of a minimum of six months of acceptable primary liner performance being submitted for department approval.

- In all cases, leachate recirculation is prohibited on areas where any soil cover has been applied unless provisions for runoff collection and containment are provided. For double-lined landfills, in no instance may the volume of leachate to be recirculated give cause to increase the primary liner systems leakage beyond the 20 gal/acre per day operational threshold based on a 30-day average and/or to increase the potential for groundwater contamination.

- All leachate recirculation proposals must be supported by an operations manual prepared in accordance with the provisions of subdivision's 360-2.9(a) and (j) of this subpart.

25.4 BIOREACTOR DESIGN

The design of bioreactor landfill components is similar to the conventional MSW landfills described in Chapters 15 through 21. The main design components of a landfill include the liner, leachate collection, gas collection, and final cover. However, one must recognize that the conventional landfills are essentially *dry landfills,* while the bioreactor landfills are *wet landfills.* As such, the performance standards for these two different landfill types are often conflicting and pose challenges to designers. Different issues that must be specifically addressed to provide satisfactory performance of bioreactor landfills are discussed in this section.

25.4.1 Cell Size

For economic and regulatory reasons, an emerging trend in traditional landfill design is to build deep cells

(or phases) that are completed within two to five years. This trend bodes well for bioreactor landfill evolution. Phase cell construction can more easily take advantage of emerging technological developments rather than committing long term to a design that might prove to be inefficient. Once closed, methanogenic conditions within the cell (phase) are optimized, and gas generation and extraction are facilitated. However, extremely deep landfills might be so dense in the lower portions that waste permeability will inhibit leachate flow. In these instances, it might be necessary to limit addition and/or recirculation to the upper levels or to develop adequate internal drainage management capability.

25.4.2 Liner and Leachate Collection System

Federal regulations prescribe a 1-ft maximum allowable leachate head on the bottom liner. The same criterion is used for conventional MSW landfills. This criterion can readily be achieved through appropriate design of liner and leachate collection and removal systems. Refer to Chapters 17 and 18 for detailed analysis and design of these systems. Adding liquid to solid waste will increase the density of the waste, which can be of critical importance in the design of the liner system and leachate collection pipes. The increase in load may be as much as 30% or more because of expected moisture uptake and settlement. Potential for clogging of leachate collection and removal systems must be carefully considered in the design, and cleanouts may be required in key areas of the landfill's LCRS. Stability of waste mass and liner slopes must be ensured during and after landfill operations as detailed in Section 25.4.6. Since leachate-head predictions are generally based on mathematical models, regulatory agencies may require monitoring to verify performance.

25.4.3 Liquid Injection System

An estimate of required liquid volumes is necessary for a liquid injection system design. Sufficient liquid supply (i.e., leachate and water) must be ensured to provide required liquid supply, and sufficient storage will be required to ensure that peak leachate-generation events can be accommodated. Methods for injecting liquids into the waste can then be properly selected.

To optimize the decomposition process, liquids should be injected to achieve moisture content close to the field capacity. Field capacity is defined to be moisture content at which the maximum amount of liquid is held in the voids against gravity drainage (in other words, at moisture contents above field capacity, gravity drainage occurs). The volume of liquid needed to reach waste field capacity can be based on prior field studies, model predictions, or landfill-specific measurement. Reinhart and Townsend (1997) and Phaneuf (2000) note that moisture contents of 25% (wet weight basis) is a minimum for bioreactor effectiveness, although better results are obtained for moisture contents of up to 40 to 70%. However, complete saturation is not conducive to methanogenesis. Phaneuf (2000) indicates that field capacity of MSW is achieved with the addition of between 25 and 50 gal/ton of solid waste; this corresponds to adding about 100 to 300 lb of water per cubic yard of waste, depending on initial density. Baker and Eith (2000) indicate that to achieve optimum anaerobic conditions requires the addition of between 40 and 80 gal/yd^3 to achieve moisture contents of 45 to 65%, assuming that the waste is already at a moisture content of 10 to 20%. Data provided by Hater (2000) indicate that a slightly higher range of liquid volumes is reasonable, which are in the range of 47 to 116 gal/yd^3 of waste, but are typically 70 gal/yd^3 of waste.

In general, achieving a moisture condition that is close to field capacity requires the addition of significant volumes of liquids. If enough leachate is not produced, injection of other approved liquids will be necessary. In addition to injection of clean water, consideration is given to existing on-site contaminated liquids, such as contaminated surface water runoff from siltation ponds or contaminated groundwater if there happens to be a pump and treat collection operation nearby. Certain municipal wastewater treatment plant effluents may also be used. The contaminant loading of these wastewaters should be evaluated to ensure that they will not increase the pollution potential of the bioreactor landfill, and they will be com-

patible with the bioreactor's microbiology (Phaneuf, 2000).

There are different methods of injecting liquids into the solid waste. A common method is *working face application,* in which the liquid is applied directly to the waste to yield target moisture supplementation levels (desired gallons per ton or cubic yard) during active landfilling. The advantages of this method are simplicity and direct access to the waste mass. The disadvantage is that this application is a one-time event. This method may also cause odor problems due to increasing gas generation shortly after application.

Different methods can be used to *inject liquids* into solid waste that is already in place, and these methods include (1) surface irrigation, (2) infiltration ponds, (3) vertical injection wells and manholes, and (4) horizontal trenches. *Surface irrigation* involves using tank trucks with an attached spray bar applying leachate to the surface of waste. On the other hand, *infiltration ponds* are constructed using waste as berms and storing leachate in it. Infiltration of leachate from the pond into the waste occurs due to gravity. As the moisture in the waste gets to field capacity, infiltration continues to lower elevations. Both surface irrigation and infiltration ponds are simple and cost-effective; however, concerns with odor, vectors, and litter may make these methods unattractive.

Vertical wells and manholes have been used on several projects and often have spacing of 100 to 200 ft. These are easy to install during landfill operation or on a retrofit basis and have relatively low cost. Odor problems can be controlled with the appropriate operation, and injection rates are high if units are not close to the waste slopes. The major shortcoming of vertical systems is their limited ability to distribute fluids laterally. Fluids tend to concentrate around the well/manhole, resulting in dry zones at shallower levels.

Horizontal trenches are often the most effective means to distribute fluids during landfill operations. Trenches are usually installed as the waste mass is constructed. Trenches typically consist of horizontal pipes embedded in pervious media, such as gravel, glass cullet, or tire chips. Recent experience has shown that certain types of tire chips compress under vertical pressure from thick waste layers (greater than 50 ft) and become less permeable (Reddy and Saichek, 1998). Pervious cover layers can be predesigned into waste bodies and integrated with constructed injection trenches or vertical wells/manholes to form a distribution system. Horizontal systems have three main advantages: (1) fluid distribution throughout the waste mass is maximized, (2) injection rates can be very high, and (3) injection ports are outside operating and traffic areas and have little adverse impact on operations. Typical trench spacings are 100 to 200 ft horizontally and 40 ft vertically. In most cases, the horizontal system is the best approach for a high-quality bioreactor project. Dual-function trenches, which add liquids and extract gas, are being used at several locations. When placed at the top of the landfill, the trench spacing is reduced horizontally, which is intended to allow greater gas collection efficiency as well as to reduce odors.

Selection of a particular liquid injection method depends on site-specific conditions such as climate, malodors, worker exposure, environmental impacts, evaporative loss, reliability, uniformity, and aesthetics. Buried trenches or vertical wells offer advantages of minimum-exposure pathways, good all-weather performance, and favorable aesthetics. However, they may be adversely affected by differential settlement. Guidance on liquid addition, alternative design, and performance can be found in Reinhart and Townsend (1997).

Water balance models are helpful to assess the distribution of liquids during bioreactor landfill operations. In the case of conventional MSW landfills, water balance models such as HELP (discussed in Chapter 18) are used to determine leachate generation. In cases of bioreactor landfills, the injection of liquids causes complex leachate flow conditions, including heterogeneous moisture distribution. Furthermore, free moisture may also be generated as a result of waste decomposition. Laboratory simulation studies reveal that the free moisture generated during decomposition can be quite significant (Bogner et al., 2001). To assess the transient changes in moisture distribution as a result of liquid injection and waste degradation progresses, advanced hydrodynamic water balance models are needed (Bogner et al., 2001).

25.4.4 Gas Extraction System

A bioreactor landfill will generate more landfill gas (LFG) in a much shorter time than a drier landfill will. To control gas efficiently and avoid odor problems, the bioreactor LFG extraction system may require installation of large pipes, blowers, and related equipment early in its operational life. Horizontal trenches, vertical wells, near-surface collectors, or hybrid systems may be used for gas extraction. Greater gas flow is readily accommodated by increased pipe diameter as capacity increases with the square of pipe diameter. Because of the higher rate of gas generation, closer spacing of collection systems may be required. Liquid addition systems should be separate from gas extraction system to avoid flow impedance. The porous leachate removal system underlying the waste should be considered for integration with the gas extraction system.

Enhanced gas production can affect side slopes and cover negatively if an efficient collection system is not installed during the active landfill phase. Uplift pressure on geomembrane covers during installation can cause ballooning of the membrane and may lead to some local instability and soil loss. Temporary venting or aggressive extraction of gas during cover installation might facilitate cover placement. Once the final cover is in place, venting should be adequate to resist the uplift force created by LFG pressure buildup. The designer should consider the pressure buildup condition on slope stability when the collection system is shut down for any significant amount of time.

25.4.5 Final Cover System

Final cover system requirements for both conventional and bioreactor landfills are the same, and the final covers are designed to minimize infiltration into the waste mass and to help contain landfill gas emissions. The final cover systems for long-term closure consist (from bottom to top) of a foundation layer, gas collection layer, barrier layer, surface water drainage layer, cover soil layer, and topsoil with vegetative growth. Refer to Chapter 19 for detailed analysis and design of landfill final cover systems.

For conventional landfills, the final cover system is constructed within one year upon the landfill attaining its final height. The timing of final cover construction for bioreactor landfills needs careful consideration. For bioreactor landfills, the final cover should not be placed until the majority of settlement occurs, which may take five years or more. Therefore, an interim closure measure may be required to optimize the disposal capacity of bioreactor landfills. If odors are a concern, a temporary geomembrane liner may be needed. If odors are not a concern, a soil cover should be such that it does not affect the gas extraction system.

25.4.6 Slope Stability

Injection of excess liquids during the bioreactor operations will affect the geotechnical properties of waste. One should take into account the changes in the properties of waste and evaluate the stability of slopes. The stability analysis procedures used in standard geotechnical engineering can also be used for the analysis of bioreactor landfill slopes. Seismic effects should also be considered during geotechnical analysis, when appropriate.

The geotechnical properties of wastes needed for stability analyses include unit weight and shear strength. Typically, the unit weight of MSW in conventional landfills ranges from 50 to 75 lb/ft^3 near the ground surface to 95 to 110 lb/ft^3 at depth, depending on the waste composition, soil content, waste compaction, and other environmental factors (see Chapter 15). However, these values represent relatively dry MSW placed at a moisture content significantly below field capacity. Injecting fluids in bioreactor landfills causes the waste to be fully saturated and the unit weight is expected to increase by 20 to 30%, with about 60 to 100 lb/ft^3 near the surface to 110 to 140 lb/ft^3 at depth (Kavazanjian et al., 2001). Hater (2000) measured the unit weight of the wastes at four landfills where leach-

ate recirculation was used. Based on these measurements, the average in-place unit weight was found to be 112 lb/ft^3 or about 66% more dense than for conditions without leachate circulation, which were measured at 67 lb/ft^3. The increase in unit weight depends on the initial moisture content of the waste, leachate generation or recirculation amounts, waste field capacity, and void ratio available for liquids. As MSW degrades chemically and biologically, the void space in a unit volume is expected to decrease, leading to further increase in the unit weight. However, no specific data are available on the effects of degradation of waste on increase in the unit weight of the waste.

The other important geotechnical property needed for stability analysis is the shear strength of the waste. Because of the low degree of saturation and relatively high permeability of MSW, drained shear strength of MSW is used in the analysis of waste mass stability in conventional landfills (see Chapter 15). In the case of bioreactor landfills, initially the waste conditions are similar to that of a conventional landfill; therefore, the drained shear strength of undegraded waste in bioreactor landfills is the same as that of waste in the conventional landfills. However, as the liquids are injected in bioreactor landfills, the waste is fully saturated and degraded waste may have relatively low permeability, and under such conditions, undrained shear strength of the degraded waste is of great engineering significance.

For stability analysis, both undrained and drained shear strengths of the degraded waste are required. Unfortunately, no data are available on undrained shear strength of degraded MSW and such data are crucial in evaluating stability of bioreactor landfills in seismic regions. Information on the drained shear strength of degraded wastes is also very limited. Kavazanjian et al. (2001) obtained representative degraded waste samples from a bioreactor landfill and conducted direct shear tests under saturated drained conditions. Figure 25.2 shows the shear stress–displacement behavior and strength envelopes for different shear deformation of 1, 2, and 3 in. in an 18-in.-diameter shear box. The results show that the degraded waste exhibits ductile stress–displacement behavior with a friction angle of approximately 39°. This indicates that drained shear strength of degraded waste is higher than the typical

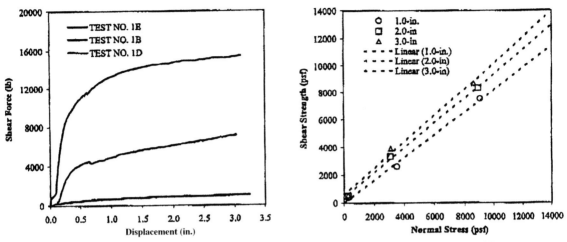

Figure 25.2 WMI bioreactor waste direct shear testing. (*From Kavazanjian et al., 2001.*)

drained shear strength of the waste in conventional landfills (Chapter 15); however, more testing is needed to determine the drained shear strengths of degraded wastes under varying environmental settings. For design purposes, the drained shear strength of degraded waste in bioreactor landfills may be assumed equal to the drained shear strength of waste in conventional landfills.

Injection of liquids in a bioreactor landfill endangers the stability of waste mass slopes, for the following reasons (Kavazanjian et al., 2001):

- Increased driving force due to the increase in unit weight of the waste mass due to liquid injection
- Decreased strength due to decreases in the effective ("intragranular") stress corresponding to the increases in pore pressure that result from liquid injection (leachate head buildup)
- Decreased strength due to transformation of waste mass by the biological and chemical processes that enhance degradation, turning the waste into an inherently weaker material

Lacking reliable data on shear strength of degraded waste, Isenberg et al. (2001) conducted sensitivity modeling using a typical landfill configuration to better understand how potential reduction in shear strength, coupled with increases in unit weight, will influence

waste mass slope stability. The landfill configuration, shown in Figure 25.3, consisted of:

- A composite bottom liner (e.g., smooth geomembrane over compacted clay soil) lying 10 ft below grade, sloping at 2% slope toward a perimeter leachate collection pipe
- 3H:1V final side slopes with benches at 40 ft vertical intervals
- 5% minimum final cover top slope
- Maximum waste depth 140 ft separated into three horizontal waste layers (upper, middle, and lower)

Four different bioreactor types were assumed as designated below:

- *Type O:* baseline condition; no liquids recirculation practiced (conventional Subtitle D landfill)
- *Type I:* limited and/or intermittent liquid recirculation and application practiced
- *Type II:* moderate and well-controlled liquid recirculation practiced; below field capacity
- *Type III:* heavy liquids recirculation and application; maximum lateral and vertical extent; approaches field capacity or beyond

The unit weights were assumed to increase with depth, and the values applied to the baseline case (type

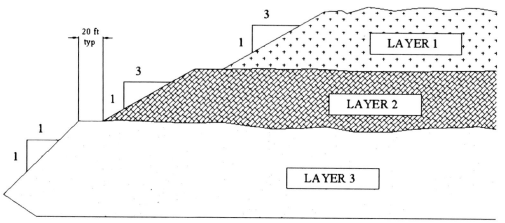

Figure 25.3 Typical landfill model configuration. (From Isenberg et al., 2001.)

O) were 45, 55, and 65 lb/ft³ for layer 1 (upper), layer 2 (middle), and layer 3 (lower), respectively. To model the combined influence of higher moisture content and waste settlement, unit weight values for type I, II, and III bioreactors were represented by increases of 25, 50, and 75%, respectively, over the baseline layer densities as shown in Table 25.2. Initially, the shear strengths were assumed to be the same for all landfill types. The friction angle values were assumed to be 26, 30, and 34° (upper, middle, and lower layers, respectively), and the cohesion values were assumed to be 200, 250 and 300 lb/ft² (upper, middle, and lower layers, respectively). It was assumed that the waste would not be saturated under controlled recirculation, and no pore pressures buildup would occur. Slope stability analyses were performed with circular and block modes of failure using the PCSTABL computer model. Circular failure planes were assumed through the waste mass, and block failure was assumed to occur along the bottom smooth geomembrane interface with interface friction angle of 8 to 10°. The factors of safety under baseline conditions for each of the landfill type are summarized in Table 25.2. These results show that waste mass by itself is stable even with lower shear strengths. The block failure mode generally indicated that the combination of low shear strength waste and low interface of the friction angle bottom liner system has a significant impact on the stability of the landfill.

Sensitivity analyses were conducted to investigate the effects of reduced shear strengths and leachate head buildup, and the results of these analyses are shown in Table 25.2. Overall, this sensitivity modeling showed that:

- By comparison to shear strength, unit weight of waste is not as critical a parameter in slope stability.
- Depending on the landfill configuration, block failure modes are likely to be more critical than circular failure modes within the waste mass.
- A buildup of liquid on the bottom liner system of 1, 5 and 10 ft reduced the FS value slightly, although the values for block failure were already less than 1.5.
- It is recommended that a variation of unit weight and shear strength with depth should be modeled with multilayers rather than using a single layer with average values.
- Slope stability analyses for bioreactor landfills need to consider site-specific conditions and the degree of bioreactor operations in order to make reasonable judgments in regard to slope stability.

25.4.7 Settlement

A bioreactor landfill will experience more rapid, total, and complete settlement than will a drier landfill. Ac-

Table 25.2 Input parameters and slope stability results

Bioreactor Type	γ_{wet} (lb/ft³)	C_u (lb/ft²)	ϕ_u (deg)	FS (circular)	FS (block)
1. Baseline Shear Strengths					
0	45	200	26		1.59[a]
	55	250	30	2.88	1.45[b]
	65	300	34		
I	56.3	200	26		1.55[a]
	68.8	250	30	2.74	1.37[b]
	81.3	300	34		
II	67.5	200	26		1.52[a]
	82.5	250	30	2.66	1.37[b]
	97.5	300	34		
III	78.8	200	26		1.50[a]
	96.3	250	30	2.59	1.35[b]
	113.8	300	34		
2. Reduced Shear Strength ($\Delta\phi_u = 2°$ and $\Delta C_u = 40$ to 60 lb/ft²)					
0	45	160	24		
	55	200	28	2.59	1.51[b]
	65	240	32		
I	56.3	160	24		
	68.8	200	28	2.46	1.48[b]
	81.3	240	32		
II	67.5	160	24		
	82.5	200	28	2.38	1.45[b]
	97.5	240	32		
III	78.8	160	24		
	96.3	200	28	2.33	1.43[b]
	113.8	240	32		
3. Reduced Shear Strength ($\Delta\phi_u = 4°$ and $\Delta C_u = 80$ to 120 lb/ft²)					
0	45	120	22		
	55	150	26	2.26	1.43[b]
	65	180	30		
I	56.3	120	22		
	68.8	150	26	2.17	1.40[b]
	81.3	180	30		
II	67.5	120	22		
	82.5	150	26	2.11	1.38[b]
	97.5	180	30		
III	78.8	120	22		
	96.3	150	26	2.07	1.38[b]
	113.8	180	30		
4. Reduced Shear Strength ($\Delta\phi_u = 6°$ and $\Delta C_u = 120$ to 180 lb/ft²)					
0	45	80	20		
	55	100	24	1.95	1.35[b]
	65	120	28		
I	56.3	80	20		
	68.8	100	24	1.89	1.33[b]
	81.3	120	28		

Table 25.2 (continued)

Bioreactor Type	γ_{wet} (lb/ft³)	C_u (lb/ft²)	ϕ_u (deg)	FS (circular)	FS (block)
II	67.5	80	20		
	82.5	100	24	1.84	1.31[b]
	97.5	120	28		
III	78.8	80	20		
	96.3	100	24	1.78	1.30[b]
	113.8	120	28		
5. Reduced Shear Strength ($\Delta\phi_u = 8°$ and $\Delta C_u = 160$ to 240 lb/ft²)					
0	45	40	18		
	55	50	22	1.52	1.26[b]
	65	60	26		
I	56.3	40	18		
	68.8	50	22	1.47	1.24[b]
	81.3	60	26		
II	67.5	40	18		
	82.5	50	22	1.43	1.23[b]
	97.5	60	26		
III	78.8	40	18		
	96.3	50	22	1.39	1.22[b]
	113.8	60	26		

6. Leachate Head Buildup (Baseline shear strengths)

Bioreactor Type	γ_{wet} (lb/ft³)	C_u (lb/ft²)	ϕ_u (deg)	Head Build-up (ft)[c]	FS (block)
III	78.8	200	26	1-C	2.59[a]
	96.3	250	30	5-C	2.59[a]
	113.8	300	34	10-C	2.51[a]
III	78.8	200	26	1-B	1.49[a]
	96.3	250	30	5-B	1.45[a]
	113.8	300	34	10-B	1.39[a]
III	78.8	200	26	1-B	1.35[b]
	96.3	250	30	5-B	1.32[b]
	113.8	300	34	10-B	1.26[b]
III	78.8	200	26	1-C	1.39[a]
	96.3	250	30	5-C	1.39[a]
	113.8	300	34	10-C	1.39[a]
III	78.8	200	26	1-B	1.42[a]
	96.3	250	30	5-B	1.39[a]
	113.8	300	34	10-B	1.31[a]

Source: Isenberg et al. (2001).

[a] Assume 10° interfacial friction angle with bottom liner.

[b] Assume 8° interfacial friction angle with bottom liner; change in density only from baseline.

[c] C indicates circular failure surface; B indicates block failure surface.

celerated settlement results from both an increased rate of solid waste decomposition and increased compression through higher specific weights. Settlement observations were made at several bioreactor landfills. At Keele Valley landfill, a settlement rate of 10 to 12 cm/month was observed in wet areas as opposed to 5 to 7 cm/month in dry areas, under identical waste characteristics (Mosher et al., 1997). At the Yolo County landfill, settlement in bioreactor cells was found to be three times higher than that of the waste in a conventional landfill (Yolo County, 1999).

Settlement during the landfilling operations will affect the performance of the final surface grade, surface drainage, roads, and gas-collection piping system. Because of the significant increase in settlement magnitude and rate, it could be very beneficial to overfill the waste above design grade before placement of the final cover. Alternatively, a significant benefit may accrue if the final cover and final site-improvement installations are postponed and the rapid settlement is used to recapture airspace. Settlement impacts can readily be accommodated by the project design. Since settlement will be largely complete soon after landfill closure, long-term maintenance costs and the potential for fugitive emissions will be avoided.

25.5 BIOREACTOR LANDFILL OPERATIONS AND MAINTAINENCE

The operation of a bioreactor landfill should be treated as operating a large biological waste digester, and all operations should be closely monitored. The operation of the bioreactor landfill should be such that biological degradation of the organic constituents of the waste is achieved in an efficient manner. To accomplish this, the operations and maintenance program should address the following issues: (1) solid waste pretreatment or segregation, (2) leachate seeps, (3) daily and intermediate cover, and (4) management of nutrients and other supplement addition.

25.5.1 Solid Waste Pretreatment or Segregation

Bioreactor operations are most efficient and effective where the waste has a high organic content and large exposed specific surface areas. For this reason, bioreactor operations should be concentrated on waste segregated to maximize its organic content and shredded, flailed, or otherwise manipulated to increase its exposed surface area. Waste segregation could include separation of construction and demolition wastes from MSW. Limited shredding can be achieved by spreading waste in thin lifts and using landfill equipment to break open plastic bags and break down containers. Mechanical shredding can be efficient and effective in reducing particle size and opening bags; however, it is an intensive, high-maintenance, and high-cost activity that might not be cost-effective. Moreover, shredded wastes may become exceedingly dense after placement, thereby limiting moisture penetration.

25.5.2 Leachate Seeps

Adding liquid to solid waste landfill increases the potential for leachate seeps or breakouts, and the landfill must be operated to minimize such possibilities. Leachate must be precluded from contaminating stormwater runoff. Monitoring for leachate seeps is mandatory, and the operation plan must include a rapid response action to correct leachate seeps as they develop. Such measures as installation of slope and toe drains, surface regrading, filling and sealing cracks as necessary to reduce surface water infiltration, and reducing the liquid addition rate are some of the standard methods used to address this condition. Managing liquid addition tare, amount, and location can limit the potential for slope seeps.

25.5.3 Daily and Intermediate Cover

Generally, 6 in. of daily soil cover material is required at conventional MSW landfills to control vectors, fire, blowing litter, odors, and scavenging. Federal and state regulations allow using alternative daily cover materials such as relatively permeable waste materials, including foundry sands, shredded tires, contaminated soils, incinerator ash residue, green waste/compost, and auto fluff. Many manufactured alternative daily cover materials have also been developed and marketed on the basis that they help to conserve air space. Examples of manufactured materials include sprayed-on slurries, polymer foams, and removable tarps.

The use of daily cover in a bioreactor landfill requires special attention (Miller et al., 1991; Phaneuf, 2000). A cover material more permeable than the waste can direct leachate to the sides, where the leachate must be collected and drained properly. A cover material less permeable than the waste can create barriers to the effective percolation of liquids and cause internal leachate mounding that can promote surface seeps and contribute to waste mass stability concerns. It can impede leachate distribution and LFG flow to collections and distribution systems. Its ability to serve as a barrier should be reduced through scarifying, or partial removal, prior to placing solid waste over it. When placed within 50 ft of the slopes, it should be graded to drain back into the landfill to preclude leachate from reaching the slope and emerging as a seep. Use of alternative covers that do not create such barriers can mitigate these effects. Therefore, the characteristics of the daily and intermediate cover materials used, or to be used, needs to be factored into the evaluation of all bioreactor landfill operations.

25.5.4 Management of Nutrients and Other Supplement Addition

Nutrients requirements are generally supplied by waste components (Barlaz et al., 1990), but research suggests that nutrients and other biological and chemical supplements may enhance biological activity. Addition of such supplements has not yet been attempted in the field. As with waste segregation, or shredding, the costs of nutrients and other additions will need to be justified.

Optimum pH for methanogens is approximately 6.8 to 7.4. Buffering of leachate to maintain pH in this range has been found to improve gas production in laboratory studies. Particular attention to pH and buffering needs should be given during early stages of leachate recirculation. Careful operation of the bioreactor landfill initially through slow introduction of liquids should minimize the need for buffering.

25.5.5 Monitoring Program

As mentioned earlier, the bioreactor landfill should be treated as a biological digester, and a monitoring program should be an integral part of the operations to monitor the performance so that the process can be managed and ultimately improved. The monitoring program should be tailored to site-specific conditions and should provide the following data:

- Liquid injection volumes
- Temperature within waste
- Moisture content of waste
- Cellulose/lignin in waste
- Leachate yield and quality
- Waste density
- Settlement
- Gas flow/quality
- Leachate levels within waste

Other parameters, such as redox, volatile solids, and biochemical methane production, that indicate biological activity may also be monitored.

25.6 CASE STUDIES

Several demonstration projects have been initiated or completed to understand technical, regulatory, public awareness, and operational issues in dealing with bioreactor landfills. A number of municipal landfills have gained approval from the USEPA to test bioreactor landfill technology. The tests are being conducted under the EPA's Project XL, which through flexibility in regulatory requirements encourage businesses and governments to experiment in areas that may benefit the environment and public health. These tests are intended to address technical issues such as biodegradation rates of waste and stability and settlement of bioreactors as well as nontechnical issues such as limited regulatory awareness and negative perception, limited performance data, and limited data on costs.

A survey conducted by Gou and Guzzone (1997) showed that approximately 130 MSW landfills are currently employing leachate recirculation. Examples of bioreactor landfill activities in selected states are:

- *California.* For three years, Yolo County has been operating a bioreactor demonstration cell that contains 9000 tons of waste. Yolo County is negotiating with concerned state regulatory agencies to permit and then operate the next 15-acre landfill cell of the Yolo County central landfill as a bioreactor.
- *Florida.* The state recently allocated more than $3.2 million to establish a demonstration bioreactor landfill.
- *Georgia.* Two aerobic bioreactor landfill projects are operational; one at the Live Oak landfill in Atlanta and the other at the Baker Road landfill in Columbia County.
- *Iowa.* The Bluestem Solid Waste Authority has received a $500,000 state grant for its bioreactor project at the Bluestem No. 2 landfill near Marion. Waste placement began in December 1998.
- *New York.* Six large-scale bioreactor demonstration projects have been conducted (Phaneuf, 2000). Four of the bioreactor projects used aerobic decomposition process, and the other two used anaerobic decomposition process.
- *South Carolina.* The State Research and Development and Demonstration Program is sponsoring an aerobic activity at the Aiken County landfill.

The main purpose of these projects is to obtain performance data to address both technical and nontechnical issues. Further information on two projects, from the list above, is presented briefly below.

Case Study 1: Yolo County Bioreactor Landfill Project. The Yolo County landfill bioreactor project consists of two demonstration cells, roughly 100 ft² and 40 ft deep, at the Yolo County central landfill, Davis, California. The first cell, the control cell, is set up in the same manner as a conventional MSW landfill, with a bottom liner and surface liner in place to prevent liquid infiltration or leakage. The second cell, the enhanced cell, is set up as a bioreactor landfill. Both cells were built using a Subtitle D composite liner system

consisting of compacted soil and a HDPE geomembrane liner. Figure 25.4 shows the control and enhanced cells schematically. Because of the liquid addition, the enhanced cell was required to have a second liner system placed below the primary liner. A liquid detection and collection system was placed between these liners to detect any possible leakage from the primary liner. Each cell was filled with about 9000 tons of MSW from April through October 1995. The waste was placed in 5-ft lifts, with 1 ft of shredded greenwaste as daily cover. Large bulky items such as couches and mattresses were excluded from the cells. Several monitoring instruments were installed into both cells during their construction, including pressure, temperature, and moisture sensors. Liquid dispersion in the enhanced cell is accomplished with 14 infiltration trenches dug into the surface of the cell. The specific objectives of the project are:

- Demonstrate substantially accelerated landfill gas generation and biological stabilization while maximizing gas capture.
- Estimate the landfill life extension that can be realized through rapid waste decomposition.
- Demonstrate that the recirculation of leachate is an effective leachate treatment strategy.
- Provide regulatory agencies with information to develop guidelines for the application of this technology.
- Better understand the movement of moisture through landfills.
- Disseminate information resulting from the continued monitoring of the project.
- Monitor the biological conditions within the landfill cells.
- Assess the performance of shredded tires as a medium for the transfer of landfill gas to collection pipes.

Both cells were highly instrumented to monitor conditions within the waste. These systems collect, measure, and record data independently from each other. Moisture and temperature sensors were placed throughout the waste in each cell. Data from these sen-

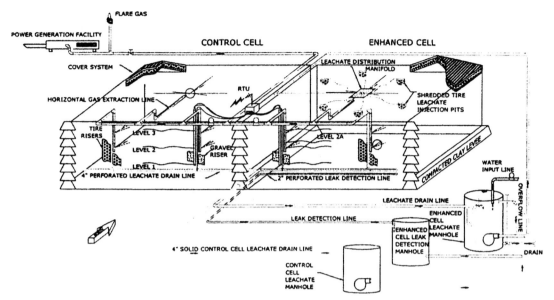

Figure 25.4 *Yolo County bioreactor landfill demonstration project. (From Yolo County, 1999.)*

sors are automatically recorded and sent to the Yolo County office via a remote telemetry unit. The landfill gas generated in the cells is collected using vertical collection wells. Some of these wells employ a conventional medium of gravel, and others use shredded tires. Each cell also has a horizontal gas collection system, consisting of a layer of shredded tires over the cell's surface. Landfill gas not captured by the vertical wells is collected in this horizontal system, thereby preventing surface emissions. The use of shredded tires is demonstrating their possible effectiveness as a gas extraction medium. The leachate monitoring system uses manholes that allow personnel to monitor leachate quantity and quality.

The monitoring data collected as of 1999 revealed the following:

1. *Air quality.* Based on air quality data, oxygen-deficient conditions were found in landfills, which results in anaerobic microbial activity. An important result of the microbial biodegradation activity is the production of landfill gas, which consists primarily of methane and carbon dioxide, both greenhouse gases. Air quality regulations exist to control landfill gas emissions from landfills that emit at a rate above a specific threshold. These regulations require the gas to be collected and destroyed. Although some surface cover systems have only 60 to 90% efficiency in preventing gas loss to the atmosphere, new impermeable surface liners, similar to the one used in this project, are becoming widely used and provide nearly 100% gas capture efficiency.

2. *Renewable energy.* As discussed previously, methane and carbon dioxide are two of the main landfill gases produced and both are greenhouse gases. Methane is 21 times more potent than carbon dioxide in its effects on the atmosphere. In a MSW landfill, the gas is commonly collected and destroyed by flaring. Flaring destroys the methane by burning. But methane is also a potential source of energy. Because the methane gas generation in a MSW landfill is far less than optimal, and even less predictable, it is not usually feasible to operate gas-to-energy conversion facilities at these landfills. Up to 50% of the gas generated from a traditional MSW landfill occurs more than 30 years after its closure, which is also beyond the mandated gas collection time period.

A bioreactor landfill greatly improves the gas generation rate, decreasing the time frame for landfill gas generation from several decades to between five and 10 years. For this project, the enhanced cell has produced over 75% more landfill gas than the control cell since June 1996. Of the landfill gas, the percent methane in the enhanced cell has averaged 54%. The control cell was not significantly different in its concentration, averaging 52%, until March 1998, when it dropped to 30%. Monitoring continues to determine if this drop is a temporary condition. Because the quantity of landfill gas generation has been much higher in the enhanced cell, the total amount of methane generated has also been significantly larger. As of September 1998, the cumulative methane volume was 1.014 ft^3 per dry pound for the enhanced cell and 0.556 for the control cell. Figure 25.5 shows the cumulative methane gas volumes for both cells.

Because the rate of methane generation is significantly higher, the feasibility of harnessing this renewable energy is greatly improved. This means that the energy market could increasingly depend on this type of renewable energy for the provision of electric generation rather than nonrenewable fossil fuels.

3. *Leachate.* There are five distinctive phases of waste decomposition: initial, transition, acid, methane fermentation, and mature phase. These phases occur sequentially and may be recognized in the leachate characteristics and composition. Leachate carries a high pollutant load during the initial phase of waste decomposition. These levels are reduced significantly once the methane fermentation phase of waste decomposition is under way. Most of the waste is stabilized, and landfill gas generation is nearly complete during the maturation phase.

In a bioreactor landfill, the methane fermentation phase is reached much faster and is completed in a shorter time frame, as discussed previously, leading to appreciably reduced pollutant levels early in the life of the landfill. By reducing the pollutant load early, the risk of groundwater contamination from seepage that could occur due to the age of the liner, defects, or other mishaps, is greatly reduced. Leachate treatment costs are tied directly to the pollutant load. Therefore, significant savings can be realized for the disposal of leachate that has been even partially decontaminated by microbial actions. Monitoring continues on this project in an effort to quantify the potential treatment effect.

4. *Landfill life extension.* Waste decomposition results in landfill settlement, which creates a volume of space that could be reused for additional waste placement. Unfortunately, the majority of landfill settlement from conventional landfill practices occur after the 30-year postclosure period, at which time it cannot be reduced. A bioreactor landfill accelerates the decompo-

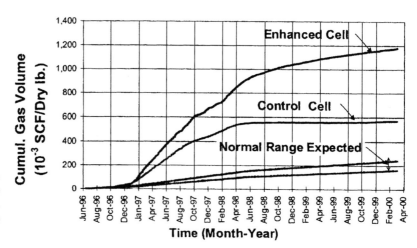

Figure 25.5 Measured methane gas volumes at Yolo County bioreactor landfill demonstration project. (From Yazdani et al., 2000.)

sition process to allow for additional waste placement during the life of the landfill. This reuse of space could potentially extend the life of the landfill by 20%. Settlement surveys for this project shown in Figure 25.6 exhibit that the settlement rate of the enhanced cell is nearly four times faster than the control cell. If the settlement rate continues, in less than four years the enhanced cell will have settled approximately 20%.

Subsequent to the demonstration project, Yolo County has proposed a full-scale 20-acre landfill module with both anaerobic and aerobic bioreactor areas (Yazdani et al., 2000). In the first phase of this project, a 12-acre module has been constructed. One 9.5-acre cell will be operated anaerobically and the other 2.5-acre cell aerobically. The liner system consists, from top to bottom, of an operations/drainage layer capable of maintaining less than 1 ft of head over the liner, a 60-mil HDPE geomembrane liner, and 2 ft of compacted clay ($k \leq 10^{-7}$ cm/s), as shown in Figures 25.7 and 25.8. The liner and leachate collection system consists, from top to bottom, of a 2-ft-thick chipped tires operations/drainage layer ($k > 1$ cm/s) over 6 in. of pea gravel, a blanket geocomposite drainage layer, a 60-mil HDPE liner, 2 ft of compacted clay ($k \leq 6 \times 10^{-9}$ cm/s), 3 ft of compacted earth fill ($k \leq 1 \times 10^{-8}$ cm/s), and a 40-mil HDPE vapor barrier layer as shown in Figure 25.8. The operation and monitoring

data will help compare the performance of bioreactors operated under anaerobic and aerobic conditions.

Case Study 2: City of Fargo Bioreactor Project. Richard et al. (2000) reported a five-year pilot-scale study conducted to investigate the effects of circulating leachate versus clean water at a specially constructed bioreactor landfill in Fargo, North Dakota. The specific objectives of the study were to:

- Evaluate the feasibility of MSW stabilization in parallel landfill bioreactors applying actual leachate to one cell and clean water in controlled amounts to the parallel test cell.
- Compare MSW stabilization rate and leachate characteristics from parallel reactors.
- Observe the time required for significant reduction in leachate characteristics.
- Observe the effects of cold weather.
- Develop an operational process that could be applied to full-scale rectors.

The parallel bioreactors constructed for this study were approximately 12 × 12 ft in area, with a depth of about 10 ft. Each reactor had a PVC-lined bottom, a leachate collection system draining to a central sump, a leachate/water distribution network at the upper 2-ft

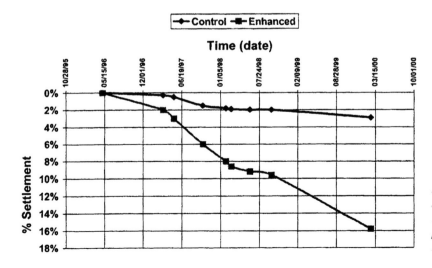

Figure 25.6 Measured waste settlement at Yolo County bioreactor landfill demonstration project. (From Yazdani et al., 2000.)

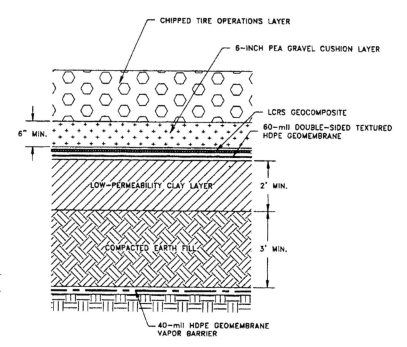

Figure 25.7 *Bottom liner for full-scale 20-acre bioreactor landfill: Yolo County. (From Yazdani et al., 2000.)*

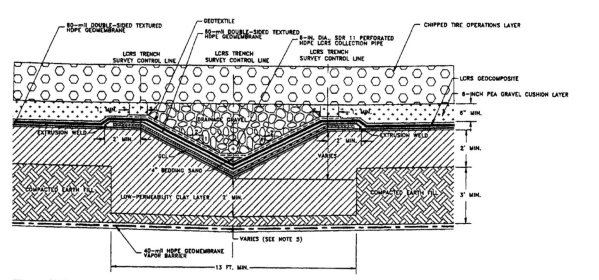

Figure 25.8 *Bottom liner and leachate collection trench for full-scale 20-acre bioreactor landfill: Yolo County. (From Yazdani et al., 2000.)*

depth, and temperature-monitoring probes stacked at three depths within the active decomposition zone.

Generally, once per week during the first years of the study, leachate was pumped out and recirculated to one reactor, while leachate in the parallel reactor was removed and an equal portion of clean water introduced. This routine was followed during the warmer months, generally April through November in North Dakota, while freezing was allowed during winter months.

Figure 25.9 shows a cross section of the bioreactor with the leachate recirculation system. The MSW used was collected from typical neighborhoods and was hand placed and compacted, and all plastic bags were opened to ensure an even liquid application throughout the waste. Leachate samples were collected over approximately five years, with most samples analyzed during the early stages when significant differences in leachate were apparent and changing dramatically. The volume of leachate and/or clean water applied and subsequently recovered was measured and analyzed for chemical oxygen demand (COD), pH, temperature,

heavy metals, and several inorganic constituents. The results of percent liquids recovered, COD, pH, and TDS are shown in Figure 25.10. These results strongly suggest that after an initial acclimation period, the rate of stabilization is not materially affected by the characteristics of the liquid applied.

25.7 RESEARCH ISSUES

Research on bioreactor landfills is being conducted by various researchers to increase the efficiency and acceptability of bioreactor landfills. The research issues include (Reinhart et al., 2000; USEPA, 2003) the following:

- *Characterization of the extent of chemical and biological degradation of waste.* An effort is being made to obtain waste samples from bioreactors and characterizing them for composition. The fate of the remaining waste constituents, such as heavy metals over the long term under varying conditions is also being explored.

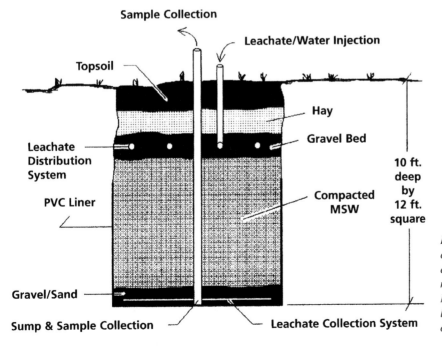

Figure 25.9 Leachate recycle in south cell and clean water addition in north cell at the Fargo, North Dakota, bioreactor landfill project. (From Richard et al., 2000.)

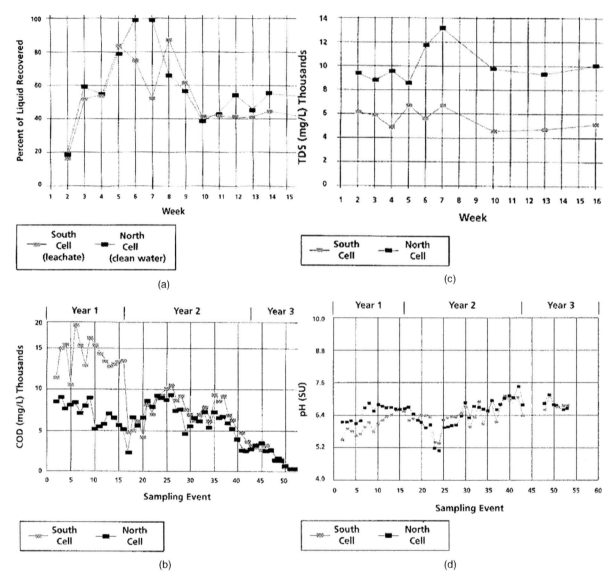

Figure 25.10 *Monitoring data at the Fargo, North Dakota, bioreactor landfill project.* (*From Richard et al., 2000.*)

- *Implication of various compaction methods on increase in unit weight and settlement and reduction in permeability.* Although settlement increases the air space, the reduction in permeability makes uniform liquid injection difficult. Compaction also contributes to anisotropic conditions within the

landfill that will promote lateral migration of injected liquids.

- *Beneficial effects of introducing nonindigeneous liquids.* The introduction of fresh water and liquids containing biosolids on biodegradation rates is being investigated at several landfills. Nonindigeneous liquids help to supplement nutrients and moisture, to dispose of liquid waste products, to compensate for insufficient leachate volumes, and/or to avoid concentration of inorganic contaminants in leachate. Risks of introducing non-indigenous liquids on potential groundwater contamination should also be investigated.

- *Leachate hydrodynamics.* Flow of injected liquids in bioreactor landfills is complicated because of the heterogeneous characteristics of waste. Laboratory testing, field monitoring, and mathematical modeling are being used to understand the flow and distribution of injected liquids under various landfill operating conditions. Mathematical modeling may be used to evaluate various operating strategies (Al-Yousfi and Pohland, 1998).

- *Geotechnical properties of degraded waste.* In particular, the undrained shear strength of degraded waste is needed for accurate stability analysis of the landfill slopes.

- *Leachate quality.* Field studies are being carried out to determine leachate quality during bioreactor operations. In general, organic concentrations are found to decrease with time; however, the fate of heavy metals under long-term conditions is unknown.

- *Aerobic/anaerobic bioreactors.* Efforts are being made to develop hybrid systems that use anaerobic and aerobic microorganisms to degrade and stabilize the wastes efficiently.

- *Cost analysis.* The analysis procedures that account for all aspects of bioreactor landfills, such as rapid gas production, recovered landfill space, reduced environmental impact, and reduced post-closure care, are being developed. Criteria for termination of postclosure monitoring are being developed (Barlaz et al., 2002).

25.8 SUMMARY

Bioreactor landfills offer several processes to achieve rapid degradation, and thus stabilization, of the relatively rapidly degrading organic waste materials within a relatively short time. Although it requires increased management and more environmental controls, the bioreactor landfill can result in enhanced performance, fewer long-term environmental risks, and higher potential revenue to help defray operational costs. Over the long term this should result in considerable environmental protection and cost savings.

QUESTIONS/PROBLEMS

25.1. State advantages and disadvantages of various liquid injection methods for bioreactor landfills.

25.2. Compare and contrast environmental and financial benefits of conventional and bioreactor landfills.

25.3. Based on published literature, prepare a list of bioreactor landfill projects that are either completed or ongoing in the United States.

25.4. Why is clogging of the leachate collection and removal system so crucial for bioreactor landfills? What are various methods for mitigating such clogging problems?

25.5. Based on expected temperature increases in bioreactor landfills, what considerations should be taken in the design of liner and leachate collection systems for bioreactor landfills?

25.6. List different hydrodynamic (water balance) models that have been used to determine moisture distribution in bioreactor landfills. Explain their main limitations.

25.7. List various models used to predict landfill gas generation and extraction in bioreactor landfills. Explain the limitations of these models.

25.8. What are the effects of pH of waste on biodegradation rates? How can pH be controlled in bioreactor landfills?

25.9. What are the effects of leachate recirculation on the fate and transport of inorganic contaminants in bioreactor landfills?

25.10. Explain why settlement is higher and occurs faster in aerobic bioreactors than in anaerobic or conventional landfills.

25.11. Explain how geotechnical properties of wastes will be affected by operation of bioreactor landfills aerobically and anaerobically.

REFERENCES

Al-Yousfi, A. B., and Pohland, F. G., Strategies for simulation, design, and management of solid waste disposal sites as landfill bioreactors, *Pract. Per. Hazard., Toxic Radioact. Waste Manag.*, Vol. 2, No. 1, pp. 13–21, 1998.

Baker, J. A., and Eith, A. W., The bioreactor landfill: perspectives from an owner/operator, pp. 27–39 in *Proceedings of the 14th GRI Conference*, Las Vegas, NV, 2000.

Barlaz, M. A., Ham, R. K., and Schaefer, D. M., Methane production from municipal refuse: a review of enhancement techniques and microbial dynamics, *Crit. Rev. Environ. Control*, Vol. 19, No. 6, pp. 557, 1990.

Barlaz, M. A., Rooker, A. P., Kjeldsen, P., Gabr, M. A., and Borden, R. C., Critical evaluation of factors required to terminate the postclosure monitoring period at solid waste landfills, *Environ. Sci. Technol.*, Vol. 36, No. 16, pp. 3457–3465, 2002.

Bogner, J., Reddy, K., and Spokas, K., Dynamic water balance aspects of bioreactor landfills, presented at the Sardinia Conference, 2001.

Campman, C., and Yates, A., Bioreactor landfills: an idea whose time has come, *MSW Manag.*, pp. 70–81, Sept./Oct. 2002.

CFR (*Code of Federal Regulations*), 40 CFR, U.S. Government Printing Office, Washington, DC, 2002.

Gou, V., and Guzzone, B., *State Survey on Leachate Recirculation and Landfill Bioreactors*, Solid Waste Association of North America, Silver Spring, MD, 1997.

Hater, G. R., Leachate recirculation, landfill bioreactor development and data acquisition at Waste Management, Landfill Methane Outreach Program, 3rd Annual LMOP Conference, Washington, DC, 2000.

Isenberg, R. H., Law, J. H., O'Neill, J. H., and Dever, R. J., Geotechnical aspects of landfill bioreactor design: is stability a fatal flaw? pp. 51–62 in *Proceedings of the SWANA Landfill Symposium*, 2001.

Kavazanjian, E., Hendron, D., and Corcoran, G. T., Strength and stability of bioreactor landfills, pp. 63–70 in *Proceedings of the SWANA Landfill Symposium*, 2001.

Mehta, R., Barlaz, M. A., Yazdani, R., Augenstein, D., Bryars, M., and Sinderson, L., Refuse decomposition in the presence and absence of leachate recirculation, *J. Environ. Eng.*, Vol. 128, No. 3, pp. 228–236, 2002.

Miller, L. V., Mackey, R. E., and Flynt, J., Evaluation of a PVC liner and leachate collection system in a 10-year-old municipal solid waste landfill, presented at the 29th Annual SWANA Solid Waste Exposition, 1991.

Mosher, F., McBean, E. A., Crutcher, A. J., and MacDonald, N., Leachate recirculation to achieve rapid stabilization of landfills: theory and practice, pp. 121–134 in *Proceedings of the SWANA 2nd Annual Landfill Symposium*, 1997.

Pacey, J., Augenstein, D., Morck, R., Reinhart, D., and Yazdani, R., The bioreactive landfill, *MSW Manag.*, pp. 53–60, Sept./Oct. 1999.

Phaneuf, R. J., Bioreactor landfills: regulatory issues, pp. 9–26 in *Proceedings of the 14th GRI Conference*, Las Vegas, NV, 2000.

Phaneuf, R. J., and Vana, J. M., Landfill bioreactors: a New York state regulatory perspective, *MSW Manag.*, pp. 46–52, May/June 2000.

Pohland, F. G., *Landfill Bioreactors: Historical Perspective, Fundamental Principles, and New Horizons in Design and Operation*, EPA/600/R-95/146, USEPA, Washington, DC, 1994.

Reddy, K. R., and Saichek, R. E., Characterization and performance assessment of shredded scrap tires as leachate drainage material in landfills, in *Proceedings of the 14th International Conference on Solid Waste Technology and Management*, Philadelphia, 1998.

Reinhart D. R., and Townsend, T. G., *Landfill Bioreactor Design and Operation*, CRC Press, Boca Raton, FL, 1997.

Reinhart, D. R., McCreanor, P. T., and Townsend, T. G., The bioreactor landfill: research and development needs, pp. 1–8 in *Proceedings of the 14th GRI Conference*, Las Vegas, NV, 2000.

Richard, D., Zimmerman, R., and Grubb, B., Leachate recirculation or clean-water addition: a comparative study, *MSW Manag.*, pp. 104–110, May/June 2000.

Sullivan, P., Getting down to cases: just what is a bioreactor landfill? *MSW Manag.*, pp. 64–67, July/Aug. 2000.

SWANA (Solid Waste Association of North America), Request for comment on bioreactor landfill definition, sent to USEPA, June 29, 2001.

USEPA (U.S. Environmental Protection Agency), *Landfill Bioreactor Design and Operation*, EPA/600/R-95/146, USEPA, Washington, DC, 1995.

———, State of the practice for bioreactor landfills, presented at the Workshop on Bioreactor Landfills, Arlington, VA, Sept. 6–7, 2000, EPA/625/R-01/012, Office of Research and Development, USEPA, Cincinnati, OH, 2002.

———, Workshop on Bioreactor Landfills, Arlington, VA, Feb. 27–28, 2003, Office of Solid Waste and Office of Research and Development, USEPA, 2003.

Warith, M., Bioreactor landfills: experimental and field results, *Waste Manag.*, Vol. 22, pp. 7–17, 2002.

Yazdani, R., Augenstein, D., and Pacey, J., U.S. EPA Project XL: Yolo County's accelerated anaerobic and aerobic composting (full scale controlled landfill bioreactor) project, pp. 77–105 in *Proceedings of the 14th GRI Conference*, Las Vegas, NV, 2000.

Yolo County, *The Yolo County Landfill Bioreactor Demonstration Project*, 1999.

26

SUBAQUATIC SEDIMENT WASTE: IN-SITU CAPPING

26.1 INTRODUCTION

Contaminated sediments are a common problem in bodies of water. Many industries have been located near a body of water to take advantage of the easy access to water in their various processes. Throughout the decades, many contaminants have been discharged into the water, where they are bound to sediments on the floor of the water body. Once the contaminants bind to the sediments, the organisms that feed off the bottom of the water bodies consume the contaminant; thus, the contaminant eventually enters the food chain. Therefore, contaminated sediments can pose serious risks to human health and the environment.

Many contaminated sediment sites have been remediated using different techniques, depending on the site characteristics. The various remediation technologies can be divided into (1) removal technologies, and (2) nonremoval technologies. *Removal technologies* consist of mechanical dredging and hydraulic dredging. Mechanical dredging uses construction equipment such as backhoes to dislodge and excavate contaminated sediments. Hydraulic dredging pumps or sucks the contaminated sediments out through a pipe in a slurry form. Dredged sediments are either disposed of in confined disposal facilities (CDFs) that are designed similar to surface impoundments/landfills or treated on- or off-site similar to contaminated soils. *Nonremoval technologies* include in-situ capping (or containment) and in-situ treatment. In-situ capping consists of placing a layer of material over the contaminated sediments to isolate them from the environment. In-situ treatment consists of either treating the contamination chemically to render it immobile or treating the con-

tamination biologically or chemically to degrade or remove it (USEPA, 1994).

The human and ecological populations are exposed to various risks when sediments in water bodies become contaminated. In assessing how to remediate these environments, it is necessary to determine what additional risks will arise when performing the remediation. One of the ways to avoid creating additional risks is not to move or treat the sediments but to contain them subaqueously by creating a physical barrier between the contaminated sediments and the aquatic (water) environment. This can be accomplished by in-situ capping (ISC). Figure 26.1 shows cross sections of typical ISC systems. The sequence of steps involved in the design of ISC systems is shown in Figure 26.2. The ISC systems are designed to perform the following functions (USEPA, 1994; Palermo, 1998; Palermo et al., 1998):

- *Physical isolation of the contaminant:* reduction of the risk of exposure by preventing the migration of contaminants up the food chain. An in-situ cap can prevent benthic organisms from consuming the contaminant and introducing the contaminant into the food chain. Source control should also be considered in achieving physical isolation of the contaminant.

- *Stabilization of the sediments:* prevention of sediments from resuspending, transporting, and/or redepositing in another area of the water body. For example, if there is contamination upstream, it would be necessary to contain those sediments

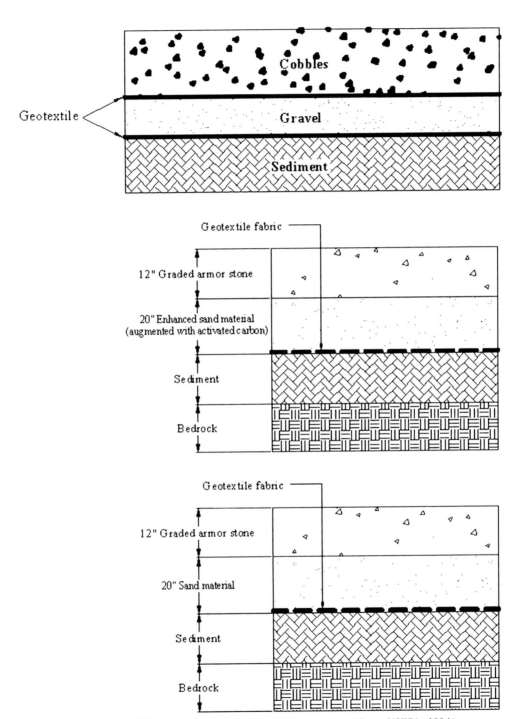

Figure 26.1 *Schematic of typical in-situ capping systems. (From USEPA, 1994.)*

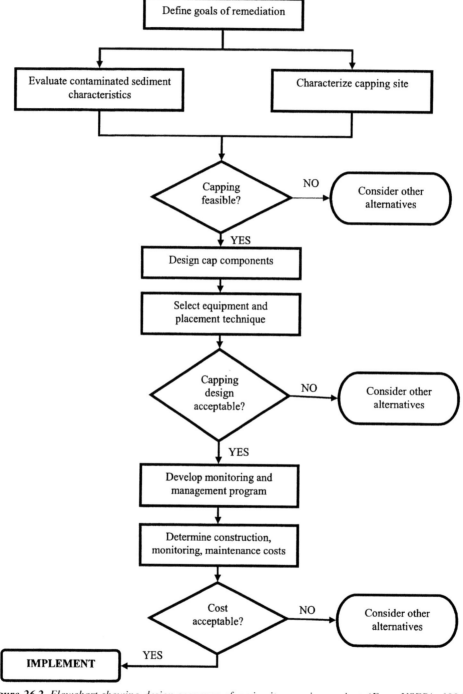

Figure 26.2 *Flowchart showing design sequence of an in-situ capping project. (From USEPA, 1994.)*

so as not to contaminate the downstream area of the water body.

- *Chemical isolation* (i.e., reduction of the flux of dissolved contaminants): containment of chemicals and nutrients from the overlying water column.

The advantage of in-situ capping is that the contaminants are not disturbed, so there is no further exposure to the contaminants. The disadvantage is that it is still a new technology and long-term performance of the cap is not known. ISC is more suitable in low-energy environments, where the cap is less likely to be disturbed; however, ISC is possible in high-energy environments if extra armoring is put in place to maintain the cap integrity. Water depth should be considered also, in both the feasibility of cap construction and the stability of the environment. Generally speaking, deeper water is more stable and therefore more conducive to capping, but it may be difficult to construct.

In this chapter, various terms used in ISC are defined. Then, the general design considerations of ISC, including the geotechnical issues, are presented, followed by a discussion of the construction, monitoring, regulatory, and economic aspects of ISC. Finally, a few case studies are presented that describe the design and construction of ISC under varying site conditions.

26.2 RELEVANT TERMINOLOGY AND DEFINITIONS

When designing structures to be built in or around water and sediments, the following water properties, sediment properties, and flux (movement) properties need to be determined at different stages of an in-situ capping project to ensure that the cap will perform the remediation goals.

26.2.1 Water Properties

- *Hydraulic conductivity of porous media:* the ease with which water flows through porous material (e.g., capping materials and sediments). It is primarily dependent on the characteristics of the porous material. Hydraulic conductivity values for various porous media can be found in the literature (see Chapters 5 and 7), or it can be determined using Darcy's law. The hydraulic conductivity of porous media can also be determined by performing laboratory tests or field tests (see Chapters 5 and 7).

- *Benthic population:* the type(s) and quantity(s) of microorganisms inhabiting the site area. These organisms feed at the bottom of water bodies and are the entryway of substances into the food chain. The amount of feeding on and penetration into the ISC depend on the local benthic population.

- *Suspended solids:* the amount of organic and inorganic solids floating in the water. These can be determined by taking a representative water sample of a known volume, evaporating the water content, and weighing the solids that remain.

- *Contaminant concentration:* the amount of contaminant contained in water and in sediment. This is measured by different analytical methods in laboratories and is usually expressed in weight per dry weight of sediment (e.g., mg/kg) and weight per volume of water (e.g., mg/L).

- *Salinity:* salt concentration in the water body, generally measured in mg/L.

26.2.2 Soil or Sediment Properties

- *Grain size distribution.* The grain size distribution of capping materials and sediments is usually determined using ASTM D 422 (see Chapter 5). *Fine-grained materials* are defined as materials with more than 50% by weight passing a No. 200 sieve. They provide a better chemical barrier than coarse-grained materials because of the high sorption capacity of fine-grained materials. Hydrophobic organic pollutants bind quickly to the abundant fresh sorption sites in fine-grained material, thus reducing the flux rate. *Coarse-grained materials* such as sandy soils, materials with grain sizes ranging from 0.075 to 4.75 mm, provide a more stable cap than fine-grained materials be-cause they are less likely to resuspend in the wa-ter, can be stable at steeper slopes, and deter bio-turbation.

- *Plasticity indices.* A series of threshold moisture contents is observed when the water content of a soil is steadily changed, indicating the consistency (plasticity) of the soil. The *liquid limit* (LL) is the threshold moisture content where the soil changes from a plastic state to a viscous fluid state; the *plastic limit* (PL) is the threshold moisture content where soil changes from a semisolid state to a plastic state. These limits, also known as *Atterberg limits,* are determined using the standard test method ASTM D 4318. See Chapter 5 for more details.

- *Organic material.* The humified and nonhumified compounds make up the organic fraction of the solid phase in soils and sediments. This is usually measured as total organic concentration (TOC). Organic material can cause high plasticity, high shrinkage, high compressibility, low permeability, and low strength (Mitchell, 1993). The method most commonly used to measure organic carbon is ASTM D 2974. See Chapter 4 for details.

- *Specific gravity.* Specific gravity is the heaviness of a substance compared to that of water. If a substance has a specific gravity greater than 1, it will sink in water; if it is less than 1, it will float in water. Specific gravity is determined using method ASTM D 854.

- *Porosity.* Porosity, denoted as n, is the volume of the voids (spaces between grains filled with either air or water) divided by the total volume of a porous medium. Porosity is calculated based on the phase diagram, as explained in Chapter 5.

- *Classification.* The Unified Soil Classification according to ASTM D 2487 classifies soils and sediments into coarse-grained, fine-grained, and peat. See Chapter 5 for details.

- *Consolidation.* Consolidation is the reduction in volume of the sediment due to gravity. For the purposes of in-situ capping, if the material is not

a fine-grained material, zero consolidation of the cap may be assumed. If the material is a fine-grained material, a consolidation analysis must be performed to determine the loss of cap thickness due to consolidation. Consolidation properties can be determined by ASTM D 2435 and can also be estimated using empirical equations. See Chapter 5 for details.

- *Shear strength.* Shear strength is the resistance of the soil to failure when shear stress is applied. *Shear stress,* which is the stress factor perpendicular to the force being applied to the object (in the case of ISC, weight), can cause the sediments to shift or move. Shear strength can be measured directly by various field and laboratory tests (see Chapter 5 for more details) and is given by the Mohr–Coulomb failure criterion:

$$\tau = c + \sigma \tan \phi \qquad (26.1)$$

where τ is the shear strength, c the cohesion parameter, ϕ the angle of internal friction, and σ the normal stress on the sediment.

- *Liquefaction.* Liquefaction generally develops in cohesionless materials such as sand and silt materials. It occurs when the material develops a high pore water pressure as a result of applied stresses and results in the transformation of material from a solid to a flowing, liquidlike state (Mitchell, 1993). The material can quickly travel thousands of feet if liquefaction occurs. Liquefaction can be caused by seismic activity, wave action, blasting, and propwash from a vessel at the surface.

26.2.3 Flux Properties or Flux-Determining Properties

- *Erosion.* The potential for erosion is determined by the streamflow or tidal velocity forces, depth, turbulence, wave-induced currents, ship/vessel drafts, engine and propeller types, maneuvering patterns, sediment particle size, and sediment cohesion. These parameters, combined with the composition of the cap and frequency of vertical erosion, can be used to develop an accurate model of future erosion.

- *Sorption.* Sorption is the absorption (diffusion into the particle) and/or adsorption (sticking to the particle surface) of contaminants to soil that retards the transport of contaminants. See Chapter 8 for details.

- *Ion exchange.* Ion exchange occurs when ionic substances sorb onto capping materials and sediments. Ion exchangers have a structure containing an excess of fixed negative or positive charges and take up ions from solution as needed to neutralize the charge. Clays are ion exchangers (Hemond and Fechner, 1994).

- *Surface complexation.* Surface complexation occurs when ionic chemicals form both electrostatic and chemical bonds with the metal or oxygen atoms of the oxyhydroxide (metal) surface of aquifer solids (Hemond and Fechner, 1994).

- *Redox-mediated flocculation.* Redox-mediated flocculation occurs when oxidation–reduction reactions cause individual small particles to aggregate into clumps to form a larger particle.

- *Bioturbation.* Bioturbation is the mixing and disturbance of sediments by benthic organisms. A site-specific study should be performed to determine a specific benthos that will recolonize the area, the type and amount of organisms, and the bioturbation depth profile to predict the amount of bioturbation that will occur. It can be assumed that bioturbation (and therefore constant mixing) will occur at the surface layer of the cap (approximately the top 3 to 4 in.). This thickness should be added to the thickness of the cap design unless the material of the cap will resist burrowing by benthic organisms.

- *Advective losses (advective flux).* Advection losses are due to movement of the contaminant by the bulk motion of fluid. The amount of contaminant that is moving per unit area, called the *flux density,* is given by

$$J = CV \qquad (26.2)$$

where J is the flux density, C the chemical concentration, and V the fluid velocity. When written for water flow, this is translated into specific discharge flux (Darcy flux), from Darcy's law:

For one-dimensional flow: $q_h = K_h i_h$

$$(26.3)$$

For three-dimensional flow: $\mathbf{q}_h = -\mathbf{K}_h \, \nabla \mathbf{h}$

$$(26.4)$$

where q_h (or $\mathbf{q}_h$) is the fluid flow or specific discharge, K_h (or $\mathbf{K}_h$) the hydraulic conductivity, and i_h (or $\nabla \mathbf{h}$) is the hydraulic head gradient.

- *Diffusive Losses.* Diffusive losses are due to movement of the contaminant resulting from the concentration gradient. Diffusion is calculated using Fick's laws and is also in units of flux density. According to Fick's first law, steady-state flux is given by:

For one-dimensional flow: $J = -D \dfrac{dC}{dx}$

$$(26.5)$$

For three-dimensional flow: $\mathbf{J} = -\mathbf{D} \, \nabla \mathbf{C}$

$$(26.6)$$

where J (or $\mathbf{J}$) is the flux density, D (or $\mathbf{D}$) the Fickian mass transport coefficient, C the chemical concentration, and x the distance over which a concentration change is being considered. Fick's second law can be used for determining nonsteady-state diffusion losses. Refer to Chapter 8 for details.

26.3 SITE EVALUATION

The site evaluation requires performing the five tasks described below.

26.3.1 Site and Surrounding Area Characterization

The first task in any remediation project is characterizing the *site and the surrounding area.* Because the cap is a long-term solution, this requires considering both current and potential future site conditions. The most obvious characterization is the physical environment, which consists of the following: (1) water body dimensions, (2) bathymetry, (3) tidal and flow patterns, (4) ice formation, (5) aquatic vegetation, (6) bridges, (7) land, and (8) human-made structures. This information will provide good insight into the underlying conditions of the site.

26.3.2 Evaluation of Hydrodynamic Conditions

The *hydrodynamic conditions* that need to be taken into consideration are currents, waves, and flows of the water body. This information can usually be gathered from federal, state, or local government agency records. Water column currents and bottom currents need to be considered to estimate the degree of dispersion of both the contaminated sediment and capping material during cap placement. Included in these analyses should be conditions during storms for a more conservative and realistic evaluation. Then modeling must be done to determine the effects of the cap on existing hydrodynamic conditions to determine how stable the in-situ cap will be when it is placed.

26.3.3 Geotechnical and Geological Conditions

Geotechnical and geological conditions are needed to estimate the consolidation and stability of the sediments once the cap is in place. The data needed are the stratification of the soils underlying the sediment, the depth to bedrock, and the physical properties of the soils. Similarly, hydrogeological conditions should be examined, specifically the groundwater–surface water interactions. The general hydraulic pressure gradients should be determined. This will indicate if there is an upward hydraulic gradient at this interface. If there is

a large upward hydraulic gradient, groundwater would be driven up through the contaminated sediment and in-situ cap, dispersing the contamination. Refer to Chapter 7 for more information on determining groundwater flow.

26.3.4 Characterization of the Sediments

A very important part of the site evaluation is the characterization of the sediments. The properties of the sediments affect cap design, equipment needed to place the cap, modeling and monitoring of the remediation site, and cap performance. The properties of the contaminated sediments that need to be determined are the organic content, plasticity indices, grain size distribution, specific gravity, and USCS classification. The physical properties of the underlying soils and the thickness of the underlying soils and contaminated layer should also be assessed. An estimate of the consolidation and shear strength properties of the contaminated sediment should be evaluated to predict cap stability.

26.3.5 Chemical Characterization

Finally, chemical properties of the site, both water and sediments, should be analyzed. The concentration of contaminants in the water and the salinity of the water should be assessed. Also, the concentrations of the contaminants, including aereal extent and vertical distribution in the sediment, should be analyzed to be used in modeling the migration of contaminants in the design stage of a project. Contaminant properties and concentrations are also helpful in developing a monitoring program after a cap has been placed.

26.4 CAP DESIGN

The composition and dimensions (thickness) of the components of a cap can be referred to as the cap design. The steps involved in the design of caps are shown in Figure 26.3 and described briefly in this section.

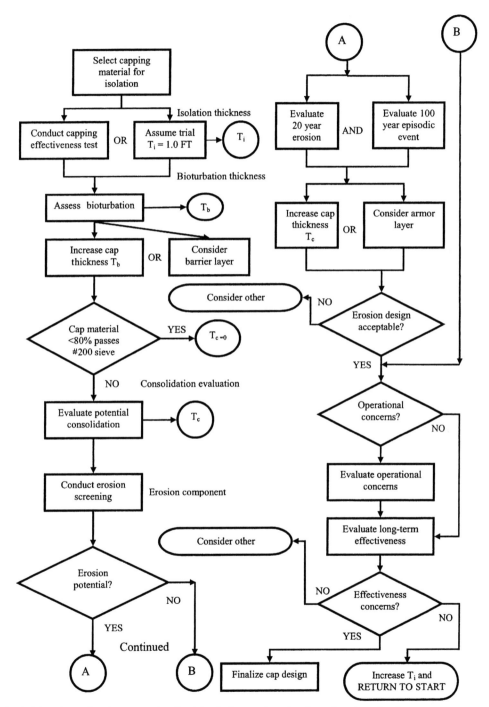

Figure 26.3 *Flowchart showing steps involved in design and evaluation of various in-situ cap components.* (*From USEPA, 1994.*)

26.4.1   Selection of Capping Materials

The first step in the design of a cap is the selection of capping materials that are available and are compatible relative to the contaminants. Two commonly used materials are (1) natural granular materials and (2) geosynthetic materials.

Natural Granular Materials. The most commonly used materials are granular materials, which can be quarry sand or other coarse-grained soils. Sandy and coarse-grained materials are often used because they are much more stable than fine-grained materials and do not resuspend as readily, which is important when placing the cap on the area of contamination. Also, benthic populations are less likely to penetrate a sandy and coarse-grained material cap. Armor stones are also used as an additional layer to provide protection from erosion and benthic penetration.

The physical properties of granular material must be analyzed to determine the proper dimensions and placement. These characteristics include density, organic content, grain size distribution, and specific gravity. The same testing methods used to determine the contaminated sediment properties should be used to determine properties of coarse-grained materials to be consistent. It is also important that the granular materials be free of contamination.

Geosynthetic Materials. Another option for capping material is the use of geosynthetic materials. Geosynthetic materials consist of geomembranes and geosynthetic clay liners, which have a very low hydraulic conductivity, and geotextiles, which are porous. As detailed in Chapter 17, geosynthetics are commonly used in other environmental control projects, such as landfills. When geotextiles are used with granular materials, the function is primarily for stabilization of the cap and the ease of constructability. Like the granular material, geotextiles must be compatible with the contaminant. The properties to be considered are the chemical barrier properties and the stability of the geotextile.

26.4.2   Cap Components

The simplest cap is a cap that provides physical isolation (i.e., it provides a buffer between the contami-

nation and the local benthic population). In this case, a 1-ft cap would be sufficient, and no further analysis or modeling would be needed. However, if stopping the contamination flux is the goal, additional components may be needed, and thorough modeling needs to be done to predict if the contaminant flux will be adequately controlled. If each component of the in-situ cap is designed for a singular function (physical isolation, sediment stabilization, or reduction of flux of dissolved contaminants), the design approach is conservative because the properties are considered additive. If one component of the cap is serving multiple functions, the design approach should not rely on the properties being additive. Protective layers and armoring should be added to the cap design to ensure integrity.

Bioturbation and consolidation of the sediment under self-weight and the weight of the overlying cap should also be taken into account when determining if contaminant flux will be stopped or sufficiently retarded. Once the bioturbation and consolidation components are modeled, extra thickness should be added to the cap to account for the thickness loss expected to be caused by these two phenomena. For any material, it is important to determine the consolidation of the underlying sediments and the advective flux of contaminants up through the cap as a result of the consolidation.

26.4.3   Erosion Potential

The potential for erosion should be carefully modeled, since the purpose of an in-situ cap is long-term containment of the contamination. If the erosion potential is expected to be too great to contain the sediments, an additional armor layer must be added to the design to ensure the integrity of the cap. There are three predominant causes of erosion in an in-situ cap:

1. *Rival/tidal-induced erosion.* This type of erosion is dominant in unnavigable portions of rivers or in areas where there is no navigation. In deeper rivers and estuaries, no additional protections are needed to prevent this type of erosion. When modeling the effects of river/tidal induced erosion, a model called the *Shields diagram* can be

used. It is important to consider the frequency and intensity of storms in the area; using a 100-year return interval is usually appropriate.

2. *Wave-induced erosion.* This type of erosion often occurs in open-water ISC sites, such as lakes, estuaries, and harbors. There are a few modeling programs to determine the effects of this type of erosion, the most often used being a model developed by the U.S. Army Corps of Engineers (USACE) called LTFATE.

3. *Propeller-induced erosion.* Propeller-induced erosion is found in urban and industrial waterways. Propeller jet wash from commercial and recreational vessels has the ability to resuspend bottom sediments. A model currently used to predict propeller-induced erosion can be found in USEPA (1994).

Stability is the most critical factor that needs to be evaluated for conditions occurring immediately after the cap placement. The reason for it is that excess pore water pressure can cause high shear stress and cause failure of the sediments and/or cap. This can be avoided by gradually placing the cap on a large area, then waiting for a sufficient amount of time for the excess pore water pressure to disperse or dissipate. More time is needed for materials with low permeability and low strength. This is important because if slope failure does occur, the contaminated sediments will mix with the cap materials and can be resuspended in the water column. This is discussed further in Section 26.4.5.

The stability and the erosion of the cap at the time of cap construction also need to be considered in the design of an in-situ cap. These components can be achieved in the same design parameter or separately, depending on the choice of capping material and site conditions. Because the cap will be placed in a dynamic environment, the movement surrounding the cap must be considered when assessing long-term stability of the cap and the rate of erosion. When modeling sediment stability, it is necessary to take into account the sediments' vertical movements. If granular material is used, pore size may be greater on the sides of the cap and be insufficient to keep the contaminated sediment in place. Analytical techniques and numerical

methods can be used to determine vertical migration of the sediments (USEPA, 1994).

For small areas with a thick cap, edge effects need to be taken into consideration. Slope failures can occur at the edge of the cap, so an overlap of the cover is needed. The length of the overlap can be estimated by using the friction versus relative density plots. If the cap is composed of a sandy material, investigations must be made to determine if liquefaction will be an issue at the site.

26.4.4 Settlement Analysis

The consolidation settlement analysis for cap depends on the type of cap material used. If the cap material is a sand, no added thickness is needed to account for consolidation. If the cap material is a silt or clay, an evaluation must be done to determine what thickness must be added to keep the cap's integrity. Consolidation analysis is also needed for postplacement monitoring to differentiate between cap elevation changes due to erosion and consolidation, and thus model more accurately for losses due to advective contaminated flux. Refer to Chapter 5 for more details on settlement analysis. Cargill (1983) recommends using the finite strain consolidation theory for evaluating the settlement of hydraulically placed dredged material caps. A finite strain computer model, MOUNDS, can be used if test data are available from self-weight consolidation tests and/or standard oedometer tests. There is also a model known as CONSOL, and it has been the experience of investigators that both models give reasonably accurate settlement predictions.

A preliminary evaluation of the settlement of the sediments due to the load from the cap system can be made using Terzaghi's small strain consolidation theory (Terzaghi et al., 1996). The average consolidation settlement of the sediment layer can be modeled assuming that the total settlement is comprised of primary settlement due to expulsion of pore water from the sediment and secondary compression due to time-dependent creep of the organic component of the sediments. Primary consolidation (S_c) of the sediment can be evaluated using the following equations (Mohan et al., 1999; 2000):

$$S_c = \frac{C_C}{1 + e_0} \left[H_s \log_{10} \left(\frac{p_0 + \Delta p}{p_0} \right) \right] \quad (26.7a)$$

$$p_0 = \gamma_s' \frac{H_s}{2} \quad (26.7b)$$

$$\Delta p = \gamma_c' H_c \quad (26.7c)$$

$$\gamma_s' = \gamma_w \frac{G_s - 1}{1 + e_0} \quad (26.7d)$$

where C_c is the compression index of the sediment, e_0 the initial void ratio of the sediment layer, H_s the height (or thickness) of the sediment layer, p_0 the initial vertical effective stress at the middle of the layer, Δp the additional vertical effective stress due to effective cap weight, γ_s' the effective unit weight of the sediment, γ_c' the effective unit weight of the cap, G_s the specific gravity of the sediment, and γ_w the unit weight of water.

Secondary consolidation settlement (S_s) is due to time-dependent deformation (creep) of the sediments under a relatively constant load. It can be evaluated using the equation.

$$S_s = H_s \frac{C_\alpha}{1 + e_p} \log_{10} \frac{t_d}{t_p} \quad (26.8)$$

where C_α is the coefficient of the secondary compression, e_p the void ratio at the end of primary consolidation, t_d the end of the design time frame, and t_p the time for the primary consolidation to occur.

The total settlement (S_t) of the sediments due to stresses induced by the proposed cap can then be computed as the sum of primary and secondary settlements:

$$S_t = S_c + S_s \quad (26.9)$$

The length of time elapsed following cap placement (t_e) and the degree of consolidation (U in %) can be expressed as

$$t_e = H_s^2 \frac{T_v}{C_v} \quad (26.10)$$

$$U(\%) = f(T_V) \quad (26.11)$$

where T_V is the dimensionless time factor (see Chapter 5) and C_v is the coefficient of consolidation.

26.4.5 Stability Analysis

As mentioned earlier, construction of the cap over soft sediments may cause stability concerns due to increased stresses in sediments. The primary stability concern during cap placement is the potential for the sediments to fail under the increased stresses and to form "mud waves" that could be trapped in subsequent layers of cap material. This concern is particularly valid since contaminated sediments typically have very high water content and low shear strength, resulting from the lack of sediment consolidation. Cap stability analysis involves the consideration of two major factors: (1) bearing capacity of the sediments underlying the cap, and (2) slope stability of the cap.

For the purpose of bearing capacity analysis, the cap can be considered as a long footing resting over the sediments with a surcharge q which is given by

$$q = \gamma_c' H_c \quad (26.12)$$

where γ_c' is the effective unit weight of the cap and H_c is the thickness of the cap. Since the sediments are soft, it can be assumed that the sediments will fail in an undrained local shear failure mode. The ultimate bearing capacity of the sediments (q_{ult}) can then be expressed as (Ling et al., 1995)

$$q_{ult} = 3.43 c_u \quad (26.13)$$

where c_u is the undrained shear strength of the sediments. Assuming a factor of safety of 3.0, the maximum thickness of the cap ($H_{c,max}$) for bearing stability is given by (Ling et al., 1995)

$$H_{c.\max} = \frac{1.14c_u}{\gamma_c'} \qquad (26.14)$$

Slope failures can occur at the edges (side slopes) of the cap layer due to single or multiple factors. These factors include the removal of soil mass by erosion, natural slope movements, overloading of the soil, transient effects (earthquakes), increase in lateral pressure, changes in soil structure, and pore pressure changes. In general, slope failures can be classified into five types: (1) translational failure, (2) planar or wedge failure, (3) circular failure, (4) noncircular failure, and (5) a combination mode of failures, consisting of one or more of the preceding types.

The protection of a slope against potential failure is typically expressed in terms of the *factor of safety,* which can be expressed in three ways, depending on the type of analysis. First, in the limit equilibrium analysis, the shear strength required to maintain stability is computed and compared to the magnitude of the available shear strength (Abramson et al., 1996). Then the factor of safety (FS) can be expressed as

$$FS = \begin{cases} \dfrac{c_u}{\tau_r} & \text{for total stress analysis} \\ & \qquad\qquad (26.15) \\ \dfrac{c' + \sigma' \tan \phi'}{\tau_r} & \text{for effective stress analysis} \\ & \qquad\qquad (26.16) \end{cases}$$

where c_u is the total available undrained shear strength,

τ_r the required shear strength, c' the effective cohesive strength, ϕ' the effective friction angle, and σ' the effective normal stress. In the other approaches, the factor of safety is defined as the ratio of the total resisting forces to the total driving forces for planar failures, and as the ratio of the total resisting moments to the total driving moments for circular failures. Cheney and Chassie (1982) suggest the following relationship for computing the factor of safety of slopes against circular failure:

$$FS = \frac{6c_u}{\gamma_f h_f} \qquad (26.17)$$

where γ_f is the unit weight of the fill material and h_f is the height of the fill.

The slope stability of an underwater cap may be evaluated using the method described by Leschinsky and Smith (1989). This method assumes a log-spiral type of failure in the sand cap and a circular failure in the sediments, as shown in Figure 26.4. Since the cap is submerged under water, the effective stress parameters should be used for analysis. The cap and sediment properties are defined as several individual layers with a corresponding thickness of cap (H_c), effective unit weight of the cap material (γ_c'), effective friction angle of the cap material (ϕ'), thickness of the sediment (H_s), and undrained shear strength of the sediment layer (c_{ui}). The analysis is typically conducted on the steep-

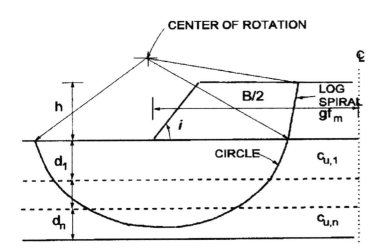

Figure 26.4 Cap slope stability analysis. (From USEPA, 1994.)

est slope (or thickest cap location) by picking an arbitrary cross section of width B (see Figure 26.4). Once
the material parameters are known, the factor of safety
can be computed for a variety of failure assumptions
using standard theories described above. Experience
indicates that typically, a factor of safety of at least 1.5
in the short term (i.e., immediately following construction) and 2.0 is desirable in the long term (Mohan et
al., 2000).

26.4.6 Contaminant Release Analysis

Simple analyses can be performed to account for different mechanisms that could potentially release contaminants through the cap system (Mohan et. al., 1999;
2000): (1) diffusion, in the absence of groundwater
flow; (2) advection and dispersion, if groundwater flow
through the cap is present; (3) release of pore water
due to sediment compression; and (4) erosion of contaminated sediments due to pore water release during
sediment compression. These analyses are described
briefly below.

Contaminant Release by Diffusion. In the absence
of groundwater flow, molecular diffusion can cause the
sediment-bound contaminants to be released into the
overlying water column. The time for 5% of the contaminants to break through the upper surface of the
cap into the water column (defined as the breakthrough
time, t_b) can be evaluated as follows (Wang et al.,
1991; Thoma et al., 1993):

$$0.95 = -2 \sum_{N=1}^{N=5} (-1)^N \exp\left(\frac{-D_t N^2 \pi^2 t_b}{H_c^2}\right) \quad (26.18)$$

$$D_t = \frac{D_e}{n + \rho_b K_{oc} f_{oc}} \quad (26.19)$$

where H_c is the cap thickness, D_t the effective transient
transport diffusion coefficient, D_e the effective diffusivity, n the sediment porosity, K_{oc} the chemical distribution coefficient, f_{oc} the organic carbon content (expressed as a fraction), and ρ_b the bulk density.

Steady-state time (t_{ss}) is defined as the time for contaminant release to reach 95% of the maximum release
rate (flux) and is expressed as

$$t_{ss} = \frac{3.69 H_c^2}{D_t \pi^2} \quad (26.20)$$

The steady-state flux (F_{ss}) can be computed by

$$F_{ss} = \frac{D_e(C_0 - C_w)}{H_c} \quad (26.21)$$

$$C_0 = \frac{C_s}{K_p} \quad (26.22)$$

where C_w is the chemical concentration in the water
column, C_0 the average chemical concentration in the
pore water, K_p the organic carbon partition coefficient
($K_p = K_{oc} \times f_{oc}$), and C_s the depth averaged sediment
chemical concentration. The long-term flux (F_l) can
then be estimated from

$$F_l = C_0 \sqrt{\frac{D_e R}{\pi t}} \quad (26.23)$$

where R is the retardation factor ($R = n + \rho_b K_p$) and
t is the time period.

***Contaminant Release by Advection and Disper-
sion.*** The contaminant release at the sand–armor interface due to the processes of advection and dispersion (expressed as C) can be evaluated based on the
works of Ogata and Bank (1961), Tchobangolous and
Schroeder (1987), and Fetter (1999). For simple
boundary and initial conditions,

$$C = 0.5 C_0 \, \text{erfc}\left(\frac{R H_c - V t}{2\sqrt{R D_H t}}\right) \quad (26.24)$$

where C_0 is the sediment–sand (lower) interface chemical concentration, H_c the cap thickness, V the groundwater velocity, t the time, D_H the hydrodynamic dispersion coefficient, and erfc is the complementary error

function (see Chapter 8). Moo-Young et al. (2001) present such contaminant transport analysis considering advection, dispersion, and liner sorption. Refer to Chapter 8 for more details on such analyses.

Contaminant Release by Pore Water Release during Sediment Compression.

Chemical migration into the sand cap layer due to pore water release during sediment compression can be estimated based on the consolidation properties of the sediment layer (Mohan et al., 2000). Assuming that pore water expulsion during sediment compression results in a short-term flux of dissolved contaminants into the cap system, the flux per area (F_c) can be estimated as

$$F_c = \frac{S_t C_t}{K_{oc} f_{oc} \times 10^{-3}} \qquad (26.25)$$

where S_t is the total settlement of the sediment, C_t is the average concentration of the contaminant in the sediment zone, K_{oc} is the partitioning coefficient of the contaminant, and f_{oc} is the total organic carbon content of the sediment layer.

Now, assuming that the contaminates in the pore water are sorbed into the bottom 1-in. (0.025 m) of the cap layer [based on the experimental results of Gunnison et al. (1987)], the concentration of contaminant in the bottom thickness of the t (generally, 0.025-m) layer of the cap (C_{bs}) can be estimated as follows:

$$C_{bs} = \frac{F_c}{t[G_s(1 - n_s)]} \qquad (26.26)$$

where n_s and G_s are the porosity and specific gravity of the cap layer, respectively. The actual thickness of the sorptive layer (L_{SSL}) can be then estimated as

$$L_{SSL} = \frac{0.025 C_{bs}}{K_{oc} f_{oc-s} C_{EPW}} \qquad (26.27)$$

where f_{oc-s} is the total organic content of the isolation layer and C_{EPW} is the contaminant concentration in expelled pore water.

Erosion of Contaminated Sediments Due to Pore Water Release during Sediment Compression.

The potential for erosion of contaminated sediment particles due to pore water release during sediment compression can be evaluated based on the results of the settling analysis. Knowing the total settlement (S_t) and the time (t_e) to achieve that from the equations in Section 26.4.4, the equivalent velocity of the release of the pore water (V_{PW}) can be computed as follows:

$$V_{PW} = \frac{S_t}{t_e} \qquad (26.28)$$

This velocity can then be used to compute the particle size of the sediments that would potentially be eroded by the pore water flux into the sand cap and the associated erosion loss volumes (Vanoni, 1977).

Comprehensive mathematical models are also available to determine the effectiveness of the cap design. The USEPA model is often used for this analysis (USEPA, 1994). This considers bioturbation, consolidation, and erosion to evaluate the effectiveness of the cap material and cap thickness. It is important to note that all properties used in this model are the properties after the cap has been placed.

In addition to mathematical modeling, various laboratory tests can be performed to determine how effective the cap will be in chemical isolation (USACE, 1991; Wang et al., 1991; USEPA, 1994). Although currently there are no standard lab tests that account for advective and diffusive properties, there are a few that are considered acceptable. The USACE uses tests that evaluate cap thickness acceptability by column testing. This test does not take into account groundwater-induced advection. Louisiana State University (LSU) has devised a test using a capping simulator cell that tests for breakthrough time and time to achieve steady-state conditions. Other tests are Environment Canada's tank test to observe the physical layering that occurs when a cap is placed and the diffusion coefficients for the material, and a USACE leach test to assess water quality through the contaminated sediment layer. These tests can be used to determine the specific diffusion and partition coefficients for the materials to be used in the cap.

26.5 CONSTRUCTION AND MONITORING

After an acceptable design for the in-situ cap has been developed, the method of construction and the monitoring plan need to be decided and implemented. The construction and monitoring of the ISC will be a significant portion of the cost, so they should be considered in the initial feasibility study. Experience has shown that no single technique is consistently more successful than another. All current methods of construction have limitations due to mobility, cap depth, and interference due to the water column. The method used will depend on the site-specific conditions.

26.5.1 Construction Methods

The methods of cap placement currently in use vary widely. For granular material placement, which is a component in almost all in-situ caps, the greatest need is for accurate placement that controls the density of the material being applied and the rate of application. If the cap is applied too fast, it can cause the underlying contaminated sediments to fail, thus releasing the contamination. The possibility of resuspending the contaminants is the primary concern; however, there is currently not an accepted method to predict this. There are two basic methods of cap construction:

1. *Land-based placement.* This technique works best in constricted areas such as narrow channels and areas close to shore. This method usually entails the capping material being trucked in and placed with backhoes, clamshells, dumped from trucks, and/or spread with bulldozers. The reach of the equipment is the primary limitation of this method. Armoring layers, used for shoreline and streambank erosion protection, are also placed using this equipment.

2. *Pipeline or barge placement.* This method works well in open and deep areas such as harbors and lakes.

Several variations of the pipeline or barge methods are currently available. In surface discharge using conventional dredging equipment, the capping material is surface-released from barges or hopper dredges and settles on the capping site. Spreading by barge movement uses a split-hull barge (barge with a bottom opening) in which the hull is gradually opened and the capping material is slowly released. In this method, *bridging* or fast release of the material can sometimes occur. Both of these techniques are used in large, deep areas.

Hydraulic washing of coarse sand is a method where the capping material is sprayed overboard a barge with a hose. This method can achieve a uniform thickness and works well for depths of 10 ft or less. A pipeline with a baffle plate or sand box uses a discharge pipe with either a baffle plate or sand box attached to the end of it, which dissipates the energy from the flowing cap material and allows the material to settle vertically with more uniformity. This prevents the disruption of the contaminated sediments and achieves a thinner cap layer. A hopper dredge pump-down is a method limited to depths of 60 to 70 ft. It uses hopper dredges that have pump-out capability (reverse-flow) to distribute the capping material.

A submerged diffuser is a diffuser placed at the end of a vertical discharge pipe, which allows the material to be discharged with low energy below the water column. This is a preferred method for large areas. The sand spread barge uses the same technique but with a slurried material; a gravity-fed downpipe (tremie) is also the same technique but with a larger-diameter pipe. These are all used in large capping areas.

The method of placement of geosynthetics and geomembranes vary significantly from one contractor to the other. Barges, cranes, divers, and anchors have all been used in the placement of a geosynthetic cap. When using geosynthetic layers, the main priority is to place the cap quickly, so as to get increased stability. For all cap placements, controls must be put in place to ensure that the cap is actually being placed at the correct location.

26.5.2 Monitoring

During and after cap placement, monitoring must be done to detect if the in-situ cap is achieving the project goals. The specific items that must be monitored are cap integrity, cap thickness, cap consolidation, benthic recolonization, and chemical migration potential. During and immediately after the cap is placed, monitoring should be frequent, while less frequent monitoring is necessary in the long term.

The monitoring plan should be a tiered plan that accounts for possible aspects of failure and strategies to correct the failure. The purpose of short-term monitoring is to determine if the process is going as planned and the cap is functioning as expected. The purpose of long-term monitoring is to determine if the remedial objectives of the cap are being met and the cap's physical integrity is being maintained. An approach that works well is to monitor for the occurrence that prompted the site to need remediation.

Construction monitoring, the first and easiest phase of monitoring, consists of two key features: (1) the materials being used for capping must undergo laboratory analysis to confirm the materials are what the project specified, and (2) the construction methods are monitored, to ensure that proper procedures are followed and an accurate placement and thickness is attained. The most commonly used technique for construction method monitoring is bathymetry, where a bathymetric survey is taken before and after the cap is placed to determine the thickness of the cap throughout the site area. Other techniques that have been used are settling plates, sediment profiling cameras, sediment resuspension modeling, and core sampling. These methods are used on a site-specific basis.

The second phase of monitoring is *cap performance monitoring.* This phase is needed to determine if the contaminant containment objectives are being or will be met. Sediment stabilization can be monitored using bathymetric surveys, especially if bathymetry was re-

corded before and immediately after cap placement. Knowing the initial depth, thickness, and area of the cap, cap thickness, and component integrity (physical integrity of the stabilization component), cap performance can be monitored both short and long term. Long-term monitoring can be time-based (specified time interval monitoring) or event-based (monitoring after erosion-inducing events such as storms or floods). Event-based monitoring requires the ability to monitor with little advance notice, which indicates that a recording gauge will be needed on-site that triggers monitoring when specified events occur. This type of monitoring is preferred because failures are detected more quickly and can be repaired before a major failure occurs.

The third phase of monitoring is *physical isolation monitoring,* which assesses whether the in-situ cap is intact, if all contaminated sediments are covered, if physical loss of contamination has been prevented, and if benthos penetration has occurred. Bathymetry can also be used in this case to determine the capped area and the armor layer placement. Sampling of the cap can determine if the benthos population has penetrated the ISC, and sediment traps can be installed in the cap to monitor all aspects of physical isolation.

The fourth phase of monitoring is related to *chemical isolation.* This is the most difficult objective to monitor in in-situ capping. Currently, no definitive method is used in this monitoring. The difficulty lies in the slowness of the flux processes and the variability in material used in capping. *Peepers* have been used somewhat successfully in this area; they are small, semipermeable bags within the cap that monitor the pore water properties.

The last consideration in construction and monitoring is the operation and maintenance of the in-situ cap. Operational concerns are managed during placement of the cap to ensure a good project. There is also routine maintenance, usually to repair or rebuild a portion of the cap after erosion has occurred, especially after storms and floods. Finally, the cap must be repaired or modified if it is found that it is not meeting the objectives and standards of the design.

26.6   REGULATORY AND ECONOMIC CONSIDERATIONS

Regulations and costs associated with in-situ capping systems vary depending on the site-specific conditions. Uses of the waterway that might be interfered with are recreational and commercial navigation, flood control, waterfront development, and sensitive or important aquatic habitats. Examples of federal regulations that need to be taken into consideration are the Rivers and Harbors Act of 1899 and the Clean Water Act of 1972.

Any activity that has the potential to affect the course, capacity, or condition of a navigable water of the United States must be permitted under Section 10 of the Rivers and Harbors Act (33 CFR 403). The permit program for Section 10 permits is managed by the U.S. Army Corps of Engineers (USACE). Federal regulations on the USACE permit program are contained in 33 CFR 320-330. For ISC projects, a Section 10 permit will be required. The permitting process requires contacting the Coast Guard and local and regional navigation users as well as assessing the potential for the cap to obstruct flows, leading to flooding or erosion.

The placement of fill materials in the waters of the United States is regulated under sections of the Clean Water Act. Specifically, Section 404 designates the USACE as the lead federal agency in the regulation of dredge and fill discharges, using guidelines developed by the USEPA in conjunction with the USACE. Federal regulations on the Corps permit program are contained in 33 CFR Parts 320–330. Therefore, for the construction of ISC, a permit is required under Section 404, and a certification of water compliance is also required from the state under Section 401.

The applicability of the Toxic Substances Control Act (TSCA), which deals with PCBs, should be assessed for sites where sediments are contaminated with PCBs. Compliance with the Resource Conservation and Recovery Act (RCRA) is also assessed based on site-specific conditions. Experience, based on a limited number of ISC projects completed, shows that both TSCA and RCRA may not necessarily be invoked because the sediments are not removed.

The major costs associated with an in-situ capping project are the capping material, the equipment needed to construct the cap (including transportation of the material), and the labor. These costs are influenced primarily by the cap design, site accessibility, and water depth. Operation and maintenance costs are secondary compared to the three costs above, but still significant enough to be included in the economic analysis. These costs are important to consider to ensure the monitoring and therefore the integrity of the cap. Finally, if barriers or enforcement is needed to keep people and vessels away from the site, it will also be a cost significant enough to be considered.

26.7   CASE STUDIES

A limited number of ISC projects have been completed under varying site conditions (USEPA, 1994; Hagerty and Trotman, 2001). USEPA (1994) summarized several ISC projects completed in the United States and abroad, as shown in Table 26.1. ISC has been applied to riverine, nearshore, and estuarine settings. Conventional dredging using construction equipment and other techniques have been used for ISC projects. The performance monitoring data collected at these sites indicate that ISC can be an effective technique for long-term containment of contaminants.

Hagerty and Trotman (2001) described three case studies where ISC was used to manage some or all of the contaminated sediments in place. The first case study described the ISC construction to contain PCB-including sediment from an embayment area of the St. Lawrence Seaway near Massena, New York. Initially, sediment was hydraulically dredged and removed, but the cleanup objective could not be met. Therefore, it was decided to install ISC to contain the remaining PCB sediment in an area of approximately 75,000 ft^2. Depth of water in the cap area varied from 0 (shoreline areas) to approximately 15 ft. The cap was designed to provide both chemical and physical isolation of the remaining sediment from the water column. The cap design, shown in Figure 26.5(a), consisted of three layers, listed from the bottom up: (1) a 6-in.-thick sand isolation layer, (2) a 6-in.-thick gravel bedding layer, and (3) a 6-in.-thick armor layer. The sand isolation layer consisted of a clean medium to coarse sand

Table 26.1 **Summary of selected in-situ capping projects**

Project Location	Contaminants	Site Conditions	Cap Design	Construction Methods
Kihama Inner Lake, Japan	Nutrients	3700 m³	Fine sand, 5 and 20 cm	Not available
Akanoi Bay, Japan	Nutrients	20,000 m³	Fine sand, 20 cm	Not available
Denny Way, Washington	PAHs, PCBs	1.2 ha near shore with depths from 6 to 18 m	0.79 m of sandy sediment	Barge spreading
Simpson–Tacoma, Washington	Creosote, PAHs, dioxins	6.88 ha near shore with varying depth	1.2 to 6.1 m of sandy sediment	Hydraulic pipeline with "sandbox"
Eagle Harbor, Washington	Creosote	22 ha within empayment	0.9 m of sandy sediment	Barge spreading and hydraulic jet
Sheboygan River, Wisconsin	PCBs	Several areas of shallow river/floodplain	Sand layer with armor stone	Direct mechanical placement
Manistique River, Michigan	PCBs	1858-m² shoal in river with depths of 3–5 m	40-mil plastic liner	Placement by crane from barge
Hamilton Harbor, Ontario	PAHs, metals, nutrients	10,000-m² portion of an industrial harbor	0.5 m sand	Triemie tube
Eitrheim Bay, Norway	Metals	100,000 m²	Geotextile and gabions	Deployed from barge
St. Lawrence River, Massena, New York	PCBs	6968 m²	6 in. sand/6 in. gravel/6 in. stone	Placed by bucket from barge

Source: USEPA (1994).

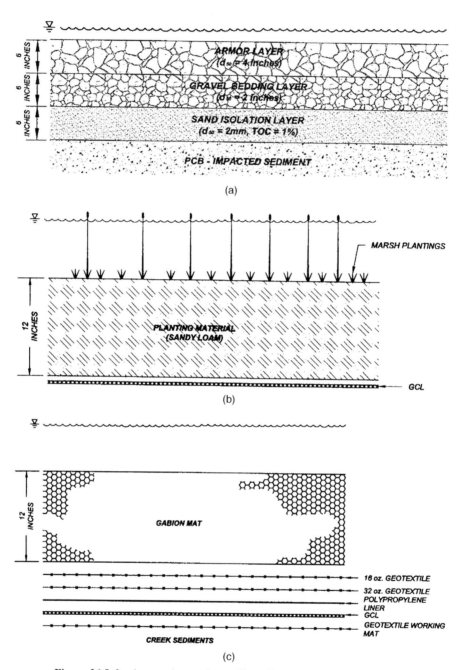

Figure 26.5 *In-situ capping projects.* (*From Hagerty and Trotman, 2001.*)

mixed with 1% granular activated carbon. The organic carbon provided the chemical isolation to minimize the upward diffusion of PCBs through the sand and into the water column. The gravel bedding layer consisted of gravel with an average grain size of 2 in., while the armor layer consisted of gravel with an average size of 4 in. Both the bedding and armor layers provided erosion protection for the isolation layer. These layers were designed according to the established USACE (1991) and USEPA (1994) guidance documents, and the design takes into account erosive forces from wave action, river flow velocities, ice scour, and propeller wash. The cap was constructed using a barge-mounted trackhoe. The capping material was dumped one bucket at a time through the water column onto the capping area. Based on postconstruction cross sections, all areas of the cap received the minimum amount of capping material; however, many areas of the cap received as much as six times the required depth of material. Such variability was attributed to the installation technique used.

The second case study is known as the Marathon Battery Remediation Project located on an inland cove directly off the Hudson River in Cold Spring, New York. The contaminants of concern at the site were cadmium and nickel in the sediments, which were the result of point-source discharges into a marsh and the Hudson River from a battery manufacturing facility. Figure 26.5(b) shows the cross section of the capping system used at this site to contain the contaminated sediments. It consisted of using a GCL and 12 in. of sandy loam planting material on top of the GCL. Sediments containing greater than 100 ppm total cadmium were removed prior to installation of the GCL and the overlying planting material. Once planting material installation and grading was complete, container-grown wetland plants were planted in the marsh. Installation of the cap system in the marsh was performed with minor complications. The marsh subgrade following sediment removal was not uniform, so some areas of the marsh required excess GCL panel seam overlaps.

The third case study is the Galaxy/Spectron Superfund site in Elkton, Maryland. The contaminants of concern were volatile organic compounds (VOCs) and dense non-aqueous-phase liquids (DNAPLs) resulting from both point- and non-point-source discharge of organic solvents from a solvent recycling facility located along Little Elk Creek. DNAPLs were present in the groundwater, sediments, and bedrock. A stream isolation concept via a liner (consisting of both a barrier layer and protective layer) with a passive groundwater collection system installed beneath the liner was designed to separate the surface flow of the creek from the groundwater discharges and creek sediments. A cross section of the liner system is shown in Figure 26.5(c). The liner system consisted of the following components, listed from the bottom up:

- *Geotextile working mat.* This mat was installed for erosion and sediment control, to reduce the release of organic vapors during construction, and to serve as a cushion between the underlying subgrade and the overlying liner system.
- *Geosynthetic clay liner* (GCL). The GCL serves as a secondary liner.
- *Scrim-reinforced polypropylene liner.* This geomembrane serves as the primary liner.
- *Geotextile cushion.* This cushion consists of both a 32- and a 16-oz nonwoven geotextile placed over the primary liner to protect the liner from the overlying gabion mat.
- *Gabion mat.* A 12-in. gabion mat was placed over the geotextile cushion to serve as the primary protective layer. The gabion stone was infilled with sand and planting material to prevent subsurface flow within the gabions and to help reestablish benthic and riparian habitats.

The entire system was constructed by diverting the creek flow via pumps. Prior to the installation of the liner system, concrete anchor/groundwater cutoff walls and a series of collection pipes were installed directly beneath the creek bed. The creek bed was then graded and prepared for liner construction. The project effectively isolated the impacted sediments and groundwater from the creek flow, and the groundwater that is removed is being treated.

26.8 SUMMARY

In-situ capping of subaqueous contaminated sediments is a remediation technology to contain the contamination from dispersing or entering the food chain by placing a layer of material over the contamination. The three remediation objectives that in-situ capping meets are physical isolation of the contaminant, stabilization of the sediments, and chemical isolation of the contaminant (reduced flux).

A technical knowledge of sediment engineering is necessary to determine site conditions, cap conditions, design parameters, and construction needs. The first step in an in-situ capping project is site evaluation. This will determine if an in-situ cap is a feasible option for remediation. If an in-situ cap is feasible to build, regulatory and economic issues must be considered. If it is decided that an in-situ cap will be used, the site evaluation will provide the information needed to design an effective in-situ cap.

In the design of the cap, many modeling techniques (relationships, equations, computer models, etc.) are employed to estimate how the contaminants, cap material, water environment, and benthic population will react in the new environment. The main concern is to keep the contaminants and cap material from becoming mobile in the environment.

After the in-situ cap has been designed and modeled for effectiveness, the in-situ cap needs to be constructed. This is usually done using typical construction machinery or barges carrying specialized equipment. During the placement of the cap and after the cap has been installed, monitoring should be implemented to ensure that the cap is containing the contamination. The monitoring plan should be prepared prior to the start of construction.

Regulations and costs associated with in-situ capping systems vary depending as the site-specific conditions. Several ISC projects have been completed under varying site conditions, and these projects indicate that ISC can be an effective technique for long-term containment of contamination.

QUESTIONS/PROBLEMS

26.1. Four basic options for remediation of contaminated sediments exist: (1) containment in place, (2) treatment in place, (3) removal and containment, and (4) removal and treatment. State the advantages and disadvantages of each option.

26.2. What are the primary functions of an in-situ capping system?

26.3. What types of site conditions limit the use of in-situ cap systems to contain sediments in waterways?

26.4. A lake near a former industrial center has been found to contain the PCB Aroclor 1242. The following ISC design has been proposed.

Water

H_{c_1} = 30 cm	Filter layer (graded armor):
	γ_{c_1} = 115 lb/ft^3
H_{c_2} = 50 cm	Isolation layer (sand):
	γ_{c_2} = 110 lb/ft^3 G_s = 2.65 n_s = 0.25
H_s = 400 cm	Contaminated sediment:
	γ_s = 100 lb/ft^3 C_c = 0.3 D_e = 4 × 10^{-6} cm^2/s
	C_u = 500 lb/ft^2 e_0 = 0.4 K_{oc} = 198,000 L/kg
	$\dfrac{C_\alpha}{C_c}$ = 0.02 f_{oc} = 0.05

(a) Determine the primary consolidation of the ISC.

(b) Perform a bearing capacity analysis for the sediment layer. Will the layer soil after the proposed cap is in place?

(c) If additional graded armor is desired for increased erosion protection, what thickness could be added before bearing capacity failure of the sediment layer? Assume a safety factor of 3.

(d) What is the breakthrough time, t_b, of the Aroclor 1242. For simplicity, assume that the filter layer does not impede the contaminant release. Also, calculate the steady-state time and the retardation factor.

(e) Estimate the future contaminant concentration near the bottom of the isolation layer before secondary consolidation begins.

REFERENCES

Abramson, L. W., Lee, T. J., Sharma, S., and Boyce, G. M., *Slope Stability and Stabilization Methods,* Wiley, New York, 1996.

Cargill, K. W., *Procedure for the Prediction of Consolidation in Soft Fine-Grained Dredged Material,* D 83-1, U.S. Army Corps of Engineers, Washington, DC, 1983.

Cheney, R. S., and Chassie, R. G., *Soil and Foundations Workshop Manual,* Federal Highway Administration, Washington, DC, 1982.

Fetter, C. W., *Contaminant Hydrogeology,* 2nd ed., Prentice Hall, Upper Saddle River, NJ, 1999.

Gunnison, D., Brannon, J. M., Sturgis, T. C., and Smith, I., *Development of Simplified Column Test for Evaluation of Thickness of Capping Material Required to Isolate Contaminated Dredged Material,* D 87-2, U.S. Army Corps of Engineers, Vicksburg, MS, 1987.

Hagerty, P. A., and Trotman, T. D., In situ contaminated sediment management: unique case studies, in *Contaminated Soils,* Vol. 6, Kostecki, P. T., Calabrese E. J., and Dragun, J. (eds.), Amhurst Scientific Publishers, Amhurst, MA, 2001.

Hemond, H., and Fechner, E., *Chemical Fate and Transport in the Environment,* Academic Press, San Diego, CA, 1994.

Leschinsky, D., and Smith, D. S., Deep seated failure of granular embankment over clay: stability analysis, *Soils Found.,* Vol. 29, No. 3, pp. 105–114, 1989.

Ling, H. I., Leshinsky, D., Gilbert, P. A., and Palermo, M. R., Geotechnical considerations related to in-situ capping of contaminated submarine sediments, in *Proceedings of the 29th TAMU & WEDA XVI Technical Conference,* New Oreleans, LA, 1995.

Mitchell, J. K., *Fundamentals of Soil Behavior,* Wiley, New York, 1993.

Mohan, R. K., Mageau, D. W., and Brown, M. P., Modeling the geophysical impacts of underwater in-situ cap construction, *MTS J.,* Vol. 33, No. 3, pp. 80–87, 1999.

Mohan, R. K., Brown, M. P., and Barnes, C. R., Design criteria and theoretical basis for capping contaminated marine sediments, *App. Ocean Res.,* Vol. 22, No. 2, pp. 85–93, 2000.

Moo-Young, H., Myers, T., Tardy, B., Ledbetter, R., Vanadit-Ellis, W., and Sellasie, K., Determination of the environmental impact of consolidation induced convective transport through capped sediment, *J. Hazard. Mater.,* Vol. 85, pp. 53–72, 2001.

Ogata, A., and Bank, R. B., *A Solution to the Differential Equation of Longitudinal Dispersion in Porous Media,* Paper 411-A, U.S. Geological Survey, Washington, DC, 1961.

Palermo, M. R., Design considerations for in-situ capping of contaminated sediments, *Water Sci. Technol.,* Vol. 37, No. 6/7, pp. 315–321, 1998.

Palermo, M., Maynord, S., Miller, J., and Reible, D. 1998. *Guidance for In-Situ Subaqueous Capping of Contaminated Sediments,* EPA 905/B-96/004, Great Lakes National Program Office, USEPA, Chicago.

Tchobangolous, G., and Schroeder, E. D., *Water Quality: Characteristics, Modeling, Modification,* McGraw-Hill, New York, 1987.

Terzaghi, K., Peck, R. B., and Mesri, G., *Soil Mechanics in Engineering Practice,* 3rd ed., Wiley, New York, 1996.

Thoma, G. J., Reible, D. D., Valsaraj, K. T., and Thibodeaux, L. J., Efficiency of capping contaminated sediments in situ: 2. Mathematics of diffusion–adsorption in the capping layer, *Environ. Sci. Technol.,* Vol. 27, No. 1, 1993.

USACE, (U.S. Army Corps of Engineers), *Design Requirements for Capping,* Dredging Research Technical Notes, Waterways Experiment Station, Vicksburg, MS, 1991.

USEPA, (U.S. Environmental Protection Agency), 1994. *ARCS Remediation Guidance Document,* EPA/905/B-94/003. Great Lakes National Program Office, USEPA, Chicago.

Vanoni, V. A., *Sedimentation Engineering,* ASCE Manuals and Reports on Engineering Practice No. 54, ASCE, Reston, VA, 1977.

Wang, X. Q., Thibodeaux, L. J., Valsaraj, K. T., and Reible, D. D., Efficiency of capping contaminated sediments in Situ: 1. Laboratory-scale experiments on diffusion-adsorption in the capping layer, *Environ. Sci. Technol.,* Vol. 25, No. 9, 1991.

INDEX